T0137318

Advances in Intelligent Systems and Computing

Volume 1095

The series "Advances in Intelligent Systems and Computing" contains publications on theory, applications, and design methods of Intelligent Systems and Intelligent Computing. Virtually all disciplines such as engineering, natural sciences, computer and information science, ICT, economics, business, e-commerce, environment, healthcare, life science are covered. The list of topics spans all the areas of modern intelligent systems and computing such as: computational intelligence, soft computing including neural networks, fuzzy systems, evolutionary computing and the fusion of these paradigms, social intelligence, ambient intelligence, computational neuroscience, artificial life, virtual worlds and society, cognitive science and systems, Perception and Vision, DNA and immune based systems, self-organizing and adaptive systems, e-Learning and teaching, human-centered and human-centric computing, recommender systems, intelligent control, robotics and mechatronics including human-machine teaming, knowledge-based paradigms, learning paradigms, machine ethics, intelligent data analysis, knowledge management, intelligent agents, intelligent decision making and support, intelligent network security, trust management, interactive entertainment, Web intelligence and multimedia.

The publications within "Advances in Intelligent Systems and Computing" are primarily proceedings of important conferences, symposia and congresses. They cover significant recent developments in the field, both of a foundational and applicable character. An important characteristic feature of the series is the short publication time and world-wide distribution. This permits a rapid and broad dissemination of research results.

** Indexing: The books of this series are submitted to ISI Proceedings, EI-Compendex, DBLP, SCOPUS, Google Scholar and Springerlink **

More information about this series at http://www.springer.com/series/11156

Rafik A. Aliev · Janusz Kacprzyk ·
Witold Pedrycz · Mo Jamshidi ·
Mustafa B. Babanli · Fahreddin M. Sadikoglu
Editors

10th International Conference on Theory and Application of Soft Computing, Computing with Words and Perceptions - ICSCCW-2019

 Springer

Editors
Rafik A. Aliev
Joint MBA Program
Azerbaijan State Oil and Industry University
Baku, Azerbaijan

Janusz Kacprzyk
Polish Academy of Sciences
Systems Research Institute
Warsaw, Poland

Witold Pedrycz
Department of Electrical
and Computer Engineering
University of Alberta
Edmonton, AB, Canada

Mo Jamshidi
Department of Electrical
and Computer Engineering
University of Texas at San Antonio
San Antonio, TX, USA

Mustafa B. Babanli
Azerbaijan State Oil and Industry University
Baku, Azerbaijan

Fahreddin M. Sadikoglu
Department of Mechatronics
Near East University
Nicosia, Turkey

ISSN 2194-5357   ISSN 2194-5365   (electronic)
Advances in Intelligent Systems and Computing
ISBN 978-3-030-35248-6   ISBN 978-3-030-35249-3   (eBook)
https://doi.org/10.1007/978-3-030-35249-3

This Springer imprint is published by the registered company Springer Nature Switzerland AG
The registered company address is: Gewerbestrasse 11, 6330 Cham, Switzerland

Preface

The 10th International Conference on Theory and Applications Soft Computing, Computing with Words and Perceptions (ICSCCW-2019) is the premier international conference organized by Azerbaijan Association of "Zadeh's Legacy and Artificial Intelligence" (Azerbaijan), Azerbaijan State Oil and Industry University (Azerbaijan), University of Siegen (Siegen, Germany), BISC—Berkeley Initiative in Soft Computing (USA), University of Texas, San Antonio (USA), Georgia State University (Atlanta, USA), University of Alberta (Canada), University of Toronto (Toronto, Canada), TOBB Economics and Technology University (Turkey), and Near East University (North Cyprus).

This volume presents the proceedings of the 10th International Conference on Theory and Applications Soft Computing, Computing with Words and Perceptions (ICSCCW-2019), held in Prague, Czech Republic, on August 27–28, 2019. It includes contributions from diverse areas of soft computing and computing with words such as uncertain computation, decision making under imperfect information, neuro-fuzzy approaches, deep learning, and natural language processing. The topics of the papers include theory and application of soft computing, information granulation, computing with words, computing with perceptions, image processing with soft computing, probabilistic reasoning, intelligent control, machine learning, fuzzy logic in data analytics and data mining, evolutionary computing, chaotic systems, soft computing in business, economics and finance, fuzzy logic and soft computing in earth sciences, fuzzy logic and soft computing in engineering, fuzzy logic and soft computing in material sciences, soft computing in medicine, biomedical engineering, and pharmaceutical sciences.

This volume covers new ideas from theories of soft computing and computing with words and their applications in economics, business, industry, education, medicine, earth sciences, and other fields. This volume will be a useful guide for academics, practitioners, and graduates in fields of soft computing and computing with words. It will allow for increasing of interest in development and applying of these paradigms in various real-life fields.

August 2019   Rafik Aliev
                                                          Chairman of ICSCCW-2019

Organization

Chairman

R. A. Aliev, Azerbaijan

Co-chairmen and Guest Editors

J. Kacprzyk, Poland
M. Jamshidi, USA
W. Pedrycz, Canada
M. B. Babanli, Azerbaijan
F. S. Sadikoglu, North Cyprus

International Program Committee

I. G. Akperov, Russia
R. R. Aliev, North Cyprus
F. Aminzadeh, USA
K. Atanassov, Bulgaria
I. Batyrshin, Mexico
H. Berenji, USA
K. Bonfig, Germany
D. Dubois, France
D. Enke, USA
B. Fazlollahi, USA
T. Fukuda, Japan
M. Gupta, Canada
H. Hamdan, France, UK
O. Huseynov, Azerbaijan
C. Kahraman, Turkey
O. Kaynak, Turkey
V. Kreinovich, USA
D. Kumar Jana, India

V. Loia, Italy
R. Abiyev, North Cyprus
E. Babaei, Iran
P. Moog, Germany
M. Nikravesh, USA
V. Niskanen, Finland
V. Novak, Czech Republic
I. Perfilieva, Czech Republic
H. Prade, France
H. Roth, Germany
A. Khashman, UK
M. Salukvadze, Georgia
T. Takagi, Japan
K. Takahashi, Japan
V. B. Tarasov, Russia
S. Ulyanov, Russia
R. Yager, USA
N. Yusupbekov, Uzbekistan

Organizing Committee

Chairman

U. Eberhardt, Germany

Co-chairmen

T. Abdullayev, Azerbaijan
L. Gardashova, Azerbaijan
E. Tuncel, North Cyprus

Members

N. E. Adilova, Azerbaijan
M. M. M. Elamin, North Cyprus
A. Alizadeh, Azerbaijan
A. Yusupbekov, Uzbekistan
A. Guliyev, Azerbaijan
S. Uzelaltinbulat, North Cyprus
B. Guirimov, Azerbaijan
G. Sadikoglu, North Cyprus
K. Jabbarova, Azerbaijan
S. Akdag, North Cyprus
M. A. Salahli, Turkey
H. Aşkaroğlu, North Cyprus

Conference Organizing Secretariat

Azadlig Ave. 20, AZ 1010 Baku, Azerbaijan
Phone: +99 412 498 15 89, Fax: +99 412 498 45 09
E-mail: icsccw.org@gmail.com

Abstracts

A Life Quality Focused Regional Development Fuzzy Dynamic Model: An Example of Cognitive Dynamic Systems Modeling

Janusz Kacprzyk[ORCID]

Polish Academy of Sciences Member, Academia Europaea Member, European Academy of Sciences and Arts Foreign Member, Bulgarian Academy of Sciences Foreign Member, Spanish Royal Academy of Economic and Financial Sciences (RACEF) Foreign Member, Finnish Society of Sciences and Letters Systems Research Institute, Warsaw, Poland
kacprzyk@ibspan.waw.pl

Abstract. We are concerned with some important issues related to dynamic systems modeling, notably related to how to include aspects of human cognition in such models. We assume as a general perspective and a point of departure Yingxu Wang's cognitive informatics which is, roughly speaking, a multidisciplinary field within informatics, or computer science, that is based on results of cognitive and information sciences, and which deals with human information processing mechanisms and processes and their applications in broadly perceived computing. The emphasis is on processes in the human brain. However, in our work we advocate an approach which assumes the brain process-centered cognitive informatics to be the foundation, but—for our purposes—it may be proper to distinguish an "outer" cognitive informatics which explicitly makes reference not what proceeds "internally" in the brain, which is an area of interest of the traditional cognitive informatics, but what proceeds. And how, "externally," i.e., what people can see, judge, evaluate, etc., and what is clearly a result of cognitive information specific processes in the brain. Cognitive informatics constitutes a foundation of cognitive computing which synergistically employs tools and techniques from, e.g., information science, data sciences, computational sciences, computer science, artificial and computational intelligence, cybernetics, systems science, cognitive science, (neuro)psychology, brain science, linguistics, etc., to just mention a few.

To show that the cognitive computing can yield an "added value," we consider—as an example of a dynamic systems modeling—a scenario-based dynamic regional development planning model. The region is characterized by seven life quality indicators related to economic, social, environmental, etc., qualities, which evolve over some planning horizon due to some investments, mostly by some regional or governmental agencies. There are investment scenarios over the planning horizon, which specify funds meant for the development of the particular life quality indexes, and some desired levels of these indexes, both objective, i.e., set by authorities, and subjective, i.e., perceived by the inhabitant groups. As a result of a particular investment scenario, the life quality indexes evolve over the planning horizon, and their temporal evolution is evaluated by the authorities and inhabitants. This evaluation has both an objective, i.e., against the "officially" set thresholds, and subjective, i.e., as perceived by various humans and their groups, aspects.

We employ Kacprzyk's fuzzy dynamic programming-based approach to the modeling and planning/programming of sustainable regional development, with soft constraints and goals, but with a more sophisticated assessment of variability, stability, and balancedness of consecutive investments. The evaluation of the development quality measures and then their optimization are then proposed so that cognitive computing, notably by inclusion of some decision making, behavioral, social, etc., biases, in particular the so-called status quo and minimal change biases. We extend the model to include a more sophisticated analysis of variability of temporal evolution of some life quality indicators and a human perception of its goodness. We also mention how to reflect elements related to fairness. We show how the new elements of the regional development model proposed can change the best development scenarios derived.

How to Process Big Data with the Help of Fuzzy Transforms

Irina Perfiljeva[iD]

Centre of Excellence IT4Innovations division of the University of Ostrava,
Institute for Research and Applications of Fuzzy Modeling
irina.perfilieva@osu.cz

Abstract. The "Big Data" processing is the main challenge of contemporary science. Experimental sciences like biology or chemistry are facing an explosion of the data available from experiments. However, much of the data is highly redundant and can be represented using a smaller number of parameters without significant loss of information. The underlying mathematical technique making possible this type of representation is called dimensionality reduction.

A modern approach is based on Laplacian eigenmaps that characterize the local neighborhood in terms of the measure of closeness. The optimization problem focused on a low-dimensional representation of the data set is formulated, and the cost function is introduced. The proposed solution is formulated in terms of eigenmaps of the graph Laplacian.

In fuzzy literature, the most relevant dimensionality reduction technique is fuzzy (F)-transform. It is based on a granulation of a domain (fuzzy partition) and gives a tractable image of an original data. The main characteristics concerning input data are size reduction, noise removal, invariance to geometrical transformations, knowledge transfer from conventional mathematics, and fast computation.

In the talk, we discuss how the F-transform can be explained in terms of Laplacian eigenmaps and combined with the PCA.

We demonstrate the efficiency of the proposed combination on the example of pattern recognition in a large database. We also compare this combination with other relevant techniques (besides other, LENET-like CNN) from the computation time and success rate points of view.

The Theory of Intermediate Quantifiers in Fuzzy Natural Logic

Vilem Novak[iD]

Institute for Research and Applications of Fuzzy Modeling, University of Ostrava, NSC IT4Innovations, 30. dubna 22, 701 03 Ostrava 1, Czech Republic
vilem.novak@osu.cz

Abstract. Fuzzy Natural Logic (FNL). This is a mathematical theory whose goal is to provide a mathematical model of natural human reasoning whose typical feature is the use of natural language. FNL includes the model of the vagueness phenomenon and follows results of the logical analysis of natural language.

Till now, FNL consists of the (1) formal theory of evaluative linguistic expressions, (2) formal theory of fuzzy/linguistic IF-THEN rules, linguistic descriptions, and approximate reasoning, and (3) the formal theory of intermediate and fuzzy generalized quantifiers. The theory is developed as a special theory of (Lukasiewicz) fuzzy-type theory (FTT; a higher-order fuzzy logic). In this talk, we will present the theory of intermediate quantifiers that were introduced in. Evaluative linguistic expressions are expressions of natural language, for example, small, medium, big, roughly one hundred, very short, more or less deep, not tall, roughly warm or medium-hot, quite roughly strong, roughly medium size, etc. They form a small, syntactically simple, but very important part of natural language which is present in its everyday use any time. In FNL, special formal theory of FTT has been constructed using which semantics of the evaluative expressions is modeled. This theory is applied in the theory of intermediate quantifiers.

Intermediate quantifiers are natural language expressions such as most, a lot of, many, a few, a great deal of, a large part of, and small part of. In correspondence with this approach they are in FNL modeled as special formulas of fuzzy-type theory in a certain extension of the formal theory of evaluative linguistic expressions. Typical intermediate quantifiers are "Most (Almost all, A few, Many) *Bs* are A."

These quantifiers occur in the generalized Aristotle syllogisms. We can prove validity of 120 syllogisms (and falsity of the other ones) and also form a generalized square of opposition in our theory.

Applications of FNL include models of commonsense human reasoning, managerial decision making, linguistic control of processes, forecasting and mining information from time series, and other ones.

Acknowledgment. The paper is supported by the project "Centre for the development of Artificial Intelligence Methods for the Automotive Industry of the region."

Granular Data Descriptors: Their Generative and Discriminative Facets

Witold Pedrycz🄬

Department of Electrical and Computer Engineering, University of Alberta, Edmonton AB T6R 2V4, Canada
wpedrycz@ualberta.ca

Abstract. Concepts constitute a concise manifestation of key features of data. As being built at the higher level of abstraction than the data themselves, they capture the essence of the data and usually emerge in the form of information granules. Data descriptors—information granules—characterize data in a concise way and are further used as building blocks of classifiers or predictors. Generative and discriminative aspects of information granules and information granularity deserve detailed investigations.

In this talk, we identify three main ways in which concepts are encountered and characterized: (i) numeric, (ii) symbolic, and (iii) granular. Each of these views come with their advantages and become complementary to some extent.

The numeric concepts are built by engaging various clustering techniques. The quality of numeric concepts evaluated at the numeric level is described by a reconstruction criterion. The symbolic description of concepts, which is predominant in the realm of artificial intelligence (AI) and symbolic computing, can be represented by sequences of labels (integers). In such a way, qualitative aspects of data are captured. This facilitates further qualitative analysis of concepts and constructs involving them by reflecting the bird's-eye view of the data. They come hand in hand with a variety of analyses concerning constructs involving symbols, namely stability, distinguishability, redundancy, and conflict. The granular concepts augment numeric concepts by bringing information granularity into the picture and invoking the principle of justifiable granularity in their construction.

At the applied end, we elaborate on granular transfer learning and granular classifiers.

Self-Organizing Maps for Interval-Valued Data. Trend to Clustering of Big Data

Hani Hamdan

Laboratoire des Signaux et Systèmes, CentraleSupélec, CNRS,
Université Paris-Sud, Université Paris-Saclay, Paris, France
Hani.Hamdan@centralesupelec.fr

Abstract. Since the beginning of this century, interval-valued data have been used more and more in data mining and soft computing. Consequently, many data analysis methods have been extended to interval data this last decade. Interval-valued data are used in many applications where they represent data imprecision, measurement inaccuracy, or measurand variability. The self-organizing map is a type of artificial neural network that is trained using unsupervised learning. It has been widely used for visualization by creating low-dimensional views of high-dimensional data. In our plenary talk, we will present the self-organizing map for interval-valued data. While considering different distances to compare two vectors of intervals, we present several self-organizing maps, and we show their use as multidimensional unsupervised classifiers. For each considered distance type, the optimization-based machine learning algorithm is set up. A particular attention is paid to the case of adaptive distances. To show the usefulness of our developed approaches, we apply them to simulated and real interval-valued data sets (Monte Carlo experiments and flaw diagnosis application using acoustic emission). An original approach to clustering of Big Data using the self-organizing map will also be proposed.

Keywords: Self-organizing maps · Estimation · Clustering · Interval-valued data · Big Data

Contents

Some Fundamental Results on Z-valued Calculus

R. A. Aliev[1,2(✉)] ⓘ

[1] Georgia State University, Atlanta, USA
raliev@asoa.edu.az
[2] Azerbaijan State Oil and Industry University,
20 Azadlig Avenue, AZ1010 Baku, Azerbaijan

Abstract. Zadeh proposed a concept of Z-number as a formal basis of handling of imprecision and partial reliability of information. A Z-number $Z = (A, B)$ is a pair of fuzzy numbers used to describe information about a value of a real-valued random variable X. A is considered as a fuzzy evaluation of X, and B is a fuzzy reliability of A. A Z-number is a special case of a Z-set, where the latter corresponds to a general case when X is not necessarily defined in the real line R. The concept of a Z-set would enhance neoclassical analysis that embraces methods to reflect imprecision, randomness and uncertainty in computation and reasoning. This concept is closely related to fuzzy set theory, set-valued analysis and its extensions.

This study for the first time presents the core of the Z-calculus. Basic constructions of the Z-calculus, such as Z-sets, Z-functions, Z-derivatives, Z-integrals, and some concepts of Z-algebra are considered. Application of z-calculus to Z-matrix equation and Z-differential equations are considered. Development of Z-calculus is important for construction of formal models which are closer to real-world problems (in realm of economics, engineering, earth sciences and other fields) than those based on classical, probabilistic or fuzzy calculus.

Keywords: Z-set · Z-valued function · Z-matrix · Z-eigenvalues and eigenvectors

1 Introduction

Zadeh proposed a concept of Z-number as a formal basis of handling of imprecision and partial reliability of information. A Z-number $Z = (A, B)$ is a pair of fuzzy numbers used to describe information about a value of a real-valued random variable X. A is considered as a fuzzy evaluation of X. In general, A is partially reliable, and this is described by a value of probability measure of A, $P(A)$. As a probability distribution of X is not exactly known, B is used as a fuzzy restriction over $P(A)$. In [1] Zadeh outlines a general approach to computation with Z-numbers. The approach is based on an extension principle that compose fuzzy arithmetic and probabilistic arithmetic. Inspired by this seminal work, during several years we conducted a research results were published in [2–11]. In [3, 4] approaches which allow to carry out basic arithmetic

© Springer Nature Switzerland AG 2020
R. A. Aliev et al. (Eds.): ICSCCW 2019, AISC 1095, pp. 1–23, 2020.
https://doi.org/10.1007/978-3-030-35249-3_1

operations over Z-numbers at a low computational cost are proposed. In [2] some basics of linear programming with Z-valued variables and parameters are proposed. The book [5] is devoted to various branches of computations over Z-numbers. The authors consider formalization of arithmetic operations, algebraic operations, optimization problems, aggregation operators and other problems of handling Z-valued information. Applications to planning, decision making, modeling are proposed. Basics of Hukuhara difference of Z-numbers are introduced in [7]. [8] is devoted to formalization of Z-number valued functions.

Other authors published a series of works of with serious theoretical basis and important practical applications. These findings are devoted to handling Z-valued information [10], Z-valued differential equations [13, 14], distance between Z-numbers [15], analysis of uncertainty in Z-numbers [16, 17] and other problems.

There is no comprehensive work on Z-calculus devoted to a general view on various existing results and new research problems. Existing works are devoted to Z-numbers which are special case of Z-sets – Z-sets used to describe information on a real-valued random variable X. However, a general consideration of a concept of Z-set is needed. In this work we try to make a first step toward a realm of Z-calculus. The rest of the paper is organized as follows. In Sect. 2 we introduce a concept of a Z-set. Operations over Z-sets and their properties are considered. Some results on partial ordering of Z-sets are given in Sect. 3. In Sect. 4 we introduce a new distance between Z-numbers. Section 5 is devoted to basics of Z-valued functions. Relation between derivative and definite integral of such a function is analyzed. An introduction to Z-valued system of linear equations is given in Sect. 6. The problems of determination of eigenvalues and eigenvectors of Z-number valued matrices are analyzed in Sect. 7. Some insights to Z-differential equations are considered in Sect. 8. Section 9 concludes.

2   Z-sets and Operations over Z-sets

2.1   A Z-set

Definition 2.1. Let X be a space of objects and $x \in X$. A Z-set in X is defined by a membership function $\mu_A : X \to [0, 1]$, probability distributions p_Z associated with X and membership function $\mu_B : [0, 1] \to [0, 1]$. $\mu_A(x)$ is a degree to which x belong to a fuzzy set A. $\mu_B(b)$ is a degree to which a value of probability measure of A, $b = P(A)$, belongs to a fuzzy set B according to a fuzzy restriction

$$P(A) \text{ is } B.$$

Accordingly, taking into account that $P(A)$ is defined as $P(A) = \int_X p_Z(x)\mu_A(x)$,

$P(A)$ *is* B induces a set G of distributions p_Z.

Thus, a Z-set is a triple $Z = (A, B, G)$. A is a fuzzy set, B is a fuzzy restriction on probability measure of A, $P(A)$, induced by a set of distributions G.

Below we consider three main operations of Z-sets which are defined as follows.

2.2 Operations over Z-sets

2.2.1 Complement of a Z-set

Let $Z = (A, B, G)$ be a Z-set and G be a set whose elements are all the probability distributions:

$$G = \left\{ p_Z(x) : \int_{-\infty}^{+\infty} p_Z(x)dx = 1, \int_{-\infty}^{+\infty} p_Z(x)\mu_A(x) \text{ is } B, \int_{-\infty}^{+\infty} xp_Z(x) = \int_{-\infty}^{+\infty} x\mu_A(x)dx / \int_{-\infty}^{+\infty} \mu_A(x)dx \right\}$$

The complement $\overline{Z}$ of Z-set $Z = (A, B, G)$ is defined as follows.

$\overline{G}$ set of probability distributions defined as

$$\overline{G} = \left\{ p_Z(x) : \int_{-\infty}^{+\infty} p_Z(x)dx = 1, \int_{-\infty}^{+\infty} p_Z(x)\mu_A(x) \text{ is } B, \int_{-\infty}^{+\infty} xp_Z(x) = \int_{-\infty}^{+\infty} x\mu_A(x)dx / \int_{-\infty}^{+\infty} \mu_A(x)dx \right\}$$

Then

$$\overline{Z} = (\overline{A}, 1 - B, \overline{G}),$$

where $\mu_{\overline{A}}(x) = 1 - \mu_A(x)$ and $1 - B$ is arithmetic operation.

2.2.2 Union of Z-sets

Union of Z-sets $Z_i = (A_i, B_i, G_i)$, $i = 1, 2$ is defined as

$$Z_{12} = Z_1 \cup Z_2 = Zor(Z_1, Z_2) = (A_{12}, B_{12}, G_{12}).$$

At first, let us consider the case of continuous Z-sets. The union $A_{12} = A_1 \cup A_2$ of the fuzzy sets A_1 and A_2 is defined as follows:

$$A_{12}(x) = (A_1 \cup A_2)(x) = \max(A_1(x), A_2(x)).$$

The set G_{12} of resulting distributions p_{12} is defined as

$$G_{12} = G_1 \cup G_2,$$

where

$$G_i = \left\{ p_{Z_i}(x) : \int_{-\infty}^{+\infty} p_{Z_i}(x)dx = 1, \int_{-\infty}^{+\infty} p_{Z_i}(x)\mu_{A_i}(x) \text{ is } B_i, \int_{-\infty}^{+\infty} xp_{Z_i}(x) = \int_{-\infty}^{+\infty} x\mu_{A_i}(x)dx / \int_{-\infty}^{+\infty} \mu_{A_i}(x))dx \right\},$$
$$i = 1, 2.$$

Convolution $p_{12} = p_{Z_1 \cup Z_2}(x) = p_{Z_1}(x) \circ_\cup p_{Z_2}(x) = p_1 \circ_\cup p_2$ of probability distributions is defined as

$$p_{12}(x) = p_{Z_1 \cup Z_2}(x) = \frac{(p_{Z_1}(x) + p_{Z_2}(x) - p_{Z_1}(x)p_{Z_2}(x))p_X(x)}{\int_X (p_{Z_1}(x) + p_{Z_2}(x) - p_{Z_1}(x)p_{Z_2}(x))p_X(x)dx},$$

p_X describes joint consideration of Z-sets $Z_i = (A_i, B_i, G_i)$, $i = 1, 2$. In special case, we can use $p_X(x) = \frac{1}{card(X)}$.

Then $Z_{12} = (A_{12}, B_{12}, G_{12})$, where

$$G_{12} = \{p_{12}(x) : p_{12}(x) = p_{Z_1 \cup Z_2}(x)$$
$$= \frac{(p_{Z_1}(x) + p_{Z_2}(x) - p_{Z_1}(x)p_{Z_2}(x))p_X(x)}{\int_X (p_{Z_1}(x) + p_{Z_2}(x) - p_{Z_1}(x)p_{Z_2}(x))p_X(x)dx}, p_i(x) \in G_i\},$$

$$B_{12} = \{(\mu_{p_{12}}(p_{12}), \mu_{A_{12}} \cdot p_{12}) : p_{12} \in G_{12}\}.$$

2.2.3   Intersection of Z-sets

Intersection of Z-sets $Z_i = (A_i, B_i, G_i)$, $i = 1, 2$ is defined as

$$Z_{12} = Z_1 \cap Z_2 = Zand(Z_1, Z_2) = (A_{12}, B_{12}, G_{12}).$$

The intersection $A_1 \cap A_2$ of the fuzzy sets A_1 and A_2 is defined as follows:

$$A_{12}(x) = (A_1 \cap A_2)(x) = \min(A_1(x), A_2(x)).$$

The set G_{12} of resulting distributions p_{12} is defined as

$$G_{12} = G_1 \cap G_2,$$

where

$$G_i = \left\{ p_{Z_i}(x) : \int_{-\infty}^{+\infty} p_{Z_i}(x)dx = 1, \int_{-\infty}^{+\infty} p_{Z_i}(x)\mu_{A_i}(x) \text{ is } B_i, \int_{-\infty}^{+\infty} xp_{Z_i}(x) = \int_{-\infty}^{+\infty} x\mu_{A_i}(x)dx \bigg/ \int_{-\infty}^{+\infty} \mu_{A_i}(x))dx \right\},$$
$$i = 1, 2.$$

Convolution $p_{12} = p_{Z_1 \cap Z_2}(x) = p_{Z_1}(x) \circ_\cap p_{Z_2}(x) = p_1 \circ_\cap p_2$ of probability distributions is defined as

$$p_{12}(x) = p_{Z_1 \cap Z_2}(x) = \frac{p_{Z_1}(x)p_{Z_2}(x)p_X(x)}{\int_X p_{Z_1}(x)p_{Z_2}(x)p_X(x)dx}.$$

Then $Z_{12} = (A_{12}, B_{12}, G_{12})$, where

$$G_{12} = \{p_{12}(x) : p_{12}(x) = p_{Z_1 \cap Z_2}(x) = \frac{p_{Z_1}(x)p_{Z_2}(x)p_X(x)}{\int_X p_{Z_1}(x)p_{Z_2}(x)p_X(x)dx}, p_i(x) \in G_i\}$$

Thus,

$$B_{12} = \{(\mu_{p_{12}}(p_{12}), \mu_{A_{12}} \cdot p_{12}) : p_{12} \in G_{12}\}.$$

2.2.4 Properties of Operations on Z-sets

Theorem 2.1. Commutative law for intersection of Z-sets is satisfied:

$$Z_{12} = Z_1 \cap Z_2 = Z_2 \cap Z_1 = Z_{21}.$$

Proof. For A_{12} one has:

$$A_{12}(x) = (A_1 \cap A_2)(x) = \min(A_1(x), A_2(x))$$

$$= \min[A_2(x), A_1(x)] = (A_2 \cap A_1)(x) = A_{21}(x).$$

The set of distributions of intersection Z_{12} can be described as follows.

$$G_{12} = \{p_{12}(x) : p_{12}(x) = p_{Z_1 \cap Z_2}(x) = \frac{p_{Z_1}(x)p_{Z_2}(x)p_X(x)}{\int_X p_{Z_1}(x)p_{Z_2}(x)p_X(x)dx}, p_i(x) \in G_i\},$$

where

$$G_i = \{p_{Z_i}(x) : \int_{-\infty}^{+\infty} p_{Z_i}(x)dx = 1, \int_{-\infty}^{+\infty} p_{Z_i}(x)\mu_{A_i}(x) \text{ is } B_i, \int_{-\infty}^{+\infty} xp_{Z_i}(x) = \int_{-\infty}^{+\infty} x\mu_{A_i}(x)dx / \int_{-\infty}^{+\infty} \mu_{A_i}(x))dx,$$

$$i = 1, 2.$$

Thus, G_{12} is a set of convolutions $p_{12}(x)$. $p_{12}(x)$ satisfies commutativity property:

$$p_{12}(x) = p_{Z_1 \cap Z_2}(x) = \frac{p_{Z_1}(x)p_{Z_2}(x)p_X(x)}{\int_X p_{Z_1}(x)p_{Z_2}(x)p_X(x)dx}$$

$$= \frac{p_{Z_2}(x)p_{Z_1}(x)p_X(x)}{\int_X p_{Z_2}(x)p_{Z_1}(x)p_X(x)dx} = p_{Z_2 \cap Z_1}(x) = p_{21}(x).$$

Therefore, G_{12} also satisfies commutativity property:

$$G_{12} = \{p_{12}(x) : p_{12}(x) = p_{Z_1 \cap Z_2}(x) = \frac{p_{Z_1}(x)p_{Z_2}(x)p_X(x)}{\int_X p_{Z_1}(x)p_{Z_2}(x)p_X(x)dx}, p_i(x) \in G_i\}$$

$$= \{p_{21}(x) : p_{21}(x) = p_{Z_2 \cap Z_1}(x) = \frac{p_{Z_2}(x)p_{Z_1}(x)p_X(x)}{\int_X p_{Z_2}(x)p_{Z_1}(x)p_X(x)dx}, p_i(x) \in G_i\} = G_{21}$$

$$G_{12} = G_1 \cap G_2 = G_2 \cap G_1 = G_{21}.$$

Then $B_{12} = \{(\mu_{p_{12}}(p_{12}), \mu_{A_{12}} \cdot p_{12}) : p_{12} \in G_{12}\} = (\mu_{p_{21}}(p_{21}), \mu_{A_{21}} \cdot p_{21}) : p_{21} \in G_{21}\} = B_{21}$.

Thus, $Z_{12} = Z_{21}$. The proof is completed.

Theorem 2.2. Commutative law for union of Z-sets holds:

$$Z_{12} = Z_1 \cup Z_2 = Z_2 \cup Z_1 = Z_{21}.$$

The proof has been omitted.

Theorem 2.3. Associative law holds for intersection of Z-sets:

$$(Z_1 \cap Z_2) \cap Z_3 = Z_1 \cap (Z_2 \cap Z_3).$$

The proof has been omitted.

Theorem 2.4. Associative law holds for union of Z-sets:

$$(Z_1 \cup Z_2) \cup Z_3 = Z_1 \cup (Z_2 \cup Z_3).$$

The proof has been omitted.

Theorem 2.5. Distribute law holds for Z-sets:

$$Z_1 \cap (Z_2 \cup Z_3) = (Z_1 \cap Z_2) \cup (Z_1 \cap Z_3),$$

$$Z_1 \cup (Z_2 \cap Z_3) = (Z_1 \cup Z_2) \cap (Z_1 \cup Z_3).$$

Proof. Let us consider $Z_1 \cap (Z_2 \cup Z_3) = (Z_1 \cap Z_2) \cup (Z_1 \cap Z_3)$.
The proof of $Z_1 \cup (Z_2 \cap Z_3) = (Z_1 \cup Z_2) \cap (Z_1 \cup Z_3)$ is analogous.
In opened notation:

$$(A_1, B_1, G_1) \cap ((A_2, B_2, G_2) \cup (A_3, B_3, G_3)) =$$
$$((A_1, B_1, G_1) \cap (A_2, B_2, G_2)) \cup ((A_1, B_1, G_1) \cap (A_3, B_3, G_3)) \cdot$$

Let us show that in terms of HMFs, the distributive law holds for any fuzzy sets A_1, A_2, A_3:

$$A_{1(23)}(x) = \min(A_1(x), \max(A_2(x), A_3(x))) =$$
$$\max(\min(A_1(x), A_2(x)), \min(A_1(x), A_3(x))) = A_{(12)(13)}(x) \cdot$$

In HMF-based representation, fuzzy sets A_1, A_2, A_3 are described as follows.

$$A_1(\mu, \alpha_1) = [a_{1L}(\mu, \alpha_1), a_{1R}(\mu, \alpha_1)] =$$
$$\{a_1(\mu, \alpha_1) : a_1(\mu, \alpha_1) = a_{1L}(\mu) + \alpha_1 \cdot (a_{1R}(\mu) - a_{1L}(\mu))\} \,'$$

$$A_2(\mu, \alpha_2) = [a_{2L}(\mu, \alpha_2), a_{2R}(\mu, \alpha_2)] =$$
$$\{a_2(\mu, \alpha_2) : a_2(\mu, \alpha_2) = a_{2L}(\mu) + \alpha_2 \cdot (a_{2R}(\mu) - a_{2L}(\mu))\} \,'$$

$$A_3(\mu, \alpha_3) = [a_{3L}(\mu, \alpha_3), a_{3R}(\mu, \alpha_3)] =$$
$$\{a_3(\mu, \alpha_3) : a_3(\mu, \alpha_3) = a_{3L}(\mu) + \alpha_3 \cdot (a_{3R}(\mu) - a_{3L}(\mu))\} \, .$$

We need to prove

$$\min(A_1(\mu, \alpha_1), \max(A_2(\mu, \alpha_2), A_3(\mu, \alpha_3))) =$$
$$\max(\min(A_1(\mu, \alpha_1), A_2(\mu, \alpha_2)), \min(A_1(\mu, \alpha_1), A_3(\mu, \alpha_3)))$$

The left part can be expressed as follows:

$$\min(A_1(\mu, \alpha_1), \max(A_2(\mu, \alpha_2), A_3(\mu, \alpha_3))) =$$
$$\min(a_{1L}(\mu) + \alpha_1 \cdot (a_{1R}(\mu) - a_{1L}(\mu)),$$
$$\max((a_{2L}(\mu) + \alpha_2 \cdot (a_{2R}(\mu) - a_{2L}(\mu))), (a_{3L}(\mu) + \alpha_3 \cdot (a_{3R}(\mu) - a_{3L}(\mu))))$$

One can see that $A_1(\mu, \alpha_1), A_2(\mu, \alpha_2), A_3(\mu, \alpha_3)$ are real numbers for any values of $\mu, \alpha_1, \alpha_2, \alpha_3$. Then one has

$$\min(a_{1L}(\mu) + \alpha_1 \cdot (a_{1R}(\mu) - a_{1L}(\mu)),$$
$$\max((a_{2L}(\mu) + \alpha_2 \cdot (a_{2R}(\mu) - a_{2L}(\mu))), (a_{3L}(\mu) + \alpha_3 \cdot (a_{3R}(\mu) - a_{3L}(\mu)))) =$$
$$\max(\min(a_{1L}(\mu) + \alpha_1 \cdot (a_{1R}(\mu) - a_{1L}(\mu)), (a_{2L}(\mu) + \alpha_2 \cdot (a_{2R}(\mu) - a_{2L}(\mu))),$$
$$\min(a_{1L}(\mu) + \alpha_1 \cdot (a_{1R}(\mu) - a_{1L}(\mu)), (a_{3L}(\mu) + \alpha_3 \cdot (a_{3R}(\mu) - a_{3L}(\mu))) =$$
$$\max(\min(A_1(\mu, \alpha_1), A_2(\mu, \alpha_2)), \min(A_1(\mu, \alpha_1), A_3(\mu, \alpha_3))).$$

Thus, $A_{1(23)} = A_{(12)(13)}$.
Let us now consider G sets:

$$G_i = \{p_{Z_i}(x) : \int_{-\infty}^{+\infty} p_{Z_i}(x)dx = 1, \int_{-\infty}^{+\infty} p_{Z_i}(x)\mu_{A_i}(x) \text{ is } B_i, \int_{-\infty}^{+\infty} xp_{Z_i}(x) = \int_{-\infty}^{+\infty} x\mu_{A_i}(x)dx / \int_{-\infty}^{+\infty} \mu_{A_i}(x))dx,$$
$$i = 1, \ldots 3\}.$$

Let us prove that

$$p_{1(23)}(x) = p_1 \circ_\cap (p_2 \circ_\cup p_3) = (p_1 \circ_\cap p_2) \circ_\cup (p_1 \circ_\cap p_3) = p_{(12)(13)}(x).$$

We recall that

$$p_{1(23)}(x) = \int_V p(x_1, x_2, x_3)dv = \int_V p(x_1, x_2, x_3)dx_1 dx_2 dx_3,$$

where $V = \{v = (x_1, x_2, x_3) : x = \min(x_1, \max(x_2, x_3))\}$ and $p(x_1, x_2, x_3)$ is a joint distribution. As independent random variables are considered, we have:

$$p(x_1, x_2, x_3) = p_1(x_1)p_2(x_2)p_3(x_3).$$

At the same time,

$$p_{(12)(13)}(x) = \int\limits_{W} p_1(x_1)p_2(x_2)p_3(x_3)dw,$$

where $W = \{w = (x_1, x_2, x_3) : x = \max(\min(x_1, x_2), \min(x_1, x_3))\}$.
As $\min(x_1, \max(x_2, x_3)) = \max(\min(x_1, x_2), \min(x_1, x_3))$, one has $V = W$. Thus,

$$p_{1(23)}(x) = p_{(12)(13)}(x).$$

Therefore, one has $G_{1(23)} = G_{(12)(13)}$,
where $G_{1(23)} = \{p_{1(23)}(x) : p_{1(23)}(x) = p_1 \circ_{\min} (p_2 \circ_{\max} p_3), p_i(x) \in G_i, i = 1, \ldots, 3\}$,

$G_{(12)(13)} = \{p_{(12)(13)}(x) : p_{(12)(13)}(x) = (p_1 \circ_{\min} p_2) \circ_{\max} (p_1 \circ_{\min} p_3), p_i(x) \in G_i, i = 1, \ldots, 3\}$.

As $A_{(12)3} = A_{(12)(13)}$, and $G_{1(23)} = G_{(12)(13)}$, then $B_{1(23)} = B_{(12)(13)}$.
Thus, $Z_{1(23)} = Z_{(12)(13)}$. The proof is completed.

Theorem 2.6. Idempotent law hold for Z-sets:

$$Z_1 \cap Z_1 = Z_1,$$

$$Z_1 \cup Z_1 = Z_1,$$

In opened notation:

$$(A_1, B_1, G_1) \cap (A_1, B_1, G_1) = (A_1, B_1, G_1),$$

$$(A_1, B_1, G_1) \cap (A_1, B_1, G_1) = (A_1, B_1, G_1).$$

Proof has been omitted.

3 Partial Ordering of Z-sets

In this section we consider partial ordering of Z-sets as one of the basic concepts of Z-valued abstract algebra.

Objects. Let X be a space of a random variable x. The objects of study are Z-sets $Z = (A, G, B)$ that describe information about values of x. These Z-sets are described as Z-objects as follows:

$$M : Z \to F_L \times F_K \times F_S$$

Here $L = [0, 1]$ is a poset describing membership function $A()$ range, K is a mapping $K : \mathcal{G} \to [0, 1]$, where $\mathcal{G}$ is a system G sets. S is a mapping $S : \mathcal{P} \to [0, 1]$, where $\mathcal{P}$ is the range of probability measure $P(A)$. Thus, $S = [0, 1]$ is a poset describing membership function $B()$ range.

Partial Ordering over Z-sets $\succcurlyeq$. Formulation of partial order requires to consider dominance of both fuzzy sets and probability measures of fuzzy sets with respect to a set of distributions. The following definitions apply.

Definition 3.1. Equality of Z-sets. A degree to which Z-sets Z_1 and Z_2 are equal, $Z_1 = Z_2$, is defined as

$$\deg_=(Z_1, Z_2) = \min(\deg_=(A_1, A_2), \deg_=(G_1, G_2)),$$

where $\deg_=(A_1, A_2)$ is defined as usual, and $\deg_=(G_1, G_2)$ is defined as

$$\deg_=(G_1, G_2) = sup_{p_1 = p_2} \min(\mu_1(p_1), \mu_2(p_2)),$$

$$p_1 = p_2 \text{ if and only if } \int_a^b p_1(x)A_1(x)dx = \int_a^b p_2(x)A_2(x)dx$$

Definition 3.2. Inclusion of a Z-set. Inclusion of Z-set Z_2 into Z_1, $Z_1 \succcurlyeq Z_2$ is defined as

$$\deg_{\succcurlyeq}(Z_1, Z_2) \geq \deg_{\succcurlyeq}(Z_2, Z_1)$$

such that

$$\deg_{\succcurlyeq}(A_1, A_2) \geq \deg_{\succcurlyeq}(A_2, A_1)$$

and

$$\deg_{\succcurlyeq}(G_1, G_2) \geq \deg_{\succcurlyeq}(G_2, G_1).$$

$$\deg_{\succcurlyeq}(G_1, G_2) \text{ is defined as}$$

$$\deg_{\succcurlyeq}(G_1, G_2) = sup_{p_1 \succcurlyeq p_2} \min(\mu_1(p_1), \mu_2(p_2)),$$

$$p_1 \succcurlyeq p_2 \text{ iff } \int_a^b p_1(x)A_1(x)dx \geq \int_a^b p_2(x)A_2(x)dx$$

Let us prove that the properties of poset of Z-sets: reflexivity, antisimmetry, transitivity.

Reflexivity. Reflexivity implies

$$Z_1 \succcurlyeq Z_1.$$

According to Definition 2, one has

$$\deg_{\succcurlyeq}(Z_1, Z_1) \geq \deg_{\succ}(Z_1, Z_1).$$

The proof is obvious.

Antisimmetry. Antisimmetry is described as

$$Z_1 \succcurlyeq Z_2 \text{ and } Z_2 \succcurlyeq Z_1 \text{ implies } Z_1 = Z_2$$

According to Definition 2, one has for $Z_1 \succcurlyeq Z_2$ and $Z_2 \succcurlyeq Z_1$:

$$\deg_{\succcurlyeq}(Z_1, Z_2) \geq \deg_{\succ}(Z_2, Z_1) \text{ and } \deg_{\succcurlyeq}(Z_2, Z_1) \geq \deg_{\succ}(Z_1, Z_2)$$

This implies

$$\deg_{\succcurlyeq}(Z_2, Z_1) = \deg_{\succcurlyeq}(Z_1, Z_2).$$

Then

$$\deg_{=}(Z_1, Z_2) = 1.$$

In other words, $Z_1 = Z_2$.

Proof. Given $\deg_{\succcurlyeq}(Z_1, Z_2) \geq \deg_{\succ}(Z_2, Z_1)$ and $\deg_{\succcurlyeq}(Z_2, Z_1) \geq \deg_{\succ}(Z_1, Z_2)$ one has:

$$\deg_{\succcurlyeq}(A_1, A_2) \geq \deg_{\succ}(A_2, A_1) \text{ and } \deg_{\succcurlyeq}(A_2, A_1) \geq \deg_{\succ}(A_1, A_2)$$

$$\deg_{\succcurlyeq}(G_1, G_2) \geq \deg_{\succ}(G_2, G_1) \text{ and } \deg_{\succcurlyeq}(G_2, G_1) \geq \deg_{\succ}(G_1, G_2)$$

Then $\deg_{\succcurlyeq}(A_1, A_2) = \deg_{\succcurlyeq}(A_2, A_1)$. In other words, $A_1 \subseteq A_2$ and $A_2 \subseteq A_1$ imply $A_1 = A_2$. Therefore, $\deg_{=}(A_1, A_2) = 1$. On the other hand, $\deg_{\succcurlyeq}(G_1, G_2) = \deg_{\succcurlyeq}(G_2, G_1)$, that is,

$$sup_{p_1 \succcurlyeq p_2} \min(\mu_1(p_1), \mu_2(p_2)) = sup_{p_2 \succcurlyeq p_1} \min(\mu_1(p_1), \mu_2(p_2)).$$

One can show that this is possible only when

$$sup_{p_1 \succcurlyeq p_2} \min(\mu_1(p_1), \mu_2(p_2)) =$$
$$sup_{p_1 = p_2} \min(\mu_1(p_1), \mu_2(p_2)) = sup_{p_2 \succcurlyeq p_1} \min(\mu_1(p_1), \mu_2(p_2)) = 1.$$

Thus, $\deg_{=}(G_1, G_2) = 1$.

Therefore, $\deg_{=}(Z_1, Z_2) = \min(\deg_{=}(A_1, A_2), \deg_{=}(G_1, G_2)) = 1$. The proof is completed.

Transitivity. Transitivity is described as

$$Z_1 \succcurlyeq Z_2 \text{ and } Z_2 \succcurlyeq Z_3 \text{ implies } Z_1 \succcurlyeq Z_3$$

In terms of a degree,

$$\deg_{\succcurlyeq}(Z_1, Z_2) \geq \deg_{\succcurlyeq}(Z_2, Z_3) \text{ implies } \deg_{\succcurlyeq}(Z_1, Z_3)$$

Proof. $\deg_{\succcurlyeq}(A_1, A_2) \geq \deg_{\succcurlyeq}(A_2, A_3)$ implies $\deg_{\succcurlyeq}(A_1, A_3)$ because inclusion operation for fuzzy sets is transitive. $\deg_{\succcurlyeq}(G_1, G_2) \geq \deg_{\succcurlyeq}(G_2, G_3)$ implies $\deg_{\succcurlyeq}(G_1, G_3)$ because $\succcurlyeq$ is transitive for p_1, p_2 and p_3:

$$\int_a^b p_1(x)A_1(x)dx \geq \int_a^b p_2(x)A_2(x)dx \geq \int_a^b p_3(x)A_3(x)dx \text{ implies}$$

$$\int_a^b p_1(x)A_1(x)dx \geq \int_a^b p_3(x)A_3(x)dx$$

The proof is completed.

Let us formulate Z-valued algebraic structure $Z = (\mathcal{Z}, u, \succcurlyeq)$. Denote u an operation over Z-sets Z_i and Z_i'. In general U is some binary operation.

Definition 3.3. Monotonicity of operation over Z-sets. u operation is monotonic if

$$Z_i \succcurlyeq Z_i', i = 1, \ldots, n$$

implies

$$u(Z_1, \ldots, Z_n) \succcurlyeq u(Z_i', \ldots, Z_n'), i = 1, \ldots, n.$$

Definition 3.4. Z-valued algebraic structure $\mathbf{Z} = (\mathcal{Z}, u = (\cup, \cap), \succcurlyeq)$ is a construct of systems of Z-sets if every its operation is monotonic (Definition 3.3) in the ordered structure induced from L (Definition 3.2).

4 Distance Between Z-sets

The first definitions for distance of Z-numbers were proposed in [5, 8]. A more comprehensive study is presented in [15]. In this section we develop the ideas in [5, 8, 15] to propose a new definition of distance of Z-sets.

Let $Z_1 = (A_1, B_1, G_1)$ and $Z_2 = (A_2, B_2, G_2)$ be two Z-sets. Denote by $A_i^{\alpha_k}$ and $B_i^{\beta_k}$ k-th α-cuts of A_i and B_i respectively, $\alpha_k \in \{\alpha_1, \alpha_2, \ldots, \alpha_n\} \subset [0, 1], \beta_k \in \{\beta_1, \beta_2 \ldots, \beta_n\} \subset [0, 1]$. Denote $a_{i\alpha_k}^L = \min A_i^\alpha, a_{i\alpha_k}^R = \max A_i^\alpha$ and $b_{i\beta_k}^L = \min B_i^{\beta_k}, b_{i\beta_k}^R = \max B_i^{\beta_k}$, $i = 1, 2$.

Assume that we investigate distance between Z-sets $Z_1 = (A_1, B_1, G_1)$ and $Z_2 = (A_2, B_2, G_2)$. Distance between A_1 and A_2 is computed as

$$D_A(A_1, A_2) = \sup_{\alpha \in (0,1]} D_A^\alpha(A_1, A_2),$$

where

$$D_A^\alpha(A_1, A_2) = \left| \frac{a_{1\alpha}^L + a_{1\alpha}^R}{2} - \frac{a_{2\alpha}^L + a_{2\alpha}^R}{2} \right|.$$

Distance between B_1 and B_2 is computed as

$$D_B(B_1, B_2) = \frac{1}{2} \left(\int_0^1 (D_B^{1\beta}(B_1, B_2) + D_B^{2\beta}(B_1, B_2)) d\alpha \right),$$

where

$$D_B^{1\alpha}(B_1, B_2) = \left| \frac{b_{1\alpha}^L + b_{1\alpha}^R}{2} - \frac{b_{2\alpha}^L + b_{2\alpha}^R}{2} \right|,$$

$$D_B^{2\alpha}(B_1, B_2) = \left| (b_{1\alpha}^R - b_{1\alpha}^L) - (b_{2\alpha}^R - b_{2\alpha}^L) \right|.$$

Now we have to find distance between underlying probability distributions. Let p_{Z_1} and p_{Z_2} be the underlying probability distributions of $Z_1 = (A_1, B_1, G_1)$ and $Z_2 = (A_2, B_2, G_2)$ respectively. The distance between p_{Z_1} and p_{Z_2} can be expressed as

$$D_p(p_{Z_1}, p_{Z_2}) = \inf_{p_{Z_1} \in G_1, p_{Z_2} \in G_2} \left\{ \left(1 - \int_R (p_{Z_1} p_{Z_2})^{\frac{1}{2}} dx \right)^{\frac{1}{2}} \right\}$$

Here we can use Hellinger and other different types of probabilistic distances (total variation distance, Bhattacharyya distance, Mahalanovis distance, Jeffreys distance etc.). Given $D_A(A_1, A_2)$ and $D_B^{total}(B_1, B_2)$, the proposed comprehensive distance for Z-sets is defined as

$$D(Z_1, Z_2) = \beta D_A(A_1, A_2) + (1 - \beta) D_B^{total}(B_1, B_2)$$

where $D_B^{total}(B_1, B_2)$ is a distance for reliability restriction computed as

$$D_B^{total}(B_1, B_2) = w D(B_1, B_2) + (1 - w) D_p(p_{Z_1}, p_{Z_2}).$$

$\beta, w \in [0, 1]$ are decision maker's preference degrees.

5   A Z-number-Valued Function

5.1   Basic Definitions

Z-number-valued functions provide a basis for handling bimodal information. Taking into account complexity of dealing with such information, we adopt conceptual framework of discrete functions theory suggested in [19, 20]. One of the key concept of this theory is a concept of approximate limit. This concept is adequate one for study of Z-valued functions in the sense that qualitative information supported by Z-numbers can be described by linguistic approximation.

Let us denote by $\mathcal{Z}^n$ the space of elements which are n-vectors of discrete Z-numbers

$$\mathbf{Z} = (Z_1, Z_2, \ldots, Z_n) = ((A_1, B_1), (A_2, B_2), \ldots, (A_n, B_n)).$$

Denote $\mathcal{Z}_{[c,d]} = \left\{ (A, B) \middle| A \in \mathcal{D}_{[c,d]}, B \in \mathcal{D}_{[0,1]} \right\}$, $[c, d] \subset \mathcal{R}$, and $\mathcal{Z}_+ = \{(A, B) \in \mathcal{Z} \middle| A \in \mathcal{D}_{[0,\infty)}\}$, $\mathcal{Z}_- = \mathcal{Z} \backslash \mathcal{Z}_+$. Let us provide an introductory definition of a Z-number valued function.

Definition 5.1. A discrete Z-number valued function of discrete Z-numbers [5, 8].
A discrete Z-number valued function of discrete Z-numbers is a mapping $f : \mathcal{Z} \to \mathcal{Z}$.

Analogously, in a continuous setting, the Z-function is defined as a mapping between spaces of continuous Z-numbers.

Consider a case $f : \Omega \to \mathcal{A}$, where Ω is a universe of discourse. Consider a sequence of Z-numbers $\mathbf{Z} = \{Z_i | Z_i \in \mathcal{A}, i = 1, 2, \ldots, n, \ldots\}$. Definitions of limit and continuity of Z-valued functions introduced in [5, 8] are given below.

Definition 5.2. An r-limit of discrete Z-number valued function of discrete Z-numbers [5, 8]. An element $Z_d \in f(\mathbf{Z})$ is called an r-limit of f at a point $Z_{a,i} \in \mathbf{Z}$ and denoted $Z_d = r - \lim_{Z_x \to Z_a} f(Z_x)$ if for any sequence $\mathbf{Z}$ satisfying the condition $Z_a = \lim \mathbf{Z}$, the equality $Z_d = r - \lim f(\mathbf{Z})$ is valid.

Definition 5.3. (q, r)-continuous Z-function [5, 8]. A function $f : \mathcal{A} \to \mathcal{A}$ is called (q, r)-*continuous* at a point $Z_a \in \mathcal{A}$ if for any $\varepsilon > 0$ there is $\delta > 0$ such that the inequality $D(Z_a, Z_x) < q + \delta$ implies the inequality $D(f(Z_a), f(Z_x)) < r + \varepsilon$, or in other words, for any Z_x with $D(Z_a, Z_x) < q + \delta$, we have $D(f(Z_a), f(Z_x)) < r + \varepsilon$.

The distance D is considered in sense of [8].

For the spaces of continuous Z-numbers the continuity of Z-valued function is defined in a traditional sense, that is, when $q = 0$ and $r = 0$.

5.2   A Derivative and a Definite Integral of a Z-number-Valued Function

The study of differential and integral calculus for Z-number-valued functions includes only few works [5, 13, 14]. As it is mentioned [14], "the first attempt towards uncertain differential equations based on Z-numbers was carried out in 2015 [5] where Hukuhara-type derivative was defined for a function of Z-numbers". In [7] conditions on existence of Hukuhara difference of Z-numbers $Z_1 -_h Z_2 = Z_{12}$ were further analyzed. [14] is devoted to Z-number-valued function calculus, where Z^+-numbers are considered instead of Z-numbers. The authors obtained strong results on differentiability and integrability of Z-number-valued function, Laplace transform of Z-number-valued function, existence and uniqueness of solution of Z-number-valued differential equation.

In this section we propose first results on differentiation and integration of Z-number-valued functions, namely, mappings to a space of Z-numbers. The results rely on the concepts of Hukuhara difference of Z-numbers and approximate limit.

Definition 5.4. Hukuhara difference of discrete Z-numbers [7]. For discrete Z-numbers $Z_1 = (A_1, B_1)$ and $Z_2 = (A_2, B_2)$ their Hukuhara difference denoted $Z_{12} = Z_1 -_h Z_2$ is the discrete Z-number $Z_{12} = (A_{12}, B_{12})$ such that $Z_1 = Z_2 + Z_{12}$.

The conditions on existence of Hukuhara difference $Z_{12} = Z_1 -_h Z_2$ are described in details in [7].

On the basis of the concept of Hukuhara difference of Z-numbers, the following definitions of derivative of Z-number-valued function were proposed in [5].

Definition 5.5. Discrete derivative of a discrete Z-number-valued function at a crisp point [5]. $f : \Omega \to \mathcal{Z}$ may have the following discrete derivatives $\Delta f(n)$ at n:

the forward right-hand discrete derivative $\Delta_r f(n) = f(n+1) -_h f(n)$.

the backward right-hand discrete derivative $\Delta_{r-} f(n) = \frac{f(n) -_h f(n+1)}{-1}$.

the forward left-hand discrete derivative $\Delta_l f(n) = f(n) -_h f(n-1)$.

the backward left-hand discrete derivative $\Delta_{l-} f(n) = \frac{f(n-1) -_h f(n)}{-1}$.

An existence of $\Delta_r f(n)$ or $\Delta_l f(n)$ implies that the length of support of the first component A_n of a Z-number $f(n) = (A_n, B_n)$ is non-decreasing. Existence of the other two derivatives implies non-increasing support of the first component of a Z-number $f(n) = (A_n, B_n)$.

Based on definition 6, the concept of a generalized Hukuhara differentiability is defined in [5] as follows.

Definition 5.6. Generalized Hukuhara differentiability [5]. Let $f : \Omega \to \mathcal{Z}$ and $n \in \Omega$. We say that f is differentiable at $n \in \Omega$ if

$$\Delta_r f(n) \text{ and } \Delta_l f(n) \text{ exist}$$

or

$$\Delta_{r-} f(n) \text{ and } \Delta_{l-} f(n) \text{ exist}$$

or

$$\Delta_r f(n) \text{ and } \Delta_{l-} f(n) \text{ exist}$$

or

$$\Delta_{r-} f(n) \text{ and } \Delta_l f(n) \text{ exist}$$

***Definition 5.7.* The definite integral** [21, 22]. The definite integral I of Z-number-valued function $f : [a, b] \to Z$ is defined as an r-limit:

$$F(x) = \int_a^b f(t)dt = \sum_{i=1}^n f(x_i)\Delta x,$$

$$d(F(x), I) \le r.$$

We have derived the following results on relation between differentiability and integrability.

Theorem 5.1 [21]. Let $f : [a, b] \to Z$ be a Z-valued function. Then the following is satisfied in terms of r-limit:

A Z-valued function $F(x_k) = \int_a^x f(t)dt = \sum_{i=1}^k f(x_i)$ is differentiable function in terms of definition and $F'(x_k) = f(x)$

A Z-valued function $G(x) = \int_x^b f(t)dt = \sum_{i=k}^n f(x_i)$ is differentiable function in terms of definition and $G'(x) = -f(x)$

For the proof, see [19].

Theorem 5.2 [22]. If a Z-number-valued function $f : [a, b] \to Z$ is differentiable on interval [a, b] then the following is satisfied in terms of r-limit:

$$\int_a^b f'(x)dx = \sum_{i=1}^{n-1} f'(x_i)\Delta x_i = f(b) -_h f(a).$$

For the proof, see [22].

The obtained results describe relation between approximate evaluations of derivative and integral of Z-number-valued function.

6   Z-valued System of Linear Equations

Statement of Problem. Consider the following system of fully Z-valued linear equations:

$$Z_{a11}Z_{x1} + \ldots + Z_{a1n}Z_{xn} = Z_{c1}$$

$$.$$
$$.$$
$$. \tag{6.1}$$

$$Z_{an1}Z_{x1} + \ldots + Z_{ann}Z_{xn} = Z_{cn}$$

$Z_{aij} = (A_{aij}, B_{aij}), Z_{xj} = (A_{xj}, B_{xj}), Z_{ci} = (A_{ci}, B_{ci}), i, j = 1, \ldots, n$ are Z-numbers.

The problem is to find a vector of variables $Z_{xj} = (A_{xj}, B_{xj}), j = 1, \ldots, n$ that satisfies the above given system of equations.

In order to solve fully Z-valued system of linear equations, we have to consider fuzzy system of linear equations, and a family of systems of linear equations with random variables X_j and random parameters X_{aij}, X_{bi}. This family is induced by sets of distributions related to each of Z-numbers $Z_{aij}, Z_{xi}, Z_{bi}, i, j = 1, \ldots, n$.

The fuzzy system of linear equations:

$$A_{a11}A_{x1} + \ldots + A_{a1n}A_{xn} = A_{b1}$$

$$\vdots$$

$$A_{an1}A_{x1} + \ldots + A_{ann}A_{xn} = A_{bn}$$

A representative of the family of probabilistic systems of linear equations:

$$p_{a11}{}^{o*}p_{x1}{}^{o+} \ldots {}^{o+}p_{a1n}{}^{o*}p_{xn} = p_{b1}$$

$$\vdots$$

$$p_{an1}{}^{o*}p_{x1}{}^{o+} \ldots {}^{o+}p_{ann}{}^{o*}p_{xn} = p_{bn}$$

This can be rewritten as $p_{a11\times\ldots\times a1nn}{}^{o}p_{x1\times\ldots\times xn} = p_{b1\times\ldots\times bn}$ where $p_{x1\times\ldots\times xn}$ and $p_{b1\times\ldots\times bn}$ are n-dimensional probability distributions.

The problem is stated as follows.

Find such vector of variables $(Z_x) = (Z_{x1}, \ldots, Z_{xn})$ that the distance between Z-number valued vector $(Z_a)(Z_x) = \left(\sum_{j=1}^{n} Z_{a_{1j}} Z_{xj}, \ldots, \sum_{j=1}^{n} Z_{a_{1j}} Z_{xj} \right)$ (the left hand side of (6.1)) and Z-number valued vector $(Z_c) = (Z_{c1}, \ldots, Z_{cn})$ (the right hand side of (6.1)) is minimized:

$$d((Z_a)(Z_x), (Z_c)) \to min \tag{6.2}$$

s.t.

$$Z_x \in Z \tag{6.3}$$

Z is a search space which is specifically defined for a problem at hand.

We propose to use evolutionary programming approach, particularly, DEO-based approach, to solve problem (6.2)–(6.3).

7 Eigenvalues and Eigenvectors of Matrices of Z-numbers

7.1 Basic Definitions

Eigenvalues and eigenvectors are widely used in various practical applications in decision making, planning, control and other fields [23]. Particularly, these concepts underlie analysis of consistency of a decision maker's (DM) knowledge. In real-world problems, DM's knowledge is naturally characterized by imprecision and partial reliability. This involves combination of fuzzy and probabilistic information. The concept of a Z-number is a formal construct to describe such kind of information. In this section we initiate study of Z-number valued eigenvalue and eigenvector of matrices, components of which are Z-numbers.

Definition 7.1. A Z-valued matrix. A Z-valued matrix (Z_{ij}) is a matrix of Z-numbers that describe partially reliable information on values of random variables $X_{ij}, i, j = 1, \ldots, n$:

$$\left(Z_{ij} = (A_{ij}, B_{ij}) \right) = \begin{pmatrix} Z_{11} = (A_{11}, B_{11}) & \ldots & Z_{1n} = (A_{1n}, B_{1n}) \\ \cdot & \ldots & \cdot \\ Z_{n1} = (A_{n1}, B_{n1}) & \ldots & Z_{nn} = (A_{nn}, B_{nn}) \end{pmatrix}$$

Let us formulate definitions of eigenvalue and eigenvector of (Z_{ij}).

Definition 7.2. A Z-eigenvalue of Z-valued matrix. A Z-valued eigenvalue of Z-valued square matrix (Z_{ij}) is such a Z-number $Z_\lambda = (A_\lambda, B_\lambda)$ that the following holds:

$$det\left(Z_{ij} - Z_\lambda I \right) = det \begin{pmatrix} Z_{11} - Z_\lambda & \ldots & Z_{1n} \\ \cdot & \ldots & \cdot \\ Z_{n1} & \ldots & Z_{nn} - Z_\lambda \end{pmatrix} = Z_0 \qquad (7.1)$$

where I is a traditional (non-fuzzy) identity matrix, Z_0 is a Z-singleton $Z_0 = (\hat{0}, \hat{1})$.

Thus, Z-valued eigenvalue $Z_\lambda = (A_\lambda, B_\lambda)$ is a root of n-th order characteristic equation

$$Z^n + Z_1 Z^{n-1} + \ldots + Z_n = Z_0,$$

where $Z_r, r = 1, \ldots, n$ are coefficients induced by the elements of Z-valued matrix (Z_{ij}). So, n Z-valued eigenvalues $Z_{\lambda s} = (A_{\lambda s}, B_{\lambda s})$ are to be found for a Z-valued square matrix (Z_{ij}). In particular case, Z^+-number $Z_{\lambda s}^+ = (A_{\lambda s}, p_{\lambda s})$ can be used instead of $Z_{\lambda s} = (A_{\lambda s}, B_{\lambda s})$.

Definition 7.3. A Z-eigenvector of Z-valued matrix. A vector of Z-numbers $(Z_Y) = (Z_{Y1} = (A_{Y1}, B_{Y1}), \ldots, Z_{Yn} = (A_{Yn}, B_{Yn}))$ is referred to as a Z-valued eigenvector of Z-valued square matrix (Z_{ij}) if it satisfies the following Z-valued linear system of equations:

$$\left(Z_{ij}\right)(Z_Y) = Z_\lambda(Z_Y) \tag{7.2}$$

where $Z_\lambda = (A_\lambda, B_\lambda)$ is Z-valued eigenvalue.

Thus, n Z-valued eigenvectors $(Z_{Ys}), s = 1, \ldots, n$ are to be found, one for each Z-valued eigenvalue $Z_{\lambda s} = (A_{\lambda s}, B_{\lambda s})$.

7.2 A General Formulation for Problems of Computation of Z-valued Eigenvalues and Eigenvectors

The generic problem of computation of Z-valued eigenvalues is as follows:

Find $Z_{\lambda s} = (A_{\lambda s}, B_{\lambda s}), s = 1, \ldots, n$ such that

$$det \begin{pmatrix} Z_{11} - Z_\lambda & \cdots & Z_{1n} \\ \cdot & \cdots & \cdot \\ Z_{n1} & \cdots & Z_{nn} - Z_\lambda \end{pmatrix} = Z_0$$

The problem of computation of $Z_{\lambda s}^+ = (A_{\lambda s}, p_{\lambda s})$ can be considered as follows.

Find $Z_{\lambda s}^+ = (A_{\lambda s}, p_{\lambda s})$ such that

$$det \begin{pmatrix} Z_{11}^+ - Z_{\lambda s}^+ & \cdots & Z_{1n}^+ \\ \cdot & \cdots & \cdot \\ Z_{n1}^+ & \cdots & Z_{nn}^+ - Z_{\lambda s}^+ \end{pmatrix} = Z_0^+ \tag{7.3}$$

This problem can be described as the problems of computation of $A_{\lambda s}$ and $p_{\lambda s}$.

Given fuzzy numbers A_{ij} find $A_{\lambda s}$ such that

$$det \begin{pmatrix} A_{11} - A_{\lambda s} & \cdots & A_{1n} \\ \cdot & \cdots & \cdot \\ A_{n1} & \cdots & A_{nn} - A_{\lambda s} \end{pmatrix} = \hat{0} \tag{7.4}$$

where $\hat{0}$ is a fuzzy singleton.

Given probability distributions p_{ij} of random variables X_{ij} and the constraint

$$det \begin{pmatrix} X_{11} - X_{\lambda s} & \cdots & X_{1n} \\ \cdot & \cdots & \cdot \\ X_{n1} & \cdots & X_{nn} - X_{\lambda s} \end{pmatrix} = 0, \tag{7.5}$$

find probability distribution $p_{\lambda s}$ of a desired random variable $X_{\lambda s}$. Symbolically, we can describe this problem as

Find $p_{\lambda s}$ such that

$$det \begin{pmatrix} p_{11} \ominus p_{\lambda s} & \cdots & p_{1n} \\ \cdot & \cdots & \cdot \\ p_{n1} & \cdots & p_{nn} \ominus p_{\lambda s} \end{pmatrix} = 0.$$

Next, we have to find Z-valued eigenvectors (Z_{Ys}) as solutions of a Z-valued linear system (7.2). (Z_{Ys}) can be found through computation of $(Z_{Ys}^+) = ((A_{Ys}, p_{Ys}))$ such that A_{Ys} and p_{Ys} are solutions of two problems associated with system (7.2).

Given $A_{\lambda s}$ find A_{Ys} that satisfies the following linear system:

$$\begin{pmatrix} A_{11} & \cdots & A_{1n} \\ \cdot & \cdots & \cdot \\ A_{n1} & \cdots & A_{nn} \end{pmatrix} \begin{pmatrix} A_{Ys1} \\ \cdots \\ A_{Ysn} \end{pmatrix} = A_{\lambda s} \begin{pmatrix} A_{Ys1} \\ \cdots \\ A_{Ysn} \end{pmatrix}. \tag{7.6}$$

Given random variable $X_{\lambda s}$ compute the random vector $(Y_s), s = 1, \ldots, n$ by solving the following linear system:

$$\begin{pmatrix} X_{11} & \cdots & X_{1n} \\ \cdot & \cdots & \cdot \\ X_{n1} & \cdots & X_{nn} \end{pmatrix} \begin{pmatrix} Y_{s1} \\ \cdots \\ Y_{sn} \end{pmatrix} = X_{\lambda s} \begin{pmatrix} Y_{s1} \\ \cdots \\ Y_{sn} \end{pmatrix} \tag{7.7}$$

Symbolically, the last problem can be formulated as given below. Find the vector of probability distributions $(p_{Ys}), s = 1, \ldots, n$ that satisfies the following system:

$$\begin{pmatrix} p_{11} & \cdots & p_{1n} \\ \cdot & \cdots & \cdot \\ p_{n1} & \cdots & p_{nn} \end{pmatrix} \begin{pmatrix} p_{Ys1} \\ \cdots \\ p_{Ysn} \end{pmatrix} = p_{\lambda s} \begin{pmatrix} p_{Ys1} \\ \cdots \\ p_{Ysn} \end{pmatrix}.$$

7.3   An Example

Consider the following 3×3 Z-valued matrix, where components of Z-numbers are triangular fuzzy numbers (TFNs):

$$(Z_{ij}) =$$

$$\begin{pmatrix} ((0.93, 0.95, 1), (0.88, 0.94, 1)) & ((2, 2.5, 3), (0.56, 0.78, 1)) & ((1.5, 2, 2.25), (0.7, 0.8, 0.9)) \\ ((0.33, 0.4, 0.5), (0.67, 0.78, 0.97)) & ((0.93, 0.95, 1), (0.88, 0.94, 0.97)) & ((1, 1.5, 2), (0.65, 0.75, 0.85)) \\ ((0.4, 0.5, 0.67), (0.65, 0.75, 0.85)) & ((0.5, 0.67, 1), (0.7, 0.72, 0.75)) & ((0.93, 0.95, 1), (0.88, 0.94, 0.97)) \end{pmatrix}$$

The obtained Z-eigenvalues are as follows:

$$Z_{\lambda 1} = (A_{\lambda 1}, p_{\lambda 1}) = ((2.5, 3, 3.7), (3.08, 0.04)),$$

$$Z_{\lambda 2} = (A_{\lambda 2}, p_{\lambda 2}) = ((-0.37, -0.01, 0.02), (-0.008 + 0.37i, 0.18)),$$

$$Z_{\lambda 3} = (A_{\lambda 3}, p_{\lambda 3}) = ((-0.37, -0.01, 0.02), (-0.008 - 0.37i, 0.18)).$$

Thus, the obtained Z-eigenvectors $(Z_{Ys}) = ((A_{Ys1}, p_{Ys1}), \ldots, (A_{Ys3}, p_{Ys3}))$ are

$$(Z_{Y1}) = \begin{pmatrix} ((0.84, 0.844, 0.844), (0.7, 0.5)) \\ ((0.41, 0.416, 0.4253), (0.3, 0.3)) \\ ((0.345, 0.349, 0.36), (0.28, 0.2)) \end{pmatrix},$$

$$(Z_{Y2}) = \begin{pmatrix} ((0.83, 0.833, 0.853), (0.51, 0.6)) \\ ((-0.27, -0.22, -0.17), (-0.13 + 0.23i, 0.32)) \\ ((-0.22, -0.16, -0.1), (-0.11 - 0.19i, 0.26)) \end{pmatrix},$$

$$(Z_{Y3}) = \begin{pmatrix} ((0.83, 0.833, 0.853), (0.51, 0.6)) \\ ((-0.27, -0.22, -0.17), (-0.13 - 0.23i, 0.3)) \\ ((-0.22, -0.16, -0.1), (-0.11 + 0.19i, 0.3)) \end{pmatrix}.$$

8   Toward Z-differential Equations

8.1   Basic Definitions

In [14] they proposed a study of Z-differential equations for the case of Z^+-numbers. In this section we are going to consider Z-differential equations in Z-numbers setting. We consider Z-initial value problem and an approach for determining an approximate solution.

Below we provide some necessary definitions and theorems, which will be used in this section. The set of all real numbers is denoted by $\mathcal{R}$, the set of all fuzzy numbers on $\mathcal{R}$ by $\mathcal{E}^1$. By $[A]^\mu$ we show the well-known μ-cut of a fuzzy set $\tilde{A}$ whose left and right end-points are indicated by $\underline{x}^\mu$ and $\bar{x}^\mu$, respectively. In this section, μ-cut is equivalent to α-cut in traditional notations of the fuzzy set theory.

Definition 8.1. **A horizontal membership function of a fuzzy number** [24]. Let $\tilde{A} : [a, b] \subseteq \mathcal{R} \to \mathcal{E}^1$. The horizontal membership function of $\tilde{A}$ is a representation of $\mu_{\tilde{A}}(x) = \tilde{A}(x)$ as $\mathcal{H}(\tilde{A}) : [0, 1] \times [0, 1] \to [\underline{x}, \bar{x}]$, $^{gr}A = {}^{gr}A(\mu, \alpha_x)$ in which "gr" stands for the granule of information included in $x \in [a, b]$, μ is the membership degree of x in $\tilde{A}$, $\alpha_x \in [0, 1]$ is called relative-distance-measure variable, and $x(\mu, \alpha_x) = \underline{x}^\mu + \alpha_x(\bar{x}^\mu - \underline{x}^\mu)$, $\alpha_x, \mu \in [0, 1]$.

In this section, α_x is not α-cut of a fuzzy number, but denotes relative position of a point within μ-cut.

Definition 8.2. The (μ, v)-cut of the Z-number $Z = (\tilde{A}, \tilde{B})$ is defined as $Z^{(\mu, v)} = ([\tilde{A}]^\mu, [\tilde{B}]^v) = ([\underline{A}^\mu, \bar{A}^\mu], [\underline{B}^v, \bar{B}^v])$, where $\mu, v \in [0, 1]$.

Moreover, using $\mathcal{H}^{-1}(A^{gr}(\mu, \alpha_x)) = [\tilde{A}]^\mu = [\inf_{\lambda \geq \mu} \min_{\alpha_x} A^{gr}(\mu, \alpha_x), \sup_{\lambda \geq \mu} \max_{\alpha_x} A^{gr}(\mu, \alpha_x)]$,

$\mathcal{H}^{-1}(B^{gr}(v, \beta_x)) = [\tilde{B}]^v = [\inf_{\eta \geq v} \min_{\beta_x} B^{gr}(v, \beta_b), \sup_{\eta \geq \mu} \max_{\beta_b} B^{gr}(v, \beta_b)]$ the (μ, v)-cut of

the vertical membership function of $(\tilde{A}, \tilde{B})$, which is in fact the span of the information granule, can be obtained.

8.2   Z Initial Value Problem and Approximate Solution

Let $\mathcal{Z}(\mathcal{R})$ be the set of all Z-numbers defined in $\Re$. Let's consider a Z-initial value problem of Z-differential equation as:

$$z' = f(t, z) \tag{8.1}$$

$$z(t_0) = z_0 \tag{8.2}$$

where $z : [t_0, t_f] \to \mathcal{Z}, f : [t_0, t_f] \times \mathcal{Z} \to \mathcal{Z}$, and $z_0 \in \mathcal{Z}$ is an initial condition.

Assume that a unique solution $z(t)$ of (8.1)–(8.2) exists. Then we can write

$$^{gr}z'(t) = {}^{gr}f(t, {}^{gr}z(t)),$$

$$^{gr}z(t_0) = {}^{gr}z_0.$$

That is,

$$z'({}^{gr}A, {}^{gr}B)(t) = {}^{gr}f(t, z({}^{gr}A, {}^{gr}B)(t)),$$

$$^{gr}z(t_0) = {}^{gr}z_0.$$

Given that ${}^{gr}A = {}^{gr}A(\mu, \alpha_x)$ and ${}^{gr}B = {}^{gr}B(v, \beta_b)$ we have

$$^{gr}z'(t, \mu, \alpha_A, v, \beta_B) = {}^{gr}f(t, {}^{gr}z(t, \mu, \alpha_A, v, \beta_B)),$$

$$^{gr}z(t_0, \mu, \alpha_A, v, \beta_B) = {}^{gr}z_0(\mu, \alpha_A, v, \beta_B).$$

Thus, we described HMF-based Z-initial value problem. Let us consider an extension of Euler method for fuzzy setting to a Z-valued problem. Symbolically, this method can be described as

$$z(t_{i+1}) = z(t_i) + h \cdot f(t_i, z(t_i)),$$

$$i = 0, 1, 2, \ldots, N - 1.$$

By using HMFs setting, we have

$$^{gr}z(t_{i+1}, \mu, \alpha_A, v, \beta_B) = {}^{gr}z(t_i, \mu, \alpha_A, v, \beta_B) + h \cdot {}^{gr}f(t_i, {}^{gr}z(t_i, \mu, \alpha_A, v, \beta_B)), \tag{8.3}$$

$$i = 0, 1, 2, \ldots, N - 1.$$

The algebraic operation on the right-hand side of (8.3) can be implemented by using the HMF-based approach to computation of Z-numbers [25]. Then, consecutive computation of (8.3) for $i = 0, 1, 2, \ldots, N - 1$ would provide the approximate solution to (8.1)–(8.2).

9 Conclusion

In this paper we describe a general framework of studies on Z-calculus. Basics of Z-sets and Z-valued functions, introductory concepts of Z-valued abstract algebra, some insights to Z-valued linear algebra and Z-differential equations are considered. Our future works will be devoted to further developments. In the realm of set-theoretic studies, axiomatic system of Z-sets, norm, distances and metrics of Z-sets, properties of Z-numbers, Z-geometry and Z-logics are to be considered, comparison of Z-sets with the other extensions of fuzzy sets are also of interest. This, in turn, may result in development of Z-approximate reasoning and Z-number based approach to computational linguistics. In the realm of analysis of Z-valued functions, Z-valued algebraic and differential equations and inequalities, and Z-preferences theory are interesting directions. This may underlie development of Z-decision tools as Z-AHP, Z-TOPSIS and other methods.

References

1. Zadeh, L.A.: A note on Z-numbers. Inf. Sci. **181**, 2923–2932 (2011)
2. Aliev, R.A., Alizadeh, A.V., Huseynov, O.H., Jabbarova, K.I.: Z-number based linear programming. Int. J. Intell. Syst. **30**, 563–589 (2015)
3. Aliev, R.A., Alizadeh, A.V., Huseynov, O.H.: The arithmetic of continuous Z-numbers. Inf. Sci. **373**, 441–460 (2016)
4. Aliev, R.A., Alizadeh, A.V., Huseynov, O.H.: The arithmetic of discrete Z-numbers. Inf. Sci. **290**, 134–155 (2015)
5. Aliev, R.A., Huseynov, O.H., Aliyev, R.R., Alizadeh, A.V.: The Arithmetic of Z-Numbers: Theory and Applications. World Scientific, Singapore (2015)
6. Aliev, R.A., Huseynov, O.H.: Decision Theory with Imperfect Information. World Scientific, Singapore (2014)
7. Aliev, R.A., Perdycz, W., Huseynov, O.H.: Hukuhara difference of Z-numbers. Inf. Sci. **466**, 13–24 (2018)
8. Aliev, R.A., Perdycz, W., Huseynov, O.H.: Functions defined on a set of Z-numbers. Inf. Sci. **423**, 353–375 (2018)
9. Aliev, R.A.: Uncertain Computation Based on Decision Theory. World Scientific Publishing, Singapore (2017)
10. Lorkowski, J., Aliev, R., Kreinovich, V.: Towards decision making under interval, set-valued, fuzzy, and Z-number uncertainty: a fair price approach. In: Proceedings of the IEEE International Conference on Fuzzy Systems, FUZZ-IEEE 2014, pp. 2244–2253, IEEE (2014)
11. Aliev, R.A., Kreinovich, V.: Z-numbers and type-2 fuzzy sets: a representation result. Intell. Autom. Soft Comput. (2017). https://doi.org/10.1080/10798587.2017.1330310
12. Yager, R.: On Z-valuations using Zadeh's Z-numbers. Int. J. Intell. Syst. **27**, 259–278 (2012)
13. Jafari, R., Razvarz, S., Gegov, A.: Solving differential equations with Z-numbers by utilizing fuzzy sumudu transform. In: Arai, K., Kapoor, S., Bhatia, R. (eds.) Intelligent Systems and Applications, IntelliSys 2018. Advances in Intelligent Systems and Computing, vol. 869. Springer, Cham (2019)
14. Mazandarani, M., Zhao, Y.: Z-differential equations. IEEE Trans. Fuzzy Syst. https://doi.org/10.1109/tfuzz.2019.2908131

15. Shen, K., Wang, J.: Z-VIKOR method based on a new comprehensive weighted distance measure of Z-number and its application. IEEE Trans. Fuzzy Syst. **26**(6), 3232–3245 (2018)
16. Jiang, W., Cao, Y., Deng, X.: A novel Z-network model based on bayesian network and Z-number. IEEE Trans. Fuzzy Syst. (2019). https://doi.org/10.1109/tfuzz.2019.2918999
17. Kang, B., Deng, Y., Hewage, K., Sadiq, R.: A method of measuring uncertainty for Z-number. IEEE Trans. Fuzzy Syst. **27**(4), 731–738 (2019)
18. Renyi, A.: Foundations of Probability. Dover Publications, San Francisco (2007)
19. Burgin, M.: Neoclassical analysis: fuzzy continuity and convergence. Fuzzy Set. Syst. **75**(3), 291–299 (1995)
20. Burgin, M.: Theory of fuzzy limits. Fuzzy Set. Syst. **115**, 433–443 (2000)
21. Huseynov, O.H.: Toward a derivative of a Z-number-valued function. In: Advances in Intelligent Systems and Computing (2019, submitted)
22. Huseynov, O.H.: Toward a derivative of a Z-number-valued function. In: Advances in Intelligent Systems and Computing (2019, submitted)
23. Edelman, A., Rao, N.R.: Random matrix theory. Acta Numerica **14**, 233–297 (2005)
24. Piegat, A., Landowski, M.: Horizontal membership function and examples of its applications. Int. J. Fuzzy Syst. **17**, 22–30 (2015)
25. Aliev, R.A., Alizadeh, A.V., Huseynov, O.H.: An introduction to the arithmetic of Z-numbers by using horizontal membership functions. Procedia Comput. Sci. **120**, 349–356 (2017)

Artificial Intelligence-Based New Material Design

M. B. Babanli[(✉)]

Azerbaijan State University of Oil and Industry,
Azadlig Avenue, 20, Baku AZ1010, Azerbaijan
mustafababanli@yahoo.com

Abstract. For a long time, design of new materials was implemented on the basis of costly and time-consuming experimental approach. Nowadays machine learning, data mining and Big Data techniques allow to develop a new approach, a so-called data-driven design of new materials. In this paper, we consider an approach to alloy discovery that is based on a synergy of deep learning and fuzzy logic methods. The approach allows to design materials with predefined characteristics by computer-aided generation of the underlying crystal structures and optimization of their parameters. An example is provided to illustrate the approach.

Keywords: Crystal structure · Material properties · Fuzzy logic · Deep learning · Big data

1 Introduction

In our previous works, we overviewed existing works devoted to material design, developed fuzzy logic-based approaches to selection of materials on the basis of vector of properties, and solved problems of material design with required characteristics by searching for an optimal composition and design temperatures [1–7]. A more important aspect is synthesis of new materials at the level of crystal structures and optimization of lattice parameters. In this realm, a series of works is devoted to application of machine learning (ML) [8], including deep learning (DL), and fuzzy logic. Let us overview some of these works. In [9] data-driven material design in additive manufacturing is considered. The authors propose using self-consistent clustering analysis for modeling process-structure-property relationships. Such factors influencing material properties as voids, phase composition and others are taken into account. The proposed approach underlie the cycle of design, properties prediction and material optimization. An approach to modeling relationship between microstructure and properties based on machine learning (ML) is proposed in [10]. The approach relies on Gaussian process regression method that searches for an optimal variant within a large database of synthetic microstructures. A new method for assessing similarity of structures by using interatomic distances is outlined in [11]. Relying on hierarchical clustering theory, the authors formulated a definition of crystal structure prototype. Crystal structure prototype database is constructed and its usefulness for material discovery problems is demonstrated through examples. Application of data-mining, ML and statistical

© Springer Nature Switzerland AG 2020
R. A. Aliev et al. (Eds.): ICSCCW 2019, AISC 1095, pp. 24–32, 2020.
https://doi.org/10.1007/978-3-030-35249-3_2

techniques to crystal structure prediction, compounds exploration and related problems can also be found in [12–19]. A neural network based system that can gather information in scientific papers to search for material recipes were proposed in [20]. In [21] the authors of [20] apply variational autoencoder neural network-based DL approach to inorganic materials synthesis. The proposed DL-based system is used to model relation between material recipes and crystal structures. A new ML-based approach to materials prediction is proposed in [22]. The approach deals with interatomic potential and uses an active learning algorithm for selection an optimal training dataset. It is illustrated that the proposed method outperforms some of the existing techniques in terms of generality and computational efficiency. In [23] prediction model based on fuzzy NN is used to model relationship between processing parameters and mechanical properties of a titanium alloy. As input variables, deformation temperature, degree of deformation, solution and aging temperatures are used. Output variables are the ultimate tensile strength, yield strength, elongation and area reduction. Testing results show that the model outperforms classical regression method. The model can be effectively used for optimization of processing parameters to achieve desired properties.

To my knowledge, existing works on application of ML and data mining techniques to prediction of material properties, relation between material properties and crystal structures parameters, optimization of parameters, and other problems are scarce. In this paper we consider a systematic approach to synthesis of new alloys on the basis of crystal parameters prediction and their relationship with desired properties. The main idea of the approach relies on the use of materials Big data, data mining and DL techniques to process these data. Big data is considered as repository of hidden knowledge for material discovery. Data mining and DL techniques are applied to extraction and use of this knowledge for designing new materials, properties prediction and crystal structure optimization.

The paper is organized as follows. Section 2 includes some preliminary material used in the sequel. In Sect. 3 we outline our approach to computational synthesis of materials and property prediction. Section 4 is devoted to illustration of application of the approach to modeling relation between lattice parameters and mechanical properties of TiNi alloy. Section 5 concludes.

2 Preliminaries

Definition 1. Fuzzy sets [24, 25]. Let U be a classical set of objects, called the universe, whose generic elements are denoted u Membership in a classical subset A of U is often viewed as a characteristic function μ_A from U to $\{0, 1\}$ such that

$$\mu_A(u) = \begin{cases} 1 & \textit{iff } u \in A \\ 0 & \textit{iff } u \notin A \end{cases}$$

where $\{0, 1\}$ is called a valuation set; 1 indicates membership while 0 - non-membership.

A is a fuzzy subset if $\mu_A : U \rightarrow [0, 1]$ is a function with values in $[0, 1]$. $\mu_A(u)$ is the degree of membership of u in A. As closer the value of $\mu_A(u)$ is to 1, so much u belongs to A.

Definition 2. A fuzzy number [24]. A continuous fuzzy number is a fuzzy subset A of the real line $\mathcal{R}$ with membership function $\mu_A : \mathcal{R} \rightarrow [0, 1]$, which possesses the following properties: (a) A is a normal fuzzy set; (b) A is a convex fuzzy set; (c) α-cut A^α is a closed interval for every $\alpha \in (0, 1]$; (d) the support of A, supp(A) is bounded.

Deep Learning [26]. Deep learning is a class of ML methods that relies on the use of artificial NNs (ANN) with multiple layers. The use of such ANN allows to extract higher level features from raw input. In material science perspective, the raw input covers lattice parameters, composition, transition temperatures, material properties data. Higher level features are reliable structure-property linkage descriptions, new generated crystal structures etc.

3   A General Outline of the Proposed Approach

In this section we outline the proposed approach to design of new materials on the basis of DL and data mining techniques. The conceptual scheme describing basic stages of this approach is shown in Fig. 1.

At *the first stage*, we have to process Big data of crystal structures that hold information about the basic properties of already known materials. *The second stage* covers choice of various material recipients for future synthesis of alloys with required properties. This will be used to study material-receptor complexes in atomic detail when predicting properties of a required alloy. A recipe is described as a vector components of which store information about features (referred to as a *descriptors* [19]) such as compounds (metals and other recipients) and temperature at which an alloy is designed. At *the third stage* we consider existing and, mainly, stable structures from big data and generate potential ones. That is, a large number of stable structures are considered which greatly narrow the space that is to be explored experimentally. These data composes a search space for application of data-driven approaches to prediction of alloys properties at the next, *fourth stage*. At this stage, we have to create a model f (NN, regression model, fuzzy rules or a hybrid SC model) by using DL to describe relation between descriptors X and properties Y hidden in big data. Construction of a model includes development of its structure (e.g. fuzzy recurrent neural network) and adjusting its parameters to achieve a predefined accuracy of modeling relationship. Initially developed model need to be tested on a basis of structures that were not used for the model construction. The *fifth stage* implements an application of the model to computational synthesis of alloys with required characteristics and/or properties prediction. This requires identification of promising target candidates for synthesis. Computational synthesis is implemented by search and optimization of compound (on base of possible candidate structures) similar to the target. Here we can use case-based similarity approach. As a result, a list of best potential candidates is created. For these candidates, the properties are computed by using the created model f. As a result, an alloy near to targeted one is synthesized computationally at the *sixth stage*. At the final, *seventh stage*, researches have to create new material in the laboratory using experimental synthesis guided by the computer calculation. The created material is tested to verify the values of properties computed at the previous stage.

The outlined approach can be used for a large class of material discovery problems. In the next section we illustrate application of the approach to modeling structure-property relationship for a TiNi alloy.

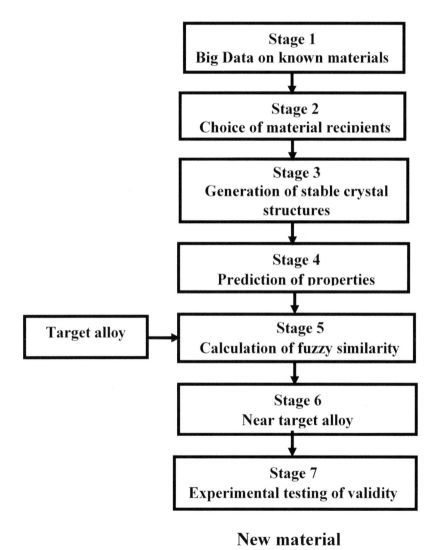

New material

Fig. 1. Structure of AI-based new materials discovery

4   An Example. Modeling of Structure-Property Relationship for a TiNi Alloy by Using Fuzzy Clustering

Let us consider construction of a model describing relationship f between the martensite lattice parameters $X = (a, b, c, \beta)$ (a, b, c – lengths of the unit cell edges, β is the angle between sides with length a and b) and mechanical characteristics $Y = (y_1, y_2, y_3)$ of TiNi alloy (y_1 – conventional ultimate strength, MPa, y_2 – conventional yield strength, MPa; y_3 – unit elongation, %):

$$Y = f(X).$$

Due to ambiguity inherent in crystal structures, we intend to construct model f as fuzzy IF-THEN rules derived from a set of experimental data (Table 1). Such a model is a universal approximator and provides a good summarization of complex set of data.

Table 1. Experimental data [27].

a	b	c	β	y_1	y_2	y_3
0.2904	0.412	0.465	97.9	983.15	235.27	49.4
0.2909	0.412	0.466	98	983.15	235.27	49.4
0.2899	0.412	0.464	97.8	983.15	235.27	49.4
0.2889	0.412	0.462	96.8	981.48	220.79	50.18
0.2894	0.413	0.464	97.1	981.48	220.79	50.18
0.2902	0.413	0.465	97.2	984	253	49
0.2916	0.4144	0.4644	97	983.145	235.2729	49.40427

For construction of fuzzy IF-THEN rules we use Fuzzy C-Means (FCM) clustering method. FCM method partitions a large set of data to several groups that represent various situations of relationship between lattice parameters and alloy characteristics. Inputs of these rules are linguistic evaluations of lattice parameters, whereas outputs describe the related characteristics. We derived the following fuzzy IF-THEN rules:

IF a is "VL" and b is "VL" and c is "L" and β is "L"
THEN y_1 is "L", y_2 is "VL", y_3 is "VH",

IF a is "H" and b is "M" and c is "VH" and β is "VH"
THEN y_1 is "VH", y_2 is "M", y_3 is "L",

...

IF a is "M" and b is "H" and c is "M" and β is "L"
THEN y_1 is "VL", y_2 is "VH", y_3 is "M",
IF a is "M" and b is "VH" and c is "M" and β is "M"
THEN y_1 is "VH", y_2 is "H", y_3 is "VL",

where VL, L, M, H, VH abbreviations in the antecedents and consequents of the rules denote linguistic evaluations *"Very Low"*, *"Low"*, *"Medium"*, *"High"* and *"Very High"* respectively. Fuzzy description of linguistic terms for some inputs and outputs of the rules are given in Figs. 2 and 3.

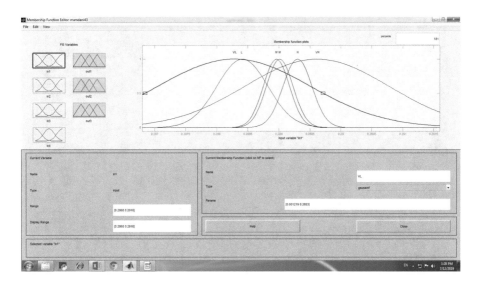

Fig. 2. Fuzzy description of linguistic terms for lattice parameter a.

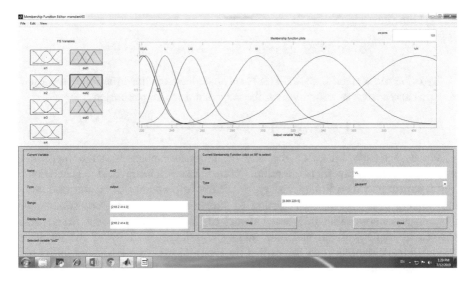

Fig. 3. Fuzzy description of linguistic terms for conventional yield strength characteristic, y_2.

By using fuzzy inference within these IF-THEN rules, we can compute values of the characteristics $Y = (y_1, y_2, y_3)$ for lattice parameters which do not exist in the experimental data. The graphical description of the fuzzy inference system is shown in Fig. 4.

Fig. 4. Fuzzy inference within the derived IF-THEN rules

For example, consider the following lattice parameters which are note presented in the experimental data (Table 1):

$$a = 0.29065 \quad b = 0.4122 \quad c = 0.4652 \quad \beta = 98$$

By using the derived fuzzy IF-THEN rules we have obtained the following values of the characteristics of material with the considered lattice parameters:

$$Y = (y_1, y_2, y_3) = (983.29, \quad 235.6, \quad 49.38)$$

We have shortly illustrated the way the proposed approach can be applied to modeling structure-property relationship. Future works will be devoted to generation of crystal structures which fit target alloy, prediction of properties of synthesized alloys and refining structures and calculation of properties. For these purposes, a more comprehensive soft computing based model can be used as recurrent fuzzy neural network to implement DL and data mining algorithms. Validity of the results of computational synthesis is to be verified by experimental testing of synthesized new alloys. Practicality of the results would rely on analysis of scientific efficiency and practical synthesizability of the alloys.

5 Conclusion

This study is the first step toward development of AI-based models of material discovery. We describe a general outline of an approach to data-driven alloy synthesis by using DL and fuzzy logic methods [28–32]. An example illustrating fuzzy If-Then rules based modeling of relationship between crystal structure parameters and properties of TiNi alloys is considered.

References

1. Babanli, M.B., Huseynov, V.M.: Z-number-based alloy selection problem. Procedia Comput. Sci. **102**, 183–189 (2016)
2. Babanli, M.B., et al.: Review on the new materials design methods. In: Aliev, R., Kacprzyk, J., Pedrycz, W., Jamshidi, M., Sadikoglu, F. (eds.) 13th International Conference on Theory and Application of Fuzzy Systems and Soft Computing—ICAFS-2018. Advances in Intelligent Systems and Computing, vol. 896. Springer, Cham (2018)
3. Babanli, M.B.: Synthesis of new materials by using fuzzy and big data concepts. Procedia Comput Sci **120**, 104–111 (2017)
4. Babanli, M.B.: Theory and practice of material development under imperfect information. In: Advances in Intelligent Systems and Computing, vol. 896, pp. 4–14. Springer (2018)
5. Babanli, M.B.: Fuzzy modeling of phase diagram under imprecise thermodynamic data. In: Proceedings of the Tenth World Conference "Intelligent Systems for Industrial Automation", pp. 265–266. b-Quadrat Verlag (2018)
6. Babanli, M.B.: Fuzzy Logic-Based Material Selection and Synthesis. World Scientific, Singapore (2019)
7. Babanli, M.B.: Fuzzy logic and fuzzy expert system-based material synthesis methods (2019). IntechOpen, https://doi.org/10.5772/intechopen.84493. Available from: https://www.intechopen.com/online-first/fuzzy-logic-and-fuzzy-expert-system-based-material-synthesis-methods
8. Constable, D.J.C.: The practice of chemistry still needs to change. Curr. Opin. Green Sustain. Chem. **7**, 60–62 (2017)
9. Yan, W., Lin, S., Kafka, O.L., et al.: Modeling process-structure-property relationships for additive manufacturing. Front. Mech. Eng. **13**, 482 (2018)
10. Jung, J., Yoon, J.I., Park, H.K., Kim, J.Y., Kim, H.S.: An efficient machine learning approach to establish structure-property linkages. Comput. Mater. Sci. **156**, 17–25 (2019)
11. Su, C., Lv, J., Li, Q., Wang, H., Zhang, L., Wang, Y., Ma, Y.: Construction of crystal structure prototype database: methods and applications. J. Phys.: Condens. Matter **29**(16), 165901 (2017)
12. Latypov, M.I., Kalidindi, S.R.: Data-driven reduced order models for effective yield strength and partitioning of strain in multiphase steels. J. Comput. Phys. **346**, 242–261 (2017)
13. Yabansu, Y.C., Steinmetz, P., Hötzer, J., Kalidindi, S.R., Nestler, B.: Extraction of reduced-order process-structure linkages from phase-field simulations. Acta Mater. **124**, 182–194 (2017)
14. Paulson, N.H., Priddy, M.W., McDowell, D.L., Kalidindi, S.R.: Reduced-order structure-property linkages for polycrystalline microstructures based on 2-point statistics. Acta Mater. **129**, 428–438 (2017)

15. Khosravani, A., Cecen, A., Kalidindi, S.R.: Development of high throughput assays for establishing process-structure-property linkages in multiphase polycrystalline metals: application to dual-phase steels. Acta Mater. **123**, 55–69 (2017)
16. Iskakov, A., Yabansu, Y.C., Rajagopalan, S., Kapustina, A., Kalidindi, S.R.: Application of spherical indentation and the materials knowledge system framework to establishing microstructure-yield strength linkages from carbon steel scoops excised from high-temperature exposed components. Acta Mater. **144**, 758–767 (2017)
17. Yabansu, Y.C., Patel, D.K., Kalidindi, S.R.: Calibrated localization relationships for elastic response of polycrystalline aggregates. Acta Mater. **81**, 151–160 (2014)
18. Ramprasad, R., Batra, R., Pilania, G., Mannodi-Kanakkithodi, A., Kim, C.: Machine learning in materials informatics: recent applications and prospects. npj Comput. Mater. **3**, 54 (2017)
19. Seko, A., Togo, A., Tanaka, I.: Descriptors for machine learning of materials data. In: Tanaka, I. (ed.) Nanoinformatics. Springer, Singapore (2018)
20. Kim, E., Huang, K., Saunders, A., McCallum, A., Ceder, G., Olivetti, E.: Materials synthesis insights from scientific literature via text extraction and machine learning. Chem. Mater. **29** (21), 9436–9444 (2017)
21. Kim, E., Huang, K., Jegelka, S., Olivetti, E.: Virtual screening of inorganic materials synthesis parameters with deep learning. npj Comput. Mater. **3**, Article no. 53 (2017)
22. Gubaev, K., Podryabinkin, E.V., Hart, G.L., Shapeev, A.V.: Accelerating high-throughput searches for new alloys with active learning of interatomic potentials. Comput. Mater. Sci. **156**, 148–156 (2019)
23. Han, Y., Yang, X., Zeng, W., Lu, W.: Nonlinear relationship between processing parameters and mechanical properties in Ti6Al4V alloy by using fuzzy neural network. In: Venkatesh, V., Pilchak, A.L., Allison, J.E., Ankem, S., Boyer, R., Christodoulou, J., Fraser, H.L., Imam, M.A., Kosaka, Y., Rack, H.J., Chatterjee, A., Woodfield, A. (eds.) Proceedings of the 13th World Conference on Titanium (2016)
24. Marwin, H.S.S., Preuss, M., Waller, M.P.: Planning chemical syntheses with deep neural networks and symbolic AI. Nature **555**, 604–610 (2018)
25. Zadeh, L.A.: Fuzzy sets. Inf. Control **8**, 338–353 (1965)
26. Yann, L., Yoshua, B., Geoffrey, H.: Deep learning. Nature **521**(7553), 436–444 (2015)
27. Prokoshkin, S.D., Khmelevskaya, I.Yu., Korotitskij, A.V., Trubitsyna, I.B., Brailovskij, V., Tyurenn, S.: On the lattice parameters of the B19' martensite in binary Ti-Ni shape memory alloys. Fiz. Met. Metalloved. **96**(1), 62–71 (2003)
28. Aliev, R.A., Perdycz, W., Huseynov, O.H.: Functions defined on a set of Z-numbers. Inf. Sci. **423**, 353–375 (2018)
29. Aliev, R.A., Alizadeh, A.V., Huseynov, O.H., Jabbarova, K.I.: Z-number-based linear programming. Int. J. Intell. Syst. **30**(5), 563–589 (2015)
30. Aliev, R.A., Pedrycz, W.: Fundamentals of a fuzzy-logic-based generalized theory of stability. IEEE Trans. Syst. Man Cybern. Part B (Cybern.) **39**(4), 971–988 (2009)
31. Aliev, R.A., Alizadeh, A.V., Huseynov, O.H.: The arithmetic of continuous Z-numbers. Inf. Sci. **373**, 441–460 (2016)
32. Aliev, R., Tserkovny, A.: Systemic approach to fuzzy logic formalization for approximate reasoning. Inf. Sci. **181**(6), 1045–1059 (2011)

Surface Roughness Modelling and Prediction Using Artificial Intelligence Based Models

Musa Alhaji Ibrahim[1,2(✉)] ⓘ and Yusuf Şahin[1] ⓘ

[1] Near East University, Nicosia, Cyprus
abbanfati2014@gmail.com, yusufsahin@neu.edu.tr
[2] Kano University of Science and Technology, Wudil, Kano, Nigeria

Abstract. Surface roughness is a significant factor in product quality. Experimental investigation of surface roughness is associated with cumbersomeness, high cost and energy consumption. The results of artificial intelligence (AI) based models namely artificial neural network (ANN) and adaptive neuro fuzzy inference system (ANFIS) were presented. Empirical data of hard turning operations containing cutting speed, feed rate and depth of cut and average surface roughness (Ra) were used as inputs and target process parameters respectively to train the AI based models. The performances of the models were evaluated using determination coefficient (DC) and root mean square error (RMSE) criteria. The results of the one-process-parameter influence on Ra donated that both models showed that feed rate was the most influential and dominant process parameter on Ra. However, the sensitivity analysis results indicated that the two different AI based models behaved differently with ANN having cutting speed and depth of cut as the most influential and dominant input-combination process parameters while that of ANFIS showed that feed rate and depth of cut were while describing and predicting the same surface roughness process under the same conditions. The results of the models were in concord with the empirical results of the average surface roughness. Both models predicted Ra but ANN performed better than ANFIS in both scenarios. The models can be used to design for the average surface roughness of hardened steel in hard turning operation thus saving cost, time and energy consumption in the experimental determination of average surface roughness, Ra.

Keywords: ANN · ANFIS · Surface roughness

1 Introduction

Surface roughness machining by hard turning operation is a complicated process in which machines are operated for longer time thus leading to more energy consumption. The process parameters namely cutting speed (V), feed rate (F) and depth of cut (D) [1] can be manipulated to address this. It is significant that the suitable cutting parameters are chosen. Improper choice of these parameters will lead to abrasive wear and deterioration of surface roughness quality. Surface roughness quality proportionally affects the life span of products [2]. This determines the quality, integrity of products and is one of the costumers' requirements [3, 4]. Poor surface roughness quality increases production cost [5]. Consequently, the modeling and prediction of surface roughness is

© Springer Nature Switzerland AG 2020
R. A. Aliev et al. (Eds.): ICSCCW 2019, AISC 1095, pp. 33–40, 2020.
https://doi.org/10.1007/978-3-030-35249-3_3

desired for saving cost, time and energy consumption for high quality surface roughness machining. Artificial intelligence (AI) based models like artificial neural network (ANN) and adaptive-neuro-fuzzy-inference system (ANFIS) have been used. These black boxes are precise, simple and speedy [6, 7]. Existing literature had shown relevant studies on surface roughness modeling and prediction using the AI based models [8–12]. The surface roughness describes the smoothness and non-smoothness of products surfaces. Therefore, in both engineering decisions and surface roughness machining, predicting and estimating the values of the surface roughness is an important fact.

In this study, ANN using Lavenberg-Marquartdt containing back propagation algorithm and ANFIS with the first order Takagi-Sugeno type of fuzzy structures were used to model and predict the average surface roughness (Ra) and determination coefficient (DC) and root mean square error (RMSE) were employed to evaluate the performances of the developed AI based models.

2   Methods

2.1   Artificial Neural Network (ANN)

Artificial neural network (ANN) is a data-driven system that mimics a human brain and it functions like the human brain by means of various interconnected nodes (neurons). Mathematically, ANN is non-linear regression and interpolation. When input signals are fed to the ANN, they are assigned different weights and biases by the neurons and the weights are continuously adjusted to reduce the error through activation function between the observed and the computed data. The Levenberg-Marquardt containing back propagation (BP) learning algorithm was chosen to update the parameters in feed forward neural network (FFNW) having single layer hidden neuron and differing number of neurons for estimation of Ra. The BP minimizes error by transmitting the errors back to the neurons iteratively until a minimal error is obtained. The data to the ANN is feed rate, cutting speed and depth of cut and the target is average surface roughness as inputs and target respectively. The sigmoidal activation transfer function was selected due to its sensitivity to small changes. The data was split into training (70%) and testing (30%). The ANN was trained with single hidden layer for Ra estimation. Number of neurons in the hidden layer is decided by trial and error. The data [13–16] was normalized between 0 and 1 using Eq. (1).

$$N = \frac{\beta i - \beta min}{\beta max - \beta min} \tag{1}$$

where $N, \beta i, \beta min, \beta max$. βmin represent normalized, actual, minimum and maximum values respectively. The normalization gives equal attention to all parameters irrespective of magnitude and takes care of dimensions.

2.2   Adaptive Neuro Fuzzy Inference System (ANFIS)

ANFIS integrates ANN and fuzzy inference system (FIS) abilities. ANFIS is known as universal estimator for dealing with complex problems using hybrid-learning algorithm. The ANN builds the suitable IF-THEN-RULE while the FIS provides membership function. A common ANFIS architecture is as shown in Fig. 1. The ANFIS is made up of five (5) layers. These layers function differently from one another but nodes of the same layer execute similar function. A brief explanation of these different layers is provided as follows:

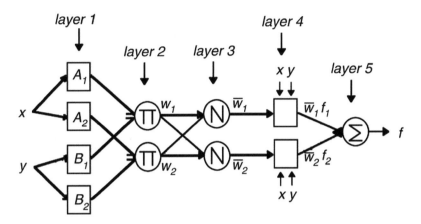

Fig. 1. ANFIS structure with two input parameters

Layer 1 deals with fuzzification of input characteristic values between the ranges 1–0. The needed values like membership functions for each i_{th} node are defined in this layer, presented in Eqs. (2) and (3):

$$O_{1,i} = \beta_{Ai}(X) \tag{2}$$

$$O_{1,i} = \beta_{Bi}(Y) \tag{3}$$

where X and Y are the inputs, Ai, and Bi are linguistic labels related to this node function.

In layer 2, nodes assign rule in the ANFIS to find the matching rule W_i. The inputs are multiplied in this layer thus producing the outputs, shown in Eq. (4):

$$Wi = \beta_{Ai} \times \beta_{Bi} \tag{4}$$

The membership values are normalized in layer 3. The formulation of the firing strengths (outputs) for node i_{th} in this layer is shown in Eq. 5:

$$W_i = \frac{wi}{w1 + w2}, \quad i = 1, 2 \tag{5}$$

Layer 4 establishes the relationship between the inputs and the outputs Eq. (6):

$$O_{4,i} = wi = (P_i X + Q_i Y + ri) \tag{6}$$

where wi is layer 3's output and Pi, Qi and r1 are regulative parameters.

In layer 5, defuzzification occurs and each node generates summation of incoming signals from preceding nodes and yields a single output value. The rule for each output is introduced into the output layer. Overall output can be computed using Eq. (7):

$$O_{5,1} = \sum_i wifi = \sum_i \frac{wifi}{wi} \tag{7}$$

The first order Sugeno type fuzzy inference system that uses hybrid-learning algorithm was used in this study. Trial-and-error approach to determine the suitable membership function was used to construct the ANFIS models.

2.3   Performance Evaluation

To investigate and determine the efficiency and performance of the models determination of coefficient (DC) and root mean square error (RMSE) were used Eqs. (8) and (9) respectively.

$$RMSE = \sqrt{\sum_{i=1}^{N} \frac{(Qo - Qp)^2}{N}} \tag{8}$$

$$DC = 1 - \sqrt{\frac{\sum_{i=1}^{N}(Qo - Qp)^2}{\sum_{i=1}^{N}(Qo - Qm)^{2)}}} \tag{9}$$

where Qo, Qp, Qm are the observed data, predicted and mean values respectively. RMSE ranges from $0 - \infty$ while DC lies between $-\infty$ and 0. When DC approaches 1 and RMSE reaches 0, it means higher efficiency and performance of the models.

3   Results

3.1   ANN and ANFIS Results

In this section, the results of the two ANN and ANFIS models are presented. Table 1 provides the result of the ANN model for Ra.

Table 1. DC and RMSE for ANN model of Ra.

Model	Input	Structure	Training		Testing	
			DC	RMSE	DC	RMSE
ANN1	V	1-12-1	0.5779	0.0804	0.5188	0.1736
ANN2	F	1-6-1	0.7948	0.0560	0.6462	0.1489
ANN3	D	1-12-1	0.5419	0.0837	0.4532	0.1851

From Table 1, it is seen that feed rate was found to be the most dominant process parameter on the Ra followed by cutting speed and depth of cut. ANN with 1-6-1 structure showed better performance and minimal error. This means that ANN2 could perform with DC (efficiency) of 79.48%, RMSE (minimal error) of 0.0560 and DC (efficiency) of 64.62%, RMSE (minimal error) of 0.1489 in training and testing data respectively better than the rest.

A comparison of the computed and observed Ra in training phase is as shown in Fig. 2. It can be seen that the observed and predicted values are close to each other.

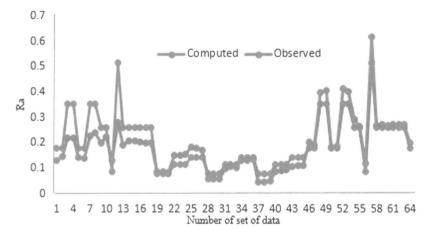

Fig. 2. Correlation between observed and computed data of Ra for ANN2 in training phase

Table 2 shows similar trend to ANN models in terms of the chronological dominance of the cutting parameters. Model ANFIS5 having a DC of 69.15% and 59.41% with RMSE of 0.0687 and 0.1595 in both training and testing respectively is more effective than the rest. Figure 3 manifests fair match with observed surface roughness values compared of the ANFIS5 model.

Table 2. DC and RMSE for the ANFIS model of Ra.

Model	Input	Membership	Training		Testing	
			DC	RMSE	DC	RMSE
ANFIS4	V	Trapezoidal	0.2330	0.1083	0.2395	0.2183
ANFIS5	F	Gaussian	0.6915	0.0687	0.5941	0.1595
ANFIS6	D	Triangular	0.3938	0.0963	0.2044	0.2233

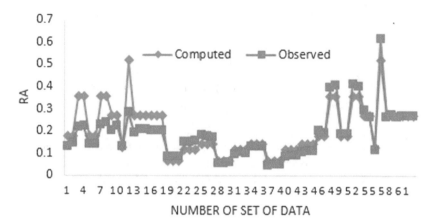

Fig. 3. Comparison of observed and computed data of Ra for ANFIS5 in training phase

3.2   Sensitivity Analysis Result

In this study, AI based sensitivity analysis was performed to determine the dominant inputs combination variables on Ra. Table 3 shows the results. From Table 3, the most dominant inputs combination process parameters were cutting speed and depth of cut for ANN13 with efficiency of 93.66% in training and in testing 90.30% phases at 15 hidden neurons and reduced error of 0.0883 and 0.1809 in training and testing phases respectively. However, the most dominant combination process parameters to the ANFIS18 model were feed rate and depth of cut with DCs of 77.60% and 71.69% and RMSEs of 0.0585 and 0.1332 in training and testing stages respectively. This shows that ANN13 described and estimated Ra better than ANFIS18 and that both models behaved differently while dealing with the same conditions. The pattern between the observed and computed data of ANN13 is as shown in Fig. 4.

Table 3. Sensitivity result of ANN and ANFIS of Ra.

Model	Combination	Structure	Training		Testing	
			DC	RMSE	DC	RMSE
ANN13	V, D	1-15-1	0.9366	0.0883	0.9030	0.1809
ANN14	F, D	1-5-1	0.7376	0.0652	0.5607	0.1659
ANN15	V, F, D	1-15-1	0.6472	0.0936	0.4105	0.1423
ANFIS16	V, F	Trap	0.6899	0.0689	0.6041	0.1575
ANFIS17	V, D	Gauss	0.3861	0.0237	0.3483	0.2021
ANFIS18	F, D	Tri	0.7760	0.0585	0.7169	0.1332
ANFIS19	V, F, D	Gauss2	0.7285	0.0644	0.8527	0.0961

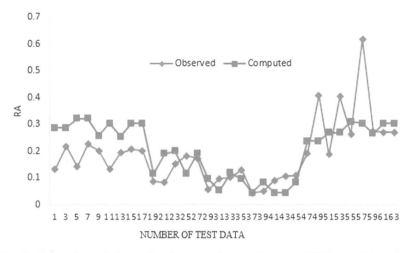

Fig. 4. Comparison of observed and computed data of Ra for ANN13 in training phase

4   Conclusion

Artificial neural network (ANN) and adaptive neuro fuzzy inference system (ANFIS) AI based models were used to model average surface roughness, Ra, of hardened steel in turning operation. For one-input-effect both models showed that feed rate was the most dominant process parameter on the average surface roughness (Ra) of the hardened steel. Sensitivity analysis showed that the models demonstrated different combinations of process parameters with ANN having cutting speed and depth of cut as dominant and influential process parameters while ANFIS exhibited feed rate and depth of cut being the dominant and influential process parameters. Both models described and predicted Ra but ANN did better than ANFIS. The models can be used to design for the average surface roughness of hardened steel in turning operation thus saving cost, time and energy consumptions.

References

1. Özel, K.Y.T.: Predictive modeling of surface roughness and tool wear in hard turning using regression and neural networks. Int. J. Mach. Tools Manuf. **45**(4–5), 467–479 (2005)
2. Sofuo, A., Orak, S.: An ANN-based method to predict surface roughness in turning operations. Arab. J. Sci. Eng. **42**, 1929–1940 (2017)
3. Kwon, Y.J., Tseng, T.-L.B., Konada, U.: A novel approach to predict surface roughness in machining operations using fuzzy set theory. J. Comput. Des. Eng. **3**, 1–13 (2016)
4. Laouissi, A., Yallese, M.A., Belbah, A., Belhadi, S., Haddad, A.: Investigation, modeling, and optimization of cutting parameters in turning of gray cast iron using coated and uncoated silicon nitride ceramic tools Based on ANN, RSM, and GA optimization. Int. J. Adv. Manuf. Technol. **101**, 523–548 (2019)

5. Rajeev, D., Dinakaran, D., Lead, G., Muthuraman, S.: Prediction of roughness in hard turning of AISI 4140 steel through artifical neural network and regression models. Int. J. Mech. Eng. Technol. **7**(5), 200–208 (2016)

6. Hossain, A.N.: Surface roughness prediction modelling for commercial dies using ANFIS, ANN and RSM. Int. J. Ind. Syst. Eng. **16**(2), 156–183 (2014)

7. Kamruzzaman, M., Rahman, S.S., Ashraf, M.Z.I., Dhar, N.R.: Modeling of chip–tool interface temperature using response surface methodology and artificial neural network in HPC-assisted turning and tool life investigation. Int. J. Adv. Manuf. Technol. **90**(5–8), 1547–1568 (2017)

8. Sarma, D.K., Dixit, U.S.: A comparison of dry and air-cooled turning of grey cast iron with mixed oxide ceramic tool. J. Mater. Process. Technol. **190**, 160–172 (2007)

9. Karayel, D.: Prediction and control of surface roughness in CNC lathe using artificial neural network. J. Mater. Process. Technol. **209**, 3125–3137 (2009)

10. Ho, S.J., Lee, S.Y., Chen, K.C., Ho, S.S.: Accurate modeling and prediction of surface roughness by computer vision in turning operations using an adaptive neuro fuzzy inference system. Int. J. Mach. Tools Manuf. **42**, 1441–1446 (2002)

11. Davim, S.R., Gaitonde, J.P., Karnikc, V.N.: Investigations into the effect of cutting conditions on surface roughness in turning of free machining steel by ANN models. J. Mater. Process. Technol. **205**, 16–23 (2008)

12. Zhong, Z.W., Khoo, L.P., Han, S.T.: Prediction of surface roughness of turned surfaces using neural networks. Int. J. Adv. Manuf. Technol. **28**(7–8), 688–693 (2006)

13. Sahin, Y., Motorcu, A.R.: Surface roughness model in machining hardened steel with cubic boron nitride cutting tool. Int. J. Refract. Met. Hard Mater. **26**, 84–90 (2008)

14. Bouacha, K., Athmane, M., Mabrouki, T., Rigal, J.: Statistical analysis of surface roughness and cutting forces using response surface methodology in hard turning of AISI 52100 bearing steel with CBN tool. Int. J. Refract. Met. Hard Mater. **28**(3), 349–361 (2010)

15. Sahin, Y., Motorcu, A.R.: The develpoment of surface roughness model when turning hardened steel with ceramic cutting tool using response methodology. Multidiscip. Model. Mater. Struct. **4**, 290–304 (2008)

16. Akkus, H.: Determining the effect of cutting parameters on surface roughness in hard turning using the Taguchi method. Measurement **44**, 1697–1704 (2011)

Soft Methods of Public Administration in the Era of Digitalization

Natalia V. Mamitova$^{(\boxtimes)}$ (iD) and Marina A. Makhotenko (iD)

Educational Institution of Higher Education "Southern University (IMBL)",
Rostov-on-Don, Russia
nvmamitova@mail.ru, mahotencko@iubip.ru

Abstract. The article is devoted to the theoretical and methodological justification of public administration in the realities of the process of digitalization of the state and society. The conceptual features of the process of digitalization and its relationship with automation on the basis of "soft models" are indicated. The role and characteristics of public administration, principles, methodology, as well as the possibility of its implementation through the achievements of information and technological progress are outlined. The analysis of the existing practical adaptations of the concept of public administration on the example of the open government project is carried out. The analysis of measures of responsibility of the Government in the digital state is carried out. The necessity of modernization of the existing institutional infrastructure is determined. As a result of the fuzzy analysis of the state of affairs in public administration formed the main practical approaches to this process.

Keywords: Public administration · NON-factors · Digitalization · Automation and soft computing · Information and communication technologies · Digital transformation of social processes

1 Introduction

The modern period of development of the Russian state is characterized by a systemic reform of the political, economic and social foundations of civil society. In this regard, the task is to ensure the transfer of public services online using remote services in the coming years, and the document flow between government agencies will be carried out exclusively in electronic form. These and other measures to digitalize the work of the authorities, based on the use of artificial intelligence, will, according to the President of the Russian Federation, increase the responsibility of state bodies, transparency of their work and fight corruption.

Traditionally, we will consider the key issues of public administration in comparison with other countries. Mathematical modeling allows us to translate the problem of improving the system of public administration and decision-making in it from the verbal-descriptive to the "engineering" plane of applied calculations and their practical use. At the same time, it is necessary to proceed from the fact of the determining influence of NON-factors at all stages of public administration and, accordingly, all its real models [1].

© Springer Nature Switzerland AG 2020
R. A. Aliev et al. (Eds.): ICSCCW 2019, AISC 1095, pp. 41–46, 2020.
https://doi.org/10.1007/978-3-030-35249-3_4

The main mathematical models of public administration and related processes and phenomena include, among others, fuzzy ("soft") models of the Federation, which will focus in more detail.

2 Materials and Method

Governance is a special phenomenon that is characteristic of absolutely to all spheres of society, be they social institutions, economic, political and spiritual. Due to its importance, the principle of federalism received a significant "soft mathematization" of the model of this control. Federalism characterizes "such a way of territorial organization of state power, in which the constituent state performs common tasks with guarantees of a certain level of independence of its individual parts" [2].

We consider a mathematical model that allows us to study the stability of the Federal management process, a phenomenon that characterizes the conflict system, which is "in a state of constant change, which is trying to find a balance between unity and diversity, centralism and regionalism, the division of power and the separation of powers, as well as between symmetry and asymmetry" [3].

Mechanisms and incentives for sustainability vary in content and degree of influence [4].

Public opinion and the threat of popular unrest can play a "tough" role (deterministic components of the model) in maintaining the integrity of the country. More "soft" include institutional arrangements described by "soft" (fuzzy) models. These include, in particular, structural (for example, the distribution of powers between subjects, the inclusion of regions in the adoption of Federal decisions), political (say, through Federal parties) and legal (legislation).

The mathematical model proposed in [1] takes into account the above "soft" and "hard" mechanisms [5]. Let the control Object (Federation) consist of N regions (subjects of Federation). The model includes a situation when each i-th of them receives a share of AI funds from the Federal Fund ("common good"), formed from the contributions of subjects, and seeks under various pretexts to leave at its disposal, not wanting to transfer the rest of their income. If the j-th subject strictly follows the constitutional Treaty, we will talk about "acceptable behavior»; otherwise—about "opportunistic behavior". Within the meaning Δu_r - characterizes the degree of such behavior (deviation from the "honest") and can apply not only to economic situations.

The utility function of the r-th region is represented as the sum of these components:

$$u_r(x_1, \ldots, x_N) = \sum_{k=1}^{N} (1 - x_k)x_r \qquad (1)$$

Each entity seeks to maximize its utility function (1). Thus, the model lays the desire to get more out of the General Fund with a decrease in contributions to it.

In the context of this work, the term "governance" will mean the context in which the subject of this process is the state (public administration), in view of the complexity and diversity of the phenomenon, as well as the lack of a generally accepted definition, it is necessary to consider this definition in more detail, thus to identify the essence, content and composition of the concept relevant in this work.

Based on the works of Russian scientists, the term "public administration" (public administration) is identified by many authors with public administration. So S. M. Dubonos believes that public administration is a kind of another form of implementation of state power – social management, that is, the category of public administration aimed at meeting public needs, but again without participation in the decision-making of the population. At the same time, the author distinguishes transparency, legitimacy and democracy as characteristic features.

Thus, in most cases, the interpretation of the concept of "public administration" loses the meaning of the concept of public administration, that is, the involvement of the General public in decision-making. Nevertheless, according to the above authors, the opinions and interests of civil society should be taken into account in the policy, and its results should be transparent. That is, the authors do not reject the need for the state to adhere to the public interest in the exercise of its power, but exclude the need for public power, often identifying it with the state, thereby distorting the content of both categories.

Obviously, at the moment, the generally accepted definition of "public administration", and moreover, existing interpretations are contrary to the content used in this term of concepts, and hence of the concept.

The main advantage of this office is the ability to work effectively with NON-factors characteristic of solving problems at the national level: with fuzzy, inaccurate and approximate data that can not be performed using classical modeling methods, often not covering all the essence of the complex system under consideration, which is public administration.

Modern control models include hybrid models that take into account the basic properties of complex systems, including:

- Neurocomputing + fuzzy logic (NF);
- fuzzy logic + genetic algorithms (FG);
- fuzzy logic + chaos theory (FCh);
- neural networks + genetic algorithms (NG);
- neural networks + chaos theory (NCh);
- neural networks + fuzzy logic + genetic algorithms (NFG) etc.

To describe vagueness, especially in public administration, probability theory and statistical methods are not enough, they are designed to work with randomness when it comes to belonging of some object to a clear set.

Fuzzy is usually taken into account by three main methods depending on the so-called degree of uncertainty of the input data. In the theory of stability at first find exact solutions, and then estimate their variation at fluctuations of initial data in limits of admissible errors. Stochastic methods consider random variables as an explication of the initial concepts corresponding to real objects and phenomena, and then use probabilistic considerations at all stages of obtaining the decision, which is also random.

For the formalized description of real situations in which there is no complete certainty and unambiguity, that is, there are so-called NON-factors [6], such mathematical apparatus as the theory of fuzzy sets is used. One of the applications of this device is the theory of decision-making [7]. The mathematical apparatus of fuzzy sets is quite complex; however, systems based on fuzzy sets have been developed and successfully implemented in such areas as process control, transport management, medical diagnostics, technical diagnostics, financial management, stock forecasting, pattern recognition. This apparatus is simply necessary in the assessment for investment analysis in the context of risk and uncertainty of public administration.

Methods of the theory of fuzzy sets and fuzzy intervals are methods of decision-making under uncertainty. Their use assumes expression of initial parameters and target criteria in the form of a vector of values, hit in each interval of which is characterized by some degree of uncertainty. Carrying out, for example, operations of addition, multiplication with such intervals by the rules of the theory of fuzzy sets, the evaluator receives effective intervals for the target criterion. The risk assessment of this method can be considered as complex (based on the fuzzy interval as a separate value) or as separate (if the probability distribution curve, in this case the uncertainty level curve, is compared to the fuzzy interval).

This trend is predominant in the modern study of socio-economic problems and has the format of collocation forecasting [8] or study of covariance functions of the system processes. To understand these patterns it is advisable to present a scheme of formation and propagation of disturbances in the system.

Formalization of the x signal propagation that changes linearly Y_1, presented as:

$$Y_1 = b_0 + b_1(b_{0x} + b_{1x}x) = b_0 + b_1 b_{0x} + b_1 b_{1x}x = b_0' + b_1'x \tag{2}$$

For prediction, using covariance functions $\tilde{C}_{yy}(\tau)$ stationary ergodic random process at $0 \leq \tau \geq N - 1$ the formula is applicable:

$$Cov(Y_{i-1}, Y_i) = \tilde{C}_{yy}(\tau) = \frac{1}{N - \tau} \sum_{\tau=1}^{n-\tau} (y_{\tau+1}) - \bar{y})(y_\tau - \bar{y}) \tag{3}$$

where $\bar{y}$ - sample average $\bar{y} = \frac{1}{N} \sum_{i=1}^{n} y_1$, at the same time, the increase in the number of iterations reduces the accuracy of the forecast and this imposes a limit $\tau \in [0, \frac{N}{2}]$.

The optimal model in the form of regression on an infinite Hilbert space is determined as a result of a search by the scheme:

$$\overline{C}_{yy}(0), \overline{C}_{yy}(1), \ldots, \overline{C}_{yy}(\tau_{Max}) \tag{4}$$

The empirical potential of this approach is due to the possibility of visual graphical representation of time series, their structure and dynamics. The most important condition for this analysis, according to [9] is the study "resilience, adaptability, catastrophes. The main question is whether the behavior of the system will change significantly as a result of changes (desirable, undesirable, unknown, unpredictable) in the management mode. That is, will the system perform its functions sustainably?".

The possibility of integration with the procedures of cenological analysis is due to the fundamental bases of these approaches, based on the classical representation of the dynamic process Lyapunov and Poincaré by means of differential equations of the form:

$$\frac{\partial x(t)}{\partial t} = f(x,t), \quad x(0) = c \tag{5}$$

where $\frac{\partial x(t)}{\partial t}$ - changing the x parameter of the system in time t; $f(x,t)$ - the law of change of parameters (phase coordinates) x; c - constant, the value of x at the initial time t = 0.

In other words, if x(t) solution starts within small distances from the origin x (0) = 0, then it must remain inside some "larger" channel – trajectory (A), (B). A stronger definition of sustainability meets the requirement $x(t) \to 0$ by $t \to \infty$ (trajectory (2) in Fig. 4), i.e. that the system would eventually return to the initial point of equilibrium is asymptotic stability, according to Lyapunov [10].

The difference of the proposed approach is the possibility of analyzing a set of trajectories, the stability characteristic of which is determined by structural (structural-topological) stability, since it is possible to use as an equilibrium point the value of the form of some canonical (standard) distribution, the possibility of determining which arises as a result of prognostic evaluation [9].

3   Conclusion

Information and technological progress today opens up extremely ambitious opportunities for the transformation of socio-economic processes, both at the business and at the state level. However, the existing conceptual and organizational contradictions, related both directly to the theoretical representation and directly to the implementation in practice of certain tools and mechanisms of the digital economy, indicate the relative "immaturity" of these ideas at this stage of development of society. Nevertheless, the theoretical and methodological Foundation has already been formed. Further implementation of the concept of digital society depends entirely on the actions of the state and business representatives, as a kind of consortium and, most importantly, on the public will and desire of citizens to build, offer or improve models of interaction and governance in society and the state. In the short term, it is difficult to talk about the possible implementation of certain principles of public administration in full, but the planned infrastructure reforms, which form the basis for their implementation, will accelerate this complex process due to objective needs. For Russia, such innovations play a strategic role as one of the elements of economic diversification.

References

1. Novikov, D.A., Belov, M.V.: Theory of complex activity. Part I. The structure of complex activities. Uncertainty and generation of integrated activities. Manag. Problems **4**, 36–45 (2018)
2. Minakov, P.A.: Public power: political science aspect. UFA: RITS Bashgu, p. 200 (2008)

3. Sevryugin, V.E.: On the doctrine of public administration in the administrative law of the Russian Federation. Bull. Tyumen State Univ. **3**, 92–97 (2011)
4. Novikov, D.A.: Models of development and development of technology of complex activity. Manag. Large Syst. **77**, 171–218 (2019)
5. Bednar, J.: The Robust Federation. Principles of Design. Cambridge University Press, Cambridge (2009)
6. Balanova, S.: https://www.forbes.ru/kompanii/350323-illyuzornyy-mir-pyat-glavnyh-mifov-cifrovoy-ekonomiki
7. Holkin, D.: https://medium.com/internet-of-energy/b7b196140c22
8. Demieva, A., Chepus, A., Mamitova, N.: The comparative analysis of the law policy of parliamentary responsibility of the executive government of the Russian Federation and foreign states. J. Soc. Sci. Res. (1), 499–502 (2018)
9. Novikov, D.A.: The analytical complexity and the error of the solution of problems of control of organizational-technical systems. Autom. Telemech. **5**, 107–118 (2018)
10. Khramov, V.V.: Theory of information processes and systems, Department of RSTU, Rostov-na-Don (2011). https://elibrary.ru/item.asp?id=32764234

Fake News Detection on Social Networks with Artificial Intelligence Tools: Systematic Literature Review

Murat Goksu and Nadire Cavus[✉]

Department of Computer Information Systems, Near East University,
99138 Nicosia, Cyprus, Mersin 10, Turkey
muratgoksu26@gmail.com, nadire.cavus@neu.edu.tr

Abstract. Rapid advances in technology have enabled print media to be published online and the emergence of Facebook, Twitter, YouTube and other social networks. Social networks have become an important way for people to communicate with each other and share their ideas. The most important feature of social networks is the rapid information sharing. In this context, the accuracy of the news or information published is very important. The spread of fake news in social networks has recently become one of the biggest problems. Fake news affects people's daily life and social order and may cause some negativity. In this study, the most comprehensive and prestigious electronic databases have been examined in order to find the latest articles about the detection of fake news in social networks by systematic literature review method. The main aim of the study is to reveal the benefits of artificial intelligence tools used in the detection of fake news and their success levels in different applications. As a result of the study, it was concluded that the success levels of artificial intelligence tools are over 90%. This study is thought to be a guide both researchers and individuals related to this field.

Keywords: Fake news detection · Social networks · Artificial intelligence

1 Introduction

Today, online social networks such as, Facebook, Twitter, and YouTube have become one of the most important ways in which people communicate with each other, share their ideas and discuss any event. Increasing technological consumption and use of social network along with the developments in information technologies have attracted the attention of malicious users who produce fake news [1]. Nowadays, almost everyone has a few social network accounts. And through these accounts one can access news information really quickly. It is difficult to check the available information published on the Internet and social networks, because there are often many conflicting sources. The diversity of resources, the fact that social networks are personal spaces and the use of online networks worldwide are important factors in the emergence of fake news [2]. It is difficult to check the information/news published on the Internet because there are a lot of resources that often conflict with each other. Unfortunately, news reached through these networks may not always be accurate. Some news sources

© Springer Nature Switzerland AG 2020
R. A. Aliev et al. (Eds.): ICSCCW 2019, AISC 1095, pp. 47–53, 2020.
https://doi.org/10.1007/978-3-030-35249-3_5

are misleading in the form of fake news, especially in social networks, to manipulate people in different ways. There are lots of websites to spread the wrong information and hide the real news. They often publish fake news, propaganda material, deceptions, conspiracy theories that conceal true news [3]. In this way, changing the perception of people away from the real agenda. The main purpose of fake news sites is to influence the public in a number of areas politically, and their widespread examples are seen throughout the world. Therefore, fake news has become a global problem that must be overcome. It is believed that the fake news problem can be solved automatically with artificial intelligence (AI) tools without human intervention with machine learning (ML) and deep learning (DL) techniques. From this point of view, artificial intelligence applications will be examined in order to detect the fake news and to reach the desired result in a short time. With this research, two research questions were put to close the gap in previous research and to contribute to the more useful of AI techniques. Firstly, what are the benefits of AI applications in fake news detection and secondly How do such benefits differ according to different types of AI practices? For the purpose of answering these questions, a systematic literature review was conducted to identify and analyze relevant publications. In addition, AI applications in clustering of ML, DL, and artificial neural networks (ANN) were investigated in the detection of fake news. Furthermore, the study identified gaps in existing research and guides future studies in this area.

2   Artificial Intelligence Tools in Fake News

Machine learning and deep learning methods are used in the detection of fake news from AI tools. As a result of literature review, it is seen that especially ML methods are used more frequently in general. In this context, more ML will be emphasized. It is possible to define AI as a whole of software and hardware systems that have many capabilities such as human behavior, digital logic execution, motion, speech and sound detection. In other words, AI makes computers think like humans [4]. Alan Turing, underlined that in 1950, raised the idea that whether the machines think [5]. With his famous Turing Test, he was able to distinguish whether a machine was intelligent or not. If a person cannot distinguish between an interaction or a human or a machine, then that machine is a clever machine that can think. The common views of the researchers were that the studies on artificial intelligence should be increased to an advanced level. For example, the winner of the quiz, IBMs' artificial intelligence Watson [6], and the GO intelligence, the Google AI that defeats the world champion can be shown. Nowadays, the rising power of personal computers and mobile devices enables the concept of AI to be applied in traditional educational environments such as schools and universities. Machine learning is considered a kind of artificial intelligence that can even expose results that are not programmed [7]. In 1959, Arthur Samuel defined ML as the machine's ability to learn the results that were not specifically programmed [8]. ML can also be expressed as solving the problem by using statistical methods on the problem-based data set [9]. For this, the machine needs to perform the learning process on the problem-based data set. In this context learning can be supervised [10], semi-supervised [11], unsupervised [12] and reinforcement [13]. Deep learning is the most current approach used to develop AI for machines to perceive and understand the world. The aim of deep learning is to prepare the software for a computer model instead of constructing the software step by step [14]. In the face of the scenarios encountered in solving the problem, the model in question can produce alternative solutions. DL is mainly used in the areas of face recognition, voice recognition, defense and security and health [15].

3   Methodology

After determining the subject of the research, academic sources in the literature were searched. A four-stage research approach was applied. Firstly, related publications were systematically identified. Then the identified publications were grouped and abandoned. In the final stage, related studies were analyzed by matching them in the framework of artificial intelligence tools.

3.1   Data Collection

A systematic literature review was conducted in the most comprehensive and prestigious databases to identify publications on AI news and social network tools. Databases related to information systems discipline such as IEEE Xplore (IEEE), and SpringerLink has included as well as more general databases like Web of Science and ScienceDirect. In order to match the search model and to reach potentially relevant articles, a search was made by title, abstract, keywords ("artificial intelligence" or "machine learning" or "deep learning" and "fake news" or "fake news" or "rumor" and "social network" or "twitter" or "Facebook"). As a result of the research, 351 articles were obtained. These articles were filtering with these terms' fake news, or false news, or rumor, or disinformation or misinformation and artificial intelligence or machine learning and detection and social media or social networks or Facebook or Twitter or YouTube. After the filtering, the remaining 23 articles were read and analyzed completely.

3.2   Data Analysis

First, after the data collection phase, the articles to be analyzed were collected regularly. These articles were then classified into AI, DL, ML and Natural Language Process (NLP) Artificial/Convolutional Neural Network (ANN-CNN). After this stage, because all the classification methods will be evaluated separately, all articles have been subjected to a thorough evaluation within their subject group. In order to increase the clarity of the research and to reach semantically consistent results, if the same subject is logically relevant, it is examined in different categories repeatedly.

4 Results and Discussion

The use of AI tools in the detection and analysis of fake news and their successful results are instrumental in increasing the confidence in AI every day. In addition to teaching and learning content of ML from AI tools, it showed that there are more willing and interested users to deal with the fake news problem [16–28]. The biggest problem in the detection of fake news is to reach the content and dataset to be able to analyze. In this context, it has been observed that ML classifiers have high level of success in reaching the dataset and content in detecting fake news [28]. As a result of research on data collected from social networks such as Twitter, it was found that ML methods can detect fake news with an accuracy of 91% [19] and 99,9% [21]. In addition, some social networks that use clickbait's and phishing methods to increase advertising revenue have achieved 99,4% accuracy [20]. The study [22], which consisted of small data compiled from news articles, resulted in an accuracy of 77.2%. In another study [23], if the data collected were lost or noisy data, data evaluation techniques were used to determine how the missing data affected the result. The focus is on data preprocessing. Accordingly, the success was determined to be 16%. Kasbe and Jain [24] study on Facebook, a popular social network, examined articles as titles and content, although the rate of detecting fake news in headlines is around 80%, this rate is 92% in content analysis. Granik&Mesyura work [25] focused on spam messages on Facebook and achieved an average success rate of 74%. The method used in [26], unlike all other studies, focused on fake news and satirical news and produced a hybrid model that predicts the human factor, the ML approach, and the classification confidence of algorithms and determines whether the task requires human input. Therefore, the results differ from other studies and the measurement of success is more specific. Zhuk et al. [27] focused on the classification of fake news by analyzing the delivery and distribution mechanisms. As a result, it is remarkable that it creates awareness in terms of adding human factor to the subject. Alom et al. [1] focused on Twitter features of Twitter users in order to improve existing spam detection mechanisms on Twitter and achieved 91% successful results. As a result of the examination and evaluation articles where ML was applied, it was observed that ML tools were extremely successful in detecting fake news. In studies using DL methods [29–31], in Deep Neural Networks [29], Recurrent Neural Network (RNN), Long Short-Term Memories (LSTM), Gated Recurrent Unit (GRU) and Convolutional Neural Networks (CNN) In [30] and in [31] Convolutional Neural Networks (CNN) algorithms are used. It was observed that the in terms of results were very close to ML algorithms. In the study [32], natural language processing method was used among AI methods and 69.4% success was obtained in terms of the result. Atodiresei et al. [33] scored Twitter and users on the Twitter data using the Forensic Asset Recognition method from the AI tools. Vijayan and Mohler [34] focused on twit data during the election period. In their studies, retweets and their distribution were examined. When the articles which are not categorized by the authors [35–37] are examined; In [35], DL methodology is applied, [36] is based on a more qualitative research where no quantitative values are used, and [37] is based on a general classification of Twitter data and the spread of fake news is measured.

The systematic literature review showed that only a literature research was conducted with the superficial data in the field of fake news [18] and that there was a large gap in this field caused us to do a literature research on this subject. The use of AI tools to detect fake news is undoubtedly a very broad topic. However, when it is seen that Artificial Intelligence technologies have started to separate their sub-branches as a state-of-the-art, it is considered that the selected topic is suitable for research. When it is seen that AI is subdivided into sub-branches such as ML, DL, NLP, and the specific studies of each field are examined, it is seen that artificial intelligence is a product of advanced technology. It is known that each sub-branch uses different methodologies that serve different purposes in their own branches. However, investigating the detection of fake news that is the most common problem in social networks and reaching a common consensus makes the research valuable. The success levels of the researches [16–28] which applied to different algorithms of ML were higher than other studies. Although the content of each research is different, it is seen that the spread of fake news is examined in different ways and each one is evaluated correctly in its own field. However, it should be noted that DL and other methodologies (e.g. NLP, ANN/CNN) are still in their maturation phase. It is considered that the success rates are not very high because the methodology applied in deep learning techniques takes longer time in ML and does not have enough data sets to fully understand the learning techniques [29–31]. Likewise, the low level of success in natural language processing and other methods stems from the lack of training of the machine [32–34].

5   Conclusion

AI is suitable for use in social networks, and many applications that successfully implement AI tools to improve fake news detection were identified in the study. However, AI should not be seen as a magic bullet in social networks. Each AI implementation is unique and therefore the benefits described may not apply in all contexts. In order to take advantage of the benefits, each application must be fully implemented to avoid drawbacks in user interaction or system success rates. It is considered that this study will fill a gap in this field and shed light on further research. This study is thought to be a guide for researchers and individuals interested in the detection of fake news.

References

1. Alom, Z., Carminati, B., Ferrari, E.: Detecting spam accounts on Twitter. In: IEEE/ACM International Conference on Advances in Social Networks Analysis and Mining, pp. 1191–1198. IEEE (2018)
2. Ehsanfar, A., Mansouri, M.: Incentivizing the dissemination of truth versus fake news in social networks. In: 12th System of Systems Engineering Conference, pp. 1–6. IEEE (2017)
3. Shu, K., Wang, S., Liu, H.: Understanding user profiles on social network for fake news detection. In: 2018 IEEE Conference on Multimedia Information Processing and Retrieval, pp. 430–435. IEEE (2018)

4. Ongsulee, P.: Artificial intelligence, machine learning and deep learning. In: 2017 15th International Conference on ICT and Knowledge Engineering, pp. 1–6. IEEE (2017)
5. Machinery, C.: Computing machinery and intelligence-AM Turing. Mind **59**(236), 433 (1950)
6. Shah, H.: Turing's misunderstood imitation game and IBM's Watson success. In: Keynote in 2nd Towards a Comprehensive Intelligence Test Symposium, AISB (2011)
7. Burkov, A.: The Hundred-Page Machine Learning Book. Hakupäivä (2019)
8. Zhang, J., Li, Z., Pu, Z., Xu, C.: Comparing prediction performance for crash injury severity among various machine learning and statistical methods. IEEE Access **6**, 60079–60087 (2018)
9. Kachalsky, I., Zakirzyanov, I., Ulyantsev, V.: Applying reinforcement learning and supervised learning techniques to play Hearthstone. In: 2017 16th IEEE International Conference on Machine Learning and Applications, pp. 1145–1148. IEEE (2017)
10. Yuan, J., Yu, J.: Semi-supervised learning with bidirectional adaptive pairwise encoding. In: 2016 15th IEEE International Conference on Machine Learning and Applications, pp. 677–681. IEEE (2016)
11. Nijhawan, R., Srivastava, I., Shukla, P.: Land cover classification using super-vised and unsupervised learning techniques. In: 2017 International Conference on Computational Intelligence in Data Science, pp. 1–6. IEEE (2017)
12. Dharani, M., Sivachitra, M.: Motor imagery signal classification using semi supervised and unsupervised extreme learning machines. In: 2017 International Conference on Innovations in Information, Embedded and Communication Systems, pp. 1–4. IEEE (2017)
13. Mukhopadhyay, S., Tilak, O., Chakrabarti, S.: Reinforcement learning algorithms for uncertain, dynamic, zero-sum games. In: 2018 17th IEEE International Conference on Machine Learning and Applications, pp. 48–54. IEEE (2018)
14. Qu, X., Wei, T., Peng, C., Du, P.: A fast face recognition system based on deep learning. In: 2018 11th International Symposium on Computational Intelligence and Design, vol. 1, pp. 289–292. IEEE (2019)
15. Akyön, F.Ç., Alp, Y.K., Gök, G., Arıkan, O.: Deep learning in electronic warfare systems: automatic intra-pulse modulation recognition. In: 2018 26th Signal Processing and Communications Applications Conference, pp. 1–4. IEEE (2018)
16. Jeong, Y., Kim, S., Yoon, B.: An algorithm for supporting decision making in stock investment through opinion mining and machine learning. In: 2018 Portland International Conference on Management of Engineering and Technology, pp. 1–10. IEEE (2018)
17. Della Vedova, M.L., Tacchini, E., Moret, S., Ballarin, G., DiPierro, M., de Alfaro, L.: Automatic online fake news detection combining content and social signals. In: 2018 22nd Conference of Open Innovations Association, pp. 272–279. IEEE (2018)
18. Cardoso Durier da Silva, F., Vieira, R., Garcia, A.C.: Can machines learn to detect fake news? A survey focused on social network. In: 52nd Hawaii International Conference on System Sciences (2019)
19. Aborisade, O., Anwar, M.: Classification for authorship of tweets by comparing logistic regression and Naive Bayes classifiers. In: 2018 IEEE International Conference on Information Reuse and Integration, pp. 269–276. IEEE (2018)
20. Aldwairi, M., Alwahedi, A.: Detecting fake news in social media networks. In: 9th International Conference on Emerging Ubiquitous Systems and Pervasive Networks, pp. 215–222 (2018)
21. Aphiwongsophon, S., Chongstitvatana, P.: Detecting fake news with machine learning method. In: 2018 15th International Conference on Electrical Engineering/Electronics, Computer, Telecommunications and Information Technology, pp. 528–531. IEEE (2018)

22. Gilda, S.: Evaluating machine learning algorithms for fake news detection. In: 2017 IEEE 15th Student Conference on Research and Development, pp. 110–115. IEEE (2017)

23. Kotteti, C.M.M., Dong, X., Li, N., Qian, L.: Fake news detection enhancement with data imputation. In: 2018 IEEE 16th International Conference on Dependable, Autonomic and Secure Computing, 16th International Conference on Pervasive Intelligence and Computing, 4th International Conference on Big Data Intelligence and Computing and Cyber Science and Technology Congress, pp. 187–192. IEEE (2018)

24. Jain, A., Kasbe, A.: Fake news detection. In: 2018 IEEE International Students' Conference on Electrical, Electronics and Computer Sciences, pp. 1–5. IEEE (2018)

25. Granik, M., Mesyura, V.: Fake news detection using Naive Bayes classifier. In: 2017 IEEE First Ukraine Conference on Electrical and Computer Engineering, pp. 900–903. IEEE (2017)

26. Shabani, S., Sokhn, M.: Hybrid machine-crowd approach for fake news detection. In: 2018 IEEE 4th International Conference on Collaboration and Internet Computing, pp. 299–306. IEEE (2018)

27. Zhuk, D., Tretiakov, A., Gordeichuk, A.: Methods to identify fake news in social network using machine learning. In: Proceedings of the 22nd Conference of Open Innovations Association, p. 59 (2018)

28. Helmstetter, S., Paulheim, H.: Weakly supervised learning for fake news detection on Twitter. In: 2018 IEEE/ACM International Conference on Advances in Social Networks Analysis and Mining, pp. 274–277. IEEE (2018)

29. Granik, M., Mesyura, V., Yarovyi, A.: Determining fake statements made by public figures by means of artificial intelligence. In: 2018 IEEE 13th International Scientific and Technical Conference on Computer Sciences and Information Technologies, vol. 1, pp. 424–427. IEEE (2018)

30. Girgis, S., Amer, E., Gadallah, M.: Deep learning algorithms for detecting fake news in online text. In: 2018 13th International Conference on Computer Engineering and Systems, pp. 93–97. IEEE (2018)

31. Seo, Y., Seo, D., Jeong, C. S.: FaNDeR: fake news detection model using media reliability. In: TENCON 2018–2018 IEEE Region 10 Conference, pp. 1834–1838. IEEE (2018)

32. Traylor, T., Straub, J., Snell, N.: Classifying fake news articles using natural language processing to identify in-article attribution as a supervised learning estimator. In: 2019 IEEE 13th International Conference on Semantic Computing, pp. 445–449. IEEE (2019)

33. Atodiresei, C.S., Tănăselea, A., Iftene, A.: Identifying fake news and fake users on Twitter. Procedia Comput. Sci. **126**, 451–461 (2018)

34. Vijayan, R., Mohler, G.: Forecasting retweet count during elections using graph convolution neural networks. In: 2018 IEEE 5th International Conference on Data Science and Advanced Analytics, pp. 256–262. IEEE (2018)

35. Gupta, S., Thirukovalluru, R., Sinha, M., Mannarswamy, S.: CIMTDetect: a community infused matrix-tensor coupled factorization based method for fake news detection. In: 2018 IEEE/ACM International Conference on Advances in Social Networks Analysis and Mining, pp. 278–281. IEEE (2018)

36. Figueira, Á., Oliveira, L.: The current state of fake news: challenges and opportunities. Procedia Comput. Sci. **121**, 817–825 (2017)

37. Vosoughi, S., Roy, D., Aral, S.: The spread of true and false news online. Science **359** (6380), 1146–1151 (2018)

Using Soft Computing Methods for the Functional Benchmarking of an Intelligent Workplace in an Educational Establishment

Gurru I. Akperov$^{(\boxtimes)}$ ⓘ, Vladimir V. Khramov ⓘ,
and Anastasiya A. Gorbacheva ⓘ

Private Educational Institution of Higher Education
"SOUTHERN UNIVERSITY (IMBL)", Rostov-on-Don, Russia
pr@iubip.ru, vxpamov@inbox.ru, asya379@yandex.ru

Abstract. The article describes a method developed for assessing the functional proximity of an intelligent workplace (IWP) of a student with the best available examples based on a fuzzy multiple attribute decision-making method. The developed method is aimed at evaluating the success of the learning process and consists in aggregating individual indicators of the reviewed functions. The proposed method demonstrates the possibility of adapting the standard integral score method used to estimate the IWP to fuzzy initial data, thus providing a number of major advantages. In particular, the proposed method allows us to change the entire complex of the studied functions and their parameters depending on the study goals and objectives without substantially reworking the model. Besides, we can adjust the weights of parameters to match the sectoral and territorial specificity of the university or to comply with the expert assessments; bring together quantitative and qualitative estimates of the IWP indicators with the corresponding estimates of several intelligent learning tools selected as templates.

Keywords: Comparative analysis of enterprise data · Integrated assessment · Aggregation · Fuzzy multiple attribute decision-making methods

1 Introduction

The selection of an intelligent workplace (IWP) for an educational establishment, the structure of its algorithmic and software tools for supporting the ongoing educational process may be viewed as a multi-parameter optimization process with fuzzy criteria. The optimization is based on the fuzzy multi-criteria decision-making method. Its mathematical model becomes available upon formalizing the optimization goal (the "ideal" variant) as a requirement vector, describing a variety of alternatives and developing qualitative optimization criteria.

© Springer Nature Switzerland AG 2020
R. A. Aliev et al. (Eds.): ICSCCW 2019, AISC 1095, pp. 54–60, 2020.
https://doi.org/10.1007/978-3-030-35249-3_6

2 Materials and Method

The generalized mathematical formulation of the selection problem [1] can be represented by the following set:

$$<V, F, P_T, L, W; T, A>$$ (1)

where the initial sets are as follows:

V is a set of alternatives;

F is a set of descriptions;

P_T is a set of possible outcomes;

L is a vectorial criterion of the outcome estimate;

W is a preference structure.

A certain solution (rule) or algorithm T must be built up, which allows to perform the required action A with the set of alternatives V.

The preference structure determines the procedure for comparing L (P) estimators, and the decision rule (or algorithm) T determines the principle of choosing elements from the set V based on the comparison results in accordance with the required action A.

In this case, the mathematical description of the IWP selection procedure and its elements consists in the interpretation of the above components (1) with due account for the peculiarities of the initial data. The homogeneous structure of the system description and of its elements allows us to view them as design alternatives.

A set of alternatives consist of a finite number of elements:

$$V = \{V_I, i = 1\}.$$ (2)

where each of the elements may be described by a fuzzy parameter vector:

$$V_i : \overline{P}_i = \left[p_{i1}, p_{i2}, \ldots, p_{ij}, \ldots, p_{im}\right].$$ (3)

The ideal variant (goal) may be described by a requirement vector:

$$\overline{P}_T = \left[p_{T1}, p_{T2}, \ldots, p_{Tj}, \ldots, p_{Tm}\right]$$ (4)

The importance of the requirement parameter must be taken into account when setting the preference vector:

$$\overline{W} = \left[w_1, w_2, \ldots, w_j, \ldots, w_m\right]$$ (5)

The above initial data sets the rank of the selection model:

$$Rang(T) = m \times n.$$ (6)

In general, the initial data may be represented in the form of a problem matrix:

$$\left\| P_{ij} \right\|; \quad i = \overline{1, n}, \quad j = \overline{1, m}, \tag{7}$$

where P_{ij} is an estimator of alternative i in accordance with parameter j;

n is a number of parameters included;

m is a number of candidate solutions.

This representation format allows us to express the initial data (see Table 1).

The decision-making procedure can be represented as a sequence of actions to assess, rank and select alternatives in accordance with the set of criteria [1].

The set-theoretic approach used in the studies to select the IWP and its elements is based on the progressive contraction of the set of possible design solutions. Selection stages can be considered as filters, which screen the wrong alternatives. The filtering process is a decision-making process for selecting alternatives under uncertainty.

Table 1. Initial data

IWP alternatives	Parameters					
	P_1	P_2	$\cdots$	P_j	$\cdots$	P_m
V_1	P_{11}	P_{12}	$\cdots$	P_{1J}	$\cdots$	P_{1m}
V_2	P_{21}	P_{22}	$\cdots$	P_{2J}	$\cdots$	P_{2m}
$\cdots$	$\cdots$	$\cdots$	$\cdots$	$\cdots$	$\cdots$	$\cdots$
V_n	P_{n1}	P_{n2}	$\cdots$	P_{nJ}	$\cdots$	P_{nm}

Standard stages in the decision-making pattern include the what-if evaluation with ranking and selecting the most preferable alternatives [1, 2]. As is evident, the initial finite set of constraints F is contracted at the first stage. As a result, a set of admissible alternatives is formed $V_d = V \cap F$.

Then, after evaluating the alternatives in accordance with the selected criteria, the set of V_d is contracted to the set of rational alternatives $V_e \in V_d$. The set V_r contains the desired solution $S \in V_r$ obtained by introducing additional preferences or by direct indication of the alternative [3].

Thus, the problem of selecting alternatives under uncertainty of the initial data for a particular application involves selecting a set of preference criteria.

An important step in the decision-making process for selecting an alternative is selecting the evaluation criteria, or the estimator L. This selection largely determines the type of the process. Taking into account the features of the fuzzy sets and the fact that in this case the mutual arrangement of the alternatives within a common hierarchy is more important than their grouping, we use the disagreement metric [4] based on the Hamming distance:

$$L_i = \bar{W} \times (\bar{\Pi}_T - \bar{\Pi}_i). \tag{8}$$

As is evident, the inclusion of high-quality fuzzy estimators will lead to the emergence of an indefinite total estimator L and, accordingly, a certain risk of selecting a non-optimal alternative. However, this uncertainty, along with the uncertainty of quantitative parameters, must be taken into account in the selection process. In case of uncertainty in the initial data, indicator (8) cannot be used explicitly, since it does not provide for operating with fuzzy values.

Therefore, we considered it necessary to adapt this disagreement (as compared to [3]) for using in conditions of fuzzy data including fuzzy description of the alternative and fuzzy requirements. Besides, we presided the procedure for setting the criterion in accordance with preferences depending on the importance of each requirement.

Technical requirements are given in the form of a vector $\bar{P}_T$ (4), each component of which represents an estimate (either qualitative or quantitative) of the desired value of the IWP parameter or its element. Characteristics of the considered alternative are introduced by vector $\bar{P}_T$. The disagreement between the description of the alternative and the requirements in general can be calculated using the following formula:

$$\bar{L}_i = \bar{P}_i - \bar{P}_T. \tag{9}$$

In order to determine the difference between fuzzy vectors, we need to describe the disagreement for each component. To calculate the disagreement for each j-th parameter of the description of the i-th alternative, we need to calculate the difference between the parameter P_{ij} and the requirement for it (the "ideal" variant) P_{Tj}.

$$l_{ij} = l(P_{ij} - P_{Tj}). \tag{10}$$

In case of using linguistic variables for estimating the requirement parameters and describing the i-th alternative of V_i, we need to determine the value (10) using a method similar to the centroid method known in the probability theory [4, 5]. It means that the difference between the two fuzzy sets, corresponding to the parameters P_{Tj} and P_{ij}, consists of two elements: deterministic component l_{ij}^d and undetermined component l_{ij}^f. The deterministic part of the estimator l_{ij}^d may be graphically interpreted as the segment connecting the two centers of gravity of the corresponding membership functions within A and B with coordinates (μ_a, u_a) and (μ_b, u_b) respectively.

Then the estimator l_{ij}^d may be calculated using the formula:

$$l_{ij}^d = \sqrt{(u_b - u_a)^2 + (\mu_b - \mu_a)^2}. \tag{11}$$

The estimator l_{ij}^f has the following value:

$$l_{ij}^f = \sqrt{S_a^2 + S_b^2 - S_a S_b} \tag{12}$$

where S_a and S_b are the areas limited by the membership functions within the sets A and B, respectively. In this case, the areas S_a and S_b estimate the uncertainty level of parameters P_{Tj} and P_{ij}. Formula (12) allows us to take into account the indefinite part of the difference between the two fuzzy sets. The fuzzy value of l_{ij} can be expressed by the interval approximation:

$$l_{ij} = l_{ij}^d \pm (l_{ij}^f)/2. \tag{13}$$

For all n alternatives, the disagreement estimators can be reduced to an $m \times n$ matrix, which can be called a solution matrix. Each i-th row of the decision matrix characterizes the i-th alternative by m-parameters.

The relative uncertainty level of the estimator l_{ij} can be determined using the formula:

$$\Delta l_{ij} = l_{ij}^f / l_{ij}^d.$$

Thus, having determined the disagreement (6), we can determine the total disagreement for the i-th alternative as the sum of the disagreements for each parameter within the m set:

$$L_i = \sum_{j=1}^{m} l_{ij} \ , \ i = \overline{1, n}. \tag{14}$$

The indicator (14) is used if the given parameters are equivalent and meet the technical specification requirements. When it comes to specifying the priority of a single parameter or a group, the weight vector is assigned $\bar{W}$.

Random assignment of weight coefficients in the estimator can lead to the problem distortion. In this case, an alternative which does not meet the problem requirements can be selected. Setting the priority for each of the estimate parameters with due account for the IWP user preferences may be viewed as a semantic orientation of the criterion (8). There are a few ways to set the priority. We can set it as follows: by the priority number $J = (1, 2, \ldots, m)$, the priority vector $\bar{V} = [v_1, v_2, \ldots, v_m]$, or the weight vector $\bar{W} = [w_1, w_2, \ldots, w_m]$.

In this article, we set the priority using a weight vector 4 $\bar{W}$ of the dimension m, according to the number of the used parameters, the components of which satisfy the condition:

$$\sum_{j=1}^{m} W_j = 1; \quad 0 \leq w_j \leq 1, \quad j = \overline{1, n} \tag{15}$$

It is recommended to build a vector $\bar{W}$ in stages, since the direct assignment of all m numbers is quite complicated. This is due to limited human ability to estimate the complex totality of parameters [4, 6].

At the front end, it is recommended to set the priority order J expressed by the ordinal scale. The components of the priority vector can be determined by pairwise ranking of the parameters ordered in accordance with J [6]. To do this, we need information about two compared parameters, since, by setting $V_m = 1$, we can calculate V_{m-1}, V_{m-2}, etc., up to V_1.

Finally, the values of the weight vector components can be determined using the formula (19):

$$W_j = \frac{\sum\limits_{k=j}^{m} V_{kj}}{\sum\limits_{j=1}^{m} \sum\limits_{k=1}^{m} V_{kj}} \tag{16}$$

The described method narrows down the multidimensional problem of the vector determination $\bar{W}$ to the repeated simple pairwise comparison and ranking. Similar methods are known as the transitive closure of a binary relation (12).

The total weighted disagreement is determined as a scalar product for all alternatives:

$$L_i^W = \bar{l}_i \times \bar{W} \tag{17}$$

The value of the deterministic (D_i) and undetermined (F_i) components for each i-th alternative can be determined using the following formulas:

$$F_d = \sum_{j=1}^{n} W_i * l_{ij}^d \tag{18}$$

$$F_i = \sum_{j=1}^{n} W_i * l_{ij}^f \tag{19}$$

The total disagreement can be expressed as follows:

$$L_i^W = D_i \pm F_i/2 \tag{20}$$

Using estimates (20) and (21), we can calculate the relative uncertainty level:

$$\Delta F_i = F_i/D_i \tag{21}$$

3   Discussion

Thus, the expression (20) can be selected as a criterion for estimating the alternatives in accordance with their requirement satisfaction degree. The proposed fuzzy multiple attribute decision-making method has such an important advantage as variability. In other words, it enables adding certain parameters; changing their weights in accordance with the study objectives and expert estimates; supplementing the estimation complex with new blocks without aligning the method with the changes made.

Besides, the proposed method, unlike the standard methods, allows us to use both the values of the studied parameters, and the dynamics of their change to perform an integrated assessment. Each IWP can be assigned with an estimate vector, which can be supplemented and expanded, depending on the study objectives, through the use of additional indicators. As a result, we may establish a basis for the IWP clustering within one specialty with the subsequent study of the dependencies existing between their characteristics.

The decision-making on the selection of the optimal alternative can be represented as a decision matrix [6]. Each table line describes a vector representing the disagreement of the alternative descriptor with the established requirements. These requirements are given in the bottom line of the table. The preference structure is defined by a string of weight coefficients established for each of the parameters.

References

1. Serdyuchenko, P.Ya., Khramov, V.V.: Principles of fuzzy aggregation in the management of complex systems. In: The Collection: Problems of Ensuring the Efficiency and Sustainability of Complex Technical Systems XIX Interdepartmental Scientific and Technical Conference, pp. 288–291 (2000)
2. Chernyshev, Yu.O., Khramov, V.V.: Features of the aggregation of qualitative features of reference points in vision systems. News TSE **3**(21), 55 (2001). https://elibrary.ru/item.asp?id=12886334
3. Khramov, V.V.: Features of the majority processing of fuzzy information. In: The Collection: Spectral Methods of Information Processing in Scientific Research Reports of the All-Russian Conference 2000, pp. 136–138. RFBR, Institute of Mathematical Problems of Biology RAS (2000). https://elibrary.ru/item.asp?id=32656899
4. Danchenko, D.P., Khramov, V.V., Tsarkov, A.N.: Aggregation of several sources of fuzzy information in the ergatic system. In: The Collection: Problems of Ensuring the Efficiency and Stability of the Functioning of Complex Technical Systems Collection of Works, pp. 441–443 (2003). https://elibrary.ru/item.asp?id=32760833
5. Khramov, V.V.: Intelligent information systems: the presentation of knowledge in information systems. Teaching aid Rostov-on-Don (2011). https://elibrary.ru/item.asp?id=32762297
6. Khramov, V.V., Gvozdev, D.S.: Intellectual information systems: intellectual data analysis, p. 152. Rostov State University of Communications, Rostov-on-Don (2016). https://elibrary.ru/item.asp?id=28322733

Examination of Computational Thinking Skill Levels of Secondary School Students: The Case of Near East College

Burak Simsek ⓘ and Sezer Kanbul$^{(\boxtimes)}$ ⓘ

Near East University, Nicosia 99138, TRNC, Mersin 10, Turkey
{burak.simsek, sezer.kanbul}@neu.edu.tr

Abstract. This study is a descriptive research conducted with the purpose of identifying the computational thinking abilities of 103 secondary school students at the ages of 13 and 14. The research data were collected using a questionnaire form which consisted of two sections, namely personal information form and Computational Thinking Skills (CTS) scale. Statistical analysis of research data; frequency analysis, definitive statistics, Shapiro-Wilk test, Mann-Whitney U test, Kruskal-Walls test and Spearman correlation analysis were conducted. At the end of the study it was found out that more than half of the students (52.43%) favored mathematics and natural science courses, that the algorithmic thinking abilities of this group were higher, that students without siblings solved problems better and that there is statistically significant and positive correlation between the scores obtained from all dimensions in the computational thinking skills.

Keywords: Computational thinking · Algorithmic thinking · Problem solving

1 Introduction

One of the most central publications which offered functional definition of elementary and secondary education levels defines Computational Thinking as an algorithmic problem-solving approach [1]. Computational Thinking uses common paths with scientific thinking in understanding such concepts as computability, intelligence, wisdom and human behaviors in engineering while designing and evaluating a complicated system at problem solving stage with mathematical thinking [2, 3].

A study which examined the Computational Thinking researches conducted between 2006 and 2016 with biometric methods and tried to identify the sub-study fields of the area, tendencies in the field, and leading publications, authors and concepts showed that Computational Thinking is being studied more widely in the fields of education and computer sciences. However, the literature on the reasons for which Computational Thinking should be taught to every child is almost non-existent [4].

Coding Education has considerable importance in teaching students Computational Thinking at early ages. Coding education is showing progress on a daily basis. Coding education which began to be included in the curricula beginning from elementary school offers success to children in analytical thinking, creativity and problem-solving as well as the field of computer. It has been displayed that children who learn coding can produce solutions to the problems they encounter, correct their errors and evaluate their results.

© Springer Nature Switzerland AG 2020
R. A. Aliev et al. (Eds.): ICSCCW 2019, AISC 1095, pp. 61–68, 2020.
https://doi.org/10.1007/978-3-030-35249-3_7

Coding Education offers an entirety of basic skills required by the literacy of the 21st century along with such skills as Creativity, Algorithmic Thinking, Critical Thinking, Problem Solving and Cooperativity that it contains [5–7].

In this context, computational thinking skills of students are important and the purpose of this study is to describe the existing computational thinking skills of secondary school students.

1.1   Sub-problems

1. What are the introductory features of students included in the study?
2. What is the computational thinking level of students?
3. Is there any significant difference between computational thinking abilities of students according to their;

    3.1. gender?
    3.2. number of siblings?
    3.3. age?
    3.4. courses they favor?

4. Is there any significant difference between sub-dimensions of students in the computational thinking abilities scale?

2   Method

2.1   Research Model

The research is a descriptive study conducted with the purpose of identifying the computational thinking skills of students. Survey method was used as research method.

2.2   Sample of the Study

The sample of the study consists of 103 secondary school students at the age of 13 and 14 at Northern Cyprus Near East College. Participants were determined based on simple random sampling. Simple random sampling is a method of selection of a sample in which each sampling unit in the population has an equal probability of being selected [8].

2.3   Data Collection Tools

Research data were collected through a questionnaire form consisting of two sections, namely personal information form and computational thinking abilities.

In the first section introductory information of the students such as gender, age, number of siblings and favorite course were included.

The second section used *"Computational Thinking Skills (CTS)" scale* which was developed with the purpose of identifying the computational thinking skills of 103 secondary school students who were included in the study.

2.4 Statistical Analysis of the Data

Statistical Package for Social Sciences (SPSS) 24.0 software was used in the statistical analysis of the research data. The distribution of students according to their introductory characteristics was determined using frequency analysis and such complementary statistics as mean, standard deviation, minimum and maximum values belonging to the scores obtained from Computational Thinking Skills Scale were provided. Shapiro-Wilk test was used in order to examine the conformity of the data set to normal distribution so that the tests which would be used in the comparison of the Computational Thinking Skills Scale scores of students according to their introductory characteristics could be performed and it was found out that they did not comply with normal distribution. Accordingly, Mann-Whitney U test and Kruskal-Walls test, which are non-parametric hypothesis tests, were used in the study. The correlations between scores obtained by students from Computational Thinking Skills Scale were identified with Spearman correlation test.

3 Findings

An examination of Table 1 shows that 49.51% of the students included in the study were males and 50.49% were females, 21.36% were 13 years old and 78.64% were 14 years old. It has been found out that 12.62% had no siblings, 71.84% had one sibling and 15.53% had two and more siblings. An examination of the favorite courses of the students showed that 52.43% favored mathematics and natural sciences, 15.53% favored Turkish and social sciences and 32.04% favored drawing-music-physical education courses.

Table 1. Distribution of students according to their introductory characteristics

	Number (n)	Percentage (%)
Gender		
Male	51	49,51
Female	52	50,49
Age		
13 years old	22	21,36
14 years old	81	78,64
Number of siblings		
None	13	12,62
One sibling	74	71,84
Two and more	16	15,53
Most favorite course		
Mathematics and natural sciences	54	52,43
Turkish and social sciences	16	15,53
Drawing-music-physical education	33	32,04
Total	103	100,00

Table 2. Scores obtained by students from computational thinking skills scale

	n	$\bar{x}$	s	Min	Max
Creativity	103	3,73	0,61	2,13	5,00
Algorithmic thinking	103	3,05	1,05	1,00	5,00
Cooperativity	103	3,74	0,97	1,00	5,00
Critical thinking	103	3,41	0,97	1,00	5,00
Problem solving	103	3,55	0,78	1,25	5,00
Entire scale	103	3,49	0,69	1,41	5,00

An examination of Table 2 shows that students received $\bar{x} = 3,73 \pm 0,61$ points from creativity sub-dimension, $\bar{x} = 3,05 \pm 1,05$ points from algorithmic thinking sub-dimension, $\bar{x} = 3,74 \pm 0,97$ points from cooperativity sub-dimension, $\bar{x} = 3,41 \pm 0,97$ points from critical thinking sub-dimension, and $\bar{x} = 3,55 \pm 0,78$ points from problem-solving sub-dimension of the computational thinking skills scale. It has been found out that students received $\bar{x} = 3,49 \pm 0,69$ points from the entire computational thinking skills scale and the minimum and maximum scores were 1.41 and 5.0, respectively.

Table 3. Comparison of scores obtained by students from computational thinking skills scale according to their genders

	Gender	n	x	s	Mean rank	Z	p
Creativity	Male	51	3,67	0,65	49,72	−0,770	0,441
	Female	52	3,79	0,57	54,24		
Algorithmic thinking	Male	51	3,19	1,00	55,73	−1,255	0,209
	Female	52	2,92	1,10	48,35		
Cooperativity	Male	51	3,74	0,99	51,92	−0,027	0,979
	Female	52	3,73	0,97	52,08		
Critical thinking	Male	51	3,41	0,97	52,21	−0,070	0,945
	Female	52	3,40	0,97	51,80		
Problem solving	Male	51	3,55	0,74	51,86	−0,046	0,963
	Female	52	3,55	0,82	52,13		
Entire scale	Male	51	3,51	0,69	53,67	−0,561	0,575
	Female	52	3,48	0,69	50,37		

An examination of Table 3 shows that there is no statistically significant difference between the scores obtained from the entire computational thinking skills scale and the sub-dimensions of the scale according to the gender of students ($p > 0,05$).

Table 4. Comparison of scores obtained by students from computational thinking skills scale according to the number of their siblings

	Number of siblings	n	$\bar{x}$	s	Mean rank	X^2	p	Difference
Creativity	No siblings	13	4,13	0,59	70,15	5,808	0,055	
	One sibling	74	3,66	0,59	48,59			
	Two and more	16	3,76	0,62	53,03			
Algorithmic thinking	No siblings	13	3,46	1,20	63,35	6,661	0,036*	1-2
	One sibling	74	2,89	1,00	47,26			2-3
	Two and more	16	3,49	1,02	64,72			
Cooperativity	No siblings	13	4,21	0,80	65,92	3,306	0,191	
	One sibling	74	3,67	0,98	50,25			
	Two and more	16	3,64	1,02	48,78			
Critical thinking	No siblings	13	3,87	1,05	66,62	4,091	0,129	
	One sibling	74	3,30	0,97	48,85			
	Two and more	16	3,50	0,83	54,69			
Problem solving	No siblings	13	4,04	0,72	71,77	6,560	0,038*	1-2
	One sibling	74	3,47	0,74	48,85			1-3
	Two and more	16	3,51	0,88	50,50			
Entire scale	No siblings	13	3,94	0,72	70,96	6,606	0,037*	1-2
	One sibling	74	3,40	0,67	48,11			1-3
	Two and more	16	3,58	0,65	54,56			

An examination of Table 4 shows that there is statistically significant difference between the scores obtained from the algorithmic thinking sub-dimension of the computational thinking skills scale according to the gender of students (p < 0,05).

It has been found out that there is statistically significant difference between the scores obtained from the entire computational thinking skills scale according to the number of siblings of students (p < 0,05).

Table 5. Comparison of scores obtained by students from computational thinking skills scale according to their age

	Age	n	$\bar{x}$	s	Mean rank	Z	p
Creativity	13 years old	22	3,73	0,47	51,23	−0,137	0,891
	14 years old	81	3,73	0,64	52,21		
Algorithmic thinking	13 years old	22	3,28	0,79	58,57	−1,165	0,244
	14 years old	81	2,99	1,11	50,22		
Cooperativity	13 years old	22	3,73	1,00	52,20	−0,036	0,971
	14 years old	81	3,74	0,97	51,94		
Critical thinking	13 years old	22	3,38	0,74	49,39	−0,464	0,642
	14 years old	81	3,41	1,02	52,71		
Problem solving	13 years old	22	3,48	0,75	50,11	−0,334	0,738
	14 years old	81	3,57	0,79	52,51		
Entire scale	13 years old	22	3,52	0,51	52,11	−0,020	0,984
	14 years old	81	3,49	0,73	51,97		

An examination of Table 5 shows that there is no statistically significant difference between the scores obtained from the entire computational thinking skills scale and the sub-dimensions of the scale according to the gender of students (p > 0,05).

Table 6. Comparison of scores obtained by students from computational thinking skills scale according to their favorite course

	Course	n	$\bar{x}$	s	Mean rank	X^2	p	d
Creativity	M	54	3,84	0,54	57,73	4,271	0,118	
	T	16	3,63	0,66	47,38			
	D	33	3,60	0,67	44,86			
Algorithmic thinking	M	54	3,56	0,82	66,17	30,753	0,000*	1-2
	T	16	2,00	0,70	22,50			1-3
	D	33	2,72	1,04	43,12			
Cooperativity	M	54	3,80	0,94	53,82	0,849	0,654	
	T	16	3,75	1,15	53,94			
	D	33	3,62	0,95	48,08			
Critical thinking	M	54	3,70	0,85	60,76	12,277	0,002*	1-2
	T	16	2,72	1,07	32,78			2-3
	D	33	3,26	0,92	46,98			
Problem solving	M	54	3,69	0,69	57,76	4,616	0,099	
	T	16	3,41	0,91	49,50			
	D	33	3,38	0,82	43,79			
Entire scale	M	54	3,72	0,58	61,64	12,594	0,002*	1-2
	T	16	3,10	0,70	35,97			2-3
	D	33	3,32	0,73	44,00			

*d: difference, *M: Mathematics and natural sciences, T: Turkish and social sciences, D: Drawing-music-physical education*

It has been found out that there is statistically significant difference between the scores obtained from the algorithmic thinking sub-dimension of the computational thinking skills scale according to the favorite courses of students (p < 0,05) (Table 6).

It has been found out that there is statistically significant difference between the scores obtained from critical thinking sub-dimension of the computational thinking skills scale according to the favorite courses of students (p < 0,05).

It has been found out that there is statistically significant difference between the scores obtained from the entire computational thinking skills scale according to the favorite courses of students (p < 0,05).

Table 7. Correlations between scores obtained by students from sub-dimensions of the computational thinking skills scale

		Creativity	Algorithmic thinking	Cooperativity	Critical thinking	Problem solving	Entire scale
Creativity	r	1,000					
	p	.					
Algorithmic thinking	r	0,388	1,000				
	p	0,000*	.				
Cooperativity	r	0,426	0,256	1,000			
	p	0,000*	0,016	.			
Critical thinking	r	0,556	0,563	0,449	1,000		
	p	0,000*	0,000*	0,000*	.		
Problem solving	r	0,678	0,546	0,671	0,679	1,000	
	p	0,000*	0,000*	0,000*	0,000*	.	
Entire scale	r	0,738	0,689	0,675	0,838	0,909	1,000
	p	0,000*	0,000*	0,000*	0,000*	0,000*	.

An examination of Table 7 shows that there are significant and positive correlations between scores obtained by students from all the sub-dimensions of the computational thinking skills scale ($p < 0,05$).

4   Discussion and Conclusion

In this research, the computational thinking skills levels of secondary school students are described and the following results are obtained.

It has been found out that the computational thinking skill levels of students are similar in terms of their genders and ages; however, an examination of their favorite course shows that more than half of the students favor mathematics and natural sciences most (52.43%).

It has been concluded that in the entire scale students who favor mathematics and natural sciences enjoy higher computational thinking skills compared to other students. As a result, another finding is that algorithmic thinking skills of students who favor mathematics and natural sciences is higher compared to other students, and that critical thinking skills of students whose favorite course is Turkish and social sciences is lower compared to other students.

It was found out that students had developed computational thinking skills, that based on the scale items "they believed that they could solve most of the problems they faced", "they liked having cooperative learning experiences with their group friends", "they more easily learned the topics covered using mathematical symbols and concepts", "they thought that trying to solve complicated problems was fun".

Another conclusion is that students who have no siblings have received higher scores in problem-solving sub-dimension. In addition, the scores obtained by students without any siblings are higher compared to students with single sibling and two and more siblings.

The study which applied Computational Thinking Skills (CTS) scale displayed similar results. The study group of this study consisted of 241 seventh and eighth grade students at secondary schools in the city center of Amasya. According to the perception of students, their Computational Thinking skills are considerably higher. However, problem-solving skills of students were found to be rather low compared to other skills [9].

Finally, it has been found out that there is statistically significant and positive correlation between scores obtained by students from all sub-dimensions in the computational thinking skills scale ($p < 0.05$). As the scores obtained by students from any sub-dimension in computational thinking skills scale increase, so do the scores they obtained from other sub-dimensions.

References

1. Barr, V., Stephenson, C.: Bringing computational thinking to K-12: what is involved and what is the role of the computer science education community? ACM Inroads **2**(1), 48–54 (2011)
2. Price, A.: Creative Maths Activities for Able Students. Ideas for Working with Children Ager 11 to 14. Paul Chapman Publishing, London (2006)
3. Wing, J.M.: Computational thinking. Commun. ACM **49**(3), 33–35 (2006)
4. Ozcinar, H.: Bibliometric analysis of computational thinking research. Educ. Technol. Theory Pract. **7**(2), 149–171 (2017). https://doi.org/10.17943/etku.288610
5. Prensky, M.: Listen to the natives. Learn. Dig. Age **4**(63), 8–13 (2005)
6. Shailaja, J., Sridaran, R.: Computational thinking the intellectual thinking for the 21st century. Int. J. Adv. Netw. Appl., 39–46 (2015). Special issue
7. Yadav, A., Hong, H., Stephenson, C.: Computational thinking for all: pedagogical approaches to embedding 21st century problem solving in K-12 classrooms. TechTrends **60**, 565–568 (2016)
8. Levy, P.S., Lemeshow, S.: Sampling of Populations: Methods and Applications. Wiley, Hoboken (2013)
9. Korkmaz, O., Cakir, R., Ozden, M.Y., Oluk, A., Sarioglu, S.: Investigation of individuals' computational thinking skills in terms of different variables. OMU J. Fac. Educ. **34**(2), 68–87 (2015). https://doi.org/10.7822/omuefd.34.2.5

Z-Differential Equations

Rafik A. Aliev[1,2]($\boxtimes$) ⓘ and Akif V. Alizadeh[3] ⓘ

[1] Azerbaijan State Oil and Industry University, 20 Azadlig Avenue,
Baku AZ1010, Azerbaijan
raliev@asoa.edu.az
[2] Georgia State University, Atlanta, USA
[3] Department of Control and Systems Engineering,
Azerbaijan State Oil and Industry University, 20 Azadlig Avenue,
Baku AZ1010, Azerbaijan
akifoder@yahoo.com, a.alizade@asoiu.edu.az

Abstract. Existing studies on Z-number valued DEs are scarce. In this paper we make a first step towards formalization of Z-number based initial value problem and its solution. An example is provided to illustrate the proposed approach.

Keywords: Z-number · Reliability · Horizontal membership functions · Differential equation

1 Introduction

Differential equations are used for modeling of processes in various fields. An important related aspect is that real-world processes are characterized by uncertainty sourced from experimental conditions, data collection, measurement etc. To account for uncertainty, two main theories are used –probability theory (for modeling of random phenomena) and fuzzy sets theory (for modeling of imprecision). Fuzzy DEs (FDEs) [2] are used to model dynamical processes with fuzzy number-valued variables in various fields – medicine, chemical engineering, economics, control theory applications etc [1, 3–6]. The theory of FDE relies on fuzzy arithmetic. The latter includes two main formalisms: standard fuzzy arithmetic and horizontal membership functions-based (HMF) fuzzy arithmetic [7, 8]. The use of standard approach to describe FDE is based on Hukuhara derivative, existence of which is related to very restrictive conditions. In view of this, the use of HMF approach is computationally more effective for dealing with FDEs [19].

Often, fuzzy and probabilistic uncertainties are considered separately. In order to account for combination of fuzzy and probabilistic information, Zadeh proposed the concept of a Z-number [9]. A Z-number is a pair of fuzzy numbers $Z = (A, B)$ used to describe a value of a random variable X. Fuzzy number A, plays the role of fuzzy restriction on the values which x can take. Fuzzy number B describes reliability of A and is formalized as a fuzzy value of probability measure of A. Fuzziness of B sourced from the fact that actual probability distribution of X is not known.

© Springer Nature Switzerland AG 2020
R. A. Aliev et al. (Eds.): ICSCCW 2019, AISC 1095, pp. 69–77, 2020.
https://doi.org/10.1007/978-3-030-35249-3_8

A series of studies is devoted to theory and applications of Z-numbers [10–13]. Few works are devoted to studies of differential calculus with Z-number valued information. The first study was proposed in [14], where Hukuhara derivative of Z-number valued function was considered. Applications of fuzzy Sumudu transform and neural networks to solving DEs with Z-numbers are considered in [15] respectively. [16] is devoted to study where Z-numbers are used to describe uncertainty only in initial condition of DE. In [16, 17] they consider extension of gH-differentiability and g-differentiability concepts for Z-number based setting. A fundamental study of DE with Z-numbers is conducted in [19]. However, Z^+-numbers are used in analysis instead of Z-numbers.

In this paper we propose a new approach to study of Z-number valued DEs. A Z-number valued initial problem is formulated and an approach to obtain its approximate solution is proposed.

The paper is structured as follows. Necessary definitions used in the paper are given in Sect. 2. In Sect. 3 we describe the proposed approach of describing and solving initial value problem of Z-valued DE. An example of the proposed approach is given in Sect. 4. Section 5 concludes.

2   Preliminaries

Let $\mathcal{E}^1$ denote the set of all fuzzy numbers (defined over real line $\mathcal{R}$). Let $[A]^\mu = [\underline{x}^\mu, \overline{x}^\mu]$ denote μ-cut of a fuzzy number A.

The trapezoidal fuzzy number A is characterized using the tuple $A = (a; b; c, d)$, $a \le b \le c < d$, its μ-cuts are

$$[A]^\mu = [a + \mu(b-a), d - \mu(d-c)] \mu \in [0, 1]. \qquad (1)$$

Definition 1. A Z-number [9]. A Z-number is an ordered pair $Z = (A, B)$, where A is a fuzzy number playing a role of a fuzzy constraint on values that a random variable X may take:

$$X \text{ is } A$$

and B is a fuzzy number with a membership function $\mu_B : [0, 1] \rightarrow [0, 1]$, playing a role of a fuzzy constraint on the probability measure of A:

$$P(A) \text{ is } B.$$

$P(A)$ is defined as follows:

$$P(A) = \int_{\mathcal{R}} \mu_A(x) p(x) dx \qquad (2)$$

Definition 2 [5, 6]. The horizontal membership function (HMF) of A is a representation of $\mu_A(x) = A(x)$ as $\mathcal{H}(A) : [0,1] \times [0,1] \to [a,b]$, $\mathcal{H}(A) = {}^{gr}A = x(\mu, \alpha_x) = x$ in which "gr" stands for the granule of information included in $x \in [a,b]$, μ is the membership degree of x in A, $\alpha_x \in [0,1]$ is called relative-distance-measure variable, and μ-cut is defined as

$$x(\mu, \alpha_x) = \underline{x}^\mu + \alpha_x(\overline{x}^\mu - \underline{x}^\mu), \quad \alpha_x, \mu \in [0,1]. \tag{3}$$

Below we introduce a definition of distance between fuzzy numbers $A_1, A_2 \in \mathcal{E}^1$ in terms of HMF.

Definition 3. The function $d^{gr} : \mathcal{E}^1 \times \mathcal{E}^1 \to \mathcal{R}^+ \cup \{0\}$ defined as

$$d^{gr}(\tilde{A}_1, \tilde{A}_2) = \sup_\mu \max_{\alpha_1, \alpha_2} [|A_1^{gr}(\mu, \alpha_1) - A_2^{gr}(\mu, \alpha_2)|], \tag{4}$$

is referred to as distance between fuzzy numbers A_1 and A_2.

Arithmetic Operations of Random Variables [18]. Let X_1 and X_2 be two dependent discrete random variables with the corresponding outcome spaces $X_1 = \{x_{11}, \ldots, x_{1i}, \ldots, x_{1n_1}\}$, $X_2 = \{x_{21}, \ldots, x_{2i}, \ldots, x_{2n_2}\}$, and the corresponding discrete probability distributions p_1 and p_2. The probability distribution of $X_1 * X_2$ is the convolution $p_{12} = p_1 \circ p_2$ of p_1 and p_2 which is determined as follows:

$$p_{12}(x) = \sum_{x = x_1 * x_2} mean([p_1(x_1), p_2(x_2)]), \text{for} * \in \{+, -\} \text{additive operations}, \tag{5}$$

$$p_{12}(x) = \sum_{x = x_1 * x_2} geomean([p_1(x_1), p_2(x_2)]), \text{for} * \in \{\cdot, /\} \text{multiplicative operations}, \tag{6}$$

for any $x \in \{x_1 * x_2 | x_1 \in X_1, x_2 \in X_2\}$, $x_1 \in X_1$, $x_2 \in X_2$.

Below we provide theorems which underlie computation of functions provided that uncertainty of arguments is described by probability distribution.

Definition 4 [19]. The (μ, v)-cut of the Z-number $Z = (\tilde{A}, \tilde{B})$ is defined as $Z^{(\mu, v)} = ([\tilde{A}]^\mu, [\tilde{B}]^v) = ([\underline{A}^\mu, \overline{A}^\mu], [\underline{B}^v, \overline{B}^v])$, where $\mu, v \in [0,1]$.

Moreover, using $\mathcal{H}^{-1}(A^{gr}(\mu, \alpha_x)) = [\tilde{A}]^\mu = [\inf_{\lambda \geq \mu} \min_{\alpha_x} A^{gr}(\mu, \alpha_x), \sup_{\lambda \geq \mu} \max_{\alpha_x} A^{gr}(\mu, \alpha_x)]$,

$$\mathcal{H}^{-1}(B^{gr}(v, \beta_b)) = [\tilde{B}]^v = [\inf_{\eta \geq v} \min_{\beta_b} B^{gr}(v, \beta_b), \sup_{\eta \geq \mu} \max_{\beta_{dx}} B^{gr}(v, \beta_b)] \tag{7}$$

the (μ, v)-cut of the vertical membership function of (A, B), which is in fact the span of the information granule, can be obtained.

Definitions of Z-number based on (μ, v)-cut concept is as follows.

Definition 5 [16, 19]. The parametric form of the Z-number $Z = (A, B) \in \mathcal{Z}$ is represented by the ordered pair of functions

$$Z = [\underline{Z}^{(\mu,v)}, \overline{Z}^{(\mu,v)}] = [([\underline{A}]^\mu, [\underline{B}]^v), ([\overline{A}]^\mu, [\overline{B}]^v)]. \tag{8}$$

Definition 6. The parametric form of Z-number in terms of HMF is represented by

$$\mathcal{H}(Z) = {}^{gr}Z = ({}^{gr}A(\mu, \alpha), {}^{gr}B(v, \beta)). \tag{9}$$

Below we introduce a definition of distance between Z-numbers in terms of HMF.

Definition 7. The function ${}^{gr}D : \mathcal{Z} \times \mathcal{Z} \to \mathcal{R}^+ \cup \{0\}$ defined as

$$^{gr}D(Z_1, Z_2) = \gamma^{gr}d(\tilde{A}_1\tilde{A}_2) + (1 - \gamma)^{gr}d(\tilde{B}_1\tilde{B}_2), \tag{10}$$

is a distance between two Z-umbers Z_1 and Z_2. $\gamma \in [0, 1]$ is a user preference degree, $d(\tilde{A}_1, \tilde{A}_2)$ is defined according to Definition 3.

Let us shortly describe the methodology of construction of functions of Z-numbers on the basis of HMF approach [20].

Operation on Z-Numbers. In order to account for relation of fuzzy and probabilistic uncertainties, Zadeh introduces the compatibility constraint:

$$\int_{\mathcal{R}} u p_X(u) du = \frac{\int_{\mathcal{R}} u \mu_{A_X}(u) du}{\int_{\mathcal{R}} \mu_{A_X}(u) du}, \int_{\mathcal{R}} u p_Y(u) du = \frac{\int_{\mathcal{R}} u \mu_{A_Y}(u) du}{\int_{\mathcal{R}} \mu_{A_Y}(u) du}, \text{(compatibility)}, \tag{11}$$

This implies that expected values of possibility and probability distributions are equal. However, sometimes this constraint is not sufficient to adequately describe close relation between two types of uncertainty in Z-number. We propose a so-called similarity constraint:

$$\int_{\mathcal{R}} (\mu_{A_X}(u) - p_X(u))^2 du \to \min, \int_{\mathcal{R}} (\mu_{B_X}(u) - p_Y(u))^2 du \to \min, \text{(similarity)}. \tag{12}$$

This implies that not only expected values, but also concentrations of two types of uncertainties are similar. Consequently, in agreement with earlier results we can write:

$$\mu_{p_Z}(p_Z) = \mu_{B_X}\left(\int_{\mathcal{R}} \mu_{A_X}(u) p_X(u) du\right) \wedge \mu_{B_Y}\left(\int_{\mathcal{R}} \mu_{A_Y}(u) p_Y(u) du\right) \to max \tag{13}$$

$$\int_{\mathcal{R}} (\mu_{A_X}(u) - p_X(u))^2 du \to \min, \tag{14}$$

$$\int_{\mathcal{R}} \left(\mu_{A_Y}(u) - p_Y(u) \right)^2 du \to \min. \tag{15}$$

subject to

$$p_Z = p_X \circ p_Y, \tag{16}$$

$$\int_{\mathcal{R}} p_X(u)du = 1, \quad \int_{\mathcal{R}} p_Y(u)du = 1, \tag{17}$$

$$\int_{\mathcal{R}} u p_X(u)du = \frac{\int_{\mathcal{R}} u \mu_{A_X}(u)du}{\int_{\mathcal{R}} \mu_{A_X}(u)du}, \quad \int_{\mathcal{R}} u p_Y(u)du = \frac{\int_{\mathcal{R}} u \mu_{A_Y}(u)du}{\int_{\mathcal{R}} \mu_{A_Y}(u)du}, \tag{18}$$

Construction of μ_{p_Z} according to (13)–(18) allows for more adequate computation of B_{12} to form a Z-number $Z_{12} = (A_{12}, B_{12})$. Moreover, we will consider dependent random variables in computation of $p_Z = f(p_X, p_Y)$ according to (5)–(6). In this paper, a discretized version of (13)–(18) is used for approximate solution of Z-valued DE.

3 Toward Formulation of Z-Differential Equations

In they proposed a study of Z-differential equations for the case of Z^+-numbers. In this section we are going to consider Z-differential equations in Z-numbers setting. We consider Z-initial value problem and an approach for determining an approximate solution. Paper [19] is one of the first studies devoted to Z-valued differential equations. In [19], not fully Z-number valued differential equation, but Z^+-number valued differential equation is considered. However, a serious question is related to separate consideration of fuzzy-valued and stochastic components of Z-number valued differential equation. The issue is that fuzzy restriction A on a value of variable of interest X and its reliability B are closely related. This issue was not taken into account in. Also in Z-valuation concept related probability distributions never is known. So there is need to investigate a fully Z-number valued differential equation.

Z-Initial Value Problem. Let $\mathcal{Z}(\mathcal{R})$ be the set of all Z-umbers defined in $\Re$. Let's consider a nonlinear Z-differential equation as:

$$z'(t) = f(t, z(t)), \tag{19}$$

$$z(t_0) = z_0, \tag{20}$$

where $z : [t_0, t_f] \to \mathcal{Z}, f : [t_0, t_f] \times \mathcal{Z} \to \mathcal{Z}$, and $z_0 \in \mathcal{Z}$ is an initial condition.

Assume that a unique solution $z(t) = (A, B)(t)$ of (19)–(20) exists. Then we can write

$$^{gr}z'(t) = {}^{gr}f(t, {}^{gr}z(t)), \tag{21}$$

$$^{gr}z(t_0) = {}^{gr}z_0. \tag{22}$$

That is,

$$z'({}^{gr}A, {}^{gr}B)(t) = {}^{gr}f(t, z({}^{gr}A, {}^{gr}B)(t)), \tag{23}$$

$$^{gr}z(t_0) = {}^{gr}z_0. \tag{24}$$

Given that $^{gr}A = {}^{gr}A(\mu, \alpha_x)$ and $^{gr}B = {}^{gr}B(v, \beta_b)$ we have

$$^{gr}z'(t, \mu, \alpha_A, v, \beta_B) = {}^{gr}f(t, {}^{gr}z(t, \mu, \alpha_A, v, \beta_B)), \tag{25}$$

$$^{gr}z(t_0, \mu, \alpha_A, v, \beta_B) = {}^{gr}z_0(\mu, \alpha_A, v, \beta_B). \tag{26}$$

An Approximate Solution. Consider approximate solution of (19)–(20) in setting of the Euler method:

$$^{gr}z(t_{i+1}, \mu, \alpha_A, v, \beta_B) = {}^{gr}z(t_i, \mu, \alpha_A, v, \beta_B) + h \cdot {}^{gr}f(t_i, {}^{gr}z(t_i, \mu, \alpha_A, v, \beta_B)), \\ i = 0, 1, \ldots, N-1. \tag{27}$$

The algebraic operation on the right-hand side of (27) can be implemented by using the HMF-based approach to computation of Z-numbers. Addition of dependent random variables in the right-hand side of (27) is considered. The reason is that, the future value, $^{gr}z(t_{i+1}, \mu, \alpha_A, v, \beta_B)$, naturally depends on its preceding value, $^{gr}z(t_i, \mu, \alpha_A, v, \beta_B)$. Then, consecutive computation of (27) would provide the approximate solution to (19)–(20).

4 Application

We will consider example on Z-differential equation given in. Example is related with modeling of cerebrospinal fluid (CSF) flow.

$$z'(t) = \left(-\frac{k}{r}\right) \cdot (z(t))^2 + (k \cdot I_f(t)) \cdot z(t) + \frac{k \cdot p_d}{r} \cdot z(t), z(t_0) = z_0, \tag{28}$$

where $z(t)$ is the CSF pressure in mm H_2O, $k > 0$ is the cerebral elasticity, $r > 0$ is the resistance of CSF to absorption, I_f is the rate of CSF formation and p_d is the pressure of the venous system which is usually equal to $z(t_0)$.

Assume that $z(t_0) = (z_A, z_B) = ((110, 115, 120), (0.5, 0.65, 1))$. Without los of generality, we consider exact values for k, r and $I_f(t)$ as $k = \frac{1}{0.6}$, $r = 700$, and $I_f(t) = 0.1$.

Solution of (28) is shown in Figs. 1 and 2.

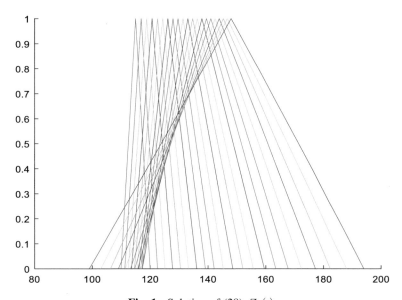

Fig. 1. Solution of (28), $Z_A(t)$.

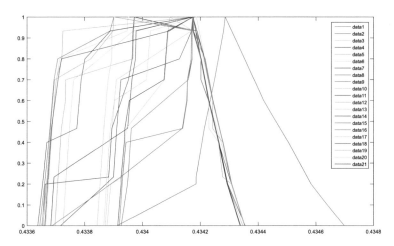

Fig. 2. Solution of (28), $Z_B(t)$.

As one can see, the fuzzy reliability $Z_B(t)$ is preserved better during the process (remains higher than 0.4, Fig. 2). Better results can be obtained if we use (μ, v)-cut threshold (for simplicity, we omit this one). Thus, the proposed approach provides an initial step toward modeling of Z-number valued DEs with improved processing of combination of two types of uncertainties.

5 Conclusions

In this paper we initiated research of Z-number valued DEs. We used HMF-based approach to computation of Z-numbers as a basis for formalization of Z-number based initial value problem and its solution. For the first time, Z-number valued DEs are considered without reduction to fuzzy and Z^+-based analogous. At the same time, we consider computation with Z-numbers which describe information on dependent random variables. This allows for a more adequate modeling of real dynamical processes, as future states naturally depend on preceding ones. Z-initial value problem is formulated and an approach to computation of its approximate solution within the Euler method is proposed. An example of Z-number valued DE that describes dynamical process in medical sphere is considered to illustrate the approach. The obtained results show that the proposed approach allows for better processing of fuzzy reliability of information as one of the key indices of bimodal information.

References

1. Ghaemi, F., Yunus, R., Ahmadian, A., Salahshour, S., Suleiman, M., Saleh, SF.: Application of fuzzy fractional kinetic equations to modelling of the acid hydrolysis reaction, Abstract and Applied Analysis (2013)
2. Ngo, V.H.: Fuzzy fractional functional integral and differential equations. Fuzzy Sets Syst. **280**, 58–90 (2015)
3. Mazandarani, M., Najariyan, M.: Differentiability of type-2 fuzzy number-valued functions. Commun. Nonlinear Sci. Numer. Simul. **19**, 710–725 (2014)
4. Najariyan, M., Farahi, M.H.: Optimal control of fuzzy linear controlled system with fuzzy initial conditions. Iran. J. Fuzzy Syst. **10**, 21–35 (2013)
5. Mazandarani, M., Zhao, Y.: Fuzzy bang-bang control problem under granular differentiability. J. Franklin Inst. **355**(12), 4931–4951 (2018)
6. Mazandarani, M., Pariz, N.: Sub-optimal control of fuzzy linear dynamical systems under granular differentiability concept. ISA Trans. **76**, 1–17 (2018)
7. Piegat, A., Landowski, M.: Horizontal membership function and examples of its applications. Int. J. Fuzzy Syst. **17**, 22–30 (2015)
8. Piegat, A., Landowski, M.: On fuzzy RDM-arithmetic. In: Hard and Soft Computing for Artificial Intelligence, Multimedia and Security, vol. 534, pp. 3–16 (2016)
9. Zadeh, L.A.: A note on Z-numbers. Inf. Sci. **181**, 2923–2932 (2011)
10. Aliev, R.A.: Operations on Z-numbers with acceptable degree of specificity. Procedia Comput. Sci. **120**, 9–15 (2017)
11. Aliev, R.A., Alizadeh, A.V., Huseynov, O.H.: An introduction to the arithmetic of Z-numbers by using horizontal membership functions. Procedia Comput. Sci. **120**, 349–356 (2017)

12. Aliev, R.A., Huseynov, O.H., Aliyev, R.R.: A sum of a large number of Z-numbers. Procedia Comput. Sci. **120**, 16–22 (2017)
13. Jafari, R., Razvarz, S., Gegov, A., Paul, S.: Fuzzy modeling for uncertain nonlinear systems using fuzzy equations and Z-numbers. In: Lotfi, A., Bouchachia, H., Gegov, A., Langensiepen, C., McGinnity, M. (eds.) Advances in Computational Intelligence Systems, UKCI 2018. Advances in Intelligent Systems and Computing, vol. 840. Springer, Cham (2019)
14. Aliev, R.A., Huseynov, O.H., Aliyev, R.R., Alizadeh, A.V.: The Arithmetic of Z-Numbers: Theory and Applications. World Scientific Publishing Co., Singapore (2015)
15. Jafari, R., Razvarz, S., Gegov, A.: Solving differential equations with Z-numbers by utilizing fuzzy Sumudu transform. In: Arai, K., Kapoor, S., Bhatia, R. (eds.) Intelligent Systems and Applications, IntelliSys 2018. Advances in Intelligent Systems and Computing, vol. 869. Springer, Cham (2018)
16. Pirmuhammadi, S., Allahviranloo, T., Keshavarz, M.: The parametric form of z-number and its application in z-number initial value problem. Int. J. Intell. Syst. **32**, 1030–1061 (2017)
17. Qiu, D., Xing, Y., Dong, R.: On ranking of continuous Z numbers with generalized centroids and optimization problems based on Z-numbers. Int. J. Intell. Syst. **33**, 3–14 (2017)
18. Jaroszewicz, S., Korzen, M.: Arithmetic operation on independent random variables: a numerical approach. SIAM J. Sci. Comput. **34**(3), A1241–A1265 (2012)
19. Mazandarani, M., Zhao, Y.: Z-differential equations. IEEE Trans. Fuzzy Syst. (2019). https://doi.org/10.1109/tfuzz.2019.2908131
20. Aliev, R.A., Perdycz, W., Huseynov, O.H.: Functions defined on a set of Z-numbers. Inf. Sci. **423**, 353–375 (2018)

Quantum Fuzzy Inference Based on Quantum Genetic Algorithm: Quantum Simulator in Intelligent Robotics

Sergey V. Ulyanov[1,2]([⊠]) [iD]

[1] Dubna State University, Universitetskaya St. 19, Dubna 141982, Russia
ulyanovsv@mail.ru
[2] INESYS LLC (EFKO GROUP), Business Centre "Central City Tower",
Ovchinnikovskaya naberezhnaya 20, Bld 1, Moscow 115035, Russia

Abstract. Successful sophisticated search solutions of intractable robotic task's as global robust intelligent and cognitive smart control in unpredicted (unconventional)/hazard control situations or multi-criteria imperfect control goal is based on quantum control principles (as quantum neural network for deep machine learning or quantum genetic optimization algorithm). It is important in these cases to choose types and kind of quantum correlations, as example, between PID-controller in coefficient gain schedule. Extracted from classical states (as example, from modeling of control coefficient gain's laws) quantum hidden correlations (that physically rigor and mathematically strong correctness, and corresponds to main qualitative properties in general of ill-defined control object) are considered as an additional physical computing and hidden quantum information resources. These information resources changes the time-dependent laws of the coefficient gains schedule of the traditional controllers as PID-controllers with guarantees the achievement of control goal in hazard situations. This article discusses the application of quantum genetic algorithm to automatically choice the optimal type and kind of correlations in the quantum fuzzy inference. Efficiency of quantum search algorithm in imperfect KB self-organization on the Benchmark system "cart – pole" demonstrated.

Keywords: Quantum computing simulator · Quantum genetic algorithm · Quantum fuzzy inference · Intelligent robotic

1 Quantum Fuzzy Inference Model

The integration of several fuzzy controllers (FC) and quantum fuzzy controller (QFC) based on quantum fuzzy inference (QFI) which allows create a new synergetic quality effect in smart management – online self-organization of imperfect knowledge base (KB) [1] on Fig. 1 is shown. Basis of industrial management systems is proportional–integral–derivative (PID) controller, which is used in 70% of the industrial automation, but often can't cope with the task of control and does not work correct in unpredicted/hazard control situations. The application of quantum computing and quantum sophisticated search algorithms, as a special example, QFI, allows increasing

© Springer Nature Switzerland AG 2020
R. A. Aliev et al. (Eds.): ICSCCW 2019, AISC 1095, pp. 78–85, 2020.
https://doi.org/10.1007/978-3-030-35249-3_9

robustness without the cost and loss of a temporal information resource in on line. Structure of corresponding QAG on Fig. 2 is shown.

In general, the structure of a quantum algorithmic gate (QAG) based on a quantum genetic algorithm (QGA) described in (1) (where Int – interference; U_F - quantum oracle (entanglement); H – Hadamard matrix; the choice of S is dependent from type of quantum algorithm [2, 3]):

$$QAG = \left[(Int \otimes {}^n I) \cdot U_F \right]^{h+1} \cdot [QGA][{}^n H \otimes {}^m S] \qquad (1)$$

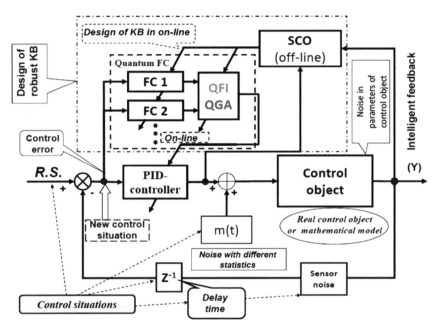

Fig. 1. Structure of robust intelligent control system based on QFI (SCO – soft computing optimizer of KB; QCO – quantum computing optimizer of KB; FC – fuzzy controller).

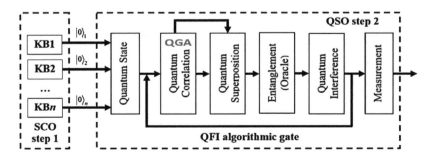

Fig. 2. QAG structure of QFI with QGA based on Eq. (1).

The first part in designing Eq. (1) is the choice of the type of operator U_F for the determination the entangled state in superposition box. Basic unit of such an intelligent control system is the quantum genetic search algorithm (QGSA) (see, Fig. 3).

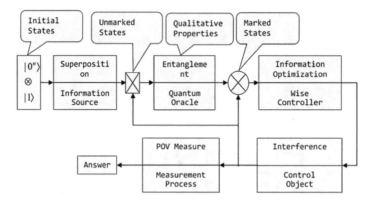

Fig. 3. Quantum search algorithm structure of intelligent self-organized control system.

Results of simulation show computing effectiveness of robust stability and controllability of (QFI + QGA)-controller and new *information synergetic effect*: from two fuzzy controllers with imperfect knowledge bases can be created robust intelligent controller (extracted hidden quantum information from classical states is the source of value work for controller [1]) in on-line. Intelligent control systems with embedding intelligent QFI-controller can be realized either on classical or on small hybrid quantum processors (as an example, on D-Wave processor type). Two classes of quantum evolution (1) are described: quantum genetic algorithm (QGA) and hybrid genetic algorithm (HGA). The sophisticated QFI algorithm for determining new PID coefficient gain schedule factors K (see, below Fig. 4) consists of such steps as normalization, the formation of a quantum bit (*qbit*), after which the optimal structure of a QAG and the state with the maximum amplitude are selected, decoding is performed and the output of quantum search algorithm is a new parameter K. At the input, the QFI obtains coefficients from the fuzzy controller knowledge bases formed in advance from the KB optimizer on soft computing. The next step is carried out normalization of the received signals [0, 1] by dividing the current values of control signals at their maximum values (max k), which are known in advance.

Formation of Quantum Bits: The superposition of the quantum system "*real state - virtual state*" as quantum bit (qbit) has the form (2):

$$|\psi\rangle = \frac{1}{2}\left(\underbrace{\sqrt{p(|0\rangle)}|0\rangle}_{real\ state} + \underbrace{\sqrt{1 - p(|0\rangle)}|1\rangle}_{virtual\ state}\right) \qquad (2)$$

Remark. Firstly, the probability density functions are determined. They are integrated and they make the probability distribution function. They allow defining the virtual state of the control signals for generating a superposition via Hadamard transform of the current state of the entered control signals. The law of probability is used as following: $p(|0\rangle) + (|1\rangle) = 1$, where $p(|0\rangle)$ is the probability of the current real state and $p(|1\rangle)$ is the probability of the current virtual state. The next step is selection of the type of quantum correlation - constructing operation of entanglement design. Three types of quantum correlation are considered: spatial, temporal and spatio-temporal. Each of them contains valuable quantum information hidden in a KB.

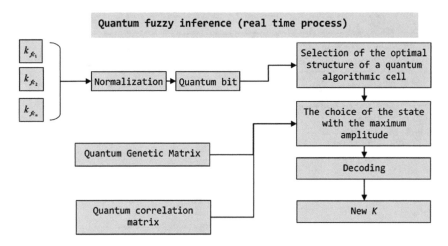

Fig. 4. Quantum fuzzy inference algorithm.

Quantum correlation considered as an additional physical computational resource, which allows increasing the successful search for solutions of algorithmically unsolvable (intractable) problems. In our case, the solution search of the problem of ensuring global robustness of the control object under conditions of unexpected control situations by designing the optimal structure and laws of time dependent changing the PID controller gains by classical control methods is an algorithmically unsolvable (intractable) problem. The solution of this problem is possible based on quantum soft computing technologies [4]. The output parameters of the PID-regulators are considered as active information-interacting agents, from which the resulting PID controlling force on the control object is formed. In a multi-agent system, there is a new synergistic effect arising from the exchange of information and knowledge between active agents (swarm synergetic information effect) [1]. There are several different types and operators of quantum genetic algorithms [4, 5]. QGA and HGA are tested on example of the roots searching task of equation as: $f(x) = |x - \frac{5}{2} + \sin(x)|$. QGA resulting performance indicates the following (see, Fig. 5). It can be seen that after about 30 populations, the value of the fitness function ceases to change. HGA shows the following results (see, Fig. 6).

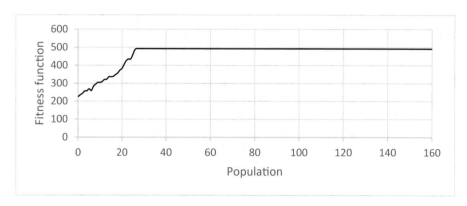

Fig. 5. Result of QGA

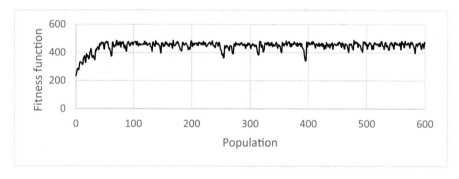

Fig. 6. Result of HGA

Remark. One of the interesting ideas was proposed in 2004, taking the first steps in implementing the genetic algorithm on a quantum computer [4]. The author proposed this quantum evolutionary algorithm, which can be called the reduced quantum genetic algorithm (RQGA). The algorithm consists of the following steps: (1) Initialization of the superposition of all possible chromosomes; (2) Evaluation of the fitness function by the operator F; (3) Using Grover's algorithm; (4) Quantum oracle; (5) Using of the diffusion operator Grover G; (6) Make an evaluation of the decision. The search for solutions in RQGA is performed in one operation. In this case the matrix form is the result of RQGA action.

2 Simulator Structure and Examples of Applications

In the development of quantum genetic algorithm [5] in this article on the model of the inverted pendulum (autonomous robot) was discovered a few problems. *Firstly*, testing a written algorithm on a robot takes a lot of time. *Secondly*, it may encounter an incorrectly working HW, and it is rather difficult to identify the malfunction itself. *Thirdly*, the GA is the selection of parameters that work best in a particular situation, but it's quite common that these parameters were very bad, which makes it difficult to set up a dynamically unstable object.

Description of the Problem. The main goal of the simulator development is SW testing, educational goals, and ability demonstration to observe the pendulum's behavior when using various intelligent control algorithms with different parameters: application only the PID controller, adding a fuzzy controller to the intelligent control system, using the GA and neural network, and apply QGA. The simulator is interesting because it covers many areas required for its implementation. There are also many different ways of development: improvement of the *2D* model or even implementation in *3D*, control of the pendulum in on line (changing the parameters of the pendulum, adding various noises), making the simulator more universal for simply creating simulations of other tasks based on results of the developed project.

Selection of Development Toolkit. Simulator access is as simple as possible and it is implemented as a non-typical web application. The diagram of the sequence of the user's work with the system and the interaction of the model, the presentation and the template are developed and demonstrated on Fig. 7.

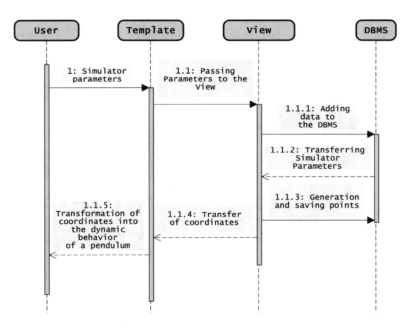

Fig. 7. Sequence diagram of system.

Most of the server-side work is math. It is necessary to calculate the position of the carriage, the angle of inclination of the pendulum in space. For this reason, Python and the Django framework, which implements the model-view-controller (MVC) approach, were chosen as the programming language (or in Django), this is the model-view-template (MVT). MySQL is used to store all data, and the architecture has been developed for adding Redis to be faster, if the MySQL operation speed is insufficient.

Figure 8 show Benchmark results of quantum intelligent control simulation of "cart - pole" system with QGA (box for the type choice of "Quantum correlation" with the application of QGA on Fig. 2).

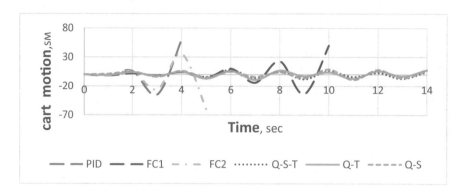

Fig. 8. Cart behavior (Q-S-T – quantum spatial-temporal correlation; Q-T – quantum temporal correlation; Q-S – quantum spatial correlation; FC – fuzzy controller).

On Fig. 9 results of simulation and experimental results comparison demonstrated.

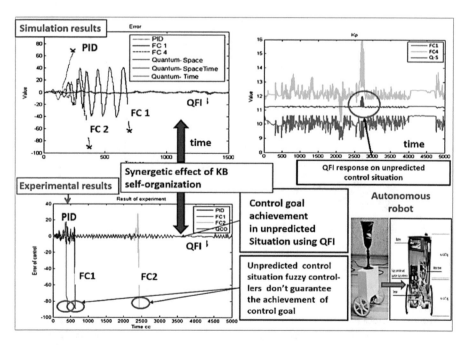

Fig. 9. Simulation & experimental results comparison for unpredicted control situation in cases of PID-controller, fuzzy controller and QFI-controller (b).

Mathematical modeling and experimental results are received for the case of unpredicted control situation and knowledge base of fuzzy controller was designing with SW of QCOPTKB^{TM} for teaching signal measured directly from control.

QGA in Fig. 4 for this case recommended the spatial quantum correlation as was early received in [1, 6].

3 Conclusion

- New circuit implementation design method of quantum gates for fast classical efficient simulation of quantum algorithms developed.
- Benchmarks of design application as Grover's QSA and QFI based on QGA demonstrated.
- Applications of QAG approach in intelligent control systems with quantum self-organization of imperfect knowledge bases described on concrete examples.
- The results demonstrate the effective application possibility of end-to-end quantum technologies and quantum computational intelligence toolkit [7–9] based on quantum soft computing for the solution of intractable classical and algorithmically unsolved problems as design of global robustness of intelligent control systems in unpredicted/hazard control situations and intelligent robotics [1, 6, 7, 9, 10].

References

1. Ulyanov, S.V.: Self-Organized Robust Intelligent Control - Quantum Fuzzy Inference in Unpredicted. Hazard Environments and Quantum Soft Computational Intelligence Toolkit in MatLab. LAP Lambert Academic Publishing, Saarbrücken (2015)
2. Ulyanov, S.V., Albu, V., Barchatova, I.: Quantum Algorithmic Gates: Information Analysis & Design System in MatLab. LAP Lambert Academic Publishing, Saarbrücken (2014)
3. Ulyanov, S.V., Albu, V., Barchatova, I.: Design IT of Quantum Algorithmic Gates: Quantum Search Algorithm Simulation in MatLab. LAP Lambert Academic Publishing, Saarbrücken (2014)
4. Ulyanov, S.V.: Quantum soft computing in control processes design: quantum genetic algorithms and quantum neural network approaches. In: Aliev, R. (ed.) 5th International Symposium on Soft Computing for Industry, WAC (ISSCI') 2004, vol. 17, pp. 99–104. Springer, Heidelberg (2004)
5. Lahoz-Beltra, R.: Quantum genetic algorithms for computer scientists. Computers $5(4)$, 31–47 (2016)
6. Litvintseva, L.V., Ulyanov, S.V.: Quantum fuzzy inference for knowledge base design in robust intelligent controllers. J. Comput. Syst. Sci. Int. 46 (6), 908–961 (2007)
7. US Patent No 8,788,450 B2: Self-organizing quantum robust control methods and systems for situations with uncertainty and risk. (Inventor: S.V. Ulyanov) (2014)
8. US Patent No 6,578,018 B1: System and method for control using quantum soft computing (Inventor: S.V. Ulyanov) (2003). US Patent No 7,383,235 B1 (2003); EP PCT 1 083 520 A2 (2001)
9. Ulyanov, S.V.: Quantum fast algorithm computational intelligence Pt I: SW/HW smart toolkit. Artif. Intell. Adv. $1(1)$, 18–43 (2019)
10. Ulyanov, S.V., Ryabov, N.V.: The quantum genetic algorithm in the problems of intelligent control modeling and supercomputing. Softw. Syst. $32(2)$, 181–189 (2019)

Solution of Zadeh's "Fast Way" Problem Under Z-Information

Rashad R. Aliyev[1]($\boxtimes$) (iD) and Nigar E. Adilova[2] (iD)

[1] Department of Mathematics, Faculty of Arts and Sciences,
Eastern Mediterranean University, Famagusta,
North Cyprus, via Mersin 10, Turkey
rashad.aliyev@emu.edu.tr
[2] Azerbaijan State Oil and Industry University, 20 Azadlig Avenue,
Baku AZ1010, Azerbaijan
adilovanigarr@gmail.com

Abstract. The concept of Z-numbers introduced by Prof. L. Zadeh created the new approach to decision-making problem. Use of Z-information for describing and calculating uncertain information was more convenient and decision making under Z-information was more adequate. Zadeh's achievements on base of Z-information became a source of motivation for other researchers.

In this paper we propose the solution method for Zadeh's "fast way" problem under Z-information. This solution method is based on Z-dynamic programming approach. To illustrate this problem under Z-information, the numerical example is analyzed.

Keywords: Z-numbers · Z-information · Z-dynamic programming · "Fast way" problem

1 Introduction

The "fast way" problem is widely used in many spheres such as routing, transportation matters, economics etc. Some researchers took a close interest for solving the "fast way" problem [1–6]. However, existing works on proposed area are related to crisp or fuzzy numbers. They investigated the shortest path problem and found a solution to this problem by using the calculation of fuzzy length.

This current study proposes the solution method to find the "fast way" tracking F from A in a city network under Z-information. The analyzed method can be applied in many areas which are interested in time-consuming problem. The investigated method presents the shortest, fast and convenient way to get from initial source to final destination by using Z-information.

For the first time, Prof. L.A. Zadeh introduced the concept of Z-numbers [7]. He described Z numbers coordinated with an uncertain variable X as a selected pair of fuzzy numbers (A, B), where A is a constraint or a restriction on values, and B is a measure of certainty for A. As compared to fuzzy information [17, 18], Z-information is describes partial reliability (B part). Based on Z-numbers, Zadeh initiated the "fast way" problem. Nowadays, a series of works devoted to Z-numbers exist [11–16].

© Springer Nature Switzerland AG 2020
R. A. Aliev et al. (Eds.): ICSCCW 2019, AISC 1095, pp. 86–92, 2020.
https://doi.org/10.1007/978-3-030-35249-3_10

Then Dubois and Prade first considered this subject with the application of fuzzy sets [8]. C.M. Klein improved this subject and used the importance of length inside integer numbers using up a dynamic-programming recursion [1].

Crucially, in [6] authors considered the new approach under fuzzy min concept. Their proposed study was inspired by the concept described in [9]. But this method was limited with the linear objective programming. Afterwards it was upgraded with the use of Bellman dynamic programming [10]. In the mentioned study, authors applied two different approaches to find the shortest path. In the first approach they proposed a triangular fuzzy number, and in the second approach they used the opportunities of trapezoidal fuzzy number.

Unfortunately, currently there is no any fundamental work based on Z-information in this field.

The proposed study consists of 5 sections. In Sect. 2 authors give some preliminary information about the analyzed problem. Section 3 includes the statement of the "fast way" problem. The solution of the problem is represented in Sect. 4. Section 5 is a conclusion section.

2   Preliminaries

Definition 1: *Z-numbers* [11–13]. As mentioned above, a continuous Z-number is organized with random uncertain variable X as an ordered pairs of (A, B) fuzzy numbers. A is a value of restriction or constraint (X is A), B is a fuzzy number with a membership function $\mu_B : [0,1] \to [0,1]$. So

$$P(A) \text{ is } B.$$

A discrete Z-number is a selected pair $Z = (A, B)$ of discrete fuzzy numbers *(A, B)*. Here A has the same meaning as it has in a continuous Z-number. B is a discrete fuzzy number with a membership function $\mu_B : \{b_1, \ldots, b_n\} \to [0,1]$, $\{b_1, \ldots, b_n\} \subset [0,1]$. Then according to a fuzzy constraint, the probability measure of A is calculated as follows:

$$P(A) = \sum_{i=1}^{n} \mu_A(x_i)p(x_i), P(A) \in \text{supp}(B).$$

$\mathcal{Z}^n$ is accepted as a space of elements which means n-vectors of discrete Z-numbers:

$$Z = (Z_1, Z_2, \ldots, Z_n) = ((A_1, B_1), (A_2, B_2), \ldots, (A_n, B_n)),$$

then $\mathcal{Z}_{[c,d]} = \left\{ (A,B) \middle| A \in \mathcal{D}_{[c,d]}, B \in \mathcal{D}_{[0,1]} \right\}$, $[c,d] \subset \mathcal{R}$, and $\mathcal{Z}_+ = \{(A,B) \in \mathcal{Z} \middle| A \in \mathcal{D}_{[0,\infty)} \}$, $\mathcal{Z}_- = \mathcal{Z} \backslash \mathcal{Z}_+$.

For two Z-numbers such as $Z_1 = (A_1, B_1)$ and $Z_2 = (A_2, B_2)$, the subset will be $\mathcal{A} \subset \mathcal{Z}$. Denoted by $A_i^{\alpha_k}$ and $B_i^{\beta_k}$ k-th α-cuts of A_i and B_i are $\alpha_k \in \{\alpha_1, \alpha_2, \ldots, \alpha_n\} \subset [0, 1]$, and $\beta_k \in \{\beta_1, \beta_2, \ldots, \beta_n\} \subset [0, 1]$, respectively.

Definition 2: *The sum of Z-numbers* [14–16]. Let us consider $Z_1 = (A_1, B_1)$ and $Z_2 = (A_2, B_2)$ to be continuous Z-numbers about values of the random variables X_1 and X_2. The result of calculation of addition $Z_{12} = Z_1 + Z_2$ is structured as follows:

Firstly, compute the addition $A_{12} = A_1 + A_2$ of fuzzy numbers. *Secondly,* calculate the discretized $R_{12} = R_1 + R_2$ as convolution $p_{12} = p_1 \circ p_2$ of p_1 and p_2. *Third,* figure out the discretized $\mu_{p_j}, j = 1, 2$. Finally, calculate μ_{p_j} as given below:

$$\mu_{p_j}(p_{jl}) = \mu_{B_j}\left(\sum_{i=1}^{n_j} p_{jl}(x_{ji})\mu_{A_j}(x_{ji})\right).$$

Through $\mu_{p_j}, j = 1, 2$, compute the discretized $\mu_{p_{12}}$ by using the formula

$$\mu_{p_{12}}(p_{12s}) = \max_{p_{1l_1}, p_{2l_2}}[\mu_{p_1}(p_{1l_1}) \wedge \mu_{p_2}(p_{2l_2})]$$

subject to $p_{12s} = p_{1l_1} \circ p_{2l_2}$

Calculate the discretized $\mu_{A_{12}}$, and $\mu_{B_{12}}$ by using the formula

$$\mu_{B_{12}}(b_{12s}) = \max \mu_{p_{12s}}(p_{12s})$$

subject to $b_{12s} = \sum_{i=1}^{n} \mu_{A_{12}}(x_i)p_{12s}(x_i)$

The obtained discretized $\mu_{B_{12}}$ can then be approximated by appropriate typical continuous membership function. So, the addition $Z_{12} = Z_1 + Z_2$ is obtained as $Z_{12} = (A_{12}, B_{12})$.

Definition 3. *The Hamming distance based metrics on* $\mathcal{Z}$ [15]. The Hamming distance based Z-metrics on $\mathcal{Z}$ is defined as follows:

$$D(Z_1, Z_2)$$
$$= \left(\frac{1}{n+1}\sum_{k=1}^{n}\left\{\left|a_{1\alpha_k}^L - a_{2\alpha_k}^L\right| + \left|a_{1\alpha_k}^R - a_{2\alpha_k}^R\right|\right\} + \frac{1}{m+1}\sum_{k=1}^{m}\left\{\left|b_{1\beta_k}^L - b_{2\beta_k}^L\right| + \left|b_{1\beta_k}^R - b_{2\beta_k}^R\right|\right\}\right)$$

$$(1)$$

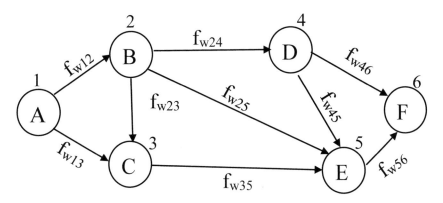

Fig. 1. Acyclic network

3 Statement of the Problem

Let's assume that we have an acyclic network G = (A, F) with 6 vertices and 9 edges where A is the source city, and F is the destination city. An acyclic network is shown in Fig. 1.

The edge weights are represented as Z-numbers (Table 1).

Table 1. Coordinates for edge weights.

$f_{w12} = [(5, 10, 15), sure];$
$f_{w13} = [(10, 15, 20), very\ sure];$
$f_{w23} = [(15, 18, 21), sure];$
$f_{w24} = [(6, 9, 12), sure];$
$f_{w25} = [(15, 20, 25), very\ sure];$
$f_{w35} = [(20, 25, 30), sure];$
$f_{w45} = [(8, 10, 12), very\ sure];$
$f_{w46} = [(20, 22, 24), sure];$
$f_{w56} = [(6, 9, 12), very\ sure]$

where *sure* and *very sure* are degrees of certainty. The values according to *sure* and *very sure* degrees are described as [0.7, 0.8, 0.9] and [0.8, 0.9, 1]. The purpose is to define minimum way to reach the destination city from the source city.

4 Solution of the Problem

Firstly, let's assume that the length of source node is

$$S_w(1) = [(0.01, 0.02, 0.03), \ (0.8, 0.9, 1)] \qquad (2)$$

To calculate the length of the way between A and B, we use the coordinates of $S_w(1)$ and f_{w12}

$$S_w(2) = \min_{i<2}\{S_w(1) + f_{w12}\}$$
$$= \{[(0.01, 0.02, 0.03), \ 0.8, 0.9, 1] + [(5, 10, 15), (0.7, 0.8, 0.9)]\}$$

The final result for $S_w(2)$ (which means for tracking way $A \rightarrow B$) will be defined with these numbers:

$$S_w(2) = \min_{i<2}\{S_w(1) + f_{w12}\} = \{[(0.01, 0.02, 0.03), \ (0.8, 0.9, 1)] + [(5, 10, 15), (0.7, 0.8, 0.9)]\}$$
$$= (5.01, \ 10.01, \ 15.03), (0.59, \ 0.73, \ 0.89)$$

$$(3)$$

The most important thing to define the "fast way" is finding the minimum between last two destinations. By using formula (1) it is possible to define the "fast way" between two destinations. The computation of the "fast way" among tracking ways $(A \rightarrow C)$ and $(B \rightarrow C)$ is carried out by using the following formula:

$$S_w(3) = \min_{i<3}\{S_w(i) + f_{w13}\} = \min\{S_w(1) + f_{w13}, \ S_w(2) + f_{w23}\} =$$

$$\min\left\{ \begin{array}{l} [(0.01, 0.02, 0.03), \ (0.8, 0.9, 1)] + [(10, 15, 20), \ (0.8, 0.9, 1)], \\ [(5.01, \ 10.01, \ 15.03), (0.59, \ 0.73, \ 0.89)] + [(15, 18, 21), (0.7, 0.8, 0.9)] \end{array} \right\} =$$

$$\min\{[(10.01, \ 15.02, \ 20.03), (0.67, \ 0.82, \ 0.95)], [(25.01, \ 33.05, 41.03), (0.5, \ 0.67, \ 0.86)]\} =$$

$$[(10.01, \ 15.02, \ 20.03), (0.67, \ 0.82, \ 0.95)] = S_w(1) + f_{w13}$$

So the "fast way" among $(A \rightarrow C)$ and $(B \rightarrow C)$ is the first one, i.e. $(A \rightarrow C)$. Then, let's calculate the length of tracking way $(B \rightarrow D)$:

$$S_w(4) = \{S_w(2) + f_{w24}\}$$
$$= [(5.01, \ 10.01, \ 15.03), (0.59, \ 0.73, \ 0.89)] + [(6, 9, 12), (0.8, 0.9, 1)]$$
$$= (11.01, \ 19.01, \ 27.03), (0.45, \ 0.61, \ 0.81)$$

In accordance with these rules, let's find the "fast way" among tracking ways $(B \rightarrow E)$, $(C \rightarrow E)$, and $(D \rightarrow E)$:

$$S_w(5) = \min_{i<5}\{S_w(i) + f_{w15}\} = \min\{S_w(2) + f_{w25},\ S_w(3) + f_{w35},\ S_w(4) + f_{w45}\}$$

$$= \min\left\{ \begin{array}{l} [(5.01,\ 10.01,\ 15.03),\ (0.59,\ 0.73,\ 0.89)] + [(15, 20, 25),\ (0.8, 0.9, 1)], \\ [(10.01,\ 15.02,\ 20.03),\ (0.67,\ 0.82,\ 0.95)] + [(20, 25, 30),\ (0.7, 0.8, 0.9)], \\ [(11.01,\ 19.01,\ 27.03),\ (0.45,\ 0.61,\ 0.81)] + [(8, 10, 12),\ (0.8, 0.9, 1)] \end{array} \right\}$$

$$= \min\left\{ \begin{array}{l} [(20.01,\ 30.01,\ 40.03),\ (0.52,\ 0.69,\ 0.86)] \\ [(30.01,\ 40.02,\ 50.03),\ (0.56,\ 0.75, 0.92)] \\ [(19.01,\ 29.01,\ 39.03),\ (0.39,\ 0.56,\ 0.79)] \end{array} \right\}$$

$$= [(19.01,\ 29.01,\ 39.03),\ (0.39,\ 0.56,\ 0.79)] = S_w(4) + f_{w45}$$

The "fast way" among three given values will be the tracking way $(D \rightarrow E)$. For the last step, let's find the "fast way" for $(D \rightarrow E)$ and $(E \rightarrow F)$.

$$S_w(6) = \min_{i<6}\{S_w(i) + f_{w16}\} = \min\{S_w(4) + f_{w46},\ S_w(5) + f_{w56}\}$$

$$= \min\left\{ \begin{array}{l} [(11.01,\ 19.01,\ 27.03),\ (0.45,\ 0.61,\ 0.81)] + [(20, 22, 24),\ (0.7, 0.8, 0.9)], \\ [(19.01,\ 29.01,\ 39.03),\ (0.39,\ 0.56,\ 0.79)] + [(6, 9, 12),\ (0.8, 0.9, 1)] \end{array} \right\}$$

$$= \min\left\{ \begin{array}{l} [(31.01,\ 41.01,\ 51.03),\ (0.36,\ 0.52,\ 0.74)] \\ [(25.01,\ 38.01,\ 51.03),\ (0.34,\ 0.52,\ 0.77)] \end{array} \right\}$$

$$= [(25.01,\ 38.01,\ 51.03),\ (0.34,\ 0.52,\ 0.77)] = S_w(5) + f_{w56}$$

The "fast way" between two given ways will be the tracking way $(E \rightarrow F)$.

Thus, the "fast way" from the source A to the destination F is summarized as below:

$$A \rightarrow B \rightarrow D \rightarrow E \rightarrow F.$$

5 Conclusion

In this study, the authors have formulated Zadeh's "fast way" problem and proposed the solution approach based on Z-information. The numerical examples are used to illustrate the experimental analysis of the suggested approach to the problem. Finally, the "fast way" between source and destination cities is obtained.

References

1. Klein, C.M.: Fuzzy shortest paths. Fuzzy Sets Syst. **39**, 27–41 (1991)
2. Lin, K., Chen, M.: The fuzzy shortest path problem and its most vital arcs. Fuzzy Sets Syst. **58**(5), 343–353 (1994)
3. Liu, S.T., Kao, C.: Network flow problems with fuzzy arc lengths. IEEE Trans. Syst. Man Cybern.: Part B **34**, 765–769 (2004)
4. Mares, M., Horak, J.: Fuzzy quantities in networks. Fuzzy Set. Syst. **10**, 135–155 (1983)

5. Okada, S., Gen, M.: Fuzzy shortest path problem. Comput. and Indust. Eng. **27**, 465 (1994)
6. Okada, S., Soper, T.: A shortest path problem on a network with fuzzy are lengths. Fuzzy Sets Syst. **109**, 129–140 (2000)
7. Zadeh, L.A.: A note on Z-numbers. Inf. Sci. **181**, 2923–2932 (2011)
8. Dubois, D., Prade, H.: Fuzzy Sets and Systems: Theory and Applications. Academic Press, New York (1980)
9. Zimmermann, H.J.: Fuzzy mathematical programming. Comput. Ops. Res. **10**(4), 291–298 (1983)
10. De, P.K., Bhincher, A.: Dynamic programming and multi objective linear programming approaches. App. Math. Inf. Sci. **5**(2), 253–263 (2011)
11. Aliev, R.A., Alizadeh, A.V., Huseynov, O.H.: The arithmetic of continuous Z-numbers. Inf. Sci. **373**, 441–460 (2016)
12. Aliev, R.A., Alizadeh, A.V., Huseynov, O.H.: The arithmetic of discrete Z-numbers. Inf. Sci. **290**, 134–155 (2015)
13. Aliev, R.A., Huseynov, O.H., Aliyev, R.R., Alizadeh, A.V.: The Arithmetic of Z-Numbers: Theory and Applications. World Scientific, Singapore (2015)
14. Aliev, R.A.: Uncertain Computation Based on Decision Theory. World Scientific Publishing, Singapore (2017)
15. Aliev, R.A., Perdycz, W., Huseynov, O.H.: Functions defined on a set of Z-numbers. Inf. Sci. **423**, 353–375 (2018)
16. Aliev, R.A., Alizadeh, A.V., Huseynov, O.H., Jabbarova, K.I.: Z-number-based linear programming. Int. J. Intell. Syst. **30**(5), 563–589 (2015)
17. Aliev, R.A., Pedrycz, W.: Fundamentals of a fuzzy-logic-based generalized theory of stability. IEEE Trans. Syst. Man Cybern. Part B (Cybern.) **39**(4), 971–988 (2009)
18. Aliev, R., Tserkovny, A.: Systemic approach to fuzzy logic formalization for approximate reasoning. Inf. Sci. **181**(6), 1045–1059 (2011)

Categorized Representations
and General Learning

Serge Dolgikh[1,2]([⊠])

[1] Solana Networks, Ottawa, Canada
sdolgikh@solananetworks.com
[2] National Aviation University, Kiev, Ukraine

Abstract. In this study we investigate the connection between general statistical properties of data and its representations with native category structures that can be formed in unsupervised observation without significant prior knowledge either in the form of ground truth data, or pre-known categories. We use a deep autoencoder model and data representing samples of Internet datagrams to produce categorized representations of input data space, with emergent density structure that was measured and visualized. We then attempt to use the emergent structure to develop a general approach to learning of new concepts in an iterative environment driven process that requires very small ground truth data and can be triggered by a single encounter. In the experiments, iterative learning applied to several categories of applications resulted in better than random accuracy of classification for most studied categories and demonstrated a direction in which discovery and early learning of new concepts can be bridged to permanent learning, with potential applications in the fields where massive amounts of prior knowledge aren't yet available, as well as in self-learning and general learning artificial systems.

Keywords: Artificial intelligence · Deep learning · Unsupervised learning · General learning · Computational neuroscience

1 Introduction

SPONTANEOUS emergence of higher-level concept sensitive features in machine learning models was reported in a number of studies previously. By applying a deep autoencoder model to image classification, Le et al. [1] observed spontaneous emergence of concept sensitive neurons activated by images in certain higher-level category, such as 'cat's face'. The effect was observed after training with very large datasets of images in a passive, unsupervised mode without any exposure to ground truth.

In [2] spontaneous formation of grid-like cells, similar to those observed in mammals was detected in a recurrent neural network with deep reinforcement learning.

The emergence of concept-sensitive structure has been observed in unsupervised training of deep autoencoder models with data representing Internet traffic in [3]. It was shown that such emergent structures can be used in "landscape-based" approach to learning that offers higher flexibility and need significantly less ground truth data.

© Springer Nature Switzerland AG 2020
R. A. Aliev et al. (Eds.): ICSCCW 2019, AISC 1095, pp. 93–100, 2020.
https://doi.org/10.1007/978-3-030-35249-3_11

As another approach to identifying new patterns in general data, applications of clustering methods in novelty detection are well known, such as OLINDDA method by Spinosa et al. and many others [4–6], see Pimentel et al. for a comprehensive review of the field [7].

In this study we attempted to look at these results in unsupervised novelty learning at a slightly different angle: what analysis can be performed, and conclusions derived about data *before* any pre-known structure or framework of knowledge has been applied? And if such analysis is indeed possible, could it shed some light on information processes that may lead to emergence of general self-learning systems capable to acquire new concepts directly from the environment without the requirement of prior knowledge in the form of known category frameworks, ground truth data, and other?

2   Methods

2.1   Native and Limit Entropy

Let's consider a general empirical dataset $X(M, N)$ with rows representing observation instances and columns, observation parameters x_i, $i = 1...N$. With instances of the parameter x_i recorded in empirical observations one can calculate the empirical probability distribution p_i for all parameters. Now, one can calculate the native empirical entropy E_i of x_i, and subsequently, of the entire dataset as:

$$Ee = \sum_{i}^{N} Ei = \sum_{i}^{N} \sum_{k}^{M} -pk \log pk \qquad (1)$$

Now, a question can be asked, how ordered (or random) is the empirical data represented by observations in X? First let's consider an example. Suppose there's a camera placed in a forest that takes a snapshot of its surroundings every hour of light time, for a year. One can calculate the native entropy of this set of images as defined above, resulting in some empirical value of E_f. Further, one can expect the result to be rather ordered than random as majority of images would fall in just a few general categories, or patterns with only comparatively minor differences within each category, for example, "forest in a warm season", "forest in winter" and so on.

These intuitive conclusions can be formalized by introducing the concept of maximum, or limit entropy of a given empirical dataset E_l, by considering p_i in (1) as uniform distributions with all states (or values) having the same probability.

Comparing E_e and E_l should then give us some ground for the conclusion about the nature of data in X: if $E_e \ll E_l$ as was intuitively expected in the earlier example, the data can be considered as rather ordered and will be of further interest for us in this study (note the ratio of E_e to E_l calculated with same scaling and histogram parameters, is less dependent on scaling and sampling of empirical distributions and remains approximately constant).

2.2   Categorized Representations

The next question that can be asked in the case of ordered data with low empirical entropy $E_e/E_1 \ll 1$ is the following: in a low entropy state one can expect probability distribution of states in the data to be dominated by relatively small number of macro states (compared to the total number of possible states, as above). Can these macro states or "patterns" that dominate observation be identified and if yes, how, with only raw data and without any prior knowledge of the nature and structure of observation?

Granted, there exist many methods and approaches to reducing noise and complexity of data such as principal component analysis, kernel density estimation [8, 9] and many others. However, while offering insights into underlying distributions, they may not necessarily help in identifying the principal patterns in the arbitrary, general data.

One possibility beyond such methods that was described in [10] is the "template strategy". As another example, if we hear the word "forest" we instantly invoke an image of a forest in the most familiar, or common to us form. This image is a template and once learned, the variety of observations of forests will be linked and relative to it. Applying information templates allows the learning system to compact data beyond just dimensionality reduction while preserving its essential information content.

Let's attempt to formalize this concept. Suppose there exists a transformation T_C from input data space X to some representation space R, such that other than certain fraction $\varepsilon(I)$, X is transformed to a finite number of compact category regions $\{R_i\}$. Also let's assume that T_C satisfies certain conditions of continuity, so that its inverse transformation is continuous too. We shall call T_C categorizing transformation, and R, a categorized representation of X.

If at least one categorized representation R_C exists, one can define categorized probability distribution as the probability function $\tilde{p}_i$ of $T(x)$ being in R_i, $x \subset I$. Comparing the entropy of the categorized distribution with native entropy of the input data may provide a measure of how closely the native categories $\{R\}$ and categorizing transformation T_C approximate or "describe" X.

2.3   Data

The data used in the study to build an example of categorized representation consisted of recordings of Internet data packets in a public Internet network (packet dumps). We used publicly available dataset (WITS, [11]) that was processed as a set of "conversations" associated to the same source and destination with parameters representing size and timing statistics of data packets in the conversations [12]. For the detailed definition of models and the data, including composition, parameters and layout of deep neural network models refer to [3].

Being a live recording in a busy Internet network, the data had a wide representation of patterns, with over 4,000 distinct applications present in the dataset. We believe it is well suited to illustrate the concepts introduced in this study because it is sufficiently diverse while on the other hand, smaller number of parameters allows for simpler categorized representations and more direct analysis thereof.

2.4   Model

The model that was used to produce an example of categorized representation in this study is a deep autoencoder neural network of near-symmetrical layout, with significant compression in the central layer. In the study, the dimension of input layer was in the range of 22–30, whereas that of the central (Encoded) layer, 3–10 with compression factor defined as the ratio of the input to encoded dimension, up to 10.

The models were first trained in an unsupervised autoencoder mode to achieve good reproduction of inputs. Two measures of the quality of reproduction i.e., average deviation of the output of the model from the input were used: (1) cost function, MSE, had starting value in the range of 0.25 dropping to 0.001–0.002 after 100 epochs of self-supervised training; and (2) accuracy, measured as the coincidence of the maximum value position in the input sample to that in the output, thus a measure related to covariance of input and output, achieved, on average, 95%.

Trained models create categorized representations of input data space via "encoding" transformation T_E of input data space to the activations of the central layer of the model, as [3]:

$$R(X) = Encoded(X) \qquad (2)$$

where X is the input sample, and Encoded, a neural network submodel that transforms the input layer to activations of the central layer of the model.

3   Results

3.1   Data Analysis

The original data, scaled to interval [0, 1] had the maximum dimension of 1.0, the mean of 0.03–0.38 and the variance of 0.001–0.14.

The order factor of the native empirical entropy to the limit entropy for the dataset was calculated as approximately 0.069. The native and the limit entropy were calculated with an empirical histogram of 1 million bins, after calculated value of the factor reached a plateau and did not change significantly with higher number of bins.

The component analysis of the dataset revealed five principal components with relative variance of 0.1 and above, and 13 above 0.01, out of the total of 22 parameters.

3.2   Unsupervised Landscape

Categorizing transformation of input data to representation space was performed by applying the encoding transformation (2) to the input data.

The resulting 3-dimensional representation space had the maximum dimension of approximately 0.02 indicating that in the process of unsupervised training the input data space had been compressed by up to two orders of magnitude in each principal dimension, or by volume, up to 10^5 in the three largest principal dimensions. Figure 1 demonstrates visualization of distribution of higher density regions in the 3-dimensional representation space.

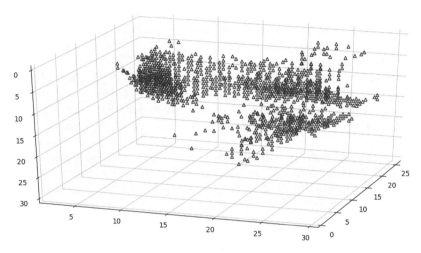

Fig. 1. Distribution of dense regions in the representation space

Density structure analysis with combination of density clustering [3, 13] and multi-dimensional histogram analysis [14] resulted in identification of 40–50 regions (clusters) with higher than average density.

With the categorized representation, we calculated categorized probability function as empirical probability of a random sample being transformed into one of the identified clusters. The resulting distribution had 12 dominant clusters with probability greater than 1% and the resulting categorized entropy of 2.33/2.94 calculated over dominant clusters vs. the entire category set, respectively, corresponding to the order factor in the range 0.006–0.0074.

In conclusion it can be noted that higher-level concept regions can be visualized directly in the representation space by plotting transformed values of samples of a given concept or class [3]. This possibility can be exploited in designing models capable of learning directly from observation of the environment with minimal ground truth.

3.3   Application to General Learning

The results in the previous sections can be applied to one of the essential tasks in general learning: developing the ability to acquire new concepts directly from the environment without prior knowledge or large amounts of ground truth data. This possibility is based on the observation that categorized representation that emerges in unsupervised observation of the environment can be used to establish an association between important (for the learner) inputs and structures in the categorized representation with minimal ground truth data, even down to a single encounter.

The general concept of the approach is as follows: starting with the set of structures or "native categories" identified in the unsupervised learning phase, at the first encounter with an instance X_n of new concept C_n one can identify the region R_n where $T(X_n)$ lands, thus representing the in-class data for a binary classifier of C_n; the rest of the category regions in the first iteration can be considered as representing the out-of-class data. This allows to create the first iteration of labeled training dataset for the binary classifier with random points generated in R_n (for in-class labeled data) and R_o as out-of-class, respectively.

In the learning iterations, the model attempts to predict the class of unlabeled samples, accumulating small sets of genuine ground truth data in a trial and error learning process. In- and out-of-class regions are then recalculated with the new data and the new iteration of binary classifier is trained, with improved classification accuracy.

The process of iterative learning was applied to several categories of Internet applications, with accuracy of the binary classifier measured after the first single encounter, and five learning iterations with 5 genuine true and false samples, each. The results in Table 1 represent the recall and false positive rate of prediction, in percent.

Table 1. Iterative learning for Internet applications.

Internet application	Initial learning	Iterative learning
DNS (network protocol)	85.6, 18.8	90.5, 6.9
Instant messenger	86.8, 44.3	91.7, 6.6
XBox (video game)	76.4, 17.7	78.8, 13.8
E-mail	72.2, 31.6	85.7, 17.9
BitTorrent (file sharing)	61.5, 43.0	66.4, 42.1

As can be observed from these results, the method of iterative learning provides successful, better than random prediction in most studied cases.

4   Conclusions

In this study we defined the concepts of categorizing transformation and categorized representation and demonstrated an example of categorized representation obtained with a deep autoencoder neural network model, with calculation of the native entropy of the data, categorized distribution and visualization of the emergent information structure in the categorized representation space.

A method of early iterative general learning based on the structure emergent in the representation space in unsupervised observation of the environment was introduced and successfully applied to learning of new higher-level concepts.

The essential question that remains is how categorized representations can be created? At this time, the answer in the general case is not known and finding the conditions under which categorized representations exist for general data with high native order would be an interesting and challenging endeavor in the information theory.

In practice, different data types yield themselves to a variety of approaches and methods resulting in effective representations, such as: reduction by principal components [15]; convolutional network representations in classification of images and manifold learning [16, 17] and others. In this wide range, self-supervising neural network models have clear strengths due to their versatility in application to different and diverse data and the power of universal approximation [18].

Specifically, feed-forward autoencoder models discussed in this study, the simultaneous constraints of compression or sparsity in the layers that create representations of input data, and high accuracy of reproduction in the output entail that the compressed representation must retain significant part of information about the input data, as discussed in Sect. 2.1. Intuitive argument may suggest that the template strategy could be the easiest way to achieve that thus "encouraging" models to create categorized representations, a hypothesis that may be related to results by Tishby et al. on the bottleneck principle [19]. This argument may also have interesting implications for parallels between machine and biologic systems studied extensively e.g. [20, 21] and the literature but is beyond the scope of this study.

In conclusion, two notes. First, there's no reason to expect that categorizing transformation, if exists, must be unique. A different transformation can result in an entirely different set of native categories and categorized probabilities. Thus, the set of native category frameworks describing the given data can be empty or contain one or more independent categorizing transformations with different properties.

And secondly, the process of initial factorization by similarity via categorized representation described above does not need to stop at the first level and can be repeated for each category space $X_k = T^{-1}(R_k)$, as well as for the non-categorized fraction. Thus, it would result in a recursive description of general data with finergrained hierarchies of native categories and is limited only by the availability of data.

References

1. Le, Q.V., Ransato, M.A., Monga, R., Devin, M., Chen, K.: Building high-level features using large scale unsupervised learning arXiv:1112.6209 (2012)
2. Banino, A., Barry, C., Kumaran, D.: Vector-based navigation using grid-like representations in artificial agents. Nature **557**, 429–433 (2018)
3. Dolgikh, S.: Spontaneous concept learning with deep autoencoder. Int. J. Comput. Intell. Syst. **12**(1), 1–12 (2018)
4. Spinosa, E., de Carvalho, A.C.P.L.F., Gama, J.: OLINDDA: a cluster-based approach for detecting novelty and concept drift in data streams. In: ACM Symposium Applied Computing (SAC), Seoul, South Korea, pp. 448–452 (2007)
5. Fanizzi, N., d'Amato, C., Esposito, F.: Conceptual clustering and its application to concept drift and novelty detection. In: ESWC 2008: The Semantic Web: Research and Applications, pp. 318–332 (2008)
6. Albertini, M.K., de Mello, R.F.: A self-organizing neural network approach to novelty detection. In: ACM Symposium Applied Computing (SAC), Seoul, South Korea, pp. 462–466 (2007)

7. Pimentel, M., Clifton, D., Clifton, L., Tarassenko L.: A review of novelty detection. Sig. Process. **99**, 215–249 (2014)

8. Halko, N., Martinsson, P.G., Tropp, J.A.: Finding structure with randomness: Probabilistic algorithms for constructing approximate matrix decompositions. arXiv:0909.4061 (2009)

9. Parzen, E.: On estimation of a probability density function and mode. Ann. Math. Stat. **33** (3), 1065–1076 (1962)

10. Bengio, Y.: Learning deep architectures for AI. Found. Trends Mach. Learn. **2**(1), 1–127 (2009)

11. WITS passive datasets, Waikato University, Waikato, New Zealand. https://wand.net.nz/wits (2018)

12. Alshammari, R., Zincir-Heywood, A.: Investigating two different approaches for encrypted traffic classification. In: 6th Annual Conference on Privacy, Security and Trust, Fredericton, pp. 156–166 (2007)

13. Comaniciu, D., Meer, P.: Mean shift: a robust approach toward feature space analysis. IEEE Trans. Patt. Anal. Mach. IntelL. **24**(5), 603–619 (2002)

14. Werman, M., Peleg, S., Rosenfeld, A.: A distance metric for multidimensional histograms. Comput. Vis. Graph. Image Process. **32**(3), 328–336 (1985)

15. Von Petersdorff, T: Example for Principal Component Analysis (PCA): Iris data, University of Maryland. https://www.math.umd.edu/~petersd/666/html/iris_pca.html. Accessed 2019

16. Kavukcuoglu, K., Sermanet, P., Boureau, Y.L., Gregor, K., Matheu, M., LeCun, Y.: Learning convolutional feature hierarchies for visual recognition. In: Proceedings of the 23rd International Conference on Neural Information Processing Systems, Vancouver, Canada, vol. 1, pp. 1090–1098 (2010)

17. Lunga, D., Prasad, S., Crawford, M., Ersoy, O.: Manifold-learning-based feature extraction for classification of hyperspectral data: a review of advances in manifold learning. IEEE Sig. Process. Mag. **31**(1), 55–66 (2014)

18. Hornik, K., Stinchcombe, M., White, H.: Multilayer feedforward neural networks are universal approximators. Neural Netw. **2**(5), 359–366 (1989)

19. Tishby, N., Pereira, F.C., Bialek, W.: The Information Bottleneck method. arXiv:physics/0004057 (2000)

20. Hassabis, D., Kumaran, D., Summerfield, C., Botvinick, M.: Neuroscience inspired artificial intelligence. Neuron **95**(2), 245–258 (2017)

21. Getting, P.A.: Emerging principles governing the operation of neural networks. Ann. Rev. Neurosci. **12**, 185–204 (1989)

Fuzzy Dynamic Programming Approach to Multistage Control of Flash Evaporator System

Shamil A. Ahmadov[1] and Latafat A. Gardashova[2(✉)]

[1] French-Azerbaijani University, 183 Nizami Street, Baku, Azerbaijan
shamilahmadov@yandex.ru
[2] Azerbaijan State Oil and Industry University,
Azadlig 35, Nasimi, Baku, Azerbaijan
latsham@yandex.ru

Abstract. This article is devoted to the fuzzy dynamic programming approach to multistage control problem of flash evaporator system. In considered problem where in each stage exist the fuzzy goal and constraints. It is clear that direct application of mathematical programming theory to solving multistage problem does not lead to aim. The basic feature of this approach lies in its capability to handle any kind of constraints. In this view of point for solving multistage optimization problem is used the discrete principle of maximum and dynamic programming method.

Keywords: Multistage control · Fuzzy condition · Flash evaporator system · Fuzzy dynamic programming

1 Introduction

Major importance of dynamic programming is that dynamic programming underlies a number of numerical algorithms for solving optimization problems. The general idea of the such algorithms based on the principle of optimality is that all interval on which it is necessary to optimize process is divided into such number of small intervals that the solution of a problem of optimization on this small interval was already simple. After that the solution of a problem begins with the end. According to the principle of optimality, optimal control on the last step does not depend on what was control on the previous steps.

In the modeling, dynamic programming converts a problem with repeated decision goals and existing limited resources into a series of interrelated subproblems arranged in stages. Each subproblem is tractable than the given original problem [1, 2].

The paper [3] is devoted the simplest problem of fuzzy optimal control. In this work is discussed the basic questions of fuzzy dynamical systems. A solution of the optimal control problem is based on the Bellman equation.

The author of article [4] offers methodical support of optimal control on fuzzy multistage processes. The problem of optimal control synthesis is formed in this paper. Obtained functional equation gives possibility to find maximum fuzzy goal. The

R. A. Aliev et al. (Eds.): ICSCCW 2019, AISC 1095, pp. 101–105, 2020.
https://doi.org/10.1007/978-3-030-35249-3_12

solution of the functional equation gives at each stage the feedback control low at this stage.

It is known that many real life problem are discrete. The discrete problems also arise due to multistage character of the process. In such cases decision is made in finite steps in advance and the optimization problem becomes multistage The basic tool for solving multistage optimization problem is dynamic programming method. The preferences and disadvantages of this method for solving optimal control problem are well known from scientific literature [5–8, 11, 12]. In this paper we use dynamic programming approach to solve multistage control problem of flash evaporator system.

This paper is organized as follows. Section 2 is devoted to the preliminaries, which contains definitions and background information related to the considered topic. In Sect. 3, we formulate the statement of the problem. In Sect. 4, we consider numerical example and results of computer simulation related to the considered problem. Section 5 presents our conclusions.

2 Preliminaries

Definition 1 [10]. A fuzzy set is a class of objects with a continuum of grades of membership. Such a set is characterized by a membership (characteristic) function which assigns to each object a grade of membership ranging between zero and one.

Definition 2. Fuzzy control problem [1]. Find such sequence of controls $\{u_0, u_1, \ldots, u_N\}$, which maximize criterion:

$$J = X_N \rightarrow \max \tag{1}$$

$$X_{t+1} = X_t \circ U_t \circ R(x_t, u_t, x_{t+1}) \tag{2}$$

$$U_t \in U_t^F, t = 0, 1, \ldots, N \tag{3}$$

where $X_t = \sum \mu_{X_t}(x_t)/x_t$, $U_t = \sum \mu_{U_t}(u_t)/u_t$,

$$X_t \in F(X_t), U_t \in F(U_t), x_t \in V, \ u_t \in W$$

Here $F(X_t), F(U_t)$ are sets of fuzzy sets, which describe the fuzzy state and controls defined on V and W.

Definition 3 [9]. Bellman equation is defined as follow:

$$x_1 = T(x_0, a_0), \ V(x_0) = \max_{a_0}\{F(x_0, a_0) + \beta V(x_1),$$

subject to $a_0 \in \Gamma(x_0), \ x_1 = T(x_0, a_0)$.

3   Statement of the Problem

We consider optimal control problem of flash evaporator system by using fuzzy dynamic programming approach. Process flow diagram of the multistage flash evaporator system is given in Fig. 1.

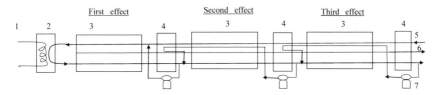

Fig. 1. Flow diagram of the multistage flash evaporator system (1-Steam; 2- Brine heater, 3- Heat recovery, 4- Heat rejection, 5- Feed, 6- Destillate, 7- Blow down)

In this system optimization factors are: number of total stage, total temperature differences between the flashing and recycle brine, exit temperature and concentration of each effect, number of stage allocated to each effect, allocation of distillate production in each effect and brine velocities of various concentrations. We will take the role of the total temperature differences and exit brine temperatures of each effect into account. In this problem as state variable x_n is chosen the temperature of the flashing brine and the decision variable u_n is chosen the recycle ratio of each effect.

The problem is to define such sequences of control variables u which maximize fuzzy goal under a fuzzy constraint on control variables.

4   Numerical Example and Simulation Results

Let's consider the following fuzzy regulator problem:
Assume that X and U is from $[-1, 1]$.

$$X_{n+1} = a_n x_n + u_n, n = 0, 1, \ldots, N - 1. \ N = 3$$

a_n-some real number, characterizing by n-th stage. At X set fuzzy relation S with membership function μ_S and in X space G set with membership grade μ_G is defined. Then fuzzy relation $S_n, n = 1, \ldots, N$ with membership function $\mu_{S_n} : X \times U \times X \to [0, 1]$ is given:

$$\mu_{S_n}(x_n, u_n, x_{n+1}) = \frac{1}{1 + (x_{n-1} - a_n x_n - u_n)^2}$$

The problem is to search $u_0, u_1, \ldots, u_{N-1}$ of points of U set membership degree of x_0 state to fuzzy set G with relations with membership functions,

$$\mu_S(x_0, u_0, x_1), \mu_S(x_1, u_1, x_2), \dots \mu_S(x_{N-1}, u_{N-1}, x_N).$$

Using $D_{N-n} = \max\limits_{x_{N-n-1}, u_{N-n}} G \circ S_{N-n+1} \circ \dots, \circ S_N, n = 1, \dots, N,$ we take $f_N(x_N) = \mu_G(x_n)$.

According to the Bellman's principle of optimality [9] we are getting:

$$f_{N-n}(x_{N-n}) = \max\limits_{x_{N-n+1}, u_{N-n}} \min\{\mu_{S_{N-n+1}}(x_{N-n}, u_{N-n+1}), f_{N-n+1}(x_{N-n+1})\}$$

By using N = 3 and

$$\mu_G(x) = \begin{cases} 0, x \in [-1,0] \\ \dfrac{1}{(1+x^{-2})} \end{cases}, x \in (0,1]; \quad f_N(x_N) = \begin{cases} 0, x_n \in [-1,0] \\ \dfrac{1}{(1+x_N^{-2})} \end{cases}, x_N \in (0,1]$$

is defined the following equations:

$$\textit{if } n = 3 : f_0(x_0) = \max\limits_{x_1, u_0} \min \left(\frac{1}{1+(x_1 - a_0 x_0 - u_0)^2}, f_0(x_1) \right);$$

$$\textit{if } n = 2 : f_1(x_1) = \max\limits_{x_2, u_1} \min \left(\frac{1}{1+(x_2 - a_1 x_1 - u_1)^2}, f_1(x_2) \right);$$

$$\textit{if } n = 1 : f_2(x_2) = \max\limits_{x_3, u_2} \min \left(\frac{1}{1+(x_3 - a_2 x_2 - u_2)^2}, f_2(x_3) \right)$$

At the end we obtain fuzzy decision as following form:

for $a_0 = -2$; $x_0 = 0.07625$ obtained u = $2.37344E - 06$; $\mu_G(x) = 0.37$

for $a_1 = -1$; $x_1 = 0.617$ obtained u = -0.9064; $\mu_G(x) = 0.98$

for $a_2 = -1.1$; $x_2 = 0.684$ obtained u = 1; $\mu_G(x) = 0.96$

for $a_3 = 1$; $x_3 = 0.463397$ obtained u = 0; $\mu_G(x) = 0.97$

for $a_4 = 2$; $x_4 = 1$ obtained u = -0.24588; $\mu_G(x) = 0.99$.

5 Conclusions

In this paper the fuzzy dynamic programming approach to multistage control problem of flash evaporation system is discussed. The solution of a problem is based on obtained functional equation. As optimal control in considering stage is chosen such value of maximizing decision on which the according state with maximal grade belongs to fuzzy goal.

References

1. Aliev, R.A., Mamedova, G.A., Aliev, R.R.: Fuzzy Set Theory and its Application. Tabriz, Iran (1993)
2. Yu, S., Gao, S., Sun, H.: A dynamic programming model for environmental investment decision-making in coal mining. Appl. Energy **166**, 273–284 (2016)
3. Egereva, I.: Methodological support of optimal control of fuzzy multistage processes. Program Syst.: Theory Appl. **6:1**(24), 11–19 (2015). (in Russian)
4. Paluh, B.V., Dzyuba, S.M., Egereva, I.A., Emelyanova, I.I.: The simplest problem of fuzzy dynamical systems optimal control. https://libeldoc.bsuir.by/bitstream/123456789/12307/1/Paluh_TheSimplest.PDF
5. Parida, P.K., Sahoo, S.K., Sahoo, K.C.: Some studies on multistage decision making under fuzzy dynamic programming. Int. J. Logic Comput. **1**(1), 52–66 (2011)
6. Chang, S.L.: Fuzzy dynamic programming and the decision making process. In: Proceedings of 3rd Princeton Conference of Information Sciences and Systems, Princeton, NJ, USA, pp. 200–203 (1969)
7. Bellman, R.E., Zadeh, L.A.: Decision-making in a fuzzy environment. Manag. Sci. **17**(4), 141–164 (1970)
8. Esogbue, A.O., Bellman, R.E.: Fuzzy dynamic programming and its extensions. TIMS/Stud. Manag. Sci. **20**, 147–167 (1984)
9. Bellman, R.E., Zadeh, L.A.: Decision making in fuzzy conditions (1970). http://ayouty.com/_Gam_Shorouq_final/5info_sys/FUZZY_DECISION.pdf
10. Zadeh, L.A.: Fuzzy sets. Inf. Control **8**(3), 338–353 (1965)
11. Aliev, R.A., Pedrycz, W.: Fundamentals of a fuzzy-logic-based generalized theory of stability. IEEE Trans. Syst. Man Cybern. Part B (Cybern.) **39**(4), 971–988 (2009)
12. Aliev, R., Tserkovny, A.: Systemic approach to fuzzy logic formalization for approximate reasoning. Inf. Sci. **181**(6), 1045–1059 (2011)

Synthesis of Fuzzy Terminal Controller for Chemical Reactor of Alcohol Production

Latafat A. Gardashova$^{(\boxtimes)}$ (iD)

Azerbaijan State Oil and Industry University,
Azadlig 20, Nasimi, Baku, Azerbaijan
latsham@yandex.ru, qardashova@asoiu.edu.az

Abstract. The most investigated area in optimal control of discrete processes under uncertain conditions is optimal control of fuzzy systems, i.e. represented by different-type fuzzy equations. In this paper, the fuzzy terminal control problem described by a fuzzy relational equation (FRE) to take into account the fuzzy state and controls is discussed. Kernel of FRE-based method is Bellman-Zadeh approach. From this point of view, a fuzzy terminal control problem is transfigured to the multistage decision making scheme. The solution of the discussed problem is defined as intersection of fuzzy goal and fuzzy constraints. At the end, obtained fuzzy decisions were maximized.

Keywords: Fuzzy relation · Fuzzy relational equation · Optimal control · Fuzzy condition · Terminal regulator · Reactor of alcohol production

1 Introduction

Real-life control processes cannot be described without considering uncertainty over parameters and variables. If information exists but is accompanied with uncertainty, then we use fuzzy uncertainty tools [1–3].

The studies devoted to solving optimal control processes have a long history. A notable part of these studies were applying various fuzzy equations (for instance, fuzzy recurrence equations, fuzzy relational equations, etc.).

FREs are without doubt the basic inverse problems emerging from fuzzy logic and fuzzy relational calculus. It is incontestable fact, the calculus of fuzzy relations is a powerful tool with applications in fuzzy control.

FREs are used as an effective tool in many areas, such as fuzzy decision making], intelligence technology [4] etc.

The first paper about fuzzy relational equations is given in scientific literature by Sanchez [5], where the max-min composition was adopted.

In scientific literature, there are a lot of researches inspecting the solvability of FREs [6–8].

In fact, FREs can be classified on many various compositions. The basic compositions include max-product [9], max-min [5], inf-α [7] max-Archimedean t-norm [8], and inf-αT [9] compositions and etc.

Authors of paper [4] discovered and explained that inf-α composition is better in fuzzy relational calculus.

© Springer Nature Switzerland AG 2020
R. A. Aliev et al. (Eds.): ICSCCW 2019, AISC 1095, pp. 106–112, 2020.
https://doi.org/10.1007/978-3-030-35249-3_13

Markovskii [10] proved that solving FREs with max-product composition is more related to the covering problem.

Lin [11] developed Markovskii's work with max-Archimedean-t-norm composition over fuzzy relational equations. Also, Lin [12] looked into FREs with u-norm and described the problem of solving a system of FREs into covering problem.

Fuzzy relational equations [13] and related problems is discussed by Xiong and Shu. Their paper is dedicated to the problem of solution method a system of relational equations based on fuzzy data with inf-implication aggregation.

In this paper, our main contribution is that we apply max-min composition to optimal control of the process, described by fuzzy relational equation.

The rest of the paper is organized as follows. Section 2 is devoted to the preliminaries, which contains definitions and background information related to the considered topic. In Sect. 3, we formulate the statement of the problem. In Sect. 4, we consider numerical example and results of computer simulation related to the considered problem. Section 5 presents our conclusions.

2 Preliminaries

Definition 1 [14]. A fuzzy relation in a set X is an arbitrary function $R: X \times X \to [0, 1]$. The family of all fuzzy relations in X is denoted by FR(X).

Definition 2 [1]. The fuzzy relation equation is an equation of the form $A \circ R = B$ where A and B are fuzzy sets, R is a fuzzy relation, and A R stands for the composition of A with R. It is known that [1] there exists three types of the fuzzy relational equation, i.e.

$$B = A \circ R$$

$$B = A * R$$

$$B = A \,\square\, R$$

where A and B are treated as fuzzy sets defined on the space universes of discourse X and Y, R denotes a fuzzy relation expressed on the Cartesian product X and Y. $\circ$, $*$ and $\square$ are max-min, sup-prod, and inf-max composition operators, respectively. Equations can be rewritten in the following form:

$$\mu_B(y) = \sup_{x \in A} \min(\mu_A(x), \mu_R(x, y)),$$

$$\mu_B(y) = \sup_{x \in X}(\mu_A(x) \times \mu_R(x, y))$$

$$\mu_B(y) = \inf_{x \in X} \max(\mu_A(x), \mu_R(x, y)),$$

$$y \in B.$$

Definition 3. Fuzzy control problem. Find such sequence of controls $\{u_0, u_1, \ldots, u_N\}$, which maximize criterion:

$$J = X_N \rightarrow \max \tag{1}$$

$$X_{t+1} = X_t \circ U_t \circ R(x_t, u_t, x_{t+1}) \tag{2}$$

$$U_t \in U_t^F, t = 0, 1, \ldots, N \tag{3}$$

where $X_t = \sum \mu_{X_t}(x_t)/x_t$, $U_t = \sum \mu_{U_t}(u_t)/u_t$,

$$X_t \in F(X_t), U_t \in F(U_t), x_t \in V, \; u_t \in W$$

Here $F(X_t)$, $F(U_t)$ are sets of fuzzy sets, which describe the fuzzy state and controls defined on V and W [17, 18].

3   Statement of the Problem

Our aim is to construct controller for control temperature in the chemical reactor of alcohol production. We consider optimal control system by using fuzzy dynamic programming on base of giving control, goal and relation between control, goal and target. The problem is to define such sequences of control variables that maximize fuzzy goal under a fuzzy constraint on control variables. Control variable is water consumption, state parameter is temperature of reaction zone. Termination time in the problem is fixed and given. Membership functions are normal and convex functions. Mathematical model of the considered problem is given in Sect. 4.

4   Numerical Example and Simulation Results

Solution of the problem is based on Bellman-Zadeh approach [15, 16]. To find solution of the problem, (1)–(3) is transformed to multistage decision making scheme. In this case maximal value of criterion (1) is defined by expert. Then the problem (1)–(3) is to achieve this value under the fuzzy constraints.

Let's consider the following fuzzy terminal regulator problem:

$$J = X_2 \rightarrow \max \tag{4}$$

$$X_{t+1} = X_t \circ U_t \circ R(x_t, u_t, x_{t+1}), \; t = 0, 1. \tag{5}$$

$$U_0 - \text{below } 45, \quad U_1 - below\ 53 \tag{6}$$

$$F(X_t) = \left\{ \frac{1}{70} + \frac{0.5}{74} + \frac{0.3}{76}, \frac{0.3}{70} + \frac{0.5}{74} + \frac{1}{76} + \frac{0.5}{78} + \frac{0.3}{80} \right\},$$

$$F(U_t) = \left\{ \frac{1}{45} + \frac{0.5}{53}, \frac{0.5}{45} + \frac{1}{53} + \frac{0.5}{62}, \frac{0.5}{53} + \frac{1}{62} + \frac{0.5}{70}, \frac{0.5}{62} + \frac{1}{70} \right\}$$

For $(u_t; x_t; x_{t+1}) = (45, 53, 62, 70; 70, 74, 76, 78, 80; 70)$ maximum value of criterion is defined by expert as $R = \begin{bmatrix} 0.3 & 0.5 & 0.5 & 0.5 & 0.3 \\ 0.3 & 0.5 & 1.0 & 0.5 & 0.3 \\ 0.3 & 0.5 & 0.5 & 0.5 & 0.3 \\ 0 & 0 & 0.3 & 0.3 & 0.3 \end{bmatrix}$,

for $(u_t; x_t; x_{t+1}) = (45, 53, 62, 70; 70, 74, 76, 78, 80; 74)$,

$$R = \begin{bmatrix} 0.3 & 0.5 & 0.5 & 0.5 & 0.3 \\ 0.3 & 0.5 & 0.5 & 0.5 & 0.5 \\ 0.3 & 0.5 & 0.5 & 0.5 & 0.5 \\ 0 & 0 & 0.3 & 0.5 & 0.5 \end{bmatrix},$$

for $(u_t; x_t; x_{t+1}) = (45, 53, 62, 70; 70, 74, 76, 78, 80; 76)$,

$$R = \begin{bmatrix} 0.3 & 0.3 & 0.3 & 0.3 & 0.3 \\ 0.3 & 0.3 & 0.3 & 0.5 & 0.3 \\ 0.3 & 0.3 & 0.3 & 0.5 & 1.0 \\ 0 & 0 & 0.3 & 0.5 & 1.0 \end{bmatrix},$$

for $(u_t; x_t; x_{t+1}) = (45, 53, 62, 70; 70, 74, 76, 78, 80; 78)$,

$$R = \begin{bmatrix} 0 & 0 & 0 & 0 & 0 \\ 0 & 0 & 0.3 & 0.5 & 0.5 \\ 0 & 0 & 0.3 & 0.5 & 0.5 \\ 0 & 0 & 0.3 & 0.5 & 0.5 \end{bmatrix},$$

for $(u_t; x_t; x_{t+1}) = (45, 53, 62, 70; 70, 74, 76, 78, 80; 80)$,

$$R = \begin{bmatrix} 0 & 0 & 0 & 0 & 0 \\ 0 & 0 & 0.3 & 0.3 & 0.3 \\ 0 & 0 & 0.3 & 0.3 & 0.3 \\ 0 & 0 & 0.3 & 0.3 & 0.3 \end{bmatrix}.$$

By using value of fuzzy goal and initial state we transform (4)–(6) problem to the form:

$$G^2 : X_2^{\max} = \frac{0.4}{74} + \frac{0.6}{76} + \frac{0.7}{78} + \frac{1}{80}, \ U_0^F = \frac{1}{45} + \frac{0.5}{53} + \frac{0.3}{62}, \ U_1^F = \frac{0.3}{45} + \frac{0.8}{53} + \frac{1}{70}.$$

$$(7)$$

Then in this case we are getting recurrence equations:

$$\begin{cases} \mu_{G^{2-V}}(x_{2-V}) = \max_{x_{2-V}}\{\mu_{G^{2-V}}(X_{2-V}) \cdot \mu_{X_{2-V}}(x_{2-V})\}, \\ \mu_{G^{2-V}}(X_{2-V}) = \max_{U_{2-V}} \min\{\mu_{U_{2-V}}(U_{2-V}) \cdot \mu_{G^{2-V+1}}(X_{2-V+1})\}, \\ \mu_{G^{2-V+1}}(X_{2-V+1}) = \max_{x_{2-V+1}} \min\{\mu_{x_{2-V+1}}(x_{2-V+1}), \mu_{x_{2-V+1}}(x_{2-V+1})\}, \\ x_{2-V+1} = X_{2-V} \circ U_{2-V} \circ R(x_{2-V}, U_{2-V+1}, X_{2-V+1}). \end{cases} \quad (8)$$

$$v = 1, 2$$

The solution begins from the last stage, $v = 1$. The set of recurrence equations for this stage has a form:

$$\begin{cases} \mu_{G^1}(x_1) = \max \mu_{G^1}(X_1) \times \mu_{X_{12}}(x_1), \\ \mu_{G^1}(X_1) = \max_{u_1} \min\{\mu_{u_1}(u_1), \mu_{G^2}(X_2)\}, \\ \mu_{G^2}(X_2) = \max_{x_2} \min\{\mu_{x_2}(x_2), \mu_{x_2^G}(x_2)\}. \end{cases}$$

Below membership value of each element of fuzzy set X_2 to fuzzy goal G^2 is constituted. Result of relational equation for $X_1 = \frac{1}{70} + \frac{0.5}{74} + \frac{0.3}{76}$ and $U_1 = \frac{1}{45} + \frac{0.5}{53}$ is defined $X_2 = \frac{0.3}{70} + \frac{0.5}{74} + \frac{0.5}{76} + \frac{0.5}{78} + \frac{0.5}{80}$; for $U_2 = \frac{0.5}{45} + \frac{1}{53} + \frac{0.5}{62}$ is defined. $X_2 = \frac{0.3}{41} + \frac{0.5}{42} + \frac{1}{43} + \frac{0.5}{44} + \frac{0.5}{45}$; for $U_3 = \frac{0.5}{21} + \frac{1}{22} + \frac{0.5}{23}$ is defined $X_2 = \frac{0.3}{41} + \frac{0.5}{42} + \frac{0.5}{43} + \frac{0.5}{44} + \frac{0.5}{45}$; for $U_4 = \frac{0.5}{62} + \frac{1}{70}$ is defined $X_2 = \frac{0.3}{70} + \frac{0.5}{74} + \frac{0.5}{76} + \frac{0.5}{78} + \frac{0.5}{80}$.

In next step we define maximal grade value belongs to intersection of fuzzy goal X_2 and X_2^G. Then we obtain for $X_1 = \frac{1}{70} + \frac{0.5}{74} + \frac{0.3}{76}$, and for all U_t and X_t is defined X_2^G as follows: $X_2^G = \{\frac{0}{70} + \frac{0.4}{74} + \frac{0.5}{76} + \frac{0.5}{78} + \frac{0}{80}, \frac{0.5}{70} + \frac{0.4}{74} + \frac{0.5}{76} + \frac{0.5}{78} + \frac{0.5}{80}, \frac{0}{70} + \frac{0.4}{74} + \frac{0.5}{76} + \frac{0.5}{78} + \frac{0.5}{80}, \frac{0}{70} + \frac{0.4}{74} + \frac{0.5}{76} + \frac{0.5}{78} + \frac{0.5}{80}\}$. For $X_1 = \frac{0.3}{70} + \frac{0.5}{74} + \frac{1}{76} + \frac{0.5}{78} + \frac{0.3}{80}$ is defined $X_2^G = \{\frac{0}{70} + \frac{0.4}{74} + \frac{0.5}{76} + \frac{0.5}{78} + \frac{0.5}{80}, \frac{0}{70} + \frac{0.4}{74} + \frac{0.5}{76} + \frac{0.5}{78} + \frac{0.5}{80}, \frac{0}{70} + \frac{0.4}{74} + \frac{0.5}{76} + \frac{0.5}{78} + \frac{0.5}{80}, \frac{0}{70} + \frac{0.4}{74} + \frac{0.5}{76} + \frac{0.5}{78} + \frac{1}{80}\}$.

According to (8) is formed the following result: for $X_1 = \frac{1}{70} + \frac{0.5}{74} + \frac{0.3}{76}$: $\mu_{G^2}(X_2) = \{\frac{0.5}{76}, \frac{0.6}{76}, \frac{0.5}{76}, \frac{0.5}{76}\}$, for $X_1 = \frac{0.3}{70} + \frac{0.5}{74} + \frac{1}{76} + \frac{0.5}{78} + \frac{0.3}{80}$: $\mu_{G^2}(X_2) = \{\frac{0.5}{76}, \frac{0.5}{80}, \frac{0.5}{80}, \frac{1}{80}\}$.

Let us consider to stage v = 2. We transform fuzzy state X_1 to fuzzy goal X_1^G by replacing its elements membership grade for membership grades to fuzzy goal. Then we are getting:

If $X_1 = \frac{1}{70} + \frac{0.5}{74} + \frac{0.3}{76}$ then are getting $X_1^G = \{\frac{0.3}{70} + \frac{0.5}{74} + \frac{0.5}{76} + \frac{0.5}{78} + \frac{0.5}{80}, \frac{0.3}{70} + \frac{0.5}{74} + \frac{0.5}{76} + \frac{0.5}{78} + \frac{0.3}{80}, \frac{0.5}{70} + \frac{0.5}{74} + \frac{0.5}{76} + \frac{0.5}{78} + \frac{0.5}{80}, \frac{0.3}{70} + \frac{0.5}{74} + \frac{0.5}{76} + \frac{0.5}{78} + \frac{0.5}{80}\}$.

If $X_1 = \frac{0.3}{70} + \frac{0.5}{74} + \frac{1}{76} + \frac{0.5}{78} + \frac{0.3}{80}$ then are getting:

$$X_1^G = \left\{ \frac{0.3}{70} + \frac{0.5}{74} + \frac{0.5}{76} + \frac{0.5}{78} + \frac{0.5}{80}, \frac{0.3}{70} + \frac{0.5}{74} + \frac{0.5}{76} + \frac{0.5}{78} + \frac{0.5}{80}, \right.$$
$$\left. \frac{0.3}{70} + \frac{0.5}{74} + \frac{0.5}{76} + \frac{0.5}{78} + \frac{0.5}{80}, \frac{0.3}{70} + \frac{0.5}{74} + \frac{0.5}{76} + \frac{0.5}{78} + \frac{0.5}{80} \right\}.$$

Next step we calculate $\mu_{G^1}(X_1)$. For $X_1 = \frac{1}{70} + \frac{0.5}{74} + \frac{0.3}{76}$: $\mu_{G^1}(X_1) = \{\frac{1}{76}, \frac{0.5}{74}, \frac{0.5}{74}, \frac{0.5}{74}\}$, for $X_1 = \frac{0.3}{70} + \frac{0.5}{74} + \frac{1}{76} + \frac{0.5}{78} + \frac{0.3}{80}$: $\mu_{G^1}(X_1) = \{\frac{0.5}{74}, \frac{0.5}{74}, \frac{0.5}{74}, \frac{0.5}{74}\}$.

At the end we obtain fuzzy decision as following form:

$$U_0 = \frac{1}{45} + \frac{0.5}{53}$$
$$U_1 = \frac{1}{45} + \frac{0.5}{53} \quad or \quad \frac{0.5}{45} + \frac{1}{53} + \frac{0.5}{62}$$

5   Conclusions

The considered problem describing by fuzzy relational equation is characterized by uncertainty state and controls. Solving such a problem by using the existing classical optimization methods is impossible. The discussed problem corresponds to Bellman-Zadeh scheme, so its solution is defined as intersection of fuzzy goal and fuzzy constaints. For solving fuzzy analogue of dynamic programming, the problem is modifed. To find solution for the given problem, it was described by sup-min fuzzy relational equations. As optimal control in considering stage is chosen such value of maximizing decision on which the according state with maximal grade belongs to fuzzy goal. After defining optimal control on last stage, calculated fuzzy decision and was maximized decision.

References

1. Aliev, R.A., Aliev, R.R.: Soft Computing and Its Application. World Scientific, New Jersey (2001)
2. Aliev, R.A., Abdikiev, N.M., Shachnazarov, M.M.: Intelligent control systems, Moscow (1990). (in Russian)
3. Gardashova, L.A., Gahramanli, Y., Babanli, M..: Fuzzy neural network based analysis of the process of oil product sorption with foam polystyrene. Int. J. Eng. Res. Appl. 7(9), 85–90 (2017)
4. Di Nola, A., Pedrycz, W., Sessa, W.S.: Fuzzy relation equations and algorithms of inference mechanism in expert systems. In: Approximate Reasoning in Expert Systems, pp. 355–367. Elsevier Science Publishers B.V., North Holland, Amsterdam (1985)
5. Sanchez, E.: Resolution of composite fuzzy relation equations. Inf. Control 30, 38–48 (1976)

6. Martino, F.D., Sessa, S.: Spatial analysis and fuzzy relation equations. Adv. Fuzzy Syst. **2011**, 1–14 (2011)
7. Li, Y.M., Wang, X.P.: Necessary and sufficient conditions for existence of maximal solutions for inf-α composite fuzzy relational equations. Comput. Math Appl. **55**, 1961–1973 (2008)
8. Belohlavek, R.: Sup-t-norm and inf-residuum are one type of relational product: unifying framework and consequences. Fuzzy Sets Syst. **197**, 45–58 (2012)
9. Xiong, Q.Q., Wang, X.P.: Solution sets of inf-αT fuzzy relational equations on complete Brouwerian lattices. Inf. Sci. **177**, 4757–4767 (2007)
10. Markovski, A.V.: On the relation between equations with max-product composition and the covering problem. Fuzzy Sets Syst. **153**, 261–273 (2005)
11. Lin, J.L.: On the relation between fuzzy max-Archimedean t-norm relational equations and the covering problem. Fuzzy Sets Syst. **160**, 2328–2344 (2009)
12. Lin, J.L., Wu, Y.K., Guu, S.M.: On fuzzy relational equations and the covering problem. Inf. Sci. **181**, 2951–2963 (2011)
13. Xiong, Q., Shu, Q.: Fuzzy relational equations and the covering problem. In: 16th World Congress of the International Fuzzy Systems Association (IFSA) and 9th Conference of the European Society for Fuzzy Logic and Technology (EUSFLAT), pp. 77–84 (2015)
14. Zadeh, L.A.: Fuzzy sets. Inform. Control **8**, 338–353 (1965)
15. Bellman, R.E., Zadeh, L.A.: Decision making in a fuzzy environment. Manag. Sci. **17**(4), 141–164 (1970)
16. Loia, V., Sessa, S.: Fuzzy relation equations for coding/decoding processes of images and videos. Inf. Sci. **171**, 145–172 (2005)
17. Aliev, R.A., Pedrycz, W.: Fundamentals of a fuzzy-logic-based generalized theory of stability. IEEE Trans. Syst. Man Cybern. Part B (Cybern.) **39**(4), 971–988 (2009)
18. Aliev, R., Tserkovny, A.: Systemic approach to fuzzy logic formalization for approximate reasoning. Inf. Sci. **181**(6), 1045–1059 (2011)

Empirical Adaptation of Control Parameters in Differential Evolution Algorithm

Petr Bujok[(⊠)] [iD]

University of Ostrava, 30. dubna 22, 70301 Ostrava, Czech Republic
petr.bujok@osu.cz

Abstract. Proper settings of two control parameters in differential evolution (DE) algorithm play a crucial role when solving various optimisation problems. There are a lot of investigated adaptive mechanisms sampling the values of these parameters differently. In the proposed adaptive approach, efficient and inefficient settings of both DE control parameters are considered. A sampling of the control parameters is performed using the performance of parameters' settings achieved in a preliminary experimental study. The proposed approach is applied to a frequently used adaptive jDE algorithm. The statistical analysis of the results provides better information about the efficiency of the proposed technique. More proper settings in the adaptation of DE control parameters provide better performance than a more general approach when solving CEC 2017 problems.

Keywords: Differential evolution · jDE · Scale factor · Crossover ratio · Experimental study · Real-world · Problems · CEC 2017 benchmark set

1 Introduction

This paper is focused on a more proper setting of two control parameters in Differential Evolution (DE) algorithm which was obtained in a huge preceding experiment. In this experiment, a lot of various settings are used and evaluated by real-world problems [1] to find more successful and poorer values of these parameters. After this analysis, a well-known adaptive DE algorithm is selected to be enhanced using the results from our previous experiment. The original and all newly proposed variants of DE algorithm are compared when solving CEC 2017 benchmark set [2]. The main goal is to show if a more specific setting of DE control parameters increases the efficiency of the DE algorithm.

2 Differential Evolution

Differential evolution was introduced by Storn and Price in 1995 [3] as a global optimiser for continuous optimisation problems with a real-value objective function. It has been one of the most frequently used evolutionary algorithms in recent years [4]. DE develops a population of N potential solutions using evolutionary operators of

R. A. Aliev et al. (Eds.): ICSCCW 2019, AISC 1095, pp. 113–120, 2020.
https://doi.org/10.1007/978-3-030-35249-3_14

mutation, crossover, and selection. Details of this simple efficient optimiser are provided in [3].

DE has very few control parameters. Apart from the size of population N common for all these techniques, it is the choice of mutation and crossover strategy, and the pair of parameters F and CR, controlling the mutation and crossover. Some recommended settings of DE control parameters are available in [3–5], but these recommendations are valid only for part of optimisation problems. The summary of the results in DE research has been recently presented in the following comprehensive papers [7–9].

2.1 Adaptive Variant of jDE

In 2006, Brest et al. proposed a simple and efficient adaptive DE variant (mostly called 'jDE' in literature) [6]. It uses mutation rand/1 and binomial crossover, where both corresponding control parameters (F and CR) are self-adapted. The values of these control parameters are encoded with each individual of the population and survive to next generation if the individual is successful. In other words, F and CR values remain the same if they generate a trial vector which outperforms the old solution.

The values of F and CR are initialised randomly for each point and are randomly mutated with given probabilities τ_1 and τ_2. Then, new values of $CR \in [0, 1]$ uniformly distributed, and F also distributed uniformly in $[F_l, F_u]$ are, where $F_l = 0.1$, $F_u = 0.9$.

3   Efficiency of DE Control Parameters

There exist many various adaptive approaches to adapt F and CR values during the search process in DE algorithm [7, 9, 10], where the values of the control parameters F and CR are typically sampled from one interval (for example $[0, 1]$). When a relatively wide sampling interval of control parameters is used, a relatively big amount of unsuccessful F, CR values is sampled and used, which can cause slow convergence of the DE algorithm.

Therefore, the same sets of equidistantly sampled values with step 0.05 are used (i.e. $\{0, 0.05, 0.1, \cdots 0.95, 1\}$) in the first stage of this experiment. The most widely used mutation rand/1 and binomial crossover were selected in this experiment. Settings were evaluated by real-world optimisation problems CEC 2011.

All combinations are statistically assessed using the non-parametric Friedman test with $p < 5 \times 10^{-6}$. Each DE setting is evaluated by a mean-rank value representing the overall performance from all selected problems (lower mean-rank means better setting).

The plots representing the efficiency of all 441 combinations of F, CR on selected problems are in Figs. 1 and 2. The most efficient combination is represented by a black square, and the least efficient is illustrated by a white square. Figure 1a represents the results for all problems. We can see that there is an interesting continuous 'dark' area where the best combination of $F = 0.45$ and $CR = 0.95$ are located. Conversely, the worst performance is provided by $F = 0$ and $CR = 1$. This fact is caused by zero 'diversity' step ($F = 0$) and high propagation of such solutions ($CR = 1$). It is possible to see that there is no value of F and CR from $[0, 1]$ which generally performs the worst. This setting performs the worst for each subset of problems. On the other hand, there are two 'bright' regions where the efficiency of DE setting is rather worse, especially *CR* is close to 1, and *F* is rather zero.

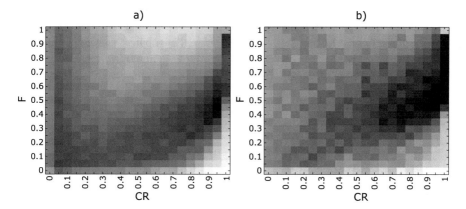

Fig. 1. Mean ranks from the Friedman test (a) all problems, (b) problems with $D < 15$.

For problems with a small dimension ($D \leq 15$), the comparison of all F, CR settings is in Fig. 1b. The 'distribution' of DE parameters efficiency is somewhat different. The best performing combination of control parameters is $F = 0.6$ and $CR = 1$. When compare these results with results of all problems (Fig. 1a), rather higher values of F are performing better, a slightly higher value of CR confirms the success of this combination because the whole mutation vector is used.

When the dimension of the problems is at the middle level ($15 < D \leq 40$), the plot of mean ranks from the Friedman test is in Fig. 2a. The best performing combination is $F = 0.45$ and $CR = 0.95$ for all test problems. The difference is visible in the 'dark' arch where $0.5 \leq CR \leq 1$ together with $0.05 \leq F \leq 0.6$. On the other hand, high F values with $CR \in [0.5, 0.9]$ achieves rather worse results.

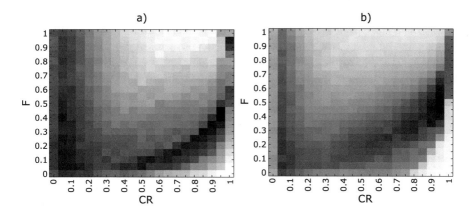

Fig. 2. Mean ranks from the Friedman test (a) problems with $15 < D \leq 40$, and (b) $D > 40$.

Figure 2b shows mean ranks from the Friedman test applied only on problems with a high dimension level ($D > 40$). This plot is very similar to Fig. 1a, where all problems are integrated. The only 'visual' difference is obvious that generally bad performing settings achieve even worse mean ranks and, similarly, best settings of all test problems perform even better when only high-dimensional problems are solved.

We can see that for higher CR values (approx. $CR > 0.25$), for all subsets there is a similar situation. An interesting fact is that with higher CR values, the efficiency and also the inefficiency is increased. High and low peaks of the lines take turns when smaller, or higher F values are applied. Besides this, it is obvious that small CR values are more efficient for problems with a low dimension, whereas in problems with a middle dimension, low CR values are evaluated poorly.

4 Newly Proposed Adaptive DE Variants

Based on the results from the previous experiment, new ideas of adaptation of F and CR parameters are proposed and studied. The adaptive DE variant introduced in 2006 by Brest et al. called jDE was selected to be enhanced by a designed mechanism. This DE variant uses only one strategy - a combination of most often used mutation rand/1 and binomial crossover. Values of CR are initialised for each individual independently from the uniformly distributed interval $[0, 1]$, and they are reinitialised with a small probability $\tau_2 = 0.1$. Values of F are also initialised for each individual independently uniformly from the interval $[0.1, 0.9]$ and reinitialised with the probability $\tau_1 = 0.1$.

In this section, four different settings of F and CR parameters in jDE algorithm are described. All these ideas are based on the results of the previous experiment. A primary goal is to show if some more specific setting of DE control parameters increases the efficiency of the jDE algorithm.

4.1 jDE$_{FCR1}$

At first, the results of the efficiency of F and CR from all problems are used. The idea is to focus on a smaller (not necessary one) area of the sampling intervals based on better performing results. Figure 1a reveals that no CR value is performed so bad. Therefore, the sampling interval of CR remains the same as in the original jDE. On the other side, when $CR < 0.15$, rather smaller values of F are efficient $F \in [0.05, 0.1]$. Conversely, when CR is maximal, $CR = 1$, F is sampled from $F \in [0.85, 0.95]$. When $0.15 \leq CR \leq 0.95$, then values of F are sampled from the linearly increased area, where lower values F_l are from 0.05 to 0.4 and the upper values F_u are computed as $F_u = F_l + 0.1$. This approach in jDE is called jDE$_{FCR1}$.

4.2 jDE$_{FCR2}$

The next idea (called jDE$_{FCR2}$) is to avoid the most inefficient 'bright' area(s) and sample the values of F, CR from combinations which are not inefficient. For $CR = 1$, F is sampled from $F \in [0.5, 1]$. When $0.8 \leq CR < 1$, then values of F are sampled from the linearly increased area, where lower values F_l are from 0.05 to 0.25 and the upper values F_u are computed as $F_u = F_l + 0.7$. If $0.25 \leq CR < 0.8$, then values of F are sampled from $F \in [0, 0.7]$ and when $CR < 0.25$ then values of F are sampled from $F \in [0, 1]$.

4.3 jDE$_{FCR3}$

Further, the results for three subsets of the problems (Figs. 1b, 2a and b) are involved (variant jDE$_{FCR3}$). The idea is to avoid inefficient parameters' combinations as is in jDE$_{FCR2}$. The sampling areas for $D \leq 15$ are as follows. When $0.75 \leq CR$, then $F \in [0.05, 0.95]$ and when $CR > 0.75$ then values of F are sampled from the linearly increased area, where lower values F_l are from 0.15 to 0.45 and the upper values F_u are from 0.85 to 0.95. For $15 < D \leq 40$ when $CR = 1$, then $F \in [0.5, 0.95]$, and when $CR = 0.95$, then $F \in [0.25, 0.65]$. When $0.6 \leq CR < 0.95$, then values of F are sampled from linearly increased area, where lower values F_l are from 0.05 to 0.2 and the upper values F_u are computed as $F_u = F_l + 0.2$. For $0.25 \leq CR < 0.6$, the values of F are sampled from $F \in [0, 0.3]$, and when $CR < 0.25$, then $F \in [0, 1]$. For problems with high dimension level, $D > 40$, F, CR are sampled as follows. When $CR = 1$, then $F \in [0.55, 0.95]$. When $0.8 \leq CR < 1$, then values of F are sampled from the linearly increased area, where lower values F_l are from 0.05 to 0.25 and the upper values F_u are computed as $F_u = F_l + 0.5$. When $0.25 \leq CR < 0.8$, then $F \in [0.05, 0.4]$, and when $CR = 0.05$ then $F \in [0, 1]$. Values of $CR = 0$ are not sampled.

4.4 jDE$_{FCR4}$

The last proposed variant is called jDE$_{FCR4}$, and it specifies a very small sampling area for the parameters. The most similar potentially efficient area for all four subsets of problems is 'dark' linearly increased stripe. Therefore here, when $0.75 \leq CR \leq 1$, then F is sampled linearly from $F \in [0.25, 0.95]$. Notice, that if the mutation of F or CR value for any individual occurs, the value of the second parameter is also resampled to be in the requested area.

5 Experimental Setting

Although the test suite of 22 real-world problems selected for CEC 2011 competition in Special Session on Real-Parameter Numerical Optimization [1] is used in the tuning phase, the test suite for a special session and competition on Single Objective Bound Constrained Real-Parameter Numerical Optimization (CEC 2017) is applied in the comparison phase. All the 30 problems of CEC 2017 are used with four dimension levels, $D = 10, 30, 50, 100$. For each algorithm and problem, 51 independent runs were carried out. The run of the algorithm stops if the prescribed number of function evaluations $maxFES = D \times 10^4$ is reached. The experiments in this paper can be divided into two independent parts. In the first part, a classic DE (rand/1/bin) with 441 settings of F and CR was applied on CEC 2011 problems. The only control parameter is the population size, and it was set to $N = 100$.

In the second part of the study, four different jDE variants were applied to the second set of CEC 2017 problems. The population size was also $N = 100$ for all four algorithms and problems. The other control parameters are set up according to the recommendation of authors in their original papers, i.e. $\tau_1 = \tau_2 = 0.1$.

6   Results

At the first stage of the experiment, all algorithms are compared on real-world problems CEC 2011 where the proposed methods provided good performance. The reason to use the second CEC 2017 set of problems is to show the applicability of the proposed advanced jDE variants. All presented results were achieved on the CEC 2017 set.

Table 1. Mean ranks from the Friedman test for $D = 10, 30, 50, 100$, CEC 2017.

D	jDE_{FCR2}	jDE_{FCR3}	jDE_{FCR4}	jDE_{FCR1}	jDE
10	**2.1**	4.1	3	**2.7**	3.1
30	3.2	**2.5**	**2.6**	3.5	3.2
50	3	**2.5**	**2.9**	3.1	3.6
100	3	**2.6**	3.1	3.1	3.2
avg	2.81	2.91	2.92	3.1	3.26

A global insight into all algorithms performance is provided by the Friedman statistical test. The zero hypothesis about the equality of the algorithms' results was rejected in each stage of the run, where the significance level was set 1×10^{-5}. For better illustration, the mean ranks are illustrated in Table 1.

Table 2. Counts of best and worst positions of algorithms from the Kruskal-Wallis test (wins/similar/losses), CEC 2017.

D	jDE	jDE_{FCR1}	jDE_{FCR2}	jDE_{FCR3}	jDE_{FCR4}
10	2/26/2	5/23/2	10/19/1	0/16/14	2/24/4
30	8/10/12	2/20/8	0/30/0	10/15/5	5/24/1
50	5/12/13	2/21/7	4/24/2	11/16/3	4/23/3
100	7/13/10	6/17/7	2/24/4	10/18/2	3/20/7
Σ	22/61/37	15/81/24	16/97/7	31/65/24	14/91/15

The worse total results are provided by the original jDE. The worst performing newly proposed method is jDE_{FCR1}, which achieves substantially good performance for $D = 10$. The best results regarding all 120 problems are achieved by jDE_{FCR2} and jDE_{FCR3}, both are based on avoiding the inefficient parameters' combinations. An interesting difference between these methods is for lower $D = 10$, where the best performing jDE_{FCR2} wins.

These results show that it is better to avoid the worse performing combinations of the control parameters than set these values more strictly regarding the best performing results.

More details of the comparison are provided by the Kruskal-Wallis non-parametric one-way ANOVA test applied to each of 120 CEC 2017 test problems. It was found out that the performance of the algorithms in the comparison significantly differs. Dunn's method was then applied for multiple comparisons. In Table 2, numbers of significant wins, similar results, and the number of the last positions are computed for each dimension level. For $D = 10$, the best results on ten problems are achieved by jDE_{FCR2}, followed by jDE_{FCR1} with five wins. The worst performance is for jDE_{FCR3}. For $D = 30$, the best results on ten problems are achieved by jDE_{FCR3}, followed by the original jDE with eight wins. The worst performance is for jDE_{FCR2}. For $D = 50$, the best results are achieved by jDE_{FCR3} (11 problems), the worst performance are achieved by jDE_{FCR1}. For high $D = 100$, the best results are provided again by jDE_{FCR3}, followed by the original jDE (7 wins). The least efficiency is for the jDE_{FCR2} variant. The best results on most number of problems are achieved by jDE_{FCR3}, the three remaining three proposed jDE variants performs similarly, only jDE_{FCR2} has the least number of losses.

Table 3. Counts of wins and losses of the proposed algorithms against jDE from the Wilcoxon test (wins/similar/losses), CEC 2017.

D	jDE_{FCR1}	jDE_{FCR2}	jDE_{FCR3}	jDE_{FCR4}
10	9/16/5	14/12/4	0/19/11	5/20/5
30	8/10/12	8/14/8	11/9/10	12/9/9
50	10/12/8	13/13/4	11/13/6	12/13/5
100	9/16/5	8/22/0	10/17/3	11/18/1
Σ	36/54/30	43/61/16	32/58/30	40/60/20

In Table 3, the results of the non-parametric Wilcoxon rank-sum test are illustrated. Each newly proposed jDE variant is compared with the original jDE on all 120 CEC 2017 problems. For each dimension, the count of significant wins, insignificant differences, and number of losses of the proposed variant are computed. The sense of this comparison is to avoid the influence of the remaining methods performance of the experiment when only two algorithms are compared. We can see that the best performance compared to jDE is provided by jDE_{FCR2}, which achieves the highest number of wins and the least number of losses. This method is more powerful for a lower dimension level, and $D = 50$. For $D = \{30, 100\}$ the count of wins of the remaining proposed methods are slightly bigger, but jDE_{FCR2} is the very rarely worse than jDE. Very good results are achieved by jDE_{FCR4}, which drops back only for low dimensional problems.

7   Conclusions

In this paper, four new ideas of the control parameter setting in adaptive differential evolution are proposed. All methods are developed on a set of 22 real-world minimisation problems and applied on the CEC 2017 benchmark set to show the independency of tuned sampling intervals. The results from a Friedman test show that all new jDE variants perform the better compared with the original jDE variant. The main difference between the two best-performing methods is that jDE_{FCR3} achieves better results for the higher dimension level, and the best performing jDE_{FCR2} is efficient even for the low dimension of the problems. This fact leads to a conclusion that it is better to avoid worse performing combinations of the control parameters than set these values more strictly regarding the best performing results.

The comparison of the algorithms on each of 120 problems separately using the Kruskal-Wallis test shows high performance of the jDE_{FCR3} variant. The results of the comparison of each proposed method with the original jDE separately confirm high performance of jDE_{FCR2} and indicates the good potential of the jDE_{FCR4} variant, especially for $D > 10$. Further research in this area will be focused on other adaptive DE variants which use sampling of F and CR values.

References

1. Das, S., Suganthan, P.N.: Problem definitions and evaluation criteria for CEC 2011 competition on testing evolutionary algorithms on real-world optimization problems. Technical report, Singapore (2010)
2. Awad, N.H., Ali, M.Z., Liang, J.J., Qu, B.Y., Suganthan, P.N.: Problem definitions and evaluation criteria for the CEC 2017 special session and competition on single objective real-parameter numerical optimization. Technical report, China (2016)
3. Storn, R., Price, K.V.: Differential evolution - a simple and efficient heuristic for global optimization over continuous spaces. J. Global Optim. **11**, 341–359 (1997)
4. Price, K.V., Storn, R., Lampinen, J.: Differential Evolution: A Practical Approach to Global Optimization. Springer, Heidelberg (2005)
5. Feoktistov, V.: Differential Evolution in Search of Solutions. Springer, Heidelberg (2006)
6. Brest, J., Greiner, S., Bošković, B., Mernik, M., Žumer, V.: Self-adapting control parameters in differential evolution: a comparative study on numerical benchmark problems. IEEE Trans. Evol. Comput. **10**, 646–657 (2006)
7. Das, S., Suganthan, P.N.: Differential evolution: a survey of the state-of-the-art. IEEE Trans. Evol. Comput. **15**, 27–54 (2011)
8. Neri, F., Tirronen, V.: Recent advances in differential evolution: a survey and experimental analysis. Artif. Intell. Rev. **33**, 61–106 (2010)
9. Das, S., Mullick, S., Suganthan, P.N.: Recent advances in differential evolution-an updated survey. Swarm Evol. Comput. **27**, 1–30 (2016)
10. Tvrdík, J., Poláková, R., Veselský, J., Bujok, P.: Adaptive variants of differential evolution: towards control-parameter-free optimizers. In: Zelinka, I., Snášel, V., Abraham, A. (eds.) Handbook of Optimization - From Classical to Modern Approach. Intelligent Systems Reference Library, vol. 38, pp. 423–449. Springer, Berlin (2012)

The Use of Multiple Correspondence Analysis to Map the Organizational Citizenship Behaviour and Job Satisfaction of Administrative Staff

Cemre S. Gunsel Haskasap[1] , Tulen Saner[2] ,
and Serife Zihni Eyupoglu[2(✉)]

[1] University of Kyrenia, Kyrenia, TRNC, Mersin 10, Turkey
cemre.gunsel@kyrenia.edu.tr
[2] Near East University, 99138 Nicosia, TRNC, Mersin 10, Turkey
{tulen.saner,serife.eyupoglu}@neu.edu.tr

Abstract. Job satisfaction itself is said to be a strong predictor of organizational citizenship behaviour the assumption being that satisfied employees will display citizenship behaviour whereas dissatisfied employees will be reluctant to display their citizenship behaviour. The aim of the study was to understand whether a relationship between job satisfaction, organizational citizenship behavior, and demographic variables exist amongst administrative staff working at a private university hospital using Multiple Correspondence Analysis (MCA). MCA was used because it is a useful tool to explore whether a relationship exist between various variables and illustrates how the variables are related as well as presenting statistical results both analytically and visually. Results of the study indicate that a relationship between job satisfaction, organizational citizenship behavior, and demographic variables is evident.

Keywords: Multiple Correspondence Analysis · Job satisfaction · Organizational citizenship behaviour · Administrative staff · Hospital · North Cyprus

1 Introduction

Work is a major part of people's lives. It is a substantial source of personal realization as well as personal and professional enhancement, and income [1]. Due to the fundamental role that work plays in most people's life, the satisfaction felt in regards to one's job is a vital element in overall well-being [2]. As a result, job satisfaction has been a frequently researched topic. The causes of job satisfaction (and dissatisfaction) are a continuing topic of research for scientists and managers. The underlying premise is that workers who are satisfied are likely be more productive as well as stay with the organization for a longer period of time, however workers who are dissatisfied with their job are more likely to be less productive and more likely to quit the organization [3].

According to [4], in order for an organization to function effectively and to ensure continuousness employees need to display three different kinds of behaviours. Firstly, employees need to be influenced to enter and stay in an organization [4]. Secondly, the employees need to perform their specific job requirements, and thirdly, the employees need to display innovative and spontaneous activity in their efforts towards the achievement of organizational goals which go ahead of their official job descriptions [4]. This third grouping of behaviour is what is known as "organizational citizenship behaviour" (OCB) [5]. The definition for OCB given by [5] is "individual behaviour that is optional but not precisely or obviously acknowledged through the official reward scheme, and that in the total fosters the valuable execution of the organization".

In today's competitive environment, organizations are become more dependent on their human resources who serve as a source of competitive advantage. Organizations now demand employees who are willing to go beyond their official roles specified in their job description. That is, organizations now demand good organizational citizens in order to gain and sustain a competitive advantage. In this sense OCB has been an essential topic of interest in the fields of management and psychology over the last 20 to 30 years. Consequently it has acknowledged a large amount of devotion in the literature [6–9]. OCB is said to contribute to the effectiveness of any organization by improving coworker and managerial productivity; facilitating the organization in become accustomed to changes in the environment; and underpinning the synchronization within and through work groups [8]. OCB enhances organizational performance as it can lubricate the social machinery of the organization; it offers the suppleness necessary to work and overcome many unanticipated possibilities; it reduces friction, and leads to increased efficiency [10]. According to [5], OCB is the total summation of OCB performed over time and across individuals in the work group, divisions and organizational levels, that adds to the whole performance of the organization. Organizations that inspire good citizenship behaviours are more exciting places of work and these organizations can employ and keep the best people. OCB include such actions as helping coworkers who have heavy workloads; consulting with others before taking action; not complaining about trivial issues in the workplace; displaying work attendance that goes beyond the norm; and getting involved in the political process of the organization. Consequently, it is central to comprehend the variables that meaningfully and positively assist in creating this constructive behaviour within the workplace.

Job satisfaction itself is said to be a strong predictor of OCB [2]. The assumption is that satisfied employees will display their citizenship behaviour whereas dissatisfied employees will be reluctant to display their citizenship behaviour. In fact, [2] found job satisfaction to be the preeminent predictor of OCB. Some behavioral scientists even go so far as to claim that job satisfaction is the sole predictor of OCB.

Non-academic professional employees/administrative staff are fundamental components for higher educational institutions today. Administrative staff have the responsibility for the various daily activities of a college or university [11]. Administrative staff working at colleges and universities are employed mainly to provide academic support, student services, and institutional support. It has been argued that administrative support staff are a necessity for all academic divisions, colleges and universities and would find it almost impossible to function without their assistance [12].

Due to the lack of research into the administrative staff in colleges and universities, and even more so in regards to empirical work related to the job satisfaction and organizational citizenship behaviour of the administrative staff an obvious gap in the literature exists. In this respect this study aims to contribute to partially filling this gap through researching the job satisfaction and organizational citizenship behaviour relationship amongst administrative staff at a private university hospital in North Cyprus. The study will also provide evidence as to whether individual/demographical characteristics are associated with this relationship.

2   Method

2.1   Sample

A total of 150 administrative staff work at the hospital in question. The data for this study was attained from a sample of 108 administrative staff working at a private university hospital in the TRNC. According to [13] a sample size of 108 is considered suitable for a population of 150. A contact person from the hospital administration was chosen in order to ease the data collection process and make it easier to access the respondents.

2.2   Study Instruments

The job satisfaction of the respondents was measured through the use of the short-form Minnesota Satisfaction Questionnaire (MSQ) [14]. It is made up of 20 items that measure 20 facets of the job. Study respondents expressed the degree of their satisfaction with each of the 20 facets of their job on a five-point Likert scale. The scale ranged from 1 (very dissatisfied) to 5 (very satisfied). An item from the MSQ is "The feeling of accomplishment I get from my job". The 24-item scale developed by [15] was used to measure OCB. Respondents expressed the degree of their organizational citizenship behaviour with each of the 24 items on a five-point Likert scale. The scale ranged from 1 (strongly disagree) to 5 (strongly agree). An item from the OCB measure is "I help others who have heavy workloads".

2.3   Procedure

Respondents completed the MSQ and OCB scale as well as the demographical questions as a single survey online. The study instrument was sent to the 108 respondents via e-mail. The completed questionnaires were returned to the authors. Of the 108 questionnaires 100 usable questionnaires were returned giving a 92% response rate. The respondents were informed that the study was conducted solely for scientific purposes and that their identities would be held confidential.

2.4   Data Analysis

The study uses multiple correspondence analysis (MCA) to analyze job satisfaction, OCB and demographic variables. MCA is suitable to map both variables and individuals, thus enabling for the construction of complex visual maps whose structuring can be interpreted. MCA can also be utilized to suggest unexpected dimensions and relationships in the tradition of exploratory data analysis.

3 Results

The application of multiple correspondence analyses (MCA) was conducted to visualize the associations between job satisfaction, OCB and demographic variables (age, gender and marital status). The association between the variables is indicated in Fig. 1 presented below. To perform the MCA job satisfaction (JS), organizational citizenship behaviour (OCB) and demographic variables were included to the multiple correspondence analysis.

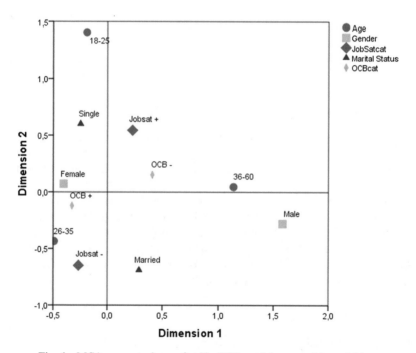

Fig. 1. MCA perceptual map for JS, OCB, and demographic variables.

Job satisfaction is located on the vertical axis of the correspondence map, while OCB is located on the horizontal axis. As can be seen from Fig. 1, in the upper left quadrant, the profile of respondents with high job satisfaction and high OCB is positioned. The common characteristics of these respondents are they are single, female and in the 18–25 years age group. This indicates that the younger female administrative staff are more satisfied with their job when compared with the other employees and they are more willing to display extra voluntary behavior in their work.

In the right upper quadrant, the profile of respondents with high job satisfaction but low OCB is positioned. The common characteristic of these respondents is that they are between the ages 36 and 60. This shows that middle aged administrative staff are satisfied with their job, however this does not inspire them to perform any extra voluntary behavior in their work.

In the left lower quadrant, the profile of respondents with low job satisfaction but high OCB is positioned. The common characteristic of these respondents is that they are between the ages 26–35. This indicates that even though the administrative staff in this age group are not satisfied with their job, they are willing to act as good organizational citizens by displaying extra voluntary behavior in their work.

In the right lower quadrant, the profile of respondents with low job satisfaction and low OCB are positioned. The common characteristics of these respondents are that they are male and married. This indicates that married male administrative staff are not satisfied with their job and are not willing to act as good organizational citizens.

4   Conclusion

High levels OCB and JS are two of the most important factors that indicate the effective functioning of any organization. This is particularly true in the case of hospitals because hospitals are organizations where people's health conditions are the major concern. When the suggestions for practice are taken into account, the outcomes from the present study indicate that demographic characteristics of employees do influence job satisfaction and OCB. Therefore, this can be taken into consideration in the recruitment and selection of new employees and to identify the kind of individuals who are more likely to display the type of performance and behavior desired and required by hospitals. In this respect in order to attract and retain needed administrative staff, it is important for hospital management to understand the vital role of OCB and JS on overall performance.

References

1. Gustainiene, L., Endriulaitiene, A.: Job satisfaction and subjective health among sales managers. Baltic J. Manag. **4**(1), 51–65 (2009)
2. Smith, C.A., Organ, D.W., Near, J.P.: Organizational citizenship behavior: its nature and antecedents. J. Appl. Psychol. **68**(4), 653–663 (1983)
3. Sarker, S.J., Crossman, A., Chinmeteepituck, P.: The relationship of age and length of service with job satisfaction: an examination of hotel employees in Thailand. J. Manag. Psychol. **18**(7/8), 745–758 (2003)
4. Katz, D.: The motivational basis of organizational behavior. Behav. Sci. **9**, 131–146 (1964)
5. Organ, D.W.: Organizational Citizenship Behavior: The Good Soldier Syndrome. Lexington Books, Lexington (1988)
6. Organ, D.W., Ryan, K.: A meta-analytic review of attitudinal and dispositional predictors of organizational citizenship behaviour. Personnel Psychol. **48**(4), 775–802 (1995)
7. Bateman, T.S., Organ, D.W.: Job satisfaction and the good soldier: the relationship between affect and employee citizenship. Acad. Manag. J. **26**(4), 587–595 (1983)

8. Podsakoff, P.M., MacKenzie, S.B., Paine, J.B., Bachrach, D.G.: Organizational citizenship behaviors: a critical review of the theoretical and empirical literature and suggestions for future research. J. Manag. **26**(3), 513–563 (2000)

9. Niehoff, B.P., Moorman, R.H.: Fairness in performance monitoring: the role of justice in mediating the relationship between monitoring and organizational citizenship behaviour. Acad. Manag. J. **36**, 527–556 (1993)

10. Podsakoff, N.P., Whiting, S.W., Podsakoff, P.M., Blume, B.D.: Individual and organizational level consequences of organizational citizenship behaviors: a meta-analysis. J. Appl. Psychol. **94**, 122–141 (2009)

11. Smerek, R.E., Peterson, M.: Examining Herzberg's theory: improving job satisfaction among non-academic employees at a university. Res. High. Educ. **48**(2), 229–250 (2006)

12. Knight, P., Trowler, P.R.: Departmental Leadership in Higher Education: New Directions for Communities of Practice. Open University Press, Buckingham (2001)

13. Sekaran, U., Bougie, R.: Research Methods for Business: A Skill-Building Approach, 7th edn. Wiley, West Sussex (2016)

14. Weiss, D.J., Dawis, R.V., England, G.W., Lofquist, L.H.: Manual for the Minnesota Satisfaction Questionnaire. The University of Minnesota Press, Minneapolis (1967)

15. Podsakoff, P.M., MacKenzie, S.B., Moorman, R.H., Fetter, R.: Transformational leader behaviors and their effects on followers' trust in leader, satisfaction and organizational citizenship behaviors. Leadersh. Q. **1**(2), 107–142 (1990)

Spatiotemporal Precipitation Modeling by AI Based Ensemble Approach

Selin Uzelaltinbulat[1]([⊠]) [iD], Vahid Nourani[2] [iD],
Fahreddin Sadikoglu[3] [iD], and Nazanin Behfar[4] [iD]

[1] Department of Computer Engineering, Near East University,
Mersin 10, North Cyprus, TRNC, Turkey
selin.uzelaltinbulat@neu.edu.tr
[2] Department of Civil Engineering, Near East University,
Mersin 10, North Cyprus, TRNC, Turkey
vnourani@yahoo.com
[3] Department of Electrical Electronic Engineering, Near East University,
Mersin 10, North Cyprus, TRNC, Turkey
fahreddin.sadikoglu@neu.edu.tr
[4] Department of Water Resources Engineering, University of Tabriz,
Tabriz, Iran
n.behtar@yahoo.com

Abstract. This study aimed at time-space estimations of monthly precipitation via a two-stage modeling framework. In temporal modeling as the first stage, three different AI models were applied to observed precipitation data from seven stations located in the Turkish Republic of Northern Cyprus (TRNC). In this way two scenarios were examined, each employing a specific inputs set. Afterwards, the outputs of single AI models were used to generate ensemble techniques to improve the performance of the precipitation predictions by the single AI models. To end this aim, two linear and one nonlinear ensemble techniques were proposed and then, the obtained outcomes were compared. In the second stage, for estimation of the spatial distribution of precipitation over whole region, the results of temporal modeling were used as inputs for the IDW spatial interpolator. The cross-validation was finally applied to evaluate the overall accuracy of the proposed hybrid spatiotemporal modeling approach. The obtained results in temporal modeling stage demonstrated that the non-linear ensemble method revealed higher prediction efficiency.

Keywords: Precipitation · Black box modelling · Artificial Intelligence · Ensemble method · Spatial interpolation · North Cyprus

1 Introduction

Precipitation is the most important component of the hydrologic cycle and accurate modeling of precipitation plays a critical role in design, planning and management of water resources and hydraulic structures. For such a spatiotemporal modeling of hydro-climatologic processes, usually a time series prediction model is linked to a spatial interpolation tool [1, 2]. Recently, Artificial Intelligence (AI) methods as such black

© Springer Nature Switzerland AG 2020
R. A. Aliev et al. (Eds.): ICSCCW 2019, AISC 1095, pp. 127–136, 2020.
https://doi.org/10.1007/978-3-030-35249-3_16

box methods showed great efficiency in modeling the dynamic precipitation process in the presence of the non-linearity, uncertainty and irregularity of the used data. One of the most commonly used AI methods for the precipitation modeling is Feed Forward Neural Network (FFNN). [3] employed Artificial Neural Network (ANN) for prediction of the monthly precipitation over 36 meteorological stations of India to estimate the monsoon precipitation of upcoming years. [4] employed ANN for real time precipitation predicting and flood management in Bangkok, Thailand. [5] applied ANNs for forecasting the rainfall time series using the temporal and spatial rainfall intensity data and pointed to the Wavelet-Elman model as the best method for rainfall forecasting.

As another type of AI model, the Least Square Support Vector Machine (LSSVM) is one of the most effective predicting methods as an alternative method of ANN. The LSSVM is capable of predicting non-linear, non-stationary and stochastic processes. The LSSVM has been used for prediction of precipitation in the recent decades. [7] forecasted the monthly precipitation over a state in China employing LSSVM method. Using the available observed data of 2 different stations from Turkey, [8] employed the LSSVM with and without wavelet based data pre-processing technique for prediction of precipitation time series. In addition to the ANN and LSSVM methods, the Adaptive Neural Fuzzy Inference System (ANFIS) model, which incorporates both the ANN learning power and fuzzy logic knowledge representation, has been considered as a robust model for precipitation prediction because of fuzzy concept ability in handling the uncertainty involved in the study processes. The ANFIS can analyze the relationship involved in the input and output data sets via a training scheme to optimize the parameters of a given Fuzzy Inference System (FIS) [9]. [11] suggested that, different ensemble approaches, compared to single techniques would provide the results with minimum error variance. Also, [12] revealed improving the forecasting accuracy by combining the results from the single models. [10] confirmed that reliability, objectivity, and accuracy of the analysis are increased by the use of ensemble systems. [13] indicated that performance of the seepage modeling can be enhanced by the ensemble method up to 20%. In addition to the temporal modeling, spatial interpolators can be useful tools to estimate the precipitation for any desired point within the study region where there is not any installed rainfall gauge. Geostatistical methods have been extensively employed in hydro-climatic modeling to estimate the missing data or peroides at points without observation instruments (e.g. see, [14]). Among several Geostatistical methods, Inverse Distance Weighting (IDW) method could gain the attention of the researchers due to its simplicity and reliable accuracy.

2   Methodology

Data from seven main stations were used in this study to predict the precipitation which are Lefkoşa, Ercan, Girne, Güzelyurt, Gazimağusa, Geçitkale, Yeni Erenköy (Table 1).

Usually, as a conventional method, linear Correlation Coefficient (CC) is computed between potential inputs and output to select most dominant input variables. To select exogenous station in second scenario, Mutual Information criterion used [1].

Table 1. The characteristics of stations and statistics of the precipitation data.

Station	Altitude (m)	Max precipitation (mm)	Mean precipitation (mm)	Std. Dev. precipitation (mm)
Ercan	123 m	71.0	25.2	0.97
Gazimağusa	1.8 m	104.7	27.9	1.27
Geçitkale	44 m	70.0	27.0	1.12
Girne	0 m	142.0	38.4	1.95
Güzelyurt	65 m	100.7	23.7	1
Lefkoşa	220 m	66.2	22.8	0.92
YeniErenköy	22 m	76.0	33.3	1.46

$$MI(A, B) = H(A) + H(B) - H(A, B) \tag{1}$$

where A and B are the probability distributions of X and Y and H(A) and H(B) show the entropies of A and B respectively, and $H(A, B)$ is their joint entropy as:

$$H(A, B) = -\sum_{a \in A} \sum_{b \in B} p_{AB}(a, b) log p_{AB}(a, b) \tag{2}$$

The MI between the observed precipitation time series of all seven stations relative to each other were calculated and tabulated in Table 2. from Table 2, overall, Ercan's

Table 2. The MI between the observed precipitation time series of stations

Station	Ercan	Gazimağusa	Geçitkale	Girne	Güzelyurt	Lefkoşa	YeniErenköy
Ercan	–	0.993	1.038	1.085	0.958	1.074	0.992
Gazimağusa	0.993	–	0.939	0.893	0.964	0.971	0.941
Geçitkale	1.038	0.939	–	0.868	0.908	0.974	0.925
Girne	1.085	0.893	0.868	–	0.911	0.949	0.876
Güzelyurt	0.958	0.964	0.908	0.911	–	0.983	0.947
Lefkoşa	1.074	0.971	0.974	0.949	0.983	–	0.967
YeniErenköy	0.992	0.941	0.925	0.876	0.947	0.967	–
Mean MI	1.02	0.950	0.942	0.931	0.945	0.986	0.941

precipitation data are more non-linearly correlated with the precipitation time series of other stations, maybe due to its central position with regard to the others.

Prior to the modeling, the monthly average precipitation data were first normalized:

$$P_{norm} = \frac{P_{(t)} - P_{\min(t)}}{P_{\max(t)} - P_{\min(t)}} \leq 1 \tag{3}$$

where P_{norm} is the normalized value of the $P_{(t)}$; $P_{\max(t)}$ and $P_{\min(t)}$ are the max and min values of the observed data. Due to the training and verification goals, about 70% of whole data were used for calibration and the rest 30% of data for verifying the trained models. The Root Mean Square Error (RMSE) and Determination Coefficient (DC) were used to evaluate the prediction efficiency of the models as [6]:

$$RMSE = \sqrt{\frac{\sum_{i=1}^{n}(P_{obs_i} - P_{com_i})^2}{n}} \tag{4}$$

$$DC = 1 - \frac{\sum_{i=1}^{n}(P_{obs_i} - P_{com_i})^2}{\sum_{i=1}^{n}(P_{obs_i} - \bar{P}_{obs})^2} \tag{5}$$

where n is the data number, P_{obs_i} is the observed data, and P_{com_i} is the predicted (computed) data. DC ranges from $-\infty$ to 1 with a perfect score of 1 and RMSE ranges from 0 to $+\infty$ with the perfect value of 0. Legates & McCabe (1999) showed that any hydro-environmental method may be adequately evaluated by DC and RMSE criteria.

2.1 Proposed Methodology

The research in this paper is separated into two parts, the temporal and spatial stages of modeling as shown in Fig. 1 and as discussed in the following sub-sections.

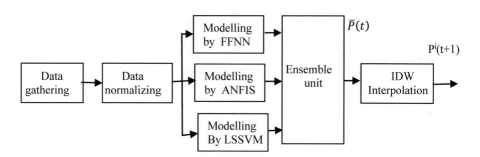

Fig. 1. Schematic of the proposed methodology (P = precipitation data)

In scenario 1, each station's own data at pervious time steps were used for predicting the same station's precipitation, while in scenario 2, another station's data in addition to each station's data were used for modelling to enhance the prediction performance. For modeling via the first scenario, the aim was to predict precipitation value using the station values at previous time steps. So, the precipitation could be patterned as:

$$P_t^i = f(P_{t-1}^i, P_{t-12}^i) \tag{6}$$

where i denotes to the station name (as Ercan, Gazimağusa, Geçitkale, Girne, Güzelyurt, Lefkoşa and Yeni Erenköy stations) and (P_{t-1}^i, P_{t-12}^i) are the precipitation values of ith station corresponding to time steps (t – 1) and (t − 12) or 1 & 12 months ago.

In scenario 2, the prediction formula (6) was modified by introducing precipitation value from Ercan station P_t^{Ercan} as exogenous input. Therefore, the general mathematical formulation of this scenario can be expressed as:

$$P_t^i = f(P_{(t-1)}^i, P_{(t-12)}^i, P_t^E rcan) \tag{7}$$

3 Ensemble Unit

Clearly the combining the outputs from several prediction methods can improve the final accuracy of a time series modeling tool. In an ensembling process the outcomes of various models are used and as so, the final outputs will not be sensitive to selection of the best methods. Therefore, predicts of ensemble method will be more safe and less risky than the results of the single best methods [16]. In current paper, 3 ensemble techniques were applied to combine of the outputs of the used AI based models to enhance the overall efficiency of the predictions as:

a. the simple linear averaging method:

$$\bar{P}(t) = \frac{1}{N} \sum_{i=1}^{N} P_i(t) \tag{8}$$

where $\bar{P}(t)$ is the output of the simple ensemble model, N shows the number of single models (in this study, $N = 3$) and $P_i(t)$ stands for the outcome of the ith method (i.e. ANN, ANFIS and LSSVM) in time step t.

b. the linear weighted averaging method:

$$\bar{P}(t) = \sum_{i=1}^{N} w_i P_i(t) \tag{9}$$

where i shows imposed weight on the output of ith method that may be computed on the basis of the performance measure of ith method as:

$$w_i = \frac{DC_i}{\sum_{i=1}^{N} DC_i} \tag{10}$$

where DC_i measures the model efficiency (such as coefficient of determination).

c. the non-linear neural ensemble method:
   For the nonlinear neural ensemble method another FFNN model is trained by feeding the outputs of single AI models as inputs to the neurons of the input layer.

4 Spatial Modeling Using IDW Method

In the IDW method, the weightings are solely a function of the distance between the point of interest and the sampling points for i = 1, 2, …, n. Considering the distance Di between these two points, the value of a point of interest point takes the form:

$$Z = \frac{\sum_{i=1}^{n} \frac{1}{D_i^q} Z_i}{\frac{1}{D_i^q}} \tag{11}$$

where Z is the interpolated value of the point of interest; Zi is the value of sampling point i; Di is the distance between the interpolated and sampled values, and q is an appropriate constant. If the parameter q takes a value of 1 or 2, the method is called respectively, inverse distance interpolation or inverse distance squared interpolation.

5 Results and Discussions

5.1 Temporal Modeling Results

For the temporal modeling firstly single AI based methods were trained and verified via two different input scenarios and then the prediction efficiency of AI based modeling was enhanced using ensemble techniques. The obtained results for single model were presented in [15].

5.2 Ensemble Modeling Results

In the ensemble modeling, the outputs of three AI based single models were combined to improve the prediction performance. In this step, only the verification dataset was employed to compute the weights of the averaging methods. Results of different ensemble methods are shown by Tables 3 and 4 respectively for scenarios 1 & 2.

The outputs obtained ensemble techniques indicate that almost all three ensemble techniques could produce reliable results in comparison to the single AI methods.

As it can be seen from Tables 3 and 4 and Fig. 2, the efficiency of neural ensemble is better than the linear ensembling methods in most cases. In terms of DC, simple linear, weighted linear and neural averaging methods enhanced the predicting performance up to 15%, 15%, 38% and 16%, 15%, 24% in calibration step for scenarios 1 and 2 and up to 35%, 35%, 54% and 12%, 12%, 28% in verification step for scenarios 1 and 2, respectively. It's obvious that the neural ensemble could lead to better results by 10% on average. Furthermore, the ensembling model could improve the efficiency of precipitation modelling of Girne station more than other stations by 21% on average

Table 3. Results of ensembles using simple (SLA), weighted (WLA) and non-linear averaging (NLA) methods for scenario 1

Station	Ensemble method[a]	Model structure[a]	DC		RMSE (normalized)	
			Calibration	Verification	Calibration	Verification
Ercan	SLA	–	0.678	0.643	0.177	0.149
	WLA	(0.357 – 0.331 – 0.312)	0.680	0.644	0.177	0.149
	NLA	(3,16,1)	0.786	0.677	0.148	0.146
Gazimağusa	SLA	–	0.560	0.520	0.182	0.121
	WLA	(0.347 – 0.320 – 0.333)	0.559	0.521	0.182	0.121
	NLA	(3,3,1)	0.702	0.540	0.155	0.126
Geçitkale	SLA	–	0.7431	0.650	0.134	0.094
	WLA	(0.337 – 0.323 – 0.340)	0.741	0.651	0.135	0.094
	NLA	(3,12,1)	0.765	0.670	0.128	0.092
Girne	SLA	–	0.753	0.516	0.111	0.173
	WLA	(0.352 – 0.302 – 0.346)	0.750	0.522	0.112	0.173
	NLA	(3,11,1)	0.825	0.678	0.095	0.157
Güzelyurt	SLA	–	0.779	0.432	0.139	0.143
	WLA	(0.337 – 0.333 – 0.330)	0.776	0.433	0.140	0.143
	NLA	(3,4,1)	0.774	0.447	0.137	0.143
Lefkoşa	SLA	–	0.594	0.561	0.171	0.138
	WLA	(0.343 – 0.324 – 0.333)	0.592	0.564	0.171	0.138
	NLA	(3,2,1)	0.706	0.585	0.150	0.138
YeniErenköy	SLA	–	0.628	0.489	0.128	0.081
	WLA	(0.340 – 0.321 – 0.339)	0.623	0.489	0.129	0.080
	NLA	(3,13,1)	0.690	0.491	0.118	0.079

Table 4. Results of ensembles using simple (SLA), weighted (WLA) and non-linear averaging (NLA) methods for scenario 2

Station	Ensemble method	Model structure[a]	DC		RMSE (normalized)	
			Calibration	Verification	Calibration	Verification
Gazimağusa	SLA	–	0.851	0.699	0.116	0.107
	WLA	(0.336-0.320-0.344)	0.847	0.699	0.118	0.107
	NLA	(3,20,1)	0.900	0.722	0.095	0.102
Geçitkale	SLA	–	0.880	0.681	0.096	0.096
	WLA	(0.345-0.321-0.334)	0.873	0.691	0.099	0.093
	NLA	(3,5,1)	0.883	0.727	0.100	0.086
Girne	SLA	–	0.889	0.734	0.079	0.144
	WLA	(0.345-0.322-0.333)	0.884	0.744	0.080	0.142
	NLA	(3,16,1)	0.947	0.813	0.090	0.122
Güzelyurt	SLA	–	0.923	0.686	0.089	0.124
	WLA	(0.338-0.328-0.334)	0.913	0.681	0.096	0.123
	NLA	(3,18,1)	0.885	0.668	0.106	0.121
Lefkoşa	SLA	–	0.895	0.627	0.101	0.127
	WLA	(0.342-0.335-0.323)	0.884	0.633	0.107	0.125
	NLA	(3,17,1)	0.953	0.691	0.064	0.123
YeniErenköy	SLA	–	0.884	0.690	0.077	0.067
	WLA	(0.350-0.312-0.338)	0.880	0.690	0.078	0.067
	NLA	(3,19,1)	0.929	0.787	0.060	0.059

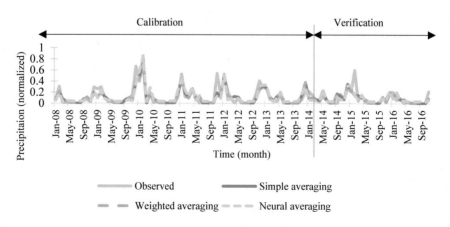

Fig. 2. Results of precipitation prediction using simple, weighted and neural averaging methods and observed precipitation.

in the verification step. On the other hand, this method could not improve the modelling efficiency of Güzelyurt station meaningfully.

6   Conclusion

In this paper, firstly the FFNN, ANFIS and LSSVM predictors were developed using the precipitation data from seven stations. Thereafter, the ensemble methods were employed to increase the temporal modeling efficiency. Secondly, the outputs of neural ensemble method were utilized in the spatial modeling stage. In this stage, through 7 steps for all stations, one station's data were individually removed from the modeling process, then it's values were estimated by the predicted values from 6 other stations for the verification period. At temporal stage 2 scenarios were considered with different input variables that in scenario 1 each station's own pervious data was used for modeling while in scenario 2, the central station's (Ercan station) data were also employed in addition to each station's own data. The results of two employed scenarios indicated that scenario 2 had better performance and could enhance the modeling efficiency up to 58%, in the verification step because of employing the observed data from the Ercan station as exogenous input in simulating other stations' precipitation. Whereas, among three single AI models, the FFNN model showed better performance in most cases in the verification step. Furthermore, the ensemble model based on the non-linear neural averaging produced better predictions than the single models and linear ensemble models up to 38% and 54% for calibration and verification steps, respectively. After temporal modeling, the IDW method was employed for the spatial estimation of precipitation. The cross-validation results showed that the model is able to produce successful estimations for monthly precipitation values over the region with average DC value of 0.7. Furthermore, IDW method was more beneficial than AI methods in cases that a station's own data contains some degrees of noises.

References

1. Nourani, V., Uzelaltinbulat, S., Sadikoglu, F., Behfar, N.: Artificial intelligence based ensemble modeling for multi-station prediction of precipitation. Atmosphere **10**(2), 80 (2019)
2. Nourani, V., Ejlali, R.G., Alami, M.T.: Spatiotemporal groundwater level forecasting in coastal aquifers by hybrid artificial neural network-geostatistics model: a case study. Environ. Eng. Sci. **28**(3), 217–228 (2010)
3. Guhathakurta, P.: Long lead monsoon rainfall prediction for meteorological sub-divisions of India using deterministic artificial neural network model. Meteorol. Atmos. Phys. **101**(2), 93–108 (2008)
4. Hung, N.Q., Babel, M.S., Weesakul, S., Tripathi, N.K.: An artificial neural network model for rainfall forecasting in Bangkok, Thailand. Hydrol. Earth Syst. Sci. **13**, 1413–1425 (2009)
5. Devi, S.R., Arulmozhivarman, P., Venkatesh, C.: ANN based rainfall prediction - a tool for developing a landslide early warning system. In: Advancing Culture of Living with Landslides-Workshop on World Landslide Forum, pp. 175–182 (2017)

6. Nourani, V., Andalib, G.: Daily and monthly suspended sediment load predictions using wavelet-based AI approaches. J. Mountain Sci. **12**(1), 85–100 (2015)
7. Lu, G.Y., Wong, D.W.: An adaptive inverse-distance weighting spatial interpolation technique. Comput. Geosci. **34**(9), 1044–1055 (2008)
8. Kisi, O., Cimen, M.: Precipitation forecasting by using wavelet-support vector machine conjunction model. Eng. Appl. Artif. Intell. **25**(4), 783–792 (2012)
9. Akrami, S.A., Nourani, V., Hakim, S.J.S.: Development of nonlinear model based on wavelet-ANFIS for rainfall forecasting at Klang Gates Dam. Water Resour. Manag. **28**(10), 2999–3018 (2014)
10. Yamashkin, S., Radovanovic, M., Yamashkin, A., Vukovic, D.: Using ensemble systems to study natural processes. J. Hydroinformatics **20**(4), 753–765 (2018)
11. Bates, J.M., Granger, C.W.J.: The combination of forecasts. Oper. Res. Q. **20**, 451–468 (1969)
12. Makridakis, S., Andersen, A., Carbone, R., Fildes, R., Hibon, M., Lewandowski, R., Winkler, R.: The accuracy of extrapolation (time series) methods: results of a forecasting competition. J. Forecasting **1**(2), 111–153 (1982)
13. Sharghi, E., Nourani, V., Behfar, N.: Earthfill dam seepage analysis using ensemble artificial intelligence based modeling. J. Hydroinformatics **20**(5), 1071–1084 (2018)
14. Caruso, C., Quarta, F.: Interpolation methods comparison. Comput. Math Appl. **35**(12), 109–126 (1998)
15. Sharifi, S.S., Delirhasannia, R., Nourani, V., Sadraddini, A.A., Ghorbani, A.: Using ANNs and ANFIS for modeling and sensitivity analysis of effective rainfall. In: Recent Advances in Continuum Mechanics, Hydrology and Ecology, pp. 133–139 (2013)
16. Kasiviswanathan, K.S., Cibin, R., Sudheer, K.P., Chaubey, I.: Constructing prediction interval for artificial neural network rainfall runoff models based on ensemble simulations. J. Hydrol. **499**, 275–288 (2013)

Consistency of Fuzzy If-Then Rules
for Control System

Nigar E. Adilova[(⊠)] [ID]

Azerbaijan State Oil and Industry University, 20 Azadlig Ave., Baku AZ1010,
Azerbaijan
adilovanigarr@gmail.com

Abstract. Fuzzy If-Then rules are frequently used to describe the conditional statements that consist of fuzzy logic. It is closely connected with fuzzy inference process which is formulated from fuzzy logic operators and fuzzy If-Then rules. To define consistency of given rules enables decision makers to select more important criteria. However, researches on consistency for fuzzy control systems are still scarce.

This paper describes some preliminary investigations on consistency of fuzzy If-Then rule based on control systems. Some numerical examples related to this subject are characterized in our research.

Keywords: Fuzzy number · Fuzzy If-Then rules · Consistency of fuzzy rules · Fuzzy control system

1 Introduction

The solution of the consistency problem in decision making systems has become widespread. The importance of consistency is that, where the proposed criteria is compared with others and the most correct and alternative one is chosen among them.

AHP (Analytic hierarchy process) being the basis of consistency matter was introduced by Saaty [1]. He emphasized the opportunities of consistency problem and its solution. Based on his ideas other authors used from pairwise comparisons which are proposed by Saaty in decision making systems. This method enables to check the consistency of the decision maker's criteria. Afterwards author suggested the calculation and application of consistency index [2].

In [3] author proposed three different approaches to consistency problem in fuzzy logic. He described this matter in wider sense and described consistency degrees for fuzzy controllers. A little later a learning algorithm based on consistency and completeness conditions was proposed. This learning algorithm united in a single process rule was tested on different databases [4].

Then authors improved consistency subject and gave an approach to this subject under fuzzy theories [5].

Based on hierarchy process pairwise comparison method was improved [6]. An approximation methodology was considered within a distance-based framework. By using goal programming authors introduced a flexible tool for computing priority weights.

© Springer Nature Switzerland AG 2020
R. A. Aliev et al. (Eds.): ICSCCW 2019, AISC 1095, pp. 137–142, 2020.
https://doi.org/10.1007/978-3-030-35249-3_17

Furthermore, in [7] author discussed the importance of AHP and demonstrated it in a step-by-step manner, at the end of this paper he determined the possible inconsistency values. Pedrycz and Zhang defined the consistency in the way of continuity [8].

We use in this paper an approach [9] how to analyze one of main properties of qualities of fuzzy rule-base consistency of rule-base. This study is considered for organizing 7 fuzzy rules and finding their inconsistency degrees.

During construction of rule-base collision such inconsistency is applied. If antecedents overlap, the consequents come to be different from given two rules, it means that these two rules are inconsistent. For more interpretation, these two following rules will be inconsistent:

$$If\ X_1\ is\ A_1\ and\ X_2\ is\ A_2\ THEN\ Y\ is\ C$$

$$If\ X_1\ is\ A_1\ and\ X_2\ is\ A_2\ THEN\ Y\ is\ D$$

This proposed study is organized as follows. Section 2 defines the preliminary information on fuzzy IF … THEN rules, consistency of fuzzy relation matrix etc. In Sect. 3 the statement of the consistency problem of IF … THEN rules is investigated. In Sect. 4 experimental analyses for consistency of fuzzy rule-base of the given control system are explored. Section 5 is dedicated to the conclusion.

2   Preliminaries

***Definition 1. Fuzzy number* [10]:** A fuzzy number is a set A on R which has the following properties: (a) A is a normal fuzzy set; (b) A is a convex fuzzy set; (c) $\alpha-$ cut of A, A^α is a closed interval for every $\alpha \in (0, 1]$; (d) the support of A, A^{+0} is bounded.

***Definition 2. Fuzzy If-Then rules* [11–14]:** Fuzzy if-then rule statements are commonly used to define the conditional statements that possess fuzzy logic.
A single fuzzy if-then rule is shown in this form:

$$If\ x\ is\ A\ then\ y\ is\ B$$

where both A and B are linguistic values defined by fuzzy sets.
   Multi-input multi-output fuzzy system is described as follows

If X_1 is A_1 and X_2 is A_2 and ….. and X_n is A_n
then Y_1 is B_1 and Y_2 is B_2 and ….. and Y_m is B_m

where A_i and B_i are information granules.

Definition 3. Inconsistency of fuzzy rules:
Two fuzzy rules are inconsistent if they have the same if-part, but different then-parts as noticed above.

Definition 4. Consistency of fuzzy rules: To calculate the consistency of two arbitrary fuzzy rules, definitions of Similarity of Rule Premise (SRP) and Similarity of Rule Consequent (SRC) are given. Two fuzzy rules R_1 and R_k are given as below:

$$R_1 : \text{ If } x_1 \text{ is } A_{i1}(x_1) \text{ and } x_2 \text{ is } A_{i2}(x_2) \text{ and} \ldots x_n \text{ is } A_{in}(x_n), \text{ then } y \text{ is } B_i(y)$$

$$R_k : \text{ If } x_1 \text{ is } A_{k1}(x_1) \text{ and } x_2 \text{ is } A_{k2}(x_2) \text{ and} \ldots x_n \text{ is } A_{kn}(x_n), \text{ then } y \text{ is } B_k(y)$$

Then Similarity of Rule Premise and Similarity of Rule Consequent between rule i and rule k are defined by using fuzzy similarity measure as following:

$$SRP(i, k) = \wedge_{j=1}^{n} S(A_{ij}, A_{kj})$$

$$SRC(i, k) = S(B_i, B_k) \tag{1}$$

where n is the total number of the input variables. The consistency between rule $R(i)$ and $R(k)$ can be defined as:

$$Cons(R(i), R(k)) = \exp\left\{ -\frac{\left(\frac{SRP(i,k)}{SRC(i,k)} - 1.0\right)^2}{\left(\frac{1}{SRP(i,k)}\right)^2} \right. \tag{2}$$

Thus, an inconsistency index of the fuzzy system can be given by this formula:

$$f_{incons} = \sum_{i=1}^{N} \sum_{\substack{1 \leq k \leq n \\ k \neq 1}} [1.0 - Cons(R(i), R(k))] \tag{3}$$

3 Statement of the Problem

Let's presume that these following 7 rules are involved in fuzzy control system [12].

1. If the error e is negative big (NB) THEN the control action u is negative big (NB).
2. If the error e is negative medium (NM) THEN the control action u is negative medium (NM).
3. If the error e is negative small (NS) THEN the control action u is negative small.
4. If the error e is zero (ZE) THEN the control action u is zero. (4)
5. If the error e is positive small (PS) THEN the control action u is positive small (PS).
6. If the error e is positive medium (PM) THEN the control action u is positive medium (PM).
7. If the error e is positive big (PB) THEN the control action u is positive big (PB).

Here, errors for fuzzy control system are defined with e, control actions are described as u. Through (4) formula membership function can be computed as follows:

Rule 1

$$\mu_{E1}(e) = 1.00/-10 + 0.73/-7 + 0.34/-3 + 0.20/0 + 0.13/3 + 0.08/7 + 0.06/10$$
$$\mu_{U1}(u) = 1.00/-1 + 0.92/-0.7 + 0.67/-0.3 + 0.50/0 + 0.37/0.3 + 0.26/0.7 + 0.20/1$$

Rule 2

$$\mu_{E2}(e) = 0.73/-10 + 1/-7 + 0.61/-3 + 0.34/0 + 0.20/3 + 0.11/7 + 0.08/10$$
$$\mu_{U2}(u) = 0.91/-1 + 1.00/-0.7 + 0.86/-0.3 + 0.67/0 + 0.50/0.3 + 0.34/0.7 + 0.26/1$$

Rule 3

$$\mu_{E3}(e) = 0.34/-10 + 0.61/-7 + 1.00/-3 + 0.73/0 + 0.41/3 + 0.20/7 + 0.13/10$$
$$\mu_{U3}(u) = 0.61/-1 + 0.86/-0.7 + 1.00/-0.3 + 0.92/0 + 0.74/0.3 + 0.50/0.7 + 0.37/1$$

Rule 4

$$\mu_{E4}(e) = 0.20/-10 + 0.34/-7 + 0.73/-3 + 1.00/0 + 0.73/3 + 0.34/7 + 0.20/10$$
$$\mu_{U4}(u) = 0.50/-1 + 0.67/-0.7 + 0.92/-0.3 + 1.00/0 + 0.92/0.3 + 0.6726/0.7 + 0.50/1$$

Rule 5

$$\mu_{E5}(e) = 0.13/-10 + 0.20/-7 + 0.41/-3 + 0.73/0 + 1.00/3 + 0.61/7 + 0.34/10$$
$$\mu_{U5}(u) = 0.37/-1 + 0.50/-0.7 + 0.74/-0.3 + 0.92/0 + 1.00/0.3 + 0.86/0.7 + 0.67/1$$

Rule 6

$$\mu_{E6}(e) = 0.08/-10 + 0.11/-7 + 0.20/-3 + 0.34/0 + 0.61/3 + 1.00/7 + 0.73/10$$
$$\mu_{U6}(u) = 0.26/-1 + 0.34/-0.7 + 0.50/-0.3 + 0.67/0 + 0.86/0.3 + 1.00/0.7 + 0.91/1$$

Rule 7

$$\mu_{E7}(e) = 0.06/-10 + 0.08/-7 + 0.13/-3 + 0.20/0 + 0.34/3 + 0.73/7 + 1.00/10$$
$$\mu_{U7}(u) = 0.20/-1 + 0.26/-0.7 + 0.37/-0.3 + 0.50/0 + 0.67/0.3 + 0.92/0.7 + 1.00/1$$

For given 7 rules and their two properties (errors and control actions) membership functions were evaluated. Problem is to test consistency of given by (4) fuzzy rule-base of the investigated control system. By using the values of membership functions the proposed problem is solved.

4 Solution of the Problem

Firstly, we are going to compute the similarity between Rule 1 and Rule 2.

e1	1	0.73	0.34	0.2	0.13	0.08	0.06
u1	1	0.92	0.67	0.5	0.37	0.26	0.2

e2	0.73	1	0.61	0.34	0.2	0.11	0.08
u2	0.91	1	0.86	0.67	0.5	0.34	0.26

Based on (1) formula similarity for rule promise and rule consequent is analyzed. In conclusion SRP and SRC are evaluated as follows.

$$\text{SRP} \quad 0.679641$$
$$\text{SRC} \quad 0.827214$$

In accordance with the values of SRP and SRC consistency index will be 0.985407 as given (2) and inconsistency index will be 0.01459316 as shown (3) formula.

By using noticed method we calculated the inconsistent index between all given rules.

f_{12}	0.01459316
f_{13}	0.020310866
f_{14}	0.017540734
f_{15}	0.193064166
f_{16}	0.009892732
f_{17}	0.196592936
f_{23}	0.022481725
f_{24}	0.022646961
f_{25}	0.018068121
f_{26}	0.013229326
f_{27}	0.009892732
f_{34}	0.019944672
f_{35}	0.022698899
f_{36}	0.018473963
f_{37}	0.013966112
f_{45}	0.018449689
f_{46}	0.144290551
f_{47}	0.195695449
f_{56}	0.024189048
f_{57}	0.117425775
f_{67}	0.01459316

We considered all possible variants to compute inconsistency index. If we have 7 rules, it means that from obtained similarity values we will have 21 inconsistency indexes. And as can be seen, inconsistency index for each rule between the next rule were noticed.

From obtained 21 inconsistency indexes for summarizing total inconsistency degree we will find the average inconsistency degree.

$$f_{incons} = 0.053716.$$

5   Conclusion

In this study consistency of IF-Then rule-base for fuzzy control system was analyzed. The total inconsistency degree was obtained from seven rules. The analysis has shown that inconsistency of the rule base of investigated fuzzy control system is 0.05 and acceptable from point of new of designer.

References

1. Saaty, T.L.: The Analytic Hierarchy Process. McGraw-Hill International, New York (1980)
2. Saaty, T.L.: Theory and Applications of the Analytic Network Process: Decision Making with Benefits, Opportunities, Costs, and Risks. RWS Publications, Pittsburgh (2005)
3. Gottwald, S.: On some types of consistency consideration in fuzzy logic. In: Fuzzy Neuro Systems, pp. 366–373 (1997)
4. González, A., Pérez, R.: Completeness and consistency conditions for learning fuzzy rules. Fuzzy Sets Syst. **96**(1), 37–51 (1998)
5. Gottwald, S., Novák, V.: An approach toward consistency degrees of fuzzy theories. Int. J.- General Syst. **29**(4), 499–510 (2000)
6. Dopazo, S., Jacinto, G.-P.: Consistency-driven approximation of a pairwise comparison matrix. Kybernetika **39**(5), 561–568 (2003)
7. Vargas, R.V.: Using the hierarchy process (AHP) to select and prioritize projects in a portfolio, Washington (2010)
8. Zhang, Z., Pedrycz, W.: A consistency and consensus-based goal programming method for group decision-making with interval-valued intuitionistic multiplicative preference relations. IEEE Trans. Cybern. 1–15 (2018)
9. Yaochu, J., von Seelen, W., Sendhoff, B.: An approach to rule-based knowledge extraction. In: IEEE International Conference on Fuzzy Systems Proceedings. IEEE World Congress on Computational Intelligence (1998)
10. Aliev, R.A.: Fundamentals of the Fuzzy Logic-Based Generalized Theory of Decisions. Springer, Heidelberg (2013)
11. Novák, V., Lehmke, S.: Logical structure of fuzzy IF-THEN rules. Fuzzy Sets Syst. **157**(15), 2003–2029 (2006)
12. Aliev, R.A., Aliev, F., Babaev, M.: Fuzzy Process Control and Knowledge Engineering in Petrochemical and Robotic Manufacturing. Verlag, Germany (1991)
13. Aliev, R.A., Pedrycz, W.: Fundamentals of a fuzzy-logic-based generalized theory of stability. IEEE Trans. Syst. Man Cybern. Part B (Cybern.) **39**(4), 971–988 (2009)
14. Aliev, R., Tserkovny, A.: Systemic approach to fuzzy logic formalization for approximate reasoning. Inf. Sci. **181**(6), 1045–1059 (2011)

Identifying Prospective Teachers' Perceived Competence Towards Use of Assessment and Evaluation Methods

Saide Sadıkoğlu$^{(\boxtimes)}$, Cigdem Hursen , and Erkan Bal

Near East University, Mersin 10, North Cyprus, Turkey
{saide.sadikoglu, cigdem.hursen, erkan.bal}@neu.edu.tr

Abstract. The purpose of this study is to identify prospective teachers' perceived competence towards using assessment and evaluation methods. In addition, the study also aims to explore the problems that prospective teachers face when using assessment and evaluation methods. A total of 100 prospective teachers at their final year of studies at the Department of Early Childhood Education of a private university participated in the study. The results of the study revealed that the prospective teachers mostly prefer to use traditional assessment and evaluation methods and face-to-face assessment methods. It was also identified that the prospective teachers' perceived competence towards using assessment and evaluation methods are at medium level.

Keywords: Product-oriented assessment methods · Process-oriented assessment methods · Prospective teachers · Early childhood education

1 Introduction

Assessment and evaluation is an integral part of the education process and analyse the quality of the education process Kilmen and Kösterelioğlu [1]. With or without realizing it, teachers make judgments about their students' performances in the learning-teaching process. Teachers' feedback on the quality of the individuals' work by using their professional judgment is explained as the focus of evaluation Jones Black and William [2], state that innovations designed to strengthen the feedback that students receive about their learning provide significant learning gains. The intersection between evaluation and learning is critical for the promotion or prevention of quality in education Hopfenbeck et al. [3]. Pointing out to the changing and developing society, Toptaş [4], emphasises that new assessment and evaluation approaches are adopted in the curricula that are updated to train the needed manpower. From a different point of view, Şad and Göktaş [5], state that the lack of adoption of modern assessment and evaluation approaches is regarded as a serious problem for teacher training. This study aims to identify the current knowledge and skill levels of prospective teachers about assessment and evaluation methods. In order to achieve this objective, answers to the following questions were sought:

© Springer Nature Switzerland AG 2020
R. A. Aliev et al. (Eds.): ICSCCW 2019, AISC 1095, pp. 143–150, 2020.
https://doi.org/10.1007/978-3-030-35249-3_18

1. What is the frequency of prospective teachers' use of assessment and evaluation?
2. How is the prospective teachers' perceived competence toward using assessment and evaluation methods?
3. What are prospective teachers' problems in using assessment and evaluation methods?
4. What are the views of prospective teachers about the implementation of assessment and evaluation methods?

2 Method

This study is conducted with the purpose to identify the perceived competence of prospective pre-school teachers towards using assessment and measurement methods and the problems they experience while implementing these methods. A quantitative method was employed in the study, in which a total of 100 prospective teachers studying at the Early Childhood Education department of a private university participated. Demographic characteristics of the study group are presented in Table 1.

Table 1. Demographic characteristics of the study group

Gender	n	%
Female	90	90,0
Male	10	10,0
Age	n	%
25–30	93	93,0
31–35	1	1,0
36–40	6	6,0

As it is presented in Table 1, 90 of the prospective teachers participated in the study were female and 10 were male. A total of 93 of the prospective teachers participated in the study belong to the 25–30 age group, one of them belong to the 31–35 age group and six of them belong to the 36–40 age group.

2.1 Data Collection Tools

This study aims to identify prospective teachers' perceived competence towards using assessment and evaluation methods. In this context, the survey for "teachers' frequency of using assessment and evaluation techniques and their perceived competence" prepared by Gelbal and Kelecioğlu [6], was employed.

The survey consists of 5 sections. In Sect. 1, while demographic characteristics of participants are identified, in Sect. 2, identification of the opinions of the participants about the frequency of use of assessment tools was aimed. In the third section of the survey, the perceived competence of teachers are questioned. Section 2 is prepared to identify the frequency of teachers' use of assessment tools, and the views of teachers are scored between "mostly", "occasionally" and "never". In Sect. 3 which was created to identify the perceived competence of teachers, the three levels are stated as "very", "medium" and "none". In the fourth section of the survey, it was aimed to identify the views of teachers about the problems they experience while implementing the assessment tools. In this section of the survey, teachers can mark multiple options for expressions. In the fifth section of the survey, identification of the views of teachers about process-oriented assessment tools was aimed. In this section, teachers' views consisted of three levels: "I agree", "I do not have any views" and "I disagree".

In this study, quantitative data obtained from Sects. 2, 3.1, 3.2 and 3.3 of the survey were analysed with frequency and percentage values, while the data obtained from Sect. 3.4 of the survey was analysed with minimum, maximum, mean and standard deviation analysis techniques. The statements in Sect. 3.4 of the survey are scored from 3 to 1, and it can be said that as the total score approaches 3, the views increase in positive direction. Cronbach's Alpha reliability coefficient for the whole survey was calculated as 0.94.

3   Results

In this section the findings obtained from quantitative results are presented.

3.1   Prospective Teachers' Frequency of Using Assessment and Evaluation Tools and Methods

The results of the quantitative data obtained in order to identify the frequency of prospective teachers to use assessment tools and methods during their practical training in schools are presented in Table 2.

Table 2. Frequency of using assessment tools and methods

Assessment methods	Frequency of use					
	Mostly		Occasionally		Never	
	f	%	f	%	f	%
Traditional methods	47	47.0	39	39.0	14	14.0
Face-to-face methods	52	52.0	31	31.0	17	17.0
Methods under new approach	23	23.0	59	59.0	18	18.0
Methods related to students' self-assessment	35	35.0	48	48.0	17	17.0

In the study it is identified that 35% of prospective teachers perceived themselves very competent to use traditional methods, while 56% perceived themselves to be competent at a medium level. Nine percent of prospective teachers do not perceive themselves competent enough to use traditional measurement methods.

While 38% of prospective teachers perceive themselves competent to use of face-to-face assessment methods, 45% perceive themselves to be competent at a medium level. On the other hand, 17% of the prospective teachers do not feel competent to implement face-to-face assessment methods. While 24% of prospective teachers perceive themselves competent for new methods, 61% perceive themselves to be competent at a medium level and 15% do not feel competent at all. When the methods of self-assessment of the students are analysed, 27% of prospective teachers perceive themselves competent, 62% perceive themselves competent at a medium level and 11% do not perceive themselves competent. The obtained findings reveal that the prospective teachers perceive themselves the most competent (38%) at using face-to-face assessment methods.

3.2   Prospective Teachers' Perceived Competence Towards Assessment Tools and Methods

The quantitative data obtained in order to identify the prospective teachers' perceived competence regarding the use of assessment tools and methods are presented in Table 3.

Table 3. Perceived competence levels towards using assessment tools and methods

Assessment methods	Perceived competence level					
	Very		Medium		None	
	f	%	f	%	f	%
Traditional methods	35	35.0	56	56.0	9	9.0
Face-to-face methods	38	38.0	45	45.0	17	17.0
Methods under new approach	24	24.0	61	61.0	15	15.0
Methods related to students' self-assessment	27	27.0	62	62.0	11	11.0

In the study it is identified that 35% of prospective teachers perceived themselves very competent to use traditional methods, while 56% perceived themselves to be competent at a medium level. Nine percent of prospective teachers do not perceive themselves competent enough to use traditional measurement methods.

While 38% of prospective teachers perceive themselves competent to use of face-to-face assessment methods, 45% perceive themselves to be competent at a medium level. On the other hand, 17% of the prospective teachers do not feel competent to implement face-to-face assessment methods. While 24% of prospective teachers perceive themselves competent for new methods, 61% perceive themselves to be competent at a medium level and 15% do not feel competent at all. When the methods of self-assessment of the students are analysed, 27% of prospective teachers perceive themselves competent, 62% perceive themselves competent at a medium level and 11% do not perceive themselves competent. The obtained findings reveal that the prospective teachers perceive themselves the most competent (38%) at using face-to-face assessment methods.

3.3 The Distribution of Problems Prospective Teachers Experience When Using Assessment Tools

The distribution of problems that prospective teachers experience when using assessment tools during their practical training are presented in Table 4.

Table 4. Distribution of problems prospective teachers experience when using assessment tools

Experienced problems	Traditional methods		Face-to-face methods		New approach		Students' self-assessment	
	f	%	f	%	f	%	f	%
No problems	51	51.0	28	28.0	17	17.0	4	4.0
Crowded class size	35	35.0	27	27.0	30	30.8	8	8.0
Lack of time	26	26.0	24	24.0	43	43.0	7	7.0
Difficulty of evaluation	19	19.0	35	35.0	30	30.0	16	16.0
Difficulty of implementation	12	12.0	46	46.0	29	29.0	13	13.0
Difficulty of preparation	24	24.0	27	27.0	36	36.0	13	13.0
Minimal parent support	30	30.0	26	26.0	30	30.0	14	14.0
Reluctance of students	48	48.0	18	18.0	24	24.0	10	10.0
Insufficient explanations in the programme	35	35.0	29	29.0	25	25.0	11	11.0
Lack of students' collaboration when needed	32	32.0	39	39.0	20	20.0	9	9.0
Students' lack of information	28	28.0	28	28.0	35	35.0	9	9.0
Complexity of the methods	26	26.0	30	30.0	40	40.0	3	3.0
Unsuitability with the class level	23	23.0	29	29.0	37	37.0	11	11.0

Table 4 presents the problems that prospective teachers experience when using the assessment methods. The findings reveal that the majority of prospective teachers (51%) so not experience any problems using traditional methods. However, 35% of the prospective teachers stated that they have problems when the classroom was crowded while implementing the traditional methods. 43% of the prospective teachers stated that the assessment methods including the new applications take time to implement. 35% of the prospective teachers mentioned the difficulty of evaluating face-to-face assessment methods, while 46% stated that this method was difficult to implement. The prospective teachers who mentioned the difficulty of preparing methods for new approaches (36.0%), also stated that students were reluctant to implement traditional measurement methods (48%). In the implementation of face-to-face assessment methods, while 39% of the prospective teachers stated that they experienced problems because the students did not collaborate, 40% explained that the assessment methods under new approaches were complex and 37% stated that they are not suitable for classroom levels. The obtained findings reveal that the majority of the prospective teachers do not experience any problems when using the traditional assessment methods. In this context, it is seen as a necessity to provide courses that will improve the competency levels of prospective teachers in the use of process-oriented assessment methods in teacher training programs.

3.4 The Views of Prospective Teachers on the Implementation of Assessment and Evaluation Methods

The average scores of prospective teachers' views on the implementation of process-oriented assessment and evaluation methods are presented in Table 5.

Table 5. The distribution of average scores of prospective teachers' views on the implementation of process-oriented assessment and evaluation methods

Assessment and evaluation methods	N	Min	Max	Mean	Sd.
Self-assessment	100	1.00	3.00	2.26	.562
Peer-assessment	100	1.00	3.00	2.01	.619
Observation	100	1.00	3.00	2.10	.500
Oral presentation	100	1.00	3.00	2.12	.516
Project	100	1.00	3.00	2.12	.581
Portfolio	100	1.00	3.00	2.08	.499

As it is presented in Table 5, the study also aims to identify the views of prospective teachers on the implementation of process-oriented assessment and evaluation methods. It is revealed that prospective teachers do not have any knowledge about methods of self-assessment method among process-oriented measurement and evaluation methods (M = 2.26, sd = .562). Similarly, the study reveals that prospective teachers do not have any knowledge about peer-assessment (M = 2.01, sd = .619), observation (M = 2.10, sd = .500), oral presentation (M = 2.12, sd = 516), project (M = 2.12, sd = .518) and portfolio (M = 2.08, sd = .499) assessment and evaluation methods. The findings indicate that there is a need for practices that will improve the prospective teachers' views about process-oriented assessment and evaluation methods in a positive way.

4   Discussion and Conclusion

In this study, which is conducted to determine the prospective teachers' frequency of using assessment and evaluation methods and their perceived competence, it is identified that traditional assessment and evaluation methods are mostly used. Similarly, it is identified that the prospective teachers mostly prefer face-to-face assessment methods. In the study conducted by Gelbal and Kelecioğlu [6], similar results were obtained. The findings support the findings obtained from this study.

Another finding obtained from the study is that the majority of the prospective teachers have a perceived competence to use assessment and measurement tools at a medium level. The majority of the prospective teachers do not perceive themselves fully competent to use of assessment and evaluation methods. In a different study conducted with teachers, it was identified that teachers' perceptions of competence for process-oriented assessment tools were above the medium level Hursen and Birkollu [7]. It is supposed that the reason why that the majority of prospective teachers' perceived competence levels to use assessment and measurement methods are at a medium level is because they have not yet started their practicing their profession. The study also aims to identify the problems experienced by prospective teachers when using assessment and evaluation methods. The study reveals that prospective teachers experience problems in implementing assessment and evaluation methods due to crowded class size, lack of time, difficulty in evaluation, reluctance of students, and lack of collaboration and difficulty of implementation. In a similar study, Öztürk and Şahin [8], attempted to identify the difficulties that prospective teachers may experience when applying process-oriented assessment and evaluation methods. The researchers concluded that while implementing assessment and evaluation methods, problems may arise from parents, teachers, students, school facilities, environment and time factors Öztürk and Şahin, [8]. The findings of the above-mentioned study; the lack of time, reluctance of students and lack of environment in particular support the results obtained from this study. Ektem, Keçici and Pilten [9], indicate that the lack of time, crowded class size and teachers' heavy workload are among the reasons of the problems experienced in implementing process-oriented assessment and evaluation methods.

Another finding obtained from the research is the views of prospective teachers about the implementation of process-oriented assessment and evaluation methods. Prospective teachers stated that they do not have knowledge about the implementation of process-oriented measurement and evaluation methods. This reveals that prospective teachers' views about process-oriented assessment and evaluation methods are not as positive as expected. In their study, In this context, it can be stated that prospective teachers are not familiar with the process-oriented assessment methods.

In line with the results obtained, it is recommended that prospective teachers participate in courses, trainings and seminars through which they can gain experience and competence in the implementation of process-oriented assessment and evaluation methods. In addition, it is recommended to include practices in teacher training programmes that provide prospective teachers experience with process-oriented assessment and evaluation methods.

References

1. Kilmen, S., Kösterelioğlu, İ.: Examination of teachers' perceptions towards alternative assessment approaches with CHAID analysis. Elementary Educ. Online **16**(1), 256–273 (2017)
2. Jones, C.A.: Assessment for learning. Vocational Learning Support Programme, 16–19 (2005). https://dera.ioe.ac.uk/7800/1/AssessmentforLearning.pdf
3. Baird, J.A., Hopfenbeck, T.N., Newton, P., Stobart, G., Steen–Utheim, A.T.: State of the Field R. Assessment and Learning. Oxford University Centre for Educational Assessment Report OUCEA, 14(2) (2014)

150   S. Sadıkoğlu et al.

4. Toptaş, V.: Classroom teachers' perceptions about the use of alternative assessment and evaluation methods in mathematics courses. Educ. Sci. **36**(159), 205–219 (2011)
5. Şad, S.N., Göktaş, Ö.: Investigation of traditional and alternative measurement and evaluation approaches among teaching staff at higher education. Ege Eğitim Dergisi **2**(14), 79–105 (2013)
6. Gelbal, S., Kelecioğlu, H.: Teachers' proficiency perceptions of about the measurement and evaluation techniques and the problems they confront. Hacettepe Üniversitesi Eğitim Fakültesi Dergisi (H. U. J. Educ.) **33**, 135–145 (2007)
7. Hursen, C., Birkollu, S.S.: Determining the relationship between teachers' attitudes towards the use of process-oriented measurement tools and self-efficacy perceptions. Folklor/edebiyat **25**(97–1), 462–477 (2019)
8. Öztürk, Y.A., Şahin, Ç.: Pre-service teachers' views on difficulties on implementation of process-oriented assessment and evaluation methods. Buca Eğitim Fakültesi Dergisi **36**, 109–129 (2013)
9. Ektem, I.S., Keçici, S.E., Pilten, G.: The opinions of primary teachers related with process oriented measurement and evaluation methods. Ahi Evran Üniversitesi Kırşehir Eğitim Fakültesi Dergisi (KEFAD) **17**(3), 661–680 (2016)

Fuzzy Ordination of Breast Tissue with Electrical Impedance Spectroscopy Measurements

Meliz Yuvalı[1]([✉]) [iD], Cemal Kavalcıoğlu[2] [iD], Şerife Kaba[3] [iD], and Ali Işın[3] [iD]

[1] Faculty of Medicine, Department of Biostatistics, Near East University, Near East Boulevard, P.O. Box: 99138, Nicosia, TRNC, Mersin 10, Turkey
meliz.yuvali@neu.edu.tr
[2] Department of Electrical and Electronic Engineering, Near East University, Near East Boulevard, P.O. Box: 99138, Nicosia, TRNC, Mersin 10, Turkey
cemal.kavalcioglu@neu.edu.tr
[3] Department of Biomedical Engineering, Near East University, Near East Boulevard, P.O. Box: 99138, Nicosia, TRNC, Mersin 10, Turkey
{serife.kaba,ali.isin}@neu.edu.tr

Abstract. Electrical impedance spectroscopy (EIS) is a useful technique which requires minimum invasion into body employed for the characterization of living tissues with the facility and low cost. EIS techniques assist diagnosis, in a different way, by providing information regarding the electrical conductivity and permittivity properties of the patient's cells and tissues. Measuring the bio-impedance of the tissues, allows scientists to take into consideration the capacitive characteristics of the tissues along with their resistive characteristics. So by effectively measuring electrical impedances through body tissues, cancerous tissues can be differentiated and diagnosed. In this study, breast tissues obtained from 106 patients have been classified via fuzzy logic according to the data accumulation in the EIS device. The device gives 9 different impedance features for each patient which are then reduced to 6 classes. These classes are; glandular tissue, connective tissue, adipose, mastopathy, fibro-adenoma and carcinoma. The aim of this study is to design a fuzzy system to classify breast tissues with EIS test results.

Keywords: Breast tissue · Electrical Impedance Spectroscopy · Fuzzy logic

1 Introduction

The human breast is a modified skin wrapped in the fibrous fascia. The surface pectoral fascia is based underneath the surface of the breast and in the retromammary space. The breast consists of three main structures which are skin, subcutaneous tissue and breast tissue [1]. The characteristics of Electrical impedance spectral evolves with the alterations of breast tissue pathological constellation [2]. The electrical impedance measurement splits up into six groups. These groups are; Glandular tissue (gla), Connective tissue (con), Adipose (adi), Mastopathy (mas), Fibro-adenoma (fad) and Carcinoma

© Springer Nature Switzerland AG 2020
R. A. Aliev et al. (Eds.): ICSCCW 2019, AISC 1095, pp. 151–157, 2020.
https://doi.org/10.1007/978-3-030-35249-3_19

(car). These groups can simply divide into two categories as the first three groups are normal tissues while the remaining ones are lesion tissues. Mastopathy (mas), Fibro-adenoma (fad) lesions tissues are the benign disease, and the Carcinoma (car) is the malignant disease [3].

As one of the most lethal diseases in the modern world, cancer is a disease where cells of the patient grow, divide and spread unnaturally without control. Early diagnosis of cancer, especially before the cancer cells spread to wide tissue areas, is key for effective treatment and survival of the patient. The method for non-invasively diag-nosing cancer early on, without imposing any adverse ionizing radiation, is the tech-nique of Electrical Impedance Spectroscopy (EIS) [4]. While conventional medical imaging methods assist cancer diagnosis by providing images of the internal structures or images of the function of internal structures of the patient, EIS techniques assist diagnosis, in a different way, by providing information regarding the electrical con-ductivity and permittivity properties of the patient's cells and tissues [5]. By using non-invasive electrical impedance electrodes an EIS system applies electrical currents to the patient's body and measures tissue resistances and capacitances by recording the voltage drops in the applied electrical signals. As cancer cells/tissues have different electrical impedance characteristics than the healthy ones, with these measurements tissues of the patient can be mapped and cancerous ones can be diagnosed [6].

The author in [7] used a method to classify breast tissues by using the electrical impedance spectral properties. Nine breast tissue characteristics were acquired by measuring the electrical impedance spectra of in vitro and freshly collected breast tissue and then breast tissue was classified using support vector machine method [7]. The author in [8] presented computerized breast cancer detection using "resonance-frequency based electrical impedance spectroscopy". This system was used to collect a data set containing 74 malignant and 215 benign tumors confirmed by biopsy [8]. The study in [4] suggested that sensitive tumor detection would increase when additional information was given to the doctor from Electrical Impedance Spectroscopy and internal intelligence. The most widely used bio-impedance model, the Cole model was then inserted into the data [4]. The approach and results can be converted into breast cancer [4].

Although important studies have been conducted in the literature classifying the results between normal tissue and cancer tissue from the electrical impedance device, in our study, we used a fuzzy-based approach to classify breast tissues. The purpose of this study is to classify breast tissues using electrical impedance device test results. In this study, section two presents the methodology, section three describes the probatory findings, and the outcome comes up in the last section.

2 Methodology

2.1 Database

In this study, UCI (California Irvine University) breast tissue data set of 106 patients obtained from electrical impedance spectral device are used [9]. The minimum-maximum values of the electrical impedance spectroscopy device specifications given in the following tables are determined by using the IBM SPSS (Statistical Package for Social Sciences, version 21) package program. The abbreviations of the impedance features as follows (Tables 1 and 2):

Table 1. Breast tissue data set 1.

	IO		PA		HFS		DA	
	Min-Max	BTC	Min-Max	BTC	Min-Max	BTC	Min-Max	BTC
L1	1600,000-2800,000	adi	0,144-0,358	car	0,032-0,468	car	175,020-640,276	con
L2	649,369-1724,090	con	0,058-0,166	gla	0,011-0,292	gla	74,635-264,805	car
L3	269,496-551,879	car	0,045-0,232	mas	0,000-0,438	mas	72,931-1063,441	adi
L4	144,000-355,000	fad	0,026-0,108	con	(-0,015)-0,163	fad	24,437-157,884	mas
L5	121,000-554,654	mas	0,020-0,189	fad	(-0,021)-0,145	con	20,589-150,224	gla
L6	103,000-502,000	gla	0,012-0,201	adi	(-0,066)-0,378	adi	19,648-89,558	fad

Table 2. Breast tissue data set 2.

	Area		A/DA		Max IP		DR		P	
	Min-Max	BTC	Min-Max	BTC	Min-Max	BTC	Min-Max	BTC	Min-Max	BTC
L1	1402,232-174480,476	adi	15,938-44,895	car	51,855-436,100	adi	143,258-632,165	con	1475,372-2896,582	adi
L2	1189,545-11888,392	car	14,635-164,072	adi	35,603-96,563	car	65,541-253,785	car	528,699-1524,609	con
L3	304,271-11852,485	con	3,161-33,601	mas	23,976-143,092	con	10,595-150,917	mas	329,091-656,769	car
L4	144,467-5305,123	mas	2,870-19,649	fad	18,226-47,561	gla	7,569-86,577	fad	160,374-385,133	fad
L5	78,258-2657,910	gla	2,757-17,693	gla	9,102-49,328	mas	5,721-142,496	gla	141,766-553,358	mas
L6	70,426-1370,838	fad	1,596-43,387	con	7,969-43,692	fad	(-9,258)-977,552	adi	124,979-544,039	gla

"L: Level, BTC: Breast fiber ordination, I0: Impedivity (ohm) at zero density, PA500: Phase angle at 500 kHz, HFS: High-frequency slope of phase angle, DA: Impedance space between spectral ends, AREA: Area under spectrum, A/DA: Area standardized by DA, MAX IP: Maximum of the spectrum, DR: Distance, between I0 and real part of the maximum density point, P: Length of the spectral curve."

3 Classification

3.1 Fuzzy Logic

Lotfi A. Zadeh launched fuzzy logic for the first time in 1965 with fuzzy sets as an elongation of the classic set theory designated by open sets. At that point, entire polynomial mathematics utilizing fuzzy sets that elucidate the established logic as an augmentation of its exact activities is characterized as fuzzy logic [10]. In a number of cases, a fuzzy logic method is a main inexact map of the scalar outcome in which an input data vector determines a linguistic explication calculated by digits. Herewith, the fuzzy system is incomparable, the fuzzy logic system can conduct quantitative data and grammar. There are a few distinct affiliations, which proposes that there are numerous conceivable outcomes that lead to logic [11].

Fuzzy Logic Preference Reasons. General features of fuzzy logic are as follows: Data Fuzzy logic is tolerant to make data unclear. Fuzzy logic has facilities such as an easy understanding of the concept and non-rigid usage. Conventional limiting methods can be blended in fuzzy logic. The fuzzy model can illustrate the complication in random nonlinear functions which can be achieved by experienced experts. Fuzzy logic uses inborn language. Fuzzy systems are not the conventional process of the control and fuzzy systems simplify applications in most cases [11].

3.2 Membership Functions (MF)

A membership function [12] determines how the input space for each point is varied between 0 and 1. Input data space between whiles refers to the universe of discourse, an ostentatious title for a plain theme. The label is entitled to every single MF with description. As an illustration, an input variant, namely, **input-1** and **input-2** involves six membership features label; Fibro-adenoma, Carcinoma, Glandular, Connective, Adipose and Mastopathy during this work. Selected MF type is:

Gaussian: Gaussian fuzzy affiliation features are fairly communal in the literary perspective of fuzzy logic since they are the essential linkage between fuzzy systems and radial basis function (RBF) neural networks system. This function computes fuzzy membership values using Gaussian membership features. A Gaussian membership function is not the identical as a Gaussian likelihood deploy.

For instance, a Gaussian membership function has the highest value of 1 all the time [13].

3.3 Linguistic Variables

System input data or outcome variables are named with respect to quantitative values in a natural or artificial language. A lingual variant is popularly collapsing into lingual denominate group; Carcinoma, Fibro-adenoma, Mastopathy, Glandular, Connective, and Adipose.

3.4   Universe of Discourse

The set of elements of fuzzy are adopted from the Universe, in short, for the universe of discourse. The universe considers all elements. Even the universe is dependent on the content [13].

IF - THEN Rules. IF-THEN rule expressions are operating in order to equip the conditional depositions implicating fuzzy logic. A sole fuzzy IF-THEN rule definition can be as follows:

"a" and "b" are variants of linguistic, fuzzy groups portrayed in X and Y in terms, separately. In the event that X is "a" at that point, Y is "b". "a" and "b" are linguistic factors, which are fuzzy groups named as X and Y terms individually collected in the universe of discourse. If-part of the rule "X is a" is the primary choice, the remaining part of the rule "Y is b" is named as the result.

Classification Results. Fuzzy Inference System (FIS) has four types of modules which are fuzzification, knowledge base, inference engine, and defuzzification. Fuzzy derivation techniques are the title given for straightforward and nonstraightforward strategies. Mamdani's and Sugeno's techniques are mostly used for direct strategies. Indirect techniques are progressively confounded. Most basic fuzzy derivation procedure is Mamdani technique and it is a common learning forecast example. The model developed by Mamdani operates with well-defined data inlet and with lingual terms and in terms which serves as an essential piece of this model and is a conviction evaluate in assessing the future utility when the real utility is obscure.

4   Experimental Results

Fuzzy Rules. In this examination, the fuzzy standards rely upon the number of information factors of the semantic IF-THEN Develops in which overall construct is "IF a, THEN b" and the Fuzzy Rules of "b" and MF. In the fuzzy standard based choice model delineated in Fig. 1 has 1 variable and 6 membership functions = 6^1 = 6 rules be acquired.

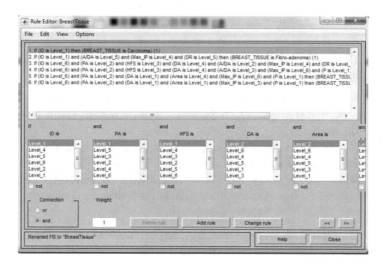

Fig. 1. Rule editor.

5 Conclusion

The pathophysiology of human breast tissue can be reflected by the electrical impedance spectral characteristics. In this paper, breast tissue classification based on electrical impedance spectral characteristics is proposed for the diagnosis of breast disease in the early period. The electrical impedance spectra of the breast tissue was collected in vitro and measured freshly and 9 breast tissue characteristics were obtained, and the breast tissue was then classified using a fuzzy logic system. Experimental results indicate that this method effectively classifies breast tissue. In order to live tissue characterization due to their ease of use and low cost, electrical impedance spectroscopy is preferred as a minimally invasive technique with significant advantages. This manuscript defines, how this technique can be implemented to breast tissue classification and breast cancer diagnosis. The Fuzzy Inference System (FIS) is proposed to classify breast tissues from Electrical Impedance Spectroscopy Measurements.

References

1. Morris, E., Liberman, L.: Breast MRI. Springer, New York (2005)
2. Sibbering, M., Courtney, C.: Management of breast cancer: basic principles. Surgery (Oxford) **34**(1), 25–31 (2016). https://doi.org/10.1016/j.mpsur.2015.10.005
3. Da Silva, J.E., De Sá, J.M., Jossinet, J.: Classification of breast tissue by electrical impedance spectroscopy. Med. Biol. Eng. Comput. **38**(1), 26–30 (2000)
4. Moqadam, S.M., Grewal, P.K., Haeri, Z., Ingledew, P.A., Kohli, K., Golnaraghi, F.: Cancer detection based on electrical impedance spectroscopy: a clinical study. J. Electr. Bioimp. **9**, 17–23 (2018)
5. Zarafshani, A., Bach, T., Chatwin, C.R., Tang, S., Xiang, L., Zheng, B.: Conditioning electrical impedance mammography system. Measurement **116**, 38–48 (2018)

6. Morimoto, T., Kimura, S., Konishi, Y., Komaki, K., Uyama, T., Monden, Y., Kinouchi, D. Y., Iritani, D.T: A study of the electrical bioimpedance of tumors. J. Invest. Surg. **6**, 25–32 (1993)
7. Liu, C., Chang, T., Li, C.: Breast tissue classification based on electrical impedance spectroscopy. In: 2015 International Conference on Industrial Technology and Management Science. Atlantis Press (2015)
8. Gao, W., Fan, M., Zhao, W., Zheng, B., Li, L.: Computerized detection of breast cancer using resonance-frequency-based electrical impedance spectroscopy. In: Medical Imaging 2017: Imaging Informatics for Healthcare, Research, and Applications, vol. 10138, p. 1013816. International Society for Optics and Photonics (2017)
9. UCI (California Irvine University). http://archive.ics.uci.edu/ml/datasets/breast+tissue. Accessed 10 Jan 2019
10. Adedeji, B., Badiru, J.Y.C.: Fuzzy Engineering expert systems with Neural Network Applications. Department of Industrial Engineering University of Tennessee Knoxville, TN. School of Electrical and Computer Engineering University of Oklahoma Norman, OK (2002)
11. Sandya, H.B., Hemanth Kumar, P., Himanshi Bhudiraja, S.K.R.: Fuzzy rule based feature extraction and classification of time series signal. Int. J. Soft Comput. Eng. (IJSCE) **3**, 2231–2307 (2013)
12. Jantzen, J.: Tutorial on Fuzzy Logic. Technical University of Denmark (2008)
13. Güler, I., Ubeyli, D.E.: Adaptive neuro-fuzzy inference system for classification of EEG signals using wavelet coefficients. A Department of Electronics and Computer Education, Faculty of Technical Education, Gazi University, Ankara, Turkey, Department of Electrical and Electronics Engineering, Faculty of Engineering (2005)

A Memetic and Adaptive Continuous Ant Colony Optimization Algorithm

Mahamed Omran[1] and Radka Polakova[2(✉)]

[1] Department of Computer Science, Gulf University for Science
and Technology, P.O. Box 7207, 32093 Hawally, Kuwait
omran.m@gust.edu.kw
[2] Faculty of Social Studies, University of Ostrava, Ostrava, Czech Republic
radka.polakova@osu.cz

Abstract. This paper proposes two new variants of the Continuous Ant Colony Optimization algorithm, ACO_R. The first variant, called the Adaptive ACO_R ($AACO_R$), uses the relative diversity of the solutions in the algorithm's archive to adapt its parameters. The second variant, called the memetic $AACO_R$ ($MAACO_R$), uses a local search operator to improve the performance of $AACO_R$. Both variants were tested on the 22 IEEE CEC 2011 real-world optimization problems and compared with ACO_R and two *state-of-the-art* optimization methods. The results demonstrate the merits of the proposed approaches.

Keywords: Optimization · Ant colony algorithm · Real-world optimization problems

1 Introduction

Ant Colony Optimization (ACO) [1] mimics real ant colony foraging behavior. Real ants begin their search for food sources randomly. When an ant finds a food source, it deposits a chemical substance called *pheromone* on its way back to the colony. Indirect communication between ants is achieved using the deposited pheromone. Pheromone attracts ants and thus the more pheromone on a path, the more ants will follow this path. This way ants can find the shortest path between a food source and its colony.

ACO was originally proposed to solve discrete optimization problems. Several attempts were made to modify ACO to solve problem in the continuous domain (e.g., [2–4]). The most important attempt is ACO_R [4] where the pheromone model is represented by an archive of solutions. This archive is used to model a probability distribution, which is used to generate new solutions. A distribution model of ant colony foraging by analyzing the relationship between the position distribution and food source during ant colony foraging was proposed in [3]. The approach [3] was compared with ACO_R on six problems, four of which have only two dimensions. The fifth function was a simple unimodal function (the sphere function). The only relatively challenging function was a 10-dimensional Rosenbrock function and in this function ACO_R performs better than the proposed method. [2] proposed the unified ACO, UACOR, which combines ACO_R with two other ACO_R-variants. However, UACOR

© Springer Nature Switzerland AG 2020
R. A. Aliev et al. (Eds.): ICSCCW 2019, AISC 1095, pp. 158–166, 2020.
https://doi.org/10.1007/978-3-030-35249-3_20

requires tuning more than 20 parameters, making it difficult to be used for practical applications.

In this paper, we propose to improve the original ACO_R by adapting one of its main parameter using the diversity of the archive and using a local search operator. The proposed ACO_R-variants are compared with the original ACO_R and two *state-of-the-art* optimization methods on 22 real-world optimization problems.

The rest of the paper is organized as follows. Section 2 describes the ACO_R algorithm. The diversity mechanism used in this study is explained in Sect. 3. Section 4 provides a discussion of the adaptive variant of ACO_R while Sect. 5 enhances the adaptive variant by using a local search operator. Experimental results are discussed in Sect. 6. Section 7 concludes the paper and presents future plans.

2 ACO_R: The Continuous Ant Colony Optimization Algorithm

The original ACO_R algorithm [4] consists of three main components.

2.1 Pheromone Representation

ACO_R maintains an archive of k solutions where each solution is a vector of n components, i.e.

$$s_i = [s_{i1}, s_{i2}, \ldots, s_{in}]^T$$

where $1 \leq i \leq k$ and n is the problem's dimension. The solutions are sorted in ascending order (for minimization problems) based on their objective functions such that,

$$f(s_1) \leq f(s_2) \leq \cdots \leq f(s_k)$$

2.2 Probabilistic Solution Construction

Each solution, s_i in the archive is assigned a weight, w_i. This weight is defined as follows.

$$w_i = \frac{1}{qk\sqrt{2\pi}} e^{\frac{-(i-1)^2}{2q^2k^2}} \tag{1}$$

where q is a user-specified parameter. A small value of q favors the selection of better solutions while larger values make selection more uniform. To generate a new solution, one solution is chosen from the archive using the following probabilities:

$$p_i = \frac{w_i}{\sum_{r=1}^{k} w_r} \qquad (2)$$

where $1 \leq i \leq k$. After choosing a solution, s_c, m solutions are generated from a normal distribution with a mean of s_c and a standard deviation of σ_j,

$$\sigma_j = \xi * D_j \qquad (3)$$

where

$$D_j = \sum_{r=1}^{k} \frac{|s_{r,j} - s_{c,j}|}{k - 1} \qquad (4)$$

and $\xi > 0$ is a parameter of ACO_R called the *pheromone evaporation rate*. Small values of ξ allow ACO_R to converge quickly while larger values slow down the convergence.

2.3   Pheromone Update

The m recently-constructed solutions are appended to the k archive solutions. The $k + m$ solutions are sorted according to their objective functions and the best k solutions are kept in the archive.

3   The Diversity Mechanism

Polakova et al. [5] proposed adapting the population size of Differential Evolution [6] based on the population diversity. In this paper, the diversity mechanism used in [5] is used to adapt the ξ parameter of ACO_R. In this section, the diversity mechanism is described in the context of ACO_R. The diversity of the solutions' archive is measured by the average squared distance between each solution in the archive and the archive mean, i.e.,

$$DI = \sqrt{\frac{1}{k} \sum_{i=1}^{k} \sum_{j=1}^{n} \left(x_{i,j} - \bar{x}_J \right)^2} \qquad (5)$$

where $\bar{x}_j$ is the jth component of the mean vector of the solutions in the archive.

The initial archive diversity is denoted by DI_{init} and it is used to compute the relative diversity,

$$RD = \frac{DI}{DI_{init}}. \qquad (6)$$

RD is kept near its required value rRD, which linearly decreases from the initial value of 1 at the beginning of the search to a value near zero at the end of the search as depicted in Fig. 1. *RD* is considered high if it is greater than 1.1*rRD, while *RD* is assumed low if it is less than 0.9*rRD.

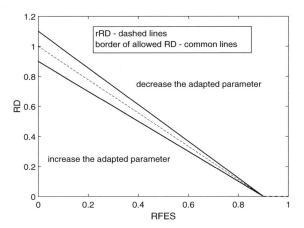

Fig. 1. The required relative diversity, rRD, and the area of the allowed relative diversity, RD. RFES = FES/MaxFES, where FES is the current number of function evaluations and MaxFES is maximum number of function evaluations.

4 AACO$_R$: The Adaptive Continuous Ant Colony Optimization Algorithm

The AACO$_R$ algorithm differs from the ACO$_R$ approach described in Sect. 2 in the following:

- The ξ parameter has a big impact on the performance of AACO$_R$. If ξ is large, the chances of getting diverse solutions increase and thus premature convergence (due to low diversity) may be avoided. A small ξ generates solutions around a chosen one and thus improves exploitation when diversity is high. Hence, in AACO$_R$, the ξ parameter is adapted in each generation based on the relative diversity, *RD* (defined in (6)), as follows,

$$\xi = 0.9 * \xi \ \ \text{if } RD > 1.1 * rRD$$

$$\xi = 1.1 * \xi \ \ \text{if } RD < 0.9 * rRD \tag{7}$$

$$\xi = \xi = \text{otherewise.}$$

If $\xi > 0.9$, it is set to 0.9 and if $\xi < 0.1$, it is set back to 0.1.

A new ξ is generated anew for each of the m solutions constructed at each generation according to the following distribution,

$$\xi' \sim \mathcal{N}(\xi, 0.1). \tag{8}$$

If $\xi' > 0.9$, it is set to 0.9 and if $\xi' < 1*10^{-2}$, it is set back to $1*10^{-2}$.

- According to [2] and to our own experiments on many problems, it seems that $q \in [0.1, 0.3]$ generally yields good results. Hence, the q parameter is generated before each new probabilistic solutions' construction from a normal distribution with a mean of 0.2 and a standard deviation of 0.1, i.e.,

$$q \sim \mathcal{N}(0.2, 0.1). \tag{9}$$

If $q > 0.9$, it is set to 0.9 and if $q < 1*10^{-4}$, it is set back to $1*10^{-4}$.
- To improve the archive's diversity, if two solutions in the archive are identical, the second one is replaced as follows [7],

$$s_{i,j} \sim \mathcal{N}\left(s_{i,j}, (b_j - a_j)/k\right) \tag{10}$$

The details of the AACO$_R$ approach is shown in Algorithm 1.

Algorithm 1: AACO$_R$
$S \leftarrow \emptyset$;
for $i \leftarrow 1$ **to** k **do**
    for $j \leftarrow 1$ **to** n **do**
        $s_{i,j} \sim U(a_j, b_j)$;
    end
    $S \leftarrow S \cup \{s_i\}; f_i \leftarrow f(s_i)$;
end
$S \leftarrow \text{Sort}(S)$;
while *stopping criterion is not satisfied* **do**
    $S' \leftarrow \emptyset$;
    for $l \leftarrow 1$ **to** m **do**
        $q \sim \mathcal{N}(0.2, 0.1)$; Choose a solution s_c using (2); $\xi' \sim \mathcal{N}(\xi, 0.1)$;
        for $j \leftarrow 1$ **to** n **do**
            $D_j = \sum_{r=1}^{k} \frac{|s_{r,j} - s_{c,j}|}{k-1}$; $\sigma_j = \xi' D_j$; $s'_{l,j} \sim \mathcal{N}(s_{c,j}, \sigma_j)$;
            if $s'_{l,j} < a_j$ **or** $s'_{l,j} > b_j$ **then** $s'_{l,j} \sim U(a_j, b_j)$ **end**
        end
        $S' \leftarrow S' \cup \{s'_l\}; f_l \leftarrow f(s'_l)$;
    end
    $S'' \leftarrow S' \cup S$; Remove duplicates in S''; $S'' \leftarrow \text{Sort}(S'')$;
    Discard the m worst solutions in S''; $S \leftarrow S''$;
    Compute RD using (6) and rRD; Adjust ξ using (7);
end

5 MAACO$_R$: The Memetic Adaptive Continuous Ant Colony Optimization Algorithm

Memetic Algorithms (MAs) [8] are population-based optimization algorithms consisting of an evolutionary or swarm-intelligent framework and a set of local search algorithms. Thus, MAs try to balance exploration and exploitation. MAs have been successfully used to solve many optimization problems. For more details about MAs, interested readers are referred to [9]. The MAACO$_R$ approach is identical to the AACO$_R$ algorithm described in Sect. 4 except that it uses a local search operator to improve solutions in the archive. Thus, MAACO$_R$ can be considered to be a memetic algorithm. The idea of using a local search operator has already been suggested by [4] where it was called the *daemon actions*.

In MAACO$_R$, after computing *RD* and *rRD* in Algorithm 1, if *RD* > 1.1*rRD* (i.e., the diversity is high) then a solution, s_p, in the archive is chosen as in Sect. 2.2 but using (9). Once the solution is selected, a local search method is applied.

Thus, the better the solution, the more probable it will be refined by the local search. The maximum allowed number of function evaluation for the local search operator, *maxFE$_{LS}$*, is set to *3kn* as in [7]. The *interior point* method (as implemented by MATLAB's `fmincon` function) is used as the local search method.

6 Experimental Results and Discussions

The two proposed approaches were tested on the 22 real-world optimization problems described in the IEEE CEC 2011 benchmark [10]. The problems have dimensions ranging from 1 to 216. For more details, please refer to [10]. The proposed approaches were compared the Genetic Algorithm with Multi-Parent Crossover (GA-MPC) algorithm [11] and the Success-History based parameter Adaptation DE (SHADE) algorithm [12], which ranked fourth (out of 21 algorithms) in the IEEE CEC competition on real-parameter, single-objective optimization [13].

30 runs were carried out for each algorithm on each test problem. The maximum number of function evaluation was set to 150,000. The Wilcoxon rank-sum test was used to check the significance of the difference between competing approaches.

For all ACO$_R$ variants, the archive size, *k*, is set to the maximum of *n* and 100. According to [4], *k* may not be smaller than *n*. For simplicity, *m* was set equal to *k*. Based on our empirical results, *q = 0.2* and ξ = 0.6 yield good results for ACO$_R$. These values are generally consistent with recommendations of [2] on two other sets of problems.

Table 1 summarizes the results of AACO$_R$, ACO$_R$, GA-MPC, and SHADE on the 22 benchmark problems. It is clear that AACO$_R$ outperforms ACO$_R$ on most of the problems. Only on three problems ACO$_R$ performs better. However, AACO$_R$ performs slightly worse than GA-MPC while SHADE outperforms AACO$_R$ on most of the problems. These results confirm those reported by [14], which show the superiority of adaptive DE variants compared to other nature-inspired, swarm intelligence techniques.

Table 1. The median values achieved by AACO$_R$, ACO$_R$, GA-MPC, and SHADE over 30 runs and 150,000 function evaluations. The symbols +, −, and ≈ denote AACO$_R$ is *significantly* better, worse, or equivalent to the competing approach.

Prob.	AACO$_R$	ACO$_R$	GA-MPC	SHADE
T_1	0.000000e+00	3.910170e−27(+)	0.000000e+00(≈)	3.911441e−03(+)
T_2	−2.650759e+01	−1.568641e+01(+)	−2.652659e+01(≈)	−2.393334e+01(+)
T_3	1.151489e−05	1.151489e−05(≈)	1.151489e−05(≈)	1.151489e−05(≈)
T_4	1.391382e+01	2.095740e+01(+)	1.387176e+01(≈)	1.412512e+01(≈)
T_5	−3.438548e+01	−2.232203e+01(+)	−3.435230e+01(≈)	−3.582279e+01(−)
T_6	−2.742973e+01	−1.652561e+01(+)	−2.300593e+01(+)	−2.913335e+01(−)
T_7	8.063723e−01	1.681343e+00(+)	8.987662e−01(+)	1.275726e+00(+)
T_8	2.200000e+02	2.200000e+02(≈)	2.200000e+02(≈)	2.200000e+02(≈)
T_9	1.412322e+05	1.755058e+03(−)	4.960875e+04(−)	2.115255e+03(−)
T_{10}	−2.143326e+01	−2.051518e+01(+)	−2.144594e+01(≈)	−2.164373e+01(−)
$T_{11.1}$	4.390609e+05	5.303599e+04(−)	5.258440e+04(−)	5.233486e+04(−)
$T_{11.2}$	2.251137e+07	2.187702e+07(−)	2.130389e+07(−)	1.732090e+07(−)
$T_{11.3}$	1.547882e+04	1.548153e+04(≈)	1.544512e+04(−)	1.544419e+04(−)
$T_{11.4}$	1.905830e+04	1.928275e+04(+)	1.847996e+04(−)	1.810009e+04(−)
$T_{11.5}$	3.295176e+04	3.303416e+04(+)	3.295739e+04(≈)	3.274038e+04(−)
$T_{11.6}$	1.354859e+05	1.383297e+05(+)	1.347830e+05(≈)	1.266317e+05(−)
$T_{11.7}$	1.947609e+06	2.366081e+06(+)	2.012955e+06(+)	1.886032e+06(−)
$T_{11.8}$	9.714528e+05	1.686460e+06(+)	9.780152e+05(≈)	9.367560e+05(−)
$T_{11.9}$	1.585215e+06	2.337204e+06(+)	1.208241e+06(−)	9.422527e+05(−)
$T_{11.10}$	9.406951e+05	1.686460e+06(+)	9.780152e+05(≈)	9.367560e+05(−)
T_{12}	1.563267e+01	1.536405e+01(≈)	1.638916e+01(+)	1.671959e+01(+)
T_{13}	2.195148e+01	2.125662e+01(≈)	2.022996e+01(≈)	1.771337e+01(−)
+/≈/−	N/A	14/5/3	4/12/6	4/3/15

Table 2 summarizes the results of MAACO$_R$, AACO$_R$, GA-MPC, and SHADE on the 22 problems. The results show that MAACO$_R$ outperforms both AACO$_R$ and GA-MPC on all the problems. Compared to SHADE, MAACO$_R$ outperforms SHADE on 12 problems while SHADE performs better on only seven problems.

Table 2. The median values achieved by MAACO$_R$, AACO$_R$, GA-MPC, and SHADE over 30 runs and 150,000 function evaluations. The symbols +, −, and ≈ denote MAACO$_R$ is *significantly* better, worse, or equivalent to the competing approach.

Prob.	MAACO$_R$	AACO$_R$	GA-MPC	SHADE
T_1	2.499025e−21	0.000000e+00(−)	0.000000e+00(≈)	1.741370e−02(+)
T_2	−2.755586e+01	−2.678641e+01(+)	−2.656204e+01(+)	−2.425601e+01(+)
T_3	1.151489e−05	1.151489e−05(≈)	1.151489e−05(≈)	1.151489e−05(≈)
T_4	1.383530e+01	1.394430e+01(≈)	1.391286e+01(≈)	1.377076e+01(−)
T_5	−3.684537e+01	−3.445934e+01(+)	−3.423079e+01(+)	−3.569183e+01(+)
T_6	−2.621796e+01	−2.742966e+01(≈)	−2.300593e+01(+)	−2.912497e+01(−)
T_7	5.968294e−01	8.581542e−01(+)	9.673012e−01(+)	1.229416e+00(+)
T_8	2.200000e+02	2.200000e+02(≈)	2.200000e+02(≈)	2.200000e+02(≈)
T_9	8.051277e+00	1.369662e+05(+)	6.075286e+04(+)	2.069244e+03(+)
T_{10}	−2.141939e+01	−2.142110e+01(≈)	−2.140138e+01(≈)	−2.164447e+01(−)
$T_{11.1}$	5.203720e+04	5.996242e+05(+)	5.260032e+04(+)	5.245549e+04(≈)
$T_{11.2}$	1.731204e+07	2.366245e+07(+)	2.114536e+07(+)	1.732029e+07(+)
$T_{11.3}$	1.544674e+04	1.547824e+04(+)	1.544554e+04(≈)	1.544419e+04(−)
$T_{11.4}$	1.808090e+04	1.900633e+04(+)	1.848990e+04(+)	1.811159e+04(+)
$T_{11.5}$	3.288041e+04	3.295353e+04(+)	3.293972e+04(+)	3.274050e+04(−)
$T_{11.6}$	1.242184e+05	1.355264e+05(+)	1.342241e+05(+)	1.266282e+05(+)
$T_{11.7}$	1.898382e+06	1.949880e+06(+)	2.063109e+06(+)	1.890031e+06(−)
$T_{11.8}$	9.281198e+05	9.503433e+05(+)	9.710117e+05(+)	9.368683e+05(+)
$T_{11.9}$	9.354230e+05	1.474837e+06(+)	1.197236e+06(+)	9.418822e+05(+)
$T_{11.10}$	9.281198e+05	9.503433e+05(+)	9.710117e+05(+)	9.368683e+05(+)
T_{12}	1.539850e+01	1.587734e+01(+)	1.612934e+01(+)	1.654397e+01(+)
T_{13}	2.134060e+01	2.198689e+01(≈)	2.044000e+01(≈)	1.723092e+01(−)
+/≈/−	N/A	15/6/1	15/7/0	12/3/7

7 Conclusions

Two new ACO$_R$-variants have been proposed in this paper. The first variant, AACO$_R$, uses the relative diversity of the solutions in the ACO$_R$'s archive to adapt the ζ parameter of ACO$_R$. A memetic variant of AACO$_R$ was then proposed, where a local search operator was used to further improve good solutions in the archive. Both methods were tested on 22 real-world engineering optimization problems with good results.

References

1. Dorigo, M., Maniezzo, V., Colorni, A.: Ant system: optimization by a colony of cooperating agents. IEEE Trans. Syst. Man Cybern. B **26**(1), 29–41 (1996)
2. Liao, T., Stützle, T., Montes de Oca, M., Dorigo, M.: A unified ant colony optimization algorithm for continuous optimization. Eur. J. Oper. Res. **234**, 597–609 (2014)

3. Liu, L., Dai, Y., Gao, J.: Ant colony optimization algorithm for continuous domains based on position distribution model of ant colony foraging. Sci. World J. 20 (2014)
4. Socha, K., Dorigo, M.: Ant colony optimization for continuous domains. Eur. J. Oper. Res. **185**, 1155–1173 (2008)
5. Polakova, R., Tvrdik, J., Bujok, P.: Adaptation of population size according to current population diversity in differential evolution. In: Proceedings of the IEEE 2017 Symposium Series on Computational Intelligence (SSCI), pp. 2627–2634. IEEE, USA (2017)
6. Storn, R., Price, K.: Differential evolution - a simple and efficient heuristic for global optimization over continuous spaces. J. Global Optim. **11**, 341–359 (1997)
7. Omran, M., Clerc, M.: APS 9: an improved adaptive population-based simplex method for real-world engineering optimization problems. Appl. Intell. **48**(6), 1596–1608 (2017)
8. Moscato, P.: On evolution, search, optimization, genetic algorithms and martial arts: towards memetic algorithms. Technical Report 826 (1989)
9. Neri, F., Cotta, C.: Memetic algorithms and memetic computing optimization: a literature review. Swarm Evol. Comput. **2**, 1–14 (2012)
10. Das, S., Suganthan, N.P.: Problem definitions and evaluation criteria for CEC 2011 competition on testing evolutionary algorithms on real world optimization problems. Technical report, Jadavpur University, Nanyang Technological University (2010)
11. Elsayed, S., Sarker, R., Essam, D.: GA with a new multi-parent crossover for solving IEEE-CEC2011 competition problems. In: Proceedings of the IEEE Congress on Evolutionary Computation, pp. 1034–1040. IEEE, USA (2011)
12. Tanabe, R., Fukunaga, A.: Success-history based parameter adaptation for differential evolution. In: Proceedings of the IEEE 2013 Congress on Evolutionary Competition, pp. 71–78. IEEE, Mexico (2013)
13. Liang, J., Qin, B., Suganthan, N.P., Hernandez-Diaz, G.: Problem definitions and evaluation criteria for the IEEE CEC 2013 special session on real-parameter optimization. Technical report, Zhengzhou University, Nanyang Technological University (2013)
14. Bujok, P., Tvrdik, J., Polakova, R.: Adaptive differential evolution vs nature-inspired algorithms: an experimental comparison. In: Proceedings of the IEEE 2017 Symposium Series on Computational Intelligence (SSCI), pp. 2604–2611. IEEE, USA (2017)

Evaluating the Efficacy of Adult HIV Post Exposure Prophylaxis Regimens in Relation to Transmission Risk Factors by Multi Criteria Decision Method

Murat Sayan[1,2] , Tamer Sanlidag[2,3] , Nazife Sultanoglu[2,5] ,
Berna Uzun[2,4(✉)] , Figen Sarigul Yildirim[6] ,
and Dilber Uzun Ozsahin[2,7,8]

[1] Faculty of Medicine, Clinical Laboratory, PCR Unit, Kocaeli University,
Kocaeli, Turkey
`sayanmurat@hotmail.com`
[2] Research Center of Experimental Health Sciences, Near East University,
99138 Nicosia, Cyprus
`{tamer.sanlidag,berna.uzun,`
`dilber.uzunozsahin}@neu.edu.tr,`
`nsultanoglu@hotmail.com`
[3] Department of Medical Microbiology, Manisa Celal Bayar University,
Manisa, Turkey
[4] Department of Mathematics, Near East University, 99138 Nicosia, Cyprus
[5] Faculty of Medicine, Department of Medical Microbiology and Clinical
Microbiology, Near East University, 99138 Nicosia, Cyprus
[6] Clinical of Infectious Diseases, Health Science University,
Antalya Educational and Research Hospital, Antalya, Turkey
`drfigensarigul@yahoo.com`
[7] Department of Biomedical Engineering, Near East University,
99138 Nicosia, Cyprus
[8] Gordon Center for Medical Imaging, Radiology Department,
Massachusetts General Hospital and Harvard Medical School, Boston, USA

Abstract. There are approximately 39 million people living with the human immunodeficiency virus (HIV) worldwide. Although antiretroviral treatment (ART) has revolutionized the treatment and management of the HIV infection, a cure still remains elusive. In addition to ART's success at prolonging the lives of the infected individuals, it also has its uses in other cases. For example, post exposure prophylaxis (PEP) ART regimens are used after potential exposure to HIV to prevent the establishment of the HIV infection by up to 80% when administered in a timely manner. There are many guidelines available for these situations; however, they have differences in terms of their content and very few guidelines suggest specific PEP ART regimens for particular risk factors. Our study uses a multi criteria decision making method called Fuzzy PROMETHEE to evaluate the effectiveness of selected PEP ART regimens according to specified transmission risk factors for HIV. The risk factors were the criteria in this study and were weighted according to experts' opinions in terms of their relative importance.

© Springer Nature Switzerland AG 2020
R. A. Aliev et al. (Eds.): ICSCCW 2019, AISC 1095, pp. 167–174, 2020.
https://doi.org/10.1007/978-3-030-35249-3_21

Keywords: Multi criteria decision method · Fuzzy PROMETHEE · HIV · Risk factors · Adult post exposure prophylaxis regimen

1   Introduction

Globally, it is estimated that 39 million people were living with HIV in 2018, of whom 13 million were receiving antiretroviral treatment (ART). ART has been used since 1996 to treat HIV infection and has successfully decreased the morbidity and mortality associated with HIV [1, 2]. There are approximately 30 antiretroviral agents, which are divided into seven groups according to their molecular mechanisms and therapeutic targets. ART is a combination of anti-HIV agents, with preferably at least three anti-HIV agents from the two groups. Mono-therapy is not an option due to increased risk of resistance to applied anti-HIV agent [3].

HIV is mainly transmitted through unprotected sex, blood transfusions, and needle sharing or needle stick injuries. Although ART is used to treat HIV infection, it is also used in cases of potential exposure to HIV to prevent the establishment of the HIV infection. This intervention is referred to as post-exposure prophylaxis (PEP), which includes the use of ART for a period of 28 days. PEP should be initiated soon after the potential exposure to HIV within 72 h. If PEP is administered on time and adherence to the ART is maintained for the 28-day course, PEP can reduce the risk of HIV infection by over 80% [4].

There are many international and national guidelines for types of PEP exposure, including occupational or non-occupational and populations (adults or children). Occupational exposure to HIV includes needle stick or mucocutaneous injuries, whereas a non-occupational transmission route involves sexual contact [5–7].

Although several studies have assessed the risk factors for the administration of PEP regimens, the existing guidelines standardize the PEP regimens according to the occupational or non-occupational exposure to HIV. This study will systematically assess the efficacy of 22 PEP ART regimens over each particular risk factor consisting of blood transfusion, needle sharing, needle stick injury, mucocutaneous injury, receptive anal intercourse, insertive anal intercourse, receptive penile-vaginal intercourse, insertive penile-vaginal intercourse, receptive oral intercourse and insertive oral intercourse.

This systemic comparison between the particular risk factors and the PEP regimens was performed through a technique called fuzzy PROMETTHE, which is a multi-criteria decision aid system. Fuzzy PROMETHEE allows decision makers to find the best alternative in a system via the calculation of the outranking amongst the alternatives. In particular, our approach is to determine the best ART regimen to be used for particular risk factors for HIV infection.

2   Methods

This study analyzed the effectiveness of selected adult PEP ART regimens by using the fuzzy PROMETHEE method. The results were obtained by ranking the PEP ART regimens through mutual comparison between the alternatives. In the fuzzy

PROMETHEE method, partial ranking is performed by PROMETHEE I, whereas complete ranking of the alternatives within the system is performed by PROMETHEE II. As a consequence, the obtained outranking scores allowed the ordering of the PEP ART regimes from best to worst in terms of their effectiveness. For the calculation of the outranking scores, two types of information are required. These are the criteria associated with their weight (relative importance) and the preference function. P represents the preference function and it is used in the evaluation of alternatives a and $a_{t'}$ in association with their criteria function. The scale from 0 to 1 is used for the preference of one alternative over the other. A variety of preference functions exist for the PROMETHEE method. These include U-shape, V-shape, linear and Gaussian. The basic steps of the PROMETHEE method are as follows:

Step 1. Determine each criterion j, with a specific preference function represented as $p_j(d)$.

Step 2. Every criterion is defined with weights denoted as $w_T = (w_1, w_2, \ldots, w_k)$. Weights of the criterion can be equal if their importance is equal. In addition, normalization can be used for the weights; $\sum_{i=1}^{k} w_k = 1$.

Step 3. $a_t, a_{t'} \in A$ evaluates the outranking relation in π for each alternative, represented as a_t and $a_{t'}$, within the system

$$\pi(a_t, a_{t'}) = \sum_{k=1}^{K} w_k \cdot [p_k(f_k(a_t) - f_k(a_{t'}))], \qquad AXA \rightarrow [0, 1] \qquad (1)$$

In this formula, π (a, b) indicates the preference index, which is a measure that defines the intensity of preference for an alternative a_t as opposed to another alternative $a_{t'}$ in a system considering all criterion at the same time.

Step 4. Evaluation of the leaving and entering outranking flows:

- Leaving (or positive) flow for the alternative a_t:

$$\Phi^+(a_t) = \frac{1}{n-1} \sum_{\substack{t'=1 \\ t' \neq t}}^{n} \pi(a_t, a_{t'}) \qquad (2)$$

- Entering (or negative) flow for the alternative a_t:

$$\Phi^-(a_t) = \frac{1}{n-1} \sum_{\substack{t'=1 \\ t' \neq t}}^{n} \pi(a_{t'}, a_t) \qquad (3)$$

In this formula, n represents the number of alternatives, and each alternative is compared to (n − 1) number of alternatives within the system. The leaving flow stated as $\Phi^+(a_t)$ shows the strength of an alternative as $a_t \in A$, whereas the entering flow stated as $\Phi^-(a_t)$ shows the weakness of the alternatives in $a_t \in A$. Following the outranking flow evaluation, the PROMETHEE I and PROMETHEE II methods are used to evaluate the partial and complete order based on the net flow order, respectively.

Step 5. Partial order of alternatives A are determined as followed:

Alternative a_t is preferred to alternative $a_{t'}$ $(a_t Pa_{t'})$ in PROMETHEE I, if it meets one of the conditions stated as follows:

$(a_t P a_{t'})$ if;

$$
\begin{cases}
\Phi^+(a_t) > \Phi^+(a_{t'}) \text{ and } \Phi^-(a_t) < \Phi^-(a_{t'}) \\
\Phi^+(a_t) > \Phi^+(a_{t'}) \text{ and } \Phi^-(a_t) = \Phi^-(a_{t'}) \\
\Phi^+(a_t) = \Phi^+(a_{t'}) \text{ and } \Phi^-(a_t) < \Phi^-(a_t, a_{t'})
\end{cases}
\tag{4}
$$

If two alternatives presented as a_t and $a_{t'}$ possess the same leaving and entering flows, a_t is equal to $a_{t'}$ $(a_t I a_{t'})$:

$$
(a_t I a_{t'}) \text{ if}: \Phi^+(a_t) = \Phi^+(a_{t'}) \text{ and } \Phi^-(a_t) = \Phi^-(a_{t'})
\tag{5}
$$

a_t is incomparable to $a_{t'}$ $(a_t R a_{t'})$ if

$$
\begin{cases}
\Phi^+(a_t) > \Phi^+(a_{t'}) \text{ and } \Phi^-(a_t) > \Phi^-(a_{t'}) \\
\Phi^+(a_t) < \Phi^+(a_{t'}) \text{ and } \Phi^-(a_t) < \Phi^-(a_{t'})
\end{cases}
\tag{6}
$$

Step 6. Net outranking flow evaluation for each alternative by the formula is presented below

$$
\Phi^{net}(a_t) = \Phi^+(a_t) - \Phi^-(a_t)
\tag{7}
$$

A complete order obtained by the net flow can be calculated through PROMETHEE II and is defined by:

$$
a_t \text{ is preferred to } a_{t'} \ (a_t P a_{t'}) \text{ if } \Phi^{net}(a_t) > \Phi^{net}(a_{t'})
\tag{8}
$$

$$
a \text{ is indifferent to } a_{t'} \ (a_t I a_{t'}) \text{ if } \Phi^{net}(a_t) = \Phi^{net}(a_{t'}).
\tag{9}
$$

In fact, better alternative possess the highest $\Phi^{net}(a_t)$ value [8–11].

The Fuzzy PROMETHEE method was used in the current study. This is an adaptation of the PROMETHEE method with fuzzy scale which contains fuzzy numbers. By this way, this allows the processing raw or fuzzy data in the actual decision-making environment. Triangular fuzzy numbers are any real numbers between zero and 1. The magnitude of the triangular fuzzy numbers $\tilde{F} = (N, a, b)$ is determined by the Yager index in relation to the center of the triangle with $YI = (3N - a + b)/3$. Adult PEP regimens were the alternatives to be ranked in this system which were selected from available guidelines and experts' opinions as presented in Table 1.

Table 1. Adult post exposure prophylaxis regimens analyzed in Fuzzy PROMETHEE.

Adult post exposure prophylaxis antiretroviral combinations:

Integrase inhibitor-based regimens:
DTG + TDF/FTC (dolutegravir + tenofovir disoproxil fumarate/emtricitabine)
EVG/c + TDF (elvitegravir/cobicistat + tenofovir disoproxil fumarate)
RAL + TDF/FTC (raltegravir + tenofovir disoproxil fumarate/emtricitabine)
RAL + TDF + 3TC (raltegravir + tenofovir disoproxil fumarate + lamivudine)
RAL + ZDV + FTC (raltegravir + zidovudine + emtricitabine)
Protease inhibitor-based regimens:
DRV/r + TDF/FTC (darunavir/ritonavir + tenofovir disoproxil fumarate/emtricitabine)
DRV/r + TDF + 3TC (darunavir/ritonavir + tenofovir disoproxil fumarate + lamivudine)
DRV/r + ZDV + FTC (darunavir/ritonavir + zidovudine + emtricitabine)
DRV/r + ZDV + 3TC (darunavir/ritonavir + zidovudine + lamivudine)
ATV/r + TDF/FTC (atazanavir/r + tenofovir disoproxil fumarate/emtricitabine)
ATV/r + TDF + 3TC (atazanavir/r + tenofovir disoproxil fumarate + lamivudine)
ATV/r + ZDV + FTC (atazanavir/r + zidovudine + emtricitabine)
ATV/r + ZDV+ 3TC (atazanavir/r + zidovudine + lamivudine)
LPV/r + TDF/FTC (lopinavir/r + tenofovir/emtricitabine)
LPV/r + ZDV + 3TC (lopinavir/r + zidovudine + lamivudine)
LPV/r + TDF/FTC (lopinavir/r +tenofovir/emtricitabine)
LPV/r + TDF + 3TC (lopinavir/r +tenofovir + lamivudine)
LPV/r + ZDV + FTC (lopinavir/r + zidovudine + emtricitabine)
LPV/r + ZDV + 3TC (lopinavir/r + zidovudine + lamivudine)
Non-nucleoside reverse transcriptase inhibitor-based regimens:
RPV + TDF/FTC (rilpivirine + tenofovir disoproxil fumarate/emtricitabine)
RPV + TDF + 3TC (rilpivirine + tenofovir disoproxil fumarate + lamivudine)
RPV + ZDV + FTC (rilpivirine + zidovudine + emtricitabine)
RPV + ZDV + 3TC (rilpivirine + zidovudine + lamivudine)

The criteria, which in this case were the HIV transmission risk factors, were linked with each particular adult PEP regimen and categorized into Linguistic scale. This scale consisted the linguistic terms very high (VH), important (I), medium (M), low (L) and very low (VL), which were associated with triangular fuzzy scale, as represented in Table 2. Here, the fuzzy numbers indicated the weights (relative importance) of the criteria that were determined by the decision maker – experts' opinions.

As a result, the Fuzzy PROMETHEE method allowed the calculation of outranking flows of each alternative in fuzzy numbers. Subsequently, defuzzification by the Yager index with Gaussian preference function was applied. In the ranking procedure, the visual PROMETHEE decision lab program was used, which defined the out ranking flow: net flow, positive flow and negative flow. Net flow calculation is made through the extraction of negative flow from the positive flow.

Table 2. Triangular fuzzy scale

Linguistic scale for evaluation	Triangular fuzzy scale	Criteria-risk factors for acquisition of HIV infection
Very high (VH)	(0.75, 1, 1)	Blood transfusion
Important (H)	(0.50, 0.75, 1)	Needle sharing, needle stick injury, insertive penile vaginal intercourse, receptive anal intercourse
Medium (M)	(0.25, 0.50, 0.75)	Mucocutaneous injury, insertive anal intercourse, receptive penile vaginal intercourse,
Low (L)	(0, 0.25, 0.50)	Receptive oral intercourse, insertive oral intercourse
Very low (VL)	(0, 0, 0.25)	–

3 Results

By using the Fuzzy PROMETHEE method, we were able to compare the effectiveness of 22 adult PEP ART regimens in regard to risk factors with their related weights and parameters. The effectiveness of each regimen calculated and ranked via the net flow, which allowed the ordering of the selected PEP regimens according to their potency. Table 3 indicates the use of PEP regimens in descending order in terms of their use in preventing the establishment of HIV infection after a potential exposure.

Table 3. Complete ranking of the post exposure prophylaxis regimens.

No	Complete ranking	Adult antiretroviral regimen	Net flow	Positive flow	Negative flow
1	1	DTG + TDF/FTC	0,0628	0,0628	0,0000
2	1	EVG/c + TDF	0,0628	0,0628	0,0000
3	1	RAL + TDF/FTC	0,0628	0,0628	0,0000
4	1	DRV/r + TDF/FTC	0,0628	0,0628	0,0000
5	1	ATV/r + TDF/FTC	0,0628	0,0628	0,0000
6	1	LPV/r + TDF/FTC	0,0628	0,0628	0,0000
7	7	RAL + TDF + 3TC	−0,0047	0,0094	0,0141
8	7	RAL + ZDV + FTC	−0,0047	0,0094	0,0141
9	7	RAL + ZDV + 3TC	−0,0047	0,0094	0,0141
10	7	DRV/r + TDF + 3TC	−0,0047	0,0094	0,0141
11	7	DRV/r + ZDV + FTC	−0,0047	0,0094	0,0141
12	7	DRV/r + ZDV + 3TC	−0,0047	0,0094	0,0141
13	7	ATV/r + TDF + 3TC	−0,0047	0,0094	0,0141
14	7	ATV/r + ZDV + FTC	−0,0047	0,0094	0,0141
15	7	ATV/r + ZDV + 3TC	−0,0047	0,0094	0,0141
16	7	LPV/r + TDF + 3TC	−0,0047	0,0094	0,0141
17	7	LPV/r + ZDV + FTC	−0,0047	0,0094	0,0141
18	7	LPV/r + ZDV + 3TC	−0,0047	0,0094	0,0141
19	19	RPV + TDF/FTC	−0,0801	0,0000	0,0801
20	19	RPV + TDF + 3TC	−0,0801	0,0000	0,0801
21	19	RPV + ZDV + FTC	−0,0801	0,0000	0,0801
22	19	RPV + ZDV + 3TC	−0,0801	0,0000	0,0801

In addition, by using the visual PROMETHEE decision lab program, we have indicated and evaluated the advantages and disadvantages of each PEP ART regimen in regard to their effectiveness at preventing the acquisition of HIV infection after a potential exposure in the presence of different risk factors - transmission routes. When the risk factors are written above the zero threshold level, this indicates that the particular regimen is effective in terms of preventing the establishment of the HIV infection in the presence of that indicated risk factor. The weight of each risk factor is added up, and forms the width of the bar. If the regimen is not effective at preventing the establishment of HIV infection in the presence of the indicated risk factors, then they are written below the zero threshold level showing the disadvantage of that particular regimen.

4 Discussion and Conclusion

We were able to evaluate the effectives of selected adult PEP ART regimens in terms of their use in the presence of different risk factors by using the Fuzzy PROMETHEE method. According to our evaluation, we have indicated that DTG + TDF/FTC, EVG/c + TDF, RAL + TDF/FTC, DRV/r + TDF/FTC, ATV/r + TDF/FTC and LPV/r + TDF/FTC regimens were more effective for use in case of potential exposure to HIV in terms of all listed risk factors.

The use of fuzzy PROMETHEE in the health sciences is increasing rapidly due its ability to evaluate rough data in the actual environment in a rapid and reliable manner. This application can be adapted or changed easily by the decision maker allowing the determination of the best regimens to be used in a patient specific manner.

References

1. WHO: The Discovery and Development of Antiretroviral Agents. http://apps.who.int/medicinedocs/en/m/abstract/Js21619en/. Accessed 16 May 2019
2. WHO: WHO Data and Statistics (2018). https://www.who.int/gho/publications/world_health_statistics/2018/en/
3. AIDSinfo: HIV Treatment: The Basics Understanding HIV/AIDS. https://aidsinfo.nih.gov/understanding-hiv-aids/fact-sheets/21/51/hiv-treatment–the-basics. Accessed 16 May 2019
4. Sultan, B., Benn, P., Waters, L.: Current perspectives in HIV post-exposure prophylaxis. HIV AIDS (Auckl) **6**, 147 (2014)
5. WHO: Guidelines on post-exposure prophylaxis for HIV and the use of co-trimoxazole prophylaxis for HIV-related infections among adults, adolescents and children. World Health Organization (2014)
6. CDC: Updated Guidelines for Antiretroviral Postexposure Prophylaxis After Sexual, Injection-Drug Use, or Other Nonoccupational Exposure to HIV, United States (2016)
7. WHO: Post-exposure prophylaxis to prevent HIV infection, WHO. https://www.who.int/hiv/topics/prophylaxis/info/en/. Accessed 13 Dec 2018
8. Sultanoglu, N., Uzun, B., Yildirim, F.S., Sayan, M., Sanlidag, T., Ozsahin, D.U.: Selection of the most appropriate antiretroviral medication in determined aged groups (≥ 3 years) of HIV-1 infected children. In: Advances in Science and Engineering Technology International Conferences (ASET), pp. 1–6. IEEE Xplorer, Dubai (2019)

9. Sayan, M., Sultanoglu, N., Uzun, B., Yildirim, F.S., Sanlidag, T., Ozsahin, D.U.: Determination of post-exposure prophylaxis regimen in the prevention of potential pediatric HIV-1 infection by the multi-criteria decision making theory. In: Advances in Science and Engineering Technology International Conferences (ASET), pp. 1–5. IEEE Xplorer, Dubai (2019)

10. Uzun, B., Yildirim, F.S., Sayan, M., Sanlidag, T., Ozsahin, D.U.: The use of fuzzy PROMETHEE technique in antiretroviral combination decision in pediatric HIV treatments. In: Advances in Science and Engineering Technology International Conferences (ASET), pp. 1–4. IEEE Xplorer, Dubai (2019)

11. Ozsahin, D.U., Ozsahin, I.: A fuzzy PROMETHEE approach for breast cancer treatment techniques. Int. J. Med. Res. Health Sci. 7(5), 29–32 (2018)

An Intelligent Guide for Ranking the Hospitals in EU Countries Effective on Pancreatic Cancer Treatment via Fuzzy Logic

Günay Kibarer[1](✉) ⓘ and Cemal Kavalcıoğlu[2] ⓘ

[1] Department of Biomedical Engineering, Near East University,
Near East Boulevard, P.O. Box 99138, Nicosia, TRNC, Mersin 10, Turkey
aysegunay.kibarer@neu.edu.tr
[2] Department of Electric and Electronic Engineering, Near East University,
Near East Boulevard, P.O. Box 99138, Nicosia, TRNC, Mersin 10, Turkey
cemal.kavalcioglu@neu.edu.tr

Abstract. Pancreatic diseases, especially the most widely recognized type, pancreatic adenocarcinoma, are exceptionally difficult to treat and mostly detected just at a phase with no reversible possibility for medical treatment, which is the main therapeutic remedy. The best treatment for pancreatic malignancy relies upon how far it has spread, or its stage. The hospitals which are listed as the best for the treatment of pancreatic cancer, specifically, are mostly found in European Union (EU) Countries, especially in Germany, due to the companies owned capable of manufacturing technologically advanced diagnostic and treatment devices at low cost facilitating the trade of these devices by hospitals. In this study, best ten hospitals founded in EU countries superior in pancreatic cancer diagnosis and treatment were selected and the necessary hospital profile was provided for each from their websites. These data were then placed in a Fuzzy Rule-Based Model ranking the hospitals in increasing effectiveness for the treatment of pancreatic cancer providing a valuable guide for the patients.

Keywords: Pancreatic cancer · Cancer therapy · Fuzzy logic

1 Introduction

Cancer is one of the most ten diseases causing the death of millions of people around the world, and the scientists are living real difficulty in finding a definite therapy since each cancerous cell has its own properties and internal composition so the treatment of an individual cell will not be able to cure the other existing cells at all [1].

Until now, several therapies have been figured out to treat the cancerous cells according to their type such as chemotherapy, surgery, radiation, immunotherapy, targeted therapy, hormone therapy, stem cell transplant and nano medicine which may result with extended life spam or complete cure of the patients if discovered at an early stage [2]. Pancreatic malignant growth is grouped by which part of the pancreas is influenced: the part that makes stomach related substances (exocrine) or the part that makes insulin and different hormones (endocrine).

© Springer Nature Switzerland AG 2020
R. A. Aliev et al. (Eds.): ICSCCW 2019, AISC 1095, pp. 175–182, 2020.
https://doi.org/10.1007/978-3-030-35249-3_22

Pancreatic adenocarcinoma is the most prevalent type covering 95% of all cases. Different less regular exocrine pancreatic tumors include; adenosquamous carcinoma, squamous cell carcinoma, and goliath cell carcinoma. Endocrine pancreatic malignant growths are exceptional, and are named by the sort of hormone created. Some pancreatic islet cell tumors do not emit hormones and are known as non-discharging islet tumors of the pancreas. The best treatment for pancreatic malignancy relies upon its stage, or how far it has spread. A stage is a term utilized in malignant growth treatment to depict the degree of spread. The phases of pancreatic disease are: Stage 0: No spread; Stage I: Local development; Stage II: Local spread; Stage III: Wider spread; Stage IV: Confirmed spread [3].

There are many hospitals dealing with various types of cancer diseases by the most advanced medical techniques around the world, however, our research is mainly focused on the best ten hospitals dealing with pancreatic cancer and treatment of this disease [4]. In our study, it was aimed to prepare a guide for ranking the hospitals existing in European Union (EU) countries most effective on the treatment of pancreatic cancer types via Mamdani Model which is one of the most widely utilized fuzzy inference techniques. Usually, psychological disorders, depression and hopelessness may be observed in patients accompanying cancer diagnosis. They are panic stricken and even lose their motivation ending up with suicides. Therefore, we believe that our study will be a valuable guide for those patients living difficulty in deciding at which hospital they should be treated concerning their cancer type.

2   Methodology

2.1   Database

In this study, detailed information about each hospital is provided on the basis of 8 variables used as input data, mainly concerning its location, doctors, employees, cleanliness, placement, treatment techniques, being a research hospital or not, and finally its survival rate. The input data in Table 1 are abbreviated as: H-hospital; L-location; MD-medical doctor; E-employee; C-cleanliness; P-place; TT-treatment therapy; RH-research hospital; SR-survival rate; chemo-chemotherapy, CS-cytological surgery; PCR-pancreatic carcinoma resection; rad-radiology [5–14].

Table 1. Input data for the selected hospitals in EU countries.

H	L	MD	E	C	P	TT	RH	SR
Clinique La Colline	Geneva	6	22	✓	Center	chemo & CS	✓	5–6%
Double Check	Zurich	4	–	✓	Center	chemo & CS	X	10–15%
Hamburg-Eppendorf	Hamburg	7	70	✓	North	chemo, PSR & CS	✓	20%
Klinik Beau-Site	Bern	7	18	✓	Center	chemo, PSR & CS	X	22%
Klinikum Rechts Der Isar	Munich	8	67	✓	Center	chemo, PSR & CS	✓	8%
Surgeon	Geneva	5	–	✓	South	chemo & CS	X	5–8%
Hirslanden Klinik Aarau	Aarau	3	–	✓	Upstate	chemo, PSR & CS	X	3%
Charité Berlin	Berlin	3	35	✓	North	chemo, PSR, rad. & CS	✓	17–25%
Klinik Im Park	Zurich	5	82	✓	Center	chemo, PSR, rad. & CS	✓	20–28%
J.W Goethe University Hospital	Frankfurt	1	–	✓	Center	chemo & CS	✓	3%

3 Classification

3.1 Fuzzy Logic

In the mid-1960's, fuzzy logic was developed to make up for the lack of two-valued logic and probability theory. In the logic of probability, the likelihood level can be known, but not the correctness or falsehood of the propositions. An event and subject cannot be judged out of the possibility. The latest theory about fuzzy logic systems has been proposed as Lutfi Alesker Zadeh's Fuzzy Logic Theory. The truth values of variables may be any real integer between 0 and 1, inclusive, in fuzzy logic. It is employed to handle the concept of partial truth, The truth value may range between completely true and completely, thus, a proposal can neither be called true nor false when fuzzy logic is concerned. The greatest advantage of using Fuzzy Logic is that it can explain the existing system with linguistic qualifiers. We have the opportunity to express a complicated system at our disposal in terms that we manage only by ourselves. Fuzzy Logic is faster, with smaller software [15].

3.2 Fuzzy Sets

Accession of an element in a set is either all or nothing in a standard set theory. Therefore maps of the characteristic function of an element can be identified either by 1 (in the set) or 0 (not in the set) [15].

3.3   Membership Functions (MF)

A membership function is a curve that describes how each point in the input space is mapped to a membership value (or degree of membership) among 0 and 1. Input space is occasionally referred to as universe of discourse, a fancy name for a simple concept. Label is a name entitled for each MF that is described. For example, in this study an input variable such as input_1, input_2, and input_3 has eight MFs labeled as location, number of doctors, number of employees, cleanliness, place of hospital, treatment techniques, research hospital inquiry and survival rate. Chosen MF type is:

- **Gaussian: MF type** for this study is **Gaussian.** Popularity of Gaussian fuzzy membership functions in literature may be explained by the relation between fuzzy systems and radial basis function (RBF) neural networks. Fuzzy membership values are computed using a Gaussian membership function which should not be confused with Gaussian probability distribution. The maximum value of a Gaussian membership function may only be 1 [16].
- **Linguistic Variables:** Instead of the numeric values, some values are natural input words or linguistic variables, and expressions are system input or output variables. A linguistic variable is generally divided into linguistic terms. In this study, variables are location, number of doctors, number of employees, cleanliness, place of hospital, treatment techniques, research hospital inquiry and survival rate.

3.4   Universe of Discourse

Fuzzy set elements have been received from the universe in order to short or universe of discourse. Universe includes all elements which can be take into account. Even the universe itself depends on context [16].

IF - THEN Rules

To formulate conditional statements including fuzzy logic subjects; operators, verbs, and fuzzy sets, these if-then rule expressions have been used. A single fuzzy if-then rule is defined as follows: a and b are called linguistic variables. If X is "a", Y becomes "b", where fuzzy sets are specified in X and Y ranges, respectively.

Classification Outcome: Mamdani method is the most widely utilized fuzzy inference technique which is a knowledge-based prediction model. FIS consists of four modules, namely, a knowledge-based module, fuzzification module, defuzzification module, and inference engine module.

The model works with open data input, linguistic intervals, and terms. A significant benefit of this model is that when the actual value is not known, it is a reliable measure to predict the future values.

4 Experimental Results

The output data and figures of the methodology employed for the classification of hospitals in increasing order of effectiveness on pancreatic cancer are demonstrated below.

Fuzzy Rules. Number of input variables, the linguistic terms and if-then constructs support fuzzy rules [16]. In the fuzzy rule-based selection model defined in Fig. 1 below, 8 variables and 6 membership functions = 6^1 = 6 rules are obtained (Figs. 2, 3 and 4).

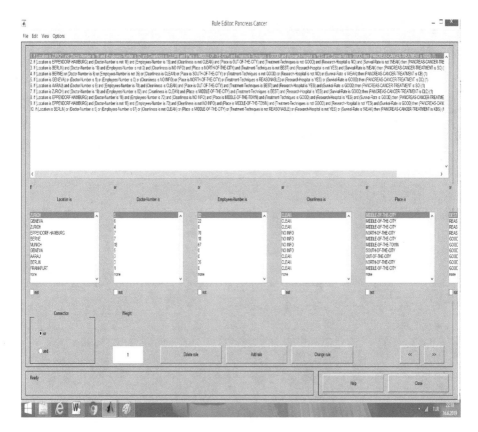

Fig. 1. Rule editor.

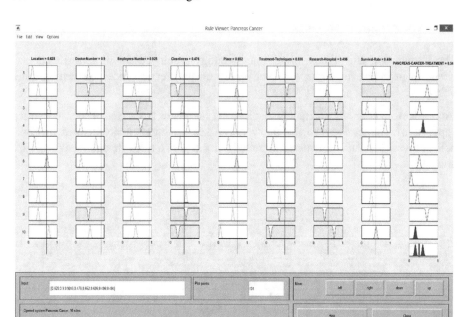

Fig. 2. Fuzzy rule-based selection process.

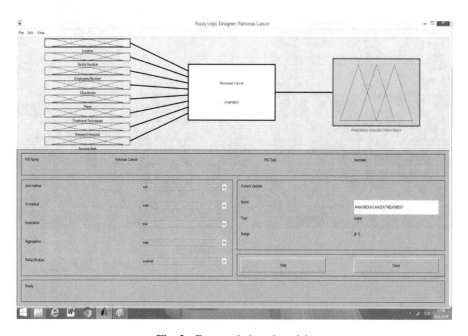

Fig. 3. Fuzzy rule-based model.

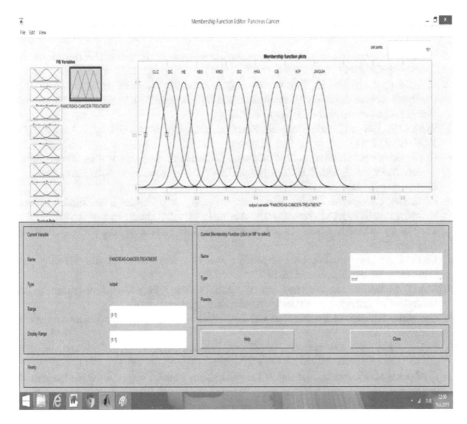

Fig. 4. Membership function of output classifier.

5 Conclusion

Until now Fuzzy Logic has been rarely applied in 'health-care organizations' like hospitals. A group of my students conducting a project on pancreatic cancer selected the best ten hospitals in EU countries and provided the related hospital profile to be used as input data in fuzzy inference technique. The Mamdani model employed in the fuzzy rule-based selection model with 8 variables and 6 membership functions was utilized to serve as a valuable tool for the classification of those hospitals. We strongly believe that the resulting study will be a state-of-the-art in this new field and yield a most promising and trustable guide for patients who live difficulties in their decisions for selecting the hospital which will provide the most successful treatment for them, extending their life time and even ending up with a complete cure of this fatal disease. In a long term study, the number of hospitals on different types of cancer may also be included within a wide variety of locations so as to enable the patients in making the right choice for the hospital concerning their health problems.

References

1. Worldwide cancer data. https://www.wcrf.org/dietandcancer/cancer-trends/worldwide-cancer-data. Accessed 2019
2. McGuigan, A., Kelly, P., Turkington, R.C., Jones, C., Coleman, H.G., McCain, R.S.: Pancreatic cancer: a review of clinical diagnosis, epidemiology, treatment and outcomes. World J. Gastroenterol. **24**(43), 4846 (2018)
3. Ryan, D.P., Hong, T.S., Bardeesy, N.: Pancreatic adenocarcinoma. N. Engl. J. Med. **371**(11), 1039–1049 (2014)
4. Hope European Hospital and Healthcare Federation. http://www.hope.be/wp-content/uploads/2015/11/79_2009_OTHER_Hospitals-in-27-Member-States-of-the-European-Union-eng.pdf. Accessed 2019
5. Hirslanden Clinique La Colline, Hirslanden Hospital group website. https://hirslanden.com/en/international/clinics/kliniken/hirslanden-clinique-la-colline.html. Accessed 2019
6. The history of Double Check. Double check hospital website https://doublecheck.ch/about/. Accessed 2019
7. UKE: University of Hamburg-Eppendorf website. https://www.uke.de/english/general/about-us/index.html. Accessed 2019
8. Klinik Beau-Site: Hirslanden Hospital group website. https://www.hirslanden.com/en/international/clinics/kliniken/klinik-beau-site.html. Accessed 2019
9. University Hospital rechts der Isar, website. https://www.mri.tum.de/international-patients. Accessed 2019
10. Patient Care, Surgeon website. http://www.myswisssurgeon.com/patient-care. Accessed 2019
11. Hirslanden Klinik Aarau, Hirslanden Hospital group website. https://www.hirslanden.com/en/international/clinics/kliniken/klinik-aarau.html. Accessed 2019
12. Charite Universitatsmedizin Berlin, Charite website. https://www.charite.de/en/. Accessed 2019
13. Hirslanden Klinik Im park. https://www.hirslanden.ch/en/klinik-im-park/home.html. Accessed 2019
14. Goethe Universitat Frankfurt. http://www.goethe-university-frankfurt.de/62971597/Doctors___hospitals. Accessed 2019
15. Jantzen, J.: Tutorial on fuzzy Logic. Technical report. Technical University of Denmark, Oersted-DTU, Automation, Bldg 326, 2800 Kongens Lyngby (No. 98-E, p. 868), Denmark (2008)
16. Sandya, H.B., Hemanth Kumar, P., Himanshi Bhudiraja, S.K.R.: Fuzzy rule based feature extraction and classification of time series signal. Int. J. Soft Comput. Eng. (IJSCE) **3**(2), 2231–2307 (2013)

Efficacy Evaluation of Antiretroviral Drug Combinations for HIV-1 Treatment by Using the Fuzzy PROMETHEE

Murat Sayan[1,2] , Dilber Uzun Ozsahin[2,3,8(✉)] ,
Tamer Sanlidag[2,4] , Nazife Sultanoglu[2,5] ,
Figen Sarigul Yildirim[6] , and Berna Uzun[2,7]

[1] Faculty of Medicine, Clinical Laboratory, PCR Unit, Kocaeli University,
Kocaeli, Turkey
sayanmurat@hotmail.com

[2] Research Center of Experimental Health Sciences, Near East University,
Nicosia, Cyprus
{dilber.uzunozsahin, tamer.sanlidag,
berna.uzun}@neu.edu.tr, nsultanoglu@hotmail.com

[3] Department of Biomedical Engineering, Near East University,
99138 Nicosia, Cyprus

[4] Department of Medical Microbiology, Manisa Celal Bayar University,
Manisa, Turkey

[5] Faculty of Medicine, Department of Medical Microbiology
and Clinical Microbiology, Near East University, 99138 Nicosia, Cyprus

[6] Clinical of Infectious Diseases, Health Science University,
Antalya Educational and Research Hospital, Antalya, Turkey
drfigensarigul@yahoo.com

[7] Department of Mathematics, Near East University, 99138 Nicosia, Cyprus

[8] Gordon Center for Medical Imaging, Radiology Department,
Massachusetts General Hospital and Harvard Medical School, Boston, USA

Abstract. The Human Immunodeficiency Virus (HIV) causes disease by damaging the immune system. If treatment is not initiated, the immune system collapses and this leads to Acquired Immune Deficiency Syndrome (AIDS). The drugs used in the treatment of HIV infection slow or stop the damage caused by the virus to the immune system. In this study, we analyzed the treatment options of HIV since there are many antiretroviral drug combinations available for the treatment and each combination has different properties. The variety of different combinations can cause confusion for physicians in practice. Based on this aim, we proposed the fuzzy PROMETHEE technique, a multi-criteria decision making technique based on mutual comparison of the options. The most common antiretroviral drug combinations used in the HIV treatment were evaluated and compared corresponding to their parameters by the PROMETHEE technique. According to our results, integrase-based inhibitor drug combinations were predominantly preferred. BIC + TAF/FTC (bictegravir + tenofoviralfenamide/emtricitabine) outranked the other antiretroviral drug combinations with a net flow of 0.0437, followed by DTG + ABC/3TC (dolutegravir + abacavir/lamivudine) then DTG + TAF/FTC (dolutegravir + tenofoviralfenamide/emtricitabine). The results obtained with the application of decision-making

© Springer Nature Switzerland AG 2020
R. A. Aliev et al. (Eds.): ICSCCW 2019, AISC 1095, pp. 183–189, 2020.
https://doi.org/10.1007/978-3-030-35249-3_23

theories on these option treatment methods will provide significant information for relevant patients, HIV treatment specialists and drug-makers.

Keywords: HIV treatment options · Multi criteria decision making · Preference ranking organization method for enrichment evaluations (PROMETHEE) · Fuzzy PROMETHEE

1 Introduction

The human immunodeficiency virus (HIV) is a member of the retrovirus family and consists of two genotypes, namely HIV-1 and HIV-2 [1]. HIV primarily targets and destroys the CD4+T cells of the immune cells. This alters the immune system of an infected individual and increases the risk of opportunistic infections and certain cancers, such as candidiasis of bronchi, trachea, esophagus, or lungs, cryptococcosis, tuberculosis, pneumocystis carinii pneumonia, Kaposi's sarcoma and invasive cervical cancer [2].

The normal range for CD4+T cells in healthy individuals is about 500–1,500 cells per mm^3 of blood. When the CD4+T cell count decreases below 200 cells per mm^3 of blood upon gradual destruction due to HIV infection, the disease progresses to an advanced stage called AIDS [3]. HIV transmission occurs when the body fluids of HIV infected individuals such as blood, semen, breast milk and vaginal or rectal fluids come into contact with the mucous membrane or damaged tissue of an uninfected individual. This can occur through unprotected anal or vaginal sex or via direct injection into the bloodstream, such as through needle sharing or a needle stick injury [4]. According to the World Health Organization (WHO) statistics, it is reported that at the end of 2017, 36.9 million people were living with HIV and 940 000 HIV-related deaths occurred. Although, there is no cure for the HIV infection, the management of the disease is maintained by the use of anti-retroviral treatment (ART). ART is a combination of anti-HIV drugs referred to as an HIV regimen and in 2017, 21.7 million people were receiving ART for their HIV infection [5, 6]. Currently, there are about 30 anti-HIV drug agents available for clinical use approved by Food and Drug Administration (FDA) [7].

Anti-HIV drugs target different steps of the replication cycle of HIV and they are divided into different groups based on their molecular mechanism, each targeting specific steps of the HIV replication. Currently, there are 7 groups, which are the nucleoside reverse transcriptase inhibitors (NRTIs) (i.e., abacavir, zidovudine), non-nucleoside reverse transcriptase inhibitors (NNRTIs) (i.e., efavirenz, nevirapine), protease inhibitors (PIs) (i.e., ritonavir, darunavir), fusion inhibitors (i.e., enfuvirtide), integrase inhibitors (INs) (i.e., dolutegravir, raltegravir), CCR5 antagonists (i.e., maraviroc) and post-attachment inhibitors (i.e., ibalizumab) [8]. Preferably, an ART regimen should combine three anti-HIV drugs from two different groups [6]. Monotherapy consisting of only one anti-HIV drug is not an option for treatment due to the increased risk of virologic failure and drug resistance [9, 10].

The use of ART has moved HIV infection from the category of being a deadly disease to a chronic manageable disease; significantly reduced the morbidity associated with the infection and prolonged the lifespan of HIV-infected individuals. This is achieved by maintaining the viral loads below detection levels with all available techniques and preventing the onset of AIDS [11]. Although there are many HIV treatment guidelines available for clinical practice, prescribing the most effective ART regimen in a patient-specific manner remains a complicated task [9, 12, 13]. According to many guidelines, there are three main classifications for initial combination treatment in antiretroviral-naive HIV positive patients: Recommended (previously preferred), alternative, and other regimens [12]. These management and treatment guidelines are published by various medical organizations, associations, group of experts and various authorities such as the World Health Organization, the US Centers for Disease Control and Prevention, the European AIDS Clinical Society, the British HIV Association, health ministry and departments, medical society and foundations, and medical research councils of various countries. The guidelines provide best practices and options developed from evidence-based medicine, although there are emerging issues, different expert opinions and experiences, as well as contradictory interpretations. However, the guides themselves must be updated periodically. Such diverse and numerous guidelines can cause confusion for physicians who use guidelines in practice.

In order to resolve the confusion in regard to choosing the most effective ART regimen by following guidelines, we propose the use of a multi-criteria decision making technique referred to as Fuzzy PROMETHEE to mutually compare between the selected ART regimens and to determine the most potent ART regimen amongst them. In this study, the PROMETHEE method is integrated with fuzzy logic to evaluate the HIV treatment options for physicians to develop more standardized and easy methods in their practices. The PROMETHEE method was presented in 1985 by Brans and Vincle [14] while fuzzy logic was defined by Zadeh in 1965 [15]. This hybrid method has been applied successfully for the ranking of options in fuzzy environment since the beginning of the 2000's [16, 17].

2   Methods

In this study, we have analyzed the effectiveness of the most commonly used and recommended HIV treatments options using the fuzzy PROMETHEE method. This method allowed the mutual comparison of the selected 17 adult HIV-1 ART Regimens by ranking the options. PROMETHEE I and PROMETHEE II are used for partial and complete ranking of the alternatives in a closed system, respectively. The differences in the calculated outranking of the alternatives allowed the ART Regimens to be ranked from the best to worst in terms of their effectiveness according to their outranking score. In order to perform the evaluation of the outranking, criteria were linked with their weights (relative importance) and preference functions. Preference function denoted as (P_j) evaluates the Alternatives a and $a_{t'}$ according to criteria functions. Preference for one of the alternatives over the others is represented with the values on the scale from 0 to 1. There are different types of preference function available for PROMETHEE method and software, including Usual (type I), U-shape (type II), V-shape (type III), linear (type V) and Gaussian (type VI). The basic steps of the PROMETHEE method can be followed in [18].

The PROMETHEE method is adapted to the Fuzzy PROMETHEE method, which involves the use of a fuzzy scale with fuzzy numbers. This enables process rough (fuzzy) data in the actual decision-making environment. The real numbers from 0 to 1 are the triangular fuzzy numbers. The Yager index determines the magnitude of the triangular fuzzy numbers $\tilde{F} = (N, a, b)$. This is obtained in relation to the center of the triangle with $YI = (3N - a + b)/3$. The alternatives to be ranked in this system were the adult ART regimens listed in Table 1, determined from the latest HIV treatment guidelines and experts' opinions. The criteria associated with each alternative regimen were categorized into Linguistic scale: Very high (VH), Important (H), Medium (M), Low (L) and Very low (VL), which were linked with a triangular fuzzy scale (presented in Table 2). The fuzzy number stated the relative importance of the criteria, in other words the weight, which were determined by the experts' opinions.

Table 1. List of HIV-1 adult anti-retroviral treatment regimens.

Recommended Antiretroviral combinations:

Integrase inhibitor-based regimens:
BIC + TAF/FTC (bictegravir + tenofovir alafenamide/emtricitabine)
DTG + ABC/3TC (dolutegravir + abacavir/lamivudine)
DTG + TAF/FTC (dolutegravir + tenofovir alafenamide/emtricitabine)
EVG/c + TDF/FTC (elvitegravir/cobicistat + tenofovir disoproxil fumarate/emtricitabine)
EVG/c + TAF/FTC (elvitegravir/cobicistat + tenofovir alafenamide/emtricitabine)
RAL + TAF/FTC (raltegravir + tenofovir alafenamide/emtricitabine)
Protease inhibitors-based regimens:
DRV/r + TDF/FTC (darunavir/ritonavir + tenofovir disoproxil fumarate/emtricitabine)
DRV/c + TAF/FTC (darunavir/cobicistat + tenofovir alafenamide/emtricitabine)
Non-nucleoside reverse transcriptase inhibitor-based regimens:
RPV/TDF/FTC (rilpivirine/tenofovir disoproxil fumarate/emtricitabine)
RPV/TAF/FTC (rilpivirine/tenofovir alafenamide/emtricitabine)

Alternative Antiretroviral Combinations:
Integrase inhibitor-based regimens:
RAL + ABC/3TC (raltegravir + abacavir/lamivudine)
Protease inhibitor-based regimens:
ATV/r + TAF/FTC (atazanavir/ritonavir + tenofovir alafenamide/emtricitabine)
ATV/r + ABC/3TC (atazanavir/ritonavir + abacavir/lamivudine)
DRV/r + ABC/3TC (darunavir/ritonavir + abacavir/lamivudine)
Non-nucleoside reverse transcriptase inhibitor-based regimens:
EFV + TDF/FTC (efavirenz + tenofovir disoproxil fumarate/emtricitabine)

Other Antiretroviral combinations:

RAL + DRV/r (raltegravir + darunavir/ritonavir)
LPV + DRV/r (lopinavir + darunavir/ritonavir)

Table 2. Linguistic fuzzy scale.

Linguistic scale for evaluation	Triangular fuzzy scale	Criteria
Very high (VH)	(0.75, 1, 1)	Side effects, plasma turnover, drug-drug interaction, false prescription, compliance, previously treatment, pregnancy, opportunistic infection, immunologic recovery, virologic recovery
Important (H)	(0.50, 0.75, 1)	Number of tablets, dose frequency, genetic barrier, time of suppression, age, working condition, serodiscordant couple, cancer, GFR, CD4+T cell count, coinfection, comorbidity, mental disorder
Medium (M)	(0.25, 0.50, 0.75)	Limitation, inefficient drug combination, genetic testing, bone density, member of key population, T-DRM, viral load
Low (L)	(0, 0.25, 0.50)	Cost, size of tablet, supplement, absorption, HIV genotype/subtype
Very low (VL)	(0, 0, 0.25)	Not determined

The Fuzzy PROMETHEE calculated the outranking flow in fuzzy numbers. This was followed by defuzzification by the Yager index with Gaussian preference function.

Then, the visual PROMETHEE decision lab program was used to rank adult ART regimens within the system, defining the outranking flows of net flow, positive flow and negative flow. There are many applications about Fuzzy PROMETHEE method shown in [19]. The difference between the positive and negative outranking flow results in the net flow.

3   Results

Complete ranking results obtained from the Fuzzy PROMETHEE method for the adult ART combinations with the positive, negative and net flow rankings are listed in Table 3. Based on the table, BIC + TAF/FTC outranks the other antiretroviral drug combinations with a net flow of 0.0437, followed by DTG + ABC/3TC then DTG + TAF/FTC. The least ranked antiretroviral combination is ATV/r + ABC/3TC with a net flow of −0.0279 followed very closely by DRV/c + TAF/FTC with a net flow rank of −0.0278.

Table 3. Complete ranking results for HIV-1 adult anti-retroviral treatment regimens obtained from Fuzzy PROMETHEE method.

Complete ranking	Adult anti-retroviral regimens	Net flow	Positive flow	Negative flow
1	BIC + TAF/FTC	0.0437	0.0490	0.0053
2	DTG + ABC/3TC	0.0383	0.0452	0.0069
3	DTG + TAF/FTC	0.0370	0.0434	0.0063
4	EVG/c + TAF/FTC	0.0270	0.0391	0.0121
5	EVG/c + TDF/FTC	0.0195	0.0341	0.0146
6	RAL + ABC/3TC	0.0054	0.0245	0.0191
7	RAL + TAF/FTC	0.0037	0.0229	0.0192
8	RPV/TAF/FTC	−0.0022	0.0301	0.0323
9	RAL + DRV/r	−0.0080	0.0190	0.0270
10	RPV/TDF/FTC	−0.0103	0.0253	0.0357
11	LPV + DRV/r	−0.0122	0.0158	0.0280
12	EFV + TDF/FTC	−0.0124	0.0240	0.0364
13	DRV/r + ABC/3TC	−0.0217	0.0120	0.0337
14	DRV/r + TDF/FTC	−0.0244	0.0115	0.0359
15	ATV/r + TAF/FTC	−0.0277	0.0073	0.0350
16	DRV/c + TAF/FTC	−0.0278	0.0088	0.0367
17	ATV/r + ABC/3TC	−0.0279	0.0081	0.0360

4   Conclusion

In this study, we have analyzed different adult ART regimens for HIV treatment using the fuzzy PROMETHEE method. A number of parameters and different ART combinations were compared, resulting in complete ranking of the ART regimens. According to our evaluation, it was indicated that integrase-based ART regimens were predominantly preferred.

With the results of this study, it is easier for decision makers to decide on a more favorable drug combination of treatment to embark upon based on the strengths and weaknesses of each combination. These parameters, weights and options can be updated easily as more data is obtained and according to the desires of the decision maker and the health condition of the patient. In addition, the results of the fuzzy PROMETHEE method can be used to inform treatment guidelines and provide feedback on the success of HIV treatment efforts.

References

1. Walker, B., McMichael, A.: The T-cell response to HIV. Cold Spring Harb. Perspect. Med. **2** (11) (2012)
2. CDC: Opportunistic Infections-Living with HIV-HIV Basics-HIV/AIDS-CDC. https://www.cdc.gov/hiv/basics/livingwithhiv/opportunisticinfections.html. Accessed 15 Apr 2019

3. U.S. Department of Veterans Affairs: CD4 count (or T-cell test) - HIV/AIDS. https://www.hiv.va.gov/HIV/patient/diagnosis/labs-CD4-count.asp. Accessed 15 Apr 2019
4. HIV/AIDS-CDC: HIV Transmission-HIV Basics. https://www.cdc.gov/hiv/basics/transmission.html. Accessed 16 Apr 2019
5. WHO: WHO-Data and statistics. https://www.who.int/hiv/data/en/. Accessed 16 Apr 2019
6. AIDSinfo: Antiretroviral Therapy (ART)-Definition-AIDSinfo. https://aidsinfo.nih.gov/understanding-hiv-aids/glossary/883/antiretroviral-therapy. Accessed 16 Apr 2019
7. U.S. Food and Drug Administration: HIV/AIDS - Antiretroviral drugs used in the treatment of HIV infection. https://www.fda.gov/ForPatients/Illness/HIVAIDS/ucm118915.htm. Accessed 16 Apr 2019
8. AIDSinfo: Drug Class-Definition-AIDSinfo. https://aidsinfo.nih.gov/understanding-hiv-aids/glossary/1561/drug-class. Accessed 16 Apr 2019
9. European AIDS Clinical Society: European AIDS Clinical Society Guidelines Version 9.0 (2017)
10. Michael, S., et al.: Antiretroviral drugs for treatment and prevention of HIV infection in adults. JAMA 320(4), 379–396 (2018)
11. Oguntibeju, O.: Quality of life of people living with HIV and AIDS and antiretroviral therapy. HIV AIDS (Auckl) 4, 117–124 (2012)
12. AIDSinfo: Guidelines for the Use of Antiretroviral Agents in Adults and Adolescents Living with HIV (2018)
13. British HIV Association: Clinical Guidelines. https://www.bhiva.org/Clinical-Guidelines. Accessed 16 Apr 2019
14. Brans, J.P., Vincke, P.H.: A preference ranking organisation method: the PROMETHEE method for multiple criteria decision-making. Manag. Sci. 31(6), 647–656 (1985)
15. Zadeh, L.A.: Fuzzy sets. Inf. Control 8(3), 338–353 (1965)
16. Geldermann, J., Spengler, T., Rentz, O.: Fuzzy outranking for environmental assessment case study: iron and steel making industry. Fuzzy Sets Syst. 115(1), 45–65 (2000)
17. Ozsahin, I., Uzun Ozsahin, D., Uzun, B.: Evaluation of solid-state detectors in medical imaging with fuzzy PROMETHEE. J. Instrum. 14 (2019)
18. Uzun Ozsahin, D., Ozsahin, I.: A fuzzy PROMETHEE approach for breast cancer treatment techniques. Int. J. Med. Res. Health Sci. 7(5), 29–32 (2018)
19. Uzun Ozsahin, D., et al.: Evaluating X-Ray based medical imaging devices with fuzzy preference ranking organization method for enrichment evaluations. Int. J. Adv. Comput. Sci. Appl. 9(3) (2018)

Fuzzy Estimation of Level of Country's Social Security

Gorkhmaz Imanov[1](✉), Malahat Murtuzaeva[1],
and Yadulla Hasanli[2] (iD)

[1] Control Systems of Azerbaijan National Academy of Science,
B. Vahabzadeh Street 9, AZ1141 Baku, Azerbaijan
korkmazi2000@gmail.com, malaxat-55@rambler.ru
[2] Scientific Research Institute of Economic Studies Under Azerbaijan State
University of Economics, Istiglaliyyat str., 6 (I building), AZ1001 Baku,
Azerbaijan
yadulla59@mail.ru

Abstract. In this article were analyzed indicators, such as unemployment rate, life expectancy at birth, Gini coefficient, research and development expenditures, poverty level, military expenditure, which formulate social security. On the stage of fuzzification of indicators social security was applied Gaussian membership function. In order to define level of country's social security formula fuzzy intuitionistic linguistic fuzzy number and average intuitionistic linguistic fuzzy number were used. By using statistical information for two periods – first (2006–2009) and second (2014–2016) years of Azerbaijan defined fuzzy aggregate index social security. Based on the calculation by formulas of average fuzzy linguistic number define results for two periods. Results of calculation level of social security in the second period were bigger than in the first period. Main indicators such as, that had an impaction quality of social security, were improving Poverty level, Gini coefficient and results gave recommendation for improving level Research & Development expenditures, level of Unemployment.

Keywords: Intuitionistic fuzzy sets · Intuitionistic linguistic fuzzy number · Aggregated index · Mean of collection of ILFN-s · Social security

1 Introduction

Social security is one of the most important directions of economic security. According to Huber et al. economic security consists of the following components: ensuring deserving conditions for living and developing the identity; the political, military capability of the society and the country and the socio-economic stability in order to eliminate internal and external threats [1]. In order to define level of country's social security econometric models were used. In this paper was proposed method on the base of intuitionistic fuzzy linguistic theory [2] to define level of country's social security.

In order to estimate the level of country's social security the following indicators, which influenced on its level were analyzed:

© Springer Nature Switzerland AG 2020
R. A. Aliev et al. (Eds.): ICSCCW 2019, AISC 1095, pp. 190–196, 2020.
https://doi.org/10.1007/978-3-030-35249-3_24

- **Unemployment rate (% of total labor force) – UNE:** Unused resources and the economy's spare capacity indicate unemployment. As is known, unemployment tends to be cyclical and decreases when the economy expands as companies hire more workers to meet growing demand, and increases when economic activity slows. Increasing the level of unemployment is already a first state, connected with the imperfection of the mechanisms of regulation and self-regulation of the economic system. The high level of unemployment is one of the big acute problems of global and national proportion [3];
- **Life expectancy at birth (years) – LEB:** Average number of years that a newborn is expected to live if current mortality rates do not change. High life expectancy and its quality indicate a stable level of social security. Low one is about dangerous state of social tension. Social tension manifests itself in a number of most significant symptoms (a significant increase in discontent among the population, mistrust of the authorities, conflict in society, anxiety, stressfulness of relations) and is determined by the influence of man-made, natural and social factors [4];
- **Gini coefficient (1-100) – GIN:** The Gini coefficient is a measure of economic inequality of households. The coefficient measures the variance of income or distribution of wealth among a population. At the same time, inequality is not dangerous by itself, it is dangerous to achieve a certain critical value, which may pose a threat to the country. It is for this reason that the Gini coefficient determines not only the level of social stratification, but also the level of social and political stability [5];
- **Research and development expenditures (% of GDP) – R&D:** this consists of the total (current and capital) expenditure on R&D carried out by resident companies, universities, research institutes, etc. It includes R&D financed from abroad but excludes residents' funds for R&D performed outside the domestic economy. R&D expenditures is measured as percentage of GDP [6];
- **Poverty level (%) – POV:** The situation in which, taking into account the country's economic and social circumstances, the person's minimum basic needs cannot be met and will be below the monetary threshold, will be called the poverty line for the country. As countries develop, the poverty line often rises: in rich countries it is higher than in poorer countries. In order to track the level of development, the number of people living below the poverty line is monitored by the governments. The poverty line is also one of the important indicator for SDG 1, ending poverty "in all its types." [7].

Military Expenditure (% of GDP) – MIL: There are two ways to measure military spending: (1) spending in real terms and (2) as a percentage of GDP. Military expenditure in real terms is important as the nominal level of expenditure impacts on the result of war. Country's military expenditure is formed by the whether country is at war or not. If the country is not in a state of war, part of the budget is spent on supporting military power. All government expenditures include opportunity costs of priorities and objectives. Military conflicts increase military expenditure that is a matter for concern among the population about the opportunity cost in terms of spending on human, economic and social development [8]. To define level of social security in many investigations were used econometric instruments. In this paper was proposed intuitionistic linguistic fuzzy approach to the challenge.

2 Fuzzy Linguistic Approach to Assess Level of Social Security

In order to assess level of social security of country were used statistical information of Azerbaijan for two periods – 2007–2009 and 2014–2016 years. Main instruments of estimation aggregate index of social security were intuitionistic fuzzy linguistic numbers. On the basis of Atanassov's researches [2] intuitionistic fuzzy set (IFS) Wang and Li proposed a linguistic intuitionistic fuzzy set [9].

$$A = \left\{ \langle x, \left[S_{\theta(x)}, \mu_A(x), v_A(x) \right] \rangle | x \in X \right\} \tag{1}$$

where $S_{\theta(x)} \in S$, $\mu_A : X \to [0, 1]$ and $v_A : X \to [0, 1]$, that satisfies the condition $\mu_A(x) + v_A(x) \leq 1$, $\mu_A(x)$ and $v_A(x)$, represent the membership and non-membership degrees, respectively, of elements x to the linguistic value $S_{\theta(x)}$.

For each intuitionistic linguistic set $A = \left\{ \langle x, \left[S_{\theta(x)}, \mu_A(x), v_A(x) \right] \rangle | x \in X \right\}$, there is $\pi_A(x) = 1 - \mu_A(x) - v_A(x)$, which is called the fuzzy intuitionistic index of the element x of the linguistic variable $S_{\theta(x)}$.

For the intuitionistic linguistic set $A = \left\{ \langle x, \left[S_{\theta(x)}, \mu_A(x), v_A(x) \right] \rangle | x \in X \right\}$, $\left(S_{\theta(x)}, (\mu_A(x), v_A(x)) \right)$ triple is called an intuitionistic linguistic fuzzy number.

For fuzzification indicators of social security were used Gaussian membership function (Eq. 2).

Table 1. Indicators of social security

Indicators	2007	2008	2009	2014	2015	2016
UNE	6.3	5.9	5.7	4.9	5.0	5.0
LEB	73.0	73.4	73.5	74.2	75.2	75.2
GIN	31.8	31.8	33.12	25.41	28.78	28.87
R&D	1.7	1.7	0.25	0.2	0.2	0.2
POV	15.8	13.2	10.9	5.0	4.9	5.9
MIL	2.864	3.291	3.325	4.555	5.468	3.69

Taking into account the four bands, the methodology of the Sustainable Development Goals Index and dashboards were used to determine the threshold of linguistic indicators. The maximum (i.e., the upper limit) indicated by the green band is reached for each variable. Three colored bands from yellow to orange and red indicate a growing distance from the target. The yellow-orange thresholds for all indicators were set as the value between the red and green thresholds, which demonstrated in Table 2.

Table 2. Supports linguistic fuzzy variables

Indicators	Green	Yellow	Orange	Red
1. UNE	(0.5; 5.1)	(4.9; 7.65)	(7.35; 10.2)	(9.8; 25.9)
2. LEB	(63.7; 77)	(61.25; 66.3)	(56.8; 62.5)	(46.1; 61.2)
3. GIN	(27.5; 30)	(30; 35)	(35; 40)	(40; 63)
4. R&D	(1.47; 3.7)	(1.274; 1.53)	(0.98; 1.326)	(0; 1.02)
5. POV	(0.5; 5.33)	(5.22; 10.93)	(10.51; 16.64)	(15.98; 25)
6. MIL	(0.54; 4.53)	(4.356; 8.52)	(8.18; 12.5)	(12.01; 16.16)

Information that was taken into account between green (best) and red (waste), which were demonstrated in Table 2, defined linguistic index:

$$Green - S_G = 3; \quad Yellow - S_Y = 2; \quad Orange - S_O = 1; \quad Red - S_R = 0$$

Gaussian membership functions are one of the popular methods for specifying fuzzy sets due to their smoothness and brief notation. The advantage of these curves is that they are smooth and nonzero at all points. Since Gaussian membership function was poor in response in all cases this proves that it is better to use the theory Gaussian membership function in data on probabilities and statistics [9]. Intuitionistic fuzzy Gaussian membership $\mu(x)$ and non-membership $v(x)$ functions are defined as [10, 11].

$$\mu(x) = e^{-\frac{1}{2}*\left(\frac{x-\bar{x}}{\sigma}\right)^2} - \varepsilon$$

$$v(x) = 1 - e^{-\frac{1}{2}*\left(\frac{x-\bar{x}}{\sigma}\right)^2} \qquad (2)$$

where $\bar{x}$ – central value and width $\sigma > 0$.

For every indicators were defined parameters ε: Unemployment rate (UR) - 0.05; Life expectancy at birth (LEB) - 0.06; Research and development expenditure (R&D) - 0.06; Poverty level (PL) - 0.06; Military expenditure (ME) - 0.09. Results of computation membership and non-membership function indicator, demonstrated in Table 3.

Table 3. Parameters Intuitionistic fuzzy linguistic numbers

Indicators	θ	μ	ν	π	θ	μ	ν	π	θ	μ	ν	π
Years	2007				2008				2009			
UNE	2	0.95	0.01	0.05	2	0.91	0.05	0.05	2	0.82	0.13	0.05
LEB	3	0.78	0.16	0.06	3	0.58	0.36	0.06	3	0.53	0.41	0.06
GIN	2	0.63	0.31	0.06	2	0.63	0.31	0.06	2	0.67	0.27	0.06
R&D	3	0.76	0.18	0.06	3	0.76	0.18	0.06	0	0.44	0.5	0.06
POV	1	0.43	0.51	0.06	1	0.85	0.09	0.06	2	0.196	0.744	0.06
MIL	3	0.88	0.04	0.09	3	0.67	0.24	0.09	3	0.65	0.26	0.09
Years	2014				2015				2016			
UNE	2	0.37	0.59	0.05	2	0.42	0.53	0.05	2	0.42	0.53	0.05
LEB	3	0.26	0.68	0.06	3	0.05	0.89	0.06	3	0.05	0.89	0.06
GIN	3	0.86	0.08	0.06	3	0.6	0.35	0.06	3	0.545	0.4	0.06
R&D	0	0.42	0.52	0.06	0	0.42	0.52	0.06	0	0.42	0.52	0.06
POV	3	0.77	0.17	0.06	3	0.83	0.11	0.06	2	0.47	0.48	0.06
MIL	2	0.45	0.46	0.09	2	0.94	0.03	0.24	3	0.35	0.56	0.09

Then on the base [12] following formulas were defined fuzzy weights of every indicators social security.

$$\lambda_k = \frac{\left(\mu_k + \pi_k\left(\frac{\mu_k}{\nu_k}\right)\right)}{\sum_{k=1}^{6}\left(\mu_k + \pi_k\left(\frac{\mu_k}{\nu_k}\right)\right)} \tag{3}$$

$$\sum_{i=1}^{6} \lambda_k = 1$$

Results of computation demonstrated in Table 4.

Table 4. Fuzzy weight of indicators social security

Indicators	2007	2008	2009	2014	2015	2016
UNE	0.551808	0.472997	0.436142	0.047559	0.043372	0.197729
LEB	0.062183	0.048578	0.106406	0.03133	0.003766	0.01717
GIN	0.02919	0.058958	0.186658	0.537766	0.08228	0.197027
R&D	0.054207	0.109487	0.077723	0.058185	0.042916	0.195648
POV	0.01429	0.231691	0.00564	0.256475	0.30277	0.233497
MIL	0.288323	0.078289	0.18743	0.068686	0.524896	0.15893
Sum	1	1	1	1	1	1

By using following formulas of intuitionistic linguistic weighted average were computed fuzzy aggregate index social security (**FASS**) for two periods.

$$FASS = \left\langle S_{\sum_{k=1}^{6} \lambda_k * \vartheta\left(a_{ij}^k\right)}, \left(1 - \prod_{k=1}^{6}\left(1 - \mu\left(a_{ij}^k\right)\right)^{\lambda_k}, \prod_{k=1}^{6}\left(v\left(a_{ij}^k\right)\right)^{\lambda_k}\right)\right\rangle \quad (4)$$

$$FASS(2007) = \langle 2.39; 0.937; 0.034; 0.029\rangle$$

$$FASS(2008) = \langle 2.005; 0.85; 0.087; 0.06\rangle$$

$$FASS(2009) = \langle 2.138; 0.72; 0.21; 0.06\rangle$$

$$FASS(2014) = \langle 2.71; 0.78; 0.146; 0.07\rangle$$

$$FASS(2015) = \langle 2.172; 0.875; 0.085; 0.04\rangle$$

$$FASS(2016) = \langle 1.98; 0.4; 0.49; 0.1\rangle$$

According to [6] if $\tilde{\alpha}_j = \left\langle S_{\theta(\tilde{\alpha}_j)}, \left(u(\tilde{\alpha}_j), v(\tilde{\alpha}_j)\right)\right\rangle$ $(j = 1, 2, \ldots n)$ be a collection of the Intuitionistic Linguistic Numbers (ILN), then defines the mean of these ILNs as

$$\tilde{m}_j = \left\langle S_{\theta(\tilde{m}_j)}, \left(u(\tilde{m}_j), v(\tilde{m}_j)\right)\right\rangle \quad (5)$$

where $\tilde{m}_1$ – *mean of FASS for* 2007–2009 *(first period)*, $\tilde{m}_2$ – 2014–2016 years (second period): [13]

$$\theta(\tilde{x}) = \frac{1}{n}\sum \theta(\tilde{\alpha}_i) \quad u(\tilde{x}) = 1 - \left(\prod\left(1 - u\left(a_j\right)\right)\right)^{\frac{1}{n}} \quad v(\tilde{x}) = \left(\prod v\left(a_j\right)\right)^{\frac{1}{n}}$$

On the base of formulas noticed above were defining $\tilde{m}_1$ and $\tilde{m}_2$:

$$\tilde{m}_1 = \langle 2.1778; 0.862; 0.086\rangle; \tilde{m}_2 = \langle 2.287; 0.745; 0.183\rangle$$

3   Results of Analysis

As is shown from results of calculation level social security in the second period have improved and was more than in the first period. Meaning of indicators of poverty level mainly have influenced on the situation, which in the first period was red - $S_r = 1, 3$ and in the second period it improved to yellow - $S_y = 3$. The value of Gin coefficients in the first period were orange - $S_o = 2$, and in the second period improved to yellow - $S_y = 3$. Level of Research & Development expenditures decrease from green - $S_g = 3$ to red - $S_r = 0$. Level of Unemployment does not change in both periods and equal to yellow - $S_e = 2$. Life expectancy at birth in both periods is equal to S_g – green.

4   Conclusion

Application intuitionistic fuzzy linguistic set instrument to analyses of development social indicators give us possibility to define not only quantity volume and also quality level of social security for two periods. It defined level of indicators, which has influenced to quality of social security. Results of analyses give possibility to person, which makes decisions in socioeconomic development to do some corrects in management process.

References

1. Huber, G., Rehm, P., Schlesinger, M., Valletta, R.: Economic Security at risk: findings from the economic security index. Rockefeller Foundation, Yale University (2010)
2. Atanassov, K.: Intuitionistic fuzzy sets. Fuzzy Sets Syst. **20**(1), 87–96 (1986)
3. Focus Economics. https://www.focus-economics.com/economic-indicator/unemployment-rate
4. World Health Organization. https://www.who.int/whosis/whostat2006DefinitionsAndMetadata.pdf
5. Corporate Finance Institute. https://corporatefinanceinstitute.com/resources/knowledge/economics/gini-coefficient/
6. Gross domestic spending on R&D (indicator) (2019). https://doi.org/10.1787/d8b068b4-en
7. Poverty and Inequality. http://datatopics.worldbank.org/world-development-indicators/themes/poverty-and-inequality.html
8. Military versus social expenditure: The Opportunity Cost of World Military Spending, Stockholm International Peace Research Institute (SIPRI), Media backgrounder, 5 April 2016
9. Wang, J.Q., Li, H.B.: Multi-criteria decision-making method based on aggregation operators for intuitionistic linguistic fuzzy numbers. Control Decis. **25**, 1571–1574 (2010)
10. Sadollah, A.: Which membership function is appropriate in fuzzy system?, Chap. 1. In: Fuzzy Logic Based in Optimization Methods and Control Systems and Its Applications (2018)
11. Radhika, C., Parvathi, R.: Intuitionistic fuzzification functions. Global J. Pure Appl. Math. **12**(2), 1211–1227 (2016)
12. Boran, F.E., Genc, S., Kurt, M., Akay, D.: A multi-criteria intuitionistic fuzzy group decision making for supplier with TOPSIS method. Expert Syst. Appl. **36**(8), 11363–11368 (2009)
13. Liu, P.: Some generalized dependent aggregation operators with intuitionistic linguistic numbers and their application to group decision making. J. Comput. Syst. Sci. **79**(1), 131–143 (2013)

Artificial Intelligence Based and Linear Conventional Techniques for Reference Evapotranspiration Modeling

Jazuli Abdullahi[1]($\boxtimes$) (iD), Gozen Elkiran[1], and Vahid Nourani[2]

[1] Department of Civil Engineering, Faculty of Civil and Environmental Engineering, Near East University, Nicosia, Cyprus via Mersin 10, Turkey
jazuli.abdullahi@neu.edu.tr
[2] Department of Water Resources Engineering, Faculty of Civil Engineering, University of Tabriz, Tabriz, Iran
vnourani@yahoo.com

Abstract. This study was aimed at investigating the potentials of Artificial Neural Network (ANN), Support Vector Regression (SVR) and Multiple Linear Regression (MLR) techniques for the estimation of reference evapotranspiration (ET_0) in a semiarid station of Nigeria. To do so, 34 years daily monthly average data including maximum, minimum and mean temperatures (T_{max}, T_{min}, and T_{mean}), relative humidity (R_H) and wind speed (U_2) were used as input parameters. Three models were developed from each technique using three different input combinations. FAO Penman Monteith method was used as the basis upon which the performances of the models were assessed. The results revealed that models developed using T_{min}, T_{max} and U_2 produced better performance. The results also depicted that with the unique capability of each technique, different results would be obtained, both ANN and SVR models could lead to efficient and reliable results, but MLR model could not produce reliable performance due to its inability to deal with nonlinear aspect of ET_0.

Keywords: Reference evapotranspiration · Penman Monteith · Nigeria

1 Introduction

Evapotranspiration (ET) refers to the transfer of water from the surface of the earth to the atmosphere through evaporation from surfaces of wet plant, water and soil, and through plant stomata by transpiration [1]. The reference surface of reference evapotranspiration or reference crop evapotranspiration (ET_0) is a hypothetical glass reference crop having an estimated crop depth of 0.12 m, 70 sm^{-1} of a fixed surface resistant and 0.23 albedo [2]. Sequel to May 1990 expert consultation meeting, FAO Penman Monteith method (called FAO-56-PM) was suggested as the sole and the standard method of calculating ET_0 [2].

Although physical-based and conceptual models are the reliable tools for investigating the actual physics of the phenomenon, they have practical limitations, and when accurate predictions are more important than the physical understanding, employing black box models can be more useful. Multiple Linear Regression (MLR) is a classic

© Springer Nature Switzerland AG 2020
R. A. Aliev et al. (Eds.): ICSCCW 2019, AISC 1095, pp. 197–204, 2020.
https://doi.org/10.1007/978-3-030-35249-3_25

black box tool for modeling linear relationship between dependent and one or more independent variables [3]. In recent years, application of Artificial Intelligence (AI) (for examples, Artificial Neural Network (ANN), Support Vector Regression (SVR) etc.) are widely adopted, through which several research papers were published. [4] applied ANN to model ET_0 and compared the results with FAO-56-PM. [5] employed ANN and empirical equations for daily ET_0 estimation using limited input climate variables. For SVR models (based on the Support Vector Machine, SVM, concept), [6] modeled daily ET_0 in a semiarid mountain area by integrating genetic algorithm and SVM. Using limited meteorological data, [7] evaluated the performances of SVM, ELM and four tree-based ensemble models for predicting daily ET_0. To the best knowledge of the authors, no study was conducted by applying AI and MLR models for ET_0 modeling in Nigeria. Therefore, the aim of this study was to investigate the potentials of ANN, SVR and MLR models for ET_0 prediction in Maiduguri, a semiarid climate station of Nigeria.

2   Materials and Method

2.1   Study Location and Data

Maiduguri is the capital city of Borno state, situated in the northeastern Nigeria at 11° 51' N latitude, 13° 05' E longitude and 354 m altitude. The climate of the station is semiarid with light annual rainfall of around 300 mm. Maiduguri contains savannah or tropical grasslands vegetation, it has daily mean temperature of around 22–35 °C but often rise above 40 °C between March and June [8]. Figure 1 shows the location of the study.

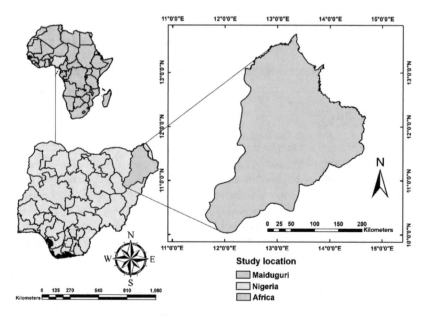

Fig. 1. Study location.

The minimum, maximum and mean temperatures (T_{min}, T_{max}, T_{mean}) (°C), relative humidity (R_H) (%), and wind speed (U_2) (m/s) data used in this study were obtained from National Aeronautics and Space Administration (NASA) (www.nasa.gov). The NASA data for U_2 were obtained at 10 m above the earth surface. The U_2 data were converted to 2 m above the earth surface using [2] equation.

Thirty four (34) years daily monthly average data were used in this study covering a period from 1983–2016 (408 number of observations). The data were divided into 75% (306 observations) for training and 25% (102 observations) for testing. Before ET_0 modeling, the data were normalized between 0 and 1 for more integrity and to reduce redundancy using [9]:

$$R_{norm} = \frac{R_i - R_{min}}{R_{max} - R_{min}} \tag{1}$$

where R_{norm} is the normalized value, R_i is the observed value, R_{max} and R_{min} are the maximum and minimum data values, respectively.

To analyze and determine the performance and efficiency of the models proposed, [10] research was endorsed, which stated that Determination Coefficient (DC or Nash-Sutcliffe efficiency criterion Eq. 2) and Root Mean Square Error (RMSE Eq. 3) can sufficiently evaluate any hydro-climatic model.

$$DC = 1 - \frac{\sum_{i=1}^{N}\left(R_i - \hat{R}_i\right)^2}{\sum_{i=1}^{N}\left(R_i - \overline{R}\right)^2} \tag{2}$$

$$RMSE = \sqrt{\frac{\sum_{i=1}^{N}(R_i - \hat{R}_i)^2}{N}} \tag{3}$$

where R_i, was defined in Eq. 1, N, $\overline{R}$, and $\hat{R}_i$ are respectively the number of observations, mean of the observed values, and predicted values. The best performance is achieved with DC and RMSE close to 1 and 0, respectively.

2.2   Proposed Methodology

In this study, two AI (ANN and SVR) and a linear conventional (MLR) models were developed to model ET_0 in a semiarid climate station of Nigeria. FAO-56-PM regarded as the sole and standard method for ET_0 estimation was used as the referenced ET_0 through which the performances of the applied techniques were assessed. Three models were developed from each technique using 3 different input combinations, given by:

$$M1: ET_0 = f(T_{min}, T_{max}, T_{mean}) \tag{4}$$

$$M2: ET_0 = f(T_{min}, T_{max}, R_H) \tag{5}$$

$$M3: ET_0 = f(T_{min}, T_{max}, U_2) \tag{6}$$

where $T_{min}, T_{max}, T_{mean}, R_H$ and U_2 were already defined. The results from each model were obtained and presented. The proposed methodology in this study is presented in Fig. 2.

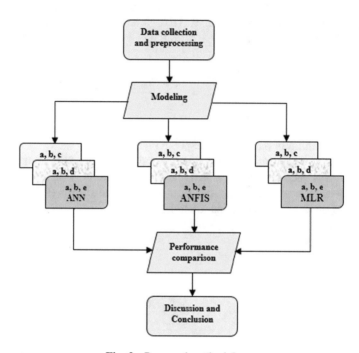

Fig. 2. Proposed methodology.

In Fig. 2, a, b, c, d and e represent $T_{min}, T_{max}, T_{mean}, R_H$ and U_2 used for 3 input combinations for the 3 developed models.

2.3   FAO Penman Monteith (FAO-56-PM) Method

[2] described FAO-56-PM method as the sole standard method for computing ET_0, as such was used in this study and the equation is given by:

$$ET_0 = \frac{0.408\Delta(R_n - G) + \gamma \frac{900}{T+273} U_2(e_s - e_a)}{\Delta + \gamma(1 + 0.34U_2)} \tag{7}$$

where ET_0 is the reference evapotranspiration (mm/day), Δ is slope vapor pressure curve (kpa/°C), R_n is net radiation at the crop surface (MJ/m²/day), G is soil heat flux density (MJ/m²/day), T is mean daily air temperature at 2 m height (°C), U_2 is wind speed at 2 m height (m/s), e_s is saturation vapor pressure (kpa), e_a is actual vapor pressure (kpa), $e_s - e_a$ is saturation vapour pressure deficit (kpa), γ is psychrometric constant (kpa/°C).

2.4   Artificial Neural Network (ANN)

ANN is an AI based approach that uses input and output datasets to simulate system altitude. It comprises of different number of neurons (nodes) processing elements which are interconnected with information processing characteristics of an adorable attribute including nonlinearity, learning, generalization capability and noise tolerance. The most widely applied ANN approach nowadays is Feed Forward Neural Network (FFNN) trained with Back Propagation (BP) algorithm [11]. Figure 3 shows the FFNN architecture.

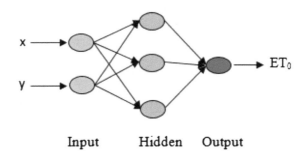

Input   Hidden   Output

Fig. 3.   The FFNN structure.

2.5   Support Vector Regression (SVR)

SVR was developed based on support vector machine (SVM). First introduced by [12] is an AI based model capable of providing satisfactory solutions to the problems of prediction, classification, regression and pattern recognitions [13]. In SVR modeling, instead of minimizing the errors between computed and observed values, the operational risk is minimized as the objective function. To catch the nonlinear data pattern, the output goes through a nonlinear kernel after linear regression is first fitted on the data.

2.6   Multiple Linear Regression (MLR)

In general, for multi-linear regression (MLR), the dependent variable y, and n regressor variables may be related by [9]:

$$y = b_0 + b_1x_1 + b_2x_2 + b_3x_3 + \ldots + b_ix_i + \xi \qquad (8)$$

where x_i is the value of the ith predictor, b_0 is the regression constant, and b_i is the coefficient of the ith predictor and ξ is the error term.

3   Results and Discussion

In this section, the results obtained by two AI based (ANN, and SVR) and one conventional (MLR) techniques using 3 sets of input combinations are presented for ET_0 estimation for Maiduguri station, Nigeria.

The ANN model was trained using Levenberg Marquardt algorithm with single hidden layer and different number of hidden neurons. The best nodes number that provides better performance was determine using trial and error procedure. Radial basis function (RBF) kernel was used for model creation for SVR technique. For better performance, reliability and efficiency achievement, the RBF kernel's parameters in SVR were tuned. Table 1 shows the results obtained from each technique for the 3 developed models. In the Table, x-y-z represent number of input parameters, hidden layer neurons and number of output for ANN models, RBF is the kernel's parameters used in SVR model construction and x-y are the number of inputs and output for MLR models.

Table 1. Results of the developed models.

Model	Model no.	Structure	Training		Testing	
			DC	RMSE	DC	RMSE
ANN	M1	3-9-1	0.903	0.078	0.857	0.073
	M2	3-12-1	0.926	0.068	0.920	0.067
	M3	3-8-1	0.946	0.045	0.929	0.055
SVR	M1	RBF	0.840	0.100	0.776	0.091
	M2	RBF	0.866	0.071	0.823	0.105
	M3	RBF	0.950	0.043	0.793	0.114
MLR	M1	3-1	0.471	0.182	0.140	0.179
	M2	3-1	0.545	0.169	0.166	0.177
	M3	3-1	0.639	0.150	0.369	0.153

As seen in Table 1 for ANN models in terms of higher DC and lower RMSE, M3 (T_{min}, T_{max} and U_2) performed better than M1 (which has inputs T_{min}, T_{max} and T_{mean}) and M2 (which has inputs, T_{min}, T_{max} and R_H) in both the training and testing steps. Though as semiarid climate, Maiduguri station is characterized by high temperatures (T_{min}, T_{max} and T_{mean}), which can have significant effect on ET_0, but this results indicate that combining different factors that affect ET_0 would be more significant than dwelling on a sole factor for ET_0 modeling. In addition, R_H being the ratio of the partial pressure to equilibrium vapor pressure of water at a given temperature implies that as the temperature increases the amount of water vapor needed for saturation increases and hence, ET_0 increases. As such, ET_0 modeling would be more effective when temperatures and R_H are combined. However, Maiduguri constitutes Sahara desert, which results in wind blowing from one direction to another, this increases the U_2 of the station. According to [14], though U_2 alone might not have much effect in the estimation of climatological parameters, but its inclusion with combination of other parameters, better estimations could be achieved. This could be why M3 with inclusion of U_2 produced the best results among the ANN models.

The performances of the models for SVR are similar to that of ANN technique; the only exception is in the testing step whereby M2 performed better than M3 and M1. This is due to unique nature of each model as different approaches were involved in solving the difficult behavior of ET_0. MLR has the least performance among the applied techniques. It underestimates ET_0 especially in the testing step where it has the lowest DC (0.140) and highest RMSE (0.179) for M1. The poor performance could be because; (i) the data in nature contain both linear and nonlinear phenomena, MLR model been developed to deal with linear aspect of the system may not have good performance, (ii) the data is more nonlinear in nature. The general results show that ANN technique produced better prediction performance than SVR and MLR techniques due to its ability to deal with uncertain behavior of ET_0. Figure 4 shows the performances of all the developed models in term of time series in the testing step for SVR technique.

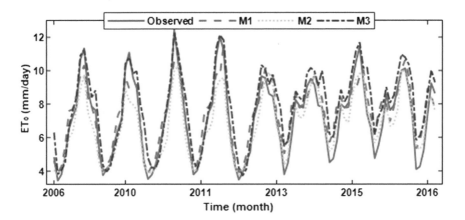

Fig. 4. Time series of the observed and predicted models for SVR technique in the testing step.

4 Conclusion

In this study, two AI based (ANN and SVR) and MLR models were applied for ET_0 modeling in Maiduguri, a semiarid climate station of Nigeria. For this purpose, 3 models were developed from each technique using 3 different kinds of input combinations to determine the optimum performance. FAO-56-PM being the standard method for ET_0 estimation was used as the referenced ET_0 upon which the performance of the other models were assessed. DC and RMSE were used as performance evaluation criteria.

The results showed that both ANN and SVR models produced reliable performances and could be employed successfully for modeling ET_0 in the study station. Meanwhile, MLR models underestimated ET_0 due to its inability to cope with nonlinear aspect of ET_0. The results also revealed that all the 3 input combinations models could be used efficiently for ET_0 modeling but M3 which has inputs combination of T_{min}, T_{max} and U_2 produced the best results in comparison to M1 and M2.

References

1. Odhiambo, L.O., Yoder, R.E., Yoder, D.C.: Estimation of reference crop evapotranspiration using fuzzy state models. Trans. ASAE **44**, 543–550 (2001)
2. Allen, R.G., Pereira, L.S., Raes, D., Smith, M.: Crop evapotranspiration-guidelines for computing crop water requirements. FAO Irrigation and drainage paper 56. FAO, Rome, 300 (9) (1998)
3. Ladlani, I., Houichi, L., Djemili, L., Heddam, S., Belouz, K.: Estimation of daily reference evapotranspiration (ET_0) in the North of Algeria using adaptive neuro-fuzzy inference system (ANFIS) and multiple linear regression (MLR) models: a comparative study. Arab. J. Sci. Eng. **39**, 5959–5969 (2014). https://doi.org/10.1007/s13369-014-1151-2
4. Abdullahi, J., Elkiran, G.: Prediction of the future impact of climate change on reference evapotranspiration in Cyprus using artificial neural network. In: 9th International Conference on Theory and Application of Soft Computing, Computing with Words and Perception, pp. 276–283. Elsevier, London (2017). https://doi.org/10.1016/j.procs.2017.11.239
5. Antonopoulos, V.Z., Antonopoulos, A.V.: Daily reference evapotranspiration estimates by artificial neural networks technique and empirical equations using limited input climate variables. Comput. Electron. Agric. **132**, 86–96 (2017). https://doi.org/10.1016/j.compag.2016.11.011
6. Yin, Z., Wen, X., Feng, Q., He, Z., Zou, S., Yang, L.: Integrating genetic algorithm and support vector machine for modeling daily reference evapotranspiration in a semi-arid mountain area. Hydrol. Res. **48**, 1177–1191 (2017). https://doi.org/10.2166/nh.2016.205
7. Fan, J., Yue, W., Wu, L., Zhang, F., Cai, H., Wang, X., Xiang, Y.: Evaluation of SVM, ELM and four tree-based ensemble models for predicting daily reference evapotranspiration using limited meteorological data in different climates of China. Agric. For. Meteorol. **263**, 225–241 (2018). https://doi.org/10.1016/j.agrformet.2018.08.019
8. Arku, A.Y., Musa, S.M., Mofoke, A.L.E.: Determination of water requirement and irrigation timing for Amaranthus hybridus in Maiduguri metropolis, north-eastern Nigeria. In: Fourth International Conference on Sustainable Irrigation, pp. 279–289. WIT Press, England (2012). https://doi.org/10.2495/si120241
9. Elkiran, G., Nourani, V., Abba, S.I., Abdullahi, J.: Artificial intelligence-based approaches for multi-station modelling of dissolve oxygen in river. Global J. Environ. Sci. Manag. **4**, 439–450 (2018). https://doi.org/10.22034/gjesm.2018.04.005
10. Legates, D.R., McCabe Jr., G.J.: Evaluating the use of "goodness-of-fit" measures in hydrologic and hydroclimatic model validation. Water Resour. Res. **35**, 233–241 (1999)
11. Abdullahi, J., Elkiran, G., Nourani, V.: Application of Artificial Neural Network to predict reference evapotranspiration in Famagusta, North Cyprus. In: 11th International Scientific Conference on Production Engineering Development and Modernization of Production, pp. 549–554 (2017)
12. Cortes, C., Vapnik, V.: Support-vector networks. Mach. Learn. **20**, 273–297 (1995)
13. Nourani, V., Elkiran, G., Abba, S.I.: Wastewater treatment plant performance analysis using artificial intelligence–an ensemble approach. Water Sci. Technol. **78**, 2064–2076 (2018). https://doi.org/10.2166/wst.2018.477
14. Nourani, V., Elkiran, G., Abdullahi, J., Tahsin, A.: Multi-region modeling of daily global solar radiation with artificial intelligence ensemble. Nat. Resour. Res. **28**, 1217–1238 (2019). https://doi.org/10.1007/s11053-018-09450-9

Z-Value Based Risk Assessment: The Case of Tourism Sector

A. M. Nuriyev and K. Jabbarova^(✉)

Azerbaijan State Oil and Industry University, Azadliq Avenue, 20,
AZ1010 Baku, Azerbaijan
a_nour@mail.ru, konul.jabbarova@mail.ru

Abstract. In this study we are analyzing potentialities of the Z-numbers in improving the quality of risk assessment. Rel0iability of relevant information is unaccounted for in many risk assessment approaches and this circumstance limits the descriptive power of the current approaches. Suggested by L. Zadeh a bi-component Z-number Z = (A, B) allows to take into account the reliability of information. Usually, A and B are perception-based and in effect are imprecise. Z-value based risk assessment framework for the tourist sector is proposed. A general and computationally effective approach suggested for computations with Z-numbers can be utilized for risk factors estimation. Efficiency of the proposed approach is illustrated by country-specific examples.

Keywords: Tourism risk factors · Risk factors importance · Risk sub-factors · Risk assessment · Linguistic terms · Z-number · Z-value-based risk evaluation

1 Introduction

Humanity lives in a world of threats and dangers. The most common threat estimation is risk. Risk serves as a measure of the danger perceived by a person.

Risk assessment uses various mathematical tools: probability theory, theory of possibilities, fuzzy approach, etc. Probability is the most common tool, but other tools also exist, including imprecise (interval) probability and representations based on the theories of possibility and evidence, as well as qualitative approaches [1, 2].

Zadeh [3] noted that "Humans have a remarkable capability to make rational decisions based on information which is uncertain, imprecise and/or incomplete. Formalization of this capability, at least to some degree, is a challenge that is hard to meet". In this note, the concept of a Z-number is introduced and methods of computation with Z-numbers are outlined. Since its introduction, the concept of Z-numbers has been successfully applied as a new direction in the analysis of uncertain and complex systems in various areas of science and technology.

Suggested by Kang et al. [4] an approach based on converting a Z-number to a fuzzy number leads to loss of original information reducing the benefit of using original Z-number-based information [5].

The work of Zadeh [6] discusses different methods, applications, and systems based on the Z-number concept. Aliyev and colleagues suggested a general and computationally effective approach to computation with Z-numbers. The approach is applied to

© Springer Nature Switzerland AG 2020
R. A. Aliev et al. (Eds.): ICSCCW 2019, AISC 1095, pp. 205–213, 2020.
https://doi.org/10.1007/978-3-030-35249-3_26

the computation of arithmetic and algebraic operations, t-norms and s-norms, and construction of typical functions [5, 7–10].

Zadeh indicated risk assessment as one of the main areas of application of Z- numbers [11].

İn this paper the using of Z-value based risk assessment on the example of the tourism sector are discussed. It is proposed to consider a tourist travel as a project and to use a project-oriented approach.

2 Preliminaries

Definition 1. A discrete fuzzy number. A fuzzy subset A of the real line $\mathcal{R}$ with membership function $\mu_A : \mathcal{R} \to [0, 1]$ is a discrete fuzzy number if its support is finite, i.e. there exist $x_1, \ldots, x_n \in \mathcal{R}$ with $x_1 < x_2 < \ldots < x_n$, such that $\text{supp}(A) = \{x_1, \ldots, x_n\}$ and there exist natural numbers s, t with $1 \leq s \leq t \leq n$ satisfying the following conditions:

1. $\mu_A(x_i) = 1$ for any natural number i with $s \leq i \leq t$
2. $\mu_A(x_i) \leq \mu_A(x_j)$ for each natural numbers i, j with $1 \leq i \leq j \leq s$
3. $\mu_A(x_i) \geq \mu_A(x_j)$ for each natural numbers i, j with $t \leq i \leq j \leq n$.

Definition 2. Random variables and probability distributions. A random variable, X, is a variable whose possible values x are numerical outcomes of a random phenomenon. Random variables are of two types: continuous and discrete.

A continuous random variable X is a variable which can take an infinite number of possible values x. A discrete random variable is a variable which can take only a countable number of distinct values.

To determine probability that a continuous random variable X takes any value in a closed interval $[a, b]$, denoted $P(a \leq X \leq b)$, the concept of probability distribution is used. A probability distribution or a probability density function is a function $p(x)$ such that for any two numbers a and b with $a \leq b$:

$$P(a \leq X \leq b) = \int_a^b p(x)dx,$$

where $p(x) \geq 0$, $\int_{-\infty}^{\infty} p(x)dx = 1$.

Consider a discrete random variable X with outcomes space $\{x_1, \ldots, x_n\}$. Probability of an outcome $X = x_i$, denoted $P(X = x_i)$ is defined in terms of a probability distribution. A function p is called a discrete probability distribution or a probability mass function if

$$P(X = x_i) = p(x_i),$$

where $p(x_i) \in [0, 1]$ and $\sum_{i=1}^{n} p(x_i) = 1$.

Definition 3. A Z-number [3]. A Z-number is an ordered pair $Z = (A, B)$, where A is a fuzzy number playing a role of a fuzzy constraint on values that a random variable X may take: X *is A* and B is a fuzzy number with a membership function μ_B: $[0, 1] \rightarrow$ $[0, 1]$, playing a role of a fuzzy constraint on the probability measure of A: $P(A)$ *is B*.

Now let us consider the suggested arithmetic operations over discrete Z-numbers.

Definition 4. Z-valued weighted arithmetic mean. Let as consider real-valued weighting vector $W = (W_1, W_2, \ldots, W_n)$ and Z-valued vector $Z = (Z_1, Z_2, \ldots, Z_n)$. A weighted arithmetic mean operator $WA()$ assigns to any two vectors W and Z a unique Z-number Z_W:

$$WA(Z_1, Z_2, \ldots, Z_n) = Z_W = (A_W, B_W)$$

Components A_W, B_W are determined as follows.

$$A_W^\alpha = [W_1 A_1 + W_2 A_2 + \ldots + W_n A_n]^\alpha \text{ for each } \alpha \in [0, 1],$$

$$R_W = W_1 R_1 + W_2 R_2 + \ldots + W_n R_n.$$

B_W is computed in accordance with methodology given in [5].

3   Risk in Tourism Sector

Traveling like any other activity is potentially dangerous and despite the protective measures taken, there is always some level of risk. The need to classify risks accompanying the tourist business is caused by the desire of decision-makers (managers, travelers and etc.) to clearly structure possible problems and prevent adverse outcomes of tourist activities. These risks are diverse, are difficult to analyze.

It should be noted that many research works are devoted to the analysis of tourism risks from the point of view of tourism firms and institutions and much less works devoted to the risk analysis associated directly with the journey. However, single tourism is becoming increasingly popular in the world, especially such types as extreme, leiser etc.

Regardless of the way travel is organized, risk analysis, especially for security, will always be relevant. Studies of recent years are aimed at improving the methodology for analyzing tourism risks. Below we provide a brief survey of the research publications on identification and study of travelers risks and threats.

Many researchers pointed natural, transport, technological, weather and climatic conditions, biological threats, and political instability, as tourist risk factors [12–16].

Björk and Kauppinen-Räisänen used functional, social, physical, psychological, financial risk types in order to explore how travellers perceive countries in term of risk and which risk types and personal safety dimensions correlate with country risk perception [17].

Kordic et al. [12] and Yun and Maclaurin [18] showed that the safety and security or risk issues have been as one of the factors in destination research and pointed as important factors of tourism destination competitiveness.

In mentioned the number of research studies related to safety, security, and risk in travel and tourism seems to be increasing.

Yang and Nair arguments the risk perception as a valid and convincing tool to investigate tourist's concerns prior to and/or when they are taking up a trip [19].

4 Tourism Risk Factors

Before assessment of tourism risks, we should pay attention to several important aspects. If we consider a tourist trip from the point of view of a traveler, then, of course, each trip should be considered as a project. According to PMBOK guide [20] "*A project is a temporary endeavor undertaken to create a unique product, service, or result.*" Today project management has moved from narrow professional spheres to all areas of human activity. Any tourist trip is aimed at achieving goals - adventure, excursions, treatment, etc. Each journey is different from the other, even if the same country was committed (different years, changing circumstances, etc.).

Therefore, it is advisable to use the project risk analysis methodology for risk analysis of a tourist trip/travel.

Project management always occurs under the influence of a large number of factors that change in the process of project implementation. Uncertainty is one of the conditions of the project. Consequently, one of the main processes in project management is risk management, which is present at all stages of its life cycle.

As important stage of study the results of research in the field of tourism risks were analyzed. Various risks/hazards of tourist trips were considered in the relevant literature and the list of tourism threats was compiled. Using Delphi method the risk register was prepared and involved experts identified risk factors and sub-factors affecting travel safety. Each factor and sub-factor was determined by their weight of importance. Five countries were selected and the risk factors and sub-factors for these five countries were assessed.

We define five important risk factors R_j, j = 1, ..., 5: *threats/potential risks associated with the country of destination, RF_1; threats/potential risks associated with natural environment of country, R_2; threats/potential risks associated with the activities of the tour operator, R_3; threats/potential risks associated with travel (transportation), R_4, threats/potential risks associated with the traveler's person, R_5.*

Each risk factor has 3–6 sub-factors. Decision-maker (manager, traveller, consultant etc.) should make decision about travel destination using risk factors and sub-factors given in Table 1. For many reasons it is not suitable to note the names of countries (alternatives). We are analyzing a problem solution on example of two countries.

Table 1. Tourist travel risk Z-evaluation

Item	Risk factors	Country A1		Country A2	
		Z-evaluation of threats	Z-evaluation of importance weights	Z-evaluation of threats	Z-evaluation of importance weights
1	**Destination country's risks**		**(VH, EL)**		**(VH, EL)**
1.1	Terrorist threats	(L, VL)	(H, VL)	(L, VL)	(H, VL)
1.2	Crime situation	(M, EL)	(H, VL)	(M, L)	(H, VL)
1.3	Cultural/mental differences	(L, VL)	(L, VL)	(M, L)	(H, VL)
1.4	Level of local sanitation	(M, VL)	(M, L)	(M, VL)	(M, L)
1.5	Level of local emergency services	(M, VL)	(M, L)	(M, L)	(M, L)
1.6	Mobile communications/Internet	(VL, L)	(M, L)	(L, VL)	(M, VL)
2	**Natural environment's risks**		**(H, VL)**		**(H, VL)**
2.1	Probability of natural disasters	(M, L)	(H, VL)	(M, L)	(H, VL)
2.2	Climatic and weather conditions	(M, L)	(H, VL)	(M, L)	(M, L)
2.3	Biological threats	(L, EL)	(M, L)	(L, VL)	(M, L)
2.4	Ecological situation	(M, EL)	(L, VL)	(M, L)	(M, L)
2.5	Terrain	(M, L)	(H, VL)	(H, VL)	(M, L)
3	**Tour operator's risks**		**(M, L)**		**(M, L)**
3.1	Unreliable partners	(L, EL)	(H, VL)	(M, L)	(H, VL)
3.2	Improper insurance package	(L, EL)	(H, VL)	(M, L)	(H, VL)
3.3	Financial problems of the tour operator	(L, EL)	(H, VL)	(M, L)	(H, VL)
3.4	Overbooking at placement	(M, L)	(L, VL)	(M, L)	(H, VL)
3.5	Unreliable information about travel	(M, L)	(L, VL)	(M, L)	(M, L)
4	**Risks associated with transportation**		**(M, VL)**		**(M, L)**
4.1	Low carrier service	(L, EL)	(H, VL)	(M, L)	(H, VL)
4.2	Overbooking	(VL, EL)	(M, L)	(M, L)	(M, L)
4.3	Crime situation at airports	(M, L)	(L, VL)	(L, VL)	(M, L)
5	**Risks associated with the traveler's person**		**(VH, EL)**		**(H, VL)**
5.1	Chronic diseases	(L, EL)	(H, VL)	(M, L)	(M, L)
5.2	Age	(L, VL)	(M, L)	(L, VL)	(L, VL)
5.3	Allergenic reactions	(L, VL)	(H, VL)	(M, L)	(M, L)
5.4	Bad habits	(L, VL)	(M, L)	(L, VL)	(L, VL)

5　Tourism Risk Factors

It should be noted than in many areas, risk experts deal with the prediction like this one *"very likely that the level of threat N is medium"* or *"extremely likely that this factor is very important"*. This prediction can be formalized as a Z-number-based evaluation X is Z (A, B). A collection of Z-valuations is referred to as Z-information.

In our work experts evaluate risk factors and sub-factors and their importance weight using Z-numbers. So we have **Z-value based risk** or risk factors for each alternative. Let's consider using Z-numbers to calculate the level of risk or risk factors. Since, using the Z-numbers, experts can evaluate each Z-valued risk factor and sub-factor, as well as their importance weights, Multi-Criteria Decision Analysis (MCDA) can be used to calculate the overall risk level for each alternative.

Arithmetic operations on Z-numbers as well as the ranking of Z-numbers [21] and aggregation of Z-information allow using Multi-Criteria Decision Analysis (MCDA) [22, 23] for the solving decision-making problem of choosing safe travel destination. Codebooks of components A and B of Z-numbers are shown in Table 2.

Table 2. The encoded linguistic terms for A and B components of Z-numbers

Scale	A		B	
	Level	Membership function value (triangle)	Level	Membership function value (triangle)
1	Very Low	1/1, 1/1, 0/0	Unlikely	1/0.05, 1/0.05, 0/0.25
2	Low	0/1, 1/2, 0/3	Not very likely	0/0.05, 1/0.25, 0/0.5
3	Medium	0/2, 1/3, 0/4	Likely	0/0.25, 1/0.5, 0/0.75
4	High	0/3, 1/4, 0/5	Very likely	0/0.5, 1/0.75, 0/1
5	Very high	0/4, 1/5, 1/5	Extremely likely	0/0.75, 1/1, 1/1

For each country (alternative) the Z-evaluations of threats and importance weights were completed. Now we compute for Z-value based risk for each country.

Step 1. We should calculate for each Z-valued risk factor Ra_{ij} its value for i-th alternative using the weighted average-based aggregation of the corresponding importance weights of the sub-factors.

$$Z_{a_{ij}} = \frac{\sum_k^{K_j} Z_{x_{ijk}} \cdot Z_{w_{jk}}}{\sum_k^{K_j} Z_{w_{jk}}}$$

$Z_{x_{ijk}}$ – Z-valued evaluation of k-th sub-factor of j-th factor for i-th alternative
$Z_{w_{jk}}$ – Z-valued evaluation of importance weight of k-th sub-factor of j-th factor

$$Z_{y_1} = \frac{\sum_{i=1}^{6} Z_{x_i} \cdot Z_{w_i}}{\sum_{i=1}^{6} Z_{w_i}} = \frac{(L,VL) \cdot (H,VL) + (M,EL) \cdot (H,VL) + (L,VL) \cdot (L,VL) +}{(H,VL) + (H,VL) + (L,VL)}$$

$$= \frac{(M,ML) \cdot (M,L) + (M,VL) \cdot (M,L) + (VL,L) \cdot (M,L)}{+ (M,L) + (M,L) + (M,L)} = ((0.54 \ 2.23 \ 8.47)(0.1 \ 0.31 \ 0.83));$$

$$Z_{y_2} = \frac{\sum_{i=1}^{5} Z_{x_i} \cdot Z_{w_i}}{\sum_{i=1}^{5} Z_{w_i}} = \frac{(M,L) \cdot (H,VL) + (M,L) \cdot (H,VL) + (L,EL) \cdot (M,L) +}{(H,VL) + (H,VL) + (M,L)}$$

$$= \frac{(M,EL) \cdot (L,VL) + (M,L) \cdot (H,VL)}{+ (L,VL) + (H,VL)} = ((0.79 \ 2.83 \ 9.02)(0.32 \ 0.6 \ 0.91));$$

$$Z_{y_3} = \frac{\sum_{i=1}^{5} Z_{x_i} \cdot Z_{w_i}}{\sum_{i=1}^{5} Z_{w_i}} = \frac{(L,EL) \cdot (H,VL) + (L,EL) \cdot (H,VL) + (L,EL) \cdot (H,VL) +}{(H,VL) + (H,VL) + (H,VL)}$$

$$= \frac{(M,L) \cdot (L,VL) + (M,L) \cdot (L,VL)}{+ (L,VL) + (M,L)} = ((0.43 \ 2.18 \ 8.1)(0.3 \ 0.59 \ 0.9));$$

$$Z_{y_4} = \frac{\sum_{i=1}^{3} Z_{x_i} \cdot Z_{w_i}}{\sum_{i=1}^{3} Z_{w_i}} = \frac{(L,EL) \cdot (H,VL) + (VL,EL) \cdot (M,L) + (M,L) \cdot (L,VL)}{(H,VL) + (M,L) + (L,VL)}$$

$$= ((0.27 \ 1.5 \ 7.2)(0.15 \ 0.36 \ 0.88));$$

$$Z_{y_5} = \frac{\sum_{i=1}^{4} Z_{x_i} \cdot Z_{w_i}}{\sum_{i=1}^{4} Z_{w_i}} = \frac{(L,EL) \cdot (H,VL) + (L,VL) \cdot (M,L) + (L,VL) \cdot (H,VL) + (L,VL) \cdot (L,VL)}{(H,VL) + (M,L) + (L,VL) + (L,VL)}$$

$$= ((0.43 \ 2 \ 7)(0.1 \ 0.34 \ 0.89))$$

Step 2. After obtaining Z-value based evaluations of risk factors the country risk computes as the weighted averaged-based aggregation of risk factors Z_{a_j}, j = 1, ... 5 obtained at previous step

$$Z_a = \frac{\sum_{1}^{5} Z_{a_j} \cdot Z_{w_j}}{\sum_{1}^{5} Z_{w_j}}$$

For the 1-st country the total risk compute as

$$Z_y^I = \frac{Z_{y_1} \cdot (VH,EL) + Z_{y_2} \cdot (H,VL) + Z_{y_3} \cdot (M,L) + Z_{y_4} \cdot (M,VL) + Z_{y_5} \cdot (VH,EL)}{(VH,EL) + (H,VL) + (M,L) + (M,VL) + (VH,EL)}$$

$$= ((0.29 \ 2.18 \ 14.40)(0.22 \ 0.46 \ 0.8))$$

Analogously we computed the overall country risk evaluations for the other country: $Z_y^{II} = ((0.34 \ 2.79 \ 21.08)(0.16 \ 0.41 \ 0.86))$

Step 3. Ranking the countries according to the risk assessment results.
For ranking Z-value based risks the approach of ranking of Z-numbers is used [21]. The result of ranking is shown below

Country II vs. Country I:

$$do\left(Z_y^{II}\right) = 1, \ do\left(Z_y^{I}\right) = 0.17;$$

Thus, the Country II is the best – more secure and safe for travel.

6 Conclusion

In this study we consider application of Z-value based project risk assessment. Suggested approach is used for risk assessment in tourism sector. At every phase of the project management process, various risks originate owing to the occurrence of uncertain events. Uncertainty of the project information is caused by nature of project. Risks directly influence objectives of the project-content, duration, cost, and quality and due to these risks should be identified, assessed and controlled. Risk assessment tools based on traditional probabilistic or possibility models have a limited capacity of description and processing project-related uncertain information and not in all cases are relevant for the risk assessment.

Risk analysis includes risks that can be evaluated based on the past statistical data and risks that are an inherent part of the given situation without past data available. Evaluation of such risks is based on unconscious experience and intuition. Recent advances in computation with Z-numbers allow to conceptualize and process uncertain information by using perception-based and linguistically expressed fuzzy numbers, describing both restriction on the value of the uncertain variable and reliability of the value. By using the Z-numbers project risk analyst can evaluate each Z-valued risk factor and sub-factor, as well as their importance weights and then Multi-Criteria Decision Analysis (MCDA) can be used to calculate the overall risk level for each alternative.

References

1. Aven, T.: Risk assessment and risk management: review of recent advances on their foundation. Eur. J. Oper. Res. **253**, 1–13 (2016)
2. Zhang, Z., Li, K., Zhang, L.: Research on a risk assessment method considering risk association. Math. Probl. Eng. 1–7 (2016)
3. Zadeh, L.A.: A note on Z-numbers. Inf. Sci. **181**(14), 2923–2932 (2011)
4. Kang, B.Y., Wei, D., Li, Y., Deng, Y.: Decision making using Z-numbers under uncertain environment. J. Comput. Inf. Syst. **8**, 2807–2814 (2012)
5. Aliev, R.A., Huseynov, O.H., Aliev, R.R., Alizadeh, A.V.: The Arithmetic of Z-Numbers: Theory and Applications. World Scientific, Singapore (2015)
6. Zadeh, L.A.: Methods and Systems for Applications for Z-Numbers, US Patent, Patent No.: US 8,311,973 B1, Date of Patent: 13 November (2012)
7. Aliev, R.A., Alizadeh, A.V., Huseynov, O.H.: The arithmetic of discrete Z-numbers. Inf. Sci. **290**, 134–155 (2015)

8. Aliev, R.A., Huseynov, O.H., Zeinalova, L.M.: The arithmetic of continuous Z-numbers. Inf. Sci. **373**, 441–460 (2016)

9. Aliev, R.A., Huseynov, O.H., Aliyev, R.R.: A sum of a large number of Z-numbers. Proc. Comput. Sci. **120**, 16–22 (2017)

10. Aliev, R.A., Huseynov, O.H., Alieva, K.R.: Z-valued T-norm and T-conorm operators-based aggregation of partially reliable information. Proc. Comput. Sci. **102**, 12–17 (2016)

11. Zadeh, L.A.: Z-numbers—a new direction in the analysis of uncertain and complex systems. In: 7th IEEE International Conference on Digital Ecosystems and Technologies, Menlo Park, 24 July (2013)

12. Kordic, N., Živković, R., Stanković, J., Gajic, J.: Safety and security as factors of tourist destination competitiveness (2015). https://doi.org/10.15308/sitcon-34-38

13. Simanavicius, A., Edmundas, J., Biruta, S.: Risk assessment models in the tourism sector. Amfiteatru Economic **17**(39), 836–846 (2015)

14. Shaw, G., Saayman, M., Saayman, A.: Identifying risks facing the South African tourism industry. S. Afr. J. Econ. Manag. Sci. **15**(2), 190–206 (2012)

15. Williams, A., Balaz, V.: Tourism, risk tolerance and competences: travel organization and tourism hazards. Tour. Manag. **35**, 209–221 (2013)

16. Ural, M.: Risk management for sustainable tourism. Eur. J. Tourism Hosp. Recreation **7**, 63–71 (2016)

17. Björk, P., Kauppinen-Räisänen, H.: Destination countries' risk image as perceived by Finnish travellers. Finnish J. Tourism Res. **9**, 21–38 (2013)

18. Yun, D., Maclaurin, T.: Development and validation of an attitudinal travel safety scale (2019). https://www.researchgate.net/publication/228637939

19. Yang, E.C.L., Nair, V.: Tourism at risk: a review of risk and perceived risk in tourism. Asia-Pac. J. Innov. Hosp. Tourism **3**(2), 239–259 (2014)

20. PMBOK Guide: A Guide to the Project Management Body of Knowledge, p. 756, 6th edn. Project Management Institute (2017)

21. Aliev, R., Huseynov, O., Serdaroglu, R.: Ranking of Z-numbers and its application in decision making. Int. J. Inf. Technol. Decis. Making **15**, 1–17 (2016)

22. Sharghi, P., Jabbarova, K., Alieva, K.R.: Decision making on an optimal port choice under Z-information. Proc. Comput. Sci. **102**, 378–384 (2016)

23. Gardashova, L.: Application of operational approaches to solving decision making problem using Z-numbers. Appl. Math. **5**, 1323–1334 (2014)

Improving Project Management of Socio-Economic Development of the Region with the Use of Soft Computing

D. I. Dinnik$^{(\boxtimes)}$ ⓘ, N. S. Grigoryeva ⓘ, and Y. E. Galoyan ⓘ

Private Educational Institution of Higher Education "SOUTHERN UNIVERSITY (IMBL)", Rostov-on-Don, Russia
iubip502@yandex.ru, grigorievans@bk.ru,
galoyan@iubip.ru

Abstract. There has been determined the option of assessing the development of the region based on the soft model using fuzzy granular approach to the formation of the results. The graph of connectivity of the main indicators of economic development is received, ways of use of the formed granules for an adequate and flexible assessment of strategic development in the main directions of the region and the Russian Federation as a whole are specified: health care; education; mortgage and rent of housing and the urban environment; international cooperation and export; labor productivity; small business and support of individual entrepreneurial initiative; reform of control and Supervisory activity; safe and high-quality roads; ecology.

Keywords: Project management · Federal regulation system · Regional development · Region · Management system · Efficiency · Priority project · Modeling

1 Introduction

In recent years, Russia has been actively creating and developing projects for its strategic development. Twelve priority areas are identified, including: health care; education; mortgage and rental housing; housing and communal services and urban environment; international cooperation and export; labor productivity; small business and support for individual entrepreneurship; reform of control and supervision activities; safe and quality roads; single-industry towns; ecology [1, 2]. Taking into account the peculiarities of the digital economy, intellectualization of its management, based on soft models and estimates, we consider the processes of designing the socio-economic development of the region in the framework of granular fuzzy approaches aimed at the concentration of financial, administrative, managerial resources, increasing responsibility for the solution of tasks, as well as increasing the impact of the use of financial resources [3].

Today, project management is a key element of improving the management efficiency of public authorities, a way of organizing activities in which the implementation

© Springer Nature Switzerland AG 2020
R. A. Aliev et al. (Eds.): ICSCCW 2019, AISC 1095, pp. 214–220, 2020.
https://doi.org/10.1007/978-3-030-35249-3_27

of important strategic tasks is structured into separate projects and programs (granules), and a set of appropriate soft tools and methods is used to manage them. One of the main institutional barriers to improving the efficiency of regional development management in the project approach is the fundamental impossibility of a clearly formulated and transparent indicative system that allows to measure the degree of compliance of the results of the activities of executive bodies with their functions and public obligations [2]. In this regard, there is a need to revise the existing methods of assessing the management of regional development, as the above reasons for the problems of implementing project management prevent the creation of new social value in the Rostov region.

2 Problem

In the context of economic instability in the world and the conduct of sanctions policy towards the Russian Federation, the question of the inclusion of such a risk factor in the project management system becomes acute. Soft computing technology has been intensively developed over the last decade. It is able to solve the problems of management of poorly structured control objects, so relevant to the General theory and practice of design of control systems. In economic administration, there is often a need to assess the efficiency of various complex economic objects on the basis of a large volume of heterogeneous statistical indicators. Quite often it is also required to rank the studied objects according to the constructed estimates of their effectiveness.

From the point of view of modeling under certainty (or fractional uncertainties) this task is difficult to formalize because it requires a complex analysis of the uncorrelated among themselves of statistical data with the subsequent Association of the received diverse individual assessments, in General, a single evaluation of the object. In addition, the evaluation of objects on various parameters involves the use of criteria that in practice often do not have a clear formulation and are based on the subjective judgments of experts.

3 Materials and Method

The use of the theory of fuzzy sets and soft computing, which allows to create models in conditions of (including subjective) uncertainty, is promising for a reliable comprehensive quantitative assessment of the functioning of complex economic objects and systems. As applied to the problem to be solved, granular mathematics can be used as an adequate tool. At the same time, we believe that "information granulation is a method of combining data arrays on the basis of merging and separation on the principle of functional similarity, as well as the properties of indistinguishability of these objects, based on the principle of points being pulled to a certain Central point". The quantitative methods of decision-making currently developed help to choose the best of the many possible solutions only under one particular type of uncertainty or under conditions of full certainty. In addition, most of the existing methods to facilitate quantitative research within the framework of specific decision-making tasks are based on extremely simplified models of reality and unnecessarily rigid restrictions, which reduces the value of research results and often leads to wrong decisions. The difference between the information granulation method used in this study and the known ones is its applicability for combining semantically related values, which are represented by fuzzy values.

3.1 Original Wording

Interpretation as two universal sets on the principle of "to be united". Let it be that P and Q are two objects, universal sets of different nature. Objects P and Q can be granulated (assigned to one granule B, cluster) by a binary relation, meaning semantic indistinguishability $(P, Q) \in B$. We can say that Q "shrinks" to P and this variant will be called P center. Information granulation is a set of $CB = \{B^k | k \in K\}$, fuzzy binary relationship B, where K is a complete set of indexes that lists indistinguishability, functionality, or similarities.

3.2 Geometric Interpretation

Information granulation for P is a family of definite and/or fuzzy subsets that are related to P, where K is the power of the subset. The center of the information granule can be viewed from the point of view of various mathematical formal representations oriented to the geometric formulation. From this point of view, the center of the information granule is a point, and the granule itself is limited in space by some neighborhood h. The boundaries of this neighborhood can be determined by both metric and non-metric methods, that is, it is possible to obtain either a clear information granule or a fuzzy one.

It is possible to classify the studied structures on the basis of several representations as global (granulation is induced by binary relations) or local (granulation is induced by the neighborhood system). An example of a global granulation is, for example, the attachment of neighborhoods of a topological space, and an example of a local pelletizing is the clarification of the boundaries of neighborhoods of the topological subset. Note that local information granulation is induced by the neighborhood of the system of spaces to obtain finite granules at each point of space. These two types of granulation are not equivalent. In many tasks of control and management of a complex system, there is no need to obtain an optimal clear solution for each moment of time, since the cost of information accumulation and rigid elimination of residuals in the system may exceed the effect achieved.

Most often, the specific content of the problem requires a given level of fuzzy solutions. The real problems contain fuzzy conditions and some indistinctness of purpose due to the fact that their formulation is carried out by a person. We should take into account that the uncertainty factor in solving problems largely changes the methods of decision-making: the principle of representation of the initial data and parameters of the model changes, the concept of optimal solution becomes ambiguous.

Management of socio-economic development of the region is a set of interrelated systems and a set of measures implemented by the Executive authorities of the Russian Federation, for systematic and balanced development and implementation of public obligations to society. A comprehensive diagnosis of the project management system of regional development on the example of the Rostov region was carried out according to the algorithm:

1. Analysis of the organizational structure of the highest permanent Executive body of the Rostov region;
2. Key macroeconomic indicators of the Rostov region;
3. Analysis of program-target management in the Rostov region;
4. Analysis of budget execution in Rostov region;
5. Evaluation of the effectiveness of the Executive authorities of the Rostov region in the framework of the presidential Decree;
6. Determination of performance indicators and construction of a model of the regional regulatory system.

Resources, including financial, are necessary for the implementation of management functions, so the work presents a vertical and horizontal analysis of the budget of the Rostov region (Fig. 1).

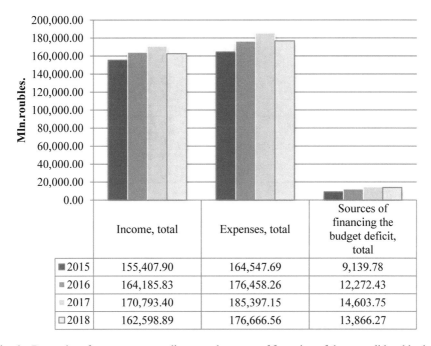

	Income, total	Expenses, total	Sources of financing the budget deficit, total
■2015	155,407.90	164,547.69	9,139.78
■2016	164,185.83	176,458.26	12,272.43
▪2017	170,793.40	185,397.15	14,603.75
☐2018	162,598.89	176,666.56	13,866.27

Fig. 1. Dynamics of revenues, expenditures and sources of financing of the consolidated budget deficit of the Rostov region for 2015–2018, in million rubles

The performance of the budget is mixed. In almost all areas of budget spending there is a situation in which in 2016 there was an increase in the performance of the budget compared to the level of 2015, and 2017, in its turn, is the best of the analyzed periods, in which the deviation of actual indicators from the planned average of 0.1–5% in the direction of underutilization of financial resources. In 2017, as in 2015–2016, insufficient development of funds took place in the national economy, which becomes noticeable when considering the overall performance indicators of the budget, which are at a high level.

The external socio-economic environment is based on the management model of Granberg [1]. Each top of the model includes a system of indicators that characterize the socio-economic situation of the region and the management system.

The developed model shows the relationship of elements (vertices) both positive and negative. It is necessary to investigate the system for changes of variables to key vertices. To do this, we conduct a scenario analysis, giving impulses (positive and negative) to the vertices of the model. For the Executive authorities of the Rostov region, this approach will be very interesting, because of the many scenarios of interaction it is possible to choose the most optimal.

Table 1. Protocol for scenario analysis (fragment)

Ways of development	1	2	3	4	5	6	7	8	9
V0	0	0	0	0	0	0	0	0	0
V01	0	0	1	0	1	0	1	0	0
V02	1	0	0	1	1	0	1	0	1
V03	1	1	0	1	1	0	1	1	1
V04	1	0	0	1	0	0	1	0	0
V05	1	0	1	1	1	0	1	0	0
...	...	...	...	...	...	...	...	...	...
V21	1	1	1	1	0	1	1	1	1
V22	1	0	0	1	0	0	0	0	1
V23	0	1	1	1	1	1	1	1	1
V24	0	0	1	1	0	1	0	1	1

The review revealed that the definition of quality indicators affecting the equilibrium functioning and development of the system, and the formation of conditions for its ultimate effectiveness entails four main groups of factors: market, macroeconomic, industry-specific, and sociotechnical.

The formation of ways modeling the impact of factors was carried out according to the prepared protocol (Table 1). In the model, activating vertices were sequentially set, into which vectors of control actions on the system were introduced.

Players in the scenarios can be any element of the system. The conditions for the study were developed on the basis of statistics of the Rostov region and the results of a survey of experts. In relation to the forthcoming stage of scenario analysis of interaction within the regional regulatory system: the main target parameters, priorities and directions of development of the system are determined.

To assess the impact of factors on each other, a survey of experts was conducted, including: representatives of Executive authorities, leading scientists involved in public administration and regional development, representatives of the business community (Table 2).

Table 2. Expert assessment of the influence of factors in the cognitive model of interaction of PPC (V0) with the environment

Connection	Rate	Connection	Rate	Connection	Rate
V01-V3	5	V17-V1	9	V9-V18	8
V03-V01	7	V2-V17	8	V22-V10	7
V05-V02	3	V23-V2	−8	V14-V10	7
V06-V02	4	V2-V12	6	V21-V10	−8
V03V010	8	V4-V2	8	V10-V19	8
V03-V04	5	V17-V3	7	V10-V18	9
V22-V03	9	V20-V3	−6	V23-V11	8
V03-V05	5	V5-V3	7	V10-V11	9
V14-V04	9	V12-V3	10	V11-V12	8
V04-V09	9	V4-V20	−5	V12-V24	9
V04-V02	−7	V22-V4	9	V10-V13	9
V04-V010	7	V4-V5	8	V15-V13	8
V06-V04	5	V4-V23	7	V16-V15	7
V06-V05	8	V5-V19	8	V18-V15	6
V06-V09	6	V5-V21	−9	V16-V18	6
V17-V06	7	V6-V14	5	V09-V17	8
V09-V07	9	V7-V8	6	V18-V24	7
V3-V07	8	V7-V15	7	V19-V24	8
V08-V07	−8	V8-V9	8	V23-V20	9
V08-V09	−8	V8-V10	8	V011-V02	−6
V3-V08	8	V8-V14	9	V011-V09	8
V09-V10	6	V8-V13	9	V03-V011	6
V09-V9	7	V15-V8	7	V04-V011	7
V09-V6	4	V14-V9	6	V09-V8	8
V13-V09	9	V9-V10	8	V012-V2	5
V4-V09	8	V10-V9	10	V2-V09	8
V010-V09	2	V9-V13	9	V03-V012	10
V7-V1	5	V9-V11	10	V012-V6	8
V1-V11	6	V21-V9	−9	V23-V9	−7
V3-V1	8	V9-V19	8	V23-V10	−4

4   Conclusion

Implementing the presented model under project management, it was possible to reduce the duration of the period of development and coordination of programs and projects, increase the speed of investment development in the region and the growth rate of GRP, achieve the following effects in the management system using cognitive models:

1. Effects for society – participation and control over the implementation of projects, initiator of projects, improving the quality of services, increasing the number of benefits.

2. Effects for business – removal of administrative barriers, increasing the investment climate in the region.
3. Effects for the Executive authorities – increasing the principle of transparency and executive discipline, improving performance.

Using the cognitive model of the Rostov region it is possible to analyze the problem of implementation of projects (programs), to isolate its essence, to reformulate the problem into a task and to formulate the final goal and result, as well as to plan steps, resources and risks.

Granular computing is a promising methodology for computing. In large-scale calculations, the solution of the problem consists of three stages: (1) the choice of the granulation formalization method in algebraic or geometric formulation; (2) the formalization in the form of a granular structure, for example, in the form of a binary granular representation; and (3) the application of granular structures to problems arising in the field of intelligent control systems of the region's economy.

References

1. Grigoryeva, N.S., Kolycheva, Z.Ya.: Strategic directions of innovational regional development. Sci. Educ.: Hous. Econ. Entrepreneurship Law Manag. 6(85), 22–25 (2017)
2. Grigoryeva, N.S.: Necessity and opportunities of stable regional development. In: Works of International Scientific-Practical Conference "Transport-2015" «Rostov State Transport University», pp. 72–73 (2015)
3. Grigoryeva, N.S.: Priority areas for improving the socio-economic efficiency of state regulation of small business. In: Problems and Prospects of Business Development in Russia Collection of Reports of the International Scientific and Practical Conference, pp. 101–107 (2011)

Evaluation of Tax Administration Effectiveness Using Brown-Gibson Model

A. F. Musayev[1]([⊠]) [ID], A. G. Aliyev[2] [ID], A. A. Musayeva[3] [ID],
and M. Kh. Gazanfarli[4] [ID]

[1] The Azerbaijan University, J. Hajibeyli st., 71, AZ1007 Baku, Azerbaijan
akif.musayev@gmail.com
[2] ANAS Institute of Information Technology, B. Vahabzadeh st., 9,
AZ1141 Baku, Azerbaijan Republic
alovsat_qaraca@mail.ru
[3] ANAS Institute of Control Systems, B. Vahabzadeh st., 9, AZ1141 Baku,
Azerbaijan Republic
aygun.musayeva@gmail.com
[4] ANAS Institute of Economy, H. Javid ave., 115, AZ1143 Baku,
Azerbaijan Republic
q.miraa@gmail.com

Abstract. One of the factors that play an important role in the formation of an effective tax system is to increase the effectiveness of tax administration. This research is implemented on the assessment of tax "tax collection per a tax employee", which is one of the characterized indicators of the administration efficiency. Evaluation is investigated as a multi-criteria decision-making problem using Delphi query method. As the survey results consist of both quantitative as well as qualitative indicators, this evaluation is conducted by 2 methods that have computing opportunity by taking into consideration these indicators: The weighted sum model and Brown-Gibson model.

Keywords: Tax system · Tax administration · Effectiveness indicators of tax system · Multi-criteria decision-making problem · Fuzzy inference system · Weighted sum model · Brown-Gibson model

1 Introduction

As the tax policy is a part of the state economic strategy, its effective implementation has a significant impact on economic development and social welfare of any country. Tax policy of the state is realized through the improvement of tax legislation, tax administration, and tax authorities' structure. As the tax system forms the basis of the state regulation mechanism of the economy, the effective activities of the country entire economic complex depends directly on its proper and efficient organization. Although many different researches and calculations have been carrying out for years, there is no system of concrete methods and indicators for assessing the effectiveness of the tax system yet. However, the following indicators are used in the research for different countries:

© Springer Nature Switzerland AG 2020
R. A. Aliev et al. (Eds.): ICSCCW 2019, AISC 1095, pp. 221–228, 2020.
https://doi.org/10.1007/978-3-030-35249-3_28

– Characteristic indicators of classical taxation principles: fairness; adequacy; simplicity; transparency; administrative ease.
– Tanzi qualification diagnostic indicators (Tanzi productivity test) [1]: dispersion index, erosion index, index of lag in tax collection, specialization index, burden index, index of collection expenditures, concentration index, objectivity index.
– Gill system in diagnostic indicators. Effectiveness of indicators is analyzed in two groups, quantitatively and qualitatively in the Gill's proposed tax system [2]:

1. Quantitative indicators: tax collection efficiency, tax burden index, tax collection deficiency, administration efficiency, voluntary tax solvency, tax concealment potential, fiscal accounting inaccuracy, and tax avoidance potential;
2. Qualitative indicators: tax culture, tax administrator work efficiency, tax administrator morality level and others.

Stallmann used the characteristic indicators of taxation principles to assess the impact of taxes on society and the economy [3]. Evaluation of Lithuania and Ireland tax systems by V. Tanzi diagnostic indicators and comparative analysis of tax revenues are reflected in Juozaitien's research [4]. Skackauskiene offered a single-evaluation method with the hierarchical system of the tax system's core, partly integrated and complex integrated indicators in its research [5]. The aim of Jakštonytė and Giriūnas' research to create a model for evaluating the effectiveness of the universal tax system, which allows evaluating monological, logical and statistical methods of comparative analysis, management philosophy [6].

As tax policy and tax administration are intensely related together, measurement of its effectiveness plays an important role in the formation of an effective tax system. Numerous studies have been realized on this issue, new guidance and econometric models have been proposed by scholars from different countries to ensure administrative efficiency [7, 8].

The effects of changes and additions in the tax administration and legislation to the tax potential were assessed by Musayev et al. [9–11] using Mamdani and Sugeno fuzzy inference methods comparatively.

One of the important stipulations for accountability and transparency in tax administration is the measurement of activities that constitute a part of strategic and operational planning processes. A part of our research has been devoted to measure the performance of tax authorities. This problem was also widely analyzed in the William Crandall's book [12]. Moreover, the professionalism level of the staff was assessed over the officials in tax system by Musayev et al. [13].

In addition, the effectiveness of tax administration also of the tax system depends on some factors such as "additional estimates for tax inspections", "tax receipts for each cost of tax authorities' work", and others.

In this paper "tax collection per a tax employee" which is one of the indicators characterizing the activities of the tax employees, that means the assessment of the amount of taxes collection during the financial year to a tax officer has been investigated as multi-criteria making-decision problem [14, 15] and has been compared by calculating for a concrete country with its two methods.

In general, the multi-criteria decision-making problem methods used for distinguishing of alternatives are based on different criteria. However, in this paper alternatives (tax authority employees) have been evaluated not only for comparisons, but also with the purpose of obtaining a final result.

2 Statement of Problem

"Tax collection per a tax employee" which is one of the tax system effectiveness indicators is evaluated by Weighted Sum model (using Mamdani inference system) and Brown-Gibson model based on query that is held among tax employees via Delphi method.

Assessment is realized as follows. Firstly, the query is held based on Delphi method among the tax employees to determine indicators that are characterized their performance level like realizing inspections, preparation of reports, level of organization of tax administration, and others. In this case, the jth answer of the ith expert c_{ij} written in quantitative and qualitative expression is as follows:

$$c_i = \left(c_{i1}, c_{i2}, \ldots c_{ij}, \ldots c_{in}\right), \ i = \overline{1, m} \ (i \neq j) \tag{1}$$

Indicators affecting the activities of taxpayers are expressed not only by quantitative, but also by qualitative indicators. Therefore, the results of the survey consist of quantitative ($t = 1$) expressed by time and qualitative indicators are conveyed by linguistic words ($t = 2$), and in this case the expression (1) is grouped according to $t = 1, \ldots, 2$ directions and is written as follows:

$$c_i^t = \left(c_{i1}^t, c_{i2}^t, \ldots c_{ij}^t, \ldots c_{in}^t\right), \ i = \overline{1, m} \ (i \neq j), \ t = 1, 2 \tag{2}$$

2.1 Evaluation via Weighted Sum Model

As it is mentioned above, two methods are used in this assessment. First, let's carry out the process with the weighted sum model [16].

As the results of the survey are quantitative and qualitative, qualitative indicators will be quantified by the Mamdani inference [17] system. For this purpose, the qualitative indicators are distinguished from the quantitative indicators as per Eq. (2), and the process is executed for each of the ith expert responses included in the system as input variables for each qualitative indicator.

As a result of quantitative expression of qualitative indicators, the expression (1) will be as follows:

$$\acute{c}_i = \acute{c}_{i1}, \acute{c}_{i2}, \ldots \acute{c}_{iJ} \ldots \acute{c}_{in}, \ i = \overline{1, m} \ (i \neq j) \tag{3}$$

Finally, the assessment of the average processing period of each tax officer during the fiscal year is accomplished by weighted sum model, one of the solving methods of multi-criterion decision-making problem by utilizing the importance of the questions (indicators) that characterize them.

$$\acute{c}_i = \sum_{j=1}^{n} \acute{c}_{ij} w_j, \ i = \overline{1,m}, \ j = \overline{1,n} \ (i \neq j) \tag{4}$$

w_j expresses the weights of obtained information, calculated with the following formulae, which depends on the significance of indicators.

$$w_j = \frac{2(n+1-r_j)}{n(n+1)}, \ j = \overline{1,n} \tag{5}$$

r_j-rank (order of importance), $\sum_{j=1}^{n} w_j^{\acute{d}t} = 1$, $w_j^{\acute{d}t} \geq 0$.

Weights are equal when the importance rates of the indicators are the same.

The final result mean that "tax collection per a tax employee" - τ is defined as following expression.

$$\tau = \frac{TR \ (tax \ receipts)}{\frac{\sum_{i=1}^{m} \acute{c}_i}{m}} \tag{6}$$

2.2   Evaluation via Brown-Gibson Model

The another used method was proposed by P. Brown and D. Gibson in 1972 to evaluate the alternatives characterized by quantitative and qualitative indicators [18, 19]. The mathematical expression of this model is as follows:

$$\acute{c}_i = CFM_i * [X * OFM_i + (1 - X)SFM_i] \tag{7}$$

$\acute{c}_i$ - Measure for an alternative i; CFM_i - Critical factor measure; OFM_i - Objective (quantitative) factor measure; SFM_i - Subjective (qualitative) factor measure; X - Objective factor decision weight; $(1 - X)$ - Subjective factor decision weight. The final result τ is calculated by taking the obtained $\acute{c}_i$ with (7) into account in (6).

3   Application of the Methodology

The tax collection per a tax employee has been evaluated in the tax system of the Azerbaijan Republic by the methods given above is as follows.

Firstly, an inquiry was conducted among 38 tax employees with the Delphi method. The survey was held based on some indicators like inspections, settlement of disputes with taxpayers, level of organization of tax administration, and others, for determining time spent related with tax employees' activities. The survey was repeated three times to obtain robust results (Table 1).

Table 1. Fragment of obtained information.

№	INDICATORS (obtained by survey)							
	Ethical behaviour	The level of tax administration	On-site inspection	...	Time spent of disputes	Time spent tax registration	Opening time of bank account	Calculating time of tax return
1	High	Middle	0	...	45	30	2	5
2	Normal	Middle	22	...	20	0	1	2
3	High	High	25	...	30	30	2	3
...	...	...	...	...	...	...	...	...
15	High	High	20	...	20	0	2	2
...	...	...	...	...	...	...	...	...
38	High	High	25	...	0	0	2	2

Next, all obtained information was grouped according to t = 1, 2 directions, that means quantitative (objective) and qualitative (subjective) indicators, based on expression (2).

3.1   Assessment with Weighted Sum Model

Prior to applying method, conversion to quantitative indicators to evaluate of the effect by expressing with qualitative information to tax employees' activities (t = 2 direction) problem was solved.

The Mamdani inference system was employed for this estimation. The process was conducted individually for each expert on three qualitative indicators such as "ethical behaviour" "tax administration level", "intervention". Consequently, the quantitative expression of the indicator "effectiveness" per time spent, was obtained. As an example, the process was evaluated as follows for one expert's appropriate indicators:

The above 3 qualitative indicators were entered into the system as input variables. "Ethical behaviour" was defined by the input variable as "high", "normal", and "low" terms, and the membership function of each term was established.

"Intervention", and "tax administration level" input variables were also expressed with "regular", "sometimes", "never", and "high", "middle", "low" terms set and membership functions of each term was established. After the fuzzification of the input variables, the set of rules were created.

This process was replicated for the remaining 38 experts' responses. The calculations were carried out using the fuzzy toolbox of the MathWorks, MATLAB Software R2018b.

The new row of indicators determined by Eq. (3) includes the quantitative expression of the qualitative indicators in terms of time expenditures. Then weights of criteria, considering their importance, were assessed with Eq. (5), as in Table 2 (Fig. 1).

Table 2. Weights of criteria

w_1	w_2	w_3	w_4	w_5	w_6	w_7	w_8	w_9	w_{10}	w_{11}
0.15	0.166	0.136	0.11	0.12	0.09	0.08	0.06	0.015	0.05	0.03

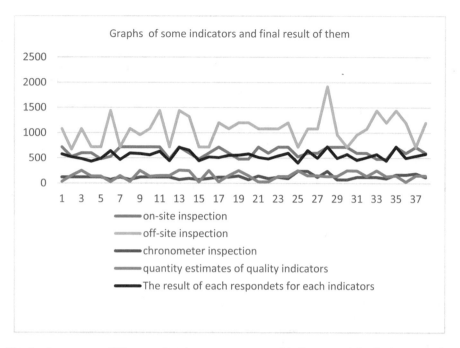

Fig. 1. A summary of 38 respondents' responses to several indicators and the final outcome for each of them.

3.2   Assessment with Brown-Gibson Model

For applying this model, first, indicators should be grouped as qualitative (subjective) and quantitative (objective) indicators as Eq. (2) and the parameters used was evaluated by the formula (7).

Table 3. Evaluation of "the tax collection for per a tax employee" with Brown-Gibson model

$X = 0.9$	$(1 - X) = 0.1$	$\tau = 1.615.674$
$X = 0.8$	$(1 - X) = 0.2$	$\tau = 1.816.817$
$X = 0.7$	$(1 - X) = 0.3$	$\tau = 2.075.162$
$X = 0.6$	$(1 - X) = 0.4$	$\tau = 2.419.159$
$X = 0.5$	$(1 - X) = 0.5$	$\tau = 2.899.866$
$X = 0.4$	$(1 - X) = 0.6$	$\tau = 3.618.990$
$X = 0.3$	$(1 - X) = 0.7$	$\tau = 4.812.390$
$X = 0.2$	$(1 - X) = 0.8$	$\tau = 7.180.106$
$X = 0.1$	$(1 - X) = 0.9$	$\tau = 14.134.184$

$$\tau = \frac{VD\ (tax\ receipts)}{\frac{\sum_{i=1}^{m} \acute{c}_i}{m}}, \quad \acute{c}_i = CFM_i * [X * OFM_i + (1 - X)SFM_i], \quad (i = \overline{1, 38})$$

These calculations were executed with MS Excel.

The productivity of tax employees (alternatives) due to tax collection was assessed by depending on the effects of objective and subjective indicators as in Table 3.

4 Conclusion

As we mentioned, the effectiveness of tax administration can be estimated by using several indicators. However, our paper emphasizes the determination of the tax administration effectiveness by evaluating the "tax collection per a tax employee" characterized by the productivity of the tax employee, which is essential for measuring. Different methods are proposed for this, calculations are implemented using these methods and suggestions are given to improve efficiency by comparing results. In addition, as some indicators, including tax registration, ethical behavior, calculation of tax return, and others which characterize the activities of tax employees are expressed in two forms as qualitative and quantitative, so quantitative expressions of qualitative indicators are calculated for their using together in the evaluation process. At the same time, which indicator (quantitative or qualitative) being much more effective is analyzed by proposed methods. As a result, it is determined the improving qualitative indicators plays a more important role than quantitative indicators for increasing of tax administration effectiveness.

References

1. Tanzi, V.: The IMF and tax reform. IMF Working Paper 90/39 (2004)
2. Gill, J.: A diagnostic framework for revenue administration. The Tax Policy and Administration Thematic Group. The World Bank technical paper 472 (2000)
3. Stallmann, J. I.: Evaluating tax systems. Missouri Legislative Academy (2004)
4. Juozaitienė, L.: Comparison of Lithuanian and Irish tax systems using V. Tanzi qualification diagnostic indicators. Ekonomika ir vadyba: aktualijos ir perspektyvos 2(15), 65–75 (2009)
5. Skackauskiene, I.: Tax system evaluation model. In: 6th International Scientific Conference Vilnius, Lithuania, Business and Management, Selected Papers, Vilnius, pp. 719–727 (2010)
6. Jakštonytė, G., Giriūnas, L.: Tax system efficiency evaluation modeling with reference to V. Tanzi criteria. Economics and management (2010)
7. Kumar, S., Nagar, A.L., Samanta, S.: Indexing the effectiveness of tax administration. Econ. Polit. Wkly. 42(50), 104–110 (2007)
8. Das-Gubta, A., Estrada, G.B., Park, D.: Measuring tax administration effectiveness and its impact on tax revenue. Working Papers DP-2016–17. Economic Research Institute for ASEAN and East Asia (ERIA) (2016)
9. Musayev, A.F., Madatova, Sh.G., Rustamov, S.S.: Evolution of the impact of the tax legislation reforms on the tax potential by fuzzy inference method. In: 12th International Conference on Application of Fuzzy Systems and Soft Computing (ICAFS), pp. 507–514. Elsevier, Vienna (2016)
10. Musayev, A.F., Madatova, Sh.G., Rustamov, S.S.: Mamdani-type fuzzy inference system for evaluation of tax potential. In: Sixth World Conference on Computing dedicated to 50th Anniversary of Fuzzy Logic and Applications and 95th Birthday Anniversary of Lotfi A. Zadeh, Berkley, USA (2016)

11. Rustamov, S.S., Musayev, A.F., Madatova, Sh.G.: Evaluation of the Impact of state's administrative efforts on tax potential using Sugeno-Type fuzzy inference method. In: ICAFS 2018, Advances in Intelligent Systems and Computing, vol. 896, pp. 352–361 (2019)
12. Crandall, W.: Revenue Administration: Performance Measurement in Tax Administration. Technical Notes and Manuals 10/11 (2010)
13. Musayev, A.F., Guliyev, H.H., Efendiyev, G.M., Gazanfarli, M.Kh.: Evaluation of the staff's professionalism level by modified weighted sum model. In: Tenth World Conference "Intelligence Systems for Industrial Automation", WCIS-2018, Tashkent, Uzbekistan (2018)
14. Zadeh, L.A.: Fuzzy sets. Inf. Control. **8**, 338–353 (1965)
15. Triantaphyllou, E.: Multi-criteria decision making methods. A Comparative Study (2000)
16. Fishburn, P.C.: Additive Utilities with Incomplete Product Set: Applications to Priorities and Assignments. Operations Research Society of America (ORSA), Baltimore, MD, USA (1967)
17. Mamdani, E.H., Assilian, S.: An experiment in linguistic synthesis with a fuzzy logic controller. Int. J. Man-Mach. Stud. **7**(1), 1–13 (1975)
18. http://bytepublish.com/B-GModel/B-G%20Model.pdf
19. https://www.revolvy.com/page/Brown%E2%80%93Gibson-model

A Semi-supervised Fuzzy Clustering Approach via Modifications of the DBSCAN Algorithm

Erind Bedalli[1,2]([✉]) [iD], Enea Mançellari[1] [iD], and Rexhep Rada[2] [iD]

[1] Epoka University, Tirana, Albania
{ebedalli,emancellari}@epoka.edu.al
[2] University of Elbasan, Elbasan, Albania
{erind.bedalli, rexhep.rada}@uniel.edu.al

Abstract. In the data mining context, semi-supervised learning is applicable in circumstances where only a scarce amount of information on the intrinsic structure of a dataset is available. This information may be in the form a few labelled instances or a relatively small set of constraints on the pairwise memberships of particular instances. In this study we are providing a semi-supervised fuzzy clustering model which modifies versions of conventional DBSCAN algorithm in order to generate soft clusters which foreclose the noise points. The employed modifications are mostly related to the control parameters of the algorithm intending to utilize the additional information (which in our case is in the form of a few labelled instances) and adaptations towards the fuzzy clustering approach. Finally, several experimental procedures have been conducted on synthetic and real-world benchmark datasets in order to assess the accuracy of our employed model and to compare it to the conventional algorithms of the respective domain.

Keywords: Semi-supervised learning · Density-based clustering · DBSCAN algorithm · Fuzzy clustering

1 Introduction

Supervised and unsupervised learning are two notable disciplines of data mining, encompassing a multitude of methodologies employed in revealing patterns in collections of data. Supervised learning operates with the proviso of the presence of additional information in the form of a training set, which typically contains instances that are properly labelled beforehand. Based on this training set, a mapping function is generated which will infer the categorization of the new instances [1]. On the other hand, unsupervised learning is entirely data-driven as no external information is provided. The data elements are clustered into groups (clusters) relying solely on the similarities/dissimilarities among them [2].

Semi-supervised learning is conceived as a blending of the supervised and unsupervised techniques. It operates in scenarios where only a scarce amount of additional information on the intrinsic structure of the datasets is available [3]. In these circumstances, due to the scantiness of the additional information, maintaining a supervised learning approach is infeasible. Nevertheless, leaving this small amount of additional

© Springer Nature Switzerland AG 2020
R. A. Aliev et al. (Eds.): ICSCCW 2019, AISC 1095, pp. 229–236, 2020.
https://doi.org/10.1007/978-3-030-35249-3_29

information unexploited and pursuing an unsupervised learning procedure would be a suboptimal alternative [4]. Semi-supervised learning relies on the principle of incorporating both labeled and unlabeled instances in order to explore the geometric structure of unlabeled data, thus addressing the synchronization of the detected structural regularities according to the patterns of the labeled instances [5].

In this work we are presenting a semi-supervised learning approach based on modifications of the conventional DBSCAN algorithm into two main directions: adapting it towards semi-supervised learning and augmenting it to produce fuzzy clusters. The conventional DBSCAN algorithm is guided by two control parameters: *Eps* (ε) representing the neighborhood radius and *Minpts* representing the minimal number of points that a region of radius ε must comprise in order to make allowance for a region to be dense [6]. The presence of additional information enables an automatic adjustment of the *Eps* (ε) parameter whose value is adapted in order to be concordant with the pairwise memberships of the available labelled instances. Thus, in our model *Minpts* is the only parameter which has to be externally provided. Furthermore our model is also augmented with a view to generating fuzzy clusters by quantitative discrimination between core and border points. Finally a series of experimental procedures have been conducted on several benchmark and artificial datasets and its accuracy is assessed mostly based on fuzzy extensions of the Rand index.

2   Related Work

In the recent decades, much progress has been made towards many forms of semi-supervised learning, especially in the domain of semi-supervised learning with instance-level constraints. The seminal notion of must-link and cannot-link constraints as two fundamental types of pairwise constraints was presented by Wagstaff et al. [7]. A must-link constraint indicates that the two instances must belong to the same cluster, while a cannot-link constraint indicates that the two instances must belong to distinct clusters. Instead of solely utilizing this information for cluster validation, the main idea is to utilize this information to provide "guidance" for the cluster generation procedure. Semi-supervised learning is also successfully involved in collaboration with the partition-based fuzzy clustering algorithms like fuzzy c-means. The additional information here is mainly utilized to adapt the objective function [8].

Lelis and Sander have shown a semi-supervised clustering approach (namely the SSDBSCAN algorithm) with pairwise constraints by carrying out modifications on the DBSCAN algorithm. A significant advantage of that methodology is that it represents a good alternative for an automatic knowledge discovery process as no user intervention is indispensable [9].

On the other hand there have been several successful approaches in extending the DBSCAN algorithm in order to generate fuzzy clusters, but these methodologies are applied only in an unsupervised way [10].

3   The Conventional DBSCAN Algorithm

DBSCAN (Density-Based Spatial Clustering of Applications with Noise) algorithm is an eminent representative of the category of density-based clustering algorithms. It relies on the principal notion that a cluster comprises a concentrated set of instances, separated by sparse (or empty) regions where the concentration of the instances is distinctly low [11]. In this context, a cluster is delineated as a maximal contiguous group of instances with density surpassing a specified threshold. During this clustering procedure, some instances may be located in the intermediate separating regions, thus they do not take part in any of the clusters and consequently are perceived as noise or outliers [12]. On the other hand, the instances which are comprised in the clusters are of two forms, which are core and border points. The first form indicates that the instance is completely involved in the cluster, while the second form indicates that the point is in the boundary of the cluster.

The DBSCAN algorithm is guided by two pre-defined control parameters which are: *Eps* (ε) representing the neighborhood radius and *Minpts* representing the minimal number of points that a region of radius ε must comprise in order to make allowance for a region to be dense. Based on the *Eps* (ε) parameter, for every instance x_i of the dataset, we define the ε – neighborhood as $N_\varepsilon(x_i) = \{x \in X | d(x_i, x) \le \varepsilon\}$. An instance x_i will be labelled as a core point if the number of elements in its ε – neighborhood is at least *MinPts* and it is labelled as a border point if the number of elements in its ε – neighborhood is smaller than *MinPts* but its ε – neighborhood contains at least one core point [13].

In this framework, a density edge is characterized as an edge linking a pair of core points. In addition, two core points p and q are delineated as density-connected, if there exists a path consisting of density edges connecting them as shown in the Fig. 1.

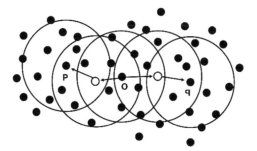

Fig. 1. An illustration of density-connected points

The notion of density-connected pairs of points is the essence of the DBSCAN algorithm, based on which the generation of the clusters is carried out. The algorithm operates by iteratively selecting an unassigned core point as the first member of a new cluster and gradually broadening the current cluster towards any possible direction picking every unassigned core point which is density-connected to the first member of the cluster. This iterative procedure is completed once that no unassigned core point remains. Finally the border points are distinguished and assigned to the clusters of the respective nearest core point. The entire procedure can be briefly depicted by the following pseudo-code [14]:

1. Construct the ε – neighborhoods for each instance of the dataset and label the instances as core, border and noise points based on the neighborhood analysis of each data point.
2. Relinquish the noise points (so hereafter they will not be involved the clustering procedure).
3. While there are core points unassigned to any cluster:
   3.1 Pick an unassigned core point (denote it as c_i).
   3.2 Generate a new cluster consisting of the point x and all the other core points which are density-connected to x.
4. Distribute the border points into the clusters of their respective nearest core point.

The DBSCAN algorithm has the desirable properties of detecting clusters of diverse shapes and densities and to automatically establish the common number of clusters. On the other hand the main handicaps of this algorithm are the sensitivity to the control parameters and the coercion in dealing with clusters of diverse densities [15]. The conventional DBSCAN algorithm is an unsupervised learning procedure as it assumes no prior available information. In the following section will be discussed on adaptations of this algorithm to exploit the presence of additional prior information about pairwise memberships of instances.

4 Modifications on the DBSCAN Algorithm for Semi-supervised Fuzzy Clustering

Our primary intention in this work is to construct a model which will be able to operate in presence of additional information and generate fuzzy clusters. The original form of the available information is a relatively small group of labelled instances, but we are not going to exploit the available information directly in this form. Instead, based on this information we can easily generate sets of pairwise constraints (must-link or cannot-link constraints) about the labeled instances. Based on these pairwise constraints, the DBSCAN algorithm can be modified in order to narrow the interval of possible values of *Eps* (ε) and then automatically pick a value within this interval (in our case the midpoint of this interval is being picked). The interval of possible values of *Eps* (ε) is narrowed by both must-link and cannot-link constraints. The must-link constraints decrease the upper bound of this interval, while the cannot-link constraints increase the lower bound of this interval. So, assuming that p_1 and p_2 are instances with a cannot-link constraint among them and let ε_1 be the value of the largest edge in the group of all density-connected paths among p_1 and p_2, then ε_1 is a lower bound on the value of *Eps*. for the semi-supervised model. On the other hand, assuming p_1 and p_2 are instances with a must-link constraint and let ε_2 be the value of the largest edge in the group of all density-connected paths among p_1 and p_2, then ε_2 is an upper bound on the value of Eps for the semi-supervised model. In this way all the pairwise constraints are inspected, updating either the lower bound or the upper bound of the interval of possible values.

Furthermore, the algorithm is augmented so that it will generate fuzzy partitions. In this way instances do not have to exclusively belong to one of the clusters in a complete way, but an instance may have partial memberships in several clusters. There are several methodologies for applying fuzzy approaches on the DBSCAN algorithm [16]. In our case the employed methodology is based on distances, where the membership of the point p_i in the cluster C_j is evaluated according to:

$$\mu(p_i, C_j) = \begin{cases} 1 & \text{if } d(p_i, p_c) \leq \varepsilon_{min} \\ 0 & \text{if } d(p_i, p_c) > \varepsilon_{max} \\ \frac{\varepsilon_{max} - d(p_i, p_c)}{\varepsilon_{max} - \varepsilon_{min}} & \text{otherwise} \end{cases} \tag{1}$$

Here ε_{min} and ε_{max} represent the lower and upper bounds in the interval of possible values of Eps, while p_c represents the core point in C_j which is closest to the point p_i. As a distance metric many options are theoretically acceptable (like Euclidean distance, Manhattan distance, max distance, Canberra distance, complements of similarity functions etc.), Nevertheless, in our model we have employed exclusively the Euclidean distance.

Finally, our model may be summarized by the following pseudocode:

1. Based on the provided labelled instances, construct the set of pairwise must-link and cannot-link constraints.
2. Inspect every pairwise constraint and apply the updates on the lower and upper bounds of the Eps.
3. Pick as ε the midpoint of the interval and construct the ε – neighborhoods for each instance of the dataset and label the instances as core, border and noise points based on the neighborhood analysis of each data point.
4. Relinquish the noise points (so hereafter they will not be involved the clustering procedure).
5. While there are core points unassigned to any cluster:
   5.1 Pick an unassigned core point (denote it as c_i).
   5.2 Generate a new cluster consisting of the point x and all the other core points which are density-connected to x.
6. Evaluate the fuzzy memberships of the border points into the clusters.

5   Experimental Setup and Results

In order to practically assess the proposed model, a multitude of experimental procedures have been conducted on several benchmark datasets of the UCI repository and also on two synthetic datasets. The benchmark datasets are Breast Cancer Wisconsin (Diagnostic), Dermatology, Diabetes, Glass, Vehicle, Wine and Yeast [17]. The first synthetic dataset is generated by mixtures of random normal distributions with varying means and standard deviations, while the second synthetic dataset is intended to be more complex and is randomly generated as a mixture of a few distinct forms of random distributions. The main characteristics of the employed benchmark and synthetic datasets are presented in the following table (Table 1):

Table 1. Main characteristics of the employed datasets

Dataset	Number of attributes	Number of instances
Breast Cancer (D)	32	569
Dermatology	33	366
Diabetes	8	768
Glass	9	214
Vehicle	18	846
Wine	13	178
Yeast	8	1484
Synth-1	9	250
Synth-2	12	320

In the experimental procedures, for comparison purposes, firstly the conventional DBSCAN is applied (in an unsupervised way) and then the semi-supervised model is applied several times with varying amounts (5%, 10%, 15%, 20%) of labelled instances which are randomly picked from the dataset. Finally the generated results are compared to the intrinsic structure of the dataset, and the accuracy is evaluated using fuzzy extensions of the Rand Index. For each dataset the models are tested with a few distinct values of the *MinPts* parameter and the average value of the accuracy is recorded for comparison reasons. The results of the experimental procedure are summarized in the following table (Table 2):

Table 2. The accuracy of the algorithms on the respective datasets

Dataset	DBSCAN	Semi-supervised fuzzy model			
		5%	10%	15%	20%
Breast Cancer (D)	0.652	0.693	0.745	0.780	0.812
Dermatology	0.615	0.644	0.687	0.752	0.748
Diabetes	0.704	0.733	0.762	0.798	0.809
Glass	0.758	0.789	0.823	0.844	0.865
Vehicle	0.703	0.738	0.759	0.757	0.786
Wine	0.731	0.750	0.772	0.803	0.831
Yeast	0.640	0.683	0.708	0.730	0.741
Synth-1	0.745	0.790	0.823	0.840	0.855
Synth-2	0.679	0.720	0.773	0.779	0.806

As it can be easily observed from the previous table, our semi-supervised model has a significantly better accuracy compared to the conventional unsupervised DBSCAN algorithm. Obviously, the accuracy of the clustering model is considerably improved while additional information is provided. While the amount of the available information increases, the accuracy of the model is also expected to increase (with very few exceptions).

6   Conclusions

In this study we have presented a semi-supervised fuzzy clustering approach based on modifications of the conventional DBSCAN algorithm. Our model operates in scenarios where scarce additional information is available in the form of a relatively small group of labelled instances. The additional information is pre-processed in order to obtain a set of must-link and cannot-link pairwise constraints. This set of constraints is utilized in controlling automatically one of the key parameters of the DBSCAN algorithm, which is the Eps value. Consequently, the minimum number of points (*MinPts*) is the only parameter which has to be controlled by external intervention. In the last stage, our model is augmented in order to generate fuzzy clusters. This is maintained based on the evaluations about the upper and lower bounds of the Eps parameter performed in the previous stage.

Experimental procedures conducted on nine datasets (including seven benchmark datasets of the UCI repository and two artificial datasets) demonstrated a significant improvement in the accuracy of our semi-supervised fuzzy model compared to the conventional unsupervised approach.

References

1. Zhu, X.: Semi-supervised learning literature survey. University of Wisconsin-Madison, Department of Computer Sciences (2005)
2. Bedalli, E., Mançellari, E., Asilkan, O.: A heterogeneous cluster ensemble model for improving the stability of fuzzy cluster analysis. Proc. Comput. Sci. **102**, 129–136.4 (2016)
3. Berkhin, P.: A survey of clustering data mining techniques. In: Grouping Multidimensional Data, pp. 25–71. Springer, Heidelberg (2006)
4. Bedalli, E., Ninka, I.: Adapting the fuzzy c-means clustering algorithm for a semi-supervised learning approach. Sci. Innov. New Technol. **1**, 61 (2014)
5. Grabocka, J., Bedalli, E., Schmidt-Thieme, L.: Supervised nonlinear factorizations excel in semi-supervised regression. In: Pacific-Asia Conference on Knowledge Discovery and Data Mining, pp. 188–199. Springer, Cham (2014)
6. Schubert, E., Sander, J., Ester, M., Kriegel, H.-P., Xu, X.: DBSCAN revisited, revisited: why and how you should (still) use DBSCAN. ACM Trans. Database Syst. (TODS) **42**(3), 19 (2017)
7. Wagstaff, K., Cardie, C., Rogers, S., Schrödl, S.: Constrained k-means clustering with background knowledge. In: ICML, vol. 1, pp. 577–584 (2001)

8. Grira, N., Crucianu, M., Boujemaa, N.: Semi-supervised fuzzy clustering with pairwise-constrained competitive agglomeration. In: The 14th IEEE International Conference on Fuzzy Systems. IEEE (2005)

9. Lelis, L., Sander, J.: Semi-supervised density-based clustering. In: Ninth IEEE International Conference on Data Mining. IEEE (2009)

10. Abir, S., Eloudi, Z.: Soft DBSCAN: improving DBSCAN clustering method using fuzzy set theory. In: 2013 6th International Conference on Human System Interactions (HSI), pp. 380–385. IEEE (2013)

11. Gan, J., Tao, Y.: DBSCAN revisited: mis-claim, un-fixability, and approximation. In: Proceedings of the 2015 ACM SIGMOD International Conference on Management of Data, pp. 519–530 (2015)

12. Bedalli, E., Enea, M., Esteriana, H.: Exploring user feedback data via a hybrid fuzzy clustering model combining variations of FCM and density-based clustering. In: International Conference on Intelligent Networking and Collaborative Systems, pp. 71–81. Springer, Cham (2018)

13. Khan, K., Rehman, S.U., Aziz, K., Fong, S., Sarasvady, S.: DBSCAN: past, present and future. In: The Fifth International Conference on the Applications of Digital Information and Web Technologies (ICADIWT 2014), pp. 232–238 (2014)

14. Amin, K., Johansson, R.: Choosing DBSCAN parameters automatically using differential evolution. Int. J. Comput. Appl. **91**(7), 1–11 (2014)

15. Carlos, R., Spiliopoulou, M., Menasalvas, E.: C-DBSCAN: density-based clustering with constraints. In: International Workshop on Rough Sets, Fuzzy Sets, Data Mining, and Granular-Soft Computing, pp. 216–223. Springer, Heidelberg, (2007)

16. Dino, I., Bordogna, G.: Fuzzy extensions of the DBSCAN clustering algorithm. Soft Comput. **22**(5), 1719–1730 (2018)

17. Arthur, A., Newman, D.: UCI machine learning repository (2007)

Decision Making Problem of a Single Product Dynamic Macroeconomic Model on Base of Fuzzy Uncertainty

Latafat A. Gardashova[1]($\boxtimes$) (iD) and Babek G. Guirimov[2]

[1] Azerbaijan State Oil and Industry University, Azadlig 20,
Nasimi, Baku, Azerbaijan
latsham@yandex.ru
[2] SOCAR Midstream Operations, Baku, Azerbaijan
guirimov@hotmail.com

Abstract. A problem of fuzzy optimal control for a single-product dynamical macroeconomic model is considered in which the gross domestic product is divided into productive consumption, gross investment, and non-productive consumption. The multi-criteria model is described by a fuzzy differential equation (FDE) to take into account the imprecision inherent in dynamics of real-world systems. We applied DEO (Differential Evolution Optimization) and fuzzy Pareto optimality (FPO) formalism to solve the considered problem that allows to softly narrow a Pareto optimal set by determining degrees of optimality for considered solutions.

Keywords: Single-product dynamic macroeconomic model · Multiobjective optimal control · Pareto optimality · Fuzzy optimality · Fuzzy differential equation

1 Introduction

A real-life control model of economy cannot be modelled without considering uncertainty or impreciseness over parameters and variables. This uncertainty may be defined in stochastic and non-stochastic (fuzzy) sense.

The studies devoted to solving optimal control problems for dynamic economic models have a long history. The models of economic growth have attracted noticeable interest in the area of mathematical economics.

The intensive research in this direction has been done to support developed countries in construction of accurate economic growth models to improve country economic development [1–8]. A number of various mathematical methods of measuring the effectiveness of economic growth have been suggested.

The information relevant to the behaviour of an economic process is uncertain. Uncertainty can be of two types: aleatory (probabilistic) and epistemic (possibilistic) [9]. In the existing works, economic uncertainty is mainly captured by using stochastic techniques. However, the use of stochastic techniques is based on crucial assumptions which notably constraint its use for adequate modelling of real-world economic

© Springer Nature Switzerland AG 2020
R. A. Aliev et al. (Eds.): ICSCCW 2019, AISC 1095, pp. 237–245, 2020.
https://doi.org/10.1007/978-3-030-35249-3_30

uncertainty. In order to deal with possibilistic data L. Zadeh suggested the fuzzy sets theory and fuzzy logic. These theories have passed a long way and provided a lot of successful applications. In particular, control systems based on fuzzy logic were successfully applied for control of various complex and uncertain objects and provided better results than their classical counterparts. Fuzzy mathematics describes relations between uncertain quantities including arithmetic operations, properties of fuzzy functions, dynamics of possibilistic uncertainty, and, in general, analysis over spaces of fuzzy sets. Fundamental approach to modelling behaviour of a dynamical system under possibilistic uncertainty is fuzzy differential equations (FDEs) [10]. Theory of FDEs [9, 11] is developed to describe uncertainty in trajectories of a dynamical system.

The existing approaches based on the classical definition of Pareto optimality create several difficulties when applied to problems with high number of criteria. As compared to classical Pareto Optimality which determines whether a considered solution is optimal, the FPO determines a degree of Pareto optimality of a considered solution. In other words, FPO is developed to differentiate "more optimal" solutions from "less optimal" solutions. The motivation to use FPO is that it is more suitable for problems characterized by uncertainty and many objective functions.

In the present paper we suggest an approach to decision making in a single-product dynamic economic model. The considered problem is represented as a multi-objective optimal control problem of dynamics of capital under uncertain relevant information. In order to model possibilistic uncertainty in dynamics of capital a linear FDE is used.

The use of FDE allows taking into account imprecision inherent in the dynamics that may be naturally conditioned by influence of various external factors, unforeseen contingencies of future and so on.

It should be noted that dynamic programming is very computationally intensive and even more for fuzzy problems. Therefore, we decided to apply a modified version [12] of the evolutionary DEO (Differential Evolution Optimization) method [13] to our problem.

The paper is organized as follows. Section 2 is devoted to the preliminaries, which contains definitions and background information related to the considered topic. In Sect. 3, we formulate the statement of the problem. The method of solution is presented in Sect. 4. In Sect. 5, we consider examples and results of computer simulation related to the considered problem. And suggests future research avenues for neoclassical macroeconomic analysis. Section 6 presents our conclusions.

2   Preliminaries

Definition 1. Fuzzy Pareto Optimality Formalism [14]**.** Fuzzy Pareto Optimality formalism extends the classical Pareto Optimality concept in order to differentiate among Pareto Optimal alternatives. Let's shortly explain Fuzzy Pareto Optimality formalism suggested in [2, 14, 15].

In [14] they introduce two-place functions nbF, neF, nwF that for each pair of alternatives $\tilde{u}_i, \tilde{u}_k$ count the number of criteria with respect to which $\tilde{u}_i$ performs "better", "equivalent", "worse" respectively, than $\tilde{u}_k$:

$$nbF(\tilde{u}^i, \tilde{u}^k) = \sum_{j=1}^{M} \mu_b^j (J_j(x, \tilde{u}^i) - J_j(x, \tilde{u}^k)), \tag{1}$$

$$neF(\tilde{u}^i, \tilde{u}^k) = \sum_{j=1}^{M} \mu_e^i (J_j(x, \tilde{u}^i) - J_j(x, \tilde{u}^k)) \tag{2}$$

$$nwF(\tilde{u}^i, \tilde{u}^k) = \sum_{j=1}^{M} \mu_w^j (J_j(x, \tilde{u}^i) - J_j(x, \tilde{u}^k)). \tag{3}$$

where $\mu_b^j, \mu_e^j \mu_w^j$ are membership functions for linguistic evaluations "better", "equivalent" and "worse" respectively for j-th criteria [15]. $\mu_b^j, \mu_e^j \mu_w^j$ are constructed such that Ruspini condition holds, which, in turn, results in the following condition:

$$nbF(\tilde{u}^i, \tilde{u}^k) + neF(\tilde{u}^i, \tilde{u}^k) + nwF(\tilde{u}^i, \tilde{u}^k) = \sum_{j=1}^{M} (\mu_b^j + \mu_e^j + \mu_w^j) = M. \tag{4}$$

On the base of $nbF(\tilde{u}^i, \tilde{u}^k)$, $neF(\tilde{u}^i, \tilde{u}^k)$, and $nwF(\tilde{u}^i, \tilde{u}^k)$ they calculate $(1 - kF)$-dominance as a dominance in the terms of its degree. This concepts suggests that $\tilde{u}^i$ $(1 - kF)$-dominates $\tilde{u}^k$ iff

$$neF(\tilde{u}^i, \tilde{u}^k) < M, nbF(\tilde{u}^i, \tilde{u}^k) \geq \frac{M - neF(\tilde{u}^i, \tilde{u}^k),}{kF + 1} \tag{5}$$

With $kF \in [0, 1]$.

In order to determine the greatest kF such that $\tilde{u}^i$. $(1 - kF)$-dominates $\tilde{u}^k$, a function d is introduced:

$$d(\tilde{u}^i, \tilde{u}^k) = \begin{cases} 0, & if, \ nbF(\tilde{u}^i, \tilde{u}^k) \leq \frac{M - neF(\tilde{u}^i, \tilde{u}^k)}{2} \\ \frac{2nbF(\tilde{u}^i, \tilde{u}^k) + neF(\tilde{u}^i, \tilde{u}^k) - M}{nbF(\tilde{u}^i, \tilde{u}^k)} \end{cases}, \ otherwise \tag{6}$$

Given d, the desired greatest kF is found as $1 - d(\tilde{u}^i, \tilde{u}^k)$. $1 - d(\tilde{u}^i, \tilde{u}^k) = 1$ means Pareto dominance of $\tilde{u}^i$ over $\tilde{f}_k$ whereas $d(\tilde{u}^i, \tilde{u}^k) = 0$ means no Pareto dominance of $\tilde{u}^i$ over $\tilde{u}^k$.

In contrast, to determine whether $\tilde{u}^*$ is Pareto optimal, in fuzzy optimality formalism they determine whether $\tilde{u}^*$ is Pareto optimal with the considered degree kF. $\tilde{u}^*$ is kF optimal if and only if there is no $\tilde{u}^i \in U$ such that $\tilde{u}^i(1 - kF)$ dominates $\tilde{u}^*$.

The main idea of fuzzy optimality concept suggests to consider $\tilde{u}^*$ in terms of its degree of optimality $do(\tilde{f}^*)$ determined as follows:

$$do(\tilde{u}^i) = 1 - \max_{\tilde{u}_i} d(\tilde{u}^i, \tilde{u}^*). \tag{7}$$

Definition 2. Fuzzy stability [12]. The solution $\tilde{x}(t, t_0, \tilde{x}_0)$ of system $\tilde{x}' = \tilde{f}(t, \tilde{x}')$ is said to be uniformly fuzzy Lipschitz stable if there exist $\tilde{M}$, $\tilde{\delta} > 0$ such that

$$\|\tilde{x}(t, t_0, \tilde{x}_0) - \tilde{x}(t, t_0 \tilde{\tilde{x}})\|_{fH} \leq \tilde{M}\|\tilde{x}_0 - \tilde{\tilde{x}}\|_{fH} \tag{8}$$

for $t \geq t_0$, $\|\tilde{x}_0 - \tilde{\tilde{x}}_0\|_{fH} \leq \tilde{\delta}$. A degree of stability is defined by a degree of inequality (8) satisfaction – by the distance between the left and the right parts of (8). The larger this distance is, i.e. the farther $\|\tilde{x}(t, t_0 \tilde{y}_0) - \tilde{x}(t, t_0, \tilde{x}_0)\|_{fH}$ it from $\tilde{M}\|\tilde{y}_0 - \tilde{x}_0\|_{fH}$, the larger a degree of stability is.

3 Statement of the Problem

Single-product macroeconomic models are the models dealing with changes of the interconnected aggregated macroeconomic indicators in the economic system producing only one product. Such indicators are: gross domestic product, final product, manpower resources, productive assets, capital investments (formation), consumption, etc.

A macroeconomic model considers factors characterizing the production: labor costs *(L)*, labor facilities (BPA – Basic Productive Assets) K, and labor subjects. Natural resources W^S and labor subjects are the components returned to the production. Gross product is the result of production activity X (designated by u_1). Gross product X is divided into production consumption W and final product Y. In turn the final product Y in the distribution block Py is divided into gross capital investments I and non-productive consumption C (or u_2). Gross capital investments are divided into depreciation charges A and net capital investments Ik expanding the productive assets.

In a single-product model the productive costs W are proportional to production volume, the rate of increase in basic productive assets is proportional to net capital investments. The capital investments or gross investments I are completely spent for a gain in BPA and depreciation charges. The depreciation charges A are proportional to the BPA.

It is required to formalize a problem under conditions of maximizing income, cutting costs of gross product, increasing the BPA at a time t, and increasing the discounted sum of direct consumption during a planning term.

Let us consider a single-product dynamic macroeconomic model that reflects interaction among factors of production when a gross domestic product (GDP) is divided into productive consumption, gross investment and non-productive consumption as the performance of production activity. In its turn productive consumption is assumed to be completely consumed on capital formation and depreciation. These processes are complicated by the presence of possibilistic, that is, fuzzy uncertainty which is conditioned by imprecise evaluation of future trends, unforeseen contingencies and other vagueness and imprecision inherent in economical processes. Under the above mentioned assumptions the considered dynamic macroeconomic model can be described by the following differential equation:

$$\tilde{K}(n+1) = \tilde{K}(n) + \frac{1}{q}\left[(1-a)\tilde{u}_1(n) + \mu\tilde{K}(n) - \tilde{u}_2(n)\right] \tag{9}$$

Here K is a variable describing imprecise value of capital, u_1 (the first control variable) is the value of GDP, u_2 (the second control variable) is the value of non-productive consumption, a is the direct cost, q is basic fund rate, μ is amortization expenses. Later three variables are coefficients related to the productive consumption, net capital formation and depreciation, respectively.

Let us consider a multi-objective optimal control problem of (9)–(11) within the period of planning $t \in [t_0, T]$ with four objective functions (criteria): J_1 – profit, J_2 – reduction of production expenditures of GDP, J_3 – the value of capital at the end of period $[t_0, T]$, J_4 – the discount sum of direct consumptions over $[t_0, T]$. The considered multiobjective optimal control problem is formulated as

$$Max(\tilde{J}_1(\tilde{u}_1, \tilde{u}_2) = \int_{t_0}^{T} p(t)\tilde{u}_2(t)dt, \quad \tilde{J}_2(\tilde{u}_1, \tilde{u}_2) = -c\int_{t_0}^{T} |\tilde{u}_1(t)|dt, \quad \tilde{J}_3(\tilde{u}_1, \tilde{u}_2) = \tilde{K}(t), \quad \tilde{J}_4(\tilde{u}_1, \tilde{u}_2) = \int_{t_0}^{T} \theta(t)\tilde{u}_2(t)dt)$$

$$\tag{10}$$

Here $p(t) = \frac{1}{(1+R)^t}$ is the price of production unit, produced at the time t, R is inflation rate, $\theta(t) = e^{rn\Delta}$ is the discount function, $c = const > 0$ is the scientific-technical progress coefficient, r is discounting coefficient.

The above is subject to the following constraints:

$$\underline{u} \le u_1 \le \overline{u}, \quad \underline{u} \le u_2 \le \overline{u}, \quad \underline{K} \le K_n \le \overline{K} \tag{11}$$

4 Method of Solution

As was stated above, in this work we solve a problem of optimal control for a single-product dynamical macroeconomic model with the help of a method of differential evolution.

According to the methodology, optimal control problem can be reduced to an approximate multi-objective linear programming problem. We used the following expanded example of the problem described by (9)–(11):

$$\tilde{K} = -\frac{\beta}{\gamma} + \left(\tilde{k}_0 + \frac{\beta}{\gamma}\right)e^{\gamma t}$$

$$\tilde{K}_n = \frac{b_1(1-a)-b_2}{(1-a)a_1-(\mu+a_2)} + \left(\tilde{k}_0 - \frac{b_1(1-a)-b_2}{(1-a)a_1-(\mu+a_2)}\right)e^{\gamma t}$$

$$\beta = b_1(1-a) - b_2, \quad \gamma = (1-a)a_1 - (\mu+a_2), \quad t = n\Delta$$

$$\tilde{J}_1 = \sum_{n=0}^{N} p(a_2 \tilde{k}_n + b_2)\Delta = \sum_{n=0}^{N} pa_2(-\tfrac{\beta}{\gamma} + (\tilde{k}_0 + \tfrac{\beta}{\gamma})e^{\gamma t})\Delta \rightarrow \max$$

$$\tilde{J}_2 = -c \sum_{n=0}^{N} \left| a_1(-\tfrac{\beta}{\gamma} + (\tilde{k}_0 + \tfrac{\beta}{\gamma})e^{\gamma n \Delta} + b_1) \right| \Delta \rightarrow \max$$

$$\tilde{J}_3 = \sum_{n=0}^{N} (-\tfrac{\beta}{\gamma} + (\tilde{k}_0 + \tfrac{\beta}{\gamma}))e^{\gamma n \Delta} \rightarrow \max$$

$$\tilde{J}_4 = \Delta \sum_{n=0}^{N-1} e^{rn\Delta}((-\tfrac{\beta}{\gamma} + (\tilde{k}_0 + \tfrac{\beta}{\gamma})e^{\gamma n \Delta})a_2 + b_2) \rightarrow \max$$

Here

$$\underline{\tilde{K}_n} \leq \tilde{K}_n \leq \overline{\tilde{K}_n}, \quad \underline{\tilde{U}_1} \leq a_1 \tilde{k}_n + b_1 \leq \overline{\tilde{U}_1}, \quad \underline{\tilde{U}_2} \leq a_2 \tilde{k}_n + b_2 \leq \overline{\tilde{U}_2}$$

5   Simulation Results

Let's consider solving problem (9–11) on the base of the method suggested in Sect. 4 under the following data:

$\tilde{K}_0 = 20\tilde{0}0, \quad 20\tilde{0}0 \leq \tilde{K}_n \leq 30\tilde{0}0, \quad 6\tilde{0}0 \leq a_1 k_n + b_1 \leq 8\tilde{0}0, \quad 5\tilde{0}0 \leq a_2 k_n + b_2 \leq 7\tilde{0}0,$
$n = \overline{0,9}, \ \Delta = 0.2, \ N = 10, \ r = 0.05, \ \mu = 0.15,$
$a_1 \geq 0, \ a_2 \geq 0.005, \ N = 9; \ q = 1.02; \ a = 0.54; \ c = 0.5; R = 0.1;$

The DE's parameters used were: population size NP = 100, crossover constant CR = 0.9, and the weighting factor F = 1.0.

At the first stage we obtained a set of the feasible solutions $\tilde{u}^i = (\tilde{u}_1^i, \tilde{u}_2^i)$, where $\tilde{u}_1^i(t, \tilde{x}) = a_1^i \tilde{x} + b_1^i$ and $\tilde{u}_2^i(t, \tilde{x}) = a_2^i \tilde{x} + b_2^i$ for the considered problem.

At the second stage we compare values of criteria for the obtained feasible solutions.

Given the values of criteria for feasible solutions we determined the corresponding set of Pareto optimal solutions. The set of Pareto optimal solutions is given in Table 1.

At the next stage we calculated values of nbF, neF, nwF. Some of the calculated values of nbF, neF, nwF are given below:

$$nbF(\tilde{u}^1, \tilde{u}^2) = 0.1, \quad nbF(\tilde{u}^3, \tilde{u}^1) = 0.00367, \quad nbF(\tilde{u}^5, \tilde{u}^3) = 0.0049$$
$$neF(\tilde{u}^1, \tilde{u}^2) = 3.9281, \quad neF(\tilde{u}^3, \tilde{u}^3) = 4, \quad neF(\tilde{u}^5, \tilde{u}^4) = 3.9996$$
$$nwF(\tilde{u}^1, \tilde{u}^2) = 0.00516, \quad nwF(\tilde{u}^3, \tilde{u}^1) = 0.1, \quad nwF(\tilde{u}^5, \tilde{u}^3) = 0.000389$$

Given the calculated values of nbF, neF, nwF for each pair of solutions $\tilde{u}_i, \tilde{u}_k$ we determined the greatest kF such that $\tilde{u}_i \ 1 - kF$ dominates $\tilde{u}_k$ as $kF = 1 - d(\tilde{u}_i, \tilde{u}_k)$, where calculated $d(\tilde{u}_i, \tilde{u}_k)$ is given below:

Table 1. Pareto optimal set (fuzzy triangular numbers). Criteria values.

Feasible solutions	J1	J2
Ex001	(1282.5, 1283.5, 1284.3)	(−736.4, −713.3, −690.3)
Ex003	(1252.9, 1286.8, 1320.7)	(−619.1, −600, −580.9)
Ex004	(1267.7, 1286.8, 1305.9)	(−617.8, −600, −582.2)
Ex005	(1252.7, 1286.8, 1320.9)	(−619.8, −600, −580.1)
Ex007	(1252.8, 1286.8, 1320.8)	(−619.5, −600, −580.5)
	J3	J4
Ex001	(2842.7, 2934.9, 3027.2)	(1307.5, 1308.1, 1308.9)
Ex003	(1781.7, 2000.0, 2218.3)	(1277.2, 1311.9, 1346.5)
Ex004	(1920.9, 2000.0, 2079.1)	(1292.2, 1311.8, 1331.5)
Ex005	(1746.1, 2000.0, 2253.9)	(1276.9, 1311.8, 1346.8)
Ex007	(1763.2, 2000.0, 2236.8)	(1277.1, 1311.8, 1346.6)

$$d(\tilde{u}_i, \tilde{u}_k) = \begin{bmatrix} 0 & 0.92269 & 0.94729 & 0.92147 & 0.92206 \\ 0 & 0 & 0.96749 & 0 & 0 \\ 0 & 0 & 0 & 0 & 0 \\ 0 & 0.89083 & 0.95589 & 0 & 0.88995 \\ 0 & 0.89165 & 0.96105 & 0 & 0 \end{bmatrix}$$

Finally, for each $\tilde{u}^*$ we calculated its degree of optimality $do()$:

$$do(\tilde{u}^1) = 1, \quad do(\tilde{u}^2) = 0.077307, \quad do(\tilde{u}^3) = 0.032506, \quad do(\tilde{u}^4) = 0.078531, \quad do(\tilde{u}^5)$$
$$= 0.07794$$

The optimal solution $\tilde{u}^1$ has the highest degree of optimality $do(\tilde{u}^1) = 1$.

However, this does not complete our research on solving of the considered problem. For each the considered fuzzy Pareto optimal solutions $\tilde{u}^1, i = 1, \ldots, 5$ we also need to investigate stability of the corresponding solutions of FDE the fuzzy stability criterion (see Preliminaries). The reason is that stability is the most important property of a control process. In order to investigate stability of trajectory $\tilde{x}^1$ defined by $\tilde{u}^1$ we choose $\tilde{V}_{(3)}(t, \tilde{x}^i, \tilde{\tilde{x}}) = \|\tilde{x}^i - \tilde{\tilde{x}}\|_{fH}$, where as condition $\tilde{V}'_{(3)}(t, \tilde{x}^i, \tilde{\tilde{x}}) \leq 0$ the following one is used: $\tilde{V}'_{(3)}(t, \tilde{x}^i, \tilde{\tilde{x}}) = \frac{1}{q}(a_1^i - aa_1^i - \mu - a_2^i) < 0$. By verifying condition for fuzzy Pareto optimal solutions $\tilde{u}^i, i = 1, \ldots, 5$ we obtained the following results: $\tilde{V}'_{(3)}(t, \tilde{x}^1, \tilde{\tilde{x}}) = -0.0439135$, $\tilde{V}'_{(3)}(t, \tilde{x}^2, \tilde{\tilde{x}}) = -0.177450922$, $\tilde{V}'_{(3)}(t, \tilde{x}^3, \tilde{\tilde{x}}) = -0.177451196$, $\tilde{V}'_{(3)}(t, \tilde{x}^4, \tilde{\tilde{x}}) = -0.177450667$, $\tilde{V}'_{(3)}(t, \tilde{x}^5, \tilde{\tilde{x}}) = -0.177451039$

Therefore, the solutions of fuzzy differential equation corresponding to solutions $\tilde{u}^i, i = 1, \ldots, 5$ are stable. Then the selected solution of the considered optimal control problem is solution $\tilde{u}^1$ which has the highest degree of optimality among the stable solutions.

6 Conclusions

The considered dynamical problem is multi-criteria and is characterized by uncertainty intrinsic to real-world economy with non-statistical parameters and fuzzy conditions. Solving such a problem by using the existing classical optimization methods is very hard (if not impossible) due to non-intuitive and inefficient problem description, inadequate crisp modeling and computational complexity. Hence thereby obtained results would be non-informative and unrealistic.

The application of fuzzy modeling and evolutionary optimization approaches has allowed us to simplify description of fuzzy multi-objective control problem for a single-product dynamical macroeconomic model. For solving it we for the first time used a modified DEO method [12, 13] with multiple fuzzy constraints and objectives. For the problem complicated by four conflicting criteria and non-stochastic uncertainty, we obtained intuitively meaningful solution after relatively small processing time. The solution has undergone the stability/sensitivity test and is proven to be sustainable.

References

1. Aliev, R.A.: Modelling and stability analysis in fuzzy economics. Appl. Comput. Math. 7(1), 31–53 (2008)
2. Aliev, R.A., Pedrycz, W.: Fundamentals of a fuzzy-logic-based generalized theory of stability. IEEE Trans. Syst. Man Cybern. Part B (Cybern.) 39(4), 971–988 (2009)
3. Kantorovich, L.V., Zhiyanov, V.I.: The one-product dynamic model of economy considering structure of funds in the presence of technical progress. Rep. Acad. Sci. USSR 211(6), 1280–1283 (1973)
4. Kozlova, O.R.: Generalized solutions in one problem of optimal control for the Tobin macroeconomic model. Autom. Remote Control 69(3), 483–496 (2008)
5. Kryazhimskii, A., Watanabe, C.: Optimization of Technological Growth. Gendaitosho, Kanagawa (2004)
6. Krasovskii, A.A., Tarasyev, A.M.: Dynamic optimization of investments in the economic growth models. Autom. Remote Control 68(10), 1765–1777 (2007)
7. Kwintiana, B., Watanabe, C., Tarasyev, A.M.: Dynamic optimization of R&D intensity under the effect of technology assimilation: econometric identification for Japan's Automotive industry. Interim report IR-04-058, IIASA, Laxenburg, Austria (2004)
8. Leontyev, V.: Input-Output Economics. Oxford University Press, New York (1966)
9. Bede, B., Gal, S.: Generalizations of the differentiability of fuzzy-number valued functions with applications to fuzzy differential equations. Fuzzy Sets Syst. 151(3), 581–599 (2005)
10. Diamond, P., Kloeden, P.: Metric Spaces of Fuzzy Sets. Theory and Applications. World Scientific, Singapore (1994)
11. Lakshmikantham, V., Gnana, T.B., Vasundhara, J.D.: Theory of Set Differential Equations in Metric Space. Cambridge Scientific Publishers, Cambridge (2005)
12. Aliev, R.A., Guirimov, B.G.: Type-2 Fuzzy Neural Networks and Their Applications. Springer, Heidelberg (2014)
13. Storn, R., Price, K.: Differential evolution - a simple and efficient heuristic for global optimization over continuous spaces. J. Global Optim. 11, 341–359 (1997)
14. Aliev, R.A., Huseynov, O.H.: Decision theory with imperfect information. World Scientific Publishing Co. Pte. Ltd., Singapore (2014)

15. Farina, M., Amato, P.: A fuzzy definition of "optimality" for many-criteria optimization problems. IEEE Trans. Syst. Man Cybern. Part A Syst. Hum. **34**(3), 315–326 (2004)
16. Aliev, R., Tserkovny, A.: Systemic approach to fuzzy logic formalization for approximate reasoning. Inf. Sci. **181**(6), 1045–1059 (2011)

Application of an Artificial Intelligence Decision-Making Method for the Selection of Maintenance Strategy

Mohammad Yazdi[1] , Tulen Saner[2(✉)] ,
and Mahlagha Darvishmotevali[2]

[1] Universidade de Lisboa, Lisbon, Portugal
mohammad.yazdi@tecnico.ulisboa.pt
[2] Department of Tourism and Hotel Management, Near East University,
99138 Nicosia, TRNC, Mersin 10, Turkey
{tulen.saner,mahlagha.darvish}@neu.edu.tr

Abstract. Multi-criteria decision-making (MCDM) methods have been extensively used in different types of engineering applications. Therefore, the viability and reliability of the methods are always considered as a requirement in order to be adopted in different situations. Best worst method (BWM) which has recently been introduced to increase the consistency and uniformity of MCDM, using a multi-objective mathematical programming, has enough capability to determine the optimum results in a decision making problem. The study aims to utilize the BWM to select and prioritize maintenance strategy in offshore operation. In this respect, a group of decision makers participated in the assessment by expressing their opinions. Typically, if the final evaluation result is close to each expert's opinion, subsequently the expert will accept it. Due to the fact that, BWM results in the optimum values; the results would be accepted by all the experts that have participated in the study.

Keywords: BWM · Multi-criteria decision making · Off-shore operation · Optimization

1 Introduction

In recent years, multi-criteria decision making (MCDM) methods are the main part of decision making science in the different engineering application [1]. In this respect many scholars have widely developed MCDMs in order to acquire much more realistic results. As an example, MCDMs are used in aerospace [2], chemical process [3, 4], construction [5], environment [6], etc. However, the significant concern of evaluators to use the MCDMs methods in different industrial sectors are always questioning the validity and reliability of MCDMs techniques. In order to solve a decision making problem, a group of experts which have relevant background and expertise are grouped together to express their opinions. Thereafter, the expert who has a better qualification profile, the final results would be close to his or her opinions. In contrast, humanity causes that all experts expect that the final results should be close to their individual

© Springer Nature Switzerland AG 2020
R. A. Aliev et al. (Eds.): ICSCCW 2019, AISC 1095, pp. 246–253, 2020.
https://doi.org/10.1007/978-3-030-35249-3_31

expressions. Subsequently, the outputs of MCDMs methods are typically failed to bring satisfaction to all employed experts. According to this fact, those methods which result in the optimal value based on mathematical programming can properly illustrate their inherent merits to the assessors as a reliable and valid technique.

In this study, MCDM method is used, which is based on mathematical programming called Best worst method (BWM) to solve a realistic decision making problem and compare the results with the such available conventional MCDM methods like as AHP (analytical hierarchy process) [7] and TOPSIS (technique for order preference by similarity to ideal solution) [8]. BWM has received considerable attention since its proposal and that is why in recent years BWM have been widely utilized by many scholars for different engineering applications [9]. The organization of the study is as follows. In Sect. 2, the methodology of using BWM is explained. An application of case study is provided in Sect. 3. Finally, the conclusion, cover the limitations of study as well as some directions for the future study are discussed in Sect. 4.

2   Method

Rezaei (2015) [10] introduced the Best worst method (BWM(as an appropriate alternative to AHP in MCDM problems. BWM method has considerable advantages such as better consistency and obtaining optimal importance weight. In addition, BWM decreases the pairwise comparisons' iteration significantly by merely performing the comparisons which means that decision-makers need to examine the preference of the best criterion over other criteria and also obtain the preference of all criteria over the worst criterion, engaging Saaty scale number which is between 1 and 9. It has received considerable attention since its proposal and this is why in recent years BWM has been widely used by many scholars for different engineering applications (see an overview paper on integrations and applications of the BWM [9]). BWM can be applied in different multi-criteria decision making problem in four different states which provided in details as follows:

Stage 1: Computing the best and the worst criteria. In this step, the best and the worst criteria between the main criteria and all alternatives are computed based on the practical circumstances and a decision maker's opinion. Main criteria are denoted in the same order as $C_1, C_2, C_3, \ldots, C_i$. The alternatives are represented as $A = \{A_1, A_2, \ldots, A_n\}$. In this accordance, the best and worst main criteria are denoted as C_{Best} and C_{Worst}, respectively. In a similar way, the best and worst alternative given by decision maker is shown as C_{Best} and C_{Worst}.

Stage 2: Determining the preference of the best criterion/alternative over the rest of the specified criteria/alternatives. According to the obtained results from stage 1, the preference of the best criterion/alternative over the rest ones is provided by decision maker utilizing Saaty scale 1 to 9 representing in [11]. The best to others (BO) obtained from decision maker illustrated as $BO = \{BO_{B1}, BO_{B2}, \ldots, BO_{Bn}\}$, where BO_{Bi} reflects the rating of the best criterion/alternative B over ideal criterion/alternative i, particularly $BO_{BB} = 1$.

Stage 3: Determining the preference of the rest of criteria/alternatives over the worst one. Similar to stage 2, the preference of all other criteria/alternatives over the worst one is also estimated by decision maker utilizing comparison scale 1 to 9. The vector of other to worst (WO) obtained from decision maker is denoted as $OW = \{OW_{1W}, OW_{2W}, \ldots, OW_{nW}\}$ where OW_{jW} reflects the preference of any criterion/alternative j to the worst criterion/alternative W, particularly $OW_{WW} = 1$.

Stage 4: Finding the optimal weight of all criteria/alternative. The optimal importance weight of all criteria is denoted as $\left(\omega_{C_1}^*, \omega_{C_2}^*, \ldots, \omega_{C_n}^*\right)$. In order to obtain the optimal importance weight of criteria, the maximum absolute differences of $|\omega_B/\omega_i - BO_{Bi}|$ and $|\omega_i/\omega_W - OW_{iW}|$ have to be minimized for all i. The aforementioned conditions can be mathematically modeled as follows:

Model 1:

$$\min \; \max_i\{|\omega_B/\omega_i - BO_{Bi}|, |\omega_i/\omega_W - OW_{iW}|\}$$

Subject to:
$$\sum_i \omega_i = 1, \omega_i \geq 0, \text{ for all } i.$$

The provided model 1 can simply be converted into model 2 as:

Model 2:

$$\min \zeta$$

Subject to:

$$|\omega_B/\omega_i - BO_{Bi}| \leq \zeta, |\omega_i/\omega_W - OW_{jW}| \leq \zeta, \sum_i \omega_j = 1, \omega_i \geq 0, \text{ for all } i.$$

By solving model 2, the optimal value of ζ as well as the optimal importance weight of criteria obtained by decision maker can be derived. The same procedure is also applied for alternatives respect to each criteria. In order to figure out the reliability and validity of the obtained results, the consistency ratio (CR) is computed by dividing the ζ over the consistency index (CI) [12, 13]. Similar to AHP, in BWM the smaller value of ζ is signifying the smaller CR. Thus, high consistency situation simply means that the value of CR is close to zero. Rezaei (2016) [14] further discussed that model 2 can resulted the multiple optimal solutions for a problem which has more than three criteria. In other words, for not fully consistent comparisons having more than three criteria/alternatives, the intervals weights can be used to obtain the priority of criteria in the multi-optimality problems. In this accordance, Rezaei (2016) [14] concluded that a single solution is preferred in most cases compare with multiple optimal solutions.

Therefore, Rezaei (2016) [14] introduced a linear model of BWM to compute the single solution inoptimality problems. Then, the problem (model 1) can be formulated into model 3 as:

Model 3:

$$\min \ \max_i \{|\omega_B - BO_{Bi} \cdot \omega_i|, |\omega_i - OW_{iW} \cdot \omega_W|\}$$

Subject to:

$$\sum_i \omega_i = 1, \omega_i \geq 0, \text{ for all } i.$$

Model 3 can also be modeled into the linear mathematical programing as:
Model 4:

$$\min \zeta^*$$

Subject to:

$$|\omega_B - BO_{Bi} \cdot \omega_i| \leq \zeta^*, |\omega_i - OW_{iW} \cdot \omega_W| \leq \zeta^*, \sum_i \omega_i = 1, \omega_i \geq 0, \text{ for all } i.$$

Model 4 can properly be used to make a single solution of the optimal value of ζ^* and optimal weights $(\omega_1^*, \omega_2^*, \ldots, \omega_n^*)$ provided by a decision maker whereas non-linearity and inconsistency problems in model 2 may result in multiple solutions. The closer value of ζ^* to zero means that the comparison obtained by the decision maker has high consistency that is while in model 2 the CI is used for this purpose. Finally, the weight of alternative with consideration of all criteria is determined by multiplying the optimal weigh of each criterion by the corresponding optimal weight of each alternative. In the next section, an application maintenance strategy selection in off-shore operation is provided to show the effectiveness and feasibility of BWM.

3   Results

The BWM is established in a decision making assessment of the selection of an on-board machinery (crane) maintenance strategy for off-shore operating according to group of independent experts. The hierarchical model of this decision making assessment procedure is as depicted in Fig. 1.

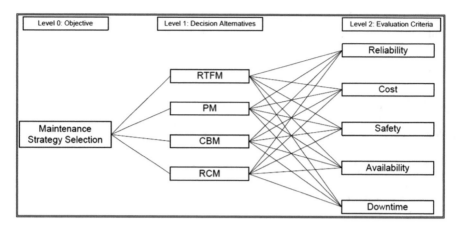

Fig. 1. Hierarchical structure of maintenance strategy selection. Note: RTFM, PM, CBM, and RCM stand for run-to-failure maintenance, preventive maintenance, condition based maintenance, and reliability centered maintenance, respectively (Adopted after [15]).

In this study, a heterogeneous group, including three independent experts are asked to express their opinions about the defined decision-making problem. The three experts who are going to make judgements have different backgrounds and expertise in their professions. The experts' quality profile is the same as the study of [1]. Let's consider the three experts as E#1, E#2, and E#3 which has the order of importance of the weight of the criteria using BWM method as 0.37, 0.33, and 0.30. Herein, it is asked from all employed experts to express the worst and the best criteria and order the criteria according to the BWM procedure explained in Sect. 2. Next, in the same procedure, the experts are evaluating the alternatives corresponding to the special criteria. Based on the BWM method, the optimum results which show the importance weight of all criteria and alternatives are provided in Tables 1 and 2, respectively.

Table 1. Weight of criteria based on importance using BWM method

Experts	Criteria	Reliability	Cost	Safety	Availability	Downtime
E#1	Best(Reliability)	1	3	3	5	7
	Worst (Down-time)	8	3	6	4	1
	Optimal value	0.458	0.189	0.189	0.114	0.050
E#2	Best (Safety)	3	9	1	4	5
	Worst (Cost)	4	2	9	5	4
	Optimal value	0.194	0.046	0.498	0.146	0.116
E#3	Best(Reliability)	1	6	2	8	4
	Worst (Availability)	7	4	5	1	5
	Optimal value	0.446	0.092	0.276	0.049	0.138
Aggregated optimal value		0.367	0.113	0.317	0.105	0.098

Let's take the first experts opinions to obtain the corresponding optimum importance weight of criteria. Using model 2, the optimal weight of the criteria derived as model 2:

Model 2:

$$\min \zeta^{E\#1}$$

Subject to:

$$\left| \omega_{\text{Reliability}}^{E\#1} - 3 \cdot \omega_{\text{Cost}}^{E\#1} \right| \leq \zeta^{E\#1},$$

$$\left| \omega_{\text{Reliability}}^{E\#1} - 3 \cdot \omega_{\text{Safety}}^{E\#1} \right| \leq \zeta^{E\#1},$$

$$\left| \omega_{\text{Reliability}}^{E\#1} - 5 \cdot \omega_{\text{Availability}}^{E\#1} \right| \leq \zeta^{E\#1},$$

$$\left| \omega_{\text{Reliability}}^{E\#1} - 7 \cdot \omega_{\text{Downtime}}^{E\#1} \right| \leq \zeta^{E\#1},$$

$$\left| \omega_{\text{Reliability}}^{E\#1} - 8 \cdot \omega_{\text{Down-time}}^{E\#1} \right| \leq \zeta^{E\#1},$$

$$\left| \omega_{\text{Cost}}^{E\#1} - 3 \cdot \omega_{\text{Down-time}}^{E\#1} \right| \leq \zeta^{E\#1},$$

$$\left| \omega_{\text{Safety}}^{C.3} - 6 \cdot \omega_{\text{Down-time}}^{E\#1} \right| \leq \zeta^{E\#1},$$

$$\left| \omega_{\text{Availability}}^{C.3} - 4 \cdot \omega_{\text{Down-time}}^{E\#1} \right| \leq \zeta^{E\#1},$$

$$\omega_{\text{Reliability}}^{E\#1} + \omega_{\text{Cost}}^{E\#1} + \omega_{\text{Safety}}^{E\#1} + \omega_{\text{Availability}}^{E\#1} + \omega_{\text{Down-time}}^{E\#1} = 1,$$

$$\omega_{\text{Reliability}}^{E\#1}, \omega_{\text{Cost}}^{E\#1}, \omega_{\text{Safety}}^{E\#1}, \omega_{\text{Availability}}^{E\#1}, \text{ and } \omega_{\text{Down-time}}^{E\#1} \geq 0.$$

Table 2. The importance of weight of alternatives based on related criterion

Criteria						
Alternatives	Reliability	Cost	Safety	Availability	Downtime	Final weight
RTFM	0.210	0.180	0.30	0.170	0.140	0.224
PM	0.230	0.280	0.250	0.250	0.250	0.246
CBM	0.280	0.450	0.340	0.200	0.290	0.311
RCM	0.280	0.090	0.110	0.380	0.320	0.219

Once, the optimum value of all alternatives related to the specific criterion are obtained, the summation of multiplication procedure have been done in order to derive the final importance weights. As it can be seen from Table 2, the priority of alternatives as maintenance strategy selection falls into the CBM ≻ PM ≻ RTFM ≻ RCM. Next, the CR is required to be computed, and in case of inconsistency, it is asked from the experts to express their opinion once more to improve the consistency. However, in our case study, the CR is on acceptable area.

In the next step, the ranking of strategy selection is compared with two different methods as AHP and TOPSIS. As it can be seen from the Fig. 1, using mathematical modeling can effectively change the final results. As an example, the best strategy using BWM is CBM whereas using TOPSIS method the best one is RTFM (Fig. 2).

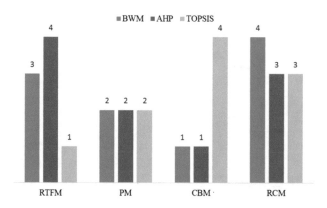

Fig. 2. The comparison between BWM, AHP, and TOPSIS

4 Conclusion

In this study, a recently introduced multi-criteria decision making method called BWM is used to select an appropriate maintenance strategy in the off-shore operation. The BWM because of its inherent features such as better consistency compared with AHP, and resulting in the optimal values, has enough advantages to solve a decision-making problem. Recommendation for future study can be suggested to overcome some challenges envisaged by this study. Firstly, the decision-makers' opinion is subjective-based, and subjectivity is partially meaning fusion and ambiguity.

Therefore, in order to get more realistic results, an extension of fuzzy BWM could be useful. Secondly, subjectivity can be reduced by time. Accordingly, it is vital to map our static system into dynamic one. For this purpose, dynamic BWM using Bayesian updating mechanism can properly deal with a dynamic environment.

References

1. Yazdi, M., Nikfar, F., Nasrabadi, M.: Failure probability analysis by employing fuzzy fault tree analysis. Int. J. Syst. Assur. Eng. Manag. 8(4), 1177–1193 (2017)
2. Di Bona, G., Silvestri, A., Forcina, A., Falcone, D.: AHP-IFM target: an innovative method to define reliability target in an aerospace prototype based on analytic hierarchy process. Qual. Reliab. Eng. Int. 33, 1731–1751 (2017)
3. Wang, C.-N., Huang, Y.-F., Cheng, I.-F., Nguyen, V.: A multi-criteria decision-making (MCDM) approach using hybrid SCOR metrics, AHP, and TOPSIS for supplier evaluation and selection in the gas and oil industry. Processes. 6(12), 252 (2018)
4. Yazdi, M., Darvishmotevali, M.: Fuzzy-based failure diagnostic analysis in a chemical process industry. In: Aliev, R., Kacprzyk, J., Pedrycz, W., Jamshidi, M., Sadikoglu, F. (eds.) 13th International Conference on Theory and Application of Fuzzy Systems and Soft Computing - ICAFS-2018. Advances in Intelligent Systems and Computing, vol. 896, pp. 724–731. Springer, Cham (2019)
5. Yazdi, M.: Risk assessment based on novel intuitionistic fuzzy-hybrid-modified TOPSIS approach. Saf. Sci. 110, 438–448 (2018)
6. Tsai, S.-B., Chien, M.-F., Xue, Y., Li, L., Jiang, X., Chen, Q.: Using the fuzzy DEMATEL to determine environmental performance: a case of printed circuit board industry in Taiwan. PLoS ONE 10, e0129153 (2015)
7. Saaty, T.L.: The analytic network process. Decis. Mak. Anal. Netw. Process. 95, 1–26 (2005)
8. Zhang, X., Xu, Z.: Extension of TOPSIS to multiple criteria decision making with pythagorean fuzzy sets. Int. J. Intell. Syst. 29, 1061–1078 (2014)
9. Liao, H., Shen, W., Tang, M., Mi, X., Lev, B.: The state-of-the-art survey on integrations and applications of the best worst method in decision making: why, what, what for and what's next? Omega 87, 205–225 (2019)
10. Rezaei, J.: Best-worst multi-criteria decision-making method. Omega 53, 49–57 (2015)
11. Saaty, T.L.: Creative thinking, problem solving and decision making. RWS Publications, Pittsburgh (2010)
12. Yazdi, M.: Footprint of knowledge acquisition improvement in failure diagnosis analysis. Qual. Reliab. Eng. Int. 35(1), 405–422 (2018)
13. Yazdi, M.: Acquiring and sharing tacit knowledge in failure diagnosis analysis using intuitionistic and pythagorean assessments. J. Fail. Anal. Prev. 19(2), 369–386 (2019)
14. Rezaei, J.: Best-worst multi-criteria decision-making method: Some properties and a linear model. Omega 64, 126–130 (2016)
15. Asuquo, M.P., Wang, J., Zhang, L., Phylip-Jones, G.: Application of a multiple attribute group decision making (MAGDM) model for selecting appropriate maintenance strategy for marine and offshore machinery operations. Ocean Eng. 179, 246–260 (2019)

Quality Assessment of Oil that Difficult to Recover Based on Fuzzy Clustering and Statistical Analysis

M. K. Karazhanova[1]([⊠]) ⓘ, L. B. Zhetekova[1], and K. K. Aghayeva[2]

[1] Caspian State University of Technology and Engineering named after Sh. Yessenov, 32 Microdistricts, Aktau 130003, Mangistau Region, The Republic of Kazakhstan
mikado_70@inbox.ru, zhetekova81@mail.ru

[2] Oil and Gas Institute of Azerbaijan National Academy of Sciences, F.Amirov 9, AZ1000 Baku, Azerbaijan
kamila.agayeva@hotmail.com

Abstract. Based on the analysis and generalization of literature data, results of the classification of difficult to recover oil selected from Uzen, Zhetibay, Kalamkas, Karakudyk and Karamandybas oil fields of Kazakhstan using fuzzy cluster analysis are presented. Classification of types of difficult to recover oil is considered according to a set of features, including content of sulfur, chlorides, oil density, oil viscosity, permeability of occurrence conditions. Analysis of the classification results of difficult to recover reserves was performed, which showed the need to divide the total sample (set) into homogeneous groups according to a set of classification criteria, for which fuzzy cluster analysis is most suitable. A parameter characterizing the quality of oil is proposed.

Keywords: Hard-to-extract reserves · Sulfur · Chlorides · Viscosity · Oil density · Quality indicator

1 Introduction

In recent years, there has been an increase in the volume of extraction of hard-to-extract oils, which creates a number of technological and economic problems in the process of their recover at various stages, transportation and processing. Inclusion in the development of hard-to-extract oil reserves with high density and viscosity, high sulfur content, wax, tar, heavy metals doesn't only worsen the chemical characteristics of the crude oil and related technological indicators, but also increases the negative impact on the environment, increasing the environmental costs of the territories. Expansion of the refining volumes of the considered oil in recent years, hardness of which is determined not only by the anomalous properties of the crude oils, but also by the complicated geological conditions of occurrence, makes it important and necessary to study qualitative features of the difficult to recover oil. Some features of the physicochemical properties of hard-to-extract oils and conditions of their occurrence are discussed in [1, 2]. However, quality indicators of such oils, inherent in particular, to the fields of

© Springer Nature Switzerland AG 2020
R. A. Aliev et al. (Eds.): ICSCCW 2019, AISC 1095, pp. 254–258, 2020.
https://doi.org/10.1007/978-3-030-35249-3_32

Kazakhstan, have not been studied enough, which makes it difficult to solve both problems of oil production and transportation of hydrocarbon raw materials, and technological problems of petrochemistry and oil refining in conditions of growth in recent years of hard-to-extract reserves. The urgency of the noted problems is especially important for the conditions of Kazakhstan. In this regard, this report presents the results of the analysis of qualitative indicators of various types of difficult to recover oil selected from fields in Kazakhstan, using their classification according to a set of features that determine their quality. The analysis of the classification results of difficult to recover reserves was performed, which showed the need to divide the total sample (set) into homogeneous groups according to a set of classification criteria, for which cluster analysis is most appropriate [3]. In this regard, in order to substantiate the chosen method, this report examines the main essence and results of the application of cluster analysis, assessment of the difficulty degree in extracting reserves and its effect on the performance indicator.

A brief analysis of the classification results for difficult to recover oil. Due to the fact that oil is a very complex natural object, researchers study it from various aspects. Thus, genesis of oil and formation of oil fields, issues of their search and exploration, study of the chemical composition of oils and development of ways of their processing are being studied.

Solution of all these and related issues is possible with a rational classification, which allows to fully and accurately characterize oil from both scientific and practical points of view. This will help choose a grounded method of exposure to the reservoir. In addition, creation of such classification is a very difficult task, which, according to a large number of researchers, has not yet found a satisfactory solution [1–9].

E.M. Khalimov proposed definition of the concept of difficult to recover reserves on the basis of the basic principles and criteria of recognition in 1987, i.e. attribution of oil difficult to recover reserves was formulated in [8]. Based on the generalization of these criteria and taking into account suggestions of other specialists, list of the main types of hard-to-extract oils is given in [2], according to which it is possible to classify hard-to-extract oils with the properties and conditions of occurrence listed in the article. In this connection, the main types of hard-to-extract oils are presented on the basis of generalization of the criteria for classifying oil reserves as hard-to-extract ones using information from the database on the physicochemical properties of the world's oils created at the Institute of Petroleum Chemistry SB RAS. These issues were discussed in a large number of scientific publications, so we limit ourselves to brief information on difficult to recover reserves.

Application of clustering methods in analyzing information. In recent years, in solving classification problems, cluster analysis methods have been widely used [3–5]. As noted in the literature, a large number of studies have been accumulated, in which more than a hundred different clustering algorithms have been proposed, hierarchical and non-hierarchical cluster analyzes, fuzzy clustering are among them.

In recent years, these methods are widely used in the tasks of analyzing information, data mining and processing. Traditional cluster analysis methods imply a clear split of the original set into subsets, at which each point (recognizable object) after splitting falls into only one cluster, i.e. characteristic function (an analogue of the membership function in ordinary sets) is equal to one for this cluster, and zero for the others. In case of missing in any cluster, a new cluster is formed. However, this restriction is not always true. Often it is necessary to produce a partition so as to determine the degree of belonging of each object to each set. In this case, it is advisable to use fuzzy methods of cluster analysis. The tasks of fuzzy cluster analysis and cluster analysis in general, or, as they are otherwise called automatic classification, have recently been widely used in economics, sociology, medicine, geology, oilfield business and other industries, i.e. wherever there are sets of objects of arbitrary nature, described as vectors x = {x_1, x_2, ..., x_n}, which must be automatically divided into groups of objects that are homogeneous in their characteristics [3]. In other words, a cluster can be characterized as a group of elements with common properties, values of classification features that are close to each other. Two signs may be called as characteristics of the cluster: internal homogeneity; external isolation, i.e. heterogeneity with other such clusters.

In [3], in order to classify difficult to recover reserves, clustering was performed using a fuzzy cluster analysis algorithm. For this purpose, data on viscosity, oil density and permeability of oil occurrence conditions for a large number of fields in Kazakhstan are collected. The cluster analysis was performed according to three characteristics-viscosity and oil density, as well as reservoir permeability. As a result of the implementation of the fuzzy cluster analysis program, four classes were obtained, each of which characterizes the degree of difficulty in extracting oil [3].

Results of cluster analysis and quality assessment of difficult to recover oil by a complex of signs. However, Kazakhstan's oils also differ in composition, we have increased the number of attributes by adding sulfur and chloride content, and cluster analysis was performed on five grounds: % content of sulfur, chloride, density, viscosity and permeability of occurrence conditions. As a result of the implementation of the fuzzy cluster analysis algorithm, the data set on the marked fields is divided into three homogeneous classes. The results are shown in the Table 1.

As shown in the table, the results of oil cluster-analysis were divided into three classes. In this case, the first cluster has shown almost clearly, the other two clusters pertain to the second and third classes with a degree close to each other. This point to the fact that according to the considered oil properties types, these clusters are close to each other as well. Therefore, at the next stage, additional data were also included in the analysis for other fields in Kazakhstan.

Assessing and analyzing the regularities of changes in oil quality indicators from the different regions in process, the possibility of using a comprehensive oil quality indicator is discussed in [2] with reference to similar operations. This indicator allows for comparative evaluation of oil quality. Similar parameter is calculating by us by summarizing all the considered oil properties using weighting coefficient:

$$K = \frac{1}{0,036C_S + 0,262\ln(C_{Cl}) + 0,557\ln(\rho) + 0,145\ln(\upsilon)},$$

where C_S - concentration of sulfur, C_{Cl} - concentration of chlorine, ρ - density of oil, υ - viscosity of oil.

Using the obtained indicator is quite convenient for analyzing the quality specifics of various types of oil that difficult to recover.

Table 1. Fuzzy cluster-analysis on oil properties.

C_S, %	C_{Cl}, mg/l	ρ, kg/m³	ν MPa	K_p, mD	Cluster	μ_1	μ_2	μ_3
1	2	3	4	5	6	7	8	9
0,1	3,8	781	4,51	43	1	1	0	0
0,18	7,9	786	3,58	42	1	1	0	0
—	—	—	—	—	—	—	—	—
0,06	41,4	779,6	4,21	80	1	1	0	0
0,05	18,1	767	2,73	38,4	1	1	0	0
—	—	—	—	—	—	—	—	—
0,15	23,4	720	1,33	5,16	1	0,9999	0	0
0,17	24	747	1,64	8,55	1	1	0	0
1,2	27,4	888	18,5	532	3	0	0,4984	0,5016
1,2	23,95	886	22	238	2	0	0,5024	0,4976
1,2	24,2	891	21,4	339	2	0	0,5007	0,4993
1,2	22,3	876	21	411	3	0	0,4972	0,5028
1,2	28,1	873	19,5	250	2	0	0,5013	0,4987
1,55	22,5	888	24,7	397	3	0	0,4999	0,5001
1,57	15,3	884	21,4	243	2	0	0,5014	0,4986
1,3	18,2	889	25,3	828	3	0,0002	0,4991	0,5007
1,12	35,2	871	26,9	228	2	0	0,5013	0,4987
1,4	16,9	894	31,5	189	2	0	0,501	0,499
0,31	329	820,3	10,57	1,6	1	0,9063	0,0468	0,0468
—	—	—	—	—	—	—	—	—
0,16	192	814,9	6,36	3,1	1	0,9987	0,0006	0,0006
0,15	3,2	822	4,94	130	1	1	0	0
—	—	—	—	—	—	—	—	—
0,2	40,4	831	4,96	90	1	1	0	0

The increase of the oil quality indicators is consistent with improving oil quality and reduce – decline in the oil quality and this to be a logical and convention on different types of oil that difficult to recover for comparison.

2   Conclusion

Classification of hard-to-recover oil reserves is important when choosing a stimulation method. The use of a fuzzy cluster analysis allows you to solve this problem successfully in the presence of a set of features that characterizes the degree of complexity of oil recovery.

A complex parameter is proposed that characterizes the degree of complexity of oil recovery from a field with hard-to-recover reserves.

References

1. Maksutov, R., Orlov, G., Osipov, A.: High-viscosity oil reserves completion in Russia. Technol. FEC **6**, 36–40 (2005)
2. Yashenko, I.G., Polishyk, U.M.: Difficult to recover oil and tits properties analysis based on the oil quality qualifications. Vestnik Russian Natural Sciences Academy (West-Siberian department), vol. 19, pp. 37–44 (2016)
3. Akhmetov, D.A., Efendiyev, G.M., Karazhanova, M.K., Koylibaev, B.N.: Classification of hard-to-recover hydrocarbon reserves of Kazakhstan with the use of fuzzy cluster-analysis. In: 13th International Conference on Application of Fuzzy Systems and Soft Computing-ICAFS 2018, pp. 865–872 (2018)
4. Efendiyev, G.M., Mammadov, P.Z., Piriverdiyev, I.A., Mammadov, V.N.: Clustering of geological objects using FCM-algorithm and evaluation of the rate of lost circulation. Proc. Comput. Sci. **102**, 159–162 (2016)
5. Efendiyev, G., Mammadov, P., Piriverdiyev, I., Mammadov, V.: Estimation of the lost circulation rate using fuzzy clustering of geological objects by petrophysical properties. Visnyk Taras Shevchenko Nat. Univ. Kyiv **2**(81), 28–33 (2018)
6. Aliev, R.A., Guirimov, B.G.: Type-2 Fuzzy Neural Networks and Their Applications (2014). http://www.springer.com/us/book/9783319090719. Accessed 19 July 2019
7. Turksen, I.B.: Full Type 2 to type n fuzzy system models. In: Seventh International Conference on Soft Computing, Computing with Words and Perceptions in System Analysis, Decision and Control, Turkey, Izmir, pp. 21–22 (2013)
8. Lisovsky, N.N., Khalimov, E.M.: About reserves difficult to recover qualifications. Vestnik CDC Rosnedra. **6**, 33–35 (2009)
9. Purtova, I.P., Varichenko, A.I., Shpurov, I.V.: Oil reserves difficult to recover. Terminology. The problems and conditions of completion in Russia, Science and FEC (technical scientific research), vol. 6, pp. 21–26 (2011)

Method of Identification of Extremist Texts in the Russian Language Based on the Fuzzy Logic

Olga Vitchenko[1] , Yuri Dashko[1] , Evgeny Tishchenko[2] ,
and Lyudmila Sakharova[2(✉)]

[1] Private Educational Institution of Higher Education
"SOUTHERN UNIVERSITY (IMBL)", Rostov-on-Don, Russia
owinf@mail.ru, dashko@iubip.ru
[2] Rostov State University of Economics, Rostov-on-Don, Russia
celt@inbox.ru, l_sakharova@mail.ru

Abstract. The article describes the author's methodology of the study of the Russian language texts in order to identify their extremist orientation. This method is based on the fuzzy logic. The identified problems in solving the task are indicated. Then the description of the method is presented. There are also main points of fuzzy logic and soft computing tools used in the development of the method. We also determined a set of indicators consisting of four groups: (1) markers of extremist orientation (by types of extremism); (2) markers of linguistic manipulation; (3) indicators of verbal aggression; (4) scientific and journalistic terms. There is quantitative and qualitative characterization for each group of indicators. After that, the process of formation of the evaluation of the studied text on the extremist orientation is described in detail. This process is described on the basis of the system being fuzzy-logical conclusions, called fuzzy multilevel [0,1] – classifiers.

Keywords: Text · Analysis · Extremist orientation · Fuzzy logic · Soft computing · Method

1 Problems of Revealing of Extremist Orientation in the Russian Texts

Analysis of various studies [1–6] allows us to formulate the following problems of solving the problems of identifying extremist orientation in the texts in Russian:

(1) psychologists and linguists have not developed criteria for extremist texts for automatic text analysis;
(2) the developed methods are based in most cases on dictionary systems, which limits their use in existing information systems that require huge costs for updating dictionaries, databases and a large number of rules;
(3) binary classification is mostly used (or divided into three classes), which does not fully solve the problem of identifying complex thematic and emotional orientation of texts in natural languages;

© Springer Nature Switzerland AG 2020
R. A. Aliev et al. (Eds.): ICSCCW 2019, AISC 1095, pp. 259–265, 2020.
https://doi.org/10.1007/978-3-030-35249-3_33

(4) the differentiating signs of definition of extremist orientation of texts at their automatic classification and rubrication are insufficiently investigated;
(5) automatic allocation of texts of illegal subjects requires serious application and fundamental research and development.

Since these problems are characterized by the probabilistic uncertainty of the input data and the uncertainty of expert assessments Fuzzy logic tools are preferred for the definition of indicators in this area. In addition, the fuzzy logic and hybrid fuzzy neural networks allow to go from binary classification and to consider in evaluating the layers of complexity of the thematic and emotional focus of the natural language. Finally, fuzzy inference systems allow the aggregation of large volumes of indicators into complex is estimated without significantly complicating the corresponding mathematical models and the software that implements them.

2   The Main Terms of Fuzzy Logic Used in the Method

The term «fuzzy logic» was introduced by Lotfi Zadeh in 1994 [5] and denotes a set of inaccurate, approximate methods for solving problems of weakly structured systems management in the field of Humanities, robust control, biology, medicine.

Initially, the mathematical apparatus of soft computing (SC – Soft Computing) was a set of models, methodologies and algorithms used in the following areas: fuzzy logic (FL – fuzzy sets, fuzzy logic, fuzzy controllers), neural calculations (NC – neural networks), genetic calculations (GC), probabilistic calculations (PC). Later in this toolkit were included reasoning on the basis of evidence (evidential reasoning), belief networks, chaotic systems and parts of the theory of machine learning. As Lotfi Zadeh notes, the guiding principle of soft computing is: "tolerance for inaccuracy, uncertainty, and partial truth to achieve ease of manipulation, robustness, low solution cost, and better agreement with reality" [6].

Models and systems in which FL, NC, GC and PC, as well as classical mathematical modeling used in some combination, are called hybrid models and systems [3].

Fuzzy inference systems are the preferred tool for fuzzy logic, which allows to aggregate large amounts of information and build on their basis comprehensive assessments of the state of systems. In particular, the most successful attempts to form such estimates are associated with the use of fuzzy multilevel [0,1] – classifiers.

3   A Set of Indicators of Content Monitoring of the Internet Using Hybrid Fuzzy Neural Networks

To identify communities with extremist orientation, as well as to assess their degree and dynamics of destructive impact on Internet users, it is proposed to carry out content monitoring of Internet information through the use of hybrid fuzzy neural networks that calculate a comprehensive assessment based on a set of indicators.

The set of indicators consists of four groups:

(1) markers of extremist orientation (by types of extremism);
(2) markers of psychological manipulation;
(3) indicators of verbal aggression;
(4) scientific and journalistic terms.

For example, the first group of markers of religious extremism includes:

(1) tokens, meaning direct calls for the initiation of religious strife; promotion of exclusivity on the basis of religious affiliation; violation of the rights, freedoms and legitimate interests of a man and citizen, depending on his religious affiliation;
(2) the list of religious extremist groups, their leaders, activists, centers;
(3) distorted geographical names, carrying ideological color;
(4) specific religious-manipulative terms of rituals, adherents and opponents of religion;
(5) religious extremist appeals and slogans;
(6) calls for monetary donations, and to perform pilgrimage, entry into a volunteer corps, etc.;
(7) subversive language;
(8) references to the materials included in a list of religious extremist materials.

The second group, a group of markers of psychological manipulation, includes:

(1) general markers of psycholinguistic manipulation (subjunctive mood, violation of the logic of assumptions and consequences, etc.);
(2) "chopped" abundance of incentive vocabulary (verbal forms of joint action in the first person; formative suffixes -and-(in the Russian language) in the 2nd person; compound forms in 3rd person; modal particles and any words of a motivating character);
(3) highlighting in the text;
(4) the so-called "dialogue with the reader";
(5) lexical ironic stamps of extremist orientation (by types of extremism);
(6) the presence in the text of emotional and psychological characteristics of the image of "enemies", including the enumeration of criminal or socially reprehensible actions, standardly attributed to them by extremists (by types of extremism);
(7) the presence in the text of emotional and psychological characteristics of the image of "heroes", opposed to the image of "enemies";
(8) the presence in the text of the image of "the people" as a passive substance, suffering from the actions of "enemies";
(9) the presence in the text the terminology characteristic for modern historical myth-making (by the types of extremism);
(10) the critical amount of references to historical events in "hot spots", military events, natural, technological and social disasters with a view to fomenting an atmosphere of fear and uncertainty;
(11) characteristic tokens, reflecting the appeal to the youth audience;
(12) the presence of verbal repetition.

The group of verbal aggression is standard and contains dictionaries of expressions characterizing the vocabulary of physical and psychological violence and destruction; the vocabulary of negative emotional evaluation; the vocabulary of negative rational evaluation, abusive and profanity, etc. In accordance with the classification of emotions by Izard K. [4], in the expressive text, there are three types of hostile emotions: anger, disgust and contempt. They are followed by primary emotions – fear, resentment, envy, pain, sadness, humiliation, neglect, and so on.

The group of scientific and journalistic terms is a standard and is intended for screening out scientific and journalistic articles on the study of extremist topics.

The dictionary of each group is supposed to be created both on the basis of already existing dictionaries of the corresponding orientation, and on the basis of statistical analysis of a sample of texts of extremist orientation.

It is obvious that different lexical units are characterized by different degrees of belonging to extremist vocabulary. Therefore, it is proposed to compare the values of fuzzy membership functions on the basis of statistical analysis of the sample of reference texts for which the degree of expression of the studied words is estimated by expert means. As an alternative, for example, pairwise expert assessments can be used.

4   Method of Formation of Estimation of an Extremist Orientation of the Text on the Basis of Systems of Fuzzy-Logical Conclusions

Formation of the evaluation of the studied text on the extremist orientation is proposed to be carried out on the basis of systems of fuzzy-logical conclusions, called fuzzy multilevel [0,1] – classifiers and applied by Nedosekin A. to assess the risk of bankruptcy of the company [4].

It is assumed that in the study of the text, four linguistic variables are introduced into consideration, the numerical values of which correspond to its estimates for each of the groups:

g_1 = "comprehensive assessment of the content of extremist markers";

g_2 = "comprehensive assessment of the content of psychological manipulation markers";

g_3 = "comprehensive assessment of the content of the words of verbal aggression";

g_4 = "comprehensive assessment of the content of scientific and journalistic terms".

Each of the assessment is a linguistic variable, with a universal set [0,1] and a term a set of five terms $G = \{G_1, G_2, G_3, G_4, G_5\}$. Terms can be assigned the following meaning (static classifiers or classifiers of the first type):

G_1 – "assessment 1, the text does not contain the studied vocabulary";

G_2 – "assessment 2, the text contains single elements of the studied vocabulary";

G_3 – "assessment 3, the text contains elements of the studied vocabulary";

G_4 – "assessment 4, the text contains the studied vocabulary";

G_5 – "assessment 5, the text contains the vocabulary under study in a critical volume".

Also the linguistic variable

g = "comprehensive assessment of checking extremist orientation"

is introduced. Its numerical value is calculated on the basis of fuzzy multilevel [0,1] – classifiers, as well as on the basis of aggregation of the values of the first three of these estimates.

The fourth assessment is used to filter out scientific articles and journalistic research: the publication is searched, as well as the output of the publication and other markers that clearly indicate the publication of the article as the reviewed material. In the case when the search gave no result, the fourth estimation is not considered.

The

g = "comprehensive assessment of the text for extremist orientation"

is used to rank the texts, as well as to study the dynamics, for example, the content of the public in time (which can serve to control the level of its extremist orientation).

The formation of each of the four estimates of the text is based on the aggregation (using fuzzy multi-level [0,1] – classifiers) of numerical values of indicators for the corresponding group.

Estimates are aggregated on the basis of sets of indicators. Thus, for example,

g_1 = "complex assessment of the content of extremist orientation markers"

is formed on the basis of numerical values of six indicators corresponding to the listed subgroups. Each of the subgroups, according to the theory of fuzzy multilevel [0,1] – classifiers has its own weight, determined by expert estimates. For each indicator the normalized numerical value is calculated. To do this, the value of the membership function of each of the encountered markers is multiplied by the relative frequency of its occurrence of the word in the text; then the products are summarized.

The linguistic variable

B_i = "level of the i-th indicator"

$(i = 1,2, ...,6)$ is introduced. A set of values that the variable B_i is the term set of the five terms $B = \{B_1, B_2, B_3, B_4, B_5\}$:

B_1 – "very low level";

B_2 – "low level";

B_3 – "average index level";

B_4 – "high level";

B_5 – "very high level".

Each studied indicator will be compared with the value of the membership functions relating it to the corresponding term of the linguistic variable. The construction of the membership function is the main problem, which is proposed to be solved with the help of hybrid fuzzy neural networks. Possible alternatives are membership functions in the form of fuzzy triangular numbers; fuzzy trapezoidal numbers; sigmoid functions; Gaussian functions, etc. It is proposed to adjust the parameters of accessory functions on the basis of hybrid fuzzy neural networks, using the method of Golub-Pereyra.

Fuzzy multilevel [0,1] – classifiers of the second type, or dynamic classifiers can also be used to assess the dynamics of extremist-oriented communities based on their content analysis. The content of the messages of each of the community members is analyzed in three ways:

g_1 = "assessment of the dynamics of the content of signal words of extremist content";

g_2 = "assessment of the dynamics of the content of psychological manipulation markers";

g_3 = "assessment of dynamics of the content of the words of verbal aggression".

Each of the estimates is a linguistic variable, with a universal set [0,1] and a term set of five terms $G = \{G_1, G_2, G_3, G_4, G_5\}$. Terms can be assigned the following meanings (static classifiers or classifiers of the first type):

G_1 – "sustainable departure from the studied vocabulary";

G_2 – "a departure from the studied vocabulary";

G_3 – "stagnation";

G_4 – "the growth of the studied vocabulary";

G_5 – "sustainable growth of the studied language".

Directions should be ranked by experts; apparently, the second direction should have the most weight. In this case, the numerical estimates of each of the groups are also calculated on the basis of aggregation of the indicators included in its subgroups. However, subgroup numbers are calculated more easily than in the previous case. Calculation of the normalized values of the studied indicators for the considered period N of some time intervals (for example, weeks) is carried out on the basis of the scheme taking into account the importance of different time periods due to weight coefficients:

$$x_i = 0,5 \left(1 + \sum_{i=1}^{N-1} k_i I_i \right) \tag{1}$$

$$k_i = \frac{2(N-i)}{(N-1)N} \tag{2}$$

k_i – weight coefficients determined by the Fishburne rule; numbering of time periods is in reverse order;

I_i – integer functions defined in such a way that the value "1" corresponds to the increase in the i-th indicator (deterioration of the situation); the value "−1" – decrease in the i-th indicator; the value "0" – stabilization, no changes.

At the same time, the term sets of indicators have the same form as above, and the membership functions can be standard uniform, for example, trapezoidal. Thus, the proposed method of monitoring information in Internet communities will allow to assess the level of their extremist dynamics and trends in their development.

5   Advantages of the Developed Method

The method has the following advantages:

(1) the ability to analyze the content of the Internet community in four groups of indicators, including several subgroups, reflecting shades of extremist orientation; also to form a final numerical assessment of the Internet community, taking into account both the level of individual indicators and their dynamics;

(2) the ability to account for the formation of the group, arbitrarily large number of records without their dimensionalization;

(3) the contribution of each of the indicators to the final assessment varies by weight, which can be changed according to the wishes of the experts;

(4) estimates given for each of the individual groups serve as material for the construction of a comprehensive evaluation of the entire system;

(5) the assessments help to rank the online community in terms of their extremist aggression;

(6) the proposed method is relatively easy to implement, easily formalized for software systems.

The method has the important advantage that the result of its application is a table of estimates, which summarizes the unified data on individual indicators. Linguistic recognition of the final complex assessment allows us to judge the situation in the Internet community, and the numerical value of the corresponding fuzzy variable – to give it a quantitative assessment ("how bad" or "how well"). The result is easily analyzed on the basis of a table of assessments, which is the material for additional sociological and psychological studies of the phenomenon of extremist communities on the Internet.

The method is the author's innovation and has not been used previously for content monitoring of Internet communities.

References

1. Ananiava, M.I., Kobozeva, M.V., Soloviev, F.N., Polyakov, I.V., Tchepovskii, A.M.: About the problem of revealing extremist orientation in texts. Vestnik NGU. Ser.: Inf. Technol. **14**(4), 5–13 (2016)
2. Izard, K.E.: Psychology of Emotions. Piter, St. Petersburg (1999)
3. Ledeneva, T.M., Moiseev, S.A.: Formalization of properties of interpreted linguistic scales and terms of fuzzy models. Appl. Inform. **4**(40), 126–132 (2012)
4. Nedosekin, A.: Fuzzy Financial Management. AFA Library, Moscow (2003)
5. Zade, L.A.: Fundamentals of a new approach to the analysis of complex systems and decision making processes. "Mathematics today. Knowledge", Moscow (1974)
6. Zade, L.A.: The concept of a linguistic variable and its application to making approximate decisions Mir. Moscow (1976)

Evaluation of Financial Stability of Azerbaijan Commercial Banks Using the Fuzzy Logic Methods

Ramin Rzayev[1,2(✉)] [iD], Sevinj Babayeva[2], Rovshan Akbarov[3], and Emin Garibli[3]

[1] Azerbaijan State Oil and Industry University,
Azadlig ave. 34, AZ1010 Baku, Azerbaijan
raminrza@yahoo.com
[2] Institute of Control Systems of ANAS, B. Vahabzadeh street 9,
AZ1141 Baku, Azerbaijan
babayevasevinj@yahoo.com
[3] Azerbaijan State University of Economics, Istiqlaliyyat street 6,
AZ1000 Baku, Azerbaijan
rovshanakperov@yahoo.com, egaribli@hotmail.com

Abstract. The paper considers two fuzzy methods of multifactor assessment of alternatives and groups of financial indicators, which are used to assess the financial stability of the four leading commercial banks of Azerbaijan. At the same time, the choice of financial ratios that have the greatest impact on the financial stability of a bank is dictated by the desire to establish a strict correlation between the financial condition of banks and these factors. Therefore, not trying to invent new ratios to assess liquidity, profitability, capital adequacy, quality of assets and liabilities, the paper employs a set of ratios that are most commonly used in various methods for selected financial indicators of bank stability.

Keywords: Financial stability of bank · Financial ratio · Fuzzy set

1 Introduction

Analysis of the activities of a commercial bank (CB) is a system of special knowledge related to the study of the financial and economic results of its activities, identifying factors that have a significant impact on its financial stability, identifying trends and justifying development directions. In [1, 2], we proposed a method for evaluating the financial indicators of CB for the current period of their financial activities, based on the use of fuzzy methods of multifactor assessment of alternatives. At the same time, the method itself was tested on the example of hypothetical CB, which in itself is not quite convincing from the point of view of practical significance. Therefore, without going into details of the essence of the problem, which we described in [1], we will proceed directly to the statement of the problem of evaluating the financial sustainability of CB, using the examples of current financial statements of the commercial banks of Azerbaijan.

© Springer Nature Switzerland AG 2020
R. A. Aliev et al. (Eds.): ICSCCW 2019, AISC 1095, pp. 266–274, 2020.
https://doi.org/10.1007/978-3-030-35249-3_34

2 Problem Statement

We selected four successful banks of Azerbaijan as the evaluated alternatives: a_1 – "Yapi Kredi Bank" CJSC; a_2 – "Capital Bank" OJSC; a_3 – "Express bank" OJSC; a_4 – "PASHA Bank" OJSC. The data on the financial indicators of their economic activity during the reporting period are summarized in Table 1. In this case, the carriers of information on the financial indicators for each CB are open access data [3–6].

Table 1. Financial data of current reports of alternative CB.

No.	Financial indicator	Value of financial indicator ($\times$1000AZN)			
		a_1	a_2	a_3	a_4
01	C - Capital	80999	379839	143963	499149
02	C1 - Tier 1 Capital	59494	191463	138345	328238
03	RWA – Risk weighted assets	312251	1621463	238355	1553166
04	CD – Citizens' deposits	116825	877393	92234	775936
05	CF – Corporate funds	226312	1578184	62073	2114025
06	TRF – Total raised funds	343137	2455577	154307	2889961
07	TL – Total liabilities	357463	2860457	171241	3459266
08	DL – Demand liabilities	49003	635583	58831	875668
09	RIL – Raised interbank loans	4317	66068	10665	399973
10	IEA – Income-earning assets	455355	3211739	298886	3923287
11	TA – Total assets	438462	3240296	315204	3958415
12	TD – Total debt	455355	3211739	298886	3923287
13	ELLP – Estimated loan loss provisions	38048	90643	9770	63324
14	RBR – Raised bank resources	347454	2783716	164972	3289934
15	LO – Loans overdue	86170	28680	18069	116090
16	TLD – Total loan debt	1589820	1826306	1964140	2006021
17	TACCMP – Total amount of credit claims for major participants	5040	4465	33649	138639
18	HLA – Highly liquid assets	40181	248290	24828	315691
19	RF – Raised funding	355073	2875035	183266	3499839
20	LA – Liquid assets	36650	757390	66903	801499
21	DTL – Demand term liabilities	289325	1819280	79137	1275797
22	P – Profit	7953	112407	1458	89092
23	NII – Net interest income	20781	176053	19411	141739
24	TAII – Total assets with interest income	191660	2126562	196414	2455848
25	GAE – General and administrative expenses	22858	124024	21754	74818
26	NOI – Net operating income	56190	888890	38150	111910
27	OE – Operating expenses	5577	11137	2160	13640
28	OI – Operating income	11196	100026	5975	24831

Thus, based on the data from Table 1, it is necessary to evaluate the levels of financial stability of CB a_k ($k = 1 \div 4$), which forms the following aim of the work. The main aim of this study is to adapt the fuzzy models proposed in [1, 2] to the evaluation of the financial sustainability of the four leading CB of Azerbaijan based on their current financial indicators given in Table 1.

3 Multifactor Assessment of Financial Sustainability of CB

To evaluate the financial sustainability of CB based on the data given in Table 1, so-called financial ratios (FR) are calculated, which act as evaluation criteria. At the same time, the choice of these FR is determined by the significant impact on the financial stability of CB, their compatibility, mutual comparability in dimension and orientation. In [7], the most versatile list of FR of CB stability was proposed, which we used in [1, 2] in the form of Table 2 below, where, along with the respective calculation formulas, their standard values are also given.

Table 2. The system of financial ratios of sustainability of commercial banks.

FR	Sustainability ratio	Calculation formula ($\times 100\%$)
F_1	Capital adequacy ratio	F_1 = C/RWA
F_2	Tier 1 Capital adequacy ratio	F_2 = C1/RWA
F_3	Customer base ratio	F_3 = (CD + CF)/TRF
F_4	Resource base stability ratio	F_4 = (TL-DL)/TL
F_5	Ratio of dependence on IL	F_5 = RIL/TRF
F_6	Asset use efficiency ratio	F_6 = IEA/TA
F_7	Loan policy aggressiveness ratio	F_7 = TD/RBR
F_8	Loan policy quality ratio	F_8 = (TD-ELLP)/TD
F_9	Overdue loans share	F_9 = LO/TLD
F_{10}	Concentration of credit risks for shareholders	F_{10} = TACCMP/C
F_{11}	Ratio of highly liquid assets and raised funds	F_{11} = HLA/RF
F_{12}	Quick liquidity ratio	F_{12} = HLA/DL
F_{13}	Current liquidity ratio	F_{13} = LA/DTL
F_{14}	Raised funds structure ratio	F_{14} = DL/RF
F_{15}	Return on assets ratio	F_{15} = P/TA
F_{16}	Return on capital ratio	F_{16} = P/C
F_{17}	Net interest margin	F_{17} = NII/TAII
F_{18}	Expenditure pattern	F_{18} = GAE/NOI
F_{19}	Operating expenses and income ratio	F_{19} = OE/OI
F_{20}	Operating expenses and assets ratio	F_{20} = OE/TA

The calculated values of FR, as the quality criteria for the investigated CB a_k ($k = 1 \div 4$), are summarized in Table 3. The standard values of these criteria are also given there. Analysis of the calculated values of FR indicates that all four CB do not fully comply with the standard values - criteria for evaluating financial stability. Nevertheless, even in the presence of such initial conditions, two fuzzy methods of multi-criteria assessment are applicable, which we considered in [1, 2].

Table 3. Calculated and standard values of bank stability criteria.

Quality criteria	Criterion value for a commercial bank				Standard value
	a_1	a_2	a_3	a_4	
F_1	25.94035	23.4257	60.39857	32.13752	10
F_2	19.05326	11.80804	58.04158	21.13348	6
F_3	100	100	100	100	80
F_4	86.29145	77.78037	65.64433	74.68631	70
F_5	1.258098	2.690529	6.911546	13.84008	≤ 15
F_6	103.8528	99.11869	94.82304	99.11257	85
F_7	131.0548	115.376	181.1738	119.2512	$60 \div 70$
F_8	91.64432	97.17776	96.7312	98.38595	$96 \div 99$
F_9	5.42011	1.570383	0.919945	5.787078	≤ 4
F_{10}	6.222299	1.175498	23.37337	27.77507	≤ 35
F_{11}	11.31626	8.636069	13.54752	9.020158	3
F_{12}	81.99702	39.06492	42.20224	36.05145	15
F_{13}	12.66742	41.6313	84.54073	62.8234	50
F_{14}	13.80082	22.10697	32.10143	25.02024	≤ 50
F_{15}	1.81384	3.469035	0.462558	2.250699	≥ 1.5
F_{16}	9.81864	29.59333	1.01276	17.84878	≥ 8
F_{17}	10.84264	8.278762	9.882697	5.771489	≥ 5
F_{18}	40.67984	13.95268	57.02228	66.85551	≤ 85
F_{19}	49.81243	11.13411	36.15063	54.93134	$50 \div 70$
F_{20}	1.271946	0.343703	0.68527	0.344582	≥ 4.75

4   Evaluation of Financial Stability by Fuzzy Inference Method

So, assuming that FR F_i $(i = 1 \div 20)$ are the criteria for evaluating the financial sustainability of CB, and their qualitative characteristics in the form of terms of the corresponding linguistic variables, we shall apply the fuzzy model proposed in [2] to evaluate the financial stability of the above-stated alternative CB a_k $(k = 1 \div 4)$ in the form of the following set of rules:

e_1: "If F_{20} = NOT LESS THAN 4.75 and F_{19} = WITHIN THE NORM and F_{18} = NOT GREATER THAN 85 and F_{17} = NOT LESS THAN 5 and F_{16} = NOT LESS THAN 8 and F_{15} = NOT LESS THAN 1.5 and F_{14} = NOT GREATER THAN 50 and F_{13} = WITHIN THE NORM and F_{12} = WITHIN THE NORM and F_{11} = WITHIN THE NORM, then Y = ACCEPTABLE";

e_2: "If F_{20} = NOT LESS THAN 4.75 and F_{19} = WITHIN THE NORM and F_{18} = NOT GREATER THAN 85 and F_{17} = NOT LESS THAN 5 and F_{16} = NOT LESS THAN 8 and F_{15} = NOT LESS THAN 1.5 and F_{14} = NOT GREATER THAN 50 and F_{13} = WITHIN THE NORM and F_{12} = WITHIN THE

NORM and F_{11} = WITHIN THE NORM and F_1 = WITHIN THE NORM and F_2 = WITHIN THE NORM, then Y = MORE THAN ACCEPTABLE";

e_3: "If F_{20} = NOT LESS THAN 4.75 and F_{19} = WITHIN THE NORM and F_{18} = NOT GREATER THAN 85 and F_{17} = NOT LESS THAN 5 and F_{16} = NOT LESS THAN 8 and F_{15} = NOT LESS THAN 1.5 and F_{14} = NOT GREATER THAN 50 and F_{13} = WITHIN THE NORM and F_{12} = WITHIN THE NORM and F_{11} = WITHIN THE NORM and F_1 = WITHIN THE NORM and F_2 = WITHIN THE NORM and F_3 = WITHIN THE NORM and F_4 = WITHIN THE NORM and F_5 = NOT GREATER THAN 15 and F_6 = WITHIN THE NORM and F_7 = WITHIN THE NORM and F_8 = WITHIN THE NORM and F_9 = NOT GREATER THAN 4 and F_{10} = NOT GREATER THAN 35, then Y = PERFECT";

e_4: "If F_{20} = NOT LESS THAN 4.75 and F_{19} = WITHIN THE NORM and F_{18} = NOT GREATER THAN 85 and F_{17} = NOT LESS THAN 5 and F_{16} = NOT LESS THAN 8 and F_{15} = NOT LESS THAN 1.5 and F_{14} = NOT GREATER THAN 50 and F_{13} = WITHIN THE NORM and F_{12} = WITHIN THE NORM and F_{11} = WITHIN THE NORM and F_3 = WITHIN THE NORM and F_4 = WITHIN THE NORM and F_5 = NOT GREATER THAN 15 and F_6 = WITHIN THE NORM and F_7 = WITHIN THE NORM and F_8 = WITHIN THE NORM and F_9 = NOT GREATER THAN 4 and F_{10} = NOT GREATER THAN 35, then Y = VERY ACCEPTABLE";

e_5: "If F_{20} = NOT LESS THAN 4.75 and F_{19} = WITHIN THE NORM and F_{18} = NOT GREATER THAN 85 and F_{17} = NOT LESS THAN 5 and F_{16} = NOT LESS THAN 8 and F_{15} = NOT LESS THAN 1.5 and F_{14} = NOT GREATER THAN 50 and F_{13} = WITHIN THE NORM and F_{12} = WITHIN THE NORM and F_{11} = WITHIN THE NORM and F_1 = WITHIN THE NORM and F_2 = WITHIN THE NORM and F_3 = NOT WITHIN THE NORM and F_4 = NOT WITHIN THE NORM and F_5 = NOT GREATER THAN 15 and F_6 = NOT WITHIN THE NORM and F_7 = NOT WITHIN THE NORM and F_8 = NOT WITHIN THE NORM and F_9 = NOT GREATER THAN 4 and F_{10} = GREATER THAN 35, then Y = ACCEPTABLE";

e_6: "If F_{20} = LESS THAN 4.75 and F_{19} = NOT WITHIN THE NORM and F_{14} = GREATER THAN 50 and F_{13} = NOT WITHIN THE NORM and F_{12} = NOT WITHIN THE NORM and F_{11} = NOT WITHIN THE NORM, then Y = UNACCEPTABLE".

Here, we select the discrete set J = {0; 0.1; 0.2;...; 1} as the universe for fuzzy subsets describing the values of the linguistic variable Y and the following: S = ACCEPTABLE – $\mu_S(j) = j$; MS = MORE THAN ACCEPTABLE – $\mu_{MS}(j) = j^{1/2}$; P = PERFECT – $\mu_P(j) = 1$ if $j = 1$, ¡and $\mu_P(j) = 0$, if $j < 1$; VS = VERY ACCEPTABLE – $\mu_{VS}(j) = j^2$; US = UNACCEPTABLE – $\mu_{US}(j) = 1-j$ as membership functions restoring these fuzzy sets.

The left-hand terms of the rules are described by fuzzy subsets of the universe {a_1, a_2, a_3, a_4} that are restored by Gaussian membership function $\mu_{Fi}(u) = \exp[-(u-u_{i0})^2/\sigma_i^2]$ (i = 1÷20), where density σ^2 = 2500 is selected for all cases.

Thus, based on the data from Table 3, we have:

- WITHIN THE NORM (F_1): A_1 = {0.9034/a_1; 0.9304/a_2; 0.3620/a_3; 0.8220/a_4};
- WITHIN THE NORM (F_2): A_2 = {0.9341/a_1; 0.9866/a_2; 0.3385/a_3; 0.9125/a_4};
- WITHIN THE NORM (F_3): A_3 = {0.8521/a_1; 0.8521/a_2; 0.8521/a_3; 0.8521/a_4};
- WITHIN THE NORM (F_4): A_4 = {0.8993/a_1; 0.9761/a_2; 0.9924/a_3; 0.9913/a_4};
- NOT GREATER THAN 15% (F_5): A_5 = {1/a_1; 1/a_2; 1/a_3; 1/a_4};
- WITHIN THE NORM (F_6): A_6 = {0.8675/a_1; 0.9234/a_2; 0.9621/a_3; 0.9234/a_4};
- WITHIN THE NORM (F_7): A_7 = {0.2251/a_1; 0.4389/a_2; 0.0071/a_3; 0.3790/a_4};
- WITHIN THE NORM (F_8): A_8 = {0.9924/a_1; 1/a_2; 1/a_3; 1/a_4};
- NOT GREATER THAN 4% (F_9): A_9 = {0.9992/a_1; 1/a_2; 1/a_3; 0.9987/a_4};
- NOT GREATER THAN 35% (F_{10}): A_{10} = {1/a_1; 1/a_2; 1/a_3; 1/a_4};

- WITHIN THE NORM (F_{11}): $A_{11} = \{0.9727/a_1; 0.9874/a_2; 0.9565/a_3; 0.9856/a_4\}$;
- WITHIN THE NORM (F_{12}): $A_{12} = \{0.1661/a_1; 0.7932/a_2; 0.7438/a_3; 0.8376/a_4\}$;
- WITHIN THE NORM (F_{13}): $A_{13} = \{0.5726/a_1; 0.9724/a_2; 0.6205/a_3; 0.9363/a_4\}$;
- NOT GREATER THAN 50% (F_{14}): $A_{14} = \{1/a_1; 1/a_2; 1/a_3; 1/a_4\}$;
- NOT LESS THAN 1.5% (F_{15}): $A_{15} = \{1/a_1; 1/a_2; 0.9996/a_3; 1/a_4\}$;
- NOT LESS THAN 8% (F_{16}): $A_{16} = \{1/a_1; 1/a_2; 0.9807/a_3; 1/a_4\}$;
- NOT LESS THAN 5% (F_{17}): $A_{17} = \{1/a_1; 1/a_2; 1/a_3; 1/a_4\}$;
- NOT GREATER THAN 85% (F_{18}): $A_{18} = \{1/a_1; 1/a_2; 1/a_3; 1/a_4\}$;
- WITHIN THE NORM (F_{19}): $A_{19} = \{1/a_1; 0.5465/a_2; 0.9261/a_3; 1/a_4\}$;
- NOT LESS THAN 4.75% (F_{20}): $A_{20} = \{0.9952/a_1; 0.9923/a_2; 0.9934/a_3; 0.9923/a_4\}$.

Then, taking into account the adopted formal descriptions, the rules will be written in the following symbolic form:

e_1: "$(X = A_{20})$ & $(X = A_{19})$ & ... & $(X = A_{11}) \Rightarrow Y = S$";
e_2: "$(X = A_{20})$ & $(X = A_{19})$ & ... & $(X = A_{11})$ & $(X = A_1)$ & $(X = A_2) \Rightarrow Y = MS$";
e_3: "$(X = A_{20})$ & $(X = A_{19})$ & ... & $(X = A_1) \Rightarrow Y = P$";
e_4: "$(X = A_{20})$ & $(X = A_{19})$ & ... & $(X = A_3) \Rightarrow Y = VS$";
e_5: "$(X = A_{20})$ & $(X = A_{19})$ &...& $(X = A_{11})$ & $(X = \neg A_{10})$ & $(X = \neg A_9)$ & $(X = \neg A_8)$ & $(X = \neg A_7)$ & $(X = \neg A_6)$ &$(X = \neg A_5)$ &$(X = \neg A_4)$ &$(X = \neg A_3)$ & $(X = A_1)$ & $(X = A_2) \Rightarrow Y = S$";
e_6: "$(X = \neg A_{20})$ & $(X = \neg A_{19})$ & $(X = \neg A_{14})$ & $(X = \neg A_{13})$ & $(X = \neg A_{12})$ & $(X = \neg A_{11}) \Rightarrow Y = US$".

Next, applying the intersection rule of fuzzy sets [8], for the left parts of the rules we get the resulting fuzzy sets: $M_1 = \{0.1661/a_1; 0.5465/a_2; 0.6205/a_3; 0.8376/a_4\}$; $M_2 = \{0.1661/a_1; 0.5465/a_2; 0.3385/a_3; 0.8220/a_4\}$; $M_3 = \{0.1661/a_1; 0.4389/a_2; 0.0071/a_3; 0.3790/a_4\}$; $M_4 = \{0.1661/a_1; 0.4389/a_2; 0.0071/a_3; 0.3790/a_4\}$; $M_5 = \{0/a_1; 0/a_2; 0/a_3; 0/a_4\}$; $M_6 = \{0/a_1; 0/a_2; 0/a_3; 0/a_4\}$.

As a result, the rules will take the following, even more compact form:

e_1: "$(X = M_1) \Rightarrow Y = S$"; e_2: "$(X = M_2) \Rightarrow Y = MS$"; e_3: "$(X = M_3) \Rightarrow Y = P$"; e_4: "$(X = M_4) \Rightarrow Y = VS$"; e_5: "$(X = M_5) \Rightarrow Y = S$"; e_6: "$(X = M_6) \Rightarrow Y = US$";

Transforming these rules by means of the Lukasiewicz implication: $\mu_H(u, j) = \min\{1, 1 - \mu_X(u) + \mu_Y(j)\}$ [9], for each pair $(u, j) \in U \times J$, we get fuzzy relations in the form of matrices: $R_1, R_2, ..., R_6$. As a result of the intersection of these fuzzy relations, we end up with a common functional solution in the form of the following matrix

$$R = \begin{array}{c|ccccccccccc} & 0 & 0,1 & 0,2 & 0,3 & 0,4 & 0,5 & 0,6 & 0,7 & 0,8 & 0,9 & 1 \\ \hline a_1 & 0,8339 & 0,8339 & 0,8339 & 0,8339 & 0,8339 & 0,8339 & 0,8339 & 0,8339 & 0,8339 & 0,8339 & 1,0000 \\ a_2 & 0,4535 & 0,5535 & 0,5611 & 0,5611 & 0,5611 & 0,5611 & 0,5611 & 0,5611 & 0,5611 & 0,5611 & 1,0000 \\ a_3 & 0,3795 & 0,4795 & 0,5795 & 0,6795 & 0,7795 & 0,8795 & 0,9795 & 0,9929 & 0,9929 & 0,9929 & 1,0000 \\ a_4 & 0,1624 & 0,2624 & 0,3624 & 0,4624 & 0,5624 & 0,6210 & 0,6210 & 0,6210 & 0,6210 & 0,6210 & 1,0000 \end{array}$$

To obtain quantitative estimates of the financial stability of the specified CB a_k ($k = 1 \div 4$), the compositional inference rule in the fuzzy environment is first applied: $E_k = G_k \circ R$, where E_k is a fuzzy interpretation of the bank's stability level, G_k is a representation of the k-th bank's financial stability in the form of the fuzzy set on U. Then, applying the compositional rule [8]: $\mu_{Ek}(j) = \max\{\min(\mu_{Gk}(u), \mu_R(u))\}$, where

$\mu_{Gk}(u) = 0$, if $u \neq u_k$, and $\mu_{Gk}(u) = 1$, if $u = u_k$, we get $\mu_{Ek}(j) = \mu_R(u_k, j))$, i.e. E_k is the k-th row of R.

According to this, the fuzzy inference about the degree of acceptability of the financial stability of the 1st bank a_1 is the fuzzy subset of the universe J with the values of the membership function located in the 1st row of the matrix R: E_1 = {0.8339/0; 0.8339/0.1; 0.8339/0.2; 0.8339/0.3; 0.8339/0.4; 0.8339/0.5; 0.8339/0.6; 0.8339/0.7; 0.8339/0.8; 0.8339/0.9; 1/1}.

A quantitative estimate of the financial stability of a_1 is obtained by defuzzifying the fuzzy inference according to the following procedures:

- for $0 < \alpha < 0.8339$: $\Delta\alpha = 0.8339$, $E_{1\alpha}$ = {0; 0.1; 0.2; ...; 0.9; 1}, $M(E_{1\alpha}) = 0.50$;
- for $0.8339 < \alpha < 1$: $\Delta\alpha = 0.1661$, $E_{1\alpha}$ = {1}, $M(E_{1\alpha}) = 1$.

Using the formula $F(C) = \alpha_{max}^{-1} \int_0^{\alpha_{max}} M(C_\alpha)d\alpha$, where α_{max} is the maximum value on C, we shall find the numerical evaluation of the financial stability of CB a_1:

$$F(E_1) = \int_0^1 M(E_{1\alpha})d\alpha = 0.5 \cdot 0.8339 + 1 \cdot 0.1661 = 0.5830.$$

For the fuzzy evaluation of the financial stability of CB a_2 E_2 = {0.4535/0; 0.5535/0.1; 0.5611/0.2; 0.5611/0.3; 0.5611/0.4; 0.5611/0.5; 0.5611/0.6; 0.5611/0.7; 0.5611/0.8; 0.5611/0.9; 1/1} respectively, we have:

- for $0 < \alpha < 0.4535$: $\Delta\alpha = 0.4535$, $E_{2\alpha}$ = {0; 0.1; 0.2; ...; 0.9; 1}, $M(E_{2\alpha}) = 0.50$;
- for $0.4535 < \alpha < 0.5535$: $\Delta\alpha = 0.1$, $E_{2\alpha}$ = {0.1; 0.2; ...; 0.9; 1}, $M(E_{2\alpha}) = 0.55$;
- for $0.5535 < \alpha < 0.5611$: $\Delta\alpha = 0.0076$, $E_{2\alpha}$ = {0.2; 0.3; ...; 0.9; 1}, $M(E_{2\alpha}) = 0.60$;
- for $0.5611 < \alpha < 1$: $\Delta\alpha = 0.4389$, $E_{2\alpha}$ = {1}, $M(E_{2\alpha}) = 1$,

$$F(E_2) = \int_0^1 M(E_{2\alpha})d\alpha = 0.5 \cdot 0.4535 + 0.55 \cdot 0.1 + 0.6 \cdot 0.0076 + 1 \cdot 0.4389$$
$$= 0.7252.$$

Fuzzy evaluations of the financial stability of others CB are obtained by similar actions, namely: $F(E_3) = 0.6132$ and $F(E_4) = 0.7541$. Thus, the 4th bank a_4 – "Pasha Bank", which has the largest total index (0.7541), is the most financially stable. Further, in descending order: a_2 – "Capital Bank" (0.7252), a_3 – "Express bank" (0.6132), a_1 – "Yapı Kredi Bank" (0.5830). At the same time, according to the scale obtained in [1, 2], all these banks have high degrees of financial stability.

5   Evaluation of the Financial Stability by Maxmin Convolution

The fuzzy subsets A_i ($i = 1 \div 20$) of the universe {a_1, a_2, a_3, a_4} constructed in the previous section describe the quality criteria using the corresponding membership functions, the values of which show the relationships between financial ratios F_{ik}

($i = 1\div20$; $k = 1\div4$) of each of the specified banks and the corresponding evaluation criterion. Assuming that these criteria have the same significance for the decision-maker, we determine the desired set of alternatives, say A, by intersecting fuzzy sets containing evaluations of alternatives by the selection criteria: $A = A_1 \cap A_2 \cap \ldots \cap A_{20}$. Then, according to the fuzzy maximum convolution method, from the point of view of financial stability, the best CB is considered to be the one that has the maximum degree of membership in the fuzzy set A. At the same time, the operation of intersection of fuzzy sets is performed according to the rule [9]: $\mu_A(a_k) = \min\{\mu_{Ai}(a_k)\}$. In our case, the desired set is formed as:

$A = [\min\{0.9034; 0.9341; 0.8521; 0.8993; 1; 0.8675; 0.2251; 0.9924; 0.9992; 1; 0.9727; 0.1661; 0.5726; 1; 1; 1; 1; 1; 1; 0.9952\}; \{0.9304; 0.9866; 0.8521; 0.9761; 1; 0.9234; 0.4389; 1; 1; 1; 0.9874; 0.7932; 0.9724; 1; 1; 1; 1; 1; 0.5465; 0.9923\}; \{0.3620; 0.3385; 0.8521; 0.9924; 1; 0.9621; 0.0071; 1; 1; 1; 0.9565; 0.7438; 0.6205; 1; 0.9996; 0.9807; 1; 1; 0.9261; 0.9934\}; \{0.8220; 0.9125; 0.8521; 0.9913; 1; 0.9234; 0.3790; 1; 0.9987; 1; 0.9856; 0.8376; 0.9363; 1; 1; 1; 1; 1; 1; 0.9923\}]$.

The ranking of alternatives is based on the resulting priority vector, which, according to [9], has the form: $\max\{\mu_A(a_k)\} = \max\{0.1661; 0.4389; 0.0071; 0.3790\}$. The latter means that from the point of view of financial stability, the best CB is a_2 – "Capital Bank", which corresponds to the highest value 0.4389. Then, in descending order: a_4 – "PASHA Bank" (0.3790); a_1 – "Yapi Kredi Bank" (0.1661); a_3 – "Express bank" (0.0071).

6   Conclusion

According to the results of using two fuzzy methods of multi-criteria evaluation of alternatives, total indexes of financial stability of four leading commercial banks of Azerbaijan were obtained on the basis of relevant data on their financial indicators from available information sources. Nevertheless, the rankings of banks based on the results of the use of these methods somewhat differ from each other. This can be explained by the fact that the data of the CB are insignificantly different in their initial financial indicators and essentially occupy leading positions in the financial market of Azer-baijan. However, we tend to consider the ranking based on the fuzzy inference method to be more reliable, because the inference engine used here is more sensitive to minor discrepancies in the source financial data.

References

1. Rzayev, R., Babayeva, S., Rzayeva, I., Jamalov, Z.: Evaluation of commercial banks using the fuzzy inference method for analyzing their financial indicators of stability. Math. Mach. Syst. **4**, 128–144 (2015). (in Russian)
2. Rzayev, R., Babayeva, S., Babayev, T.: Automated information system for the comprehensive assessment of the financial Stability of commercial banks. Prob. Control Inform. **3**, 71–86 (2017). (in Russian)

3. Yapı Kredi Bank: Azerbaijan CJSC. https://www.yapikredi.com.az/Files/Annual%20Audit% 20Report%20-%20AZE%202017.pdf. Accessed 18 April 2019
4. Capital Bank: OJSC. https://kapitalbank.az/useruploads/reports/FS_AZE_2017.compressed. pdf. Accessed 18 April 2019
5. Express Bank: OJSC. https://www.expressbank.az/assets/files/Annual_report_2017.pdf. Accessed 18 April 2019
6. Pasha Bank: OJSC. https://www.pashabank.az/about_us/lang,az/#!/financial_reports/. Accessed 18 April 2019
7. Lotobaeva, G., Nasonova, A.: The system of key indicators of stability of commercial bank. Banking **3**, 76–79 (2006). (in Russian)
8. Zadeh, L.: The Concept of a Linguistic Variable and its Application to Approximate Reasoning. Elsevier Publishing, New York (1974)
9. Andreichikov, A. Andreichikova, O.: Analysis, synthesis, planning decisions in the economy. Finance and Statistics, Moscow (2000) (in Russian)

Fuzzy-Multiple Modification of the Spectrum-Point Methodology for Assessing the Financial Condition of the Company (Based on the Audit-IT)

Natalia G. Vovchenko[1]([✉]) [iD], Luydmila V. Sakharova[1] [iD],
Tatyana V. Epifanova[1] [iD], and Victoriya S. Kokhanova[2] [iD]

[1] Rostov State University of Economics, Rostov-on-Don, Russia
nat.vovchenko@gmail.com, L_Sakharova@mail.ru,
rostovshell@mail.ru
[2] Private Educational Institution of Higher Education "SOUTHERN
UNIVERSITY (IMBL)", Rostov-on-Don, Russia
kohanovavs@yandex.ru

Abstract. A methodology has been developed for assessing the financial condition of an enterprise on the basis of a fuzzy-plural modification of the integral scoring method Audit-IT (the program "Your Financial Analyst"). The modification was carried out using the approach developed by Nedosekin A. for assessing the risk of bankruptcy of an enterprise and consisting in aggregating the indicators of coefficient analysis by means of standard five-point classifiers. The proposed methodology demonstrates the possibility of transforming a standard integral score method for assessing the financial condition of an enterprise into a corresponding fuzzy-multiple methodology, which has several important advantages. In particular, it allows you to change the complex of the studied parameters, depending on the goals and objectives of the study without significant processing of the model; adjust the weights of the parameters depending on the sectoral and territorial specifics, as well as expert assessments; bring together quantitative estimates of indicators with estimates of the dynamics of their change.

Keywords: Analysis of financial statements · Integrated assessment · Aggregation · Fuzzy-Multiple method · JEL classification: G21; J21; J24; O32; O33

1 Introduction

Methods of analysis of financial and economic activities of the enterprise are the basis of theoretical studies of financial analysis [1]. They are based on the practice of managing the country's economy as a whole and its sectors [2].

As a tool for the analysis most often are financial ratios They make it possible to identify both the dynamics of indicators and the limits of permissible values (restrictions) and ratios of indicators. Developed criteria, based on which the coefficients give

© Springer Nature Switzerland AG 2020
R. A. Aliev et al. (Eds.): ICSCCW 2019, AISC 1095, pp. 275–283, 2020.
https://doi.org/10.1007/978-3-030-35249-3_35

a qualitative assessment of the financial condition of the organization. Most often, these criteria are based on integral point estimates [3]. Such assessments are of great importance for government agencies that form investment and tax policies, potential investors and partners of the company, as well as managers of the company itself. The method of integral estimation takes into account all interrelations between indicators, and also allows to trace the possible dynamics with accuracy and to reveal deviations [4]. At the same time, integrated assessment methods have several disadvantages. First of all, it is a subjective approach to the determination of expert assessments and the neglect of the specific features of the enterprise and industries [5]. Scoring is automatic. Most indicators have equal or arbitrarily established weight. When adding new indicators, you have to change the whole order of counting, etc. Integral techniques are difficult to modify. A significant problem in them is to take into account the opinions of experts, as well as the uncertainty of external conditions.

However, the preferred tool for financial analysis is fuzzy logic. The Nedosekin A. model of bankruptcy risk of an enterprise based on standard five-point [0,1] - classifiers [6] became widely known. As the author of the model himself showed, it can be generalized to a set of indicators that go beyond the financial statements and accounted for in the Argenti quantitative scale [7]. In addition to the traditional block of the level of finance, the author reviewed the block of enterprise management. This unit includes the level of top management, financial management, marketing and advertising units, development of a distribution network and branches, etc. The model showed a high level of prediction accuracy, flexibility and modifiability. Its novelty lies in the possibility of aggregating a large number of heterogeneous indicators. The applied approach indicates the way to build similar fuzzy-logical models based on the existing integral scoring models for assessing the financial condition of the company.

This article presents a fuzzy-plural modification of the Methodology for Analyzing the Financial Condition of an Organization Audit-IT (Your Financial Analyst software). To implement the methodology, we used the author's methods for assessing the state of systems based on fuzzy-plural aggregation of sets of indicators [8, 9].

2   Materials and Method

In accordance with the methodology of Audit-IT [10], the analysis of the financial condition is carried out according to the accounting (financial) statements of organizations - forms No 1 "Balance sheet" and No 2 "Report on financial results" ("Profit and loss report"). For the purpose of qualitative assessment of financial indicators, a scale of five basic gradations is used. At the same time, the intervals of values "excellent", "good", "satisfactory", "unsatisfactory", "critical" are set. Numerical intervals of indicators corresponding to each qualitative value are established on the basis of expert assessments and reflect the standards adopted for a particular industry.

Summarizing (integral) assessment of the financial condition of the organization consists of assessing the financial situation and assessing the effectiveness of the organization, Table 1.

Table 1. The scale gradations summative evaluation scoring methodology Audit-IT.

Mark		Symbol	Qualitative characteristics of financial condition
From	Before	(rating)	
2	1.6	AAA	Excellent
1.6	1.2	AA	Very good
1.2	0.8	A	A good
0.8	0.4	BBB	Positive
0.4	0	BB	Normal
0	−0.4	B	Satisfactory
−0.4	−0.8	CCC	Unsatisfactory
−0.8	−1.2	CC	Bad
−1.2	−1.6	C	Very bad
−1.6	2	D	Critical

The following indicators are involved in the generalized assessment of the financial condition (the weight of the indicator is shown in parentheses): autonomy coefficient (0.25); the ratio of net assets and share capital (0.1); the ratio of own working capital (0.15); current (total) liquidity ratio (0.15); quick ratio (intermediate) liquidity (02); absolute liquidity ratio (0.15). In a generalized assessment of the effectiveness of activities, the following indicators are involved: return on equity (0.3); return on assets (0.2); return on sales (0.2); revenue dynamics (0.1); turnover of working capital (0.1); the ratio of profit from other operations and revenues from core activities (0.1).

Based on the scores of the financial position and the effectiveness of the activity, a summary assessment is calculated - a score of financial condition. This score is obtained as the sum of the financial position score, multiplied by 0.6, and the score of financial results, multiplied by 0.4. That is, the indicators are taken in the proportion of 60% and 40% respectively, since the indicator of the financial position largely characterizes the financial condition of the organization. Depending on the value of the financial status score (in accordance with the table above), an organization is assigned one of 10 financial rating values - from AAA (best) to D (worst). Thus, in the course of the analysis, the block of analysis of the financial situation falls into four blocks: (1) the structure of the property and the sources of its formation (12 indicators); (2) assessment of the value of the organization's net assets (3 indicators); (3) analysis of the financial sustainability of the organization (13 indicators); (4) liquidity analysis (7 indicators). Analysis of the effectiveness of the organization includes: (1) an overview of the organization's performance (10 indicators); (2) analysis of profitability (10 indicators); (3) the calculation of indicators of business activity (turnover) (6 indicators). However, to build an integrated assessment for each of the two blocks, only 6 main indicators are used, which is apparently due to the inefficiency of the proposed aggregation algorithm for a large number of parameters. Since the scores of the indicators can be both positive and negative, when summed into the final assessment, the terms of opposite signs can compensate each other. As a result, significant deviations of the two parameters in the direction of the "critical" and "excellent" states can lead to a final assessment of "satisfactory". As a result, with the aggregation of a large number of parameters, some uninformative "average value" of the integral estimate can be obtained.

It should also be noted that the methodology includes too many classes (10 pieces), which makes the results difficult to foresee. We have proposed a method for calculating the integral assessment, based on the aggregation of parameter values by means of a system of fuzzy-logic conclusions - standard five-level [0,1] - classifiers. The construction of the methodology was carried out in two stages. The analysis of the financial position and the effectiveness of the activity of OAO "Zarya" for the period from 01/01/2016 to 12/31/2018 (3 years) was taken as the source material. At the first stage, the estimations of the financial condition and efficiency of the enterprise, as well as the generalizing assessment (financial score) were made on the basis of the parameter sets used in the Audit-IT methodology. At the second stage, a method was developed for calculating comprehensive assessments of the financial and economic activity of an enterprise in four blocks: (1) financial stability; (2) liquidity; (3) profitability; (4) business activity. Each assessment is calculated on the basis of a set of indicators appearing in the analytical report Audit-IT. When calculating a comprehensive assessment based on a set of indicators, two main problems arise: (1) aggregation of time series into numerical values of indicators; (2) an assessment of numerical values of indicators and their aggregation in a complex assessment. For most of the parameters used to build the financial score in the Audit-IT methodology, qualitative assessments are known, including by industry (given in the "Methods for analyzing the financial condition of organizations" on the Audit-IT website). In this case, the aggregation of time series of values for N years was made on the basis of the formula [9].

$$x_i = \sum_{i=1}^{N} k_i P_i, \qquad k_i = \frac{2(N+1-i)}{(N+1)N}$$

where P_i are the parameter data; k_i are the weights determined by the rules given Fishburne "weight" of years; the numbering of years is carried out in reverse order.

If there are no qualitative assessment parameters, and in principle is the dynamics of their changes, a simplified diagram of the aggregation. The account of the dynamics of the period under review N years is based on the scheme:

$$x_i = 0,5\left(1 + \sum_{i=1}^{N-1} k_i I_i\right). \qquad k_i = \frac{2(N-i)}{(N-1)N}.$$

ki are weighting coefficients determined by Fishburne rule; Numbering time periods is carried out in reverse order. I_i are integer functions determined so that the value "1" corresponds to an increase of i-th index (situation worsened); value "−1" corresponds decrease of i-th parameter; value "0" corresponds stabilization, no change. Aggregation of indicators in a comprehensive evaluation suggested exercise based on the standard five-level [0.1] - classifiers. Each of the indicators assigned to a linguistic variable term - many of which consists of five terms G = {G1, G2, G3, G4, G5}: G1 is "critical condition"; G2 is "unsatisfactory condition" G3 is "satisfactory condition"; G4 is "good status"; G5 is "excellent condition". Each linguistic variable has a standard trapezoidal membership function, the definition of four numbers (a_1, a_2, a_3, a_4). For each of the indicators in accordance with the intervals qualitative assessment provided in "Method financial analysis organizations" (Audit-IT), are set corresponding to the four numbers obtained starting fuzzification intervals (Tables 2 and 3). The table shows the four groups.

Table 2. Fuzzy trapezoidal numbers defining the functions of terms belonging.

No	G1	G2	G3
1.1.	(∞; ∞; −0,1; 0,1)	(−0.1, 0.1, 0.545, 0.555)	(0.545; 0.555; 0.570; 0.574)
1.2.	(∞; 0), (∞; ∞; −0.02; 0.02)	(0.80, 0.84; ∞; ∞)	(−0.02, 0.02, 0.104, 0.108)
1.3.	(∞; ∞; −0.22; −0.18)	(−0.22; 0.18; 0.08; 0.12)	(0.08; 0.12; 0.102; 0.106)
1.4.	(∞; ∞; 0.45; 0.55)	(0.45; 0.55; 0.84; 0.86)	(0.84; 0.86; 0.882; 0.886)
1.5.	(∞; ∞; −0.25; −0.15)	(0.25; 0.15; 0.045; 0.055)	(0.045; 0.055; 0.052)
1.6.	(∞; ∞; −0.1; 0.1)	(−0.1, 0.1, 0.495, 0.505)	(0.495; 0.505; 0.515; 0.525)
1.7.	(0; 0; 0.15; 0.25)	(0.15; 0.25; 0.35; 0.45)	(0.35; 0.45; 0.55; 0.65)
1.8.	(0; 0; 0.15; 0.25)	(0.15; 0.25; 0.35; 0.45)	(0.35; 0.45; 0.55; 0.65)
2.1.	(∞; ∞; 0.9; 1.1)	(0.9; 1.1; 1.78; 1.82)	(1.78; 1.82; 1.865; 1.875)
2.2.	(∞; ∞; 0.45; 0.55)	(0, 0.45; 0.55; 0.79; 0.81)	(0.79; 0.81; 0.83; 0.834)
2.3.	(∞; ∞; 0.045; 0.055)	(0.045; 0.055; 0.148; 0.152)	(0.148; 0.152; 0.154; 0.158)
2.4.	(0; 0; 0.15; 0.25)	(0.15; 0.25; 0.35; 0.45)	(0.35; 0.45; 0.55; 0.65)
3.1.	(∞; ∞; −0.05; 0.05)	(0.05; 0.05; 0.119; 0.121)	(0.119; 0.121; 0.124; 0.126)
3.2.	(∞; ∞; −0.02; 0.02)	(0.02; 0.02; 0.059; 0.061)	(0.059; 0.061; 0.062; 0.064)
3.3.	(∞; ∞; 0.9; 1.1)	(0.9; 1.1; 1.49; 1.51)	(1.49; 1.51; 1.55; 1.57)
3.4.	(∞; ∞; −0.02; 0.02)	(0.02; 0.02; 0.049; 0.051)	(0.049; 0.051; 0.051; 0.052)
3.5.	(∞; ∞; −0.02; 0.02)	(0.02; 0.02; 0.118; 0.122)	(0.118; 0.122; 0.124; 0.126)
3.1.	(280; 320; ∞; ∞)	(158; 162; 280; 320)	(152; 156; 158; 162)
3.2.	(36; 40; ∞; ∞)	(16; 18; 36; 40)	(14; 16; 16; 18)
3.3.	(162; 166; ∞; ∞)	(70; 72; 162; 166)	(67; 69; 70; 72)
3.4.	(960; 972; ∞; ∞)	(480; 500; 960; 972)	(310; 314; 480; 500)

Table 3. Extension

No	G4	G5
1.1	(0.570; 0.574; 0.645; 0.655), (0.85; 0.95; ∞; ∞)	(0.645; 0.655; 0.85; 0.95)
1.2.	(0.104, 0.108; 0.109; 0.113), (0.52, 0.56; 0.80; 0.84)	(0.109, 0.113; 0.52; 0.56)
1.3.	(0.102; 0.106; 0.13; 0.17)	(0.13; 0.17; ∞; ∞)
1.4.	(0.882; 0.886; 0.895; 0.905)	(0.895; 0.905; ∞; ∞)
1.5.	(0.055; 0.052; 0.14; 0.16)	(0.14; 0.16; ∞; ∞)
1.6.	(0.515; 0.525; 0.595; 0.605)	(0.595; 0.605; ∞; ∞)
1.7.	(0.55; 0.65; 0.75; 0.85)	(0.75; 0.85; 1; 1)
1.8.	(0.55; 0.65; 0.75; 0.85)	(0.75; 0.85; 1; 1)
2.1.	(1.865; 1.875; 1.95; 2.05)	(1.95; 2.05; ∞; ∞)
2.2.	(0.83; 0.834; 0.95; 1.05)	(0.95; 1.05; ∞; ∞)
2.3.	(0.154; 0.158; 0.18; 0.22)	(0.18; 0.22; ∞; ∞)
2.4.	(0.55; 0.65; 0.75; 0.85)	(0.75; 0.85; 1; 1)
3.1.	(0.124; 0.126; 0.229; 0.231)	(0.229; 0.231; ∞; ∞)
3.6.	(0.062; 0.064; 0.11; 0.13)	(0.11; 0.13; ∞; ∞)
3.7.	(1.55; 1.57; 2.9; 3.1)	(2.9; 3.1; ∞; ∞)
3.8.	(0.051; 0.052; 0.08; 0.12)	(0.08; 0.12; ∞; ∞)
3.9.	(0.124; 0.126; 0.155; 1.65)	(0.155; 1.65; ∞; ∞)
3.5.	(80; 84; 152; 156)	(∞; ∞; 80; 84)
3.6.	(7; 9; 14; 16)	(∞; ∞; 7; 9)
3.7.	(33; 37; 67; 69)	(∞; ∞; 33; 37)
3.8.	(302; 306; 310; 314)	(∞; ∞; 302; 306)

3   Results

The calculation carried out on the basis of the data in Table 4.

Table 4. Numerical values of indicators

N	Significance				The aggregated value
	31.12.2015	31.12.2016	31/12/2017	31.12.2018	
Weight	0.1	0.2	0.3	0.4	1
1.1.	0.66	0.51	0.48	0.5	0.512
1.2.	0.52	0.95	1.1	1.01	0.976
1.3.	0.28	−0.3	−0.58	−1.13	−0.658
1.4.	0.73	0.67	0.61	0.54	0.606
1.5.	0.21	−0.22	−0.4	−0.53	−0.355
1.6.	10.33	−11.7	−1 568.99	−3 171.06	−1 740.428
1.7.	231729255	252735675	289544237	313802383	1
1.8.	231707991	252714411	289522973	313781119	1
2.1.	1.75	1.14	0.86	0.51	0.865
2.2.	1.53	1.04	0.84	0.51	0.817

(continued)

Table 4. (*continued*)

N	Significance				The aggregated value
	31.12.2015	31.12.2016	31/12/2017	31.12.2018	
2.3.	0.23	0.16	0.12	0.21	0.175
Weight	–	1/6	2/6	3/6	1
3.1.	–	8	11	8.4	9.2
3.2.	–	8	10.7	10.8	10.3
3.3.	–	31.1	14.7	20.5	20.33
3.4.	–	12.6	11.4	10.5	11.15
3.5.	–	22	23.2	21.5	22.15
3.6.	–	693032679	609821837	623979575	0.67
4.1	–	96	119	106	108.67
4.2.	–	3	2	<1	1.67
4.3.	–	73	97	77	83
4.4.	–	132	168	183	169.5

Stage 1. Table 5 shows the calculation of the weights of terms summative evaluation of the financial status of the indicators used in the Audit-IT methodology. Assessment of the financial condition:

$$G = 0.125 * 0.45 + 0.3 * 0.25 + 0.5 * 0.2 + 0.7 * 0 + 0.885 * 0.1 = 0.32 \ (G2, \text{"bad"}).$$

Similarly, the estimate *effectiveness of the company:* 0.85 (G5, "excellent").

Finally, the aggregation of the estimates *the financial condition and performance of the enterprise on the basis of* standard five-level fuzzy [0,1] - classifiers obtained integral evaluation (score) of financial and economic activity of the enterprise: 0.53 (G3, "satisfactory").

Stage 2. Designed four integrated enterprise evaluation: stability, liquidity, profitability and turnover. Calculation generalizing evaluation enterprise stability is presented in Table 6.

Table 5. Payment summarizing assessment of the financial status of multiple fuzzy procedure for a set of indicators used in the Audit-IT

N	Hotspots		0.125	0.3	0.5	0.7	0.885
	Weight	x_i	G1	G2	G3	G4	G5
1.1.	0.25	0.512	0	1	0	0	0
1.8.	0.1	1	0	0	0	0	1
1.3.	0.15	−0.658	1	0	0	0	0
2.1.	0.15	0,865	1	0	0	0	0
2.2.	0.2	0.817	0	0	1	0	0
2.3.	0.15	0,175	1	0	0	0	0
Weight	1		0.45	0.25	0.2	0	0.1

Table 6. Payment summative evaluation of the enterprise sustainability on fuzzy-multiple method

N	Hotspots		0,125	0.3	0.5	0.7	0,885
	Weight	xi	G1	G2	G3	G4	G5
1.1.	0.25	0.512	0	1	0	0	0
1.2.	0.1	0.976	0	1	0	0	0
1.3.	0.15	−0.658	1	0	0	0	0
1.4.	0.1	0.606	0	1	0	0	
1.5.	0.1	−0.355	1	0	0	0	0
1.6.	0.1	−1 740.428	1	0	0	0	0
1.7.	0.1	1	0	0	0	0	1
1.8.	0.1	1	0	0	0	0	1
Weight			0.45	0.45	0	0	0.2

$$G = 0.125 * 0.45 + 0.3 * 0.45 + 0.5 * 0 + 0 + 0.7 * 0 + 0.885 * 0.2 = 0.37;$$
$$\mu_2(0.37) = 0.8; \mu_3(0.37) = 0.2 \ (G2, \ \text{``bad''}).$$

Similarly calculated summarizing assessment of liquidity based on the aggregation of the three standard ratios (current, quick and absolute liquidity). It is equal to 0.25, which corresponds to the term G2, "bad". Summative Evaluation of profitability of the enterprise is calculated on the basis of aggregation of the six indicators (return on sales, return on sales net profit margin, return on assets, return on equity, earnings dynamics) and is equal to 0.87, which corresponds to the term G5, "excellent". Finally, summarizing the assessment of business activity (turnover) of the enterprise is based on the aggregation of four indicators (turnover of working capital, inventories, receivables, home equity) and is equal to 0.69, which corresponds to the G4, "good".

Based on the above four evaluations based aggregate estimate of financial and economic activity of the enterprise; it is equal to 0.51, which corresponds to the term G3, "satisfactory". Thus, the proposed fuzzy-multiple method, on the one hand, agrees with the point "method of analysis of the financial condition of the organization", based on which it is built. On the other hand, it has such an important advantage, as the variation, i.e. consideration of the possibility to introduce additional parameters; modify their weights in accordance with the objectives of the study and expertise; complement the complex estimation of the new units without any additional "tuning" method under the changes being made. Furthermore, the proposed technique, unlike the conventional techniques can be used to form the complex evaluation value not only studied parameters, but also the dynamics of their changes. Each company as a result can be assigned to the vector evaluation (g1, g2, g3, g4), which may be supplemented and expanded, depending on the purposes of the study, due to the study of additional indicators: the number of employees, the tax burden, relative volumes of state and investment, etc. As a consequence, it is possible to create a basis for the clustering of companies within the same industry, followed by the study of relationships between the characteristics of the enterprise.

4   Conclusion

A technique illustrating the process of modifying standard integrated point estimation model enterprise financial condition on the basis of fuzzy logic conclusions systems - fuzzy five-level [0.1] - classifiers. Compared with standard techniques, fuzzy multiple methods of financial statement analysis has a number of new features to change the set of tested parameters depending on the goals and objectives of the study with no significant processing model; adjust the weight parameters depending on the sectoral and territorial specificities, and expert assessments; bring together quantitative estimates of parameters with estimates of the dynamics of their changes.

References

1. Artyukhov, A.V., Litvin, A.A.: Analysis of the financial condition of the enterprise: the nature and the need for. Young Sci. **11**, 744–747 (2015)
2. Fursova, M.H., Ilyin, A.A., Moses, L.V.: Analysis of economic activity: a tutorial. Publisher VSUES Voronezh, Russia (2011)
3. Hriplivy, F.P., Hriplivy, A.F.: Comparative analysis of methods for evaluating the financial condition of the organization. Polythemat. Netw. Electron. Sci. J. Kuban State Agrarian Univ. **81**, 22 (2012)
4. Pitchers, M.S.: Innovative instruments forecasting evaluation of the financial condition of the company. Bull. S. Ural State Univ. Ser. Econ. Manage. **30**, 56 (2012)
5. Smelova, T.A., Merzlikina, G.S.: Evaluation of the economic viability of the enterprise in crisis management. Volgograd State Technical University, Volgograd, Russia (2003)
6. Nedosekin, A.O.: Financial management in vague terms. (FUZZY FINANCIAL MANAGEMENT). Russia, Moscow, AFA Library (2003)
7. Nedosekin, S.A.: Business risk assessment based on fuzzy data: Monograph. St. Petersburg, Russia (2005)
8. Sakharova, L.V., Stryukov, M.B., Akperov, G.I.: Optimization of agricultural land use on the basis of mathematical methods of financial analysis and the theory of fuzzy sets. Adv. Intell. Syst. Comput. **896**, 790–798 (2019)
9. Vovchenko, N.G., Stryukov, M.B., Sakharova, L.V., Domokur, O.V.: Fuzzy-logic analysis of the state of the atmosphere in large cities of the industrial region on the example of Rostov region. Adv. Intell. Syst. Comput. **896**, 709–715 (2019)
10. Audit-IT.: Financial Analysis. Auditing firm "Avdeyev and K": audit and accounting services, 1999–2019 (2019). https://www.audit-it.ru

Evaluation of Efficacy of Integration Processes at the Manufacturing Enterprises of Industry Sector on the Regions of a Country

Tural Suleymanli Namig[(✉)]

Odlar Yurdu University, 13, Koroglu Rahimov Street,
Baku, Azerbaijan Republic
suleymanli.tural@gmail.com

Abstract. An integration shall be considered as a strategic alliance of enterprises, which allow to strengthen the competitive posture of participants of an integration and to obtain financial and other rewards of business growth. In our work, based on the indices calculated we derived a unified indicator - Integral Parameter of Efficacy of integration of manufacture enterprises of industry on the regions of a republic.

Keywords: Efficacy of an integration · Synergy effect · Integral parameter of efficacy

1 Introduction

On observing the integration processes as an object it is expedient to evaluate the efficiency of this object both at the stage of determination of it's realization and at the stage of assessment of results obtained. In addition to that, the evaluation shall be carried out both on posture of each participant and on the resultant synergetic effect. Thus, in the process of evaluation, the approach is directed to different levels of efficacy and to the time horizon of efficacy.

During the integration process both expenditures and yields are considered not depending on the assessment methods employed. The set of objects, subjects and processes is being involved to the mechanism of realization of an integration process and namely this set forms the expenditures and the yields parts of this mechanism. On observing the options of implementation of the project, on selection of the realization mechanism it is necessary to head towards the maximum economic gain achievable.

1.1 Integration Processes with Synergetic Effect

It is possible to relate the following integration processes to the class of integration processes having the capability to pose mutual impact with the positive synergetic result:

- the integration processes which complement each other as an outcome of realization of a multitude of separate results;

R. A. Aliev et al. (Eds.): ICSCCW 2019, AISC 1095, pp. 284–291, 2020.
https://doi.org/10.1007/978-3-030-35249-3_36

– the integration processes that are interdependent and yield positive result only when act together, otherwise yield negative result if they act independently.

There are several types of integration synergy and some of them and listed below:

- *Functional* (is provided by improvement of the management system and by decrease of conditional fixed costs on expansion of business scale).
- *Purposeful* (reveals itself in an increase of financial and other possibilities as a result of the engineering of a product with a cutback of costs related to interrelation with the external environment, of it's production and of it's sale).
- *Complex* (gives an additional opportunity because of improvement of provisioning of resources and improvement of sales of a product).
- *Conglomerate* (consists of distribution of risks and diminishing of their impact).

In addition, it is necessary to note the gains, which are possible but are often lost at the evaluation of the efficacy of integration projects. These cases happen when companies – participants of the integration process – failed to be fully loaded with a work.

1.2 Defining Temporary Boundaries for Expected Synergetic Effect

During the process of integration and as the result of it on the event of provision of additive, synergetic effect for all system on account of orchestrated transmutations of local races the system dynamic shall arise. This is controlled transformation of parameters of overall system in the result mutual communication (alignment over reciprocal connections) of dynamic processes of it's subsystems. It is necessary to define this dynamic, to measure it quantitatively and, on the base of it, to determine the efficacy of this integration and the level of efficiency. So that, because of the change of scale of business it is necessary to locate the equipoise of relative and absolute indicators during the design of system of evaluation indices.

The advantage of this approach is that the temporary boundaries of theirs are defined and this gives a capability to precisely determine the slices at the evaluation of efficacy.

2 Methods

2.1 Overview of Methodology

The potential efficacy of the integration process is to be determined at the stage of it's substantiation from economic point of view and it is possible to utilize the system of indices applicable to the evaluation of an investment project, namely *net profit, net present value, index of profitability*. The methodics to determine these indices are well known, but taking into account the specificity of the projects under the consideration it is expedient to direct attention to the following: what shall be the reporting timeframe for project's indices and how shall profitability degree of current value of cash flows before and after integration be selected?

It is reasonable to set reporting timeframe equal to the duration of the integration project. In turn the duration of the project shall cover not only the compound duration of organizational, economical and other activities stipulated in -the project, but also the duration of institutionalization of a new structure, of a new business model, the duration of accommodation of the staff, the suppliers and the consumers (one year or more depending on the scale of changes happening).

The efficacy can be evaluated both with an expenditure and total indices system and with an integrated index. The latter shall reflect the essence of an integration project, it's ultimate target.

The decision of integration can be regarded as the most important strategic decision taken by every one of participating production enterprises. In such a condition, many ones think that a market capitalization can be taken as an integrated index reflecting the purpose and the scopes of changes taking place during the integration process. This approach is quite widespread in the countries possessing developed market economy. In such an approach an effect of synergy can be determined as a surplus, as a remainder of deduction of the sum of participating enterprises' capitalizations before the integration from the capitalization of unified enterprise. Alas, there are certain methodic problems existing on realization of that approach. Additionally it is necessary to take into consideration the organizational-juridical form of project participants. It is possible to utilize the determination of stock prices on the stock market at the calculations of market values of those integrating enterprises, which are open joint stock ones. Also it is possible that advantages of an unified enterprise will not be reflected on it's stock price, or be reflected after a some time lag.

The methodic of calculating of a market value also depend on profitability of uniting companies. It is possible that for the sake of strategic purposes enterprises of operating with a loss or of operating with a low profit kind and completely successful ones could unite, because they depend on each other on resources, goods' production or sale. In such a circumstances, it is expedient to select the appropriate approach and the method of profit normalization for those enterprises that operate with a loss, and to model the cash flows for various options before integration and after it.

Concerning value-based approach to the evaluation of the efficacy, one's note the positive and the negative dynamic on variance of market capitalization criterion. The positive system dynamic satisfy the common interest of participants.

The composition of business processes are subject to change as the result of integration, and the impact of theirs on enterprise's activity indices shall not be weakened, on the contrary, it shall be strengthened, one mean the quality of the processes shall increase. The quality of the processes could be evaluated by the means of a group of indicators, such as, for example, resource indicators, process indicators, product indicators. The indicator of results' achievement in a timely manner and the indicator of degree of reach of goals set forth also can be appertained to the product indicator class. That make us believe that qualitative evaluations do not always provide complete conception about variances and there is a need for an expert estimations.

When revising the processes of restructuration of enterprises and holding a supposition that the ultimate goal of these processes is to keep or improve the strategic sustainability, in order to evaluate the efficacy, it is allowed to employ indices bound to these components of a strategic sustainability, such as production-technological sustainability, financial, market position and personnel sustainability. The subsystems mentioned could be represented by the system of indicators. In order to evaluate the efficacy of restructuration the etalon values compatible with the goals of restructuration are to be defined and to be matched with actual achieved indicator values.

The carried out researches demonstrated that to devise scientifically justified methods of an evaluation of the efficacy of an integration there is a necessity for more profound research of the economic essence of "efficacy" category (including with the regards to the peculiarities of reciprocal communications between agricultural industry entities). So that on a view of the essence and economic connotation of efficacy problem we have spotted the following based on studying the proceedings of local and foreign scientists devoted to that problem.

There is no terminological accord in the usage of the term "efficacy" at the analysis of production-economic systems. As a rule the term "efficacy" is used as a synonym of productivity, optimality, sustainability (see Table 1). Within the framework of process approach the evaluation of efficacy is limited to the comparison of characteristics of "inputs" and "outputs". From the system point of view efficacy represents disparate characteristics of a system's quality.

Table 1. The modern approaches to the term "efficacy"

Categories	Substance
1. Economic approach	
Productivity, effectiveness	Overall parameter of production system's design
2. Production-technological approach	
System optimality, system equilibrium	Internal characteristics of the resource potential, the proportions between its components
3. Systematic approach, complex approach	
Responsiveness, durability stability	The state of the system, its capability to withstand the extraneous influence
Synergy capability	The outcome of the cooperative reciprocal connections of subsystems which change the quality of the system during self-organization of it
4. Commercial approach	
Competitive capacity	The competence to preserve the stable position at the competitive struggle
Use	The degree of fulfillment of user's requirements
Innovativeness	The competence to utilize the modern ICT

2.2   The Characteristics of Approaches to the Assessment of Efficacy

Thus, it should be noted that the qualitative aspect of efficacy can be based on the "productivity" and the "optimality" categories. We have systemized the essential characteristics of presented approaches and are demonstrating them on the Table 2.

Table 2. The characteristics of approaches to the assessment of efficacy

Comparison criteria	Distinctive features	
	Productivity	Optimality
Economic synonym	Productivity, effectiveness	Optimality, equilibrium
Content of the category	Production changes, effect	The process of integration
Direction of the diagnostic	Retrospective (temporary result yielded)	Prediction
Content of the diagnostic	The characteristics of the summation of activities, the measurement of productivity of usage of resources	Internal content of the production
Essential fundament	Process approach	Complex approach
Evaluation of the system	The system of quantitative and qualitative indices	Critical and median quantifiers' system

The specific calculations of different components of the efficacy category based on the approaches mentioned above are performed. Amongst them is the determination of degree of integration of industry and trade capitals at production enterprises of industry sector on the regions of the republic (see Table 3).

Calculations demonstrates that the more high indices of sale of the product over the own distribution channels are peculiar to Baku city, Absheron, Ganja-Qazakh and Lankaran economic regions. This indicates the more high degree of integration in this regions (see Table 3).

Table 3. The determination of degree of integration of industry and trade capitals at production enterprises of industry sector on the regions of the republic (in million manats) [1].

$I_{sodc} = \frac{V_{sodc}}{V_s}$ Economic Regions	The index of sale of the product over the own distribution channels (I_{sodc})	The volume of sales of the product over the own distribution channels (V_{sodc})	Compound volume of the sales (V_s)
1. Baku city	0,676	18551,4	27432,4
2. Absheron	0,649	634,7	977,3
3. Ganja-Qazakh	0,557	342,7	615,7
4. Shaki-Zakatala	0,463	89,5	193,2
5. Lankaran	0,555	56,7	102,0
6. Quba-Khachmaz	0,507	47,8	94,3
7. Aran	0,524	600,7	1145,6
8. Upper-Karabakh	0,427	12,5	29,3
9. Mountanous Shirvan	0,520	10,2	19,6
10. Nakhchivan AR	0,600	565,5	942,5

Afterwards we have performed the calculations to determine the degree of connections' cooperation at production enterprises of industry sector on the regions of the republic - the coverage index (Icov) and the power index (Ipow) are considered.

Calculations demonstrates that the most high coverage indices (Icov) are peculiar to Baku city, Absheron, Shaki-Zakatala, Upper-Karabakh and Nakhchivan AR (see Column 3 of Table 4).

When it comes to the indices of production power (Ipow), the more high values are peculiar to Baku city, Absheron and Ganja-Qazakh economic regions (see Column 4 of Table 4).

An important stage of this research is the process of determination the effectiveness of the management within the corporation at production enterprises of industry sector on the regions of the republic. The index of share of mother enterprise in the charter capital of a given enterprise (Iregsh) and the index of share of mother enterprise in the overall property of a given enterprise (Iovrlsh) are the indices to be considered. Calculations have shown that usually these indices have values more than 80% and thus affirm the leading role of a mother enterprise having the capability to amplify the integration processes of daughter production enterprises in an industry sector in the future (see Column 5 and Column 6 of Table 4).

Table 4. Indices for input to fuzzy integral parameter generation system

1 Economic Regions	Indices for input to fuzzy integral parameter generation system				
	2	3	4	5	6
	Isodc	Icov	Ipow	Iregsh	Iovrlsh
1. Baku city	0,676	0,741	0,818	0,900	0,900
2. Absheron	0,649	0,763	0,922	0,975	0,867
3. Ganja-Qazakh	0,557	0,698	0,894	0,911	0,846
4. Shaki-Zakatala	0,463	0,716	0,782	0,894	0,727
5. Lankaran	0,555	0,631	0,756	0,860	0,631
6. Quba-Khachmaz	0,507	0,642	0,665	0,851	0,920
7. Aran	0,524	0,759	0,718	0,939	0,882
8. Upper-Karabakh	0,427	0,552	0,624	0,733	0,700
9. Mountanous Shirvan	0,520	0,523	0,633	0,818	0,818
10. Nakhchivan AR	0,600	0,754	0,772	0,905	0,826

3   Results

A fuzzy integral parameter generating system based on neural networks with the specified inputs and output [2–5] is designed. Output value is the Integral Parameter of Efficacy of integration of manufacture enterprises of industry at regions of a republic (IPE).

Figure 1 shows the rule editing window for generating the resulting fuzzy value for the IPE of Efficacy of integration of manufacture enterprises of industry at regions of a republic.

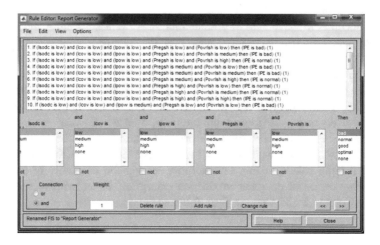

Fig. 1. The rule editing window for generating the resulting fuzzy value for the IPE of Efficacy of integration of manufacture enterprises of industry

4   Conclusion

Defuzzification of the resulting fuzzy values for the various economic regions of the republic gives the values shown in Table 5.

Table 5. The resulting fuzzy and defuzzified values for Integral Parameter of Efficacy of integration of manufacture enterprises of industry at regions of a republic (IPE)

Economic Regions	IPE	
	Fuzzy value	Exact value
1. Baku city	Optimal	0,854
2. Absheron	Optimal	0,882
3. Ganja-Qazakh	Good	0,793
4. Shaki-Zakatala	Good	0,735
5. Lankaran	Normal	0,701
6. Quba-Khachmaz	Good	0,735
7. Aran	Good	0,798
8. Upper-Karabakh	Bad	0,633
9. Mountanous Shirvan	Bad	0,682
10. Nakhchivan AR	Good	0,789

The results of the calculations show that the more high values of Integral Parameter of Efficacy of manufacture enterprises of industry at regions of a republic are peculiar to Baku city and Absheron economic region; medium values are peculiar to Ganja-Qazakh, Shaki-Zakatala, Quba-Khachmaz, Aran and Nakhchivan AR economic regions; low values are peculiar to Lankaran; most low values are peculiar to Upper-Karabakh and Mountanous Shirvan economic regions.

References

1. The industry of Azerbaijan Republic: Statistical almanach. ARDSK, Baku (2018)
2. Jang, J.S.R., Sun, C.T., Mizutani, E.: Neuro-Fuzzy and Soft Computing. Prentice Hall, Upper Saddle River (1997)
3. Zadeh, L.A.: Fuzzy sets as a basis for a theory of possibility. Fuzzy Sets Syst. **1**, 3–28 (1978)
4. Aliev, R.A., Fazlollahi, B., Aliev, R.R.: Soft Computing and Its Applications in Business and Economics. Springer, Heidelberg (2004). https://doi.org/10.1007/978-3-540-44429-9
5. Beale, M.H., Hagan, M.T., Demuth, H.B.: Neural Network Toolbox. User's Guide Math Works Inc., Natick (2014)

Methods of Assessing the Risk of Bankruptcy of an Enterprise Based on a Set of MDA-Models and the Theory of Fuzzy Sets

V. Alekseychik Tamara[1] , A. Vasilenko Alla[1] ,
B. Stryukov Michael[1] , and S. Kokhanova Victoriya[2](✉)

[1] Rostov State University of Economics,
B. Sadovaya Street, 69, 344002 Rostov-on-Don, Russia
alekseychik48@mail.ru, allvasilenko@yandex.ru,
mstryukov@mail.ru
[2] Sothern University (IMBL), M. Nagibina Pr., 33a/47,
344068 Rostov-on-Don, Russia
kohanovavs@yandex.ru

Abstract. The aim of the work is to develop a methodology for comprehensive assessment of the risk of bankruptcy of an enterprise based on fuzzy-multiple aggregation of estimates obtained through the use of a set of classical models. The technique is based on the use of fuzzy multi-level classifiers and allows the aggregation of estimates for the three groups of models. In each group, an enterprise is assessed according to several conditions. For example, for Altman and Taffler-Tishou models, the assessment is carried out in two states ("medium risk - high risk"). For the so-called Irkutsk model and the Savitskaya model, the assessment is performed according to five conditions ("very low risk—low risk—medium risk—high risk—very high risk"). It is significant that the analysis applies only those indicators that most reflect the possibility of bankruptcy of the enterprise. At the final stage, the normalized bankruptcy risk estimates obtained in each of the groups are aggregated into a final assessment. This assessment is an integral indicator of the risk of bankruptcy of an enterprise. The novelty of the proposed methodology consists in the possibility of combining the conclusions obtained on the basis of various non-standardized methods using different evaluation criteria. In addition, the technique allows to take into account in the model weights, reflecting the reliability of the models for the studied group of enterprises.

Keywords: Complex assessment · Aggregation · Fuzzy-Set methodology · Systems of fuzzy-logical conclusions · Risk of bankruptcy

1 Introduction

The task of determining the risk of bankruptcy of an enterprise is relevant for all persons interested in the financial position of the enterprise, that is, its management, owners, investors, creditors, tax and administrative authorities, etc. In the financial analysis is well known a number of indicators characterizing certain aspects of the current financial situation of the company.

© Springer Nature Switzerland AG 2020
R. A. Aliev et al. (Eds.): ICSCCW 2019, AISC 1095, pp. 292–300, 2020.
https://doi.org/10.1007/978-3-030-35249-3_37

At present, a large number of integral coefficients have been developed, which characterize the general situation and the probability of an enterprise bankruptcy on the basis of these indicators of its financial position. First of all, these are the so-called MDA models, or models built on the basis of multiple discriminant analysis. These are statistical models, and their construction is carried out on the basis of statistical data of financial statements. The MDA-models include the Altman Model, the Fox Model, the Tuffler Model, the Irkutsk Model (IEEE), the Fulmer Model, etc. Among the most famous works in which models were created using this method, the following can be noted: Altman [1], Deakin [3], Taffler [4], Altman [4], Zaitseva [5], Davydova & Belikov [6].

The algorithm for constructing all MDA models includes five points: (1) the selection of bankrupt enterprises; (2) the formation of a sample of non-bankrupt enterprises; (3) calculation of financial ratios for both groups; (4) building a regression equation that classifies all enterprises into bankrupts and non-bankrupts using the Multiple Discriminant Analysis Tool (MDA); (5) checking the adequacy of the constructed model.

The paper [7] analyzed the results of applying foreign and Russian models to the financial statements of Russian enterprises. It has been established that for the prediction of bankruptcy, the so-called "classic" western models (Altman, Taffler, Springgate) are more effective than models originally built by Russian companies. In general, the results of using foreign models are ambiguous: models with high overall reliability are not very good for bankrupt and non-bankrupt groups, and models that work well for a group of bankrupts are not very effective for "healthy" companies and have low overall reliability. It is impossible to say that at the present time there are absolutely reliable MDA - models, on the basis of which one could get reliable information about the risk of bankruptcy of an enterprise. As a result, the credibility of the study increases with the simultaneous use of several models at once, both foreign and domestic.

The problem is that when several of these models are used in practice at the same time, the result is often scattered: for some models, the probability of bankruptcy is high, and for others - low. Therefore, an important task is to combine the results obtained and the formation of quantitative estimates based on a set of estimates calculated using different models. At the same time, it is impossible to obtain estimates as "average values" based on several traditional models of bankruptcy risk assessment. This is due to the fact that the models are not unified, have different criteria, and also allocate a different number of enterprise conditions.

The most promising solution is to aggregate estimates based on fuzzy logic methods [8–10]. Fundamentals of the theory of fuzzy sets are described in the works of Lofti Zade [10]. The greatest effect when using the method of fuzzy sets is typical for the evaluation of processes that are based on subjective judgments. This applies to both estimates of the probability of bankruptcy, and to the audit in general.

Audit as a process based on subjective judgments of an auditor also implies an assessment of the reliability of financial statements in the face of uncertainty. The possibilities of using the tools of the theory of fuzzy sets for the purpose of taking optimal actions in the process of planning audit procedures were also investigated [11].

Fuzzy-logical inference systems (first of all, matrix aggregation systems) allow building system estimates based on a set of indicators, taking into account their weights. In addition, the systems of fuzzy-logical conclusions allow to take into account expert assessments, which is an indisputable advantage in adapting the mathematical apparatus of research to solve the problems facing the domestic economy.

2 Materials and Methods

We have developed a methodology for aggregating estimates of the risk of bankruptcy of an enterprise, calculated on the basis of a number of well-known MDA-models. The study was based on 13 foreign and domestic methods implemented on the basis of the service afdanalyse.ru/Example/Analiz_Ball_1.xls. Below we provide a brief description of the techniques available on the same site.

Altman Two-Factor Model:

$$Z = -0.3877 - 1.0736 * Clr + 0.0579 * (Bc/P),$$

where: Clr is current liquidity ratio; Bc is borrowed capital; P is liabilities. Conclusion: if Z < 0 is less than 50%, and decreases with decreasing Z, if Z = 0, it is approximately 50%, if Z > 0 is more than 50%, and increases with Z (3 states).

Modified Five-Factor Altman Model:

$$Z = 0.717X1 + 0.847X2 + 3.107X3 + 0.42X4 + 0.998X5,$$

where: X1 is working capital to the amount of assets, estimates the amount of the company's net liquid assets in relation to total assets. X2 is retained earnings to total assets of the company, reflects the level of financial leverage of the company. X3 is profit before tax to total assets; reflects the effectiveness of the company's operating activities. X4 is the carrying value of equity/borrowed capital (liabilities). X5 is sales to the total value of the assets of the enterprise, characterizes the profitability of the assets of the enterprise. Conclusion: if Z < 1.23 – high risk; if Z is from 1.23 to 2.89 – there is no certainty; if Z is over 2.9 – low risk (3 states).

Altman Model for Non-production Companies:

$$Z = 6.56X1 + 3.26X2 + 6.72X3 + 1.05X4,$$

where: X1 = Working capital/Assets; X2 = Retained earnings/Assets; X3 = Profit before tax/assets or EBIT/assets; X4 = Equity/Liabilities. Conclusion: if Z ≤ 1.1 – high risk; if Z is in the range of 1.1 to 2.6, the situation in the enterprise is stable; if Z ≥ 2.6, the situation is unstable (3 states).

Altman Model for Emerging Markets:

$$Z = 3.25 + 6.56X1 + 3.26X2 + 6.72X3 + 1.05X4,$$

where: X1 = Working capital/Assets; X2 = Retained earnings/Assets; X3 = Profit before tax/assets or EBIT/assets; X4 = Equity/Obligations. Conclusion: if Z is equal to or less than 1.1 – the situation is critical, the organization is highly bankrupt; if the value of Z is equal to or exceeds 2.6 – an unstable situation, the probability of an organization going bankrupt is small, but it is not excluded; if the indicator Z is within the range from 1.10 to 2.6 – low probability of bankruptcy of the organization (3 states).

Model by Taffler-Tishou:

$$Z = 0.53X1 + 0.13X2 + 0.18X3 + 0.16X4,$$

where: X1 is the ratio of profit before tax to the amount of current liabilities; X2 is the ratio of the sum of current assets to total liabilities; X3 is the ratio of the amount of current liabilities to total assets; X4 is the ratio of revenue to total assets. Conclusion: when Z > 0.3, the probability of bankruptcy is low; at 0.2 < Z < 0.3, the state of uncertainty; at Z < 0.2 is high probability of bankruptcy (3 states).

Fulmer Model:

$$H = 5,528X1 + 0,212X2 + 0,073X3 + 1,270X4 - 0,120X5 + 2,335X6 + 0,575X7 + 1,083X8 + 0,894X9 - 6,075,$$

where: X1 = retained earnings of previous years [1]/Balance sheet [1]; X2 = sales Revenue/Balance sheet [1]; X3 = Profit before taxes/Equity; X4 = Cash flow/long-Term and short-term liabilities [1]; X5 = long-Term liabilities [1]/Balance sheet [1]; X6 = short-Term liabilities/Total assets [1]; X7 = log (tangible assets); X8 = Working capital [1]/long-Term and short-term liabilities [1]; X9 = log (profit before tax + interest payable/interest paid). [1] = average value for the period (value at the beginning + value at the end of the period)/2. Conclusion: insolvency is inevitable at H < 0(2 States).

Springate Model:

$$Z = 1,03X1 + 3,07X2 + 0.66X3 + 0,4X4,$$

where: X1 = Working capital/Balance sheet; X2 = EBIT/Balance sheet; X3 = EBIT/Current liabilities; X4 = sales Revenue (net)/Balance sheet. Conclusion: at Z < 0.862 the company is a potential bankrupt (2 States).

Irkutsk Model:

$$P = 8.38X1 + X2 + 0.054X3 + 0.63X4,$$

where: X1 = the net working (operating) capital/assets; X2 = net profit/equity; X3 = net profit/balance sheet total; X4 = net profit/total costs. Conclusion: if R is less than 0 —the probability of bankruptcy is maximum (90%–100%); if R 0–0.18—high (60%–80%); if R 0.18–0.32—average (35%–50%); if R 0.32–0.42—low (15%–20%); if R is more than 0.42—minimum (up to 10%) (5 States).

Liss Model:

$$Z = 0,063X1 + 0,092X2 + 0,057X3 + 0,0014X4,$$

where: X1 = working capital/amount of assets; X2 = profit from sales/amount of assets; X3 = retained earnings/amount of assets; X4 = equity/debt capital. Conclusion: if Z < 0.037 high probability; if Z > 0.037 low probability (2 States).

Model by O. P. Zaitseva:

$$K = 0,25X1 + 0,1X2 + 0,2X3 + 0,25X4 + 0,1X5 + 0,1X6$$

where: X1 = the loss ratio of the enterprise; X2 = the ratio of accounts payable and receivables; X3 = the ratio of short-term liabilities and the most liquid assets; X4 = the loss ratio of sales, characterized by the ratio of net loss to the volume of sales of these products; X5 = the ratio of financial leverage (financial risk); X6 = the load factor of assets as the inverse ratio of the turnover of assets.

We must compare the actual value (AV) with the normative value (NV), which is calculated by the formula:

$$NV = 0,25 * 0 + 0,1 * 1 + 0,2 * 7 + 0,25 * 0 + 0,1 * 0,7 + 0,1 \\ * X6 \, (last \, year).$$

Conclusion: if the actual coefficient is greater than the normative Factor > KP, then the probability of bankruptcy of the enterprise is extremely high; if less—then the probability of bankruptcy is insignificant (2 States).

Model of R.S. Saifullin and G.G. Kadykov:

$$R = 2X1 + 0,1X2 + 0,08X3 + 0,45X4 + X5,$$

where: X1 = equity ratio; X2 = current liquidity ratio; X3 = the intensity of turnover of advanced capital; X4 = management ratio; X5 = return on equity. Conclusion: if R < 1 – unstable state; if R ≥ 1 – stable state (2 States).

Model by V.V. Kovalev:

$$N = 25\,N1 + 25\,N2 + 20\,N3 + 20\,N4 + 10\,N5,$$

where N1 is the inventory turnover ratio; N2 is the current liquidity ratio; N3 is the capital structure ratio; N4 is the return on assets: profit before tax/assets; N5 is the efficiency ratio: profit before tax/sales revenue. Conclusion: if N ≥ 100—good financial position; if N < 100 - negative financial position (2 States).

Model by G.V. Savitskaya:

$$Z = 0,111X1 + 13,239X2 + 1,676X3 + 0,515X4 + 3,80X5,$$

where X1 = the share of working capital in the formation of current assets; X2 = the ratio of working capital to the main; X3 = the turnover ratio of total capital; X4 = return on assets of the enterprise; X5 = the coefficient of financial independence. Conclusions: when z Is greater than 8, the risk of bankruptcy is small; when Z is from 8 to 5, the risk of insolvency is small; when Z is from 5 to 3, the average risk of bankruptcy; when Z is lower than 3 – the risk is small; when Z is lower than 1, the company is bankrupt (5 States).

As can be seen from the descriptions of the models, five of them (formally) distinguish two states of the enterprise in relation to the risk of bankruptcy; six distinguish three states and two distinguish five states.

So-called matrix schemes of data aggregation, fuzzy multilevel [0,1] - classifiers, two -, three-and five-point are used for information aggregation.

As a carrier of a linguistic variable, we define a segment of the real axis [0,1]. We introduce a linguistic variable "BR (bankruptcy risk)" with a term-set of G values consisting of three terms: G1 – "BR low"; G2 – "BR medium"; G3 – "BR high".

Matrix scheme of data aggregation based on three-level fuzzy classifiers is based on the formula:

$$g = \sum_{i=1}^{N} p_i \sum_{j=1}^{3} \alpha_j \mu_{ij}(x_i)$$

where α_j are nodal points of the standard classifier (centers of gravity of terms), p_i is the weight of the i-th factors in the convolution, $\mu_{ij}(x_i)$ is the value of the membership function of the j-th qualitative level relative to the current value of the i-th factor

The value of g is then recognized on the basis of a standard fuzzy classifier, according to the specified membership functions.

If the linguistic variable "BR (bankruptcy risk)" is described by a term-set of five terms: (G1 – "BR very low"; G2 – "BR low"; G3 – "BR medium"; G4 – "BR high"; G5 – "BR very high"), we get a standard five-point [0,1] – classifier.

Finally, if the linguistic variable "BR" is described by a term-set of two terms (G1 – "BR low"; G2 – "BR high"), we obtain the simplest binary classifier.

Systems of multipoint classifiers allow us to calculate a comprehensive assessment of the risk of bankruptcy of the enterprise by rationing estimates and aggregating them on the basis of matrix schemes.

3   Results

A study of the risk of bankruptcy of Open Society "Donskoye" on the basis of financial statements for 2016–2017 was carried out. At the first step, 6 models with two terms were aggregated using two-point classifiers; at the second step, 5 models with three terms were aggregated; at the third step, 2 models with five terms were aggregated. Finally, at the fourth step, the final comprehensive assessment of bankruptcy risk is built on the basis of three groups of models, using standard three-point classifiers. It was believed that all models of equilibrium (weights can be varied). The process of aggregation of results is presented in Tables 1, 2, 3, and 4 (introduced reduction of RB – "bankruptcy risk"). The results of direct calculation of bankruptcy risk by models are given in Table 2.

V. A. Tamara et al.

Table 1. Aggregation of BR estimates obtained on the basis of MDA – models distinguishing between two states (G1 is "BR low"; G2 is "BR high").

Model	Term	2016		2017	
		G1	G2	G1	G2
1. Fulmer model		1	0	1	0
2. Springite Model		1	0	1	0
3. Liss model		1	0	1	0
4. Model Zaitseva		1	0	0	1
5. Saifullin-Kadykov model		1	0	1	0
6. Model Kovalev		1	0	1	0
Term weights		1	0	5/6	1/6
Final grades		g_2 (2016) = 0,26*1 + 0,74*0 = 0,26		g_2 (2017) = 0,26*5/6 + 0,74*1/6 = 0,34	
Integrated value		$g_2 = g_2(2016)*1/3 + g_2(2017)*2/3 = 0,26*1/3 + 0,34*2/3 = 0,31$			

Table 2. Aggregation of BR estimates obtained on the basis of MDA – models that distinguish three states (G1 is "BR low"; G1 is "BR medium"; G1 is "BR high").

Model	2016			2017		
	G1	G2	G3	G1	G2	G3
1. Two-factor Altman model	1	0	0	1	0	0
2. Modified five-factor Altman model	1	0	0	0	1	0
3. Altman model for non-production companies	0	1	0	0	1	0
4. Altman model for emerging markets	1	0	0	1	0	0
5. Model Taffler-Tishou	1	0	0	1	0	0
Term weights	4/5	1/5	0	3/5	2/5	0
Final grades	g_3 (2016) = 0,155*4/5 + 0,5*1/5 + 0,845*0 = 0,224			g_3 (2017) = 0,155*3/5 + 0,5*2/5 + 0,845*0 = 0,293		
Integrated value	$g_3 = g_3$ (2016)*1/3 + g_3 (2017)*2/3 = 0,224*1/3 + 0,293*2/3 = 0,27					

Table 3. Aggregation of estimates of BR, obtained on the basis of MDA-models, distinguishing five states (G1 is "BR is very low"; G2 is "BR is low"; G3 is "BR is medium"; G4 is "BR is high"; G5 is "BR is very high").

Model	2016					2017				
	G1	G2	G3	G4	G5	G1	G2	G3	G4	G5
1. Irkutsk model	1	0	0	0	0	1	0	0	0	0
2. Model Savitskaya	0	1	0	0	0	0	0	1	0	0
Term weights	1/2	1/2	0	0	0	1/2	0	1/2	0	0
Final grades	g_5 (2016) = 0,125*1/2 + 0,3*1/2 + 0,5*0 + 0,7*0 + 0,875*0 = 0,21					g_5 (2017) = 0,125*1/2 + 0,3*0 + 0,5*1/2 + 0,7*0 + 0,875*0 = 0,31				
Integrated value	$g_5 = g_5$ (2016)*1/3 + g_5 (2017)*2/3 = 0,21*1/3 + 0,31*2/3 = 0,28									

Table 4. Aggregation of BR estimates from Tables 2, 3, and 4 into the final estimate.

Estimates	Terms		
	G1	G2	G3
g1 = 0,31	0,4	0,6	0
g2 = 0,27	0,65	0,35	0
g3 = 0,28	0,6	0,4	0
Term weights	0,55	0,45	0
Integrated value	g = 0,155*0,55 + 0,5*0,45 + 0,845*0 = 0,31		

4   Discussion

Thus, the final aggregated estimate based on the considered models has a numerical value of 0.31 (in accordance with the theory of fuzzy sets, it can be considered that it is likely that the expert will refer the enterprise to the corresponding term). The value of the membership functions:

$$\mu(0,31) = \mu_2(0,31) = 0,45; \ \mu(0,31) = \mu_1(0,31) = 0,55.$$

Thus, we can assume that the company can be assigned to the first term ("RB low") with a probability of 0.45 and to the second term ("RB average") with a probability of 0.55.

Therefore, the analysis of financial condition of the enterprise on the basis of thirteen different models allowed to calculate the aggregated value giving an assessment of risk of bankruptcy on an interval [0;1]. Conventionally, this value can be considered as the risk of bankruptcy, calculated taking into account the views of thirteen independent experts.

5   Conclusion

The technique, the novelty of which is the ability to aggregate the results of the analysis of the risk of bankruptcy of the enterprise, resulting from the use of a complex of different models of bankruptcy. In this case, models can use different criteria and classify the state of the enterprise in different ways. As follows from the description of the methodology, the complex of models used may vary depending on the objectives of the study.

References

1. Altman, E.I.: Financial ratios, discriminant analysis and the prediction of corporate bankruptcy. J. Financ. **4**, 589–609 (1968)
2. Deakin, E.: Discriminant analysis of predictors of business failure. J. Account. Res. **10**, 167–179 (1972)

3. Taffler, R.J.: The assessment of company solvency and performance using a statistical modeling. Account. Bus. Res. **13**(52), 295–307 (1983)
4. Altman, E.I.: Further empirical investigation of the bankruptcy cost question. J. Financ. **39** (4), 1067–1089 (1984)
5. Zaitseva, O.P.: Crisis Management in Russian company. Aval, 11–12, 66–73 (1998)
6. Davydova, G.V., Belikov, AYu.: Methods of quantitative assessment of bankruptcy risk of enterprises. Risk Manag. **3**, 13–20 (1999)
7. Fedorova, E., Gilenko, E., Dovzhenko, S.: Bankruptcy prediction for Russian companies: Application of combined classifiers. Exp. Syst. Appl. **18**(40), 7285–7293 (2013)
8. Nedosekin, A.O.: Fuzzy Financial Management. AFA Library, Moscow (2003)
9. Nedosekin, A.O., Kozlovsky, A.N., Abdulaeva, Z.I.: Analysis of branch economic stability by fuzzy-logical methods. Econ. Manage. Prob. Solut. **5**, 10–16 (2018)
10. Zadeh, L.A.: Concept of a Linguistic Variable and its Application to Making Approximate Decisions. Mir, Moscow (1976)
11. Yakimova, V.A.: Optimization of audit actions on the basis of an assessment of sufficiency of auditor proofs and labor input of process of their collecting. Int. Account. **43**, 25–36 (2012)

Fuzzy Multiple Methods of Diagnosis and Credit Risk of Bankruptcy of the Agricultural Enterprises of the Region on the Basis of Score and MDA-Models

B. Stryukov Michael[1]([✉]) [iD], V. Domakur Olga[2] [iD],
K. Medvedskaya Tatyana[3] [iD], D. Kostoglodova Elena[1] [iD],
and V. Martynov Boris[4] [iD]

[1] Rostov State University of Economics, Rostov-on-Don, Russia
mstryukov@mail.ru, kafedra_finance@mail.ru
[2] Belarusian State Academy of Communication, Minsk, Belarus
domakur@tut.by
[3] Don State Technical University, Rostov-on-Don, Russia
medvedskaya7l@mail.ru
[4] Private Educational Institution of Higher Education,
"Southern University (IMBL)", Rostov-on-Don, Russia
martynov@iubip.ru

Abstract. The aim of the work is to develop a methodology for comprehensive assessment of the creditworthiness and risk of bankruptcy of industry enterprises in the region based on fuzzy-multiple aggregation of estimates obtained through the use of a combination of classical MDA and scoring models. The technique is based on the use of fuzzy three-level classifiers and allows the aggregation of estimates for a group of models and a set of enterprises in the industry. Initially, bankruptcy risk estimates are calculated for a set of regional enterprises based on classical models (for example, Altman, Taffler, Seyfullin-Kadikov). Then they are aggregated across the aggregate of enterprises, with the result that we get an aggregated assessment of the risk of bankruptcy of enterprises in the industry according to each of the methods. Finally, the obtained aggregate estimates are used to form a comprehensive assessment of the risk of bankruptcy of enterprises in the industry in the region. The novelty of the proposed methodology consists in the possibility of combining the conclusions obtained on the basis of various non-standardized methods using different evaluation criteria. In addition, the method allows to take into account in the model weights, reflecting the importance of enterprises in the final assessment and the reliability of the models for the studied group of enterprises.

Keywords: Creditworthiness · Bankruptcy risk · Aggregation · Fuzzy-Multiple method

© Springer Nature Switzerland AG 2020
R. A. Aliev et al. (Eds.): ICSCCW 2019, AISC 1095, pp. 301–308, 2020.
https://doi.org/10.1007/978-3-030-35249-3_38

1   Introduction

Analysis of the creditworthiness of the borrower, as well as the risk of its bankruptcy is an important task for potential creditors of the company, as well as its partners. An even more serious task is to assess the attractiveness of a given industry of the region for potential lenders.

To solve the problem of assessing the risk of bankruptcy of an individual enterprise, a number of methods have been developed and used, such as the Altman [1], Taffler [2] etc. Most of the techniques are based on the so-called MDA models. In MDA-models, a certain final parameter is considered, which is a regression dependence on a set of well-known coefficients (such as autonomy ratio, liquidity ratios, profitability, etc.). Despite the practicality and wide applicability in modern financial analysis, these models have several drawbacks. For example, MDA models are not able to provide a reasonable quantitative estimate of the probability of bankruptcy; they are able to indicate only its qualitative degree—as low, high, very high, etc.

In addition, in all MDA-models there is a so-called "zone of uncertainty", if it falls into which the calculated total indicator can not make a clear conclusion about the probability of bankruptcy [3]. Different models of bankruptcy risk assessment often give very different conclusions, and their aggregation in the framework of existing methods is not possible. Finally, even with assessing the risk of bankruptcy of individual enterprises of a given cluster, it is extremely difficult to calculate a kind of comprehensive assessment, based on which potential lenders could assess the attractiveness of the region for lending.

The creditworthiness of an enterprise is the ability of a company to repay its short-term liabilities on time and in full. Scoring models that use bankruptcy risk assessment models (Altman, Beaver, Lisa, Taffler, Savitskaya, Kadyrov, Zhdanov, etc.) are most often used to assess the creditworthiness of enterprises in the industry. These models give an assessment of the class of creditworthiness depending on the level of risk of non-repayment of debts. After calculating the credit rating, its value is compared with the levels of bankruptcy risk.

In addition, there are scoring methods for assessing the creditworthiness of enterprises (for example, analysis of the creditworthiness of the borrower according to the method of Sberbank). As their common drawback, one should also note the impossibility of using them for a comprehensive assessment of the credit attractiveness of enterprises in the industry of a given region.

This problem can be solved by applying the theory of fuzzy sets [4], which is currently being successfully used in financial analysis [5]. This article presents a method for diagnosing creditworthiness and bankruptcy risk of industry enterprises based on fuzzy inferences systems, so-called fuzzy multilevel [0, 1] - classifiers.

2   Materials and Methods

2.1   Analysis of the Creditworthiness of the Borrower (Sberbank Method)

As a source material for the construction of an integrated assessment of the creditworthiness of enterprises in the agrarian region, the Sberbank method, implemented in the Audit-IT software, was used.

In accordance with the Sberbank methodology, the borrower's creditworthiness is determined based on the values of six indicators: (1) absolute liquidity ratio; (2) intermediate (fast) liquidity ratio; (3) current ratio; (4) the coefficient of availability of own funds (except for trade and leasing organizations); (5) product profitability; (6) the profitability of the enterprise. Depending on the value of each of the indicators, it can be assigned to one of three categories (Table 1). To calculate the total score of the borrower, you need to multiply the category number of the indicator by its weight coefficient (Table 1), and then sum the values obtained.

Borrowers are divided, depending on the amount of points received, into three classes: (1) first-class - lending of which is beyond doubt (the amount of points is up to 1.25 inclusive); (2) second class - lending requires a balanced approach (more than 1.25, but less than 2.35 inclusive); (3) the third class - lending is associated with increased risk (more than 2.35). The calculation scheme for the category to which the borrower can be attributed is given in Table 1.

Table 1. Distribution of indicators by categories according to their numerical value (Savings Bank methodology)

Indicator	Designation	Weight	Reference: category index		
			Category 1	Category 2	Category 3
Absolute liquidity ratio	R1	0.05	0.1 or higher	0.05–0.1	Less than 0.05
Coefficient of intermediate (fast) liquidity	R2	0.1	0.8 or higher	0.5–0.8	Less than 0.5
Current ratio	R3	0.4	1.5 or higher	1.0–1.5	Less than 1.0
Coefficient presence equity	R4	0.2	0.4 or higher	0.25–0.4	Less than 0.25
Product profitability	R5	0.15	0.1 or higher	Less than 0.1	Unprofitable
The profitability of the company	R6	0.1	0.06 and above	Less than 0.06	Unprofitable
Score	S	1	1.25	From 1.25 to 2.35	Over 2.35

2.2 Methods for Assessing the Risk of Bankruptcy of an Enterprise

To assess the risk of bankruptcy of an enterprise, three standard methods were used, also implemented in the Audit-IT software: Altman's Z-score (Altman's model), Taffler's model, and Seifullin-Kadykov's model. Below is a brief description of the techniques.

Altman's Z-score is calculated using the following formula (4-factor model for private non-production companies):

$$Z - score = 6.56T1 + 3.26T2 + 6.72T3 + 1.05T4,$$

where T1 is the ratio of working capital to the value of all assets; T2 is the ratio of retained earnings to the value of all assets; T3 is the ratio of EBIT to the value of all assets; T4 is the ratio of equity to debt.

The estimated probability of bankruptcy depending on the value of the Altman Z-account is: (1) 1.1 or less – a high probability of bankruptcy; (2) from 1.1 to 2.6 – the average probability of bankruptcy; (3) from 2.6 and higher – low probability of bankruptcy.

The Tuffler model includes four factors:

$$Z = 0.53X1 + 0.13X2 + 0.18X3 + 0.16X4,$$

where X1 = Sales Profit/Short-term Liabilities; X2 = Current assets/Liabilities; X3 = Short-term liabilities/Assets; X4 = Revenue/Assets.

The probability of bankruptcy according to the Taffler model is determined as follows: if Z is greater than 0.3, then the probability of bankruptcy is low; if Z = 0.3, then the probability is average; if Z is less than 0.2, then the probability of bankruptcy is high.

The five-factor model of the Saifullin-Kadykov method: The final indicator is determined by the formula:

$$R = 2K1 + 0.1K2 + 0.08K3 + 0.45K4 + K5,$$

where K1 is the ratio of own funds; K2 is the current ratio; K3 is the asset turnover ratio; K4 is a commercial margin (profitability of sales); K5 is the return on equity.

According to the Saifullin-Kadykov model, with the value of the final indicator R < 1, the probability of bankruptcy of an organization is considered high, with R = 1 medium, with R > 1 low.

2.3   Math Tools

The segment of the real axis [0,1] is defined as the carrier of the linguistic variable. The linguistic variable "bankruptcy risk" (of each enterprise and region) has a term-set of G values, consisting of three terms: G1 is "Low bankruptcy risk"; G2 is "Average bankruptcy risk"; G3 is "High risk of bankruptcy". The linguistic variable "borrower's creditworthiness" (both of an individual enterprise and a region) also has a term-set G consisting of three terms: G1 is "The crediting of which is beyond doubt"; G2 is "Lending requires a balanced approach"; G3 is "Lending is associated with increased risk".

The matrix scheme of data aggregation based on three-level fuzzy classifiers is based on the formula:

$$g = \sum_{i=1}^{N} p_i \sum_{j=1}^{3} \alpha_j \mu_{ij}(x_i)$$

where α_j are standard nodal points classifier (terms centroids), p_i is i-th weight factors in the contraction, $\mu_{ij}(x_i)$ is membership function value j-th level of quality with respect to the current value of i-th factor (the standard used trapezoidal). Then the index g is subjected to recognition based on the standard fuzzy classifier in accordance with the membership functions.

3   Results

Diagnostics of creditworthiness and risk of bankruptcy of enterprises in the region was carried out on the basis of a random sample of ten enterprises of the agroindustrial complex of the Rostov Region: (1) Litvinenko Ltd; (2) Aksai Land Ltd; (3) Manych-Agro Ltd; (4) JSC "Friendship"; (5) "Mutilinskoe" Ltd; (6) Red October Company; (7) "Light" Ltd; (8) Rassvet Ltd; (9) SEC named after Shaumyan; (10) Agrofirma Celina Ltd.

Calculations of the creditworthiness of each of the enterprises on the basis of the Sberbank methodology, as well as an assessment of the probability of the risk of its bankruptcy (based on the Altman, Taffler and Seifullin-Kadykov methods) were made using the Audit-IT software. For the financial analysis used the accounting statements of enterprises for the years 2015–2017.

Let us briefly consider the stages of the implemented methodology for diagnosing creditworthiness and bankruptcy risk of regional enterprises.

3.1   Credit Rating

At the first stage, an analysis of the financial condition of each of the enterprises was carried out using Audit-IT software.

At the second stage, some calculation results are summarized in Table 2.

At the third stage, the obtained estimates were aggregated based on fuzzy three-level [0, 1] classifiers (Table 3). At the same time, the weights of enterprises are the shares of each of them in the total revenue for 2017. It was established that the numerical value of the aggregated variable "the creditworthiness of enterprises in the region" is $G = 0.22$, which corresponds to the first class, "Lending is not in doubt".

Table 2. Creditworthiness of the borrower: 1 - the actual value of the parameter; 2 - category; S - total points

Indicator	Weight	Litvinenko Ltd		Aksai Land Ltd		Manych-Agro Ltd		JSC "Friendship"		"Mutilinskoe" Ltd	
		1	2	1	2	1	2	1	2	1	2
R1	0.05	29.23	1	0.02	3	<0.01	3	<0.01	3	0.04	3
R2	0.1	68.74	1	1.14	1	0.24	3	2.48	1	0.04	3
R3	0.4	190.46	1	2.51	1	0.7	3	11.42	1	3.97	1
R4	0.2	0.93	1	0.58	1	0.33	2	0.92	1	0.51	1
R5	0.15	0.31	1	0.15	1	0.15	1	0.33	1	0.04	2
R6	0.1	0.29	1	0.1	1	0.13	1	0.34	1	0.03	2
S	1		1		1.1		2,3		1.1		1.55
Category			1		1		2		1		2

Table 3. The calculation of the aggregated assessment of the creditworthiness of agricultural enterprises

N	Company Name	Revenue for 2017	Weight coefficient	Terms		
				1 class	2 class	3 class
1.	Litvinenko Ltd	79 570	0.012	1	0	0
2.	Aksai Land Ltd	150 894	0.023	1	0	0
3.	Manych-Agro Ltd	414 200	0.062	0	1	0
4.	JSC "Friendship"	268 102	0.040	1	0	0
5.	"Mutilinskoe" Ltd	41 310	0.006	0	1	0
6.	Red October Company	664 244	0.100	1	0	0
7.	"Light" Ltd	1 334 894	0.201	1	0	0
8.	Rassvet Ltd	767 386	0.116	1	0	0
9.	SEC named after Shaumyan	778 727	0.117	1	0	0
10.	Agrofirma Celina Ltd	2 143 745	0.323	1	0	0
		6 643 072	1	0.932	0.068	0

$G = 0.2*0.932 + 0.5*0.068 + 0.8*0 = 0.22$ (1 class, "Lending is not in doubt")

3.2   Bankruptcy Risk Assessment of Regional Enterprises

The study included the following steps.

First stage. Financial analysis of each of the enterprises on the software Audit-IT.
Second stage. Formation of summary tables for ten enterprises, respectively, for methods of Altman, Taffler, Seifullin-Kadykov (Table 4 for methods of Altman).
Third stage. Aggregation of estimates for enterprises calculated on the basis of the Altman model into an integral assessment of the risk of bankruptcy of enterprises in the region (Table 5). It is established that G (Altman) = 0.237.
Fourth stage. The same for the Taffler model. It is established that G (Taffler) = 0.219.
Fifth stage. The same for the Seifullin-Kadykov model. It is established that G (Seifullin-Kadykov) = 0.239
Sixth stage. Aggregation of the obtained integral assessments of the bankruptcy risk of regional enterprises based on the Altman, Taffler, Seifullin-Kadykov models into the final assessment (the methods are considered equilibrium, Table 6).

Determined that G = 0.247

$$\mu(0.247) = \mu_1(0.247) = 0.765, \mu(0.247) = \mu_2(0.247) = 0.235$$

that is, there is a more low degree of bankruptcy than the average, according to estimates of all three models.

Table 4. Calculation of the risk of bankruptcy of agricultural enterprises based on the Altman model

Indicator	Litvinenko Ltd	Aksai Land Ltd	Manych-Agro Ltd	JSC "Friendship"	"Mutilinskoe" Ltd
T_1	0,37	0,35	-0,18	0,63	0,44
T_2	0,87	0,42	0,15	0,76	0,51
T_3	0,16	0,12	0,06	0,15	0,06
T_4	6,93	1,36	0,49	11,73	1,06
Z-score:	**13,62**	**5,88**	**0,24**	**19,9**	**6,06**
The degree of risk of bankruptcy	Low	Low	High	Low	Low

Table 5. Calculation of the aggregated risk assessment of bankruptcy of agricultural enterprises of the region based on the Altman model

N	Company Name	Weight coefficient	Terms		
			G1	G2	G2
1.	Litvinenko Ltd	0.012	1	0	0
2.	Aksai Land Ltd	0.023	1	0	0
3.	Manych-Agro Ltd	0.062		0	1
4.	JSC "Friendship"	0.040	1	0	0
5.	"Mutilinskoe" Ltd	0.006	1	0	0
6.	Red October Company	0.100	1	0	0
7.	"Light" Ltd	0.201	1	0	0
8.	Rassvet Ltd	0.116	1	0	0
9.	SEC named after Shaumyan	0.117	1	0	0
10	Agrofirma Celina Ltd	0.323	1	0	0
		1	**0.938**	**0**	**0.062**

$G = 0.2*0.938 + 0.5*0 + 0.8*0.062 = 0.237$

Table 6. Calculation of the aggregated risk assessment of bankruptcy of agricultural enterprises in the region based on three models

Model	Weight model	Evaluation region	Terms		
			G1	G2	G3
Altman Z-Score:	1/3	0.237	0.815	0.185	0
Z-score Taffler:	1/3	0.219	0.905	0.095	0
Saifullina-Kadykov Model	1/3	0.239	0.805	0.195	0
Term weight			0.842	0.158	0

$G = 0.2*0.842 + 0.5*0.158 + 0.8*0 = 0.247$

Thus, on the basis of the analysis carried out, it was established that the value of the aggregated variable "the solvency of regional enterprises" is G = 0.22, which corresponds to the first class, "lending is not in doubt".

It was also established that the integral assessment of the bankruptcy risk of enterprises in the region, built on the basis of Altman, Taffler, Seyfulli-na-Kadykov models under the assumption that the methods are equilibrium, is G = 0.247. That is, there is a greater degree of bankruptcy than the average, according to estimates of all three models.

Thus, a methodology has been built up that allows diagnostics of creditworthiness and bankruptcy risk of regional enterprises based on accounting reports from Internet sources. It is significant that the analysis is based on open data sources.

Obviously, a complete picture can be obtained only on the basis of a similar analysis of all enterprises in the relevant sector of the region. This requires the creation of software.

However, in general, the experiment showed the practical feasibility of the analysis based on the proposed method.

4   Conclusion

A technique has been developed, the novelty of which lies in the ability to aggregate the results of the analysis of the solvency and bankruptcy risk of an enterprise, obtained as a result of applying a complex of various bankruptcy models, to construct the corresponding integral assessments of the region. Another important advantage of the methodology is the ability to build estimates based on open Internet data. In addition, the weights of various standard methods of assessing bankruptcy risk and creditworthiness may vary depending on expert assessments and research objectives.

References

1. Altman, E.I.: Financial ratios, discriminant analysis and the prediction of corporate bankruptcy. J. Financ. **4**, 589–609 (1968)
2. Taffler, R.J.: The assessment of company solvency and performance using a statistical modeling. Account. Bus. Res. **13**(52), 295–307 (1983)
3. Agarwal, V., Taffler, R.: Comparing the performance of market-based and accounting based bankruptcy prediction models. J. Bank. Financ. **32**, 1541–1551 (2008)
4. Zadeh, L.A.: Concept of a Linguistic Variable and its Application to Making Approximate Decisions. Mir, Moscow (1976)
5. Nedosekin, A.O., Kozlovsky, A.N., Abdulaeva, Z.I.: Analysis of branch economic stability by fuzzy-logical methods. Econ. Manag. Prob. Solut. **5**, 10–16 (2018)

Toward Eigenvalues and Eigenvectors of Matrices of Z-Numbers

R. A. Aliev[1,2(✉)], O. H. Huseynov[3] ⓘ, and K. R. Aliyeva[4]

[1] Joint MBA Program, Georgia State University, Atlanta, USA
raliev@asoa.edu.az
[2] Azerbaijan State Oil and Industry University, 20 Azadlig Ave.,
AZ1010 Baku, Azerbaijan
[3] Research Laboratory of Intelligent Control and Decision
Making Systems in Industry and Economics, Azerbaijan State Oil
and Industry University, Baku, Azerbaijan
oleg_huseynov@yahoo.com
[4] Department of Instrument Making Engineering, Azerbaijan State Oil
and Industry University, 20 Azadlig Avenue, AZ1010 Baku, Azerbaijan
kamalann@gmail.com

Abstract. Eigenvalues and eigenvectors are widely used in various practical applications in decision making, planning, control and other fields. Particularly, these concepts underlie analysis of consistency of a decision maker's (DM) knowledge. In real-world problems, DM's knowledge is naturally characterized by imprecision and partial reliability. This involves combination of fuzzy and probabilistic information. The concept of a Z-number is a formal construct to describe such kind of information. In this paper we initiate study of Z-number valued eigenvalue and eigenvector of matrices, components of which are Z-numbers. A statement of problem and a solution approach for computation of Z-number valued eigensolutions are proposed. An example is provided to prove validity of the proposed approach.

Keywords: Z-number · Fuzzy matrix · Random matrix · Eigenvalue · Eigenvector · Decision making

1 Introduction

A wide class of real world decision problems in realm of economics, management, business and other fields are formalized as problems of linear algebra. Particularly, one can mention problems of consistency of judgement-based pairwise comparison of alternatives in multiattribute decision making. Classical techniques of linear algebra are well developed. However, one has to account for uncertainty inherent in information relevant to real-world problems [1, 2]. A series of works are devoted to account for various types of uncertainties in linear algebra-based models. A large class of studies devoted to handling probabilistic uncertainty in such models is based on the theory of random matrices [3, 4]. A random matrix is a matrix elements of which are random variables. These studies are devoted to computation of eigenvalues, condition numbers and eigenvectors, Jacobian of random matrix, and other problems. Nowadays random

© Springer Nature Switzerland AG 2020
R. A. Aliev et al. (Eds.): ICSCCW 2019, AISC 1095, pp. 309–317, 2020.
https://doi.org/10.1007/978-3-030-35249-3_39

matrices are applied in machine learning, signal processing, computer graphics, econometrics and other fields. A series of works are devoted to fuzzy linear algebra. These works include studies of fuzzy matrices, fuzzy eigenvalues and eigenvectors, fuzzy systems of linear equations and other problems [5–21].

A class of real-world problems is characterized by combination of fuzzy and probabilistic uncertainties. One of the important formal constructs developed to deal with such type of uncertainty is the Z-number concept introduced by Zadeh [29]. A series of works devoted to theory and applications of Z-numbers exist [22–34].

There are no works on matrices with Z-valued information in existence. In this paper we proceed with research of Z-valued matrices as matrices elements of which are Z-numbers. This implies that we consider a synergy of fuzzy and probabilistic uncertainties, but not their 'mechanical sum'. We formulate definitions of Z-matrices, Z-valued eigenvalue and Z-valued eigenvector of a Z-matrix. The problems of computation of Z-valued eigenvalues and Z-valued eigenvectors are formulated and the solution approach is proposed. An example is provided to illustrate the proposed method

The paper is structured as follows. Section 2 includes necessary definitions used in the sequel. New concepts as Z-matrix, Z-valued eigenvalues and eigenvectors are formulated. In Sect. 3, the problems of computation of Z-valued eigenvalues and eigenvectors of Z-matrix are stated. The solution approach for the considered problems is given in Sect. 4. In Sect. 5, we provide an example of computation of Z-valued eigenvalues and eigenvectors. Section 6 is conclusion.

2 Preliminaries

Definition 1 Continuous Z-number [25–28]. A continuous Z-number is an ordered pair $Z = (A, B)$ where A is a continuous fuzzy number playing a role of a fuzzy constraint on values that a random variable X may take: X *is A*. B is a continuous fuzzy number with a membership function $\mu_B : [0, 1] \rightarrow [0, 1]$, playing a role of a fuzzy constraint on the probability measure of A: $P(A)$ *is B*.

Definition 2 A fuzzy square matrix [8]. A fuzzy square matrix $\left(A_{ij}\right)$ is a matrix elements of which are fuzzy numbers $A_{ij}, i, j = 1, \ldots, n$ that describe fuzzy restrictions on values of random variables $X_{ij}, i, j = 1, \ldots, n$:

$$\left(A_{ij}\right) = \begin{pmatrix} A_{11} & \ldots & A_{1n} \\ . & \ldots & . \\ A_{n1} & \ldots & A_{nn} \end{pmatrix}.$$

Definition 3 A random square matrix [3]. A random square matrix $\left(X_{ij}\right)$ is a matrix of random variables $X_{ij}, i, j = 1, \ldots, n$:

$$\left(X_{ij}\right) = \begin{pmatrix} X_{11} & \ldots & X_{1n} \\ . & \ldots & . \\ X_{n1} & \ldots & X_{nn} \end{pmatrix}.$$

Each random variable X_{ij} is governed by pdf $p_{ij}, i, j = 1, \ldots, n$.

Definition 4 A Z-valued matrix. A Z-valued matrix (Z_{ij}) is a matrix of Z-numbers that describe partially reliable information on values of random variables $X_{ij}, i, j = 1, \ldots, n$:

$$
\left(Z_{ij} = \left(A_{ij}, B_{ij} \right) \right) = \begin{pmatrix} Z_{11} = (A_{11}, B_{11}) & \cdots & Z_{1n} = (A_{1n}, B_{1n}) \\ \cdot & \cdots & \cdot \\ Z_{n1} = (A_{n1}, B_{n1}) & \cdots & Z_{nn} = (A_{nn}, B_{nn}) \end{pmatrix}
$$

Let us formulate definitions of eigenvalue and eigenvector of (Z_{ij}).

Definition 5 A Z-eigenvalue of Z-valued matrix. A Z-valued eigenvalue of Z-valued square matrix (Z_{ij}) is such a Z-number $Z_\lambda = (A_\lambda, B_\lambda)$ that the following holds:

$$
det\left(Z_{ij} - Z_\lambda I \right) = det \begin{pmatrix} Z_{11} - Z_\lambda & \cdots & Z_{1n} \\ \cdot & \cdots & \cdot \\ Z_{n1} & \cdots & Z_{nn} - Z_\lambda \end{pmatrix} = Z_0 \tag{1}
$$

where I is a traditional (non-fuzzy) identity matrix, Z_0 is a Z-singleton $Z_0 = (\hat{0}, \hat{1})$.

Thus, Z-valued eigenvalue $Z_\lambda = (A_\lambda, B_\lambda)$ is a root of n-th order equation

$$
Z^n + Z_1 Z^{n-1} + \ldots + Z_n = Z_0,
$$

where $Z_r, r = 1, \ldots, n$ are coefficients induced by the elements of Z-valued matrix (Z_{ij}). So, n Z-valued eigenvalues $Z_{\lambda s} = (A_{\lambda s}, B_{\lambda s})$ are to be found for a Z-valued square matrix (Z_{ij}). In special case, Z^+-number $Z_{\lambda s}^+ = (A_{\lambda s}, p_{\lambda s})$ can be used.

Definition 6 A Z-eigenvector of Z-valued matrix. A vector of Z-numbers $(Z_Y) = (Z_{Y1} = (A_{Y1}, B_{Y1}), \ldots, Z_{Yn} = (A_{Yn}, B_{Yn}))$ is referred to as a Z-valued eigenvector of Z-valued matrix (Z_{ij}) if it satisfies the following Z-valued system of equations:

$$
\left(Z_{ij} \right) (Z_Y) = Z_\lambda (Z_Y) \tag{2}
$$

where $Z_\lambda = (A_\lambda, B_\lambda)$ is Z-valued eigenvalue.

Thus, n Z-valued eigenvectors $(Z_{Ys}), s = 1, \ldots, n$ are to be found, one for each Z-valued eigenvalue $Z_{\lambda s} = (A_{\lambda s}, B_{\lambda s})$.

3 Statement of the Problem

We consider a problem of computation of Z-valued eigenvalues and Z-valued eigenvectors. The generic problem of computation of Z-valued eigenvalues is as follows:

Find $Z_{\lambda s} = (A_{\lambda s}, B_{\lambda s}), s = 1, \ldots, n$ such that (1) holds.

Note that in $Z_{ij} = (A_{ij}, B_{ij}), A_{ij}$ is a fuzzy number and B_{ij} is a fuzzy constraint over $P(A_{ij}) = \int_{\mathcal{R}} \mu_{A_{ij}}(x) p_{ij}(x) dx$. This implies that a fuzzy set of possible pdfs p_{ij} is induced by B_{ij}. Thus, given $Z_{ij} = (A_{ij}, B_{ij})$, we have to compute $A_{\lambda s}$ and to construct fuzzy restriction $B_{ij}:P(A_{\lambda s}) = \int_{\mathcal{R}} \mu_{A_{\lambda s}}(x) p_{\lambda s}(x) dx$ is $B_{\lambda s}$. In other words, a fuzzy set of distributions p_{ij} induces a fuzzy set of distributions $p_{\lambda s}$. According to the general framework of computation of Z-numbers [29], computation of $Z_{\lambda s} = (A_{\lambda s}, B_{\lambda s})$ is implemented through computation of a Z^+-number $Z_{\lambda s}^+ = (A_{\lambda s}, p_{\lambda s})$ based on Z^+-numbers $Z_{ij}^+ = (A_{ij}, p_{ij})$ as follows.

Given fuzzy numbers A_{ij} find $A_{\lambda s}$ such that

$$det(A_{ij} - A_{\lambda}I) = \hat{0} \tag{3}$$

$\hat{0}$ is a fuzzy singleton.
Given pdfs p_{ij} of random variables X_{ij} and the constraint

$$det(X_{ij} - X_{\lambda}I) = 0 \tag{4}$$

find pdf $p_{\lambda s}$ of a desired random variable X_{λ}.

Next, we must find Z-valued eigenvectors (Z_{Ys}) as solutions of a Z-valued system (2). (Z_{Ys}) can be found through computation of $(Z_{Ys}^+) = ((A_{Ys}, p_{Ys}))$ such that A_{Ys} and p_{Ys} are solutions of two problems associated with system (2).

Given $A_{\lambda s}$ find A_{Ys} that satisfies the following system:

$$(A_{ij})(A_{Y_s}) = A_{\lambda_s}(A_{Y_s}). \tag{5}$$

Given random variable $X_{\lambda s}$ compute the random vector $(Y_s), s = 1, \ldots, n$ by solving the following system:

$$(X_{ij})(Y_s) = X_{\lambda_s}(Y_s). \tag{6}$$

4   Solution of the Problem

In this section we propose an approach to solve the problems formulated in Sect. 3.

At the first step, we consider a Z^+-matrix $\left(Z_{ij}^+ = (A_{ij}, p_{ij})\right)$ related to Z-matrix (Z_{ij}). For simplicity, pdf p_{ij} is to be found as normal distribution $p_{ij} = (m_{ij}, _{ij})$. Formally, we have to find σ_{ij} by solving the optimization problem:

$$\sum_{k=1}^{N} \mu_{A_{ij}}(x_{ijk}) \frac{1}{\sqrt{2\pi}_{ij}} e^{-\frac{(m_{ij}-x_{ijk})^2}{ij}} \Delta x \to b_{ij} \tag{7}$$

$$m_{ij} = \sum_{k=1}^{N} \frac{\mu_{A_{ij}}(x_{ijk})x_{ijk}}{\mu_{A_{ij}}(x_{ijk})}.(compatibility\ constraint[29]) \tag{8}$$

At the second step, given fuzzy matrix (A_{ij}) and the related random matrix (p_{ij}), we compute $Z_{\lambda s}^{+} = (A_{\lambda s}, p_{\lambda s})$, where $A_{\lambda s}$ is a fuzzy eigenvalue of (A_{ij}), and $p_{\lambda s}$ is a probabilistic eigenvalue of (p_{ij}). $A_{\lambda s}$ is computed by using a decomposition approach [14] where matrices of triangular fuzzy numbers (TFNs) $(A_{ij}) = ((A_{ijl}, A_{ijm}, A_{iju}))$ for computational efficiency purposes. Thus, eigenvalues $A_{\lambda s} = (\lambda_{sl}, \lambda_{sm}, \lambda_{su})$ as TFNs are obtained.

In order to compute pdf $p_{\lambda s}$, we consider a random variable $X_{\lambda s}$ as a root of characteristic polynomial equation:

$$X^n + c_1 X^{n-1} + \ldots + c_n = 0$$

The coefficients c_i are random variables with pdfs $p_{ci}, i = 1, \ldots, n$ induced by pdfs p_{ij}. The computation of the roots $X_{\lambda s}$ are difficult even for non-probabilistic case and is based on a numerical approach [3]. Thus, we have to compute numerically the pdf $p_{\lambda s}$. We will find $p_{\lambda s}$ as normal pdf $p_{\lambda s} = (m_{\lambda s, \lambda s})$ given p_{ci} as $p_{ci} = (m_{ci, ci})$. For each discretized numerical realization x_{ij} of values of random matrix elements Xij governed by pdfs p_{ij}, the corresponding numerical realization of random eigenvalue $x_{\lambda s}, s = 1, \ldots, n$ is computed. Thus, a discretized set of random matrix elements induces a discretized set $\{x_{\lambda sk}, k = 1, \ldots, N\}$. For the corresponding set, mean and standard deviations $m_{\lambda s, \lambda s}$ are computed. That is, we find $p_{\lambda s} = (m_{\lambda s, \lambda s})$.

Given eigenvalue $Z_{\lambda s}^{+} = (A_{\lambda s}, p_{\lambda s})$, we have to compute Z-valued eigenvalue $Z_{\lambda s} = (A_{\lambda s}, B_{\lambda s})$. We recall that all the possible distributions $p_{\lambda s}$ are induced by all the possible distributions p_{ij}. Thus, for each $Z_{ij} = (A_{ij}, B_{ij})$ we can compute a finite (discretized) set of distributions p_{ij}. B_{ij} induces a fuzzy restriction over distributions p_{ij}:

$$\mu_{p_{ij}}(p_{ij}) = \mu_{B_{ij}}\left(\int_{\mathcal{R}} \mu_{A_{ij}}(x) p_{ij}(x) dx\right).$$

In turn, $\mu_{p_{ij}}$ will induce a fuzzy restriction over set of distributions $p_{\lambda s}$, described by possibility distribution $\mu_{p_{\lambda s}}$:

$$\mu_{p_{\lambda s}}(p_{\lambda s}) = \sup_{(p_{ij})} \min_{i=1,\ldots,n, j=1,\ldots,n} \mu_{p_{ij}}(p_{ij}),$$

Thus, $B_{\lambda j}$ can be constructed as follows:

$$\mu_{B_{\lambda s}}(b_{\lambda s}) = \sup_{p_{\lambda s}} \mu_{p_{\lambda s}}(p_{\lambda s}), b_{\lambda s} = \int_{\mathcal{R}} \mu_{A_{\lambda s}}(x_{\lambda s}) p_{\lambda s}(x_{\lambda s}) dx_{\lambda s}.$$

At the *third step*, given n eigenvalues $Z_{\lambda s}^{+} = (A_{\lambda s}, p_{\lambda s})$, we have to compute the corresponding n eigenvectors $(Z_{Ys}^{+}) = ((A_{Ys}, p_{Ys})), s = 1, \ldots, n$. For this purpose, two systems of Eqs. (5) and (6) related to (2) are considered. (A_{Ys}) is fuzzy vector as a solution of (5), (p_{Ys}) is a vector of pdfs as a solution of (6).

(A_{Ys}) is computed based on the decomposition approach [14]. In order to compute random vector (Y_s) as a solution of system (6), we have to consider a discretized set of realizations of (6). Each realization is a crisp system with matrix (x_{ij}), vector (y_{sj}) and a scalar value $x_{\lambda j}$ (numerical realizations of random matrix X_{ij}, random vector Y_{ij} and random variable $X_{\lambda s}$). The solution of this system is a vector $(y_{sj}), i = 1, \ldots, n$. For a discretized set of realizations of (6), we will have a discretized set of vectors $\{(y_{is})_k, i = 1, \ldots, n, k = 1, \ldots, N\}$. For this set a vector of mean values and standard deviations $((m_{s1}, s1), \ldots, (m_{sn}, sn))$ can be computed. Thus, we can find pdfs of elements of (Y_s) as $(p_{s1} = (m_{s1}, s1), \ldots, p_{sn} = (m_{sn}, sn))$.

Thus, $(Z_{Ys}^+) = ((A_{Ys}, p_{Ys})), s = 1, \ldots, n$ eigenvectors are found. Given eigenvector $(Z_{Ys}^+) = ((A_{Ys1}, p_{Ys1}), \ldots, (A_{Ysn}, p_{Ysn}))$, we have to compute Z-valued eigenvector $(Z_{YS}) = (Z_{Ys1} = (A_{YS1}, B_{YS1}), \ldots Z_{Ysn} = (A_{Ysn}, B_{Ysn}))$. Any component of $(p_{Ys1}, \ldots, p_{Ysn})$ vector is only one possible distribution. It is needed to construct all possible distributions p_{Ysj} to compute B_{Ysj} for any component of (Z_{Ys}). We recall that a vector of all the possible distributions $p_{Ys1}, \ldots, p_{Ysn}$ is induced by all the possible distributions $p_{\lambda s}$. Thus, for each $Z_{\lambda s} = (A_{\lambda s}, B_{\lambda s})$ we have a finite (discretized) fuzzy set of distri-

butions $p_{\lambda s}$: $\mu_{p_{\lambda s}}(p_{\lambda s}) = \mu_{B_{\lambda s}}\left(\int_{\mathbb{R}} \mu_{A_{\lambda s}}(x_{\lambda s})p_{\lambda s}(x_{\lambda s})dx_{\lambda s}\right)$. In turn, $\mu_{p_{\lambda s}}$ induces a fuzzy

restriction over set of distributions p_{Ysj}, described by possibility distribution $\mu_{p_{Ysj}}$:

$$\mu_{p_{Ysj}}(p_{Ysj}) = \sup_{p_{\lambda s}} \min_{s=1,\ldots,n} \mu_{p_{\lambda s}}(p_{\lambda s}), j = 1, \ldots, n.$$

p_{Ysj} is a component of eigenvector $(p_{Ys1}, \ldots, p_{Ysn})$ computed on the basis of $p_{\lambda s}$.

Thus, for any component $Z_{Ysj} = (A_{Ysj}, B_{Ysj})$ of Z-valued eigenvector (Z_{Ys}), B_{Ysj} can be constructed as follows: $\mu_{B_{Ysj}}(b_{Ysj}) = \sup_{p_{Ysj}} \mu_{p_{Ysj}}(p_{Ysj}), b_{Ysj} = \int_{\mathbb{R}} \mu_{A_{Ysj}}(x_{Ysj})p_{Ysj}(x_{Ysj})dx_{Ysj}$

5   An Example

Consider the following 3×3 Z-valued matrix, where components of Z-numbers are triangular fuzzy numbers (TFNs):

$$(Z_{ij}) = \begin{pmatrix} ((0.93, 0.95, 1), (0.88, 0.94, 1)) & ((2, 2.5, 3), (0.56, 0.78, 1)) & ((1.5, 2, 2.25), (0.7, 0.8, 0.9)) \\ ((0.33, 0.4, 0.5), (0.67, 0.78, 0.97)) & ((0.93, 0.95, 1), (0.88, 0.94, 0.97)) & ((1, 1.5, 2), (0.65, 0.75, 0.85)) \\ ((0.4, 0.5, 0.67), (0.65, 0.75, 0.85)) & ((0.5, 0.67, 1), (0.7, 0.72, 0.75)) & ((0.93, 0.95, 1), (0.88, 0.94, 0.97)) \end{pmatrix}$$

Let us shortly describe the procedures of computation of Z-valued eigenvalues and Z-valued eigenvectors for this matrix (Sect. 4). For A parts of Z-numbers we use decomposition approach [14]. The fuzzy eigenvalues $A_{\lambda 1} = (\lambda_{1l}, \lambda_{1m}, \lambda_{1u})$ computed by using this approach are $A_{\lambda 1} = (2.5, 3, 3.68), A_{\lambda 2} = (-0.37, -0.01, 0.02)$, $A_{\lambda 3} = (-0.37, -0.01, 0.02)$. The following pdfs p_λ of eigenvalues are obtained: $p_{\lambda 1} = (3.05, 0.035), p_{\lambda 2} = (-0.02 + 0.344i, 0.15), p_{\lambda 3} = (-0.02 - 0.344i, 0.15)$. The obtained Z-eigenvalues: $Z_{\lambda 1} = (A_{\lambda 1}, p_{\lambda 1}) = ((7.97, 18.12, 10.32), (3.05, 0.035)), Z_{\lambda 2} = (A_{\lambda 2}, p_{\lambda 2}) = ((-0.73 - 0.46, 0.22), (-0.02 + 0.344i, 0.15)), Z_{\lambda 2} = (A_{\lambda 2}, p_{\lambda 2}) = ((-0.73 - 0.46, 0.22), (-0.02 - 0.344i, 0.15))$.

At the *third step*, we compute eigenvectors $Z_{Ys} = (A_{Ys}, p_{Ys})$ for the obtained eigenvalues $Z_{\lambda s} = (A_{\lambda s}, p_{\lambda s})$. The fuzzy eigenvectors (A_{Ys}), $s = 1, \ldots, 3$ obtained on the basis of the approach proposed in [14] are as follows:

$$(A_{Y1}) = \begin{pmatrix} (0.44, 0.52, 0.64) \\ (0.22, 0.26, 0.32) \\ (0.18, 0.22, 0.27) \end{pmatrix}, \ (A_{Y2}) = \begin{pmatrix} (-9.68, -0.26, 0.52) \\ (-4.8, -0.13, 0.26) \\ (-4, 0.1, 0.2) \end{pmatrix},$$

$$(A_{Y3}) = \begin{pmatrix} (-9.68, -0.26, 0.52) \\ (-4.8, -0.13, 0.26) \\ (-4, 0.1, 0.2) \end{pmatrix}.$$

At the same time, for the probabilistic eigenvalues computed at the previous step, the corresponding eigenvectors are found as:

$$(p_{Y1}) = \begin{pmatrix} (0.74, 0.025) \\ (0.365, 0.21) \\ (0.3, 0.17) \end{pmatrix}, \ (p_{Y2}) = \begin{pmatrix} (0.51, 0.025) \\ (-0.12 + 0.21i, 0.35) \\ (-0.12 - 0.21i, 0.35) \end{pmatrix},$$

$$(p_{Y3}) = \begin{pmatrix} (0.3, 0.17) \\ (-0.10 - 0.17i, 0.28) \\ (-0.10 + 0.17i, 0.28) \end{pmatrix}.$$

Thus, the obtained Z-eigenvectors $(Z_{Ys}) = ((A_{Ys1}, p_{Ys1}), \ldots, (A_{Ys3}, p_{Ys3}))$ are

$$(Z_{Y1}) = \begin{pmatrix} ((0.84, 0.844, 0.844), (0.7, 0.5)) \\ ((0.41, 0.416, 0.4253), (0.3, 0.3)) \\ ((0.345, 0.349, 0.36), (0.28, 0.2)) \end{pmatrix},$$

$$(Z_{Y2}) = \begin{pmatrix} ((0.83, 0.833, 0.853), (0.51, 0.6)) \\ ((-0.27, -0.22, -0.17), (-0.13 + 0.23i, 0.32)) \\ ((-0.22, -0.16, -0.1), (-0.11 - 0.19i, 0.26)) \end{pmatrix},$$

$$(Z_{Y3}) = \begin{pmatrix} ((0.83, 0.833, 0.853), (0.51, 0.6)) \\ ((-0.27, -0.22, -0.17), (-0.13 - 0.23i, 0.3)) \\ ((-0.22, -0.16, -0.1), (-0.11 + 0.19i, 0.3)) \end{pmatrix}.$$

6 Conclusion

Problems of computation of eigenvalues and eigenvectors under combination of fuzzy and probabilistic information are important for real-world applications in decision making, economics and other fields. Existing works devoted to eigenvalues and eigenvectors of fuzzy matrices are scarce. No works are devoted to eigenvalues and eigenvectors of Z-valued matrices. For the first time, we developed an approach for computation of eigensolutions of Z-valued matrix. An example provided here shows validity of the proposed approach.

References

1. Aliev, R.A., Pedrycz, W.: Fundamentals of a fuzzy-logic-based generalized theory of stability. IEEE Trans. Syst. Man, Cybern. Part B (Cybern.) **39**(4), 971–988 (2009)
2. Aliev, R., Tserkovny, A.: Systemic approach to fuzzy logic formalization for approximate reasoning. Inf. Sci. **181**(6), 1045–1059 (2011)
3. Edelman, A., Rao, N.R.: Random matrix theory. Acta Numer. **14**, 233–297 (2005)
4. Anderson, G.W., Guionnet, A., Zeitouni, O.: An Introduction to Random Matrices. Cambridge University Press, Cambridge (2009)
5. Bhowmik, M., Pal, M.: Generalized intuitionistic fuzzy matrices. FarEast J. Math. Sci. **29**(3), 533–554 (2008)
6. Gavalec, M.: Computing matrix period in max-min algebra. Discret. Appl. Math. **75**, 63–70 (1997)
7. Hashimoto, H.: Canonical form of a transitive fuzzy matrix. Fuzzy Sets Syst. **11**, 157–162 (1983)
8. Thomson, M.G.: Convergence of powers of a fuzzy matrix. J. Math. Anal. Appl. **57**, 476–480 (1977)
9. Hashimoto, H.: Transitivity of fuzzy matrices under generalized connectedness. Fuzzy Sets Syst. **29**(2), 229–234 (1989)
10. Hashimoto, H.: Transitivity of generalized fuzzy matrix. Fuzzy Sets Syst. **17**, 83–90 (1985)
11. Hemasinha, R., Pal, N.R., Bezdek, J.C.: Iterates of fuzzy circulant matrices. Fuzzy Sets Syst. **60**, 199–206 (1993)
12. Khan, S.K., Pal, M.: Interval-valued intuitionistic fuzzy matrices. Notes Intuit. Fuzzy Sets **11**(1), 16–27 (2005)
13. Behera, D., Huang, H.-Z., Chakraverty, S.: Solution of fully fuzzy system of linear equations by linear programming approach. Comput. Model. Eng. Sci. **108**(2), 67–87 (2015)
14. Praščević, N., Praščević, Ž.: Application of fuzzy AHP method based on eigenvalues for decision making in construction industry. Tech. Gaz. **23**, 57–64 (2016)
15. Allahviranloo, T., Salahshour, S., Khezerloo, M.: Maximal- and minimal symmetric solutions of fully fuzzy systems. J. Comput. Appl. Math. **235**, 4652–4662 (2011)
16. Laarhoven, P.J.M.V., Pedrycz, W.: A fuzzy extension of Saaty's priority theory. Fuzzy Sets Syst. **11**, 199–227 (1983)
17. Allahviranloo, T., Salahshour, S., Khezerloo, M.: Maximal and minimal symmetric solutions of fully fuzzy systems. J. Comput. Appl. Math. **235**, 4652–4662 (2011)
18. Behera, D., Chakraverty, S.: Solution method for fuzzy system of linear equations with crisp coefficients. Fuzzy Inf. Eng. **5**, 205–219 (2013)
19. Behera, D., Chakraverty, S.: New approach to solve fully fuzzy system of linear equations using single and double parametric form of fuzzy numbers. Sadhana **40**, 35–49 (2015)
20. Ishizaka, A.: Comparison of fuzzy logic, AHP, FAHP and hybrid fuzzy AHP for new supplier selection and its performance analysis. Int. J. Integr. Supply Manag. **9**, 1–22 (2014)
21. Krejcí, J.: Fuzzy eigenvector method for obtaining normalized fuzzy weights from fuzzy pairwise comparison matrices. Fuzzy Sets Syst. **315**, 26–43 (2017)
22. Zadeh, L.A.: A note on Z-numbers. Inform. Sci. **181**, 2923–2932 (2011)

23. Aliev, R.A., Alizadeh, A.V., Huseynov, O.H.: The arithmetic of discrete Z-numbers. Inform. Sci. **290**, 134–155 (2015)
24. Aliev, R.A., Zeinalova, L.M.: Decision making under Z-information. In: Guo, P., Pedrycz, W. (eds.) Human-Centric Decision-Making Models for Social Sciences (Studies in Computational Intelligence), pp. 233–252. Springer, Heidelberg (2014)
25. Aliev, R.A., Huseynov, O.H.: Decision Theory with Imperfect Information. World Scientific, Singapore (2014)
26. Aliev, R.A., Huseynov, O.H., Aliyev, R.R., Alizadeh, A.V.: The Arithmetic of Z-Numbers: Theory and Applications. World Scientific, Singapore (2015)
27. Aliev, R.A., Alizadeh, A.V., Huseynov, O.H.: The arithmetic of continuous Z-numbers. Inform. Sci. **373**, 441–460 (2016)
28. Aliev, R.A., Perdycz, W., Huseynov, O.H.: Functions defined on a set of Z-numbers. Inform. Sci. **423**, 353–375 (2018)
29. Aliev, R.A., Alizadeh, A.V., Huseynov, O.H., Jabbarova, K.I.: Z-number based linear programming. Int. J. Intell. Syst. **30**, 563–589 (2015)
30. Yager, R.R.: On a view of Zadeh's Z-numbers. Adv. Comput. Intell. Commun. Comput. Inf. Sci. **299**, 90–101 (2012)
31. Yager, R.: On Z-valuations using Zadeh's Z-numbers. Int. J. Intell. Syst. **27**, 259–278 (2012)
32. Kang, B., Deng, Y., Hewage, K., Sadiq, R.: A method of measuring uncertainty for z-number. IEEE Trans. Fuzzy Syst. **27**, 731–738 (2018)
33. Kang, B., Chhipi-Shrestha, G., Deng, Y., Hewage, K., Sadiq, R.: Stable strategies analysis based on the utility of Z-number in the evolutionary games. Appl. Math. Comput. **324**, 202–217 (2018)
34. Banerjee, R., Pal, S.K.: Z*-numbers: augmented Z-numbers for machine-subjectivity representation. Inf. Sci. **323**, 143–178 (2015)

Intelligent Search of Spatial Data Analysis Context

Stanislav Belyakov[1] , Marina Belyakova[1] ,
Alexander Bozhenyuk[1(✉)] , and Janusz Kacprzyk[2]

[1] Southern Federal University, Nekrasovsky 44, 347922 Taganrog, Russia
beliacov@yandex.ru, mlbelyakova@sfedu.ru,
avb002@yandex.ru
[2] Systems Research Institute Polish Academy of Sciences,
Newelska 6, 01-447 Warsaw, Poland
janusz.kacprzyk@ibspan.waw.pl

Abstract. The paper deals with the task of controlling the course of interactive analysis of spatial data using geographic information systems. The main problem of analysis is considered a significant amount of data that the analyst receives on requests to the information database of the geographic information system. The goal of the management is to provide the user with a set of cartographic images that are as useful as possible for the analysis. The concept of semantic content of images and analysis contexts is introduced. A conceptual knowledge model of the analysis context is proposed. Its distinguishing feature is the presence of two components. The first component is the core of the context. It includes fundamental knowledge of the method of analysis. Knowledge determines the spatial, temporal and semantic boundaries of the field of analysis. The second component is valid context conversions that preserve its meaning. A condition is introduced to preserve the meaning of the analysis process as the correspondence of the requested data to the limit of permissible transformations. An algorithm for searching the context that preserves the semantic content of the analysis is given. The algorithm uses an archive of use cases of sequences of change of contexts. The results of an experimental study of the effectiveness of the method are given on the example of a corporate geographic information system.

Keywords: Intellectual Geo-information systems · Context · Analysis management

1 Introduction

The analysis of spatial data is needed in many areas of production, planning, design and business. Modern means of analysis are geographic information systems (GIS) and network services. Information base of GIS includes electronic maps, databases of semantic data and knowledge, links to external data sources. This enables users to interactively solve complex application problems by studying this data. The complexity of the tasks is due to the incompleteness, uncertainty and ambiguity of spatial data. The process of constructing solutions is accompanied by a large amount of analytical work

© Springer Nature Switzerland AG 2020
R. A. Aliev et al. (Eds.): ICSCCW 2019, AISC 1095, pp. 318–324, 2020.
https://doi.org/10.1007/978-3-030-35249-3_40

necessary for searching in GIS data, which will allow you to choose the option of problem statement, compare possible alternative solutions, and evaluate ways to implement them. The interactive nature of the analysis is an important feature of the analyst's work with GIS. Starting a session of analysis, the user proceeds to form a mental image of a spatial situation in his mind. Visual study of cartographic images in different scales, types, angles and thematic representations is an external manifestation of the intellectual process in the analyst-GIS system. Here, the effectiveness of GIS is determined by how useful the user will be the flow of visual information. To describe the spatial objects and relations that generate the meaning of the image, use a special information structure - context. The context includes any information that enables the user to adequately assess the situation [1]. GIS includes many different contexts. Each of them reflects a special subjective vision of reality by its creator. Therefore, without taking special measures to select the analysis context, the effectiveness of the inter-active interaction turns out to be low.

This paper discusses an approach to determining the context of analysis using a special intelligent procedure. It is assumed that the client part of the GIS provides the ability to either explicitly or automatically select the context from the set described in the system. In the first case, the user receives a list of "most appropriate" contexts and selects any of them to continue the analysis. In the second case, the context is implicitly changed by the system that monitors the actions of the user analyzing cartographic images. In both cases, the achievement of the necessary semantic content does not require the user to study the existing set of contexts.

2 Search Methods

2.1 Known Approaches to Improving the Quality of Spatial Data Analysis

A traditionally used method for analyzing spatial data is the use of a previously developed template [2, 3]. The template includes a list of classes of objects, layers, communication templates with external databases, types and presentation formats of cartographic images. Having loaded spatial data in the selected template, the user receives a thematic map of the analyzed area. Its study is implemented in a conven-tional manner on or off the active layers, change of scales, types and angles. Map conversion to an acceptable level of visual perception is performed by the user man-ually. Changing the semantic content of the field of analysis requires a transition to another pattern, which is also implemented manually and adversely affects the user's understanding of the situation. GIS and analyst work independently.

Problems of perception of visual information and dialogue management are explored in the field of User experience design (UX, UXD, UED). The results obtained are used in the development of interfaces to the GIS software core, but they are not sufficient for manipulating cartographic representations. At the same time, a focus on studying user behavior stimulated the emergence of analysis contexts as a necessary information structure. The well-known Jacob's law of Internet UX [4] indicates that professionally oriented user groups have a certain common intuitive understanding of interfaces. Accordingly, there is knowledge of best practice analysis. However, the question of the form of presentation of this knowledge and the mechanism of its application remains little studied.

Geographic information systems and services are essentially positioned as Big Data systems. The reason is that any spatial data has an explicit or implicit time reference and never expires. The continuous growth of the information base as it exploits is natural, which raises the problem of studying large amounts of data. Problems of cartographic analysis, therefore, are adjacent to the problems of visual analytics for Big Data. An example is the work [5], which studies the role of visual analysis in the modern sense of "digital creativity". The authors explore the process of forming the mental image of the problem as a cycle of viewing the image and rethinking the formulation of an applied problem. The works in this area offer a variety of specialized analysis management models.

Situational awareness of the user-analyst is the main condition for the construction of effective solutions to the set applied problem in the GIS-analyst system [6]. Awareness of the goal, its structuring, the formulation of the subtasks corresponding to the set goal, the search for their solutions through the perception of information about the situation in this area of space, forecasting the development of the situation - these are the components of situational awareness [7]. Achieving comprehensive situational awareness has been the subject of many years of research in engineering, psychology, engineering, and design [8]. The study of the means of dialogue with artificial systems concentrates on the analysis of cognitive processes. The goal of many studies is to find the best distribution of cognitive load in human-machine systems through the use of artificial intelligence and cognitive systems [9]. Existing results should be adapted to the specificity of the cartographic analysis.

Research in the field of visual analytics allowed us to formulate the concept of guidance in the human – machine analytics process [10]. As noted in this paper, the visual analysis of complex situations ceases to be an auxiliary decision-making tool. It becomes the main component that assumes guidance in the process of solving an applied problem [10, 11]. The authors draw attention to intelligent guidance, controlling the global goal of the analysis. It should be noted that the generalization of the considered approaches does not allow to directly implement the guidance mechanism in the GIS.

2.2   Proposed Method for Searching of Context of Spatial Data Analysis

The concept of the "semantic content" of a cartographic image cannot be strictly defined for obvious reasons. The term "meaning" in the task in question reflects the user's subjective awareness of the situation being studied. The user gives the image a certain meaning. However, one has to assume that the semantic content carries the context. If the context established by the geographic information system does not correspond to the meaning of the analyst's understanding, then the data flow becomes of little use. As practice shows, this happens with a fairly high probability. The probability is the higher, the more uncertain the task facing the analyst, the more noticeable the "knowledge gap" in assessing the situation [12]. We suggest using the concept of context transformations to display and evaluate the meaning that the context carries. It consists in setting a set of permissible context transformations that do not change the meaning contained in it.

Accordingly, if the limits of permissible transformations are violated, then we can talk about the loss of meaning of the analyzed data.

The context is as follows:

$$c_m = <c_m*, H(c_m)>, \tag{1}$$

Where c_m* is the core of the context, and $H(c_m)$ is the set of its admissible transforms. The core includes fundamental knowledge of the objects and relationships inherent in the context of analysis c_m. In essence, they reflect, for example, a scientifically based or heuristic method for analyzing a problem. The context core contains the rules for determining the spatial, temporal and semantic boundaries of the scope of analysis. Any response to a user request to the GIS information base is processed in the established context by the rules of the kernel. Thus, the redundancy of cartographic images is eliminated. All objects and relationships that are not relevant in the given context are deleted. As the analysis showed, knowledge concentrated in c_m* is necessary but not sufficient to reflect the semantic orientation of the analysis. This is explained by the following:

(1) The similarity of nuclei does not completely determine their semantic proximity. For example, the core contexts of manual and automotive transportation seem closer to each other than to the context of the tourist infrastructure of the city. However, it is the latter context that is more useful in solving the problem of moving certain loads. Moving into this context is not obvious, but it does provide useful data on transportation options in a much shorter way;

(2) Core contexts do not reflect the deep knowledge of expert analysts about the transfer of existing experience in analyzing situations. The expert's well-known advantage is his ability to apply experience and build a reliable solution in previously unexplored conditions. It can be assumed that behind this is operating with "reasonable" (permissible) differences in situations and objects. The lack of such information leads to incorrect transfer of experience.

In order to increase the reliability of the comparison of the context meanings, a set $H(c_m)$ of admissible transformations of the boundaries of the domain of analysis is introduced in formula (1), which preserve the essence of context c_m. Set $H(c_m)$ includes objects, relations and functions, the presence (or absence) of which in the working area of the analysis indicates that the context has retained its meaning.

The task of finding an adequate context is as follows. GIS includes a description of a variety of contexts $C = \{c_1, c_2, \ldots c_M\}$, each of which can be set by the analyst in a session an arbitrary number of times.

The set of contexts used in the session forms a sequence:

$$C^{(m)} = \left\{ c_i^{(m)} \right\}, C^{(m)} \subseteq C,$$

where m is the session number. Let at the moment of time t_a, with the established context $C(t_a)$, objects and relations are observed that violate the permissible change in the spatial, semantic or temporal boundary. It is necessary to choose the context $c_p \in C$ that ensures the continuation of the analysis in the same semantic direction.

Analyzing the formulation of the problem, it should be noted that using the history of analysis precedents is a reasonable solution. Technically, history is a sequence of commands that a client sent to a GIS server in a session. This sequence is automatically saved in a log-file. From it, the sequences of changing contexts are uniquely restored.

The simplest solution to the context search problem would be to select the most frequently encountered context. However, this solution is not satisfactory, since it does not reflect the semantic orientation of the analysis process. We assume that the semantic orientation is reflected by a sequence of contexts. Then, having a set of data about the chains for different sessions, it can be argued that:

(1) Two sessions x and y were conducted in the same semantic direction, if the corresponding sequences coincide:

$$\left|C^{(x)}\right| = \left|C^{(y)}\right|, \forall 0 < a \le \left|C^{(x)}\right| : c_a^{(x)} = c_a^{(y)}$$

(2) Two sessions x and y were partially conducted in the same sense direction, if they include one or several matching fragments of sequences:

$$\tilde{c}^{(x)} \subset C^{(x)}, \tilde{c}^{(y)} \subset C^{(y)}, \left|\tilde{c}^{(x)}\right| = \left|\tilde{c}^{(y)}\right|,$$
$$\forall c_a^{(x)} \in \tilde{c}^{(x)}, c_a^{(y)} \in \tilde{c}^{(y)} : c_a^{(x)} = c_a^{(y)}$$

These statements allow us to formulate the following algorithm for searching for a promising context:

1. $c_p = \varnothing$, to consider the current session sequence as **Specified Sequence**.

2. To sort known from the history of the analysis of the sequence of contexts in ascending order of frequency.

3. Cycle through the list of sorted sequences:

4. If **Current Sequence** is equivalent to **Specified Sequence**, then go to step 6.

5. If **Current Sequence** is partially consistent with **Specified Sequence**, AND there is context c_{next} in **Current Sequence**, which is after the equivalent fragment of the sequence, then $c_p = c_{next}$.

6. End.

If $c_p = \varnothing$ at the end of the algorithm, then the current context is saved. Otherwise, the analyst will be asked to continue working in the new context.

3   Results

Since the effect of the proposed method should be evaluated by analyst satisfaction, a software model of an intelligent context search module was developed. The module was implemented in the corporate GIS in AutoLisp language of AutoCad Map system, the information base of which contains about 10^6 objects. The knowledge base includes 27 contexts that are used to manage the engineering communications of the enterprise, transport logistics, process equipment accounting, video surveillance and protection, emergency response, and others. A group of 23 users was given the opportunity to use the system for a month. Users solved the following types of tasks:

1. Accounting and maintenance of process equipment;
2. Construction and repair of buildings and structures;
3. Engineering communications management;
4. Transport logistics in the territory of the enterprise and in the adjacent territory;
5. Energy management;
6. Ensuring surveillance and territorial security;
7. Emergency response;
8. Transport network design;
9. Property management;
10. Placement and transportation of hazardous waste.

Each user rated the following quality indicators on a 10-point scale:

(A_1) the ability to formulate a task and ways of finding solutions with significant uncertainty of the original goal;

(A_2) the ability to find an acceptable solution for high-risk problems;

(A_3) the ability to design a satisfactory solution within a tight time frame for decision-making;

(A_4) reduction of cognitive load arising during the search and selection of cartographic materials.

The estimation range [−5, +5] was supposed to be used as follows: −5 corresponds to the most negative attitude, 0 corresponds to the neutral attitude, +5 corresponds to the most positive attitude.

Table 1 shows the survey results:

Table 1. Evaluation results.

	Task type number									
	1	2	3	4	5	6	7	8	9	10
A_1	2	3	4	4	2	3	0	1	1	3
A_2	4	3	4	5	1	3	4	3	0	5
A_3	4	4	4	4	5	3	4	5	4	5
A_4	3	1	4	4	−1	3	4	1	0	2

The columns correspond to the types of tasks, and the rows correspond to the estimates of the quality indicators.

Results can be assessed as follows. In general, satisfaction with the work of the intellectual module is assessed positively. Despite the subjectivity, such an assessment suggests the usefulness of managing the course of cartographic analysis. The greatest effect was observed in terms of A_3, i.e. in conditions of limited time for making decisions.

You can notice the types of tasks for which satisfaction is low enough, and sometimes clearly negative (columns 5 and 9). It can be assumed that the reason could be insufficient "training" of the intellectual module. As follows from the above search algorithm, its satisfactory work requires an archive of various analysis precedents. It was noted that for these tasks this number is small.

4   Conclusion

The method of searching the context for analyzing spatial data considered in this paper uses a special understanding of the meaning of contexts. It consists in the explicit selection of the core component, which reflects objective knowledge, and the set of permissible transformations that reflect the experts' subjective understanding of the deep content of the context. The proposed method of presentation allows one to reliably estimate the semantic proximity of contexts through admissible transformations.

Further studies on the problem considered should be carried out in the direction of using machine learning methods to recognize semantic areas of analysis, as well as combining contexts.

Acknowledgments.. This work has been supported by the Russian Foundation for Basic Research, Projects #19-07-00074, #18-01-00023 and #17-01-00060

References

1. Dey, A., Abowd, G.: Towards a better understanding of context and context-awareness. In: Gellersen H.-W. (ed.) HUC 1999, LNCS, vol. 1707, pp. 304–307. Springer, Heidelberg (1999). https://doi.org/10.1007/3-540-48157-5_29
2. Longley, P., Goodchild, M., Maguire, D., Rhind, D.: Geographic Information Systems and Sciences, 3rd edn. Wiley, Hoboken (2011)
3. Shashi, S., Hui, X.: Encyclopedia of GIS. Springer, New York (2008)
4. Nielsen, J.: Law of Internet User Experience. https://www.nngroup.com/videos/jakobs-law-internet-ux/. Accessed 02 Dec 2018
5. Cybulski, J., Keller, S., Nguyen, L., Saundage, D.: Creative problem solving in digital space using visual analytics. Comput. Hum. Behav. **42**, 20–35 (2015). https://doi.org/10.1016/j.chb.2013.10.061
6. Endsley, M.: Design and evaluation for situation awareness enhancement. In: Proceedings of the Human Factors Society 32nd Annual Meeting, pp. 97–101. CA: Human Factors Society, Santa Monica (1988). https://doi.org/10.1177/154193128803200221
7. Endsley, M., Bolte, B., Jones, D.: Designing for Situation Awareness: An Approach to Human-Centered Design. Taylor & Francis, London (2003)
8. Ziemke, T., Schaefer, K., Endsley, M.: Situation awareness in human-machine interactive systems. Cognit. Syst. Res. **46**, 1–2 (2017). https://doi.org/10.1109/JSYST.2019.2918283
9. Nilsson, M., Van Laere, J., Susi, T., Ziemke, T.: Information fusion in practice: A distributed cognition perspective on the active role of users. Inf. Fusion **13**(1), 60–78 (2012). https://doi.org/10.1016/j.inffus.2011.01.005
10. Collins, C., Andrienko, N., Schreck, T., Yang, J., Choo, J., Engelke, U., Jena, A., Dwyer, T.: Guidance in the human–machine analytics process. Vis. Inform. **2**(3), 166–180 (2018). https://doi.org/10.1016/j.visinf.2018.09.003
11. Andrienko, N., Lammarsch, T., Andrienko, G., Fuchs, G., Keim, D., Miksch, S., Rind, A.: Viewing visual analytics as model building. Comput. Graph. Forum **37**(6), 275–299 (2018). https://doi.org/10.1111/cgf.13324
12. Berndtsson, M., Mellin, J.: Active database knowledge model. In: Liu, L., Özsu, M.T. (eds.) Encyclopedia of Database Systems. Springer, New York (2018)

Using Data Mining with Fuzzy Methods for the Development of the Regional Marketing Geospace Information System

G. Akperov Imran[(⊠)] [iD]

Private Educational Institution of Higher Education
"Southern University (IMBL)", Rostov-on-Don, Russia
rector@iubip.ru

Abstract. The article reviews the use of data mining specifics based on applying models with a combined architecture in the fuzzy marketing. The use of digital economy requires changing criteria and priorities. This includes the artificial intelligence deployment, which enables monitoring of a huge number of parameters of economic, political and social processes in general and in relation to a relevant region or to an economic cluster, and implementing adequate executive decision-making procedures. The review of regional economic processes reveals a connection and interaction between their individual parameters, such as the ability to operate and interact with other systems without any restrictions, with reference to geographic location, or the so called geo-interoperability. This concept is closely related to the theory of geo-information space (GIS), a phenomenon formed by System of Systems, which includes components of natural and artificial origin and has certain synergistic properties. The article reviews a number of features of the GIS approach in the study of distributed social, ecological, economic and marketing processes, including their versatility, interdisciplinarity and the possibility of using the advantages of a systematic approach in these studies. There are various approaches to assessing such properties of the marketing geospace as adaptability, self-learning, self-adjustment, and sustainable development. The article reveals the connection between the regional marketing space and emerging economic clusters.

Keywords: Geo-information marketing space · Marketing systems architecture · Marketing information systems · Interoperability · Combined models

1 Introduction

The digital economy, which is currently growing and developing, has its own characteristic features and priorities. The most important of them is the world-wide deployment of the artificial intelligence, which enables monitoring of a huge number of parameters of economic, political and social processes in general and in relation to a relevant region, and implementing adequate executive decision-making procedures.

The review of regional processes reveals a connection and interaction between their individual parameters, such as interoperability, or "the ability to operate and interact with

© Springer Nature Switzerland AG 2020
R. A. Aliev et al. (Eds.): ICSCCW 2019, AISC 1095, pp. 325–330, 2020.
https://doi.org/10.1007/978-3-030-35249-3_41

other processes or systems without any restrictions, and geo-interoperability when it is connected it to a certain geographic location. A number of researchers [1, 2] speak of geo-information space as a phenomenon formed by System of Systems, which includes components of natural and artificial origin and possessing certain synergistic properties.

The GIS approach used in the study of distributed social, ecological, economic and marketing processes possesses such features as versatility, interdisciplinarity and the possibility of using the advantages of a systematic approach in these studies.

In [2], the term "geo-information space" is described as "a complex of coordinated territorial information models. GIS is a digital description of a combination of particular geospace representations, created in a computing environment and intended for computer-based use in solving spatial problems and developing spatial solutions".

2   Unified Geo Data Space as an Information System

The unified geo data space of a given region contains the following ordered information: geoinformation, spatial objects, geo data model arrays, displaying their main physical, chemical, economic and other properties.

Unlike the objective reality, the unified geo data space is a virtual object, designed and developed by man in a computing environment only. Its representations in such applications as construction, economics, geology, ecology, etc., are largely predetermined by the problem reviewed and are called digital layouts.

The regional unified geo data space is intended for solving spatial problems by means of data mining, "identifying geographically distributed patterns, developing spatial solutions for managing life support, preserving the natural environment, developing public production, providing information and reference support to the community, as a basis for establishing and operating territorial information management systems".

A unified geo data space is a geo-information system in terms of technical and informational aspects. It includes a knowledge base in the territory of the constituent entity or in the Federal District of the Russian Federation.

As of now, a considerable theoretical basis has been created for modeling the spatio-temporal pattern of the economy and social aspects. Besides, the basic patterns of individual GIS layers have been established, including transformation of their individual components in a competitive economic environment. P.A. Minakir wrote that "Spatial economics is a form of economy existing as an integrity of interacting economic entities, somehow distributed within a geographic space. An economic entity means an operator participating in at least one of the production processes, exchange and consumption".

The system analysis revealed "methods of lexico-semantic modeling of cognitive knowledge structures among the number of the studied methods. These methods allow us to apply the peculiarities of the geographic information environment. The proposed data mining structure is based on hybrid forms of knowledge representation regarding the marketing data: this includes a combination of hierarchical classification with the possibility of establishing and modifying several types of thesaurus and semantic relationships" [2].

This approach actively uses such concepts as a "cognitive geospace" and "cognitive geo-interoperability". It is aimed at coordinating the activities of experts from various subject matter areas in information systems (IS) operated in automatic/semi-automatic mode. This approach is also used in studying ways and methods for creating a digital generation of geo-information technologies.

3 Fuzzy Regional Marketing Space

Analyzing the results of marketing activities of individual market participants ultimately reveals a certain behavioral pattern, which is common to the entire marketing system and depends on the active environment, internal and external relations that form the marketing GIS. It is worth mentioning that "spacial marketing requires the use of innovation-oriented methods in production and management system, which means continuous improvement of intellectual activity, which becomes necessary due to the increasing complexity of studying the real and virtual interaction of businesses" and their expedient interaction in solving economic problems.

System studies of the marketing GIS suggest the further optimization of the conditions under which it develops, and of its effective functioning and development in the future. Marketing system studies should pay particular attention "to the study of the causal relationships of the marketing system's behavior and the identification of its structure and properties that will ensure the effective implementation of marketing activities" [2].

System marketing study must include the following main areas required for the development of spatial economics:

Theory issues, including big data mining and model evaluation, study and development of marketing concepts, tools and technologies; measurements and evaluation of the marketing properties and marketing space properties specific to a particular region, development and actual application of synergetic modeling methods.

Identification of fuzzy certainty factors in the adopted marketing models and assessment of their influence on the architecture and manageability of the marketing geo-information space.

Marketing policy at all levels: strategies, programs, and implementation mechanisms.

Pricing and tariff setting rules in regional marketing systems; external general and local marketing interoperability.

The spatio-temporal approach used in the study of marketing geo information systems (GIS) is based on the principle of consistency. It provides for viewing a marketing GIS as a system consisting of various types of correlating elements (which is actually defined as System of Systems [1]).

A marketing GIS can be described as a complex of socio-economic entities (subspaces) of the market space involved in the continuous spatial interactions regarding the demand generation and profit earning. All types of these subspaces possess a number of common properties, such as dimensionality, mutual spatial arrangement of elements, availability of nodes (centers), networks, etc.

The key advantage of the spatial approach is the possibility of a "multidimensional (multi-layered) arrangement of a spatial marketing system, within which certain economic, social, geographical, political and technological factors interact. These factors determine the equilibrium functioning and development of the entire spatial marketing system, and ensure the maximum efficiency of its functioning".

Thus, the marketing GIS may be viewed as a complex system of systems, an integrity of various physical subsystems and their multi-dimensional connections within the framework of the spatial concept. This includes social, industry, territorial and other connections.

4   Complex Models in the Marketing GIS Study

The marketing geo-information space is a complex multi-dimensional fuzzy and non-linear system. To study this system, adequate mathematical tools are required, typical of standard open dynamic systems, which, as a rule, significantly depend on initial conditions. Thus, in addition to traditional fuzzy certainty factors that impede the evaluation of the dynamics and its predictability, there may be the so-called deterministic randomness caused by bifurcations and strange attractors. This suggests a possible emergence of chaos in the incomplete predictability of the economic process behavior.

As was shown above, the virtual geo-information space may be used as a basis for studying the marketing space. It provides for expanding the scope of research (by studying and identifying the possible interoperability of various GIS subspaces), as well as developing nomenclature and methodologies of numerous sciences. We mean here "social, humanitarian, sociological and technical sciences, modeling and forecasting the interaction and correlation of these subspaces, generalizing theoretical results and creating interdisciplinary knowledge bases".

The review of [1] revealed that the definition of quality indicators affecting the equilibrium functioning and development of the marketing spatial system, and the formation of conditions for its ultimate effectiveness entails four main groups of factors: market, macroeconomic, industry-specific, and sociotechnical.

The use of hybrid intelligent system models is required to solve this kind of multi-criteria problems, including the following:

- fuzzy logic systems,
- genetic algorithms and artificial neural networks,
- imitation systems and analytical systems,
- expert systems and chaotic systems.

The main advantage of using these systems includes the possibility of effective dealing with fuzzy certainty factors in these systems, such as fuzzy, uncertain and approximate data. No analysis can be performed using standard modeling patterns, as they do not cover all the features of the reviewed complex system, such as fuzzy marketing.

The current hybrid research methods include the following:

- neuro computing + fuzzy logic (NF);
- fuzzy logic + genetic algorithms (FG);
- fuzzy logic + chaos theory (FCh);

- neural networks + genetic algorithms (NG);
- neural networks + chaos theory (NCh);
- neural networks + fuzzy logic + genetic algorithms (NFG), etc.

Hybrid methods can be applied to develop sustainable adaptive architectures of marketing information systems in the framework and under the conditions of the marketing GIS.

5 Method of Assessing the Sustainability of the Marketing GIS as a Complex Information System

The complexity of mathematical modeling used in marketing research is determined by the following factors:

- the complexity of the study subject itself, fuzziness and non-linearity of marketing processes, threshold and bifurcation effects;
- the effect of interoperability and interaction of marketing variables, which are mostly fuzzy, underdetermined and interdependent;
- the difficulty of direct measurements of fuzzy marketing variables;
- fuzzy and unstable marketing interactions;
- significant influence of the human factor on all marketing processes.

The term "soft computing" appeared as a result of combining (hybridizing) intellectual data processing methods and combining various artificial intelligence techniques. The term was introduced by L. Zadeh in 1994.

Soft computing is a system of computational methods providing the basis for understanding, designing and developing intelligent systems. These methods complement each other and are used in various combinations to create hybrid intelligent systems.

A model intended for studying such important properties of a spatial marketing system as stability and adaptability of its information architecture can be an example of the above approach [2]. The stability of the GIS marketing architecture is determined by its structural stability, status parameters, and, above all, stability of its functioning and development process.

A number of economists consider that the end of the XX century and in the early XXI century were characterized by the active clustering processes in their economies [1]. This included the development of groups of integrated companies with common territorial or industry-based features, and of associated entities. The advantages of clusters as of certain integral entities are primarily determined by their new properties and characteristics that are not inherent in their individual elements. This fact provides for a whole new level of production scale, market activity and competitiveness. Therefore, clusters can be considered as new system entities. Cluster elements, just like cluster systems (product manufacturers, sales networks, scientific and educational institutions, service providers, etc.), form close correlations, thus developing new properties and characteristics different from those of individual cluster elements. As we can see, there is a certain logic in correlating and adapting the regional marketing space and the regional cluster. The adaptability of the marketing system means, first of all, its flexibility and self-adjustment.

Adaptive architecture is a "smart" architecture able to adjust to the needs of a particular enterprise and allowing businesses to respond quickly to the market changes and IT flows. For this purpose, the sustainability of the marketing information system architecture must be regularly reviewed. To this end, the above-mentioned hybrid models combining the advantages of simulation, optimization, and soft computing can be applied. We can emphasize certain "methods that can demonstrate such properties as adaptability and the ability to learn and to self-adjust. These are, first of all, neural networks and fuzzy logic. Both methods are modeling tools and are applied after the stage of learning or knowledge extraction. Neural networks are used when dependent and independent variables are linked by complex non-linear relations".

When solving multi-criteria problems of studying the behavior of spatial marketing systems, models based on the advantages of imitation, optimization and expert cognitive systems should be preferred. Functions of the indicators affecting the equilibrium behavior and the development of the marketing spatial system, and forming the conditions for maximizing its effectiveness, can be defined and developed using hybrid data mining methods.

6 Conclusion

Thus, the following results have been obtained in these studies:

(1) An intellectual knowledge base has been identified as a problem class within formation of the marketing information space;
(2) the following problems of optimizing the inter-correlations within human-computer systems are formulated for solving problems of estimating and monitoring complex objects under conditions of a prior uncertainty:

- the problem of assessing the enterprise monitoring status using fuzzy data from various sources;
- the problem of choosing possible ways for aggregating the information in fuzzy marketing systems.

(3) Options for combined soft computing models are proposed for the study of practically significant marketing applications.

References

1. Akperov, I., Khramov, V., Lukasevich, V., Mityasova, O.: Fuzzy methods and algorithms in data mining and formation of digital plan-schemes in earth remote sensing. Proc. Comput. Sci. **120**, 120–125 (2017)
2. Karpik, A.P., Lisitsky, D.V.: Electronic geo-space - the essence and conceptual framework. Geodesy cartogr. **5**, 41–44 (2009)

Assessment the Attractiveness of Countries for Investment by Expert Knowledge Compilation

Ramin Rzayev[1,2(✉)], Sabina Aliyeva[1], Galib Hajiyev[3], and Tarana Karimova[3]

[1] Azerbaijan State Oil and Industry University, Azadlig ave. 34, AZ1010 Baku, Azerbaijan
raminrza@yahoo.com, sabinaliyeva2017@mail.ru
[2] Institute of Control Systems of ANAS, B. Vahabzadeh str. 9, AZ1141 Baku, Azerbaijan
[3] Azerbaijan State University of Economics, Istiqlaliyyat str. 6, AZ1000 Baku, Azerbaijan
galib.haciyev@gmail.com, taranakarimov@gmail.com

Abstract. To compile the initial expert knowledge regarding influences the fuzzy methods for assessing the investment attractiveness of countries are considered. The proposed approaches were tested to assess the investment attractiveness of 10 hypothetical countries characterized by similar in values consolidated expert estimates of economic, socio-political, and other influences.

Keywords: Investment attractiveness · Expert evaluation · Fuzzy set

1 Introduction

Creation of a favorable investment climate, as combination of political, socio-economic, and legal conditions in the territory of a particular country, is impossible without taking into account the factors that in a varying degree influence the investment decision-making realized by foreign investors. In fact, the factors of investment attractiveness of a country (FIAC) are the criteria used by international rating agencies for the integral assessment of the investment attractiveness of the country. The most general list of FIAC is following: x_1 – political; x_2 – social; x_3 – economic; x_4 – ecological; x_5 – criminal; x_6 – financial; x_7 – resource and raw materials; x_8 – labor; x_9 – production; x_{10} – innovative; x_{11} – infrastructure; x_{12} – consumer; x_{13} – institutional; x_{14} – legislative. All existing approaches to the calculation of integral indicators of investment attractiveness have their limitation, generated by the "blurring" and ambiguity of estimates of weakly structured FIAC $x_i (i = 1 \div 14)$. To overcome them, some expert groups compile ordinal ratings of country for investment attractiveness, determining its place among the rest. However, this approach is not perfect either, because it does not create the visibility for a potential investor: how much one country is more attractive and riskier than another, how much difference between countries occupying consecutive ordinal positions are significant or insignificant. Therefore,

R. A. Aliev et al. (Eds.): ICSCCW 2019, AISC 1095, pp. 331–339, 2020.
https://doi.org/10.1007/978-3-030-35249-3_42

based on these prerequisites, it becomes obvious the importance and actuality of the study of methods for transparent assessment of weakly structured indicators of the country's investment attractiveness (CIA) using a fuzzy analysis of expert evaluations obtained at an early stage of the subject area.

2   Expert Evaluation of CIA

As noted above, CIA is a multifactorial category characterized by system of the FIAC $x_i (i = 1 \div 14)$, which has a significant impact on the modern market of foreign investments [1]. The expert assessment of SIA implies: (1) ranking of factors x_i according to their priority; (2) identification the weights of x_i, based on their relative impact on the level of CIA; (3) assessment the FIPS using consolidated expert evaluations; (4) calculation of the integral index reflecting the overall level of the CIA.

Suppose the rank scores of FIAC x_i were obtained by independent survey of fifteen experts. At the same time, each expert was asked to arrange x_i according to the principle: the most important factor should be designated by the number "1", the next less important one – by the number "2" and further in descending order of expert preference. Thus, obtained rank scores are ordered in the form of Table 1.

Table 1. Expert rank scores of FIAC priority.

Expert	Evaluated factors and their rank scores (r_{ij})													
	x_1	x_2	x_3	x_4	x_5	x_6	x_7	x_8	x_9	x_{10}	x_{11}	x_{12}	x_{13}	x_{14}
1	2	4	1	10	14	3	5	13	6	12	7	11	8	9
2	1	3	2	13	14	4	5	10	9	11	8	12	7	6
3	3	4	1	12	13	2	6	8	7	10	5	9	11	14
4	3	4	2	13	14	1	5	6	7	9	8	10	12	11
5	2	5	1	13	14	4	3	6	8	9	7	10	11	12
6	1	13	2	14	12	3	7	8	9	10	6	11	5	4
7	3	10	2	9	11	1	4	12	13	14	8	7	6	5
8	1	13	2	14	12	3	5	4	6	11	7	8	10	9
9	2	13	1	14	12	4	3	6	5	8	7	9	11	10
10	3	4	2	12	13	1	5	7	6	14	8	11	10	9
11	4	5	2	13	12	3	1	14	6	10	7	11	8	9
12	2	4	1	9	13	3	10	11	12	14	5	8	7	6
13	3	14	2	13	12	4	1	11	5	10	7	9	8	6
14	3	14	1	12	13	2	4	9	10	11	5	7	8	6
15	1	14	2	13	12	3	4	11	9	10	5	8	6	7
$\sum$	34	124	24	184	191	41	68	136	118	163	100	141	128	123

To establish the degree of consistency of expert opinions, it is necessary to apply the Kendall's coefficient of concordance, which demonstrates the multiple rank correlation of expert rank scores. According to [2], this coefficient is calculated by the formula: $W = 12 \cdot S / [m^2 (n^3 - n)]$, where m is the number of experts; n is the number of

FIAC x_i; S is the deviation of expert rank scores from the average value of the ranking, which is calculated as [2]: $S = \sum_{i=1}^{n} \sum_{j=1}^{m} [r_{ij} - m(n+1)/2]^2$, where $r_{ij} \in \{1, 2, \ldots, 14\}$ is the rank of the i-th FIAC, which is established by the j-th expert. For our case, $S = 36945.50$ and, accordingly, $W = 12 \cdot 36945.50/[15^2(14^3 - 14)] = 0.7218 > 0.6$, which indicates the fairly strong consistency of expert rank scores relative to the degrees of importance x_i. At the preliminary stage of the independent survey, each expert was also instructed to establish the values of the normalized estimates of the weights of FIAC x_i ($i = 1 \div 14$). The results of this survey are summarized in Table 2.

Table 2. Expert normalized estimates of the weights of FIAC.

Expert	Evaluated factors and normalized estimates of their weights													
	x_1	x_2	x_3	x_4	x_5	x_6	x_7	x_8	x_9	x_{10}	x_{11}	x_{12}	x_{13}	x_{14}
1	0.125	0.105	0.135	0.045	0.010	0.115	0.095	0.020	0.085	0.030	0.075	0.040	0.065	0.055
2	0.135	0.115	0.125	0.020	0.010	0.105	0.095	0.045	0.055	0.040	0.065	0.030	0.075	0.085
3	0.115	0.105	0.135	0.030	0.020	0.125	0.085	0.065	0.075	0.045	0.095	0.055	0.040	0.010
4	0.115	0.105	0.125	0.020	0.010	0.135	0.095	0.085	0.075	0.055	0.065	0.045	0.030	0.040
5	0.125	0.095	0.135	0.020	0.010	0.105	0.115	0.085	0.065	0.055	0.075	0.045	0.040	0.030
6	0.135	0.020	0.125	0.010	0.030	0.115	0.075	0.065	0.055	0.045	0.085	0.040	0.095	0.105
7	0.115	0.045	0.125	0.055	0.040	0.135	0.105	0.030	0.020	0.010	0.065	0.075	0.085	0.095
8	0.135	0.020	0.125	0.010	0.030	0.115	0.095	0.105	0.085	0.040	0.075	0.065	0.045	0.055
9	0.125	0.020	0.135	0.010	0.030	0.105	0.115	0.085	0.095	0.065	0.075	0.055	0.040	0.045
10	0.115	0.105	0.125	0.030	0.020	0.135	0.095	0.075	0.085	0.010	0.065	0.040	0.045	0.055
11	0.105	0.095	0.125	0.020	0.030	0.115	0.135	0.010	0.085	0.045	0.075	0.040	0.065	0.055
12	0.125	0.105	0.135	0.055	0.020	0.115	0.045	0.040	0.030	0.010	0.095	0.065	0.075	0.085
13	0.115	0.010	0.125	0.020	0.030	0.105	0.135	0.040	0.095	0.045	0.075	0.055	0.065	0.085
14	0.115	0.010	0.135	0.030	0.020	0.125	0.105	0.055	0.045	0.040	0.095	0.075	0.065	0.085
15	0.135	0.010	0.125	0.020	0.030	0.115	0.105	0.040	0.055	0.045	0.095	0.065	0.085	0.075
Σ	1.835	0.965	1.935	0.395	0.340	1.765	1.495	0.845	1.005	0.580	1.175	0.790	0.915	0.960

Starting from the data presented in Table 2, let us conduct preliminary calculations for the subsequent identification of the weights of the FIAC in the form of averages over the groups of normalized estimates. In particular, averaging over the i-th group of normalized estimates of FIAC weights is performed iteratively according to [3] as:

$$\alpha_i(t+1) = \sum_{j=1}^{m} w_j(t)\alpha_{ij}, \qquad (1)$$

where $w_j(t)$ is the weight characterizing the degree of competence of the j-th expert ($j = 1 \div m$) at the time t. In this case, the averaging process is completed under

$$\max_i \{|\alpha_i(t+1) - \alpha_i(t)|\} \leq \varepsilon, \qquad (2)$$

where ε is the allowable accuracy of calculations. Assuming $\varepsilon = 0.0001$ and the same competences among experts, characterized at the initial stage $t = 0$ as $w_j(0) = 1/m$, the average values for groups of normalized estimates of FIAC weights in the 1st approximation can be obtained from the special case of (1): $\alpha_i(1) = \sum_{j=1}^{15} \alpha_{ij}/15$ ($i = 1 \div 14$), as following numbers: $\alpha_1(1) = 0.1223$; $\alpha_2(1) = 0.0643$; $\alpha_3(1) = 0.1290$;

$\alpha_4(1) = 0.0263$; $\alpha_5(1) = 0.0227$; $\alpha_6(1) = 0.1177$; $\alpha_7(1) = 0.0997$; $\alpha_8(1) = 0.0563$; $\alpha_9(1) = 0.0670$; $\alpha_{10}(1) = 0.0387$; $\alpha_{11}(1) = 0.0783$; $\alpha_{12}(1) = 0.0527$; $\alpha_{13}(1) = 0.0610$; $\alpha_{14}(1) = 0.0640$.

It is not difficult to notice that the requirement (2) for the 1st approximation is not satisfied. Therefore, to proceed to the next stage it is necessary to compute the normalizing multiplier $\eta(1) = \sum_{i=1}^{14} \sum_{j=1}^{15} \alpha_i(1)\alpha_{ij} = 1.2992$. Then, according to [3] and

$$\begin{cases} w_j(1) = \frac{1}{\eta(1)} \sum_{i=1}^{14} \alpha_i(1) \cdot \alpha_{ij} \ (j = \overline{1, 14}), \\ w_{15}(1) = 1 - \sum_{j=1}^{14} w_j(1), \ \sum_{j=1}^{15} w_j(1) = 1, \end{cases} \quad (3)$$

indicators of competence of experts in the 1st approximation will be the corresponding numbers: $w_1(1) = 0.0672$; $w_2(1) = 0.0670$; $w_3(1) = 0.0666$; $w_4(1) = 0.0668$; $w_5(1) = 0.0672$; $w_6(1) = 0.0664$; $w_7(1) = 0.0658$; $w_8(1) = 0.0666$; $w_9(1) = 0.0666$; $w_{10}(1) = 0.0672$; $w_{11}(1) = 0.0668$; $w_{12}(1) = 0.0654$; $w_{13}(1) = 0.0665$; $w_{14}(1) = 0.0670$; $w_{15}(1) = 0.0670\}$. Then, according to (1) or, more precisely, $\alpha_i(2) = \sum_{j=1}^{15} w_j(1)\alpha_{ij}$, the average values over the groups of normalized estimates of the FIAC weights in the 2nd approximation are the following numbers: $\alpha_1(2) = 0.3148$; $\alpha_2(2) = 0.1494$; $\alpha_3(2) = 0.3121$; $\alpha_4(2) = 0.1020$; $\alpha_5(2) = 0.1170$; $\alpha_6(2) = 0.2914$; $\alpha_7(2) = 0.2361$; $\alpha_8(2) = 0.1834$; $\alpha_9(2) = 0.1848$; $\alpha_{10}(2) = 0.1471$; $\alpha_{11}(2) = 0.2241$; $\alpha_{12}(2) = 0.1563$; $\alpha_{13}(2) = 0.2160$; $\alpha_{14}(2) = 0.2283$.

Checking the obtained values for the fulfillment of (2) and making sure that: max $\{|\alpha_i(2) - \alpha_i(1)|\} = \max\{|0.3148 - 0.1223|$; $|0.1494 - 0.0643|$; $|0.3121 - 0.1290|$; $|0.1020 - 0.0263|$; $|0.1170 - 0.0227|$; $|0.2914 - 0.1177|$; $|0.2361 - 0.0997|$; $|0.1834 - 0.0563|$; $|0.1848 - 0.0670|$; $|0.1471 - 0.0387|$; $|0.2241 - 0.0783|$; $|0.1563 - 0.0527|$; $|0.2160 - 0.0610|$; $|0.2283 - 0.0640|\} = 0.1924 > \varepsilon$, it is necessary to find the normalizing multiplier as $\eta(2) = \sum_{i=1}^{14} \sum_{j=1}^{15} \alpha_i(2)\alpha_{ij} = 3.5028$. In this case, the indicators of competence of experts in the 2nd approximation will be the following numbers: $w_1(2) = 0.066324$; $w_2(2) = 0.06683$; $w_3(2) = 0.06541$; $w_4(2) = 0.06584$; $w_5(2) = 0.06614$; $w_6(2) = 0.068159$; $w_7(2) = 0.06662$; $w_8(2) = 0.06711$; $w_9(2) = 0.06674$; $w_{10}(2) = 0.06626$; $w_{11}(2) = 0.06595$; $w_{12}(2) = 0.06585$; $w_{13}(2) = 0.06712$; $w_{14}(2) = 0.06772$; $w_{15}(2) = 0.06792$.

In the 3rd approximation, the averaging for the groups of normalized estimates of the FIAC weights calculated for the special case of (1): $\alpha_i(3) = \sum_{j=1}^{15} w_j(2)\alpha_{ij}$, are the following numbers: $\alpha_1(3) = 0.3163$; $\alpha_2(3) = 0.1507$; $\alpha_3(3) = 0.3136$; $\alpha_4(3) = 0.1037$; $\alpha_5(3) = 0.1188$; $\alpha_6(3) = 0.2930$; $\alpha_7(3) = 0.2377$; $\alpha_8(3) = 0.1851$; $\alpha_9(3) = 0.1864$; $\alpha_{10}(3) = 0.1488$; $\alpha_{11}(3) = 0.2257$; $\alpha_{12}(3) = 0.1581$; $\alpha_{13}(3) = 0.2178$; $\alpha_{14}(3) = 0.2302$. For this case, condition (4) is also not satisfied: max$\{|0.3163 - 0.3148|$; $|0.1507 - 0.1494|$; $|0.3136 - 0.3121|$; $|0.1037 - 0.1020|$; $|0.1188 - 0.1170|$; $|0.2930-0.2914|$; $|0.2377 - 0.2361|$; $|0.1851 - 0.1834|$; $|0.1864 - 0.1848|$; $|0.1488 - 0.1471|$; $|0.2257-0.2241|$; $|0.1581 - 0.1563|$; $|0.2178 - 0.2160|$; $|0.2302 - 0.2283|\} = 0.0019 > \varepsilon$. Therefore, it is necessary to proceed to the next iteration step by appropriate multiplier $\eta(3) = \sum_{i=1}^{14} \sum_{j=1}^{15} \alpha_i(3)\alpha_{ij} = 3.5271$. Then according to (3) the experts' competence indicators in the 3rd approximation will be following numbers: $w_1(3) = 0.06632$;

$w_2(3) = 0.066826$; $w_3(3) = 0.065413$; $w_4(3) = 0.065841$; $w_5(3) = 0.066133$; $w_6(3) = 0.068157$; $w_7(3) = 0.06663$; $w_8(3) = 0.067108$; $w_9(3) = 0.066742$; $w_{10}(3) = 0.066259$; $w_{11}(3) = 0.06595$; $w_{12}(3) = 0.065858$; $w_{13}(3) = 0.06712$; $w_{14}(3) = 0.067717$; $w_{15}(3) = 0.06792$. Substituting them into the (1), we obtain the average values of the normalized estimates of the FIAC weights for the i-th groups ($i = 1 \div 14$): $\alpha_1(4) = 0.31635$; $\alpha_2(4) = 0.15071$; $\alpha_3(4) = 0.31362$; $\alpha_4(4) = 0.10375$; $\alpha_5(4) = 0.11882$; $\alpha_6(4) = 0.29296$; $\alpha_7(4) = 0.23774$; $\alpha_8(4) = 0.18507$; $\alpha_9(4) = 0.18636$; $\alpha_{10}(4) = 0.14879$; $\alpha_{11}(4) = 0.22573$; $\alpha_{12}(4) = 0.15814$; $\alpha_{13}(4) = 0.21782$; $\alpha_{14}(4) = 0.23021$. In this case, condition (4) is satisfied, i.e. $\max\{|\alpha_i(4) - \alpha_i(3)|\} = 0.0000025 < \varepsilon$. This means that $\alpha_i(4)$ ($i = 1 \div 14$) are the final weights of FIAC x_i.

The method of expert evaluations involves the discussion of FIAC by another group of specialists. Each expert is provided with a list of evaluation criteria and it is proposed to individually obtain the independent estimation of the CIA level by the following 5-point scale: 1 – UNSATISFACTORY, 2 – SATISFACTORY; 3 – MORE THAN SATIS-FACTORY; 4 – VERY SATISFACTORY; 5 – PERFECT. Further, expert estimations are analyzed for their consistency (or inconsistency) according to the rule: the maximum allowable difference between two expert estimates by any evaluation criterion relative to FIAC x_i ($i = 1 \div 14$) should not exceed 3. This rule allows filtering unacceptable deviations in expert estimates for each FIAC. The total indexes can be calculate by:

$$r_1 = 100 \times \left[\sum\nolimits_{i=1}^{14} \alpha_i e_i\right] \Big/ \left[\max_i \sum\nolimits_{i=1}^{14} \alpha_i e_i\right], \quad r_2 = \left[\sum\nolimits_{i=1}^{14} \alpha_i e_i\right] \Big/ \left[\sum\nolimits_{i=1}^{14} \alpha_i\right] \quad (4)$$

where α_i is a weight reflecting the relative significance of the i-th FIAC; e_i is the consolidated expert estimation of the CIA relative to i-th influence; the maximum values of r_1 and r_2 means the maximum level of CIA, and vice-versa. Further, suppose that the experts was offered to test 10 alternative countries a_k ($k = 1 \div 10$) using the 5-point system for assessing the degree of influence of FIAC on their investment attractiveness. As a result of the consolidation of expert estimates relative to influence of FIAC in the form of arithmetic averages and the use of formulas (4) for the declared countries, aggregated estimates of the CIA levels are obtained and summarized in the form of Table 3.

Table 3. Aggregated expert estimates of the CIA levels.

Country	The weights of FIAC														r_1	r_2
	α_1	α_2	α_3	α_4	α_5	α_6	α_7	α_8	α_9	α_{10}	α_{11}	α_{12}	α_{13}	α_{14}		
a_1	4.25	2.25	4.75	1.55	4.75	2.65	3.35	2.70	3.25	2.65	4.45	4.85	2.85	4.25	71.7	3.58
a_2	4.85	4.65	3.55	4.60	3.95	4.15	3.25	2.85	4.85	3.75	3.65	3.85	3.00	4.25	78.6	3.93
a_3	0.75	4.25	1.55	0.55	4.10	2.35	2.00	3.75	4.65	1.35	4.35	1.55	0.85	3.45	48.8	2.44
a_4	4.30	3.15	0.45	4.75	1.45	3.15	1.10	3.30	2.15	0.15	2.00	1.15	1.25	0.35	40.5	2.03
a_5	3.25	2.95	1.10	4.15	3.35	4.45	1.45	1.35	0.25	4.15	3.95	4.50	2.25	2.10	54.0	2.70
a_6	2.65	2.85	2.15	2.40	3.95	0.95	1.85	3.05	2.20	4.45	1.45	1.35	0.55	0.35	39.9	1.99
a_7	2.45	3.75	1.00	1.05	4.25	2.15	3.25	4.65	1.00	1.80	2.65	0.25	0.85	2.65	44.5	2.22
a_8	2.50	4.55	4.25	2.25	2.00	2.75	1.85	4.00	1.65	1.95	0.15	3.15	3.95	3.65	56.1	2.81
a_9	0.75	0.75	3.55	1.50	0.45	2.85	3.50	3.85	2.35	3.75	3.10	2.25	0.65	2.10	46.6	2.33
a_{10}	1.00	2.15	1.45	1.65	4.85	2.95	1.65	3.85	0.85	1.25	4.55	3.45	0.20	1.55	42.5	2.12

3   Assessment of the CIA Using Fuzzy Inference System

Based on the "blurring" and ambiguity of estimates of weakly structured FIAC, as a qualitative criterion of assessment we consider the evaluation concept "FAVORABLE", which being one of the terms of the linguistic variable "*environmental level*" (political, social, economic, etc. according to the list) can be used to assess the CIA level. To reflect FAVORABLE criterion for evaluating each FIAC, the appropriate fuzzy subset of the discrete universe $U = \{a_1, a_2, ..., a_{10}\}$ are constructed on the basis of the consolidated expert estimates of the CIA relative to each FIAC (see Table 3) and the Gaussian function $\mu(e) = \exp\{-(e-5)^2/\sigma^2\}$, where $\sigma^2 = 4$ is the density. For example, the following appropriate fuzzy set C_1 is a criterion for assessing the favorableness of the political environment in a state in terms of its investment attractiveness:

- FAVORABLE (political environment) – $C_1 = \{0.8688/a_1;\ 0.9944/a_2;\ 0.0109/a_3;\ 0.8847/a_4;\ 0.4650/a_5;\ 0.2514/a_6;\ 0.1968/a_7;\ 0.2096/a_8;\ 0.2096/a_9;\ 0.0183/a_{10}\}$.

As evaluation concepts characterizing the CIA level we choose the terms of the output linguistic variable y, which are described by following fuzzy subsets of the universe $J = \{0;\ 0.1;\ ...;\ 1\}$ [3]: $S = $ ACCEPTABLE, $\mu_S(j) = j$; $MS = $ MORE THAN ACCEPTABLE, $\mu_{MS}(j) = j^{1/2}$; $P = $ PERFECT, $\mu_P(j) = 1$ if $j = 1$ and $\mu_P(j) = 0$ if $j < 1$; $VS = $ VERY ACCEPTABLE, $\mu_{VS}(j) = j^2$; $US = $ UNACCEPTABLE, $\mu_{US}(j) = 1-j$. Then, to describe the cause-effect relations between FIAC and CIA-levels the following fuzzy implicative rules are applied:

$$d_1 : (x_1 = C_1)\&(x_3 = C_3)\&(x_7 = C_7)\&(x_9 = C_9)\&(x_{11} = C_{11})\&(x_{12} = C_{12})\&(x_{14} = C_{14}) \Rightarrow (y = S);$$
$$d_2 : (x_1 = C_1)\&...\&(x_3 = C_3)\&(x_7 = C_7)\&...\&(x_9 = C_9)\&(x_{11} = C_{11})\&(x_{12} = C_{12})\&(x_{14} = C_{14}) \Rightarrow (y = MS);$$
$$d_3 : (x_1 = C_1)\&(x_2 = C_2)\&...\&(x_{14} = C_{14}) \Rightarrow (y = P);$$
$$d_4 : (x_1 = C_1)\&(x_3 = C_3)\&(x_6 = C_6)\&...\&(x_{14} = C_{14}) \Rightarrow (y = VS);$$
$$d_5 : (x_1 = C_1)\&...\&(x_6 = C_6)\&(x_7 = \neg C_7)\&(x_9 = C_9)\&(x_{10} = \neg C_{10})\&(x_{11} = C_{11})\&(x_{12} = \neg C_{12})$$
$$\&(x_{13} = C_{13})\&(x_{14} = C_{14}) \Rightarrow (y = S);$$
$$d_6 : (x_1 = \neg C_1)\&(x_3 = \neg C_3)\&(x_6 = \neg C_6)\&(x_{11} = \neg C_{11})\&(x_{14} = \neg C_{14}) \Rightarrow (y = US).$$

As a result of rule transformation by fuzzy operation "&" and Lukasiewicz's implication, the general solution is obtained in the form of the following matrix

	0	0.1	0.2	0.3	0.4	0.5	0.6	0.7	0.8	0.9	1
a_1	0.5350	0.6350	0.7350	0.8350	0.9086	0.9490	0.9490	0.9490	0.9490	0.9490	0.9845
a_2	0.5350	0.6350	0.6851	0.6851	0.6851	0.6851	0.6851	0.6851	0.6851	0.6851	0.9944
a_3	0.9891	0.9929	0.9929	0.9929	0.9929	0.9929	0.9929	0.9929	0.9929	0.9929	0.8998
a_4	0.9955	0.9972	0.9972	0.9972	0.9972	0.9972	0.9972	0.9972	0.9972	0.9847	0.8847
$R = a_5$	0.9964	0.9964	0.9964	0.9964	0.9964	0.9964	0.9964	0.9964	0.9964	0.9964	0.9272
a_6	0.9955	0.9955	0.9955	0.9514	0.8514	0.7514	0.6514	0.5514	0.4514	0.3514	0.2514
a_7	0.9865	0.9964	0.9964	0.9514	0.8514	0.7514	0.6514	0.5514	0.4514	0.3514	0.2514
a_8	0.9972	0.9972	0.9972	0.9972	0.9972	0.9972	0.9972	0.9972	0.9972	0.9688	0.8688
a_9	0.9891	0.9943	0.9943	0.9943	0.9943	0.9943	0.9912	0.8912	0.7912	0.6912	0.5912
a_{10}	0.9865	0.9968	0.9968	0.9968	0.9968	0.9968	0.9968	0.9968	0.9968	0.9968	0.9506

According to the composition rule in the fuzzy environment [3], the k-th row of the matrix R defines the fuzzy conclusion relative to CIA level of the k-th country ($k = 1 \div 10$). For numerical estimation of CIA level, the defuzzification rule is applied [3], which implies forming the level sets $E_\alpha = \{j | \mu_E(j) \geq \alpha\}$ and calculation of their cardinal numbers in the form of $M(E_\alpha) = \sum_{i=1}^{n} j_i / n$. Particular, for the fuzzy conclusion on CIA level of the a_1 (1st row of the R) $E_1 = \{0.535/0;\ 0.635/0.1;\ 0.735/0.2;\ 0.835/0.3;\ 0.9086/0.4;\ 0.949/0.5;\ 0.949/0.6;\ 0.949/0.7;\ 0.949/0.8;\ 0.949/0.9;\ 0.9845/1\}$, we have:

- for $0 < \alpha < 0.535$: $\Delta\alpha = 0.535$, $E_{1\alpha} = \{0;\ 0.1;\ 0.2;\ ...;\ 0.9;\ 1\}$, $M(E_{1\alpha}) = 0.5$;
- for $0.535 < \alpha < 0.635$: $\Delta\alpha = 0.1$, $E_{1\alpha} = \{0.1;\ 0.2;\ ...;0.9;\ 1\}$, $M(E_{1\alpha}) = 0.55$;
- for $0.635 < \alpha < 0.735$: $\Delta\alpha = 0.1$, $E_{1\alpha} = \{0.2;\ 0.3;\ ...;\ 0.9;\ 1\}$, $M(E_{1\alpha}) = 0.60$;
- for $0.735 < \alpha < 0.835$: $\Delta\alpha = 0.1$, $E_{1\alpha} = \{0.3;\ 0.4;\ ...;\ 0.9;\ 1\}$, $M(E_{1\alpha}) = 0.65$;
- for $0.835 < \alpha < 0.9086$: $\Delta\alpha = 0.0736$, $E_{1\alpha} = \{0.4;\ 0.5;\ ...;\ 0.9;\ 1\}$, $M(E_{1\alpha}) = 0.70$;
- for $0.9086 < \alpha < 0.9490$: $\Delta\alpha = 0.0404$, $E_{1\alpha} = \{0.5;\ 0.6;\ 0.7;\ 0.8;\ 0.9;\ 1\}$, $M(E_{1\alpha}) = 0.75$;
- for $0.949 < \alpha < 0.9845$: $\Delta\alpha = 0.0355$, $E_{1\alpha} = \{1\}$, $M(E_{1\alpha}) = 1$.

Then the numerical estimate of the fuzzy output E_1 can be obtained as follows [3]:

$$F(E_1) = \alpha_{\max}^{-1} \int_0^{\alpha_{\max}} M(E_{1\alpha})d\alpha = (0.5 \cdot 0.535 + 0.55 \cdot 0.1 + ... + 1 \cdot 0.0355)/0.9845$$
$$= 0.5737.$$

The numerical estimates of fuzzy conclusions for CIA-levels of other countries are obtained similarly: $F(E_2) = 0.6656$; $F(E_3) = 0.4955$; $F(E_4) = 0.4938$; $F(E_5) = 0.4965$; $F(E_6) = 0.3417$; $F(E_7) = 0.3419$; $F(E_8) = 0.4921$; $F(E_9) = 0.4492$; $F(E_{10}) = 0.4982$.

4 Assessment of the CIA Using Fuzzy Maximin Convolution

Constructed above the fuzzy subsets C_i ($i = 1 \div 14$) of the universe $\{a_1, a_2, ..., a_{10}\}$ describe the qualitative assessment criteria by Gaussian membership function, through the values of which the relations of each FIAC x_i to the corresponding evaluation criterion are revealed. Assuming that the criteria C_i have the same significance, we define the desired set of alternatives C by intersecting of fuzzy sets containing estimates of alternative countries by the selection criteria: $C = C_1 \cap C_2 \cap ... \cap C_{14}$. Then, according to the fuzzy maximin convolution method, from the point of view of investment attractiveness, the best is a country, which belongs to the fuzzy set C with the maximum degree. In this case, the desired set is formed as: $C = [\min\{0.869; 0.151;$ 0.985; 0.051; 0.985; 0.2514; 0.506; 0.267; 0.465; 0.251; 0.927; 0.994; 0.315; 0.869\}; \{0.994; 0.970; 0.591; 0.961; 0.759; 0.835; 0.465; 0.3149; 0.994; 0.677; 0.634; 0.719; 0.368; 0.869\}; \{0.011; 0.869; 0.051; 0.0071; 0.817; 0.173; 0.105; 0.677; 0.969; 0.036; 0.899; 0.051; 0.0135; 0.549\}; \{0.885; 0.425; 0.0057; 0.985; 0.043; 0.425; 0.022; 0.486; 0.131; 0.0028; 0.105; 0.025; 0.0297; 0.005\}; \{0.465; 0.350; 0.0223; 0.835;

0.506; 0.927; 0.043; 0.036; 0.0036; 0.835; 0.759; 0.939; 0.151; 0.122}; {0.251; 0.315; 0.131; 0.185; 0.759; 0.017; 0.084; 0.387; 0.141; 0.927; 0.043; 0.036; 0.007; 0.0045}; {0.197; 0.677; 0.018; 0.020; 0.869; 0.131; 0.465; 0.969; 0.018; 0.077; 0.251; 0.0036; 0.0135; 0.251}; {0.2096; 0.951; 0.869; 0.151; 0.105; 0.282; 0.084; 0.779; 0.061; 0.098; 0.0028; 0.425; 0.759; 0.634}; {0.011; 0.011; 0.591; 0.047; 0.006; 0.315; 0.5698; 0.719; 0.173; 0.677; 0.406; 0.151; 0.0088; 0.122}; {0.018; 0.131; 0.043; 0.061; 0.994; 0.3497; 0.061; 0.719; 0.0135; 0.0297; 0.951; 0.549; 0.0032; 0.051}]. The ranking of countries by the CIA-level is based on the resulting vector of priorities max $\{\mu_C(a_k)\}$ = max{0.051; 0.3149; 0.0071; 0.0028; 0.0036; 0.0045; 0.0036; 0.0028; 0.0057; 0.0032}. In terms of CIA, the best country is a_2, which corresponds to the highest value 0.3149. Following: a_1 – 0.0510, a_3 – 0.0071, etc. descending.

5 Conclusion

According to the approach the FIAC weights α_i ($i = 1 \div 14$) were identified on the base of agreed expert conclusions relative to the priority of the FIAC. It became the basis for the forming of the integral estimates of the CIA-levels by formulas (4). The methods of fuzzy inference and maximin convolution, which using different modes of aggregation compile obtained expert knowledge relative to influences of FIAC. A comparison of the results obtained by all methods is presented in Table 4.

Table 4. Integral estimates of the CIA-levels by applying of all methods.

Country	Expert evaluation methods				Fuzzy assessment methods			
	Criterion r_1	Rank	Criterion r_2	Rank	Fuzzy inference	Rank	Maxmin	Rank
a_1	71.67	2	3.58	2	0.5737	2	0.0510	2
a_2	78.57	1	3.93	1	0.6656	1	0.3149	1
a_3	48.79	5	2.44	5	0.4955	5	0.0071	3
a_4	40.51	9	2.03	9	0.4938	6	0.0028	9, 10
a_5	53.95	4	2.70	4	0.4965	4	0.0036	6, 7
a_6	39.89	10	1.99	10	0.3417	10	0.0045	5
a_7	44.48	7	2.22	7	0.3419	9	0.0036	7, 6
a_8	56.11	3	2.81	3	0.4921	7	0.0028	10, 9
a_9	46.63	6	2.33	6	0.4492	8	0.0057	4
a_{10}	42.47	8	2.12	8	0.4982	3	0.0032	8

The ranks of countries in terms of CIA obtained by using all methods coincided with the first, second and eighth countries. In other cases, the estimates obtained by fuzzy maximin convolution method resonate strongly with others, since this method is relatively "crude" and it is focused on choosing the best alternative. The remaining estimates obtained by fuzzy inference are also somewhat different from the expert ones. This can be explained by the fact that the initial consolidated expert data on the relative influence of FIAC do not differ significantly from each other. However, we consider ranking based on fuzzy inference as more reliable, because it is more sensitive to minor discrepancies in the source data and it is flexible for decision makers.

References

1. Aksenova, N., Prikhodko, E.: Modern approaches to assessing the investment attractiveness of the country. Finance Credit **32**(416), 56–64 (2010). (In Russian)
2. Lin, A., Wu, W.: Statistical Tools for Measuring Agreement. Springer, New York (2012)
3. Rzayev, R.R., et al.: Two approaches to country risk evaluation. Adv. Inf. Commun. **1**, 793–812 (2019)

A Fuzzy Semantic Data Triangulation Method Used in the Formation of Economic Clusters in Southern Russia

Gurru I. Akperov and Vladimir V. Khramov[✉]

Sothern University (IMBL), M. Nagibina pr., 33a/47,
344068 Rostov-on-Don, Russia
rector@iubip.ru, vxpamov@inbox.ru

Abstract. The article describes the formation of clusters as evolving polymorphic systems in algorithms of systems analysis based on the international experience. The authors pay special attention to the range of cluster development patterns and the necessary conditions for their formation. The given range of patterns is required to develop algorithms for a comparative analysis of their reflection in the national and regional regulatory documents in order to accelerate the clustering process in the Russian economy.

Keywords: Cluster · Polymorphic systems · Clustering patterns

1 Introduction

A number of economists consider that the accelerated socioeconomic development of a number of European and Southeast Asian countries at the end of the XX century and in the early XXI century depend on the clustering processes in their economies [1]. This includes the development of groups of integrated companies with common territorial or industry-based features, and of associated entities. The advantages of clusters as of certain integral entities are primarily determined by their new properties and characteristics that are not inherent in their individual elements. This fact provides for a whole new level of production scale, market activity and competitiveness. Therefore, clusters can be considered as new system entities. Cluster elements, just like cluster systems (product manufacturers, sales networks, scientific and educational institutions, service providers, etc.), form close correlations, thus developing new properties and characteristics different from those of individual cluster elements.

2 Cluster Formation Patterns

To study the characteristic (distinctive) features and patterns of cluster composition, we need to define the basic premises for the further theoretical conclusions. We accept the fact that clusters are a kind of polymorphic system as the first basic premise of the developed approach. This will enable us to use the system modeling analysis and the developing tools of the general systems theory.

© Springer Nature Switzerland AG 2020
R. A. Aliev et al. (Eds.): ICSCCW 2019, AISC 1095, pp. 340–344, 2020.
https://doi.org/10.1007/978-3-030-35249-3_43

The ratio between commercial investments in tangible and intangible assets is one of the markers indicating the rates of these changes in the knowledge building for the needs of economic development.

Being powerful tools of competition in the fast- developing world markets, clusters actually act as concentrators of market-oriented intellectual resources accelerating their capitalization and expansion to economic rotation in promising combinations with other production factors.

It is worth mentioning that a high formalization levels and the expected economic results enable the mathematical modeling of these processes and transition to the formulation and solution of mathematically rigorous optimization problems.

As far as clusters are concerned, we should mention systems with highly integrated and actively interacting elements. In this sense, active interaction means interaction that is significantly more active than the background interaction.

Clusters are sometimes treated as networks, e.g. business networks [1]. However, taking into account the above mentioned feature (high integration of cluster elements), which is not always inherent in networks, we should not confuse these concepts, since clusters have clustering centers (their distinguishing feature), which attract surrounding elements with their powerful potential field, while the potential field of networks is more uniform.

Another basic premise includes the following: Clusters are subsystems within national or regional innovation systems.

The development of cluster initiatives as specific institutional structures of national innovation systems was comprehended in European countries gradually.

Clusters evolve as integrated system entities and become independent elements of the emerging regional innovation environment (open non-linear dynamic system that interchanges matter, energy and information with other systems).

If we analyze the source of a significant flow of resources in innovation systems in order to actually influence the regional budgeting, we can see that the required effect can be achieved either through a sufficiently large number of small and medium-sized innovative companies, such as were in the USA and in a number of other developed countries at the end of the last century, or through a concentration of forces and resources with the further formation of powerful cluster complexes covering large-scale sales markets. Thus, budget-forming function is one of the distinguishing features of cluster systems.

3   Clusterization: The Study of the Sources of Economic Integration and Self-organization

Clusterization of the regional economy, or formation of regional industrial clusters, irrespective of the initiating force (either the government or the business) is followed, first of all, by restructuring the main types of correlations of the key elements initiated by the diverse, though directed management. New types of correlations lead to the formation of certain polymorphic system entities with specific properties of their key elements, which are not characteristic of each separate entity being integrated. Thus, clustering processes are supposed to create a certain complex of stable system entities corresponding to the relevant principles of composition with certain properties and strong internal relations. Counting on the fact that many diverse elements participate in clustering processes, and that each of them influences the process components, we can hardly believe that the necessary interrelations will randomly develop in such multi-component environments. Therefore, defining patterns characteristic of the formation of the required regulating influence on the developing clustering processes is of certain interest.

Analyzing the experience of European countries in the development of clustering process we can see that the national cluster policy has developed, first of all, in small states. In federal states, a significant emphasis was transferred to the regions. Quite interesting, that only 5 of the 31 European countries did not have national cluster development programs. These are federal countries where cluster development programs are decentralized. In 2008, regional cluster development programs were established in 57% of European countries. The regional cluster development policy focused on interaction, cooperation and building a framework for the further network localization. It was aimed at establishing rules and selecting the appropriate cluster forms and objectives. Based on the arguments mentioned above, we can define the following pattern: in federal countries, clustering policies are formed by the federal subjects.

The following fact is worth mentioning when we speak about the transition to the cluster-centered management: with an increase in the number of participants in any system entities, their integration ability decreases; in other words, their intersecting areas are reduced. This feature is more like a consistent pattern, arising from the analysis of the developing complex polymorphic systems. Still, synergistic processes ensuring the system controllability and emergence play a significant part.

Besides, we should note one more consistent pattern in the development of complex polymorphic systems. This pattern is significant for clustering processes, but is seldom reflected in the regulatory documents of cluster development policy: The system operator that deals with cluster entities claiming sustainable competitiveness must deal with the cooperation interests of the cluster entity only.

Cluster management must clearly understand the role of participants and shareholders in order to maintain the balance between continuance, on the one hand, and changes, on the other hand.

In the view of some European analysts, cluster managers maintaining contact with other elements can access the production through their contacts with firms. These interaction priorities, though, are determined by the production itself, as are the priorities of interaction with other clusters, as well as with the global market and financial institutions. Since cluster parameters are constantly changing, their managers must constantly interact with the public sector, research and educational institutions, other cluster entities, the international market and financial institutions.

Strategic issues relating to the development processes, including those defining the general vector for developing cluster initiatives and key figures, play the most important role in the full-scale development of cluster initiatives. The involvement of companies in the strategic analysis is mandatory. The so-called third mission of the university as a coordinator of the regional clustering processes can be recognized here. A decent and well-prepared strategy and strong ties between cluster participants form the basis for developing and implementing a range of measures that would satisfy the needs of cluster participants in the most successful (systemically synergistic) form.

Design and operation of cluster systems must comply with principles of convergence. It is the key condition which ensures that the management effect achieved in the innovation systems correlates with the selected objectives.

We should also mention that the study of the cluster development dynamics involves modeling methods and mechanisms for describing cluster components similar to self-organizing maps or artificial two-dimensional neural networks, such as algorithms describing hex grids. Such a method is used to represent cluster components. It enables studying the formation of the cluster architecture depending on its self-organization, and the relationship of this architecture with the unified geo data space of the region.

The following pattern can be clearly recognized: if a system lacks some of its essential elements and their interrelations, it may retain its system-like appearance, but will fundamentally differ from the one we plan to establish.

Thus, it can be argued that combining the same system elements, but using different types of connections will form entirely different systems with different properties, parameters and characteristics.

The issue of identifying cluster development patterns itself has already been raised by quite a few authors and is in no way revolutionary. However, algorithms of the systemic analysis imposed to it and combined with the identification and systematization of the main patterns of arranging the analyzed systemic entities suggest taking a different look at the processes occurring today in the country and its regions in terms of the cluster based economic development. It seems quite obvious, that a brief look at this analysis reveals serious contradictions with the regulatory initiatives of the national and regional levels.

The initial data available in regard to the cluster status and structural processes are incomplete, fuzzy and contradictory. To control the cluster formation process, a special research (methodological, cross-check) triangulation method should be used. It includes an integrated process of opposing or comparing methods with each other, and is aimed at increasing the validity of field achievements.

A cross-check triangulation method may be used as follows:

- to compare different interpretations of various studies;
- to compare interpretations using different methods;
- to compare the current scenario results with relevant results obtained in other studies;
- to repeat the results obtained with the same object after a certain period.

4   Conclusions

Triangulation method improves the data reliability by identifying the aggregation of controversial data and forming a so-called "group judgement" about a single empirical object. Mathematically, the procedure is reduced to selecting the majority voting of favorable results m from all possible results n. It looks as follows if applied to the soft computing methods:

If $\mu 1$, $\mu 2$, $\mu 3$, ... μn are the values of the fuzzy set adjective of the corresponding (equally significant) properties of an economic entity which claims its position within the cluster, then the value of the membership function of the entire object shall be determined in accordance with the following procedure: all values of μi shall be arranged in descending order, and the utmost left m-th shall be selected. This shall be the value of the resulting membership degree of the entire object required to participate in the cluster formation.

Reference

1. Akperov, I., Khramov, V., Lukasevich, V., Mityasova, O.: Fuzzy methods and algorithms in data mining and formation of digital plan-schemes in earth remote sensing. Proc. Comput. Sci. **120**, 120–125 (2017)

Application of Advanced Process Control Technologies for Optimization of Polymers Production Processes

Nodirbek Yusupbekov[1], Fakhritdin Abdurasulov[2], Farukh Adilov[3], and Arsen Ivanyan[3(✉)] (iD)

[1] Tashkent State Technical University, Tashkent, Uzbekistan
dodabek@mail.ru
[2] "Oltin Yo'l GTL" Company, Tashkent, Republic of Uzbekistan
info@oltinyolgtl.com
[3] LLC "XIMAVTOMATIKA", Tashkent, Republic of Uzbekistan
{Farukh.Adilov, Arsen.Ivanyan}@himavtomatika.uz

Abstract. This article describes latest advanced process control technologies in application of polymers' production. Polymers production are complex non-linear processes which require specific multi-direction approach in process control methodology to achieve maximization of required polymer's yield and real economic result. There are several mathematical modelling techniques exist and new advanced techniques being developed for both process and control modelling and further programming them in special software environments.

Keywords: Advanced process control (APC) · Polymers · Polymerization process · Model predictive control (MPC) · Non-linear controller (NLC) · Optimization · Melt index · Differential/algebraic equations (DAEs)

1 Introduction

Polymer production is one of the most important areas of applied chemistry due to its significant economic and social impact. Polymers as materials are present in almost every field of human activity. They range from commodity materials, such as polyethylene or PVC (poly vinyl chloride), up to highly specialized and expensive materials for drug release or space-craft related applications. Polymers are long molecules or "macromolecules" produced from simple small chemical components or monomers. The chemical reaction by which monomers are transformed into polymers is called polymerization and its control presents serious challenges for the chemical engineer, due to the fact that these reactions are usually highly exothermic and often proceed in very viscous media that render mass and heat transport difficult. Also, these reactions are famous for behaving in non-linear fashion and several instances of multiplicities and sustained oscillations have been reported in the literature for even industrial scale reactors [1]. In order to aid in the design, operation, control and optimization of this kind of reactors and reactions, standard mathematical modeling of the polymerization process is an invaluable technique.

© Springer Nature Switzerland AG 2020
R. A. Aliev et al. (Eds.): ICSCCW 2019, AISC 1095, pp. 345–351, 2020.
https://doi.org/10.1007/978-3-030-35249-3_44

To describe polymerization processes it is required to divide these processes on many parts and apply different mathematical modelling principles to each of part, then to summarize them into one dynamic process model which should be monitored and researched further for development of process control techniques and creating production benefits.

This article introduces the advanced process control methodology for polymers production processes which is based on nonlinear multivariable model-based controller/optimizer.

2   Methods and Results

Advanced process control (APC) is now widely used industrial practice of implementation of model-based software that is used to direct the process operation using different mathematical techniques [2–4]. Industrial APC applications require that the process model created accurately represents the process dynamics. Improved economics of the operation or production improvements are typical driving forces for using these applications. Advanced process control software includes:

- Offline and online model building, model verification, and data analysis tools that ensure model accuracy
- Driver software to collect historical data from control systems or online process historian systems
- Operator guidance to advise where the APC is taking the process
- Prediction of controlled variables based on future planned moves of manipulated variables
- Run-time APC tuning to adjust for changes in process dynamics without having to rebuild the model
- APC controller monitoring and performance analysis software, and not the proportional-integral-differential (PID) or process controller tuning and related performance

Online optimization is designed to continuously monitor the state of the process and, through model reference, predict an optimum operation path. Online optimization software typically employs technology for solving simultaneous equations. The model is a part of an application package that presents the best direction for the process operation to ensure that the objectives are met. The output of the optimization software can be presented in an advisory fashion to the operator, or can be set as a new target to a lower-level control strategy or APC control strategy. Improved economics, business performance, safety, or environmental concerns are the typical driving forces for using APC applications and can be summarized as following objectives:

- honor all operating constraints and environmental and safety limits
- operate process close its constraints while maintaining operational safety
- maintain products quality specifications
- maximize unit throughput
- maximize production/recovery of most valuable products

- minimize off-spec products and give-away of valuable products
- optimize energy usage (minimize energy consumption, maximize heat recovery)
- optimize utilities, chemicals, catalysts and equipment utilization
- stabilize plant operation and minimize variability of key control and quality variables
- mitigate/reject the effect of process, utilities and environmental disturbances
- guarantee smooth transitions between different operating conditions
- reduce operator interventions
- achieve high control system operating factor

As already stated above APC for processes of polymers production cannot be implemented based on standard mathematical modelling principles. Nonlinear multi-variable model-based controller (NLC) uses a dynamic process model specified as a set of differential/algebraic equations (DAEs). Models can be based on first principles or they can be empirically derived. Nonlinear dynamic model of the system is incorporated within the controller structure. This strategy is in contrast with linear model predictive control (MPC) algorithms, which assume a linear model of the process dynamics. This is particularly important in polymerization processes where different grades result in significantly different gains and dynamics between manipulated variables (MVs) and calculated variables (CVs) - e.g. Melt Index and manipulated variables like temperature or Hydrogen flow.

Typical examples of processes that could benefit from this controller include polymerization reactors where grade changes are necessary, distillation towers with variable feed composition and flow, or other processes that have highly nonlinear characteristics.

Nonlinear controller (NLC) is most widely used with a first principles-based engineering model for polymer processes. This methodology has been developed based on scientific researches of different scientists worldwide [5–7]. The model uses well-established reaction kinetic mechanisms for polymers to estimate fundamental physical properties of polymers such as production rate, melt index and density. It also uses the actual physical plant equipment data (such as vessel volumes) to establish dynamics in the unit. This ensures that key quality control is very predictable over a broad range of operating regions even if the initial data used to estimate model parameters did not include these regions. This is unlike many other solutions where data-driven models must have been trained on data from each region where control is required.

Some of the inferential calculations that are provided as part of the model and hence can be used directly by NLC are:

- Polymer Production Rate
- Number Average Molecular Weight
- Weight Average Molecular Weight
- Instantaneous and Bed-average Melt Index
- Instantaneous and Bed-average Density
- Polymer Density calculations
- Reactor Dew point temperatures
- Reactor Monomer Conversions
- Reactor Superficial Gas velocities

- Reactor Minimum Fluidization velocities
- Catalyst productivity
- Space Time yield
- Recycle Gas compositions
- Actual concentrations for reactor feeds: H_2, C_3, C_2, etc.

NLC has the ability to control the polymer process all the way through the grade transitions. The user can set the desired response for a CV such Melt Index in terms of time to reach one time constant (t) and the controller will drive the unit to respond. Limitations on speed of response become related to the reaction kinetics rather than the controller architecture.

The current prediction of a given CV is controlled by manipulating MVs such that it is maintained between the high and low reference trajectory which move out over time towards the CV's SPHI and SPLO limits.

NLC will also control the trajectory of a key CV during the transition. This minimizes the amount of off spec produced and, in some cases, can keep the transition product within a certain specification thereby classifying it as a higher value product. Once on the new grade the NLC will continue to control to specification without a change in the model being necessary.

The NLC system is an implementation of a true constrained multivariable optimal control algorithm. Constraint violations are predicted and handled dynamically, as opposed to linear MPC system that only address constraint violations at the predicted steady state. Over-constrained conditions (insufficient manipulated variables available to meet all control objectives) are handled through constraint prioritization.

NLC is capable of performing economic optimization in a dynamic sense when control objectives are met. This scenario exists when the number of manipulated variables exceed the number of CV's with set points, or when set point high and low ranges are used.

NLC is a model predictive controller utilizing dynamic process models. These models are represented by a set of differential/algebraic equations (DAEs). A DAE model is a collection of variables and differential and/or algebraic equations. The reaction kinetics in a polymer reactor are a good example of differential equations found in the DAE model. The general form of the DAE constraints is shown as follows:

$$g\left(\frac{dy}{dz}, y, u, z\right) = 0.0 \tag{1}$$

where:

y is the dependent variables, u the independent variables and z is the integration variable e.g. time, length, etc.

y and u may have upper and lower bounds on their values and the system also solves boundary conditions such as values when z = 0.0 (initial conditions).

NLC DAE system is based on orthogonal collocation on finite elements, which is a technology suitable for the solution of a wide range of differential/algebraic equation models. Orthogonal collocation methods provide the mechanism for the equation-based solution of the dynamic models used in NLC. The equation-based technology provides flexibility to handle many different specifications of a model, including those for state estimation, state feedback, dynamic data reconciliation and closed loop dynamic optimization and control.

As a separate part of polymers production process polymer reactor dynamic model is considered to be developed and inter-linked with NLC DAE system. The reactor models are integrated with a general kinetic model for the transition metal catalyzed polymerization of olefins. In reactor model the dependence of rate constants used in all of the kinetic schemes on temperature and pressure is represented by the following equation:

$$k = k_0 e^{\frac{-(E_A + V_A P)}{RT}} \tag{2}$$

where k_0, E_A, V_A, P, T and R are the frequency factor, activation energy, activation volume, pressure, temperature, and ideal gas constant, respectively.

The important measurable product properties, such as melt flow index and density are computed by the reactor models.

Reaction mechanisms are defined for each of the reaction classes:

- Site Activation Reactions. Site activation reactions model the creation of vacant active sites from potential active sites available in the catalyst.
- Initiation Reactions. Initiation reactions model the creation of active polymer chains of length one from a vacant active site.
- Propagation Reactions. Propagation reactions model the growth of active polymer chains by combination with a monomer unit.
- Chain Transfer Reactions. Chain transfer reactions model the conversion of a live polymer chain to a dead polymer chain which also produces an active polymer chain of length one.
- Site Transformation Reactions. Site Transformation reactions model the transformation of live polymer chains with active sites of one type into vacant active sites of another type. Dead polymer is also produced in this reaction.
- Site Deactivation Reactions. Site deactivation reactions model the conversion of live polymer chains into dead sites and also produces dead polymer.

Reaction rates are integrated over time to produce the values for the total number of all types of reactions occurring. The relative reaction rates and concentrations of monomer, polymer and initiator determine the likelihood of each reaction class occurring and hence can be used to calculate the polymer distribution chain for both live polymer (with active sites) and dead polymer (no active sites).

As a result of these reactions, at any given time there will be a distribution of polymer chains from short chains that have quickly terminated to long chains that have propagated. This distribution curve of polymer chains can be characterized by moments describing the distribution curve.

Once the process model developed, the control model structure has to be defined. The variables in the model that will be accessed have to be classified, which is the process of categorizing each variable in the problem according to its properties. The controlled state variables (or CVs) are controlled over the control horizon using reference trajectories. These reference trajectories relate the rate of response of the controlled variables to the current error. Violation variables are used to allow deviations from the desired trajectory. These violation variables are scaled internally to be relative violations. The user sets the value of τ (Tau) which may be interpreted as the time constant for the response of a controlled variable moving from its current position to its setpoint high or low limit.

NLC includes an objective function that determines to a large degree the manipulated variable moves and is described below:

$$\Psi = \mu_1 F_1(e, w) + \mu_2 F_2(y, u, v, c) + \mu_3 F_3(u, c) \tag{3}$$

$F_1(e, w)$ is the controller error objective function term and is the most heavily weighted term in the composite objective function. It minimizes the integrated average weighted error over the time horizon. $F_2(y, u, v, c)$ is the economic objective function term and minimizes this function according to cost coefficients applied to both controlled and manipulated variables. $F_3(\Delta u, c^{\Delta u})$ is the minimum movement objective function term and is designed to minimize the movement of manipulated variables when the controller error and economic objective function terms have been minimized. The term ensures a unique solution to the control problem and prevents large manipulated variable moves being made to achieve small economic improvements.

3   Conclusion

It is already proven in practice that implementation of NLC technique as APC system for polymers production plants improves polymer profitability in the following areas:

- Improve plant throughput by 2 to 8% during periods of high polymer demand by pushing the unit up against multiple plant constraints.
- Minimize the amount of off-spec material produced during grade transitions. NLC can reduce grade transition times by as much as 30% which means as much as 2% more on-spec polymer can be produced without increasing feed-rate.
- Minimize the amount of wide-spec material produced during normal operations by minimizing process upsets that can cause reactor excursions and reducing product quality variations by as much as 50%.
- Minimize the amount of off-spec material produced during unplanned shutdowns.

Moreover, NLC increases operation stability, which reduces process upsets while improving safety.

There is also further investigation activity going on related to the integration of NLC models with simulation "digital twin" models [8] of polymers plants and incorporation this structure to cloud environment with getting more enterprise operation profits in real-time mode [9].

References

1. Flores-Tlacuahuac, A., Saldívar-Guerra, E., Guerrero-Santos, R.: Dynamic modelling, nonlinear parameter fitting and sensitivity analysis of a living free-radical polymerization reactor. Comput. Aided Chem. Eng. **16**, 21–39 (2003)
2. Howes, S., LePore, J., Mohler, I., Bolf, N.: Implementing Advanced Process Control for Refineries and Chemical Plants. https://www.picontrolsolutions.com/wp-content/uploads/2015/10/Implementing-Advanced-Process-Control-for-Refineries-and-Chemical-Plants-DOC-Version.pdf
3. Lee, J.S., Na, S.S., Lee, D.E.: A step-by-step approach toward advanced process control system in petrochemical industry. In: Proceedings of the 17th World Congress the International Federation of Automatic Control, Seoul, Korea, pp. 10600–10601 (2008)
4. Garcia, M.R.S., Pitta, R.N., Fischer, G.G., Neto, E.R.N.: Optimizing Diesel Production Using Advanced Process Control and Dynamic Simulation. https://www.academia.edu/14092181/Optimizing_Diesel_Production_Using_Advanced_Process_Control_and_Dynamic_Simulation
5. Hiroya, S., Morimasa, O., Satoshi, O., Kouji, A., Masahiro, O., Wang, Y.: Industrial application of a nonlinear model predictive control to polymerization reactors. Control Eng. Pract. **9**, 819–828 (2001)
6. Rose, T.P., Devadhas, G.G., Rex, S.R.: Invention of a suitable controller for a non-linear chemical process. In: 2014 International Conference on Control, Instrumentation, Communication and Computational Technologies (ICCICCT), pp. 1462–1467 (2014)
7. Ibrehem, A.S., Hussain, M.A., Ghasem, N.M.: Decentralized advanced model predictive controller of fluidized-bed for polymerization process. Iran. J. Chem. Chem. Eng. **31**(4), 91–117 (2012)
8. Yusupbekov, N.R., Abdurasulov, F.R., Adilov, F.T., Ivanyan, A.I.: Application of cloud technologies for optimization of complex processes of industrial enterprises. In: Aliev, R.A., et al. (eds.) 13th International Conference on Theory and Application of Fuzzy Systems and Soft Computing ICAFS-2018. AISC, vol. 896, pp. 1–7. Springer, Cham (2019)
9. Yusupbekov, N.R., Jurayev, T.T., Abdurasulov, F.R., Sattarov, Sh.B., Adilov, F.T., Ivanyan, A.I.: Investigation of further improvement of integrated intelligent control system for Ustyurt Gas-Chemical Complex. In: Proceedings on International Scientific-Technical Conference on Actual Problems in Processes and Productions Automation, Karshi, Uzbekistan, pp. 43–47 (2017)

Soft Models of Law Enforcement in the Realities of Russian Society and Their Use in Intellectual Decision Support Systems

A. Sergey Grigoryan⬤, A. Pasikova Tatyana⬤,
and A. Levitskaya Elena(✉)⬤

Sothern University (IMBL), M. Nagibina pr., 33a/47, Rostov-on-Don, Russia
tatyana-plotko@yandex.ru, civilistiubip@yandex.ru

Abstract. Based on the categories of uncertainty and certainty in law, their nature is studied and "soft" models are formed. Special attention is paid to the role of abstraction and formalization in law education. Uncertainty is considered from two sides of the properties of law: positive and negative. The basic concepts are specified by examples from the Russian practice of law enforcement. The possibilities of using artificial intelligence to improve the efficiency and validity of legal decisions are shown.

Keywords: Legal regulation · Certainty in the law · Uncertainty in the law · Soft models legal processes · The abstractness and formalism of the law · The evaluation of the concept

1 Introduction

In the study of uncertainty in law, the concepts of "uncertainty" and "abstraction" are usually distinguished. "Uncertainty is a concept more capacious, characteristic of any matter, knowledge, including scientific, which is impossible without such a technique as abstraction. As you know, the concepts reflect the General certainty of any class of objects. This is achieved by abstraction as a specific form of cognitive process. Legal norms are the result of human cognitive activity. Their content is also characterized by generalizations and abstractness of content as a condition of legal regulation. At the same time we have to deal with the phenomenon of uncertainty in the form of generalizations and abstractions" [1, 2], which in the theory of artificial intelligence are divided into various "NON-factors" (fuzziness, inaccuracy, incompleteness, underdeterminacy, incorrectness, etc.) [3, 4], with the help of which legal concepts are formulated.

Thus, uncertainty (as well as certainty) is inherent in any system, including legal. "These qualities are interrelated to the opposite extreme, which, of course, peculiar to the law. Legal science the focus has traditionally given such grounds of law, as a certainty" [5].

In the study of uncertainty (caused by a set of NON-factors [3]) in the law should not be confused with the concept of "abstraction". Obviously, "uncertainty is a concept more capacious, characteristic of any matter, knowledge, including scientific, which is

R. A. Aliev et al. (Eds.): ICSCCW 2019, AISC 1095, pp. 352–358, 2020.
https://doi.org/10.1007/978-3-030-35249-3_45

impossible without such a technique as abstraction. As you know, the concepts reflect the General certainty of any class of objects. This is achieved by abstraction as a specific form of cognitive process. Legal norms are the result of human cognitive activity. Their content is also characterized by generalizations and abstractness of content as a condition of legal regulation" [2].

2   Materials and Method First Section

Consider some model M defined by the set of X variables and the set of R relations on these variables [4]. Each of the variables $x \in X$ is associated with a range of values, which is a subdomain of the universe U.

Let the model $M = (X, R)$ be a description of some fragment of reality, within the legal field, in which the variables from X represent specific values that are characteristics of real objects. "The set of vectors of variables of an arbitrary subset $X' \subseteq X$ will be called a generalized vector (denotate) of this subset, and the denotate of the whole set X—denotate of the model M. Thus, in the model M the real vector quantity-denotate, i.e. constant, is described by the variable" [3].

Let us consider further an arbitrary Legal System (LS), which displays information about the fragment of reality Q (in the process of LS fragment Q does not change in the sense that all its actual properties – legal acts - remain constant).

In General, the LS has incomplete information about Q. moreover, this incompleteness may be the result of a combination of several reasons (NON-factors), for example [6, 7]:

- lack of information about the existence of some quantities in Q included in the model;
- lack of certainty about the values of some denotates to select the type of data representing their variables;
- the absence or approximation (limited accuracy) of information on the value of a denotate within the range of values defined by the data type;
- lack of information about the presence of part of the relationship;
- approximate data on the nature of a relationship R.

In turn, sanctions on the object of exposure to the LS may also include a significant share of uncertainty. The degree of certainty of such sanctions can also vary, and depending on this, they can be divided into absolutely-defined, relatively-defined and alternative. It should be noted.

3   Uncertainty in Law and Principles of Law First Section

The greatest level of uncertainty in the law is seen in the analysis of the principles of law as its main ideas (for example, the principles of criminal law enshrined in Art. 3, 4, 6, 7 of the criminal code). The principles of law in the most General form are designed to reflect the laws of public life, transforming them into the basis of the content of law. However, without typification and generalization of specific life circumstances, their transfer to the level of General and uncertain regulations it is impossible to do. This circumstance gives the principles of law the quality of "centers" of legal regulation. Analysis of the current Russian legislation shows that it has a sufficient number of regulatory legal acts, characterized by a high degree of generalization of the regulatory material. Thus, the Constitution of the Russian Federation has a high degree of normative generalization, and the level of generalization in the document itself is different, Federal constitutional laws regulating the most important social relations, for example, key institutions of democracy, federalism, state structure, etc.

Of particular interest to researchers are the so-called framework laws. This method of legal regulation is often used in the joint jurisdiction of the Federation and its subjects. According to the Constitution of the Russian Federation in the field of joint jurisdiction Federal laws and laws of subjects of the Russian Federation are adopted. Federal laws on the subjects of joint jurisdiction and the laws of the subjects of the Russian Federation adopted in their development are of great positive importance for the development of Russian legislation and the improvement of Federal relations.

On the one hand, by defining the General basis of legal regulation, the Federal authorities ensure the unity and integrity of the Russian legal system, the uniform functioning of legal principles in the field of joint jurisdiction. On the other hand, this method of Federal legislative regulation provides for the possibility of the subjects of the Russian Federation to adopt normative legal acts taking into account the specifics of a particular region, which do not contradict the basic provisions of the Federal law. Uncertainty of framework laws provides flexibility and balance of legal regulation at the level of the Federation and its subjects. Such legal regulation is a form of transition from legal uncertainty to legal certainty. The practice of application of framework laws is widely used in regional legislation: as in Federal laws, key legal concepts are often formulated, the competence and powers of regional state bodies, the rights and obligations of participants in legal relations, etc. are determined. Framework laws often act as a kind of program for the development and adoption of thematic laws and other normative legal acts at the level of subjects of the Russian Federation. Uncertainty in law and valuation concepts. The uncertainty of Russian law and the level of regulatory legal generalizations are directly expressed in the rules of law, containing evaluative concepts and terms.

The standard-setter uses them consciously in order to fully and consistently cover certain social relations and take into account the dynamics of their development by legal regulations, their range.

Evaluation constructions give the subjects of law enforcement a certain freedom in the interpretation of the legal norm by means of the possibility of filling the evaluation term with its own content depending on the actual situation. Interpretation, for example, judicial, in the course of enforcement of evaluative terms in most cases is the process of their replacement by more formalized concepts that fix the legal meaning of phenomena and objects. At the heart of this process is the same pattern—the transition from legal uncertainty to legal certainty, which is an important feature of law enforcement. The value of evaluation terminology in judicial law enforcement is great, when in the process of interpretation the function of binding legal formalism to specific life situations is carried out. However, the role of evaluative concepts in judicial practice, especially their application and development—little studied problems.

The most important role in the study of this issue is the explanation of the content of normative evaluation concepts in such acts of interpretation of the Supreme judicial bodies as the decisions of the Plenums of the Supreme Court and the Supreme Arbitration Court of the Russian Federation. As a result of the disclosure of the content of evaluation concepts, specific factual circumstances are determined. It follows from this another, no less important conclusion that the evaluative concepts of the so-called cross-cutting type (used in various branches of law) should be interpreted in the law itself with the help of terminology, to some extent clarifying the content and scope of the assessment in legal regulation. Judicial practice in recent years has been actively explaining valuation concepts.

4 Digitalization in Jurisprudence

One of the rather traditional problems of legal proceedings both in Russia and in a number of other States is the workload of courts and judges, which in turn entails an increase in the time for consideration and resolution of cases, and as a rule, does not leave the judges the opportunity to allocate separate time for the study of particularly complex cases [7].

At present, digitalization is widely introduced in the world space, the fourth industrial revolution is widely spoken about, since the level of development and diversity of artificial intelligence systems (AIS) already allows to introduce cyber-systems into production, logistics and other spheres [8, 9].

If we talk about the law, the "Watson" from the global company IBM in 2015 was presented to users [10]. It is a super computer that is capable of self-learning and self-development. Its main function, emphasized by the name – help, it summarizes and analyzes a huge database of legal acts, court decisions and other legal documentation in a moment to answer the question, prepare a statement of claim, draft contract or conclusion.

The digitalization of the Russian legal space has already begun and is developing at a fairly rapid pace. The introduction of the state automated system "Justice" has already made it possible to significantly speed up the document flow, familiarize the participants with the movement of the case, simplify the calculation of the state fee, etc.

Already now it is possible to automate the decision-making process in the simplest cases, determined by a soft algorithm. So on divorce proceedings, in the absence of children and disputes over property, it turns branching algorithm that does not require the intervention of a judge, on the principle of building "Yes", "no", "insufficient data".

This is the simplest category of cases, which, however, currently requires the attention of the judge.

Decisions in most cases of writ proceedings can be made on the basis of a linear algorithm.

So, now, according to Art. 122 of the Civil procedure code of the Russian Federation the court order is taken out if:

– the claim is based on a notarized transaction;
– the claim is based on a transaction made in simple written form;

- -the claim is based on the notary's protest of the bill in non-payment, non-acceptance and undated acceptance;
- the declared requirement about collecting of the alimony for minor children not related to the establishment of paternity, contest of paternity (motherhood) or necessity of attraction of other interested persons;
- the requirement about collecting the salary added, but not paid to the worker, the sums of payment of leave, payments at dismissal and (or) other sums added to the worker is declared;
- declared by territorial authority of Federal Executive authority on ensuring the established order of activity of courts and execution of judicial acts and acts of other bodies the requirement about collecting the expenses made in connection with search of the Respondent, or the debtor, or the child;
- the requirement about collecting the added, but not paid monetary compensation for violation by the employer of the established term respectively of salary payment, vacation payment, payments at dismissal and (or) other payments due to the worker is declared;
- the requirement about collecting debt on payment of premises and utilities, and also services of telecommunication is declared;
- the declared requirement about collecting of obligatory payments and contributions with members of the homeowners or building society.

Given the fact that a court order is issued by a judge only on the basis of examination of the document, in the absence of calling and hearing the parties, witnesses, investigation of material evidence, it seems that the introduction of AI systems to resolve such cases is quite possible.

Further automation and digitalization of accounting and data recording systems for all state and other resources, further implementation and dissemination of electronic document management should also contribute to this.

So according to the requirements of the notarized transaction, the claimant may apply to the court, the system automatically generates and sends a request to the unified database of the Russian Federation the notarial action, where specific notary listed as attesting to the transaction from its database in a certain pattern sends a confirmation document in the attachment and denial of the existence of such a notarial action.

On the basis of the answer received from the notary, the AI forms either a refusal to issue a court order due to lack of grounds, or goes to the next stage of access verification – the presence of payment of a fee, after passing this stage, analyzes the submitted document and makes a court order.

The same algorithm is possible with the requirements based on the notary's protest of the bill in non-payment, non-acceptance and undated acceptance.

When applying for alimony claims for minor children, not complicated by any additional requirements, if the applicant is based only on the official income of the parent from whom alimony is collected, such information can also be provided automatically, in a certain template, by the Federal tax service.

The same applies to claims for the recovery of accrued, but not paid to the employee wages, amounts of vacation pay, payments upon dismissal and (or) other amounts accrued to the employee, as well as claims for the recovery of accrued, but not paid monetary compensation for violation by the employer of the established period, respectively, payment of wages, vacation pay, payments upon dismissal and (or) other payments due to the employee.

Requirements of bailiffs for recovery of expenses incurred in connection with the search for the defendant, or the debtor, or the child can also be confirmed by template documents recorded in their database.

The sums of debts on payment of premises and utilities, and also services of telecommunication, obligatory payments and contributions from members of partnership of owners of housing or construction cooperative are also fixed in special programs and in the established template can be sent to court.

Thus, by automating the routine part of justice, it is possible to achieve a number of advantages – the release of time and effort of judges to deal with more complex cases where the intellectual, moral, professional potential of an experienced person – judge is really needed, the acceleration of the consideration of cases where human intervention is not required or can be minimal.

The introduction of THESE in the justice sector is consistent with the plot of the implementation of its basic principles – the rule of law, equality, accessibility.

The legislator recognizes the importance of this area, for example, the Order of the Government of the Russian Federation dated 23.03.2018 № 482-R "On approval of the action plan ("road map") to improve legislation and eliminate administrative barriers in order to ensure the implementation of the National technological initiative in the direction of "TechNet" (advanced production technologies), is fixed as one of the points of the development of a long-term plan of standardization in the field of advanced production technologies, including advanced information technologies and technologies of cyberphysical systems, for 2018–2025 years, including the long-term plan of Rosstandart on standardization in the field of advanced production technologies, the creation of a regulatory framework for technologies underlying the creation and application of advanced production technologies - cyberphysical systems (industrial Internet of things, big data, artificial intelligence, etc.).

Further development should be the study of the use of artificial intelligence to make decisions on a wider range of cases under the control of a human judge. In the process of training, the system will determine the weight of each factor, aligning the existing judicial practice. The main issue of the study should be a set of factors required for training and decision-making system, and the main advantage – no need for mathematical formalization of existing legislation.

5   Conclusion

Thus, legal uncertainty, firstly, a sign of law, manifested in the incompleteness of the form and content of legal phenomena; secondly, the property of law, which has both positive and negative value, and, thirdly, the pattern of transition from the quality of uncertainty to the quality of certainty inherent in law as well as other social phenomena. Uncertainty as an attribute of law is a complex phenomenon that requires further research.

References

1. Bogdanov, E.V.: Categories "certainty" and "uncertainty" as elements of contractual regulation of public relations. Legis. Econ. **4**, 21–28 (2012)
2. Diev, V.S.: Uncertainty as an attribute and decision-making factor. Vestnik NSU Ser.: Philos. **8**, 3–8 (2010)
3. Narinyani, A.S.: Nudepreteens in systems of representation and processing of knowledge. Proc. USSR Acad. Sci. Tech. Cybern. **5**, 3–28 (1986)
4. Nariniani A.S.: NON-FACTORS: Inaccuracy and Underdeterminacy - distinction and interrelation. Russ. Acad. Sci., Teor. Syst. UPR **5**, 44–56 (2000)
5. Vlasenko, N.: Uncertainty in law: nature and forms of expression. J. Russ. Law **2**, 3–43 (2013)
6. Rybina, G.V., Borozdin, D.S., Rudakov, A.A.: Some approaches to modeling of non-knowledge factors in training integrated expert systems. Artif. Intell. Decis.-Making **1**, 22–46 (2008)
7. Khramov, V.V., Gvozdev, D.S.: Intelligent information systems: data mining. Rostov State University of Railway Engineering, Rostov-on-Don (2016)
8. Zade L.: The Concept of a linguistic variable and its application to making approximate decisions. Mir, Moskva (1976)
9. Khramov, V.V.: Basics of computer application and programming. Moskva (1995)
10. Watson. https://ru.wikipedia.org/wiki/IBM_Watson. Accessed 07 July 2019

Multiattribute Evaluation of Weapon Systems Under Z-Information

K. I. Jabbarova$^{(\boxtimes)}$ (ID)

Department of Computer-Aided Control Systems,
Azerbaijan State Oil Academy,
20 Azadlig Ave., AZ1010 Baku, Azerbaijan
konul.jabbarova@mail.ru

Abstract. Weapon systems are complex technical systems characterized by imperfect relevant information. Zadeh introduced the Z-number concept to formalize imprecision and partial reliability of information. In this paper, we consider a hierarchical decision making based on an evaluation of weapon systems under Z-number-based information. Z-number-valued weighted average aggregation operator is used for solving of this problem.

Keywords: Weapon system · Discrete Z-number · Partial reliability

1 Introduction

A weapon system is a complex system [1]. In [1] they proposed a general approach for weapon systems evaluation under fuzzy information. Weapon system comparison relies on ranking of fuzzy numbers.

In [2] approximate reasoning-based model is applied to evaluate the weapons systems. The model is represented by fuzzy rules. Finally, the obtained results are compared with other fuzzy evaluation approaches.

In [3] a new method for evaluating weapon systems under fuzzy information is presented.

In this paper we consider a hierarchical decision problem of weapon systems ranking under Z-valued information. The paper is structured as follows. Section 2 introduces an outline for arithmetic of Z-numbers. In Sect. 3, decision making problem of weapon system comparison under Z-valued information is formulated. Soluton of the problem is described in Sect. 4. Section 5 concludes.

2 Preliminaries

Definition 1. A discrete Z-number [1–7]: A discrete Z-number is an ordered pair $Z = (A, B)$ where A is a discrete fuzzy number playing a role of a fuzzy constraint on values of a random variable X: X *is* A. B is a discrete fuzzy number with a membership function $\mu_B : \{b_1, \ldots, b_n\} \rightarrow [0,1], \{b_1, \ldots, b_n\} \subset [0,1]$, playing a role of a fuzzy constraint on the probability measure of A: $P(A) = \sum_{i=1}^{n} \mu_A(x_i) p(x_i)$ *is* B.

© Springer Nature Switzerland AG 2020
R. A. Aliev et al. (Eds.): ICSCCW 2019, AISC 1095, pp. 359–365, 2020.
https://doi.org/10.1007/978-3-030-35249-3_46

Operations over Discrete Z-numbers: Let $Z_1 = (A_1, B_1)$ and $Z_2 = (A_2, B_2)$ be discrete Z-numbers describing information about values of X_1 and X_2. Consider computation of $Z_{12} = Z_1 * Z_2$, $* \in \{+, -, \cdot, /\}$. The first stage is computation of $A_{12} = A_1 * A_2$ [4].

The second stage involves construction of B_{12}. We realize that in Z-numbers Z_1 and Z_2, the 'true' probability distributions p_1 and p_2 are not exactly known. In contrast, fuzzy restrictions represented in terms of the membership functions are available

$$\mu_{p_1}(p_1) = \mu_{B_1}\left(\sum_{k-1}^{n_1} \mu_{A_1}(x_{1k})p_1(x_{1k})\right), \mu_{p_2}(p_2) = \mu_{B_2}\left(\sum_{k=1}^{n_2} \mu_{A_2}(x_{2k})p_2(x_{2k})\right).$$

Probability distributions $p_{jl}(x_{jk}), k = 1, \ldots, n$ induce probabilistic uncertainty over $X_{12} = X_1 * X_2$. Given any possible pair p_{1l}, p_{2l}, the convolution $p_{12s} = p_{1l} \circ p_{2l}$ is computed as

$$p_{12s}(x) = \sum_x p_{1l}(x_1)p_{2l}(x_2), \forall x \in X_{12}; x_1 \in X_1, x_2 \in X_2.$$

Given p_{12s}, the value of probability measure of A_{12} is computed:

$$P(A_{12}) = \sum_{k=1}^n \mu_{A_{12}}(x_{12k})p_{12s}(x_{12k}).$$

However, p_{1l} and p_{2l} are described by fuzzy restrictions which induce fuzzy set of convolutions:

$$\mu_{p_{12}}(p_{12}) = \max_{\{p_1, p_2 : p_{12} = p_1 \circ p_2\}} \min\{\mu_{p_1}(p_1), \mu_{p_2}(p_2)\}$$

Fuzziness of information on p_{12s} induces fuzziness of $P(A_{12})$ as a discrete fuzzy number B_{12}. The membership function $\mu_{B_{12}}$ is defined as

$$\mu_{B_{12}}(b_{12s}) = \sup(\mu_{p_{12s}}(p_{12s})) \tag{2}$$

subject to

$$b_{12s} = \sum_k p_{12s}(x_k)\mu_{A_{12}}(x_k) \tag{3}$$

As a result, $Z_{12} = Z_1 * Z_2$ is obtained as $Z_{12} = (A_{12}, B_{12})$.

Ranking of Discrete Z-numbers. According to R. Aliev's approach [4], Z-numbers are ordered pairs, for ranking of which there can be no unique approach. For purpose of comparison, the authors suggest to consider a Z-number as a pair of values of two attributes – "one attribute measures value of a variable, the other one measures the associated reliability" [4]. Then it will be adequate to compare Z-numbers as multiattribute alternatives. Basic principle of comparison of multi-attribute multi-criteria alternatives in this case is the Fuzzy Pareto optimality principle.

3   Statement of Problem

Selection among tactical missile systems of three companies is considered. These companies are evaluated by using five criteria $c_j, j = 1, \ldots, 5$ c_1 - tactics, c_2 - port services, c_3 - maintenance, c_4 - economy, c_5 - advancement (Table 1) [4]. Decision relevant information is expressed by Z-numbers.

Table 1. Evaluation criteria

Item Criteria & Sub-criteria	Weighting	Company A	Company B	Company C
C1: Tacits	*(VH, VL)*			
1. Effective range (km)	(HA, VL)	(VG, EL)	(P, VL)	(A, VL)
2. Flight height (m)	(VL, VL)	(L, VL)	(H, EL)	(M, VL)
3. Flight velocity (M, No)	(VH, VL)	(P,VL)	(VG, VL)	(A, EL)
4. Reliability (%)	(VH, VL)	(G, VL)	(VG, VL)	(P, VL)
5. Firing accuracy (%)	(VH, VL)	(G,VL)	(VG, EL)	(P, VL)
6. Destruction rate (%)	(HA, VL)	(P, VL)	(VG, EL)	(A, VL)
7. Kill radius	(AA, VL)	(A, VL)	(P, L)	(VG, VL)
C2: Technology	*(LA, VL)*			
8. Missile scale (cm) (1xd-span)	(BA, VL)	(L, VL)	(H, EL)	(M, VL)
9. Reaction time (min)	(VH, VL)	(H, VL)	(L, L)	(M, VL)
10. Fire rate (round/min)	(VH, VL)	(P, VL)	(P, L)	(VG, EL)
11. Anti-jam (%)	(H, VL)	(P, VL)	(VG, VL)	(A, VL)
12. Combat capability	(VH, VL)	(VG, EL)	(G, EL)	(G, EL)
C3: Maintenance	*(VL, VL)*			
13. Operation condition requirement	(A,VL)	((1-H), VL)	((1-L), VL)	((1-L), VL)
14. Safety	(AA, VL)	(VG, EL)	(G, EL)	(G, EL)
15. Defilade	(L, VL)	(G, VL)	(VG, VL)	(G, L)
16. Simplicity	(LA,VL)	(G, VL)	(G, VL)	(G, L)
17. Assembly	(LA, VL)	(G, VL)	(G, VL)	(P, L)
C4: Economy	*(A, VL)*			
18. System cost (10,000)	(H, VL)	(L, VL)	(H, EL)	(M, VL)
19. System life (years)	(H, VL)	(VG,VL)	(VG,EL)	(P, VL)
20. Material limitation	(A, VL)	((1-H), VL)	((1-L), VL)	((1-L), EL)
C5: Advancement	*(HA, VL)*			
21. Modularization	(A, VL)	(A, VL)	(G, VL)	(A, VL)
22. Mobility	(HA, VL)	(P,VL)	(VG, EL)	(G, VL)
23. Standardization	(LA, VL)	(G, VL)	(G, VL)	(VG, EL)

There are subcriteria which have both have a negative and positive connotations in the problem. Flight height, missile scale, reaction time, system cost, operation condition requirement, material limitation have a negative connotation.

Codebooks for fuzzy numbers A and B (components of Z-numbers) are shown in Tables 2, 3 and 4.

Table 2. The encoded linguistic terms for A components of Z-numbers (positive connotation)

Scale	Level	Linguistic value
1	Poor	$\{1/0.05, 0/0.35\}$
2	Average	$\{0/0.05, 1/0.35, 0/0.7\}$
3	Good	$\{0/0.35, 1/0.7, 0/1\}$
4	Very Good	$\{0/0.7, 1/1\}$

Table 3. The encoded linguistic terms for A components of Z-numbers (negative connotation)

Scale	Level	Linguistic value
1	Low	$\{1/0.1, 0/0.5\}$
2	Medium	$\{0/0.1, 1/0.5, 0/1\}$
3	High	$\{0/0.5, 1/1\}$

Table 4. The encoded linguistic terms for B components of Z-numbers

Scale	Level	Linguistic value
1	Unlikely	$\{1/0.05, 1/0.05, 0/0.25\}$
2	Not very likely	$\{0/0.05, 1/0.25, 0/0.5\}$
3	Likely	$\{0/0.25, 1/0.5, 0/0.75\}$
4	Very likely	$\{0/0.5, 1/0.75, 0/1\}$
5	Extremely likely	$\{0/0.75, 1/1, 1/1\}$

Table 5. The encoded linguistic terms for A parts of weights under Z-numbers

Scale	Level	Linguistic value
1	Very Low	$\{1/1, 0/2\}$
2	Low	$\{0/1, 1/2, 0/3\}$
3	Low Average	$\{0/2, 1/3, 0/4\}$
4	Below Average (BA)	$\{0/3, 1/4, 0/5\}$
5	Average	$\{0/4, 1/5, 0/6\}$
6	Above Average (AA)	$\{0/5, 1/6, 0/7\}$
7	High Average (HA)	$\{0/6, 1/7, 0/8\}$
8	High	$\{0/7, 1/8, 0/9\}$
9	Very High	$\{0/8, 1/9, 1/9\}$

Codebooks of the A component of Z-number valued importance weights of each criterion and sub-criterion are shown in Table 5.

4   Solution of the Problem

Let us solve the considered problem. At first we should compute overall evaluation of each company.

At the first stage, we obtain the Z-valued criteria evaluations by using weighted average-based aggregation of subcriteria values. The weighted average relies on operations over Z-numbers (Sect. 2).

$$
Z_{y_{ij}} = \frac{\sum\limits_{k}^{K_j} Z_{x_{ijk}} \cdot Z_{w_{jk}}}{\sum\limits_{k}^{K_j} Z_{w_{jk}}}, \tag{4}
$$

where $Z_{x_{ijk}}$ is a Z-number-valued evaluation w.r.t c_{jk} criterion, $Z_{w_{jk}}$ is a Z-number-valued importance weight of c_{jk}. The results for company B are as follows.

$$
Z_{y_{11}} = \frac{\sum\limits_{k=1}^{7} Z_{x_{11k}} \cdot Z_{w_{1k}}}{\sum\limits_{k=1}^{7} Z_{w_{1k}}} = \frac{(P, VL) \cdot (HA, VL) + (H, EL) \cdot (VL, VL) + (VG, VL) \cdot (VH, VL)}{(HA, VL) + (VL, VL) + (VH, VL)}
$$

$$
\frac{+ (VG, VL) \cdot (VH, VL) + (VG, VL) \cdot (VH, VL) + (VG, EL) \cdot (HA, VL) + (P, L) \cdot (AA, VL)}{(VH, VL) + (VH, VL) + (HA, VL) + (AA, VL)}
$$

$$
= ((0.42 \quad 0.74 \quad 1)(0.11 \quad 0.32 \quad 0.78));
$$

$$
Z_{y_{12}} = \frac{\sum\limits_{k=1}^{5} Z_{x_{12k}} \cdot Z_{w_{2k}}}{\sum\limits_{k=1}^{5} Z_{w_{2k}}} = \frac{(H, EL) \cdot (BA, VL) + (L, L) \cdot (VH, VL) + (P, VL) \cdot (VH, VL)}{(BA, VL) + (VH, VL) + (VH, VL)}
$$

$$
\frac{+ (VG, VL) \cdot (H, VL) + (G, EL) \cdot (VH, VL)}{+ (H, VL) + (VH, VL)} = ((0.25 \quad 0.5 \quad 0.9)(0.27 \quad 0.61 \quad 0.9));
$$

$$
Z_{y_{13}} = \frac{\sum\limits_{k=1}^{5} Z_{x_{13k}} \cdot Z_{w_{3k}}}{\sum\limits_{k=1}^{5} Z_{w_{2k}}} = \frac{((1 - L), VL) \cdot (A, VL) + (G, EL) \cdot (AA, VL) + (VG, VL) \cdot (L, VL)}{(A, VL) + (AA, VL) + (L, VL)}
$$

$$
+ \frac{(G, VL) \cdot (LA, VL) + (G, VL) \cdot (LA, VL)}{(LA, VL) + (LA, VL)} = ((0.16 \quad 0.55 \quad 1.5)(0.28 \quad 0.58 \quad 0.9));
$$

$$Z_{y_{14}} = \frac{\sum_{k=1}^{3} Z_{x_{14k}} \cdot Z_{w_{4k}}}{\sum_{k=1}^{3} Z_{w_{4k}}} = \frac{(H, EL) \cdot (H, VL) + (VG, EL) \cdot (H, VL)}{(H, VL) + (H, VL)}$$

$$+ \frac{((1 - L), EL) \cdot (A, VL)}{(A, VL)} = ((0.35 \quad 0.84 \quad 1.5)(0.3 \quad 0.59 \quad 0.99));$$

$$Z_{y_5} = \frac{\sum_{k=1}^{3} Z_{x_{15k}} \cdot Z_{w_{5k}}}{\sum_{k=1}^{3} Z_{w_{5k}}} = \frac{(A, VL) \cdot (A, VL) + (G, VL) \cdot (HA, VL) + (VG, EL) \cdot (LA, VL)}{(A, VL) + (HA, VL) + (LA, VL)}$$

$$= ((0.35 \quad 0.84 \quad 1.5)(0.3 \quad 0.59 \quad 0.99));$$

At the second stage, we compute the overall company evaluation Z_y as the weighted average-based aggregation of the criteria evaluations $Z_{y_j}, j = 1, \ldots, 5$ obtained at First stage:

$$Z_y = \frac{\sum_{j=1}^{5} Z_{y_j} \cdot Z_{w_j}}{\sum_{j=1}^{5} Z_{w_j}} = ((0.27 \quad 0.78 \quad 1.69)(0.3 \quad 0.59 \quad 0.91)).$$

Analogously, we computed the overall company evaluations Z_y for the other companies:

$$Z_y^{Company\ A} = ((0.14 \quad 0.41 \quad 1.3)(0.3 \quad 0.59 \quad 0.88));$$

$$Z_y^{Company\ C} = ((0.1 \quad 0.48 \quad 1.56)(0.3 \quad 0.59 \quad 0.88));$$

At the third stage we rank these evaluations by using the approach described in Sect. 2:

Company B vs. Company C:

$$do\left(Z_y^B\right) = 1, do\left(Z_y^C\right) = 0;$$

Company B vs. Company C:

$$do\left(Z_y^B\right) = 1, do\left(Z_y^A\right) = 0;$$

Thus, the best one is Company B.

5 Conclusion

A decision making problem on selection of weapon system under Z-number valued information is considered. The solution is based on weighted average of Z-numbers and comparison of Z-numbers. The obtained results provide ranks of alternatives.

References

1. Cheng, C.: Evaluating weapon systems using ranking fuzzy numbers. Fuzzy Sets Syst. **107**, 25–35 (1999)
2. Othman, M., Shaiful, A., Mohammad, I., Norshimah, R., Mohamad, F.B.Y.: Fuzzy evaluation of weapons system. Comput. Inf. Sci. **2**(3), 24–31 (2009)
3. Chen, S.: A new method for evaluating weapon systems using fuzzy set theory. IEEE Trans. Syst. Man Cybern.-Part A: Syst. Hum. **26**(4), 493–497 (1996)
4. Wu, D., Mendel, J.M.: Computing with words for hierarchical decision making applied to evaluating a weapon system. IEEE Trans. Fuzzy Syst. **18**(3), 441–460 (2010)
5. Aliev, R.A., Alizadeh, A.V., Huseynov, O.H.: The arithmetic of discrete Z-numbers. Inf. Sci. **290**, 134–155 (2015)
6. Aliev, R.A., Zeinalova, L.M.: Decision making under Z-information. In: Pedrycz, W., Guo, P. (eds.) Human-Centric Decision-Making Models for Social Sciences, pp. 233–252. Springer, Heidelberg (2014)
7. Zadeh, L.A.: A note on Z-numbers. Inf. Sci. **181**, 2923–2932 (2010)

Evaluation of Desirable Isothermal Reactions Rate in Uncertain Environment

Mahsati Jafarli$^{(\boxtimes)}$ (iD)

Azerbaijan State Oil and Industry University,
Azadlig 20, Nasimi, Baku, Azerbaijan
mehseti.babanli@gmail.com

Abstract. In this paper, we are focusing on the evaluation of an first order isothermal reaction rate in uncertain environment. The dependence between reaction rate and concentration of substance is analyzed. By using Fuzzy c-means clustering method, a set of rules is extracted from data. By determining and applying fuzzy clusters, a corresponding fuzzy inference system is designed and applied in the approximate reasoning of chemical reactions. By using fuzzy C-means approach fuzzy IF…THEN rule is constructed, which describes the relation between the reaction rate and decomposition of substance.

Keywords: Isothermal reaction · Fuzzy c-means · Fuzzy inference

1   Introduction

In scientific literature, chemical reactions are carried out at different reaction rates. Chemical reaction rates are analyzed in [1] by using fuzzy logic and approximate reasoning. By using this method, the approximate value of the reaction rate is determined, and, in this case, temperature and particle size are taken as the key factors defining reaction rates.

In [2], a noniterative method is described to solve fuzzy reaction equation. The advantage of this work is the ability to obtain different-quality solutions and to choose the ones that are better to describe the behavior of the real-word system.

Taking into consideration that differential equations are important tools to describe and understand real-world processes, it is important to analyze the behavior of a chemical reaction by using differential equations. First-order differential equations in chemistry are discussed in [3]. Authors of [4] have discussed the kinetics of the chemical reaction by using isothermal calorimetric data. A package on reaction kinetics is presented and described in [5].

A steady state approximation approach to solve rate equations is presented in [6]. In accordance with this approach, coupled differential equations are converted into a system of algebraic equations, one for each species in the reactions.

In our previous work we suggest the approach to investigate dynamics of first-order isothermal reactions and its stability, described by fuzzy differential equation.

Review of the related papers on modeling of chemical reactions was defined that analyzing of solutions of fuzzy differential equations problem by using fuzzy inference

© Springer Nature Switzerland AG 2020
R. A. Aliev et al. (Eds.): ICSCCW 2019, AISC 1095, pp. 366–372, 2020.
https://doi.org/10.1007/978-3-030-35249-3_47

method are very rarely. From this point of view in this paper we try to investigate evaluation of desirable isothermal reactions rate in uncertain environment The paper is structured as follows. In Sect. 2, we present some prerequisite material on fuzzy number, fuzzy stability etc. In Sect. 3 statement of the problem is given. In Sect. 4, we describe solution of the fuzzy differential equation described dynamics of investigated isothermal reaction. Section 5 offers some conclusions.

2   Preliminaries

Definition 1 Fuzzy number [7, 8]. A fuzzy number is a fuzzy set A on R which possesses the following properties: (a) A is a normal fuzzy set; (b) A is a convex fuzzy set; (c) α-cut of A, A^{α} is a closed interval for every $\alpha \in (0, 1]$; (d) the support of A, A^{+0} is bounded.

Definition 2 Fuzzy C-means algorithm [9]. This algorithm works by assigning membership to each data point corresponding to each cluster center on the basis of distance between the cluster center and the data point. More the data is near to the cluster center more is its membership towards the particular cluster center. Clearly, summation of membership of each data point should be equal to one. After each iteration membership and cluster centers are updated.

Main objective of fuzzy c-means algorithm is to minimize:

$$J(U, V) = \sum_{i=1}^{n} \sum_{j=1}^{c} (\mu_{ij})^m \|x_i - v_j\|^2$$

where, "$\|xi{-}vj\|$" is the Euclidean distance between ith data and jth cluster center.

Definition 3 Mamdani reasoning [10]. Let's assume that list of the rules contains the following If-then rules:

$$\text{IF } x_1 \text{ is } A_{11} \text{ and } \dots\dots x_n \text{ is } A_{1n} \text{ THEN y is } B_1$$
$$\text{IF } x_1 \text{ is } A_{21} \text{ and } \dots\dots x_n \text{ is } A_{2n} \text{ THEN y is } B_2$$
$$\dots\dots \dots\dots \dots\dots$$
$$\text{IF } x_1 \text{ is } A_{m1} \text{ and } \dots\dots x_n \text{ is } A_{mn} \text{ THEN y is } B_m$$

where $x_j, j = 1\dots n$ - input variables, y - is output variable. A_{ij} and B_i are linguistic values in accordance with input and output variables of the rules.

We can describe fuzzy inference process by using list of the rules as follow:

1. For each rule level of validity preconditions is defining as follows:

$$\alpha_i = \min_{j=1}^{n}[\max_{X_j}(A_j'(x_j) \wedge A_{ij}(x_j))]$$

where $A_j'(x_j)$ - are new independent values of input variables

2. For each rule calculate individual outputs:

$$B_i'(y) = \min(\alpha_i, B_i(y))$$

3. Calculate aggregative output: $B'(y) = \max\left(B_1'(y), B_2'(y), \ldots, B_m'(y)\right)$

3 Statement of the Problem

Data on relationship between rate coefficient and decomposition of substance for first order isothermal reaction is given in Table 1.

Table 1. Data on the coefficient rate and concentration of substance

Number	Decayed concentration	Coefficient rate
1	0,381217	8,33E−05
2	0,472708	0,000111
3	0,550671	0,000139
4	0,617107	0,000167
5	0,67372	0,000194
...	...	...
15	0,934125	0,000472
16	0,943865	0,0005
...	...	...
34	0,996849	0,001
35	0,997315	0,001028
36	0,997712	0,001056
37	0,99805	0,001083
38	0,998338	0,001111

The problem is to design fuzzy IF...THEN rules-based fuzzy model which describes relationship between rate coefficient and substance decomposition of this type reactions.

Then on base of derived model to find desirable rate coefficient of reaction by using fuzzy approximate reasoning.

4 Solution of the Problem

For design of IF-THEN fuzzy model fuzzy C-means(FCM) approach is used. By using data which described in Table 1 is defined centers of clusters. Dataset contains 38 records: input - amount of concentration, output - coefficient rate.

For simulation FCM based clustering initial data are: Cluster numbers = 5, Max iteration = 1000, exponent = 2, Min. Improvement = 0,000001.

Fragment of clusters and their membership functions is given in Tables 2 and 3.

Table 2. Center of clusters

Center of clusters	
0.6152	0.0002
0.7768	0.0003
0.9867	0.0008
0.9036	0.0004
0.4271	0.0001

Table 3. Fragments of values membership function

Cluster 1	Cluster 2	Cluster 3	Cluster 4	Cluster 5
0.0361	0.0126	0.0054	0.0072	0.9387
0.0897	0.0197	0.0069	0.0098	0.8739
0.6761	0.2183	0.0237	0.0438	0.0381
0.1839	0.6985	0.0299	0.0636	0.0241
0.0084	0.9771	0.0037	0.0093	0.0016
	...	...		...
0.0007	0.0021	0.9853	0.0116	0.0003
0.0008	0.0023	0.9841	0.0125	0.0003
0.0008	0.0024	0.9830	0.0134	0.0004

To describe extracted rules from data the codebook is used (Table 4).

Table 4. Codebook

Decayed concentration	Fuzzy number	Rate constant	Fuzzy number
Small	(0.3812, 0.6152, 0.998)	Very low	(0.00011, 0.0002, 0.0011)
Middle	(0.3812, 0.7768, 0.998)	Low	(0.00011, 0.0003, 0.0011)
Very large	(0.3812, 0.9867, 0.998)	High	(0.00011, 0.0008, 0.0011)
Large	(0.3812, 0.9036, 0.998)	Middle	(0.00011, 0.0004, 0.0011)
Very small	(0.3812, 0.427, 0.998)	Very very low	(0.00011, 0.0001, 0.0011)

Constructed fuzzy model is the following form:

If Decayed concentration is small Then coefficient of rate is very low;
If Decayed concentration is middle Then coefficient of rate is low;

If Decayed concentration is very large Then coefficient of rate is high;
If Decayed concentration is large Then coefficient of rate is middle;
If Decayed concentration is very small Then coefficient of rate is very very low.

Fuzzy inference is performed by using Mamdani Fuzzy inference system.

Test.1. IF amount of decayed Concentration is 0.6897 Then coefficient of rate is equal to 0.000528 (Fig. 1).

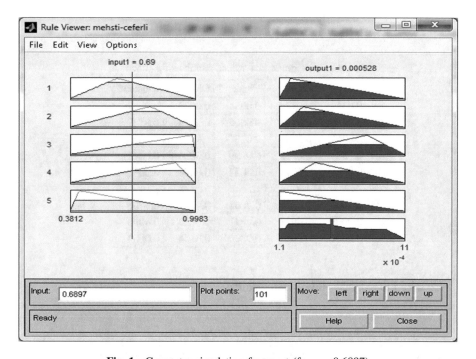

Fig. 1. Computer simulation fragment (for c = 0.6897)

Test.2. IF amount of decayed Concentration is 0.98 Then coefficient of rate is equality 0.000662 (Fig. 2).

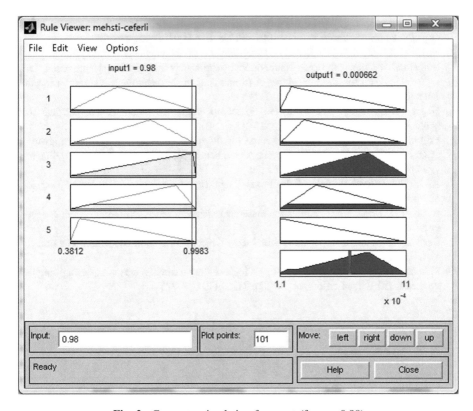

Fig. 2. Computer simulation fragment (for c = 0.98)

5   Conclusion

The article analyzes the behavior of dependence process for the rate of reaction and the concentration of species is first order isothermal reaction.

Designed fuzzy IF…THEN model gives ability to calculate relationship between given decayed concentration and rate coefficient of reaction. All calculation were made in Matlab environment.

The model presented in the article can be used in the analysis of different first order isothermal chemical processes.

References

1. Khaliq, A., Amais A.: Approximate reasoning in chemical reactions, 59 p. Master thesis, Blekinge Institute of Technology (2010). http://www.diva-portal.org/smash/get/diva2: 832139/FULLTEXT01.pdf
2. Can, E., Bayrak, M.A.: A new method for solution of fuzzy reaction equation. Commun. Math. Comput. Chem. **73**, 649–661 (2015). Claire Vallance. Reaction kinetics. 31 p. http:// vallance.chem.ox.ac.uk/pdfs/KineticsLectureNotes.pdf

3. Scholz, G., Scholz, F.: First-order differential equations in chemistry. Chem Text **1**, 1 (2015). https://link.springer.com/article/10.1007/s40828-014-0001-x
4. Nowicki, L., Siuta D., Godala, M.: Determination of the chemical reaction kinetics using isothermal reaction calorimetry supported by measurements of the gas production rate. A case study on the decomposition of formic acid in the heterogeneous Fenton reaction. Thermochimica Acta **653**, 62–70 (2017)
5. Nagya, A.L., Papp, D., Tóth, J.: ReactionKinetics—A mathematica package with applications. Chem. Eng. Sci. **83**(1), 12–23 (2012)
6. Erturk, V.: Differential transform method for solving a boundary value problem arising in chemical reactor theory. In: Conference Mathematical Modelling and Analysis (2017). inga.vgtu.lt/*art/konf/programme_detailed.php
7. Aliev, R.A.: Fundamentals of the Fuzzy Logic-Based Generalized Theory of Decisions, 322 p. Springer, Heidelberg (2013)
8. Aliev, R.: Uncertain Computation-Based Decision Theory, 521 p. World Scientific Publishing, Singapore (2018)
9. Bezdek, J.C.: Pattern Recognition with Fuzzy Objective Function Algorithm. Plenum, New York (1981)
10. Mamdani, E.H.: Application of fuzzy logic to approximate reasoning using linguistic synthesis. IEEE Trans. Comput. **26**(12), 1182–1191 (1977)

Vehicle Detection and Tracking Using Machine Learning Techniques

Kamil Dimililer[1,3,4](✉) ⓘ, Yoney Kirsal Ever[2,3,4] ⓘ,
and Sipan Masoud Mustafa[5] ⓘ

[1] Department of Electrical and Electronic Engineering, Faculty of Engineering,
Near East University, Nicosia, Mersin 10, North Cyprus, Turkey
kamil.dimililer@neu.edu.tr
[2] Department of Software Engineering, Faculty of Engineering,
Near East University, Nicosia, Mersin 10, North Cyprus, Turkey
yoneykirsal.ever@neu.edu.tr
[3] Research Center of Experimental Health Sciences, Near East University,
Nicosia, Mersin 10, North Cyprus, Turkey
[4] Applied Artificial Intelligence Research Centre, Near East University, Nicosia,
Mersin 10, North Cyprus, Turkey
[5] Duhok Polytechnic University, 61 Zakho Road, 1006 Mazi Qr, Duhok, Iraq
sipanmmustafa@gmail.com

Abstract. More than two decades machine learning techniques have been applied in multidisciplinary fields in order to find more accurate, efficient and effective solutions. This research tries to detect vehicles in images and videos. It deploys a dataset from Udacity in order to train the developed machine learning algorithms. Support Vector Machine (SVM) and Decision Tree (DT) algorithms have been developed for the detection and tracking tasks. Python programming language have been utilized as the development language for the creation and training of both models. These two algorithms have been developed, trained, tested, and compared to each other to specify the weaknesses and strengths of each of them, although to present and suggest the best model among these two. For the evaluation purpose multiple techniques are used in order to compare and identify the more accurate model. The primary goal and target of the paper is to develop a system in which the system should be able to detect and track the vehicles automatically whether they are static or moving in images and videos.

Keywords: Vehicle detection · Vehicle tracking · SVM · Decision tree · Image detection · Object detection and tracking

1 Introduction

Since the population and transport system increase day by day, the demand for managing them increase at the same time. The world is getting populated so fast. Therefore, the number of machines from any types including vehicles increased at the same time [1, 2]. That being said, new topics like traffic, accidents and many more issues are needed to be managed. It is hard to manage them with the old methods, new trends and technologies have been found and invented to handle each and every

© Springer Nature Switzerland AG 2020
R. A. Aliev et al. (Eds.): ICSCCW 2019, AISC 1095, pp. 373–381, 2020.
https://doi.org/10.1007/978-3-030-35249-3_48

milestone that human kind is trying achieve. One of these challenges is traffic in highways and cities. Many options like traffic light, sign, etc. deployed in order to deal with these phenomena [3]. It seems that these options are not enough or not so efficient alone. New technologies like object detection and tracking are invented in order to utilize automated camera surveillance to produce data that can give meanings for a decision-making process. These phenomena have been used for different kind of issues. The new trend Intelligent Transport System (ITS) has many elements which object detection and tracking are one of them. This system is used to detect vehicles, lanes, traffic sign, or vehicle make detection. The vehicle detection and classify ability gives us the possibility to improve the traffic flows and roads, prevent accidents, and registering traffic crimes and violations.

There are different techniques and methods for vehicle detection and classification. The variety of these techniques are in types of algorithms like Support Vector Machine (SVM) [3], Convolutional Neural Network (CNN) [4], Decision Tree, Recurrent Neural Network (RNN) etc. The field is constantly evolving since the industry is focused on this system or Computer visionary [6–9]. In this work we investigate two algorithms SVM and Decision Tree to identify how they can apply in the field and which one works better than the other. Machine learning is used to detect objects. There are many techniques doing this job but in order to identify the best model among the suggested models, this work is going to explore the topic on how to detect the objects while at the same time compare two suggested models and suggest the best one which yields highest accuracy and performance.

Object detection attracted the attention in research industry lately. Researchers are trying to explore the topic to reach to an accepted accuracy level. Machine learning is used to detect objects. There are many techniques doing this job but in order to identify the best model among the suggested models, this work is going to explore the topic on how to detect the objects while at the same time compare two suggested models and suggest the best one which yields highest accuracy and performance.

The aim of this paper is to take two algorithms, SVM and Decision Tree to detect vehicles in images and videos. It compares the two algorithms with the same dataset and pre-processing methods.

2   Related Works

In the past few decades researchers had a great interest in vehicle detection and tracking. The topic attracted the attention quit much. Different sensing modalities have been used for detecting the objects or specifically vehicles. These modalities are LIDAR, radar, and computer vision. The attraction caused by immense progress of image processing that various methods and techniques have been invented and proposed [7–15].

Vehicle detection also called computer vision object recognition, basically the scientific method that machines recognize and adapt human sight properties. The main duty of a vehicle detection system is to localize one or more vehicles in input images. There are two methods in vehicle detection system: sliding window method [15] and local features method [14]. In the local features-based method, the system usually finds the features of one object or the group of the objects in the very beginning. After that it tries to categories the founded features into different classes via classification models. This is usually the final step where the system decides category of the object it belongs to. It claims that the strongest point of this method is the geometry and known features of the targets before the detection. While it can be a weakness or limitation as well because it can only detect pre-learned objects. However, the second method is known as sliding window-based system. It works differently. It scans input image with a number of windows of different sizes. Afterwards, it analyzes if the target is in the window or not [16].

A method proposed by Wang et al. [13] for detecting objects using edge detection in UAV footage as demonstrated in Fig. 1. Researcher presents an example of how to use edge detection to recognize objects. In order to identify the straight lines on a vehicle an edge detection algorithm has been used. Before doing so, the algorithm removes noises from the background via and by the help of higher threshold in the process.

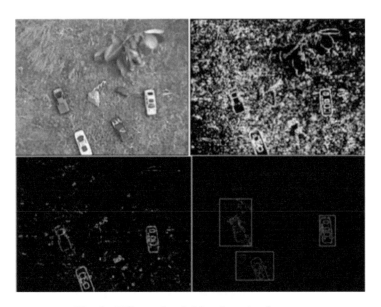

Fig. 1. Different thresholds using edge features.

Another method suggested by Sokalski et al. [18] uses color feature and edge detection to separate and distinguish artificial and natural objects. However, during color extraction of the original image, the nine features of the original image are used. Although these features are uses to specify the edges and changes to separate artificial and natural objects like illustrated in Fig. 2. It summarizes local descriptors into texture features and color features. Color feature usually works fast in order to detect objects but in scenarios or somehow generally the accuracy is challenged. But texture feature works better with objects in images due to having more information about object in textures.

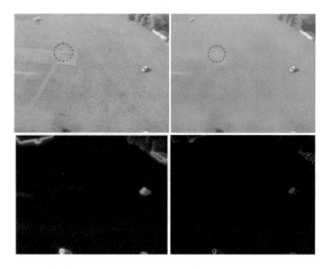

Fig. 2. Edge and color feature for artificial objects.

Susaki et al. [12] suggested a two-arrange way to deal with the programmed location of vehicles inside aeronautical symbolism. Their methodology depends on the utilization of different fell Haar classifiers [18] for vehicle arrangement and an auxiliary check organize that endeavors to dispose of non-vehicle hopefuls dependent on UAV elevation and vehicle measure limitations. To accomplish vehicle introduction invariance, they utilize four separate fell Haar classifiers. The four classifiers are then assessed utilizing a question picture at various sizes and positions utilizing a sliding window approach and numerous classifier location, and identification covers are settled utilizing a spatial combining strategy. This can be seen in Fig. 3 below. Berni et al. [5] later broadened this work with the utilization of extra warm symbolism for warm mark affirmation, which improved execution extensively.

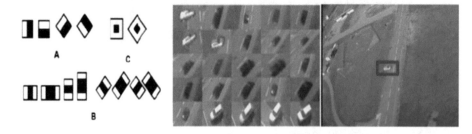

Fig. 3. Features, training samples and detection results.

Chen et al. [7] concentrated on programmed vehicle location in provincial conditions. Their approach comprises of a course location calculation, which is included two phases. In the first phase, a Harris corner indicator is utilized to distinguish highlights of enthusiasm for the pictures; the creators contend that "vehicles specifically have an expansive number of edges and corners looked at to characteristic articles". Next, a productive sliding window approach is utilized to decide districts with an element thickness higher than a foreordained edge. These creators' exploration analyzed the execution of two picture fix descriptors, an altered Histogram of Oriented Angles (HoG) highlight and Histogram of Gabor Coefficients highlights. The execution of three factual grouping strategies; K-Nearest Neighbors (k-NN), Random Forests (RF) and SVM are suggested. Starting with the first phase of calculations, a normal location rate of 85% is acquired and it is found that the top performing classifier is Random Forests by utilizing Histogram of Gabor Coefficients highlights; this was equipped for characterizing 98.9% of vehicles and 61.9% of foundation pictures effectively.

Sahli et al. [11] proposed a nearby component-based methodology for programmed vehicle identification in low-goals elevated symbolism. Their methodology was created with the point of being free from the imperatives identified with location strategies dependent on a vehicle's visual appearance. Their approach depends on the extraction of Scale-Invariant Feature Transform (SIFT) highlights from vehicle and foundation pictures. These highlights are utilized to prepare a SVM classifier to characterize a model that can be utilized to order SIFT highlights separated from the autos and foundation in a question picture. The creators' bunching strategy depends on an altered liking propagation (ALP) calculation that is bound by the spatial limitations identified with the geometry of vehicles at the given goals. They got a grouping exactness of 95.2% in aeronautical symbolism of a parking garage containing 105 vehicles, with no false-positive recognitions.

3   Methodology

In this paper some applications and tools for creating models and experiments are utilized and applied. For example, python is chosen as the programming language of the model development, for training the dataset with a lot of vehicle and non-vehicle images, to create ML models.

3.1   Tools Used and Dataset

Python is a general-purpose programming language that is used for different algorithms and applications. Especially, in the past few years' it became popular in in artificial intelligence and machine learning applications due to having many efficient and handy libraries. Apart from this other open source libraries such as Numpy, Matplotlib, and Jupyter Notebook are used for working faster with large amounts of data, creating different kind of graphs and figures for variety of aims is used to extract color features and create histograms, and a web-based application which makes us able to create and modify live codes, equations, plaintexts and visualizations respectively.

The experiments are carried out with the PC that is being used to train and test the models has 8 GB, Quad Core i7 processor with Intel HD 4 GB graphics card.

Udacity website provides great resources of images for training the classifiers. Vehicles and non-vehicles samples of the KITTI vision benchmark suite have been used for training as shown in Fig. 4. These sample images are combination of the GTI vehicle images, the KITTI vision benchmark suite, and extracted images from the project video.

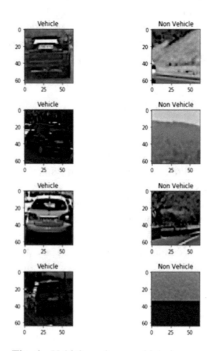

Fig. 4. Vehicle and non-vehicle images

3.2   Implementation

Since self-driving technologies have been increased recently, importance of vehicle detection and tracking are become crucial. In this project, a software pipeline is developed to detect vehicles in a video. In order to proceed with proposed algorithm, the following tasks are needed to be achieved.

- Histogram of Oriented Gradients (HOG) feature extraction on a labeled training set of images will be performed and a Linear SVM classifier will be trained.
- A sliding-window technique will be implemented and trained classifier to search for vehicles in images will be used.
- Pipeline will be run on a video stream and a heat map of recurring detections will be created for frame by frame in order to reject outliers and follow detected vehicles.
- A bounding box for vehicles detected will be estimated.

4   Results and Discussions

Many libraries have been utilized to perform various tasks like Numpy is used for importing and processing data, Matplotlib is used for visualization of the extracted color features. There are two algorithms, SVM and Decision Tree, are used to classify the images and a pipeline to track images in the videos. Data is divided into two parts training and testing. The system can detect the cars but cannot track them in a video. In order to do that, a pipeline needed to be defined. SVC and Decision Tree results comparison using evaluation matrices is given in the following figure (Fig. 5).

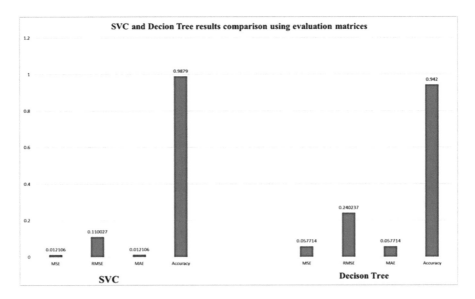

Fig. 5. SVC and Decision Tree comparison result

Three evaluation matrices with accuracy percentage level Since the evaluation matrices are the error levels, as much as the values for MSE, RMSE, and MAE are low the accuracy level is high and if they are high then the accuracy level is low. As you can see in the Fig. 5, all MSE (Mean Squared Error), RMSE (Root Mean Squared Error), and MAE (Mean Absolute Error) are lower in SVC at the same time accuracy level is 98% for SVC but for Decision Tree the Error levels are high and the accuracy level is 94%. This shows that SVC works better for vehicle detection and tracking tasks.

5   Conclusions and Future Works

This paper developed two classifier algorithms by using SVM and DC to detect and track vehicles in order to find efficient and effective one.

The models are trained with the same dataset and the evaluation results showed that SVM performs better than Decision Tree. The dataset can be improved, for testing these models using a better and larger dataset with a massive number of vehicle and non-vehicle images from different places, angels, cars, roads, cameras, distances etc. Furthermore, few other models can be added to the comparison list of models in order to make the comparison more reliable and more predictable.

References

1. Baha, N.: Real-time obstacle detection approach using stereoscopic images. Int. J. Inf. Eng. Electron. Bus. **6**(1), 42–48 (2014)
2. Bahrepour, M., Akbarzadeh-T, M.R., Yaghoobi, M., Naghibi-S, M.B.: An adaptive ordered fuzzy time series with application to FOREX. Expert Syst. Appl. **38**(1), 475–485 (2011)
3. Bambrick, N.: Support Vector Machines for dummies; A Simple Explanation. AYLIEN—Text Analysis API—Natural Language Processing, pp. 1–13 (2016). http://blog.aylien.com/support-vector-machines-for-dummies-a-simple/. Accessed 28 Apr 2018
4. Bush, I.J., Dimililer, K.: Static and dynamic pedestrian detection algorithm for visual based driver assistive system. J. ITM Web Conf. **9**, 03002 (2017). EDP Sciences
5. Berni, J.A., Zarco-Tejada, P.J., Suárez, L., Fereres, E.: Thermal and narrowband multispectral remote sensing for vegetation monitoring from an unmanned aerial vehicle. IEEE Trans. Geosci. Remote Sens. **47**(3), 722–738 (2009)
6. Khashman, A., Sekeroglu, B., Dimililer, K.: ICIS: a novel coin identification system. Lecture Notes in Control and Information Sciences, vol. 345, pp. 913–918. Springer, Heidelberg (2006)
7. Chen, X., Meng, Q.: Robust vehicle tracking and detection from UAVs. In: 7th International Conference of Soft Computing and Pattern Recognition (SoCPaR), pp. 241–246. AIRCC Publishing Corporation (2015)
8. Chen, X.: Automatic vehicle detection and tracking in aerial video, Doctor of Philosophy Thesis in Loughborough University (2015)
9. Kalghatgi, M.P., Ramannavar, M., Sidnal, N.S.: A neural network approach to personality prediction based on the big-five model. Int. J. Innov. Res. Adv. Eng. **2**(8), 56–63 (2015)
10. Kanistras, K., Martins, G., Rutherford, M.J., Valavanis, K.P.: Survey of unmanned aerial vehicles (UAVs) for traffic monitoring. In: Handbook of Unmanned Aerial Vehicles, pp. 2643–2666 (2015)
11. Sahli, S., Ouyang, Y., Sheng, Y., Lavigne, D.A.: Robust vehicle detection in low-resolution aerial imagery. In: Airborne Intelligence, Surveillance, Reconnaissance (ISR) Systems and Applications VII, vol. 7668, pp. 76680G. International Society for Optics and Photonics (2010)
12. Susaki, J.: Region-based automatic mapping of tsunami-damaged buildings using multi-temporal aerial images. Nat. Hazards **76**(1), 397–420 (2015)
13. Wang, X., Zhu, H., Zhang, D., Zhou, D., Wang, X.: Vision-based detection and tracking of a mobile ground target using a fixed-wing UAV. Int. J. Adv. Rob. Syst. **11**(9), 156 (2014)
14. Noh, S., Shim, D., Jeon, M.: Adaptive sliding-window strategy for vehicle detection in highway environments. IEEE Trans. Intell. Transpt. Syst. **17**, 323–335 (2015)
15. Ofor, E.: Machine learning techniques for immunotherapy dataset classification (Master dissertation, Near East University) (2018)
16. Yu, X., Shi, Z.: Vehicle detection in remote sensing imagery based on salient information and local shape feature. Optik - Int. J. Light Electron Opt. **126**(20), 2485–2490 (2015)

17. Su, A., Sun, X., Liu, H., Zhang, X., Yu, Q.: Online cascaded boosting with histogram of orient gradient features for car detection from unmanned aerial vehicle images. J. Appl. Remote Sens. **9**(1), 096063 (2015)

18. Sokalski, J., Breckon, T.P., Cowling, I.: Automatic salient object detection in UAV imagery. In: Proceedings of 25th International Unmanned Air Vehicle Systems, pp. 1–12. IEEE (2010)

Fuzzy Logic-Based Approach to Electronic Circuit Analysis

K. M. Babanli[(⊠)] and Rana Ortac Kabaoglu

Istanbul University, Cerrahpasa, 34320 Avcilar, Istanbul, Turkey
kenanbabanli@gmail.com, rana@istanbul.edu.tr

Abstract. Electronic circuit analysis is an important research field. One of the key application fields is communication. An important problem in the field of the communication is related to the use of chaotic signals under uncertainty. In this paper we consider a fuzzy logic-based approach for analysis of chaotic electronic circuit. Fuzzy differential equations are used to model chaotic dynamics of an electronic circuit under uncertainty.

Keywords: Fuzzy number · Fuzzy differential equation · Electronic circuit · Chaotic signals

1 Introduction

Chaos is a fundamental phenomenon and has important practical applications due to such properties as ergodicity, aperiodicity, and sensitive dependence on initial conditions. Chaos applications can be found in spacecraft control, radar imaging, signal processing and communication. The perspectives of applications are based on such criteria as security, broadband, low cost, and the use of small size electronic circuits. In contrast to applications of conventional methods, chaos applications allow to integrate source encoding, channel encoding, and noise robustness. At its initial stage, research devoted to applications of chaos was related to the feasibility, security, and weakness to attacks in an ideal channel. The physical constraint of the channel was conditioned by additional white noise. The recent increase of research is due to chaos application in the modern communication channels as optical, radio frequency and underwater acoustic channels.

Several problems of communication can be modeled by different types of differential equations. Application of chaos theory in secure communication was analyzed in [1, 2]. In order to account for uncertainty in communication channels, fuzzy differential equations (FDEs) can be used to describe dynamics of systems under uncertainty in the initial conditions and parameters. Chen and Lee reported a new chaotic system developed on the basis of the Euler equation for the motion of rigid body [3]. This system is referred to as Chen-Lee system. This system is described by ordinary nonlinear differential equations. An implementation of the system by using an electronic circuit is proposed.

R. A. Aliev et al. (Eds.): ICSCCW 2019, AISC 1095, pp. 382–389, 2020.
https://doi.org/10.1007/978-3-030-35249-3_49

In [4] authors analyze an FDE that describes dynamics of the RL-type electric circuit. A dynamic behavior of system, the control and anti-control of chaos are investigated.

Chen-Lee system control based on fuzzy logic and Lyapunov direct method is considered in [2].

In [5] a system of fuzzy linear equations is used to model an electrical circuit characterized by uncertainty. A linear programming-based method is used to solve the considered system of equations.

A fractional-order Chen-Lee system is implemented by using chain fractance and tree fractance circuits in [6]. The authors claim that this system can be applied for secure communication.

In [7], they study control of chaos and chaotification in the Chen–Lee system. The efficiency of time delay constants was analyzed numerically. The considered system exhibits such dynamic behaviors as fixed points and chaotic motion.

A fuzzy expert system for diagnosis of parametric faults of analog electronic circuits is proposed in [8, 9]. In contrast to a classical system, the developed system is also able to locate faults.

Stability of equilibrium points of Chen-Lee system is considered in [10]. Implementation of this system by several electronic circuits is proposed. The authors claim that the proposed implementation can be used for encryption purposes.

Circuit analysis is needed for telecommunication purposes[11]. In this realm, consideration of numerous factors characterized by uncertainty is challenge. In this paper we consider Chen-Lee system modeled by FDEs for electronic circuit analysis. The model presented in the article can be used in analysis of electronic circuits.

The paper is structured as follows. In Sect. 2 we present some prerequisite material used in the paper. Statement of the problem is formulated in Sect. 3. Section 4 is devoted to solution of the problem. Section 5 concludes.

2   Preliminaries

Definition 1 [9]. A fuzzy number is a fuzzy set A on R which possesses the following properties: (a) A is a normal fuzzy set; (b) A is a convex fuzzy set; (c) α-cut of A, A^{α} is a closed interval for every $\alpha \in (0, 1]$; (d) the support of A, A^{+0} is bounded.

Definition 2. Chen-Lee System [10]. Chen-Lee system is described by the following system of DEs:

$$\dot{y}_1 = -y_2 y_3 + a_1 y_1$$
$$\dot{y}_2 = y_1 y_3 + b_1 y_2$$
$$\dot{y}_3 = y_1 y_2 + c_1 y_3$$

where a_1, b_1 and c_1 are system parameters. The necessary conditions for this system to generate chaos are as follows: $a_1 > 0$, $b_1 < 0, c_1 < 0$ and $0 < a < -(b+c)$.

3   Statement of the Problem

Our aim is to analyze an electronic circuit for chaotic fuzzy Chen-Lee system. We consider the following FDEs as a model of the considered system:

$$
\begin{aligned}
\hat{\tilde{y}}_1 &= -\tilde{y}_2\tilde{y}_3 + a_1\tilde{y}_1 \\
\tilde{\dot{y}}_2 &= \tilde{y}_1\tilde{y}_3 + b_1\tilde{y}_2 \\
\dot{\tilde{y}}_3 &= \tilde{y}_1\tilde{y}_2 + c_1\tilde{y}_3
\end{aligned}
\tag{1}
$$

where a_1, b_1 and c_1 are crisp parameters, y_1, y_2, y_3 are the fuzzy-valued angular velocities about principal axes fixed at the center of mass. In electronic circuit representation of (1), y_1, y_2, y_3 represent voltage on circuit outputs.

The problem is to find parameters and initial conditions such that the solution of (1) will be represented as a fuzzy chaotic signal.

4   Solution of the Problem

Assume that following parameters and initial conditions are used for system (1):

$$d_1 = 40; \ e_1 = 30; f_1 = 20; \ yi10 = (y_{iL0}, yi0, yiU0) = (-20.0, \ 0.20, \ 20.0);$$

$$y_{U0} = [20.0, \ 15.0, \ 10.0]; \ y0 = [0.20, \ 0.15, \ 0.10]; \ y_{L0} = [-20.0, \ -15.0, \ -10.0];$$

$a_1 = 5.2; \ b_1 = -10.3; \ c_1 = -38.0.$, here y_{L0} and y_{U0} are lower and upper bound.

In this case, system (1) exhibits chaotic behavior. We determined that if $y_1 \in [-d_1, d_2], y_2 \in [-e_1, e_2]$ and $d_1 > 0, e_1 > 0$ then the fuzzy model of the considered Chen-Lee system can be represented as follows:

IF y_1 is positive and y_2 is positive THEN the system behavior is described by dynamics shown in Fig. 1.
IF y_1 is positive and y_2 is negative THEN the system behavior is described by dynamics shown in Fig. 2.
IF y_1 is negative and y_2 is positive THEN the system behavior is described by dynamics shown in Fig. 3.
IF y_1 is negative and y_2 is negative THEN the system behavior is described by dynamics shown in Fig. 4.

The results of computer simulation of the considered model by using Hukuhara derivative and generalized derivative are shown below (Figs. 1, 2, 3, 4, 5, 6 and 7). As one can see, the solutions are represented as chaotic signals. Projection of phase portraits for the considered fuzzy chaotic Chen-Lee system is described in Fig. 7.

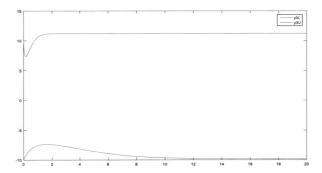

Fig. 1. Solution by using rule 1 (fragment)

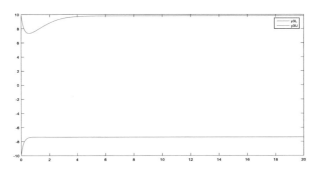

Fig. 2. Solution by using rule 2 (fragment)

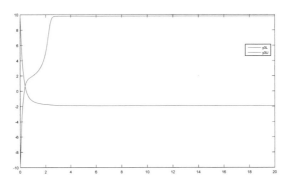

Fig. 3. Solution by using rule 3 (fragment)

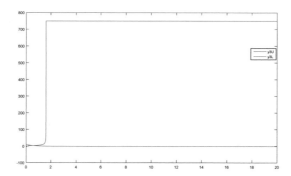

Fig. 4. Solution by using rule 4 (fragment)

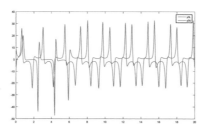

Fig. 5. Solution of fuzzy differential equation by using Hukuhara derivative (fragment)

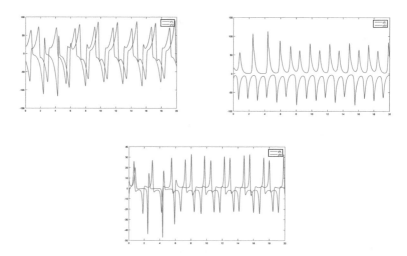

Fig. 6. Solution of fuzzy differential equation by using generalized derivative

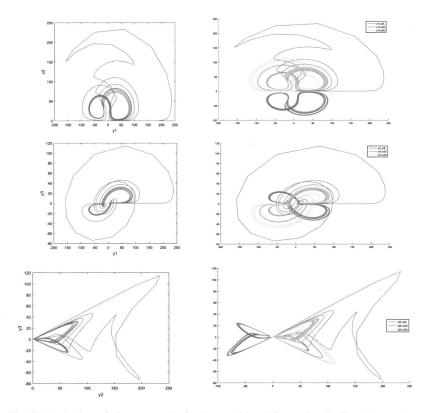

Fig. 7. Projection of phase portraits for the considered fuzzy chaotic Chen-Lee system.

The solution of (1) shown in Figs. 1, 2, 3, 4, 5 and 6 is obtained numerically by using Runge-Kutta method. For this purpose, ode87 solver of Matlab software is used. This solver is an accurate integrator of 8th-order (the local error normally expected is O (h^9), for more information, see [12, 13]).

The following code in Matlab m-file is used to obtain solution of FDEs system (1):

```
fuzzyODE_ChenLeeSystem2.m
function dydt = fuzzyODE_ChenLeeSystem2(~,y)
global a1 b1 c1 d1 e1 r

P1=1/2.*(1+y(1)./d1);
P2=1/2.*(1-y(1)./d1);
Q1=1/2.*(1+y(2)./e1);
Q2=1/2.*(1-y(2)./e1);


I1=(P1.*Q1./(P1+P2+Q1+Q2)).*r(1);
I2=(P1.*Q2./(P1+P2+Q1+Q2)).*r(2);
I3=(P2.*Q1./(P1+P2+Q1+Q2)).*r(3);
I4=(P2.*Q2./(P1+P2+Q1+Q2)).*r(4);


dydt = zeros(3,1); % a column vector

dydt=[I1;I1;I1] .* [-e1*y(3)+a1*y(1);d1*y(3)+b1*y(2);d1*y(2)/3+c1*y(3)]+...
   [I2;I2;I2] .* [e1*y(3)+a1*y(1);d1*y(3)+b1*y(2);d1*y(2)/3+c1*y(3)]+...
   [I3;I3;I3] .* [-e1*y(3)+a1*y(1);-d1*y(3)+b1*y(2);-d1*y(2)/3+c1*y(3)]+...
   [I4;I4;I4] .* [e1*y(3)+a1*y(1);-d1*y(3)+b1*y(2);-d1*y(2)/3+c1*y(3)];
```

fuzzyODE_ChenLeeSystemResults2.m

```
clc; clear;
global a1 b1 c1 d1 e1 r
% a1 = 5.0; b1= -10.0; c1 = -38.0;
a1 = 5.2; b1= -10.3; c1 = -38;
d1=40.0; e1=30.0; f1=20.0;
% yR0 = [d1,e1,f1];
yR0 = [20.0,15.0,10.0];
y0 = [0.20,0.15,0.10];
yL0 = [-20.0,-15.0,-10.0];
% yL0 = [-d1,-e1,-f1];


r = [1;1;1;1];
t0=0;tf=20.0;
options = odeset('RelTol',1e-6,'AbsTol',[1e-6],'InitialStep',0.01,'MaxStep',0.01);
[tR,yR] = ode87(@ fuzzyODE_ChenLeeSystem2,[t0,tf],yR0,options);
[t, y] = ode87(@ fuzzyODE_ChenLeeSystem2,[t0,tf],y0,options);
[tL,yL] = ode87(@ fuzzyODE_ChenLeeSystem2,[t0,tf],yL0,options);
```

5 Conclusion

Chaotic signals simulation is one of the basic problems considered in different engineering applications. In this paper, a fuzzy differential equation-based model of chaotic Chen-Lee system is considered. Behavior signals are analyzed by using If-Then rules. This means that the considered chaotic signals can be represented by electronic circuits. Our future work will be related to design of an electronic chip that generates chaotic signals under fuzzy uncertainty.

References

1. Mata-Machuca, J.L., Martínez-Guerra, R., Aguilar-Lopez, R., Aguilar-Ibanez, C.: A chaotic system in synchronization and secure communications. Commun. Nonlinear Sci. Numer. Simul. **17**(4), 1706–1713 (2012)
2. Li, S.Y., Ge, Z.M.: Generalized synchronization of chaotic systems with different orders by fuzzy logic constant controller. Expert Syst. Appl. **37**(3), 1357–1370 (2011)
3. Chen, H.-K., Lee, C.-I.: Anti-control of chaos in rigid body motion. Chaos Solitons Fractals **21**(4), 957–965 (2004)
4. Silvio, A.B.S., Ferreira, L., Pires, D.M., Velozo, F.A.: Fuzzy differential equation in a RL type circuit. Int. J. Adv. Appl. Math. Mech. **6**(1), 43–48 (2018)
5. Srinivas, B., Rao, B.S.: A new innovative method for solving electrical circuit analysis. Int. J. Comput. Appl. Math. **12**(2), 311–323 (2017)
6. Wang, S.-P., Lao, S.-K., Chen, H.-K., Chen, J.-H., Chen, S.-Y.: Implementation of the fractional-order Chen–Lee system by electronic circuit. Int. J. Bifurcat. Chaos **23**(02), 1350030 (2013)
7. Chen, G.H.: Controlling chaos and chaotification in the Chen-Lee system by multiple time delays. Chaos Solitons Fractals **36**(4), 843–852 (2008)
8. Grzechca, D.E.: Construction of an expert system based on fuzzy logic for diagnosis of analog electronic circuits. Int. J. Electron. Telecommun. **61**, 77–82 (2015)
9. Aliev, R.A., Aliev, R.R.: Soft Computing and Its Application. World Scientific, Singapore (2001)
10. Tam, L.-M., Chen, H.-K., Sheu, L.J., Lao, S.-K.: Alternative implementation of the chaotic Chen-Lee system. Chaos Solitons Fractals **41**(4), 1923–1929 (2009)
11. Li, S.-Y., Yang, C.-H., Lin, C.-T., Ko, L.-W., Chiu, T.-T.: Chaotic motions in real fuzzy electronic circuits. Abstr. Appl. Anal. **2013**, 1–14 (2013)
12. Prince, P.J., Dorman, J.R.: High order embedded Runge-Kutta formulae. J. Comp. Appl. Math. **7**, 67–75 (1981)
13. Hairer, E., Nørsett, S.P., Wanner, G.: Solving Ordinary Differential Equations. Nonstiff Problems. Springer Series in Computational Mathematics, vol. 8, 2nd edn. Springer, Heidelberg (1993)

Understanding Behavior of Biological Network via Invariant Computation

Rza Bashirov$^{(\boxtimes)}$ ⓘ and Guy Romaric Yemeli Ngandjoug ⓘ

Eastern Mediterranean University, Famagusta, North Cyprus, Mersin-10, Turkey
rza.bashirov@emu.edu.tr, yemeliguy@gmail.com

Abstract. It is of practical interest to know whether a biological network contains mass-preserving and state-preserving subnetworks. In a mass-preserving subnetwork, the total mass is constant and, therefore, bounded. In any state-preserving subnetwork biochemical reactions bring the subnetwork back to a given state. For instance, any reversible reaction forms a state-preserving subnetwork. In a large intricate biological network, it is rather cumbersome task to determine mass-preserving and state-preserving subnetworks.

Akt and MAPK are important pathways regulating cell proliferation, differentiation, senescence and apoptosis. In the present research, we derive information from Reactome and KEGG databases as well as from existing literature, to date, to create rather detailed Petri net model of Akt and MAPK pathways and crosstalk one pathway to another and perform their qualitative analysis with computation of P-invariants and T-invariants to determine mass-preserving and state-preserving subnetworks. We also make deductions regarding the temporal behavior of both pathway.

Keywords: Hybrid Petri nets · Transition invariant · Place invariant

1 Introduction

Akt and MAPK pathways are central to many diseases and are often de-regulated in many cancers. It was observed in many cancers that one of the proteins in these pathways is mutated. Components of the Akt and MAPK pathways were found in cancer cells. Akt and MAPK pathways were investigated separately by several authors. This is the main motivation behind the present research to perform qualitative analysis of the whole network composed of these two important signaling pathways and their crosstalk. With this in mind we explore Petri nets as modelling framework to create detailed model of underlying biological network and then use P-invariants and T-invariants as well as siphon and trap properties to predict structural and behavioral characteristics of underlying network.

In terms of metabolic networks a P-invariant represents substrate conservation, while in signal transduction or gene regulatory networks P-invariant often describes different states of a protein, a protein complex or a gene. A T-invariant has two interpretations in biochemical context. The T-invariant corresponds to a multi-set of transitions occurrence of which in some order reproduces a specified marking. Study of

© Springer Nature Switzerland AG 2020
R. A. Aliev et al. (Eds.): ICSCCW 2019, AISC 1095, pp. 390–396, 2020.
https://doi.org/10.1007/978-3-030-35249-3_50

T-invariants may contribute to understanding of the characteristics of biological networks. Given a marking, if T-invariant does exists then it is called feasible. Each entry in a T-invariant can be interpreted as relative firing rate of specified transition.

This paper is organized as follows. In the following section we outline P-invariants and T-invariants, the research methods, and explain why these methods sound in context of qualitative analysis of metabolic, signaling and genetic networks. Next we provide and analyze simulation results; overview structural and behavioral characteristics of underlying network derived from simulation results, provide minimal semipositive P-invariants and T-invariants and discuss their biological meaning. Finally, the paper ends up with conclusions.

2 Materials and Methods

In this section, we briefly discuss the materials and methods behind the present research [1–16]. For more information on Petri nets interested readers are referred to [7–10], while for biological context concerning Akt and MAPK pathways we recommend [3, 4, 6, 11, 12].

Let A be adjacency matrix of a Petri net. A place vector x (transition vector y) is called P-invariant (T-invariant) if it is a non-trivial and non-negative integer solution of the linear equation $x \cdot A = 0(A \cdot y = 0)$. A P-invariant stands for a set of places over which the weighted sum of tokens is constant and independent of firing sequence, i.e. if M_0 is an initial marking then $x \cdot M_1 = x \cdot M_2$ holds for $M_1, M_2 \in R(M_0)$. A place belonging to a P-invariant is definitely bounded, and CPI causes structural boundedness. Similarly, a T-invariant y is a non-zero and non-negative integer solution of $C \cdot y = 0$. A T-invariant has no total effect on a marking.

To calculate a P- or T-invariant, technically, we need to solve a homogeneous linear equation system over nonnegative integers. Validation is required to systematically check the model and increase our confidence in it. A P-invariant is a set of places over which regardless of firing sequence the weighted sum of tokens remains unchanged, so that occurrence of any transition has no effect on P-invariant. This means that P-invariant conserves the number of tokens. A T-invariant represents a set of transitions, whose firing returns Petri net to the initial state.

Let x and y be two P-invariants. If x and y have the same support, then x and y are linearly dependent, meaning that there exist two positive integers α and β such that $\alpha \cdot x = \beta \cdot y$. If α and β are nonnegative integers, then $\alpha \cdot x + \beta \cdot y$ is a P-invariant. If all components of vector $x - y$ are nonnegative integers, then $x - y$ is a P-invariant.

Biological material related to Akt and MAPK pathways and their crosstalk is derived from biological databases Kyoto Encyclopedia of Genes and Genomes, and Reactome, that are respectively described in [5, 10], and from literature in the field, to date.

Snoopy [8] and Charlie [7] software tools are powerful frameworks for modelling and simulation of complex systems with Petri nets. Snoopy tool is, indeed, effective platform for quantitative analysis of biological networks. For example, in [1] and [2] Snoopy tool was respectively applied to stochastic modelling and quantitative analysis of p16-mediated pathway and fetal-to-adult hemoglobin switch network. In current work, we use Snoopy as for model development and Charlie for quantitative analysis.

In [9], it was reported that MAPK signalling pathway is usually initiated by activation RAF protein by Ras. Activated RAF (RAFP) in turn induces a cascade of mitogen-activated protein kinase/ERK kinase (MEK) and extracellular signal regulated kinase (ERK). RAFP phosphorylates MEK to MEKP and then to MEKPP. In [15], it was showed that when activated, the extracellular-signal-regulated kinases (ERK) plays important role in the induction of certain processes including cell proliferation, differentiation, development ERK is regulated by phosphorylation mediated by MEKP and MEKPP.

Akt and MAPK pathways and their crosstalk is schematically illustrated in Fig. 1. In this figure phosphatases, activators of direct reactions, and dephosphatases, activators of reverse reactions, are renamed as PHASE, for short.

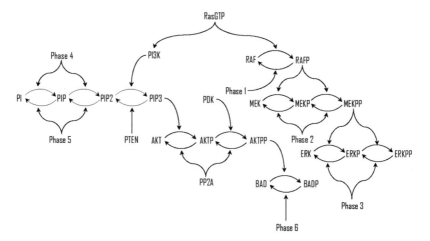

Fig. 1. Schematic illustration of cascade of the phosphorylation/de-phosphorylation reactions in Akt and MAPK pathways and their crosstalk.

The following observations regarding Akt pathway are described in [3, 6, 12, 14, 16]. Akt pathway is activated through a cascade of phosphorylation/dephosphorylation reactions. Phosphatidylinositol (PI) molecules are important players of intracellular signalling. Phosphorylation of these molecules produces phosphatidylinositol 3,4-bisphosphate (PIP2) and phosphorylation of PIP2 creates phosphatidylinositol 3,4,5-trisphosphate (PIP3). Phosphoinositide 3-kinase (PI3K) phosphorylates PIP2 to PIP3 and PTEN dephosphorylates PIP3 to PIP2. PIP3 in turn recruits AKT phosphorylation to AKTP and PDK induces dephosphorylation of AKTP to AKTPP.

3   Results

3.1   Structural Characteristics of Akt and MAPK Pathways and Their Crosstalk

Simulation results revealed that the underlying Petri net is pure and, therefore, has loop-free structure. This excludes use of read arcs and double arcs in the model. In terms of biological context, this means that a component cannot be produced and consumed by single reaction, that is, enzymatic reactions need to be formulated in more detail. We observed that all arcs have the same weight that is equal to 1, meaning that the model is represented by ordinary Petri net and, thereby, each stoichiometric parameter has the same numeric value equal to one. Simulations showed that all outgoing arcs of a place have the same weight, which means this is homogenous Petri net. This particularly implies that each component is equally consumed by multiple reactions (in terms of molecules). This Petri net has connected structure. The implication of this fact is that all biological components are directly or indirectly connected with each other via biomolecular reactions. Moreover, the Petri net has strongly connected structure, implying that all components are directly connected with each other via biomolecular reactions. Simulations also determined that the maximum of the weights of outgoing arcs is not greater than the minimum of the weights of the incoming arcs for a place. This is called non-blocking multiplicity property – the fact indicating a balance between amounts of produced and consumed molecules for a biological component. Named Petri net is not conservative, clearly indicating that the reactions in the model involve association/dissociation of different components. We also found that the Petri net does not have static conflict-free nature. This means that there do not exist reactions sharing the same components as reactants. All transitions have pre-places called no input transitions property, from which we conclude that there is a finite source for each biological component. We also observed that all places have pre-transitions, meaning that a component cannot be produced by any reaction, thus, such components are limiting. The Petri net also holds no output places property since all its places have post-transitions, indicating that a component can be infinitely accumulated and, therefore, this component cannot be fully consumed by any reaction.

3.2   Behavioral Characteristics of Akt and MAPK Pathways and Their Crosstalk

Simulation results showed that the Petri net is structurally bounded. Hence, accumulation of a component depends on initial state of the system and for each biological component there is an upper bound. The fact that Petri net is live reveals that due to the cyclic reactions that restore each other, all reactions contribute forever to the signaling. This Petri net is also reversible. This implies that there is a sequence of reactions that reproduces the initial state from any state. We found that the Petri net has not dynamically conflict free structure. Applied to biological context this means if a protein e.g., MEK gets dephosphorylated it loses the possibility to phosphorylate ERK. The number of dead states is 0 indicating that at least one reaction can always take place (Table 1).

Table 1. Minimal semi-positive P-invariants and their biological interpretations.

Minimal semi-positive P-invariants	Biological meaning
X_1 = (PI_PHASE4, PIP_PHASE4, PHASE4)	States of PHASE4
X_2 = (PIP_PHASE5, PIP2_PHASE5, PHASE5)	States of PHASE5
X_3 = (PI, PIP, PIP2, PIP3, PIP_PHASE5, PIP2_PHASE5, PIP3_PTEN, PI_PHASE4, PIP_PHASE4, PIP2_PI3K, AKT_PIP3)	States of PI
X_4 = (AKTP_PP2A, AKTPP_PP2A, PP2A)	States of PP2A
X_5 = (AKT, AKT_PIP3, AKTP, AKTP_PP2A, AKTP_PDK, AKTPP, AKTPP_PP2A, BAD_AKTPP)	States of AKT
X_6 = (BAD, BAD_AKTPP, BADP, BADP_PHASE6)	States of BAD
X_7 = (RAS_GTP, PIP2_PI3K, PI3K, RAF_RAS_GTP)	States of RAS_GTP
X_8 = (RAF_RASGTP, RAF, RAFP, RAFP_PHASE1, MEK_RAFP, MEKP_RAFP)	States of RAF
X_9 = (MEKP_PHASE2, PHASE2, MEKPP_PHASE2)	States of PHASE2
X_{10} = (MEK_RAFP, MEK, MEKP, MEKPP, MEKP_RAFP, MEKP_PHASE2, MEKPP_PHASE2, ERK_MEKPP, ERKP_MEKPP)	States of MEK
X_{11} = (ERKP_PHASE3, ERKPP_PHASE3, PHASE3)	States of PHASE3
X_{12} = (ERK, ERK_MEKPP, ERKP, ERKPP, ERKP_MEKPP, ERKP_PHASE3, ERKPP_PHASE3)	States of ERK

The Petri net contains no dead transition. This revels that for each reaction it is possible to reach a state, where the reaction can occur. The Petri net is covered by transition invariants, from which we conclude all reactions in the model are contained in circles of reactions. Each circle can restore its initial state. The Petri net is also covered by place invariants. This means that network made of Akt and MAPK pathways and their crosstalk has mass preserving structure (Table 2).

Table 2. Minimal semi-positive T-invariants and their biological interpretations.

Minimal semi-positive T-invariants	Biological interpretation
Y_1 = (R33, R36, R31, R34)	Binding of PHASE4 to PI, phosphorylation of PI to PIP and its release, binding of PHASE5 to PIP and dephosphorylation of PIP and its release reactions
Y_2 = (R39, R42, R37, R40)	Binding of PHASE4 to PIP, phosphorylation of PIP to PIP2 and its release, binding of PHASE5 to PIP2 and dephosphorylation of PIP2 and its release reactions

(*continued*)

Table 2. (*continued*)

Minimal semi-positive T-invariants	Biological interpretation
Y_3 = (R45, R48, R43, R46, R67)	Activation of PI3K, binding of PI3K to PIP2, phosphorylation of PIP2 to PIP3 and its release, binding of PTEN to PIP3 and dephosphorylation of PIP3 and its release reactions
Y_4 = (R51, R54, R49, R52)	Binding of PIP3 to AKT, phosphorylation of AKT to AKTP and its release, binding of PP2A to AKTP and dephosphorylation of AKTP and its release reactions
Y_5 = (R57, R60, R55, R58)	Binding of PDK to AKTP, phosphorylation of AKTP to AKTPP and its release, binding of PP2A to AKTPP and dephosphorylation of AKTPP and its release reactions
Y_6 = (R63, R66, R61, R64)	Binding of AKTPP to BAD, phosphorylation of BAD to BADP and its release, binding of PHASE6 to BADP and dephosphorylation of BADP and its release reactions
Y_7 = (R3, R6, R1, R4)	Binding of RAS_GTP to RAF, phosphorylation of RAF to RAFP and its release, binding of PHASE1 to RAFP, dephosphorylation of RAFP and its release reactions.
Y_8 = (R9, R18, R7, R16)	Binding of RAFP to MEK, phosphorylation of MEK to MEKP and its release, binding of PHASE2 to MEKP and dephosphorylation of MEKP and its release reactions
Y_9 = (R12, R15, R10, R13)	Binding of RAFP to MEKP, phosphorylation of MEKP to MEKPP and its release, binding of PHASE2 to MEKPP and dephosphorylation of MEKPP and its release reactions
Y_{10} = (R21, R30, R19, R28)	Binding of MEKPP to ERK, phosphorylation of ERK to ERKP and its release, binding of PHASE3 to ERKP and dephosphorylation of ERKP and its release reactions
Y_{11} = (R24, R27, R22, R25)	Binding of MEKPP to ERKP, phosphorylation of ERKP to ERKPP and its release, binding of PHASE3 to ERKPP and dephosphorylation of ERKPP and its release reactions

4   Conclusion

In the paper, a Petri net based model of the biological network made of Akt and MAPK pathways and their crosstalk has been presented. We performed qualitative analysis of underlying biological network in terms of Petri nets. By exploring P-invariants we determined 12 cycles in the biological network over each of which concentration is preserved. By exploding T-invariants we detected 11 fragments in the biological network in each of which due to the cyclic reactions that restore each other, the initial state can be reproduced. We also determine structural and behavioral characteristics through analysis of siphon-trap property. Overall, such an analysis supports the understanding of structural and behavioral characteristics of biological networks.

References

1. Bashirov, R., Akçay, Nİ.: Stochastic simulation-based prediction of the behavior of the p16-mediated signaling pathway. Fundam. Inform. **160**, 167–179 (2018). https://doi.org/10.3233/FI-2018-1679

2. Bashirov, R., Mehraei, M.: Identifying targets for gene therapy of β-globin disorders using quantitative modeling approach. Inform. Sci. **397–398**, 37–47 (2017). https://doi.org/10.1016/j.ins.2017.02.053

3. Castellano, E., Downward, J.: RAS interaction with PI3K: more than just another effector pathway. SAGE J. **2**(3), 261–274 (2011)

4. Chang, F., Steelman, L.S., Lee, J.T., Shelton, J.G., Navolanic, P.M., Blalock, W.L., Franklin, R.A., McCubrey, J.A.: Signal transduction mediated by the Ras/Raf/MEK/ERK pathway from cytokine receptors to transcription factors: potential targeting for therapeutic intervention. Leukemia **17**, 1263–1293 (2003)

5. Croft, D., O'Kelly, G., et al.: Reactome: a database of reactions, pathways and biological processes. Nucleic Acids Res. **39**, D691–D697 (2011)

6. Faes, S., Formond, O.: PI3K and AKT: unfaithful partners in cancer. Int. J. Mol. Sci. **16**(9), 21138–21152 (2015)

7. Heiner, M., Schwarick, M., Wegener, J.T.: Charlie – an extensible Petri net analysis tool. In: Devillers, R., Valmari, A. (eds.) PETRI NETS 2015, LNCS, vol. 9115, pp. 200–211. Springer, Heidelberg (2015)

8. Heiner, M., Herajy, M., Liu, F., Rohr, C., Schwarick, M.: Snoopy – a unifying Petri net tool. In: Haddad, S., Pomello, L. (eds.) PETRI NETS 2012, LNCS, vol. 7347, pp. 398–407. Springer, Heidelberg (2012)

9. Heiner, M., Gilbert, D., Donaldson, R.: Petri nets for systems and synthetic biology. In: Bernardo, M., Degano, P., Zavattaro, G. (eds.) Formal Methods for Computational Systems Biology, LNCS, vol. 5016, pp. 215–264. Springer, Heidelberg (2008)

10. Kanehisa, M., Goto, S., Kawashima, S., Okuno, Y., Hattori, M.: The KEGG resource for deciphering the genome. Nucleic Acids Res. **32**(1), D277–D280 (2008)

11. Koch, I., Chaouiya, C.: Discrete modelling: petri net and logical approaches. In: Sandun, C. (ed.) Systems Biology for Signaling Networks, pp. 821–855. Springer, New York (2010)

12. Malek, M., Kielkowska, A., Chessa, T., Clark, J., Hawkins, P.T., Stephens, L.R.: PTEN regulates PI(3,4)P2 signalling downstream of class I PI3K. Mol. Cell **68**(3), 566–588 (2017)

13. McCubrey, J.A., Steelman, L.S., Chappell, W.H.: Roles of the Raf/MEK/ERK pathway in cell growth, malignant transformation and drug resistance. BBA **1773**(8), 1263–1284 (2007)

14. Osaki, M., Oshimura, M., Ito, H.: PI3K-AKT pathway: its functions and alterations in human cancer. Apoptosis **9**(6), 667–676 (2004)

15. Shaul, Y.D., Seger, R.: The MEK/ERK cascade: from signalling specificity to diverse functions. BBA **1773**(8), 1213–1226 (2007)

16. Zhang, X., Majerus, P.W.: Phosphatidylinositol signalling reactions. Semin. Cell Dev. Biol. **9**(2), 153–160 (1998)

Fuzzy Controlled Robot Platform Tracking System

Mohamad Alshahadat[1] , Bülent Bilgehan[2](✉) ,
and Hisham Salim Alomsafer[1]

[1] Department of Computer Engineering, Faculty of Engineering,
University of Kyrenia, Girne, Mersin 10, Turkey
m.alshahadat@gmail.com, heshaam3000@yahoo.com
[2] Department of Electrical and Electronic Engineering, Faculty of Engineering,
Near East University, 99138 Nicosia, TRNC, Mersin 10, Turkey
bulent.bilgehan@neu.edu.tr

Abstract. The paper aims to introduce a fuzzy controlled tracking method using video structure analysis in MATLAB. The method uses a single overhead camera to record the trajectory and the distance traveled by the object under the test. The input is the video/image data, which is obtained by the camera and interfaced to the computer. The structure analysis decomposes the video into X-Y coordinate system divided into nine frames which includes information about their colors, positions, shapes, and movements. As the object moves vertically and horizontally, a control signal is generated. The generated signal is passed to the controller via RS232. The method uses MATLAB software for image processing.

The computer software enables to detect and track the object by moving the camera in the direction of the detected object. The object detection is based on the color or shape. The computer software employed performed four main functions: (1) image correction, (2) localization through the identification process, (3) user-defined "terminal points", and (4) user-conducted conversion (converts pixel values to match the real-world distances). The precision and accuracy analysis of the method provided. A real-world application test produced accurate results.

Keywords: Object tracking · Color detection · Robot platform · Fuzzy control

1 Introduction

Object tracking has many applications, such as video surveillance, human-computer interface, vehicle navigation, and robot control. The object tracing robots are very much used in commercial, academic, and governmental development. The improvements in such systems are to be used as an aid both in robot purchasing decisions and in understanding robot capabilities before deployment. The overall processing is based on estimating the position of an object over a sequence of images. The method becomes very complex in particular, applications, such as varying illumination levels, change of appearance, reformatted shapes. One method of object tracking platform is to process by identifying and tracking some specific feature of the moving object such as color that belongs to the moving object. Therefore, trajectories of moving object can be traced

© Springer Nature Switzerland AG 2020
R. A. Aliev et al. (Eds.): ICSCCW 2019, AISC 1095, pp. 397–404, 2020.
https://doi.org/10.1007/978-3-030-35249-3_51

through this process over time. The main aim is to track the object based on the data obtained from the video sequence. The introduced system initially identifies the region of interest (ROI) of the moving target. The next step is to use an adaptive color filter to extract the color information, which enables to track the object under the test. The main contribution of this paper is the method introduced to extract the color feature that belongs to the moving object. Such process requires highly advanced processing operation. The paper [1] introduced a video-toning approach as an alternative to introduced process. Another research paper [2] introduced motion layer based object removal in videos. In general, video analysis and segmentation process considered the most time consuming factors. However, there are some papers [3, 4] presenting better algorithms to overcome the time consumption. Alternatively, some papers introduce parallel processing technique to minimize processing time [5–7]. Feature extraction [8] and localization [9] methods are some other alternative processes to decrease the execution time. The important point is to implementation both, feature extraction and localization at the same processing interval. Building a parallel system usually introduces a problem due to large amount of data. Usually, such applications make a prime assumption for each object. This improves the performance of the overall system [10, 11].

2   Methodology

A platform carrying a camera has two degrees of freedom (2 rotations: Azimuth, Elevation) and burning laser beam, the two degrees of freedom of the platform is to be controlled in order to keep a moving object (the target) at the same axis of laser beam and burn it. Figure 1 represents of the proposed system. We developed a program under Matlab that performs two tasks: object detection and object tracking.

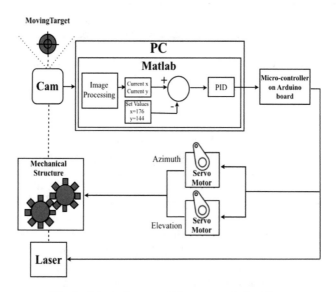

Fig. 1. Block diagram of the proposed algorithm.

2.1   Object Detection

Object detection method uses the image-processing techniques. Detection of the target consists of several steps. At the first step, a grayscale image and the blue component image was created. At the second step, the grayscale image subtracted from the blue component image. At the third step, the results are converted into a binary image, based on a threshold. At the fourth step, the small objects were removed from a binary image. The last step used to label all the connected components in the image, set properties for each labeled region and display calculated properties on the original image (see Fig. 2).

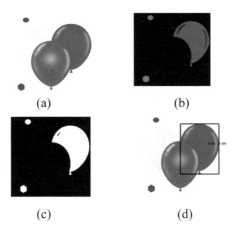

(a)   (b)

(c)   (d)

Fig. 2. Detection of the target (a) Original image (b) Subtraction result (c) Binarization, (d) Remove the small objects and Label connected components in the 2-D binary image.

2.2   Object Tracking

In this section, we developed an algorithm to control the platform to track the object. The deviation of the centroid from the center of the image (Δx, Δy) and the driving parameters of the servo motors calculated according to [12] the PID formula as:

$$\theta t + \Delta t = \theta t + P \, \varepsilon x + I \int \varepsilon x + D \, \varepsilon \dot{x} \tag{1}$$

$$\beta t + \Delta t = \beta t + P \, \varepsilon y + I \int \varepsilon y + D \, \varepsilon \dot{y} \tag{2}$$

If the target is halfway through the circular movement in the burning laser the controller sends the appropriate command to Arduino (serial communication via USB).

Figure 3 shows the flowchart of the proposed algorithm. The communication starts between Matlab and Arduino serially then the setting procedure of the camera and streaming video (serial) takes place. After the initialization process, it follows a loop to get a snapshot from the video. The snapshots are processed to determine the center of the target. Using a PID control system enables to evaluate the error. The error is then transmitted to Arduino that move a servomotor to set position (track the target).

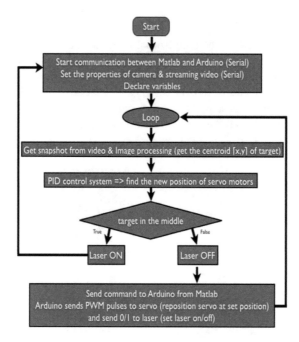

Fig. 3. Flow chart of platform tracks.

3 Error Evaluation Procedure

We designed a new calibration test, which is organized such that the performance of the system with different configuration parameters can be compared with each other. Using Lumion software a video is created in which a target is moving along a trajectory with controlled speed (Fig. 4). The platform is placed in front of the screen displaying this video. The system is required continuously to track the target as it moves along the x and y-axis. An error is considered when the centroid of the target [x or y] exceeds a set threshold. The number of errors is summed over the duration of the test.

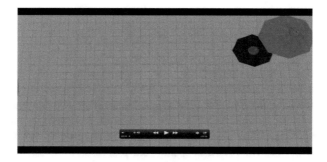

Fig. 4. Calibration test.

Figure 5 shows the process of error detection. Two sets of errors are detected during the whole cycle with four snapshots. The test results in a total mark of 2/4 = 50%. The aim of a calibration test is to tune the system with the least possible number of errors (approximately 0%) at the high speed as possible.

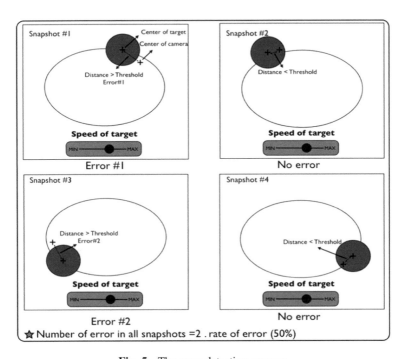

Fig. 5. The error detection process.

3.1 Calibration the Factor of the Proposed Control Algorithm

We made many tests to calibrate the two factors the proportional factors x, y, and the prediction factor (see Fig. 6).

$$\text{Prediction value} = |x2 - x1| \tag{3}$$

$$\text{Predection factor} = \text{prediction value} \times \text{K derivative} \tag{4}$$

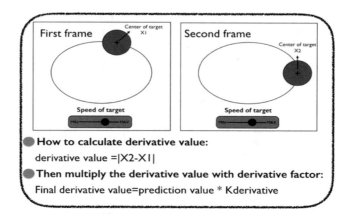

Fig. 6. Explain how to calculate the prediction factor.

We calibrated the proportional factors first then the prediction factor. The choosing of factors depends on a number of errors (minimum errors). The speed of objects that tracked in the video is (Calibration test) = 1.7 m/s. The distance between the platform and screen is 2 m or 3 m. We considered the small square at the center of the camera view to calculate the error in the tests. If the target is out of the square the errors counter counts the error else no error is produced. The square in test took 9% of the all pixels (see Fig. 7).

In our test, we calculated the errors in 1000 frames so the percentage of error depends on the number of frames. The Tables 1, 2, 3 and 4 show the calibration of the proportional and prediction factors with different distance between the target and the platform.

$$\text{Error}\% = \frac{Error}{No.frame} \times 100 \tag{5}$$

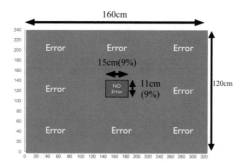

Fig. 7. Small square 9% pixels of all pixels.

Table 1. Calibration results of the proportional factor at a distance of 2 m.

Proportional factor x	Proportional factor y	Number of error	Prediction factor
1/14	1/16	overshot	0
1/18	**1/20**	**14.8%**	**0**
1/19	1/21	18%	0
1/21	1/23	26.9%	0

Table 2. Calibration the prediction factor while the distance is 2 m.

Proportional factor x	Proportional factor y	Number of error	Prediction factor
1/18	1/20	12%	1
1/18	**1/20**	**10.2%**	**1.5**
1/18	1/20	11.1%	2

Table 3. Calibration the proportional factor while the distance is 3 m.

Proportional factor x	Proportional factor y	Number of error	Prediction factor
1/14	1/16	Overshot	0
1/17	1/19	16.1%	0
1/18	**1/20**	**13.3%**	**0**
1/21	1/23	23%	0

Table 4. Calibration the prediction factor while the distance is 3 m.

Proportional factor x	Proportional factor y	Number of error	Prediction factor
1/18	1/23	8.6%	1
1/18	**1/20**	**8.9%**	**1.5**
1/18	1/20	9.6%	2

4   Conclusion

This paper presents a new fuzzy controlled method for a tracking system for robots. The performance of the system can easily be tested and does not require high-cost benchmark software program. The introduced system uses an optimization method based on fuzzy controlled feedback. Initially, the area without the object recorded as a reference. This is a basic operation with no additional assumptions. The difference between the reference and the object images allows extracting the tracking parameters.

The algorithm is tested in the laboratory live and the success rate is 100%. The algorithm works well under all conditions and the time taken to detect and track the object is <8 ms. The tracking system can further be improved by modeling the reference images. This improves the measurement values through the camera. The location of an object is based on the precise values of the distance and angle parameters. The high level of accurate parameter values is obtained from the introduced optimization algorithm. Further improvement to the system can be made by enabling it to track objects in the night. We believe that the introduced tracking platform system offers a low-cost product for accurate object location.

References

1. Wang, J., Xu, Y., Shum, Y.: Video tooning. ACM Trans. Graph. **23**(3), 574–583 (2004)
2. Zhang, Y., Xiao, J., Shah, M.: Motion layer based object removal in videos. In: Proceedings of the Seventh IEEE Workshops on Application of Computer Vision, pp. 516–521. IEEE Computer Society, Washington DC (2005)
3. Ishiguro, T., Miyamoto, R.: An efficient prediction scheme for pedestrian tracking with cascade particle filter and its implementation on Cell/B.E. In: Proceedings International Symposium on Intelligent Signal Processing and Communication Systems, pp. 29–32. IEEE, Kanazawa (2009)
4. Medeiros, H., Gao, X., Kleihorst, R., Park, J., Kak, C.: A parallel implementation of the color based particle filter for object tracking. In: Proceedings. ACM SenSys Workshop Applicat. Syst Algorithms Image Sensing (Image Sense), pp. 80–87. IEEE, Anchorage (2008)
5. Cherng, D., Yang, S., Shen, C., Lu, Y.: Real time color based particle filtering for object tracking with dual cache architecture. In: Proceedings on 8th IEEE International Conference, pp. 148–153. IEEE, Klagenfurt (2011)
6. Lu, X., Ren, D., Yu, S.: FPGA-based real-time object tracking for mobile robot. In: Proceedings on International Conference on Audio, Language and Image Processing, pp. 1657–1662. IEEE, Shanghai (2010)
7. Liu, S., Papakonstantinou, A., Wang, H., Chen, D.: Real-time object tracking system on FPGAs. In: Proceedings on Symposium on Application Accelerators in High-Performance Computing, pp. 1–7. IEEE, Knoxville (2011)
8. Lin, M., Yeh, H., Yen, H., Ma, H., Chen, Y., Kuo, C.: Efficient VLSI design for SIFT feature description. In: Proceedings on International Symposium on Next Generation Electronics, pp. 48–51. IEEE, Kaohsiung (2010)
9. El, H., Halym, I., Habib, D.: Proposed hardware architectures of particle filter for object tracking. EURASIP J. Adv. Signal Process. **1**(4), 17–36 (2012)
10. Garabini, M., Passaglia, A., Belo, F., Salaris P., Bicchi, A.: Optimality principles in variable stiffness control: The VSA hammer. In: Proceedings on IEEE/RSJ International Conference on Intelligent Robots and Systems, pp. 3770–3775. IEEE, San Francisco (2011)
11. Braun, D., Howard, M., Vijayakumar, S.: Optimal variable stiffness control: formulation and application to explosive movement tasks. Auton. Robot. **33**(3), 237–253 (2012)
12. Dorf, R., Bisghop, R.: Modern Control System, 12th edn. Upper Saddle River (2006)

A Novel DOA Estimation Method
for Wideband Sources Based on Fuzzy Systems

Bülent Bilgehan$^{(\boxtimes)}$ (iD) and Amr Abdelbari (iD)

Department of Electrical and Electronic Engineering, Near East University,
Nicosia, TRNC, Mersin 10, Turkey
{bulent.bilgehan,amr.abdelbari}@neu.edu.tr

Abstract. The Direction of Arrival (DOA) methods has great applications in
the field of wireless communication. The case may be to locate an illegal
transmission or to identify the target in the battlefield. This conference article
presents a new DOA method. The new method estimates DOAs by using Fuzzy
logic system to classify the detected peaks in the spatial spectrum and identify
the correct DOAs using an antenna array operating at 2050 MHz. The rest
detected peaks discard as it is a false peaks. The new method reduces the
computational complexity of super-resolution algorithms such as Incoherent
Multiple Signal Classification (Incoherent-MUSIC) and Test of Orthogonality of
Project Subspaces (TOPS). The performance of the current method is compared
to that of the IMUSIC and TOPS algorithms.

Keywords: Direction-of-Arrival · Spatial analysis · MUSIC · Wideband
signals · Sub-band · Array signal processing

1 Introduction

The method of estimating the direction of the incoming signal gains much interest in
recent years [1]. Some applications can be stated as mobile communication and radar
systems. The important point in array signal processing is the DOA estimation [2].
Different types of algorithms are generated to estimate the DOA for a wideband signal.
There are two main purposes for the algorithm. Firstly, determines the number of
incident signals. Secondly, it estimates the direction of the receiving signal [3]. Some of
the different algorithms are in [4]. The most popular algorithm is classified to be
MUSIC. The conventional estimation algorithms have different capabilities and limitations [5]. Mainly, the number of the signals must be smaller than the antenna array
elements, and large numbers of signal snapshots are required to achieve high accuracy.
The DOA estimation methods evolved after the time series analysis, linear prediction
methods and different processing of array methods.

The algorithms for the DOA estimation method can be categorized under two major
headings. The first category covers methods that represent the power and position of
the transmitted signal [6]. The second category can be considered as high-resolution
method. Such a method requires knowledge of the signal under consideration [7]. The
early DOA estimation methods aim to achieve the number of sources in processing [8,
9]. Once the number of sources is revealed, high-resolution method applied to estimate

© Springer Nature Switzerland AG 2020
R. A. Aliev et al. (Eds.): ICSCCW 2019, AISC 1095, pp. 405–412, 2020.
https://doi.org/10.1007/978-3-030-35249-3_52

the angular position of the sources. The methods under the classification of high-resolution are known to be more accurate than other methods. The processing of all methods involves subspace analysis. The subspace analysis follows a defined structure of the correlation matrix defining the signal under consideration. The correlation matrix includes details of the signal propagation. Another important point is that all the real signal analysis requires the addition of a noise factor. Therefore, the data space should be considered as signal and noise subspaces. The initial step in such processing is to compose the covariance matrix based on a good estimation as:

$$\hat{C}_{xx} = \frac{1}{N} \sum_{i=1}^{N} X(t_i) * X^H(t_i), \tag{1}$$

where N is the number of snapshots. $X(t_i)$ and $X^H(t_i)$ the Hermitian conjugate are the sensors data that represents the received signal which expressed in the time domain by:

$$x(t) = \sum_{k=1}^{D} s_k(t) a_m(\theta_k) + n_m(t), \tag{2}$$

where S(t) is the impinging signal and n(t) is the Additive White Gaussian Noise (AWGN) add to the mth sensor. $a(\theta_k)$ is the steering vector of the mth sensor and the kth source. To apply the ordinary subspace DOA methods, the wideband received signal is divided using Discrete Fourier Transform (DFT) to an L number of subbands in the frequency domain represented by

$$X(f_j) = \sum_{j=1}^{L} A_m(f_j, \theta_k) S_k(f_j) + N_m(f_j), \tag{3}$$

where the bold uppercase symbols denote the matrix in the frequency domain.

Conventional wideband DOA methods differ at this stage. IMUSIC calculates the spatial spectrum directly by applying the narrowband MUSIC to each subband [10]. TOPS method applies a transformation matrix first to transform all the subbands into one subband called the reference subband. Then, using the Singular Value Decomposition (SVD) to determine the final spatial spectrum where the DOAs occur at the minimum singular values [11]. These methods achieve high-resolution at high SNR values. However, at low SNR values, their performance suffers from the issues of false peaks that appear in the final spatial spectrum and the high computational costs. Therefore, new DOA estimation algorithms are needed.

In this paper, a novel DOA method uses Fuzzy system to accurately detect the sources and estimate their DOAs. The introduced method overcomes other methods issues such as false detection and higher complexity. In Sect. 2, the proposed method is presented, followed by the simulation results in Sect. 3, and finally the conclusions.

2   Methodology

In this section, the proposed method is presented. The proposed method distinguishes between the valuable subbands for further processing while still benefit from other subbands as well. The number of peaks detected in the spatial spectrum of each frequency subband varies from one frequency subband to another. The newly introduced fuzzy system applied to select the correct peaks related to the signal DOA. The proposed method determines the accurate DOA at high SNR values while achieving the least bias at the extreme low SNR values. Finally, the proposed method studies the spatial spectrum itself without any modification to the applied narrowband DOA method.

2.1   The Mathematical Foundation

Like IMUSIC method, the proposed method applies the conventional narrowband MUSIC method and hence, the spatial spectrum for each subband can be obtained by:

$$F(f_j, \theta) = argmin\{a^H(f_j, \theta) E_n^H(f_j) E_n(f_j) a(f_j, \theta)\}. \tag{4}$$

where $E_n^H(f_j)$ is the eigenvectors matrix of the noise subspace for the jth subband. θ is the hypothesis search angle varying from $-90°$ with $0.1°$ step to $90°$.

First of all, the difference between the arrays of sensors is a phase shift in frequency with respect to the reference sensor. As a matter of fact, different frequency experiences different channel path loss variations. It is assumed that the signal follows a Gaussian distribution [10]. Thus, the less affected frequencies are likely to contain DOA information with higher accuracy. Secondly, due to the fact that each covariance matrix is estimated using (1), the orthogonality between the steering vector corresponding to the signal subspace and the eigenvectors of the noise subspace is not the same in the subbands [12]. As a result, false peaks appear in the spatial spectrum of some subbands and some of them are good enough to have only the estimated DOAs with high resolution as shown in Fig. 1(a).

The fuzzy system consists of three stages: the inputs, the processing, and the outputs. The input to the proposed system is the angle values of all detected peaks for all subbands as illustrated in Fig. 1(a). The outputs of the fuzzy system are angle clusters contain all the peaks that have been detected classified into a number of clusters arranged in a descending order. Figure 1(b) shows the output of the system where the clusters contains the highest number of peaks is the estimated DOA. Other clusters with less number of peaks are corresponding to noise and false detection and discarded from the final spatial spectrum. The processing stage includes the rules base that grows as the inputs increase. The rules follow the following arguments:

Rule a. IF $p_u < $ mean $\{R_z\} + \theta$ AND $p_u > $ mean $\{R_z\} - \theta$, THEN $p_u \in R_z$ AND Add p_u to R_z.

Rule b. IF NOT $p_u < $ mean $\{R_z\} + \theta$ AND NOT $p_u > $ mean $\{R_z\} - \theta$, THEN p_u R_z AND Construct a new rule for new $R_{\beta+1}$ with its mean $\{R_{\beta+1}\}$.

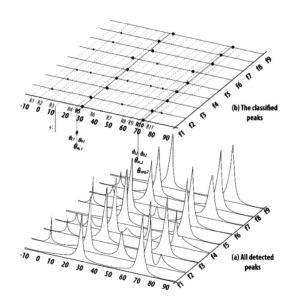

Fig. 1. Visual illustration of the proposed algorithm. (a) The inputs to the proposed system PESO are the angle values of all detected peaks. (b) The outputs of the system are angle clusters.

where θ is the variance around the mean of the zth angle cluster. The mean is calculated every time a new peak adds to the cluster; therefore, the angle cluster follows estimated DOA with more accuracy.

The number of fuzzy rules supposed to grow to double the number of R angle clusters. The maximum number of the rules in this system is limited to $2*(M + L)$ where M is the number of sensors and L is the total number of subbands. The result of the fuzzy system will be a set of angle clusters as following:

$$R = \{R_1 \quad R_2 \ldots R_z \ldots R_\beta\}, \tag{5}$$

where β is the total number of angle clusters and ($\beta \leq$ L). The DOAs belongs to the cluster with the highest number of peaks. Finally, the total spatial spectrum for the proposed method is represented by:

$$F(\theta) = \frac{1}{\alpha}\sum_{u=1}^{\alpha} F(f_u, \varphi), \quad where \quad \varphi \quad \acute{\in} \quad \{\theta_u \pm \Theta\} : \theta_i \in R_z, \tag{6}$$

$$F(\theta) = \frac{1}{L}\sum_{i=1}^{L} min\{F(f_j, \varphi)\}, \quad where \quad \varphi \quad \in \{\theta_i \pm \Theta\}. \tag{7}$$

where α is the total number of peaks in the zth angle cluster.

2.2   The Proposed Algorithm

The algorithmic summary of the new proposed method follows:

Step 1: DFT is applied to the wideband and subdivided into L subbands.
Step 2: Using (4), the spatial spectrum of each subband obtained.
Step 3: Apply the fuzzy system to arrange the observed peaks into angle clusters.
Step 4: Calculate the number of peaks per angle cluster.
Step 5: Choose clusters with the highest number of peaks as the DOA and discard other clusters.
Step 6: Plot the final spatial spectrum using (6) and (7).

3   Simulation Results

The newly introduced method that called Probabilistic Evaluation of Subspace Orthogonality (PESO) was tested with 3 transmitting simulation sources located at 30, 50° and 65°. The performance of the introduced method has been compared with both IMUSIC and TOPS methods. The Monte-Carlo simulation with 200 rounds has been tested using MATLAB with a number of subbands L = 22 and 30 snapshots for each 128 DFT points. The array of sensors are assumed to be a Uniform Linear Array (ULA) with a number of sensors M = 8.

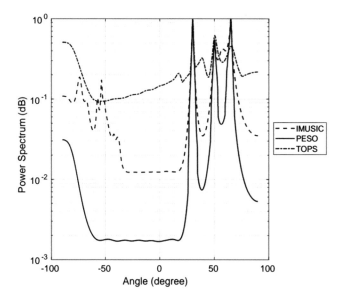

Fig. 2. The spatial spectrum at −2.5 dB for the DOA methods IMUSIC, TOPS and PESO for the sources located at 30°, 50° and 65°.

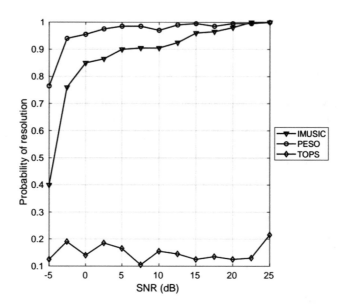

Fig. 3. The probability of resolution for the DOA methods IMUSIC, TOPS and PESO for the sources located at 30°, 50° and 65°.

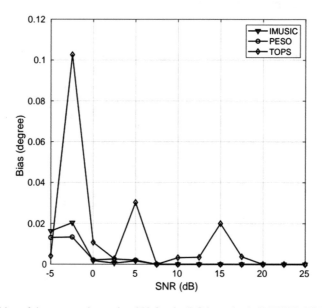

Fig. 4. The bias of the sources located at 30° for the DOA methods IMUSIC, TOPS and PESO.

The plot in the Fig. 2 proves the superiority of the newly defined PESO method. There are no false peak in the spatial spectrum and estimates the direction of arrival accurately even at a very low SNR value while TOPS and IMUSIC method shows false peaks.

Figure 3 gives the probability of resolution for three methods under the simulation test. The probability is considered as 1 if all DOAs detected within a small range. Otherwise, it is considered as 0. The suggested PESO method maintains the highest probability among other DOA methods for the complete range of SNR values.

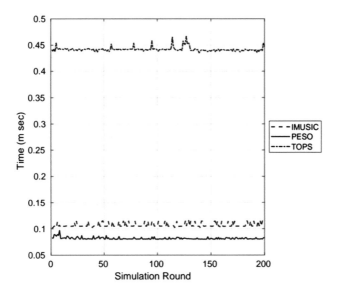

Fig. 5. The mean elapsed time of the DOA methods IMUSIC, TOPS and PESO.

The localization of the three methods is tested in Fig. 4. The PESO method has the minimum bias at the low SNR values and almost no bias above 7.5 dB while other methods suffer from high bias. TOPS performance is the worst with high bias even in high SNR values.

Finally, the newly introduced method (PESO) has a simple mathematical representation. The overall average execution time for the snapshots recorded and plotted in Fig. 5. The PESO has very low computational cost for all simulation rounds compared with other methods. IMUSIC is close to PESO due to its simplicity as well while TOPS still suffer from high complex calculations that led to high computational costs.

4 Conclusions

This paper presents a detailed comparative study of three high-resolution methods for the Direction-of-Arrival (DOA) of RF signals. We presented a comparative study of the performances of different techniques for the DOA estimation. The receiving system used is composed of a linear array of five antenna sensors. The newly introduced algorithm of processing implemented in MATLAB. Simulations carried out with MATLAB provides interesting results in terms of source localization accuracy when signals are assumed uncorrelated. The MUSIC method is known to be robust and accurate in some applications. However, in some applications, Root-MUSIC or ESPRIT observed to perform better than MUSIC. This research work proposed a new method that performs better than existing methods on the merits of accuracy even at very low SNR values, less snapshots, and computational cost.

References

1. Carlin, M., et al.: Directions-of-arrival estimation through Bayesian compressive sensing strategies. IEEE Trans. Antennas Propag. **61**(7), 3828–3838 (2013)
2. Bilgehan, B., Abdelbari, A.: Fast detection and DOA estimation of the unknown wideband signal sources. Int. J. Commun Syst **32**(11), e3968 (2019)
3. Zhang, D., Wang, F., Burgos, R., Boroyevich, D.: Common mode circulating current control of interleaved three-phase two-level voltage-source converters with discontinuous space-vector modulation. In: Proceedings of IEEE Energy Conversion Congress and Exposition, pp. 2801–2807. IEEE, San Jose (2009)
4. Liberti, J.C., Rappaport, T.S.: Smart Antennas for Wireless Communications: IS-95 and Third Generation CDMA Applications, 2nd edn. Prentice Hall, New York (1999)
5. Arja, H.E., Huyart, B., Begaud, X.: Joint TOA/DOA measurements for UWB indoor propagation channel using MUSIC algorithm. In: Proceedings of the 2nd European Wireless Technology Conference, pp. 28–29. EuMA, Rome (2009)
6. Chen, J.C., Hudson, R.E., Yao, K.: Maximum-likelihood source localization and unknown sensor location estimation for wideband signals in the near-field. IEEE Trans. Signal Process. **50**(8), 1843–1854 (2002)
7. Shahrokh, V., Champagne, B., Kabal, P.: Parametric localization of distributed sources. IEEE Trans. Signal Process. **43**(9), 2144–2153 (1995)
8. Wu, H.T., Yang, J.F., Chen, F.K.: Source number estimator using transformed gerschgorin radii. IEEE Trans. Signal Process. **43**(6), 1325–1333 (1995)
9. Wong, K.M., Zhang, Q.T., Reilly, J.P.: On information theoretic criteria for determining the number of signals in high resolution array processing. IEEE Trans. Acoust. Speech Signal Process. **38**(11), 1959–1971 (1990)
10. Wax, M., Shan, T.-J., Kailath, T.: Spatio-temporal spectral analysis by eigenstructure methods. IEEE Trans. Acoust. Speech Signal Process. **32**(4), 817–827 (1984)
11. Yoon, Y.-S., Kaplan, L.M., McClellan, J.H.: Tops: new DOA estimator for wideband signals. IEEE Trans. Signal Process. **54**(6), 1977–1989 (2006)
12. Sharman, K., Durrani, T., Wax, M., Kailath, T.: Asymptotic performance of eigenstructure spectral analysis methods. In: ICASSP 1984. IEEE International Conference on Acoustics, Speech, and Signal Processing, vol. 9, pp. 440–443. IEEE, San Diego (1984)

Application of Fuzzy Logic in Selection of Best Well for Hydraulic Fracturing in Oil and Gas Fields

T. Sh. Salavatov[1]([✉]) [ID] and Khurram Iqbal[2]

[1] Department of Petroleum Engineering,
Azerbaijan State Oil and Industrial University, Baku, Azerbaijan
petrotech@asoiu.az
[2] Dewan Petroleum Limited, Islamabad 9583, Pakistan
khurram@dewanpetroleum.com

Abstract. To meet high demand of hydrocarbons, innovative techniques are imperative, therefore hydraulic fracturing become popular to extract hydrocarbons from shale and tight formations. Designing of treatment and selection of most appropriate well for hydraulic fracturing plays a vital role to achieve maximum benefit from this expensive technology. Designing hydraulic fracturing job initiates with identification of best candidate well for job which includes understanding geological factors of area, well location, lithology, selection of proppant volume and understanding of created fracturing geometry, proppant volume. Other main constituents are fracture geometry which includes fracturing length, height and width.

Fuzzy Logic Systems application is vastly used in research area of petroleum engineering. This paper is focused on using fuzzy logic technique to decide best well for best well for hydraulic fracturing. Selection of most suitable well for hydraulic fracturing, among many zones/layers within many numbers of producing wells is reflected makes it difficult, especially when the selection process depends upon on a group of parameters having different variables, attributes and features. This process becomes multifaceted, nonlinear and advocate with uncertainties. This technique is proved to reduce uncertainties in selection of most suitable well for stimulation and hydraulic fracturing.

In the end of this paper example is also provided where fuzzy logic was used to reduce the uncertainties and by selecting the best candidate well, hydrocarbons (gas) production of candidate well was increased four times of its natural ability by using fuzzy logic.

Keywords: Hydraulic fracturing · Fuzzy logic · Well selection

1 Introduction

Petroleum industry comprises range of different events that includes exploration, drilling and production of hydrocarbons. To achieve best economical production from gas wells by removing restrictions in flow around the well bore and achieve broader drainage area is by hydraulic fracturing. Hydraulic fracturing is one of the important

© Springer Nature Switzerland AG 2020
R. A. Aliev et al. (Eds.): ICSCCW 2019, AISC 1095, pp. 413–419, 2020.
https://doi.org/10.1007/978-3-030-35249-3_53

activities whose main purpose is to improve ultimate economic recovery [1]. When reservoir production is below economic limit, the most preferable and viable solution is conducting stimulation job or hydraulic fracturing to enhance production. Therefore, successful hydraulic fracturing job is based on selection of the best well for fracturing.

Fuzzy logic application in different spheres of petroleum engineering and earth sciences is very helpful [2], in petrophysics, drilling [3], reservoir characterization [4], permeability and rock type estimation [5, 6], petroleum separation [7], stimulation [8] and optimizing formation layers and wells for fracturing [9, 10].

As the uncertainties for selecting the best candidate well for hydraulic fracturing are countless; having comparatively limited data set therefore estimates created on existing methods for which great degree of uncertainty exists. Any such system where only uncertain and vague information may exist, fuzzy logic delivers a method to evaluate system behavior by providing us to incorporate rational relationship between perceived input and output situations.

During recent years, intelligent quantitative algorithms such as artificial neural networks, fuzzy C-mean clustering, C4.5 decision tree and fuzzy logic are applied vastly to estimate reservoir quality [11–16], these estimations are considered as the best tool to selecting best candidate for hydraulic fracturing well.

2 Fuzzy Logic; Selecting Candidate Well for Hydraulic Fracturing and Dealing with Uncertainties

Fuzzy logic provides allowance in conventional Boolean logic developed to estimate perception of partial truth [17–22]. Zadeh explained that in place of regarding fuzzy theory as a single theory, one must look the process of fuzzification as an approach to oversimplify any specific theory from an independent to a continuous or "Fuzzy" form. According to Zadeh [18], uncertainty is an inevitable feature of information.

In each stage (Fuzzifier, interference engine and deffusizfier) membership functions do apply before providing output. The membership function μ (x) designates membership of elements x for base set X in the fuzzy set A. Intersection and union operators are based on minimum or maximum operations:

$$\mu_{A \cap B}(x) = \min\{\mu_A(x), \mu_B(x)\}, \ x \in X \tag{1}$$

$$\mu_{A \cup B}(x) = \max\{\mu_A(x), \mu_B(x)\}, \ x \in X \tag{2}$$

As the success of hydraulic fracturing job is solely dependent on the best selection of candidate well but group of constraints that having different fields, characteristics and structures on which this selection is based on. As every parameter is having complex, nonlinear and adherent with uncertainties therefore the only tool to deal with these uncertainties is Fuzzy Logic. As each parameter have its own nature of uncertainty therefore the volume of data increases which goes out of capability of human intellect to untangling important information by conventional methods.

Selecting operation in Fuzzy logic depends on the problems of applications. Such as membership function for fuzzy set $A \cap B$, which consequences from intersection of fuzzy sets of A and B. Where μA and μB are the members of fuzzy sets A and B, respectively. Where μA can be membership function of fuzzy set for any risk variable i.e., skin value, reservoir damage or even water saturation etc. Similarly selecting target reservoir or production zone from a group which is considered to a difficult task due to number of constraints having different fields, characteristics and structures become convenient. The selection of best nominee well for hydraulic fracturing can be predicted to be nonlinear, un-equilibrium, and adherent due to uncertainties which can be solved easily by type-2 Fuzzy sets and systems.

Fuzzy logic helps in guarantying accuracy and excellence to avoid discrepancies and uncertainties. This is used in reservoir characterization which is the initial basic step of selecting best candidate well for hydraulic fracturing which comprises of assigning reservoir its properties such as porosity, permeability, relative permeability, fluids, fluid contacts, saturation, mineral composition, particle size distribution, rock strength and many other properties to understand reservoir which includes uncertainties in geological information and spatial variability. Although oil industry is having various means to evaluate reservoir e.g., by core, logs, tests, but this data is expensive and only for one point. Each well cannot be achieved to avoid such uncertainties fuzzy logic provides a handy tool to overcome by fuzzy logic. As during hydraulic fracturing due to these uncertainties or variability from point to point may impact rock strength which effect the execution during hydraulic fracturing job.

3 Discussion

According to Xiong and Holditch, [8]; Yang, [9]; Yin and Wu, [10] initially fuzzy logic was established and used to cover the complications in conventional method for selection of candidate well for hydraulic fracturing. As Xiong and Holditch, [8] used permeability of <1 mD, skin <5.5, water saturation >37.5, porosity 6–26%, net pay thickness >100 ft, drainage area >110acres, formation pressure 2,000–3,000 psi and formation depth <10,000 ft; whereas Yang 2000 used permeability of 0–0.3 mD, skin 5–25, water saturation 0–40%, porosity 0–10%, net pay thickness 33–164 ft, formation pressure 1,450–1,741 psi but no sensitivity was used for drainage area and formation depth were used, similarly Yin and Wu [10] used permeability of 0–30 mD, skin 4–5, water saturation 75–80%, net pay thickness 10–16 ft, formation pressure 1,450–2,030 psi but no sensitivity was used for drainage area, formation depth and even porosities were used. The comparison among these authors are briefed in Table 1 below:

Table 1. Summarized comparison between different authors

Authors	Xiong and Holditch [8]	Yang [9]	Yin and Wu [10]
Fuzzy variables	9	12	7
Levels	−1, 0.5, 0.7 & 1	0.01, 0.1, 0.25 & 0.64	Not available, unsuitable, second and first priority
Constants (m & n)	m & n	$a = \frac{dup + ddown}{2}$ $b = \frac{|dup + ddown|}{2\sqrt{ln2}}$	$aij = \frac{diju + dijd}{2}$ $bij = \frac{\sqrt{(diju-dijd)2}}{\sqrt{4ln2}}$
Membership function (Gaussian function)	$f(x) = e^{-\left(\frac{x-m}{n}\right)2}$	$rij = e^{\left[-\left(\frac{x-a}{b}\right)2\right]}$	$uij(di) = e^{-\left[\left(\frac{di-aij}{bij}\right)\right]2}$
Matrix name	Relationship matrix	Fuzzy analogues matrix	Fuzzy relationship matrix
Operator	Min $bj = min\left\{1, \sum_{i=1}^{9} IiFij\right\}$	Max $bj = \sum_{K=1}^{B} AKbki$	Max $V = AR = (V_1, V_2, V_3, ..., V_m)^T$

Initial type-1 fuzzy set system was not able to deal completely with linguistic uncertainties in terms of inconsistency for decision making. As range of vagueness in any situation involves, result may vary due to some absence or ignorance in information such deficiency of information, profusion in information, conflict in evidences and any ambiguity in measurement and believes [23]. These number of deficiencies resulted further variety of uncertainties [24]. Therefore, Type-2 fuzzy set and systems were used to handle such uncertainties for a better control.

Further development of Type 2 fuzzy logic system for selection of candidate well for hydraulic fracturing and other stimulation may provide better capacity of handling linguistic uncertainties than previous methodology as the uncertainties regarding values of membership function and the variables involved in each set of selection of well more refined fuzzy sets will develop. Fuzzy logic is a limitless tool which have only one limitation which is depending on human brain, as long as any uncertainty can be brought to sensitivity type-2 fuzzy logic can be used as a best tool and can be further developed according to need.

By using above mentioned uncertainties of well location, petrophysical properties of each well, drilling conditions and issue, well deliverability of each well, reservoir thickness encountered in each well, effected permeability of each zone in each well, porosity of each well, drainage area of each well and rock mechanics, best candidate well was selected to perform a hydraulic fracturing job in which resulted into four folds of production and flowing wellhead pressure enhancement. The results of hydraulic fracturing job is presented in Fig. 1 below, where green line is showing natural ability for production of reservoir.

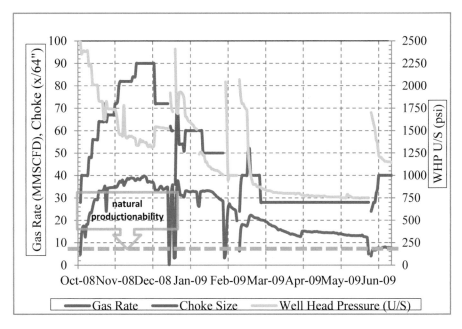

Fig. 1. Results of hydraulic fracturing after selection of candidate well through Fuzzy Logic

4   Conclusion

Nature of selection of the most suitable well for stimulation is non-linear due to involvement of many variable parameters, which are different, qualitatively as well as quantitatively. Fuzzy set theory provides flexibility in a convenient framework to evolve, and analyze all data even independent of source or characteristics. Dual concepts related to the fuzzy logic are; fuzzy membership functions and second is linguistic variables, can especially be improved to take applications to solve the intrinsic nature of selection of well for hydraulic fracturing.

   Well selection of wells, formations and sub layers for hydraulic fracturing, the procedure determines the best fit from number of wells and formations/sub-layer variables whose effect on the performance can be best suitable. Typically different types of data and information will be used in different areas and properties in different formats and configurations such as well location, drilling, petrophysical properties, well deliverability, reservoir thickness, permeability, porosity, drainage area and rock mechanics etc. to filter, final selection candidate/s. Selection process to handle more uncertainties and determining the best candidate for hydraulic fracturing treatment by considering complex, nonlinear, and imprecise nature of data. This process of decision-making is including more variable. It is obvious that ability to manage such uncertainties turns out to be a critical issue.

   Difference in other artificial intelligence methods flaws with the advantage of fuzzy logic is that fuzzy logic enables an appropriate framework associates and analyze those uncertainties even independently from their source or characteristics that helps in deep investigation on the main works that utilization of fuzzy logic in HF candidate-well selection. After reviewing literature available on artificial intelligence and fuzzy logic for selection of most suitable well for hydraulic fracturing it is recommended that type-2 fuzzy set and systems can be adopted.

Further to this different kind of algorithms i.e., artificial neural network, fuzzy C-mean and C4.5 decision tree can be used for selection of candidate well for hydraulic fracturing [22]. As making predictions on available approaches depends on high degree of uncertainties. However, it is unlikely to develop accurate mathematical model that can predict the most suitable well for stimulation but fuzzy logic leads a way to understand system behavior by allowing us to interpolate between observed input and output situations.

Fuzzy logic reflects the degree of membership for each parameter in each evaluation criteria system level and calculate the weight of each parameter affecting the well production. Further to this performance of fuzzy logic with respect to predictions able to sufficiently estimate the production of the gas well with high correlation coefficient.

Fuzzy logic not only can be used as a tool for selecting best candidate well but can also be used for optimizing hydraulic fracturing job by type of damage effecting formation, evaluating rock layers that could provide barrier during fracture propagation and may also be used for selecting best treatment fluids and additives for performing hydraulic fracturing job.

References

1. Economides, M.J., Nolte, K.G.: Reservoir stimulation, 3rd edn, p. 856. Wiley, London (2000)
2. Finol, J., Guo, Y.K., Jing, D.Y.: A rule based fuzzy model for the prediction of petrophysical rock parameters. J. Pet. Sci. Eng. **29**, 97–113 (2001)
3. Garrouch, A.A., Lababidi, H.M.S.: Development of an expert system for underbalanced drilling using fuzzy logic. J. Pet. Sci. Eng. **31**, 23–39 (2001)
4. Quenes, A.: Practical application of fuzzy logic and neural networks to fractured reservoir characterization. Comput. Geosci. **26**, 953–962 (2000)
5. Kadkhodaie, I.A., Rezaee, M.R., Moallemi, S.A.: A fuzzy logic approach for estimation of permeability and rock type from conventional well log data: an example from the Kangan reservoir in the Iran offshore gas field. J. Geophys. Eng. **3**, 356–369 (2006)
6. Khademi, H.J., Shahriar, K., Rezai, B., Bejari, H.: Application of fuzzy set theory to rock engineering classification systems: an illustration of the rock mass excavability index. Rock Mech. Rock Eng. **43**, 335–350 (2010)
7. Liao, R.F., Chan, C.W., Hromek, J., Huang, G.H., He, L.: Fuzzy logic control for a petroleum separation process. Eng. Appl. Artif. Intel. **21**, 835–845 (2008)
8. Xiong, H., Holditch, S.A.: Using a fuzzy expert system to choose target well and formations for stimulation. In: Braunschweig, et al. (eds.) Artificial İntelligence in the Petroleum İndustry: Symbolic and Computational Applications, pp. 361–379. Editions Technip, Paris (1995)
9. Yang, E.: Selection of target wells and layers for fracturing with fuzzy mathematics method. In: Sixth International Conference on Fuzzy Systems and Knowledge Discovery, pp. 366–369 (2009)

10. Yin, D., Wu, T.: Optimizing well for fracturing by fuzzy analysis method of applying computer. In: 1st IEEE International Conference on Information Science and Engineering, pp. 286–290 (2009)
11. Zadeh, L.A.: Is there a need for fuzzy logic? Inf. Sci. **178**, 2751–2779 (2008)
12. Saeedi, A., Camarda, K., Liang, J.: Using neural networks for candidate selection and well performance prediction in water-shutoff treatments using polymer gels-a field-case study. SPE Prod. Oper. **22**, 417–424 (2007)
13. Paasche, H., Tronicke, J., Holliger, K., Green, A.G., Maurer, H.: Integration of diverse physical-property models: subsurface zonation and petrophysical parameter estimation based on fuzzy c-means cluster analyses. Geophysics **71**, H33–H44 (2006)
14. Quinlan, J.R.: Improved use of continuous attributes in C4.5. arXiv preprint arXiv:cs/9603103 (1996)
15. Zadeh, L.A.: Probability measures of fuzzy events. J. Math. Anal. Appl. **23**, 421–427 (1968)
16. Guo, J., Xiao, Y.: A new method for fracturing wells reservoir evaluation in fractured gas reservoir. Math. Prob. Eng. (2014)
17. Zadeh, L.A.: Fuzzy sets. Inf. Control **8**, 338–353 (1965)
18. Zadeh, L.A.: Generalized theory of uncertainty (GTU) principal concepts and ideas. Comput. Stat. Data Anal. **51**, 15–46 (2006)
19. Zadeh, L.A.: Fuzzy logic, neural networks, and soft computing: one-page course announcement of CS 294-4. The University of California at Berkeley (1992)
20. Jang, J.S.R., Sun, C.T., Mizutani, E.: Neuro-fuzzy and Soft Computing. A Computational Approach to Learning and Machine İntelligence, p. 614. Prentice-Hall, Englewood Cliffs (1997)
21. Tinkir, M.: A new approach for interval type-2 by using adaptive network based fuzzy inference system. Int. J. Phys. Sci. **6**(19), 4502–4518 (2011)
22. Yong, X., Guo, J., Songgen, S.: A comparison study of utilizing optimization algorithm and fuzzy logic for candidate-well selection. In: SPE/IATMI Asia Pacific Oil & Gas Conference and Exhibition, Nusa Dua, Bali, Indonesia, 20–22 October 2015
23. Zimmermann, H.J.: An application-oriented view of modeling uncertainty. Fuzzy Sets Syst. **122**, 190–198 (2000)
24. Klir, G.J., Wierman, M.J.: Uncertainty-Based Information: Elements of Generalized Information Theory, p. 185. Springer, Heidelberg (1999)

Concept of Risk-Management of International Tour Operator on the Base of Fuzzy Cognitive Model

Anar Y. Rzayev and Inara R. Rzayeva[✉]

Azerbaijan State University of Economics, Istiqlaliyyat Str. 6,
1000 Baku, Azerbaijan
a.rzayev@list.ru, ina3r@mail.ru

Abstract. The article proposes a concept for developing an integrated system of typical fuzzy models for a total assessment of the level of acceptable tour operator risk based on expert opinions regarding internal and external factors of influence. The article proposes a concept for developing an integrated system of typical fuzzy models for a total assessment of the level of acceptable tour operator risk based on expert opinions regarding internal and external influences.

Keywords: Tour operator risk · Fuzzy cognitive model · Fuzzy inference

1 Introduction

Since the publication of the book by Neumann and Morgenstern, Game Theory and Economic Behavior [1], risk assessment has become one of the main applied areas in modeling risk situations in economics and business. In the economic literature, this problem has been studied quite well with examples of the economic activities of production, financial, insurance and other institutions. However, according to a few publications, the theoretical and methodological aspects of the analysis of tourist risks or risk management in the field of tourist services practically fell out of the field of vision of economists. Any activity of tour operators occurs in conditions of uncertainty, therefore, in the process of implementing its functions and providing tourist services, the tour operator (TO) is forced to take into account a large variety of risks that encompass all sorts of threats against participants in the tourist market. Each TO tries to determine for itself the degree of acceptable risks of financial and/or reputational losses. At the same time, the potential loss of the TO is inversely proportional to the volume of its working capital. That is why in the tourist sector of the economy it is necessary to pay special attention to the study of risk management skills of possible losses. Thus, it can be argued that the basis of a successful tourism business is the presence of risk management in the structure of the TO.

The main task of risk management is to determine the best (or optimal) strategy for concluding transactions that ensures the maximum growth of the TO's profit at the expense of the correct multi-criteria choice from all potential transactions with high profitability and reliability. To solve this task and related tasks in relation to all tourist operations (first of all international), in framework risk management employs tools of the

© Springer Nature Switzerland AG 2020
R. A. Aliev et al. (Eds.): ICSCCW 2019, AISC 1095, pp. 420–428, 2020.
https://doi.org/10.1007/978-3-030-35249-3_54

theory of statistical solutions that, setting a quantitative measure of tourist risk, in each case allow us to evaluate and compare the consequences and feasibility of certain transactions. Moreover, statistical analysis tools allow to formalize various tourist operations and, thus, to ensure the accumulation of experience of the TO. At the same time, to form heuristic knowledge from the history of transactions, it is necessary to use and accumulate expert opinions, and, as a result, fuzzy methods of analysis and decision-making regarding tourist operations.

The use of the apparatus of fuzzy logic in tourism management becomes possible due to the fact that along with quantitative measures of reliability of international tourist operations, it is increasingly necessary to apply their qualitative characteristics. It is proposed to use the fuzzy inference mechanism as such an apparatus [2], which is able to combine and aggregate statistical processing of the results of completed transactions with expert opinions regarding various conditions for their conclusion.

2 Problem Definition

The main functions of risk management are: prevision, prevention, localization and elimination of tourist decisions with excessive risk. At the same time, the definition and assessment of tourist risks are always relative, and the desire to attribute a numerical value to them is not always acceptable from the point of view of further interpretation of complex results. The acceptable level of risk that a TO can consider acceptable is a complex concept and cannot be considered as a simple combination of its interrelated and/or interdependent components, since each one is critically significant. Moreover, when assessing the total tourist risk the numerical averaging of the results for all types of tourist operations is not always acceptable. So, it is necessary to develop an adequate system of integrated fuzzy models for the estimate of the total risk.

3 Fuzzy Cognitive Model for Evaluating the Aggregated Risk

According to [3], risk is the probability of unfavorable aftermath or events. This is a situation that has an uncertainty of its outcome or the probability of a possible unde-sirable loss of something under the unfavorable confluence of certain conditions. In the field of tourism and especially of international tourism, risks as the probability of displays of undesirable (sometimes, dangerous) factors, manifest themselves not in isolation, but in aggregate. As a rule, one risk is displayed in the composition of another, or is an effect or a cause. Therefore, the hierarchy that is optimal in terms of risk management strategy should demonstrate the interrelation and interdependence between individual groups and types of tourist risks. As in other sectors of the econ-omy, in the field of tourist services risk is considered as a category of business and institutional activities of the TO. It reflects hidden causal-effect relationships between factors and results. Therefore, criteria for differentiation and criterions of classification of tourism risk that can be identified should be based on the reasons for their occur-rence in the business case, and, in the institutional case, on the differentiation of objects at risk on the base of which the effects of risk realization can be directly observed.

Based on these considerations, a concept and mechanisms for managing of tourism risks are being formed that satisfied the requirements of risk-management. In particular, these requirements imply the need to take into account the essential features of individual risks that allow forming the ways to influence on them. Thus, to regulate the large number of tourist risks, taking into account their interdependence by areas of their occurrence and influence, nature of influence, factors of influence, and localization areas, the logical detailization of tourist risks based on a fuzzy cognitive map (FCM) [4] (Fig. 1), which is formed on the classification of TO risks presented in [5].

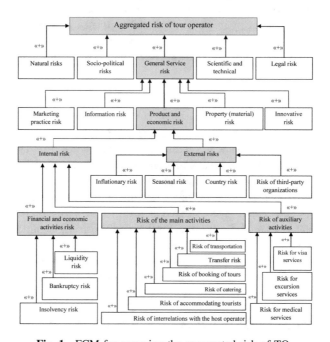

Fig. 1. FCM for assessing the aggregated risk of TO.

In practice, the "interaction" of even two elements of the FCM occurs according to more complex functional regularities, which in the usual traditional mathematical manner are very difficult to formalize. Therefore, there is a need to apply the mechanism of fuzzy inference to describe the cause-and-effect relations among the components of the aggregated tourist risk, and the corresponding analysis is based on the so-called fuzzy cognitive model (FCM) [4]. At the same time, the nodal factors of FCM are interpreted as fuzzy sets, and the cause-and-effect relations among them are established on the basis of a limited set of fuzzy linguistic rules (FLR), which are formed as

"If x_{k1} is A_{k1} and x_{k2} is A_{k2} and .... and x_{kn} is A_{kn}, then y is B_k", where y is output linguistic variable (LV), characterizing the level of consolidated risk; ($j = 1 \div n$; $k = 1, 2, \ldots$) are input LVs, characterizing risk factors; A_{kj} and B_k are the terms (values) of the corresponding input and output LVs described by appropriate fuzzy sets.

4 Verbal Description and Formalization of Cause and Effect Relationships into the FCM

Presented in Fig. 1 cause and effect relationships can be described with the help of a sufficient set of typical consistent FLR. To do this, for description of each node elements of the FCM, we choose the following consistent statements as a basis.

Risks of Financial and Economic Activities

r_{11}: "If there are no liquidity, insolvency and bankruptcy risks, then the risk level of the financial and economic activities of the TO is too low";

r_{12}: "If the liquidity risk situation most probably will not occur, there is nothing definite about the possibility of insolvency risk and there are no bankruptcy risks, then the risk level of the financial and economic activities of the TO is very low";

r_{13}: "If nothing definite can be said about the possibility of liquidity risk, the risk situation most probably does not occur due to insolvency and bankruptcy, then the risk level of the financial and economic activities of the TO is more than low";

r_{14}: "If the liquidity risk situation most probably will not occur, the possibility of insolvency and bankruptcy risk cannot be said for anything certainly, then the risk level of the financial and economic activities of the TO is low";

r_{15}: "If there is nothing definite to say about the possibility of risk in terms of liquidity and insolvency, and in bankruptcy the risk situation most probably comes, then the risk level of the financial and economic activities of the TO is high";

r_{16}: "If risk situations of liquidity and insolvency most probably come, and there is nothing definite to say about the possibility of risk of bankruptcy, then the risk level of the financial and economic activities of the TO is more than high";

r_{17}: "If it is impossible to say anything definitive about the possibility of liquidity risk, the risk situation definitely comes about insolvency, and the risk situation most probably comes about bankruptcy, then the risk level of the financial and economic activities of the TO is very high";

r_{18}: "If the risk situations of liquidity, insolvency, and bankruptcy come exactly, then the risk level of the financial and economic activities of the TO is too high."

Risk of Auxiliary Activities of the Tour Operator

r_{21}: "If there are no risks for visa, excursion, and medical services, then the risk level from the auxiliary activities of the TO is too low";

r_{22}: "If the risk situation by visa services most likely will not occur, about likelihood of risk on excursion services cannot be said for certain and there are no risks for medical services, the level of risk from the auxiliary activities of the TO is very low";

r_{23}: "If nothing can be said definitely about the possibility of the risk on visa services, risk situation most probably will not occur on excursion and medical services, then the level of risk from auxiliary activities of the TO is more than low";

r_{24}: "If the risk situation for visa services most probably will not occur, it is impossible to say anything definitely about the possibility of risk on excursion and medical services, then the level of risk from the auxiliary activities of the TO is low";

r_{25}: "If there is nothing to say definitely about the possibility of risk on visa and excursion services, and on medical services the risk situation most probably to occur, then the level of risk from auxiliary activities of the TO is high";

r_{26}: "If the risk situations for visa and excursion services are most probably to occur, and it is impossible to say anything about the risk of medical services, then the level of risk from auxiliary activities of the TO is more than high";

r_{27}: "If it is impossible to say anything definitively about the possibility visa service, a risk situation will necessarily occur about excursion services, and the risk situation on medical services is most probably to occur, then the level of risk from auxiliary activities of the TO is very high";

r_{28}: "If the risk situations for visa, excursion, and medical services come necessarily, then the level of risk from auxiliary activities of the TO is too high."

Risk of the Main Activities of the Tour Operator

r_{31}: "If the risks of accommodating tourists, their transportation and booking of tours will for sure occur, the risk level of the main activities of the TO is high";

r_{32}: "If, in addition to the listed conditions, we consider that the risk of catering will necessarily occur, then the risk level of the TO main activities is more than high";

r_{33}: "If everything that is noted in r_{32} takes place, and also if the transfer risk and the risk of interrelations with the host operator come for sure, then the risk level of the main activity of the TO is too high";

r_{34}: "If everything that is noted in r_{33} takes place, except for the risk of catering, then the risk level of the main activities of the TO is very high";

r_{35}: "If the risks of tourists accommodating, their transportation, booking of tours and relations with the host operator come for sure, but there are no transfer risks and risks of catering, then the risk level of the main activities of the TO is high";

r_{36}: "If the risks of tourists accommodating and transportation, as well as the risks of relationships with the host operator are absent, then the risk level of the main activities of the TO is low."

Internal Risk

i_{11}: "If the risks of financial-economic and main activities do not exist, and there are no risks of auxiliary activities, then the internal risk of the TO is too low";

i_{12}: "If the risk situation on financial and economic activities most probably will not occur, there is nothing definite about the possibility of the main activities risk and there are no risks of auxiliary activities, then the internal risk of the TO is very low";

i_{13}: "If it is impossible to say for sure about the possibility of risk in financial and economic activities, the risk situation is most probably not to occur in the main and auxiliary activities, then the internal risk level of the TO is more than low";

i_{14}: "If the risk situation in financial and economic activities most probably will not occur, the possibility of risk in the main and auxiliary activities cannot be said anything definite, then the internal risk level of the TO is low";

i_{15}: "If there is nothing definite to say about the possibility of risk in financial, economic and main activities, and in an ancillary activity the risk situation is most probably to occur, then the internal risk level of the TO is high";

i_{16}: "If risk situations in financial, economic and main activities most likely will occur, and it's impossible to say anything about the possibility of risk in supporting activities, then the internal risk of the TO is higher than high";

i_{17}: "If nothing definite can be said about the possibility of risk in financial and business activities, the risk situation definitely will come about in the main activity, and the risk situation in the auxiliary activity probably will come, then the internal risk level of the TO is very high";

i_{18}: "If risk situations in financial and economic, main and auxiliary activities come for sure, then the internal risk level of the TO is too high."

External Risk

e_{31}: "If inflationary and seasonal risks come for sure, then the external risk level of the TO is high";

e_{32}: "If we add to the listed conditions that the risk of third-party organizations will come for sure, then the external risk of the TO is more than high";

e_{33}: "If it takes into account all that is noted in e_{32}, and also if the country risk comes for sure, then the external risk of the TO is too high";

e_{34}: "If it takes into account all that is noted in e_{33}, except for seasonal risk, then the external risk of the TO is very high";

e_{35}: "If inflationary, seasonal and country risks come for sure, but at the same time, the risks of third-party organizations do not exist, then the external risk is still high";

e_{36}: "If inflationary and country risks do not exist, then the external risk is low."

Product and Economic Risk

p_{71}: "If there are no internal and external risks, then product-economic risk of the TO is too low";

p_{72}: "If internal risk is most probably not to occur, and external risk is absent, then product-economic risk of the TO is very low";

p_{73}: "If nothing definite can be said about the possibility of internal risk, and external risk is most probably not to occur, then product-economic risk is more than low";

p_{74}: "If internal and external risks are most probably not to occur, then product-economic risk of the TO is low";

p_{75}: "If internal and external risks are probably to occur, then product-economic risk of the TO is high";

p_{76}: "If internal risk is most probably to occur, and nothing definite can be said about the possibility of external risk, then product-economic risk is more than high";

p_{77}: "If internal risk take place for sure, and external risk is more probably to occur, then product-economic risk of the TO is very high";

p_{78}: "If internal and external risks are certain, then product-economic risk of the TO is too high."

General Service Risk

b_{31}: "If product-economic and marketing practice risks, as well as innovative risk come for sure, then the General Service risk level of the TO is high";

b_{32}: "If we add to the listed conditions that the information risk will come for sure, then the General Service risk level of the TO is more than high";

b_{33}: "If it takes into account all that is noted in b_{32}, and also if the property risk comes for sure, then the General Service risk level of the TO is too high";

b_{34}: "If it takes into account all that is noted in b_{33}, except for information risk, then the General Service risk level of the TO is very high";

b_{35}: "If product-economic and property risks, as well as marketing practice risks come for sure, but at the same time, there are no information and innovative risks, then the General Service risk level of the TO is still high";

b_{36}: "If there are no product-economic and property risks, then the General Service risk level of the TO is low."

Aggregated Risk of TO

a_{31}: "If General Service, natural and legal risks come for sure, then the aggregated risk of the TO is high";

a_{32}: "If we add to the listed conditions that the scientific-technical risk will come for sure, then the aggregated risk of the TO is more than high";

a_{33}: "If it takes into account all that is noted in a_{32}, and also if the socio-political risk comes for sure, then the aggregated risk of the TO is too high";

a_{34}: "If it takes into account all that is noted in a_{33}, except for scientific-technical risk, then the aggregated risk of the TO is very high";

a_{35}: "If General Service, natural and socio-political risks come for sure, but at the same time, there are no scientific-technical and legal risks, then the aggregated risk of the TO is still high";

a_{36}: "If there are no General Service and socio-political risks, then the aggregated risk of the TO is low."

The above reasoning in the form of verbal models allow easily to form a set of LVs and FLR to build appropriate fuzzy inference systems (FIS) relative to the TO risk, as a whole, and its components, in particular. In this case, two type of FIS consisting of eight and six FLR are used. For example, in the case of reasoning $r_{11} \div r_{18}$, all LVs and their terms are summarized in Table 1.

Table 1. Input and output LVs of FIS relative to the risks of financial and economic activities.

LV	Name	Set of terms	Universe
x_{11}	Liquidity risk	$\{A_{11}$ = BE ABSENT, B_{11} = PROBABLY WILL NOT OCCUR, C_{11} = INDEFINITE, D_{11} = PROBABLY COME, E_{11} = COME EXACTLY$\}$	$U = \{0; 0.25;$ $0.5; 0.75; 1\}$
x_{12}	Insolvency risk	$\{A_{12}$ = BE ABSENT, B_{12} = PROBABLY WILL NOT OCCUR, C_{12} = INDEFINITE, D_{12} = PROBABLY COME, E_{12} = COME EXACTLY$\}$	$U = \{0; 0.25;$ $0.5; 0.75; 1\}$
x_{13}	Bankruptcy risk	$\{A_{13}$ = BE ABSENT, B_{13} = PROBABLY WILL NOT OCCUR, C_{13} = INDEFINITE, D_{13} = PROBABLY COME, E_{13} = COME EXACTLY$\}$	$U = \{0; 0.25;$ $0.5; 0.75; 1\}$
y_1	Risk level	$\{TL$ = TOO LOW, VL = VERY LOW, ML = MORE THAN LOW, L = LOW, H = HIGH, MH = MORE THAN HIGH, VH = VERY HIGH, TH = TOO HIGH$\}$	$[0, 1]$

Symbolically, FIS can be represented by the following implicative rules:

r_{11}: $(x_{11} = A_{11})$ & $(x_{12} = A_{12})$ & $(x_{13} = A_{13}) \Rightarrow (y_1 = TL)$;
r_{12}: $(x_{11} = B_{11})$ & $(x_{12} = C_{12})$ & $(x_{13} = A_{13}) \Rightarrow (y_1 = VL)$;
r_{13}: $(x_{11} = C_{11})$ & $(x_{12} = B_{12})$ & $(x_{13} = B_{13}) \Rightarrow (y_1 = ML)$;
r_{14}: $(x_{11} = B_{11})$ & $(x_{12} = C_{12})$ & $(x_{13} = C_{13}) \Rightarrow (y_1 = L)$;
r_{15}: $(x_{11} = C_{11})$ & $(x_{12} = C_{12})$ & $(x_{13} = D_{13}) \Rightarrow (y_1 = H)$;
r_{16}: $(x_{11} = D_{11})$ & $(x_{12} = D_{12})$ & $(x_{13} = C_{13}) \Rightarrow (y_1 = MH)$;
r_{17}: $(x_{11} = C_{11})$ & $(x_{12} = E_{12})$ & $(x_{13} = D_{13}) \Rightarrow (y_1 = VH)$;
r_{18}: $(x_{11} = D_{11})$ & $(x_{12} = D_{12})$ & $(x_{13} = D_{13}) \Rightarrow (y_1 = TH)$.

To estimate of initial internal and external factors exerting influence on risk levels by areas of localization of TO operations, including evaluating the risk components of financial and economic activities, qualitative criteria are used, which are described by the following fuzzy subsets of the universe $U = \{0; 0.25; 0.5; 0.75; 1\}$: BE ABSENT: $A = \{1/0; 0.5/0.25; 0/0.5; 0/0.75; 0/1\}$; PROBABLY WILL NOT OCCUR: $B = \{0.5/0; 1/0.25; 0.5/0.5; 0/0.75; 0/1\}$; INDEFINITE: $C = \{0/0; 0.5/0.25; 1/0.5; 0.5/0.75; 0/1\}$; PROBABLY COME: $D = \{0/0; 0/0.25; 0.5/0.5; 1/0.75; 0.5/1\}$; COME EXACTLY: $E = \{0/0; 0/0.25; 0/0.5; 0.5/0.75; 1/1\}$ [3]. Terms from the right-hand parts of the above rules are described in the form of fuzzy subsets of the discrete universe $J = \{0; 0.1; 0.2; ...; 1\}$, restored by appropriate membership functions [3], namely, $\forall u \in J$: TL = TOO LOW: $\mu_{TL}(u) = 1$, if $u < 1$ and $\mu_{TL}(u) = 0$, if $u = 1$; VL = VERY LOW: $\mu_{VL}(u) = (1-u)^2$; ML = MORE THAN LOW: $\mu_{ML}(u) = (1-u)^{1/2}$; L = LOW: $\mu_L(u) = 1-u$; H = HIGH: $\mu_H(u) = u$; MH = MORE THAN HIGH: $\mu_{MH}(u) = u^{1/2}$; VH = VERY HIGH: $\mu_{VH}(u) = u^2$; TH = TOO HIGH: $\mu_{TH}(u) = 1$ if $u = 1$ and $\mu_{TH}(u) = 0$ if $u < 1$.

At the next stages of processing, the outputs of each integrated fuzzy model form inputs for the next. Thus, obtained data from internal and external sources of risk form the input characteristics for assessing the aggregate level of TO risk.

5 Conclusion

It is obvious that the typical models proposed in the article for assessing the constituent factors of TO risk need structural and parametric training in order to qualify for the necessary degree of adequacy to the problem. Moreover, it is not a fact that the proposed in Fig. 1 FCM has absorbed the absolute majority of factors affecting the aggregate level of TO risk. However, the proposed approach is a certain sense flexible relative to possible additions and/or clarifications that may be presented by experts and/or managers of TOs. Even in the proposed *"imperfect"* version, FCM, without giving absolute values for assessing risk levels, is able to respond to possible changes in the concepts of FCM and form the basis for risk-management of international TO.

References

1. von Neumann, J., Morgenstern, O.: The Theory of Games and Economic Behavior. Princeton University Press, Princeton (1944)
2. Zadeh, L.: Outline of a new approach to the analysis of complex systems and decision processes. IEEE Trans. Syst. Man Cybern. $3(1)$, 28–44 (1973)
3. Rzayev, R., et al.: Two approaches to country risk evaluation. In: Advances in Information and Communication. Future of Information and Communication Conference, vol. 1, pp. 793–812 (2019)
4. Kosko, B.: Fuzzy cognitive maps. Int. J. Man-Mach. Stud. **1**, 65–75 (1986)
5. Ovcharov, A.: Risk management in the sphere of tourist services. Manag. Pract. Bull. St. Petersburg Univ. **8**(2), 138–160 (2008). (in Russian)

Fuzzy Logic-Based User Scheduling Scheme for 5G Wireless Networks and Beyond

Amr Abdelbari⬤ and Huseyin Haci$^{(\boxtimes)}$⬤

Department of Electrical and Electronic Engineering, Near East University,
Nicosia, TRNC, Mersin 10, Turkey
{amr.abdelbari,huseyin.haci}@neu.edu.tr

Abstract. A novel Fuzzy Logic-based user scheduling scheme called "Fuzzy-based User Scheduling Evaluation (FUSE)" for ultra-dense wireless networks is proposed. The proposed scheme aims to overcome a drawback of the state of the art scheduling schemes - Maximum Signal-to-Noise Ration (Max-SNR), Proportional Fairness (PF), Exponential/Proportional Fairness (EXP/PF) and Modified Largest Weighted Delay First (M-LWDF) by providing a better balance between delay and throughput related parameters at scheduling decision making. The performance results demonstrate that the proposed scheme outperforms EXP/PF and M-LWDF.

Keywords: Fuzzy logic · User scheduling · Ultra-dense networks · Resource allocation · 5G wireless networks

1 Introduction

The number of devices connected to the internet is growing with an exponential speed. Cisco expects the number of connected devices will exceed 50 billion by 2020 [1]. With the Internet-of-Things (IoT) innovation, these devices are expected to provide smart applications and services to humans [2]. The data generated by smart applications and services will be conveyed by the fifth generation (5G) mobile networks. [3] predicts that the 5G mobile networks will require up to one thousand times (1000x) more area capacity than the current fourth generation Long-Term Evolution (LTE). Today's wireless technologies cannot support such a demand. Thus, there is need for much research to increase wireless network capacity in magnitudes.

User scheduling scheme and Radio Resource Management (RRM) methods are the major technologies that affect the capacity of a network [4]. Many research available in the literature is based on RRM methods that provide optimal data rate performance in various scenarios for future wireless networks, such as Multiple-Input Multiple-Output (MIMO) antenna arrays, small cells, distributed antenna systems, and Device-to-Device (D2D) communications [5–7].

The literature on user scheduling schemes for future ultra-dense wireless networks is limited [8]. Two popular schemes are EXP/PF and M-LWDF [9, 10]. Although very valuable research, EXP/PF and M-LWDF schemes have the following drawbacks at ultra-dense networks [8]. They provide much of the weight on delay related parameters at the scheduling decision making. In an ultra-dense network with a large number of

© Springer Nature Switzerland AG 2020
R. A. Aliev et al. (Eds.): ICSCCW 2019, AISC 1095, pp. 429–435, 2020.
https://doi.org/10.1007/978-3-030-35249-3_55

real-time traffic users, this makes the scheduler too conservative and limit the network capacity [11]. This paper proposes a novel scheduling scheme for ultra-dense wireless networks to overcome the drawbacks of today's popular schemes. The proposed scheduler is based on a Fuzzy logic control system [12], which aims to provide a high network capacity while preserving fairness among users and minimizing Quality of Service (QoS) violations.

2   System Model

A single cell system is considered with a BS serving N number of geographically distributed users. Orthogonal Frequency Division Multiplexing (OFDM) is assumed to be the multiplexing technique employed, since OFDM can transform a broadband frequency selective channel into M number of narrowband flat-fading sub-channels [13]. For the multiple access technique, Orthogonal Frequency Division Multiple Access (OFDMA) is assumed due to its advantages in providing flexibility and high system capacity [14]. 5G and beyond networks are expected to be ultra-dense networks, where N can be much larger than M [15]. This is due to IoT innovation and smart world concept. [2] envisioned for future network applications. For the channel model, simplified path loss model that includes average shadowing is used for slow fading [16]. Also, multiple fading that follows Rayleigh distribution independently and identically for different users is used for fast fading [17, 18]. The channel coherence time is assumed to be much larger that the scheduling period. Also, Additive White Gaussian Noise (AWGN) on the receiver side.

3   Methodology

The fuzzy logic system consists of three stages: the inputs, the processing, and the outputs [19]. The new method uses four classes corresponding to the priority state of the user. Depending on the related parameters of each user, Fuzzy logic system is used to classify each user and enroll it in the suitable class. Once all the users are classified according to their scheduling priority, the very high priority class is scheduled first. Then, other classes are scheduled accordingly.

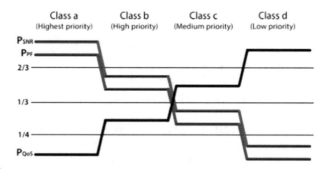

Fig. 1. The inputs to the proposed system FUSE is the values of the channel state information (CSI) parameter denoted as P_{SNR}, QoS parameter denoted as P_{QoS}, and Proportional Fairness (PF) parameter denoted as P_{PF}.

3.1 Mathematical Foundation

The fuzzy logic system consists of three stages: the inputs, the processing, and the outputs [19]. The inputs to the proposed system are the values of the channel state information (CSI) parameter, QoS parameter, and Proportional Fairness (PF) parameter for served users. The system outputs are four classes contain all users that have been classified according to their scheduling priority. The processing stage includes the rules base. Figure 1 illustrates this classification according to the priority status of each input parameter.

Let u_i is the ith user and $C_k(j)$ is the kth priority class of the jth input parameter. The CSI parameter is calculated as:

$$P_1(u_i) = \log(|SNR_i|) \tag{1}$$

where the SNR is the Signal-to-Noise Ratio for ith user. The QoS parameter is calculated as:

$$P_2(u_i) = \exp(\tau_{rt} - D_i) \tag{2}$$

where D_i is the delay of the first packet to transmit and τ_{rt} is the delay threshold of the real-time traffic. The PF parameter is calculated as:

$$P_3(u_i) = \frac{s_i}{Y_i} \tag{3}$$

where s_i is the instant channel state and Y_i is the average channel state for the ith user. Each class have a specific value for each parameter that if the inputs of any user does not exceed it, then it will be enrolled in that class. These values are arranged by the following function:

$$P_j(C_k) = \gamma_k * \frac{1}{N} * \sum_{i=1}^{N} P_j(u_i), \quad \text{where } \gamma_k \in \left\{ \frac{2}{3}, \frac{1}{3}, \frac{1}{4}, 0 \right\} \tag{4}$$

The fuzzy rule set follow these arguments:

Rule a. IF $P_j(u_i) < P_j(C_k)$ AND $P_j(u_i) > P_j(C_{k-1})$, THEN $u_i \in C_k$
Rule b. IF $P_j(u_i) < P_j(C_{k+1})$ AND $P_j(u_i) > P_j(C_k)$, THEN $u_i \in C_{k+1}$
Rule c. IF $P_j(u_i) < P_j(C_{k+2})$ AND $P_j(u_i) > P_j(C_{k+1})$, THEN $u_i \in C_{k+2}$
Rule d. IF $P_j(u_i) < P_j(C_{k+3})$ AND $P_j(u_i) > P_j(C_{k+2})$, THEN $u_i \in C_{k+3}$

where $P_j(u_i)$ is the jth input parameter value of the ith user. $P_j(C_k)$ is the jth parameter value of the kth class. For our system here, the total number of rules is $(J*K)$ where J is the total number of parameters and K is the total number of classes. The rules are not sorted in the same way for all parameters. For instance, rule set of CSI parameter is raising its values while rule set of QoS parameter is in descending order. Thus, the user with the highest SNR and shortest time till QoS violation is classified as very high priority and enrolled in class a. The final decision of scheduling is taken by

$$H(RB_v) = max\{P_1(u_i) * P_3(u_i) + P_2(u_i)\} \tag{5}$$

where RB_v is the vth Resource Block (RB).

3.2 The Proposed Algorithm

To sum up, the algorithmic approach of the new proposed method follows:

Step 1: Determining the CSI, QoS and PF for ith user at the vth RB.
Step 2: Using Fuzzy-based system, the users are classified to classes.
Step 3: Scheduling the maximum user with the most urgent priority by (5).
Step 4: Repeat steps 1-3 until all RB are occupied.

4 Simulation Results

The simulations done using MATLAB software (version 2018a) with 1000 Monte-Carlo trials. The total bandwidth is 10 MHz where each sub-carrier has 15 kHz band. The noise is assumed to be AWGN. The RT streams delay threshold is assumed to be 10 ms which will be in the 5G networks. The user's movement is set to be random within a range of 500 m. The RT traffic is forecast to be demanded more than nRT traffic. Therefore, a ratio of 70:30 RT to nRT traffic is considered. A number of 50 RBs are assumed to be available at the BS while the number of users is varying from 20 to 80. This is to simulate the scenarios from lightly to very densely populated networks.

The performance of the proposed scheduling scheme FUSE has been measured using three important performance metrics; spectrum efficiency ratio, QoS delay violations ration, and Jain's fairness ratio by comparing it with well-known schemes.

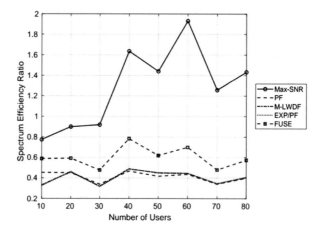

Fig. 2. The spectrum efficiency ratio of the proposed scheduling method FUSE compared with MAX-SNR, PF, EXP/PF and M-LWDF method.

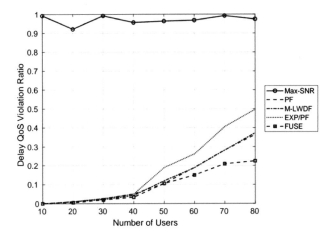

Fig. 3. The QoS delay violation for the proposed scheduling method FUSE compared with MAX-SNR, PF, EXP/PF and M-LWDF method.

Figure 2 shows the spectrum efficiency ratio verses the number of users of the total allocated throughput to the overall bandwidth for FUSE scheme compared with Max-SNR, PF, M-LWDF, and EXP/PF schemes. FUSE scheme overcomes both delay and fairness care schemes. Max-SNR achieves the highest throughput among the schemes because it gives the priority only to the good users.

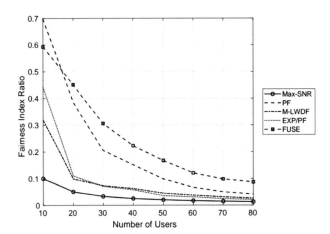

Fig. 4. The Jain's Fairness Ratio for the proposed scheduling method FUSE compared with MAX-SNR, PF, EXP/PF and M-LWDF method.

Figure 3 shows the QoS delay violations of the compared scheduling schemes where it is measuring how many packets have been dropped because its delay exceed the delay threshold. It shows that FUSE scheme has the lowest delay violations with ratio under 0.22 while other schemes' ratio is over 0.35. Max-SNR scheme performance is the worst where almost the good users only served while other users' packets left until it violate the delay limits without transmission.

In Fig. 4, Jain's fairness ratio used as the performance metric to measure the equality of sharing resources among the served users. It shows that both PF and FUSE schemes achieve higher fairness than other schemes where FUSE perform better than PF as the number of users rise higher than 20 served users. The Max-SNR scheme suffer from large bias towards the good users which can be seen on it performance on the Jain's fairness ratio.

5   Conclusions

A Fuzzy Logic-based user scheduling scheme called as FUSE is proposed. FUSE can overcome a maser drawback of both EXP/PF and M-LWDF schemes and significantly enhances the channel capacity, user fairness and packet delay metrics at ultra-dense wireless networks. It has been shown that Fuzzy Logic rules are developed to let the scheduler be more opportunistic in scheduling decisions, especially at congested scenarios.

References

1. Piao, Z., Peng, M., Liu, Y., Daneshmand, M.: Recent advances of edge cache in radio access networks for internet of things: techniques, performances, and challenges. IEEE Internet Things J. **6**(1), 1010–1028 (2019)
2. Li, W., Song, H., Zeng, F.: Policy-based secure and trustworthy sensing for internet of things in smart cities. IEEE Internet Things J. **5**(2), 716–723 (2018). https://doi.org/10.1109/jiot. 2017.2720635. ISSN 2327-4662
3. Osseiran, A., Boccardi, F., Braun, V., Kusume, K., Marsch, P., Maternia, M., Queseth, O., Schellmann, M., Schotten, H., Taoka, H., Tullberg, H., Uusitalo, M.A., Timus, B., Fallgren, M.: Scenarios for 5G mobile and wireless communications: the vision of the metis project. IEEE Commun. Magazine **52**(5), 26–35 (2014)
4. Li, C., Xia, B., Shao, S., Chen, Z., Tang, Y.: Multi-user scheduling of the full-duplex enabled two-way relay systems. IEEE Trans. Wirel. Commun. **16**(2), 1094–1106 (2017)
5. Chen, S., Ma, R., Chen, H., Zhang, H., Meng, W., Liu, J.: Machine-to-machine communications in ultra-dense networks: a survey. IEEE Commun. Surv. Tutor. **19**(3), 1478–1503 (2017)
6. Morsi, R., Michalopoulos, D.S., Schober, R.: Multi-user scheduling schemes for simultaneous wireless information and power transfer. In: 2014 IEEE International Conference on Communications (ICC), pp. 4994–4999. IEEE (2014)
7. Gao, H., Yuen, C., Ren, Y., Lv, T., Long, W.: Distributed user scheduling for mimo-y channel. IEEE Trans. Wirel. Commun. **14**(12), 7123–7139 (2015)
8. Haci, H., Abdelbari, A.: A novel scheduling scheme for Ultra-Dense networks. In: 2019 International Symposium on Networks, Computers and Communications (ISNCC): Wireless and Mobile Networks (ISNCC-2019 WMN). Springer, Istanbul (2019)
9. Xulu, S., Aiyetoro, G.: Cross-layer design approach based packet scheduling in next generation wireless networks. In: 2018 14th International Wireless Communications Mobile

Computing Conference (IWCMC), pp. 757–761. IEEE (2018). https://doi.org/10.1109/iwcmc.2018.8450439

10. Hu, Y., Wu, G.: The stability analysis for predictive M-LWDF scheduling Alg, in M-WIMAX. In: 2015 International Conference on Computational Intelligence and Communication Networks (CICN), pp. 579–584. IEEE (2015). https://doi.org/10.1109/cicn.2015.119

11. Liu, Y., Li, X., Yu, F.R., Ji, H., Zhang, H., Leung, V.C.M.: Grouping and cooperating among access points in user-centric ultra-dense networks with non-orthogonal multiple access. IEEE J. Sel. Areas Commun. **35**(10), 2295–2311 (2017). https://doi.org/10.1109/jsac.2017.2724680. ISSN 0733-8716

12. Trillas, E., Eciolaza, L.: Fuzzy Logic: An Introductory Course for Engineering Students. Springer, Switzerland (2015). https://doi.org/10.1007/978-3-319-14203-6

13. Zhou, X., Ho, C.K., Zhang, R.: Wireless power meets energy harvesting: A joint energy allocation approach in OFDM-based system. IEEE Trans. Wirel. Commun. **15**(5), 3481–3491 (2016). https://doi.org/10.1109/TWC.2016.2522410

14. Nam, C., Joo, C., Bahk, S.: Joint subcarrier assignment and power allocation in full-duplex ofdma networks. IEEE Trans. Wirel. Commun. **14**(6), 3108–3119 (2015). https://doi.org/10.1109/TWC.2015.2401566

15. Shaozhen, G., Chengwen, X., Zesong, F., Gui, Z., Xinge, Y.: Distributed chunk-based optimization for multi-carrier ultra-dense networks. China Commun. **13**(1), 80–90 (2016). https://doi.org/10.1109/CC.2016.7405706

16. Cheffena, M.: Performance evaluation of wireless body sensors in the presence of slow and fast fading effects. IEEE Sens. J. **15**(10), 5518–5526 (2015). https://doi.org/10.1109/JSEN.2015.2443251

17. Taricco, G.: On the convergence of multipath fading channel gains to the rayleigh distribution. IEEE Wirel. Commun. Lett. **4**(5), 549–552 (2015). https://doi.org/10.1109/LWC.2015.2456066

18. Wang, H., Zheng, T., Xia, X.: Secure miso wiretap channels with multi-antenna passive eavesdropper: artificial noise vs. artificial fast fading. IEEE Trans. Wirel. Commun. **14**(1), 94–106 (2015). https://doi.org/10.1109/twc.2014.2332164

19. Mohammadi, A., Dehghani, M.J.: Spectrum allocation using fuzzy logic with optimal power in wireless network. In: 2014 4th International Conference on Computer and Knowledge Engineering (ICCKE), pp. 532–536. IEEE (2014). https://doi.org/10.1109/iccke.2014.6993403

About One Approach to the Description of Semi-structured Indicators on a Given Data Sample

Araz R. Aliev[1,3(✉)] and Ramin R. Rzayev[1,2,3]

[1] Azerbaijan State Oil and Industry University, Azadlig Avenue 34,
AZ1010 Baku, Azerbaijan
alievaraz@yahoo.com, raminrza@yahoo.com
[2] Institute of Control Systems, B. Vahabzadeh Street 9,
AZ1141 Baku, Azerbaijan
[3] Institute of Mathematics and Mechanics, B. Vahabzadeh Street 9,
AZ1141 Baku, Azerbaijan

Abstract. The approach to the description of semi-structured socio-economic indicators is proposed, which is formulated on the example of the indicator "Average income" in a hypothetical company based on the weighted average estimation of employees' salaries. In the framework of this approach, the distribution function of the weights of employees' salaries is identified as a self-similar solution of a partial differential equation with initial and boundary conditions, which also initiates the membership function of a fuzzy set describing the indicator under consideration.

Keywords: Semi-structured indicator · Weighted average estimate · Fuzzy set

1 Introduction

Recent advances in solving problems of forecasting and decision-making have been achieved mainly by neuro-fuzzy data processing technologies. In particular, standard algorithms for solving problems of forecasting dynamics of various kinds of indicators in the logical basis of neural networks perform exclusively with "clean" and/or structured data, i.e. data presented as averaged numbers. Therefore, averaging the results of measuring indicators is one of the most common operations in data acquisition and control systems. In particular, in technical systems the achievement of the required accuracy in the average process is achieved by multiple measurements, where the results of individual measurements are partially compensated by positive and negative deviations from the exact value. At the same time, the accuracy of their mutual compensation improves with an increase in the number of measurements, since the mean value of negative deviations module converges to the mean value of positive deviations.

However, what shall we do with statistical data samples with the status of "historical"? As a rule, such samples characterize the behavior of economic systems for a certain period, or represent a set of indicator values that reflect a particular phenomenon, as given, which cannot be re-measured. Nevertheless, it is necessary to be

© Springer Nature Switzerland AG 2020
R. A. Aliev et al. (Eds.): ICSCCW 2019, AISC 1095, pp. 436–444, 2020.
https://doi.org/10.1007/978-3-030-35249-3_56

able to analyze correctly and assess adequately such indicators, reflecting economic processes and/or phenomena in past, present and future. At the same time, the use of quantitative methods of analysis and processing of relevant information allows to obtain the most adequate assessment of enterprises activities, micro- and macroeconomic indicators, as well as their future values. Therefore, unlike mechanical systems, where multiple measurements of the indicator are possible, in case of reflection of the generalized value of the indicator with the historical status, multiple adjustments of its averages are possible. This is the subject of this paper.

2 How to Achieve the Necessary Adequacy of Averaging?

Over the past few centuries, by efforts of many well-known scientists such as Laplace, Euler, D'Alambert, Hadamard, and others, there has been developed the universal *linear theory* of processes in various environments, in which basic models can be distinguished that reflect the amazing array of phenomena of different nature [1]. In particular, these are *the heat conduction equation*, reflecting the heat distribution in a limited area, *the wave equation* describing the vibration of the string, and *the Laplace equation*, which determines the potential of the field created by the system of electric charges.

Now, suppose that [a, b] is a long interval of length l, and it covers the set of data $U = \{u_k\}_{k=0}^{s}$. Each of the data in its own way reflects the desired value of the economic indicator P under consideration for concrete moment of time. For general reflection and next processing of the values of such indicators, there are various methods of averaging are used in mathematical statistics. We'll go the other way.

For each data of the given interval [a, b] we associate the spatial coordinate u ($a \leq u \leq b$ or $0 \leq u \leq l$). Let's analyze it at the initial stage, when $t = 0$, the value of the indicator P is reflected by the positive function $w(u, 0) = w_0(u)$, which determines the initial distribution of weights between the data u_k. The following question arises: how should the distribution of weights (or the function $w(u, t)$) be changed in order to improve the adequacy of indicator averaging?

First, based on the concept of weights, the desired function must take its values on an unit segment, i.e. $w(u, t^*) \in [0, 1]$. Secondly, its greatest value should be one, which implies the greatest weight of the average value u_{avr} of the indicator P. Actually, the search process of the function $w(u, t^*)$ is completed and, as a result, the averaging of the indicator P can be determined by the following formula of the weighted mean estimation

$$u_{\text{avr}} = \frac{\sum_{k=1}^{s} w(u_k, t^*) \cdot u_k}{\sum_{k=1}^{s} w(u_k, t^*)}, \tag{1}$$

where s is the number of measurements of the indicator P.

We will grope for the answer to the above question in a continuous medium, which in our case is characterized by continuous functions of the spatial and temporal coordinates. The existing notion of the continuous medium has become the key paradigm of modern natural science based on the theory of heat conduction and new mathematical methods laid down by the French mathematician Fourier. For this purpose, we select around a certain point $x \in [a, b]$ the neighborhood with radius $\Delta u/2$: $(x-\Delta u/2; x + \Delta u/2)$. Let $w(x, \tau)$ be a weight distribution function at the moment of time τ. After the short time Δt, this function will be equal to $w(x, \tau+\Delta t)$. Obviously, it can be changed only due to the difference between the "flows" of positive (from the right) $D(x + \Delta u/2, \tau)$ and negative (from the left) $D(x-\Delta u/2, \tau)$ deviations from the value of x, namely,

$$w(x, \tau + \Delta t) - w(x, \tau) = [D(x + \Delta u/2, \tau) - D(x - \Delta u/2, \tau)]\Delta t. \tag{2}$$

In fact, equality (2) reflects the law of conservation of matters. On an intuitive level, it is clear that the greater the difference between the weights of the historical data x that vary over a short period Δt, the greater the difference between positive and negative deviations, which means that it is necessary to find an even more adequate averaging of the value of indicator P. This understanding will help us to further reasoning.

Now we consider the limit in the continuous medium, i.e., assuming

$$D(u, t) \cong [w(u + \Delta u/2, t) - w(u - \Delta u/2, t)]/\Delta u, \tag{3}$$

we will assume that

$$D(u, t) \cong \frac{\partial w(u, t)}{\partial u}. \tag{4}$$

Substituting expression (3) into equality (2) and carrying out the limiting process for $\Delta u \to 0$ and $\Delta t \to 0$, one can obtain the following partial differential equation:

$$\frac{\partial w(u, t)}{\partial t} = \frac{\partial^2 w(u, t)}{\partial u^2} \tag{5}$$

under appropriate initial and boundary conditions

$$w(u, 0) = w_0(u); \ w(0, t) = w(l, t) = 0. \tag{6}$$

In essence, Eq. (5) is the analogue of the heat conduction equation describing heat transfer with the unit value of the thermal diffusivity in the right part of the differential equation. In some sources this coefficient is called as heat conductivity factor, although the thermal diffusivity is calculated as the ratio $a = \lambda/(\rho c)$, where λ is the heat conductivity factor, ρ is a density, c is a heat capacity. In general, as it is known, an equation of the form (5) with the corresponding coefficient in the right part of the differential equation is used to describe the particle diffusion, the penetration of a magnetic field into plasma, and many other processes. Moreover, defining the operator B in a certain Hilbert space H by the equality $Bw = -\frac{d^2 w}{du^2}$ under the conditions $w(0) = w(l) = 0$, it becomes clear that B is a self-adjoint positively defined operator. Then the problem (5), (6) can be reduced to an operator-differential equation of the form

$$\frac{dw(t)}{dt} + Bw(t) = 0 \tag{7}$$

with initial condition

$$w(0) = w_0. \tag{8}$$

where $w_0 \in D(B^{1/2})$ $(D(B^{1/2})$ is the domain of the operator $B^{1/2})$. At the same time, the solution of the Cauchy problem (7), (8) is found in the form of $w(t) = \varphi \cdot e^{-Bt}$, where e^{-Bt} is a strongly continuous semigroup generated by operator $-B$ (see [2]), and φ is an unknown vector determined from the initial condition (8): $w(0) = \varphi = w_0$.

Thus, the solution of the Cauchy problem (7), (8) is represented as $w(t) = w_0 \cdot e^{-Bt}$. It should be noted that the results on solvability of initial-boundary problems for parabolic operator-differential equations and problems without initial conditions for inverse parabolic operator-differential equations can be found in [3–6] and in the references there.

Returning to the problem (5), (6), we note that to solve it, it is necessary to know (or specify) the initial state of the unknown function $w (u, 0) = w_0(u)$. The presence of the boundary conditions $w(0, t) = w(l, t) = 0$ means that zero values of weights are maintained at the boundary of the segment $[0, l]$. In the presence of initial and boundary conditions, the computer step-by-step solution of problem (5), (6) looks like it is shown in Fig. 1. Here, the initial solution $w(u, 0) = w_0(u)$ is a sufficiently "narrow" symmetric curve with the maximum amplitude (or vertex) at the point with abscissa – the arithmetical mean of all measurements of the indicator P under consideration, which in the context of the problem (5), (6) is the middle of the segment $[0, l]$.

Let us follow the evolution of the weight distribution curve. Figure 1 shows the profiles of the weight distribution functions at different steps of the iterative search for the desired solution of the problem (5), (6). It can be seen that the maximum amplitude A of the curve $w(u, t)$ decreases, and the width of the profile L at the level $A/2$ increases. At the same time, based on the above considerations and assuming that the sum of the weights in (1) remains invariable, it becomes obvious that $A \cdot L \cong \text{const}$.

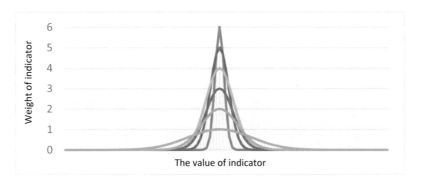

Fig. 1. Evolution of the indicator weight distribution

The decision of two variables function can be found as the product of two functions, for example, as

$$w(u, t) = f(u) \cdot g(t), \tag{9}$$

where, in the process of evolution the function $f(u)$ assigns the spatial configuration (solution form) $w(u, t)$, and $g(t)$ demonstrates the change in the amplitude of the distribution function of the weights of the indicators. Solutions of this kind do not change their forms in the process of evolution (or step-by-step iteration) and are called as *self-similar*, i.e. similar to themselves. At the same time, functions in the form of (9) have a very significant part in the study of nonlinear processes and phenomena.

Suppose, that a self-similar solution of problem (5), (6) is found in the form of (9). Then $f(u)$ is called as an *eigenfunction*, which depends very closely on the boundary conditions. In particular, in the self-similar solution of problem (5), (6) those are the infinite number of functions of the form $f_k(u) = \sin[\pi k u/l]$, $k = 1, 2, \ldots$. At the same time, each of them has its own function of amplitude change on time $g_k(t) = (A-t) \cdot \exp[-(\pi k/l)^2 \cdot t]$, $k = 1, 2, \ldots$, where A is the value of peak amplitude.

3 Identification of the Distribution Function of Measurement Weights

Let us consider a very trivial task of determining the value of the economic indicator "Average wage" in the scale of a small firm, whose staff list consists of one manager and five ordinary employees. Suppose that the monthly income of each of the five ordinary employees is $200, and that of a manager is $2,000. Then, according to the traditional method of calculation, the average income of an employee of the firm is the arithmetical mean of the income of all employees, i.e. $500. Obviously, this value does not quite adequately reflect the real level of the average income in the given firm and does not take into account the relativity of the influence of its components.

To apply the formula of the weighted average (1) as the solution to problem (5), (6), we identify the appropriate distribution function of the weights of the wage indicators of all firm employees. To this purpose let us consider Eq. (5) with the following initial and boundary conditions: $w(u, 0) = w_0(u)$, $w(0, t) = w(2050, t)$, on the interval [0, 2050] covering the statistical sample of data on wages of firm employees.

The desired function is found in the form of a self-similar solution (9), where the function $w(u, t_0) = 7 \cdot \sin[\pi u/2050] \exp[-(\pi/2050)^2 \cdot t_0] = 7 \cdot \sin[\pi u/2050]$ is chosen as the initial condition at $t_0 = 0$. The computer solution that is presented in Fig. 2, generate the final solution of the assigned task at $t = 6$ in the form of the following weight distribution function:

$$w(u, 6) = \sin[\pi u/2050] \cdot \exp[-(\pi/2050)^2 \cdot 6], \tag{10}$$

the largest value of which, as it is not difficult to notice, is 1. At the same time, the weights of salaries of two types of employees of the firm will be the corresponding numbers: $w(200, 6) = 0.3017$ and $w(2000, 6) = 0.0765$.

Calculations of the average income of firm employees are performed taking into account the identified weight distribution function (10) and the criterion of weighted mean estimation (1), or more specific factors

$$u_{avr} = \frac{\sum_{k=1}^{6} w(u_k, 6) \cdot u_k}{\sum_{k=1}^{6} w(u_k, 6)},$$

where $u_1 = u_2 = u_3 = u_4 = u_5 = 200$, $u_6 = 2000$, and $w(u_k, 6) = \sin[\pi u_k/2050] \cdot \exp[-(\pi/2050)^2 \cdot 6]$, which are presented in Table 1.

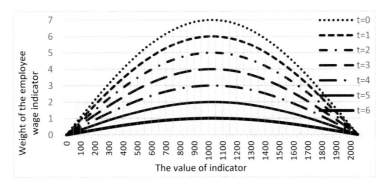

Fig. 2. Evolution of the distribution function of the weights of the components of the "Average income" indicator in the small firm.

Table 1. Estimation of the "Average salary" on the example of the staff list of small firm.

Subdivision	Position	Staff units	Salary ($)	Salary weight	Averaging by formula of:	
					Arithmetical mean ($)	Weighted mean ($)
Administration	Director	1	2,000	0.0765	2,000	153.10
Staff	Office-manager	5	200	0.3017	1,000	301.72
Total		6	×	×	500	286.92

Obviously, the averaging obtained for the indicator "Average income" in the form of the weighted mean value \$286.92 more objectively reflects the level of average salaries for the firm as a whole, rather than the arithmetic mean value \$500.

Now, let us adapt this approach for the company with a much larger number of staff units. In particular, let us consider the staff list of the limited company, which is presented in Table 2. Assuming at $t_0 = 0$ the function

$$w(u, t_0) = 7 \cdot \sin[\pi u/3550] \exp[-(\pi/3550)^2 \cdot t_0] = 7 \cdot \sin[\pi u/3550]$$

as the initial condition of the problem (5), (6), the appropriate self-similar solution can be obtained at the stage $t = 6$ as the following function (see Fig. 3):

$$w(u,\ 6) = \sin[\pi u/3550] \cdot \exp[-6(\pi/3550)^2].$$

Calculations of the average income of the employees of company, taking into account this function and the criterion of weighted mean (1), are summarized in Table 2.

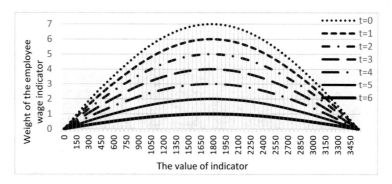

Fig. 3. Evolution of the distribution function of the weights of the components of the "Average income" indicator in the limited company.

Table 2. Estimation of the "Average salary" on the example of the staff list of ltd company.

Subdivision	Position	Staff units	Salary ($)	Salary weight	Averaging by the formula of:	
					Arithmetical mean ($)	Weighted mean ($)
Administration	Director	1	3,500	0.0442	3,500	154.82
	Secretary	2	1,200	0.8733	2,400	2,095.92
Accounting	Chief accountant	1	2,500	0.8011	2,500	2,002.85
	Payroll accountant	1	1,800	0.9998	1,800	1,799.55
Sale	Division head	1	2,700	0.6833	2,700	1,844.78
	Sales manager	20	2,000	0.9802	40,000	39,209.50
	Secretary	2	1,000	0.7739	2,000	1,547.77
Back office	Office-manager	1	1,500	0.9705	1,500	1,455.79
	Forwarding agent	5	1,300	0.9129	6,500	5,934.11
Total		34	×	×	1,850	1,810.08

The averaging obtained for the indicator "Average income" in the limited company in the form of the weighted mean estimation $1,810.08 is not significantly different from the arithmetic mean value $1,850. Nevertheless, the results of the previous example incline towards the selection of the weighted mean estimation, as the most adequate reflection of the "Average income" indicator of the company employee.

4   Conclusion

In the overwhelming majority of real socio-economic problems (and not only) the relevant data, as indicators values should be considered weakly structured and/or even unstructured, deep down [7, 8]. For example, the interval $[P_{\min}, P_{\max}]$ including all measurements can be presented as the adequate reflection of the semi-structured data of the socio-economic indicator P.

Another more adequate reflection of semi-structured data is the terms of linguistic variables of the form "ABOUT 70", which can formally be reflected by fuzzy sets [9]. For example, on the base of these reasoning the descriptions of the linguistic variable "*Salaries of a company employees*" that possess their value in the form of terms: LOW, MIDDLE, and HIGH, can be constructed by appropriate fuzzy formalisms. In particular, in the context of the "Average income" problem in the limited company with the staff list presented in Table 2, the value of indicator "Average income" can be reflected:

- in the continuous case, i.e. when the universe is the continuation $[0, 3550]$ as the membership function $w(u, 6) = \sin[\pi u/3550] \cdot \exp[-6 \, (\pi/3550)^2]$;
- in the discrete case, i.e. when the universe is the discrete set $U = \{u_1, u_2, ..., u_9\}$, consisting of the values of the salaries of company employees (see 4th column of the Table 2), in the form of fuzzy set $A = \{w(u_1, 6)/u_1; w(u_2, 6)/u_2; ...; w(u_9, 6)/u_9\}$, or more specifically: $A = \{0.7739/1000; 0.8733/1200; 0.9129/1300; 0.9705/1500; 0.9998/1800; 0.9802/2000; 0.8011/2500; 0.6833/2700; 0.0442/3500\}$.

The given procedure for describing the value of the indicator in terms of fuzzy logic is called fuzzification, i.e. its representation in the form of a fuzzy set, where the main point is the construction of the appropriate membership function [9]. Actually, the reasoning declared in the previous sections are aimed at forming a new approach to the fuzzification of semi-structured data.

References

1. Kurdyumov, S.P., Malinetsky, G.G., Potapov, A.B.: Synergetic – new directions. New in life, science, technology. Ser. "Math. Cybern. (11), 3–48 (1989). (in Russian)
2. Krein, S.G.: Linear differential equations in a Banach space. Science, Moscow (1967). (in Russian)
3. Aliev, A.R., Lachinova, F.S.: On the solvability in a weighted space of an initial-boundary value problem for a third-order operator-differential equation with a parabolic principal part. Math. Rep. **93**(1), 85–88 (2016)
4. Aliev, A.R., Mirzoev, S.S., Soylemezo, M.A.: On solvability of third-order operator-differential equation with a parabolic principal part in weighted space. J. Funct. Spaces **2017**, 1–8 (2017). Article ID 2932134
5. Aliev, A.R., Soylemezo, M.A.: Problem without initial conditions for a class of inverse parabolic operator-differential equations of third order. Math. Rep. **97**(3), 199–202 (2018)
6. Aliev, A.R., Soylemezo, M.A.: Solvability conditions in weighted Sobolev type space for one class of inverse parabolic operator-differential equations. Azerbaijan J. Math. **9**(1), 59–75 (2019)

7. Rzayev, R.R., Agayev, F., Agamaliyev, M.A., Gasanov, V.I.: Overcoming of semantic uncertainty in criterion concepts of a procedural law based on using fuzzy inferences. Procedia Comput. Sci. **102**, 209–216 (2016)
8. Yazenin, A., Rzayev, R.R., Suleymanova, A.N.: Two approaches to assessing the level of confidentiality of data to be included in a future document. Fuzzy Syst. Soft Comput. **13**(1), 59–80 (2018)
9. Zadeh, L.A.: The Concept of a Linguistic Variable and Its Application to Approximate Reasoning. Elsevier Publishing, New York (1974)

Classification of Heart Diseases Using Fuzzy Inference System (FIS) with Adaptive Noise Cancellation (ANC) Technique for Electrocardiogram (ECG) Signals

Berk Dağman$^{(\boxtimes)}$

Department of Electrical and Electronic Engineering, Near East University,
Near East Boulevard, P.O. Box: 99138, Nicosia, TRNC, Mersin 10, Turkey
berk.dagman@neu.edu.tr

Abstract. Today, the use of computer technology is increasing in the fields of medical diagnosis and treatment of illnesses and patients. The electrocardiogram is the bio-electrical signal that records the activity of the heart against the electrical time. An electrocardiogram is a significant diagnosis device in order to detect the functions of the heart. Electrocardiography is the electrical activity record in the heart to examine the operation of the cardiac muscle and the neural transmission system. In clinical practice, ECG is a very substantial diagnostic device. It is especially useful in the diagnosis of rhythm diseases, changes in electrical conduction and diagnosis of myocardial ischemia and infarction. Fuzzy logic is a combination of people's experiences, by using the obtained values with certain algorithms, depending on each rule that will be created, certain mathematical with the help of functions. ANC (Recursive Least Squares algorithm) is proposed to remove artifacts that preserve the low-frequency components and tiny properties of the electrocardiogram. When new arriving signal samples are received at each iteration, the least-squares problem solution can be computed in a recursive form resulting in RLS algorithms. It is known that RLS algorithms maintain fast convergence. The aim of this manuscript is to detect the heart diseases in the person by using Fuzzy Logic Inference System. This manuscript suggests filtering method Adaptive Noise Cancellation (RLS Algorithm) that detect the heart diseases in the person by using Fuzzy Logic Inference System to help decrease noise interference in Electrocardiogram signals and better diagnose outcomes.

Keywords: Electrocardiogram · Adaptive noise cancellation · Recursive Least-Squares (RLS) algorithm · Common diseases · Fuzzy inference system (FIS) · Fuzzy logic

1 Introduction

Electrocardiography is a method utilized to record cardiac electrical activity to examine the functioning of the heartbeat and nerve conduction system. These electrodes indicate decreasing electrical change on the skin resulting from an electrophysiological pattern of depolarization during each heartbeat of the heart muscle [1]. Our body usually

© Springer Nature Switzerland AG 2020
R. A. Aliev et al. (Eds.): ICSCCW 2019, AISC 1095, pp. 445–454, 2020.
https://doi.org/10.1007/978-3-030-35249-3_57

reports data about our health. These data include heart rate, oxygen saturation, blood pressure, neurotransmission, blood sugar, brain movements, and so on. It can be obtained by measuring physiological materials. One of the issues related to the processing of biomedical data, such as electrocardiography, is desirability. Noise is caused by external electromagnetic fields, electrical line interactions, high-frequency interference, and random body movements. Various digital filters are used to remove signal components from unwanted frequency ranges. It is not easy to use filters with constant coefficients to reduce random noise because human behavior is not fully known within the time interval. In some biomedical signal processing applications, various useful signals may overlap with various components [2]. The intervention from these resources, for example:

- **Interference of Power line:** This interference generally includes 50 Hz harmonics from the sinusoidal signal.
- **The noise of electrode contact:** This noise is caused by the loss of contact between the skin and the electrode, which affects the signal measurement.
- **The noise of muscle contraction:** The baseline of the electromyogram is generally in the microvolt range, which allows it to be insignificant.
- **The noise of electrosurgical:** This noise completely destroys the ECG signal. It can be indicated with a wide amplitude.
- **Movement of the patient:** For this reason, patient movements are transient in basal changes with electrode-skin impedance changes.

The ANC is an alternative predictor signal that is distorted by additional noise or interference. The benefit is that it is difficult or impossible for other noise processing techniques to reject the noise levels without any possible signal or noise estimates [3]. This paper has been designed as follows. Section 2 defines a brief summary of the problems identified by the ECG and the importance of the Electrocardiogram signal. Section 3 indicates common Diseases of heart. Section 4 gives details of the Filtering Technique for study. Section 5 gives information about fuzzy inference system (FIS) and classification, Sect. 6 indicates Methodology of Research Sect. 7 illustrates experimental results for the study was done and the last section will be the conclusion of this study done.

2 Problems Identified by ECG and Importance of ECG

Importance of ECG signal can be described as:

- It can be utilized to determine heartbeat speed.
- Any abnormality in the heart rhythm such as irregularities, discomfort or stability can be detected.
- Electrical signals strength and timing can be detected as they pass through each part of the heart (Fig. 1).

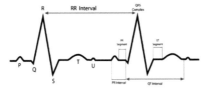

Fig. 1. Electrocardiogram of a heart in normal sinus rhythm

3   Heart Common Diseases

Electrocardiogram records can be used as a diagnostic tool to identify abnormalizations in cardiac function or to visualize the effects of cardiac tissue damage. Normalities in the QRS complex show problems in the ventricular or ventricular conjunctiva. In a normal electrocardiogram, the "T" wave has a positive amplitude, but an abnormal "T" wave in the abnormal electrocardiogram may have a negative amplitude. Various abnormal electrocardiograms and some electrocardiograms will be discussed here:

3.1   Sinus Bradycardia

Bradycardia is a term that defines heartbeats slower than normal. Sinus bradycardia, well-equipped athletes and relaxation during sleep can be. In the case of athletes, heart-pumping is powerful and effective in pumping blood, so less contraction is required. The body requires less oxygen consumption than normal activity and rests. During deep relaxation, the heart slows down at speed. However, sinus bradycardia can also occur as heart disease or as a response (Fig. 2).

3.2   Sinus Tachycardia

Excessive heartbeat over 100 beats per minute (bpm) caused by the sinoatrial node. The reasons contain, fear, stress, disease, and exercise. For each QRS it is necessary to define a "P". It may be difficult to distinguish a sinus tachyarrhythmia from an atrial tachyarrhythmia (Fig. 3).

3.3   Atrial Flutter

For each ventricular contraction, atrial flutter is a status with multiple atrial contractions. This is caused by a single large electrical signal radiating around the atrium. The rate of atrial contraction may be between 200 and 350 bpm. Due to the loosening of one side of the atrium and the contracture of one side, the amount of blood pumped by the atrium may be too small. Electrical signals enter the atrioventricular node at a very rapid rate to create ventricular contraction for each atrial contraction. Consequently, for every QRS-T complex, the Electrocardiogram has P-wave folds (Fig. 4).

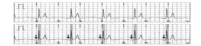

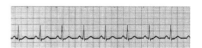

Fig. 2. Sinus Bradycardia   **Fig. 3.** Sinus Tachycardia

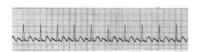

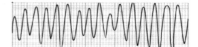

Fig. 4. Atrial flutter   **Fig. 5.** Ventricular flutter

3.4   Ventricular Flutter

Vertices can be accelerated more than 200 bpm, in this cardiac arrhythmia. This can be triggered by an extrasystole or ectopic pacemaker that occurs in the ventricles. Blood pumping is extremely inefficient. There is no visible "P" wave in the electrocardiogram record and the QRS complex and the "T" wave are regularly combined with the oscillating frequency of 180 to 250 bpm (Fig. 5).

3.5   Common Diseases of Heart

- Irregular heartbeat (palpitation).
- Sac of heart inflammation.
- The pain of chest which could be caused by a heart attack.
- Breathing shortness.

4   Filtering Technique for Study

Digital filtering methods are used to improve the signal quality and reduce the random error noise component [4]. Equation (1) shows the relationship between the actual signal from the electrocardiogram and the signal received at the output:

$$O(t) = I(t) + n(t) \tag{1}$$

Where I(t) is a real signal of Electrocardiogram measured the signal at a time **t**, n(t) is random noise affecting it, which is presumed to be additive and O(t) is received a signal from Electrocardiograph. It is not easy to remove the noise signal n (t) from the output signal O (t) without causing any damage to the input signal I (t) because normally the overlapping signal and noise spectra have been overlapped for this reason. The purpose of this manuscript is to introduce the Recursive Least Squares Algorithm method for filtering the electrocardiogram signal. Noisy electrocardiogram signals were

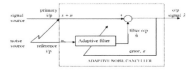

Fig. 6. ANC method

produced by adding noise distributions according to the reference signal. Adaptive Noise Cancellation (Recursive Least Squares Algorithm) technique is applied.

An Adaptive Noise Cancellation has two inputs primary and reference as illustrated in Fig. 6. Suppose that **x, b₀, b₁,** and **y** have zero means and are statistically stationary. Signal **s** is uncorrelated with **b₀** and **b₁**, and **b₁** is correlated with **b₀**.

$$s = x + b - n \Rightarrow s^2 = x^2 + (b - n)^2 + 2x(b - n) \tag{2}$$

Taking expectancy of both sides and realizing that **s** is uncorrelated with **b₀** and n,

$$E[s] = E[x^2] + E[(b - n)^2] + 2E[x(b - n)] = E[x^2] + E[(b - n)^2] \tag{3}$$

Signal power E [x²] will be unaffected as the filter is adjusted to minimize E[s²].

$$\Rightarrow minE[s^2] = E[x^2] + minE[(b - n)^2] \tag{4}$$

5   Classification

5.1   Fuzzy Logic

In the mid-1960s, fuzzy logic was developed to make up for the lack of two-valued logic and probability theory. In the logic of probability, the likelihood level can be known, but not the correctness or falsehood of the propositions. An event and subject cannot be judged out of the possibility. The last point about fuzzy logic systems has been the Lutfi Alesker Zadeh's Fuzzy Logic Theory [6]. The basic idea of fuzzy logic is that a proposition is **right, wrong, very right, very wrong, very very wrong**. In other words, the **rightness** in this way encompasses the truthfulness values in an infinite number of false and true values in classical meaning. For this reason, a proposal in fuzzy logic cannot be called **Either true or false** [7].

5.2   Fuzzy Sets

In the concept of the fuzzy set, there are expressions that are not so elementary or so elementary instead of elementary or not elementary in classical clusters. An element is represented by a membership level between 1 and 0 and not by 0 and 1 for an element [8].

5.3   Membership Functions (MF)

How each point in the input field is matched with the membership value between 0 and 1 has been indicated by An MF that is known as a curve. The entrance space is sometimes described as the concept of the discourse universe. The label is a name for each Membership Function defined. For example, In this study an input variable such as **input_1, input_2,** and **input_3** has three MFs labeled as **Normal Heart Rate = 60–100, Bradycardia Heart Rate < 60**, and **Tachycardia Heart Rate > 100**:

- **Triangle:** This function calculates fuzzy membership values using the triangle membership function.

5.4   Variables of Linguistic

Instead of the numeric values, some values are natural input words or linguistic variables, and expressions are system input or output variables. A linguistic variable is generally divided into linguistic terms. In this study, variables are **Normal Heart Rate = 60-100, Bradycardia Heart Rate < 60** and **Tachycardia Heart Rate > 100**.

5.5   Discourse Universe

Fuzzy set elements have been received from the universe for short or a discourse universe. The universe contains all elements that it can be taken account of and related to the context [9].

IF - THEN Rules

To formulate conditional statements including fuzzy logic subjects operators, verbs, and fuzzy sets used, these if-then rule expressions have been used. A single fuzzy if-then rule has been defined as follows:

Where **a** and **b** are called linguistic variables, If **x** is **a**, it becomes b, fuzzy sets are specified in **X** and **Y** ranges, respectively. If a part of the **x** rule is called a priority or a premise, a part of rule **b with y** is named a result.

6   Research Methodology

Nowadays, Biomedical Signal Processing goals to perform objective or quantitative analysis by signal analysis of physiological systems. The basic ECG range of frequency is between 0.5 Hz–100 Hz. Figure 7 illustrates the filtering system block diagram.

Adaptive Noise Cancellation (Recursive Least in order to add suitable signal distributions with the reference signal the noisy ECG signal has been composed. Squares Algorithm) is tested with white noise. Figure 8 illustrates an actual Electrocardiogram signal and the noise variance added to Electrocardiogram signal with from nvar = 1.0, 3.0.

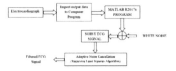

Fig. 7. The filtering system block diagram

7 Experiment and Outcomes

The outcomes of this implementation will be shown below; firstly, white noise is added at various levels by generating the true signal of the Electrocardiogram measured at a time **t**. The noisy ECG Signal is applied to Adaptive Noise Cancellation (Recursive Least Squares Algorithm) then the resultant output is compared with required filter response (Fig. 9).

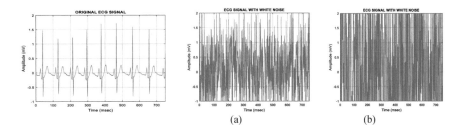

(a)   (b)

Fig. 8. Actual ECG Signal and white noise levels (a) nvar = 1.0, (b) nvar = 3.0

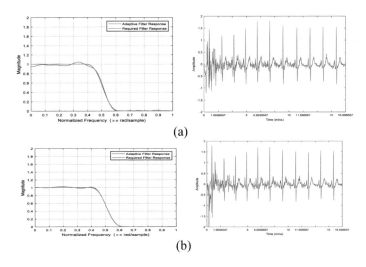

(a)

(b)

Fig. 9. Adaptive and required filter response with (a) nvar = 1.0 (b) nvar = 3.

7.1   Classification Outcomes

Mamdani method is the most widely utilized fuzzy inference technique. In the Mamdani model, four modules of FIS is defined as knowledgebase module, Fuzzification module, Defuzzification modules, and Inference engine module. The Mamdani model works with open data entries, linguistic intervals, and terms. A significant benefit of this model is that when the actual value is not known, it is a reliable measure to predict future value (Fig. 10).

Fig. 10. The fuzzy inference system structure

7.2   Fuzzy Rules

Fuzzy rules depend on the input variables number of the linguistic if-then constructs in which general form is if a then b and the Fuzzy Rules of b and MF. In the fuzzy rule-based selection model defined in Fig. 11, has 3 variables and 1 membership functions = $3^1 = 3$ rules be obtained.

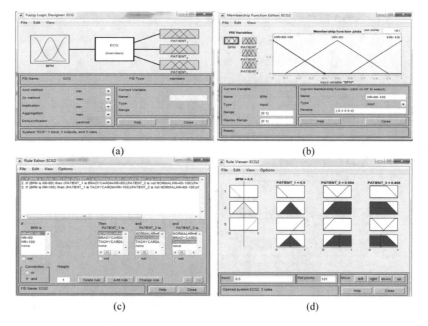

(a)   (b)

(c)   (d)

Fig. 11. (a) Fuzzy rule-based model (b) Membership function for output classifier (c) Rule editor (d) Fuzzy rule-based selection process

8 Conclusion

Nowadays, heart diseases in the world are increasing and becoming the primary cause of death. ECG is an important tool in diagnosing heart problems and the price is additionally low and readily available. However, the ECG signal is affected by many types of sounds that affect the recognition and give inappropriate data. A number of filters have been developed to clean and soften the noises found in the ECG. In this manuscript, extensive and important denoising technique ANC (RLC Algorithm) is offered and implemented on actual ECG signals corrupted with distinct amounts of noise. The proposed study, including ANC (RLC Algorithm), verifies the success of denoising the Electrocardiogram Signal with simulated sets of information. Our suggested study is to classify the type of heart diseases for ECG signals as noisy and filtered utilizing fuzzy logic control with the ANC (RLS Algorithm) has verified its success in denoising electrocardiogram signal with actual data sets. Future study will involve a separate wavelet transform (universal and local threshold) of the common and important noise reduction method. In this study, a variety of error types in the ECG signal and a solution that can be applied in electrocardiograph instruments were analyzed with white noise and the above-mentioned outcomes were obtained. To get clear, preferential standard output signals for efficient debates is the primary aim of the whole system.

References

1. Hussian, S., Babitha, S.M.: Noise removal from cardiac signals using various adaptive algorithms. In: Proceedings of International Academic Conference on Electrical, Electronics and Computer Engineering (2013)
2. Berbari, J.E.: Principles of Electrocardiography. CRC Press LLC, Boca Raton (2000)
3. Kavalcıoğlu, C., Dağman, B.: Filtering maternal and fetal electrocardiogram (ECG) signals using Savitzky-Golay filter and adaptive least mean square (LMS) cancellation technique. Bull. Transilvania Univ. Braşov – Ser. III: Math. Inf. Phy. 9(58)(2), 109–124 (2016)
4. Anderson, B.D., Moore, J.B.: Optimal Filtering. Dover Publications Inc., Mineola (2005)
5. Widrow, B., Glover, J., McCool, J., Kaunitz, J., Williams, C., Hearn, H., Zeidler, J., Dong, E., Goodlin, R.: Adaptive noise canceling : principles and applications. Proc. IEEE 63(12), 1692–1716 (1975)
6. Adedeji, B., Badiru, J.Y.C.: Fuzzy engineering expert systems with neural network applications. Department of the Industrial Engineering University of Tennes see Knoxville, TN. School of Electrical and Computer Engineering University of Oklahoma Norman, OK (2002)
7. Sandya, H.B., Hemanth Kumar, P., Himanshi Bhudiraja, S.K.R.: Fuzzy rule-based feature extraction. Int. J. Soft Comput. Eng. 3(2), 2231–2307 (2013)
8. Jantzen, J.: Tutorial on Fuzzy Logic. Technical University of Denmark (2008)
9. Pedrycz, W.: Fuzzy Control and Fuzzy Systems, 2nd edn. Research Studies Press Ltd., Baldock (1993)
10. Nayak, S., Soni, K.M., Bansal, D.: Filtering techniques for ECG signal processing. Ijreas 2(2), 671–679 (2012). ISSN: 2249–3905

11. Kavalcioglu, C., Dagman, B.: Filtering of noisy continuous glucose monitoring (CGM) signal with Savitzky - Golay Filter. In: International Biomedical Engineering Congress 2015. (IBMEC 2015). Nicosia, North Cyprus, 12–14 March 2015

12. Sadıkoğlu, F., Kavalcioglu, C.: Filtering continuous glucose monitoring (CGM) signal using Savitzky - Golay filter and simple multivariate thresholding. In: 12th International Conference on Application of Fuzzy Systems and Soft Computing (ICAFS 2016), Vienna, Austria, 29–30 August 2016

13. Güler, I., Ubeyli, D.E.: Adaptive neuro-fuzzy inference system for classification of EEG signals using wavelet coefficients. a Department of Electronics and Computer Education, Faculty of Technical Education, Gazi University, 06500 Teknikokullar, Ankara, Turkey b Department of Electrical and Electronics Engineering, Faculty of Engineering, TOBB Ekonomi ve Teknoloji Universitesi (2005)

Decision Making and Obstacle Avoidance for Soccer Robots

Rahib H. Abiyev⬤, Irfan Gunsel, Nurullah Akkaya,
Ersin Aytac$^{(\boxtimes)}$⬤, Sanan Abizada, Gorkem Say, Tolga Yirtici,
Berk Yilmaz, Gokhan Burge, and Pavel Makarov

Applied Artificial Intelligence Research Centre, Near East University,
P.O. Box 670, Nicosia, North Cyprus, Mersin-10, Turkey
{rahib.abiyev,irfan.gunsel,nurullah.akkaya,
ersin.aytac,sanan.abizada,gorkem.say,tolga.yirtici,
berk.yilmaz,gokhan.burge,pavel.makarov}@neu.edu.tr

Abstract. Robot-soccer game is characterized by a rapidly changing dynamic environment with moving obstacles. In this environment making a decision in a very short time, on-line adaptation and flexibility are very important parameters used in decision-making. Based on the decision made by a robot, the collision-free navigation of the mobile robot in this dynamic environment is also important. In the paper, the development of decision-making and navigation algorithms of a soccer robot are considered. Behavior-tree algorithm is presented and implemented for control of the mobile robots. At the same time, the type-2 fuzzy logic system is used for solving the obstacle avoidance problem. The use of the presented algorithm allows the decreasing of the path length of the robot. The efficiency of the presented algorithms has been tested with simulations and practical experiments. The obtained results demonstrate the suitability of using presented algorithms in the navigation of holonomic robots.

Keywords: Holonomoic robot · Behavior trees · Type-2 fuzzy system · Navigation

1 Introduction

The environment where soccer robot moving is characterized by the fast-moving dynamic densely cluttered objects. Decision-making module is an essential part of the system that analyzes the current state of the field and makes decisions for the future state of the robot. One of the basic criterion presenting for DM module is making a decision in a short time and on-line adaptation to a new rapidly changing environment. In literature, different approaches were considered for the development of DM module of robot soccer. These are finite state machine [1, 2], Petri nets topological map, behavior-based control [3], behaviour tree [4, 5]. The use of FSM for a soccer robot needs to determine the states of the FSM from their vision system. But the presence of many complicated situations in robot soccer game increases the number of states and also the number of transition between the states that complicate the modelling of the situations. The Petri nets can model the states with smaller graphs. But the modelling of

© Springer Nature Switzerland AG 2020
R. A. Aliev et al. (Eds.): ICSCCW 2019, AISC 1095, pp. 455–462, 2020.
https://doi.org/10.1007/978-3-030-35249-3_58

all states complicates the network. The paper presents a behavior tree approach for decision-making. BT consists of high and low-level layers including action nodes. Using BT we can model the complex states and their transition easily.

The soccer robot is moving densely cluttered, complex environments. The environment is changed every time. In such condition avoiding from the obstacles and moving to the final position is very important. For this aim, many obstacle avoidance algorithms have been designed [6–8]. In the paper, type-2 fuzzy sets are used for obstacle avoidance.

2   Structure of the Control System of Robot Soccer

The main goal is to find a solution to operate in such an environment without making crashes with other objects in real-time. The vision system is designed with four cameras at the top of the field. Information coming from these cameras are standardized with the Cartesian coordinate system. The x and y-axes respectively indicate horizontal and vertical axes. Each team receives the same information from the SSL vision system. That is an image representation of the field. This information includes the coordinates of each robot on the soccer field, the soccer field itself and the position of the ball. Each team have their own computers to analyze the information sent from the SSL vision system for decision-making for each robot. Each team has its own decision-making mechanism and create a strategy, pathfinding and speed command for motors to reach the destination coordinate for each soccer robot. The commands are sent via wireless signals to the microcontrollers on the robots. The robot after receiving the control signals sent by host computer regulates the rotational velocities of the four omni-wheels that guide the robot the destination position [9].

Figure 1 shows how the architecture of the robot soccer control system. SSL vision system is responsible for tracking the soccer robots coordinates along with the ball. SSL vision system sends this data to each team. DM algorithm uses this data to find new coordinates for soccer robots. The developed DM mechanism is based on BTs, and it is essential to make use of the DM algorithm before the pathfinding algorithm. Velocities of each soccer robots wheels are calculated after the pathfinding algorithm and sent to wheels. The output control signal then converts these values to become understandable for robot microcontroller via Wi-Fi.

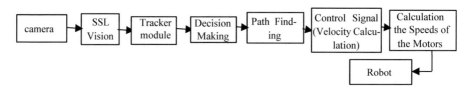

Fig. 1. Structure of the control system

3   Decision-Making Based on Behavior Trees

BT is objective oriented and each tree has its own objective that is to be accomplished. The behavior of the robot is a control law that fulfills a lot of limitations to accomplish a specific objective. Every behavior is characterized by the predefined set of actions.

They BT components are action, decorator and composite nodes. To control the flow within the tree action nodes composite and decoder nodes are used to change the state's actions. The sequence is designed such that if any of its own nodes fails it automatically fails or if all the nodes executed successfully it returns success. Selector node returns success if one of its own nodes returns success such that if the first node should fail it tries to execute the other node until success or all nodes fails.

With regard to BT, a decoder node has only one child. Using these basic BT elements other behaviors are constructed.

4   Type-2 Fuzzy System

In the paper, type-2 fuzzy logic is applied for the construction of the system for obstacle avoidance. The input for the system is the angle and distance measures from the left and right corners of the obstacle. Using input- left angle (la), right angle (ra), left distance (ld) and right distance (rd) and output turn angle (ta) the fuzzy rule base is developed. To design rule base six linguistic terms were used for each input variable and nine- for the output variable. The linguistic values defined for distance, angle and turn angle variables are presented in Fig. 2. For simplicity, the input and output variables are scaled in the rule base.

The use of type-2 fuzzy sets allows handling the uncertainty in the system. In the rule base, type-2 fuzzy membership functions are applied to represent the linguistic values of the variables. In the paper, triangle type-2 membership functions with uncertain mean are used to describe the fuzzy sets. The rule base has multiple inputs and one output. Below type-2 fuzzy rule is presented.

$$\text{IF } x_1 \text{ is } A_{1j} \text{ and } x_2 \text{ is } A_{2j} \text{ and} \ldots \text{ and } x_m \text{ is } A_{mj}\text{THEN } y_j \text{ is } B_j \tag{1}$$

where $x_1, x_2, \ldots, x_m$ are the input variables, $y_j(j = 1, \ldots, n)$ is the output variables, A_{ij} and A_j are interval type-2 fuzzy sets (triangular membership functions) of the premise and consequent parts correspondingly.

The basic issue is finding the accurate definition of the premise and consequent parts. The interval type-2 membership functions used in premise and consequent parts are given in Fig. 2.

(a)   **(b)**

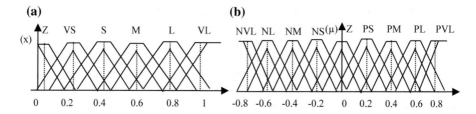

Fig. 2. Type-2 interval membership functions for- (a) input and (b) output variables

The triangle formula is applied for the designing of lower $\bar{\mu}_{\tilde{F}}(x)$ and upper $\mu'_{\tilde{F}}(x)$ membership functions. The upper and lower membership functions are denoted as $\bar{\mu}_{\tilde{F}}(x)$ and $\mu'_{\tilde{F}}(x)$, or $A'(x)$ and $\bar{A}(x)$.

$$\mu_{A_k^{\sim i}}(x_k) = \left[\underline{\mu}_{A_k^{\sim i}}(x_k), \bar{\mu}_{A_k^{\sim i}}(x_k)\right] = \underline{\mu}^i, \overline{\mu^i},\tag{2}$$

In the fuzzy inference system, "min" or "prod" t-norms operations are used to determine the firing strength of the rules. This operation is applied for lower and upper membership functions separately.

$$\underline{f} = \underline{\mu}_{\tilde{A}_1}(x_1) * \underline{\mu}_{\tilde{A}_2}(x_2) * \ldots * \underline{\mu}_{\tilde{A}_n}(x_n); \bar{f} = \underline{\mu}_{\tilde{A}_1}(x_1) * \underline{\mu}_{\tilde{A}_2}(x_2) * \ldots * \underline{\mu}_{\tilde{A}_n}(x_n);\tag{3}$$

After the determination of the firing strengths of the rules, the type reduction and defuzzification operations are applied to the resulting sets. For this purpose, the center of sets algorithm is applied.

$$Y = [y_l, y_r] = \int \ldots \int \int \ldots \int / \frac{\sum_{i=1}^{M} f^i y^i}{\sum_{i=1}^{M} f^i}\tag{4}$$
$$\quad\quad\quad\quad y^l \quad\quad y^m y^l \quad\quad y^m$$

where Y is an interval set computed by y_l and y_r, f^i is $\left[\underline{f^i}, \bar{f^i}\right] y^i = [y_l^i, y_r^i]$ is centroid of the type-2 interval fuzzy set in consequent part.

Karnik and Mendel [10] have shown that the endpoints y_l and y_r depends on $\bar{f^i}$ and $\underline{f^i}$ values.

$$y_l = y_l\left(\overline{f^1}, \ldots, \overline{f^M}, \underline{f^{L+1}}, \ldots, \underline{f^M}, y_l^1, \ldots, y_l^M\right)$$

$$y_r = y_r\left(\underline{f^1}, \ldots, \underline{f^R}, \overline{f^{R+1}}, \ldots, \overline{f^M}, y_r^1, \ldots, y_r^M\right)$$

where $\bar{f^i}$ and $\underline{f^i}$ are computed by formula (5). Karnik and Mendel presented an iterative algorithm for finding the values of y_l and y_r.

$$y_l = \frac{\sum_{i=1}^{M} f_i y_i^l}{\sum_{i=1}^{M} f_i}; y_r = \frac{\sum_{i=1}^{M} f_i y_i^r}{\sum_{i=1}^{M} f_i} \tag{5}$$

where $f^i \in F^i = \left[\underline{f_i}, \overline{f_i}\right]$, $y_i = \left[y_i^l, y_i^r\right]$. The output signal is calculated as:

$$y_l = \frac{\sum_{i=1}^{L} \overline{f_i} y_i^l + \sum_{i=L+1}^{M} \underline{f_i} y_i^l}{\sum_{i=1}^{L} \overline{f_i} + \sum_{i=L+1}^{M} \underline{f_i}}; y_r = \frac{\sum_{i=1}^{R} \underline{f_i} y_i^r + \sum_{i=R+1}^{M} \overline{f_i} y_i^r}{\sum_{i=1}^{R} \underline{f_i} + \sum_{i=L+1}^{M} \overline{f_i}}; y = \frac{y_l + y_r}{2} \tag{6}$$

The Karnik-Merndel procedure is time-consuming, but it is efficient for the design of type-2 fuzzy logic system.

5 Experimental Studies

The presented algorithms are used for navigation of soccer robots that are manufactured in our research center. The simulations and real-life implementations of presented algorithms for the soccer robots have been presented.

5.1 Simulation of Type-2 Fuzzy Obstacle Avoidance Algorithm

The presented obstacle avoidance algorithm has been tested through simulation. The avoidance from the obstacle is based on rule base that uses left angle (la), right angle (ra), left distance (ld) and right distance (rd) in order to determine the direction and value of turn angle of the robot. The safe angle is added to the determined angle in order to find the turn angle for obstacle avoidance.

The left and right angles and also left and right distance variables were used as input for the rule base.

$$\textit{If ra} = \textit{S and rd} = \textit{M and la} = \textit{M and ld} = \textit{L then ta} = \textit{PS}$$
$$\textit{If ra} = \textit{M and rd} = \textit{M and la} = \textit{S and ld} = \textit{S then ta} = \textit{NS}$$
$$\cdots$$
$$\textit{If ra} = \textit{VL and rd} = \textit{VL and la} = \textit{L and ld} = \textit{VL then ta} = \textit{NVL}$$

where la is left angle, ra is right angle, ld is left distance and rd is the right distance. Z is Zero, NS is Negative small, NM is negative medium, PS is positive small, PM is positive medium, S is small, M is medium, L is large, VL is very large. The robot uses the rule base and inference engine mechanism for finding the turn angle in order to avoid the obstacles. Based on the resulting angle, the robot alter its course in order do not hit any obstacle-object.

The comparisons with other obstacle avoidance algorithms have been done to see the type-2 fuzzy algorithms effectiveness among the others and the results of the simulations were depicted in Fig. 3. The A*, APF, RRT-Plan and RRT-smooth algorithms were used for comparison purpose. The obtained simulation results were averaged for 1000 runs. The comparisons were provided using run-time of the algorithm and the length of the path. Results of the simulation were depicted in Table 1. From the simulation results, the suitability of using type-2 fuzzy obstacle avoidance algorithm is concluded.

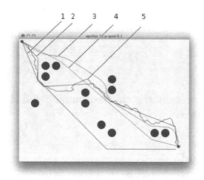

Fig. 3. Obstacle avoidance, 1- A star, 2- RRT-Smooth, 3- APF, 4- Fuzzy, 5- RRT

Table 1. Simulation results of different algorithms

Methods	Time	Length
A* (grid-size = 10)	22.535	792.548
APF	102.478	732.00
RRT Plan	8.262	849.9
RRT Smooth	14.870	748.336
Type-2 Fuzzy	1.827	701.142

5.2   Real Life Implementation

In the research center, the soccer robots with four omnidirectional wheels were designed. Each robot has 3 degrees of freedom. The omnidirectional robot and its control structure designed is given in Fig. 5. The robot soccer has four omni-wheels that are connected to the brushless DC motors. By controlling the speed of the robot the control of the speeds of the omi-wheels is carried out. The control circuit includes the control of forward and rear motors of wheels, a control circuit of dribbler mechanism and a control circuit of the kicking mechanism. The microcontroller is connected to the computer via a wireless link (robotics.neu.edu.tr).

Robots are able to turn on the spot and move in any direction regardless of orientation. The kicking mechanism of the designed robot is used for kicking the ball. The dribbling mechanism is used to catch the ball. The block diagram of a soccer robot is given in Fig. 4. The robot has a microcontroller that controls the moving mechanisms of the robot. Each of them has control circuits. These are control circuits of rear wheels, control circuits of the front wheel, control circuits of the kicker, control circuits of the dribbler, whiles link.

These algorithms are tested in a simulation program called "grSim" and implemented in real robots. As for the real-life application, RRT, DM, and Type-2 Fuzzy obstacle avoidance algorithms are tested on real soccer robots of NEUIslenders team. Running time of different hierarchy levels of BT have been observed. In complex sequences run time was 0.2 ms, for non-complex trees run time was 0.003 ms and the average run time for decision-making was 0.02 ms. Figure 5 depicts football game of the NEUIslanders robot soccer team that uses DM and fuzzy obstacle avoidance algorithms in RoboCup competition.

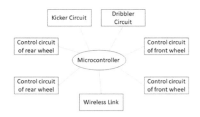

Fig. 4. Control structure of the robot-soccer

Fig. 5. The real game of NEUIslanders soccer robot team in RoboCup competition

6   Conclusions

The decision-making and obstacle avoidance algorithms have been designed for soccer robots. The decision-making has been implemented using the behavior tree that allows making a decision in a very short time. The presented BTs describe different complex states with more structured traversal defining the approach. The BT algorithm has a modular architecture that allows extending the DM for complicated states. Also, the type-2 fuzzy obstacle avoidance algorithm has been designed for guiding the robot. The algorithm is based on an estimation of left and right angles and distances and the fuzzy rule base. The obtained results demonstrate the applicability of the presented obstacle-avoidance algorithm. The experimental results obtained from the real-life application demonstrate the effectiveness of the proposed algorithms for control of soccer robots in dynamic environments.

References

1. Damas, B., Lima, P.: Stochastic discrete event model of a multi-robot team playing an adversarial game. In: Proceedings of 5th IFAC/EURON Symposium on Intelligent Autonomous Vehicles, vol. 37, no. 8, pp. 974–979. Elsevier (2004)
2. Dadios, E.P., Park, S.H.: Real time robot soccer game event detection using finite state machines with multiple fuzzy logic probability evaluators. Int. J. Comput. Games Technol. **2009,** Article ID 375905 (2009)
3. Rusu, P., Petriu, E.M., Whalen, T.E., Cornell, A., Spoelder, H.J.W.: Behavior-based neuro-fuzzy controller for mobile robot navigation. IEEE Trans. Instrum. Meas. **52**(4), 1335–1340 (2003)
4. Abiyev, R.H., Akkaya, N., Aytac, E., Ibrahim, D.: Behaviour tree based control for efficient navigation of holonomic robots. Int. J. Robot. Autom. **29**(1), 44–57 (2014)
5. Abiyev, R.H., Günsel, I., Akkaya, N., Aytac, E., Çağman, A., Abizada, S.: Fuzzy control of omnidirectional robot. Procedia Comput. Sci. **120**, 608–616 (2017)
6. Borenstein, J., Koren, Y.: The vector field histogram - fast obstacle avoidance for mobile robots. IEEE J. Robot. Autom. **7**, 278–288 (1991)
7. LaValle, S.M.: Planning Algorithms. Cambridge University Press, Cambridge (2006)
8. Abiyev, R.H., Akkaya, N., Aytac, E., Günsel, I., Çağman, A.: Improved path-finding algorithm for robot soccers. J. Autom. Control Eng. **3**(5), 398–402 (2015)
9. Abiyev, R.H., Akkaya, N., Gunsel, I.: Control of omnidirectional robot using Z-number-based fuzzy system. IEEE Trans. Syst. Man Cybern.: Syst. **49**(1), 238–252 (2019). https://doi.org/10.1109/TSMC.2018.2834728
10. Mendel, J.M.: Uncertain Rule-Based Fuzzy Logic System: Introduction and New Directions, 576 p. Prentice Hall, Upper Saddle River (2001)

Radio Wave Propagation Model for Enhancing Wireless Coverage in Elevator of Buildings

Jamal Fathi$^{(\boxtimes)}$ ⓘ and Fahreddin Sadikoglu ⓘ

Near East University, Nicosia, Turkish Republic of Norther Cyprus, Turkey
jamalfathi2004@gmail.com,
fahreddin.sadikoglu@neu.edu.tr

Abstract. The huge increase and the importance of smart phones, almost parallel with the modest technology cars. This daily growing in the numbers of mobile phones requires the increasing of the capacities of base stations in order to support the coverage area. All this, forces the manufacturers of mobile communications to rearrange the positions of base stations. This paper is studied the wave propagation loss inside the elevators, at the buildings of the Near East University (NEU) of Turkish Republic of North Cyprus. Moreover, several existing models were compared with proposed mathematical model; as much as these issues are solved according to the available locations of the base stations. Results show that proposed mathematical model is more accurate by comparison with published works.

Keywords: Mathematical model · Enhancement · Splitters · Base stations

1 Introduction

Many published studies focused their attentions to the wave propagations through free spaces, plastic fibers, and buildings in the range of 1–2000 dB/km [1–6].

The propagation loss is defined as:

$$a(dB) = 10 \tag{1}$$

where a is Loss in dB/meter, P_i and P_o are input and output powers respectively. In the available cases, scattering generally confines the recurrence at which signals can be located [1, 2]. This yields to the disconnection of the signal, or its weaknesses [3]. So blocking and reflecting surfaces in the region of the reception devices had a generous impact on the qualities of the transmission way [3, 4].

An equation that symbolized to join the log-domain and the radio signal is:

$$P_{rx} = P_{tx} + G_{tx} + G_{rx} - PL \tag{2}$$

Where, P_{tx} and G_{rx} are the received power at the receiver and the gain of the antenna in the transmitter's direction, while PL is the path loss. This paper is systemized into segments as: firstly radio wave propagation and its performance. A new,

© Springer Nature Switzerland AG 2020
R. A. Aliev et al. (Eds.): ICSCCW 2019, AISC 1095, pp. 463–469, 2020.
https://doi.org/10.1007/978-3-030-35249-3_59

related research, then the proposed model is described fourthly, the acquired effects with rapprochement researchers' results. Finally, display the acquired conclusion.

2 Related Research

The advance of exchange the connection out of radio waves signals employs continuous WiFi among transmitter and the receiver [5]. The rearrangements of the base stations are investigated in order to avoid overlap process in the cases of applicability as country, residential, and urbanized areas using 3.5 GHz [6, 7]. While in [8], recommended an accurate model to decrease the path loss of the propagated signal from outside to inside doors, depending on inferior angle penetrating the indoor situations, so these loses of the propagated signals are treated separately. In [9] an improvement aiming the minimization to propagation loss is studied, and the calculations done to improve and maximize the communication process between WiFi s and inner loops. Where in [10] unavoidable propagation losses through buildings are evaluated to the required power in different rates. While in [11] Offered a suggestion of wireless networks depending on the path between transmitted-received signals from both transmitter and receiver and then compared the propagation in multi cases including WLAN in Urban and sub-Urban areas. The case in [12] is suggested an ensured new outline which obviates networks in different cases of propagation in urban and indoor, also rated the signal's strength and capacity in Multiple-input Multiple-output (MIMO). Where, [13] Managed gauge frequencies with ranges between 203.15 and 584.15 MHz by using several channels in Ilorin City. In [14] presented a sawing inside 5.725–5.825 GHz through the UNII employed in the United States of America. Finally, gained terrific termination perpetually, once propagation that supposed for each including low band channels to have all systems using low frequencies.

3 Proposed Model

The proposed models [1–6] are used in order to improve the case of propagation lose in the transmitted signal, as much as proved a high accuracy in the path loss calculations, rearranging the locations of the base stations, and the measurements of the handover. The case inside the elevators including the separation distance between each floor made of cement. The condition for the signal (in dB) utilized as a part of this investigation is then given in the main Eq. 3.

$$Y = 24.5 + 33.8 \log(d) + 4.0 K_{floor} - 16.6 S_{win} - 9.8 G_{G_{/1}} - 0.25 A_{elv} \qquad (3)$$

Considering the transmitter to receiver technique materialized in Fig. 3, the intimate signal traverse the Hospital of the University in [1, 2, and 3] accompanied by 7 floors with the base floor scaffoldings as appeared in (see Fig. 1).

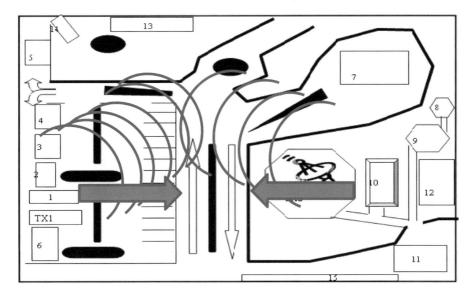

Fig. 1. Hospital scene [1–6]

Using Eq. (3) to compare the obtained results from the proposed model and with the results obtained in Free-space, Cost 231 WI, SUI, Ericsson, and Cost 231 Hata models, so that, accurate results obtained using different heights of antennas, where for rural areas. Table 1 shows the detected results.

4   Simulation Results

Table 1 includes the obtained results for Suburban area within a 3 m antenna height, while (see Fig. 2) shows the obtained results while changing the area from Urban to Suburban are. It's noticed that the proposed model has almost a stable result 110.1617 which yields to an accurate and stable system.

Table 1. Obtained results in Suburban Area 3 m antenna, 15 m antenna height, and 8 floors building

Free space	COST 231 W I	ECC-33	SUI	Ericsson	Cost 231 Hata
92.4500	107.0736	279.8598	81.2374	99.8038	121.4968
98.4706	118.5127	291.7166	94.4075	120.5982	132.1005
118.0251	155.6662	330.2266	137.1829	188.1371	166.5407
118.4706	156.5127	331.1040	138.1575	189.6759	167.3254

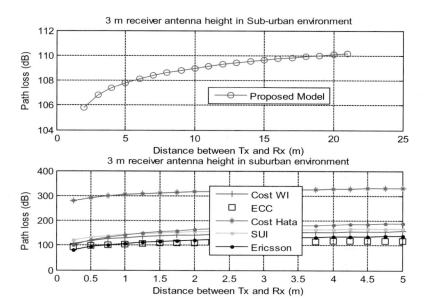

Fig. 2. Comparison results in 3 m antenna height in Suburban Area

While 810 synoptic readings are examined for path-loss estimations that have been combined into the test from 35 base stations surrounding the area of study.

Fig. 3. Laborious base-stations [1–5]

Table 2 shows the obtained results.

Table 2. Obtained results in Urban Area 3 m antenna, 15 m antenna height, and 8 floors building

Free space	COST 231 W I	ECC-33	SUI	Ericsson	Cost 231 Hata
92.4500	131.4844	279.8598	109.7492	116.1215	127.0987
98.4706	142.9236	291.7166	122.1416	125.2571	137.7024
101.9924	149.6150	298.6524	129.3907	130.6011	143.9052
104.4912	154.3627	303.5734	134.5340	134.3927	148.3062
112.4500	169.4844	319.2472	150.9159	146.4693	162.3236
113.2779	171.0574	320.8775	152.6199	147.7254	163.7816

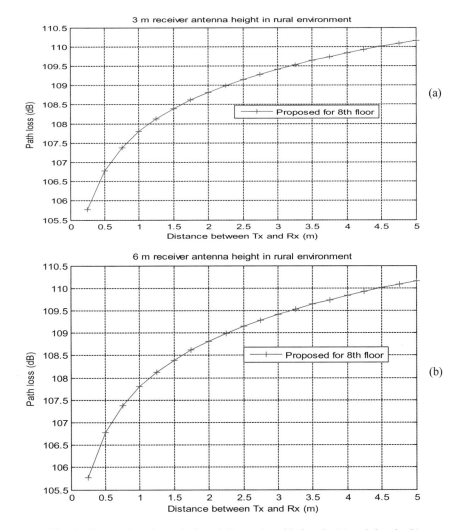

Fig. 4. Proposed mathematical model's results with 3 m in (a) and 6 m in (b)

The arrangement of base stations locations in order to minimize the overlapping process as in [1–6]. The removal procedure is done by attaching splitters, that saves time in state skipping all obtainable base stations, supplementing the protraction region in order be no disconnections.

The whole studies done earlier are done to minimize the effects of overlapping, reduce the path loss propagation, and offer full coverage inside and outside the buildings including elevators in all-weather cases. As seen in Fig. 4, the proposed mathematical model is providing stable results of propagation loss in all cases.

5   Conclusion

The results obtained in Ericsson model is 170.0844 dB with 10 m height station comparing with other models. Award the acquired results in this paper, ECC-33 used antenna with 3 m height 321.8768 dB, whilst SUI system obtained inconstant results as 109.7492–152.6199 dB, however Ericsson system exercised changeful mobile antenna height varies in the midst 6–10 m.

Finally, the propositioned mathematical model proved its validity in all cases including the elevators. Concentrating on the cement area between each two floors, this area is made of cement and covered by the material of the elevator. The propositioned model can render more events than all former models, so that the propositioned model is taking into account inside the elevators as an extra condition. Even though, the gained outcome is more accurate.

References

1. Zyad, N., Julius, D., Jamal, F.: A new algorithm to enhance radio wave propagation strength in dead spots for cellular mobile wifi downloads. 978-1-4577-1343-9/12. IEEE (2014)
2. Zyad, N., Juliusm, D., Jamal, F.: A new selected points to enhance radio wave propagation strength outside the coverage area of the mobile towers in the dead spots of cellular mobile wifi downloads. In: Long Island Section Systems, Applications and Technology Conference. IEEE (2015)
3. Zyad, F.N., Julius, D., Navarun, G., Rami, A.: New mathematical model for wireless signal path loss inside building. In: 14th International Symposium on Pervasive Systems, Algorithms and Networks, Long Island. USA (2017)
4. Jamal, F.A.B.: Estimating coverage of radio transmission into and within buildings for line of sight visibility between two points in terrain by linear prediction filter. In: Third Mosharaka International Conference on Communications, Signals and Coding. MIC-CSC, pp 1–5 (2009)
5. Jamal, F.: Improvement in strength of radio wave propagation outside the coverage area of the mobile towers for cellular mobile wifi. In: Aliev, R., Kacprzyk, J., Pedrycz, W., Jamshidi, M., Sadikoglu, F. (eds.) 13th International Conference on Theory and Application of Fuzzy Systems and Soft Computing ICAFS. Advances in Intelligent Systems and Computing, vol 896, pp 464–471. Springer, Cham (2019)

6. Jamal, F.A.B., Firudin, Kh.M: Direction prediction assisted handover using the multilayer perception neural network to reduce the handover time delays in LTE networks. Procedia Comput. Sci. **120**, 719–727 (2017)
7. Abhay, A.V.S., Wassell, I.J., Crosby, D., Sellars, M.P.: Comparison of empirical propagation path loss models for fixed wireless access systems. BT Mobility Research Unit, Rigel House, Adastral Park, Ipswich IP5 3RE, UK, pp 1–5 (2004)
8. Bose, A., Chuan, H.F.: A practical path loss model for indoor wifi positioning enhancement, 1-4244-0983-7/07. IEEE (2007)
9. Miura, Y., Yasuhiro, O.D.A., Tokio, T.: Outdoor-To-Indoor propagation modeling with the identification of path passing through wall openings, Wireless Laboratories, NTT DoCoMo, Inc. 3-5 Hikari-no-oka, Yokosuka-shi, Kanagawa, 239-8536, Japan, 0-7803-7589-0/02. IEEE (2002)
10. Durgin, G., Theodore, S.R., Xu, H.: Measurements and models for radio path loss and penetration loss in and around homes and trees at 5.85 GHz. IEEE Trans. Commun. **46**(11), 1484–1496 (1998)
11. Iskandar, Shigeru, Sh.: Prediction of propagation path loss for stratospheric platforms mobile communications in urban site LOS/NLOS environment, pp. 5643–5648, 1-4244-0355-3/06. IEEE (2006)
12. Wolfle, G., René W., Pascal, W., Philipp, W: Dominant path prediction model for indoor and urban scenarios. AWE Communications GmbH, Otto-Lilienthal-Str. 36, 71034 Boeblingen, Germany (2003)
13. Stabler, O., Reiner, H., Gerd, W., Thomas, H., Timm, H.: Consideration of MIMO in the planning of LTE networks in urban and indoor scenarios. AWE Communications GmbH Otto-Lilienthal-Straße 36, 71034. Böblingen, Germany (2011)
14. Faruk, N., Ayeni, A.A., Adediran, Y.A.: Characterization of propagation path loss at VHF/UHF bands for Ilorin City, Nigeria. Nigerian J. Technol. (NIJOTECH) **32**(2), 253–265 (2013)

Z-number Based Fuzzy System for Control of Omnidirectional Robot

Rahib H. Abiyev$^{(\boxtimes)}$ (iD), Irfan Günsel, and Nurullah Akkaya

Applied Artificial Intelligence Research Centre, Near East University,
Nicosia, North Cyprus, Turkey
{rahib.abiyev,irfan.gunsel}@neu.edu.tr,
nurullah@nakkaya.com

Abstract. This paper presents the development of a fuzzy inference system using Z-number for omnidirectional robot control. The fuzzy inference system is constructed for control of the linear and angular speed of the robot. For this aim, using expert knowledge the fuzzy Z-rules are designed for robot-soccer control. Using the designed rule base and interpolative reasoning the development of the fuzzy control system based on Z-number is carried out. The simulation of developed fuzzy control algorithm based on Z-number has shown satisfactory results at runtime. The experimental results demonstrate the efficiency of using a Z-number in the design of control system.

Keywords: Omnidirectional robot · Z-number · Fuzzy control · Z-rules

1 Introduction

The control of omnidirectional robots in environment that are densely clattered with moving obstacles is an important problem in robotics [1, 2]. Set of algorithms have been developed for this aim. Some researches are using classical PI, PD and PID controllers [2]. The performances of PI, PD and PID control systems depend on the coefficients of controllers that are based on the process model and system characteristics. However, the formulation and analysis of the system model is not easy and sometimes is not available. The soccer robots are moving in an environment that is characterized by the fast-moving obstacles and goal. In this environment, the next obstacle's position is unknown to the robot-soccer. In addition, the nonlinearities existing in dynamics of the robot has an impact on the velocity of the robot as well as the behaviour of the robot. These nonlinearities are caused by friction, vibrations, payload variation and disturbance, slippage between the wheels and terrain. For these reasons, the robots' dynamics have nonlinearities of high-order and it became too difficult to drive an accurate mathematical model. In these conditions, the development of the control system with sufficient accuracy becomes an important problem.

There are set of research works used to control the system characterizing with the nonlinearity and uncertainty. These are fuzzy control [4, 5, 8, 10–12], neural networks based control [6, 7], control based on feedback-linearization [8], sliding–mode control [9]. Fuzzy theory one of efficient approach that can be used for control of the mobile robots characterizing with nonlinearity and uncertainty existing in dynamics [13]. The

© Springer Nature Switzerland AG 2020
R. A. Aliev et al. (Eds.): ICSCCW 2019, AISC 1095, pp. 470–478, 2020.
https://doi.org/10.1007/978-3-030-35249-3_60

type-1 fuzzy sets used to characterize these linguistic variables allow handling imprecise and uncertain information to some degree. Zadeh, in his paper [14], shows that the reliabilities of the linguistic values are important information and they can be used for evaluating the values of fuzzy variables. Zadeh proposed Z-number that includes two components – constraint and reliability parameters for the design of fuzzy systems [14]. The first component of Z-number represents uncertain information, and the next component evaluates the reliability of this information. The correctness of the decision made by the rules depends on the reliability of the fuzzy values. Considering the uncertainties existing in the dynamics of mobile robots, Z-number can be a valuable alternative for designing a controller for mobile robots.

The theory of the Z-number based system are extended [14, 15] and used to solve different practical problems. These research works are the solution of the multi-criteria decision-making (MCDM) [15–17], evaluations of the educational achievement [15], estimation of food security risk level [18], dynamic plant control [19, 20]. These research works present the efficiency of using Z-number based systems in problem solutions.

This paper proposes the Z-number based controller for the omnidirectional robot control. Using expert knowledge the development of Z-number based fuzzy rule base for robot control is carried out. Based on this rule base the development of Z-number based fuzzy control system (ZNFC) for the omnidirectional robot is performed using the α-level procedure and interpolative reasoning.

2   Omnidirectional Robot Control

The design of omnidirectional robot-soccers and their control systems are carried out in our research center. The designed robot is depicted in Fig. 1(a). It has three degrees of freedom- X, Y, and rotation. As shown in Fig. 1(b) the robot-soccer has four omni-wheels that move in any direction regardless to orientation. Each omni-wheels has a diameter of 61 mm. The orientations of front wheels and rear wheels with the horizontal axis are 30 and 45° correspondingly (Fig. 1(b)). The wheels are connected to the brushless DC motors through gear mechanisms. The motors control the velocities of the robot wheels'. The microcontroller that setup on robots through motor drivers controls the DC motors [13]. The microcontroller has 2.4 GHz wireless communication with the main computer. The control system of robot contains six basic modules: vision module, data pre-processing-, decision-making, path finding-, obstacle avoidance and motion control modules [3, 10]. The decision-making module plan the behaviors and compute the new coordinates [3, 20]. The robot-soccer uses this information in order to move to the new coordinates. The kinematics and dynamics of used omnidirectional robots are described in [20].

In the paper, the ZNFC is designed for controlling the orientation (rotational) speed of the robot-soccer. The control of the robot is carried out through control of wheel speeds. The structure of the control system is given in [20]. Assume that we need to find the robot velocity that will steer the mobile robot from the start state $[x\ y\ \phi]$ to the target(reference) state $[x_r\ y_r\ \phi_r]$. The fuzzy controller's output is a function of the controller input variables- error and change-in-error. Using the reference positions

$[x_r \; y_r \; \phi_r]^T$ and current(or start) positions $[x \; y \; \phi]^T$ the errors and then change-in-error values are determined. Based on these input values, the wheels' velocities that steer the robot to the end-reference point are determined. The above mentioned can be formulated using the following formulas.

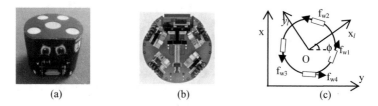

(a)   (b)   (c)

Fig. 1. (a) Robot-soccer, (b) wheels, (c) translational forces

Table 1. The rule base of the controller.

Control *(u)*		Change-in-error (e˙)						
		NL,U	NM,U	NS,U	Z,U	PS,U	PM,U	PL,U
Error *(e)*	NL,U	PL,U	PL,U	PL,U	PL,U	PM,U	PS,U	Z,U
	NM,U	PL,U	PL,U	PL,U	PM,U	PS,U	Z,U	NS,U
	NS,U	PL,U	PL,U	PM,U	PS,U	Z,U	NS,U	NM,U
	Z,U	PL,U	PM,U	PS,U	Z,U	NS,U	NM,U	NL,U
	PS,U	PM,U	PS,U	Z,U	NS,U	NM,U	NL,U	NL,U
	PM,U	PS,U	Z,U	NS,U	NM,U	NL,U	NL,U	NL,U
	PL,U	Z,U	NS,U	NM,U	NL,U	NL,U	NL,U	NL,U

$$\begin{bmatrix} v_x \\ v_y \\ v_\phi \end{bmatrix} = f\left(\begin{bmatrix} e_x & \dot{e}_x \\ e_y & \dot{e}_y \\ e_\phi & \dot{e}_\phi \end{bmatrix} \right), \quad \begin{bmatrix} e_x \\ e_y \\ e_\theta \end{bmatrix} = \begin{bmatrix} x \\ y \\ \phi \end{bmatrix} - \begin{bmatrix} x_r \\ y_r \\ \phi_r \end{bmatrix} \tag{1}$$

As mentioned the robot has three degrees of freedom. The control of the robot is realised using the angle tracking error and position tracking errors in X and Y directions and also their variations. The human (experts) knowledge is used to implement the association between the inputs and control output signals. For the design of the rule base, Mamdani's type rules have been adopted [20]. Table 1 represents the if-then control rules. As shown the rules are described by constraint and reliability parameters. The triangle membership function is utilised to represent the fuzzy values of the position tracking error, change-in-error, control signal and their reliabilities (Fig. 2). In the table and figure PL is "Positive Large", PM is "Positive Medium", PS is "Positive Small", Z is "Zero", NS is "Negative Small", NM is "Negative Medium", NL is "Negative Large", "S" is "Sometimes", "U" is "Usually", "A" is "Always".

The designed fuzzy control rules are applied to control linear, angular speeds of the wheels of the omnidirectional robot. According to the 7 linguistic terms of the input variables, the 49 rules were designed for the controller. Fuzzy rule interpolation, given in Sect. 4, is utilized for development of an inference engine of the control system.

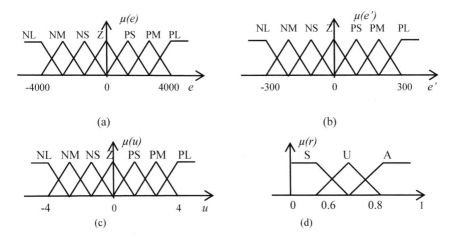

Fig. 2. Membership functions (a) tracking error, (b) change-in-error, (c) control signal and (d) reliability.

3 Interpolative Reasoning Using Z-numbers

Z-number is described by a pair of restriction A and reliability B components of fuzzy variable X as $Z = (A, B)$. In the paper, using Z-number the If-Then rule-base of the controller is designed. The interpolative reasoning mechanism is adapted for finding the output control signal. Let's consider the reasoning mechanism of the Z-number base fuzzy rule-based system. Using Table 1 the If-Then control rules are written as

$$\text{If } x_1 \text{ is } (A_{i1}, R_{i1}) \text{ and} \dots \text{and } x_m \text{ is } (A_{im}, R_{im}) \text{ Then } y \text{ is } (B_i, R_i) \qquad (2)$$

where A_{ij} and R_{ij} are constraints and reliabilities values for input variables, B_i and R_i constraints and reliabilities defined for the output variable.

The interpolation approach is used for inference mechanism. In the paper, the inference mechanism is developed using α- level procedure. Based on α-level, Koczy and Hirota [21] presented the distance measure between two fuzzy sets. Using the distance measure, for the SISO rule, Koczy and Hirota in [21] show that

$$\frac{d(A^*, A_1)}{d(A^*, A_2)} = \frac{d(B^*, B_1)}{d(B^*, B_2)} \qquad (3)$$

Here $A_1 < A^* < A_2$ and $B_1 < B_2$.

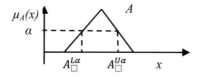

Fig. 3. Membership function, lower and upper α-levels.

The α-level of a fuzzy set of A_1 and A_2 will be denoted as A_1^α and A_2^α (Fig. 3). The lower d_L and upper d_U distances between two fuzzy sets are determined using α-level. Assume that, a fuzzy If-Then control rules of the controller is given, and we find that the current input for the controller is X_j. The lower d_L and upper d_U distances between X_j input and A_{ij} linguistic term in the rule base are determined as

$$da_{ij}^L = da^L\left(A_{ij}^\alpha, X_j^\alpha\right) = \left|A_{ij}^{L\alpha} - X_j^{L\alpha}\right|; da_{ij}^U = da^U\left(A_{ij}^\alpha, X_j^\alpha\right) = \left|A_{ij}^{U\alpha} - X_j^{U\alpha}\right| \quad (4)$$

Here $da_{ij}\left(A_{ij}^\alpha, X_i^\alpha\right) = \left\{da_{ij}^L\left(A_{ij}^\alpha, X_i^\alpha\right), da_{ij}^U\left(A_{ij}^\alpha, X_i^\alpha\right)\right\}$, is distance between X_j and A_{ij}. The interval arithmetic is utilized in finding the distances between two fuzzy sets. After sum of the distances is computed.

$$Da_i^\alpha = \sum_j^m da_{ij}\left(A_{ij}^\alpha, X_j^\alpha\right) \quad (5)$$

where $Da_i^\alpha = \{Da_i^L, Da_i^U\}$ are expressed by the lower Da_i^L and upper distances Da_i^U, n and m are the number of rules and number of input signals, correspondingly. The formula (4) and (5) are also adapted to compute a distance Dr_i^α for reliability. The total distance will be computed

$$D_i^\alpha = Da_i^\alpha + Dr_i^\alpha \quad (6)$$

where Da_i^α and Dr_i^α are distances for the constraint and reliability. $D_i^\alpha = \left(D_i^L, D_i^U\right)$. Using the following equation the output fuzzy set of the rules is calculated.

$$Y^\alpha = \sum_{i=1}^n \frac{1}{D_i^\alpha} B_{Y_i}^\alpha / \frac{1}{\sum_{i=1}^n \frac{1}{D_i^\alpha}}; R_Y^\alpha = \sum_{i=1}^n \frac{1}{D_i^\alpha} R_{Y_i}^\alpha / \frac{1}{\sum_{i=1}^n \frac{1}{D_i^\alpha}} \quad (7)$$

where Y is the constraint, R_Y is the reliability variables. The formula (7) is used to determine the Z-number output of the rule base system. To find crisp output we are using α = 0 and α = 1 levels. For the α = 0 level, the left (Y_l, R_{Yl}) and right (Y_r, R_{Yr}) values are determined. The middle (Y_m, R_{Ym}) value which is highest value is corresponding to the α = 1 level. Using the formula $Y = ((Y_l + 4 * Y_m + Y_r)/6) * ((R_{Yl} + 4 * R_{Ym} + R_{Yr})/6)$ the crisp output value is determined.

4   Experimental Studies

The developed control system based on Z-number is applied for robot-soccer control. At the first stage, the test of the fuzzy control algorithm has been performed using a software package grSim. After testing, the ZNFC is used for control of omnidirectional soccer robots that are designed in our research centre.

In the control system design, one of the basic problem was to reach the target location and to get minimum error between reference and current paths of the robot. The robot moves from the start position $[x_o \ y_o \ \phi_o]$ to target position $[x_r \ y_r \ \phi_r]$ avoiding obstacles. The results of simulations for the error and change-in-error are presented in Fig. 4. In the next simulation, the values of errors (Fig. 5(a)) and change-in-error (Fig. 5(b)) are plotted using different values of orientation angles.

The velocity control of the robot is implemented in the next simulation. The control is carried out for different values of desired reference signals v_{Ref} = (1.8; 3.6; 1.8). The simulation results of velocity control for different reference signal is given in Fig. 6. The obtained results demonstrate that ZNFC can reach the target position with the desired velocity and desired angle.

A set of simulations have been performed for testing the performance of the ZNFC and for comparative analysis. For this purpose as a reference signal (path), the rectangular trajectory has been taken (Fig. 7). This path is also used to test the performances of the PID, classical fuzzy- and type-2 fuzzy controllers. At first, proposed ZNFC is applied to steer the robot by the rectangular path. The mean absolute error (MAE) formula denoted as $E = \frac{1}{N} \sum_{k=1}^{N} |P^d - P|$ is used to measure the tracking performance. Here P is the current position, P^d is the desired position. The experiments were performed using different velocity values, when v = 3 m/s and v = 4 m/s. The experimental tests have shown that the mobile robot has more slippage in high values of velocities. The simulation results have been depicted in Fig. 7. The solid line presents the desired path, a dashed line presents the path of the robot controlled with ZNFC. The comparative results of PID-, ZNFC-, fuzzy- and type-2 fuzzy control system have been presented in Table 2. As shown, the performance of ZNFC is better than PID- and fuzzy control methods. The performances of the control systems based on type-2 fuzzy controller and ZNFC are nearly the same. For evaluation of the performances of the controllers, a maximum value of error (MVE) and MAE were used.

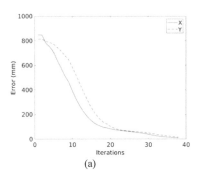

(a)

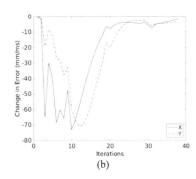

(b)

Fig. 4.   Error (a) and change in error (b) plots in x (solid line) and y (dashed line) directions.

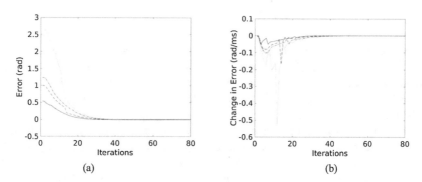

Fig. 5. The plots of errors (a) and change-in-errors (b) for orientation angles of 30° (solid line), 60° (dashed line), 70° (dash-dotted line) and 135° (dotted line)

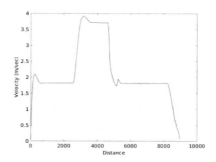

Fig. 6. The plot of robot velocity for different reference signals $v_{Ref} = (1.8; 3.6; 1.8)$

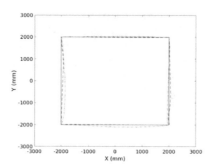

Fig. 7. Trajectory tracking when $v = 3$ m/s (dashed line), and $v = 4$ m/s (dash-dotted line)

Table 2. Comparative results of different control systems

Velocity	PID controller		Fuzzy controller		Type-2 fuzzy controller		Z-number based fuzzy controller	
	MVE (mm)	MAE	MVE (mm)	MAE	MVE (mm)	MAE	MVE (mm)	MAE
$v = 1.5$ m/s	52.217	3.736	14.252	2.279	12.054	1.352	12.465	1.198
$v = 2$ m/s	86.577	7.127	20.418	3.663	16.964	3.104	17.386	2.986
$v = 2.5$ m/s	170.737	11.849	74.821	11.644	43.558	9.885	43.764	9.846
$v = 3$ m/s	181.607	13.174	70.975	14.341	58.238	13.037	56.865	13.163
$v = 3.5$ m/s	202.652	29.000	219.158	33.409	193.48	32.284	189.158	30.485
$v = 4$ m/s	404.005	52.607	345.677	59.952	301.50	53.684	298.456	55.432

5 Conclusions

In the paper, a fuzzy inference system based on Z-number is designed for omnidirectional robot control. α-level procedure and interpolative reasoning are applied for the design of the controller. The developed Z-rule base and fuzzy inference scheme are used for control of omnidirectional soccer robots. The values of the mean absolute error and maximum absolute error of the control system are used to evaluate the control system's performance. The simulations and experimental results have shown the good tracking performance of ZNFC system. Comparisons of ZNFC, PID-, fuzzy-, and type-2 fuzzy controllers' performances demonstrate the efficiency of using the ZNFC in a mobile robot control application.

References

1. Liu, Y., Zhu, J., Williams II, R.L., Wu, J.: Omni-directional mobile robot controller based on trajectory linearization. Robot. Auton. Syst. **56**, 461–479 (2008)
2. Watanabe, K., Shiraishi, Y., Tzafestas, S.G., Tang, J., Fukuda, T.: Feedback control of an omnidirectional autonomous platform for mobile service robots. J. Intell. Robot. Syst. **22**(3–4), 315–330 (1998)
3. Abiyev, R.H., Akkaya, N., Aytac, E., Ibrahim, D.: Behaviour tree based control for efficient navigation of holonomic robots. Int. J. Robot. Autom. **29**(1), 44–57 (2014)
4. Hashemi, E., Ghaffari, J.M., Ghaffari, J.N.: Model-based PI–fuzzy control of four-wheeled omni-directional mobile robots. Robot. Auton. Syst. **59**, 930–942 (2011)
5. Masmoudi, M.S., Krichen, N., Masmoudi, M., Derbe, N.: Fuzzy logic controllers design for omnidirectional mobile robot navigation. Appl. Soft Comput. **49**, 901–919 (2016)
6. Bugeja, M.K., Fabri, S.G., Camilleri, L.: Dual adaptive dynamic control of mobile robots using neural networks. IEEE Trans. Syst. Man Cybern. Part B (Cybern.) **39**(1), 129–141 (2009)
7. Chen, C.P., Liu, Y.J., Wen, G.X.: Fuzzy neural network-based adaptive control for a class of uncertain nonlinear stochastic systems. IEEE Trans. Cybern. **44**(5), 986–993 (2014)
8. Byrnes, C.I., Isidori, A.: Output regulation for nonlinear systems: an overview. Int. J. Robust Nonlinear Control **10**(5), 323–337 (2000)

9. Chang, Y.-H., Chang, C.-W., Chen, C.-L., Tao, C.-W.: Fuzzy sliding-mode formation control for multirobot systems: design and implementation. IEEE Trans. Syst. Man Cybern. Part B (Cybern.) **42**(2), 444–457 (2012)

10. Abiyev, R.H., Akkaya, N., Aytac, E.: Navigation of mobile robot in dynamic environments, In: Proceedings of the IEEE International Conference on Computer Science and Automation Engineering (CSAE), vol. 3, pp. 480–484. IEEE, Zhangjiajie (2012)

11. Abiyev, R.H., Günsel, I., Akkaya, N., Aytac, E., Çağman, A., Abizada, S.: Robot soccer control using behaviour trees and fuzzy logic. In: ICAFS 2016. Procedia Computer Science, vol. 102, pp. 477–484 (2016)

12. Park, S., Ryoo, Y.-J., Im, D.-Y.: Fuzzy steering control of three-wheels based omnidirectional mobile robot. In: Proceedings of International Conference on Fuzzy Theory and Its Applications. IEEE, Taichung (2016)

13. Abiyev, R.H., Günsel, I., Akkaya, N., Aytac, E., Çağman, A., Abizada, S.: Fuzzy control of omnidirectional robot. In: ICSCCW 2017. Procedia Computer Science, vol. 120, pp. 608–616 (2017)

14. Zadeh, L.A.: A note on a Z-number. Inf. Sci. **181**(14), 2923–2932 (2011)

15. Aliev, R.A., Huseynov, O.H., Aliyev, R.R., Alizadeh, A.V.: The Arithmetic on Z-numbers. Theory and Application, 316 p. World Scientific Publishing (2015)

16. Aliev, R.A., Pedrycz, W., Huseynov, O.H., Eyupoglu, S.Z.: Approximate reasoning on a basis of Z-number valued If-Then rules. IEEE Trans. Fuzzy Syst. **25**(6), 1589–1600 (2017)

17. Kang, B., Wei, D., Li, Y., Deng, Y.: Decision making using z-numbers under uncertain environment. J. Comput. Inf. Syst. **8**(7), 2807–2814 (2012)

18. Abiyev, R.H., Uyar, K., Ilhan, U., Imanov, E., Abiyeva, E.: Estimation of food security risk level using Z-number based fuzzy system. J. Food Qual. **2018**, 1–9 (2018)

19. Abiyev, R.H.: Z-number based fuzzy inference system for dynamic plant control. Adv. Fuzzy Syst. **2016**, 1–9 (2016). https://doi.org/10.1155/2016/8950582

20. Abiyev, R.H., Akkaya, N., Gunsel, I.: Control of omnidirectional robot using Z-number-based fuzzy system. IEEE Trans. Syst. Man Cybern. Syst. **49**(1), 238–252 (2019). https://doi.org/10.1109/tsmc.2018.2834728

21. Koczy, L.T., Hirota, K.: Interpolative reasoning with insufficient evidence in sparse fuzzy rule bases. Inf. Sci. **71**(1), 169–201 (1993)

Development of a Vision-Based Feral Vertebrate Identifier Using Fuzzy Type II

Hamit Altiparmak$^{(\boxtimes)}$ ⓘ

Near East University, Nicosia, Turkey
hamit.altiparmak@neu.edu.tr

Abstract. Discrimination between feral vertebrates and livestock is necessary to control the population of the unwanted group of creatures in the large rangelands in Australia. This study proposes an algorithm for feral animal identification in the free environment required prior to a control action to limit the feral animal numbers compete with the livestock on the same natural resources. The inherent uncertainties of an imagery over an arbitrary scene in the nature demands a robust segmentation method. Hence, a set of methods are investigated which turned out that fuzzy logic using fuzzy membership functions is not only more realistic but also produces more reliable animal segmentation.

Keywords: Feral vertebrate · Identification · Interval type II fuzzy logic system · Animal segmentation

1 Introduction

Feral vertebrates are widespread across Australia, occupying majority of habitats [1]. Proliferation control of feral animal brings the control of natural resources required for livestock industry. They use the same natural assets that are used by the livestock. On many occasions, they may hunt or damage even larger livestock if they outnumber. Hence, there is a high demand from Australian land managers for a device that can help them to control the number of the feral vertebrates enter and live on their rangelands. For controlling the feral anima the first step is to identify them. Such identification will be carried out in the environment in which neither of the circumstances is under control. To bring some as example lighting condition is varying constantly, the imagery scene (the image frames' background) can change rapidly and on top of all, both the object (animal) under surveillance and the camera might be on motion. Dealing with such huge amount of uncertainties require robust algorithms that withstand against variations in the image.

Indeed, uncertainty is an inevitable feature of information [2]. Fuzzy logic system enables dealing with a large part of the uncertainty, in which type-1 fuzzy sets corresponds imprecision over numerical values between [0, 1]. However, in many cases it is not possible to establish the exact value of a quantity, using traditional sets [2]. Moreover, for dealing with a high degree of uncertainty or complexity, interval type-2 fuzzy can be of a remarkable aid [3]. Such methods have been utilized in the literature several times for application such as clustering, pattern recognition, intelligent control,

© Springer Nature Switzerland AG 2020
R. A. Aliev et al. (Eds.): ICSCCW 2019, AISC 1095, pp. 479–486, 2020.
https://doi.org/10.1007/978-3-030-35249-3_61

etc. [3–5]. In the present study, the edge detection for satisfactory image segmentation will be the objective. Variety of vision-based algorithms has been presented recently. Nevertheless, such methods malfunction when dealing with inaccuracy and uncertainty. Developing FLS-based image processing using IT2 has been reported numerous times in the literature such as: image segmentation developing an interval fuzzy clustering method [5] and edge detection [6].

This study focuses on the edge detection methods that can be used for a satisfactory image segmentation and in turn correct object detection. The challenge is dealing with lots of uncertainties and inaccuracy in image attributes. A significant advantage of the application is that there are numerous frames captured in sequence and frequently, that can significantly contribute more accurate animal detection. This paper organized as follows; in the first section an introductory preface is given on the short history of interval fuzzy type II especially when address image prospecting. In the second section as methodology, the circumstances are investigated and the contribution of each is discussed. Moreover, the design of the overall high level algorithm and fuzzy sets and their degree of membership is presented. The resultant images are presented in the results and discussion section to provide a clue regarding how and which method can more conveniently and more reliably contribute the animal detection. The remarks and final clues are given in the conclusion to certify the clue to be used for a feral animal and livestock classifier as a next step in the future works.

2   Methodology

2.1   Attributes of Feral Vertebrate Identification in a Wild Environment Scene

Animals are different in size shape and color. Moreover, animal species share wide range of appearance attributes. Frankly speaking, there is no method that can define each type and classify them as human does. Indeed, the reason why human-beings can successfully discriminate animal species and computers encounter huge difficulties to do so is simply long time of trainings. This is the reason why a child or a person whose occupation keep him/her busy in the city and his/her rate of animal recognition is far lower than a rangeland owner. As Einstein once recommended let us simplify the problem but not making it unreasonable, we have attempted to maintain the core of the study however consider a specific problem. It is assumed that it is dealing with a vertebrate fox which is a common case in Australia [7]. Even though animal species do not share a very comment attributes as non-living creatures do such cars which profoundly eases the vision-based identification task, there are certain number of appearance features that are useful for such purpose. A preliminary examination of a fox image shows that unlike some pests that use color features to hid in the environment they are fairly distinguishable from their environment background. On top of all their color and shape are remarkably different from their ambient objects.

2.2 The Overall Algorithm for a Vertebrate Segmentation on a Wild Environmental Scene

As earlier stated there are certain appearance attributes of foxes that can initialize the feral vertebrate (fox) detection. On top of that there are number of image pre-processing tasks required prior to the segmentation since the objective is determined. Figure 1 demonstrates a summary of the identification algorithm. However, in the present study the initial main steps of such algorithm is paved which are method comparison and selection.

2.2.1 Pattern Recognition

There are some attributes of a common pattern when dealing with a fox image. Figure 2 demonstrates a two dimensional view of such feral vertebrate (fox) recognition which can be fairly obviously distinguishable in the very left column of Table 1. The size and color contrast of each fox in each image in the left side sheds light on the same approach as a human takes to make an approximation over fox identification. One may notice that, such recognition is not performed based upon crisp values and rather it was done using a set of range of membership values.

2.3 Type II Fuzzy Set Applied for Image Segmentation and Image Processing

As mentioned in the introductory section type II fuzzy systems provides the possibility of modeling inaccuracy present on a feral fox image but in a more convenient and reliable and more straightforward approach.

2.3.1 Image Segmentation

Segmentation of an image is carried out to determine the region of interest in an image to enhancer post-object recognition. It should recognize the regions delimited by pixels sharing common properties, e.g. textures, colors etc. One efficient method reported in the literature recently proposed n algorithm using thresholding for segmentation with high emphasis on enhancing the binarization accuracy by merging type II fuzzy sets and spatial-based cutting for membership degrees.

Developing similar approach is considered in the present study and thus in the demonstration it is summarized and we refer the interested reader to that article [8].

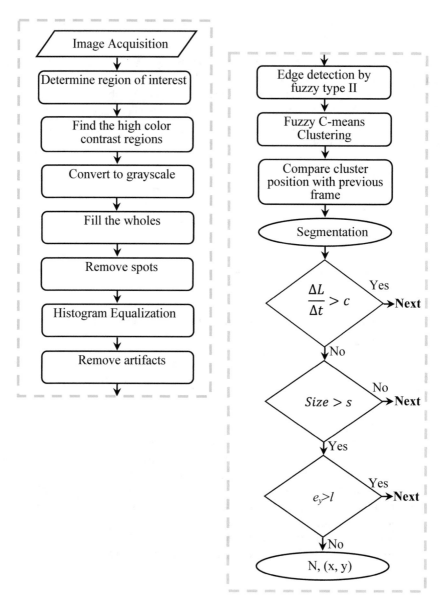

Fig. 1. The proposed algorithm for feral vertebrate identification in a sample rangeland scene in Australia; Preprocessing using human mind (Left), (Right) animal segmentation apply twice fuzzy type II

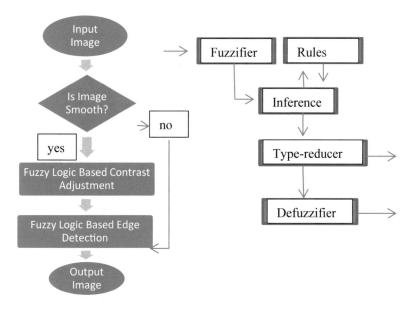

Fig. 2. Fuzzy logic image processing; Lower level algorithm embedded in the overall algorithm depicted in Fig. 1. Left: First segmentation of the region of interest; Right: pixel assessment over the segmentation required for identification

3 Experimental Results

In order to find out which edge detection method is more efficient and also verify if the proposed method is comparable with the commercially available image segmentation algorithms a set of experiments were carried out and tabulated in the Table 1 for the sake of straightforward comparison. The right column in the Table 1 shows a different foxes in different states on a variety of backgrounds. Moreover, the last two depict a scene in which two feral vertebrate is present. The edge detection and the associated segmentation of the region in which the fox (vertebrates) are present were carried out on platform of a common software mathematical package over a desktop computer which poses a processing unit as a core-$i7$, 7700 CPU having 3.6 + 3.6 GHz. The time required for the processing is very different for each applied method.

4 Discussion of the Results

Comparing the results obtained in the Sect. 3, for an unprofessional observer, it is not clear that which method is better to be used for object identification. But one may notice that the challenge in the edge detection occurs among the fox legs or say the bottom side of the image since the back side of the vertebrate is obviously almost straight and thus east to segment.

Table 1. The results of applying s set of edge detection and segmentation approaches reported recently in the literature as efficient methods on the vertebrate animal image, case study fox, captured on a different rangeland backgrounds

Original Image	Texture	Sobel	Canny	Fuzzy C-means	Fuzzy Type II

A rough lock on the second column of Table 1 from the left indicates that the texture-based segmentation of the professional software package does the job however it cannot approach the fox and most importantly it fails to cluster the animals when there are multiple of them (two last rows of the same column). Moreover, it takes min 7 s to perform the task on a very low quality image frame which means that it is not a good candidate for a real-time segmentation or on a dynamic scene.

As expected, Sobel and then Canny are good choices for edge detection however they over-focus on the very image details. It is worthwhile to mention that reducing the focus rate will severely reduce the edge detection around the fox for both aforementioned methods. Fuzzy c-means clustering performed slightly better than Sobel and Canny however busy background damages its recognition in particular images in the last three rows.

Comparing fuzzy type II edge detection with the fuzzy c-means indicates that even though it was not yet specialized for a very fox identification task, it rather focuses on the overall edges and thus less segments the legs from the body. It is a very remarkable advantage. Most importantly, the timing is big advantage however still this method in not mature if the number if its corresponding research works presented in the literature is compared with other method presented in Table 1. One significant aspect of this research is working on the black and white after the pre-processing the remarkably enhance the time-to-process. Indeed, the major advantage of the proposed algorithm using interval type II fuzzy image segmentation will emerge when multiple frames get captured by the moving camera that allows incorporating the last two steps in the algorithm in Fig. 1 dealing with fox passed trajectory and its speed which is far distinguishable from a ship or goat.

5 Conclusion

Feral animal identification was not a straightforward task especially when the background and foreground conditions are nod under the control. A set of professional methods were carried out and compared over their reliability using a human observer judgment and their correct edge detection. It is concluded that interval type II fuzzy could slight better performed in the single vertebrate image scene. It is expected to achieve better performance when an objective-specific used over a dynamic camera since Canny and Sobel cannot filter the noise as well as the fuzzy does. However the main advantage of the proposed interval type II fuzzy image segmentation will be when multiple frames are captured by the surveillance camera which can incorporate the last two steps of the proposed algorithm dealing with animal motion.

References

1. Jurdak, R., Elfes, A., Kusy, B., et al.: Autonomous surveillance for biosecurity. Trends Biotechnol. **33**(4), 201–207 (2015)
2. Zadeh, L.A.: The concept of a linguistic variable and its application to approximate reasoning-I. Inf. Sci. **8**, 199–249 (1975)
3. Castillo, O., Melin, P.L.: Type-2 Fuzzy Logic: Theory and Applications. Springer, Berlin (2008)
4. Hagras, H.A.: A hierarchical type-2 fuzzy logic control architecture for autonomous mobile robots. IEEE Trans. Fuzzy Syst. **12**, 524–539 (2004)
5. Mendoza, O., Melín, P., Castillo, O.: Interval type-2 fuzzy logic and modular neural networks for face recognition applications. Appl. Soft Comput. **9**, 1377–1387 (2009)

6. Melin, P., Gonzalez, C.I., Castro, J.R., Mendoza, O., Castillo, O.: Edge-detection method for image processing based on generalized type-2 fuzzy logic. IEEE Trans. Fuzzy Syst. **22**, 1515–1525 (2014)
7. Meek, P.D., Ballard, G.A., Fleming, P.J.S.: The pitfalls of wildlife camera trapping as a survey tool in Australia. Aust. Mammal. **37**(1), 13–22 (2015)
8. Dawoud, A.: Segmentation of dermoscopic images by the fusion of type-2 fuzziness measure in graph cuts image binarization. Int. J. Imaging Robot **15**, 73–87 (2015)

Artificial Neural Network for the Left Ventricle Detection

Elbrus Imanov[1](✉) ⓘ and Anwar A. Ibra[2] ⓘ

[1] Department of Computer Engineering, Near East University, Nicosia,
North Cyprus via Mersin 10, Turkey
elbrus.imanov@neu.edu.tr
[2] Adalil Department of IT and Telecom Technology, Tripoli, Libya
1240.anwarman1990@gmail.com

Abstract. Machine learning has proved its effectiveness through its application in medicine. Neural networks have been used for solving different dilemma in the medical field such as image analysis and diagnosis. The object detection is a common field in computer vision practices, for detecting some important objects in images or videos. In medicine, there is also a need for object detection such as, organ detection. Thus, different types of deep networks have been used in medicine for detecting organs or some parts of organs such as ventricles in heart images. This paper introduces a system based on neural network, used for automated detection in MRI cardiac images of left ventricles. Our work relies upon backpropagation neural network with a sliding window used to go through the images to find and detect the left ventricle. Firstly, network is operated in order to classify left ventricle and non-left ventricle images using a backpropagation techniques neural network. MATLAB program is run and furthermore, BPNN is automatically exercised several times before the detection process is carried out. The trained network is then validated by testing it on unseen data to gain a good generalization capability. The recognition rate achieved by BPNN is 100% on training data, 89.23% on testing data. The trained network is then used to determine if a left ventricle is contained or not in the sampled region. Moreover, the developed system seems to perform effectively in detecting left ventricles with up to 6% salt and pepper noise.

Keywords: Machine learning · Magnetic resonance imaging · Backpropagation · Artificial neural network · Fully convolution · Intelligent system

1 Introduction

Through using a neural network we can reach compact knowledge base, learning and operative processing as well [1] Machine networks have been extensively used in medicine to solve different conditions in various areas: tumor analysis, image categorization, and image segmentation [2]. Machine learning is constantly developed and improved to be served and exploited for the Artificial Intelligence model.

Object identification is an errand that generally has a place with the class of PC vision issues. It is critical that while people are exceptionally proficient in distinguishing

© Springer Nature Switzerland AG 2020
R. A. Aliev et al. (Eds.): ICSCCW 2019, AISC 1095, pp. 487–494, 2020.
https://doi.org/10.1007/978-3-030-35249-3_62

different complex systems regardless of scene limitations like fluctuating backward, machines endeavor to accomplish close expert results on object identification [3]. Artificial neural network concept comprises neural which is a crucial area of interest in biological researches to analyze the activities of the human body [4]. ANN computer is designed based on the structure of the brain [5]. Moreover, it is focused on that object location is very all the more trying for machines when contrasted with object acknowledgment. In system discovery, the system important to be distinguished can be situated in any district of a picture. While, in system acknowledgment, the systems important to be perceived is typically officially portioned; consequently, making the acknowledgment less difficult. One of the obvious choices is to consider an intelligent model such as artificial neural network. Better results for complicated relationships between dependent and independent variables can be achieved through ANN approach. Furthermore, ANN approach performs well with nonlinear problem analysis, as well [1]. ANN's are popular for its capability to learn various tasks using collected training data. Here with, lots of entirely different principles are adapted such as heuristics and elemental machine learning methods [6]. Some of the intelligent decision capabilities include tolerance to constraints such as object translation, object rotation, object scale, object illumination, and noise. Feed-forward network model regards the perception back propagation technique and the function network model as representatives, that are commonly used in prediction and pattern recognition area of interests [7]. ANN is a technique used for functional pattern classification and it is trained all the path through the error back-propagation model [8]. Backward error propagation model used is the most widespread technique that is generally known as back propagation approach [4]. Artificial neural networks have been successfully used in many important tasks such as face recognition, speaker identification, natural language processing, document segmentation, etc. Fields such as health care applications and medical diagnosis implications usually rely upon ANN's predictive and powerful classifier [9]. In this research, a type of artificial neural network known as back-propagation neural network is used to achieve the task of left ventricles detection in images obtained from Sunnybrook-cardiac database [10]. The developed left ventricle detection system using back propagation neural network is found to be quite efficient and effective for the left ventricle detection task. Left ventricle, which is the largest chamber of the heart, is considered as the first important chamber due to its activity of pumping the blood to the whole body [11]. This study has been focused on tackling the problem of automated left and right ventricle segmentation through the application of FCN, and were able to show that the FCN achieves state of the art semantic segmentation in short-axis cardiac MRI acquired at multiple sites and from other scanners [12]. Convolution Neural Network to perform semantic segmentation on images obtained from cardiac MRI scan in order to localize the left ventricle and influence the system to measure the capacity of the ventricle during the course of a heartbeat [13]. The goal of this research work is to develop a vision system for the detection of left ventricles in images. In this paper, we show a simple and new approach for detecting the left ventricle through a sliding window and a trained back propagation neural network. Sliding window machine learning approach, a window of reasonable size, e.g. m × n, is played out a pursuit over the objective data [14]. It is important to note the scope and application of this project is more and broader than the detection of left ventricles only.

2   Methodology and Techniques

Our aim is to set up a framework for artificial vision, which is able to identify the left ventricle involved in MR slicers. In this work, considering difficulties, for example, protest light, scale, interpretation, revolution, and so forth which make the recognition an intricate issue for such an open identification issue, we make plans to actualizing a wise framework which can to some degree generous adapt to the previously mentioned discovery limitations. By using back-propagation neural network we consider it as a 'cerebrum' behind recognition process. The process is achieved completed by two phases. For the first stage, prepared BPNN is used to acknowledge the left ventricle part. In the second phase BPNN is utilized for the object recognitions in the left ventricle. Example framework is shown in Fig. 1.

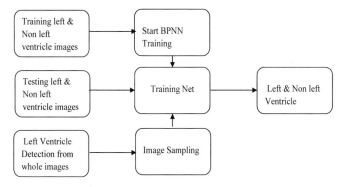

Fig. 1. Flowchart for developed system.

During the left ventricle recognition stage, left ventricle objects and non-objects is detected by the BPNN. In order to analyze left ventricle objects and non-objects (errand) testing and training data are gathered through the web. Framework, pictures containing objects of left ventricle are considered as positive illustrations or tests.

During image processing stage, collected positive and negative examples are altered from coloured to grey scale images. This converts the three RGB colour channels to a single channel representing the brightness or intensities of pixels at different locations in the images. There are many using areas for ANN, such as image analysis, speech recognition, adaptive control and so forth [4]. Hardware based red, green, blue modeling is a type of colour space model with the compounded red, green and blue channels that forms a colour vector of pixel [15]. All the images are resized accordingly to 1600 pixels.

3 Design, Training, Testing of the System

The BPNN is utilized to analyze both negative and positive examples on the gathered samples 50 samples are collected using database for positive examples means left ventricle objects: and 60 samples are randomly collected from the internet for the negative examples or in other terms, non-left ventricle object. Example positive and negative results frame our training and testing of BPNN system. The pictures are altered to gray scale with 40 × 40 pixels. Here while testing data permits to implement the prepared BPNN system on the recent data; and enable performance testing of the BPNN on invisible or a new data. It is crucial for trained ANN system to perform well with the unseen or invisible data. In our research, training uses 50 left ventricle and the 60 non-left ventricle objects; and for testing BPNN 45 left ventricle and 33 non-left ventricle objects are used. To sum up 110 training and 78 testing images are used. Note that the left ventricles are collected by cropping the full heart MRI images from the Sunny-Brook database [10]. While the non-left ventricle are obtained from the Image-Net [16]. Designed BPNN model is shown below in Fig. 2.

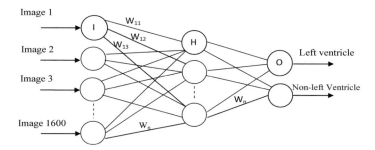

Fig. 2. Designed back propagation neural network (BPNN).

The designed BPNN model has 1600 input neurons, every input element or in other words, pixel is affiliated with an input neuron. Input neurons are non-processing, in other terms, input neuron generally conducts input pixels that were received on the

Table 1. Training parameters BPNN.

Training images	110
Input neurons	1600
Hidden neurons	70
Function	Log-Sigmoid
Rate of Learning	0.23
Rate of Momentum	0.70
MSE mean square error	0.0099
Maximum number of Epochs	3000
Training time	42 secs

processing or hidden layer neurons. Input data is extracted in the second layer that enables input data to accordingly the target points. The BPNN has two output neurons in the output layer as the purpose is to classify all the given images either into left ventricle object or non-left ventricle object; left ventricle object is stated as (1,0); while non-left ventricle object is (0,1) The training parameters BPNN are shown in Table 1. The aforementioned recognition rates achieved are shown in Table 2.

Table 2. The BPNN recognition rates

Parameter	Training	Testing
Samples	110	78
Samples correctly classified	110	70
Recognition	100%	89.23%

As the Table 2 shows, used BPNN has obtained the 100% recognition rate on training, and 89.23% on testing data. It is important to notice that 89.23% recognition rate is enough for the generalization of BPNN on unseen images or in other terms, separating new images as whether left ventricle objects or non-left ones. Cross-validation can be considered as a reliable way used in order to test the classification or the generalization power of a neural network while it is getting trained. Hence, in some cases and application cross-validation is a useful tool for creating a powerful, intelligent, and machine learning system. All the presented results are obtained using this technique of splitting. However, for more experiments, we used cross-validation by splitting the data into 70%, 20%, and 10%, on training, validation, testing, and recognition rate of 99.3%, 86.2%, 88.97%, respectively. The performance cross-validation used has slightly changed. As seen in the results, the network recognition rate during the testing has slightly decreased when cross-validation which is possibly due to the small training data used in training the network.

4   System Performance and System Evaluation

With the images involving different objects, scales or background, and so forth, to detect the left ventricle objects the BPNN model is exploited. New images or data are sampled in a non-overlapping characteristic by a mask for the purpose of detecting objects of left ventricle in the given new images. All of the images analyzed are firstly resized accordingly to 120×120 pixels with gray-scale, therefore the number of required samples and accordingly computations are reduced. On the other hand, without removing the image edges, such images containing left ventricle object with 40×40 pixels as in previously trained BPNN model can easily match with the model given for the detection process. The sliding window is shown in Fig. 3.

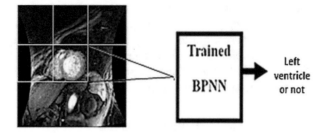

Fig. 3. Detection of left ventricle with image sampling using trained BPNN [17].

1600 pixels input field is used for the designed BPNN to achieve the sampling outcomes. The predicted output of BPNN is (1,0) for left ventricle involved windows, as it was coded in BPNN training. In contrast, the predicted output of BPNN is (0,1) for left ventricle object absent windows. As the desired output for left-ventricle object is (1,0), the BPNN with the nearest matched output to the desired output is consider as involving a left-ventricle object involving a left ventricle object. The sample images of left ventricle detection process formed by the developed model **(a)** and the incorrect left-ventricle object detection **(b)** are shown in Fig. 4.

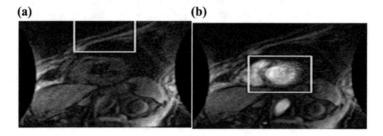

Fig. 4. (a) Correct detection outcome. (b) Wrong detection outcome using the develop system.

In order to show the effectiveness of the developed system, we perform experiments using noisy target images. The idea is to intentionally add noise to the target images for left-ventricle detection; and then task the developed system to scan the noisy

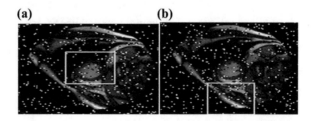

Fig. 5. (a) Correct detection outcome. (b) Wrong detection outcome develop system.

target images for left-ventricle objects. To obtain noisy target images, salt and pepper noise at different noise levels are added to the original target images. It is also the aim to observe at what noise level that the developed system begins to significantly begin to detect non-left ventricle objects in the target images. Samples of some obtained noisy target images and detection outcomes are shown in Fig. 5.

In Fig. 5(a), the true detection outcome with 6% salt & pepper noisy target image after training the networks for two more runs is shown. However, in Fig. 5(b), the trained network was not able to detect the left ventricle in the first run as it has 6% of salt and pepper noise. The developed system can perform effectively the task of cub object detection even in the face of target images with up to 6% salt & pepper noise.

5 Conclusion

The detection of objects in images is an important task required for efficient and effective use of data, where it is the aim to develop artificial systems or machines which are capable of locating an object of interest in images. Although, such tasks are traditionally formulated as computer vision problems, the problem of object detection is now being reformulated as that of a machine learning problem.

In our research, object detection of left ventricles the artificial vision system is exercised. Furthermore it is crucial to realize that, the practice can be applied with same insight and approach to detect other required necessary objects, since the same insight and approach are applicable to other detection. We have reformulated the object detection process as a machine leaning model problem to achieve more robustness in the developed system's detection practices for images with scale. In this paper we have used the back-propagation neural network as the machine learning model. The samples used of non-left and left ventricle objects are gathered from the appropriate source of data in accordance with the exercised BPNN system. In the detection processes 1600 pixels windows size in accordance with the input size of the trained BPNN is operated for the sampling of target images in a non-overlapping model. The system developed is trained for the several randomly selected images with left-ventricle objects that do not include the scene constraints. Furthermore, the developed network has shown a quite well performance ability in cub object detection even in the images of partially blocked left-ventricle objects It is seen that the developed artificial vision system is capable of performing the left ventricle object detection task with up to 6% salt and pepper noise.

References

1. Aliev, R.A., Fazlollahi, B., Aliev, R.R.: Soft Computing and Its Application in Business and Economics, p. 110. Springer, Heidelberg (2004)
2. Huang, S., Liu, J., Lee, L., Venkatesh, S., Teo, L., Au, C., Nowinski, W.: Segmentation of the left ventricle from cine MR images using a comprehensive approach. MIDAS J-Card MR Left Ventricle Segm. Challenge **27**(5), 593–599 (2009)

3. Marak, L., Cousty, J., Najman, L., Talbot, H.: 4D morphological segmentation and the MICCAI LV-segmentation grand challenge. In: Proceedings of the Workshop on Cardiac MR Left Ventricle Segmentation Challenge, USA, Chicago, pp. 1–8 (2009)
4. Sharma, V., Rai, S., Dev, A.: A comprehensive study of artificial neural networks. Int. J. Adv. Res. Comput. Sci. Softw. Eng. 2(10), 278–284 (2012)
5. Sonali, B.M., Priyanka, W.: Research paper on basic of artificial neural network. Int. J. Recent. Innov. Trends Comput. Commun. 2(1), 96–100 (2014)
6. Hadhoud, M.M., Eladawy, M.I., Farag, A., Montevecchi, F.M., Morbiducci, U.: Left ventricle segmentation in cardiac MRI image. Am. J. Biomed. Eng. 2(3), 131–135 (2012)
7. Ni, X.: Research of data mining based on neural networks. World Acad. Sci. Eng. Technol. 39, 381–384 (2008)
8. Kottaimalai, R., Rajasekaran, M.P.: EEG signal classification using principal component analysis with neural network in brain computer interface applications. In: International Conference 2013 on Emerging Trends in Computing, Communication and Nanotechnology, ICECCN, pp. 227–231. IEEE, India (2013)
9. Gharehchopogh, F.S., Khalifelu, Z.A.: Neural network application in diagnosis of patient. A case study. In: International Conference 2011 on Computer Networks and Information Technology, pp. 245–249 (2011)
10. Radau, P., Lu, Y., Connelly, K., Paul, G., Dick, A., Wright, G.: Evaluation framework for algorithms segmenting short axis cardiac MRI. The MIDAS J.-Card. MR Left Ventricle Segm. Chall. 49(2) (2009)
11. Arseny, K., Andrey, S.: An overview of techniques for cardiac left ventricle segmentation on short-axis MRI. In: International Conference 2016 on Big Data and its Applications, ICBDA 2016, vol. 801003, pp. 1–7. EDP Sciences (2016)
12. Tran, P.V.: A fully convolutional neural network for cardiac segmentation in short-axis MRI. Computer Vision and Pattern Recognition, arXiv preprint arXiv:1604.00494 (2016)
13. Zotti, C., Luo, Z., Humbert, O., Lalande, A., Jodoin, P.M.: GridNet with automatic shape prior registration for automatic MRI cardiac segmentation. arXiv preprint arXiv:1705.08943 (2017)
14. Sudowe, P., Leibe, B.: Efficient use of geometric constraints for sliding-window object detection in video. In: Proceedings of the International Conference 2011 on Computer Vision Systems, Germany, Berlin, pp. 11–20 (2011)
15. Vassilis, S., Kodogiannis, M., Boulougoura, E., Wadge, J., Lygouras, N.: The usage of soft-computing methodologies in interpreting capsule endoscopy. Eng. Appl. Artif. Intell. 20(4), 539–553 (2016)
16. Deng, J., Dong, W., Socher, R., Li, L.J., Li, K., Fei-Fei, L.: Imagenet: a large-scale hierarchical image database. In: Proceedings of the International Conference 2009 on Computer Vision and Pattern Recognition, pp. 248–255. IEEE, Florida (2009)
17. Helwan, A., Uzun Ozsahin, D.: Sliding window based machine learning system for the left ventricle localization in MR cardiac image. In: Applied Computational Intelligence and Soft Computing, pp. 1–9 (2017)

An Approach to Analysis of Useful Quality Service Indicator and Traffic Service with Fuzzy Logic

Bayram G. Ibrahimov[1] and Almaz A. Alieva[2]([⊠]) [iD]

[1] Azerbaijan Technical University, Huseyn Javid, 25, Yasamal, Baku,
Azerbaijan
i.bayram@mail.ru
[2] Mingechaur State University, Zahid Khalilov, 23, Mingechaur, Azerbaijan
almaz40@gmail.com

Abstract. The indicators of the multimedia traffic quality service and the tasks managing the distribution resources in multiservice telecommunication networks are analyzed in the presence of the incoming load self-similarity property. The aim of the work is to consider an approach of controlling the quality of service (QoS) indicator in terms of the self-similar traffic based on the prediction of the Hurst coefficient using fuzzy logic. Based on the network traffic structure study, a new approach of analyzing the quality of useful service index and service traffic has been proposed, which is based on the theory of fuzzy sets.

Keywords: Fuzzy logic · Self-similar traffic · QoS · Information and network resource · Hurst coefficient · Fuzzy set · Membership function

1 Introduction

The development of digital economy and the use of strategic plans for the "Digitalization Roadmap" requires the construction of multiservice telecommunication networks (MTN) based on the concept of the Next Generation Network (NGN) and Future Network (FNs).

In MTN construction, an important and effective direction is sharing SDN (Software-Defined Networking), NFV (Network Functions Virtualization) and IMS (Internet Protocol Multimedia Subsystem) technologies that provide [1–3]:

- acceleration of the introduction of new services and applications such as "Triple Play services" & "Bandwidth on Demand";
- an increase in the efficiency by using limited network bandwidth resources and capacity buffer memory;
- improving the quality of service (QoS), information security and network reliability.

Moreover, when using SDN & NFV and IMS technologies, a characteristic feature of modern MTN is the heterogeneous traffic, which consists of useful and service traffic possessing the self-similarity properties [2, 4]. The estimation of such parameters of

© Springer Nature Switzerland AG 2020
R. A. Aliev et al. (Eds.): ICSCCW 2019, AISC 1095, pp. 495–502, 2020.
https://doi.org/10.1007/978-3-030-35249-3_63

QoS as the average delay time of packet transmission, the probability of loss and the communication channels capacity are among the most urgent tasks today.

Due to the fact that the "quality" concept is vague and subjective, there is no possibility of describing the telecommunication processes of its formation by classical logic [5].

The existing works are based on methods for improving the indicators of quality service, such as the operational distribution of information and network resources of multiservice networks [1, 3], the method of calculating the Hurst coefficient using fuzzy models [4, 5], and modeling complex multiservice systems and telecommunication processes. Our research is devoted to solve the problem creating a new approach of analyzing the quality of useful indicator service and QoS traffic service using fuzzy logic apparatus, which allows for predicting the value of the Hurst coefficient.

2 General Formulation of Problem

Based on the study, it was revealed that one of the technical problems in the transmission of multimedia traffic over MTN packet networks is to ensure the QoS for class-differentiated traffic. A poor adaptability of packet-switched networks hinders the widespread development of MTN based on NGN and FN technologies.

The analysis of parameters describing MTN multimedia traffic with packet switching [2, 4] shows that, in mathematical modeling of service processes, in addition to traditional load parameters, it is also necessary to take into account the Hurst coefficient characterizing the self-similarity degree of a random process.

On the basis of the study, it was established [2, 3] that the test for self-assumption and the evaluation of the Hurst index are complex tasks. Moreover, the values of the Hurst coefficient obtained by using different methods, distinguish significantly. Therefore, it is necessary to propose a new technique in choosing the method for calculating the Hurst coefficient, H, where $0 < H < 1$.

In the majority of the considered papers devoted to the study of the properties in self-similar traffic and the calculation of the main indicators functioning in telecommunication networks [1–5], the Hurst coefficient value is considered to be a constant, time-independent parameter. This assumption can lead to a deterioration in the QoS indicators under the influence of the self-similarity effect in the case of a change in the Hurst coefficient at small time intervals.

According to the results of the telecommunication processes analysis the task was formulated. Namely, it is to investigate the methods of operational resource allocation of the MTN link in the presence of the self-similarity property incoming load and calculate the indicators of the Hurst coefficient based on fuzzy logic.

3 The Mechanism of Analysis and the Choice of the Approach Predicting the Hurst Coefficient Value Using Fuzzy Logic

Considering the self-similarity properties of the incoming load in MTN, the channel load factor $\rho(H_i)$ when servicing i – the useful packet is determined as follows [3]:

$$\rho(H_i) = \frac{\lambda_i}{N_k \cdot C_{i.\max}} \cdot L_{i.n} \cdot f(H_i) \leq 1 , i = \overline{1, n} \tag{1}$$

where $f(H_i)$ – a function that takes into account the self-similarity property of the incoming load and is equal to $f(H_i) = 2H_i$; H_i – Hurst coefficient for flow i-th package; λ_i – i-th packet flow rate; $L_{i,n}$ – length of the transmitted stream of i-th packet; $C_{i.\max}$ – maximum network bandwidth when streaming i-th package, $i = \overline{1, n}$.

The expression (1) determines the effective use of the MTN communication channel and the quality of traffic control. For the analysis of the parameters of multimedia traffic, the invariant value of the Hurst coefficient $H \in \{0, 1\}$ is also required (in addition to the traditional load parameters) [2–5]. Consequently, in MTN, the values of the Hurst coefficient obtained using various methods differ significantly. Such vagueness is expressed by fuzzy sets [6–9]. So, we introduce membership function $\mu_A(H_j)$, that measures degree of membership of an element H_j to a fuzzy set A. Assume that $H_e = \{H_j\}$ is a set of expected values or variations of the Hurst coefficient. Then a fuzzy set is described as follows:

$$A = \{H_j, \mu_A(H_j)\}, H_j \in H_e, \tag{2}$$

Considering (1) and (2), it is necessary to propose a new mechanism in choosing the method for calculating the Hurst coefficient. To choose the calculating method of the Hurst coefficient, it is adequate to use a fuzzy integral [6–8]. Based on the examination performed, the distribution of the fuzzy density weights of these values can be determined as a priori $\varphi(H_j), H_j \in H_e$. Considering the latter assumption and the method of calculating the Hurst coefficient, the membership function for a given random process is described by the expression:

$$\mu_A(H) = \sum_j \frac{\mu_A(H_j)}{H_j}, H_j \in H_e \tag{3}$$

For the useful self-similarity property and QoS, calculation of service quality indicator (as the most expected value H_j) is based on the fuzzy integral:

$$H_{mev} = \arg \int_j y(H_j) \cdot o \cdot \varphi(F_j), \tag{4}$$

where $y(H_j)$ – ordered by decreasing powers from membership function $\mu_A(H_j)$; $\varphi(F_j)$ – fuzzy measure of set $F_i = \ <H_{j1}, H_{j2}, \ldots, H_{ji}> $; o – composition mark.

4 The Evaluation of Self-similar Traffic Based on Fuzzy Logic

To estimate the property of self-similar traffic, some methods for calculating the Hurst coefficient based on the fuzzy logic are considered and analyzed. These methods are as follows: Variance-Time Plot, Rescaled adjusted range, R/S, the method for estimating spectral functions and the method for estimating the coefficient correlation.

As a result of the experiment and with the help of various methods for calculating indicators of the self-similar traffic property, the following values of the Hurst coefficient are obtained, H_j, $j = 1, 2, 3, 4$: $H_2 = 0,77$; $H_2 = 0,87$; $H_3 = 0,95$; $H_4 = 0,74$.

In this case, the indices in the notation correspond to the following methods for determining the value of the Hurst coefficient: H_1 – methods Variance-Time Plot; H_2 – methods rescaled adjusted range, R/S; H_3 – method for estimating spectral functions; H_4 – correlation coefficient estimation method.

Given the above methods for calculating the Hurst coefficient based on fuzzy sets, the numerical values of the membership function of the process under study can be determined as follows:

$$\mu_A(H_j) = \frac{0,40}{0,77} + \frac{0,50}{0,87} + \frac{1,0}{0,95} + \frac{0,90}{0,74}. \tag{5}$$

Taking into account the numerical values of the membership function, the membership function is ordered by decreasing powers $y(H_j)$, H_j, $j = \overline{1,\,4}$, described as follows:

$$y(H_4) = 0,90; y(H_3) = 1,0; y(H_2) = 0,50; y(H_1) = 0,40.$$

Therefore, it is assumed that the weighting coefficients of the truth of the values of the Hurst coefficient, determined by experts, have the following values for H_j, $j = 1, 2, 3, 4$:

$$\varphi(H_1) = 0,20; \quad \varphi(H_2) = 0,10; \quad \varphi(H_3) = 0,50 \; ; \varphi(H_4) = 0,40.$$

Given the last expression under given conditions, fuzzy measure $\varphi(F_j)$, $j = \overline{1,\,4}$ takes the following numeric values: $\varphi(F_2) = 0,94$; $\varphi(F_3) = 0,50$; $\varphi(F_4) = 0,81$.

Thus, numerical calculations are performed for calculating the Hurst coefficient on the basis of the proposed approach. As a result, all necessary obtained numerical values are summarized in Table 1.

From Table 1 it is seen that the minimal expected value of the Hurst parameter is $H_j = 0,73$ at $j = 4$. This is obtained by using the method of estimating the correlation coefficient with a fuzzy measure, where $\varphi(F_j) = 0,81$. The last numerical value means that, in MTN, in case of useful self-similarity property and traffic service, it is advisable to calculate the most expected value of the Hurst parameter.

Table 1. The calculated values of the Hurst coefficient in the presence of the traffic self-similarity property

H_j	$H_1 = 0,79$	$H_2 = 0,86$	$H_3 = 0,95$	$H_4 = 0,73$
$y(H_j)$	0,40	0,50	1,00	0,90
$\phi(H_j)$	0,20	0,10	0,50	0,40
$\phi(F_j)$	1,00	0,94	0,50	0,81
G	0,40	0,50	0,50	0,81
G_{max}	0,81			

5 Analysis of Methods for Predicting the Value of the Hurst Coefficient

In order to use the proposed method for estimating the correlation coefficient, an expert commission is needed. On the basis of the proposed approach, a method for predicting the value of the Hurst coefficient is determined, which allows you to distribute network and information resources adaptively. To assess the prediction of the value of the Hurst coefficient, the following methods of analytical forecasting were used [2]:

- simple predictor method and the least squares method;
- 1-st order autoregressive predictor (AR-1); 2-nd order autoregression $N_{bs}(H_j)$ of regressive predictor (AR-2).

System analysis showed that in the process of using the above methods of analytical forecasting, there are errors of underestimating and overestimating the value of the Hurst coefficient in the reference k.

1. For a case of an error in underestimation, the following expression is used:

$$\delta^+(k) = \begin{cases} \delta(k), & if \quad \delta(k) \geq 0 \\ 0, & if \quad \delta(k) < 0 \end{cases}, \tag{6}$$

2. In case of error during revaluation, the expression is as follows:

$$\delta^-(k) = \begin{cases} |\delta(k)|, & if \quad \delta(k) < 0 \\ 0, & if \quad \delta(k) \geq 0 \end{cases}, \tag{7}$$

where $\delta(k)$ – absolute forecast error determined by the expression:

$$\delta(k) = |H(k) - H^*(k)|, \tag{8}$$

where $H(k)$ and $H^*(k)$ – the real and predicted values of the Hurst coefficient of the k-th reference respectively.

Studies have shown [2, 5] that with an increase in the error of underestimating, the value of the Hurst coefficient $\delta^+(k)$ will reduce the resources allocated to the traffic service ($N_{6n}(H_j)$, bandwidth $\Delta F_k(H_j)$ and $C_{max}(H_j)$ (consequently, an increase in the probability of packet loss and a deterioration in the QoS).

Next, the second error $\delta^-(k)$ leads to an excessive overestimation of the value of the Hurst parameter and reflects the excessive allocation network and information resources of the network.

In order to select a criterion for assessing the quality of useful forecasting and service traffic, it is necessary to use underestimation coefficient $D^+(k)$ and revaluation factor $D^-(k)$:

- the underestimation coefficient is computed as follows

$$D^+(k) = \sum_k \delta^+(k) / \sum_k H(k), \qquad (9)$$

- revaluation coefficient is computed as follows

$$D^-(k) = \sum_k \delta^-(k) / \sum_k H(k) \qquad (10)$$

From the expressions (9) and (10), it can be seen that the closer the prognostic estimates of the Hurst parameter to real values, the closer to zero are the estimates under consideration $D^+(k)$ and $D^-(k)$. First, one should choose the method of predicting the value of the Hurst coefficient of the k-th reference, for which the underestimation coefficient is the smallest. Taking into account (9) and (10), numerical results of the prediction of the Hurst coefficient by selected methods are summarized in Table 2.

Table 2. Hurst parameter prediction estimates

Prediction method	$D^+(k)$	$D^-(k)$
Simple prediction	0.0016	0.0017
Least square method	0.0014	0.0015
First order autoregressive predictor method AR-1	0.0019	0.0017
Second order autoregressive predictor method, AR-2	0.0013	0.0012

From Table 2 it can be seen that the smallest errors of underestimation and overestimation are those of the most complex of the prediction methods, that is the method with the autoregressive predictor of the second order (AR-2).

To assess the effectiveness of the allocation of MTN resources based on the predicted values of the Hurst coefficient, we calculate the relative error of the system parameters both for the channel resource and the volume of the buffer store.

Further, the indicators of QoS considered as the average waiting time for the start of service, and the probability of packet loss when using the predicted value H_j are compared with values of these indicators obtained by the average value of the Hirst coefficient, $E[H_j] = 0,660$, where $0 < E[H_j] < 1$.

Now, we define the relative error of the traffic forecast:

$$\beta_i = \frac{N_{i.av} - N_i}{N_{i.av}} \cdot 100\%, \ i = \overline{1, n}, \tag{11}$$

where $N_{i.av}$ – i-th indicator of the QoS, calculated by the average value of the Hirst coefficient, $E[H_j]$; N_i – i-th indicator of the QoS, calculated by the predicted value of the Hurst coefficient.

Here, the resulting positive error shows that the allocated network resources are not sufficient for the incoming load quality service. This characterizes the excess of the probability of packet loss and the waiting time of the start service over the indicators calculated by the average value $E[H_j]$.

Further, a negative relative error indicates the percentage of MTN resources utilization. Predicting a decrease in the Hurst coefficient value, one can increase the number of simultaneously serviced connections while preserving the QoS packets.

Thus, determination of the value of the Hurst coefficient based on fuzzy set theory allows to improve the efficiency of the MTN functioning in terms of the throughput value, the average transmission delay time, the probability of packet loss, BS capacity, and the utilization rate of network resources under the influence of self-similarity effect.

6 Discussion and Conclusion

On the basis of the fuzzy set theory, a new approach to the analysis of the QoS (for heterogeneous traffic and adaptive resource allocation) in terms of the self-similarity influence is proposed. The approach is based on the Hurst coefficient prediction.

As a result of the study, the proposed approach allows eliminating ambiguity in determining the Hurst coefficient and improving the efficiency using network resources.

Estimates of the Hurst parameter value prediction are also given, and the relative traffic prediction error is determined.

References

1. Efimushkin, V., Ledovskikh, T., Ivanov, A., Shalaginov, V.: The role SDN/NFV technologies in the infrastructure of the digital economy. Exp. Test. Implement. Telecommun. 3(8), 27–36 (2018)
2. Shelukhin, O.: Modeling Information Systems. Telekom, Moscow (2011)
3. Ibrahimov, B., Humbatov, R., Ibrahimov, R.: Analysis performance multiservice telecommunication networks with using architectural concept future networks. T-Comm. 12(12), 84–88 (2018)
4. Bitner, V., Lizneva, Y.: Using the theory of neural networks in forecasting signal traffic. Inf. Space 2(3), 36–39 (2008)
5. Mammadov, H., Ibrahimov, B.: Efficiency methods forecasting of the office traffic signaling systems with use technologies of neural networks. In: Proceedings 4th World Conference on Soft Computing, Berkeley, vol. 296, pp. 241–245 (2014)
6. Zadeh, L.: Fuzzy sets. Inf. Control 8, 338–353 (1965)

7. Belman, R., Zadeh, L.: Decision making in a fuzzy environment. Manag. Sci. **17**, 141–164 (1970)
8. Aliev, R., Gurbanov, R., Aliev, R., Huseynov, O.: Investigation of stability of fuzzy dynamical systems. In: Proceedings of the Seventh International Conference on Applications of Fuzzy Systems and Soft Computing, pp. 158–164. Quadrat-Verlag, Siegen (2006)
9. Bellman, R., Giertz, M.: On the analytic formalism on the theory of fuzzy sets. Inf. Sci. **5**, 149–157 (1974)

Modeling and Evaluation of Rock Properties Based on Integrated Logging While Drilling with the Use of Statistical Methods and Fuzzy Logic

G. M. Efendiyev[1(✉)] ⓘ, P. Z. Mammadov[2] ⓘ, and I. A. Piriverdiyev[1]

[1] Oil and Gas Institute of the Azerbaijan National Academy of Sciences,
F.Amirov 9, Baku AZ1000, Azerbaijan
galib_2000@yahoo.com, igorbaku@yandex.ru
[2] Azerbaijan State Oil and Industry University, Azadlig Av. 20,
Baku AZ1010, Azerbaijan
parviz08@list.ru

Abstract. The quality of complex geological, geophysical and technological information is largely determined by the quality of information obtained during the drilling of wells. However, as the analysis shows, the quality of the information obtained during drilling does not always meet the design requirements, which in turn complicates the decision-making process. Numerous studies indicate the need for research to improve the quality of the information obtained during drilling, which requires the use of appropriate data processing and analysis methods.

Based on this, the paper is devoted to the analysis of the information obtained during drilling, the assessment of its quality, ways of applying various methods of improving the quality of information when making decisions at various stages of drilling. The possibility of using probabilistic-statistical methods and fuzzy logic when analyzing complex information and predicting the characteristics of rocks is considered. Calculations were carried out that allow dividing the section into homogeneous intervals, predicting the lithology of the rocks, and comparing the obtained results with the design ones. The latter is achieved by calculating the weighting functions for the five rock characteristics considered; as a result, their average harmonic values were determined. These values were determined for each type of rock and, when forecasting, preference was given to the type corresponding to the maximum value among the average harmonic values of weighting functions. This approach increases the reliability of the lithological forecast.

Keywords: Lithology · Rock · Fuzzy probability · Fuzzy possibility · Fuzzy logic · Uncertainty · Homogeneity · Forecast

1 Introduction

Improved hole drilling quality and efficiency are largely dependent on better data acquisition. At the same time, the quality of integrated well logging and mud logging data depends largely on the quality of data acquired during well drilling. However, the

© Springer Nature Switzerland AG 2020
R. A. Aliev et al. (Eds.): ICSCCW 2019, AISC 1095, pp. 503–511, 2020.
https://doi.org/10.1007/978-3-030-35249-3_64

analysis shows that the quality of data acquired during drilling does not always meet the design requirements, which impedes decision-making. Poor quality of acquired data is one of the reasons why wrong decisions are made, which in its turn leads to complications and incidents and affects drilling performance in general. The above supported by well drilling experience and numerous studies demonstrates the need for studies to improve the quality of drilling data and requires relevant data processing and analysis methods. To that end, this paper discusses the analysis and quality assessment of drilling data, as well as ways to use various methods to improve data quality for decision-making at different drilling stages.

2   Role of Integrated Data in Drilling Performance Improvement

Over the past few years, a large number of studies have focused on the interaction between rock cutting tools and rocks, suggesting methods and means to determine physical and mechanical properties and abrasiveness of rocks. As a whole, the analysis of past research shows that today it is possible to enhance decision-making processes only if integrated well logging and mud logging data are used which serve as the basis for process design solutions. This kind of information may be obtained in different ways. At the same time, advanced data processing and analysis methods shall be employed to obtain and use this information. Moreover, it is essential to consider drilling conditions, namely: heterogeneity, fuzziness, and randomness of factors at play. In this regard, methods of the management and decision-making theory with insufficient information, which have recently been extensively developed, can provide a sound basis for such consideration. Well log analysis methods used during well drilling are also of great importance. Mud logging while drilling that has been widely used over the past few years on a global scale addresses a variety of drilling challenges when log information for a drilled well is either absent or limited. The combination of mud logging and well logging data provides a deeper insight into the well and thus improves decision-making quality.

3   Modeling and Evaluation of Rock Properties Based on Well Logs

The analysis of mud logging data on well drilling in particular and measurement data in general is sometimes fraught with errors, uncertainty, and unstable correlations between parameters under study. It is increasingly difficult to compare the values of the same parameter measured using different methods. For example, simple measurements of process parameters, rates of penetration, and log-based rock properties may be affected by various disturbances that cannot be taken into account; these may include weight-on-bit variations, rock mineralogy and lithology, saturating fluids, mud filtrate invasion, etc. Unlike conventional methods aimed at minimizing potential errors, error estimation and analysis suggest that this error contains useful information [1, 3, 4, 7]. Error information may be used to create a reliable tool to complement conventional methods necessary for process engineers, geologists, and geophysicists who deal with forecasting challenges.

For bit performance analysis, the log can be divided into homogeneous intervals to study variation patterns in drilling parameters within each interval. Different classification methods are available for this purpose. The Rodionov method as known from geology is one of the simple methods that can be used for this operation. According to this method, at first the mass is assumed to be homogeneous throughout its depth; the Rodionov criterion is then calculated for each interval from the expression suggested by the author.

$$V\left(r_0^2\right) = \frac{n_1 + n_2 - 1}{(n_1 + n_2)n_1 n_2} \sum_{j=1}^{m} \frac{\left(n_2 \sum_{t \in A_1} x_{tj} - n_1 \sum_{t \in A_2} x_{tj}\right)^2}{\sum_{t \in T} x_{tj}^2 - \frac{1}{n_1 + n_2} \left(\sum_{t \in T} x_{tj}\right)^2} \tag{1}$$

where A_1 and A_2 are the multitudes into which space T is divided; n_1 and n_2 are the number of observations in these populations.

According to the analysis [8], the values of the Rodionov criterion are distributed following Pearson's χ^2 law. This is why the program compares each calculated value with the tabulated value for a predetermined level of significance on an interval-by-interval basis. Intervals where the calculated value exceeds the tabulated value for the Pearson criterion χ^2 are the boundary between two homogeneous units which are heterogeneous in relation to each other. We have used the above criterion to divide the log in question into homogeneous intervals in terms of rock hardness and abrasiveness and compare variations in drilling parameters within each of these intervals. These properties are in turn influenced by many factors, including grain shape, salinity, and pore-saturating fluids. Figure 1 shows how rock hardness and abrasiveness change with depth vs. the rate of penetration for different bits. It also shows homogeneous intervals drilled with the same bit types. The data have been processed using the moving average method to narrow the data scatter range and define the variation pattern for the parameter in question more clearly. Note that noise shall be taken into account when well drilling data are used in actual practice. In this context, certain random processes shall be studied against other processes, e.g. impulse noise. The automatic selection method used in [9] for valid signal assessment under operating conditions is one of the most efficient methods. The program we used includes noise filtering for processing the d-exponent also shown in the Fig. 1. A change in L statistics allows us to assess whether the system in question is homogeneous. The above figure also shows the change in the statistics. Its changing nature on the plot indicates that rock homogeneity with regard to drillability is disturbed.

Today, fuzzy logic has been successfully used to assess and characterize reservoir properties [2, 4–7, 10, 11]. Early investigators of natural science noticed that many seemingly random events follow certain patterns. These patterns or distributions were closely approximated by continuous curves referred to as 'normal curves of errors' and attributed to the laws of chance [5–7]. A normal distribution is fully specified by two parameters: its mean and variance. As a consequence, core plugs from a number of a certain rock types may have several variables that affect, let us say, their porosity, but porosity distribution will tend to be normal in shape and defined by two parameters, namely: their average value or mean and their variance or the width of the distribution. The same could be said of other parameters. In this case, the variance depends on the hidden underlying parameters and measurement error. The variance about the average value is one of key prerequisites that causes fuzziness; in this regard, the authors [7] attempt to justify why this parameter describes fuzziness and requires the use of fuzzy logic.

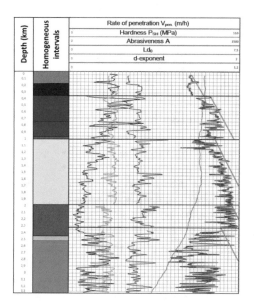

Fig. 1. Well drilling data analysis (Karabagly Field).

As a whole, the analysis shows that process simulations are greatly hindered by uncertainty related to both random and fuzzy variables. Random variables convey that variables under study may take on different values, with different probabilities. On the other hand, fuzzy variables convey the approximate nature of determination of the values of such variables. Besides, fuzzy variables may be more preferable when there is a lack of statistical data and related information necessary for more robust and reliable assessments. Such assessments of mechanical rock properties may be based on the information about physical properties of rocks evaluated from well logs using probabilistic statistical methods and the fuzzy set theory. The study of the relations in question allowed us to develop a valid calculation model for geology characterization.

Let us say that there are two types of rock. The average hardness of one rock type is 1200 MPa and the root-mean-square deviation is ±120 MPa. The other type of rock has an average hardness of 700 MPa and the root-mean-square deviation (the square root of variance) of ±80 MPa. If we assume that the hardness of some unknown rock is 850 MPa, it appears that it may be classified as either rock category in terms of hardness. (Note that there is a generally accepted twelve-point classification of rocks by their hardness.) Yet, at the same time it is more likely that this rock falls in a higher hardness category because the distribution of values for such rocks is much more 'dense', despite the fact that its hardness is equidistant from the 'most likely', or average hardness, proposed for a certain category of rocks. As noted in the literature, the fuzzy logic forecast is based on the assertion that a particular type of rock can be represented by different values of the parameters determined by different methods, although, of course, some values will be more likely than others.

To propose a single approach which includes both probabilistic and fuzzy modeling as special cases, the literature reviews what is called 'fuzzy random variables' [4, 12]. According to research, a fuzzy random variable is a random variable whose values are not ordinary real numbers, but fuzzy numbers. At the same time the author notes that the details of the definition are of the essence here.

Let us consider the application of the above provisions to forecasting the rock lithology for a drilled well based on a set of features such as hardness, abrasiveness, lithology index, porosity, and permeability, obtained as a result of well logging and mud logging while drilling. Distribution analysis showed that for each of these features, with the exception of permeability, distribution is subject to the normal law, whereas permeability is subject to the lognormal law, so their logarithms are taken as its values.

Density of the normal distribution is known to be described by the following expression:

$$P(x) = \frac{e^{-(x - \mu)^2 / 2\sigma^2}}{\sigma\sqrt{2\pi}} \tag{2}$$

where $P(x)$ is the probability density function of observation x measured in the dataset described by the arithmetic mean μ and the standard deviation σ. In this particular case, the values of hardness, abrasiveness, lithology index, porosity, permeability, and, accordingly, their arithmetic means and standard deviations will serve as a variable.

It is commonly known that in conventional statistics, the area below the curve described by the normal distribution shows the probability that the variable x will lie in the range between, say, x_1 and x_2. The curve itself represents the relative probability that the variable x will be present in this distribution. In other words, it is more likely to encounter an average value than values that are 1–2 standard deviations away from it. This curve is used to estimate the relative probability (or fuzzy possibility [7]) that a value is part of a particular dataset. If a rock has a distribution of hardness (and/or abrasiveness, lithology index, porosity, permeability) values, with an average value of μ and a standard deviation σ, then there is a fuzzy probability that the value of each of these features corresponding to the rock can be calculated using expression (2). The average value and the standard deviation are taken directly from the dataset (based on mud logging while drilling, core, and well logging data) for each rock lithotype.

A well log is typically composed of several different rock interlayers; in this case, the value of each of these features x may belong to any of these types, some of which are more likely than others. Each of these types of rocks has its average hardness (as well as other rock characteristics) and its standard deviation, resulting in pairs of μ_f and σ_f for each type of rock f_i. If it is assumed that the measured value of a rock characteristic belongs to type f, we can calculate the fuzzy probability that this characteristic *x* was measured (derived experimentally or from well logging data) using expression (2), replacing μ_f and σ_f. Such fuzzy probabilities will apply only to one particular type of rock. Therefore, a method to compare these probabilities is needed. In this case, we shall try to define our challenge once again. We have a lot of rock lithotypes, according to the lithology column in the well program. The column has to be updated using a set of features describing a particular lithological variety of rocks. To this end, we have a lot of features describing rock types, such as hardness, abrasiveness, lithology index, porosity, and permeability values. Using these values, we need to attribute the rock in question to a specific lithology type. To do so, let us turn to expression (1) again.

It is advantageous to know the relation between the fuzzy possibility for each type of rocks and the fuzzy possibility of the mean or most probable value, as mentioned in [7]. This is achieved by denormalizing Eq. (2).

The probability of the measured mean value μ will correspond to the density calculated from expression (2) and depend only on the standard deviation, i.e. the density for the case when the value of a feature is equal to its mean value:

$$P(\mu) = \frac{e^{-(\mu - \mu)^2/2\sigma^2}}{\sigma\sqrt{2\pi}} = \frac{1}{\sigma\sqrt{2\pi}} \tag{3}$$

The membership of rock, for example, by hardness *x* to type *f*, is estimated using the so-called relative weighting function, which is determined as the ratio of expression (2) to expression (3). Next, this parameter is calculated for all other rock types. To assess the membership of a rock with indicators of properties (porosity, permeability, hardness, abrasiveness, lithology indicator), it is necessary to take into account the relative frequency of occurrence of each type of rock in a well. This is achieved by multiplying the relative weighting function by the square root of the assumed occurrence of a particular type of rock *f*. If we denote it by n_f, then we can calculate the weighting function that, according to the estimated characteristic, rock *x* belongs to type *f*, using the formula:

$$F(x_f) = \sqrt{n_f}e^{-(x - \mu_f)^2/2\sigma_f^2} \tag{4}$$

The weighting function $F(P_{Sh})$ is based solely on the measured hardness (using core, mud logging or well logging data), P_{Sh}. The process is repeated for another parameter, such as porosity, $K_{por.}$. This step allows estimating the membership of the measured porosity $K_{por.}$ to lithotype *f* ($F(K_{por.})$). This process is repeated for other features describing this rock and then for each rock lithotype. At this stage, we have five weighting functions which indirectly characterize the possibility of estimation of different characteristics of rocks. They suggest that lithotype *f* is the most probable one. These weighting functions are then harmonically averaged in this form:

$$C_f = \frac{5}{\frac{1}{F(K_{por.})} + \frac{1}{F(K_{perm.})} + \frac{1}{F(P_{Sh})} + \frac{1}{F(A)} + \frac{1}{F(\Delta_\sigma)}} \tag{5}$$

This process is repeated for each lithotype *f*. The lithotype associated with the highest C_f is taken as the most probable for a given set of features. This approach makes the lithological forecast more reliable. Using the well discussed above as an example, calculations were done under this model and the results are shown in Fig. 2. The figure shows two lithology columns: the one included in the well program and the one updated under the calculation model. The identified lithology types of rocks are shown in comparison with the values of parameters, characterizing their weights. The methodology discussed in this paper was developed by analyzing large amounts of data across multiple wells; it is different from the conventional methods because the fuzzy logic combines weighting functions harmonically, based on the comparison of the evaluation results for several features. When equally possible lithologies with similar probabilities are compared, the harmonic combination selects any indicator showing that the choice of a certain lithology is unlikely.

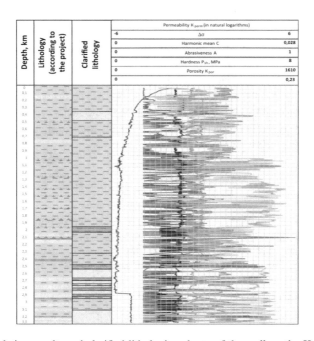

Fig. 2. Calculation results and clarified lithologic column of the well on the Karabagly field.

The lithological forecast that uses fuzzy logic is based on the assertion that a specific type of lithology can result in any source data readings, especially well log readings, although some readings are more likely than others. Thus, it is most likely that clean sandstones will be highly porous, permeable, abrasive, and with a high sigma log parameter, although there is a finite probability that logging could record values different from those generally accepted for the rock in question. It is therefore essential to have access to integrated well logging and mud logging data; in this case, the quality of the forecast information will be high.

Thus, the development and widespread introduction of state-of-the-art equipment and technology for the information support of the drilling process can improve the quality of the information and requires its appropriate analysis. Live data acquired during drilling are of great importance for drilling wells, especially in poorly studied regions with complex geological and environmental conditions. However, as can be seen in Fig. 1, there is a possibility to make a comparative analysis of changes in the properties of rocks and the rate of penetration, identify homogeneous intervals based on rock properties and drillability in general, monitor the intervals where complications are possible, etc.

4   Conclusions

The ways of obtaining and using complex information about the well section on the basis of a comparative analysis of changes in rock properties and rate of penetration are shown, which allow identifying homogeneous intervals in terms of rock properties and drillability as a whole, monitoring the intervals of possible complications, etc. An improved method for predicting the characteristics of a geological section based on complex geological, geophysical and technological information using probabilistic-statistical methods and fuzzy logic is proposed. The use of fuzzy logic provides more accurate recognition of the lithological differences of the rocks of the well section and the construction of a lithological column. Using an improved algorithm, predictive calculations were performed for well sections.

References

1. Sementsov, G., Gorbiychuk, M.: Some aspects of the studies of the geological sections of wells during drilling. News of the Higher Institutions. Min. J. **7**, 79–83 (1986)
2. Fang, J.H., Chen, H.C.: Fuzzy modeling and the prediction of porosity and permeability from the compositional and textural attributes of sandstone. J. Petrol. Geol. **20**(2), 185–204 (1997)
3. Khismetov, T.V., Efendiyev, G.M., Kirisenko, O.G., et al.: Drilling performance evaluation and decision-making based on integrated mud logging data. Oil Ind. **10**, 42–44 (2006)
4. Zadeh, L.: Fuzzy sets. Inf. Control **8**, 338–353 (1965)
5. Freund, J.E., Walpole, R.E.: Mathematical Statistics. Prentice-Hall International, New Jersey (1980)
6. Brown, D.F., Cuddy, S.J., Garmendia-Doval A.B., McCall, J.A.W.: The prediction of permeability in oil-bearing strata using genetic algorithms. In: Third IASTED International Conference Artificial Intelligence and Soft Computing (2000)
7. Cuddy, S.J., Glover, P.W.J.: The application of fuzzy logic and genetic algorithms to reservoir characterization and modeling. Soft Comput. Reservoir Charact. Model. **1**, 219–241 (2002)
8. Rodionov, D.A.: Statistical Decisions in Geology. Nedra Publishing House, Moscow (1981)
9. Mirzadzhanzade, A.Kh., Sidorov, N.A., Shirinzade, S.A.: Analysis and Design of Drilling Performance. Nedra Publishing House, Moscow (1981)

10. Aliev, R.A., Guirimov, B.G.: Type-2 Fuzzy Neural Networks and Their Applications. Springer International Publishing, Switzerland (2014)
11. Turksen, I.B.: Full type 2 to type n fuzzy system models. In: Seventh International Conference on Soft Computing, Computing with Words and Perceptions in System Analysis, Decision and Control, Turkey, Izmir, p. 21 (2013)
12. Shvedov, A.S.: On fuzzy random variables. National Research University Higher School of Economics. HSE Publishing House, Moscow (2013)

Evaluation of the Tax Policy Efficiency Under Uncertainty

Aygun A. Musayeva[1,2] , Shahzada G. Madatova[3,4(✉)] ,
Yashar A. Mammadov[5] , Alislam I. Gasimov[6] ,
and Kamala N. Ahadova[6]

[1] ANAS Institute of Control Systems, B. Vahabzadeh str. 9,
Baku AZ1141, Azerbaijan
aygun.musayeva@gmail.com
[2] The Azerbaijan University, J. Hajibeyli str. 71, AZ1007 Baku, Azerbaijan
[3] Training Centre of Ministry of Taxes of the Republic of Azerbaijan,
Agha Nematullah str. 44, Baku AZ1033, Azerbaijan
shahzademedetova@gmail.com
[4] Azerbaijan State University of Oil and Industry, Azadlig ave. 34,
Baku AZ1000, Azerbaijan
[5] ANAS Institute of Economy, H. Javid ave. 115, Baku AZ1143, Azerbaijan
mammadov60@mail.ru
[6] Azerbaijan State University of Economics (UNEC), Istiglaliyyat str. 6,
Baku AZ1001, Azerbaijan
alislam.qasimov123@gmail.com, kamalaahadova@gmail.com

Abstract. The tax burden factor of the economy is of great importance for the formation of an effective economy policy. At the same time, as it is known, when the state implements its economic functions, it has to change some of social and economic factors. This, in some cases, this may cause the tax burden to change negatively so that, it can have a significant impact on economic activity. The consumption norm, average tax burden, economic growth rate and dependence of the GDP on the structural effectiveness of the state tax policy efficiency have been evaluated through the interval analysis in the paper.

Keywords: Tax policy efficiency · Interval mathematics · Samuelson model · Macroeconomic indicators

1 Introduction

Depending on some issues of state economic policy efficiency, there are different approaches. There are such norms as population living standards, population satisfaction, social inclusiveness and others which are protected on the basis of the state economy policy, so that the state permanently controls these norms. At the same time, quantitative expressions of these norms cannot be determined by equal value. Usually, the consumption norm is determined as a result of the housekeeping survey during population census and by taking into account the physiological needs of the consumer. Other above-mentioned norms are determined by taking into account the dynamic

© Springer Nature Switzerland AG 2020
R. A. Aliev et al. (Eds.): ICSCCW 2019, AISC 1095, pp. 512–518, 2020.
https://doi.org/10.1007/978-3-030-35249-3_65

sequences created by economic targets and previous experiences and they are expressed by exact cost coefficients in the inter-sectoral balance model of the economy (Leontev's model) [1, 2]. Another approach to determine these norms is an expert evaluation. Especially, the Delphi method is more widely used [3]. The above-mentioned norms were investigated in Samuelson [4], Solow [5, 6], Harrod - Domar [7] models. For the first time, in [8], the single-factor correlations of the norms were evaluated in the real number abundance. In [9], the single-factor correlations of the norms were investigated by applying interval mathematics, which is one of the means taking into account both real number abundance and uncertainty. Davudova obtained mathematical relations expressing multifactorial correlations of the norms and these dependencies were evaluated both by the real number abundance and interval analysis [10].

In some cases, the state has to increase public expenditures in order to implement its economic functions. This requires increasing the state budget revenues, i.e. tax revenues. Increasing tax revenues raises the tax burden of the economy. This weakens economic activity and thus, a closed chain is formed. Therefore, the same logic can be said about the other norms mentioned above. Hence, economy policy institutes of the state have to take into account the changing legitimacy of the other norms mentioned above during the measure which is being implemented connected with the change of each norm.

2 Problem Statement

In this paper consumption norm, average tax burden, economic growth rate of the tax policy efficiency and dependence of the GDP on the structural effectiveness have been studied by applying interval mathematics based on the open economy model in which P. Samuelson's balance of payments is not equal to zero. The lower and upper boundaries of the macroeconomic indicator norms have been determined by the Delphi method and computational experiments were conducted in Matlab environment. The visual graphics of both points and interval values of the results are presented (Table 1).

Let's consider Samuelson's GDP expense structure model in order to write a standard model of the open economy system:

$$Y = C + I + G + \varepsilon(Ex - Im) = C + I + G + \varepsilon B \tag{1}$$

is known to be in this form, so, here Y is gross domestic product, C is consumption, I is investment costs, G is state expenditures, Ex is export, Im is import, ε is price and $B = X = Ex - Im$ is net export – payment balance. If $c = {}^{C}/_{Y}e$ housekeeping consumption rate, $s = {}^{I}/_{Y}$ investment norm, $g = {}^{G}/_{Y}$ state expenditures rate, $b = {}^{B}/_{Y}Y$ payment balance norm are marked

$$c + s + g + \varepsilon b = 1 \tag{2}$$

Let's express tax revenues with T, and average tax burden with $\theta = {}^{T}/_{Y}$. Let's choose the efficiency indicator of the state tax policy as $\tau = \frac{\Delta Y}{T} = \frac{F}{\theta}$, efficiency of the consumption costs policy as $\xi = \frac{\Delta Y}{C}$, efficiency of investment policy as $S = \frac{\Delta Y}{I}$, the efficiency of the state expenditures policy as $R = \frac{\Delta Y}{G}$ and the efficiency of foreign trade policy as $\beta = \frac{\Delta Y}{\varepsilon B}$.

The following formula [10] can be used for the purpose of evaluating:

$$
\begin{cases}
\tau = \frac{\Delta Y}{T} = \frac{F}{\theta}; \; \xi = \frac{\Delta Y}{C} = \frac{F}{c}; \; S = \frac{\Delta Y}{I} = \frac{F}{s}; \; R = \frac{\Delta Y}{G} = \frac{F}{g}; \; \beta = \frac{\Delta Y}{\varepsilon B} = \frac{F}{\varepsilon b} \\[2mm]
\tau(\theta) = F(\theta)\big/\theta = \left[\theta\big/(1+\theta) - \varepsilon b\right]^2 \big/ \theta \\[2mm]
\tau(c) = F(c)\big/\theta(c) = [1 - c - \varepsilon b]^2(1 + c - \varepsilon b)\big/4\,(1 - c + \varepsilon b) \\[2mm]
\tau(F) = F\big/\theta(F) = F\left(1 - \sqrt{F} - \varepsilon b\right)\big/\left(\sqrt{F} + \varepsilon b\right) \\[2mm]
\tau = F(\psi)\big/\theta(\psi) = \left[\frac{1}{\psi} - \varepsilon b\right]^2 \times (\psi(2 + \varepsilon b) - 1)\big/4(1 + \psi \varepsilon b)
\end{cases} \tag{3}
$$

$$
\varepsilon = \frac{1}{b}\left[\frac{\theta}{1+\theta} - \sqrt{\tau(\theta)\theta}\right] \tag{4}
$$

As exchange rate policy and fiscal policy are important policies of the state, determining the interrelationship between them is one of the important problems. In this paper, the dependence between the exchange rate policy and fiscal policy has been determined with formulation (4). The interrelationship between exchange rate policy and consumption cost policy, investment policy and public expenditure policy, respectively, were also defined (Table 2). Practical calculations related to the analysis of these dependencies will be implemented in our next studies.

Table 1. The values of c, θ, F and ψ determined by Delphi survey method.

Years	c		θ		F		ψ	
	Lower	Upper	Lower	Upper	Lower	Upper	Lower	Upper
2001	0.49	0.77	0.12	0.20	0.10	0.16	1.69	2.60
2002	0.50	0.78	0.12	0.20	0.11	0.17	1.40	2.15
2003	0.48	0.75	0.14	0.23	0.14	0.22	1.21	1.89
2004	0.45	0.70	0.14	0.24	0.16	0.24	1.05	1.61
2005	0.34	0.53	0.13	0.22	0.38	0.58	0.94	1.45
2006	0.30	0.46	0.17	0.28	0.40	0.61	0.91	1.40
...	...	...	...	...	...	...	...	...
2015	0.45	0.71	0.26	0.43	0.06	0.10	1.10	1.68
2016	0.47	0.73	0.23	0.39	0.09	0.14	1.21	1.86
2017	0.46	0.72	0.19	0.32	0.13	0.20	1.29	1.98

Source: The table was compiled by the authors.

3 The Results of Calculations and Their Graphic Descriptions

The change dynamics in consumption norms, tax burden, economic growth rates and generalized structural efficiency and in accordance with these efficiency of the state tax policy are presented according to the following years (Figs. 1, 2, 3 and 4):

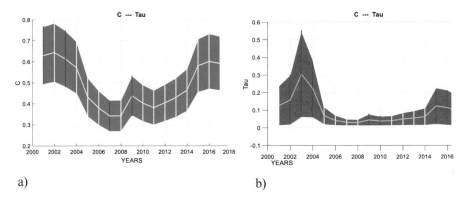

a)   b)

Fig. 1. Dependences of the state tax policy efficiency on the consumption norms (a) Consumption rate of interval of change (b) The interval of change of tax policy efficiency.

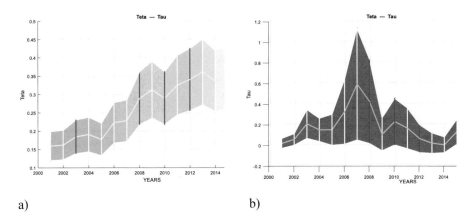

a)   b)

Fig. 2. Dependences of the state tax policy efficiency on the average tax burden (a) Interval of change of the average tax burden (b) Interval of change of tax policy efficiency.

Table 2. The values calculated based on the efficiency of the state tax policy (τ), consumption norm (c), average tax burden (θ), economic growth rate (F) and GDP structure efficiency (ψ).

Years	τ		τ(c)		τ(θ)		τ(F)		τ(ψ)	
	Lower	Upper	Lower	Upper	Lower	Upper	Lower	Upper	Lower	Upper
2001	0.51	1.31	0.005	0.23	−0.03	0.07	0.09	0.23	0.03	0.25
2002	0.56	1.43	0.008	0.29	0.00	0.11	0.11	0.26	0.05	0.34
2003	0.63	1.60	0.051	0.54	0.07	0.35	0.17	0.39	0.10	0.50
2004	0.66	1.67	0.048	0.38	0.04	0.26	0.15	0.34	0.10	0.48
2005	1.71	4.36	0.007	0.11	0.00	0.30	−0.05	0.14	0.03	0.37
2006	1.45	3.68	0.001	0.06	0.01	0.64	−0.16	0.05	0.01	0.32
...	...	...	...	...	...	...	...	...	...	...
2013	0.11	0.29	0.003	0.09	−0.08	0.12	0.03	0.09	0.02	0.39
2014	0.03	0.07	0.004	0.10	−0.08	0.08	0.01	0.04	0.03	0.38
2015	0.15	0.37	0.011	0.22	0.00	0.25	0.08	0.19	0.05	0.40
2016	0.23	0.59	0.006	0.21	0.00	0.20	0.08	0.21	0.04	0.36
2017	0.41	1.05	0.003	0.17	−0.04	0.11	0.07	0.21	0.03	0.31

Source: The table was compiled by the authors.

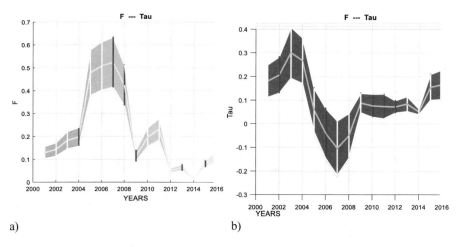

a)   b)

Fig. 3. Dependences of the state tax policy efficiency on GDP growth rates (a) Interval of change of GDP growth rate (b) Interval of change of tax policy efficiency.

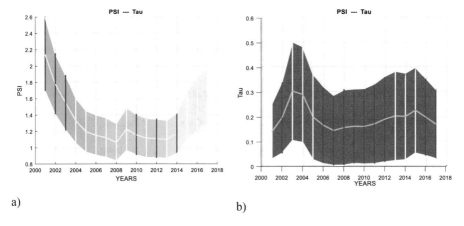

Fig. 4. Dependence of the state tax policy efficiency on GDP structure efficiency (a) Interval of change of GDP structure efficiency (b) Interval of change of tax policy efficiency.

4   Conclusion

The interval analysis of the state tax policy efficiency has been carried out through interval mathematics by applying the relations and the values of macroeconomic indicators for the Republic of Azerbaijan in 2001–2017s using the Delphi survey method in order to investigate the dependence of the state tax policy efficiency on some macroeconomic indicators. As a result of these calculations, it was determined that,

- there is a direct dependence between housekeeping consumption rates (c) and the state tax policy efficiency (τ), i.e. the increase (decrease) in c leads to the increase (decrease) in τ. In addition, the state tax policy efficiency obtained its largest interval value in 2003, and the narrowest one in 2007.
- the average tax burden (θ) decreased in 2004–2005, 2009–2010, 2013–2014, 2015–2017, and increased during the rest of the research period. Using the dependence between the average tax burden and the state tax policy efficiency, the decrease in the state tax policy efficiency was observed in 2003–2005, 2007–2009, 2010–2014 and 2015–2017;
- There is an inverse dependence between the economic growth rate (F) and the state tax policy efficiency (τ);
- The structural efficiency of GDP increased in 2008–2009 and 2013–2017, while it decreased during the remaining period. Depending on this, the state tax policy efficiency decreased in 2003–2007 and 2015–2017. The structural efficiency of the GDP and the interval of change in the state tax policy efficiency almost did not change.

As a result, it should also be mentioned the interrelationship formulations between exchange rate policy and fiscal policy, consumption cost policy, investment policy, public expenditure policy, respectively, that have been defined in this paper.

References

1. Wassily, L.: Quantitative input and output relations in the economic system of the united states. Rev. Econ. Stat. **18**(3), 105–125 (1936)
2. Wassily, L.: The input-output approach in economic analysis. In: Input-Output Relations: Proceedings of a Conference on Interindustrial Relations (held at Driebergen, Poland, pp. 1–23. (1953)
3. Alexandra, T.: Investopedia, investing, financial analysis, 13 April 2019. https://www.investopedia.com/terms/d/delphi-method.asp
4. Samuelson, P., Nordhaus, W.: Economics, 16th edn. Mass Irwin/McGraw-Hill, Boston (1998)
5. Nobelprize.org, The Prize in Economics 1987 – Press Release, 21 October 1987. https://www.nobelprize.org/nobel_prizes/economic-sciences/laureates/1987/press.html
6. Solow, R.: Contribution to the theory of economic growth. Q. J. Econ. **70**(1), 65–94 (1956)
7. Sato, R.: The harrod-domar model vs the neo-classical growth model. Econ. J. **74**(294), 380–387 (1964)
8. Vladimirov, S.: On the macroeconomic essence of the strategic development objectives of effective, balanced macroeconomic systems. Scientific works of the North-West Institute of Management, branch of RANEPA. vol. 6, 4 (21), 105–116 (2015)
9. Musayev, A., Davudova, R., Musayeva, A.: Application of interval analysis in evaluation of macroeconomic impacts of taxes. In: Proceedings of ICAFS-2018, Advances in Intelligent Systems and Computing, vol. 896, pp. 627–635 (2019)
10. Moore, R., Kearfott, R., Cloud, M.: Introduction to Interval Analysis. SIAM, Philadelphia (2009)

Knowledge Base Intelligent System of Optimal Locations for Safe Water Wells

Elbrus Imanov[1](✉) 📵 and Ezekiel Daniel[2] 📵

[1] Department of Computer Engineering, Near East University,
North Cyprus via Mersin 10, Mersin, Turkey
elbrus.imanov@neu.edu.tr
[2] Department of Physics, Daystar Christian Academy,
New Extension, Kaduna, Nigeria
tusmart4love@gmail.com

Abstract. The union of expert systems with the advanced technology elements is constantly giving us more developed and improved values or results, and indeed exact results with some human decision-making practices. As a consequence, this area of study is continuing to be the center of interest. In this paper an expert system using VP-Expert was designed to help improve the siting of water wells. The paper explains major factors to consider before siting water wells in Africa. According to World Health Organization about 900 million of the world population do not have access to sustainable safe drinking water, 84% of this estimated population dwell in rural areas. In fact, Africa has the lowest potable water coverage when compared to other continents of the world. In this paper a system was developed to identify optimal locations to site water wells. The procedure of the knowledge acquisition in the design of this system was done through interviewing geology experts, text books, journals and various related sources, and the knowledge base intelligent was represented in the rule-based procedure. These rules identify the quantity and quality of water that a targeted location is capable of producing. VP-Expert software was used for the design of the system and the system was validated by geologists from the Kaduna State Government of Nigeria. The developed system can be used effectively by governments, non- governmental organizations and individuals to improve the supply of safe drinking water.

Keywords: Artificial Intelligent · Knowledge base · Expert system · Groundwater · VP-Expert · Aquifer · Rural Water Supply Network

1 Introduction

An Expert System (ES) is a subclass of Artificial Intelligence developed to solve a specific problem in a particular domain. The designed computer system is able to simulate the conduct of a human expert to solve a problem in a particular domain. An ES is computer system that copycats human expert. As a nature of human being we are perpetually subject to decision making or selecting one or more alternatives accordingly [1]. Artificial Intelligence (AI) is related to computer programming that enables the computers to imitate or accomplish the tasks that are currently achieved

© Springer Nature Switzerland AG 2020
R. A. Aliev et al. (Eds.): ICSCCW 2019, AISC 1095, pp. 519–526, 2020.
https://doi.org/10.1007/978-3-030-35249-3_66

better by the human expertise [2]. Artificial Intelligent uses Meta, heuristics, procedural, and structural type of knowledge that allows, acting and thinking as human beings [3]. AI is developed in order to achieve smart computers that act and think as human experts, and it is achieved though reading the way a human brain thinks [4]. The ES is a division of Artificial Intelligent, and involves a set of programs that are able to reason, justify and answer the given queries in a specific domain as a human expertise [5]. The aim of this research is to develop an Expert System using VP-Expert that can be used in data deficient developing regions of the world like Africa to identify locations that will produce high yield and safe drinking water. When used, it will offer recommendations for optimal water well locations that minimize the risks to pollution and dry wells and thus maximize the benefits for people. Human experts provide the major sources of expert knowledge, and other sources are provided through journals, article, texts and databases [6]. Automated advisory programs of ES are acting as an imitator of knowledge of experts for the reasoning processes, in order to achieve a particular problem's goal [7]. Obviously, using information about groundwater is preferable when assessing locations for water wells. However, quality data on groundwater and sub-surficial geology are rarely available in Africa. To ensure that the system could be replicated globally, only publicly available datasets with global coverage were used. Hand dug and drilled water wells are the major sources of drinking water in Africa. Though these types of wells are not ideal, when properly installed they can be effective. Because hand dug wells are often open, they require daily sanitation which unfortunately is not always regulated [8]. These source of water supplies many with drinking water because they are the closest source of water from residential homes. Groundwater is the most common natural resource on earth and is also cheap to develop. Many people prefer groundwater to surface water as a source of drinking water because of its high degree of natural protection from pollution and drought [9]. Identifying the appropriate location to site water wells requires consideration of some very important factors. Water wells located near riverbeds are known to be reliable [10]. The hydrogeology of the site is a critical factor to consider since it determines the yield of the aquifer. The use and the user of the water is another vital factor because it determines the place the water will be needed. Expert system offers great opportunities to help agencies improve clean water resource planning. While Geographical Information System based models generally require large amounts of high-resolution data, the system developed in this paper was designed to be applied in data-deficient areas in Africa. VP-Expert system's design tool operates base on the production rules. VP-Expert assists only rule-base knowledge illustration or representation, which uses easy English similar to rule building [11]. To execute recommended implementation efficiently, a cautious choice of an ES shell for the precise domain purpose is very vital. VP-Expert was finally selected as the development shell for executing Water Well Location Site Expert System because of the virtuous performance of the interface and command menu of the shell. Finally to increase performance in achieving greater access to safe drinking water, the system developed here can be used to easily identify locations where there is high yield and safe groundwater. The result is an affordable solution to help individuals, governments and NGOs improve water wells siting and hence increase the quality of life of the communities.

2   Description of Knowledge Base Intelligent System

An expert is someone having comprehensive or authoritative knowledge in a particular field. Therefore, a computer system created to act as an expert to provide a solution to a problem in a particular domain is called an expert system. There are three individuals who take part in the design of the system; knowledge engineer, domain expert and the user. The domain expert is the person that has the required expertise to solve the problem that the proposed expert system is pre-planned to solve. The knowledge engineer acquires knowledge from the expert and transforms it into a format suitable for the system to use. The user consults the system to solve problem by responding to questions from the designed system. Knowledge Base System is used to refer to a computer system that has the same as a human expert in its knowledge base. Expert systems are the greatest common types of artificial intelligence application. In an expert system, the area which human intellectual endeavor to apprehend is identified as the task domain [12]. Building an Expert System requires a combination of many components that result into the decision making viz. with goals, facts, rules, inference engine, etc. [13]. Thus, we describe an expert system as a system, and not as an ordinary computer program. The components of knowledge base expert system are shown in Fig. 1.

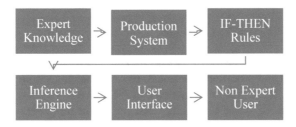

Fig. 1. Expert system architecture.

Knowledge base is the heart of an expert system. The human expertise is used as knowledge through the particular set of programs, or in other words, an ES [6]. Engineering problem solving uses heuristic knowledge, recognized scientific ideologies and computational algorithms. The domain knowledge of an expert system is saved in its KB and this module is very important that the successful application of the system relies on the excellence and dependability of the knowledge confined in its [11]. The Knowledge Base (KB) consists of stationary knowledge and dynamic knowledge that deals with the information about the sequence of action. The system solves problems by using rules from its knowledge base. The If-THEN production rules are used to represent the knowledge. Representing the domain expert's knowledge in the knowledge base is not enough and there must be an extra component that guides the execution of the knowledge. This component of the expert system is recognized as the control structure, the rule translator or the inference engine. The kind of inference mechanism relies equally on the nature of the problem domain and the technique in which knowledge is represented in the knowledge base. There are two search methods broadly used in rule-based systems, these are forward chaining and backward chaining. As the name implies search in forward chaining proceeds in the forward direction. In backward chaining the system backs a goal state or suggestion by examining the known information in the framework. The system searches the state space working from goal state to the preliminary state by applying the inverse operators. Backward chaining is a goal driven or ambitious search [14]. The working memory aims at the gathering of symbols or reliable information that mirrors the present condition of the problem which comprises the data gathered during problem implementation. The success of expert system mainly depends upon the superiority, comprehensiveness, and accurateness of the information stored in its knowledge base. This permits one to obtain more knowledge about the problem realm from the expert [15]. User interface is the component of the system which permits the user to interact with the expert system. The communication, between the user of the given system and the Inference engine, and also, any interaction between the ES programs and the user through natural language, is defined as user interface [16]. Expert System is unique and special in the sense that it has the ability to explain to a user how a conclusion was reached and this is achieved through the explanation facility. VP-Expert is a design tool that works base on production. It is made up of inference engine, the user interface, and every component needed to fully design an expert system.

3   Methodology of the Problem

The designed system employs the acquisition of knowledge from domain expert's textbooks, articles and other relevant sources to create the production rules using expert system methodology for the actualization of VP-Expert system for identifying the optimal location to site water wells. The procedure for the development of the system is categorized into two stages; knowledge acquisitions and knowledge representation. The knowledge acquisitions involve the collection of data from domain experts, papers, books and other relevant sources. Knowledge representation explains how the knowledge acquired or data collected has been transformed into a format that is suitable for use by the computer and the coding in terms of IF-THEN statement. Running the system on VP-Expert design tool and finally loading the program for consultations. The designed system is a rule-based expert system. The IF...THEN rules were used to represent the knowledge acquired, where IF demonstrate the condition and THEN provides the solutions. Development of optimal water well location expert system's workflow highlights four components namely: groundwater availability, impacts and risks of a new well, water use and access to a source [17]. The system developed in this study is intended to be a precursor to the site surveying work. The first component of the site selection process deals with the availability of groundwater. The second component of the model addresses the potential impact of a well on a groundwater source. The third component evaluates the water at a site. The value added by the framework is the ability to evaluate a site for the contamination risk, and there is a need for a macro level decision support tool that is able to identify high-risk areas more broadly. Components of Well site Selection is shown in Fig. 2.

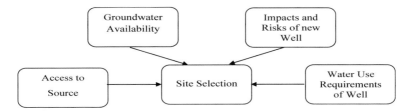

Fig. 2. Components of Well site Selection.

Table 1 shows the relation between Porosity and Permeability; the capability of water to move within the empty pores that exist between the earth formations is called permeability.

Table 1. Porosity and permeability table of sediments.

Sediment type	Porosity	Permability
Uniform size send or gravel	25–50%	High
Mixed size sand and gravel	20–35%	Medium
Glacial Till	10–20%	Medium
Silt	35–50%	Low
Clay	33–60%	Low

While the permeability of gravel is very high, its porosity is relatively low. Groundwater moves within the empty pores of earth materials. Water moves more freely between earth materials that have high permeability. In Table 2 recommended distances between water sources and contamination sources, distance to existing well when a well is pumped is shown, the groundwater in the aquifer normally moves toward the direction of the point at which the water is being tapped.

Table 2. Water sources and contamination sources.

Dist. (m)	Different types of contamination sources
100	Piles of Garbage, fuel stations, industrial waste
50	Cesspool
30	Latrines, animal pens, barns, fertilizer
15	Surface water, septic tank
7	Ditch, drain

Wells should be sited up gradient from contamination sources. The optimal water well location system was coded through the use of a VP-Expert design tool; the shell is a precise tool for designing expert systems thus only expert's systems developers are acquainted with it. The inference engine is used to cruise around the knowledge base in order to answer questions, the rules of the knowledge base are written on the editor and a user interfaces for supervising the questions. The production rules of the designed expert system consist of 8 attribute questions which serve as the input of the system and 6 possible outcomes for the location which describes the suitability of the targeted location to site water well. These input questions are shown below:

Is the subsurface soil permeable? Is the rock type cracked and fractured? What is the nature of the topography? Is the location near surface water? Is there a water indicator tree? Is there an ant mound? Is the distance to existing well greater than or equal to 30 m? Is the distance to source of contamination greater than or equal to 100 m? The possible variables of the location in the designed Expert System include:

High_Yield_Safe: This means that the location will supply high yield and safe drinking water.

High_Yield_Unsafe: This means that the location will produce high yield but unsafe water.

Moderate_Yield_Safe: This implies the location has a moderate yield and safe drinking water.

Moderate_Yield_Unsafe: This implies the location has a moderate yield and unsafe water supply.

Low_Yield_Safe: This means that though the location will have a low yield of water supply, but the water is safe for drinking.

Low_Yield_Unsafe: This implies that the location will have a low yield and unsafe water for drinking.

4   Presentation of the Developed System

During the process of developing this system, all the production rules contained in the knowledge base of the system were tested and modifications were done where necessary. The designed system was validated by geologists from Kaduna State Government of Nigeria and all relevant recommendations were included in the final designing stage. The system operates according to the backward chaining method of inference; this checks the memory to verify if the goal has been added, this is necessary because another knowledge base might have already proved the goal. The system searches the rules in its knowledge base, and if the goal has not yet been previously proved, it then looks for a rule or rules that contain the goal in its THEN part. The system then searches the memory to verify whether the memory contains the rule's goal premises. If the system cannot find the premises in its memory, then the premises are referred to as a sub-goal. The system now proves both the original goal and the sub-goals using the information it gets from the user. It was confirmed by the experts that this system can be used as a very important tool to improve the supply of potable water in Africa since the region has limited number of experts who are always working under pressure because of the overwhelming workloads. From below illustration we can see that the system has performed intelligently and with a good performance like a human expert in that domain. It has performed intelligently since it was able to answer queries and give recommendations as though it was a human expert in the field of hydrogeology. The system was validated by domain experts. The responding to questions and results of consultation are represented in Figs. 3 and 4.

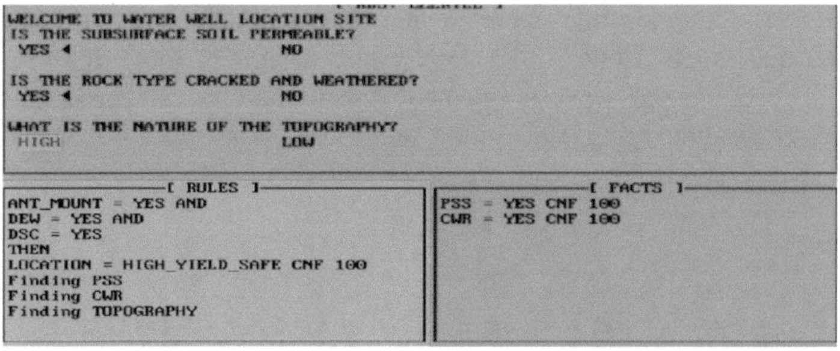

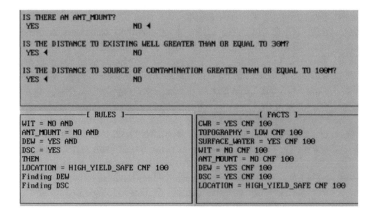

Fig. 3. The user responding to questions.

Fig. 4. Result of consultation.

5 Conclusion

In this research, system was designed to identify optimal locations to site water wells where risks to failure and waste of resources are minimized and productivity and sustainability of clean water wells are maximized in rural Africa. The purpose of the paper is twofold: improving the current research on the supply of potable water and designing a suitable system that could be implemented by individuals, governments and non-governmental organizations throughout Africa to improve the supply of safe drinking water. The knowledge base of the designed system consists of relevant production rules on factors to consider before drilling water wells. The results obtained does not only show the feasibility of drilling successful safe drinking water wells through the application of expert system, but the results also shows that the subclasses of AI can be applied in different drilling projects. The design of optimal water well location system has been presented. The knowledge acquisition and representation steps were adequately explained. The IF-THEN rules were chosen for the decision making in the developed expert system. The knowledge translated into the IF-THEN rules were collected from experts and other relevant sources, and from this acquired knowledge 134 production rules were created. The optimal water well location expert system was developed using VP-Expert System designing tool. In conclusion, the developed expert system when implemented can increase the production of safe drinking water by offering advices on the best locations to site water wells.

References

1. Aliev, R., Huseynov, O.H.: Decision theory with imperfect information. Word Scientific, River Edge (2005)
2. Ajlan, A.A.: The Comparison between forward and backward chaining. Int. J. Mach. learn. Comput. 5(2), 106–113 (2015)
3. Studer, R., Benjamins, R.V., Fensel, D.: Knowledge engineering, principles and methods. Data Knowl. Eng. 25(1–2), 161–197 (1998)
4. Kapoor, N., Bahl, N.: Comparative study of forward and backward chaining in artificial intelligence. Int. J. Eng. Comput. Sci. 5(4), 16239–16242 (2016)
5. Roseline, P., Tauro, C.J., Ganesan, N.: Design and development of fuzzy expert system for integrated disease management in finger millets. Int. J. Comput. Appl. 56(1), 31–35 (2012)
6. Patel, M., Virparia, P., Patel, D.: Web based fuzzy expert system and its application a survey. Int. J. Appl. Inf. Syst. 1(7), 11–14 (2012)
7. Bobillo, F., Delgado, M., Gómez-Romero, J., López, E.: A semantic fuzzy expert system for a fuzzy balanced scorecard. J. Expert Syst. Appl. 36(1), 423–433 (2009)
8. Awuah, E., Nyarko, K., Owus, P., Osei-Bonsu, K.: Small town water quality. Desalination 248(1), 453–459 (2008)
9. Macdonald, A.M., Davies, J., Dochartaigh, B.É.: Simple methods for assessing groundwater resources in low permeability areas of Africa. British Geological Survey Commissioned Report, CR/01/168 N (2002)
10. Dijon, R.: Groundwater Exploration in crystalline rocks in Africa. In: Proceedings of the American Society of Civil Engineers, pp. 77–81 (1981)
11. Tabibi, S.T., Zaki, T.S., Ataeepoor, Y.: Developing an expert system for diabetics' treatment advices. Int. J. Hosp. Res. 2(3), 155–162 (2013)
12. Mishkoff, H.C.: Understanding Artificial Intelligence. Instrument Learning Centre, Dallas (1985)
13. Merritt, D.: Building Expert System in prolog. Springer-Verlag, USA (1989)
14. Saghed, M.: Expert system development: some issues of design process. ACM SIGSOFT Software Eng. Notes 30(2), 1–2 (2005)
15. Patel, T., Thakkar, S.: Knowledge models, current knowledge acquisition techniques and developments. Orient. J. Comput. Sci. Technol. 6(4), 467–472 (2013)
16. Mcduffie, R., Eugene, P.: Tax expert system and future development. The CPA J. 64(1), 73 (1994)
17. Carter, R., Chilton, J., Danert., K., Olschewski, A.: Siting of drilled water wells. A guide for project managers, Rural Water Supply Network, Report number: RWSN Field Note Affiliation 5, pp. 1–15 (2010)

Scale Development for Institutional Perception Management Components Using Spss Software

Emete Yağcı ⓘ, Gülsün Başarı(✉) ⓘ, Ali Aktepebaşı ⓘ,
Şahin Akdağ ⓘ, and Rojda Kılınçaslan Akdağ

Near East University, 99138 Nicosia, North Cyprus, Turkey
{emete.yagci,gulsun.basari,sahin.akdag}@neu.edu.tr,
aliaktepebasi47@hotmail.com, rjd.kaslan@gmail.com

Abstract. The aim of this study is to develop a valid and reliable attitude scale in order to identify the attitudes on perception management with regard to the institutional perception management components. The scale development process was conducted with a total number of 969 participants among the students from TRNC, TR and foreign students. In the scale development phase, the existing attitude scales were reviewed and the professionals were consulted concerning the established items. The scale developed in the likert type was tested in terms of validity and reliability. As a result of Confirmatory Factor Analysis, the scale was comprised of 33 items and 6 dimensions. The scale is structured with 6 factors and the content and nature of items under factors were organised as *quality image, programme, sports, general view and infrastructure, accommodation and nutrition.* The Stratified Cronbach α value of the scale was generated as 0.99. This coefficient is at the acceptable level for whole scale indicating that the scale has internal consistency reliability. Consequently, this scale can be considered as valid and reliable that can be used in the identification of attitudes towards institutional perception management.

Keywords: Perception · Perception management · Institutional Perception Management Scale

1   Introduction

The concept of perception is adapted to our language through deriving from the term of perceive. This concept, which was adapted into the Western literature through a similar way, is originated from the word "capare" in Latin [1]. In general, perception is defined as a process where a person establishes a cognitive contact with the surroundings and makes of the surrounding stimulants [2]. Our perceptions that have such strong features are described as "real" by many psychologists [3].

Some people consider the perception management as convincing the target individuals or communities to think in the desired way [4], while others believe that it is a communication space to convince the target groups based on their own desires and interests and to transform them into a component to be used in accordance with their own objectives [5]. It is reflected as the activity of intentional planning, organization and control [3].

© Springer Nature Switzerland AG 2020
R. A. Aliev et al. (Eds.): ICSCCW 2019, AISC 1095, pp. 527–534, 2020.
https://doi.org/10.1007/978-3-030-35249-3_67

Pursuant to various studies [6], the use of perception management in the existing institutions as a rather effective management technique is crucially significant and necessary.

Therefore, the perception management is considered as one of the concepts that are potentially used in individual oriented institutions; hence such concept is required to be analysed at the level of educational institutions in terms of the relationship between the administrators and institution members. This study analyse the issue of perception management at the higher education institutions in detail, and a number of recommendations were proposed for better functioning at educational institutions. The most important contribution of this study would be the interpretation of theoretical and practical implementation with the results of empirical studies, and the introduction of something new to the literature through the creation of "Institutional Perception Management" Scale. The creation of a scale as a result of research can be considered as a scientific innovation.

2   Methodology

2.1   Research Pattern

This research, which was performed to evaluate whether the Institutional Perception Management (IPM) Scale measures the indicated features, is a quantitative research with a survey model identifying the perception management. This study was carried out with the descriptive survey model with the aim of describing the usability of IPM scale that was used during the scale development process, and generating information to be used for the practical research.

2.2   Research Population and Sample

This study is limited with a total number of 969 participants, who are students at a Private University located within the territories of Nicosia, TRNC during the academic years of 2017–2018, randomly selected by random stratification sampling method as TRNC (164), TR (725) and other foreign students (80).

2.3   Data Collection Tools

In terms of the scale development, *"Perceptions"* comprised by the organizational images, organizational reputations, organizational identities; *"Actions"* comprised by verbal accounts, symbolic behaviours and physical markers; *"Spokespersons"* with leaders and employees; *"Audiences"* as internal audiences and external audiences", all of which are included in the model developed by Elsbach [7, 8] constitute the components of "Institutional Perception Management" Scale and the scale was developed accordingly. The scale was performed following the determination of points, consultation with experts and pilot implementation phases respectively [9].

2.4   Data Analysis

Statistical Package for Social Sciences (SPSS) 24.0 data analysis software was utilised in the statistical analysis of data generated under the research. The model, which was proposed as 6-factor model, was executed with five models in consideration with the whole sample together with the confirmatory factor analysis. The consistency of model was tested with the Confirmatory Factor Analysis with the use of chi-square goodness of fit test, RMSEA, GFI and CFI fit indices. Confirmatory Factor Analysis (CFA) is a widely used analysis method that brings significant convenience [10]. Additionally, the model was tested whether it is invariable for the foreign students. All factors were found as statistically significant and as a result of fit indices, the model was indicated to have a good fit and suitable construct.

3   Findings

This section provides the findings generated as a result of validity and reliability tests of scale.

3.1   Scale Validity

Validity "is the correct measurement level of a measurement tool for the specific feature without confusing it with any other feature" [11]. The construct validity of scale is "the correct measurement level of scale towards an abstract concept (factor) within the context of behaviour that is desired to be measured" [12]. In terms of this study, the construct validity of this scale was assessed with (1) confirmatory factor analysis and (2) measurement invariance analysis.

Confirmatory Factor Analysis Results
The model, which was proposed as 6-factor model, was executed with five models in consideration with the whole sample together with the confirmatory factor analysis processes. Firstly, the model was checked whether there is any incomplete data, which the results showed that there is not any. Then each item was looked into whether they include any extreme values. It is considered that there is not any extreme situation as the analyses indicated that the standard z values have a distribution between -3 and $+3$. *Mahalanobis* distances were calculated whether there are any multiple extreme situations [13]. Finally, the variance inflations were evaluated and concluded to have any problems. All VIF values were found less than 30 [14].

Table 1 shows the fit indices of model generated as a result of analyses. Pursuant to Table 1, CFI and GFI fit indices were less than 0.95, yet the value of CFI as 0.90 shows that the model is consistent with the data. On the other hand, the error (misfit) indices of model are foreseen to be between 0.05–0.08. When the value is below 0.05, the model is considered to have no issues. Particularly, where the value of Root Mean Square Error of Approximation is close to 0.00, it indicates good fit; while a value less than 0.05 shows that there is minimum error between the matrixes observed to be low and generated matrixes (Browne and Cudeck 1993). Since RMSEA values calculated for the models are between 0.08 and 0.05, the error rate between the observed and

Table 1. Measurement model fit values after correction

Fit measurement	Original measurement value	Model 1	Model 2	Model 3	Model 4	Model 5
		e35 − e36	S2 − model	e14 − e15	e23− e24	e1− e4
		e37 − e38	S3 − model	S16 − model	S29- model	S12- model
			S11 − model	S34 − model	S30 − o41e41	S19 − model
					S28 − model	S27 − model
					S41 − model	S45 − model
χ^2	6168.39	5586.89	4413.96	2180.37	1963.06	1801.05
p value	0.000	0.000	0.000	0.000	0.000	0.000
χ^2/df	7.587	6.914	6.77	4.030	3.872	3.792
RSEA	0.08	0.08	0.08	0.06	0.05	0.05
GFI	0.757	0.776	0.818	0.870	0.880	0.887
CFI	0.819	0.838	0.862	0.934	0.941	0.945

generated matrixes for the model is at the acceptable level. As a result of corrections shown under Model 5, the 6-factor construct proposed theoretically provided a scale with 33 items with good construct validity (Table 2). Figure 1 shows the fifth level graphic and analytical presentation of model. The Chi-square Statistic given under Table 1 is a technique that tests the hypothesis regarding that the covariance structures of observed variables and model are fit [15]. Considering this study, we can consider that the third model is fit for the observed construct since Chi-square/degree of freedom is less than 5 [16, 17]. In terms of goodness of fit indices, the values between 0.90–0.95 show an acceptable fit while a value above 0.95 indicates a high fit [18].

The Stratified Cronbach α value of scale was calculated as 0.99. Since none of the items have a correlation value less than 0.2 between corrected item-total items, none of the items were removed after the verification analysis [19]. The reliability coefficients of scale sub-tests are given under Table 3.

Table 2. Factor distribution of items after correction

Factor 2	Factor 3	Factor 4	Factor 5	Factor 6
15	25	31	35	42
17	26	32	36	43
18		33	37	44
20			38	46
21			39	
22			40	
23				
24				

Table 3. Cronbach α reliability coefficients of sub-scales.

Sub-scale	α value	No. of items
Factor 1	0.92	10
Factor 2	0.93	8
Factor 3	0.91	2
Factor 4	0.81	3
Factor 5	0.90	6
Factor 6	0.84	4

Measurement Invariance Analysis Results

The model given in Fig. 1 was tested whether it is invariable for the students from TRNC, TR or OTHER COUNTRIES. Table 4 shows the findings regarding the tested invariance phases. Model A given in the model represents the model with free factor loads, factor correlations and error variances; Model B represents the model with fixed factor loads and free factor correlations and error variances; Model C represents the model with fixed factor loads and factor correlations, and free error variances and Model D represents the model with free factor loads, factor correlations and error variances.

More limited models and formal model were compared in order to determine the measurement invariance between groups during the phases given under Table 4, and the difference values of ΔCFI fit coefficients were evaluated (Byrne 2013; Hooper et al. 2008).

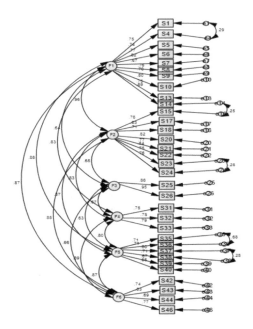

Fig. 1. Fifth level graphic and analytical presentation of model

Table 4. Fit statistics regarding the measurement invariance phases

Phases	χ^2	sd	CFI	RMSEA	ΔCFI
Model A Formal invariance	3325.32	1425	0.919	0.037	–
Model B Metric invariance	3402.35	1479	0.918	0.037	−0.001
Model C Scale invariance	3456.63	1521	0.917	0.036	−0.001
Model D Strict invariance	3615.21	1597	0.914	0.036	−0.003

As a result of multi-group DFA conducted to test the formal invariance, the fit indices show that this phase is present. The presence of formal invariance indicates that the groups separately use the same conceptual perspectives in answering the scale items as well as the students from TRNC, TR or OTHER COUNTRIES. The fit indices as a result of multiple group DFA conducted to test the metric invariance, and ΔCFI value generated as a result of CFI difference test were interpreted ($-0.01 \leq$ ΔCFI ≤ 0.01, Byrne 2013; Hooper et al. 2008). This finding indicates that it is possible to establish correlations between factors within the context of TRNC, TR or OTHER COUNTRY students. From this perspective, it is possible to compare factor scores (mean values) between groups.

A similar result was also observed in the scale invariance (ΔCFI = −0.001). Such results reflect that there is no bias based on the items between groups, which may indicate that the groups can be compared by items. Pursuant to the strict invariance results (ΔCFI = −0.003), all comparisons that would be conducted between groups regarding the model would be significant.

4   Discussion and Conclusion

This section includes the results that were obtained based on the research findings and discussion pursuant to such results.

The aim in this study with regard to the evaluation of scale structure is to identify whether the proposed Institutional Perception Management model is practical and it has a good level of construct validity. Pursuant to the aim of this research, the "Institutional Perception Management" scale was established to identify the perception management strategies and techniques used by the administrators at the educational institutions and the associated levels respectively. In the development of scale, the study by Elsbach [7, 8] was taken as a reference in the provision of contribution to the literature.

The institutional perception management scale developed in the context of institutional perception management components under the quantitative research was provided to the participants in order to develop a valid and reliable "Institutional Perception Management" scale following a trial. In terms of the reliability and validity checks of scale, the scale was firstly developed as 46 items and then was ended up with

33 items upon removing 13 statements that are not found as suitable by the Confirmatory Factor Analysis. The scale includes "Personal Information" and 6 dimensions were identified as a result of Confirmatory Factor Analysis. The scale was structured as 6-factor scale and organized as "Quality image" (10 items), "Program" (8 items), "Sports" (2 items), "General view and infrastructure" (3 items), "Social environment" (6 items), "Accommodation and nutrition" (4 items) with regard to the content and structures of items under the factors. It is a 5-likert scale as "Strongly Disagree", "Disagree", "Slightly Disagree", "Agree", and "Strongly Agree". The score range of five-rating scale is between 0–4 with the limitations of 0.00–0.78, 0.79–1.57, 1.58–2.35, 2.36–3.14, 3.15–4.00. In the test of significance, the level of $\alpha \leq .05$ was taken as a basis. The Stratified Cronbach α value of scale was calculated as 0.99. Since all of the items do not have less than 0.2 of corrected item-total correlation, none of the items were removed as a result of confirmatory analysis [19]. The result of Confirmatory Factor Analysis indicates that the scale is suitable. The ranges for the average values taken as a basis in the evaluation of perception level regarding the Institutional Perception Management were used as 0–4 and the scale was identified to have 0.99 value.

Considering the literature, there has not been a strict distinction between the concepts of "perception management", "management of perception" and "perceptional management". A number of authors defined and presented the concepts that are perceived with similar contents. The perception management, which is called "soft power" under the related literature, can be perceived as a strategical action. As a result of literature review, the studies by Elsbach [7, 8] conducted on "Institutional Perception Management" was mainly taken as a basis by the various authors, and the number of studies conducted on the perception management as the educational institutions is found as insufficient. However; where the educational institutions are considered as a whole, it can be seen that the phases at such institutions as input-procedure-output and feedback are comprised of individuals. The personal relations are significant in the realization of objectives at the educational institutions while the perception management is very crucial for the successful execution of management processes and acquisition of the desired outputs. This scale can also be used for the teachers, students and staff of educational institutions as well as the other national and international educational institutions. Additionally; the scales towards the identification of institutional perception management for different domains should be developed in accordance with the literature.

References

1. Özer, M.A.: Perception management and internal security services as a modern management techniques. Black Sea Res. **33**(33), 147–180 (2012)
2. Efron, R.: What is perception? In: Proceedings of the Boston Colloquium for the Philosophy of Science 1966/1968, pp. 137–173. Springer, Dordrecht (1969)
3. Johansson, L.R.M., Xiong, N.: Perception management: an emerging concept for information fusion. Inf. Fusion **4**(3), 231–234 (2003)
4. Özdağ, Ü.: Perception management, Propaganda. Psychological warfare, The bass crypto, Ankara (2015)

5. Korkmazyürek, H., Hazır, K.: Organizational Behavior. Beta, Istanbul (2013)
6. Karasar, N.: Scientific research method (19th bs). The Nobel prize distribution, Ankara (2009)
7. Byrne, B.M.: Structural Equation Modeling with Mplus: Basic Concepts, Applications, and Programming. Routledge, Milton Park (2013)
8. Hooper, D., Coughlan, J., Mullen, M.: Structural equation modelling: guidelines for determining model fit. Electron. J. Bus. Res. Methods **6**, 53–60 (2008). Articles, 2
9. Tekin, H.: n education Measurement and evaluation. Yargı Yayınevi, Ankara (2008)
10. Büyüköztürk, Ş.: Data Analysis Handbook for Social Sciences. Pegem Academy, Ankara (2010)
11. Kline, R.B.: Methodology in the social sciences (2005)
12. O'Brien, R.M.: A caution regarding rules of thumb for variance inflation factors. Qual. Quant. **41**(5), 673 (2007)
13. Özdamar, K.: Statistical Data Analysis by Using Software Tools. Kaan Publications, Eskisehir (2002)
14. Byrne, B.M.: Structural equation modeling with LISREL, PRELIS, and SIMPLIS, London (1998)
15. Dickey, D.: Testing the fit of our models of psychological dynamics using confirmatory methods: an introductory primer. Adv. Soc. Sci. Methodol. **4**, 219–227 (1996)
16. Everitt, B.S.: The Cambridge Dictionary of Statistics, 2nd edn. Cambridge University Press, Cambridge (2002)
17. Field, A.: Discovering Statistics Using SPSS. Sage Publication, Thousand Oaks (2005)
18. Agarwal, B.: Rule making in community forestry institutions: the difference women make. Ecol. Econ. **68**(8–9), 2296–2308 (2009)
19. Bakan, I., İlker, K.E.: Perception and perception management from an institutional perspective. J. Fac. Econ. Adm. Sci. **2**(1), 19–34 (2012)

Sea Turtle Detection Using Faster R-CNN for Conservation Purpose

Mohamed Badawy and Cem Direkoglu[(⊠)]

Center for Sustainability, Department of Electrical and Electronics Engineering,
Middle East Technical University - Northern Cyprus Campus, Kalkanli,
Guzelyurt, North Cyprus, via Mersin 10, Turkey
mu.badawy@gmail.com, cemdir@metu.edu.tr

Abstract. Automatically monitoring see turtles over extensive coastlines is an important task for environmental research and conservation nowadays. Unfortunately, some of the sea turtle species have become endangered today and this is why there is a need for search-and-rescue. Computer vision algorithms can be used for sea turtle detection and monitoring. Recently, due to the powerful Convolutional Neural Networks (CNNs), computer vision crucial applications came to reality. Although such algorithms are computationally expensive, they have proved promising results where real-time applications can be feasibly implemented given high-capability GPUs. In this paper, we present a system of sea turtles detection using a Faster R-CNN algorithm. This system performs the sea turtles' detection on a cloud (off-board). Our detection algorithm can be performed using a static camera, or a moving camera that is mounted on UAVs for surveillance, search-and-rescue purposes.

Keywords: Computer vision · Object detection · Sea turtles · Faster R-CNN

1 Introduction

Most of the sea turtle species have become endangered. There is a need for automatic coastal detection and monitoring for sea turtles. Recently, UAVs have emerged strongly and have been widely used in a lot of different applications; mainly they are used in object-detection based tasks such as surveillance, search-and-rescue and similar tasks. However, using drones to monitor and detect moving objects is a pretty difficult and tough process. It is really challenging due to the fact that the on-board camera resolution is quite low and thus most of these images are blurry and noisy. Since the target objects are quite small, the detection becomes more difficult to be achieved. The need of real-time detection also adds another challenge to the detection process. In this paper, we propose a computer vision based sea turtle detection system using a static camera, or a camera mounted on a Drone. Detecting sea turtles is never easy as it might sound due to the scarcity of sea turtles' data set as well as their diversity. Taking all these factors in consideration, we built our approach to create this system, incorporating the technical standards and trying to minimize the gap between the application requirements and the technical capabilities.

© Springer Nature Switzerland AG 2020
R. A. Aliev et al. (Eds.): ICSCCW 2019, AISC 1095, pp. 535–541, 2020.
https://doi.org/10.1007/978-3-030-35249-3_68

Sea turtle detection is an object detection problem. Recently, the Convolutional Neural Networks (CNNs) has been employed for object detection. Convolutional neural networks have two main key features: First, these networks take the advantage of local receptive fields, since image data within local spatial regions is likely to be related [1, 2]. Second, these networks are deep that means they comprise many layers to perform better performance in comparison to the traditional artificial neural networks. Thus, they are robust and prove reliability in object detection. In our work, we use the Faster R-CNN algorithm [7] to build a sea turtle detection system on a cloud (off-board). Our detection algorithm can be performed using a static camera, or a moving camera for search-and-rescue purposes. The proposed system is fast and effective.

2   Faster R-CNN

Evaluated on the VOC2007 [9] dataset, the Faster R-CNN algorithm achieves the highest accuracy, denoted as mAP, mean average precision, which is 73.2%. The main motivation beyond using faster R-CNN is its high precision in comparison to other object detection algorithms. Although the main downside of Faster R-CNN is its low FPS rate compared to YOLO [8] and SSD [12], Faster R-CNN is better when it comes to detecting objects of small size. In general, objects captured by the drone's camera have a small size and the YOLO algorithm potentially fails to detect objects of small size. We decided to use the Faster R-CNN because of its high precision which greatly surpasses other algorithms, and ability to detect objects of small size. Our main objective is correctly detecting sea turtles with the highest precision possible. Thus, we compromised the FPS rate as we are satisfied by the Faster R-CNN rate which is 7 FPS [8]. Table 1 also illustrates the performance of object detection algorithms on aerial images collected by a drone where we can see the highest precision is solely achieved by the Faster R-CNN algorithm [4].

Table 1.   Object detection results on aerial images collected by a Drone [4].

Method	Car	Cat	Dog	Person	mAP
Yolo	87.2	100.0	81.0	88.7	79.4
SSD 300	100.0	100.0	66.7	92.9	21.6
SSD 500	93.2	100.0	69.6	81.7	82.6
Faster R-CNN	**89.7**	**100.0**	**77.3**	**81.7**	**83.9**

Faster R-CNN is mainly comprised of two modules. The First module is the region proposal network (RPN). The second module is the detector that makes use of the proposed regions by the first module. Thus the entire system forms a single, unified network for object detection. Thus the RPN module directs the Fast R-CNN module where to look for the desired objects [6, 7, 10].

3   Implementation

Using Tensorflow [5] and Anaconda python 3.6, we developed the code and started the training. This process includes three important steps. First step is building the code to implement the proposed algorithm. We have used python as our developing environment and we could develop codes for building the neural network taking in consideration the weights function and the loss function as well as the number of layers. Thus, our project can substantially detect see turtles in images or videos as well as in live webcam. Second step is gathering a huge and considerable amount of data and starting the training process. For the training part, we have done a lot of experiments to optimize our training. It is of a great importance to mention that the model should be trained enough, but at the same time it shouldn't be over trained. Finally, third step is testing the module.

3.1   Training the Faster R-CNN for Sea Turtle Detection

The process of collecting datasets was never easy. There is a scarcity of sea turtles' datasets on known datasets websites such as Kaggle and Google Images. However, we could collect datasets of green and loggerhead sea turtles of approximately 1000 images of which we used around 500. It is worthy to mention that some of these images was retrieved from ImageNet [3]. We then labelled these images and annotated them one by one using LabelImg software [11]. Then, we generate the .xml files which contain the dimensions of the label box (Xmin, Xmax, Ymin, Ymax). We used around 20% of our data to test our network and 80% to train the CNN. Our first training wasn't successful since the training period was more than the required period, thus we over trained the CNN. The training reaches a number of epochs when the loss function is ultimately reduced to its minimum value after which the loss function start to rise again which in this case produces potentially wrong training and thus no detection was observed.

The next training of the iteration was done carefully by closely observing the loss function such that the neural network is not over trained. The loss function is expected to have a graph where the loss function decrements to reach levels below 0.1 where we have our CNN training completed and saturated. Thus, we shouldn't exceed the training epochs of this saturation stage. Accordingly, we did the training in a proficient manner where the loss function decrements till it reaches a value below 0.05 and that is when we shall stop the training and proceed to the next step which is testing our model.

4   Testing the Faster R-CNN for Sea Turtle Detection

Since we have built three different programs to work with our detector which are image, video, and webcam test models, we tried three of them to detect the sea turtles which worked perfectly fine. We tested our module also on different images and challenging ones and it showed promising results. Although there were some inaccuracies in detecting some of the sea turtles provided in a given photo, the network showed impressive result overall. Below, in Figs. 1 and 2, the network has been tested by giving images, videos and by also using live webcam. Images in Fig. 2 have been captured by Drones.

Fig. 1. Sea turtle detected with high confidence.

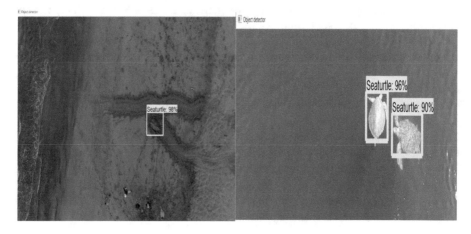

Fig. 2. Sea turtle is detected from drone video.

5 Evaluations

In this section the specifications of our machine, the precision calculations and the timing analysis are explained in detail. So we have implemented our sea turtles training and detection on a Lenovo ideapad 520 laptop with a Nvidia Card Geforce MX150 (4 GB) and intel core i7 8th generation processor. As for the hardware requirements as well as the R-CNN algorithm code can be found in https://github.com/rbgirshick/py-faster-rcnn.

5.1 Precision, Recall and F-Score Values

Precision can be defined as how accurate the predications made by the detection algorithm are. Thus, precision is defined by Eq. (1) as the ratio between the true positive results and the actual positive results of the algorithm. On the other side, recall is more meant to express with the total number of misses made by the algorithm, so as shown by Eq. (2), recall is the ratio between the true positive results and the summation of the misses and hits represented by true positive and false negative. Finally, F-score is a measure of accuracy that combines precision and recall results as given in Eq. (3). In Table 2, precision and recall related variables are listed. Table 3 shows the confusion matrix after evaluation, and Table 3 shows the computed Precision, Recall and F-Score values at threshold = 0.8 (confidence factor).

$$Precision = TP/(TP + FP) * 100\% \tag{1}$$

$$Recall = TP/(TP + FN) * 100\% \tag{2}$$

$$F1score = 2 * Precision * Recall/(Precision + Recall) \tag{3}$$

Table 2. Precision variables definition.

Variable	Definition	Explanation
TP	True Positive	Equivalent with hit
TN	True Negative	Equivalent with correct rejection
FP	False Positive	Equivalent with false error
FN	False Negative	Equivalent with miss

Table 3. Confusion matrix at threshold = 0.8.

Actual	Predicted		
	Predicted = No	Predicted = Yes	
Actual = No	TN = 270	FP = 4	274
Actual = Yes	FN = 26	TP = 90	116
	296	94	

Precision, Recall and F-Score values at threshold = 0.8:

$$Precision = 95.7\%$$

$$Recall = 77.6\%$$

$$F1score = 85.7\%$$

5.2 Time Performance Analysis

As for timing, we have an overhead of nearly 10 s when all necessary libraries and packages are imported afterwards the detection starts. Apart from the overhead, once detection starts, we are able to process and record the result for more than 3 frames per second. As mentioned in [7], the frame per second rate of Faster R-CNN reaches 5 FPS on K40 GPU using VGG model. It also reaches 17 FPS using ZF model on K40 GPU. Thus it proves it is a reliable model especially if precision is more favored than the rate of detection which is the case for our work.

6   Conclusion

In this work, we have proposed an intelligent system for sea turtles detection where the Faster R-CNN algorithm is employed in an impressive manner, and giving promising results. We have used Tensorflow-GPU deep learning framework to train our Faster-RCNN. Our work is fast and effective and it is a practical solution to object detection for drone based images. This work is really deployable and scalable as it has also been integrated in a complete system of detecting and classifying sea turtles by retrieving the videos of sea turtles from an on-board camera mounted on drone, where the user gets notifications on finding turtles in a friendly user interface. In this project we implemented the detection system on a cloud which receives the videos from the drone and automatically detects the turtles and display them in UI to notify the user about the whereabouts of the turtle by giving the frame number and the bounding box dimensions where the turtle is detected. It effectively contributes to ecosystem solutions and environmental research in general and particularly sea turtle conservation projects as sea turtles are known to be endangered species. The complete project, with all necessary files to run it, is open source and can be accessed via this link https://github.com/MuBadawy/Tensorflow-based-SeaTurtle-Detection.

Acknowledgements. This work is sponsored by Cyprus Wildlife Research Institute (CWRI). Our project team would like to acknowledge CWRI for supporting us to buy the hardware components required in the project. We also would like to thank Chantal Kohl and Stefanie Kramer from Humboldt University for supplying us some sea turtle images and videos for experimentation in our project. This project has been implemented on a server which belongs to Hossam Ahmed and Mohamed Badawy, we would like to thank them as well.

References

1. Erhan, D., Szegedy, C., Toshev, A., Anguelov, D.: Scalable object detection using deep neural networks. In: CVPR14 Proceedings of the 2014 IEEE Conference on Computer Vision and Pattern Recognition, pp. 2155–2162 (2014)
2. Girshick, R.: Fast R-CNN. https://arxiv.org/abs/1504.08083 (2015)
3. Image-net.org: ImageNet Tree View. http://image-net.org/synset?wnid=n01664065. Accessed 11 June 2019

4. Lee, J., Wang, J., Crandall, D.J., Sabanovic, S., Fox, G.C.: Real-time, cloud-based object detection for unmanned aerial vehicles. In: 2017 First IEEE International Conference on Robotic Computing (IRC), pp. 36–43 (2017)
5. Abadi, M., Agarwal, A., et al.: Tensorflow: large-scale machine learning on heterogeneous distributed systems. arXiv:1603.04467 (2016)
6. Gao, H.: Faster R-CNN explained. https://medium.com/@smallfishbigsea/faster-r-cnn-explained-864d4fb7e3f8. Accessed 3 Jan 2019
7. Ren, S., He, K., Girshick, R., Sun, J.: Faster R-CNN: towards real-time object detection with region proposal networks. https://arxiv.org/pdf/1506.01497.pdf (2016)
8. Redmon, J., Divvala, S., Girshick, R., Farhadi, A.: You only look once: unified, real-time object detection. https://www.cv-foundation.org/openaccess/content_cvpr_2016/papers/Redmon_You_Only_Look_CVPR_2016_paper.pdf (2016)
9. Sermanet, P., Eigen, D., Zhang, X., Mathieu, M., Fergus, R., LeCun, Y.: Overfeat: integrated recognition, localization and detection using convolutional networks. https://arxiv.org/pdf/1312.6229.pdf (2014)
10. Szegedy, C., Reed, S., Erhan, D., Anguelov, D.: Scalable, high-quality object detection. https://arxiv.org/abs/1412.1441 (2015)
11. Tzutalin LabelImg: Git code. https://github.com/tzutalin/labelImg (2015)
12. Liu, W., Anguelov, D., Erhan, D., Szegedy, C., Reed, S.: SSD single shot multibox detector. https://arxiv.org/abs/1512.02325 (2016)

Application of the Fuzzy Optimality Concept to Decision Making

Akif V. Alizadeh$^{(\boxtimes)}$ (iD)

Department of Control and Systems Engineering, Azerbaijan State Oil
and Industry University, 20 Azadlig Avenue, Baku AZ1010, Azerbaijan
akifoder@yahoo.com, a.alizade@asoiu.edu.az

Abstract. In Multi Criteria Decision Making with linguistic variables, the DMs may have vague information, limited attention and different information processing capabilities. This paper proposes a new fuzzy decision making method which allows fuzzy preferences in linguistic terms for alternative selection. The approach is computationally simple and its underlying concept is logical and comprehensible, thus facilitating its implementation in a computer-based system.

Keywords: Decision making · Fuzzy optimality · Pareto optimality · Degree of optimality

1 Introduction

In the realm of formal methods of decision-making, the general approach is the use of the utility theories. The main disadvantage of the utility theories is that they are based on an evaluation of vector-valued alternatives by means of a scalar-valued quantity. This transformation always leads to loss of information and is counterintuitive. Human being never separately transforms vector of attributes' values to a scalar value for comparison in reasoning or decision-making. Although approaches based on a vector-valued utility function exist, a fundamental axiomatic theory is absent. From the other side, preferences as human judgments are often vague and cannot be described by exact numerical values. However, the existing works on vector-valued utility approaches are devoted to the situations characterized by perfect decision-relevant information, which are rarely met in real-life decision making.

The other main counterargument against utility theories is that there also exist situations for which utility function cannot be applied.

The above mentioned circumstances mandate necessity to develop new decision approaches which, from one side may be based on direct pairwise comparison of vector-valued alternatives. From the other side, new approaches should be based on linguistic comparison of alternatives to be able to deal with vague vector-valued alternatives because real-life alternatives are almost always matter of a degree. Linguistic modeling of preferences will help to reduce Pareto optimal set of alternatives arriving at one optimal alternative or a narrowed subset of optimal alternatives when all relevant information is described in NL. For this purpose, a fuzzy optimality concept

© Springer Nature Switzerland AG 2020
R. A. Aliev et al. (Eds.): ICSCCW 2019, AISC 1095, pp. 542–549, 2020.
https://doi.org/10.1007/978-3-030-35249-3_69

[19] can be used as obtained from the ideas of CW-based redefinitions of the existing scientific concepts.

The layout of the rest of the paper as follows. Section 2 includes state of the art. In Sect. 3 we give preliminaries. Section 4 includes statement of the problem and method of solution. In Sect. 5 an example is presented. In Sect. 6, conclusion are given.

2 State of the Art

The existing classical decision theories, especially in economics, use utility function approach which is based on transformation of a vector to an appropriate scalar value. For many real-life decision making problems, information is judgment-based, it is vague, incomplete, imprecise, partially true etc. Preferences of a DM in this case are too inconsistent or ambiguous to permit a utility function. From the other side, in the existing theories there are so much restrictive assumptions that facilitate decision making where, really, utility function does not exists. For such cases, scoring of alternatives may be conducted by direct ranking of alternatives. Then, as a rule, instead of a utility function it is used binary relations that provide finding an optimal or near to optimal alternative(s). There exists a spectrum of works in this area.

For the performing outranking of alternatives, some methods had been developed for the traditional multiattribute decision making (MADM) problem [1, 2]. One of the first among these methods was ELECTRE method [3–5]. The general scheme of this method can be described as follows. A DM assigns weights for each criterion and concordance and discordance indices are constructed. After this, a decision rule is constructed including construction of a binary relation on the base of information received form a DM. Once a binary relation is constructed, a DM is provided with a set of non-dominated alternatives and a DM chooses an alternative from among them as a final decision.

Similar methods like TOPSIS [6–8], VIKOR [8–11] are based on the idea that optimal vector-valued alternative(s) should have the shortest distance from the positive ideal solution and farthest distance from the negative ideal solution. Optimal alternative in the VIKOR is based on the measure of "closeness" to the positive ideal solution. TOPSIS and VIKOR methods use different aggregation functions and different normalization methods. TOPSIS and VIKOR methods were also extended for interval and fuzzy numbers-valued attributes [1, 12–16].

In [17] it is suggested fuzzy balancing and ranking method for MADM. They appraise the performance of alternatives against criteria via linguistic variables as TFNs.

Main drawbacks of the approaches mentioned above are the following: (1) detailed and complete information on alternatives provided by a DM and intensive involving the latter into choosing an optimal or suboptimal alternative(s) are required; (2) TOPSIS, ELECTRE, VIKOR and similar approaches require to pose the weights of effective decision criteria that may not be always realizable; (3) Almost all these methods are based on the use of numerical values to evaluate alternatives with respect to criteria; (4) Classical Pareto optimality principle which underlies these methods makes it necessary to deal with a large space of non-dominated alternatives that sufficiently complicates the choice; (5) All these methods are developed only for MADM problems and not for problems of decision making under risk or uncertainty. Although there exists parallelism between these problems, these methods cannot be directly applied for decision making under risk or uncertainty.

3 Preliminaries

For a triangular fuzzy number $\tilde{a} = (a_1, a_2, a_3)$, the generalized mean value $gmv(\tilde{A})$ given by Lee and Li [18]:

$$gmv(\tilde{A}) = (a_1 + a_2 + a_3)/3. \tag{1}$$

Fuzzy optimality concept. In [19] they suggested the notion of fuzzy optimality for MADM problems as an extension of classical notion of Pareto optimality by grading optimality of each alternative within Pareto optimal set. This helps to differentiate more optimal alternatives from less optimal ones. Below in this section we give the main notions of the formalism in [19] to be used in the sequel.

Denote $\mathcal{A} = \{A_1, A_2, \dots, A_n\}$ the set of alternatives and $\mathcal{C} = \{C_1, C_2, \dots, C_m\}$ the set of criteria.

Definition 1 (Pareto Dominance). For any two points (candidate solutions) $A_1, A_2 \in \mathcal{A}$, A_1 is said to dominate A_2 in the Pareto sense (P-dominate) if and only if the following conditions hold:

$$C_i(A_1) \le C_i(A_2) \qquad \textit{for all } i \in \{1, 2, \dots, M\}, \tag{2}$$

$$C_j(A_1) < C_j(A_2) \qquad \textit{for at least one } j \in \{1, 2, \dots, M\}. \tag{3}$$

Definition 2 (Pareto Optimality). $A^* \in \mathcal{A}$ is PO if there is no $A \in \mathcal{A}$ such that A P-dominates A^*.

Definition 3 (Pareto Set and Front). We call PO set $\mathcal{S}_P$ and PO front $\mathcal{F}_P$ the set of PO solutions in design domain and objective domain respectively.

In [19], by directly comparing alternatives, they arrive at total degrees to which one alternative is better than, is equivalent to and is worse that another one. In order to do this, for each criterion they determine difference of between its values for considered alternatives. For obtaining the mentioned total degrees, these relative improvements are graded by membership degrees of predefined fuzzy sets "greater than 0", "equal to 0", "less than 0" and summed. Formally, these total degrees denoted nbF, neF, and nwF are determined as follows:

$$nbF(A_1, A_2) = \sum_{i=1}^{M} \mu_b^i(f_i(A_1) - f_i(A_2)), \tag{4}$$

$$neF(A_1, A_2) = \sum_{i=1}^{M} \mu_e^i(f_i(A_1) - f_i(A_2)), \tag{5}$$

$$nwF(A_1, A_2) = \sum_{i=1}^{M} \mu_w^i (f_i(A_1) - f_i(A_2)). \tag{6}$$

The membership functions $\mu_b^i, \mu_e^i, \mu_w^i$ are constructed such that Ruspini condition holds, which, in turn, results in the following condition:

$$nbF + neF + nwF = \sum_{i=1}^{M} (\mu_b^i + \mu_e^i + \mu_w^i) = M \tag{7}$$

Definition 4 $((1 - kF)$-dominance$)$. A_1 is said to $(1 - kF)$-dominates A_2 if and only if:

$$neF < M, nbF \geq \frac{M - neF}{kF + 1}, \tag{8}$$

where $0 \leq kF \leq 1$.

A function d with the following property is defined: for any couple of alternatives $A_1, A_2 \in \mathcal{A}, 1 - d(A_1, A_2)$ is the greatest kF value such that A_1 $(1 - kF)$-dominates A_2:

$$d(A_1, A_2) = \begin{cases} 0, & \text{if } nbF \leq \frac{M - neF}{2} \\ \frac{2 \cdot nbF + neF - M}{nbF}, & \text{otherwise} \end{cases} \tag{9}$$

$d(A_1, A_2) = 1$ shows that A_1 Pareto dominates A_2; $d(A_1, A_2) = 0$ shows that A_1 does not Pareto dominates A_2.

Definition 5 $(kF$-optimality$)$. A^* is kF-optimum if and only if there is no $A \in \mathcal{A}$ such that A $(1 - kF)$-dominates A^*.

A function $do : \mathcal{A} \to [0, 1]$ is defined as a degree of optimality of $A \in \mathcal{A}$ as follows:

$$do(A^*) = 1 - \max_{A \in \mathcal{A}} d(A, A^*). \tag{10}$$

In [19] it is mentioned that function do can be considered as the membership function of a fuzzy set describing the notion of kF-optimality.

Definition 6 $(kF$-Optimal Set and Front$)$. We call kF-Optimal Set $\mathcal{S}_{kF}$ and kF-Optimal Front $\mathcal{F}_{kF}$ the set of kF-Optimal solutions in design domain and objective domain respectively.

Definition 7 (Fuzzy dominance). Let $\mu_D(A_1, A_2)$ be a membership function defined as follows:

$$\mu_D(A_1, A_2) = \varphi_{\mu_D}(nbF(A_1, A_2), neF(A_1, A_2), nwF(A_1, A_2)) \tag{11}$$

Then $\mu_D(A_1, A_2)$ is a fuzzy dominance relation if for any $\alpha \in [0, 1]$ $\mu_D(A_1, A_2) > \alpha$ implies that A_1 $(1 - kF)$-dominates A_2.

In [19], particularly, they suggest to define φ_{μ_D} as follows:

$$\varphi_{\mu_D} = \frac{2 \cdot nbF(A_1, A_2) + neF(A_1, A_2)}{2M}. \tag{12}$$

Definition 8 (Fuzzy optimality). A membership function $\mu_D(A_1, A_2)$ represents the fuzzy optimality relation if for any $0 \leq kF \leq 1$ A^* belongs to the kF-cut of μ_D if and only if there no $A \in \mathcal{A}$ such that

$$\mu_D(A, A^*) > kF. \tag{13}$$

4   Statement of the Problem and Method of Solution

Vague preferences over a set of imprecise alternatives is modeled by linguistic preference relation over A. For this purpose it is adequate to introduce a linguistic variable "degree of preference" [12, 20] with term-set $T = (T_1, \ldots, T_n)$. Terms can be labeled as, for example "equivalence", "little preference", "high preference", and each can be described by a fuzzy number defined over some scale, for example [0, 1] or [0, 10] etc. The fact that $\tilde{f}$ is linguistically preferred to $\tilde{g}$ is written as $\tilde{f} \succsim_l \tilde{g}$. The latter means that there exist some $T_i \in T$ as a linguistic degree $Deg(\tilde{f} \succsim_l \tilde{g})$ to which $\tilde{f}$ is preferred to $\tilde{g}$: $Deg(\tilde{f} \succsim_l \tilde{g}) \approx T_i$.

The solution of the considered problem consists in determination of a linguistic degree of preference of $\tilde{f}$ to $\tilde{g}$ for all $\tilde{f}, \tilde{g} \in A$ by direct comparison of $\tilde{f}$ and $\tilde{g}$ as vector-valued alternatives. In the suggested approach, as a fundamental basis for determination of $\succsim_l$ a fuzzy optimality concept [19] will be used. However, in that work fuzzy optimality relation is formalized for perfect information structure, i.e. provided that all the decision relevant information is represented by precise numerical evaluations.

Direct comparison of vector-valued alternatives $\tilde{f}$ and $\tilde{g}$ may, particularly, supposes to determine the total degrees of statewise superiority, equivalence and inferiority of $\tilde{f}$ with respect to $\tilde{g}$. We suggest to determine the total degrees of statewise superiority, equivalence and inferiority, denoted nbF, neF, nwF respectively, as follows:

$$nbF(\tilde{f}, \tilde{g}) = \sum_{i=1}^{n} \mu_b^i(gmv((\tilde{f}(\tilde{s}_i) - \tilde{g}(\tilde{s}_i)) \cdot \tilde{P}_i)), \tag{14}$$

$$neF(\tilde{f}, \tilde{g}) = \sum_{i=1}^{n} \mu_e^i(gmv((\tilde{f}(\tilde{s}_i) - \tilde{g}(\tilde{s}_i)) \cdot \tilde{P}_i)), \tag{15}$$

$$nwF(\tilde{f}, \tilde{g}) = \sum_{i=1}^{n} \mu_w^i(gmv((\tilde{f}(\tilde{s}_i) - \tilde{g}(\tilde{s}_i)) \cdot \tilde{P}_i)), \tag{16}$$

where $\mu_b^i, \mu_e^i, \mu_w^i$ are membership functions for linguistic evaluations "better", "equivalent" and "worse" respectively, determined as in [19]. By following that approach, on the base of values of nbF, neF, and nwF, the value of degree of optimality $do(\tilde{h})$, as a degree of membership to fuzzy Pareto optimal set, is determined for all $\tilde{h} \in A$. $do()$ allows for justified determination of linguistic preference relation $\succsim_l$ over A.

On the base of $do()$ the degree of preference of $\tilde{f}$ to $\tilde{g}$ for any $\tilde{f}, \tilde{g} \in A$ can be determined from $do()$ as follows:

$$Deg(\tilde{f} \succsim_l \tilde{g}) = do(\tilde{f}) - do(\tilde{g}).$$

5 An Example

Here, we work out a numerical example, taken from (Chen, 2000), to illustrate the proposed methods for decision making problems with fuzzy data. Suppose that a software company desires to hire a system analysis engineer. After preliminary screening, three candidates (alternatives) $A = \{\tilde{f}_1, \tilde{f}_2, \tilde{f}_3\}$ remain for further evaluation.

Five benefit criteria are considered: C1: Emotional steadiness, C2: Oral communication skill, C3: Personality, C4: Past experience, and C5: Selfconfidence (Table 1).

Table 1. The fuzzy decision matrix and criteria weights

	C_1	C_2	C_3	Cs_4	C_5
f_1	trimf([5, 7, 9])	trimf([7, 9, 10])	trimf([3, 5, 7])	trimf([9, 10])	trimf([3, 5, 7])
f_2	trimf([7, 9, 10])	trimf([9, 10])	trimf([9, 10])	trimf([9, 10])	trimf([9, 10])
f_3	trimf([9, 10])	trimf([5, 7, 9])	trimf([7, 9, 10])	trimf([7, 9, 10])	trimf([7, 9, 10])
W	0.2	0.22	0.21	0.22	0.15

By applying fuzzy optimality concept-based approach, the considered problem can be solved as follows. At first, according to (14), (15), (16), we calculated nbF, neF, nwF:

$$nbF = \begin{bmatrix} 0 & 0 & 0.058 \\ 0.223 & 0 & 0.117 \\ 0.185 & 0.020 & 0 \end{bmatrix}, \quad neF = \begin{bmatrix} 5 & 4.777 & 4.756 \\ 4.777 & 5 & 4.863 \\ 4.756 & 4.863 & 5 \end{bmatrix},$$

$$nwF = \begin{bmatrix} 0 & 0.223 & 0.185 \\ 0 & 0 & 0.020 \\ 0.059 & 0.11 & 0 \end{bmatrix}.$$

Next we calculated μ_D according to (12), (9):

$$\mu_D = \begin{bmatrix} 0.500 & 0.523 & 0.513 \\ 0.478 & 0.500 & 0.490 \\ 0.487 & 0.510 & 0.500 \end{bmatrix}, \quad d(v_i, v_j) = \begin{bmatrix} 0 & 0 & 0 \\ 1.0 & 0 & 0.83 \\ 0.68 & 0 & 0 \end{bmatrix}.$$

Finally we calculated degree of optimality for each of the considered alternatives:

$$do = [0.0, \ 1, \ 0.17]'.$$

So, the preferences obtained are: $\tilde{f}_2 \succsim \tilde{f}_3 \succsim \tilde{f}_1$. The degrees of preferences are the following: $Deg(\tilde{f}_2 \succsim_l \tilde{f}_3) = 0.83$.

6 Conclusion

In Multi Criteria Decision Making with linguistic variables, the DMs may have vague information, limited attention and different information processing capabilities.

This paper proposes a new fuzzy decision making method which allows fuzzy preferences in linguistic terms for alternative selection.

The approach is computationally simple and its underlying concept is logical and comprehensible, thus facilitating its implementation in a computer-based system.

References

1. Lu, J., Zhang, G., Ruan, D., Wu, F.: Multi-Objective Group Decision Making. Methods, Software and Applications with Fuzzy Set Techniques. Series in Electrical and Computer Engineering, vol. 6. Imperial College Press, London (2007)
2. Zeleny, M.: Multiple Criteria Decision Making. McGraw-Hill, New York (1982)
3. Roy, B.: Multicriteria Methodology for Decision Aiding. Kluwer, Dordrecht (1996)
4. Roy, B., Berlier, B.: La Metode ELECTRE II. In: Sixieme Conference on Internationale de Rechearche Operationelle, Dublin (1972)
5. Huang, W.-C., Chen, C.-H.: Using the electre ii method to apply and analyze the differentiation theory. In: Proceedings of the Eastern Asia Society for Transportation Studies, vol. 5, pp. 2237–2249 (2005)
6. Hwang, C.L., Yoon, K.: Multiple Attribute Decision Making. Lecture Notes in Economics and Mathematical Systems, vol. 186, pp. 20–41. Springer, Berlin (1981)
7. Yoon, K.: A reconciliation among discrete compromise solutions. J. Oper. Res. Soc. **38**(3), 272–286 (1987)
8. Opricovic, S., Tzeng, G.H.: Compromise solution by MCDM methods: a comparative analysis of VIKOR and TOPSIS. Eur. J. Operat. Res. **156**(2), 445–455 (2004)
9. Opricovic, S.: Multicriteria optimization of civil engineering systems, Faculty of Civil Engineering, Belgrade (1998)
10. Opricovic, S., Tzeng, G.H.: Compromise solution by MCDM methods: a comparative analysis of VIKOR and TOPSIS. Eur. J. Oper. Res. **156**(2), 445–455 (2004)
11. Opricovic, S., Tzeng, G.H.: Extended VIKOR method in comparison with outranking methods. Eur. J. Oper. Res. **178**(2), 514–529 (2007)

12. Liu, W.J., Zeng, L.: A new TOPSIS method for fuzzy multiple attribute group decision making problem. J. Guilin Univ. Electron. Technol. **28**(1), 59–62 (2008)
13. Jahanshahloo, G.R., Lotfi, F., Hosseinzadeh, I.M.: An algorithmic method to extend TOPSIS for decision-making problems with interval data. Appl. Math. Comput. **175**(2), 1375–1384 (2006)
14. Zhu, H.P., Zhang, G.J., Shao, X.Y.: Study on the application of fuzzy TOPSIS to multiple criteria group decision making problem. Ind. Eng. Manag. **1**, 99–102 (2007)
15. Wang, Y.M., Elhag, T.M.S.: Fuzzy TOPSIS method based on alpha level sets with an application to bridge risk assessment. Expert Syst. Appl. **31**(2), 309–319 (2006)
16. Peide, L., Minghe, W.: An extended VIKOR method for multiple attribute group decision making based on generalized interval-valued trapezoidal fuzzy numbers. Sci. Res. Essays **6**(4), 766–776 (2011)
17. Vahdani, B., Zandieh, M.: Selecting suppliers using a new fuzzy multiple criteria decision model: the fuzzy balancing and ranking method. Int. J. Prod. Res. **18**(15), 5307–5326 (2010)
18. Lee, E.S., Li, R.-J.: Comparison of fuzzy numbers based on the probability measure of fuzzy events. Comput. Math Appl. **15**(10), 887–896 (1988)
19. Farina, M., Amato, P.: A fuzzy definition of "optimality" for many-criteria optimization problems. IEEE Trans. Syst. Man Cybern. Part A Syst. Hum. **34**(3), 315–326 (2004)
20. Borisov, A.N., Alekseyev, A.V., Merkuryeva, G.V., Slyadz, N.N., Gluschkov, V.I.: Fuzzy information processing in decision making systems. Moscow, "Radio i Svyaz" (in Russian) (1989)

Robust Designing of the PSS and SVC Using Genetic Algorithm

Fahreddin Sadikoglu[1(✉)] and Ebrahim Babaei[1,2]

[1] Engineering Faculty, Near East University, 99138 Nicosia, North Cyprus,
Mersin 10, Turkey
fahreddin.sadikoglu@neu.edu.tr,
e-babaei@tabrizu.ac.ir
[2] Faculty of Electrical and Computer Engineering,
University of Tabriz, Tabriz, Iran

Abstract. In recent years, flexible alternative current transmission systems (FACTS) devices as well as static VAr compensator (SVC) have been used increasingly for dynamic voltage control. Lack of coordination between SVC and power system stabilizer (PSS) parameters can oppose bad effects and generator voltage as well as angular oscillation generator, so simultaneous coordination of stabilizer parameters and SVC is of great interest. In this paper, the problem of stability of a model power system is presented with SVC and PSS block diagrams modeling. Then, the objective function of optimal performance of the model network is to provide stability in various faults. Genetic algorithm (GA) is used to determine the parameters of control systems. Finally, by simulation in MATLAB software, analysis will be reviewed and approved.

Keywords: Static VAr compensator · Power system stabilizer · Dynamic stability · Genetic algorithm

1 Introduction

With the increased electrical energy demands and need to performance power systems within the sustainability boundaries, power systems easily and with the smallest disturbance can be exposed to extreme conditions. This leads to oscillation of power systems with poor damping or even instability of power systems [1]. Problems of dynamic stability of power systems are studied in term of angular stability and voltage stability. Generator load angle oscillations as well as oscillations in the voltage are of the most important issues is the dynamic power system [2, 3]. Generator load angle oscillations are due to a mismatch between power production and power consumption of plants. PSS is used extensively for dumping these oscillations. More papers have paid on design of parameters of PSS and their proper place [4–6].

Although, PSS is very powerful in generation of stabilizer signals of power systems, but their presence alone cannot play a large role in voltage stability and could even lead to an increase in voltage oscillations [2–7]. With the advancement of power electronics and nonlinear control theory of, FACTS devices as well as SVC has been used increasingly for dynamic voltage control [7]. In addition, SVC can enhance

© Springer Nature Switzerland AG 2020
R. A. Aliev et al. (Eds.): ICSCCW 2019, AISC 1095, pp. 550–556, 2020.
https://doi.org/10.1007/978-3-030-35249-3_70

voltage oscillations, and increase dynamic and transient stability [8]. On the other hand, the lack of coordination between SVC and PSS parameters can have bad effects on generator angular oscillations and voltage oscillations, so simultaneous coordination of stabilizer parameters and SVC is of great interest of many papers [7–9].

Different methods have been used to design the controlling parameters. These methods are classified into two main Classical methods and Heuristic algorithms methods. Heuristic algorithms can be applied to the whole range of problems without any restrictions of linearization and simplification. These algorithms also can escape from local optimal points. In [10] a comprehensive comparison was done between the heuristic and classic methods. In [11] searches for locations and amounts of PSS and SVC controllers in the power networks. In that paper, location and values of controller are specified so that the network is resistant in certain predefined events. In [12] discussed optimal design of SVC and PSS parameters for three-phase faults various places. Reference [13] also examined the proper values of network controller and used GA. Reference [14] has considered resistant design of PSS parameters against all kinds of faults, but it in this paper presence of SVC is not considered.

Most of references and papers try to determine the optimal values of controller parameters in the presence of a particular disturbance. If the type of event is changed there is a probability of loss of efficiency of controllers and even instability of the system.

This paper determines the optimal values of SVC and PSS so that network against a range of events is necessary. Due to the ability GA to find optimal, this algorithm is used for optimization and problem solving. In this paper, we first formulated and introduced the prototype network. Then GA and its parameters are explained and, in the end, after the simulation results the conclusion part is presented.

2 Formulation of the Problem

SVC block diagram is shown in Fig. 1. Input of SVC is measured voltage of phase and output is the capacitive susceptance for reactive power compensation. The PSS block diagram is shown in Fig. 2. Input of PSS is speed changes of the generator from nominal value and output is stabilizer signal of excitation voltage. The power grid synchronous generator connected to an infinite bus provides share of the available loads by transformer. Generator output is equipped to SVC. The PSS is also placed at the entrance of the excitation system.
Where:

K_{PSS}: Gain of PSS
T_{1n}: Time constant for the first compensator nominator,
T_{1d}: First compensator denominator time constant,
T_{2n}: Time constant for the second compensator nominator,
T_{2d}: Time constant for the second compensator denominator,
$K_{P,SVC}$: Voltage regulators gain of SVC,
$K_{I,SVC}$: Coefficient of voltage regulator integrator of SVC
V_{ref}: Reference voltage of SVC,
t_{stop}: Final time of simulation,

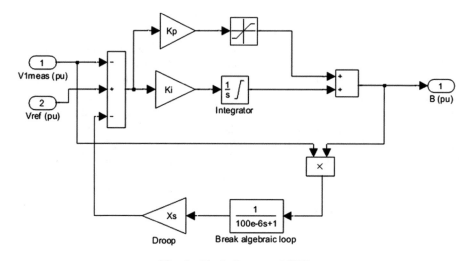

Fig. 1. Block diagram of SVC.

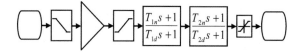

Fig. 2. Block diagram of PSS.

In addition, restrictions of the control variables are shown in Table 1.

Table 1. Restrictions of control variable.

Parameter	Min. value	Max. value
K_{PSS}	2.5	25
T_ω	1.5	15
T_{1n}	0.005	2
T_{1d}	0.001	1
T_{2n}	0.1	10
T_{2d}	0.005	15
$K_{P,SVC}$	0	100
$K_{I,SVC}$	0	750

Objective function

The objective function of studied network is expressed in the following equation:

$$\sum_{fault\ type} \int_0^{t_{stop}} t \cdot |d\omega| + \int_0^{t_{stop}} t \cdot |V_{ph} - V_{ref}| \tag{1}$$

t: Current time of simulation,

$d\omega$: Oscillation of change of the speed of the generator,

V_{ph}: Phase Voltage,

In this paper, three types of fault such as phase-to-ground short circuit, short circuit of two-phases to each other and ground, short circuit of three-phases to each other and short circuit to ground are examined.

3   Genetic Algorithm and Problem Solving

GA is a random search method based on natural selection. This algorithm is composed of a population that each person in population (chromosomes) show a sample answer and each component of chromosomes (genes) represent certain variables of problem under investigation. New generation is produced with regard to individual's fitness function and applying genetic operators (selection, coupling and mutation) and fitness function improves during repetition of algorithm.

4   Simulation Results

In this section, the proposed algorithm is examined to reduce phase voltage oscillations and oscillations of generator speed changes in case of power system and results will be presented. To assess the performance of control system in power network, faults of single-phase to ground, two-phase to ground and three-phase to ground are intended. Simulation and analysis are done by MATLAB software. GA values for control variables are presented in Table 2.

Table 2.   Investigation values of GA for controlling variables.

Parameter	Value
K_{PSS}	13.190
T_ω	7.147
T_{1n}	1.371
T_{1d}	0.012
T_{2n}	0.479
T_{2d}	10.799
$K_{P,SVC}$	1.189
$K_{I,SVC}$	0.685

And phase voltage oscillations at the single-phase to ground fault in both cases of without SVC and PSS and in the presence of them with proposed values of GA. As shown in Fig. 3, the generator voltage oscillations without SVC and PSS is more compared to time that both control systems are available; in addition, the range of variation is also greater. With SVC and PSS, oscillations are quickly damped and the system quickly reaches a steady state.

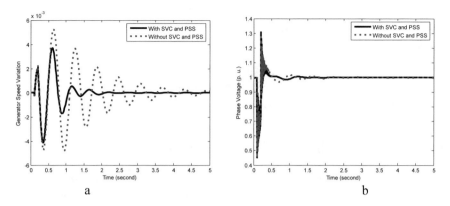

Fig. 3. Results of single-phase to ground fault simulation, (a) oscillations in generator speed change; (b) phase voltage oscillations.

Oscillations of speed change and generator phase voltage oscillations, at two-phase to ground fault in both cases of without SVC and PSS as well as in their presence, with proposed values of GA is shown in Fig. 4.

Considering the values presented in Table 2 and along with SVC and PSS, the maximum range of the generator speed oscillations reaches from 8×10^{-3} to 5×10^{-3} and maximum phase voltage oscillations range reaches from 3.0 V to 1.05 V.

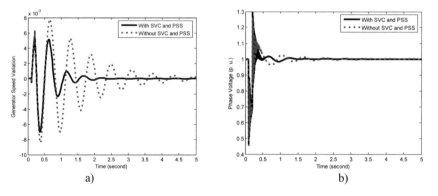

Fig. 4. Results of two-phase to ground fault simulation, (a) oscillations in generator speed change; (b) phase voltage oscillations.

From Fig. 5, it can be noted that the presence of SVC and PSS with parameters provided with GA technique system quickly reaches to operating point.

Figure 5 shows changes of phase voltage oscillations and oscillations of generator speed at three-phase to ground fault in both cases of without SVC and PSS and in the presence of them along with a GA values. As can be seen, the maximum oscillations of generator speed reduce from 0.015 to 0.006 and damping speed increases. Also, the range of maximum phase voltage oscillations and damping speed are reduced. Considering Figs. 5(a) and (b), it can be noted that with the presence of the SVC and PSS, system quickly reaches to operating point.

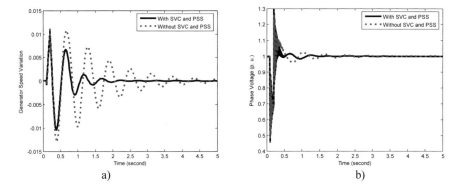

a)   b)

Fig. 5. Results of three-phase to ground fault simulation, (a) oscillations in generator speed change; (b) phase voltage oscillations.

5   Conclusion

In this paper, simultaneous optimization of stabilizer parameters of power system and SVC are discussed. In order to strengthen the power network against a wide range of events, possible short-circuits in the power network were considered in the optimization problem. Considering the capabilities of GA to find the optimal solution, the algorithm was used for optimization and problem solving. As it was seen, by increasing the number of short-circuited phases generator speed oscillations and phase voltage oscillations increased. The simulation results show that the power grid with SVC and PSS had the least oscillations against possible types of short-circuits. Control system adjusted with the proposed parameters of GA well damped oscillations and stable zed the system at a certain operating point.

References

1. Ma, J., Wang, S., Wang, Z., Qiu, Y., Thorp, J.S.: Power system energy stability region based on dynamic damping theory. IET Gener. Transm. Distrib. **10**, 2907–2914 (2016)
2. Cao, G., Chen, L., Aihara, K.: Power system voltage stability assessment based on branch active powers. IEEE Trans. Power Sys. **30**, 989–996 (2015)
3. Fallahi, H., Aghamohammadi, M.R., Parizad, A., Mohamadi, A.: Enhancing power system oscillation damping using coordination between PSS and SVC. In: Proceedings of the EPECS, pp. 1–7 (2009)
4. Zhu, Z., Geng, G., Jiang, Q.: Power system dynamic model reduction based on extended krylov subspace method. IEEE Trans. Power Sys. **31**, 4483–4494 (2016)

5. Babaei, E., Bolhasan, A.M., Sadeghi, M., Khani, S.: Power system oscillations damping using UPFC based on an improved PSO and genetic algorithm. In: Journal of International Conference of Electrical Machines and Systems, vol. 1, pp. 135–142 (2012)
6. Mohammadzadeh Shahir, F., Babaei, E.: Evaluating the dynamic stability of power system using UPFC based on indirect matrix converter. J. Autom. Control Eng. 1, 279–284 (2013)
7. Khodabakhshian, A., Hooshmand, R., Sharifian, R.: Power system stability enhancement by designing PSS and SVC parameters coordinately using RCGA. In: Proceedings of the Canadian Conference on Electrical and Computer Engineering, pp. 579–582 (2009)
8. Bian, X.Y., Geng, Y., Lo, K.L., Fu, Y., Zhou, Q.B.: Coordination of PSSs and SVC damping controller to improve probabilistic small-signal stability of power system with wind farm integration. IEEE Trans. Power Syst. 31, 2371–2382 (2015)
9. Lei, X., Lerch, E.N., Povh, D.: Optimization and coordination of damping controls for improving system dynamic performance. IEEE Trans. Power Syst. 16, 473–480 (2001)
10. Alrashidi, M.R., El-Hawary, M.E.: A survey of particle swarm optimization in electric systems. IEEE Trans. Eval. Comput. 13, 913–918 (2009)
11. Chang, Y.C.: Multi-objective optimal SVC installation for power system loading margin improvement. IEEE Trans. Power Syst. 27, 984–992 (2011)
12. Abido, M.A., Abdel Magid, Y.L.: Power system stability enhancement via coordinated design of a PSS and an SVC based controller. In: Proceedings of the IEEE ICECS, vol. 2, pp. 850–853 (2003)
13. Zhijian, L., Hongchun, S., Jilai, Y.: Coordination control between PSS and SVC based on improved genatic-tabu hybrid algorithm. In: Proceedings of the International Conference on Sustainable Power Generation and Supply, pp. 1–5 (2009)
14. Babaei, E., Galvani, S., Ahmadi Jirdehi, M.: Design of robust power system stabilizer based on PSO. In: Proceedings of the IEEE Symposium on Industrial Electronics & Applications, vol. 1, pp. 325–330 (2009)

Diagnosis of Functioning of Gas Pipelines by Cluster Analysis Method

G. G. Ismayilov⬛, E. K. Iskandarov$^{(\boxtimes)}$⬛, and F. B. Ismayilova⬛

Azerbaijan State Oil and Industry University, Azadliq 20,
AZ1010 Baku, Azerbaijan
asi_zum@mail.ru, e.iskenderov62@mail.ru,
fidan.ismayilova.2014@mail.ru

Abstract. In the article, gas pipelines were analyzed taking into account clustering of initial data on daily fluctuations in temperature, pressure and flow rates. It was found that various cases of clustering of experimental data of gas pipelines on the parameters of temperature, pressure and gas flow could be observed. Clusters do not always form these factors. For example, in gas pipelines with thermal insulation, clustering does not occur.

Keywords: Cluster analysis · Temperature mode · Gas pipeline mode · Daily temperature fluctuations · Flow rate · Pressure

1 Introduction

Analysis of the operation of gas pipelines in the territory of Azerbaijan shows that daily fluctuations in temperature affect the mode of their operation.

The influence of the temperature change of the pumped gas on its mode of operation is explained by the fact that the coefficient of hydraulic resistance of gas movement λ, in general, is related to the Reynolds number Re, which, in turn, is directly proportional to the density ρ and inversely proportional to the viscosity μ of the pumped gas. Both, density and viscosity are functions of pressure and temperature. Below in the graphs it will be seen that daily changes in temperature cause corresponding fluctuations in the gas pressure in the pipe. Therefore, daily fluctuations in the magnitude of the hydraulic resistance and changes in the operating mode of the pipeline are not excluded. However, direct calculations of λ and Re require a large amount of work. In particular, it is necessary to organize gas sampling at different times of the day to determine its composition on a chromatograph, etc. [1–4].

If the pipelines are insulated, then the effect of daily temperature fluctuations can be minimized.

The question arises from the analysis of the functioning of gas pipelines, which shows that a change in the structural forms of the flows are observed due to fluctuations in the modes of operation of pipelines that do not have heat insulation, because of daily changes in ambient temperature. In this case, direct calculations prove the existence of two daily regimes of their work, conventionally called "day" and "night."

© Springer Nature Switzerland AG 2020
R. A. Aliev et al. (Eds.): ICSCCW 2019, AISC 1095, pp. 557–564, 2020.
https://doi.org/10.1007/978-3-030-35249-3_71

2 Solution and Discussion of the Problem

The way to solve the problem is separation of modes from each other. For this purpose, you can use cluster analysis, which allows you to organize the observed data into visual structures (classes), when there are not any priori hypotheses about the classes, and the researchers are only at the descriptive stage of the study [3, 5, 6]. It also shows that cluster analysis determines the "most likely significant solution", and therefore there is no need to check the statistical significance of the studied variables. For calculation were used hourly measurements of temperatures, pressures and gas flow rates in pipelines Azadkend, Azadkend-AzerImishli, Azadkend-Bedilli, Azadkend-Bilasuvar2, Azadkend-Ovchuberekend, Azadkend-Jalilabad, Azadkend-Tukle QPS, Azadkend-RVU, Azadkend-Lerik150, Azadkend-Astara1. The Figs. 1 and 2 accordingly show dependence of the change in temperature and pressure of the pumped gas over time in the Azadkend-Bilasuvar2 pipeline. It should be noted that in all considered pipelines, temperature fluctuations are identical.

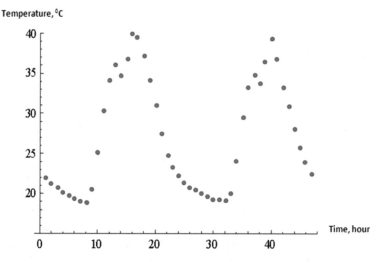

Fig. 1. Dynamics of temperature change of the gas pumped through Azadkend-Bilasuvar2 gas pipeline.

It is seen that temperature changes uniquely determine pressure changes. The same pattern characterizes the work of other gas pipelines. It follows from the figures that lower pressures correspond to greater temperatures and on the contrary, which indicates an intense mass transfer.

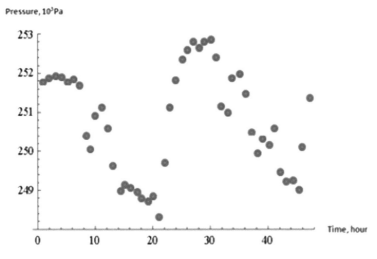

Fig. 2. Dynamics of pressure change of gas pumped through Azadkend-Bilasuvar2 gas pipeline

Cluster analysis of objects for which the values of quantitative attributes are set, begins with the calculation of differences for all pairs of objects. As a measure of difference, the distance between objects in the P-dimensional space is chosen, most often the Euclidean distance or its square. In this case, $P = 2$ and the Euclidean distance between the objects í and γ are determined by the formula:

$$d = \left\{ \sum_i (x_i - y_j)^2 + (y_i - y_j)^2 \right\}^{1/2}$$

After calculating all distances, the distance matrix d_{ij} is compiled:

$$\begin{pmatrix} o & d_{11} & d_{12} & \cdots & d_{1i} \\ d_{21} & o & d_{22} & \cdots & d_{2i} \\ d_{31} & d_{32} & o & \cdots & d_{3i} \\ \cdots & \cdots & \cdots & \cdots & \cdots \end{pmatrix}$$

Here the points of distance are zeroed on themselves. In this case, all data must have the same dimensions; otherwise, you need to go to dimensionless quantities. After that, the distances are compared with each other. Usually it is used the equation.

$$l = min \left| d_o - d_{ij} \right|$$

Also, it can be used the next equation

$$l = max \left| d_o - d_{ij} \right|$$

All statements give the same result. The results of the calculations are grouped by the similarity of the distances, which represent a cluster [7, 8]. In practice, dispersions are calculated, and the group with the lowest dispersion is selected. This is the first cluster. Then, the procedure is repeated. It is essential that the number of clusters is determined by the physical essence of the problem [9–11].

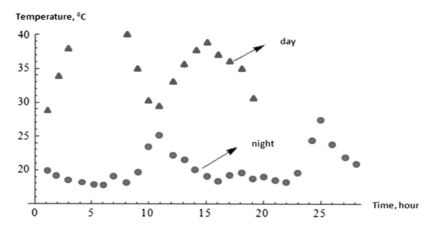

Fig. 3. Clustering the change in temperature of the gas pumped through Azadkend-Bilasuvar2 gas pipeline.

The carried out calculations divided the initial data into clusters. Figures 3 and 4 illustrate, accordingly changes in temperature, and pressure as a function of time. As it is seen, in the first case, there are two clusters, conventionally called "night" - blue, and "day" - red. On the temperature graph, the data "night" are located in the zone of 15–220 °C, the data of the "day" are located in the zone of 28–400 °C.

The clustering of data on changes in pressure over time also divides them into two classes (see Fig. 4). Nighttime data are blue, refer to the zone of higher pressures, lower temperatures and less intensive mass transfer, while daytime data are red, on the contrary, lie in the zone of less high pressures, higher temperatures and more intensive mass transfer.

For more deeply studying of the issue, it was also constructed a three-dimensional vector "temperature-pressure-flow rate" (see Fig. 5). It is seen that the breakdown into classes is related to temperature, since the "cut" of the graph passes along the temperature axis.

The processing of vectors compiled according to the series of observations of other gas pipelines shows that the following cases can be observed:

All vectors are divided into clusters, forming visual structures of the effect of temperature on the operation of the pipeline.

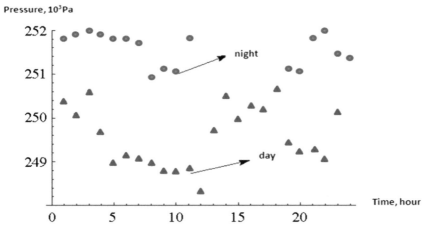

Fig. 4. Clustering the change in pressure of the gas pumped through Azadkend-Bilasuvar2 gas pipeline

Data processing in the Azadkend -RVU-2 Lankaran gas pipeline, reflects this case.

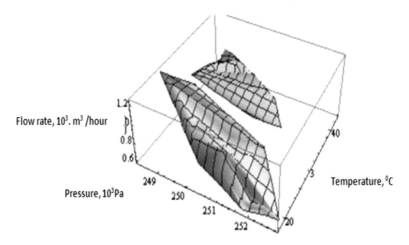

Fig. 5. Clustering of the vector "Temperature-pressure-flow rate" through Azadkend-Bilasuvar2 gas pipeline

Not all vectors form clusters. Such a case was observed when processing the results of observations through the Azadkend-Astara1 gas pipeline. The surface of pipeline is covered with insulating material (See Figs. 6, 7, 8 and 9).

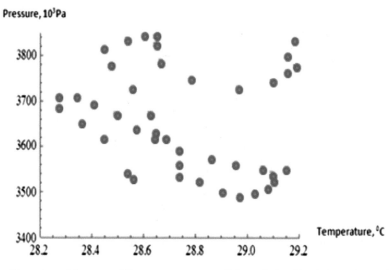

Fig. 6. Clustering of the vector "Temperature-pressure" through Azadkend-Astara1 gas pipeline

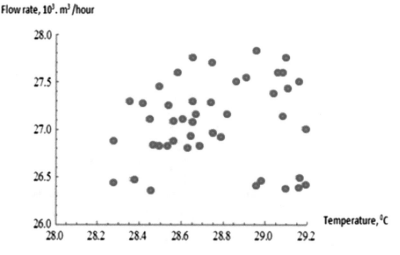

Fig. 7. Clustering of the vector "Temperature-flow rate" through Azadkend-Astara1 gas pipeline

Flow rate, 10^3. m³ / hour

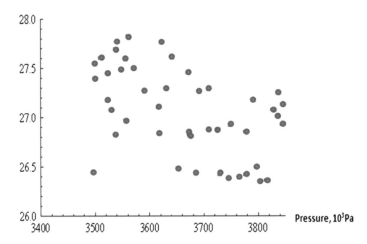

Fig. 8. Clustering of the vector "Pressure-flow rate" through Azadkend-Astara1 gas pipeline

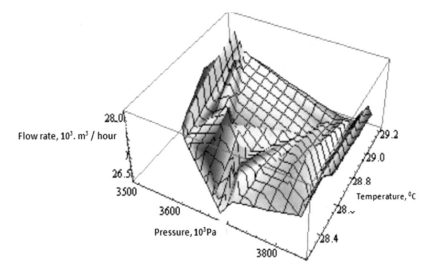

Fig. 9. Clustering of the vector "Temperature-Pressure-Flow rate" through Azadkend-Astara1 gas pipeline

During the processing of the initial data for the vectors "pressure - temperature", "flow rate - temperature" and "flow rate - pressure" data clustering does not occur (see Figs. 6, 7 and 8), also, the clustering of the vector "temperature - pressure - flow rate" in the Azadkend – Astara1 gas pipeline also does not occur (there is no "cut" of the graph) (see Fig. 9).

3   Conclusion

Thus, a series of observations of changing temperature, pressure and flow rate of pumped gas of some gas pipelines of Azerbaijan depending on time were processed using the cluster analysis method. Daily fluctuations in temperature can affect the mode of operation of gas pipelines. This effect is detected by processing the vectors "temperature-pressure", "temperature-flow rate", "pressure-flow rate" and "temperature - pressure - flow rate".

Analysis of the results of calculations shows that there are three cases of clustering of the initial data and not all vectors form clusters. In particular, for gas pipelines with thermal insulation, clustering does not occur.

References

1. Aliyev, R.A., Belousov, V.D.: Pipeline Transport of Oil and Gas. Nedra, Moscow (1988)
2. Mirzadzanzade, A.Kh., Khasanov, M.M., Bakhtizin, R.N.: Modeling of oil and gas production processes (nonlinearity, irregularity, uncertainty), Moscow (2004)
3. Ismayilov, G.G., Khudiyev, M.S., Seyfullayev, G.K.: Analysis of operation of gas pipelines based on the clustering of initial data. In: Proceedings of the XI Russian Scientific and Technical Conference Actual Problems of the Development of the Oil and Gas Complex of Russia, Moscow, p. 157 (2016)
4. Ismayilov, R.A., Seyfullaev, G.K.: Diagnosing structural changes in gas flows based on fractal analysis. In: Proceedings of Anniversary 70th International Youth Scientific Conference, Oil and Gas 2016, Moscow, vol. 1 (2016)
5. Mendel, I.D.: Cluster Analysis. Finance and Statistics, Moscow (1988)
6. Shumetov, V.G., Shumetova, L.V.: Cluster analysis - computer-assisted approach. Gos. Tekh. Univercity, Orel (2000)
7. Leonov, V.P.: The short review of cluster analysis methods. http://www.biometrica.tomsk. ru/cluster_2.htm
8. Leonov, V.P.: Literature and sites on cluster analysis. http://www.biometrica.tomsk.ru/ cluster_3.htm
9. The site of Space Research Institute of Russian Academy of Sciences. http://www.iki.rssi.ru/ magbase/REFMAN/STATTEXT/modules/stclaun.html#general
10. Baran, O.I., Grigorjev, Yu.A., Zhilina, N.M.: The algorithms and criteria of clusterization quality. In: Public Health and Public Health Service: Materials of the XLV Scientific-Practices Conference with International Participation of «Hygiene», Kemerovo, pp. 21–26 (2010)
11. Blizorukov, M.G.: Statistical methods of market analysis: manual. The Institute of Management and Entrepreneurship of Ural State University, Ekaterinburg (2008). http:// elar.usu.ru/bitstream/1234.56789/1671/6/1334937_schoolbook.pdf

Investigation of Unbalanced Open Economy System by Using Interval Mathematics for the Case of Azerbaijan Republic

Revana I. Davudova[1,2,3](✉) 📵

[1] ANAS Institute of Economics, H. Javid ave., 115, AZ1143 Baku,
Azerbaijan Republic
revanadavudova@gmail.com
[2] ANAS Institute of Control Systems, B. Vahabzadeh st., 9, AZ1141 Baku,
Azerbaijan Republic
[3] The Azerbaijan University, J. Hajibeyli st., 71, AZ1007 Baku,
Azerbaijan Republic

Abstract. This paper is devoted to the study of the relations between economic growth rate, consumption, investment, public expenditure norms, economic growth rate, general index of structural effectiveness and other macroeconomic indicators of the unbalanced open economy (UOE). Macroeconomic indicators' interrelations of the UOE have been investigated on the basis P. Samuelson model in the fields of interval mathematics with interval values, taking into account the uncertainties that can arise in the evaluation indicators. MathWorks MATLAB Software 2017a M-file has been created to implement the calculations and graphic descriptions, simulations and evaluations perform the using interval values of the macroeconomic indicators, determined by the Delphi method.

Keywords: Unbalanced perfect open economy · Macroeconomic indicators · P. Samuelson model · Interval mathematics · MATLAB M-file

1 Introduction

Improving the competitiveness of the country's national economy is a major and complex task. The result of entire economic system of the country depends on it necessarily. Improving the efficiency of the country's entire economic system, ensuring its national security and national interests are the important goal of setting problem as this form.

The main purpose of the government macroeconomic policy is to bring the current state of the economic system (ES) closer to its potential values. Minimizing the deviation between the current and potential positions of production should be a priority under any circumstances.

The purpose of this research is to compare the current and ideal-potential state of UES in the example of Azerbaijan Republic by using interval analysis.

İn this research [1] the relationships between macroeconomic indicators, such as consumption norm, economic growth rates, public spending norm, investment norm

R. A. Aliev et al. (Eds.): ICSCCW 2019, AISC 1095, pp. 565–572, 2020.
https://doi.org/10.1007/978-3-030-35249-3_72

and average tax burden, efficiency of GDP structure and others have been investigated for a Balanced Open System (BOS) when Ex = Im. It should be noted that the normative values of consumption expenditures, investment, government expenditures and other indicators sometimes could not define uniquely. These values depend on uncertain factors and are determined by various social surveys and studies. So macroeconomic norms are not unique and it may be much more appropriate to use their interval values when assessing the state of the economic system to account for uncertainty.

These studies [2, 3] are devoted to the relationships between macroeconomic indicators of the balanced open economy. Based on the Samuelson model using interval analysis the interaction of the average tax burden with macroeconomic indicators was analyzed, as the norms of consumption, public expenditures, investments and rates of economic growth and were estimated on the interval values of the macroeconomic indicators of Azerbaijan Republic defined by the method Delphi during 2005–2017. The purpose of this research is to compare the current and ideal-potential state of UES in the example of Azerbaijan Republic by using interval analysis [5–9].

2 Methodology

Samuelson model is a main macroeconomic identity, reflecting the equality of income and expenditure and linking the systems of basic macroeconomic indicators [4]. An income is considered as an aggregate income from the final production of goods and services – GDP and expenditures are considered as calculation's components of the GDP by the production method:

$$Y = C + I + G + \varepsilon B \tag{1}$$

where Y – the volume of GDP or national income forecasted (planned) by government agencies; G – public expenditures (respectively, $g = G/Y$ – public expenditure norm); I – investments (respectively, $i = I/Y$ – investing norm); T - forecasted tax revenues by the state authorities (respectively, $\theta = T/Y$ – average tax burden norm); C – public consumption (respectively, $c = C/Y$ – public consumption norm); ε - exchange rate; B – the balance of the country's balance of payments (respectively, b – balance of payments norm. $\varepsilon B = X = Ex - Im$ is net exports, which is the difference between exports (Ex) and imports (Im)). Consumption expenditures and, respectively, consumption expenses norm can be written as follows,

$$C = (1 - s)(1 - \theta)Y \quad \text{and} \quad c = \frac{C}{Y} = (1 - s)(1 - \theta).$$

We are getting the Samuelson model expressed with norms by dividing of Eq. (1) to Y:

$$c + i + g + \varepsilon b = 1 \tag{2}$$

Table 1. Utilized indicators for assessment of effectiveness of UPOES

№	Mathematical expression	Sign and name of the macroeconomic indicator
1	$F = \Delta Y/Y$	F – Economic growth rate
2	$\psi = 1/(s+g+\varepsilon b)$	ψ – General index of structural effectiveness
3	$R = \Delta Y/I = F/i$	R – Social effectiveness of investment policy
4	$\breve{g} = \Delta Y/G = F/g$	S – Social effectiveness of public expenditures policy
5	$\delta = \Delta Y/T = F/\theta$	δ – Social effectiveness of tax policy

Table 2. Mathematical expressions of the macroeconomic norms' interrelationships of UPOES

№	Impacts of average tax burden θ, balance of payments norm b and exchange rates ε on the macroeconomic norms	Impacts of public consumption norm c, balance of payments norm b and exchange rates ε on the macroeconomic norms
1	$i = g = \breve{g} = \theta/(1+\theta) - \varepsilon b$	$i = g = \breve{g} = (1 - c - \varepsilon b)/2$
2	–	$\theta = (1 - c + \varepsilon b)/(1 + c - \varepsilon b)$
3	$c = (1 - \theta)/(1+\theta) + \varepsilon b$	–
4	$F = [\theta/(1+\theta) - \varepsilon b]^2$	$F = [1 - c - \varepsilon b]^2/4$
5	$\psi = (1+\theta)/(2\theta - \varepsilon b(1-\theta))$	$\psi = 1/(1-c)$
6	$R = \theta/(1+\theta) - \varepsilon b$	$R = (1 - c - \varepsilon b)/2$
7	$\delta = \frac{F(\theta)}{\theta} = [\theta/(1+\theta) - \varepsilon b]^2/\theta$	$\delta = F(c)/\theta(c)$

Table 3. Mathematical expressions of the macroeconomic norms' interrelationships of UPOES

№	Impacts of economic growth rate F, balance of payments norm b and exchange rates ε on the macroeconomic indicators	Impacts of general index of structural effectiveness ψ, balance of payments norm b and exchange rates ε on the macroeconomic indicators
1	$i = g = \breve{g} = \sqrt{F}$	$i = g = \breve{g} = \left(\frac{1}{\psi} - \varepsilon b\right)/2$
2	$\theta = \left(\sqrt{F} + \varepsilon b\right)/\left(1 - \sqrt{F} - \varepsilon b\right)$	$\theta = (1 + \psi\varepsilon b)/(\psi(2 + \varepsilon b) - 1)$
3	$c = 1 - 2\sqrt{F} - \varepsilon b$	$c = 1 - 1/\psi$
4	—	$F = \left[\frac{1}{\psi} - \varepsilon b\right]^2/4$
5	$\psi = 1/(2\sqrt{F} + \varepsilon b)$	—
6	$R = \sqrt{F}$	$R = \left(\frac{1}{\psi} - \varepsilon b\right)/2$
7	$\delta = F/\theta(F) =$ $F - \left(1 - \sqrt{F} - \varepsilon b\right)/\left(\sqrt{F} + \varepsilon b\right)$	$\delta = \frac{F(\psi)}{\theta(\psi)} = \left[\frac{1}{\psi} - \varepsilon b\right]^2$ $\times (\psi(2 + \varepsilon b) - 1)/4(1 + \psi\varepsilon b)$

Table 4. Interval values of macroeconomic indicators

Years	Y		C		G		m	
	Lower	Upper	Lower	Upper	Lower	Upper	Lower	Upper
2005	11145.03	13774.75	4588.90	6065.79	1905.22	2526.03	0.90	0.95
2006	16684.12	20620.82	6051.20	7998.71	3373.19	4472.32	0.85	0.90
2007	25240.85	31196.55	8241.34	10893.72	5416.72	7181.72	0.82	0.87
2008	35722.11	44150.92	11675.40	15433.00	9589.04	12713.56	0.79	0.82
2009	31685.34	39161.65	13243.14	17505.30	9348.47	12394.60	0.79	0.82
2010	37793.85	46711.50	14542.05	19222.25	10471.65	13883.76	0.79	0.82
2011	46352.98	57290.20	16882.35	22315.75	13703.78	18169.05	0.78	0.81
2012	48721.89	60218.07	18798.96	24849.20	15500.69	20551.47	0.77	0.80
2013	51781.98	64000.20	21210.60	28037.00	17037.72	22589.33	0.77	0.80
2014	52522.55	64915.51	23329.05	30837.25	16651.01	22076.62	0.77	0.80
2015	48398.20	59818.00	26820.36	35452.20	15828.21	20985.71	1.03	1.08
2016	53778.43	66467.72	30863.25	40796.25	15798.66	20946.53	1.75	1.84
2017	62420.24	77148.61	35239.35	46580.75	15653.68	20754.31	1.69	1.77
2018	71019.60	87777.03	36922.80	48806.00	20219.82	26808.30	1.67	1.75

Years	I		B		T	
	Lower	Upper	Lower	Lower	Lower	Upper
2005	4774.26	6130.59	1829.13	1829.13	1905.22	2526.03
2006	5247.97	6738.87	3443.23	3443.23	3373.19	4472.32
2007	5961.82	7655.52	5345.87	5345.87	5416.72	7181.72
2008	7984.77	10253.17	9578.80	9578.80	9589.04	12713.56
2009	6475.66	8315.33	9190.05	9190.05	9348.47	12394.60
2010	8549.38	10978.18	10148.67	10148.67	10471.65	13883.76
2011	11243.23	14437.33	13973.62	13973.62	13703.78	18169.05
2012	13497.88	17332.51	15380.54	15380.54	15500.69	20551.47
2013	15727.45	20195.47	17351.71	17351.71	17037.72	22589.33
2014	15501.90	19905.85	16376.53	16376.53	16651.01	22076.62
2015	14042.16	18031.41	15573.22	15573.22	15828.21	20985.71
2016	13114.64	16840.39	15580.07	15580.07	15798.66	20946.53
2017	13684.70	17572.40	14637.74	14637.74	15653.68	20754.31
2018	15169.62	19479.17	19945.97	19945.97	20219.82	26808.30

Our paper emphasizes the study of the macro indicators' relationships of an unbalanced perfect open economy system (UPAES) within interval mathematics based on Samuelson model. Thus, $i = I/Y$ – investing norm, $g = G/Y$ – public expenditures norm and $\breve{g} = \Delta Y/G$ – public expenditures effectiveness.

For the unbalanced perfect open economy system (UPOES), with the balance of payments is not zero, are displayed the following interconnections of macroeconomic indicators. Inferring mathematical equations which given in Tables 1, 2 and 3 were implemented by us.

In perfect economies, consumption norm cannot be lower than economic growth rate, since consumption is the main cause of production.

3   Computational Experiments

The relationships between macroeconomic norms and system performance indicators of Azerbaijan Republic were estimated with interval mathematics using Delphi survey information (Table 4) based on Tables 1, 2 and 3 for 2005–2018.

Computational results are presented in Tables 5, 6 and 7:

Table 5. Impact of public consumption norm c on macroeconomic indicators

Years	c		$\theta = \frac{1-c+\varepsilon b}{1+c-\varepsilon b}$		$F = \frac{(1-c)^2}{4}$		$R = \frac{1-c-\varepsilon b}{2}$		$\psi = \frac{1}{1-c}$	
	L	U	L	U	L	L	L	U	L	U
2005	0.34	0.53	0.50	0.95	0.007	0.055	0.083	0.234	1.51	2.11
2006	0.30	0.46	0.70	1.38	0.002	0.043	0.040	0.207	1.42	1.86
2007	0.27	0.42	0.90	1.89	0.000	0.034	0.004	0.185	1.36	1.71
2008	0.27	0.42	0.91	1.93	0.000	0.033	-0.001	0.182	1.37	1.72
2009	0.34	0.53	0.57	1.16	0.001	0.039	0.025	0.197	1.52	2.14
2010	0.31	0.49	0.68	1.38	0.000	0.037	0.018	0.193	1.46	1.96
2011	0.30	0.46	0.71	1.42	0.001	0.041	0.033	0.203	1.42	1.87
2012	0.32	0.49	0.62	1.19	0.003	0.046	0.052	0.215	1.46	1.97
2013	0.34	0.52	0.55	1.06	0.003	0.047	0.056	0.217	1.50	2.09
2014	0.36	0.57	0.46	0.91	0.003	0.047	0.057	0.218	1.57	2.31
2015	0.45	0.71	0.24	0.53	0.006	0.052	0.075	0.228	1.83	3.41
2016	0.47	0.73	0.22	0.52	0.003	0.046	0.055	0.215	1.89	3.73
2017	0.46	0.72	0.25	0.57	0.002	0.044	0.044	0.209	1.86	3.58
2018	0.43	0.66	0.34	0.73	0.001	0.041	0.035	0.203	1.74	2.97

Note. In Tables 5, 6 and 7 symbols L and U mean lower and upper bounds of the interval variables, respectively.

Table 6. Impact of average tax burden θ on macroeconomic indicators

Years	θ		$c = \frac{1-\theta}{1+\theta} + \varepsilon b$		$F = \left[\frac{\theta}{1+\theta} - \varepsilon b\right]^2$		$R = \frac{\theta}{1+\theta} - \varepsilon b$		$\psi = 2\theta - \varepsilon b(1-\theta)$	
	L	U	L	U	L	L	L	U	L	U
2005	0.13	0.22	0.83	1.07	0.000	0.040	−0.200	0.001	−53.09	4.20
2006	0.17	0.28	0.85	1.17	0.003	0.107	−0.327	−0.050	−26.85	3.67
2007	0.17	0.29	0.92	1.28	0.014	0.195	−0.441	−0.119	−9.64	4.12
2008	0.22	0.36	0.84	1.23	0.005	0.181	−0.425	−0.071	−57.34	2.79
2009	0.23	0.39	0.70	1.04	−0.013	0.062	−0.249	0.053	1.99	9.29
2010	0.22	0.36	0.77	1.12	0.000	0.099	−0.314	−0.001	2.28	21.35
2011	0.24	0.41	0.72	1.08	−0.009	0.088	−0.297	0.030	1.95	10.63
2012	0.26	0.43	0.66	1.00	−0.019	0.050	−0.224	0.085	1.78	6.77
2013	0.27	0.45	0.61	0.94	−0.022	0.032	−0.179	0.125	1.64	5.26
2014	0.25	0.42	0.61	0.92	−0.019	0.020	−0.141	0.135	1.73	5.34
2015	0.26	0.43	0.49	0.73	0.001	0.065	0.039	0.254	1.54	3.46
2016	0.23	0.39	0.54	0.78	0.000	0.047	0.011	0.217	1.71	4.00
2017	0.19	0.32	0.64	0.87	−0.007	0.021	−0.047	0.145	2.16	5.86
2018	0.23	0.38	0.62	0.90	−0.014	0.020	−0.102	0.140	1.88	5.56

Table 7. Impact of economic growth rate F on macroeconomic indicators

Years	F		$c = 1 - 2\sqrt{F} - \varepsilon b$		$\theta = \frac{\sqrt{F} + \varepsilon b}{1 - \sqrt{F} - \varepsilon b}$		$R = \sqrt{F}$		$\psi = \frac{1}{2\sqrt{F} + \varepsilon b}$	
	L	U	L	U	L	L	U	L	U	L
2005	0.38	0.58	−0.83	−0.43	−15.41	5.63	0.62	0.76	0.66	0.81
2006	0.40	0.61	−1.03	−0.56	−5.14	16.11	0.63	0.78	0.64	0.79
2007	0.41	0.63	−1.17	−0.65	−21.96	−2.71	0.64	0.80	0.63	0.78
2008	0.34	0.51	−1.02	−0.53	−4.32	25.09	0.58	0.72	0.70	0.86
2009	0.09	0.14	−0.17	0.13	1.30	3.79	0.30	0.37	1.34	1.65
2010	0.16	0.24	−0.45	−0.09	2.26	25.17	0.39	0.49	1.02	1.27
2011	0.18	0.28	−0.53	−0.15	2.63	14.04	0.43	0.53	0.95	1.17
2012	0.04	0.06	0.09	0.34	0.84	1.90	0.20	0.25	1.99	2.46
2013	0.05	0.08	0.08	0.32	0.84	1.81	0.23	0.28	1.79	2.22
2014	0.01	0.02	0.42	0.58	0.45	0.82	0.11	0.13	3.76	4.65
2015	0.06	0.10	0.23	0.41	0.52	0.83	0.25	0.31	1.60	1.98
2016	0.09	0.14	0.10	0.30	0.67	1.12	0.30	0.37	1.35	1.67
2017	0.13	0.20	−0.08	0.16	0.93	1.75	0.36	0.45	1.12	1.39
2018	0.11	0.17	−0.09	0.16	1.01	2.12	0.33	0.41	1.21	1.50

We can mention that based on the meadpoints of the interval values given in Table 5, calculated on the consumption norm of macroindicators, dependencies among them are $F < R < c < \theta < \psi$ appropriately in 2005–2014.

It can be mentioned that based on the meadpoints of the interval values given in Table 6, calculated on the average tax burden of macroeconomic indicators, dependencies among them are $F < R < \theta < c < \psi$, $\psi < R < F < \theta < c$, and $R < F < \theta < c < \psi$ appropriately in 2015–2018, in 2005–2014, in 2009–2014.

According to Table 6, calculated on the economic growth rate of macroindicators, it can be mentioned the following statements are true: the relationships are $F < c < R < \theta < \psi$ in 2012, 2014–2016, $\theta < c < F < R < \psi$ in 2005 in 2006, $c < F < R < \psi < \theta$ in 2008–2011, $\theta < c < F < \psi < R$ in 2007, $c < F < R < \theta < \psi$ in 2013, $F < R < c < \psi < \theta$ and $F < c < R < \psi < \theta$, respectively in 2017 and 2018.

4   Conclusion

M-file was created on purpose the evaluation of relationships between macroindicators and effectiveness of BOE by using interval analysis in MathWorks MATLAB Software 2017a environment. The results of calculations are given in Tables 5, 6 and 7.

Let us denote through $\nearrow$, $\searrow$ and $\otimes$ respectively, the increasing and decreasing of values, the assertion "It is impossible to determine the one-sided dynamics of change". Using the signs, dynamics of changes of macroeconomic indicators, can be presented as a Table 8.

Table 8. The dynamics of changes of macroeconomic indicators

	$\bar{c}$	$\bar{\theta}$	$\bar{F}$	$\bar{R}$	$\bar{\psi}$
$\bar{c}\nearrow$	$-$	$\searrow$	$\nearrow$	$\nearrow$	$\nearrow$
$\bar{\theta}\nearrow$	$\searrow$	$-$	$\searrow$	$\nearrow$	$\nearrow$
$\bar{F}\nearrow$	$\searrow$	$\otimes$	$-$	$\nearrow$	$\searrow$
$R\nearrow$	$\nearrow$	$\nearrow$	$\nearrow$	$-$	$\otimes$
$\psi\nearrow$	$\nearrow$	$\searrow$	$\otimes$	$\otimes$	$-$

The results of the calculations can be summarized as follows:

according to calculations in respect of c, the relationships were: $\theta < c$ in 2015–2018, $c < \theta$ in 2005–2014 and $F < R < c$ during the study period;

according to calculations in respect of θ, the relationships were: $F < R$ in 2015–2018, the values ψ was almost stabile during the entire research period in 2005–2014, with the exception of 2005–2006;

according to calculations in respect of ψ during the study period were $F < \theta, c < \psi, F < \psi$.;

according to calculations in respect of F, the relationships were $c < F < R$ in 2005–2012, $F < R < c$ in 2013–2015, $c < F < R$ in 2015–2018.

References

1. Vladimirov, S.A.: On the macroeconomic essence of the strategic development objectives of effective, balanced macroeconomic systems. Sci. Works North-West Inst. Manag. Branch RANEPA 6(4(21)), 105–116 (2015)
2. Musayev, A.F., Davudova, Rİ., Musayeva, A.A.: Application of interval analysis in evaluation of macroeconomic impacts of taxes. In: Aliev, R., Kacprzyk, J., Pedrycz, W., Jamshidi, M., Sadikoglu, F. (eds.) 13th International Conference on Theory and Application of Fuzzy Systems and Soft Computing - ICAFS-2018. Advances in Intelligent Systems and Computing, vol. 896, pp. 627–634. Springer, Cham (2019)
3. Musayev, A.F., Davudova, R.I., Musayeva, A.A.: Evaluation of interrelations of macroeconomic indicators by using interval analysis. Mag. ANAS "News" Econ. Ser. 4, 5–18 (2018)

4. Samuelson, P., Nordhaus, W.: Economics, 19th edn. Irwin/McGraw-Hill, Boston (2009)
5. Moore, R.: Interval Analysis. Prentice-Hall, Upper Saddle River (1996)
6. Moore, R., Kearfott, R., Cloud, M.: Introduction to Interval Analysis. SIAM, Philadelphia (2009)
7. Mayer, G.: Interval Analysis and Automatic Result Verification. De Gruyter, Berlin (2017)
8. Nguyen, H.T., Kreinovich, V., Wu, B., Xiang, G.: Computing Statistics under Interval and Fuzzy Uncertainty. Springer, Heidelberg (2012)
9. Alefeld, G., Herzberger, J.: Introduction to Interval Computations. Academic Press, New York (1983)

Deep Learning Based Analysis in Oncological Studies: Colorectal Cancer Staging

Abubaker Faraj Khumsi[1], Khaled Almezhghwi[2] iD,
and Khaled Adweb[2](✉)

[1] College of Electronics Technology - Tripoli, Tripoli, Libya
Rainb6755@gmail.com
[2] Near East University, Mersin-10, North Cyprus, Turkey
khaldalmezghwi84@gmail.com, khaledadwep@gmail.com

Abstract. The introduction of deep learning in the medical sector for the diagnosis and classification of tissue images has proven to have the best accuracy level over the years. Hence in this study, the utilization of deep learning in colorectal cancer staging as well as the use of AlexNet Architecture in the training of datasets were used in the prediction and detection of colorectal cancer. Moreover, the novelty of the technique used in this study involved a direct analysis of the dataset. Colorectal cancer was analyzed using deep learning technique and AlexNet architecture (CNN) from a dataset of 7180 from 50 different patients having colorectal cancer, which eventually produced 900 images. The results revealed that deep learning was analyzed on several classes: adipose tissue, back tissue, debris tissue, lymphocytes tissues, mucus tissue, smooth muscle, normal tissue, cancer associated stroma and adenocarcinoma epithelium, which had an accuracy of 97.8% for all the classes, (1, 1, 0.952, 0.952, 1, 1, 0.952, and 0.952) respectively, other classes had a precision of 1 except adenocarcinoma epithelium with 0.95, (1, 1, 0.975, 0.975, 1, 1, 0.975, and 0.95) respectively for the F1-score. Hence the result revealed that deep-learning technique for colorectal cancer is very effect.

Keywords: Deep learning · AlexNet architecture · Convolutional neural network (CNN) · Colorectal cancer · Debris tissue

1 Introduction

Colorectal cancer (CRC), which is also referred to as bower or colon cancer has been recorded as at 2012 to have affected an approximate amount of 1.4 million individuals with a high occurrence in men than women. The mortality rate of colorectal cancer is ranked as the number four, with an estimated death of 700 thousand per annual [1]. In the year 2010, about 32 billion dollars were estimated for the medical and non-medical cost of CRC [2]. Moreover, the introduction of colonoscopy devices has really been beneficial and effective in the early diagnosis and screening of CRC. This has improved the chances of survival rate of patient's with CRC, hence detecting the presence of the tumor growth before the occurrence of symptoms [3].

© Springer Nature Switzerland AG 2020
R. A. Aliev et al. (Eds.): ICSCCW 2019, AISC 1095, pp. 573–579, 2020.
https://doi.org/10.1007/978-3-030-35249-3_73

The occurrence and prevalence of colorectal cancer was not common in the year 1950. Over the years this form of carcinogen has become rampant in western regions with a rough estimation of about 10% recorded mortality rate. The reason for the increase is as a result of poor diet, smoking, aging as well as poor and bad lifestyle that involves the lack of physical activities that can prevent obesity. Despite the effort put in place by medical practitioners and scientists to eliminate this deadly disease through surgery and other forms of invasive therapeutics, little or no progress has been observed over the years. The use of screening techniques has been really beneficial over the years in the diagnosis of colorectal cancer. Moreover, it has been reported that the utilization of screening for tumor will create a great impact in colorectal cancer in about 15 years to come [4].

The microscopic imaging of CRC can be very effective in describing the architecture, composition as well as the complexity of colon tumor tissues. A general assessment of the tumor characteristics as well as other parameters around the tumor such as tumor necrosis, tumor budding and glandular formation helps to give information about the tumor tissue [4–6]. The use of deep learning and other forms of machine learning algorithms has been very useful and effective in the preprocessing of images through the treatment of several datasets. Most of the visual features utilize textures, colors and shape to gain a perfect and precise recognition of a specific image [7, 8]. Biswas et al. (2016) utilized a cross wavelet transform to classify colon polyps, sigmoid colon tumor as well as Crohn's disease with a total of 1185 images. The results obtained revealed the accuracy to be 98.46% for the diseases and 98.83% for the sigmoid cancer [9].

However, this study was focused on the use of deep learning in the staging of 7180 image patches derived from 50 patience with colorectal cancer (Fig. 1). Alex net architecture was also used in this study to train the tumor data samples, which revealed high level of accuracy.

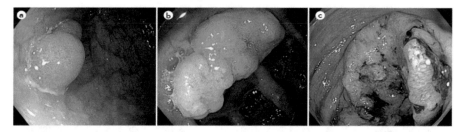

Fig. 1. Different stages of colorectal cancer. (a) Small sessile adenoma. (b) Large adenoma sessile. (c) Advanced sessile adenoma.

2   Methodology

This section explained the image preprocessing, the Alex network and different measured metric deployed to check the performance of the learner. For result evaluation of the colorectal image classification, different quantitative evaluation like accuracy, precision, recall are computed.

2.1   Dataset

The 900 images used for this research was extracted from a set of 7180 image patches which was derived from 50 patients with colorectal adenocarcinoma [10] (Fig. 2).

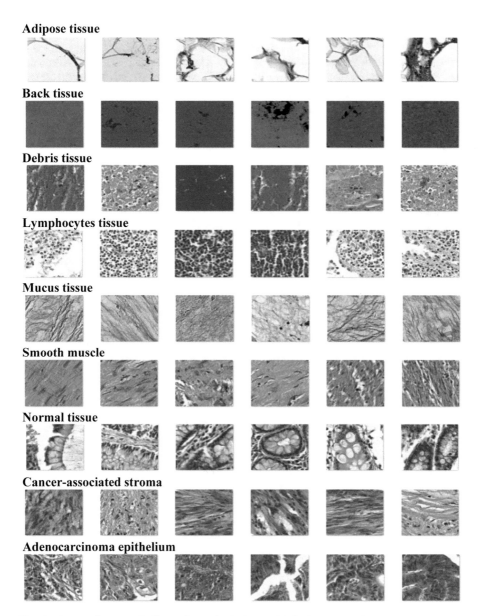

Fig. 2. Dataset images of adipose tissue, back tissue, debris tissue, lymphocytes tissues, mucus tissue, smooth muscle, normal tissue, cancer associated stroma and adenocarcinoma epithelium.

2.2   Preprocessing

All the images are color-normalized using Macenko method [11]. The normalized images are further resize from 224 by 224 to 227 by 227 which is the input size of the Alex network.

2.3   AlexNet Architecture

This neural network architecture was created by Alex Krizhevsky, Geoffrey Hilton and Ilya Sutskever who won the image classification challenge (ILSVRC) in 2012. The network architecture consists of five convolutional layers some of which are followed by maximum pooling layers and then three fully-connected layers and finally a 1000-way softmax classifier (Fig. 3).

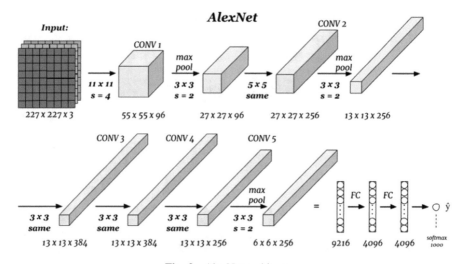

Fig. 3. AlexNet architecture

In this paper we made use of transferred learning from the Alex network. AlexNet consist of 25 layers but we only modified the 23rd and 25th layers. This reduced the training time and avoid the use of large dataset.

2.4   Performance Parameters

$$Accuracy = \frac{True\ Positive + True\ Negative}{Total} \tag{1}$$

$$Precision = \frac{True\ Positive}{True\ Positive + False\ Positive} \tag{2}$$

$$\text{Recall} = \frac{\text{True Positive}}{\text{True Positive} + \text{False Negative}} \tag{3}$$

$$\text{F1 Score} = 2 * \frac{\text{Precision} * \text{Recall}}{\text{Precision} + \text{Recall}} \tag{4}$$

3   Results and Discussion

The result from this study revealed several image readings which include: To evaluate the performance of different architectures, 80% of images of each class are randomly chosen for training and the remaining 30% for test. The initial learning rate chosen is 0.001 and the maximum epoch is 20 (Figs. 4 and 5). Several classes were evaluated in this study which include: adipose tissue, back tissue, debris tissue, lymphocytes tissues, mucus tissue, smooth muscle, normal tissue, cancer associated stroma and adenocarcinoma epithelium, which had an accuracy of 97.8% for all the classes, (1, 1, 0.952, 0.952, 1, 1, 0.952, and 0.952) respectively, other classes had a precision of 1 except adenocarcinoma epithelium with 0.95, (1, 1, 0.975, 0.975, 1, 1, 0.975, and 0.95) respectively for the F1-score (see Table 1).

Table 1. Model evaluation of dataset

SN	Class	Accuracy (%)	Precision	Recall	F1 score
1	Adipose tissue	97.8	1	1	1
2	Back tissue	97.8	1	1	1
3	Debris tissue	97.8	0.952	1	0.975
4	Lymphocytes tissue	97.8	0.952	1	0.975
5	Mucus tissue	97.8	1	1	1
6	Smooth muscle	97.8	1	1	1
7	Normal tissue	97.8	0.952	1	0.975
8	Cancer associated stroma	97.8	1	0.85	0.919
9	Adenocarcinoma epithelium	97.8	0.95	0.95	0.95

```
|==============================================================================|
|  Epoch   |  Iteration  | Time Elapsed | Mini-batch | Mini-batch | Base Learning|
|          |             | (seconds)    | Loss       | Accuracy   | Rate         |
|==============================================================================|
|       1 |         1 |      386.93 |    2.3976 |     12.50% |     0.0010 |
|       5 |        50 |    10204.90 |    0.0315 |    100.00% |     0.0010 |
|      10 |       100 |    41650.14 |    0.0092 |    100.00% |     0.0010 |
|      14 |       150 |   129801.46 |    0.0013 |    100.00% |     0.0010 |
|      19 |       200 |   139615.42 |    0.0118 |    100.00% |     0.0010 |
|      20 |       220 |   167159.67 |    0.0015 |    100.00% |     0.0010 |
|==============================================================================|
```

Fig. 4. Iteration on training dataset

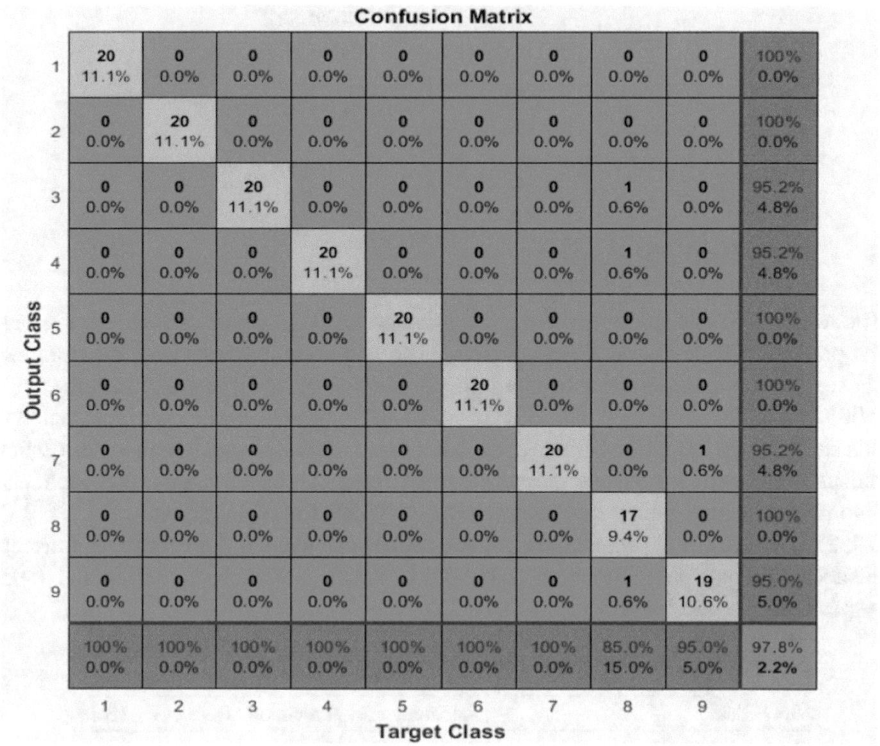

Confusion Matrix

Fig. 5. Confusion matrix

4 Conclusion

In this study, we presented deep learning approach to detect colorectal cancer from histological image of colorectal cancer and healthy tissue. The evaluation was conducted on 180 histological colorectal cancer images of 9 different classes of colorectal cancer. The experiment was carried out on a single CPU and a total of 97.8% accuracy was derived. Oncology, which has been a major focal point in the medical research sector has received a high level of attention in research over the years. The introduction of computer aided technology in the field of medicine has opened doors of possibilities towards providing medical diagnosis and therapy. The use of computer aided devices such as machine language in the detection and diagnosis of medical cases is a major step in the breakthrough of the next generation of science and computer. Despite the current limitations and setbacks observed in these techniques, it has proven to be more efficient and accurate than the previous techniques used in medical diagnosis. Predictions has reported that in the next century machine language will not only help for medical diagnosis, but with the incorporation of nanotechnology combined with artificial intelligence, several diagnosis and therapeutic functions can be performed with less invasive measures. Each class gives a prediction of 1 except debris, lymphocytes, normal and adenocarcinoma which precision of 0.952. The result for the confusion matrix in this study consist of both an output class and a target class. A better result can be achieve if the training dataset is increased but on a GPU to speed the training process. Moreover, the use of AlexNet architecture was very effective in training the dataset in this study and also functioned as an effective classifier. Furthermore, different forms of classifiers can be used in the diagnosis of cancer for future studies, such as the use of VGG and other technique combinations that can improve the quality of imaging.

However, giving to the fact that this line of study has been in trend for a long period of time and several upgrades and development are constantly been impacted in the computer aided diagnosis, there is a great possibility that deep learning as well as other computer aided machine learning diagnostic technique will take over the medical sector soonest.

References

1. Ferlay, J., Soerjomataram, I., Dikshit, R.: Cancer incidence and mortality worldwide: sources, methods and major patterns. In: GLOBOCAN International (2012)
2. Bloom, D., Cafiero, E., Jané-Llopis, E.: The global economic burden of non-communicable diseases. World Economic Forum, Geneva (2011). http://www.weforum.org/reports/global-economicburdennon-communicable-diseases
3. Wiegering, A., Ackermann, S., Riegel, J.: Improved survival of patients with colon cancer detected by screening colonoscopy. Int. J. Colorectal Dis. 31(4), 1039–1045 (2016)
4. Djemal, K., Cocquerez, J., Precioso, F.: Visual feature extraction and description. In: Benois-Pineau, J., Precioso, F., Cord, M. (eds.) Visual Indexing and Retrieval, 1st edn, pp. 5–20. Springer, New York (2012)
5. Igelnik, B.: Computational Modeling and Simulation of Intellect: Current State and Future Perspectives, 1st edn. IGI Global, Hershey (2011)
6. Biswas, M., Bhattacharya, A., Dey, D.: Classification of various colon diseases in Colonoscopy video using Cross-Wavelet features. In: 2016 International Conference on Wireless Communications, Signal Processing and Networking (WiSPNET), Chennai, pp. 2141–2145 (2016)
7. Balkwill, F.R., Capasso, M., Hagemann, T.: The tumor microenvironment at a glance. J. Cell Sci. 125, 5591–5596 (2012)
8. Schneider, N., Langner, C.: Prognostic stratification of colorectal cancer patients: current perspectives. Cancer Manag. Res. 6(1), 291–300 (2014)
9. Lanza, G., Messerini, L., Gafà, R., Risio, M.: Colorectal tumors: the histology report. Dig. Liver Dis. 43(2), 344–355 (2011)
10. Kather, N., Halama, N., Marx, A.: 100,000 histological images of human colorectal cancer and healthy tissue. https://zenodo.org/record/1214456#.XQSWlIhKjIX. Accessed 10 June 2019
11. Macenko, M., Niethammer, M., Marron, J., Borland, D., Woosley, J., Guan, X., Thomas, N.: A method for normalizing histology slides for quantitative analysis. In: 2009 IEEE International Symposium on Biomedical Imaging: From Nano to Macro, pp. 4–28. Springer, New York (2009)

Sea Turtle Species Classification for Environmental Research and Conservation

Zafer Attal$^{(\boxtimes)}$ ⓘ and Cem Direkoglu ⓘ

Center for Sustainability, Department of Electrical and Electronics Engineering,
Middle East Technical University - Northern Cyprus Campus, Kalkanli,
Guzelyurt, North Cyprus via Mersin 10, Turkey
zafer.attal@gmail.com, cemdir@metu.edu.tr

Abstract. We present two different computer vision algorithms for automatic sea turtle species classification. This study effectively contributes to environmental research, and particularly sea turtle conservation projects since sea turtles are endangered species. We classify two different sea turtle species belong to Mediterranean region, namely loggerhead and green sea turtles. The first method we experiment is the Bag of Feature (BoF) method that is integrated with Support Vector Machine (SVM) for classification. The BoF method mainly extracts color, texture and shape information of the sea turtle. The second method we employ is the Convolutional Neural Networks (CNN) that is a very effective algorithm in classification tasks. The dataset used in this work contains more than a thousand of images for each class. Results show that CNN performs maximum 70.14%, while the BoF method performs 69%. Extensive evaluation with respect to different parameter settings is presented in this paper.

Keywords: Computer vision · Feature extraction · CNN · Sea turtles

1 Introduction

There is a need for automatic classification of sea turtle species particularly in conservation projects since sea turtles are endangered animals nowadays. Recently, an artificial neural network has been used to recognize 8 different species of sea turtles in [1]. They employed traditional back propagation neural network (PBNN) algorithm as the base of their system. Also, in [7], they used artificial neural networks for pattern recognition where it has been applied on different species of fish, plant and butterfly. In this paper, we focus on sea turtles in Mediterranean region. There are two types of sea turtle species in this region: Loggerhead and Green sea turtles. We experiment two different computer vision algorithms for this purpose: Bag of Features (BoF) with Linear SVM, and Convolutional Neural Networks (CNN). Each method is explained in detail together with extensive evaluation conducted on a sea turtle dataset. These methods are the state of art methods in image classification. Each of them is trained and tested with different parameter settings to obtain the best results. We also perform the time evaluation for training and testing.

Our work is different from other published works in this domain [1, 7] since this is the first time the CNN and BoF methods have been applied for sea turtle species

© Springer Nature Switzerland AG 2020
R. A. Aliev et al. (Eds.): ICSCCW 2019, AISC 1095, pp. 580–587, 2020.
https://doi.org/10.1007/978-3-030-35249-3_74

classification. Particularly we have studied different CNN architectures, and find the optimal architectures that achieve the best results. It is very important to study sea turtle species, in environmental research and conservation projects, because sea turtles are endangered animals nowadays. In this work, we show that the state-of-art image classification (CNN and BoF) methods can achieve around 70% accuracy in this domain. We need intelligent surveillance systems to detect and classify sea turtle species where the work studied in this paper concerns about the classification problem. This will help to understand the presence and dynamics of sea turtle species in Mediterranean region.

2 Bag of Features (BoF) with Linear SVM

The algorithm goes through different stages to end up with a classification that differentiate between the two species of sea turtles. The algorithm mainly consists of four stages [3, 6]. Starting with the feature extraction phase, then the codebook creation and encoding of the features, after that the feature pooling phase, and finally the learning and classification phase. All of these phases are illustrated in Fig. 1.

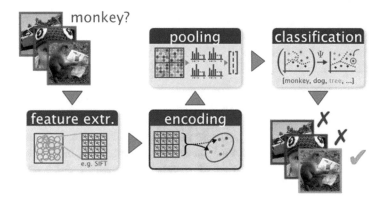

Fig. 1. BoF pipeline [8].

2.1 Feature Extraction Phase

One of the early phases of the BoF algorithm is extracting the features from the given images. Each single image is splintered to small pieces of regions where each region contains a collection of features in it. For turtles the most dominant features [11] are the beak, the carapace, the front flippers, the shell, the scutes and the rear. The feature extractor detects the interest points and forms the SIFT descriptor [3, 4]. As it is shown in Fig. 2, the given image has got multiple features that are being extracted from it. The grid step used for feature extraction in this program (8x8) and the block width is (32 64 96 128). The program filters the collected features to take the strongest 80% of them and annotate them to the tested image where the bounding box is specifying the region of interest.

Fig. 2. Turtle image before and after features extraction.

2.2   Codebook Creation and Encoding of the Features

After having the features of each image extracted, each one of them will have a code that identifies it from other images. The number of features that have been extracted from each image is very big since each interest point has its own feature as it has invariance in either scale, rotation or illumination, so this phase compresses all these features in multiple of visual vocabulary by using clustering methods such as the K-means [5] clustering to minimize sum of squared Euclidean distances between points x_i and their nearest cluster centers m_k. The mathematical expression of the K-means clustering is given below.

$$D(x,m) = \sum_{cluster\ k} \sum_{point\ i\ in\ cluster\ k} (x_i - m_k) \tag{1}$$

To improve the clustering and increase its efficiency, the number of features across all image categories is balanced before starting the process. After having many cluster centers, each one of them becomes a code vector that contains the nearest features to the center of the cluster where each vector is called word. In the end, a collection of the code vectors which represents the feature vector of the image is written in the codebook (called by visual vocabulary) in a process called encoding. The codebook is used for quantizing the features where the vector quantizer takes a feature vector and maps it to the index of the nearest visual word in the visual vocabulary (codebook).

2.3   Feature Pooling Phase

For this phase, the codewords are used to identify the images according to the distribution of the frequencies of occurrence for each codeword in the image and store it like a histogram representation as it is shown in Fig. 3(a), where these codewords are pieces of the original image. After extension to include all the images in the training set, more codewords will appear which will make the distribution of these frequencies more accurate in the corresponding histogram since for example the codewords that are shown in Fig. 3(a) will have higher frequencies compare to other codewords as they are not close to the image of the turtle given in Fig. 2. As it is shown, Fig. 3(b) shows this distribution and the relation of these different codewords.

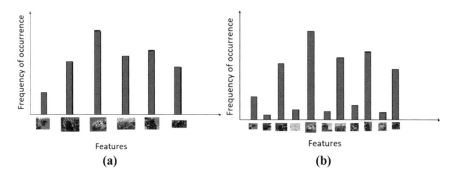

Fig. 3. (a) Frequencies of the codewords. (b) Extension of the frequency's distribution

2.4   Learning and Classification Phase

In the last phase, the training set will be used for training the classifier. At this stage, the bag of features is ready, and the training phase comes where the classifier is being trained by which features refer to which class according to the collected data from bag of feature stage which is the last three phases. Then to ensure the accuracy of the trained classifier, evaluation of the classifier is important which shows in percentage how many times did the classifier miss the correct answer and how many did it hit by using the test set. The program already knows the class of each image in the test set but not the classifier and according to the result from the classifier, the accuracy is cal-culated. While evaluating the classifier, the feature extraction phase and encoding the cluster k point i in cluster k features is going to be applied on a given image from the test set then frequency representation of the codewords will be built for the image. The histogram that will be built is going to be compared with the other histograms in the codebook to find the closest histogram where the class of the closest one is the class of the given image. The method to find the difference between the histograms is by using Euclidean distance (Eq. 2).

$$d = \sqrt{\sum |X_i - Y_i|^2} \tag{2}$$

The next step is using a SVM for classification [2, 9]. In SVM, linear kernel function (i.e. dot product) is used to map training data into kernel space. The SVM increases the gap between the positive and negative values of a given set of values by generating a linear function that separates the classes.

3   Convolutional Neural Networks (CNN) Classifier

The other method that has been tested is CNN that is the state of art method in image classification. To implement this algorithm, we constructed the default format of the layers in which it consists of input layer where all the images should have the same size to use the algorithm. Also, the size should be small, otherwise it will not work which can be counted as a disadvantage for CNN algorithm, since resizing the images will cause them to lose a lot of features and details. The original size of the images is large in comparison to the resized ones, and that explains why BoF takes a lot of time in training the classifier in comparison to the CNN as the input images are not resized for BoF algorithm. Next layer is the convolutional layer, where it is used to extract some specific features from the given images by using multiple filter layers. Also, whenever there is a convolutional layer there will batch normalization layer and RELU (Linear rectifier Unit) layer where it is used to increase the non-linearity of the system. Then the previous three layers are repeated as much as needed to increase the number of extracted features in each phase which will influence the result of classification. Max pooling layer is used next to reduce the size of the representation which will reduce the computation in the network. Fully connected layer is the next layer that is used as a representation of the flattening process of all the previous layers. The final layer is the softmax and classification layer which they are used to find the class with the highest probability and specify it as the output result of the network [10]. Figure 4 shows the layers of the CNN.

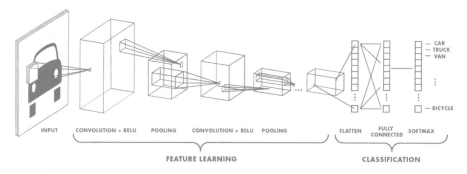

Fig. 4. CNN layers [10].

4   Classification Results

To test the accuracy of BoF + SVM and CNN classifiers, we change the number of images provided in the training set and as expected, after increasing the amount in the training set the accuracy increases, but there was not sudden change in the accuracy. Table 1 shows the confusion matrix of BoF + SVM for training set consists of 30% of the whole dataset. The average accuracy is 66% for the confusion matrix shown in Table 1. The average accuracy has been calculated by finding the mean of the accuracies (diagonal numbers on confusion matrix) where the predicted matches the known in the confusion matrix which represents the correct classification or hit (Eq. 3).

$$Accuracy = \frac{(Green,\ Green) + (Loggerhead,\ Loggerhead)}{2} \qquad (3)$$

Where $(Green, Green)$ represents Green classified as Green, and $(Loggerhead, Loggerhead)$ represents Loggerhead classified as Loggerhead. Table 2 also shows the average accuracy with respect to different percent of training set.

Table 1. Accuracy of the classifier using Linear SVM

Accuracy according to known and predicted data		Predicted	
		Green	Loggerhead
Known	Green	69%	31%
	Loggerhead	38%	62%

Table 2. Average accuracy for different percent of training set

Training set/dataset	Number of images	Average accuracy
30%	499	66%
50%	830	67%
70%	1164	69%
90%	1496	69%

Regarding CNN classifier, the parameters that have been changed are number of conv layers, number of epochs, size of filters and Mini Batch size. Table 3 shows the result of using different number of convolution layers with the specified number of filters for each convolution layers, where the first conv layer will have 8 filter layers while the second one will have 16 filters if there is and so on with fixed variable such as the size of the filters is 3×3, number of Epochs is 10 and minimum Batch size equal to 128 which is the default. From the results we can conclude that the accuracy is proportional to the number of used convolutional layers. However, increasing the number of Epochs while fixing all other parameters with 5 conv layers seems does not have a significant effect on the accuracy as shown in Table 4 while changing the filter size to 5×5 filter size, indeed increased the overall accuracy to reach 70.14%.

Table 3. Accuracy of using different number of conv layers

# of conv layers	# of filters per conv layer	Accuracy
1	8	59.95%
2	8,16	62.73%
3	8,16,32	64.35%
4	8,16,32,64	66.44%
5	8,16,32,64,128	67.13%

Table 4. Accuracy of using different number of Epochs for different filter size

# of Epochs	Accuracy with 3 × 3 filter	Accuracy with 5 × 5 filter
10	67.13%	68.00%
15	68.52%	65.65%
20	68.52%	70.14%

5   Conclusion

We have experimented two different computer vision algorithms for sea turtle species classification in Mediterranean region. The CNN classifier needs fixed size of input images where it is preferred to be small in size to make the computations possible and as a result, the required time for training is approximately 10 min on our computer. The accuracy of CNN based classification reaches up to 70.14% after experimenting with different parameter settings. On the other hand, the Bag of Features (BoF) method can deal with different size of images to get more of features. Training of BoF+SVM method can reach up to 30 min. The BoF+SVM method can achieve up to 69% accuracy. It is also noteworthy to mention that the background makes the classification more challenging, and there is a need for elimination of features of the background.

Acknowledgements. This work is sponsored by Cyprus Wildlife Research Institute (CWRI). Our project team would like to acknowledge CWRI for supporting us to buy the hardware components required in the project. We also would like to thank Chantal Kohl and Stefanie Kramer from Humboldt University for supplying us with some sea turtle images and videos for experimentation in our project. This project has been implemented on a server which belongs to Hossam Ahmed and Mohamed Badawy, we would like to thank them as well.

References

1. Adnan, K., Oyebade, O., Fahreddin, S.: Intelligent recognition of chelonioidea sea turtles. Proc. Comput. Sci. **102**, 617–623 (2016)
2. Burges, C.: A tutorial on support vector machines for pattern recognition. Data Mining Knowl. Discov. **2**, 121–167 (1998)
3. Hiba, C., Hamid, Z., Omar, A.: Bag of features model using the new approaches: a comprehensive study. Int. J. Adv. Comput. Sci. Appl. **7**(1), 226–234 (2016)
4. Lowe, D.: Distinctive image features from scale-invariant key points (2004). https://www.cs.ubc.ca/~lowe/papers/ijcv04.pdf
5. Duda, R., Hart, P., Stork, D.: Pattern Classification, 2nd edn. Wiley, New York (2001)
6. Csurka, G., Dance, C.R., Fan, L., Willamowski, J., Bray, C.: Visual categorization with bags of keypoints (2004). https://www.cs.cmu.edu/~efros/courses/LBMV07/Papers/csurka-eccv-04.pdf
7. Hernández-Serna, A., Jiménez-Segura, L.F.: Automatic identification of species with neural networks. Peer J. **2**, e563 (2014). https://doi.org/10.7717/peerj.563
8. Chatfield, K., Lempitsky, V., Vedaldi, A., Zisserman, A.: The devil is in the details: an evaluation of recent feature encoding methods (2011). http://www.bmva.org/bmvc/2011/proceedings/paper76/paper76.pdf

9. Cristianini, N., Shawe-Taylor, J.: An Introduction to Support Vector Machines: And Other Kernel-Based Learning Methods. Cambridge University Press, New York (2000)
10. Prabhu and Prabhu: Understanding of Convolutional Neural Network (CNN) - Deep Learning. Medium (2018). https://medium.com/@RaghavPrabhu/understanding-of-convolu tional-neural-network-cnn-deep-learning-99760835f148
11. Sea Turtles, SeaWorld Parks & Entertainment. https://seaworld.org/animals/facts/reptiles/ sea-turtles/

Simulation of Chemical-Technological Complexes

N. R. Yusupbekov$^{(\boxtimes)}$ ⓘ, Sh. M. Gulyamov ⓘ, A. N. Yusupbekov ⓘ, and N. A. Kabulov ⓘ

Tashkent State Technical University, Tashkent, Uzbekistan
dodabek@mail.ru, gulyamov4l@mail.ru,
abekhimprom@mail.ru, kabulov69@mail.ru

Abstract. A conceptual model of complex chemical-technological systems consisting of interacting technological and storage nodes with a continuous production pattern is proposed. An algorithm has been developed to simulate the functioning of a chemical-technological hub, which makes it possible to predict the changing characteristics of its units, taking into account perturbing influences.

Keywords: Conceptual model of the object · Simulation modeling · Technological units and hubs · Technological and storage nodes

1 Introduction

At present, the field of application of computers in the management of chemical enterprises has expanded and deepened, in particular, it has become possible to predict the results of work of individual departments and technological hubs under various conditions of their operation, using simulation methods [1, 2]. The specifics of the continuous technology chemical complex are: the continuity of the supply of raw materials and intermediate products to the process units; a large number of technological and storage nodes (measured in tens); the interconnectedness of technological and cumulative nodes material flows.

2 Conceptual Model of Sophisticated Chemical-Technological Complex Without Recycles

The main purpose of operational dispatching production management is the implementation of shift production tasks. Assume that the considered technological complex consist of m technological units $\Pi_i(i = 1, ..., M)$ and N storage nodes $C_j(j = 1, ..., N)$. At each moment of time, the technological node n_i will be characterized by its current productivity $g_i(t)$ and maximum possible productivity g_i^{max}. At the same time, the value of the latter depends on the state of raw materials available to the dispatcher, as well as the state of the equipment and the process regulation. The state of the cumulative node C_j at the current time t will be characterized by the value of its fill level $S_j(t)$.

R. A. Aliev et al. (Eds.): ICSCCW 2019, AISC 1095, pp. 588–595, 2020.
https://doi.org/10.1007/978-3-030-35249-3_75

Technological and storage nodes are viewed as dynamic links that convert input signals into output ones. Control actions are applied to the input of the process units and represent new load values, when the transition to new loads occurs gradually, over a certain time. We assume that the technological units that produce or consume several types of products, produce or consume them in a certain ratio that persists over the entire time period under consideration. With this assumption, the structural links between technological and storage nodes can be specified as a matrix

$$D = (dij) \ (i = 1, \ldots, M; \ j = 1, \ldots, N). \tag{1}$$

The d_{ij} ratio will be positive if the output of the node is the input of the C_j drive (i.e., the product is produced by the Π_i node), and d_{ij} is negative when the output of the C_j drive is the input of the technological node Π_i (that is, this product is consumed by the Π_i node). In the absence of a direct connection between the nodes Π_i and C_j, the corresponding element $d_{ij} = 0$. The values of nonzero entries in the row of the matrix D reflect the existing relationship between the products of the input or output streams of the corresponding technological node.

The current value of the unbalance $B_j(t)$ between product arrival at the consumption point with the entered values is given by the following equation:

$$B_j(t) = \sum_{i=1}^{M} d_{ij} g_i(t). \tag{2}$$

The activity of any enterprise is always planned. The program that determines the number of products is issued to the dispatcher in the form of planned tasks A_i ($i = 1, \ldots, M$) for each technological node for the interval of the dispatching shift $[D, T]$. In the current time t_1 to examine the progress of the planned tasks can be $G_i(t_1)$ the magnitude of the operating time of technological nodes.

$$G_i(t_1) = \int_0^{t_1} g_i(t) dt. \tag{3}$$

The production process is subject to perturbations. In case of perturbations associated with the displacement of the optimum of the production process or in the presence of a quickly entry of reserves to replace failed equipment, the management of the complex means maintaining the previously specified loads in the process units, that is, the control actions interpreted as a change in previously established loads. In case of disturbances, accompanied by a decrease in the value of the maximum possible production of g_i^{max} to a certain value g_i^B (lack of raw materials, equipment failures, which require some time t_i^B to restore), it becomes necessary to carry out control actions. This need may be associated with the output level of the drive interacting with the perturbed technological node to one of its borders S_j^{max} (overflow) or S_j^{min} (emptying) as a result of an increase in the unbalance caused by the disturbance between the receipt and consumption of the product of this drive. In this case, the control action is carried out in order to reduce the imbalance of the overflowing or emptying drive to "zero" and is to reduce the loads of the corresponding technological units associated with the drives in question.

Thus, the interconnectedness of the nodes of the complex can lead to a situation where a change in the program of operation of one technological node requires the restructuring of the modes of other nodes connected with it. After recovery time t_i^B, all technological units can either be returned to the operating modes specified before the disturbance occurs, that is, there are no control actions in the above-mentioned sense, or they can be brought to new values of the loads that compensate for the losses. In this case, it is necessary to calculate the necessary control actions for this purpose.

It should be noted that the absence of control actions at any time interval $[t_0, t_1]$ means maintaining the established load values of technological nodes $g_i(t_0)$ during this entire time, except for periods of restoration or forced work at a lower load associated with maintaining accumulation levels within acceptable limits. We will consider only those disturbances that entail a decrease in the value of the maximum possible production of technological units. Other types of disturbances will not be considered in the model, since they do not require the introduction of control actions in the sense we adopted above. So, the failures of the technological node Π_i can be characterized by the parameters: $\tau_i, g_{i,k_i}^B, t_{i,k_i}^B (k_i = 1, \ldots, K_i)$, where τ_i is the moment of failure; g_{i,k_i}^B - the value limiting the value of the maximum possible generation during the recovery time t_{i,k_i}^B. The k_i index indicates that the technological node may be subjected to K_i types of disturbances associated with failures of various types of equipment. The trouble-free operation of the technological node Π_i on the interval of the dispatching shift $[D, T]$ corresponds to the value $\tau_i = T$. High requirements for the reliability of chemical units involved in the continuous process, as well as redundancy of equipment, make it possible to substantiate the assumption that at the considered time interval in each technological node no more than one failure is possible out of K_i types of failures of this node. The results of practical studies of the reliability $G_i(t_1)$ of chemical equipment [3, 4] also speak in favor of accepting this hypothesis. Nevertheless, the assumption about the possibility of the occurrence of an arbitrary number of perturbations on the interval $[D, T]$ for each of the technological nodes seems more natural. However, frequent failures do not allow the dispatcher to effectively manage the course of a continuous technological process, since in this case he/she would be constantly busy eliminating the consequences of disturbances. In such situations, the problem of solving any other control problems is meaningless. With rare failures, the occurrence of several disturbances in a row in a single node within a short time seems unlikely. Therefore, the results of calculations based on the assumption of the possibility for each technological node on the dispatcher shift interval of no more than one failure should differ little from the results of the same calculations obtained taking into account an arbitrary number of failures.

Thus, on the interval dispatching shift $[D, T]$ in the technological complex is implemented a certain set of failures of technological nodes, which can be set using a set of x characteristics of these failures

$$x = \left\{ \tau_1, g^B_{1,k_1}, t^B_{1,k_1}, \ldots, \tau_M, g^B_{M,k_M}, t^B_{M,k_M} \right\}. \tag{4}$$

Let use to X denote the set of x characteristics, describing the possible set of failures of technological nodes during one dispatch shift. When modeling dispatcher actions, we proceed from the need to maximize the values of $G_i(T)$, provided that all previously accepted assumptions about the implementation of control actions are met. This can be illustrated by the example of reducing the load on the process node when the drive of the product consumed by this node is empty. In the described case, the consumer load of the limiting product is reduced to the level of production of this product by its supplier. This is equal to reducing the imbalance in the limiting product to «zero».

The solution of the same issue is more complicated in the presence of the empty drive of several consumers. Reducing the amount of workings of a group of techno-logical units can be achieved in several ways, in particular, by reducing the production of one of the nodes of the group, reducing the production of several or all technological nodes of the group in a certain proportion. The choice of a particular method and the determination of the proportions are based on information about the process imple-mented by a particular technological complex. At the same time, there is some decisive rule (or set of decisive rules possibly changing in time), which allow the dispatcher to manage the chemical-technological complex when the accumulators reach the emer-gency limits (S^{min}_j or S^{max}_j) [6, 7].

In accordance with the previously adopted designations, we will characterize the state of the technological node with its production values $g(t)$ and operating time from the beginning of the shift $G(t)$, and the state of the storage tank with the values of its filling level and unbalance between the input and output streams $B(t)$. The change in the level $S_j(t)$ of the stocks of the cumulative node C_j over the time interval $[t_0, T]$ is described by the relation

$$S_j(t_1) = S_j(t_0) + \int_{t_0}^{t_1} B_j(t)dt, \tag{5}$$

where $B_j(t)$ is the amount of imbalance between the input and output flow of the cumulative node.

The change over time of the technological node output values is determined by the control actions, disturbance characteristics, and the state of the drives directly con-nected by material flows to the given technological node. In accordance with the approach proposed in [5, 6], the change in the workings $g_i(t)$ of the process node Π_i during transient processes will be modeled using a piecewise linear function of the form

$$\bar{g}_i(t) = \begin{cases} g^0_i & at\ \ t_0 \leq t \leq t_0 + B_i, \\ g^0_i + (t - B_i) & at\ \ t_0 + B_i < t_0 + t^g_i, \\ g^1_i & at\ \ \ \ \ \ t \geq t_0 + t^g_i, \end{cases} \tag{6}$$

where t_0 – is the load change moment; g_i^0 – is the output value before the beginning of the transition process; g_i^1 – the value of output established after the end of the transition process; B_i - pure lag time; t_i^g is the duration of the transition process, i.e.

$$t_i^g = B_i + \frac{|g_i^1 - g_i^0|}{P_i^g}, \tag{7}$$

where P_i^g - the rate of change of the output value during the transition process.

The values of B_i and P_i^g are determined on the basis of two conditions. The first of these is the equality of the amount of production time during the transition process, calculated using the function $\bar{g}_i(t)$, to the real quantity of products produced during this period, i.e.

$$\int_{t_0}^{t_0 + t_i^g} \bar{g}_i(t)dt = \int_{t_0}^{t_0 + t_i^g} g_i(t)dt. \tag{8}$$

The second condition is the simultaneous achievement of the functions $\bar{g}_i(t)$ and $g_i(t)$ established after the end of the transition process of the output value g_i^1, i.e.

$$t_i^g = g_i^{-1}\left(g_i^1\right). \tag{9}$$

As a result of solving the approximation problem (6)–(9), the following expressions were obtained for the values B_i and P_i^g:

$$B_i = \frac{\left(g_i^0 + g_i^1\right)g^{-1}\left(g_i^1\right) - 2\int_{t_0}^{t_0 + t_i^g} g_i(t)dt}{g_1 - g_0}, \tag{10}$$

$$P_i^g = \frac{|g_i^1 - g_i^0|}{g_i^{-1}\left(g_i^1\right) - B}. \tag{11}$$

The values of the function g_i required for determining the approximation parameters B_i and P_i^g are proposed in [6] to calculate the load change rate P_i^R by the dispatcher using the weight function presented as a weight function-development.

The identification of the weight functions $W_i(t)$, based on the use of data from the normal functioning of the object, is carried out according to the algorithms given in [7]. At the end of the transient process, the technological node retains the steady-state value of output over the entire remaining interval, except for the duration of the perturbation. In this case, perturbations can occur both in this technological node and be transmitted to it from other technological nodes through empty or overflowing accumulative nodes directly connected to this node. During the action of the disturbances, the technological node retains the greatest possible output in the current situation. The value of the technological node is calculated by the formula (3). The unbalance between the input and output for the cumulative node is calculated by the formula (2). The time variation of the fill level of the accumulator C_j is expressed in terms of the calculated unbalance

values according to expression (1). To reproduce (simulate) the changing characteristics of the state of a model on a computer, it is necessary to proceed to finite-
difference equations describing the functioning of the model elements in discrete time.
In this case, an assumption is made about the constancy of the values of the workings
of technological nodes during each step of the simulation. Due to this, the dependences
between the characteristics of the states of the elements of the considered continuous
model are represented as fairly simple finite-difference equations.

Let be $g_i(t), G_i(t), (i = 1, 2, \ldots, M)$ и $S_j(t), B_j(t), (j = 1, \ldots, N)$ – current values of
the characteristics of the state of the elements of the model. Then their values at the
next moment $(t + \Delta t)$ of model time can be determined from the following first order
finite difference equations. The change in the production of g_i in the absence of any
effects is:

$$g_i(t + \Delta t) = g_i(t). \tag{12}$$

The change in the production of g_i in the transition process due to the control action

$$g_i(t + \Delta t) = \begin{cases} g_i(t) + P_i^g \Delta t, & \text{if } (t + \Delta t) > t_0 + B_i, \\ g_i(t), & \text{if } (t + \Delta t) \leq t_0 + B_i, \end{cases} \tag{13}$$

where B_i and P_i^g – are the characteristics of the transition process calculated by the
formulas (10) and (11) t_0 – is the initial moment of the control action.

Change of production time G_i technological node:

$$G_i(t + \Delta t) = G_i(t) + \frac{g_i(t + \Delta t) + g_i(t)}{2} \Delta t. \tag{14}$$

The change in the magnitude of the unbalance B_j. In accordance with the expression (2)
we have

$$B_j(t + \Delta t) = \sum_{i=1}^{M} d\, i\, j\, g_i(t + \Delta t). \tag{15}$$

S_j level change is:

$$S_j(t + \Delta t) = S_j(t) + \frac{B_j(t) + B_j(t + \Delta t)}{2} \Delta t. \tag{16}$$

Let the initial values of the characteristics of the state of the technological complex and
the characteristics of technological units expected on the interval of the dispatcher shift
be set $(\tau_1, g_1^B, t_1^B, \ldots, \tau_M, g_M^B, t_M^B)$, where τ_i – moments of failure, g_i^B – the maximum
possible value of generation during the time of the disturbance t_M^B.

The algorithm for calculating the characteristics of the state of the technological
complex includes the following main blocks:

Block 1. Calculation of balance values for drives by the formula (15). The input
parameters are the current values of the workings $g_i(t), (i = 1, \ldots, M)$, the result of the
block operation is the magnitudes of the offsets $B_j(t), (j = 1, \ldots, N)$.

Block 2. Determination of the magnitude of the control actions to maintain the levels of drives in the permissible limits

$$t_j^\Gamma = \frac{S_j^{ext} - S_j(t)}{B_j(t)}, \quad (j = 1, \ldots, N), \tag{17}$$

where

$$S_j^{ext} = \begin{cases} S_j^{max}, & \text{if } B_j(t) > 0, \\ S_j^{min}, & \text{if } B_j(t) < 0, \end{cases}$$

t_j^Γ – the time at which the drive level reaches one of its permissible limits.

The control actions on technological units consist in reducing the workings of suppliers of overflowing storage devices and reducing the workings of consumers of emptying storage devices. The magnitudes of these effects are determined from the condition

$$B_j\left(t + t_j^\Gamma\right) = 0. \tag{18}$$

The transition time t^g is determined by its approximation parameter from expression (7). If a $\left(t + t_j^\Gamma - t_j^g\right) > T$, then you do not need to make controls. Block input parameters are values $S_j(t), S_j^{max}, S_j^{min}, B_j(t)$, output - g_i^Γ, those. production to which it is necessary to move technological nodes, t_i^g – transition time.

Block 3. Calculation of the characteristics of transients during control actions. This calculation is carried out in accordance with expressions (10) and (11). Input parameters are values $g_i(t), g_i^y$, output - approximation parameters P_i^g, θ.

Block 4. Determining the state of the technological node. The technological node may be in one of the following states: no effects; the beginning of the transition process and the transition process itself.

The state of «the beginning of the transition process» is identified in cases where: $t + \Delta t = \tau_i$ – the moment of the start of the action of the perturbation (at this moment the transition process from producing $g_i(t)$ to produce g_i^B begins, at this value g_i^y assigned value g_i^B); $t + \Delta t = \tau_i + t_i^B$ – the moment of termination of the restrictions caused by the perturbation (the transitional process from producing g_i^B to produce $g_i(t_0)$, at this value g_i^y assigned value $g_i(t_0)$); $t + \Delta t = t_i^\Gamma$ – the moment when the control action begins when the drive is overfilled or empty, while the g_i^y assigned to g_i^Γ.

The state of «transition» is identified over time t_i^g from the beginning of the transition process. In all other cases, the «no impact» situation is identified. Block input parameters are values $\tau_i, t_i^B, t_i^\Gamma$, output parameters are k- identified situation number and value g_i^y, t_i^g - for the situation «the beginning of the transition process».

Block 5. Calculation of changes in the values of workings technological nodes. In this block, the values of workings $g_i(t)$ new values are calculated according to the identified situation $g_i(t + \Delta t)$ for the next model time. In situations of "no impact" and «beginning of the transition process» values $g_i(t + \Delta t)$ calculated by the formula (12).

In a situation of «transition process» values $g_i(t + \Delta t)$ calculated by the formula (13). Block input parameters are k - situation number, i – technological node number and size $g_i(t), g_i^y, t_i^g$, output parameter is output value $g_i(t + \Delta t)$. In one step of the simulation algorithm, a transition occurs from the state of the technological complex at the moment t to its state at the moment $(t + \Delta t)$. The current values of balances and levels of filling drives determine the need to make control actions to maintain their levels within acceptable limits. If the impact is necessary, the amount of output g_i^Γ is calculated, time to start the transition process t_i^Γ and duration t_i^g (blocks 2 and 3).

Then, in the cycle for each technological node, its state is identified (block 4), the value of $g_i(t + \Delta t)$ (block 5) and calculation of the value $G_i(t + \Delta t)$ according to the formula (14). At the end of the cycle, balance values are calculated for all drives first. $B_j(t + \Delta t)$ (block 1) and then the level values $S_j(t + \Delta t)$ according to the formula (16). The accuracy of the calculations and the time spent on all the calculations depend on the size of the simulation step Δt. Therefore, the choice of this value should be carried out for each specific case.

3   Conclusion

The approach to solving the problem of compensating losses from disturbances of the chemical complex, which consists of calculating control actions based on the results of forecasting the consequences of the expected disturbances and realizing these effects before the occurrence of the disturbance, while the control object still has sufficient resources to ensure that possible losses do not lead to disrupting targets.

References

1. Zverev, V.D., Postelin, B.V., Fomin, B.F.: Conceptual production management. Electronic equipment, Moscow, vol. 2, no. 23, pp. 69–80 (1977). (in Russian)
2. Buslenko, V.I.: Automation simulation of complex systems. Science, Moscow (1977). (in Russian)
3. Kafarov, V.V., Meshalkin, V.P.: Analysis and Synthesis of Chemical technology system. Chemistry, Moscow (1993). (in Russian)
4. Minulina, A.R.: Development of a method for predicting and assessing the reliability of industrial vertically integrated systems. Abstract of dissertation, Moscow (2015). (in Russian)
5. Kabak, I.S.: Ensuring the reliability of complex software based on artificial neural networks. Abstract of dissertation, Moscow (2015). (in Russian)
6. Prangishvili, I.V., Ambarcumyan, A.A., Poletykin, A.G., Grebenjuk, G.G., Yadykin, I.B.: The state of the level of automation of energy facilities and circuitry solutions, management for its development. In: Automation Problems, Russia, vol. 2, pp. 11–26 (2013)
7. Yusupbekov, N.R., Gulyamov, SH.M., Temerbekova, B.M., Ataullaev, A.O.: Algorithm for assessing the stability and premises of the production process in technological technologies and complexes. In: Problems of Informatics and Energy, Tashkent, vol. 3, pp. 3–12 (2014). (in Russian)

Student Performance Classification Using Artificial Intelligence Techniques

Nevriye Yılmaz[1($\boxtimes$)] (ID) and Boran Sekeroglu[2] (ID)

[1] Department of Classroom Teaching, Near East University,
Nicosia, TRNC, Mersin 10, Turkey
nevriye.yilmaz@neu.edu.tr
[2] Information Systems Engineering, Near East University, Nicosia,
TRNC, Mersin 10, Turkey
boran.sekeroglu@neu.edu.tr

Abstract. Education has vital and increasing importance almost for all countries in order to accelerate their development. Well-educated persons provide more benefits to their countries and for that reason, classification of students' performance before they enter exams or taking courses is also gained an importance. Improvement of education quality must be performed during the active semester to improve students' personal performance to response this expectation. To provide this, some of the main indicators are students' personal information, educational preferences and family properties. In this paper, artificial intelligence techniques are applied to the questionnaire results that consists these main indicators, of three different courses of two faculties in order to classify students' final grade performances and to determine the most efficient machine learning algorithm for this task. Several experiments are performed and results suggests that Radial-Basis Function Neural Network can be used effectively for this and helps to classify student performance with accuracy of 70%–88%.

Keywords: Artificial Intelligence · Education · Radial-Basis Function NN

1 Introduction

Many cases we encountered and decisions taken in our lives are passing through complicated processes in human mind and brings some uncertainties. Common uncertainties are not randomly characterized but the information sources of solutions for the problems can include varieties. At that phase, Artificial Intelligence (AI) and related technologies are offered as ideal solutions. Artificial Intelligence can solve the problems such as complexity and uncertainty which are caused by several reasons and lack of information [1].

Machine Learning, which is a part of AI studies is used in every field of our lives. Effective decision making and classification is performing not only in engineering fields, as well as in sociological fields. Scientists can predict and classify sociological events by approximation of results instead of explaining by certain and template reasons [2, 3].

© Springer Nature Switzerland AG 2020
R. A. Aliev et al. (Eds.): ICSCCW 2019, AISC 1095, pp. 596–603, 2020.
https://doi.org/10.1007/978-3-030-35249-3_76

Similarly, principles of AI has been used in sociological questionnaire which considers fuzzy properties of language in order to obtain more realistic and reliable results. In this context, it can be mentioned that AI has robust relation between the fields of language, religion, sociology, education and law researches. In addition, it supports students and instructors to build strong decision making mechanisms in education field. It is obvious that there are several related factors that effects academic success directly or indirectly, thus there are different studies to examine academic success thoroughly and to determine the cognitive-sensual components that effects this success. Some of these researches are about control algorithms, medical tests, economy, environment, psychology and education fields. But, AI is implemented generally in engineering field [4–6].

In academic success assessment process, usually traditional evaluation systems are used. They represent objective and subjective assessment evaluation measures. Quantifiable properties are generally pointed by certain assets. The subjective assets like leadership, representation and problem solving are less quantifiable. For this reason, educational serve and assessment criteria are subjective as in other qualitative researches [7]. According to this, AI principles and techniques are important to perform classification, decision making, prediction and to assess various sensual components instead of traditional assessments and also to effectively and prevalently used in researches of education field.

Considering the several factors related to the academic success of persons, effective assessment gains more importance. There are several researches in the literature about the components that effects academic success and effects of several variables [8–12]. During the determination of academic success, written exams, tests and oral examinations are considered in cognitive success of students and scales are generally used for sensual components. Therefore, these methods that are not based on certain criteria and weights cause subjectivity during the evaluation process and thus false evaluations can be performed. AI and machine learning applications provide flexible and wide perspective for modeling any problem.

Different family information, personal characteristics and preferences cause complicated and non-linear correlation between the instances (questions) and the academic success of students. Therefore, humans are not able to perform analysis to determine success of student by observing the data.

In this research, student performance analysis according to the final grades of Atatürk Faculty of Education and Faculty of Engineering of Near East University student is performed using AI and machine learning algorithms.

The rest of the paper is organized as follows; Sect. 2 gives introduction about the data collection process and dataset characteristics, Sect. 3 briefly explains considered machine learning algorithms, Sect. 4 presents performed experiments, obtained results and discussions in details and finally Sect. 5 concludes the experimental results of this research.

2   Data Collection and Dataset Characteristics

Questionnaire that consists totally 30 questions is conducted to 101 students of Turkish program of 3 courses of 2 faculties at Near East University. These courses are 'Assessment and Evaluation in Education' of Faculty of Education, 'General Physics II' and 'Electronics' of Faculty of Engineering.

Questionnaire of 'Assessment and Evaluation in Education' course is conducted for 66 students, and 21 and 14 for 'General Physics II' and 'Electronics' courses respectively.

Questions are divided into 3 categories as Personal Questions, Family Questions and Educational Preferences. The aim was to provide separate analysis of the effect of different kinds of question types on the student performance. Personal Questions part consists 10 questions, Family Questions part includes 6 questions and Educational Preferences part consists 14 questions. Table 1 summarizes the questions considered in the questionnaire.

Table 1.   Summary of questionnaire.

Personal questions	Family questions	Educational questions
Age	Mothers' Education	Weekly study hours
Sex	Fathers' Education	Reading (non-scientific)
High School Type	Number of Brother/Sister	Reading (scientific)
Scholarship Type	Parents Relationship	Attendance to Seminar/Conference
Additional Job	Mothers' Job	Effect of Projects and Activities
Sports/Arts	Fathers' Job	Attendance to Lectures
Relationship		Study type I (Group/Alone)
Salary		Study type II (Regular/Last week)
Transportation		Taking notes
Accommodation		Writing/Listening
		Effect of in-class Discussions
		Effect of Flip Classroom
		GPA of Last semester
		Expected CGPA at graduation

3   Classification Methods

Several classification algorithms can be applied in order to perform classification tasks. However, characteristics of dataset reduce the number of algorithms that can be suitable for specific kind of application. This section will briefly introduce the considered machine learning algorithms. In this research, Backpropagation Neural Network (BPNN), Radial-Basis Function Neural Network (RBFNN), Decision Tree Classifier (DT) and Logistic Regression (LOGR) is considered because of the non-linearity of data and the number of attributes and instances.

BPNN is the most commonly used supervised classification and prediction algorithm [3]. It propagates back the error which is calculated by the difference between actual and target output and updates weights using Gradient-Descent algorithm in each iteration. In this research, three hidden layers are decided to be used in order to obtain optimal convergence after several experiments thus it becomes Deep BPNN. Activation function was Sigmoid Activation Function with maximum iteration number of 1000.

RBFNN is another kind of supervised learning method similar to BPNN but it consists only single hidden layer. It uses Radial Basis Functions to calculate output of hidden neurons instead of any activation function. It can be used both for classification and prediction problems [13]. After performing several experiments, learning rate is decided to be used as 0.09, maximum iterations set to 3000 and cluster number set to the total number of classes in the dataset.

DT uses divide and conquer strategy to classify attributes [14]. Starting node, internal nodes and leaf nodes represents starting point, attributes and class labels respectively. Main drawback of DT is determining the starting node or the sequence of internal nodes during the organization. For this reasons, generally entropy method or Gini index is considered to organize tree in optimal way. In this research, Gini index is considered for tree organization. It has been applied in several classification problems [15].

Logistic Regression is a basic statistical method that performs classification tasks [16]. It is similar to Linear Regression but for classification. It uses Logistic Function (Sigmoid) instead of linear function in order to solve non-linear problems.

4 Experiments and Results

This section presents performed experiments and obtained results in details. In this research, several experiments were conducted in order to obtain optimal classification results of students grades.

4.1 Experiments

Experiments divided into four categories for each course as Personal, Family, Educational and All Questions. Considered algorithms which are BPNN, RBFNN, LOGR and DT, are trained to classify eight student grades which are AA, BA, BB, CB, CC, DC, DD and FF in decreasing order respectively based on Near East University grading policy. Finally, all courses are combined and an experiment is performed in order to determine general accuracy.

Experiments are performed according to the Hold-out method which is based on dividing dataset into pre-determined percentage of training and test set. In this research, 70% and 30% of training and testing sets are considered respectively.

Evaluation are performed according to the Accuracy of each algorithm which is calculated by dividing correctly classified instances by total instances.

4.2   Results

In Course 1, RBFNN achieved highest accuracy rates than other algorithms for each questionnaire type and all questions. 84.44%, 93.33%, 73.33% and 80.00% of classification rates are obtained for Personal, Family, Educational and All Questions experiments. It is followed by BPNN and LOGR but DT was unable to make expected classifications in this experiment. Table 2 presents details of obtained results of Course 1 experiments.

Table 2. Accuracy results of Course 1 experiments

Experiment	RBFNN	BPNN	DT	LOGR
Personal	84.44%	75.00%	30.00%	60.00%
Family	93.33%	50.00%	30.00%	80.00%
Educational	73.33%	65.00%	45.00%	50.00%
All Questions	80.00%	65.00%	42.00%	60.00%

In Course 2, obtained rates for Personal and All questions are not high as in Course 1 but highest rates were achieved by RBFNN similar to Course 1. However, LOGR achieved 100% of accuracy rates in Family and All Questions experiments which are the optimum results in this experiment. Table 3 demonstrates details of obtained results of Course 2 experiments for each algorithm. Figure 1 shows the real-valued classification graph of BPNN for All Questions experiment of Course 1.

Table 3. Accuracy results of Course 2 experiments

Experiment	RBFNN	BPNN	DT	LOGR
Personal	77.70%	75.00%	50.00%	75.00%
Family	77.00%	50.00%	25.00%	100.00%
Educational	88.80%	77.00%	75.00%	100.00%
All Questions	77.33%	76.90%	75.00%	75.00%

Table 4. Accuracy results of Course 3 experiments

Experiment	RBFNN	BPNN	DT	LOGR
Personal	50.00%	16.66%	33.33%	16.66%
Family	64.28%	16.66%	33.33%	0.00%
Educational	78.57%	50.00%	33.33%	33.33%
All Questions	85.00%	33.33%	50.00%	16.66%

In Course 3, RBFNN produced highest classification rates when other algorithms are considered however, lowest classification rates are obtained within all courses Table 4 presents obtained results for Course 3 in details.

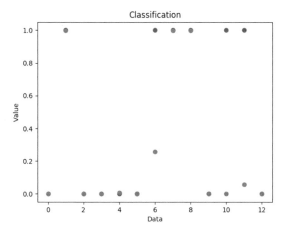

Fig. 1. BPNN real-valued classification (76.90%) graph for All Questions experiment of Course 2.

As a last experiment of this research, three courses are combined together in order to classify student performance without considering faculty or course information. Obtained results showed that neural based methods RBFNN and BPNN achieved superior results than DT and LOGR with 70% and 66.6% of classification rates respectively. Table 5 presents obtained results for combined courses considered in this research.

Table 5. Accuracy results of combined courses experiment

Experiment	RBFNN	BPNN	DT	LOGR
Combined	70.00%	66.66%	43.33%	53.33%

4.3   Discussions

Experimental results shows that neural based machine learning algorithms perform superior results than Decision Tree and Logistic Regression. However, it can be noticed that, Backpropagation produced unstable results in some experiments especially those consist minimum attributes. Increment of attributes cause more stable and increased accuracy in classification. Decision Tree produced lowest results almost in all experiments that is because of the complexity and non-linearity of data that complicates making relationship between attributes and classes. Logistic Regression produced unstable results which are sometimes optimal or lowest accuracy rates. Variety of questions and non-linearity of within sections may cause this unstable accuracy rates. Radial-Basis Function Neural Network generally achieved superior results than other algorithms with more stable and accurate results. This proves the positive effect of usage of the radial-basis functions in hidden neurons instead of considering traditional hidden neurons.

In the curriculum, the methods and techniques to be used in the teaching process should be oriented towards the application instead of traditional teaching. In addition, it is necessary that more experiments and practices are required, the importance of active participation of students and prerequisite courses. It is considered that the prerequisite for the courses is to express the course skills correctly and it will be beneficial for the effective and efficient passing of the teaching process.

In this context, when the effect of students' motivation on academic achievement is evaluated, it is recommended to organize activities aimed at the objectives of the course before starting the teaching process regarding the prejudices of the students towards these courses. During the learning and teaching process, methods and techniques (lecturing, question-answer, experiment, observation, problem solving, group study) should be applied according to the nature of the subject. As much as possible, students should be actively involved in laboratory studies. Moreover, the fact that the content does not contain abstract concepts and that the information can be used in daily life will help to learn more meaningfully and permanently. In this context, the process of learning and teaching should be made more meaningful. It is important to have the courses in the laboratories during the learning - teaching process or to include the learning environments outside the school depending on the subject covered.

Thus, it can be suggested that neural network based classification methods can effectively be used for students' performance classification before taking corresponding courses and provide more effective personal learning and teaching process.

5 Conclusions

Education and related performance evaluation is a challenging task for human beings. Therefore computer assisted systems are developed in order to improve this evaluation. One of the most important approaches in this performance evaluation is Artificial Intelligence which can classify or predict huge data. In this research, three courses of two faculties were considered for questionnaire which has three sections, in order to classify students' final grade performances. Several experiments were conducted and different machine learning algorithms were implemented.

Experimental results showed the classification of student performances according to any questionnaire type as Personal, Family or Educational Preferences, is possible with machine learning algorithms. Considering general results suggests the usage of Radial-Basis Function Neural Network in this task. It produced 70%–88% of classification rate. The other considered algorithms performed unstable results which may cause false classification for future classifications.

Radial Basis Function Neural Network can effectively be used to improve the quality of education by considering personal information of student and to guide instructors to prepare personalized education processes according to students characteristics.

Future work will include the increment of dataset instances, implementation of more machine learning algorithms especially deep learning. Also the main factors of students' failure will be determined using machine learning methods.

References

1. Uysal, M.P.: Modeling learning styles with fuzzy logic (in Turkish). In: 4th International Computer and Instructional Technologies Symposium, Konya, Turkey, pp. 1040–1045 (2010)
2. Öcal, Ö.: Student modeling with fuzzy logic approach in adaptive intelligent teaching system (in Turkish). Yüksek Lisans Tezi. Marmara Üniversitesi (2016)
3. Sekeroglu, B., Dimililer, K., Tuncal, K.: Student performance prediction and classification using machine learning algorithms. In: 8th International Conference on Educational and Information Technology, Cambridge, UK, pp. 7–11 (2019)
4. Sunter, Z., Altun, H., Sunter, S.: A new approach for harmonic elimination in single-pulse modulated single-phase inverter drive system. J. Fac. Eng. Archit. Gazi Univ. **30**(2), 237–247 (2015)
5. Dai, S., Li, L., Li, Z.: Modeling vehicle interactions via modified LSTM models for trajectory prediction. IEEE Access **7**, 38287–38296 (2019)
6. Onat, N.C., Gumus, S., Kucukvar, M., Tatari, O.: Application of the TOPSIS and intuitionistic fuzzy set approaches for ranking the life cycle sustainability performance of alternative vehicle Technologies. Sustain. Prod. Consumption **6**, 12–25 (2016)
7. Akandere, M., Özyalvaç, N.T., Duman, S.: Examination of secondary school students' attitudes towards physical education course and their academic success motivation. Konya Anatolian High School (in Turkish). Selçuk Univ. J. Soc. Sci. **24**, 1–10 (2010)
8. Cox, R.H.: Sport Psychology: Concepts and Applications, 2nd edn. McGraw-Hill Education, New York (1990)
9. Şen, Aİ., Koca, S.A.: Attitudes and reasons of secondary school students towards mathematics and science (in Turkish). Educ. Res. **18**, 236–252 (2005)
10. Chen, L.S., Cheng, C.H.: Selecting IS personnel use fuzzy GDSS based on metric distance method. Eur. J. Oper. Res. **160**(3), 803–820 (2005)
11. Dede, C.: Comparison of frameworks for 21st century skills. In: 21st Century Skills: Rethinking How Students Learn, vol. 20, pp. 51–76 (2010)
12. Demirtaş, Z.: The relationship between school culture and student achievement. Educ. Sci. **35**(158), 3–13 (2010)
13. Adnan, R., Samad, A.M., Tajjudin, M., Ruslan, F.A.: Modeling of flood water level prediction using improved RBFNN structure. In: 2015 IEEE International Conference on Control System, Computing and Engineering (ICCSCE), George Town, pp. 552–556 (2015)
14. Dougherty, G.: Pattern Recognition and Classification. Springer, Heidelberg (2013)
15. Yuan, Z., Wang, C.: An improved network traffic classification algorithm based on Hadoop decision tree. In: 2016 IEEE International Conference of Online Analysis and Computing Science (ICOACS), Chongqing, pp. 53–56 (2016)
16. Mason, C., Twomey, J., Wright, D., Whitman, L.: Predicting engineering student attrition risk using a probabilistic neural network and comparing results with a backpropagation neural network and logistic regression. Res. High. Educ. **59**(3), 382–400 (2018)

Fuzzy Set Theory Tools in Models
of Uncertainties in Economics

A. D. Chistyakov[1][(⊠)] and N. D. Eletsky[2]

[1] Don State Technical University, Gagarin Sq. 1, Rostov-on-Don, Russia
andrey.chist@gmail.com
[2] Southern University, Nagibin Av. 33A/47, Rostov-on-Don, Russia
nde527@yandex.ru

Abstract. The object of analysis is the applicability of the fuzzy sets theory tools for study the uncertainty of contemporary socio-economic structures. It's revealed the role of fuzzy set theory as formalizing instrument of macro- and microlevel interactions in uncertain economy conditions and decision-making taking into consideration the conflict of purposes. The article also illustrates the instrumental abilities of fuzzy set theory on the examples of the task adjusting the national budget's social component in conditions of modernization of the economy and the consumer choice optimization model. The budgetary opportunities of the consumer in this case will be reborn to "a budgetary surface". In its turn, the optimal consumer choice, based on the fuzzy set methodology, is formalized as a fuzzy set of options for the actual choice, taking into account fuzzy restrictions on income, time and information, and in relation to a fuzzy set of variable consumer sets of close integral utility. The equilibrium state of the market parameters reflected by the model acts as a "balance of uncertainties".

Keywords: Economy of uncertainties · Macro-and microfactors
of uncertainty · Fuzzy set theory · Formalization of uncertainties · Multifactorial
decision-making algorithms · Balance of uncertainties

1 Introduction

In the system of methodological axiomatics of approaches of limited rationality, "Homo economicus" from the subject of rational (in the meaning – abstract optimal) choice is increasingly transformed into the subject of real choice, making economic decisions in a plurality and uncertainty of restrictions and the influence of random factors, including – non-economic nature. Similarly, the firm as "point moving along the curves of supply and demand" and as the atomized agent maxing the profit in the spirit of the "Robinson Crusoe" methodology, becomes a complex social-economic Institute. Its mission and the "tree of goals", although conditioned by the market "genes" of profitability, are developing on the ground and in the atmosphere of a multicomponent system of modern mixed economy. It is characterized by the inter-action of the regularities of the free and regulated market, the attributes of the social state, the transition to the post-industrial information system, greening, globalization and other modern trends.

© Springer Nature Switzerland AG 2020
R. A. Aliev et al. (Eds.): ICSCCW 2019, AISC 1095, pp. 604–612, 2020.
https://doi.org/10.1007/978-3-030-35249-3_77

One of the consequences of the multiplicity of factors of interaction of these trends was the historical limit of models with unambiguous solutions. These models, of course, played a crucial role in the development of the methodology and theoretical body of economic science; on their basis it was possible to identify, explain and graphically interpret the key laws of the market mechanism and the behavior of economic agents. However, the further development of economic practice and theory required by the middle of the twentieth century more specific, and therefore more complex formalized, including mathematical, interpretations of economic processes. Among the prerequisites for the emergence of institutionalism, and then Keynesianism, as we know, dissatisfaction with the simplification of the used models occurred. That was initially reflected in the non-mathematical, verbal, "qualitative" nature of the analysis and presentation, and then – the synthesis of the ideas of these areas with the complicated mathematical heritage of Neoclassicism, with models of econometrics and mathematical Economics within the "mainstream". Neo- and post-Keynesianism, neo-institutionalism, cliometrics, consumer behavior theory, decision theory, economic applications of game theory and probability theory – these and other emerging trends of modern economic theory are increasingly inducing impulses of methodological and substantive synthesis, and to a large extent – on the basis of mathematical formalization. The complex and multiple character of interactions within the framework of modern economic processes, high dynamism of transformation of preconditions and external conditions, internal factors and emergent impulses of activity have generated the phenomenon of "uncertainty economy" [1–4], the essential specificity of which can be reflected by means of models using the tools of fuzzy set theory [5].

2   Problem Statement and Methods

The essential attribute of "uncertainty economy" is the need to make economic decisions on the basis of current incomplete, inaccurate and limited information, and taking into account the fact that the consequences of these decisions will be manifested in the process of constant changes in external and internal conditions of activity, so that the final effect may differ significantly from the original goals, and in some cases – contradict them. Even the formal achievement of the initial goals may lose its meaning and lead to a substantially opposite effect if the external economic and social environment has undergone substantial qualitative modifications during the period of time required to achieve the goal.

This requires continuous monitoring and adjustment of the goals of economic activity and the means to achieve them, which significantly distinguishes the modern economy not only from traditional economic systems with the constant reproduction of the same prerequisites and results of management, but also from the routine algorithms of the traditional market. The patterns of the "uncertainty economy" are manifested at the micro -, meso - and macro-levels, and nowadays – at the global level.

At present, along with the traditional problem of correlation of uncertainty and risk, there intensified discussions about the nature, structure and forms of economic uncertainties, and therefore discusses the relationship of such concepts as "uncertainty economy", "uncertainties in economics", "uncertain economics", "economy of uncertainties" and other related [6–8]. In our opinion, one of the directions of more complete and adequate formalization and modeling of economic phenomena and processes just in aspect of "uncertainties in economics" can be the involvement of tools of fuzzy set theory. These tools allow us to approach the generalized characteristics of the multidimensional and multilevel nature of socio-economic interactions in the world with an increasing degree of uncertainty [9, 10]. Let us note some approaches to the possible solution of this complex prospective problem.

In the economy of uncertainty in general, it is customary to distinguish three directions:

– Uncertainty of the type of probability – at which the values of the parameters of the studied systems X are matched with the probability $p(X)$ of obtaining (occurrence) of such parameter values. Probability distributions $P(X)$ is described by sample moments of the 1st, 2nd, 3rd... order, often called mathematical expectation, expected value, dispersion, variance, kurtosis ... Tools for describing uncertainty as "probability" were well developed in the late XIX-early XX centuries. In the economy, they are successfully used, for example, in the problems of investment risk assessment;

– The uncertainty of the type of incompleteness - in which it is taken into account that the efficiency F of the system activity, besides the controlled parameters $X = \{x_1, x_2, ...x_l, ...x_L\}$ and unmanageable, but adaptable parameters $Y = \{y_1, y_2, ... y_m, ... y_M\}$, is influenced by the unmanaged and unadaptable component $\xi(t)$, the value of which increases with the implementation time of management actions – $F = f(X) + \phi (Y) + \xi(t)$. The tools for describing the uncertainty of the type of incompleteness are in the process of formation. The beginnings of such tools can be considered the explorations of Mut, Lucas, Sargent [11];

– Uncertainty of the type of fuzzy - in which the parameters of the studied systems X, in addition to the values $(X = \{x_1, x_2, ...x_i, ...x_M\})$, determined by the degree of $\mu(x_i)$ belonging to the fuzzy set of i-s parameters. The tools for describing the uncertainty of the type of fuzzy are mainly developed in the second half of the twentieth century [12], but the use of these tools in the economy is fragmented, weakly affecting the development of economic science and especially practice.

However, the widespread introduction of methods of fuzzy set theory to the practice of economic entities will provide sustainable competitive advantages: by increasing the validity of decisions by formalizing procedures that traditionally use emergent strategies; by improving the accuracy (validity) of the solutions in comparison with traditional deterministic and stochastic statements. Herewith, methods of fuzzy set theory allow to approach optimal (or at least acceptable) solutions even in the face of conflicting goals, conflict of limitations, conflict of objectives and constraints.

3 Fuzzy Model of Social Components of the National Budget in the Context of Economic Modernization

As an example of increasing the validity of decisions by formalizing the tasks of decision-making in a conflict of goals that are traditionally used emergent strategy, let us consider the problem of adjusting the social component of the national budget under conditions of the modernization of economy.

In this task the "multi-attribute" formulation of the choice problem is realized with usage of linguistic variables. For this there are formulations: attributes of choice (goals and limitations); many alternatives to management decisions (quantitative, structural, structural-quantitative); bank of verbal → formal linguistic statements of the expert community on the impact of each alternative on attributes (the expert community in this case may include as specifically experts, as the statements in the media, in the blogosphere,… etc., approaching to the scopes of the electorate); obtaining a comprehensive assessment of each alternative based on the aggregation of linguistic information; ranking of alternatives based on the evaluation results. As attributes of the choice of alternatives, we consider a wide range of aspects (criteria) of managerial decision-making and restrictions on the adoption of these decisions.

In this example, the following attributes are used: the level of conformity of the alternative to the proclaimed mission; the potential for development under the elected alternative; compliance of the alternative to previous decisions; the impact of the adopted alternative on the stability of economic development; the impact of the adopted alternative on the satisfaction of the electorate; the impact on the transparency (expectations, predictability) of economic policy; the response time of the economic system (time lag) on the management alternative; the impact of the alternative on the overall growth rate of the economy; influence of the chosen management alternative on the exchange rate of the national currency; impact on inflation; impact on export potential; evaluation of the alternative from the standpoint of political correctness.

Alternatives to management decisions for multi-attribute selection can be formed: in the form of quantitative changes, for example, % increase in allocations, with a given step corresponding to small, medium or significant changes; in the form of structural changes, for example, allocation options; in the form of specified "step-by-step" combinations of structural and quantitative changes; in the form of other "rules" for producing a finite set of alternatives, for example, iterative search process.

Universal (for all attributes) term-set of linguistic estimates of alternatives we will accept:

$$T(X) = T_1(X) = \ldots T_i(X) = \{verygood, good, almostgood, satisfactory, almostbad,$$

$$bad, verybad\}$$

All verbal statements with the help of the "dictionary of inclusions and correspondences" are reduced to linguistic variables of the term-set **T(X)** and receive an appropriate assessment of the degree of correspondence $\mu(x)$ - Table 1.

Table 1. Linguistic variables of the term-set with degree of correspondence

Variable term-sets $T(X)$	$\mu(x)$ value
Very good	0,99
Good	0,833
Almost good	0,667
Satisfactory	0,5
Almost bad	0,33
Bad	0,167
Very bad	0,01

Formal presentation of management alternatives Al_1–Al_p assessments at At_1–At_n attributes based on the statements (opinions) of m experts involved is:

	At_1	At_2	...	At_n
Al_1	$r_{111},...r_{11j},...r_{11m}$	$r_{121,...,}r_{12j,...,}r_{12m}$	...	$r_{1n1,...,}r_{1nj,...,}r_{1nm}$
	$\omega_{111,...,}\omega_{11j,...,}\omega_{11m}$	$\omega_{121,...,}\omega_{12j,...,}\omega_{12m}$		$\omega_{1n1,...,}\omega_{1nj,...,}\omega_{1nm}$
Al_2	$r_{211,}...,r_{21j,}...,r_{21m}$	$r_{221,...,}r_{22j,...,}r_{22m}$	...	$r_{2n1,...,}r_{2nj,...,}r_{2nm}$
	$\omega_{211,...,}\omega_{21j,...,}\omega_{21m}$	$\omega_{221,...,}\omega_{22j,...,}\omega_{22m}$		$\omega_{2n1,...,}\omega_{2nj,...,}\omega_{2nm}$
...	...	...	...	...
Al_p	$r_{p11,}...,r_{p1j,}...,r_{p1m}$	$r_{p21,...,}r_{p2j,...,}r_{p2m}$		$r_{pn1,...,}r_{pnj,...,}r_{pnm}$
	$\omega_{p11,...,}\omega_{p1j,...,}\omega_{p1m}$	$\omega_{p21,...,}\omega_{p2j,...,}\omega_{p2m}$		$\omega_{pn1,...,}\omega_{pnj,...,}\omega_{pnm}$

where: r_{pnm} - linguistic evaluation of the m-th expert Advisor, the influence of the p-th alternative on the n-th attribute; ω_{pnm} - evaluation of the weight of the n-th attribute when choosing the p-th alternative according to the m-th expert.

The linguistic estimates r_{pnm} of the influence of alternatives on the attributes correspond to one of the values of the universal term-set $T(X)$

$$r_{pnm} \in T(X)\,\forall p,\, n,\, m$$

And, accordingly, can be represented (replaced) by the corresponding values of the degree of belonging $\mu(x)$

$$r_{pnm} \leftrightarrow \mu(x)|\mu(x) \in \{0.01,\ 0.167,\ 0.33,\ 0.5,\ 0.667,\ 0.833,\ 0.99\}.$$

As a complex evaluation of Al_1–Al_p alternatives, we'll use a convex combination of fuzzy estimates for At_1–At_n attributes of the m involved experts

$$\mu_{pn}(x) = \omega_{\widetilde{pn1}}\, \mu_{pn1}(x) + \omega_{\widetilde{pn2}}\, \mu_{pn2}(x) + \ldots + \omega_{\widetilde{pnm}}\, \mu_{pnm}(x)$$

$$\text{where: } \omega_{\widetilde{pni}} = \frac{\omega_{pni}}{\sum_{i=m}^{i=1} \omega_{pni}}, \rightarrow \sum_{i=m}^{i=1} \omega_{\widetilde{pni}} = 1.$$

The sought rank of an alternative management decision becomes

$$\mu_{\widetilde{p}}(x) = \omega_{\widetilde{p1}}{}^{\sim}\, \mu_{p1}(x) + \omega_{\widetilde{p2}}{}^{\sim}\, \mu_{pn2}(x) + \ldots + \omega_{\widetilde{pn}}{}^{\sim}\, \mu_{pn}(x)$$

$$\text{where: } \omega_{\widetilde{pi}}{}^{\sim} = \frac{\omega_{\widetilde{pi}}}{\sum_{i=m}^{i=1} \omega_{\widetilde{pi}}}, \rightarrow \sum_{i=m}^{i=1} \omega_{\widetilde{pi}}{}^{\sim} = 1.$$

It is obvious that management decisions taken with regard to heterogeneous, diverse attributes, based on the opinion of a wide expert community, taking into account the ranking of alternatives, are much more justified.

4 Modification of the Consumer Choice Model Under Fuzziness of Initial Parameters

As an example of using the provisions of the theory of fuzzy sets to improve the accuracy (validity) of the obtained solutions, we compare the solution of the classical (deterministic) consumer choice problem with the solution obtained based on the "fuzzy" formulation of this problem.

Classical formulation of the "consumer choice" problem implies the achievement of the optimal ("rational") use of the budget at the point of contact (tangent) of the "budget line" characterizing the possibilities (demand) of the consumer, and the "indifference curve" describing a set of equal in utility options of consumer choice (Fig. 1).

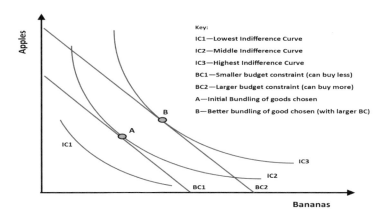

Fig. 1. Classical representation of the "consumer choice" model.

In the "fuzzy" formulation of the problem of consumer choice, the classic "budget line" will correspond to the middle of the zone of fuzzy transition from absolute, guaranteed belonging to the budget (with the indicator of belonging $\mu(x) = 1$) to the zone of complete non-compliance to the given budget (with the indicator of belonging $\mu(x) = 0$).

Assuming, for clarity, that the transition of the indicator of belonging to the budget $\mu(x)$ from 1 to 0 will be linear, the "law" of such a transition can be shown in Fig. 2.

The budgetary opportunities of the consumer in this case will be reborn to "a budgetary surface".

Since the remove from the curve of "indifference" entails a decrease in the index of belonging, the "law" of the change of belonging of this curve will be represented as a triangle, the vertex of which will correspond to the index of belonging $\mu(x) = 1$, and the indicators of belonging of the values x outside the base of the triangle will have the value of the indicator $\mu(x) = 0$.

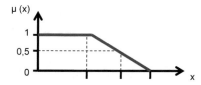

Fig. 2. The law of change of the index of belonging $\mu(x)$ for the budget of the consumer.

The intersection of the surfaces of budgetary possibilities and the surface of the fuzzy representation of the "indifference curve", corresponding to the intersection of fuzzy sets of coordinate values x, for which $\mu(x) \neq 0$, forms a new surface, covering the area of permissible values of consumer choice c "height" point above the base, characterizing the attractiveness of such a choice.

The most interesting solutions with the attractiveness indices $\mu(x)$ 25% higher than the classical solution at point K are located to the left and below this point, in the middle of the zone of acceptable consumer choice highlighted in Fig. 3. In addition, in case of fuzzy statement, we operate not with points, but with zones that allow significant freedom of decision-making. This also applies to other consumer choice constraints, which are generally not addressed in the traditional model. To an even greater extent, uncertainty is manifested in relation to the information parameters of consumer choice, especially taking into account the information features of the modern society of mass consumption.

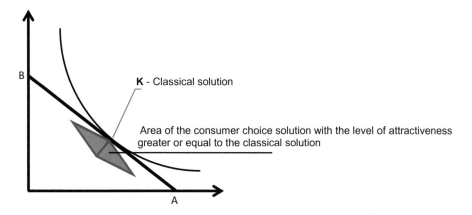

Fig. 3. Fuzzy solution for the "consumer choice" model.

It is characterized not by the absence or limitation of information about the goods, but by its overabundance, which is why the direct consumer choice of the buyer of goods and the corresponding market transaction is first preceded by the selection of necessary and sufficient information about these goods from the information and advertising flow, which includes many similar, as well as inaccurate and false signals and "white noise" [13]. This inevitably leads to a significant blurring of the boundaries of information restrictions.

5 Conclusion

Thus, the optimal consumer choice, based on the fuzzy set methodology, is formalized as a fuzzy set of options for the actual choice, taking into account fuzzy restrictions on income, time and information, and in relation to a fuzzy set of variable consumer sets of close integral utility. Herewith, the equilibrium state of the market parameters reflected by the model acts as a "balance of uncertainties". Algorithms for the analysis of the "balance of uncertainties" on the basis of the axioms and models of theory of fuzzy sets provide the tools of iterative approximation of the formalized abstractions of "ideal market" to the more complete and accurate theoretical reflection of the realities of multifactor, multilevel and multidimensional uncertain economic reality. Besides that, the instrumental importance of iterative models of fuzzy set theory increases qualita- tively as the transition from the analysis of the "equilibrium of uncertainties" to the modeling of "nonequilibrium of uncertainties". It is obvious that equilibrium is a special case (and in theoretical models – a simplified and limiting aspect) of the nonequilibrium state of the system; but nonequilibrium is a fundamental attribute of economic reality.

References

1. Knight, F.H.: Risk, Uncertainty, and Profit. Cosimo Classics, New York (2006)
2. Dequech, D.: Fundamental Uncertainty and Ambiguity. East. Econ. J. **26**(1), 41–60 (2000)
3. Kliesen, K.L.: Uncertainty and the economy. https://www.stlouisfed.org/publications/regional-economist/april-2013/uncertainty-and-the-economy. Accessed 28 July 2019
4. Pindyck, R.S.: Uncertainty in environmental economics. Rev. Environ. Econ. Policy **1**(1), 45–65 (2007)
5. Zadeh, L.A.: Fuzzy sets. Inf. Control **8**(3), 338–353 (1965)
6. Chappe, R.: Choice under uncertainty: a misnomer. https://www.ineteconomics.org/perspectives/blog/choice-under-uncertainty-a-misnomer. Accessed 28 July 2019
7. Beck, U.: Living in the world risk society. Econ. Soc. **35**(3), 329–345 (2006)
8. Moffat, M.: What uncertainty means in economics? The meaning of "uncertainty" in economics. https://www.thoughtco.com/meaning-of-uncertainty-in-economics-1147302. Accessed 28 July 2019
9. Hilsenrath, J.: The economic forecast for 2019: less growth and more uncertainty. The Wall Street Journal, 16 June 2019. https://www.wsj.com/articles/the-economic-forecast-for-2019-less-growth-and-more-uncertainty-1543892700. Accessed 16 June 2019
10. Zimmermann, H.-J.: Fuzzy set theory. Adv. Rev. **2**, 317–332 (2010)
11. Lucas, R.E., Sargent, T.J.: After Keynsian macroeconomics. Fed. Reserv. Bank Minneap. Q. Rev. **3**(2), 1–16 (1979)
12. Kaufmann, A.: Introduction to the Theory of Fuzzy Subsets. Fundamental Theoretical Elements, vol. 1. Academic Press, New York (1975)
13. The economics of uncertainty and information. https://epdf.pub/the-economics-of-uncertainty-and-information.html. Accessed 28 July 2019

Investigation of Preference Knowledge of Decision Maker on Consumer Buying Behaviour

Gunay Sadikoglu[1] and Khatira J. Dovlatova[2]($\boxtimes$)

[1] Faculty of Economics and Administrative Sciences, Near East University,
POBOX: 99138, Nicosia, TRNC, Mersin 10, Turkey
gunay.sadikoglu@neu.edu.tr
[2] Joint MBA Program, Azerbaijan State Oil and Industry University,
20 Azadlig Ave., Baku AZ1010, Azerbaijan
xdovlatova@gmail.com

Abstract. Consumer buying behaviour is characterized by such factors as fashion- consciousness, conservatism, hedonism, shopping experience, brand perception, and purchasing intention. This paper aims to obtain priority weights from consumer preference information included in a pairwise comparison matrix which does not exhibit consistency property. For solving this problem a linear programming-based method for correction of the pairwise comparison matrix to achieve consistency property was used. This paper, offers the application of crisp AHP to determine the consistency of the obtained matrix.

Keywords: AHP method · Consumer buying behaviour · Consistency index · Goal programming · Pairwise comparison

1 Introduction

Some problems of consumer buying behaviour modelling may be considered within a multi-criteria decision-making framework. Different quantitative multi-criteria decision-making (MCDM) methods as TOPSIS, AHP, and others are widely applied for decision-making [1]. The AHP method is based on pairwise comparison matrices which describe a decision maker's (DM) preferences related to alternatives and criteria. The consumer buying behaviour modelling problem can be formulated as the multi-criteria decision-making problem. Pairwise comparison matrix $M = (m_{ij})$ is obtained based on a DM's judgments. The aim is to determine a set of numerical weights $(w_1, \ldots, w_n)$ of n criteria given in a decision matrix. A matrix of (w_{ij}) the ratio of weights specified by $w_{ij} = w_i/w_j$ is computed by using matrix (m_{ij}) [2].

In the AHP process goal, criteria and alternatives are chosen and formulated with hierarchy, from goal to criteria and alternatives. In each level factors affect the result of decision-making are evaluated by specialists [3]. Computation of fuzzy AHP was first introduced by Van Laarhoven and Pedrycz followed by Csutora and Buckley offering 'Lambda-Max' method [4]. Massa et al. computed both eigenvalue problems to a universe one analyse of fuzzy data [1].

© Springer Nature Switzerland AG 2020
R. A. Aliev et al. (Eds.): ICSCCW 2019, AISC 1095, pp. 613–620, 2020.
https://doi.org/10.1007/978-3-030-35249-3_78

The structure of this paper is as follows: Sect. 2 describes preliminary information on fuzzy sets and numbers, the optimization problem in a distance-based framework, AHP method. In Sect. 3 statement of the problem and its solution are given. Section 4 states the conclusion.

2 Preliminaries

Definition 1 Fuzzy Sets and Numbers [5]. A fuzzy set A in R (real line) is defined to be a set of ordered pairs $A = \{(x, \mu_A(x))/x \in R\}$, where $\mu_A(X)$ is called the membership function for the fuzzy set. A fuzzy number is a fuzzy set on the real line that satisfies the conditions of normality and convexity.

Let us assume that the membership function of any fuzzy number $\tilde{f}$

$$
\tilde{f}(x) = \begin{cases} \dfrac{x}{a^\alpha} + \dfrac{a^\alpha - a^m}{a^\alpha}, & x \in [a^m - a^\alpha, a^m] \\[2mm] \dfrac{-x}{a^\beta} + \dfrac{a^m + a^\beta}{a^\beta}, & x \in [a^m, a^m + a^\beta] \end{cases}
$$

Definition 2 The Optimization Problem in a Distance-Based Framework [2]. Let $M = (m_{ij})_{ij}$ be the pairwise comparison matrix given by the specialist. When M confirms suitable properties (reciprocity and consistency), there exists a set of positive numbers, $\{w_1, \ldots, w_n\}$, such which $m_{ij} = w_i/w_j$ for every $i, j = 1, \ldots, n$.

Nevertheless, M does not normally verify these properties because of the existence of noise, imperfect judgements, and for other psychological causes. Consequently, the challenge is to search a set of priority weights that synthesize preference information maintained in a standard pairwise comparison matrix.

The classical Euclidean distance is now generalized by an ℓ^p-distance. So, the approximation problem can be confirmed as follows [6]:

$$
\min \left[\sum_{i=1}^{n} \sum_{j=1}^{n} \left| m_{ij} - w_{ij} \right|^p \right]^{1/p},
$$

$$
w_{ij} w_{ji} = 1 \quad \text{for all } i, j
$$

$$
w_{ij} w_{jk} = w_{ik} \text{for all } i, j, k \tag{1}
$$

$$
w_{ij} > 0 \text{ for all } i, j,
$$

$$
\sum_{i=1}^{n} w_i = 1
$$

$$n_{ij} = \frac{1}{2}\left[\left|w_i - w_i m_{ij}\right| + (w_i - w_i m_{ij})\right]$$

$$p_{ij} = \frac{1}{2}\left[\left|w_i - w_i m_{ij}\right| - (w_i - w_i m_{ij})\right],$$

Definition 3 AHP improves the graphical description of a problem in terms of the general goal, criteria, and decision alternatives. Pairwise comparisons are essential building blocks of the AHP. The AHP employs an underlying scale with linguistic values from "equally important" to "absolutely important" to rate the relative advantages for two items. Element d_{ij} of the matrix is the measure of the benefit of an item in row i when to fin contrast between the item in column j [7].

$$D^k = \begin{pmatrix} d_{11} & d_{12} & d_{1n} \\ d_{21} & \dots & d_{23} \\ d_{n1} & d_{n2} & d_{nn} \end{pmatrix},$$

The formula below indicates between criteria $C_i C_j$ in the hierarchy:

$$\tilde{d}_{ij} = (d_{ij,l}, d_{ij,m}, d_{ij,u}), \quad \tilde{d}_{ij} = (1, 1, 1) \quad d_{ij,l} > 0$$
$$i, j = 1, 2 \dots, k \qquad i \neq j, \tag{2}$$

That $\tilde{A}$ fuzzy comparison matrix consist of the following elements [8]:

$$\tilde{f}_{ij} = \frac{1}{\tilde{f}_{ij}} = \left(\frac{1}{f_{ij,u}}, \frac{1}{f_{ij,m}}, \frac{1}{f_{ij,l}}\right), \qquad i, j = 1, 2, \dots, k, \qquad i \neq j, \tag{3}$$

3 Problem Statement and Solution

In brand choosing process, different factors influence consumer buying behaviour [9]. In this study, input variables such as fashion-consciousness, conservatism, and output variables such as hedonism, shopping experience, and brand perception and their impact on purchasing intention were considered. The problem statement comprises measuring and analyzing output variables. MCDM problem involves 3 criteria C_1, C_2, C_3. C_1- hedonism, C_2- shopping experience, C_3-brand perception.

The problem is to determine the consistency index and the consistency ratio of the given fuzzy matrix.

For solving this problem, the main aim is to minimize the objective function, find weight vectors, and calculate consistency index and ratio of $\tilde{A}$ fuzzy matrix expressed in Table 1. $\tilde{A}$ fuzzy matrix divided by three crisp A_1, A_m, A_u matrices are defined by (1) and (2) as below indicated in Table 2, 3, and 4:

Table 1. Fuzzy matrix of the criteria $(\tilde{A})$

Criteria	Hedonism (C_1)	Shopping experience (C_2)	Brand perception (C_3)
Hedonism (C_1)	(1 1 1)	(7 8 9)	(5 6 7)
Shopping Experience (C_2)	(0.11 0.13 0.14)	(1 1 1)	(3 4 5)
Brand Perception (C_3)	(0.14 0.17 0.2)	(0.2 0.3 0.3)	(1 1 1)

Table 2. A_1 matrix

	C_1	C_2	C_3
C_1	1	7	5
C_2	0.11	1	3
C_3	0.14	0.2	1

Achievement function is to minimize objective function:

$$\min\left[\sum_{i=1}^{4}\sum_{j=1}^{4}(n_{ij}+p_{ij})\right],$$

Goals and constraints:

$$1w_1 - w_1 + n_{11} - p_{11} = 0$$
$$7w_2 - w_1 + n_{12} - p_{12} = 0$$
$$5w_3 - w_1 + n_{13} - p_{13} = 0$$
$$0.11w_1 - w_2 + n_{21} - p_{21} = 0$$
$$1w_2 - w_2 + n_{22} - p_{22} = 0$$
$$3w_3 - w_2 + n_{23} - p_{23} = 0$$
$$0.14w_1 - w_3 + n_{31} - p_{31} = 0$$
$$0.2w_2 - w_3 + n_{32} - p_{32} = 0$$
$$1w_3 - w_3 + n_{33} - p_{33} = 0$$
$$w_1 + w_2 + w_1 = 1$$
$$w_1 > 0$$
$$w_2 > 0$$
$$w_3 > 0$$

By using Linear programming, bellowed results are obtained for A_1:

$$w_1 = 0,745, \ w_2 = 0.106, \ w_3 = 0.149$$

Objective function

$$Z = min((n_{11} + p_{11}) + (n_{12} - p_{12}) + (n_{IJ} - p_{IJ})) = 0.856$$

The same calculation for A_m and A_u side is considered and vectors, objective functions are obtained as:

Table 3. Am matrix

	C_1	C_2	C_3
C_1	1	8	6
C_2	0.13	1	4
C_3	0.17	0.3	1

Objective function and vectors by using Table 3 is obtained: $Z = 0.526$, $w_1 = 0,774$, $w_2 = 0.097$, $w_3 = 0.129$

Table 4. A_u matrix

	C_1	C_2	C_3
C_1	1	9	7
C_2	0.14	1	5
C_3	0.2	0.3	1

Objective function and vectors by using Table 4 is obtained:

$$Z = 0,637, \quad w_1 = 0,797, \quad w_2 = 0,088, \quad w_3 = 0,114$$

By using weight vectors 3 crisp matrices are obtained and results are shown in Table 5,

Table 5. Ratio –matrix

Weight vector	Ratio-matrix
$\begin{bmatrix} 0.745 \\ 0.106 \\ 0.149 \end{bmatrix}$	$\begin{bmatrix} 1,00 & 7,02 & 4,99 \\ 0,14 & 1,00 & 0,71 \\ 0,20 & 1,41 & 1,00 \end{bmatrix}$
$\begin{bmatrix} 0.774 \\ 0.967 \\ 0.129 \end{bmatrix}$	$\begin{bmatrix} 1,00 & 0,80 & 6,00 \\ 1,25 & 1,00 & 7,50 \\ 0,17 & 0,13 & 1,00 \end{bmatrix}$
$\begin{bmatrix} 0.797 \\ 0.088 \\ 0.114 \end{bmatrix}$	$\begin{bmatrix} 1,00 & 9,06 & 6,99 \\ 0,11 & 1,00 & 0,77 \\ 0,14 & 1,30 & 1,00 \end{bmatrix}$

For 3 obtained crisp A_l, A_m, A_u matrices CI and CR are calculated as follows. After normalization new matrix is defined:

$$\begin{bmatrix} 1.00 & 7.02 & 4.99 \\ 0.14 & 1.00 & 0.71 \\ 0.20 & 1.41 & 1.00 \end{bmatrix} \rightarrow \begin{bmatrix} 0.75 & 0.74 & 0.74 \\ 0.10 & 0.11 & 0.11 \\ 0.15 & 0.15 & 0.15 \end{bmatrix}$$

Eigenvector and scale for RI are indicated in Tables 6 and 7.

Table 6. Eigenvector calculation

Eigenvector	Total sum	Max	
0.75	1.34	2.99	$A_1 = 0.75$
0.11	9.43		$A_2 = 0.11$
0.15	6.7		$A_3 = 0.15$

Using the following formula from definition 3 for the calculation of the consistency Index:

$$CI = \frac{\lambda_{max} - n}{n - 1},$$

$$CI = \frac{2.99 - 3}{3 - 1} = -0.003,$$

$$CR = \frac{CI}{RI},$$

Table 7. Average random consistency RI

n	RI
1	0.0
2	0.0
3	0.58
4	0.9
5	1.12
6	1.24
7	1.32
8	1.41
9	1.45
10	1.49

$$CI = \frac{CI}{RI} = \frac{-0.003}{0.58} = -0.005 < 0.1 \, ,$$

CI and CR is obtained for this matrix: $\begin{bmatrix} 1,00 & 0,80 & 6,00 \\ 1,25 & 1,00 & 7,50 \\ 0,17 & 0,13 & 1,00 \end{bmatrix}$

$$CI = -0.0008$$
$$CR = -0.001 < 0.1$$

CI and CR is obtained for this matrix: $\begin{bmatrix} 1,00 & 9,06 & 6,99 \\ 0,11 & 1,00 & 0,77 \\ 0,14 & 1,30 & 1,00 \end{bmatrix}$

$$CI = -0.004$$
$$CR = -0.007 < 0.1$$

4 Conclusion

The AHP method is a suitable method for MCDM process in which hierarchies between the goal, criteria and alternatives are calculated. Consisting of the comparison matrices of criteria and alternatives make an AHP method preferable. The main advantage of this paper is using a new method as a consistency-driven approximation of a pairwise comparison matrix by using goal programming. By using goal programming weight vectors and ratio matrices (3 crisp matrices) are obtained. And CI, CR is calculated consistency ratio is less than 10%. The offered method can be used for different spares of Marketing Management to a find solution to decision-making problems.

References

1. Saaty, T.L.: The Analytic Hierarchy Process. McGraw-Hill, New York (1980)
2. Dovlatova K.J.: Decision-making in investment by application of the analytic hierarchy process (AHP). WCIS-2018, Tashkent, Uzbekistan (2018)
3. Ayhan, M.B.: A fuzzy AHP approach for supplier selection problem: a case study in a gear motor company. Int. J. Manag. Value Supply Chains (IJMVSC) 4(3) (2013)
4. Van Laarhoven, P.J.M., Pedrycz, W.: A fuzzy extension of Saaty's priority theory. Fuzzy Sets Syst. 11, 229–241 (1983)
5. Aliev, R.A: Uncertain Computation Based on Decision Theory. World Scientific Publishing, Singapore (2017)

6. Dopazo, E., González-Pachón, J.: Consistency-driven approximation of a pairwise comparison matrix Kybernetika. Praha **39**(5), 561–568 (2003)
7. Oyserman, D., Schwarz, N.: Conservatism as a situated identity: Implications for consumer behavior. J. Consum. Psychol. **27**(4), 532–536 (2017)
8. Buckley, J.J., Feuring, T., Hayashi, Y.: Fuzzy hierarchical analysis revisited. Eur. J. Oper. Res. **129**, 48–64 (2001)
9. Ramya, N., Mohamed, S.A.A.: Factors affecting consumer buying behaviour. Int. J. Appl. Res. **2**(10), 76–80 (2016)

Computing with Words in Natural Language Processing

Farida Huseynova[(✉)] [iD]

Azerbaijan State Oil and Industry University, Baku, Azerbaijan
farida_hus@hotmail.com

Abstract. Today's natural language processing system is based on a system of Aristotle's binary logic and in this case, the applied methods usually cannot take the semantics into account in processing a language. The main issue of the paper is the problem of processing.

Fuzzy logic and its extension computing with words has an ability to process the data in more abstract level without taking into consideration the semantics of sentences, texts, materials, etc. in all aspects.

In this paper, we tried to use an approach to analyze a set of terms given in a natural language.

Keywords: Natural language processing · System · Method · Logic · Semantic · Fuzzy set theory · Computing with words (CWW)

1 Introduction

Zadeh [13] pointed out that humans have many remarkable abilities; one of them is the main ability "to converse, communicate, reason and make rational decisions in an environment of imprecision, uncertainty, incompleteness of information and partiality of truth."

Since, natural language processing (NLP) emerged with the aim of creating a balance between computer science and linguistics, for decades it fulfilled a mission for understanding and interpreting the languages. Some authors view NLP as the gateway between machine language and the natural language of human speech [5]. Different researches have been devoted to this field, and it is often contrasted to computational linguistics [4, 8].

Human language is a system of complex elements for expressing thoughts and ideas in communication process. The human mind can easily accomplish this task, but the computer programs face the serious difficulties in understanding a human communication.

Alain Colmerauer and his colleagues [2] note that computers try to understand what the language is really trying to say, notwithstanding for its meaning, as human analyze it. Copestake Ann [3] believes that, NLP embraces a wide range of applications and it is usually used for searching information including the restoration of meaning and sometimes exclusion of details for selection of information in machine translation.

© Springer Nature Switzerland AG 2020
R. A. Aliev et al. (Eds.): ICSCCW 2019, AISC 1095, pp. 621–625, 2020.
https://doi.org/10.1007/978-3-030-35249-3_79

In recent decades, many researches have been devoted to the understanding of structures of the languages, especially the phonetic and syntactic aspects of the languages that were mainly designed on grammar formalisms and parsing procedures. NLP reinforces manipulation in the processing of a language, because it faces a serious difficulty in understanding the semantics of natural language.

There were the earliest attempts to implement computational natural language understanding, and it is evident that, these traditional models for achieving understanding of a language were insufficient. Novák [6] states that, "there are many linguistic systems, often based on set theory and logic, attempting to grasp the meaning, (at least some phenomena) of the natural language, and none them is satisfactory in all respects."

A serious obstacle may appear in the vagueness of meaning of separate lexical units of texts preventing or hindering progress in understanding systems. However, there is no way than to deal with the vagueness in the models of natural language semantics. İt is required to develop a methodological approach for semantic interpretation in NLP systems, which may be of great importance not only theoretically, but also have a practical significance.

It is required to develop a computable semantic approach to resolve the ambiguity of the meaning and the essential connection with the grammatical level using an adequate semantic interface. It is known that, NLP, is a system, basically used for describing perceptions. It should be remembered that, classical logical systems such as prepositional logic and predicate logic are used for natural language processing and knowledge representation. And when there is uncertainty, incomplete information and limitation inherent in natural language semantic, fuzzy logic approach is expected to handle the problem.

The aim of the research is to describe application of fuzzy logic in natural language processing and computation. Foundation of computation with information described in natural language is paradigm of computing with words suggested by prof. Zadeh [11, 13].

2 NLP by Computing with Words (CWW)

Essentially, CWW is a system of computation where the objects of computation are words, phrases and propositions and they are drawn from a natural language. The propositions are the main conveyers of information. It must be remarked that CWW is the only system of computation which offers a capability to compute with information described in a natural language.

In CWW, the information is delivered by constraining the value of variable and by assumed information consisting of a collection of prepositions expressed in a natural language.

Natural language processing (NLP) and computation based on CWW paradigm consist of three steps and take place by this way; Precisation, Protoform-based reasoning, Converting to language(Fig. 1).

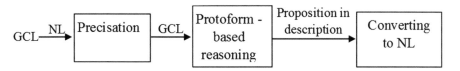

Fig. 1. Schematic description of natural language processing and computation

Precisation Module. In Zadeh's theory, it is called Generalized Constraint language. Here, the collection of propositions expressed in natural language are explicated in a formal computer language and they may be impractical and called a manipulative language. It is required to take essential steps in revealing the new ways for expressing various types of natural language statements within the general constraint language.

Protoform-Based Reasoning Module. Zadeh [10] calls the protoform-based reasoning as one of the particularly favorable and promising directions in the process. The main aim of this module is the manipulation of the propositions.

Converting to NL. By using linguistic approximation procedure, a statement in Generalized Constraint language (GCL) is converted into appropriate statement in natural language (NL).

All the distinctive features of each of these modules [1, 7, 12] here are represented and characterized according to Zadeh's theory. Zadeh's approach to the representation of fuzzy constraints is based on test-score semantics. Network of fuzzy constraints, according to this semantics is regarded as the formation from the collection of propositions of natural language.

Network of fuzzy constraints can be represented as X is R, where R is a constraining fuzzy relation and X is the constrained variable. As mentioned in [11], the expression in question is the canonical form of p. The canonical form is required for explication of fuzzy constraints in evidence which is implicit in proposition. It should be noted that a constraint expressed in a canonical form may be conditional.

3   Linguistic Approximation

The main investigation of this work is connected with converting module based on retranslation process of the third module [11]. Actually, due to primary task of the third module, the proposition V is A, where A is fuzzy subset of the universe X. The main difficulty appears in expressing this proposition in a natural language statement.

Let us consider collection of terms from a natural language i.e. a natural language vocabulary.

The converting process into natural language is substituting the proposition V is A by V is L, where L is one of the elements exposed from the natural language vocabulary. Each fuzzy subset is appropriate to a unique natural language term from vocabulary.

The process of conversion is then one of replaced proposition V is F for V is A, where F is some element from fuzzy sets collection and then expressing the output as V is L, where L is the linguistic term associated with F. The main challenge in this issue is the substitution of V is F for V is A [11].

A degree to which the statement V is A induces the statement V is F (D) can be defined as

$$D(A\acute{I}F) \;=\; Min[I(A(x), F(x))], \quad x \in X$$

where the operator I is an implication operator.

As operator I we can use [7]

$$Zadeh : I(a,b) = max(min(a,b), 1-a)$$

$$Ali-2 : I(a,b) = \begin{cases} 1, & a \le b \\ min(1-a,b), & a > b \end{cases}$$

The closeness of F to A should be related to the distance between be related to the distance between these two sets.

The closeness of A and F may be measured by difference $\Delta_j = |A(x_j) - F(x_j)|$. In particular, Hamming distance is determined as $D_1(A, F) = \Sigma^n_{j=1} \Delta_j$

Suppose that F is fuzzy set corresponding to some natural language term from vocabulary.

Consider a collection of criteria C 1, ...CN for closeness F and A. If F is a possible converting term and C1 is criteria, for example specificity, then for alternative F we have $U_i = C_i(F)$, it is the degree of the specificity of F. We associate with each U_i. Finally, we use a variable satisfaction degree SD. The universe of (SD) is the unit interval.

To determine satisfaction degree SD we use

$$if \; U_i \; is \; D_{11} \; and \; U_2 is \; D_{21}.\,...$$
$$then \, D \, is \, S_1$$
$$if \; U_1 \; is \; D_{12} \; and \; U_2 is \; D_{22}.\,...$$
$$then \, D \, is \, S_2$$
$$.\,...\,...\,...\,...\,...\,...$$
$$if \; U_1 \; is \; D_{1m} \; and \; U_2 is \; D_{2m}$$
$$then \, D \, is \, S_m$$

Having $U_i = C_i(F)$ we can use the techniques of fuzzy modeling to determine SD (F). We then select F with the maximal value of satisfaction.

4   Conclusion

It was shown in this paper that natural language processing problems are mainly based on binary logic, while a more adequate approach is based on fuzzy logic. In this paper general scheme and detailed modules of computing with words based system for processing of information given in NLP were considered.

Acknowledgement. The author would like to express deep and sincere gratitude to prof. R.A Aliev for providing invaluable guidance throughout this research and his helpful comments.

References

1. Aliev, R.A., Fazlollahi, B., Aliyev, R.R.: Soft Computing and its Applications in Business and Economics. Springer, Heidelberg (2004)
2. Colmerauer, A.: Metamorphosis grammars. In: Bolt, L. (ed.) Natural Language Communication with Computers, pp. 133–187. Springer, New York (1978)
3. Copestake, A.: Natural Language Processing, 8 Lectures (aac@cl.cam.ac.uk) (2004). http://www.cl.cam.ac.uk/users/aac
4. Desikan, B.S.: Natural Language Processing and Computational Linguistics (2018)
5. Natural language processing.: From Wikipedia, the free encyclopedia
6. Novák, V.: An introduction to fuzzy logic applications in intelligent systems. Fuzzy Sets Nat. Lang. Process. 185–200 (1989)
7. Wang, P.P.: Computing With Words, pp. 35–67. Wiley, New York (2001)
8. Tsujii, J.: Computational Linguistics and Natural Language Processing. Lecture Notes in Computer Science, pp. 52–67 (2011)
9. Yager, R.R.: On the retranslation process in Zadeh's paradigm of the computing with words. IEEE Trans. Syst. Man Cybern. - Part B: Cybern. **34**(2), 1184–1195 (2004)
10. Zadeh, L.A.: Toward a perception-based theory of probabilistic reasoning with imprecise probabilities. J. Stat. Plann. Infer. **10**, 233–264 (2002)
11. Zadeh, L.A.: From computing with numbers to computing with words - from manipulation of measurements to manipulation of perception. In: Wang, P. (ed.) Computing with Word, pp 35–67. Wiley (2001)
12. Zadeh, L.A.: Test-score semantics for natural languages and meaning representation via PRUF, empirical semantics. In: Rieger, B. (ed.) Germany: Brockmyer, pp. 198–211 (1981)
13. Zadeh, L.A.: What computing with words means to me. IEEE Comput. Intell. Mag. **5**(1), 20–26 (2010)

Synthesis of Intelligent Control System of Quadrotor-Multidimensional Dynamic Object

A. S. Alieva$^{(\boxtimes)}$ (iD)

Institute of Control Systems of ANAS,
Bakhtiyar Vahabzadeh Street 9, Baku, Azerbaijan
c_adile@yahoo.com

Abstract. The paper has been dedicated to the formulation of mathematical model of a multiconnected and multidimensional dynamic object for one of the modern mechatronic devices—a flying robot. The hierarchical architecture of two-level intelligent control (ICS) system of flying robot was suggested and its technical realization has been developed. As well as in the paper, in order to regulate the coordinates of quadrator - $x, y, z, \phi, \theta, \psi$ (for example, altitude-z value) the architecture of adaptive fuzzy PDPI controller has been proposed and the knowledge base has been synthesized. According to experimental research, it has been shown that the synthesized intelligent control system is adapted to change of disturbances, mass and air resistance.

Keywords: Multiconnected and multidimensional dynamic object ·
Flying robot-quadrotor · Fuzzy controller · Adaptation

1 Introduction

Currently in many industrial fields:-atomic energy, machinery, oil-extracting; - in scientific researches, in agriculture; in conducting military and antiterrorist operations different robotics-mechatronic devices and unmanned aerial vehicles are used in a large scale [1–4]. To ensure safety in the complex technological processes (measuring in accessible conditions, shooting, pulverizes and firefighting) and in human life creating of unmanned aerial vehicles (UAV) have been spread [2–4].

Remote control mechatronic device like quadrotor is used to carry out mentioned operations-technological processes, [3, 4]. Most of these unmanned aerial vehicles are controlled automatically via the remote control [1–4]. However, it is required to plan the flight trajectory and to control these drones according to intelligent automatic control algorithm. Structural and parametric identification of mathematical model as a control object, like quadrator as well as synthesis of intelligent control system is a key issue for the high-quality implementation of this demand.

© Springer Nature Switzerland AG 2020
R. A. Aliev et al. (Eds.): ICSCCW 2019, AISC 1095, pp. 626–631, 2020.
https://doi.org/10.1007/978-3-030-35249-3_80

2   Identification of Flying Robot-Multidimensional Dynamic Object and the Synthesis of the Control Systems

Flying robot - quadrotor is unmanned aerial vehicle (UAV), consisting of fans rotating with direct current motors (DCM) fixed at the ends of two intersection girder-frames ("plus" shaped).

Geometric model of quadrotor can be described. This type of rotation provides compensation of impulse moments generated on DCM. In the result of rotation of motor fans (F_1, F_2, F_3, F_4) each motor has vertical pulling forces. On the basis of these, quadrotor rises to the air and flys on the planned trajectory [1, 2].

While the modeling of quadrotor, we accept the change of its mass and ambient condition (resistant force) as an external disturbance impact. Thus, in several operations (such as pulveraizing of plants in agriculture, firefighting, rescue of people in the sea-beaches) initial mass of quadrotor changes – reduces sufficiently depending on the time. In the result, these factors will have an impact on the quality indicators of control. To remove mentioned influences, its control system must have automatical adaptation or rather intelligence feature.

Some features of quadrotor- technical capabilities are as follows:

Quadrotors have ultrasound, height measuring (altimeter), 3° hiroskop for measuring angular velocity and direction, accelerometers and magnitometers. Batteries are the main source of electrical power, DCM are able to provide flights from several minutes to ten- minutes with 0.010–10 kW and 5–50 m/h velocities. They have Wifi connection system for contacting with earth station.

İt is known that, mathematical model of object is important for installing of control system with quality indicator [1, 3]. When mathematical model of quadrotor is determined as a multiconnected automatic control object, its dynamics and movement modes shoud be revealed. As seen from schematic diagram, rotation direction for two rotors are clockwise and for the other two are counterclockwise.

One of the reasons of being difficult to create flying robot control system is that, as a multi connected and multidimensional control object, its mathematical model is sufficiently nonlinear.

Quadrotor, as a rigid mechanical model, consists of m- mass homogenous material. It is a flying device-robot which gravity force has an impact on its geometrical centre. It has six degrees of freedom in space. Basically, a dynamic model of quarator is defined basically on the basis of the Euler-Lagrang or Newton-Euler approach [1–4].

When mathematical model of quadrotor is determined, two coordinate systems are used:- first system is used to identify s quadrotor movement relative to earth (inert coordinate system) R^b; - second is a mobile coordinate system (R^m.) and used to identify movement of quadrotor relative to its centre point.

Let's say that the generelized coordinates of quadrator are defined as $q = [P, r]^T \in R^6$. Here $P = [x, y, z]^T \in R^b$ are the positioning coordinates of quadrator in the absolute coordinate system, and $r = [\phi, \theta, \psi]^T$ - is a vector characterized by Euler's angles.

According to Newton-Euler approach the movement of quadrator (positioning - x, y, z and turn , ϕ, θ, ψ) can be written by the follow differential equations [2–4]:

$$\ddot{x} = \left(c_\phi s_\theta c_\psi + s_\phi s_\psi\right) \frac{1}{m} U_1,$$

$$\ddot{y} = \left(c_\phi s_\theta s_\psi - s_\phi c_\psi\right) \frac{1}{m} U_1, \tag{1}$$

$$\ddot{z} = -g + \left(c_\phi c_\theta\right) \frac{1}{m} U_1,$$

$$\ddot{\phi} = \dot{\theta}\dot{\psi}\left(\frac{I_{yy} - I_{zz}}{I_{xx}}\right) - \frac{J_r}{I_{xx}}\dot{\theta}\Omega_d + \frac{1}{I_{xx}}U_2,$$

$$\ddot{\theta} = \dot{\phi}\dot{\psi}\left(\frac{I_{zz} - I_{xx}}{I_{yy}}\right) + \frac{J_r}{I_{yy}}\dot{\phi}\Omega_d + \frac{1}{I_{yy}}U_3, \tag{2}$$

$$\ddot{\psi} = \dot{\theta}\dot{\phi}\left(\frac{I_{xx} - I_{yy}}{I_{zz}}\right) - \frac{1}{I_{zz}}U_4$$

In (1) equation $s(.), c(.)$ are respectively substituents of $\sin(.)$ and $\cos(.)$ functions of Euler's angles, and are derived from matrix of complex rotation transforming mobile coordinates into the location coordinate system:

$$R = \begin{bmatrix} c\phi c\theta & s\phi s\theta c\psi - s\psi c\phi & c\phi s\theta c\psi + s\psi s\phi \\ s\phi c\theta & s\phi s\theta s\psi + c\psi c\theta & c\phi s\theta s\psi - s\phi c\psi \\ -s\theta & s\phi c\theta & c\phi c\theta \end{bmatrix} \tag{3}$$

In (1) equation $\Omega_d = \omega_2 + \omega_4 - \omega_1 - \omega_3$; I_{xx}, I_{yy}, I_{zz} are respectively inertial moments relative to main axes, I_r - a summary inertial moment of rotors relative to z axe; $U = [U_1 U_2 U_3 U_4]^T$ - are virtual control inputs determined depending on quadrate of angular velocities of relative DCM shafts $\omega_i (i = 1, 2, 3, 4)$, as well as coefficients - k_1, k_2 of pulling and resistant forces, e.g. $U = [U_1 U_2 U_3 U_4]^T = K[\omega^2]$, or:

$$\begin{bmatrix} u_1 \\ u_2 \\ u_3 \\ u_4 \end{bmatrix} = \begin{bmatrix} k_1 & k_1 & k_1 & k_1 \\ 0 & -lk_1 & 0 & lk_1 \\ -lk_1 & 0 & lk_1 & 0 \\ k_2 & -k_2 & k_2 & -k_2 \end{bmatrix} \begin{bmatrix} \omega_1^2 \\ \omega_2^2 \\ \omega_3^2 \\ \omega_4^2 \end{bmatrix} \tag{4}$$

Here $l-$ is a distancebetween the rotor shaft and gravity center of quadrator $K-$ are the constant coefficients matrix with 4×4 dimension. According to the last (4) equation, dependence of rotational velocities from control inputs can be defined as follow:

$$\omega_1^2 = \frac{u_1}{4k_1} + \frac{u_3}{2k_1 l} - \frac{u_4}{4k_2}, \qquad \omega_2^2 = \frac{u_1}{4k_1} - \frac{u_2}{2k_1 l} + \frac{u_4}{4k_2} \tag{5}$$

$$\omega_3^2 = \frac{u_1}{4k_1} - \frac{u_3}{2k_1 l} - \frac{u_4}{4k_2}, \qquad \omega_4^2 = \frac{u_1}{4k_1} + \frac{u_2}{2k_1 l} + \frac{u_4}{4k_2} \tag{6}$$

Intelligent control system is proposed as a structure with two-level architecture. At the top level the initial $x_0, y_0, z_0, \phi_0, \theta_0, \psi_0$ and final $x_s, y_s, z_s, \phi_s, \theta_s, \psi_s$ values of $x, y, z, \phi, \theta, \psi$ - coordinates define $x_{tap} = x_{opt}(t), y_{tap} = y_{opt}(t), z_{tap} = z_{opt}(t)$ and $\phi_{tap} = \phi_{opt}(t), \theta_{tap} = \theta_{opt}(t), \psi_{tap} = \psi_{opt}(t)$ values providing optimal movement trajectory and at the bottom level enter "assignment" (positive) entry of appropriate fuzzy PDPI type controllers.

Control error respect to appropriate coordinates is defined as below:

$$e_x = x_{tap} - x \,, e_y = y_{tap} - y \tag{7}$$

$$e_z = z_{tap} - z \,, e_\psi = \psi_{tap} - \psi$$

Here e_i - is a control error respect to appropriate coordinates.

Current values of x, y, z and ϕ, θ, ψ – coordinates of quadrator are regulated via fuzzy PDPI controllers, z-coordinate is regulated fuzzy PDPI controller adopted to the change of mass. Substantiation algorithm of "Fuzzy adaptive controller"- adaptation described on scheme is proposed in the Appendix 1, table of "Fuzzy PD controller"- linguistic rules are given in the Appendix 2. k_i - parameters [5] formed on Fuzzy adaptation blocks are implemented on the basis of the proposed algorithms.

Let's note that, if quality comparison of controllers is needed, then an ordinary PD and PID rules are used:

$$U_{1x} = k_{px}e_x + k_{dx}\dot{e}_x,$$

$$U_{1y} = k_{py}e_y + k_{dy}\dot{e}_y, \tag{8}$$

$$U_{1z} = k_{pz}e_z + k_{dz}\dot{e}_z + k_{iz}\int e_z dt,$$

$$U_4 = k_{p\psi}e_\psi + k_{d\psi}\dot{e}_\psi,$$

In this case, $k_{ij} = [k_{ij}^L, k_{ij}^R]$ - are interval values of the setting parameters defined on the basis of experimental modeling or analytically defined by the method given in [6]. U_{1x} and U_{1y} values of Euler –orientation angles ϕ_{tap}, θ_{tap} –(rool and pitch) determined by the follows equations depending on managerial impacts:

$$\phi_{tap} = \frac{1}{g}(U_{1x}\sin \psi_{tap} - U_{1y}\cos \psi_{tap})$$

$$\theta_{tap} = \frac{1}{g}(U_{1x}\cos \psi_{tap} - U_{1y}\sin \psi_{tap})$$

3 Technical Realization

Technical realization –simulation of intelligent control system of quadrotor is implemented on "Robot Operating System", "Gazebo" and Matlab/Simulink on Linux platform.

The S-model, based on generalization of intelligent control system of quadrotor by the subsystems, is provided in MATLAB. Time charts of trajectory are described for quadrotor movement-flight in interval of 0–80 s. Here the desired values of relevant coordinates and angles (for example, ψ_{tap}, z_{tap} and etc.) have been illustrated.

Transition processes (such as Roll, Pitch, Yaw, Altitude) from initial values $x_0, y_0, z_0,\ \phi_0, \theta_0, \psi_0$ to final values $x_s, y_s, z_s,\ \phi_s, \theta_s, \psi_s$ of $x, y, z,\ \phi, \theta, \psi$ - coordinates for quadrotor has been given.

4 Result

The proposed intelligent control algorithm is adapted to the load change of quadrotor-UAV and regulates its movement with required accuracy.

Appendix1: It is possible to estimate the mass change of quadrotor by using (1) equation. For this purpose, in case of $\cos\phi\cos\theta = \eta \cong$ cons is in the third expression of (1), it is possible to estimate approximately the mass m(t) (change) of quadrotor by following: $m(t) = {}^{\eta U_1}/_{(\ddot{z}(t) + g)}$

According the expression it is possible to conclude, that a reduction of m(t) – could be resulted in increment of $\ddot{z}(t)$. According to this fact, reduction of U_1 - could be again resulted in reduction of $\ddot{z}(t)$. On the bases of this fact, adaptation of z(t) height regulator could be implemented according to estimation of $\dot{z}$ or $\ddot{z}$ variation.

Appendix2: Linguistic rules table of PD part of fuzzy PDPI controller have been proposed as below (Table 1):

Table 1. Linguistic rules table of fuzzy controller

$\dot{E}^L$	E^L								
	NB	NM	NS	NZ	Z	PZ	PS	PM	PB
NB	PB	PM		NB	NB	NM			Z
NM			NB	NB	NM	NS		Z	PZ
NS		NB	NB	NM	NS	NZ	Z	PZ	PS
NZ	NB	NB	NM	NS	NZ	Z	PZ	PS	PM
Z	NB	NM	NS	NZ	Z	PZ	PS	PM	PB
PZ	NM	NS	NZ	Z	PZ	PS	PM	PB	PB
PS	NS	NZ	Z	PZ	PS	PM	PB	PB	
PM	NZ	Z			PS	PM	PB	PB	
PB	Z			PM	PB	PB		PB	PB

where U^L - Linguistic term set of control

Here N-neqative, P-pozitive, Z- conditional zero (the nominal value of U is to keep the flight in a position), S-small, M- medium, B-is big. Membership (affliation) functions of fuzzy term set have been shown as follow.

References

1. Jafarov, S.M., Zeynalov, E.R., Mustafaeva, A.M.: Synthesis of robust controller-regulators for omnidirectional mobile robot with irregular movement. Proc. Comput. Sci. **102**, 469–476 (2016)
2. Yushenko, A.S., Lebedev, K.R., Zabikhafar, S.Kh.: Control system of quadrotor on the bases of the adaptive neural network. In: Science and Education, MSTU N.E Baumana, vol. 7, pp. 262–277 (2017)
3. Aliyev, R.A., Jafarov, S.M.: Control in Robotic Systems, Baku (2004)
4. Oualid, D., Abdur, R.F., Deok, J.L.: Intelligent controller design for quadrotor stabilization in presence of parameter variations. J. Adv. Transp. 1–10 (2017)
5. Aliyev, R.A., Jafarov, S.M.: Principles of Construction and Design of Intelligent Systems, Baku (2005)
6. Zeynalov, E.R., Jafarov, P.S., Jafarov, S.M., Mustafaeva, A.M.: The methods of analitic synthesis of controllers for dynamic objects described by fuzzy differential equations. In: Tenth–ICAFS and SC, Lisbon, Portugal, 29–30 August, pp. 85–94 (2012)

Analysis of Indicators of the State of Regional Freight Traffic by Method of Fuzzy Linear Regression

Taras Bogachev$^{(\boxtimes)}$ ⓘ, Tamara Alekseychik ⓘ, and Olga Pushkar ⓘ

Rostov State University of Economics, Bolshaya Sadovaya Street, 69, 344002 Rostov-on-Don, Russia
bogachev73@yandex.ru, alekseychik48@mail.ru,
olga-pushkar@yandex.ru

Abstract. In this paper on the basis of statistical data 1996–2017 an analysis was made of the dependence of the volume of goods transported by road in the Rostov region on the density of public roads with hard surface, gross regional product per capita and tariff indices for freight traffic. When constructing regression equations on the basis of economic data, there is often an uncertainty that is associated with incomplete and vague information about the process being studied. In this reason, a fuzzy linear regression method is proposed for the analysis of this dependence. The linear regression coefficients are fuzzy symmetric triangular numbers. To find them, the corresponding optimization problem is solved. Two models were constructed, corresponding to the degrees of the fitting of the fuzzy linear model $h = 0.4$ and $h = 0.5$. The most adequate of the constructed fuzzy models is the model corresponding to the degree of the fitting of the fuzzy linear model $h = 0.4$, which is confirmed by the analysis of the control sample. At $h = 0.5$, the fuzziness coefficient increases sharply and the application of the model has no practical value. According to the results of the analysis, it was found that the index of tariffs for freight transportation has a decisive influence on the volume of cargo transportation.

Keywords: Fuzzy set theory · Analysis of transportation systems · Fuzzy linear regression model

1 Introduction

Transport is a backbone industry, an essential component of the region's industrial and social infrastructure. The economic specialization of industry and agriculture forms the specifics of interregional transport links.

The transport and road complex of the Rostov region combines all types of long-distance transport: rail, road, river, sea, air, intracity (electric) transport.

The possibilities and pace of socio-economic development of the territory of the Rostov region are mainly determined by the roads, which are the most important component of the transport infrastructure of the region.

© Springer Nature Switzerland AG 2020
R. A. Aliev et al. (Eds.): ICSCCW 2019, AISC 1095, pp. 632–638, 2020.
https://doi.org/10.1007/978-3-030-35249-3_81

The development of markets for goods and services, small and medium businesses in the region objectively expands the scope of application of freight road transport. In this regard, this paper proposes the task of analyzing the indicators of the carriage of goods by road on the base of fuzzy linear regression. This method was considered in [1–4].

2   Purpose of the Study

In this paper continues the research of the transport systems of southern Russia using the methods of the theory of fuzzy sets [5, 6]. The information base for the study is the annual data on the main indicators of the characteristics of vehicles of the Rostov region for the 1996–2017 [7]. To do this, we introduce the following notation:

Y – carriage of goods by road transport organizations of all activities (million tons);
X_1 – density of public roads with hard surface (km of roads per 10000 km^2 of territory);
X_2 – gross regional product per capita (thousand rubles);
X_3 – tariff indices for freight traffic (December to December of the previous year, in percent).

The aim of the exploration is to analyze the indicators of freight in the Rostov region from 1996 to 2017.

3   Description of a Fuzzy Linear Regression Model

As noted in [8], when constructing regression equations based on economic data, there is often uncertainty, which is associated not only with random measurement errors, but also with incomplete and unclear information about the process being studied. With this in mind, there are various approaches to describing uncertainty. In particular, one approach is the use of classical probabilistic methods in combination with fuzzy linear regression.

Fuzzy values of A, described by the expressions "approximately a," are often represented by so-called triangular fuzzy numbers. A triangular fuzzy number A is given by a triple of numbers $(a_m; a; a_M)$, such that

$$a_m \leq a \leq a_M \tag{1}$$

The number a is called the modal value of the fuzzy number A.

Under the condition of strict inequality (1), fuzzy number A has the following membership function:

$$\mu_A(x) = \begin{cases} 0, & x \notin [a_m; a_M] ; \\ \dfrac{x - a_m}{a - a_m}, & a_m \leq x \leq a; \\ \dfrac{a_M - x}{a_M - a}, & a \leq x \leq a_M. \end{cases}$$

In fuzzy linear regression symmetric triangular numbers are commonly used. A triangular fuzzy number $A = (a_m; a; a_M)$ is called symmetric if the condition $a - a_m = a_M - a = r$, is satisfied, $r \geq 0$. It follows $a_m = a - r$, $a_M = a + r$. The number r is usually called the fuzziness coefficient of a triangular symmetric fuzzy number.

Then the fuzzy linear regression problem in the general case can be posed as follows. Let there be k results of observations of the dependent variable Y on n factors X_i, $i = 1,...,n$. It is necessary to find fuzzy coefficients A_0, A_1, ..., A_n such that the conditions $\mu_j(y_j) \geq h$ are fulfilled, where μ_j is the membership function of the fuzzy set Y_j, h – the degree of the fitting of the fuzzy linear model, and the uncertainty associated with these coefficients would be minimal.

Let the parameters of the model $A_i = (a_i - r_i, a_i, a_i + r_i)$ are symmetric fuzzy numbers, where $a_i \in R$ and $r_i \geq 0$, $i = 0...n$. At the same time, Y also becomes a fuzzy value. In accordance with [1] to find a_i and r_i we obtain the linear programming problem:

$$f = ka_0 + \sum_{j=1}^{k}\sum_{i=1}^{n} r_i x_{ij} \rightarrow \min \tag{2}$$

$$\begin{cases} y_j \geq a_0 + \sum_{i=1}^{n} a_i x_{ij} - (1-h)\left(r_0 + \sum_{i=1}^{n} r_i x_{ij}\right), \\ y_j \leq a_0 + \sum_{i=1}^{n} a_i x_{ij} + (1-h)\left(r_0 + \sum_{i=1}^{n} r_i x_{ij}\right), \\ r_i \geq 0, \ j = 1...k. \end{cases}$$

Solving this problem for different values of the degree of the fitting of the fuzzy linear model h, we obtain the objective function in the form of fuzzy triangular numbers $Y = <Y_d, Y_m, Y_u>$, where Y_m– model the value of the index Y, Y_d – constraint on the left of the indicator Y, Y_u – constraint on the right of the indicator Y.

4 Application of a Fuzzy Linear Regression Model for the Explore of Goods Carriage by Road Transport in the Region

Table 1 shows the data for the selected indicators.

To conduct a more complete analysis of the dependence of the studied indicator on the chosen factors, we construct a model of fuzzy linear regression. Of the 22 observations, 18 (1998–2013) are attributed to the explanatory sample, 4 (2014–2017) to the control. Assume that the coefficients of the model are triangular fuzzy numbers A_0, A_1,

Table 1. The values of the selected indicators 1996–2017

Year	Y	X_1	X_2	X_3
1996	128.4	103	7088.7	131.1
1997	101.2	105	8062.9	102.9
1998	90.3	104	9163.7	76.9
1999	92.0	109	15672.2	112.0
2000	83.1	114	20004.0	170.3
2001	98.7	118	28470.0	129.1
2002	99.4	120	31942.0	115.7
2003	119.0	123	39225.0	137.0
2004	120.7	123	50843.0	113.7
2005	113.6	132	61142.0	111.5
2006	66.8	141	78642.0	105.3
2007	64.7	143	104603.0	103.8
2008	61.9	142	134137.0	113.0
2009	60.1	140	129626.0	110.7
2010	58.9	139	154128.0	109.3
2011	57.7	139	179470.0	108.8
2012	60.1	202	198129.0	104.4
2013	61.4	258	215923.0	104.0
2014	60.3	260	237466.0	106.8
2015	60.0	261	280522.0	113.1
2016	60.4	262	300186.0	106.7
2017	54.6	263	318001.0	103.6

A_2, A_3 of the form $A_i = <a_i - r_i, a_i, a_i + r_i>$, where $a_i \in R$ and $r_i \geq 0$, $i = 0...3$. In this case, Y will also become a fuzzy value. In accordance with (2) to find a_i and r_i, we obtain the linear programming problem.

$$f = 18a_0 + \sum_{j=1}^{18} \sum_{i=1}^{3} r_i x_{ij} \rightarrow \min.$$

$$\begin{cases} y_j \geq a_0 + \sum_{i=1}^{3} a_i x_{ij} - (1-h)\left(r_0 + \sum_{i=1}^{3} r_i x_{ij}\right), \\ y_j \leq a_0 + \sum_{i=1}^{3} a_i x_{ij} + (1-h)\left(r_0 + \sum_{i=1}^{3} r_i x_{ij}\right), \\ r_i \geq 0, \ j = 1...18. \end{cases}$$

Solving this problem with the values of degree of the fitting of the fuzzy linear model $h = 0.4$ and $h = 0.5$, we obtain the following values of fuzzy coefficients (Table 2).

Table 2. Values of fuzzy coefficients

h	a_0	a_1	a_2	a_3	r_0	r_1	r_2	r_3
0.4	5.866	−0.059	−0.0002	0.996	0	0	0	0.292
0.5	15.711	−0.191	0.0007	0.254	0	0	0	1.440

Note that in both cases the numbers A_0, A_1 and A_2 are not fuzzy. The equation of non-clear regression with $h = 0.4$ has the form

$$Y = 5.8666 - 0.059X_1 - 0.0002X_2 + \langle 0.704; 0.996; 1.288 \rangle X_3.$$

Considering, that $Y = <Y_d, Y_m, Y_u>$, we get a fuzzy linear regression graph (Fig. 1).

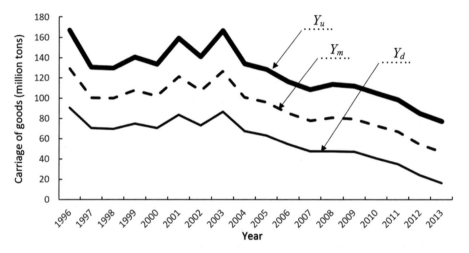

Fig. 1. Fuzzy linear regression for carriage of goods by road transport organizations (Y), $h = 0.4$, axis X – year, axis Y – carriage of goods.

Evaluation of results using the control sample is given in Table 3.

Table 3. Evaluation of results

Year	Y controlling	Left border of fuzziness interval	Right border of fuzziness interval	Indicator Y controlling
2014	60.3	13.723900	76.10600	+
2015	60.0	8.650097	74.71204	+
2016	60.4	−0.231830	62.09186	+
2017	54.6	−6.384020	54.12895	−

Therefore, the constructed model is adequate, as evidenced by the corresponding values of the explored indicator (Table 3).

The fuzzy regression equation with $h = 0.5$ is

$$Y = 15.711 + 0.191X_1 + 0.0007X_2 + \langle -1.186; 0.254; 1.694 \rangle X_3.$$

In this case, the fuzziness coefficient increases dramatically and the value of the fuzzy number Y becomes too blurred.

According to both constructed models, the gross regional product per capita has practically no effect on the volume of cargo transportation, the density of public roads with a hard surface has little effect. This suggests that the main factor affecting the volume of cargo transportation is a fuzzy indicator – the index of tariffs for freight transportation.

5 Conclusion

The task of analyzing the indicators of freight traffic in the Rostov region was considered. The model, built using fuzzy linear regression, made it possible to investigate the dependence of the carriage of goods by road transport organizations in this region on the density of public roads with hard surface, gross regional product per capita and tariff indices for freight traffic. Verification of the model corresponding to the degrees of the fitting of the fuzzy linear model $h = 0.4$ with the help of the control sample confirmed its adequacy for prediction. With an increase in the reliability threshold even by a small amount (for example, up to 0.5), the use of the model has no practical sense.

References

1. Tanaka, H., Uejima, S., Asai, K.: Linear regression analysis with fuzzy model. IEEE Trans. Syst. Man Cybern. **12**(6), 903–907 (1982)
2. Tanaka, H.: Fuzzy data analysis by possibilistic linear models. Fuzzy Sets Syst. **24**(3), 363–375 (1987)
3. Kim, K.J., Moskowitz, H., Koksalan, M.: Fuzzy versus statistical linear regression. Eur. J. Oper. Res. **92**(2), 417–434 (1996)
4. Hojati, M., Bector, C.R., Smimou, K.: A simple method for computation of fuzzy linear regression. Eur. J. Oper. Res. **166**(1), 172–184 (2005)
5. Alekseychik, T., Bogachev, T., Bogachev, V., Bruhanova, N.: The choice of transport for freight and passenger traffic in the region, using econometric and fuzzy modeling. In: 9th International Conference on Theory and Application of Soft Computing, Computing with Words and Perception, ICSCCW 2017, 22–23 August 2017, Budapest, Hungary, Procedia Computer Science, vol. 120, pp. 830–834 (2017)

6. Alekseychik, T., Bogachev, T., Bogachev, V.: Comparative assessment of the transport systems of the regions using fuzzy modeling. In: 13th International Conference on Application of Fuzzy Systems and Soft Computing (ICAFS) 2018, 27–28 August, Warsaw, Poland. Advances in Intelligent Systems and Computing, vol. 896, pp. 651–658 (2019)
7. The ROSSTAT. Region of Russia. Socio-economic indicators. Statistical compendium. http://www.gks.ru/wps/wcm/connect/rosstat_main/rosstat/ru/statistics/publications/catalog/doc_1135087342078. Accessed 21 June 2019
8. Hansen, L.P.: Uncertainty consequences for the economic. Bull. Financ. Univ. **2**(86), 6–12 (2015)

Bifurcation Analysis and Synergetic Control of a Dynamic System with Several Parameters

V. Bratishchev Alexander[1], A. Batishcheva Galina[2]([⊠]) [iD],
Y. Denisov Mikhail[2], and I. Zhuravleva Maria[2]

[1] Don State Technical University, Rostov-on-Don, Russia
avbratishchev@spark-mail.ru
[2] Rostov State University of Economics, Rostov-on-Don, Russia
gbati@mail.ru, lunatikl957@mail.ru, zhurmari@mail.ru

Abstract. The article provides a complete bifurcation analysis of the mathematical model of the dynamic system "Emergence of planned regulation" proposed by V. P. Milovanov. The behavior of trajectories at infinity is studied using the Poincare transform. With the help of theoretical analysis and numerical experiment the phase portrait of the system is obtained in Matlab package. The system turned out to be a lip in the open first quarter of the phase plane. The system of additive control of both cash and commodity flows to achieve a given dynamic equilibrium from an arbitrary initial state is constructed by the method of analytical design of aggregated regulators. Dedicated class a valid reachable States. The numerical experiment shows the stability of this state as a whole. This model allows you to predict the development of the process for any predetermined initial state of the system, as well as to control the parameters of the system to design a predetermined dynamic equilibrium.

Keywords: Cash and commodity flows · Autonomous system · Equilibrium state · Phase portrait · Stability · Aggregated variable · Invariant variety · Synergetic regulator

1 Introduction

In the article [1] with the help of qualitative theory of dynamic systems [2] bifurcation analysis was carried out, and with the help of the theory of synergetic control [3] designed synergetic regulator of the dynamic system "Intermediary activity". In this article, the same problems are solved for the dynamic system "the Emergence of planned regulation", a mathematical model of which is proposed in the monograph [4].

$$\begin{cases} x'_t = a_1 x + a_2 y - a_3 xy \\ y'_t = -b_1 xy + b_2 y^2 \end{cases}, \quad a_i, b_j > 0$$

$$\begin{cases} x'_t = a_1 x + a_2 y - a_3 xy \\ y'_t = -b_1 xy + b_2 y^2 \end{cases} \tag{1}$$

© Springer Nature Switzerland AG 2020
R. A. Aliev et al. (Eds.): ICSCCW 2019, AISC 1095, pp. 639–646, 2020.
https://doi.org/10.1007/978-3-030-35249-3_82

Here, the phase variable x(t) – the funds at the disposal of organizations, monopolies, cooperatives; y - generalized goods, the amount of which on the market at a given time is equal to the phase variable y(t). The rate of growth of funds x'_t consists of money a_1x, placed in the Bank, the purchase of goods a_3xy and payment by consumers of the goods of its moral and physical deterioration a_2y. The speed of growth in the quantity of goods y'_t is determined by its sale b_1xy and its purposeful production b_2y^2.

2 Complete Bifurcation Analysis of the System

It involves the determination of the topological structure of all equilibrium states of the system.

Note that the OX axis of the phase plane of system (1) is composed of two trajectories defined by the equations

$$\begin{cases} x'_t = a_1x \\ y = 0 \end{cases}.$$

It separates the desired elementary cells [2] of the upper and lower phase half-planes, regardless of the values of the parameters a_i, b_j.

System (1) has 2 equilibrium states

$$S_1 = (0,0), S_2 = \left(\frac{a_1b_2 + a_2b_1}{a_3b_1}, \frac{a_2b_1 + a_2b_1}{a_3b_2} \right)$$

$S_1 = (0,0)$ it is a multiple state of equilibrium [2]. Bringing the system to the canonical form and applying the theorem on the classification of multiple States of equilibrium, we establish that S_1 it is a saddle-node. Its nodal sector is unstable and makes up the upper half-cross, and the two saddle sectors make up the lower half-cross S_1.

By means of Lyapunov's theorem about stability on the first approximation [5] it is established that the state S_2 which is in the first quarter is a saddle.

For Fig. 1 the results of a computational experiment on the S-model [6] of the considered system with parameters $a_i = b_j = 1$ are shown.

Below the saddle $S_1 = (0,0)$ node there are two saddle sectors $S_2 = (2,2)$ - saddle. The equilibrium States of the system (1) at infinity are investigated in two steps. State different from the points $(0, \pm\infty)$, are studied using the Poincare transformations of the original variables: $x = 1/u, y = v/u$. Have

$$\begin{cases} u'_t = u(-a_1u + a_3v - a_2uv) \\ v'_t = v(-b_1 - a_1u + (a_3 + b_2)v - a_2uv) \end{cases} \tag{2}$$

This system has three States of equilibrium.

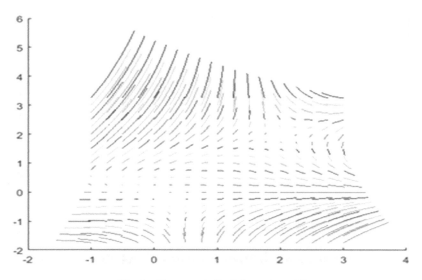

Fig. 1. Phase portrait of the system (1)

The state $S_1' = (0,0)$ has is a multiple. Since the system already has a canonical form, according to the theorem on the classification of multiple States, it is a saddle-node with a stable nodal sector in the right half-cross S_1' u, v-plane. Since the latter is displayed by the inverse transformation to the neighborhood of the point $(+\infty, 0)$ of the initial phase space, the trajectories of the system (1) are contracted to $(+\infty, 0)$.

The left half-cross of the state consists of two saddle sectors. They are displayed in the saddle sectors of the neighborhood of the point $(-\infty, 0)$.

The state $S_2' = \left(0, \frac{b_1}{a_3 + b_2}\right)$ is an unstable node of the system (2).

Its prototype is two infinitely remote points of the line $y = \frac{b_1}{a_3 + b_2}x$ of the initial phase space. Right unstable poliocretes point S_2' moves in an unstable neighborhood the top right of an infinitely distant point. The left unstable half-cross of the point S_2' passes, as the numerical experiment with the S-model of the system (2) shows, into a stable neighborhood of the left lower infinitely remote point.

The third state of equilibrium is the image of the state S_2 of the original system. State different from the points $(\pm\infty, 0)$, are investigated using the Poincare transformations of the original variables $x = v/u, y = 1/u$. Have

$$\begin{cases} u_t' = u(-b_2 + b_1 v) \\ v_t' = a_2 u - (a_3 + b_2)v + a_1 uv + b_1 v^2 \end{cases} \tag{3}$$

This system also has three equilibrium States.

The state $S_1'' = (0,0)$ is a stable node. Its right half-crossness u, v-plane is displayed by inverse Poincar \' e transformation into a stable neighborhood of the point $(0, +\infty)$.

The left stable half-cross passes, as the numerical experiment with the S-model of the system (3) shows, into the unstable neighborhood of the point $(0, -\infty)$.

Thus, the system (1) has two finite and three infinite equilibrium States. The behavior of trajectories in the vicinity of these States is established. Together with the numerical experiment on the S-model, this allows us to obtain a scheme of the phase portrait of the system (1) (Fig. 2).

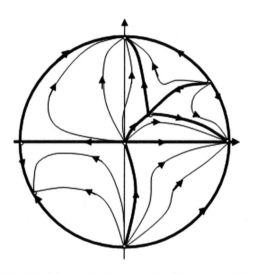

Fig. 2. Scheme of phase portrait of the system (1)

The portrait consists of six elementary cells. The system has no bifurcations of the open first quarter of the phase plane. Phase portraits for any positive parameter sets will be topologically equivalent.

Comment. All trajectories of the first quarter except two separatrix are contracted to points $(+\infty, 0)$ or $(0, +\infty)$. From an economic point of view, either the quantity of goods (overproduction) increases indefinitely or the quantity of money (inflation) increases indefinitely. In this regard, there is a problem of controlling these processes so that the trajectories of the controlled system are contracted to a predetermined final state.

3   Construction of Synergetic Regulator of the System (1)

We apply the method of ACAR (analytical design aggregated) regulators [4] to answer two questions:

(1)   how to manage the size of the cash flow x(t),
(2)   how can you manage the production of generalized goods y(t)

to any initial state the corresponding trajectories were drawn over time to a single advance to a given state (x_0, y_0)? Control is sought as an additive summand of the rate of change of the corresponding value and is a function of the current state of the system.

The equation of the regulator is as follows

$$\begin{cases} x'_t = a_1 x(t) + a_2 y(t) - a_3 x(t) y(t) + u_1(t) \\ y'_t = -b_1 x(t) y(t) + b_2 y^2(t) \end{cases} \tag{4}$$

According to the method of ACAR [3] the government is constructed through the selection of a suitable function (called aggregate variable $\psi_1(x, y)$ that defines the attracting invariant manifold $\psi_1(x, y) = 0$ of the designed controller. It is required that the derivative of this function by force (4) satisfy on the trajectories the differential equation $\psi'_{1,t} = (-1/T)\psi$. The latter has a solution $\psi_1(x(t), y(t)) = C\exp\{-t/T\} \to 0$ at $t \to +\infty$. It is in this sense that the attraction of the manifold for all trajectories of the regulator is understood. This property of the variety can be called stability in the sense of Kolesnikov [7]. Note that it is not equivalent to the stability of an invariant set in the sense of Zubov [8].

From the equation $\psi'_{1,t} = \psi'_{1,x} x'_t + \psi'_{1,y}(b_1 xy + b_2 y^2) = (-1/T)\psi$ we derive the equation of the regulator

$$\begin{cases} x'_t = \frac{-1}{\psi'_{1,x}} \left(\psi'_{1,y}(-b_1 xy + b_2 y^2) + \frac{1}{T}\psi_1 \right) \\ y'_t = -b_1 xy + b_2 y^2 \end{cases}$$

and management. $u_t = -(a_1 x + a_2 y - a_3 xy) - \frac{1}{\psi'_{1,x}} \left(\psi'_{1,y}(-b_1 xy + b_2 y^2) + \frac{1}{T}\psi_1 \right).$

According to [9] the state of equilibrium is the solution of the functional system

$$\begin{cases} \psi_1(x, y) = 0 \\ y(b_1 x - b_2 y = 0 \end{cases}$$

and therefore lies on the lines $y = 0, y = (b_1/b_2)x$. Select the aggregated variable $\psi_1(x, y) := x - cy^3$ with the parameter $c > 0$. The solution of the system with nonzero coordinates has the form

$$\begin{cases} x_0 = \frac{b_2}{b_1} \sqrt{(b_1/b_2 c)} \\ y_0 = \sqrt{(b_1/b_2 c)} \end{cases}$$

We check the asymptotic stability condition [9]:

$$\left. f'_{2y} - \frac{\psi'_{1y}}{\psi'_{1x}} f'_{2x} \right|_{(x_0,y_0)} = -2b_2 \sqrt{(b_1/b_2 c)} < 0$$

That (x_0, y_0) is stable.

Numerical experiments with the S-model of the constructed regulator under different sets of parameters and initial conditions from the first quarter show that the state (x_0, y_0) has the stability property as a whole.

Rules:

(1) With fixed parameters b_1, b_2 the selected aggregated variable allows you to design the management of the quantity of goods $x(t)$ to achieve only the States with coordinates $\left(\frac{b_2}{b_1}\sqrt{(b_1/b_2c)}, \sqrt{(b_1/b_2c)}\right), c > 0$, that is, lying on the branch of the cubic parabola $x = cx^3$.

(2) If we want to synthesize the control system to achieve an arbitrary predetermined state (x_0, y_0), $x_0, y_0 > 0$, it is necessary to change the parameters of the goods b_1, b_2: they must be linked by equality $b_1/b_2 = x_0 y_0$.

2. The equation of the regulator is as follows

$$\begin{cases} x'_t = a_1 x(t) + a_2 y(t) - a_3 x(t) y(t) \\ y'_t = -b_1 x(t) y(t) + b_2 y^2(t) + u_2(t) \end{cases}$$

We are looking for an aggregated variable $\psi_2(x, y)$ that satisfies the equation $\psi'_{2,t} = (-1/T)\psi$ on the trajectories of the regulator. From the equation

$$\psi'_{2,t} = \psi'_{2,x}(a_1 x + a_2 y - a_3 xy) + \psi'_{2,y} y' = (-1/T)\psi$$

we derive the equation of the regulator

$$\begin{cases} x'_t = a_1 x + a_2 y - a_3 xy \\ y'_t = \frac{-1}{\psi'_{2,y}}\left(\psi'_{2,x}(a_1 x + a_2 y - a_3 xy) + \frac{1}{T}\psi_2\right) \end{cases}$$

and management $u_t = b_1 xy - b_2 y^2 - \frac{1}{\psi'_{2,y}} \cdot \left(\psi'_{2,x}(a_1 x + a_2 y - a_3 xy) + \frac{1}{T}\psi_2\right)$.

The equilibrium state of the regulator is the solution of the functional system

$$\begin{cases} a_1 x + a_2 y - a_3 xy = 0 \\ \psi_2(x, y) = 0 \end{cases} \tag{5}$$

and therefore lies on the hyperbola $y = a_1 x/(a_3 x - a_2)$, passing through the origin. Select the aggregated variable $\psi_2 := x - cy^2$ with the parameter $c > 0$. The solution of the system with positive coordinates is unique (Fig. 3) and has the form

$$\begin{cases} x_0 = \dfrac{a_1^2 + 4a_2 a_3/c + a_1\sqrt{a_1^2 + 4a_2 a_3/c}}{2a_3^2} \cdot c \\ y_0 = \dfrac{a_1 + \sqrt{a_1^2 + 4a_2 a_3/c}}{2a_3} \end{cases} \tag{6}$$

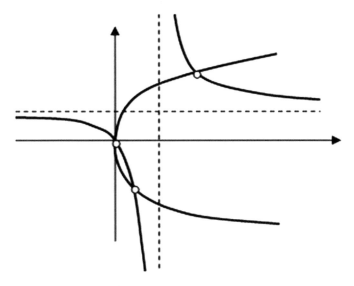

Fig. 3. System schedule (5)

We check the asymptotic stability condition [9]:

$$\left. f'_{1x} - \frac{\psi'_{2x}}{\psi'_{2y}} f'_{1y} \right|_{(x_0, y_0)} = -\frac{\sqrt{a_1^2 + \frac{4a_2a_3}{c}}}{2} < 0.$$

That (x_0, y_0) is stable.

Numerical experiments with the S-model of the constructed regulator under different sets of parameters and initial conditions from the first quarter show that the state (x_0, y_0) has the stability property as a whole.

4 Conclusion

(1) With fixed parameters a_1, a_2, a_3, the selected aggregate variable allows to synthesize the control of the quantity of goods x(t) to achieve only the States with coordinates (6), that is, lying on the parabola $x = cy^2$.

(2) If we want to synthesize the control system to achieve an arbitrary predetermined state $(x_0, y_0), x_0, y_0 > 0$, it is necessary to change the parameters of the goods a_1, a_2, a_3: they must be linked by equality $a_1 x_0 + a_2 y_0 = a_3 x_0 y_0$, a_3. That is known a_1, a_2 from this equation we find a_3, or given a_3 from this equation we select the positive a_1, a_2.

References

1. Bratishchev, A.V., Batishcheva, G.A., Zhuravleva, M.I.: Bifurcation analysis and synergetic management of the dynamic system "intermediary activity". In: Proceedings of 13th International Conference on Theory and Application of Fuzzy Systems and Soft Computing, ICAFS-2018. Advances in Intelligent Systems and Computing. AISC, vol. 896, pp. 659–667 (2018)
2. Bautin, N.N., Leontovich, E.A.: Methods of qualitative research of dynamic systems on a plane. Ed.2, Moscow, Science (1990)
3. Kolesnikov, A.A.: Synergetic Methods of Control of Complex Systems. The Theory of System Analysis. Komkniga, Moscow (2006)
4. Milovanov, V.P.: Synergetics and Self-organization. Economy. Biophysics. Komkniga, Moscow (2005)
5. Demidovich, B.P.: Lectures on Mathematical Theory of Stability. LAN Publishing House, St. Petersburg (2008)
6. Lazarev, Yu.: Modeling Processes and Systems in MATLAB. Publishing Group BHV, Peter, Kiev (2005)
7. Bratishchev, A.V.: On the characteristic polynomial of the equilibrium state of an Autonomous system having an attractive invariant variety. Differ. Equ. Control Process. **2**, 15–23 (2017)
8. Zubov, V.I.: Stability of Motion. Vysshaya SHKOLA, Moscow (1973)
9. Bratishchev, A.V.: The Mathematical Theory of Controlled Dynamical Systems. Introduction to Concepts and Methods. Publishing Center DGTU, Rostov-on-Don (2015)

Predicting Dynamics of the Consumer Commodity Market Based on Fuzzy Neural Network

Rana Mikayilova Nuru[(✉)] [ID]

Azerbaijan State Economic University (UNEC), Baku, Azerbaijan Republic
rana.mikayilova@unec.edu.az

Abstract. An important place in the diagnosis of the consumer goods market is occupied by the issues of analyzing methods for forecasting retail turnover. When forecasting retail turnover in the consumer market, an extrapolation method is partially applied. Since trade is a stable development system, their performance is relatively stable. In our work, based on the indicators obtained by extrapolating the turnover of various components of the consumer goods market, we derived a unified indicator - the ratio of commercial activity of the retail product market.

Keywords: Market · Commodity market · Statistical methods · Dynamic methods · Forecasting methods · Modeling methods · Neural network · Fuzzy logic · Coefficients and indices

1 Introduction

In the study of the consumer commodity market, the task of forecasting and determining the main parameters and contours of prospective development, usually for a short-term period, is highly relevant. Predictive assessment allows managers of subjects of the commodity market to identify the positive and negative aspects of their activities, and thus develop the main areas of work for the future.

Although consumer commodity market forecasting is short-term, these studies are a prerequisite for a further medium-term, or even long-term, forecast. This is explained by the fact that the indicators obtained from the short-term forecast enrich the information base of the subjects of the commodity market and thus create the prerequisites for their development for a longer period. The significance of short-term forecasts of the consumer goods market is expressed in the fact that for a short period no significant changes are observed and thus the possibility of obtaining relatively reliable indicators of the forecast is created, which is the starting point for medium-term and long-term forecasts of the commodity market.

© Springer Nature Switzerland AG 2020
R. A. Aliev et al. (Eds.): ICSCCW 2019, AISC 1095, pp. 647–653, 2020.
https://doi.org/10.1007/978-3-030-35249-3_83

2 Methods

2.1 Overview of Methodology

The economic literature offers a variety of approaches and methods for forecasting the consumer goods market [1–6].

Short-term projected assessment of the subjects of the consumer commodity market is the basis for the formation of a marketing strategy that is regulated and coordinated by taking into account external and internal factors. The results of the initial forecast are the basis for planning production and commercial activities. The obtained initial forecast data are systematically checked on the subject of their adequacy to the real sales dynamics. If significant deviations are observed, the forecast estimate is given again and further adjusted.

An important step in forecasting the consumer commodity market is the selection of methods and techniques that will help increase the level of predictive evaluation. At the same time, it is necessary to take into account both external factors and internal factors, including the following: a targeted approach, time approach, conjuncture approach, identification of regional-territorial delimitation, etc.

Evaluation methods are based on expert assessments, opinions, suggestions. In particular, expert assessments are composed of: the Delphi method; statistical processing of personal customer ratings; probabilistic models; simulation modeling; graphic methods, etc.

2.2 Classic Approaches for Forecasting the Consumer Commodity Market

The economic literature [7, 8] classifies two distinct approaches for the methods of forecasting of the consumer commodity market: economic-analytical and mathematical.

The economic-analytical approach is a relatively well-known method of research and forecasting of the situation in the consumer goods market, which consists in the application of the intuitive-logical reasoning of economists involved in the diagnosis of markets. The forecast is based on a subjective consideration of trade and commodity factors and represents the opinion of experts about the development of the situation in the time frame. Along with the positives of this approach, there are also negatives, among which there is a high level of subjectivity, the inability to objectively analyze, absence of the possibility of cumulatively accounting for a small number of conjunctive factors.

The economic-mathematical approach is the forecasting of the commodity market based on a number of mathematical models or systems of equations with varying degrees of adequacy of the studied processes. The content of these formations is fully consistent with the properties of ordinary extrapolation or multifactor models. The advantage of these forecasts can be considered the clear logical compatibility of the results.

Thus, for the formation of the forecast, it is necessary to study the properties of the predicted object in retro and perspective. As a mechanism for forecasting, a system of methods is used, on the basis of which the causal parameters of the retro trend are diagnosed in the activities of firms and the results of the diagnosis reflect shifts in the perspective activity of business structures. The resulting analysis of existing forecasting methods helps to systematize them into two groups: factual and heuristic.

2.3   Statistical Research Methods

Statistical research methods are applicable in almost all types of commodity market. They are used when it is necessary to obtain data on a group with a certain internal heterogeneity.

Defining the content of prognostic and managerial models in business, it is necessary to pay attention to the tasks of the statistical diagnostics of the mechanism of functioning of the business structure and forecasting their activities. Statistical methods of forecasting are based on the basis of extrapolation, which is understood as certain regularities, links and relationships that operate in the study period for a certain point in time in the future.

In scientific and applied activities in the field of statistical methods of diagnostics of information (according to the degree of particular features of the techniques associated with immersion in specific problems) there are three approaches:

1. the development and study of general-purpose methods, without taking into account the peculiarities in the scope of application;
2. the development and study of statistical models of reliable phenomena and procedures in accordance with the needs of the commodity market;
3. the use of statistical models for the statistical diagnosis of practical data.

The main phases of statistical diagnostics are: definition of diagnostic goals; optimal assessment of the initial data; correlative assessment and provision of comparative indicators; development of a system of generalized indicators; registration of existing features, properties and differences, relationships and patterns of the studied phenomena and procedures; formulation of conclusions and recommendations on the development of the commodity market.

Forecasting the conjuncture of the commodity market, based on statistical diagnostics of the market, and relies on probabilistic-statistical models built in accordance with the characteristics of the field of application.

2.4   Consumer Commodity Market and Suggested Grouping of Forecasting Methods for It

A distinctive feature of the consumer commodity market, in contrast to the industrial and agrarian commodity markets, is the fact that the processes involved in selling goods and meeting the needs of the population are carried out in this market. Therefore, the approach to the predictive assessment of the consumer commodity market should have a slightly different character. The coefficients and indicators used should be consistent with the profile of the consumer commodity market.

It should be noted that in the consumer commodity market a significant place is occupied by the retail goods market, which accounts for more than 80% of the total turnover (hereinafter referred to as the turnover of paid services and public catering).

So far, our research suggests the following grouping of forecasting methods for predicting consumer commodity market:

– the group of basic methods for forecasting consumer commodity markets includes: extrapolation by the average level of a number of dynamics; extrapolation by the average increase in the series; extrapolation by the average growth rate of the series.
– the group of methods for forecasting the consumer commodity market on the basis of econometrics.
– the group of fuzzy methods for forecasting the consumer commodity market with the use of fuzzy neural network includes: study of commercial activity (passivity) of the retail product market.

2.5   Study of Commercial Activity (Passivity) of the Retail Product Market the Use of Fuzzy Neural Network

Studies have shown that not all criteria and indicators of forecasting methods can be mathematically formulated, and in most cases they require their inclusion in the forecasting, estimation and decision-making models as the most desirable and important factors. In this regard, fuzzy logic is the most popular method that combines the human thinking style of fuzzy systems [12, 13] through the use of fuzzy sets and a linguistic model consisting of a set of fuzzy IF-THEN rules. On the other hand, having a large amount of information in hand, one should analyze their retrospectiveness and further predict their dynamics of change based on learning algorithms based on neural networks [10–13].

Particular attention was paid to forecasting methods for the consumer goods market, which combines the methods of the theory of fuzzy logic and neural network. In particular, the study of commercial activity (passivity) of the retail product market was carried out by us using the fuzzy neural network [11, 14].

Samigulin [9] introduced into scientific circulation such an indicator as the coefficient of commercial activity or commercial inactivity of the retail commodity market. This coefficient can have values below −1 to indicate the level of commercial inactivity of the retail commodity market or values above +1 to indicate the level of commercial activity of the retail commodity market. Both extremes are equally unacceptable for the commodity market - both a high coefficient of commercial passivity and a high coefficient of commercial activity.

With a high rate of commercial passivity of the retail commodity market (below − 1), which characterizes a significant excess of demand over product supply, the economy of the territory is estimated as underdeveloped, with limited resources of commodity production, with a predominance of imported goods in the market. The critical value of the passivity of the retail market is in the range from −1.2 to −1.25.

With a high indicator of the coefficient of commercial activity of the market (above +1), which characterizes the significant excess of product supply over demand, the economy of the territory is estimated as developed, with decent resources of its own commodity production. A significant excess of commodity supply over market demand is fraught with the accumulation of excessive stocks, which is unacceptable for producers. The critical value of the activity ratio of the retail market should not exceed the rate of +1.3. In Table 1 is used the following notations:

Table 1. Indicators of the components of the consumer product market

Indicators	2013	2014	2015	2016	2017	Average
FixSituationDyn	1,096	1,095	1,097	1,099	1,022	1,081
MerchSeqGain	1,099	1,100	1,109	1,015	1,025	1,069
CateringSeqTempo	1,160	1,182	1,140	1,001	1,034	1,083
ServiceSeqTempo	1,082	1,072	1,051	0,989	1,012	1,041
Cact_of_ret_commd_market	1,7073	1,4524	1,5424	1,3979	1,3874	1,4975

FixSituationDyn - extrapolation according to the average level of a number of dynamics for an unchanged situation (total turnover, in% to the previous year);

MerchSeqGain - extrapolation according to the average growth of the series (retail turnover, in% to the previous year); CateringSeqTempo - extrapolation at the average growth rate (catering turnover, in% to the previous year); ServiceSeqTempo - extrapolation at the average growth rate (turnover of paid services, in% of the previous year) Cact_of_ret_commd_market – the ratio of commercial activity of the retail market;

ServiceSeqTempo - extrapolation at the average growth rate (turnover of paid services, in% of the previous year).

As it is seen from Table 1 the commercial activity ratio of the retail market in 2013 has a value +1,7073, which exceeds the critical value. This entails that in 2013 the demand was totally satisfied by the offer and the offer of the product surplus the demand significantly. But in 2017 this ratio decreased up to value of +1,3874 which is close to the maximum critical value +1,3. This entails the steady recovery of the economy of the republic.

Using the mathematical and mathematical tools Matlab [14], a model of a fuzzy-neural forecasting system was developed.

As input values in the generation system based on fuzzy logic, we used the average of the various components of the consumer product market, averaged over 2003–2017 (see Table 1). In turn, these indicators were obtained by extrapolation, based on the main indicator of this market – turnover [12].

3   Results

Figure 1 shows the rule editing window for generating the resulting fuzzy value for the commercial activity ratio of the retail commodity product market.

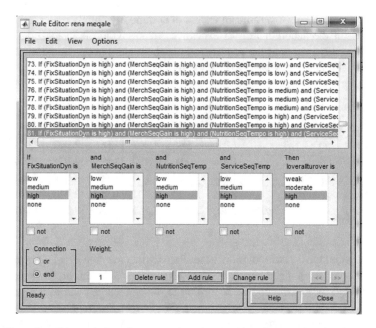

Fig. 1. The rule editing window for generating the resulting fuzzy value for the commercial activity ratio of the retail commodity product market.

4   Conclusion

Prediction of the commercial activity of the retail market for commodity products was carried out for 2020. Defuzzification of the resulting fuzzy value for the coefficient of commercial activity of the retail commodity product market gives us a value of Cact_of_ret_commd_market = 1,206. This means that by 2020 there will be an increase of goods and services of domestic production in the consumer goods market of the Republic of Azerbaijan, customer demand will be fully satisfied and an improved business environment and trading methods will function in the country.

References

1. Abryutina, M.S.: Economic analysis of the commodity market and trading activities. Business and Service, Moscow (2010)
2. Babich, T.N.: Forecasting and planning in market conditions. INFRA-M, Moscow (2012)
3. Zharikov, A.V., Goryachev, R.A.: Forecasting demand and sales. NSU, Novgorod (2017)
4. Zinnurov, U.G.: Strategic marketing planning and management. MAI, Moscow (2018)
5. Kuznetsova, G.V.: The conjuncture of world commodity markets. Yurayt, Moscow (2019)
6. Kirishneva, I.R., et al.: Forecasting and planning in market conditions. RUPS, Rostov n/A (2001)
7. Lychkina, N.N.: Simulation of economic processes. INFRA-M, Moscow (2012)

8. Mashunin, Yu.K.: Theory, mathematical modeling and forecasting of market development. J. Far Eastern Federal Univ. (2. Market structure), 3–20 (2017)
9. Samigulin, E.V.: Economic law and the parameters of the retail product market. AUCA Academic Review, pp. 117–125 (2008)
10. The trade in Azerbaijan Republic. Statistical almanac Azərbaycanda ticarət. Statistik məcmuə. AR DSK. Baku (2018)
11. Jang, J.S.R., Sun, C.T., Mizutani, E.: Neuro-Fuzzy and Soft Computing. Prentice Hall (1997)
12. Zadeh, L.A.: Fuzzy sets as a basis for a theory of possibility. Fuzzy Sets Syst. **1**, 3–28 (1978)
13. Aliev, R.A., Fazlollahi, B., Aliev, R.R.: Soft Computing and its Applications in Business and Economics. Springer, Heidelberg (2004)
14. Beale, M.H., Hagan, M.T., Demuth, H.B.: Neural Network Toolbox. User's Guide: Math Works, Inc., Natick (2014)

Properties of Set-Theoretical Operations over Z-Sets

Akif V. Alizadeh$^{(\boxtimes)}$ (iD)

Department of Control and Systems Engineering,
Azerbaijan State Oil and Industry University, 20 Azadlig Avenue,
Baku AZ1010, Azerbaijan
akifoder@yahoo.com, a.alizade@asoiu.edu.az

Abstract. In this study we initiate research on set-theoretical operations over Z-sets. We formalize set theoretic operations as complement, union and intersection of Z-sets. Some important properties of these set-theoretic operations are studied.

Keywords: Z-arithmetic · Probabilistic arithmetic · Associativity law · Commutativity law · Z-sets

1 Introduction

Set theory is one of the fundamental parts of mathematics. Theories of crisp sets, fuzzy sets and random sets are well-developed. However, no works exist on theory of Z-sets. The existing works in the scope of Z-number concept are mainly devoted to problems of computation of Z-numbers, ranking of Z-numbers, and various applications. An overview of these works is given in the preceding chapters. Let us mention some other works that are important in the light of Z-valued sets study. In [1] the authors introduce a comprehensive measure for distance between Z-numbers. They consider a Z-number in a form of fuzzy numbers A, B and a set of distributions that relate B to A. As a result, in this measure, they use distance between sets of distributions underlying two Z-numbers.

In [2] they consider Z-number as a pair of fuzzy sets following the Zadeh's interpretation, and propose a new way to formalization. Z-numbers are viewed in the frameworks of possibility theory, imprecise probabilities and belief functions. The authors show that it is easier to use random sets than convex sets of probabilities for Z-numbers. In general, the authors propose a new fundamental study of Z-numbers.

In this study, we initiate research on set-theoretical operations over Z-sets. We formalize set theoretic operations as complement, union and intersection of Z-sets. Some important properties of these set-theoretic operations are studied. Examples are included to illustrate the proposed approach.

The paper is organized as follows: Sect. 2 includes basic concepts used in the paper. In Sect. 3 we define operations over Z-sets. Section 4 concludes properties of set theoretical operations under Z-Sets. Section 5 conclusion are given.

© Springer Nature Switzerland AG 2020
R. A. Aliev et al. (Eds.): ICSCCW 2019, AISC 1095, pp. 654–661, 2020.
https://doi.org/10.1007/978-3-030-35249-3_84

2 Preliminaries

Definition 1. Let X be a space of objects and $x \in X$. A Z-set Z in X is defined by membership function $A(x) \in [0, 1]$, probability distributions P_X associated with X, membership function $B(P_x) \in [0, 1]$. Where $A(x)$ denotes that element x belongs to A with membership degree $A(x)$, $B(P(x))$ denotes that probability measure value belongs to B with degree $B(P_x)$, P_X is the underlying probability density of X. There is restriction $P_X = \int_X A(x)p(x)dx$.

Taking to accounting that the degree to which P_X satisfies Z-set $\{\mathrm{Prob}_P(A(x)), \forall x \in X\}$ is B we can get $G_P = B(\int_X A(x)p(x)dx)$.

So Z-set in X can be expressed as triple $Z = (A, B, G)$, where A is fuzzy subset of X, B is fuzzy set (restriction) on probability measure of A, G is set of probability distribution over the space P.

$P(A)$, induced by a set of distributions G:

$$G = \left\{ p_Z(x) : \int_X p_Z(x)dx = 1, \int_X p_Z(x)A(x)dx \text{ is } B, x \in X \right\}. \tag{1}$$

3 Set Theoretical Operations over Z-Sets

We consider three main operations of Z-sets $Z_i = (A_i, B_i, G_i)$ which are defined as follows.

3.1 Complement of Z-Sets

Let $Z = (A, B, G)$ be a Z-set, where G be a set whose elements are all the probability distributions, defined as (1).

The complement $\bar{Z}$ of Z-set $Z = (A, B, G)$ is defined as follows.

$\bar{G}$ set of probability distributions defined as

$$\bar{G} = \left\{ p_Z(x) : \int_X p_Z(x)dx = 1, \int_X p_Z(x)A(x)dx \text{ is } B, x \in X \right\}. \tag{2}$$

Then

$$\bar{Z} = (\bar{A}, 1 - B, \bar{G}), \tag{3}$$

where $\bar{A}(x) = 1 - A(x)$ and $1 - B$ is fuzzy arithmetic operation.

3.2 Union of Z-Sets

Union of Z-sets $Z_i = (A_i, B_i, G_i)$, $i = 1, 2$ is defined as
$$Z_{12} = Z_1 \cup Z_2 = Zor(Z_1, Z_2) = (A_{12}, B_{12}, G_{12}).$$
At first, let us consider the case of continuous Z-sets.
The union $A_{12} = A_1 \cup A_2$ of the fuzzy sets A_1 and A_2 is defined as follows:

$$A_{12}(x) = (A_1 \cup A_2)(x) = \max(A_1(x), A_2(x)). \tag{4}$$

The set G_{12} of resulting distributions p_{12} is defined as $G_{12} = G_1 \cup G_2$, where

$$G_i = \left\{ p_{Z_i}(x) : \int_X p_{Z_i}(x) dx = 1, \int_X p_{Z_i}(x) A_i(x) dx \text{ is } B_i, x \in X \right\} \tag{5}$$

for $i = 1, 2$.

Convolution $p_{12} = p_{Z_1 \cup Z_2}(x) = p_{Z_1}(x) \circ_\cup p_{Z_2}(x) = p_1 \circ_\cup p_2$ of probability distributions is defined as

$$p_{12}(x) = p_{Z_1 \cup Z_2}(x) = \frac{(p_{Z_1}(x) + p_{Z_2}(x) - p_{Z_1}(x) p_{Z_2}(x)) p_X(x)}{\int_X (p_{Z_1}(x) + p_{Z_2}(x) - p_{Z_1}(x) p_{Z_2}(x)) p_X(x) dx}, \tag{6}$$

where $p_X(x) = \frac{1}{card(X)}$.

Then $Z_{12} = (A_{12}, B_{12}, G_{12})$, where

$$G_{12} =$$
$$\left\{ p_{12}(x) : p_{12}(x) = p_{Z_1 \cup Z_2}(x) = \frac{(p_{Z_1}(x) + p_{Z_2}(x) - p_{Z_1}(x) p_{Z_2}(x)) p_X(x)}{\int_X (p_{Z_1}(x) + p_{Z_2}(x) - p_{Z_1}(x) p_{Z_2}(x)) p_X(x) dx}, \right. \tag{7}$$
$$\left. p_i(x) \in G_i, x \in X \right\}$$
$$B_{12} = \left\{ (\mu_{p_{12}}(p_{12}), \mu_{A_{12}} \cdot p_{12}) : p_{12} \in G_{12} \right\}.$$

3.3 Intersection of Z-Sets

Intersection of Z-sets $Z_i = (A_i, B_i, G_i)$, $i = 1, 2$ is defined as
$$Z_{12} = Z_1 \cap Z_2 = Zand(Z_1, Z_2) = (A_{12}, B_{12}, G_{12}).$$
The intersection $A_1 \cap A_2$ of the fuzzy sets A_1 and A_2 is defined as follows:

$$A_{12}(x) = (A_1 \cap A_2)(x) = \min(A_1(x), A_2(x)). \tag{8}$$

The set G_{12} of resulting distributions p_{12} is defined as

$$G_{12} = G_1 \cap G_2,$$

Where G_i defined as (5).

Convolution $p_{12} = p_{Z_1 \cap Z_2}(x) = p_{Z_1}(x) \circ_\cap p_{Z_2}(x) = p_1 \circ_\cap p_2$ of probability distributions is defined as

$$p_{12}(x) = p_{Z_1 \cap Z_2}(x) = \frac{p_{Z_1}(x)p_{Z_2}(x)p_X(x)}{\int_X p_{Z_1}(x)p_{Z_2}(x)p_X(x)dx}, \tag{9}$$

where $p_X(x) = \frac{1}{card(X)}$.

Then $Z_{12} = (A_{12}, B_{12}, G_{12})$, where

$$G_{12} = \{p_{12}(x) : p_{12}(x) = p_{Z_1 \cap Z_2}(x) = \frac{p_{Z_1}(x)p_{Z_2}(x)p_X(x)}{\int_X p_{Z_1}(x)p_{Z_2}(x)p_X(x)dx}, p_i(x) \in G_i, x \in X\}. \tag{10}$$

Thus,

$$B_{12} = \{(\mu_{p_{12}}(p_{12}), \mu_{A_{12}} \cdot p_{12}) : p_{12} \in G_{12}\}.$$

4 Properties of Operations on Z-Sets

Theorem 1. Commutative law for intersection of Z-sets is satisfied:
$Z_{12} = Z_1 \cap Z_2 = Z_2 \cap Z_1 = Z_{21}$.
Proof. For A_{12} one has:

$$A_{12}(x) = (A_1 \cap A_2)(x) = \min(A_1(x), A_2(x))$$
$$= \min[A_2(x), A_1(x)] = (A_2 \cap A_1)(x) = A_{21}(x).$$

The set of distributions of intersection Z_{12} can be described as follows. $G_{12} = \{p_{12}(x) : p_{12}(x) = p_{Z_1 \cap Z_2}(x) = \frac{p_{Z_1}(x)p_{Z_2}(x)p_X(x)}{\int_X p_{Z_1}(x)p_{Z_2}(x)p_X(x)dx}, p_i(x) \in G_i\}$, where G_i defined as (5).

Thus, G_{12} is a set of convolutions $p_{12}(x)$. $p_{12}(x)$ satisfies commutativity property [3]:

$$p_{12}(x) = p_{Z_1 \cap Z_2}(x) = \frac{p_{Z_1}(x)p_{Z_2}(x)p_X(x)}{\int_X p_{Z_1}(x)p_{Z_2}(x)p_X(x)dx}$$
$$= \frac{p_{Z_2}(x)p_{Z_1}(x)p_X(x)}{\int_X p_{Z_2}(x)p_{Z_1}(x)p_X(x)dx} = p_{Z_2 \cap Z_1}(x) = p_{21}(x).$$

Therefore, G_{12} also satisfies commutativity property:

$$G_{12} = \{p_{12}(x) : p_{12}(x) = p_{Z_1 \cap Z_2}(x) = \frac{p_{Z_1}(x)p_{Z_2}(x)p_X(x)}{\int_X p_{Z_1}(x)p_{Z_2}(x)p_X(x)dx}, p_i(x) \in G_i\}$$
$$= \{p_{21}(x) : p_{21}(x) = p_{Z_2 \cap Z_1}(x) = \frac{p_{Z_2}(x)p_{Z_1}(x)p_X(x)}{\int_X p_{Z_2}(x)p_{Z_1}(x)p_X(x)dx}, p_i(x) \in G_i\} = G_{21}$$

$$G_{12} = G_1 \cap G_2 = G_2 \cap G_1 = G_{21}.$$

Then

$$B_{12} = \{(\mu_{p_{12}}(p_{12}), \mu_{A_{12}} \cdot p_{12}) : p_{12} \in G_{12}\} = (\mu_{p_{21}}(p_{21}), \mu_{A_{21}} \cdot p_{21}) : p_{21} \in G_{21}\} = B_{21}.$$

Thus, $Z_{12} = Z_{21}$. The proof is completed.

Theorem 2. Commutative law for union of Z-sets holds:

$Z_{12} = Z_1 \cup Z_2 = Z_2 \cup Z_1 = Z_{21}$.

Proof. For A_{12} one has:

$$A_{12}(x) = (A_1 \cup A_2)(x) = \max(A_1(x), A_2(x))$$
$$= \max[A_2(x), A_1(x)] = (A_2 \cup A_1)(x) = A_{21}(x)$$

The set of distributions $G_{12} = G_1 \cup G_2$ of union Z_{12} can be described as follows.

$$G_{12} = \{p_{12}(x) : p_{12}(x) = p_{Z_1 \cup Z_2}(x)$$
$$= \frac{(p_{Z_1}(x) + p_{Z_2}(x) - p_{Z_1}(x)p_{Z_2}(x))p_X(x)}{\sum_X (p_{Z_1}(x) + p_{Z_2}(x) - p_{Z_1}(x)p_{Z_2}(x))p_X(x)}, p_i(x) \in G_i\},$$

where G_i defined as (5).

$p_{12}(x)$ satisfies commutativity property [3]:

$$p_{12}(x) = p_{Z_1 \cup Z_2}(x) = \frac{(p_{Z_1}(x) + p_{Z_2}(x) - p_{Z_1}(x)p_{Z_2}(x))p_X(x)}{\int_X (p_{Z_1}(x) + p_{Z_2}(x) - p_{Z_1}(x)p_{Z_2}(x))p_X(x)dx}.$$

Therefore, G_{12} also satisfies commutativity property:

$$G_{12} = \{p_{12}(x) : p_{12}(x) = p_{Z_1 \cup Z_2}(x) = \frac{(p_{Z_1}(x) + p_{Z_2}(x) - p_{Z_1}(x)p_{Z_2}(x))p_X(x)}{\sum_X (p_{Z_1}(x) + p_{Z_2}(x) - p_{Z_1}(x)p_{Z_2}(x))p_X(x)}, p_i(x) \in G_i\}$$

$$= \{p_{21}(x) : p_{21}(x) = p_{Z_2 \cup Z_1}(x)$$
$$= \frac{(p_{Z_2}(x) + p_{Z_1}(x) - p_{Z_2}(x)p_{Z_1}(x))p_X(x)}{\sum_X (p_{Z_2}(x) + p_{Z_1}(x) - p_{Z_2}(x)p_{Z_1}(x))p_X(x)}, p_i(x) \in G_i\} = G_{21}$$

$$G_{12} = G_1 \cup G_2 = G_2 \cup G_1 = G_{21}.$$

Then

$$B_{12} = \{(\mu_{p_{12}}(p_{12}), \mu_{A_{12}} \cdot p_{12}) : p_{12} \in G_{12}\} = (\mu_{p_{21}}(p_{21}), \mu_{A_{21}} \cdot p_{21}) : p_{21} \in G_{21}\} = B_{21}.$$

Thus, $Z_{12} = Z_{21}$. The proof is completed.

Theorem 3. Associative law holds for intersection of Z-sets:

$(Z_1 \cap Z_2) \cap Z_3 = Z_1 \cap (Z_2 \cap Z_3)$.

Proof. For A parts of Z-sets one has:

$$A_{(12)3}(x) = (A_{12} \cap A_3)(x) = \min[A_{12}(x), A_3(x)] = \min[\min[A_1(x), A_2(x)], A_3(x)].$$

As min operation satisfies associativity condition, then

$$\min[\min[A_2(x), A_3(x)], A_1(x)] = \min[A_1(x), A_{23}(x)] = A_{1(23)}(x).$$

Then, for any fuzzy sets A_1, A_2, A_3 the following property holds $((A_1 \cap A_2) \cap A_3)) = (A_1 \cap (A_2 \cap A_3))$, so $A_{(12)3} = A_{1(23)}$.
Let us now consider G sets:

$$G_{(12)3} = \{p_{(12)3}(x) : p_{(12)3}(x) = (p_1 \circ_\cap p_2) \circ_\cap p_3, p_i(x) \in G_i, i = 1, \ldots, 3\},$$

$$G_{1(23)} = \{p_{1(23)}(x) : p_{1(23)}(x) = p_1 \circ_\cap (p_2 \circ_\cap p_3), p_i(x) \in G_i, i = 1, \ldots, 3\}.$$

As $\circ_\cap$ in operation for random variables satisfies associativity property $p_{(12)3}(x) = p_{1(23)}(x)$ [3], we have $G_{(12)3} = G_{1(23)}$.
As, $A_{(12)3} = A_{1(23)}$, and $G_{(12)3} = G_{1(23)}$, then $B_{(12)3} = B_{1(23)}$.
Therefore, $Z_{(12)3} = Z_{1(23)}$. The proof is completed.

Theorem 4. Associative law holds for union of Z-sets:
$(Z_1 \cup Z_2) \cup Z_3 = Z_1 \cup (Z_2 \cup Z_3)$.
Proof. For A parts of Z-sets one has:

$$A_{(12)3}(x) = (A_{12} \cup A_3)(x) = \max[A_{12}(x), A_3(x)] = \max[\max A_1(x), A_2(x)], A_3(x)].$$

As max operation satisfies associativity condition, then

$$\max[\max[A_2(x), A_3(x)], A_1(x)] = \max[A_1(x), A_{23}(x)] = A_{1(23)}(x).$$

Then, for any fuzzy numbers A_1, A_2, A_1 the following property hold $((A_1 \cap A_2) \cap A_3)) = (A_1 \cap (A_2 \cap A_3))$, so $A_{(12)3} = A_{1(23)}$.
Let us now consider G sets:

$$G_{(12)3} = \{p_{(12)3}(x) : p_{(12)3}(x) = (p_1 \circ_\cup p_2) \circ_\cup p_3, p_i(x) \in G_i, i = 1, \ldots, 3\},$$

$$G_{1(23)} = \{p_{1(23)}(x) : p_{1(23)}(x) = p_1 \circ_\cup (p_2 \circ_\cup p_3), p_i(x) \in G_i, i = 1, \ldots, 3\}.$$

As $\circ_\cup$ operation for random variables satisfies associativity property $p_{(12)3}(x) = p_{1(23)}(x)$ [3], we have $G_{(12)3} = G_{1(23)}$.
As $A_{(12)3} = A_{1(23)}$, and $G_{(12)3} = G_{1(23)}$, then $B_{(12)3} = B_{1(23)}$.
Therefore, $Z_{(12)3} = Z_{1(23)}$. The proof is completed.

Theorem 5. Distribute law holds for Z-sets:

$$Z_1 \cap (Z_2 \cup Z_3) = (Z_1 \cap Z_2) \cup (Z_1 \cap Z_3),$$

$$Z_1 \cup (Z_2 \cap Z_3) = (Z_1 \cup Z_2) \cap (Z_1 \cup Z_3).$$

Proof. Let us consider $Z_1 \cap (Z_2 \cup Z_3) = (Z_1 \cap Z_2) \cup (Z_1 \cap Z_3)$. The proof of $Z_1 \cup (Z_2 \cap Z_3) = (Z_1 \cup Z_2) \cap (Z_1 \cup Z_3)$ is analogous.
In opened notation:

$$(A_1, B_1) \cap ((A_2, B_2) \cup (A_3, B_3, G_3))$$
$$= ((A_1, B_1, G_1) \cap (A_2, B_2, G_2)) \cup ((A_1, B_1, G_1) \cap (A_3, B_3, G_3)).$$

The distributive law holds for any fuzzy sets A_1, A_2, A_3:

$$A_{1(23)}(x) = \min(A_1(x), \max(A_2(x), A_3(x)))$$
$$= \max(\min(A_1(x), A_2(x)), \min(A_1(x), A_3(x))) = A_{(12)(13)}(x).$$

Let us now consider G_i sets: by (5) for $i = 1, 2, 3$.
For any probability distributions p_1, p_2, p_3 the distributive law holds:

$$p_{1(23)}(x) = p_1 \circ_\cap (p_2 \circ_\cup p_3) = (p_1 \circ_\cap p_2) \circ_\cup (p_1 \circ_\cap p_3) = p_{(12)(13)}(x).$$

Thus,
$$p_{1(23)}(x) = p_{(12)(13)}(x).$$
Therefore, one has $G_{1(23)} = G_{(12)(13)}$,
where $G_{1(23)} = \{p_{1(23)}(x) : p_{1(23)}(x) = p_1 \circ_{\min} (p_2 \circ_{\max} p_3), p_i(x) \in G_i, i = 1, \ldots, 3\}$,

$$G_{(12)(13)} = \{p_{(12)(13)}(x) : p_{(12)(13)}(x) = (p_1 \circ_{\min} p_2) \circ_{\max} (p_1 \circ_{\min} p_3), p_i(x) \in G_i, i = 1, \ldots, 3\}.$$

As $A_{(12)3} = A_{(12)(13)}$, and $G_{1(23)} = G_{(12)(13)}$, then $B_{1(23)} = B_{(12)(13)}$.
Thus, $Z_{1(23)} = Z_{(12)(13)}$. The proof is completed.

5 Conclusion

It is proved that the basic laws of set theoretical operations holds for Z-sets. The proofs are based on the analogous properties of fuzzy sets and probabilistic arithmetic. The obtained results are necessary for strong formulation of such important concepts as similarity of Z-sets, relation of Z-sets and other concepts.

References

1. Aliev, R.A., Alizadeh, A.V., Huseynov, O.H.: The arithmetic of discrete Z-numbers. Inform. Sci. **290**, 134–155 (2015)
2. Aliev, R.A., Alizadeh, A.V., Huseynov, O.H.: The arithmetic of continuous Z-numbers. Inform. Sci. **373**, 441–460 (2016)
3. Piegat, A., Plucinski, M.: Computing with words with the use of Inverse RDM Models of Membership Functions. Appl. Math. Comput. Sci. **25**(3), 675–688 (2015)

Tumor Classification Using Gene Expression and Machine Learning Models

Kubra Tuncal$^{(\boxtimes)}$ (ID) and Cagri Ozkan$^{(\boxtimes)}$ (ID)

Information Systems Engineering, Near East University, Nicosia,
TRNC, Mersin 10, Turkey
kubra.tuncal@neu.edu.tr, cagri.ozkan@neu.edu.tr

Abstract. Cancer is the most fatal cause of death and determination of the reasons, making early diagnosis and correct treatment reduces the loss of lives but humans are still far away to produce a complete and permanent solutions to this problem. Nowadays, RNA and gene researches try make this solutions step by step more effective to defect cancer and to improve these researches. However, the number of the genes and complexity of the data makes analysis and experiments more challenging for humans thus, computerized solutions such as machine learning models are needed. This paper presents preliminary results of five types of tumor classification on RNA-Seq. Three machine learning models, Support Vector Machine, Backpropagation neural network and Decision Tree is implemented and various experiments are performed for this task. Obtained results show that machine learning models can effectively be used for tumor classification using gene information and Support Vector Machine achieved superior results than other considered models.

Keywords: RNA-Seq · Backpropagation · Support Vector Machine · Decision Tree

1 Introduction

Cancer is called the structure that occurs with the accumulation of disorders in the DNA in the cells and at the same time it increases irregularly. Deoxyribonucleic acid (DNA) has an inherited structure both in humans and almost all other living things. Many of the DNA cells are in the nucleus and the remaining part is located in the mitochondria. Ribonucleic acid (RNA) is a polymeric molecule involved in coding, decoding, editing and expression of genes. Cellular organisms need messenger RNA to transmit genetic data that sends the synthesis of specific proteins. At the same time, most of the viruses use the RNA genome to encode their genetic information. Therefore, DNA and RNA must coexist and work together to accomplish tasks within a cell [1].

Gene is both a basic physical and functional part of heredity. It consists of DNA and each gene contains the information necessary to produce the specific proteins an organism needs. Each gene performs its function by coding different proteins from the other gene. At the same time, they are responsible for carrying hereditary information. When a problem occurs in the genes that are composed of DNA, the accumulation and

© Springer Nature Switzerland AG 2020
R. A. Aliev et al. (Eds.): ICSCCW 2019, AISC 1095, pp. 662–667, 2020.
https://doi.org/10.1007/978-3-030-35249-3_85

irregular proliferation of these disorders reveal the tumor structure, that is cancer. Therefore, regular functioning of genes is of great importance.

According to the Human Genome Project; there are a total number of genes between 29,000 and 36,000 in humans. Because of the role of genes in the diagnosis of cancer there are several studies on this subject.

Machine Learning is a sub-branch of artificial intelligence. Machine learning algorithms are modeling and analyze the existing data and make the best results from these data. Thus, researches performed several researches about RNA, DNA, gene and cancer prediction and classification tasks.

In the study conducted by Xiao and Wu [2], RNA-Gene expression data were studied. Three RNA sequence data sets were considered and comparison also performed with Stacked sparse auto-encoder (SSAE), Support Vector Machine (SVM), Random Forest (RF), Neural Network (NN) and Auto-encoder (AE) The results of the data used with the SSAE method achieved superior results than the other methods. Weinstein et al. [3] considered twelve types of tumors. In this study, each of the 12 types of cancer data included 5.074 total tumor specimens, 93% of which were evaluated by genome, epigenomic, gene and protein expression data on a minimum platform. It also includes multiple tumor types together and prepares a common data set. This is a study conducted to ensure that data is analyzed and interpreted by ensuring consistency across all platforms.

In the research of Danaee and Ghaeini [4], Rna-seq Expression was used as dataset. Deep learning was used to identify genes that are important in the diagnosis of breast cancer. Stacked Denoising Autoencoder (SDAE) method achieved optimal results on this dataset. Tarek and Abd Elwahab [5], used Gene expression data that consists Leukemia Dataset, Colon Dataset and Breast Cancer Datasets. Ensemble Module, produced highest results in the classification in their study. Huang and Cai [6] used Genomic and epigenomic data with support vector machine (SVM) for the classification of cancer types.

In this paper, three machine learning models as Backpropagation, Support Vector Machine and Decision Tree, are considered in order to classify gene expression cancer RNA-Seq Data Set [3] which includes 20531genes and 801 instances as preliminary experiments. Five classes as Breast Cancer Susceptibility, Kidney Renal Clear Cell Carcinoma, Colon Adenorcarcinoma, Lung Adenocarcinoma and Prostate Adenocarcinoma are included to the dataset.

The rest of the paper is organized as follows; Sect. 2 introduces the considered algorithms in this research. Section 3 presents performed experiments and obtained results in details. Finally, Sect. 4, concludes the achieved results within this research.

2   Classification Models

Classification differs from prediction by finite number of output classes. Real or binary outputs are classified as corresponding outputs according to some figure of merit. Several models which uses different algorithms are proposed in order to classify objects or region of interest correctly.

In this research, three different kinds of models are considered in order to perform classification task. These models are Backpropagation neural network, Decision Tree Classifier and Support Vector Machine.

Backpropagation is considered because of its efficiency in classification tasks of non-linear data [7] and considering one of the fundamental neural network to compare obtained results with other models.

Decision Tree Classifier is a classification algorithm based on divide and conquer [8]. It is quietly different than other classification models but it is also an effective model for the data that has non-linear relationship between attributes and instances.

Initially, Support Vector Machines proposed for binary classification but then it was modified for multi-class tasks [9]. It uses support vectors to optimize decision on hyper-plane.

These models are frequently used both in classification and prediction tasks in several researches [10–13].

3   Experiments and Results

This section presents the dataset which is the domain of the problem, performed experiments and obtained results in details. In this research, two kinds of experiments have been performed by using considered three models in order to obtain optimal classification rates.

3.1   Dataset

RNA-Seq Data Set [3] includes 20531 genes as attributes and 801 instances to classify 5 different cancer types namely Breast Cancer Susceptibility (BRCA), Kidney Renal Clear Cell Carcinoma (KIRC), Colon Adenorcarcinoma (COAD), Lung Adenocarcinoma (LUAD) and Prostate Adenocarcinoma (PRAD). In this paper, 1022 attributes (genes) are considered to classify these cancer types as preliminary research.

3.2   Experiments

Two kinds of experiments includes 50% and 70% of training ratio respectively by using Hold-out method which is based on randomly selection of training and testing instances. For all algorithms, inputs and outputs are the number of considered attributes and cancer types which are 1022 and 5 respectively.

Evaluation of each model is performed by considering the accuracy of obtained results, main indicator of classification problems, which the formula is given in Eq. 1.

$$AC = \frac{CCI}{TNI} \tag{1}$$

where AC is the accuracy, CCI and TNI is correctly classified instances and total number of instances respectively.

3.3 Parameters of Models

Each model has its unique parameter that affects the learning or testing phases. After several experiments common parameters was chosen to each experiment of each model.

In the considered Backpropagation neural network, there are 3 hidden layers with 500 neurons for each. Learning rate and momentum factor was 0.0009 and 0.90 respectively. Maximum epochs was decided to be used as 250.

In Decision Tree Classifier, determination of initial node which is the root and the sequence of the internal nodes is a challenging tasks and one of the main algorithm that solve this problem is Gini algorithm. Thus, Gini algorithm is considered to build the tree.

In Support Vector Machine, different kernel types were proposed but generally Radial-Basis Function kernel is preferred to be used in big data which has non-linear relationship between attributes and instances. Therefore, Radial-Basis Function kernel is used in this research.

3.4 Experimental Results

In the first experiment, models are trained by considering 50% of dataset. Experimental results shows that Backpropagation Neural Network and Support Vector Machine produced sufficient results however, Support Vector Machine achieved 99.75% while accuracy rate but Backpropagation was 98.70%. Accuracy rate of Decision Tree was 95.51%.

In the second experiment, models are trained by considering 70% of dataset. Similar to Experiment 1, Backpropagation Neural Network and Support Vector Machine produced superior results however, Support Vector Machine classified all untrained instances correctly with 100% of accuracy rate but Backpropagation achieved 99.17%. Accuracy rate of Decision Tree was 91.28%. Table 1 shows the obtained results in details.

Table 1. Obtained accuracy results for all experiments

Training Ratio	Backpropagation	Support Vector Machine	Decision Tree
50%	98.70%	99.75%	95.51%
70%	99.17%	100%	91.28%

3.5 Discussions on Experimental Results

Experimental results show that Support Vector Machine and Backpropagation Neural Network can be used effectively to classify cancer types or classes using gene dataset with high accuracy. However, Decision Tree even it achieved more than 90% of accuracy in both experiments, is not effective enough as other two considered models.

Efficiency of these two models can be explained by the characteristics of dataset which allows machine learning algorithms to build a relationship between instances and attributes. But the main drawback of Decision Trees is the sequence of leaves during the build of tree which may suddenly increase or decrease the accuracy of the model.

Increment of training data is caused more effective convergence in Backpropagation Neural Network and Support Vector Machine but because of its structure, Decision Tree increased its accuracy while decreasing the training data which cause minimal instances to build tree.

4 Conclusions

Genes are responsible and also indicators for several diseases and tumors. Considering the information of these thousands of genes and making prediction and classification is challenging task for human beings. Thus, implementation and application of machine learning techniques are inevitable.

In this paper, three machine learning algorithms, Support Vector Machines, Backpropagation neural network and Decision Tree is implemented to RNA-Seq dataset to classify five tumor types.

Obtained preliminary results shows that both Support Vector Machine and Backpropagation can be used efficiently for this classification task with 100% and 99.17% of accuracy. Decision Tree can also be used in this classification but not efficient as well as others.

Future work will include the consideration of all genes and implementation of more machine learning models to classify all tumor types.

References

1. Erdemir, F., Gülzade, U.: Genetik, genomik bilimi ve hemşirelik. Dokuz Eylül Üniversitesi Hemşirelik Yüksekokulu Elektronik Dergisi 3(2), 96–101 (2010). (in Turkish)
2. Xiao, Y., Wu, J., Lin, Z.: A semi-supervised deep learning method based on stacked sparse auto-encoder for cancer prediction using RNA-seq data. Comput. Meth. Programs Biomed. 166, 99–105 (2018)
3. Weinstein, J.N., et al.: The cancer genome atlas pan-cancer analysis project. Nat. Genet. 45(10), 1113–1120 (2013)
4. Danaee, P., Ghaeini, R., Hendrix, D.: A deep learning approach for cancer detection and relevant gene identification. Pacific Symp. Biocomput. 2017, 219–229 (2017)
5. Tarek, S., Abd Elwahab, R., Shoman, M.: Gene expression based cancer classification. Egypt. Inf. J. 18, 151–159 (2017)
6. Huang, S., et al.: Applications of support vector machine (SVM) learning in cancer genomics. Cancer Genomics Proteomics 15, 41–51 (2018)
7. Khashman, A., Sekeroglu, B.: Global binarization of document images using a neural network. In: Third International IEEE Conference on Signal-Image Technologies and Internet-Based System, pp. 665–672, Shanghai (2007)
8. Dougherty, G.: Pattern recognition and classification. Springer, New York (2013). https://doi.org/10.1007/978-1-4614-5323-9

9. Tong, S., Koller, D.: Support vector machine active learning with applications to text classification. J. Mach. Learn. Res. **2**, 45–66 (2001)
10. Senturk, Z.K., Senturk, A.: Yapay sinir agları ile göğüs kanseri tahmini. El-Cezeri J. Sci. Eng. **3**(2), 345–350 (2016). (in Turkish)
11. Yuan, Z., Wang, C.: An improved network traffic classification algorithm based on Hadoop decision tree. In: 2016 IEEE International Conference of Online Analysis and Computing Science (ICOACS), Chongqing, pp. 53–56 (2016)
12. Ge L, Shi, J., Zhu, P.: Melt index prediction by support vector regression. In: 2016 International Conference on Control, Automation and Information Sciences (ICCAIS), Ansan, pp. 60–63 (2016)
13. Polaka, I., Igar, T., Borisov, A.: Decision tree classifiers in bioinformatics. J. Riga Tech. Univ. (42), 118–123 (2010)

Assessment of Environmental Management in the Region on the Basis of Fuzzy-Plural Analysis of Statistical Data

Elizabeth A. Arapova$^{(\boxtimes)}$ ⑩, Sergey V. Rogozhin ⑩,
Anatoly F. Chuvenkov ⑩, and Svetlana A. Batygova ⑩

Rostov State University of Economics, B. Sadovaya str., 69,
344002 Rostov-on-Don, Russia
dist_edu@ntti.ru, sergeyvr@yandex.ru,
chuvenkovaf@mail.ru, batygova@yandex.ru

Abstract. The method of assessment of the degree of economic activity in the region the basic principles of environmental management. The assessment of the region is formed on the basis of fuzzy-multiple aggregation of the corresponding estimates of its municipalities – large cities and districts. The technique is based on the use of standard fuzzy five-level [0,1] classifiers. As a statistical material, standard data on the state and protection of the environment in the municipalities of the Rostov region for five years – 2012–2016 were used. The proposed methodology includes the following stages: (1) the formation of five assessments for each municipality: "assessment of the dynamics of polluting emissions into the air"; "assessment of the dynamics of the load on the water system"; "the degree of clutter of the territory"; "the share of protected natural areas"; "the level of financing environmental measures"; (2) aggregation of the received estimates into the final assessment of the municipality: "assessment of compliance of environmental principles"; (3) ranking of municipalities in accordance with the received assessments; (4) aggregation of assessments of municipalities into a comprehensive assessment of the region, which allows to judge the level of compliance of economic activities in the region with the basic principles of environmental management.

Keywords: Environmental management · Integrated assessment · Regional · Municipal · Multi-level fuzzy classifiers

1 Introduction

The ecological condition of the region largely depends on the planning and implementation of environmental activities at the level of municipalities - cities and regions. Environmental damage caused by human activity may be significantly reduced due to events such as: treatment of wastewater and pollutant emissions into the atmosphere; remediation of illegal landfills and the monitoring of waste disposal sites, included in the state register; expansion of protected areas; increase funding for environmental measures.

© Springer Nature Switzerland AG 2020
R. A. Aliev et al. (Eds.): ICSCCW 2019, AISC 1095, pp. 668–674, 2020.
https://doi.org/10.1007/978-3-030-35249-3_86

The control parameters describing the listed destinations, performed in Rostov region at the municipal level [1] and [2]. Edition [1] annually a detailed report on the environmental performance in big cities and districts of the Rostov region on a fixed set of indicators of diverse indicators. The volume of the information is constantly increasing, making hardly visible complex analysis of ecological environmental management in the region as a whole. Another more serious problem is a similar analysis of individual municipalities and ranking them in order of formation of ecological regional development strategy at the municipal level.

This paper proposes a method of estimating the degree of compliance with environmental management in the region, the basic principles of ecological environmental management. Assessment of the region is based on fuzzy multiple aggregation of relevant assessments of its municipalities - cities and regions. Qualification of each municipality is generated based on the aggregation time series statistics in the following areas: the dynamics of polluting emissions into the atmosphere; dynamic load on the water system; the degree of clutter territory; the proportion of protected areas; the level of funding of environmental measures.

The technique is based on applying standard fuzzy five-level [0.1] classifiers. The material used as the statistical data on the standard condition and the environment in municipalities Rostov region (cities and districts) annually given in [1]. Data compiled over five years – 2012–2016 years [2–7].

The proposed method includes the following steps:

(1) forming for each municipality five ratings: "Evaluation of the dynamics of polluting emissions into the air"; "Assessment of the dynamics of the load on the water system"; "Degree of clutter territory"; "The share of protected areas"; "The level of funding of environmental activities";

(2) aggregation of the estimates in the final assessment of the municipality, "Conformity assessment of environmental management principles of ecological environmental management".

(3) ranking municipalities in accordance with the received estimates;

(4) The aggregate estimates of municipalities in a comprehensive assessment of the region.

Consider in more detail the method of forming each of the estimates.

2   General Principles of Operation of Standard Five-Point [0,1] – Classifiers

The algorithm of formation of a complex assessment of the object on the basis of a set of indicators consists of the following stages [8], [9].

Phase 1. Introduction to the consideration of the linguistic variable "complex assessment of the state of the object on the basis of a set of indicators", the definition of its universal set, term-set, membership functions of terms.

Phase 2. Formation of a list of significant indicators, calculation of normalized values of indicators (belonging to the interval [0,1]).

Phase 3. Introduction to the consideration of the linguistic variable "index level", the definition of its universal set, term-set, membership functions of terms.

Phase 4. Ranking of indicators, indication of their weight coefficients in the final evaluation.

Phase 5. Aggregation of normalized values of indicators in a complex assessment of the state of the object, linguistic recognition of the numerical evaluation. In this study, each of the five estimates corresponds to a linguistic variable with a universal set in the form of a numerical segment. Term-set of each linguistic variable consists of five terms $G = \{G_1, G_2, G_3, G_4, G_5\}$. In each case, each of the terms carries its meaning; however, from the point of view of environmental management, we can assume that: term G_1 – "excellent"; G_2 – "good"; G_3 – "satisfactory"; G_4 – "bad"; G_5 – "very bad". The membership functions have a standard trapezoidal form (with an increase in the numerical value of the estimate corresponds to an increase in the number of the term). Each of the indicators involved in the formation of estimates is associated with a linguistic variable with a universal set in the form of a numerical segment [0,1]. Term-the set of each linguistic variable also consists of five terms $G = \{G_1, G_2, G_3, G_4, G_5\}$, and: term G_1 – "very low level of the indicator"; G_2 – "low level of the indicator"; G_3 – "average level of the indicator"; G_4 – "high level of the indicator"; G_5 – "very high level of the indicator".

Comprehensive evaluation of the state of nature is based on the aggregation of the above five ratings: g_1 = "assessment of the dynamics of polluting emissions into the air"; g_2 = "evaluation of load dynamics on water system"; g_3 = "clutter areas"; g_4 = "share protected natural territories"; g_5 = "the level of funding of environmental activities".

The linguistic variable g (municipality) = "assessment of compliance with the principles of environmental management" is introduced. Term-set consists of five terms $G = \{G_1, G_2, G_3, G_4, G_5\}$, conditionally assessing the state of the system: G_1 – "full compliance with the principles of environmental management"; G_2 – "compliance with the principles of environmental management in general"; G_3 – "partial compliance with the principles of environmental management"; G_4 – "non-compliance with the principles of environmental management"; G_5 – "complete non-compliance with the principles of environmental management".

3 Study of Large Cities of Rostov Region for Compliance with the Principles of Environmental Management

A study of major cities of the Rostov region in compliance with the principles of ecological wildlife.

Based on statistical data of [1] complex research large cities Rostov region with use of the foregoing techniques. For the calculations created software that allows you to aggregate statistical data from tables based on the standard five-point [0.1] - classifiers.

Grade Rostov region on big cities is: g = 513, which corresponds to the term "partial matching of environmental nature principles".

As can be seen from Tables 1 and 2, the highest place in the ranking of environmental wildlife in the Rostov region took Rostov-on-Don, due to the relatively high level of funding for environmental programs. The lowest rank takes Novocherkassk, which is caused, primarily, the tendency to increased pressure on the aqueous system: increasing water loss during transport, the increase of untreated waste water and decreasing the amount of recycled water used sequentially. As noted in [1], the situation Novocherkassk associated with increased development pressure on the aqueous system: increasing water consumption for technological needs Novocherkasskaya TPP due to increased power production. In Novocherkassk also no protected areas.

Place in the rating of all other cities based on the state of nature in which the first two indicators, as they are characterized relatively low (compared to the Rostov-on-Don), the level of funding of environmental activities, the almost complete absence of protected areas, with virtually no litter areas (exception - Kamensk-Shakhtinsky). In particular, a sufficiently low position of Azov, Taganrog, Volgodonsk and explained by the increase in emissions in the period 2014–2016 years. In accordance with the estimates given in [1] are marked trend: the increase in Azov air pollution particulate matter; Volgodonsk - formaldehyde and benzapiren; Taganrog - suspended solids, nitrogen dioxide and hydrogen chloride.

Table 1. Results of calculations on big cities Rostov area (fragment)

	City	G1	Term	G2	Term	G3
1.	Rostov-on-Don	0.313	G2	0.407	G3	0.061
2.	Azov	0.505	G3	0.460	G3	0.000
3.	Bataysk	0.125	G1	*	*	0.000
4.	Volgodonsk	0.459	G3	0.420	G3	0.394
5.	Gukovo	0.263	G2	0.440	G3	0.030
6.	Donetsk	0,125	G1	0.217	G2	0.030
7.	Zverevo	0.313	G2	0.497	G3	0.061
8.	Kamensk-Shakhtinsky	0.212	G2	0.577	G3	0.424
9.	Novocherkassk	0.362	G2	0.619	G4	0.030
10.	Novoshahtinsk	0.125	G1	0.420	G3	0.030
11.	Taganrog	0.362	G2	0.480	G3	0.000
12.	mine	0.125	G1	0.620	G4	0.000

* - city Bataisk a subscriber water utility of the city of Rostov-on-Don

A study of major cities of the Rostov region in compliance with the principles of ecological wildlife. The results of calculations by area Rostov region shown in Table 2 and the results of their ranking in Table 3.

Ust-Donetsk region is characterized by a high level of environmental financing (funding in 2016 amounted to 88,699.98 thousand. Rubles from the local budget). Matveyevo Kurgansky-area, the final rating is characterized by a tendency to increase the use of water, increasing losses during transport, and the reverse zero volume of water used in series; lack of protected areas, very low funding costs nature conservation events. Just as in the ranking of cities in the rest of the space is determined by the state of nature in which the first two indicators, because for all of them characterized by a relatively low level of funding of environmental activities, the almost complete absence of protected areas, the substantial absence of clutter areas.

Table 2. Ranking of major cities of the Rostov region with a linguistic recognition evaluation

City	Numerical evaluation value
G_1 - "full compliance with environmental principles of nature	
–	–
G_2 - "Environmental compliance with the principles of nature as a whole"	
1. Rostov-on-Don	0.370
G_3 - "partial compliance with environmental principles of nature"	
2. Bataysk	0.411
3. Donetsk	0.452
4. Novoshahtinsk	0.492
5. Mine	0.532
6. Kamensk-Shakhtinsky	0.536
7. Zverevo	0.539
8. Taganrog	0.544
9. Volgodonsk	0.567
10. Azov	0.579
11. Gukovo	0.588
G_4 - "non-compliant environmental principles of nature	
12. Novocherkassk	0.601
G_5 - "a complete mismatch of ecological environmental management principles	
–	–

Table 3. Ranking areas Rostov region with recognition evaluation linguistic(fragment)

Area	Numerical evaluation value	Area	Numerical evaluation value
G1		22. Morozov	0.513
–	–	23. Kuibyshev	0.515
G2		24. Martynovsky	0.515
1. Ust-Donetsk	0.348	25. Veselovsky	0.523
2. Remontnenskiy	0.365	26. Kagalnitsky	0.524
3. Tsimlyansky	0.374	27. Bagaevsky	0.539
G3		28. Bokovskaya	0.539
4. Krasnosulinskaya	0.459	29. Zernogradskiy	0.539

(*continued*)

Table 3. (*continued*)

Area	Numerical evaluation value	Area	Numerical evaluation value
5. Milyutinski	*0.464*	30. Neklinovskiy	*0.539*
6. Tarasovskiy	*0.476*	31. October	*0.543*
…	…	…	…
15. Semikarakorsk	*0.504*	40. Belokalitvenskyi	*0.579*
16. Soviet	*0.504*	41. Millerovskiy	*0.579*
17. Azov	*0.505*	42. Tatsinskaya	*0.582*
18. Chertkovsky	*0.506*	**G4**	
19. Dubovskii	*0.508*	43. Matveyev-Kurgan	*0.654*
20. Oblivskaya	*0.508*	**G5**	
21. Peschanokopskiy	*0.511*	–	–

4 Conclusion

The method is developed, which allows to build an assessment of the degree of compliance of environmental management in the region with the basic principles of environmental management. The assessment of the region is formed on the basis of fuzzy-multiple aggregation of the corresponding estimates of its municipalities-large cities and districts. The assessment of each municipality is formed on the basis of aggregation of time series of statistical data on five directions: dynamics of polluting emissions into the air; dynamics of load on the water system; degree of clutter of the territory; share of protected natural areas; level of funding for environmental protection. The method is based on the use of standard fuzzy five-level [0,1] classifiers. Data on the Rostov region are used as a statistical material. The developed method allowed to assess the state of nature management in the Rostov region and to conclude that there is a "partial compliance with the principles of environmental management".

References

1. Environmental Bulletin of the Don: On the state of the environment and natural resources of the Rostov region in 2011/2016. Government-in the Rostov region, Rostov-on-Don. http://минприродыро.рф/state-of-the-environment/ekologicheskiy-vestnik/. Accessed 18 July 2018
2. Report on the state of sanitary and epidemiological welfare of the population in the Rostov region in 2016. Federal service for supervision of consumer rights protection and human welfare, Rostov-on-Don. http://61.rospotrebnadzor.ru/index.php?Itemid=116&catid=96:2009-12-30-08-03-55&id=6813:-q-2016q&option=com_content&view=article. Accessed 30 Apr 2018
3. Arustamov, E.A., Levakova, I.V., Barkalova, N.B.: Environmental Bases of Use. Dashkov and K Publishing House, Moscow (2008)
4. Petrishchev, V.P., Dubrovskaya, S.A.: The method of integrated assessment of the ecological state of urban areas. News Samara Sci. Center Russ. Acad. Sci. **15**(3), 234–238 (2013)
5. Nedosekin, A.O.: Fuzzy Sets and Financial Management. AFA Library, Moscow (2003)

6. Sakharova, L.V., Stryukov, M.B., Akperov, I.G., Alekseychik, T.V., Chuvenkov, A.F.: Application of fuzzy set theory in agro-meteorological models for yield estimation based on statistics. In: 9th International Conference on Theory and Application of Soft Computing, Computing with Words and Perception, Budapest, Hungary, 24–25 August 2017, pp. 820–829 (2017). Procedia Computer Science, vol. 120
7. Stryukov, M.B., Sakharova, L.V., Alekseychik, T.V., Bogachev, T.V.: Methods of estimation of intensity of agricultural production on the basis of the theory of fuzzy sets. Int. Res. J. **7**(61), 123–129 (2017). Part 3
8. Kramarov, S., Temkin, I., Khramov, V.: Principles of formation of united geo-information space based on fuzzy triangulation. In: 9th International Conference on Theory and Application of Soft Computing, Computing with Words and Perception, Budapest, Hungary, 24–25 August 2017, pp. 835–843 (2017). Procedia Computer Science, vol. 120
9. Kramarov, S.O., Sakharova, L.V.: Soft computing in IU-segment: management of complex multivariate systems based on fuzzy analog controllers. Sci. Bull. South. Univ. Manag. **3**(19), 42–51 (2017)

Automated Voice Recognition of Emotions Through the Use of Neural Networks

A. V. Kurbesov$^{(\boxtimes)}$ ⓘ, D. V. Ryabkin ⓘ, I. I. Miroshnichenko ⓘ,
N. A. Aruchidi ⓘ, and K. Kh. Kalugyan ⓘ

Rostov State University of Economics, B. Sadovaya st., 69,
344002 Rostov-on-Don, Russia
akurbesov@yandex.ru, d.riabkin@yandex.ru,
kalugyan@yandex.ru, iimo2@ya.ru, bnatalya2000@mail.ru

Abstract. The paper describes the development of a software product, based on the architecture of the neural network, which allows to handle record telephone conversations and identify their emotional color. The approaches to the training of the neural network. Based on neural network developed a software product that allows you to assess the emotional state of the negotiators when contacting the emergency services to the quality of more than 60%, which was successfully tested. Carried out the integration of the software with other informational systems specified services.

Keywords: Neural networks · Automated recognition · Voice · Emotions · Algorithms · Models

1 Introduction

The urgency of the problem of voice recognition of emotion when transmitting an audio signal through the channels of various means, determines the adequacy of the behavior of a human operator when the emergency services [1, 3–6]. These services should first carry brigade ambulance and emergency services, the Ministry of Emergency Situations (MES).

Sound communication system in conjunction with a human operator must address the following key challenges:

- receiving a call and processing the call also if necessary, providing psychological support to the caller person;
- analysis of the incoming event information;
- automatic identification of caller ID, confirmation from the service provider, the subscriber location data available with this number, and other information necessary to ensure the call is answered;
- recording and documentation of all incoming and outgoing calls;
- maintaining a database of the main characteristics of the events, the beginning and the end of the response to the calls made and the main results;
- the ability to receive calls in various languages.

R. A. Aliev et al. (Eds.): ICSCCW 2019, AISC 1095, pp. 675–682, 2020.
https://doi.org/10.1007/978-3-030-35249-3_87

Software has been designed for the task, which is provided with the following functions:

- processing of audio calls and recognition of their emotional color;
- loading the data into a corporate database;
- creation of weekly reports on the emotional state of the caller.

The main objectives, are subjected to automation are:

- automatic recognition of raising the level of anxiety in the audio recordings of calls;
- the creation of a report on the level of anxiety changes in calls per week.

Anxiety - this individual psychological peculiarities, manifested in the human tendency often experience intense anxiety on relatively small or significant occasions. This is seen either as a sign of the alarming situation, or as an indication of the nature of the operator associated with the weakness of the nervous processes, or with both of them [10, 11].

Operation of the developed information system is carried on a dedicated server. The database is stored on own separate gray-belief, the information system must communicate with the database and perform the functions necessary data.

In recent years, the study of emotions to move quickly through the various smart technologies and a wide interest from researchers in the field of neuroscience, psychology, psychiatry, audiology and Informatics [2, 9]. An integral part of these studies is the availability of proven and reliable expressions of emotions. To meet these needs, it has become more and more available data sets emotional. Most kits contain either static facial expressions or voice recordings. Few contain audio-visual recording media in the North American English. And yet there are few confirmed sets of emotional expressions. To meet these needs, it was developed RAVDESS data set – large set of certified media in North American English.

2 Statement of the Problem and Solution

RAVDESS - is a marked multimodal database with recorded voices. The database is balanced on the floor and is composed of 24 voices of professional actors, voicing lexically selected statements with a North American accent. It includes a tranquil, happy, sad, angry, scared, surprised and aversive expression and peaceful, happy, sad, angry and fearful emotions. Each expression is pronounced on two levels of emotional intensity. All entries are available in the "face and voice" format "only face" and "voice only". A set of 7356 records evaluated by 10 times from the standpoint of emotional-term reliability and intensity. Estimates were provided by 247 persons, untrained study participants from North America.

A further set of 72 participants provided data test-retest. Co-communicated to the high level of emotional authenticity and reliability of pro-Werke. All recordings are available for free under a Creative Commons license.

SAVEE base data was recorded by four native speakers of English (designated as DC, JE, JK, KL) and contains the voices of graduate students and is-investigators of the University of Surrey in age from 27 to 31 years. Emotions were divided into separate categories: anger, disgust, fear, happiness, sadness and surprise. This is confirmed by cross-cultural studies of Ekman. Text material is composed of 15 sentences TIMIT divided into emotions: 3 common 2 and 10 common specific proposals that have been different for each emotion and phonetically balanced. Total 3 general and 12 emotionally-specific proposals have been recorded as neutral, to give 30 neutral sentences. This led to the 120 utterances per speaker, which reveals the following emotions: anger; disgust; fear; happiness; sadness; surprise; indifference.

For training the developed neural network we have two sets of data, RAVDESS and SAVEE were used. RAVDESS on the site of the proposed archive file containing the data set was selected from 1,000 audio recordings of 24 votes cast. The site SAVEE selected an archive containing the data set with 500 audio recordings of votes cast 4.

As a result, 1500 has been prepared audio files in wav format, and their names are given to a single mask. The file name consists of a 7-digit numeric identifier type:

$$XX\text{-}XX\text{-}XX\text{-}XX\text{-}XX\text{-}XX\text{-}XX.wav,$$

where

1 item - a record type (01 = Audio + Video, 02 = video only, audio only = 03);
2 item - the type of voice activity (= 01 speech, singing = 02);
3 item - emotions (neutral = 01, 02 = calm, 03 = happy, sad = 04, 05 = angry, violent 06 = 07 = disgust, surprise = 08);
4 item - emotional intensity (01 = normal, 02 = severe);
5 item - phrase (00 = "Another phrase" 01 = "Children talk at the door," 02 = "The dogs sit at the door");
6 item - repetition number (01 = 1st repetition, 02 = 2nd repetition);
7 item - actor ID.

For teaching Google Colaboratory remote server has been used. Google Colaboratory - this cloud service designed to simplify in the field of machine learning research. Using Colaboratory can gain remote access to the machine with the Graphics processing unit or Tensor Processing Unit is absolutely free, which greatly simplifies the training time-leg type of classifiers and neural networks. We can say that this is a kind of analogue of Google docs for Jupyter.

To test the performance of libraries, as well as learn the characteristics of the audio file was built waveform graphs and spectrograms (Figs. 1 and 2).

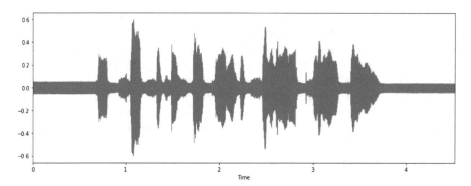

Fig. 1. Form record signal

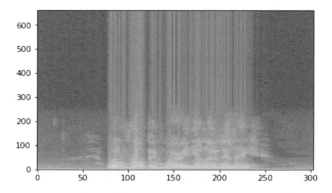

Fig. 2. Record spectrogram

Each audio file has a unique name, the elements of which can be used to determine specific emotions represented in the audio file.

Library Librosa (Python library designed to analyze audio) was used for processing and extracting audio features. Using Librosa library, it became possible to extract the characteristics of the voice, that is, MFCC (mel-frequency cepstral coefficients [10]). MFCC is a representation of the signal spectrum (Fig. 3).

On the positive side of this method include the use of an expansion in a basis of orthogonal sinusoidal (cosinusoidal) functions (signal range), allowing you to take into account the wave nature of the signal for further analysis. Spectrum projected onto the mel-scale, allows to allocate the frequencies that are best perceived by man [7, 8].

To reduce the amount of information processed quantity calculated coefficients may be limited to any desired value, which enables to produce an adequate and correct compression frame.

```
In [60]:  train[255:265]
```

```
Out[60]:
```

	4	5	6	7	8	9	...	121	122	123	124	125	126	127	128	129	0
)582	0.243815	0.234133	0.220812	0.222221	0.232087		...	0.248799	0.253912	0.260256	0.257698	0.258209	0.256242	0.255648	0.255648	0.255701	angry
)521	0.285065	0.291352	0.303514	0.308232	0.328804		...	0.234485	0.228035	0.216631	0.214859	0.212437	0.213037	0.218348	0.223208	0.224450	fearful
)765	0.108862	0.103840	0.101478	0.107730	0.103912		...	0.066940	0.036635	0.027208	0.036532	0.053178	0.065569	0.057186	0.039764	0.021314	angry
)141	0.074467	0.089486	0.088280	0.092139	0.093846		...	0.054423	0.053604	0.055540	0.058426	0.060729	0.068808	0.088886	0.098216	0.090357	sad
)724	0.281591	0.296421	0.285957	0.260214	0.257237		...	0.299710	0.291853	0.291916	0.299710	0.299710	0.299710	0.287766	0.252755	0.243608	happy
)779	0.330779	0.330779	0.330779	0.330779	0.330779		...	0.288739	0.287423	0.283312	0.291878	0.305482	0.321055	0.327999	0.301280	0.300456	calm
)433	0.169379	0.171645	0.179289	0.190308	0.182795		...	0.149075	0.147707	0.159900	0.184663	0.187635	0.168762	0.149145	0.130382	0.120786	neutral
)036	0.238554	0.242728	0.229463	0.228398	0.243454		...	0.223064	0.207814	0.210600	0.210909	0.202713	0.192792	0.192630	0.195298	0.187149	happy
)079	0.326079	0.305091	0.284397	0.274060	0.266039		...	0.156601	0.185422	0.202734	0.204833	0.213753	0.221158	0.222267	0.185138	0.151496	sad
)975	0.172604	0.173216	0.167372	0.168891	0.178888		...	0.205757	0.200951	0.197044	0.193599	0.208915	0.228052	0.219472	0.205900	0.201549	surprised

Fig. 3. MFCC voice signals group

Each audio file is converted to a list of factors. Remedy the missing factors for some audio files, which recognize the lack of din. Subsequently, the sampling frequency is doubled, which provides the possibility of allocating each fixed emotions. A further increase in the sampling rate is not appropriate, as there is noise affecting the results. Subsequent steps involve shuffling of data, the division into training and testing, and then store the neural network model.

We analyzed the use of multilayer perceptron type MLP, LSTM and CNN. Established perceptron MLP model (Fig. 4) had a very low accuracy – about 25% with 8 layers and softmax function of the output, with learning cycle in 550 epochs.

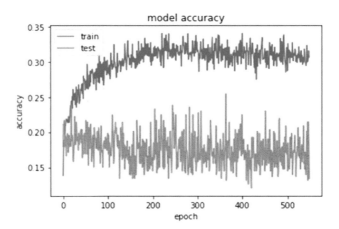

Fig. 4. Schedule network training MLP

LSTM training model had the lowest accuracy – about 15% with 5 layers, activation function, with learning cycle in 50 periods (Fig. 5).

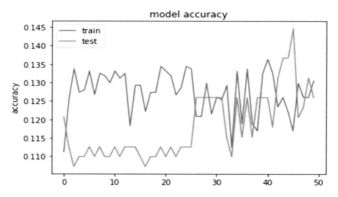

Fig. 5. Schedule network training LSTM

CNN was the best model for this classification. After training, the best accuracy of 60% was obtained numerous models. The model has 18 layers, activation function softmax and rmsprop, and learning cycle 1000 epochs (Figs. 6 and 7).

In [51]: `model.summary()`

Layer (type)	Output Shape	Param #
conv1d_1 (Conv1D)	(None, 216, 128)	768
activation_1 (Activation)	(None, 216, 128)	0
conv1d_2 (Conv1D)	(None, 216, 128)	82048
activation_2 (Activation)	(None, 216, 128)	0
dropout_1 (Dropout)	(None, 216, 128)	0
max_pooling1d_1 (MaxPooling1	(None, 27, 128)	0
conv1d_3 (Conv1D)	(None, 27, 128)	82048
activation_3 (Activation)	(None, 27, 128)	0
conv1d_4 (Conv1D)	(None, 27, 128)	82048
activation_4 (Activation)	(None, 27, 128)	0
conv1d_5 (Conv1D)	(None, 27, 128)	82048
activation_5 (Activation)	(None, 27, 128)	0
dropout_2 (Dropout)	(None, 27, 128)	0
conv1d_6 (Conv1D)	(None, 27, 128)	82048
activation_6 (Activation)	(None, 27, 128)	0
flatten_1 (Flatten)	(None, 3456)	0
dense_1 (Dense)	(None, 10)	34570
activation_7 (Activation)	(None, 10)	0

```
Total params: 445,578
Trainable params: 445,578
Non-trainable params: 0
```

Fig. 6. Summary model

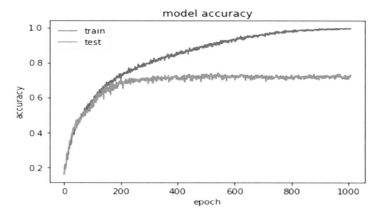

Fig. 7. Schedule CNN network training

CNN for the task of classification of emotions, using a precision of 70% was reached – after the creation of a variety of different models of the best model was found.

As a result of analysis of the input data processing algorithm was designed and built by the architecture of the neural network, which has been successfully tested. The resulting neural network satisfies the training/accuracy of the algorithm. Since learning of the neural network is an extremely long and complicated process, this factor has played a key role in the course of work. The important elements of the study was an empirical choice of network parameters that defined the ultimate accuracy of the results. It should be noted that the construction of the architecture of the neural network is quite insufficient for the integrated operation of the software. The original architecture of the information system has been developed which includes a database that stores the results, prompt release of abnormal cases (in real time), and a retrospective analysis of the information received, with the formation of a wide range of reports. The developed software is embedded in the information system of a higher level and provide interaction and integration with subsystems such as: ERA GLONASS; Sirena MES; Geographic information systems; database operators. To demonstrate the logical relationship between the processes used diagram IDEF0 (designation analogous SADT), which reflects the functional principles simulation.

In the diagram shown the main incoming data - phone number and operator ID, management processes - the top part of the arrow - Charter interdistrict situational center to ensure system call emergency services on a single number 112 and the Russian Federation legislation, the processing mechanisms - automated module for the recognition of emotions.

3   Conclusion

As a result of the work performed following results were obtained:

- developed the original architecture of the neural network, which provides recognition of emotion in his voice;

- configured neural network selection parameters for which the accuracy of the results obtained was about 60%;
- developed the original architecture of the information system database;
- the resulting program product embedded in a test operation;
- made the integration of the developed software product in the information systems of the Ministry of Emergency Situations and emergency departments.

Implementation of the results raised the adequacy of the human response - the operator and increased its speed by 23%; temporarily secured reduction in labor costs of data processing required during the formation of accounting documents 31%.

References

1. Marichal, X., Wei-Ying, M., HongJiang, Z.: Blur determination in the compressed domain using DCT information. In: Proceedings of the International Conference: Image Processing, ICIP 1999, vol. 2 (1999)
2. Evdokimov Alexey, V., Kovalenko Vasiliy, A., Kurbesov Alexandr, V., Shabanov Alexey, V.: Patients identification in medical organization using neural network. In: Aliev, R., Kacprzyk, J., Pedrycz, W., Jamshidi, M., Sadikoglu, F. (eds.) 13th International Conference on Theory and Application of Fuzzy Systems and Soft Computing – ICAFS-2018, ICAFS 2018. Advances in Intelligent Systems and Computing, vol. 896. Springer, Cham (2019)
3. Kalalembang, E., Usman, K., Gunawan, I.P.: DCT-based local motion blur detection. In: Conference: Instrumentation, Communications, Information Technology, and Biomedical Engineering (ICICI-BME), International Conference (2009)
4. Huynh-Thu, Q., Ghanbari, M.: The accuracy of PSNR in predicting video quality for different video scenes and frame rates. Telecommun. Syst. **49**(1), 35–48 (2012)
5. Determination of the pitch frequency of speech. http://ainews.ru/2018/07/pitch_tracking_ili_opredelenie_chastoty_osnovnogo_tona_v_rechi_na_primerah.html
6. Automatic recognition of human emotions on the basis of reconstruction of attractors. https://www.nbpublish.com/library_get_pdf.php?id=21913
7. Gelig, A.H., Matveev, A.S.: Introduction to the mathematical theory of the trainees recognition systems and neural networks. Tutorial: Monograph, St. Petersburg, St. Petersburg State University Press (2014)
8. Haykin, S.: Neural Networks. Full Course, Moscow, Williams (2019)
9. Redko, V.G.: Evolution, neural networks, intelligence: models and concepts of evolutionary cybernetics, Moscow, SINTEG (2017)
10. Rutkovska, D., Pilinsky, M., Rutkowski, L.: Neural networks, genetic algorithms and fuzzy systems, Moscow, Hotline – Telecom (2013)
11. Tarhov, D.A.: Neural network models and algorithms. Directory, Moscow, Radio Engineering (2014)

Non-linear Ensemble Modeling for Multi-step Ahead Prediction of Treated COD in Wastewater Treatment Plant

S. I. Abba[1,2(✉)] , Gozen Elkiran[1,2] , and Vahid Nourani[3]

[1] Department of Physical Planning Development and Maintenance,
Yusuf Maitama Sule University, Kano, Nigeria
saniisaabba86@gmail.com, gozen.elkiran@neu.edu.tr
[2] Faculty of Civil and Environmental Engineering, Near East University,
Near East Boulevard, 99138 Nicosia, North Cyprus, Turkey
[3] Department of Water Resources Engineering, Faculty of Civil Engineering,
University of Tabriz, Tabriz, Iran
vahid.nourani@neu.edu.tr

Abstract. The paper proposes the application of data-driven models, including wavelet neural network (WNN) and multilayer perceptron (MLP), for multi-step ahead modeling of treated chemical oxygen demand ($COD_{Treated}$) using neuro-sensitivity input variables selection approach. Afterward, two non-linear ensemble techniques were applied to increase the prediction performance of the single models. Daily measure data obtained from new Nicosia wastewater treatment are used in this study, the performance efficiency of the models was determined in terms of Nash–Sutcliffe efficiency (NSE) and root mean squared error (RMSE). The obtained results of single models showed that WNN increased the performance accuracy up to 7% and 8% over MLP in both calibration and verification. The results also revealed the reliability of non-linear ensemble models in multi-step ahead prediction of $COD_{Treated}$, hence, ensemble modeling could efficiently improve the performance of WNN and MLP models.

Keywords: Chemical oxygen demand · Ensemble technique · Multi-layer perceptron · Wavelet neural network · Wastewater

1 Introduction

Chemical oxygen demand (COD) is one of the key parameters for the determination of size and efficiency of wastewater treatment plant's (WWTP) facilities and process [1]. Under normal condition, COD is used to measures and determine the strength of wastewater. In addition, COD serves as the indicator for clarification of water before the disposal into receiving body [1]. Wastewater treatment plant (WWTP) is highly complicated and dynamic system due to its intricacy of the treatment process.

For monitoring the environmental and ecological health, suitable operation and control of WWTPs are crucial. The generation of a consistent prediction using several traditional models may not be achievable due to the dynamic nature non-stationarity of historical data [2]. This has made it necessary that researchers should develop stronger

© Springer Nature Switzerland AG 2020
R. A. Aliev et al. (Eds.): ICSCCW 2019, AISC 1095, pp. 683–689, 2020.
https://doi.org/10.1007/978-3-030-35249-3_88

and efficient models using the available historical data in order to improve the reliability and performance of the plant [3]. Moreover, the benefit of involving data-driven algorithms is plain clear due to their reliability inaccuracy of modeling methodologies and less time-consuming. Several modeling approaches including classical and non-linear models were used to develop reliable models for the WWTP, nevertheless, most of these approaches have issues such as determining the suitable structure, time, require much expertise to design and costly [4]. Recently, ensemble approaches have been applied for modeling several hydro-environmental problems and reported superiority in term of accuracy with regards to nonlinear single techniques.

Hence, designing a universally-capable data-driven model with a wider range of implementation become necessary in WWTP. As such, the objective of this research is (i) To explore a neuro-sensitivity approach for non-linear input variables selection (ii) To develop and compare the potential of hybrid Wavelet neural network (WNN) and Multi-layer perceptron (MLP) neural network models for multi-step ahead modeling of treated COD in new Nicosia WWTP (iii) To improve the prediction performance using two different nonlinear ensemble techniques.

2   Applied Methodology

2.1   Wavelet Neural Network (WNN)

Wavelet transform (WT) has been a powerful tool in signal processing and recently begun to explore in the field of hydrology predominantly in time series of streamflow forecasting due to its hybrid nature, localization characteristics and powerful approximation which allow hierarchical multiresolution learning of input-output data mapping [5]. Wavelet neural network (WNN) derived from the basic concept of combining the advantage of neural network and wavelet analysis, in this way the wavelet function substitutes the role of the activation function in the hidden layer node and the wavelet parameters are considered and replaces weights and threshold values [6]. The WNN model adopts a three-layer topology of NN namely: input, hidden and output layer (see, Fig. 1a). For more information about wavelet functions, [6] can be referred. The wavelet function (or mother wavelet) in its discrete form can be represented as:

$$\psi_{a,b(t)} = \frac{1}{\sqrt{a}} \psi \left(\frac{t-a}{b} \right) \qquad (1)$$

where t is time, a is position parameter, and b is scaling (or dilation) factor of the mother wavelet.

2.2   Multi-layer Perceptron (MLP) Neural Network

Multi-layer perceptron (MLP) neural network, as one of the most common kind ANN, has the capability to handle non-linear system and describe by numerous literature as a universal approximator among the different categories of ANNs [7]. As like the other traditional ANN, MLP consists of input, one or more hidden and output layers in its architecture (see, Fig. 1b). Referred to [7] for more information about MLP.

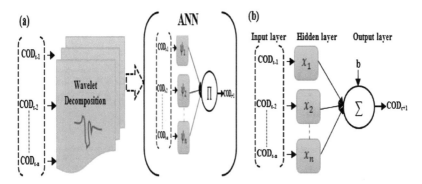

Fig. 1. Structural topology of (a) WNN and (b) MLP models

2.3   Used Data and Model Development

The daily 360 records data obtained from new Nicosia MWWTP which includes (TP_{raw}, $Conductivity_{raw}$, TSS_{raw}, T_{Nraw}, and pH_{raw}) as the input variables and ($COD_{Treated}$) as the corresponding output respectively. The normalized data were divided into 75% and 25% for both calibration and verification, respectively. According to several hydro-environmental studies, different approaches of selected appropriate input variables were reported such as correlation analysis, autocorrelation function (ACF) and the partial autocorrelation function (PACF), principal component analysis, etc., however, these methods are attributed to the linear input-output relationship [8]. As such neuro-sensitivity input variables selection was explored in this study to handle the nonlinear-correlation pattern between the input-output variables. neuro-sensitivity input variables selection approach is an approach that involved the determination of dominants variable in a non-linearity manner among the different input combinations (see Fig. 2a). For multi-step ahead of $COD_{Treated}$, two different models of input combinations were generated based on non-linear neuro-sensitivity analysis (see Eq. 2). In order to do multi-step ahead modeling of $COD_{Treated}$, the target output at an n-time step ahead could be presented in Eq. 2.

$$\left. \begin{array}{l} M1 = COD_{t+1}^{treated} = f\left(TP_{t-n}^{raw}\right), \left(Cond_{t-n}^{raw}\right), \left(TSS_{t-n}^{raw}\right) \\ M2 = COD_{t+1}^{treated} = f\left(TP_{t-n}^{raw}\right), \left(Cond_{t-n}^{raw}\right), \left(TSS_{t-n}^{raw}\right), \left(TN_{t-n}^{raw}\right), \left(pH_{t-n}^{raw}\right) \end{array} \right\} \quad (2)$$

In Eq. 2, the lower index stands for the time step. According to the autoregressive property of the process and the effluent concentration, the treated COD value may be considered as a function of raw TP, Conductivity, TSS, TN, and pH. After the calibration process, the model, the efficiency of the models was determined and analyzed in terms of Nash–Sutcliffe efficiency (NSE) and root mean squared error (RMSE) as:

$$NSE = 1 - \frac{\sum_{i=1}^{n}\left(COD_{obs_i} - COD_{pred_i}\right)^2}{\sum_{i=1}^{n}\left(COD_{obs_i} - \overline{COD}_{obs}\right)^2} \tag{3}$$

$$RMSE = \sqrt{\frac{\sum_{i=1}^{n}\left(COD_{obs_i} - COD_{pred_i}\right)^2}{n}} \tag{4}$$

where COD_{pred_i} is the predicted, COD_{obs_i} is the observed data, n is the data number and data. NSE ranges from $-\infty$ to 1 with a perfect score of 1 and RMSE ranges from 0 to $+\infty$ with a perfect value of 0.

2.4 Non-linear Ensemble Technique

Several studies proved the promising success of ensemble techniques, according to [4] ensemble technique was employed by three different approaches (single, weighted and neural) and among all the approaches non-linear ensemble reported to outperform the other types of ensemble in terms of performance improvement. Therefore, this study employed non-linear neural ensemble approaches, the outputs of the single models are considered as the new inputs to the neural ensemble model. The illustration of the general ensemble concept employed in this study is similar to that of [4].

2.5 Multi-step Ahead (MSA)

In recent years, various researchers put emphasis on time series predicting, but most of them focused on one-step-ahead predictions, which is not practical in daily life [9]. The multi-step ahead (MSA) prediction comprises of forecasting the x next values of a time-series and this can be accomplished by either independent (direct-based method) value prediction or iterative approach. The former, involved the calibration of a direct model to forecast y $(K + x)$ while the latter consist of reiterating one-step-ahead prediction to a desire horizon, this method need too much computing. The iterative prediction only uses one model to forecast all the horizons needed; the objective is to analyze a short sequence of data and try to predict the rest of the data sequence until a predefined time-step is reached. The main drawback of this approach is that the error in nearest horizons can be transmitted to others horizons (see, Fig. 2b).

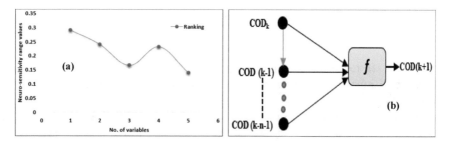

Fig. 2. (a) Nonlinear neuro-sensitivity analysis of the input-output variables (b) Basic principle for one-step-ahead time series prediction $(k = 1)$

3   Application Results and Analysis

In any AI modeling, finding the optimal architecture is the main problems due to the fact that, there is no standard pattern for selecting the desired architecture prior to the calibration phase. As such, a different number of hidden neurons ranging from 1 to 6 were observed in MLP and WNN models by trial and error procedure. It can be seen from Table 1 that the WNN model outperforms the MLP in terms of all the performance criteria. The Table 1 also depicted that, M1 with four input combinations, NSE = 0.7955 and RMSE = 0.0708 served as the best model for MLP model while for WNN model, M2 with 6 input combinations, NSE = 0.87669 and RMSE = 0.0546 proved to be the best performing model in the verification phase. Further examination and comparison of the results indicated that, WNN model superior and more reliable than MLP for all the multi-step ahead model combination. With regard to percentage variation and accuracy, WNN increases by up to 7% and 8% over MLP in both calibration and verification. The forecasts of each model are represented in Fig. 3 in the form of a scatterplot. It is seen from the scatterplots that the WNN prediction are closer to the corresponding observed COD values than those of the MLP model. As seen from the fit intersection.

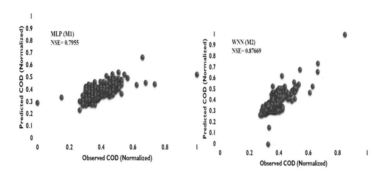

Fig. 3. Observed vs predicted scatter plots for the best model of MLP and WNN

As mentioned above ensemble modeling was performed in order to overcome the weakness of the single models, as reported by [4] single models may perform differently even with the same scenario of data. In this study, MLP and WNN ensemble (MLP-E and WNN-E) were employed and the results presented in Table 1 shows that employing ensemble techniques could improve the single model by considerable accuracy.

Table 1. Results of the single and ensemble models

Single model	Model types	No. of hidden nodes	Calibration		Verification	
			NSE	RMSE[a]	NSE	RMSE[a]
MLP	M1	4	0.8226	0.0412	0.7955	0.0708
	M2	6	0.6903	0.0480	0.6149	0.0693
WNN	M1	4	0.8584	0.0417	0.8302	0.0689
	M2	8	0.8940	0.0365	0.87669	0.0546
Ensemble models	MLPE	3	0.9258	0.0106	0.9264	0.0110
	WNNE	3	0.9665	0.0246	0.9652	0.0213

[a]Since all data are normalized, the RMSE has no dimension.

Among the ensemble models, WNN-E outperforms MLP-E and the results indicate the reliability of both the ensemble model in multi-step ahead prediction of treated COD. Figure 4 shows the graphical representation of MLP-E and WNN-E based on the Taylor diagram.

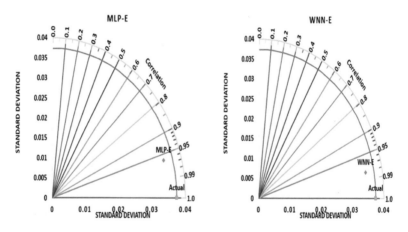

Fig. 4. Taylor diagram showing the degree of prediction in terms of R and SD for treated COD

Taylor is a two-dimensional method that exhibits how closely a model or different model matches the observed and corresponding computed values.

4   Conclusion

The performance accuracy of data-driven models i.e. wavelet neural network (WNN) and multilayer perceptron (MLP), for multi-step ahead modeling of treated chemical oxygen demand (COD$_{Treated}$). The obtained results confirmed the superiority of WNN with regards to the MLP model. The outcomes also proved that WNN outperformed MLP in term of ensemble approach, therefore justified that ensemble approaches could increase the prediction performance of the single models in both calibration and verification. The study used only two data-driven algorithms hence it was suggested that employing more data-driven models may be required to strengthen these conclusions.

References

1. Abba, S.I., Elkiran, G.: Effluent prediction of chemical oxygen demand from the wastewater treatment plant using artificial neural network application. Procedia Comput. Sci. **20**, 156–163 (2017)
2. Dogan, E., Ates, A., Yilmaz, E.C.: Application of Artificial Neural Networks to Estimate Wastewater Treatment Plant Inlet Biochemical Oxygen Demand, vol. I. Wiley InterScience, Hoboken (2015)
3. Yaseen, Z.M., et al.: Novel hybrid data-intelligence model for forecasting monthly rainfall with uncertainty analysis. Water (Switzerland) **11**(3), 502 (2019)
4. Nourani, V., Elkiran, G., Abba, S.I.: Wastewater treatment plant performance analysis using artificial intelligence – an ensemble approach. Water Sci. Technol. **78**(10), 2064–2076 (2018)
5. Wang, B., Hao, W.N., Chen, G., He, D.C., Feng, B.: A wavelet neural network forecasting model based on ARIMA. Appl. Mech. Mater. **347–350**, 3013–3018 (2013)
6. Danandeh Mehr, A., Kahya, E., Olyaie, E.: Streamflow prediction using linear genetic programming in comparison with a neuro-wavelet technique. J. Hydrol. **505**, 240–249 (2013)
7. Kim, S., Singh, V.P.: Modeling daily soil temperature using data-driven models and spatial distribution. Theor. Appl. Climatol. **118**(3), 465–479 (2014)
8. Hadi, S.J., Tombul, M.: Forecasting daily streamflow for basins with different physical characteristics through data-driven methods. Water Resour. Manag. **32**(10), 3405–3422 (2018)
9. Liu, Y., Hou, D., Bao, J., Qi, Y.: Multi-step ahead time series forecasting for different data patterns based on LSTM recurrent neural network. In: Proceedings of the 2017-th Web Information Systems and Applications Conference, WISA 2017, vol. 018, pp. 305–310 (2018)

A New Model of Movement of Liquids in Porous Medium

Ramiz S. Qurbanov$^{(\boxtimes)}$ and Aynur J. Jabiyeva◉

Azerbaijan State Oil and Industry University, Azadlig 20, Nasimi,
Baku, Azerbaijan
Raniz.gurbanov@yahoo.com, aynur.Jabiyeva@outlook.com

Abstract. The study of the movement of liquids with different rheological properties in a porous medium is of practical importance in the fields of oil production, chemical technology, hydrogeology [1]. At the same time, other approaches to the description of processes are distinguished. The study of motion are described mainly on various models. In this article, to describe the motion of rheologically stationary fluids in a porous medium, we propose a model, which we call the "hypothetical channel". In this case, the hydraulic radius and the quasi-Newtonian approach given in the work are used [5]. The essence of this approach is as follows: the movement of rheological parameters of time, foreign fluids in a porous medium are represented by the same fluid of various shapes in the channel, the hydraulic radius and shape coefficient of which are equal to the hydraulic radius and shape coefficient of the medium under consideration and at the same pressure gradients and the average velocity in the channel equal to the filtration rate in a porous medium.

Keywords: Empiric · Semi-empirical approaches · Rheological stationary fluids · Oil production · Fuzzy values · Fuzzy logic-based method

1 Introduction

There are different models of flow in porous media [2, 3]. First, H. Darcy, based on results of experiments, proposed the empirical filtering law

$$U_m = k\frac{\Delta P}{\mu L}, \tag{1}$$

Um is average speed, μ - is the dynamic liquid, $\frac{\Delta P}{L}$ – is the modulus of the pressure modulus, k – constant porous medium coefficient.

The study of motion are described mainly on various models. The filtration rate of a viscous dream of an ideal model is expressed by formula (1), instead of k the value is taken $\frac{mR^2}{8}$, in the case of a fictitious model instead of K in (1) would be as $\frac{n^2d^2}{96(1-m)}$,

Above n - is the transparency of the porous medium, d - is the effective particle diameter, m - is the coefficient of porosity.

© Springer Nature Switzerland AG 2020
R. A. Aliev et al. (Eds.): ICSCCW 2019, AISC 1095, pp. 690–696, 2020.
https://doi.org/10.1007/978-3-030-35249-3_89

Among the semi-empirical filtration models, the Kozen-Karman formula is often used.

$$U_m = \frac{n^3 d^2}{72 c'(1-m)^2} \frac{\Delta P}{\mu L},\qquad(2)$$

here c' is the coefficient that takes into account the geometry of the porous channels (usually, they take $c' \approx 2.5$).

To note that the use of these models on the one hand greatly simplifies the filtering process, and on the other hand, creates inconvenience due to the difficulties in determining the linear dimensions d, R and the coefficients m, n. The model considered as a combination of ideal and fictitious models is also not without its drawbacks.

2 A "hypothetical channel" Model

In this article, to describe the motion of rheologically stationary fluids in a porous medium, we propose a model that we called the "hypothetical channel". In this case, the hydraulic radius and the quasi-Newtonian approach given in [4] are used.

The essence of this approach is as follows: the movement of rheological parameters of time, foreign fluids in a porous medium are represented by the same fluid of various shapes in the channel, the hydraulic radius and shape coefficient of which are equal to the hydraulic radius and shape coefficient of the medium under consideration and at the same pressure gradients and the average velocity in the channel equal to the filtration rate in a porous medium according to the proposed approach, the filtration rate is expressed by the formula

$$U_m = \frac{r_n^2}{2\xi_m} \frac{\Delta P}{\mu L},\qquad(3)$$

r_n, ξ_m is the hydraulic radius and the shape factor of the channel shape factor, respectively.

For non-Newtonian fluids, the dynamic viscosity μ in (3) is replaced by the equivalent viscosity.

The dependence of the shape coefficient on for various forms of the porous medium is shown, from which it can be seen that similar dependences for the fictitious Slichter model and other models have extremes, and for the rest they are characterized by a decrease in the value in ξ_m with m. The exact value of the shape factor is determined by comparing formulas (1) and (3)

$$\xi_m = \frac{r_n^2}{2k}.\qquad(4)$$

The works show the flow of viscous liquids in channels, often found in heat exchange devices. Hydraulic radii and shape coefficients of some regions characteristic of various channel shapes. The values of ξ_m were calculated by correlating the empirical dependences λ (Re) with the generalized law of resistance [4].

The average value of the shape factor for the specified areas is $\xi_m \approx 0.416$. Then, in (4) we find out that $r_n \approx 0{,}912\sqrt{k}$.

In [7, 8], experimental studies were carried out to determine the parameters m, k, Scp for core samples taken from reservoirs of various fields, the results of which are presented (see Fig. 1).

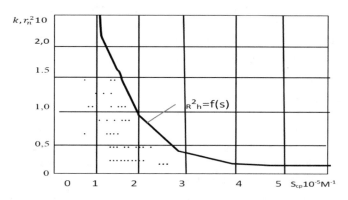

Fig. 1. Dependence of permeability coefficient and hydraulic radius on the specific surface of porous medium

3 Statement of the Problem and Solution

Based on the above correlation, it can be assumed that for porous medium, the values of r_n and $\sqrt{k}$ have the same order. Therefore, based on considerations of convenience, in the first approximation, $r_n \approx \sqrt{k}$. $\sqrt{k}$ as a linear dimension is the average characteristic of the medium, and in the case of a viscous regime, the physicochemical properties of the soil, etc.

Then the consistent variables for the filtering case are represented by the expressions

$$\tau_m = \frac{\Delta P r_n}{L} = \frac{\Delta P \sqrt{k}}{L}, \tag{5}$$

$$\gamma_m = 2\xi_m \frac{U_m}{r_n} = \frac{U_m}{\sqrt{k}}, \tag{6}$$

τ_m, γ_m are the values of stress and shear rate, respectively, reduced to the scale of the porous medium

$$\tau_m = \mu \gamma_m \tag{7}$$

with the Darcy-Weisbach formula presented as

$$\tau_m = \frac{1}{8} \lambda \rho U_m^2 \tag{8}$$

we obtain the generalized law of resistance

$$\lambda = \frac{64}{R_{e_m}^*} \tag{9}$$

$R_{e_m}^* = \frac{4\rho U_m \sqrt{k}}{\mu}$ - is the generalized Reynolds number.

The experimental data from [9, 10] and the line corresponding to the generalized law of resistance, from which it is clear that in the region of viscous flow, the data lie near dependence (9). It is not possible to unambiguously determine the boundary of the viscous regime, because, the critical value of the Reynolds number depends on the porosity and structure of the porous medium. In view of this, it would be even more adequate to use the fuzzy set theory to describe imprecision related o boundaries. However, it should be noted that the violation of the viscous regime for cemented sands and natural samples occurs at smaller R_{e_m} (approximately 10 times) than for non-cemented samples.

A viscoplastic fluid filtration model first proposed by academician A.H. Mirzadzhanzade's Buckingham formula for viscous-plastic fluids in a circular pipe can be generalized for the filtration case by replacing the mean velocity with the filtration velocity and the radius of the tube with a double hydraulic radius [11]. In this case, which takes the form:

$$U_m = \frac{k\Delta P}{L} \left[1 - \frac{4}{3}\frac{\Delta P_0}{\Delta P} + \frac{1}{3}\left(\frac{\Delta P_0}{\Delta P}\right)^4 \right] \tag{10}$$

η - is the structural viscosity.

The initial pressure gradient, in this case, is represented by the expression

$$\Delta P_0 = \alpha \frac{\tau_0}{\sqrt{k}} \tag{11}$$

$\alpha = \sqrt{m/2\beta}$ – is a coefficient that depends on the porosity and shape of the pore channels; as result of the conducted experiments, $\alpha = (155 \div 180) \cdot 10^{-4}$ was found.

In [12] they suggested the ratio for the consistent variables of the porous medium:

$$\tau_m = \frac{\Delta P r_n}{L}, \gamma_m = \frac{U_m k_1}{r_n m}, \tag{12}$$

k_1 – is the coefficient describing the shapes of the channels of the porous medium.

In the review article [13] for description of the process of filtering abnormal fluids, J. Savins uses the Kozeni-Karman model, which results as the following expressions for consistent variables

$$\tau_\varepsilon = \frac{md}{6c'(1-m)}\frac{\Delta P}{\mu L},$$ (13)

$$\gamma_\varepsilon = \frac{12(1-m)}{m^2 d}U_m.$$ (14)

Summarizing the above, on the one hand, one can see that consistent variables (5) and (6) and law of opposition (9) arising from the "hypothetical channel" model are simple and versatile, on the other hand, this model allows predicting the rheological behavior of abnormal fluids in a porous medium based on capillary rheometry results.

Numerous parallel experiments were carried out in a capillary viscometer and simulated porous medium with various abnormal systems (oil vapor, vapor and bit-stocks, resinous oil), the results of which indicate a better correlation of the data for both medium.

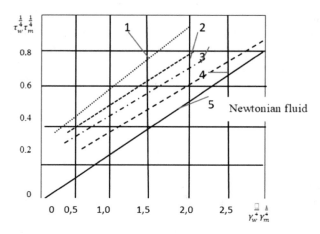

Fig. 2. Rheological curves of vapor oil at temperatures: 1–28 C, 2–34.50 C, 3–40 C, 4–50 C, 5–55 C (symbols +, • refers to the capillary results rheometry and porous medium, respectively).

Figure 2 shows the rheological curves in the coordinates $\tau_m^{1/4} - \gamma_m^{1/4}$ for the vaporizer, based on the viscometer data (d = 1.5 mm, L = 750 mm) and the porous medium (k = 8, 81D and k = 137D) at different temperatures.

The interval of temperatures is [28, 55]. The temperatures from this interval are used in further computations.

As we can see from the figure, at the same temperatures the data of capillary rheometry and filtration belong to the same curves, and at a temperature of 550 °C the test liquid becomes Newtonian.

The Kesson-Schulman was adopted as the rheological model

$$\tau_m^{1/n} = \tau_0^{1/n} + (\eta\gamma_m)^{1/n} \tag{15}$$

and the choice of a specific model was made on the basis of the Fisher criterion. The results of the calculations are presented in Table 1.

Table 1. The results of calculations

t^0, C	n	τ_0, Pa	η, Pa•C
28	4	0, 1304	5, 10
34, 5	4	0, 0270	2, 07
40	2	0, 0104	1, 76
50	2	0, 0084	0, 64
55	–		0, 43

Imprecision related to the values of τ_0, η and γ_m can be described as $\tau_0 = [0.0084, 0.13]$, $\eta = [0.43, 5.1]$ and $\gamma_m = [0.5, 2.5]$.

4 Conclusions

1. A new model has been developed to describe the movement of rheologically stationary fluids in a porous medium using the concept of an equivalent radius.
2. The developed model allows predicting the rheological behavior of abnormal fluids in a porous medium based on the results of capillary rheometry.
3. By comparing the universal rheological equation with the Darcy-Weisbach formula, a generalized law of resistance and a generalized Reynolds number are obtained.

References

1. Leibenzon, L.S.: The Movement of Natural Liquids and Gases in a Porous Medium. Gostekhizdat, Moscow (1947)
2. Scheidegger, A.E.: Physics of Fluid Flow Through Porous Medium. Gostoptekhizdat, Moscow (1960)
3. Ber, J., Zaslavsky, D., Irmey, S.: Physical and Mathematical Bases of Water Filtration. Mir, Moscow (1971)
4. Gunn, D.J., Darling, C.W.: Fluid flow. Trans. Inst. Chem. Eng. **41**(4), 163–173 (1963)
5. Abdulvagabov, A.I.: The study of the modes of motion of liquids and gases in a porous medium. Cand. Diss., Azineftehim, Baku (1961)
6. Savins, J.G.: Non-newtonian flow through porous medium. Ind. Eng. Chem. **61**(10), 18–47 (1969)
7. Trebin, G.F.: Filtration of Liquids and Gases in Porous Medium. Gostoptehizdat, Moscow (1959)

8. Kozıcki, W., Hsu, C.J., Tiu, C.: Non-newtonian flow. Chem. Eng. Sci. **22**(4), 487–502 (1967)
9. Malak, J., Hejna, J., Schmid, J.: Pressure losses and heat transfer in non-circular channels with hydraulically smooth walies. Int. J. Heat Mass Transfer **18**(1), 139–149 (1975)
10. Mirzadzhanzade, A.K.: Problems with hydrodynamics viscoplastic and viscous liquids as applied to oil production. Azernefteshr, Baku (1959)
11. Kravchenko, I.I., Kovalenko, E.K., Genkina, B.I.: To the issue of specific surface. Tr. Phew Research Institute, vol. 17, pp. 66–73 (1967)
12. Fomenko, I.E.: On the value of the specific surface, tortuosity and the pore radius of the Romashkinskoye reservoirs. Tr. Phew Institute **17**, 73–77 (1971)
13. Gurbanov, P.C., Abdınov, E.T.: On unification of rheological and hydraulic calculation. Trans. Acad. Sci. Azerb. Ser. Phys. Techn. Math. Sci. **19**(5), 173–183 (1967)

Development of CNN Model for Prediction of CRISPR/Cas12 Guide RNA Activity

Sa'id Zubaida Amee$^{(\boxtimes)}$ ⓘ, Auwalu Saleh Mubarak ⓘ,
Aşır Süleyman ⓘ, and Ozsoz Mehmet$^{(\boxtimes)}$ ⓘ

Near East University, 99138 Nicosia, Northern Cyprus via Mersin 10, Turkey
{zubaida.saidameen, suleyman.asir,
mehmet.ozsoz}@neu.edu.tr, mubarakauwal@gmail.com

Abstract. CRISPR technology is a powerful gene modifying tool that was derived from bacteria and archea. This CRISPR tool recognizes and cleaves any target DNA with the usage of a guide RNA and Cas endonuclease enzyme. Cpf1 endonuclease enzyme is a type of Cas enzyme from class 2 in the CRISPR-Cas system used for cutting DNA. The success of this genetic engineering tool depends on design of effective single guide RNA (sgRNA) to help Cas endonuclease enzyme carry out DNA cleavage. Different computational tools have been created to help with design of sgRNA however, it still remains a challenge. The application of deep learning has been extended to CRISPR technology. Here, we developed a deep learning algorithm using convolutional neural network which predicts activity of Cpf1 guide RNA. We built different models using different hyperparameters and identified the best model capable of predicting Cpf1guide RNA activity. ConvCpf1 will help in design of sgRNAs that will be highly active for genome editing using Cpf1 endonuclease enzyme, thereby saving time as well as experimental cost.

Keywords: CRISPR · sgRNA activity · CNN model

1 Introduction

1.1 CRISPR-Cas System and Deep Learning

The CRISPR-Cas system (Clustered Regularly Interspaced Short Palindromic Repeats - CRISPR associated protein) is a prokaryotic immune system that was first discovered in bacteria [1] and archea [2], which has been embraced and utilized for editing eukaryotic genomes [3]. The principle of editing gene with CRISPR-Cas9 is the use of a single guide RNA (sgRNA) that directs Cas9 to DNA target sequence of interest, and Cas9 cuts the two strands of DNA with the help of two nuclease domains generating a double strand breaks. This DNA double-strand breaks can be repaired either by non-homologous end joining (NHEJ), generating gene knockout, or by homology directed repair (HDR), leading to a gene knock-in [4]. The most common CRISPR-Cas systems are Class I and Class II with different subtypes, however, Class II systems are the most widely used because they are very efficient and simple as they require only one Cas enzyme and guide RNA to cut any gene of interest [5].

© Springer Nature Switzerland AG 2020
R. A. Aliev et al. (Eds.): ICSCCW 2019, AISC 1095, pp. 697–703, 2020.
https://doi.org/10.1007/978-3-030-35249-3_90

The application of CRISPR-Cas9 system for editing genes from eukaryotes was first demonstrated in 2013 [6–8]. Afterwards, many researches have efficaciously utilized CRISPR-Cas9 to engineer genomes of diverse organisms. Similarly, researchers discovered that CRISPR-Cas9 may be used to correct genetic illnesses and generate cancer models. But the CRISPR-Cas9 system has some drawbacks, including excessive unintended gene cutting (off-target) impact [9, 10].

Machine learning techniques has been carried out on different kind of issues in genomics and genetics [11]. The application of machine learning is now utilized in CRISPR technology. Many studies were carried out to determine the features responsible for sgRNA activity. In a research for prediction of sgRNA activity it was found that sgRNA activity was associated to certain sequences motifs [12]. Specific sequences responsible for improved sgRNA was also discovered including their PAM sequences [13]. Guanine rich and adenine deprived sgRNAs were observed to have high activity using zebra fish embryos [14].

Deep learning is a subfield of machine learning, it addresses the issue of feature extraction needed by conventional machine learning algorithms because it consist of multiple layers that can understand important features from raw data necessary for a prediction task [15]. Each hidden layer output is transformed by a non-linear function such as sigmoid function or rectified linear unit, therefore, neural networks use hidden layers to learn these non-linear transformations automatically. Altogether, these hidden layers transform input feature into complex patterns that can be used to produce output [16]. Several studies have applied deep learning algorithm to develop models for prediction of off-target mutations in CRISPR-Cas9 [17], to predict the activity of Cpf1 based on DNA target sequences [18], to predict CRISPR sgRNA activity for CRISPR/Cas9 [19], including platform design for prediction of sgRNA on-target and off-target activities [20].

Here, our aim is to develop a deep learning algorithm which can without delay predict Cpf1 activity from sgRNA sequences beyond the need for feature extraction. We developed a model utilizing a convolutional neural network to detect features from Cpf1 sgRNA sequences and make predictions about their activities.

2   Materials and Methods

2.1   Datasets

The main dataset used in this study are from [21]. Kim together with his colleagues tried to discover sgRNA features that are associated with Cpf1 efficiency. In this study, this public data was used for prediction of Cpf1 activity using Convolutional Neural Network (CNN).

2.2   Sequence Encoding

For encoding of sgRNA sequences which contains A, T, G, C, one-hot vectors was used as follows: = 'A':[1, 0, 0, 0], 'T':[0, 1, 0, 0], 'G':[0, 0, 1, 0], 'C':[0, 0, 0, 1]. Therefore, every sgRNA sequences was considered as an image and was presented as a matrix with the shape 4×23. Where 23 is the length of the sequence and 4 is the number of nucleotides as previously described by [17].

2.3   Neural Network Model

This work was implemented using Tensorflow [22] and Keras [23] Libraries. A CNN model was used, and the architecture of the model is summarized as follows:

Input is a code matrix with shape of 23×4 where 23 represent the length of sgRNA sequence and 4 as any of the 4 nucleotides A, T, G and C. Convolutional layer is the first in this network, which is capable of feature extraction from sgRNA sequences using filters.

Next was a batch normalization (BN) layer, that minimizes internal covariate shift and improve learning rate. ReLU activation function was applied on every neuron of this layer. After batch normalization (BN) layer there is a third layer (max-pooling) for global max-pooling. Max pooling (MP) is carried out to obtain maximum values of output that represent the most important feature from BN layer outputs.

Then another convolutional layer was used followed by three fully connected dense layers with the sizes of 128, 64 and 64 neurons respectively. In last dense layer, a dropout layer was used to prevent overfitting by randomly switching off some neurons using a drop unit of 0.30 as this mechanism will help the model to understand new patterns of data flow. The final output layer consists of one neuron that is correlated to the Cpf1 activity. The neuron in this final layer is fully connected to all neurons present in previous layers. This layer performs linear regression and predicts scores corresponding to activity.

3   Results and Discussion

Here, a CNN model was used to predict Cpf1 guide RNA activity. CNN is a kind of feed-forward artificial neural network. The crucial aspect approximately of CNNs is that they can identify spatial representations in data, in preference to laborious feature engineering. CNN model encompasses three varieties of layers- convolution layers, pooling layers, and fully connected layers. Inside the convolution layers, weight vectors known as filters are used to multiply all the data regions. In this manner CNN is able to identify locally correlated patterns regardless of their places in the data [20].

Indel frequencies for 1251 sgRNA sequences obtained from this dataset were used to build a deep learning-based regression model. The sgRNA sequence was used as a feature for the convolutional neural network. First, the dataset was split into training and testing. To evaluate the generalization of the model, the training set was used to find the minimum value of mean squared error for the best model selection. The training was to develop a model with the best prediction of the test set. Pairs of input (sgRNA sequences) and output (Indel frequencies) sets were used to update the model parameters, by trying to minimize the loss function between predicted value and actual output.

We developed sixteen models in which we choose the best model that will predict Cpf1 guide RNA based on minimum MSE values. The sixteenth model possess the least MSE of 24.919.

The models were built by changing some hyperparameters like the number of filters, filter length and number of convolution.

As seen from Fig. 1, we obtained MSE values after performing one convolution (Fig. 1a and b), and two convolutions (Fig. 1c and d). Using different number of filters 23, 32 and 64, to compare performance of the models, the least MSE values was obtained with 64 filters (Fig. 1a and c) which indicates good performance while 23 filters shows large MSE value, hence poor performance. Therefore, model with 64 filters has better performance. Next, we tried to see the effect of filter length on model performance. Since models with 64 filters gave better performance, we compared the performance of models with 64 filters but with different filter length (Fig. 1b and d) and we observed that model with filter length 7 gave the best results.

Two convolutional layers gave better performance, also as the length of filters increases model performance was improved. Other studies were also performed to determine the activity of Cpf1 guide RNAs. Application of machine learning for Cpf1 activity prediction was carried out using SVM model [10]. Unlike CNN, here feature extraction was first carried out using Random Forest before training the SVM model, features for the sgRNAs like the position-specific composition of nucleotide, position-nonspecific composition of nucleotide, content of GC, least free energy, as well as melting temperature were determined.

This process requires high computational task and takes time when compared with CNN where the feature extraction is done automatically.

Recently, a research previously reported an algorithm that was able to predict the activity of Cpf1 by utilizing only target sequence of DNA sequence. They used the measured activities of 1,251 guide RNA–target sequence pairs to build a model by training a logistic regression classifier. Here Elastic Net regularization technique was used for both feature selection and regularization. This process requires feature extraction which is not an easy task. Features such as Position-independent nucleotides and dinucleotides, Position-dependent nucleotides and dinucleotides, required melting temperature, amount of GC, as well as Free energy were determined [23]. Here in this research prediction of Cpf1 activity using CNN model is based on only sgRNA sequences and does not require feature extraction.

One of the limitations of deep learning in CRISPR technology is the lack of available datasets. But this research shows that a CNN can be developed based on sgRNA sequence for prediction of Cpf1 activity because it is needed since most computational tools predicts Cas9 activity.

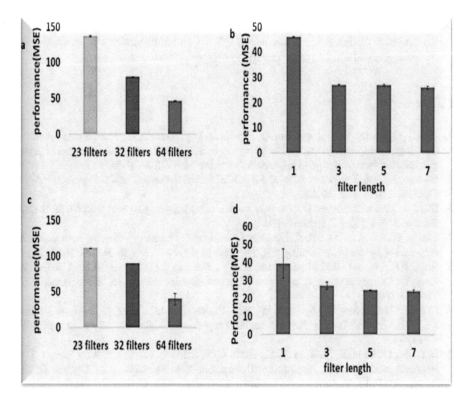

Fig. 1. MSE values comparing different model performance. **a** One convolutional layer with different number of filters. **b** One convolutional layer with 64 filters but different filter length. **c** Two convolutional layers with different number of filters. **d** Two convolutional layers with 64 filters but different filter length

4   Conclusion

In this research we developed 16 models with different hyperparameters and compared their performances using their MSE values. We used different convolutional layers and we observed that model performance was better with two convolutional layers. Here, all the computational task of feature extraction has been overcome by hidden layers of the CNN. In CRISPR technology only peculiar fragments of DNA are targeted but sgRNA is able to bind sometimes to other unintended DNA sequences causing off targets effects which can be harmful to the cell, it is therefore necessary to predict the activity of each sgRNA sequence and deep learning is applicable for accurate predictions and minimization of off-targets.

We have shown that the activity of Cpf1 sgRNA can be predicted from raw data using deep CNN model. Based on our knowledge, our model is the first deep learning model capable of predicting Cpf1 activity directly from sgRNA raw data without using feature extraction mechanism. The application of deep learning has been extended to

CRISPR technology for guide RNA activity prediction and this will go a long way to assist scientist to design efficient and active sgRNA and to minimize any unwanted side effects after gene editing, save time and experimental cost.

References

1. Ishino, Y., Shinagawa, H., Makino, K., Amemura, M., Nakatura, A.: Nucleotide sequence of the iap gene, responsible for alkaline phosphatase isoenzyme conversion in Escherichia coli, and identification of the gene product. J. Bacteriol. **169**(12), 5429–5433 (1987)
2. Davies, K., Mojica, F.: Crazy about CRISPR: an interview with Francisco Mojica. CRISPR J. **1**(1), 1–5 (2018)
3. Doudna, J.A., Charpentier, E.: The new frontier of genome engineering with CRISPR-Cas9. Science **346**(6213), 12580961-9 (2014)
4. Lin, S., Staahl, B.T., Alla, R.K., Doudna, J.A.: Enhanced homology-directed human genome engineering by controlled timing of CRISPR/Cas9 delivery. elife **3**(e04766), 1–13 (2014)
5. Rojo, F.P., Nyman, R.K.M., Johnson, A.A.T., Navarro, M.P., Ryan, M.H., Erskine, W., Kaur, P.: Crispr-cas systems: ushering in the new genome editing era. Bioengineered **9**(1), 214–221 (2018)
6. Church, G.M., Esvelt, K.M., Guell, M., Mali, P., Norville, J.E., Yang, L., Aach, J., DiCarlo, J.E.: RNA-guided human genome engineering via Cas9. Science **339**(6121), 823–826 (2013)
7. Hille, B., De Camilli, P., Murray, D., Burd, C.G., Schiavo, G., Irvine, R.F., Rameh, L.E., Nahorski, S.R., Challiss, R.A., Balla, T., Jackson, T.R., Hawkins, P.T., Thyagarajan, B., Rohacs, T., Balla, T., Stolz, L.E., Lemrow, S.M., York, J.D., Levine, T.P., Balla, A., Kim, Y.J., Lemmon, M.A., Balla, T., Aguet, F., Danuser, G., Schmid, S.L., Balla, T.: Multiplex genome engineering using CRISPR/Cas systems. Science **337**(80), 730–735 (2012)
8. Mussolino, C., Cathomen, T.: RNA guides genome engineering. Nat. Biotechnol. **31**(3), 208–209 (2013)
9. Zhang, X.H., Tee, L.Y., Wang, X.G., Huang, Q.S., Yang, S.H.: Off-target effects in CRISPR/Cas9-mediated genome engineering. Mol. Ther. - Nucleic Acids **4**(11), 1–8 (2015)
10. Zhu, H., Liang, C.: CRISPR-DT: designing gRNAs for the CRISPR-Cpf1 system with improved target efficiency and specificity. Bioinformatics **2019**(269910), 1–7 (2019)
11. Libbrecht, M.W., Noble, W.S.: Dissecting the genetics of complex traits using summary association statistics. Nat. Rev. Genet. **16**(6), 321–332 (2015)
12. Wang, T., Wei, J.J., Sabatini, D.M., Lander, E.S.: Genetic screens in human cells using the CRISPR-Cas9 system. BMJ Support. Palliat. Care **343**(6166), 256–263 (2012)
13. Doench, J.G., Hartenian, E., Graham, D.B., Tothova, Z., Hegde, M., Smith, I., Sullender, M., Ebert, B.L., Xavier, R.J., Root, D.E.: Rational design of highly active sgRNAs for CRISPR-Cas9-mediated gene inactivation. Nat. Biotechnol. **32**(12), 1262–1267 (2014)
14. Moreno-Mateos, M.A., Vejnar, C.E., Beaudoin, J.D., Fernandez, J.P., Mis, E.K., Khokha, M.K., Giraldez, A.J.: CRISPRscan: Designing highly efficient sgRNAs for CRISPR-Cas9 targeting in vivo. Nat. Methods **12**(10), 982–988 (2015)
15. Rusk, N.: Deep learning. Nat. Methods **13**(1), 35 (2015). https://doi.org/10.1038/nmeth.3707
16. Eraslan, G., Avsec, Ž., Gagneur, J., Theis, F.J.: Deep learning: new computational modelling techniques for genomics. Nat. Rev. Genet. **20**(7), 389–403 (2019)
17. Lin, J., Wong, K.C.: Off-target predictions in CRISPR-Cas9 gene editing using deep learning. Bioinformatics **34**(17), 656–663 (2018)

18. Kim, H.K., Min, S., Song, M., Jung, S., Choi, J.W., Kim, Y., Lee, S., Yoon, S., Kim, H.: Deep learning improves prediction of CRISPR-Cpf1 guide RNA activity. Nat. Biotechnol. **36**(3), 239–241 (2018)
19. Xue, L., Tang, B., Chen, W., Luo, J.: Prediction of CRISPR sgRNA activity using a deep convolutional neural network. J. Chem. Inf. Model. **59**(1), 615–624 (2019)
20. Chuai, G., Ma, H., Yan, J., Chen, M., Hong, N., Xue, D., Zhou, C., Zhu, C., Chen, K., Duan, B., Gu, F., Qu, S., Huang, D., Wei, J., Liu, Q.: DeepCRISPR: optimized CRISPR guide RNA design by deep learning. Genome Biol. **19**(1), 1–18 (2018)
21. Kim, H.K., Song, M., Lee, J., Menon, A.V., Jung, S., Kang, Y.M., Choi, J.W., Woo, E., Koh, H.C., Nam, J.W., Kim, H.: In vivo high-throughput profiling of CRISPR-Cpf1 activity. Nat. Methods **14**(2), 153–159 (2017)
22. Abadi, M., Agarwal, A., Barham, P., Brevdo, E., Chen, Z., Citro, C., Corrado, G.S., Davis, A., Dean, J., Devin, M., Ghemawat, S., Goodfellow, I., Harp, A., Irving, G., Isard, M., Jia, Y., Jozefowicz, R., Kaiser, L., Kudlur, M., Levenberg, J., Mane, D., Monga, R., Moore, S., Murray, D., Olah, C., Schuster, M., Shlens, J., Steiner, B., Sutskever, I., Talwar, K., Tucker, P., Vanhoucke, V., Vasudevan, V., Viegas, F., Vinyals, O., Warden, P., Wattenberg, M., Wicke, M., Yu, Y., Zheng, X.: TensorFlow: Large-Scale Machine Learning on Heterogeneous Distributed Systems. arXiv preprint arXiv:1603.04467 (2016)
23. Keras. https://keras.io. Accessed 20 June 2019

Fuzzy Model of Functioning of Educational-Laboratory and Production Capacities of the Educational Cluster in the Information Security Field

E. V. Zhilina$^{(\boxtimes)}$, L. K. Popova , N. A. Rutta ,
and N. E. Sheydakov

Rostov State University of Economics, B. Sadovaya Street, 69,
344002 Rostov-on-Don, Russia
black-2@mail.ru, popova_plk@mail.ru, rutic79@mail.ru,
sheidakov@mail.ru

Abstract. In this paper, the fuzzy model of functioning of shared educational-laboratory and production capacities on the example of a cluster in the information security field is developed. The proposed model helps to conduct an integrated accounting of quality factors, given their uncertainty (the results of equipment operation, the results of compliance with regulatory and technical documentation, the results of research work, the results of repairing the equipment, the results of the final state certification of students). This approach allows accurate assessing the functioning of shared educational-laboratory and production capacities to make reliable decisions on the management of educational clusters in the information security field.

Keywords: Fuzzy model · Educational-laboratory and production capacities · Cluster · Mamdani algorithm · Matlab · The linguistic variable · Membership function

1 Introduction

One of the major particularities of educational activity in the information security sphere consists in the need for expensive equipment. This fact makes it important to analyze the problem of providing the educational and research processes with modern educational-laboratory and production equipment as well as with improved approaches for using that equipment.

The study of the subject area made it possible to determine the need to develop an integrated approach to improve the management for shared educational-laboratory and production capacities (hereinafter - ELPC) on the example of information security clusters.

The analysis of the organizational system of educational clusters in Russia shows an aging problem of educational and research equipment, as well as lagging of research and analytical equipment base from the modern level of technology development. There is a need to use modern educational and experimental base to achieve the

© Springer Nature Switzerland AG 2020
R. A. Aliev et al. (Eds.): ICSCCW 2019, AISC 1095, pp. 704–711, 2020.
https://doi.org/10.1007/978-3-030-35249-3_91

qualitative level of the educational process in the information security sphere. This will allow to expand the possibilities of experimental research and to intensify the process of their implementation [5]. To ensure the efficient use of research complexes it is necessary to provide conditions for high load use. That task can be successfully carried out in the ELPC cluster.

Such shared ELPC was built inside the cluster of organizations and universities of Rostov Region (RSUE, MTUCI, TIT SFEDU, etc.). List of equipment, available for members of cluster helps to combine three types of activities, as it includes equipment for research, education and production.

However, it is necessary to note that comprehensive studies on the functioning of shared ELPC clusters have not been carried out yet. Known methodologies are not universal, as the activity of each cluster depends critically on the concentration of users and their relationships, the nature of the educational and scientific environment. There are no published scientific results describing how to assess the effectiveness of shared ELPC in the information security field.

There are several methodological approaches to build shared ELPC. A certain class of fuzzy logic tasks can be used to build the functional model of shared ELPC.

The methodology of fuzzy modelling does not replace or exclude the methodology of system modelling, but specifies it in relation to the process of building and use of fuzzy models in complex systems. The fuzzy modelling process itself represents a sequence of interrelated steps, similar to the stages of system modelling. In addition, each of the steps is performed in order to build and use a fuzzy model of the system to solve the original problem [4].

Generally, the fuzzy model is an information-logical model of the system, built on the basis of the theory of fuzzy sets and fuzzy logic [4].

The use of fuzzy models for calculating the functioning of shared ELPC in the information security field will expand the possibility of using clusters in higher education and will allow to accurately assess the output parameters of the models.

2   Main Part

2.1   A General Mathematical Model

The mathematical fuzzy model for evaluating the functioning of ELPC in the information security field can be represented in general form as follows:

$$x_i \rightarrow BP \rightarrow \text{FYLiPM}, \tag{1}$$

where

xi – input variable (input) and i $\in$ [1..6];
x1 – the results of equipment operation (REO);
x2 – results of compliance with regulatory and technical documentation (R_NTD);
x3 – results of research work (R_NIR);
x4 – results of used equipment capacity (RIM);
x5 – the results of repair of the equipment (RO);

x6 – results of the final state certification of students (IGA);
BR – the base of the rules necessary for the functioning of shared ELPC;
FYLiPM – functioning of shared ELPC (output).

2.2 Fuzzy Modelling

Modelling of shared ELPC functioning was carried out using specialized package Fuzzy Logic Toolbox from MATLAB tools. Fuzzy inference execution is implemented on the basis of the Mamdani algorithm [5].

Each of the linguistic input variables has triangular membership functions, which can be defined in general form by expressions:

$$\mu_\Delta(x_i, a, b, c) = \begin{cases} 0, & x \leq a, \\ \frac{x-a}{b-a}, & a \leq x < b, \\ \frac{c-x}{c-b}, & b \leq x \leq c, \\ 0, & c \leq x \end{cases} \tag{2}$$

where a, b, c are some numerical parameters characterizing the base of the triangle (a, c) and its vertex (b), and the following condition must be satisfied: $a \leq b \leq c$.

The figures below show graphs of membership functions of fuzzy term sets of incoming linguistic variables.

Variable x1 - "REO" (results of equipment operation) consists of three main term sets defined in the interval [0,1]: "low results" [0, 0,4], "average results" [0,2, 0,8], "high results" [0,6, 1] and two additional term sets: "below average" [0,1, 0,4], "above average" [0,6, 0,9] (see Fig. 1).

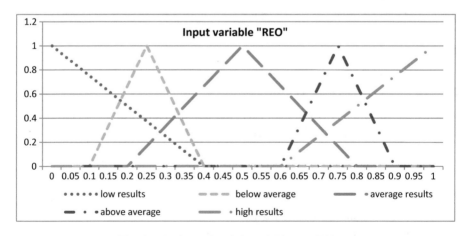

Fig. 1. The input linguistic variable "REO" (x1).

Variable x2 - "R_NTD" (results of compliance with the terms of normative and technical documentation) also consists of three-term sets defined in the interval [0,1]:

"critical" [0, 0,4], "acceptable" [0.1, 0.9], "the most appropriate" [0.6, 1] (see Fig. 2), which determines the number and quality of grants won and their implementation during the reporting period.

The composition of the linguistic variable X3 - "R_NIR" [2] (results of research) term sets is similar to the term sets of the linguistic variable "REO". It consists of three main term sets defined on the interval [0, 1]: "low results" [0, 0.4], "average results" [0.2, 0.8], "high results" [0.6, 1] and two additional term sets: "below average" [0.1, 0.4], "above average" [0.6, 0.9].

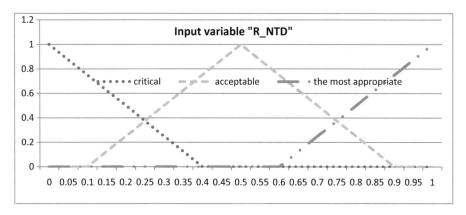

Fig. 2. The input linguistic variable "R_NTD" (x2).

The variable x4 - "RIM" (results of the used equipment capacity) consists of three main term sets defined in the interval [0,1]: "low load" [0, 0.4], "sufficiently optimal load" [0.1, 0.9], "optimal load" [0.6, 1] (see Fig. 3).

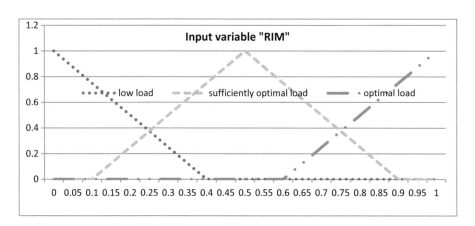

Fig. 3. The input linguistic variable "RIM" (x4).

Variable X5 - "RO" (equipment repair results) consists of two term sets defined on the interval [0,1]: "inefficient" [−0.6, 0, 0.6], "effective" [0.4, 1, 1.6] (see Fig. 4).

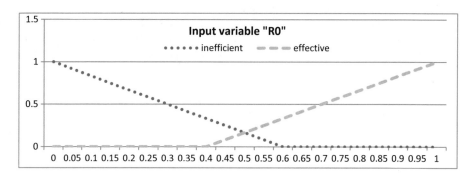

Fig. 4. The input linguistic variable "RO" (X5).

Variable X6 - "IGA" (the results of the final state certification of students) consists of four main term sets defined in the interval [0,1]: "unsatisfactory" [0, 0.2], "satisfactory" [0.1, 0.6], "good" [0.5, 0.9], "excellent" [0.8, 1], which determines the level of training of university students [2, 3] graduating from the educational program on information security [1, 3] (see Fig. 5).

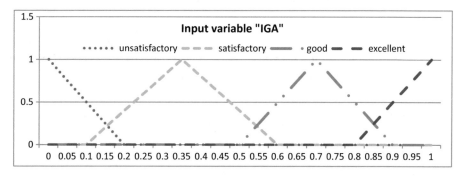

Fig. 5. The input linguistic variable "IGA" (X6).

Figure 6 shows a graph of the membership functions of fuzzy term sets of the linguistic variable "FYLiPM", given in MATLAB (output in the model).

Given the membership functions of term sets of linguistic variable "FYLiPM" will be as follows: $\mu^H_{FYLiPM}(x, 0, 0, 0.4)$ – unsuccessful, $\mu^O_{FYLiPM}(x, 0.1, 0.5, 0.9)$ – optimal, $\mu^Y_{FYLiPM}(x, 0.6, 1, 1)$ – successful.

As a result of the subject area analysis of the base of rules (BR) of ELPC functioning in the information security field is formed. In Table 1 a fragment of the rule base is given (+ - "true").

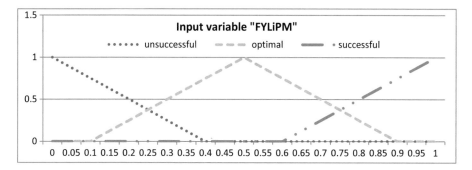

Fig. 6. The output variable "FYLiPM".

Table 1. A fragment of fuzzy production rules for assessing the functioning of shared ELPC.

Parameter	Description	Term	BR					
			1	2	3	4	5	n
x1 - REO	Results of compliance with the terms of regulatory and technical documentation	Low	+	+		+	+	
		Below the average			+			
		Average						
		Above average						
		High						+
x2- R_NTD	Results of research	Critical	+	+	+	+		
		Allowed					+	
		Most appropriate						+
x3 - R_NIR	Results of research	Low	+	+	+		+	
		Below the average				+		
		Average						
		Above average						
		High						+
x4 - RIM	The results of the used capacity of the equipment	Low load	+	+	+	+	+	
		Optimal efficiency						
		Optimal load						+
x5 - RO	The results of repair of equipment	Inefficient	+	+	+	+	+	
		Effective						
x6 - IGA	The results of the final state certification of students	Unsatisfactory	+		+	+	+	
		Satisfactory		+				
		Good						
		Excellent						+
FYLiPM	Functioning of shared ELPC	Unsuccessful	+	+	+	+	+	
		Optimal						
		Successful						+

Accumulation of conclusions according to all the rules was carried out using the max-disjunction operation [3]. In defuzzification, the method of the center of gravity for a discrete set of the membership function values is used.

The following implementations of the fuzzy model of ELPC functioning, demonstrate the output of the fuzzy inference rules of the resulting linguistic variable "FYLiPM". It shows the direct dependence of the output variable from the values of REO and IGA (see Table 2):

Table 2. The implementation of the fuzzy inference rules of the ELPC operation in the information security field.

Results	Implementation of a fuzzy model (fragments)
FYLiPM =0,508	REO=0.645 R_NIR=0.645 RO=0.664 FYLiPM=0.508 R_NTD=0.827 RIM=0.609 IGA= 0.645
the term "optimal" level of confidence μ^{O}_{FYLiPM} = 0,98.	
FYLiPM =0,47	REO=0.245 R_NIR=0.773 RO=0.809 FYLiPM=0.47 R_NTD=0.791 RIM=0.79 IGA= 0.573
the term "optimal" level of confidence $\mu^{O}_{FYLiPM} = 0,8.$	
FYLiPM =0,195	REO=0.645 R_NIR=0.645 RO=0.664 FYLiPM=0.195 R_NTD=0.827 RIM=0.609 IGA= 0.1
term "not successful" confidence level μ^{H}_{FYLiPM} = 0,75.	

3 Conclusion

1. The model of shared ELPC functioning on the example of a cluster in the information security field is developed. It was made on the basis of the theory of fuzzy sets.
2. The proposed model helps to conduct an integrated accounting of quality factors, under uncertainty (the results of equipment operation, the results of compliance with regulatory and technical documentation, the results of research work, the results of repairing the equipment, the results of the final state certification of students). Setting the value of the criterion of the confidence level significance, it is possible to change results depending on the group level of capacity shared utilization.
3. It was experimentally determined that the application of this approach allows accurate assessing of the ELPC functioning. This proves the validity of its application in the development of an integrated approach to the improvement of the organization and management for existing shared educational-laboratory and production capacities of clusters in the information security field.

References

1. Tishchenko, E.N., Zhilina, E.V., Sharypova, T.N., Palyutina, G.N.: Fuzzy models of the results of the mastering the educational programs in the field of information security. In: Aliev, R.A., Kacprzyk, J., Pedrycz, W., Jamshidi, M., Sadikoglu, F.M. (eds.) ICAFS 2018. AISC, vol. 896, pp. 694–701. Springer, Cham (2019). https://doi.org/10.1007/978-3-030-04164-9_91
2. Tishchenko, E., Sharypova, T., Zhilina, E., Cherkezov, S.: Economic and mathematical modelling of complex cooperation of academic staff of educational cluster on the basis of fuzzy sets theory. J. Appl. Econ. Sci. (JAES) 5(11), 905–907 (2016)
3. Zhilina, E.V.: Fuzzy models of student evaluation of discipline learning. Econ. Syst. Manag.: E-J. 35(11) (2011). https://uecs.ru/instrumentalnii-metody-ekonomiki/item/820-2011-11-30-11-05-39
4. Wang, L.-X., Mendel, J.M.: Generating fuzzy rules by learning from examples. IEEE Trans. Syst. Man Cybern. 22(6), 1414–1427 (1992)
5. Olishevskiy, D.P., Serbinovski, B.Y.: Modelling and Analysis of Organization and Management of Collective Use Center. Southern Federal University, Novocherkassk: SRSTU (NTI), p. 135 (2009)

About One Mathematical Model of Reliability and Safety of Complex Systems

Araz R. Aliev[1,2](✉) (ID), Vagif M. Mamedov[1],
and Mammad I. Seyidov[1]

[1] Azerbaijan State Oil and Industry University, Baku, Azerbaijan
alievaraz@yahoo.com, vaqifmammadoqlu@gmail.com,
mamed_seyidov@mail.ru
[2] Institute of Mathematics and Mechanics,
Azerbaijan National Academy of Sciences, Baku, Azerbaijan

Abstract. A generalization of existing results of application of the fuzzy set theory to problems of reliability and safety of systems is considered. On the basis of the introduced fuzzy measure the mathematical model of reliability and safety of complex systems is developed.

Keywords: Reliability · Probability · Fuzzy measure · Technical operation · Fuzzy-probabilistic event · Safety · Fuzzy model · Fuzzy-probabilistic model

1 Introduction

Nowadays, intelligent systems design [2] and development of emergent information technologies require generation of new paradigms and information processing methods to cope with uncertainty and lack of experimental data, higher demands for system functions, complexity of real-world operation conditions etc. In this regard, the use of fuzzy sets theory in the realm of problems of reliability and safety of information systems provides ability of system failures prediction, aging process monitoring, durability evaluation, optimal maintenance problem. In this work, results of long term research in field of reliability and safety of complex control systems are analyzed and mathematically validated.

2 The Principle of Methodology and the Mathematical Basis of the Approach

Let E be an arbitrary set. We assume that $\mathcal{F}$ is a ring or an algebra of subsets E_i of set E ($\forall i, E_i \subset E, i = 1, 2, \ldots$). Recall that a real-valued function $g : \mathcal{F} \to [0, +\infty)$ is referred to as a measure if:

1. $g(\varnothing) = 0$, where $\emptyset$ is an empty set.
2. $g(\bigcup_{i=1}^{n} E_i) = \sum_{i=1}^{n} g(E_i)$, where $E_i \subset E$, $E_i \cap E_j = \emptyset$ for any $i \neq j, i = 1, 2, \ldots,$ $j = 1, 2, \ldots$.

© Springer Nature Switzerland AG 2020
R. A. Aliev et al. (Eds.): ICSCCW 2019, AISC 1095, pp. 712–719, 2020.
https://doi.org/10.1007/978-3-030-35249-3_92

Note that condition $E_i \cap E_j = \emptyset$ is very restrictive and is not often practically fulfilled in problems of reliability and safety of complex systems. Thus, on the basis of Zadeh definitions of basic operations over fuzzy sets, the concept of fuzzy measure was introduced [1, 3, 7]. A fuzzy measure satisfies the conditions given below.

Let Ω be a universal set and $E_n, n = 1, 2, \ldots$, be all its possible subsets, i.e. for $\forall n, E_n \subset \Omega$. Then fuzzy measure $g : E_n \to [0, 1]$ satisfies the following axioms:

1. $g(E_n) \geq 0$ for $\forall E_n \subset \Omega$ – non-negativity axiom;
2. $g(\bigcup_n E_n) = \max_n g(E_n) \leq 1$ – boundedness axiom;
3. $g(\bar{E}_n) = 1 - g(E_n)$ – complement axiom.

Note that condition $E_i \cap E_j = \emptyset, \forall i \neq j$ is not necessary for fuzzy measure.

In [1] it is proven that, condition 3 (complement axiom) can be omitted, if $\exists E_k$ in E_n, such that $\max_n g(E_n) = g(E_k) = 1$. In other words, at least one normal fuzzy subset in E_n necessarily exist. It is clear that if fuzzy set is subnormal, it can be normalized as $G(E_n) = g(E_n)/\sup_n g(E_n)$.

Investigation and analysis of reliability and safety of complex systems (by using fuzzy и and probability measures) shows that four main types of real-world events exist:

I.   Deterministic event – it is a priori exactly known whether an event will occur or not (complete certainty);
II. Random event – it is not exactly known a priori whether an event will occur;
III. Fuzzy event – it is partially known a priori whether an event will occur (partial certainty or partial uncertainty);
IV. Fuzzy random event – event characterized by fuzziness and randomness (combination of fuzziness and probabilistic uncertainty).

For deterministic and random events a quite amount of examples exist, such well developed theories as probability theory, mathematical statistics and discrete analysis are used as formalism to deal with such events. Let us consider some examples on 3rd and 4th types. In [13] fuzzy event is defined as a fuzzy subset of a universal set, whose membership function is Borel measurable. In system reliability theory, «beginning of an aging period» of techware or control systems can be considered as fuzzy event. Indeed, operating experience shows that aging process of techware is a natural phenomena and beginning of such process is rather fuzzy than random. In [7] this issue is investigated in detail and a technique for approximate calculation of techware durability is proposed. Let us note that durability problems are still not fully resolved.

As another example, consider fuzzy events: "the system *often* fails" ("*rarely*", "*very rarely*", *very often*", etc.). Assume that these fuzzy events are described in graphical form (Fig. 1), where $\mu_{often}(\mu_{very\ seldom}, \mu_{seldom})$ are fuzzy measures N – number of failures.

It is clear that the point $\arg\max_N \mu_{often} = N_1 (\arg\max_N \mu_{seldom} = N_0)$ provides partial information, reflecting "approximate" number of times the system fails in a given period. Thus, having fuzzy (or non-numeric) information, one can judge the behavior of the system from the position of reliability and safety.

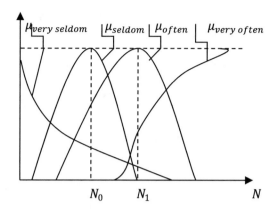

Fig. 1. Graphical description of fuzzy events

However, in the process of designing information systems, their "normal functioning" cannot always be specifically classified within some group of events (fuzzy or random) and studied using the appropriate mathematical apparatus. The issue is that maintenance factors (maintenance, spare parts supplies, etc.) improves the condition of the system, whereas destabilizing factors (temperature, vibration, humidity, etc.) worsens its condition. Therefore, depending on the complex influence of these factors, the system may be in different states. These conditions can "approximately" ("fuzzy") be predicted [8, 9, 12]. Indeed, given the causal relationships, imprecise values of parameters affecting the state of the system, one can imagine how the system failure emerges, and to assume that it will happen "soon". However, this fuzzy event may not occur due to the fact that it is not mature enough, or to occur earlier than the estimated time due to the fact that not observed and unforeseen factors (random) contributed to premature failure. As a result, we are confronted with a phenomenon characterized by a combination of fuzziness and randomness. Let it be referred to as "fuzzy-random" (or "F - P") event.

Fuzzy-random events [7, 10, 11] include: "there will be a failure in the system", "the system will function normally", "the safety of the carriage of passengers by the aircraft is sufficient", etc. It is clear that problems related to events of types II – IV require using a specific approach. However, as it is known, II and III types of problems are investigated and solved by the probabilistic theory and the fuzzy set theory, respectively. Concerning a problem of type IV, we have to deal with information characterized by fuzziness and randomness. Three cases are possible: fuzziness dominates, i.e. event is fuzzy, but its occurrence depends on random factors (F-P event); randomness dominates (P-F event), i.e. the class of probability measures is described by possibility distribution; and fuzziness and randomness equally present in an event. The measures for these tree cases are determined as [5, 10, 11]:

I. $P_E = \int g(\cdot)dP$;

II. $\rho = \{P/\forall E, N(E) \leq P(E) \leq \prod(E)\}$;

III. $G_{F-P} = \int \sum_{\oplus} g(E_i)dP$,

P_E - Lebesgue-Stieltjes integral [4], $\prod(E)$, $N(E)$ – possibility measure and necessity measure respectively [5] $\oplus$ denotes an algebraic sum.

Note that in this paper we consider case I, whereas case III is considered partially (i.e. when the operation of the maximum (axiom 2) is replaced by the algebraic sum). We show that, taking into account the above, and using Kolmogorov's axiomatic theory, one can similarly construct fuzzy-probability spaces.

Let be Ω_F an arbitrary set, where each subset describes an elementary fuzzy event. An elementary fuzzy event is considered as an event that can be represented in at least one way as the union of a finite number of pairwise disjoint fuzzy subsets. For example, "a system will fail due to a high electrical load or an operator error" is a fuzzy event that can be viewed as a union of elementary events: "a system will fail due to a high electrical load" and "a system will fail due to an operator error".

A set of events, for which fuzzy measure is defined, will be referred to as a field of fuzzy events. Analogously to the classical theory of measure, a field of fuzzy events $\Im_F$ has the following properties [10, 11]:

(1) If fuzzy measure $g(E_i)$ is defined for $E_i \subset \Im_F$, then it is also defined for the complement of E_i, i.e. $g(\bar{E}_i)$ is defined for $\bar{E}_i \subset \Im_F$.
(2) If fuzzy measure $g(E_i)$ is defined for $\forall E_i \subset \Im_F, i = 1, 2, \ldots$, then it is also defined for the union $\bigcup E_i$, $g(\bigcup E_i)$.

Axioms I–V, given in [1] verify these conditions. Then a set of fuzzy events that satisfies these conditions is referred to as algebra of fuzzy events. σ-algebra $\Im_F$ will be defined as follows:

(1) Any subset $E_{F_n} \subset \Im_F$, n = 1,2,..., is a fuzzy event;
(2) $E_{F_n} \subset \Im_F \Leftrightarrow \bar{E}_{F_n} \subset \Im_F$;
(3) $\forall n, E_{F_n} \subset \Im_F \Leftrightarrow \bigcup_{n \geq 1} E_{F_n} \subset \Im_F$;
(4) $\varnothing, \Omega_F \subset \Im_F$.

We will call quadruple $\{\Omega_F, \Im_F, g, P\}$ a fuzzy-probabilistic space where probability P is defined as Lebesgue-Stieltjes integral: $P(E_{F_n}) = \int\limits_{E_{F_n} \subset \Im_F} g(E_{F_n})dP$.

Let us prove that P is a probability measure, i.e. it satisfies Kolmogorov's axioms for the cases of fuzzy and random elementary events. At first, consider:

I. $P(E_{F_n}) \geq 0$ - non-negativity axiom.
As $P(x)$ is probabilistic measure defined over probabilistic space and fuzzy measure $g(E_{F_n}) \geq 0$ is a positive real-valued function, then the antiderivative of function $g(E_{F_n}) \cdot P'(x)$ ($P'(x)$ is a derivative) is an increasing function in Ω_F. Therefore, using a Newton-Leibnitz formula, one can claim that $P(E_{F_n}) \geq 0$. Next, consider

II. 6 – additivity.

Let for any $n, E_{F_n} \subset \Im_F$. Given axiom II for fuzzy measure, one can write $P(\bigcup_{n \geq 1} E_{F_n}) = \int g(\bigcup_{n \geq 1} E_{F_n}) d \sum_{n \geq 1} P_n$.

Assume that $\max_n g(E_{F_n}) = g(E_{F_k})$, then by using the property of differential and integral, one has: $P(\bigcup_{n \geq 1} E_{F_n}) = \int g(E_{F_k}) d \sum_{n \geq 1} P_n = \int \sum_{n \geq 1} g(E_{F_k}) dP_n = \sum_{n \geq 1} \int g(E_{F_k}) dP_n$.

This is nothing but additivity condition.

Finally, consider.

III. $P(\Omega_F) = 1$ – normality axiom.

Theorem. Normality axiom is true if and only if at least one normal fuzzy subset exist in Ω_F.

The proof of the sufficiency is obvious.

Let us prove the necessity. Let there be at least one normal fuzzy subset E_{F_n} in Ω_F. Then $\max_n g(E_{F_n}) = 1$. Taking into account that elementary events are only considered and using axiom II, one has $P(\Omega_F) = \int_{\Omega_F} g(\bigcup_{n \geq 1} E_{F_n}) dP = \int_{\Omega_F} \max_n g(E_{F_n}) dP = \int_{\Omega_F} dP = 1$.

This completes the proof. Note that, the second condition (axiom) for fuzzy мере can be written as $g(\bigcup_{n \geq 1} E_n) = \sum_{\oplus} g(E_{F_n})$. Then we call the obtained measure as fuzzy probabilistic measure. The following proposition applies.

Proposition. If $\bigcap_{i=1}^{n} E_i$ is a non-empty set and fuzzy probabilistic measure is defined over Ω_F, then one has $g(\bigcap_{i=1}^{n} E_i) = \prod_{i=1}^{n} g(E_i) = g(E_1) \cdot g(E_2) \cdots g(E_n)$.

Proof. Let $n = 2$. By using de-Morgan law one has $\overline{E_1 \cap E_2} = \overline{E_1} \cup \overline{E_2}$. Then $g(\overline{E_1 \cap E_2}) = g(\overline{E_1} \cup \overline{E_2})$. By using the second condition for fuzzy probabilistic measure in the right hand side of the equality, one has: $g(\overline{E_1} \cup \overline{E_2}) = g(\overline{E_1}) + g(\overline{E_2}) - g(\overline{E_1}) \cdot g(\overline{E_2})$. Then $1 - g(E_1 \cap E_2) = 1 - g(E_1) + 1 - g(E_2) - (1 - g(E_1)) \cdot (1 - g(E_2))$ holds and one has $g(E_1 \cap E_2) = g(E_1) \cdot g(E_2)$. The proof is completed.

By using the recurrence formula for the sum $\sum_{\oplus} g(E_n)$ the proposition can be proved for $n > 2$ by using the mathematical induction.

In the next section we tried to develop an F-P (Fuzzy-Probability) model of reliability and safety of systems taking into account the abovementioned results.

3   F-P Model of Reliability and Safety of Systems

In reliability theory, a probability that a system will perform its intended function (no-failure operation) $P(t)$ is defined as an expected value of a functional $\{\varphi[x(t)]\}$ [6], i.e.

$$P(t) = M\{\varphi[x(t)]\},$$

where $x(t)$ is stochastic process in state space and φ is some functional defined as:

$$\phi : x(t) \rightarrow \begin{cases} 0, & \text{if a system fails} \\ 1, & \text{otherwise} \end{cases}, t \text{ is time.}$$

Let $\alpha = (\alpha_1, \alpha_2, \ldots, \alpha_n), \beta = (\beta_1, \beta_2, \ldots, \beta_m)$ vectors of parameters describing the quantitative values of factors that worsen and improve the state of the system, respectively. The complex effect of these parameters on the state of the system is denoted by $(\alpha \otimes \beta)$, state space $G = \{x\}$, i.e. states X which differ in terms of reliability of a system is denoted by a vector $X[\alpha(t) \otimes \beta(t)]$.

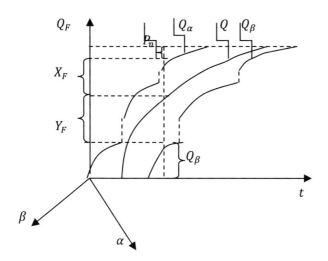

Fig. 2. Failure state described as "fuzzy – stochastic" process

In accordance with what is proposed [9, 12] $\alpha(t), \beta(t)$-fuzzy measurable parameters, that can easily be described by using fuzzy measure g. Then taking into account the abovementioned (Sect. 2), we can state that $X[\alpha(t) \otimes \beta(t)]$ vector is «fuzzy-stochastic process», i.e. it is characterized by both randomness and fuzziness. Such processes can be described by smooth, zigzag or increasing piecewise functions (Fig. 2). It can be seen that worsening factors α increase probability of failure (Q_α) at t_0 by X_F value, whereas positive factors decrease probability of failure (Q_β) by Y_F value.

In accordance with Sect. 2, we define g as $g_\varphi : X[\alpha(t) \otimes \beta(t)] \rightarrow [0, 1]$, i.e. instead of a classical measure, fuzzy measure is used, and it is assumed that the transition from a workable state to a non-functional state is not abrupt, but smooth and is described by a continuous (or discrete) interval $[0, 1]$. As an example, one can consider the case when the failure of several elements (functions) does not allow to definitely reason about the failure of the whole system. In contrast, one can reason that the violation of technology reduces the quality of functioning, but does not lead to failure of functions.

Then, using the proposed methodology and taking into account (1) one has:

$$M[g_\varphi] = \int g_\varphi[\alpha(t), \beta(t)]dP$$

This is a well known Bellman–Zadeh formula [4]. It generalizes the results in the reliability theory of complex systems. Indeed, for $g_\varphi \equiv 1$, one has Gnedenko mathematical model of reliability [6].

4 Conclusion

The obtained FP model of reliability and safety allows to deal with such problems as predicting and evaluating reliability of information system in the early stages of designing, taking into account external influencing factors and technical operation factors, a priori determining the economic feasibility of developed systems, solving an optimal multi-parameter technical maintenance problem, optimizing the maintenance period, and evaluating reliability of multi-functional systems, develop fuzzy algorithms and software for problems of reliability and safety of intelligent systems.

The introduced concept of fuzzy measure can also be useful in protecting data and securing information in global networks when considering the reliability of software and operating personnel, where, along with objective factors, subjective factors play a significant role, including man and society.

References

1. Aliev, A.R., Mamedov, V.M., Gasimov, G.G.: Analysis and processing of information in economic problems. Crisp and fuzzy technologies. In: 13th International Conference on Theory and Application of Fuzzy Systems and Soft Computing, ICAFS 2018. Advances in Intelligent Systems and Computing, vol. 896, pp. 65–72. Springer, Cham (2019)
2. Aliev, R.A., Abdikiev, N.M., Shakhnazarov, M.M.: Production systems with artificial intelligence. Radio and Communication, Moscow (1990). (in Russian)
3. Aliev, R.A., Aliev, R.R.: Soft Computing and its Application. World Scientific, London (2001)
4. Bellman, R.E., Zadeh, L.A.: Decision-making in a fuzzy environment. Manag. Sci. **17**, 141–164 (1970/71)
5. Dubois, D., Prade, A.: Theory of possibilities. In: Application to Knowledge Representation in Computer Science. Radio and Communication, Moscow (1990). (in Russian)
6. Gnedenko, B.V., Belyaev, Yu.K., Soloviev, A.D.: Mathematical Methods in the Theory of Reliability. Nauka, Moscow (1965). (in Russian)
7. Kiyasbeili, Sh.A., Mamedov, V.M.: On the evaluation of the durability of technical systems using the theory of fuzzy sets. Electronnoe Model. (1), 59–61 (1981). (in Russian)
8. Kiyasbeili, Sh.A., Mamedov, V.M.: On possible use of the theory of fuzzy sets for solving the problem of estimating and insuring the reliability of complex systems. Autom. Remote Control **43**(12), 1617–1623 (1982)

9. Kiyasbeili, Sh.A., Shishonok, N.A., Mamedov, V.M.: Some questions of the application of the theory of fuzzy sets to the problems of reliability. In: Reliability of Automated Process Control System, pp. 35–41. Naukova Dumka, Kiev (1981). (in Russian)

10. Mamedov, V.M.: F-reliability model. Trans. NAS Azerbaijan **22**(2–3), 3–9 (2002). (in Russian)

11. Mamedov, V.M.: Fuzzy and soft measurements in reliability problems of complex systems. In: Proceedings of the International Conference an Soft Computing and Measurements, Saint Petersburg, vol. 2, pp. 16–18 (2003). (in Russian)

12. Shishonok, N.A., Kiyasbeili, Sh.A., Mamedov, V.M.: Using the theory of blurred sets to assess the complex influence of technical operation factors on the reliability of the ACS. Electronnoe Model. (1), 97–99 (1982). (in Russian)

13. Zadeh, L.A.: Probability measures of fuzzy events. J. Math. Anal. Appl. **23**(2), 421–427 (1968)

Synthesis of the Two-Level Intelligent Hierarchical Control System of the Mobile Robot

E. R. Zeynalov[1] and A. S. Aliyeva[2(✉)] [iD]

[1] Azerbaijan State Oil and Industry University, Baku, Azerbaijan
zeynalelchin@gmail.com
[2] Institute of Control Systems of ANAS, Bakhtiyar Vahabzadeh st. 9, Baku,
Azerbaijan
c-adile@yahoo.com

Abstract. The actual problems related to the operation of high-quality intelligent control systems under the conditions of autonomy of two-wheeled mobile robots have been investigated in the paper. The synthesis method of the two-level intelligent control system of two-wheeled mobile robot for a dynamic object written by a multiply connected nonlinear model is proposed. According to the proposed method, an algorithmic support of the synthesis problem of a two-level intelligent control system is developed on the bases of fuzzy logic and finite automata. One of the advantages of the proposed in the paper method is to provide the activity of a two-wheel mobile robot, quickly, with high dynamic accuracy and without collision with barriers in the environment of several obstacles.

Keywords: Two-wheeled mobile robots · Intelligent two-level hierarchical control · Fuzzy logic T-S controller · Finite automata · Stateflow

1 Introduction

Some medical aid, air and sea exploration, military researches, agriculture, security and etc. included application of mobile robots moving in all directions. Currently in many fields only wheeled mobile robot (MR) is used [4, 10]. Mobile robots are applied on the purification of dangerous substances, rescue operations, security, investigation of planets and searching of mines in a large scale. For an improvement of autonomy of mobile robots the following actual issues, such as positioning of robot, planning of trajectory (path planning), motion tracking on trajectory, collision with barriers ought to be solved [1–11].

The motion of mobile robot without collision with barriers could be solved by artificial potential of the field [1–4], however it could not be accepted as an optimal solution. In the method of artificial potential field the goal is an artificial potential of the field characterizing the motion of mobile robot and obstacles.

In the result of attractive and repulsive forces the third force is generated and it enables effective control of mobile robot. We also note that, when values of attractive

© Springer Nature Switzerland AG 2020
R. A. Aliev et al. (Eds.): ICSCCW 2019, AISC 1095, pp. 720–727, 2020.
https://doi.org/10.1007/978-3-030-35249-3_93

and repulsive forces, dissemination of obstacles and reverse orientations are equal, it could be formed a situation in what the assignment is not implemented towards obstacle avoidance of mobile robot [4].

According to critical analysis of scientific literature the high- quality intelligent control methods are not satisfied [1–10] under condition of autonomy of mobile robots. Considering the latest, the synthesis method of intelligent two-level control system of dynamic object written by the multi related non-linear model has been suggested in the proposed paper and its virtual realization issues have been solved in MATLAB.

2   Synthesis of Intelligent Two-Level Control Systems of Mobile Robot

MR is the dynamic multi related object with three control outputs x, y, φ and three control inputs (ω_r, ω_l) and $(\omega_r + \omega_l)/2$ variables. For two-wheeled mobile robot the structural scheme of intelligent two-level hierarchical control system is shown in Fig. 1.

In block 1 an initial (x_0, y_0, φ_0) coordinate of two-wheeled mobile robot and final coordinate of goal (x_m, y_m, φ_m) s are formalized. In the 2-nd block current state coordinates $(x, y, \varphi, \omega_r, \omega_l)$ are determined on the bases of signals obtained from eight sensor devices.

In order to not clash the mobile robots with barriers "attractive", "repulsive", "sum" forces $(F_{attractive}(q), F_{repulsive}(q), F_{sum}, f_x, f_y)$ are determined according to artificial field tension method in the 3-rd block and transferred to the 4-th block. In the block 4 $\omega_r^{opt}, \omega_l^{opt}, \varphi^{opt}$ of intelligent low-level hierarchic automatic control system (ACS) are formalized on the bases of finite automata assignment effects via intelligent system identifying motion trajectory mobile robot without collision with barriers.

On the two-dimensional plane on the fixed coordinate system the coordinate points related to wheels of Mobile robot is determined as follows: $q = (x_1, y_1)$, but $q_0 = (x_0, y_0)$ for obstacles, $q_m = (x_m, y_m)$ for goal. Let an obstacle and goal create F_{sum} force individually by applying force on the wheel of mobile robot

$$F_{sum} = F_{attractive}(q) + F_{repulsive}(q)$$

$F_{attractive}(q)$ – is an attractive force created byaim, $F_{repulsive}(q)$ – is repulsive force created by repulsive force. Total force created by artificial potential field is determined as follows [1–4]:

$$U(q) = U_{attractive}(q) + U_{repulsive}(q),$$

$$F_{attractive}(q) = -\nabla U_{attractive}(q),$$

$$F_{repulsive}(q) = -\nabla U_{repulsive}(q),$$

$U_{attractive}(q)$ and $U_{repulsive}(q)$ express artificial potential field related to obstacles and goals respectively, operator ∇ is shown as $\frac{\partial}{\partial x}$, $\frac{\partial}{\partial y}$. In the equations of (3) and (4) attractive and repulsive forces are reversal direction.

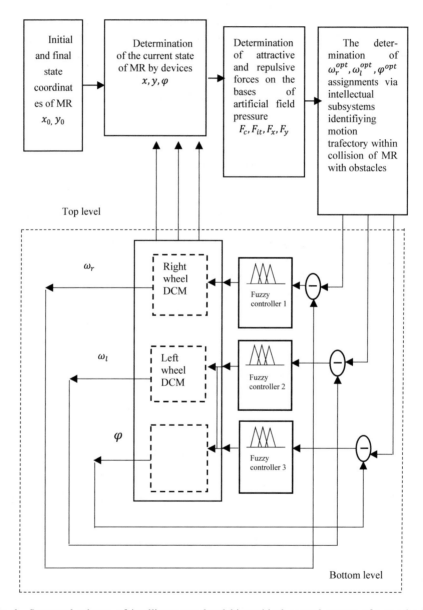

Fig. 1. Structural scheme of intelligent two-level hierarchical control system of two-wheeled mobile robot.

Continuously depending on location of obstacle and goal, total force would take mobile robot to the final point without any collision. A description of the artificial potential field depend on the location of obstacles that would repulse without collision. Beside, for motion without collision of MR it is necessary to identify the coordinates of robot, obstacle and goal $q = (x_1, y_1)$, $q_{obstacle} = (x_0, y_0)$, və $q_g = (x_g, y_g)$. In the presented case, it is assumed these coordinates are known, this suggests that mobile robots own eight sensors enabling to obtain these coordinates in real time.

For motion of differential-driven MR without collision with obstaclet two-level hierarchical control system suggested in the Fig. 1 is used. Mobile robot is as a transport tool as a control object having right and left pulling forces. The wheels of robot are parallel and do not distort and could orient in both directions. In the states space the kinematic model of mobile robot can be written as follows [3, 10]:

$$\dot{x} = \frac{(\omega_r + \omega_l)r}{2} \cos \varphi,$$

$$\dot{y} = \frac{(\omega_r + \omega_l)r}{2} \sin \varphi, \tag{1}$$

$$\dot{\varphi} = \frac{(\omega_r - \omega_l)r}{2},$$

x, y - coordinates of middle point of the axis jointed by wheels (output variables), φ - an angle generated by symmetry axis of the mobile robot with positive X axis (direction of MR), ω_r and ω_l are control effects of MR of angular velocities relatively to the right and left wheels, r is wheel coefficient (rate), and $2l$ is a distance between wheels. The assumed motion of MR is limited on XY plane and there is no any wheel error.

Coordinates of front part $-q = (x_1, y_1)$. Coordinates of q are determined depending on x, y and φ:

$$x_1 = x + L \cos \varphi, \tag{2}$$

$$y_1 = y + L \cos \varphi,$$

Here, L – is a distance between middle points of wheel axises, $(x, y)q$ point is perpendicular to the axis. Considering the latest expressions, in the states space the kinematic model of MR could be written as follows [3, 9]:

$$\begin{pmatrix} \dot{x}_1 \\ \dot{y}_1 \end{pmatrix} = A(\varphi) \begin{pmatrix} \omega_r \\ \omega_l \end{pmatrix}, \qquad A(\varphi) = \begin{pmatrix} \cos \varphi - L \sin \varphi \\ \sin \varphi \quad L \cos \varphi \end{pmatrix} \begin{pmatrix} \frac{r}{2} \quad \frac{r}{2} \\ \frac{r}{2l} \quad -\frac{r}{2l} \end{pmatrix} \tag{3}$$

Since $detA(\varphi) = -\frac{Lr^2}{2l} \neq 0$, it is clear that input-output linearization schematically can be as $(\omega_r, \omega_l) - (x_1, y_1)$.

Note that, in order to avoid collision the values of control effects of MR (initially by reverse kinematic method) ω_r and ω_l can be determined as:

$$\begin{pmatrix} \omega_r \\ \omega_l \end{pmatrix} = \frac{v_d}{\sqrt{f_x^2 + f_y^2 + \varepsilon}} \frac{1}{Lr} \begin{pmatrix} L\cos\varphi - l\sin\varphi & l\cos\varphi + L\sin\varphi \\ L\cos\varphi + l\sin\varphi & -(l\cos\varphi - L\sin\varphi) \end{pmatrix} \begin{pmatrix} f_x \\ f_y \end{pmatrix}, \quad (4)$$

v_d is constant velocity, ε is a very small constant quantity, but f_x and f_y are components of F_{sum} - relevant to X and Y directions. In the case of obstacle f_x and f_y n is identified as:

$$\begin{pmatrix} f_x \\ f_y \end{pmatrix} = \begin{pmatrix} f_{x\,attr} + f_{x\,1\,repul} + f_{x\,2\,repul} + \cdots + f_{x\,n\,repul} \\ f_{y\,attr} + f_{y\,1repul} + f_{y\,2\,repul} + \cdots + f_{y\,n\,repul} \end{pmatrix} \quad (5)$$

$F_{x\,repul}$ is an attractive force related to X, Y, direction and $f_{x\,attr}, f_{x\,k\,repul} f_{y\,attr}, f_{y\,k\,repul.}, f_{y\,attr.}, f_{y\,k\,repul} k = 1, 2, ..n$ are repulsive force components regarded to $1, 2, \ldots, n$, obstacles relatively on X direction. During the motion of MR $f_{x\,attr}, f_{x\,k\,repul} f_{y\,attr}, f_{y\,k\,repul.}, f_{y\,attr.}, f_{y\,k\,repul} (k = 1, 2)$ regarding to the field potential is obtained from follows equations [1, 4, 6]:

$$U_{attr}(q) = \frac{1}{2}\xi\left[(x_1 - x_g)^2 + (y_1 - y_g)^2 \right], \quad (6)$$

$$f_{x\,attr} = -\xi(x_1 - x_g),$$

$$f_{y\,attr} = -\xi(y_1 - y_g).$$

$$U_{k\,repul}(q) = \begin{cases} \frac{1}{k}\eta\left[\frac{1}{\rho(q,q_{obstacle.1})} - \frac{1}{\rho_{0k}} \right]^2, & if \quad \rho(q, q_{obstacle.1}) \leq \rho_{0k,} \\ 0, & if \quad \rho(q, q_{obstacle.1}) > \rho_{0k}, \end{cases} \quad (7)$$

$$f_{xk\,repul} = \begin{cases} \frac{1}{k}\eta\left[\frac{1}{\rho(q,q_{obstacle.1})} - \frac{1}{\rho_{0k}} \right]\left[\frac{1}{\rho^3(q,q_{obstacle.k})} \right](x_1 - x_{0k}), & if\ \rho(q, q_{obstacle.k}) \leq \rho_{0k}, \\ 0, & if\ (q, q_{obstacle.k}) > \rho_{0k}, \end{cases}$$

$$f_{yk\,attr} = \begin{cases} \frac{1}{k}\eta\left[\frac{1}{\rho(q,q_{obstacle.k})} - \frac{1}{\rho_{0k}} \right]\left[\frac{1}{\rho^3(q,q_{obstacle.k})} \right](y_1 - y_{0k}), & if\ \rho(q, q_{obstacle.k}) \leq \rho_{0k}, \\ 0, & if\ (q, q_{obstacle.k}) > \rho_{0k}, \end{cases}$$

$$k = 1, 2$$

Here, η – positive scaling coefficient, but $\rho(q, q_{obstacle}) = \| q - q_{obstacle} \|$ is the shortest distance between mobile robot and obstacle, $k = 1, 2$ – the number of obstacles and ρ_0 is positive coefficient characterizing influence distance of obstacle.

The fulfilment of the MR motion without collision with obstacles is conducted experimentally by fixed current gears. Each wheel of MR is controlled by Direct

Current Motor (DCM), in other words, the direction of angle - φ of robot is also controlled by changing of DCM angular velocities ω_r and ω_l of right and left wheels. Note that, DCM applied on the robots is mainly written by two compiled models [9]:

$$W(s) = \frac{\theta(s)}{U(s)} = \frac{\omega(s)}{U(s)} = \frac{K_a}{L_a J_{ef} s^2 + \left(L_a f_{ef} + R_a J_{ef}\right)s + \left(R_a J_{ef} + K_a K_b\right)}$$

Here u, L_a, R_a – respectively control voltage given to the anchor circuit of DCM is inductivity and resistance of anchor circuit, as well as, K_a, K_b – are the electrical parameters, J_{ef}, f_{ef} – effective inertial moment and friction coefficients, and it reflects mechanical properties of DCM. Since few interactions in DCM, also electrical time coefficient of gear is smaller than mechanical time coefficient, in some respects it is possible do not consider inductivity of anchor, i.e. $L_a \cong 0$, so it is possible to consider a simplified approximate mathematical model of DCM of right and left wheels [3, 9]:

$$\frac{d\omega_r}{dt} = -a_r \omega_r + k_r u_r, \tag{8}$$

$$\frac{d\omega_l}{dt} = -a_l \omega_l + k_l u_l, \tag{9}$$

here, ω_r, ω_l and u_r, u_l - angular velocities and control input tensions of right and left gears. Parameters of right and left DCM have values as $a_r = a_l = a, k_l = 1.09 k_r$, $a = 10.1, k_r = 5.5$

For high quality management of angular velocities of axises of right and left gears of MR an intelligent fuzzy T-S type regulators are suggested as:

Q_j: **If** control error is $\varepsilon_r - \tilde{E}_{rj}$ **and** $\dot{\varepsilon}_r - \tilde{\dot{E}}_{rj}$ -, **THEN** output of the right gear regulator:

$$u_r = K_{rP}^j \varepsilon_r(t) + K_{rI}^j \int_0^t \varepsilon_r(t)dt + K_{rD}^j \frac{d\varepsilon_r}{dt}, \dots j = \overline{1, Q}$$

Q_j: **If** control error is $\varepsilon_l - \tilde{E}_{lj}$ **and** $\dot{\varepsilon}_l - \tilde{\dot{E}}_{lj}$ -, **THEN** output of the left gear regulator:

$$u_l = K_{lP}^j \varepsilon_r(t) + K_{lI}^j \int_0^t \varepsilon_l(t)dt + K_{lD}^j \frac{d\varepsilon_l}{dt}, \dots j = \overline{1, Q}, (Q = 3)$$

Here, $\varepsilon_r = \omega_r^{opt} - \omega_r$ and $\varepsilon_l = \omega_l^{opt} - \omega_l$ - control error of the right and left angular velocities of the gear, $K_{rP}^j, K_{rI}^j, K_{rD}^j$ – are respectively proportionality, integration and differentiation ratios in the j-th linguistic rule of the right and left fuzzy T-S regulators, $\tilde{E}_{rj}$ – and $\tilde{\dot{E}}_{rj}$ – are fuzzy term sets of control error and its derivative.

The major aim of the bottom level is to provide tracking trajectory $\omega_r^{opt}(t)$, $\omega_l^{opt}(t)$ identified with high precision on the top level, i.e. $\varepsilon_r \to 0$ and $\varepsilon_l \to 0$.

3   Implementation of Intelligent Control System of Mobile Robot in MATLAB

For simulation of hierarchical control system of two-wheeled MR, which architecture has been described in Fig. 1, "S-model" is compiled according to Simulink, Fuzzy Logic Toolbox, Stateflow packets in MATLAB.

Firstly via "Solid Works" program 3D model of mobile robot is exported in **xml** expanding for MATLAB. This file via **simimport** command from commands windows of MATLAB is imported into the Simulink, Fuzzy Logic Toolbox. Initially subsystems have been installed in accordance with Fig. 1. When simulation starts, an information obtained from sensors of mobile robot is transferred to the subsystem installed by **Stateflow**. There on the basis of the finite state automata the direction of mobile robot motion (trajectory, in other words $\omega_r^{opt}(t), \omega_l^{opt}(t), \varphi^{opt}$) is identified by following algorithm [1–4, 10, 11].

4   Conclusion

The synthesis method of intelligent two-level hierarchical control system of dynamic object written by nonlinear model has been suggested in the paper. Algorithmic support of the synthesis problem of intelligent two-level control system has been processed according to fuzzy set and finite state machines. One of the advantages of the proposed method is a quick motion and with high dynamic accuracy without collision with obstacles of two-wheeled mobile in the environment of several obstacles

References

1. Krogh, B.H.: A generalized potential field approach to obstacle avoidance control. In: International Robotics Research Conference, Bethlehem, PA (1984)
2. García Sánchez, J.R.: Diseño y construcción de un robot móvil, aplicando el método de campospotenciales en la evasión de obstáculos. Tesis de Maestría. CIDETEC delInstitutoPolitécnicoNacional, Mexico City, Mexico (2008)
3. Sira-Ramirez, H., Agrawal, S.K.: Differentially Flat Systems. Marcel Dekker, New York (2004)
4. Silva-Ortigoza, R., Marcelino-Aranda, M., Silva-Ortigoza, G., Hernández-Guzmán, V.M., Molina-Vilchis, M.A., Saldaña-González, G., Herrera-Lozada, J.C., Olguín-Carbajal, M.: Wheeled mobile robots: a review. IEEE Latin Am. Trans. **10**(6), 2209–2217 (2012)
5. Silva-Ortigoza, R., Márquez-Sánchez, C., Marcelino-Aranda, M., Marciano-Melchor, M., Silva-Ortigoza, G., Bautista-Quintero, R., Ramos-Silvestre, E.R., Rivera-Díaz, J.C., Muñoz-Carrillo, D.: Construction of a WMR for trajectory tracking control: experimental results. Sci. World J. **2013**, 1–17 (2013)
6. Ge, S.S., Cui, Y.J.: New potential functions for mobile robot path planning. IEEE Trans. Robot. Autom. **16**(5), 615–620 (2000)
7. Vidal Calleja, T.A.: Generalización del método de campospotencialesartificiales para unvehículoarticulado. Tesis de Maestría. Sección de Mecatrónicadel Departamento de IngenieríaEléctrica del CINVESTAV-IPN, Mexico City, Mexico (2002)

8. https://www.mathworks.com/matlabcentral/fileexchange/47208-mobile-robot-simulation-for-collision-avoidance-with-simulink
9. Aliyev, R.A., et al.: Robotic Control Systems. Nargiz, Baku (2004)
10. Jafarov, S.M., Zeynalov, E.R., Mustafayeva, A.M.: Synthesis of robust controller-regulators for omnidirectional mobile robot with irregular movement. Proc. Comput. Sci. J. **102**, 469–476 (2016)
11. Jafarov, S.M., Zeynalov, E.R., Mustafayeva, A.M.: Synthesis of the optimal fuzzy T-S controller for the mobile robot using the chaos theory. Proc. Comput. Sci. J. **102**, 302–308 (2016)

Absenteeism Prediction: A Comparative Study Using Machine Learning Models

Kagan Dogruyol[1]($\boxtimes$) and Boran Sekeroglu[2]

[1] Department of Industrial Engineering, Eastern Mediterranean University, Famagusta, TRNC, Mersin 10, Turkey
kagan.dogruyol@emu.edu.tr
[2] Information Systems Engineering, Near East University, Nicosia, TRNC, Mersin 10, Turkey
boran.sekeroglu@neu.edu.tr

Abstract. Solidity of companies or institutions is related to several factors but mostly to absenteeism. Taking annual leave or pre-determined absent days of personnel may be covered by others however, unexpected absenteeism causes irredeemably poor results. Prediction of the correlation between this pre-determined and unexpected absenteeism is a challenging task and includes non-linear relationship. Neural Network based Machine Learning models are built to solve this kind of non-linear problems by using their non-deterministic nature. In this research, three neural network models; Backpropagation, Radial Basis Function and Long-Short Term Memory neural networks, are implemented to solve prediction problem of absenteeism. In addition, a comparative study is conducted between these models. Two experiments with different training ratios and three evaluation criteria are considered and implemented. The experimental results suggested that Long-Short Term Memory neural network has very high prediction rates as 99.9% in prediction problems that consists complex data and it produced superior results than other two neural network models.

Keywords: Long-Short Term Memory Network · Backpropagation · Radial basis function neural network

1 Introduction

Companies try to maximize profits and minimize costs to stay competent in the market. In the pursuit of goal achievement, employees are vital elements for the companies from bottom to top. Thus, absenteeism is considered as one of the most important complications for the companies that it can increase their costs and become an obstacle to achieve organizational goals and objectives. The absence of a worker in the workplace is called absenteeism. When it happens, either costs to increase the shortened labor force or an overhead among the other employees occur where it can also lead to demotivation among the other employees [1].

© Springer Nature Switzerland AG 2020
R. A. Aliev et al. (Eds.): ICSCCW 2019, AISC 1095, pp. 728–734, 2020.
https://doi.org/10.1007/978-3-030-35249-3_94

The absence of workers does not only affect the misuse of resources, but also the gross income of them. The reduction in employees' gross income can lead to lower purchasing power, health, and increased physical and psychological burden on them. If the absenteeism of the employees can be reduced, this will result in higher purchasing power and possibly an increase in the gross domestic product rate (GDP) [2].

Especially, Large companies with many employees are facing a major problem in absenteeism. Hence, it is essential to create and implement prediction tools for absenteeism in such companies that are heavily dependent on human resources [3]. In addition, absenteeism rate in companies is an important indicator where it can create negative consequences at the end. As an indicator, it may signify that there is a lack of employee commitment and underline the necessity for the company to take corrective measures against this problem [4].

Several surveys on how absenteeism affect the routine of conducting productive activities in companies have been conducted in recent years. Therefore, absenteeism prediction at work can assist managers to take protective measures against absence of labour to minimize financial costs [1].

Workplace absenteeism plays a crucial role in representing the productivity and profitability of a company. Therefore, the knowledge of employee absenteeism becomes the basis of an organization in its various proportions. In addition, lack of employees might decrease credibility of a company where a stopped service affects the delivery times of a product to customers. Consequently, the reasons and triggering patterns of absenteeism are important for a company where excessive occurrence of certain illnesses and injuries can be identified at an early stage. However, prediction of workplace absenteeism has not been widely researched for the last decade. Machine learning and data mining has become popular tools in human resource management in organizations [5]. In addition to this, Dogruyol, Aziz and Arayici [6] indicated that lack of knowledge is the most crucial challenge in sustainable planning. Consequently, an absenteeism prediction model to increase the knowledge related with absenteeism is necessary [7].

In the following, first, the machine learning models that will be used to predict absenteeism at work will be highlighted. Then, the experimental data and results of analysis will be underlined. Finally, the research highlights are concluded.

2 Prediction Models

In the literature, different types of prediction models have been developed and proposed. Support Vector Regression (SVR), Backpropagation (BP) neural network, Radial-Basis Function neural network (RBFNN), Long-Short Term Memory Network (LSTM) and Linear Regression (LR) are the most widely used models in prediction-based researches. In this research, neural network-based models, RBFNN, BP and LSTM are used because of the reliable prediction results of neural networks in non-linear and huge datasets.

2.1   Long-Short Term Memory Neural Network

LSTM is a recurrent neural network that is defined as a network with a memory. It remembers previous actions and states that effects next phases during the learning. Like other recurrent networks, it has a repeating module but distinctively it has four neural network layers instead of one in this module. Process through and between these layers provides LSTM to remember long-term information. It has been improved and modified in several researches for different kind of applications [8–10].

2.2   Backpropagation Method-Based Neural Network

One of the traditional neural networks is BP neural network which can be used both for classification and prediction tasks efficiently [11, 12]. Its' algorithm is still the basis for different kinds of neural networks even in deep neural networks. Shallow type includes one or two hidden layers and single input and output layers. Weight adjustments are propagated back through the layers after comparing actual and target outputs.

2.3   Radial-Basis Function-Based Neural Network

RBFNN is another kind of neural networks that can be used both for classification and prediction effectively [13, 14]. It is limited to have only a single hidden layer and it uses Radial Basis Functions within this layer.

3   Experiments and Results

In this section, the dataset used to perform experiments and obtained results will be presented in detail. In this research, two different experiments are conducted by using considered models in order to obtain optimal prediction results and to perform comparison of neural based models in prediction tasks.

3.1   Dataset

The dataset named "absenteeism at work" is obtained from UCI Machine Learning Repository. Absenteeism dataset [15] consists 20 attributes, 1 target and 740 instances. Data was collected between 2007 and 2010 in private company of Brazil. Table 1 shows the summary of attributes that included in the dataset.

3.2   Experiments

Experiments are divided into two categories according to the Hold-out method as 60% and 70% of training ratio respectively. Evaluation criteria for each experiment is based on R^2 score, Mean Squared Error (MSE) and Explained Variance (EV) Score which are the main indicators of prediction success.

Table 1. Attributes of absenteeism at work dataset

Attribute	Attribute	Attribute
ID	Service time	Social drinker
Reason of absence	Age	Social smoker
Month of absence	Work load ave/day	# of pet
Day of the week	Hit target	Weight
Seasons	Disciplinary failure	Height
Transportation expense	Education	Body mass index
Distance to work	# of children	

EV Score is defined as the regression sum of squares as shown in Eq. 1.

$$EV_s = \sum_{i=1} (f_i - \hat{y})^2 \tag{1}$$

where f_i is the predicted values and $\hat{y}$ is the real sample.

R2 score is defined as variance sample from the independent one to be predicted. Its' formula is given in Eq. 2.

$$R^2 = \frac{EV_s}{UV_s} \tag{2}$$

where EV is defined in Eq. 1 and UV is unexplained variations of samples.

MSE is based on the sum of squares of the difference between predicted and real values. Formula of MSE is given in Eq. 3.

$$MSE = \frac{1}{n}\sum_{i=1}^{n} (Y_i - \hat{Y}_i)^2 \tag{3}$$

where n is the total number of samples and Y_i and $\hat{Y}_i$ are the predicted and expected outputs of estimator respectively.

3.3   Experimental Results

First, models are trained by considering 60% of dataset. Experimental results showed that BP neural network is unable to make predictions on test set even several parameters, hidden layer numbers and neuron numbers in these layers were applied. Obtained highest prediction results of BP is 0.060, 0.063 and 0.011 for R^2 Score, EV Score and MSE respectively.

RBFNN produced more effective prediction results than BP while achieving 0.875, 0.887 and 0.0101 for R^2 Score, EV Score and MSE respectively.

LSTM network achieved highest and the most effective prediction results in this experiment. Minimized error and highest consistency is obtained by LSTM with 0.00095 MSE, 0.9903 R^2 and 0.9960 EV Scores. Table 2 shows the obtained results in Experiment 1 in details.

Table 2. Experimental results for 60% of training ratio

Criteria	RBFNN	BPNN	LSTM
MSE	0.0101	0.011	0.000956
EV	0.887	0.063	0.996
R^2	0.875	0.060	0.990

In second experiment, 70% of dataset is used to train models in order to compare the behavior of neural networks with higher training data. Similar results as in Experiment 1 are obtained for all models. BP is unable to make predictions even with higher training data and different hyper-parameters.

Increment of training data is caused more effective convergence in RBFNN. MSE is reduced to 0.0079 and EV and R^2 scores and increased to 0.907 and 0.901 respectively.

As in Experiment 1, LSTM network is the model that make more effective predictions. Increment of training data helped to reduce MSE and to increase R^2 Score but small decrement is observed in EV Score. MSE was calculated as 0.00081 and R^2 and EV Scores were calculated as 0.9917 and 0.992 respectively (Figs. 1 and 2).

Obtained results of Experiment 2 is presented in Table 3 in details.

Table 3. Experimental results for 70% of training ratio

Criteria	RBFNN	BPNN	LSTM
MSE	0.0079	0.014	0.00081
EV	0.907	0.055	0.992
R^2	0.901	0.030	0.991

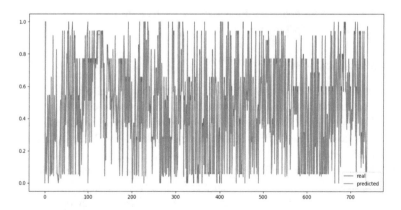

Fig. 1. Prediction graph for LSTM using 60% of training ratio.

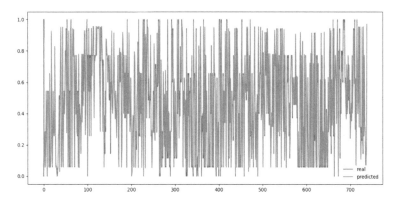

Fig. 2. Prediction graph for LSTM using 70% of training ratio.

3.4 Discussions

It is noticeable that the results obtained from neural network models are difficult to be interpreted because of their internal mechanisms especially hidden layers. Principles of neural network is to simulate human perception during the convergence. However, sometimes they fail to make efficient classification or prediction as it has happened in BP prediction model in this research. Adding more hidden layers, tuning hyper-parameters may either increase the obtained results or decrease.

On the other hand, RBFNN is more adaptable in hidden layer because of radial-basis functions and this provided more effective results than BP. Even though, MSE results were close to each other, R^2 and EV scores were quite different and RBFNN made valuable predictions on absenteeism dataset.

In addition, structure of LSTM brings an advantage on dataset that is based on time-series. While it remembers the previous actions in long and short terms, it provides more effective and superior results than other neural network models on this dataset in this research. Finally, in their recent study, Wahid et al. (2019) applied 4 tree-based machine learning algorithms to the same data set used in this study resulting with a highest accuracy level of 82% and LSTM prediction model used in this study proves to produce significantly better results than the recent studies.

4   Conclusions

Absenteeism is an essential problem for companies during their business life cycle. Unexpected absenteeism causes additional workload to other personnel and decreases the efficiency in the work. In this research, prediction of absenteeism at work is performed using three neural network models. Besides, the comparison has been performed between these models in order to determine optimal model for this kind of applications and simulations.

Obtained results show that LSTM performs at utmost level prediction of absenteeism at work problems by correlate the non-linear attributes within the dataset.

Comparisons also showed that backpropagation neural network is not useful for time-series prediction problems. Even though, Radial-Basis Function neural network has produced good prediction levels, it was way lower than the results obtained by LSTM.

Future work will include the implementation of more machine learning algorithms such as Support Vector Regression, Decision Trees etc. and compare the results with this research.

References

1. Araujo, V.S., et al.: A hybrid approach of intelligent systems to help predict absenteeism at work in companies. SN Appl. Sci. **1**(6), 536 (2019)
2. Nanjundeswaraswamy, T.: An empirical study on absenteeism in Garment industry. Manage. Sci. Lett. **6**(4), 275–284 (2016)
3. de Oliveira, E.L., Torres, J.M., Moreira, R.S., de Lima, R.A.F.: Absenteeism prediction in call center using machine learning algorithms. In: World Conference on Information Systems and Technologies, pp. 958–968. Springer, Cham (2019)
4. Cohen, A., Ronit, G.: Predicting absenteeism and turnover intentions by past absenteeism and work attitudes: an empirical examination of female employees in long term nursing care facilities. Career Dev. Int. **12**(5), 416–432 (2007)
5. Wahid, Z., et al.: Predicting absenteeism at work using tree-based learners. In: Proceedings of the 3rd International Conference on Machine Learning and Soft Computing, pp. 7–11. ACM (2019)
6. Dogruyol, K., Aziz, Z., Arayici, Y.: Eye of sustainable planning: a conceptual heritage-led urban regeneration planning framework. Sustainability **10**(5), 1343 (2018)
7. Ajit, P.: Prediction of employee turnover in organizations using machine learning algorithms. Algorithms **4**(5), C5 (2016)
8. Li, S., Wang, Q., Chen, J.: Low cost LSTM implementation based on stochastic computing for channel state information prediction. In: 2018 IEEE Asia Pacific Conference on Circuits and Systems (APCCAS), Chengdu, pp. 231–234 (2018)
9. Mirza, A.H., Cosan, S.: Computer network intrusion detection using sequential LSTM neural networks autoencoders. In: 2018 26th Signal Processing and Communications Applications Conference (SIU), Izmir, pp. 1–4 (2018)
10. Dai, S., Li, L., Li, Z.: Modeling vehicle interactions via modified LSTM models for trajectory prediction. IEEE Access **7**, 38287–38296 (2019)
11. Sekeroglu, B., Dimililer, K., Tuncal, K.: Student performance prediction and classification using machine learning algorithms. In: 8th International Conference on Educational and Information Technology, Cambridge, UK, pp. 7–11 (2019)
12. Adali, T., Sekeroglu, B.: Analysis of microRNAs by neural network for early detection of cancer. Procedia Technol. **1**, 449–452 (2012)
13. Yadav, V., Nath, S.: Daily prediction of PM10 using radial basis function and generalized regression neural network. In: 2018 Recent Advances on Engineering, Technology and Computational Sciences (RAETCS), Allahabad, pp. 1–5 (2018)
14. Adnan, R., Samad, A.M., Tajjudin, M., Ruslan, F.A.: Modeling of flood water level prediction using improved RBFNN structure. In: 2015 IEEE International Conference on Control System, Computing and Engineering (ICCSCE), George Town, pp. 552–556 (2015)
15. Martiniano, A., Ferreira, R.P., Sassi, R.J., Affonso, C.: Application of a neuro fuzzy network in prediction of absenteeism at work. In: 7th Iberian Conference on Information Systems and Technologies (CISTI), Madrid, pp. 1–4 (2012)

Stable Algorithms for Adaptation of Objects with Control Delay

H. Z. Igamberdiev[1(✉)] ⓘ, A. N. Yusupbekov[1] ⓘ, D. R. Karimov[1] ⓘ,
and O. P. Shukurova[2] ⓘ

[1] Tashkent State Technical University, str. Universitet 2,
100095 Tashkent, Uzbekistan
ihz.tstu@gmail.com, abekhimprom@mail.ru,
davron2003@icloud.com
[2] Karshi Engineering and Economic Institute, str. Mustaqillik,
180100 Karshi, Uzbekistan
oysara_shukurova@mail.ru

Abstract. An approach to solve the problem of synthesizing stable algorithms for adaptation of dynamic objects with a control delay is proposed. It was demonstrated that the problem of calculating a pseudo-inverse matrix for object control is generally unstable with respect to matrix perturbations. To solve equation for control law, A.N. Tikhonov's methods of regularization and singular decomposition are used based on the minimal pseudo-inverse matrix. The proposed computational schemes enable the synthesis of object adaptation stable algorithms with control delay and provide high quality control processes.

Keywords: Objects with control delay · Adaptation · Stable algorithms · Minimal pseudo-inverse matrix method · Regularization method · Singular decomposition

1 Introduction

Well-known works on the synthesis of systems with delays mainly use state space extension method to account delays, which resulted in significant increase of synthesis problem dimension. For example, instead of Riccati's matrix differential equation, which is the result of linear system synthesis problem with no delay, the systems with delay require a whole system of matrix partial differential equations. Number of these equations and the state space dimension quickly increase with increasing of delays. However, it is possible to propose more efficient methods of synthesis under general conditions based on structural features of systems with delays.

The control of systems with delay is quite difficult even under the causal conditions in addition to the cases, when the parameters of the system are not determined or drift slowly in control process. Since the problem of such systems control is difficult and its solution requires explicit or implicit forecast for the system's output for delay in control channel, the system model uncertainty entails further complication of control device operation algorithms, which in some cases become cumbersome and complex for engineering implementation.

© Springer Nature Switzerland AG 2020
R. A. Aliev et al. (Eds.): ICSCCW 2019, AISC 1095, pp. 735–740, 2020.
https://doi.org/10.1007/978-3-030-35249-3_95

Systems with control delays are most considered, since the component delay has been studied in the literature in details. At first, the system with similar delays in all control channels finds control appropriate in terms of quadratic functional based on the well-known identity theorem for optimal stochastic and deterministic controllers and the principle of control separation by random input signal. After that, the most difficult and interesting case of delays different in control channels is considered. Here, the control is synthesized based on delays compensation principle.

2 Problem Definition

Let's consider the problem of adaptation algorithm synthesis for objects with delays only in control. Such objects have their own parameters, which cannot be covered by a deep negative association, since the system becomes unstable. It is especially difficult to control such objects under conditions of uncertainty, when the object is unstable at zero controlling action. However, number of objects suit for adaptation algorithm construction, although some parameters have to be selected by enumeration and modeling.

Let's assume that the control object to be described by the equation

$$x_{k+1} = A^* x_k + B^* u_{k-l} + w_k,$$
$$y_k = L x_k,$$

$$u_t = \varphi_t, \quad t \in [k_0 - l,\, k_0],$$

where, $x \in R^n$; $u \in R^m$; $y \in R^s$; w_k – n-disturbance dimensional vector with limited components; A^*, B^*, L– numerical matrices depending on the vector of unknown parameters; $\xi \in \Xi$, Ξ – known possible values of vector, which will be defined later; $n \geq k \geq m$. After some time, the vector of output parameters is required to be close to some k-dimensional vector of setting actions g_0. Considering this fact, we set the target conditions in the form of the following inequality:

$$\eta_k = (y_k - g_0)^T K (y_k - g_0) \leq \varepsilon, \tag{1}$$

where, $K = K^T > 0$, ε-a positive number that determines the required accuracy of mastering the setting action.

Use the following formula to get the structure of the control device

$$y_{k+1} = \sum_{i=0}^{n-1} A_i^* y_{k-i} + \sum_{i=0}^{z} B_i^* u_{k-i-l} + \sum_{i=0}^{p} D_i w_{k-i}. \tag{2}$$

After recording this equation for a time point $k + l$ and expressing the values of vectors y_{k+l-1}, y_{k+l-2} through vectors y and u at the preceding time points according to (2), we get

$$y_{k+i} = \sum_{i=0}^{n-1} A_i y_{k-i} + \sum_{i=0}^{l+z} B_i u_{k-i} + v_k,$$

where, $v_k = \sum_{i=1}^{l-1} w_{k+l-i}$, A_i and B_i are matrices depending on the equation parameters matrix (2). Equation (2) was recorded for cases, when $n \geq l$.

Suppose that setting action g_0 is limited and vector v_k does meet the limits:

$$\|v_k\| < \varepsilon_1, \ \varepsilon_1 > 0.$$

Let's assume that the initial function φ_t was selected for interval $[k_0, l]$ with vector y inside the zone determined by inequality:

$$\|y_t\| \leq \delta, \ \delta > 0, \ t \in [k_0, l]. \tag{3}$$

We assume that there is some control u_* providing the value of vector y_k being under the zone

$$\|y_k\| \leq \delta_1, \ 0 < \delta_1 < \delta, \tag{4}$$

if $u_k = u_*$ vector y_{k+l} does not go outside the zone (3) of l steps, and then returns to zone (4).

Let's assume that the control vector to be limited as follows $\|u_k\| \leq \phi, \ \phi > 0$.
Record the target conditions (1) for a time point $k + l$ [3, 7]:

$$(y_{k+l} - g_0)^T K (y_{k+l} - g_0) \leq \varepsilon \tag{5}$$

and input the matrix C_0 and vector v_k:

$$C_0 = [A_0 \vdots \cdots \vdots A_{n-1} \vdots \cdots \vdots B_{l+z} \vdots - 1],$$

$$v_k = col[y_k, \ldots, y_{k-n+1}, \ldots, u_{k-1}, \ldots, u_{k-z-i} g_0],$$

then the target conditions (5) became:

$$[B_0 u_k + C_0 v_k + v_k]^T K [B_0 u_k + C_0 v_k + v_k] \leq \varepsilon, \tag{6}$$

and the control action enforcing (6) at $v_k \equiv 0$, became

$$B_0 u_k = f_k, \tag{7}$$

or $u_k = B_0^+ f_k = C_* v_k$, where, B_0^+ – pseudoinverse matrix, $f_k = -C_0 v_k$.

3 Solution

The system (7) generally may be incompatible. Assume, that $\bar{u}_k$ – is a rated pseudo-solution for this system of linear algebraic equations for $f_k = \bar{f}_k \in R^m$, i.e. such element from the set of $Z_0 \equiv \{u_k \in R^n : \|\bar{B}_0 u_k - \bar{f}_k\| - \inf\}$, for which $\|\bar{u}_k\| = \inf\{\|u_k\| : u_k \in Z_0\}$. It is known that the rated pseudo-solution of problem (7) is unique and can be determined with exact data $(\bar{B}_0, \bar{f})$ as follows: $\bar{u}_k = \bar{B}_0^+ \bar{f}_k$, where, $\bar{B}_0^+$ – matrix pseudo-inverse for $\bar{B}_0$ [1, 2]. However, the problem of calculating $\bar{B}_0^+$ is generally unstable with respect to matrix disturbance. Therefore, if we assume that instead of the exact data $(\bar{B}_0, \bar{f}_k)$ of the problem (7) we know approximations $(B_{0,h}, f_{k,\delta})$ meeting the conditions: $B_{0,h} \in \cup$, $\|B_{0,h} - \bar{B}_0\| \le h$, $f_{k,\delta} \in R^m$, $\|f_{k,\delta} - \bar{f}_k\| \le \delta$, where h and δ – are required accuracy of approximations, then the element $u_{k,h\delta} \equiv B_{0,h}^+ f_{k,\delta}$ will not generally comply with $\bar{u}_k$ at $h, \delta \to 0$.

To solve Eq. (7), use minimum pseudo-inverse matrix intended to solve extremum problem: find a matrix $\tilde{B}_{0,h} \in \cup$ such as

$$\left\| \tilde{B}_{0,h}^+ \right\|_* = \inf\left\{ \left\| B_0^+ \right\|_* : B_0 \in \cup, \ \left\| B_0 - B_{0,h} \right\| \le h \right\},$$

and to construct the element $u_{k,h} \equiv \tilde{B}_{0,h}^+ f_{k,\delta}$ accepted as approximation to $\bar{u}_k$. Here, $\|\cdot\|_*$ – is Euclidian norm under $\cup^*$ matrix $n \times m$.

The algorithm for searching a stable approximation to $\bar{B}_0^+$ according to set $B_{0,h}$ is appropriate to be synthesized using A.N. Tikhonov's regularization method selecting regularization parameter according to general residual principle. Regarding the algorithmic implementation of general residual principle to find pseudo-inverse matrix, it should be noted that according to A.N. Tikhonov's functional minimization theory, extremum Z_h^α does comply with the required minimum condition

$$(\alpha I + B_{0,h}^T B_{0,h}) Z = B_h^T. \tag{8}$$

Matrix Eq. (8) is equivalent to system of linear equations like

$$(\alpha I + B_{0,h}^T B_{0,h}) z_k = (b_{0,h}^T)_k, \quad k = 1, 2, \ldots, m, \tag{9}$$

where, z_k and $(b_{0,h}^T)_k$ are the main points of k-e columns of Z and $B_{0,h}^T$ matrices.

Systems solving (9) is rationally organized by method. In this case, first of all, the matrix $B_{0,h}$ is transformed to a two-diagonal form $G_h : B_{0,h} = U G_h V^T$, where U and V are orthogonal matrices. Then, the Eq. (8) transforms to

$$(\alpha I + G_h^T G_h) X = W_h, \tag{10}$$

where $W_h = V^T B_{0,h} = G_h^T U^T$ and $X = V^T Z$.

Using set α we assume m of three diagonal system of linear equations generated by the matrix Eq. (10), which are solved by sweep method. Generalized residual is calculated by equality

$$\rho_\eta(\alpha) = \left\| G_h X_h^\alpha - U^T \right\|_{m \times m} - Ch \left\| X_h^\alpha \right\|_* - \mu_h^\beta,$$

where, X_h^α – is a solution (10). After determination of generalized residual radical $\alpha(\eta)$, as well as matrix $X_h^{\alpha(\eta)}$ we can calculate the approximation $Z_h^{\alpha(\eta)}$ to $A^+ : Z_h^{\alpha(\eta)} = V X_h^{\alpha(\eta)}$.

In addition, a very effective approach to solve the interested problem is the singular value decomposition method. Introduce singular decompositions of matrices $\bar{B}_0$ and $B_{0,h}$ [16–18]: $\bar{B}_0 = \bar{U} \bar{R} \bar{V}^T$, $B_{0,h} = U_h R_h V_h^T$, where, $\bar{U}$, U_h and V_h, $\bar{V}_h$ are dimension orthogonal matrices $m \times m$ and $n \times n$ respectively, $\bar{R} = diag(\bar{\rho}_1, \ldots, \bar{\rho}_M) \in \cup$, $R_h = diag(\tilde{\rho}_1, \ldots, \tilde{\rho}_M) \in \cup$ are orthographic diagonal matrices including singular values $\bar{\rho}_k, \tilde{\rho}_k \geq 0$ $(k = 1, \ldots, M \equiv \min(m, n))$ of matrices $\bar{B}_0$ and $B_{0,h}$ arranged by lack of growth. The [17] states that any matrix having singular decomposition $B_0 = URV^T \in \cup$, $R = diag(\rho_1, \ldots, \rho_M)$, the following ratios are appropriate:

$$\sum_{k=1}^M (\rho_k - \bar{\rho}_k)^2 \leq \| B_0 - \bar{B}_0 \|^2,$$

$$\left\| B_0^+ \right\|_*^2 = \sum_{k=1}^M \Theta(\rho_k^2) = \sum_{k=1}^{r(B_0)} \rho_k^{-2}; \quad r(B_0) \equiv rang B_0,$$

where, $\Theta(\rho) \equiv \{ \rho^{-1} \text{ as } \rho \neq 0; 0 \text{ as } \rho = 0 \}$.

According to [1], we can demonstrate that the problem (1) may be solved as $\hat{B}_{0,h} = U_h \hat{R}_h V_h^T$, where, U_h and V_h to be determined from singular decomposition of the matrix $B_{0,h}$, and $\hat{R}_h = diag(\hat{\rho}_1, \ldots, \hat{\rho}_M) \in \cup$, with $\hat{\rho}_k (k = 1, \ldots, M)$ being the solution of the following extremum problem: search such $\hat{\rho}_1, \ldots, \hat{\rho}_M \geq 0$, that

$$\sum_{k=1}^M \Theta(\hat{\rho}_k^2) = \inf \left\{ \sum_{k=1}^M \Theta(\rho_k^2) : \sum_{k=1}^M (\rho_k - \tilde{\rho}_k)^2 \leq h^2; \rho_1, \ldots, \rho_M \geq 0 \right\}. \quad (11)$$

According to the solution (11) we can obtain the approximation $u_{k,h}$ to $\bar{u}_k : u_{k,\eta} = V_h \hat{R}_h^+ U_h^T f_{k,\delta}$, $\hat{R}_h^+ = diag[\Theta(\hat{\rho}_1), \ldots, \Theta(\hat{\rho}_M)] \in \cup^*$.

We assume that there is some control u_* providing that $u_k = u_*$ always does comply with the condition $\| y_{k+1} \| \leq d_*$, $d > 0$, i.e. there is some emergency control u_*, which is used, if the monitored parameters are out the acceptance limits.

Considering the fact that the matrix C_* is unknown, determine the structure of control device by the following formula:

$$u_k = u_*, \text{ if } \| y_k \| > \delta_1; \quad u_k = C_k v_k, \text{ if } \| y_k \| \leq \delta_1,$$

where, C_k is the matrix of configurable parameters.

In this case, the structure of device controlled will be selected as follows:

$$u_k = \tilde{C}_k \tilde{v}_k, \text{ if } \|y_k\| \le \delta_1, \; u_k = u_*, \text{ if } \|y_k\| > \delta_1,$$

where, $\tilde{C}_k$ is the matrix of configurable parameters.

Construction of adaptation algorithm requires the method. First of all, verify the sharpened target conditions $\eta_{k+i} \le \rho^2 \varepsilon, \; 0 < \rho < 1$ as $C_k = C_*$.

Factorize the matrix K to multipliers: $K = (B_0^+)^T H B_0^+$, where $H = H^T > 0$, then at $v_k \equiv 0$, we get:

$$(-B_0 B_0^+ A_0 v_k - f_k)^T (B_0^+)^T H B_0^+ (B_0 B_0^+ f_k - f_k) =$$
$$= (B_0^+ f_k - B_0^+ f_k)^T H (B_0^+ f_k - B_0^+ f_k) \le \rho^2 \varepsilon.$$

Pursuant to hypothesis made, the vector v_k is limited. Therefore, the following condition will be fulfilled:

$$\gamma \|C_*\| \le q_* < q, \; \gamma > 0.$$

Then, use the method of recurrent target inequalities to construct the following setting procedure for matrix C_k:

$$C_{k+l} = C_{k+l-1}, \text{ if } \eta_{k+l} \le \varepsilon, \; \|C_{k+l-1} v_{k+l}\| \le \phi;$$

$$C_{k+l} = q_* C_{k+l-1} / \|v_{k+l}\| \|C_{k+l-1}\|, \text{ if } \eta_{k+l} \le \varepsilon, \; \|C_{k+l-1} v_{k+l}\| > q$$

4 Conclusion

The above algorithms found practical use when solving problems of automation and control of multi-connected technological processes with a delay under uncertain conditions and enabled increase of production processes efficiency.

References

1. Lawson, C.L., Hanson, R.J.: Solving Least Squares Problems. Siam, Philadelphia (1995)
2. Horn, R.A., Johnson, C.R.: Matrix Analysis. Cambridge University Press, Cambridge (2012)
3. Yusupbekov, N.R., Igamberdiev, H.Z., Mamirov, U.F.: Algorithms of sustainable estimation of unknown input signals in control systems. Autom. Mob. Rob. Intell. Syst. **12**(4), 83–86 (2015). https://doi.org/10.14313/JAMRIS_4-2018/29

Assessment of the Stability of the Agricultural Production of the Region on the Basis of the Matrix of Data Aggregation Schemes, as Well as Financial, Social and Environmental Performance

Imran G. Akperov[2]($\boxtimes$) , Elizabeth A. Arapova[1] ,
Galina A. Batishcheva[1] , and Galina V. Lukyanova[1]

[1] Rostov State University of Economics, Rostov-on-Don, Russia
rector@iubip.ru
[2] Private Educational Institution of Higher Education "Southern University
(IMBL)", Rostov-on-Don, Russia
dist_edu@ntti.ru, gbati@mail.ru, lukyanova.g@yandex.ru

Abstract. The aim of the work is to develop a methodology for the integrated assessment of the sustainability of production in an agricultural region based on indicators of three groups: the financial condition of enterprises in the region, indicators of compliance of agricultural production with environmental requirements in agricultural areas of the region, and indicators of social sustainability. The possibility of calculating the assessment of the sustainability of agricultural production on the basis of aggregation of parameters characterizing specific enterprises and agricultural areas is indicated. This is a significant advantage over the techniques that operate with certain average values. As a mathematical apparatus, systems of fuzzy logical conclusions are used, allowing to aggregate heterogeneous indicators in arbitrarily large volumes without significantly complicating the model. The assessment of the financial sustainability of individual enterprises was made on the basis of the classical spectrum-point model implemented in the Audit-IT software. The novelty of the methodology lies in the ability to assess the sustainability of agricultural production in the region on the basis of open Internet data: accounting reports of enterprises and indicators of the ecological status of the regions in the region. The methodology was tested on the assessment of the sustainability of agricultural production in the Rostov region.

Keywords: Creditworthiness · Bankruptcy risk · Aggregation · Fuzzy-multiple method

1 Introduction

Currently, one of the important tasks of economic research in the field of the agro-industrial complex is the construction of techniques that allow to assess the sustainability of the development of agricultural production in the regions [1]. The concept of

© Springer Nature Switzerland AG 2020
R. A. Aliev et al. (Eds.): ICSCCW 2019, AISC 1095, pp. 741–748, 2020.
https://doi.org/10.1007/978-3-030-35249-3_96

sustainability of agricultural production includes an assessment of a set of indicators reflecting the financial and economic state of individual agricultural enterprises, the state of the social sphere, the impact of agricultural production on the region's ecology [2, 3].

At the same time, the mathematical apparatus for constructing such estimates is currently rather poorly developed. Preferably, different point models are used, including estimates of a limited number of indicators. The weights are calculated arbitrarily, and their change requires processing the entire methodology.

The most successful mathematical apparatus for carrying out such studies is a system of fuzzy-logical conclusions that allow one to measure, ration and aggregate large data arrays without significantly complicating the models [4–6].

This article presents a methodology for building a comprehensive assessment of the sustainability of agricultural production in the region based on three groups of indicators: the financial status of agricultural enterprises in the region, indicators of the ecological status of agricultural areas and their environmental management, as well as data on wages in areas of the region. The technique is based on the use of fuzzy-logical inference systems for aggregating information - standard fuzzy multilevel [0,1] - classifiers.

2   Methods

2.1   Analysis of the Financial Sustainability in the Region

Methods of assessing the financial sustainability of agricultural production in the region is based on the aggregation of the relevant assessments of individual agricultural enterprises, calculated by Audit-IT standard software. Generalizing (cumulative) assessment of the financial condition of the organization consists of the evaluation of the financial situation and evaluate the effectiveness of the organization. This uses the following gradation given in Table 1.

Table 1. The financial condition of the gradation

Mark		Symbol (rating)	Qualitative characteristics of financial condition
From	Before*		
2	1.6	AAA	Excellent
1.6	1.2	AA	Very good
1.2	0.8	A	A good
0.8	0.4	BBB	Positive
0.4	0	BB	Normal
0	–0.4	B	Satisfactory
–0.4	–0.8	CCC	Unsatisfactory
–0.8	–1.2	CC	Bad
–1.2	–1.6	C	Very bad
–1.6	2	D	Critical

In summarizing financial assessment involved the following figures (in parentheses the weight (importance) index): equity ratio (0.25); the ratio of net assets and authorized capital (0.1); a ratio of own circulating means (0,15); current ratio (total) ratio (0.15); coefficient fast (intermediate) ratio (0.2); cash ratio (0.15).

In summarizing performance assessment involved the following parameters (in brackets are the weight (importance) index): return on equity (0.3); return on assets (0,2); ROS (0,2); revenue dynamics (0,1); turnover of capital (0.1); of profits from other operations and revenues from the main activity (0,1). On the basis of points of the financial position and performance is calculated summarizing the assessment of - the financial condition score. This score is the sum of the financial position of points, multiplied by 0.6, and points of financial results, multiplied by 0.4. That is, the indicators are taken at a ratio of 60% and 40%, respectively, as indicator of the financial situation to a greater extent characterizes the financial status of the organization. Depending on the score values of the financial status (as shown in the above table) organization is assigned one of 10 values -from AAA financial rating (best) to D (worst).

For aggregating region farms evaluations used the so-called matrix circuit aggregation, fuzzy five-level [0.1] classifiers.

As a carrier of a linguistic variable is defined segment of the real axis [0,1]. Linguistic variable "bankruptcy risk" (and every company, and region) has a term set values of the G, consisting of three terms: G1 – "low risk of bankruptcy"; G2 – "average risk of bankruptcy"; G3 – "high risk of bankruptcy". Linguistic variable "creditworthiness of the borrower" (and the individual enterprise and the region) also has a term set of the G, consisting of three terms: G1 – "which no doubt lending"; G2 – "lending requires a balanced approach"; G3 – "lending is associated with an increased risk".

Matrix aggregation scheme based on three-level fuzzy classifiers formula:

$$g = \sum_{i=1}^{N} p_i \sum_{j=1}^{3} \alpha_j \mu_{ij}(x_i).$$

where α_j – standard nodal points classifier (terms centroids), p_i – i-th weight factors in the contraction, $\mu_{ij}(x_i)$ – membership function value j-th level of quality with respect to the current value of i-th factor (the standard used trapezoidal). Then the index g is subjected to recognition based on the standard fuzzy classifier in accordance with the membership functions.

2.2 Evaluation of Agricultural Nature of the Region

Evaluation of agricultural nature of the region is based on fuzzy multiple aggregation of relevant assessments of its districts. Qualification of each area is generated based on the aggregation time series statistics in the following areas: the dynamics of polluting emissions into the atmosphere; dynamic load on the water system; degree of contamination of the territory of solid waste; the proportion of protected areas; the level of funding for environmental protection and nature activities.

The technique is based on applying standard fuzzy five-level [0.1] classifiers.

Each of the figures involved in the formation of estimates assigned to a linguistic variable with the universal set as a numerical interval [0,1]. Therm-set of each linguistic variable and consists of five terms $G = \{G1, G2, G3, G4, G5\}$, meaning that there is a level indicator: G1 – "very low"; G2 – "low"; G3 – "average"; G4 – "high";. G5 – "very high".

A linguistic variable g1 = «evaluation of the dynamics of polluting emissions into the air" is also a term - the set consisting of five terms: G1 – "steady trend towards a decrease in emissions"; G2 – "tendency to reduce emissions"; G3 – "stabilization of the situation"; G4 – "tendency to increase emissions"; G5 – "steady tendency to increase emissions".

Linguistic variable: g2 = "load estimation of the dynamics on the water system" consists of five terms: G1 – "stable tendency to reduce the burden on the water system"; G2 – "the tendency to reduce the load on the water system"; G3 – "stabilization of the situation"; G4 – "tendency to increase the load on the water system"; G5 – "stable tendency of increase the burden on the water system".

Linguistic variable: g3 = "Degree of contamination of the territory of solid waste" consists of five terms: G1 – "contaminated territories is practically no"; G2 – "low level contamination"; G3 – "soiled"; G4 – "high degree of pollution"; G5 – "a very high degree of contamination".

Linguistic variable: g4 = "the share of protected areas" consists of five terms: G1 – "a very high proportion of protected areas"; G2 – "high proportion of protected areas"; G3 – "the average percentage of protected areas"; G4 – "low share of protected areas"; G5 – "protected areas are virtually absent".

Linguistic variable: g5 = "Level of funding of environmental activities" consists of five terms, which determine the level of environmental financing: G1 – "very high"; G2 – "high"; G3 – "average"; G4 – "low"; G5 – "very low".

Comprehensive assessment nature condition is generated based on the aggregation obtained above five ratings: g1 = "evaluation of the dynamics of polluting emissions into the air"; g2 = "evaluation of the dynamics of load on the water system"; g3 = "degree of contamination of the territory of solid waste"; g4 = "the share of protected areas"; g5 = "the level of funding of environmental measures".

Let us consider the linguistic variable: g (area) = "estimate agricultural nature of the region". Therm - set consists of five terms, characterizing the level of compliance with the principles of nature: G1– "fully compliant"; G2 – "is broadly in line"; G3 – "partially compliant"; G4 – "as a whole does not match"; G5 – "does not fully comply".

3   Results

3.1   Assessment of the Financial Condition of Enterprises in the Region

Implemented diagnosis creditworthiness and the risk of bankruptcy of enterprises of the region on the basis of a random sample of ten agricultural enterprises of the Rostov area: (1) Litvinenko Ltd; (2) Aksai Land Ltd; (3) Manych-Agro Ltd; (4) JSC

"Friendship"; (5) "Mutilinskoe" Ltd; (6) Red October Company; (7) "Light" Ltd; (8) Rassvet Ltd; (9) SEC named after Shaumyan; (10) Agrofirma Celina Ltd.

Calculation of estimates for individual companies is shown in Table 2, and their aggregation in Table 3. Aggregation conducted on the basis of fuzzy five-level [0.1] classifiers showed that the evaluation of the financial companies in the region elastase derived for the considered sample is 0.158, which is equivalent to the term "very good condition".

Table 2. Assessment of the financial condition of enterprises

No	Weight	Litvinenko Ltd	Aksai Land Ltd	Manych-Agro Ltd	JSC "Friendship"	"Mutilinskoe" Ltd
1.1.	0.25	0.25	0.35	−0.25	0.25	−0.175
1.2.	0.1	0.2	0.2	0.2	0.2	0.2
1.3.	0.15	0.188	0.3	−0.3	0.3	0.263
1.4.	0.15	0.3	0.3	−0.3	0.3	0.3
1.5.	0.2	0.35	0.4	−0.4	0.4	−0.4
1.6.	0.15	0.3	0.3	−0.3	−0.06	−0.3
Group 1	**0.4**	**1,588**	**1.85**	**−1.35**	**1.39**	**−0.112**
2.1	0.3	0.6	0,015	−0.075	0.6	−0.345
2.2.	0.2	0.4	0.28	−0.2	0.4	−0.2
2.3.	0.2	0.4	0.2	0.23	0.4	−0.2
2.4.	0.1	0.2	0.2	0.2	0.2	0.2
2.5.	0.1	0.1	0.05	−0.115	−0.175	0.16
2.6.	0.1	0.2	0.2	0.16	0.2	0.2
Group 2	**0.6**	**1.9**	**+945**	**0.2**	**1,625**	**−0.185**
Final grade	**One**	**1.71**	**1.49**	**−0.73**	**1.48**	**-0.14**
Term		AAA - excellent	AA - very good	CCC - unsatisfactory	AA - very good	B - satisfactory
No.	Weight	Red October Company	"Light" Ltd	Rassvet Ltd	SEC named after Shaumyan	Agrofirma Celina Ltd
1.1.	0.25	0.313	0.25	0,275	0.25	0.25
1.2.	0.1	0.2	0.2	0.2	0.2	0.2
1.3.	0.15	0.3	0.3	0.188	0.3	0.3
1.4.	0.15	0.3	0.3	0.3	0.3	0.3
1.5.	0.2	0.23	0.4	−0.05	0.4	0.4
1.6.	0.15	−0.24	0.3	0.3	0.3	0.3
Group 1	**0.4**	**1,103**	**1.75**	**1,213**	**1.75**	**1.75**
2.1	0.3	0.42	0,015	0.6	0.42	0.6
2.2.	0.2	0.4	0.04	0.4	0.4	0.4
2.3.	0.2	0.4	0.01	0.23	0.25	0.4
2.4.	0.1	0.2	0.2	0.2	0.2	0.2
2.5.	0.1	0.04	−0.175	0,175	0.1	0.05
2.6.	0.1	0.2	0.2	0.2	0.2	0.2
Group 2	**0.6**	**1.66**	**0.29**	**1,805**	**1.57**	**1.85**
Final grade	**1**	**1.33**	**1.17**	**1.45**	**1.68**	**1.79**
Term		AA - very good	A - good	AA - very good	AAA - excellent	AAA - excellent

Table 3. The calculation of the aggregate assessment of the financial condition of enterprises in the region

No	Company name	Revenues for 2017	Weight coefficient
1.	Litvinenko Ltd	79570	0.012
2.	Aksai Land Ltd	150894	0.023
3.	Manych-Agro Ltd	414200	0.062
4.	JSC "Friendship"	268102	0.040
5.	"Mutilinskoe" Ltd	41310	0.006
6.	Red October Company	664244	0.100
7.	"Light" Ltd	1334894	0.201
8.	Rassvet Ltd	767386	0.116
9.	SEC named after Shaumyan	778727	0.117
10.	Agrofirma Celina Ltd	2143745	0.323
		6643072	1

No	Therms									
	AAA	AA	A	BBB	BB	B	CCC	CC	C	D
1.	1	0	0	0	0	0	0	0	0	0
2.	0	1	0	0	0	0	0	0	0	0
3.	0	0	0	0	0	0	1	0	0	0
4.	0	1	0	0	0	0	0	0	0	0
5.	0	0	0	0	0	1	0	0	0	0
6.	0	1	0	0	0	0	0	0	0	0
7.	0	0	1	0	0	0	0	0	0	0
8.	0	1	0	0	0	0	0	0	0	0
9.	1	0	0	0	0	0	0	0	0	0
10.	1	0	0	0	0	0	0	0	0	0
	0.452	**0.279**	**0.201**	**0**	**0**	**0.006**	**0.062**	**0**	**0**	**0**

G = 0.05 * 0.452 + 0.15 * 0.279 + 0.25 * 0.201 + 0.35 * 0 + 0 + 0.45 * 0.55 * 0.006 + 0.65 * 0.062 + 0.062 + 0.75 * 0.85 * 0.95 * 0 + 0 = 0.158 (A term AA - "very good condition")

3.2   Evaluation of Agricultural Nature of the Region

The material used as the statistical data on the standard condition and the environment in areas Rostov region (cities and districts) annually given in [1]. Data compiled over five years - 2012–2016 years.

The calculation results on 36 regions Rostov region shown in Table 3 and the results of their ranking in Table 3, extension. Here, numerals indicate the parts: 1 - Azovskiy; 2 - Aksay; 3 - Bagaevsky; 4 - Belokalitvenskyi; 5 - Bokovskaya; 6 - the Upper; 7 - Veselovsky; 8 - Volgodonsk; 9 - Dubovskii; 10 - Egorlyksky; 11 - Zavetinsky; 12 - Zernogadsky; 13 - Zimovnikovsky; 14 - Kagalnitsky; 15 - Kamensky; 16 - Kasharsky; 17 - Constantine; 18 - Krasnosulinskaya; 19 - Kuibyshev; 20 - Martynovsky; 21 - Matveyevo-Kurgansky; 22 - Millerovo; 23 - Milyutinski; 24 - Morozov; 25 - Myasnikovsky; 26 - Neklinovskiy; 27 - Oblivskaya; 28 - The October; 29 - Orel;

30 - Peschanokopskiy; 31 - Proletarian; 32 - Remontnenskiy; 33 - Rodionovo-Nesvetaiskaya; 34 - MPP; 35 - Semikarakorsk; 36 - Soviet; 37 - Tarasovskiy; 38 - Tatsinskaya; 39 - Ust-Donetsk; 40 - Tselinskiy; 41 - Tsimlyansky; 42 - Chertkovsky; 43 - Sholokhov.

Aggregation of the estimates by region showed that "the assessment of the agricultural nature management in the region" for the Rostov region is 0.502, which corresponds to the term G3 - "partially compliant".

Table 4. The results of calculations by districts of the Rostov region

	G1	Term	G2	Term	G3	Term	G4	Term	G5	Term	G	Term
1.	0.313	G2	0,500	G3	0,029	G1	0.757	G4	0.971	G5	**0,505**	**G3**
2.	0.412	G3	0,482	G3	0,014	G1	1,000	G5	1,000	G5	**0.564**	**G3**
3.	0.263	G2	0.540	G3	0,019	G1	1,000	G5	0.981	G5	**0.539**	**G3**
4.	0.313	G2	0.694	G4	0,040	G1	0,992	G5	0.988	G5	**0,579**	**G3**
5.	0.313	G2	0.540	G3	0,005	G1	1,000	G5	0.930	G5	**0.539**	**G3**
6.	0.412	G3	0.634	G4	0,021	G1	0,992	G5	0.993	G5	**0.597**	**G3**
7.	0.313	G2	0,410	G3	0,014	G1	1,000	G5	0,999	G5	**0.523**	**G3**
8.	0,125	G1	0.465	G3	0,015	G1	1,000	G5	0,955	G5	**0.504**	**G3**
9.	0,125	G1	0,560	G3	0,040	G1	1,000	G5	0,987	G5	**0,508**	**G3**
10.	0,125	G1	0,443	G3	0,001	G1	0.993	G5	0.995	G5	**0,501**	**G3**
11.	0,125	G1	0.480	G3	0,022	G1	1,000	G5	0,992	G5	**0.504**	**G3**
12.	0.263	G2	0,484	G3	0,014	G1	0,990	G5	1,000	G5	**0.539**	**G3**
13.	0,125	G1	0,460	G3	0,014	G1	1,000	G5	0,999	G5	**0.504**	**G3**
14.	0,362	G2	0,400	G3	0,021	G1	1,000	G5	0.984	G5	**0.524**	**G3**
15.	0.169	G1	0,425	G3	0,002	G1	0.971	G5	0.993	G5	**0,501**	**G3**
35.	0,125	G1	0,500	G3	0,014	G1	0.984	G5	1,000	G5	**0.504**	**G3**
36.	–	–	0,460	G3	0,008	G1	0.997	G5	1,000	G5	**0.504**	**G3**
37.	0,125	G1	0,380	G2	0,000	G1	0,992	G5	0.995	G5	**0.476**	**G3**
38.	0,505	G3	0,600	G3	0,015	G1	1,000	G5	0.805	G5	**0.582**	**G3**
39.	0,125	G1	0.518	G3	0,014	G1	0.840	G5	0,000	G1	**0.348**	**G2**
40.	0.459	G3	0,440	G3	0,014	G1	0,999	G5	0,978	G5	**0,575**	**G3**
41.	0.212	G2	0.520	G3	0,001	G1	0,147	G1	0.995	G5	**0,374**	**G2**
42.	0.213	G2	0,400	G3	0,024	G1	0.987	G5	0.851	G5	**0.506**	**G3**
43.	0.459	G3	0,380	G2	0,000	G1	0.995	G5	0,999	G5	**0.551**	**G3**

3.3   Evaluation of Social Sustainability

To assess the social development of the Rostov region used the data of the average wage level of analysis results in the agricultural sector organizations on the monitoring results the Ministry of Agriculture of Rostov region as of 04.01.2018, by districts.

The analysis carried out in the same way as it was done in Sect. 3.2. As indicators of performance are taken, the corresponding percentage of the total number of people involved in the monitoring and reflecting the proportion of working with: (1) the level of wages below the average regional; (2) Salary/board at a subsistence level and above; (3) the level of/board below subsistence; (4) Salary/fees below the tripartite (regional) agreements for 2017–2019 years.

Based on the data using the standard five-level [0.1] - classifiers built aggregate score g = "social sustainability score based on financial indicators", g = 0.27, "good condition".

3.4   Integral Assessment of the Sustainability of Agricultural Production in the Region

As the evaluation of stability of the integrated farming take the arithmetic average of the obtained three ratings. Linguistic recognition evaluation made by standard five-level [0.1] - classifiers. It is found that the numerical evaluation value is equal to 0.31, which corresponds to the term "good stability".

Thus, on the basis of the analysis revealed that the complex evaluation of stability in agricultural production region constructed based on the complex financial and environmental indicators corresponds to the term "good stability". It contains the ability to calculate estimates of the stability of agricultural production parameters based aggregation characterizing specific enterprise and agricultural areas. This is a significant advantage over methods that operate by certain averages.

It should be noted that in the further set of parameters (primarily social) must be enlarged. It is necessary to include in the set of parameters of the level of medical care, access to education, social infrastructure, and others.

4   Conclusion

The technique is proposed, the novelty of which is the ability to calculate the sustainability of agricultural production in the region on the basis of financial and environmental performance of the individual companies and agricultural areas. Put in its basis the mathematical apparatus based on a system of fuzzy logic conclusions and allows the aggregation of heterogeneous information in large volumes without loss of flexibility and simplicity of the model.

References

1. Ryabov, I.: Assessment of sustainability of agricultural production in the territorial food safety system. Bull. NGIEI 9(64) (2016)
2. Vinnichek, L., Yu, Q.: Factors sustainable agricultural production 4 (45) (2017). Volga Niva
3. Knuhova, M., Topsahalova, F.: A factor of sustainable agricultural production of region. Mod. Probl. Sci. Educ. 2 (2014)
4. Nedosekin, A.: Fuzzy financial management. AFA Library, Moscow, Russia (2003)
5. Nedosekin, A.: Application of the fuzzy sets to the problems of financial management. Audit and financial analysis, No. 2 (2000). https://www.cfin.ru/press/afa/2000-2/08.shtml
6. Nedosekin, A., Kozlovsky, A., Abdulaeva, Z.: Analysis of branch economic stability by fuzzy-logical methods. Econ. Manage.: Probl. Solutions 5, 10–16 (2018)

Fuzzy-Multiple Modeling for the Analysis and Forecasting of Economic Cenosis

Alexander N. Kuzminov[1]([⊠]) [iD], Natalia G. Korostieva[1] [iD],
Ahmed I. Khazuev[3] [iD], and Oleg A. Ternovsky[2] [iD]

[1] Rostov State University of Economics, Rostov-on-Don, Russian Federation
akuzminov@sfedu.ru, nata_korostieva@mail.ru
[2] South-Russian State Polytechnic University (NPI), Novocherkassk,
Russian Federation
terol2005@mail.ru
[3] Chechen State University, Grozny, Russian Federation
docent@inbox.ru

Abstract. The article considers a new approach to the formation of a model of fuzzy-multiple analysis and forecasting the development of socio-economic systems with cenosis's features. The authors of the article proceed from the consideration that the use of fuzzy logic for analysis, forecasting and modeling economic phenomena and processes is justified by high performance and the prospect of using in the conditions of increasing uncertainty, but it is constrained by the lack of special research methods. The existing problem of the qualitative development of models in the field of fuzzy sets, soft calculations and approximate reasoning, used for dealing with numerous applied problems, can be solved by means of an interdisciplinary synthesis of related academic disciplines' achievements, including a new scientific direction - cenology. The key point of such integration is the possibility of expert-analytical support of key procedures, namely: the description of the probability distribution function for possible values; operations on fuzzy numbers within the bounds of the calculated confidence interval of such a function; soft computing based on using regularities of the distribution of prime numbers; dynamic modeling in the form of fuzzy cognitive models (Fuzzy Cognitive Maps). The resulting model demonstrates the practical implementation of fuzzy sets and soft computing for economic and financial tasks, which is confirmed by the results of empirical research.

Keywords: Economic cenosis · Cognitive modeling · Fuzzy-multiply method

1 Introduction

Economic applications of the fuzzy sets theory form an independent scientific research area, which indicates a stock of knowledge on the application of fuzzy sets and soft computing for economic and financial tasks [1]. However, the degree of qualitative implementation of the theoretical material is limited by a low level of approbation. We assume that this is due to the lack of technologies for the qualitative implementation of fuzzy-multiple analysis methods, because of the narrowness of local tools, the high complexity and uncertainty of the majority of current operations in the modern

© Springer Nature Switzerland AG 2020
R. A. Aliev et al. (Eds.): ICSCCW 2019, AISC 1095, pp. 749–757, 2020.
https://doi.org/10.1007/978-3-030-35249-3_97

economy. An interdisciplinary model is proposed that combines a number of special methods for the study of complex systems, the model characterized by the possibility to analyze and forecast under uncertainty conditions.

The key point of such integration is the possibility of expert-analytical support of key procedures, namely: the description of the probability distribution function for possible values; operations on fuzzy numbers within the bounds of the calculated confidence interval of such a function; soft computing based on using regularities of prime numbers distribution; dynamic modeling in the form of fuzzy cognitive models (Fuzzy Cognitive Maps). The logic of the proposed analytical model for complex economic systems is a combination of all the procedures mentioned. They are considered through regularities and dependencies from the point of the probabilistic approach.

This approach is a combination of techniques of methods of formalized representation of systems (MFRS) and methods based on mobilization of professional experience and intuition of specialists (MMPEI) (graphical and set-theoretic description of the systems), and it is based on the results of cognitive modeling of semi-structured systems, as well as statistical and cenological research tools [13].

Thus, developing the methodology of fuzzy-multiple modeling and economic cenosis forecasting, it is necessary to use the logical and substantive integration of three methods, which will be based on the results of imitational cognitive modeling. It will ensure that these restrictions are taken into account through the use of specific research techniques of cenological analysis. It should be noted that the proposed approach was not used as an integral tool for the study of complex socio-economic systems.

2 Method

2.1 Dynamic Modeling in the Form of Fuzzy Cognitive Models

Let us give a formalization of the cognitive approach to the construction of a model within the framework of the theory of systems as a system of equations that describe a set of multitude in set-theoretic terms [12]:

$\overline{V} = \{Vi : i \in I\}$, where I - is a set of indices, where the system is represented in the space $\overline{V}$ as a subset of the Cartesian product $\times \overline{V}$:

$$S \subset \times \{Vi : i \in I\}$$

i = 1,2,...,11 = A,B,...,K - the number of submodels.

All elements Vi: $i \in I$, of the Cartesian product $\times \overline{V}$ are called objects of system S, having input X and output Y objects:

$$S \subset X \times Y$$

Vi - submodels respectively;
X and Y - groups of observable parameters,

The commonality of the system under consideration is determined by the relation [12]:

$$S \subset X \times Y,$$

where $\times$ - is a symbol of the Cartesian product;
I - a set of indices;
Vi - elements of the system;
$X = \times \{ Vi : i \in Ix \}$ –input parameters;
$Y = \times \{ Vi : i \in Iy \}$ –output parameters;

$Ix \subset I$ and $Iy \subset I$, еслиIx ∩ Iy = ∅ (the intersection is empty) and $Ix \cup Iy = I$ (the combination gives I).

It is obvious that the system is associated both with transformations in time and state transitions, so it can be marked as dynamic (temporary), i.e. determined on temporary objects whose elements are temporary functions (functions of time):

$v : T \rightarrow A$, where T - is an ordered set of time moments;
A - is the alphabet of the object V; an element is $vi \in Vi$.

Cognitive analysis is used to describe and forecast when solving semi-structured problems, in other words, to formulate and refine hypotheses about the functioning of the object under study is considered as a complex system.

In order to understand and analyze the behavior of a complex system, a structural scheme of causal relationships is built.

Description of situations is carried out by the method of cognitive maps, which is a combination of studies was described by Gorelova:

(a) "causal pathways and the propagation of perturbations through the map;
(b) system connectivity;
(c) system complexity;
(d) system stability and definition of stable and unstable variables;
(e) sensitivity of a cognitive map to various perturbations" [14].

The effectiveness of this approach is due to the possibility of making expert decisions, guided by the cognitive presentation of analytical and forecast information, which is ensured by the use of graph theory and topology methods. A feature of cognitive simulation is the ability to intelligently assess the effects of perturbations arising in dynamics, when the values of variables at the vertices of the map can change throughout the development vector, i.e. increase, decay or remain unchanged, as well as spread within cause-effect chains affecting other vertices.

The mathematical interpretation of such a process looks like an assessment of the dependence of the values of the vertex $Ui(t + 1)$ on $Ui(t)$ and on the vertices adjacent to Vi vertices, i.e. united by cause-effect chains, when $p_j(t)$ – is the change in the vertex Vj at the time t affecting the Vi at the moment $t + 1$, which is described by the function $\pm p_j(t)$, dependant on vertices relationship nature.

In the general case, if there are several vertices Vj, adjacent to Vi, the process of the propagation of perturbations along a graph is determined by the rule:

$$Ui(t+1) = Ui(t) + \sum_{j=1}^{n} f(Vj, Vi)p_j(t) \tag{1}$$

It is modified after introducing such elements as the weighted value e_{ji} of the target vertices function f(Vj, Vi) and the mutual influence characteristic Wji, determined statistically:

$$Y_i(t+1) = Y_i(t) + \sum_{j=1}^{n} e_{ji} \cdot W_{ji}[Y_j(t+1) - Y_j(t)] \tag{2}$$

where Yj(t), Yj(t + 1) – is the value of the vertex Vj at the imitation steps at the moment t and the next one t + 1.

Further parametric study of the changes that are caused by the accumulation of **perturbations** at the vertices as a result of moving along chains and cycles is provided by means of the CogniMap 2.0 imitation modeling program [4].

The creation of possible analytical dependencies is carried out according to the results of expert estimation, static assessments or regularities of the cenological analysis of the structure [8], which makes the preliminary system clustering of groups of parameters possible. Clustering in terms of cenosis is taken into account as well. Changes of the parameters at the vertices in the process of imitation will be carried out in accordance with the rules of formulas (1) and (2).

For the considered concept of simulation estimation, a polynomial representation of the regression is applicable, reflecting the nature of the relationship between the vertices Vi and Vj, which can be represented by the formula:

$$\hat{Y} = b_0 + \sum_{j=1}^{k} b_j x_j + \sum_{\substack{u, j = 1 \\ u \neq j}}^{k} b_{uj} x_u x_j + \sum_{j=1}^{k} b_{jj} x_j^2 + \ldots \tag{3}$$

$\hat{Y}$ – a calculated value of the parameter Y.

Nonparametric analysis, which determines the general trends of the system, is formalized by linear regressions:

$$\hat{Y} = b_0 + b_j x_j, \quad j = 1, 2, \ldots k \tag{4}$$

The positive or negative impact of each of the factors Xj on the indicator Y is indicated by the value of the coefficient bj (>0, <0).

2.2 Cenological Analysis

2.2.1 Standard Distribution Probability Function

In case when any element of the system represented by the vertex of Vj is distinguished and clearly classified by its type, structural mathematical models based on hyperbolic H-distributions according to species, ranks and rank-by-parameter forms are used. The most important postulate formulated by Kudrin B. is used: "for stable distributions of unstable frequencies, almost all stable densities are not expressible in in terms of elementary functions (by standard formulas). But all stable densities (except Gaussian) decrease at large values of the argument, like hyperboles, which constitutes a formal representation of the cenological approach" [8].

For such kind of cenosis, a structural description is applicable, which is the basis for evaluating the equivalence of a certain ideal distribution, where each structural element is characterized as a separate individual, which can be identified in the categories of mathematical formalization:

$$u_i \in s_j \equiv u_k \in s_j; i \neq k, s_j \neq s_m, \tag{5}$$

where each unique element is marked with a pair of numbers representing a number in the list $u_i = 1, 2,..., U$, where U is a number of elements from the generalized group, forming an aggregate of length T, and with a number of the following type: $s_j = 1, 2,...,$ $S,$. where S is the number of species forming the system by V volume. Individuals of the same species are indistinguishable and form the population. The species, each of which is represented by an equal number of individuals, form a caste $k_k = 1, 2,..., K,$ i.e., each of the castes is a set formed by populations of the same size.

The distribution of w_i species in such a system is a certain sequence described by the functional regularity $\Omega(w_i) = \Omega(i) = \Omega(x)$:

$$\Omega(x) = \frac{W_0}{x^{1+\alpha}} \tag{6}$$

where $x \in [1, \infty]$ is the characteristic of the number of elements i = [x]; $\alpha > 0$ is the characteristic distribution index - $\gamma = 1 + \alpha$; $W_0 = AS$, $W_1 = [W_0]$, where W_0 is a theoretical value, and W1 is the actual (empirical) value of the maximum point of the distribution graph; A is the distribution constant determined by experts.

The aggregate of all cenosis elements, ranked by the leading indicator, will be formed in hyperbolic rank distribution, the form which is fundamental for most real systems and for the sequence of ranks r is described by the function:

$$\Lambda(r) = B/r^\beta; \omega(r) = u_r/U; U = u_r, \tag{7}$$

where B is a certain absolute value, β is a characteristic distribution index in the $0 > \beta > 2$ range.

The distribution curve calculated form is the required benchmark and provides an opportunity to evaluate in terms of cenosis stability, its dynamics, structural balance, etc.

2.2.2 Cenological Computations Using Regularities in the Distribution of Prime Numbers

As mentioned above, cenosis systems structure management is based on the possibility of comparing it with a certain reference form, which can be calculated mathematically using the prime numbers model [6]. For this purpose, the calculation of the number of the predominant group of elements is used in accordance with formula:

$$W_1 = N/2\ln N \tag{8}$$

The remaining numbers of the series are calculated analytically, which ensures the correctness of the result even if the sample is insufficient. The result is a canonized distribution in the form of primes, which reflects the pattern of distribution, the distance between the elements, the non-linearity of the number of local groups and allows us to predict the dynamics of movement along the curve of individual elements.

2.2.3 Operations with Fuzzy Numbers Within the Bounds of the Cenological Confidence Interval

Another example of subtle procedures for evaluating socio-economic cenosis is GZ-analysis, which indicates the limits of the confidence interval, basing on generalizing cenological indicators as on systemic ones. The heuristic stage of such an analysis deals with data on the a long period indicators dynamics and is used for preliminary evaluation calculations, forming the basis for criterial analysis [3, 7]. To forecast the elements of the socio-economic cenosis system structure using GZ-methods V. Gnatyuk proposed "models of autoregressive moving average, time series decomposition, as well as various variations of methods based on the analysis of the singular spectrum of the trajectory matrix of the time series" [3]. This stage of the research is based on information about the dynamics of rank distributions of the object under consideration from the standpoint of its structural stability, where the synchronicity of the regularity of the development of rank surfaces and changes in the characteristic index are considered according to the formula:

$$W_G = \left(\int_{r_1}^{r_2} W^g(r)dr \right) - ((r_2 - r_1)W_2) \tag{9}$$

where W(r) – is the rank parametric distribution; $W^g(r)$ is the Gaussian distribution corresponding to the cluster distribution of parameters. W_2–the value of the parameter corresponding to the characteristics of the cluster's ranks (r_2).

Then for each of the ranks considered it is possible to determine with high accuracy the confidence interval of the rank parametric cenosis distribution, or the combination of the maximum and minimum values of the parameters stable over a longer time interval. The sweep width of the confidence interval is considered in terms of quartile values.

$$\Delta W_Z = W_{0,75}^q - W_{0,25}^q \tag{10}$$

where $W_{0,75}^q$ – is the upper quartile of the parameter values distribution (of 0.75 order)
$W_{0,25}^q$ – is the lower quartile (of 0.25 order)

$$W_Z = \int_{r1}^{r2} (W(r) - W^g(r))dr. \tag{11}$$

$$\frac{\Delta W_Z/2}{\sigma} = \Phi^{-1}(p_d/2)$$

where $\Delta W_Z/2$ is the width of the system confidence interval one way from the mathematical expectation; σ is the standard root-mean-square deviation of the experimental points from the mathematical one; Φ^{-1} – is the inverse Laplace function p_d a priori accepted confidence probability (0.95).

The following system value determines the stability of the parameter characteristics from the point of view of the width of the Gaussian confidence interval and reflects the range of stable values for each specific rank over time, regardless of what real object of cenosis it contains. The interquartile range is used:

$$\Delta W_G = W_{0,75}^q - W_{0,25}^q \tag{12}$$

$$\frac{\Delta W_Z/2}{\sigma} = \Phi^{-1}(p_d/2) \tag{13}$$

Combination of these estimations provides a very accurate characteristics of the socio-economic cenosis stability, which corresponds with the general physical concept of coherence, i.e. characteristics of a complex system to entropy reduce and organization improve

$$K_{GZ} = \lim_{KK \to KO} \frac{W_Z}{W_G} \cong \frac{\Delta W_Z}{\Delta W_G} \tag{14}$$

where KK is the number of cenosis clusters; KO - the number of cenosis objects.

3   Results

Thus, a synthetic model of fuzzy-multiple analysis and forecasting of the development of socio-economic systems with cenosis features, based on the description of the probability distribution function for possible values, was developed; as well as operations with fuzzy numbers within the bounds of the calculated confidence interval of such a function; soft computing using regularities of prime numbers distribution; dynamic modeling in the form of fuzzy cognitive models. The effectiveness of the proposed integration is confirmed by the effectiveness of the analysis of various disciplines and analytical support in the form of empirical studies of economic systems of various sizes [9, 10].

4 Conclusion

A new technique has been developed which enables the study of complexly structured large-scale socio-economic objects and provides a qualitative fuzzy-multiple analysis based on the possibility of describing semi-structured problems, their evaluation and cognitive representation. The considered approach was proposed for the first time in scientific practice, it increases the accuracy of the assessment of the structural stability of systems, especially in the conditions of a wide variation of the economic values under consideration.

References

1. Gil Aluja, J.: Towards a new concept of economic research. Fuzzy Econ. Rev. (1995). https://doi.org/10.25102/fer.1995.01.01
2. Casti, J.: Large systems. Connectivity, Complexity and Disaster. Mir, Moscow (1982)
3. Gnatyuk, V.: Intellectual technologies of monitoring electric consumption of objects of the transport electrotechnical complex. Marine Intell. Technol. Sci. J. 3(37) V 1, 130–135 (2017)
4. Gorelova G.V., Kalinichenko A.I., Kuzminov A.N.: Program for cognitive modeling and analysis of regional socio-economic systems. Certificate of computer program registration RUS (2018). 2018661506. 07.09.2018
5. Gorelova, G.V., Dzharimov, N.K.: Regional Education System, Integrated Research Methodology. Printing House, Krasnodar (2002)
6. Kaczorowski, J.: On sign-changes in the remainder-term of the prime-number formula II. Acta Arith. 45, 65–74 (1985)
7. Kostrikova, N.A.: Synthesis technology of distributed intelligent control systems as a tool for sustainable development of territories and complex objects. In: Kostrikova, N.A., Merkulov, A.A., Ya Yafasov, A. (eds.) Marine Intellectual Technologies, vol. 3, no. 37 - 1 T, pp. 135–141 (2017)
8. Kudrin, B.I.: Introduction to technology. Tomsk State University, Tomsk (1993)
9. Kuzminov, A.N., Korostieva, N.G., Dzhukha, V.M., Ternovsky, O.A.: Economic coenosis stability, methodology and findings. In: Contemporary Issues in Business and Financial Management in Eastern Europe, vol. 100, pp. 61–70 (2018)
10. Kuzminov, A.N., Dzhukha, V.M., Ternovsky, O.A., Mikhnenko, T.N.: Cenological Measurement of Productive Efficiency. Eur. Res. Stud. J. XXI(2), 27–36 (2018)
11. Lavretsky E., Wise K.A.: Lyapunov stability of motion. In: Robust and Adaptive Control. Advanced Textbooks in Control and Signal Processing. Springer, London (2013)
12. PlotinskyYu, M.: Theoretical and empirical models in social processes. Uch. Logos, Moscow (1998)
13. Poelmans, J., Ignatov, D.I., Kuznetsov, S.O., Dedene, G.: Fuzzy and rough formal concept analysis: a survey. Int. J. Gen. Syst. 43(2), 105–134 (2014). https://doi.org/10.1080/03081079.2013.862377

14. Zadeh, L.: The Concept of a Linguistic Variable and its Application to Approximate Reasoning. Mir, Moscow (1975)
15. Zakharova, E.N.: On the cognitive modeling of sustainable development of socio-economic systems//Bulletin of Adygea State University. Series 1: Regionology: philosophy, history, sociology, jurisprudence, political science, cultural studies. N1 (2007). https://cyberleninka. ru/article/n/o-kognitivnom-modelirovanii-ustoychivogo-razvitiya-sotsialno-ekonomicheskih-sistem. Accessed 23 Jun 2019
16. NCS. http://www.springer.com/lncs. Accessed 21 Nov 2016

Eigensolution of 2 by 2 Z-Matrix

Kamala Aliyeva$^{(\boxtimes)}$ (iD)

Azerbaijan State Oil and Industry University, Azadlig Ave.20,
AZ1010 Baku, Azerbaijan
kamalann64@gmail.com

Abstract. In decision making problems preferences of decision maker (DM) as usual is expressed in matrix form. Main problem in DM preference formalization is testing of consistency of DM preference knowledge expressed in matrix form. This problem is related with eigensolution of a given matrix. Investigation of eigensolution of numerical and fuzzy matrix is well known. Unfortunately, up today there are works in existing scientific literature on investigation eigenvalues and eigenvectors of Z-number valued matrices. In this paper for the first time we investigate 2 by 2 decision matrix, elements of which are Z-numbers, expressing fuzzy and probabilistic uncertainty of DM preference.

Keywords: Z-number · Z matrix · Eigensolution · Fuzzy numbers · Fuzzy and probabilistic uncertainty

1 Introduction

Computation of fuzzy eigenvalues of a fuzzy matrix is a challenging problem. Determining the maximal and minimal symmetric solution can help to find the eigenvalues. Initially, Buckley found fuzzy eigenvalues when matrix has fuzzy positive elements in [1], but this method was limited. He used the founded fuzzy eigenvalues to solve a economic model. After, Chiao has studied generalized fuzzy eigenvalues and his method is consistent to Buckley method, [2]. Thoedorou, Drossosb, and Alevizosb studied fuzzy eigenvalues of fuzzy corresponding analysis; they utilized a two-step method with triangular fuzzy numbers, in [3].

Tian founded fuzzy eigenvectors of a real matrix in [4]; he studied the structure of fuzzy eigenspaces and relationships between real eigenspaces and fuzzy eigenspaces of real matrix. To date, few studies, if any, have been dealt with finding general fuzzy eigenvalues and its fuzzy eigenvectors of a fuzzy matrix classical fuzzy arithmetic. It is needed to mention that existing studies on operations over Z-numbers are based on classical fuzzy arithmetic. FFLSs have been studied by many authors, a solution method for finding fuzzy eigenvalues and fuzzy eigenvectors of a fuzzy matrix with fuzzy idempotent has not been given yet. Recently, Allahviranloo et al. [5] proposed a novel method to solve a fully fuzzy linear system based on the 1-cut expansion. In that method, some spreads and, therefore, some new solutions are derived.

The other methods introduced by other researchers have many demerits [2, 3, 6, 10] in that their method was not usable for large n but their methods can make fuzzy vector these cannot find fuzzy eigenvalue [1]. In this paper, practical method to find fuzzy

© Springer Nature Switzerland AG 2020
R. A. Aliev et al. (Eds.): ICSCCW 2019, AISC 1095, pp. 758–762, 2020.
https://doi.org/10.1007/978-3-030-35249-3_98

eigenvalues and eigenvectors is applied. In Sect. 2, the most important notations used are mentioned. Then, in Sect. 3, we present our new method to obtain eigenvalues and eigenvectors. Also, Sect. 4 gives a numerical example. The final section ends this paper with a brief conclusion.

2   Preliminaries

Definition 1 Z-number. A Z-number is an ordered pair $Z = (A, B)$ where A is a continuous fuzzy number playing a role of a fuzzy constraint on values that a random variable may take X is A and B is a continuous fuzzy number with a membership function $\mu_B : [0, 1] \rightarrow [0, 1]$ playing a role of a fuzzy constraint on the probability measure of A, P(A) is B.

A fuzzy matrix. A fuzzy square matrix (A_{ij}) is a matrix elements of which are fuzzy numbers $A_{ij}, i, j = 1, \ldots, n$ which describe fuzzy restrictions on values of random variables $X_{ij}, i, j = 1, \ldots, n$:

$$\left(A_{ij}\right) = \begin{pmatrix} A_{11} & \cdots & A_{1n} \\ . & \cdots & . \\ A_{n1} & \cdots & A_{nn} \end{pmatrix}.$$

A Random Square Matrix. A random square matrix X is a matrix of probability distributions $X_{ij}, i, j = 1, \ldots, n$ of random variables $X_{ij}, i, j = 1, \ldots, n$:

$$\left(x_{ij}\right) = \begin{pmatrix} x_{11} & \cdots & x_{1n} \\ . & \cdots & . \\ x_{n1} & \cdots & x_{nn} \end{pmatrix}$$

Each random variable $X_{ij}, i, j = 1, \ldots, n$ is governed by probability distribution $p_{ij}, i, j = 1, \ldots, n$. For simplicity of notation, we describe a random square matrix $X_{ij}, i, j = 1, \ldots, n$,

$$\left(p_{ij}\right) = \begin{pmatrix} p_{11} & \cdots & p_{1n} \\ . & \cdots & . \\ p_{n1} & \cdots & p_{nn} \end{pmatrix}.$$

Definition 2 A Z-valued matrix. A Z-valued matrix (Z_{ij}) is a matrix of Z-numbers that describe partially reliable information on values of random variables $X_{ij}, i, j = 1, \ldots, n$:

$$\left(Z_{ij} = \left(A_{ij}, B_{ij}\right)\right) = \begin{pmatrix} Z_{11} = (A_{11}, B_{11}) & \cdots & Z_{1n} = (A_{1n}, B_{1n}) \\ . & \cdots & . \\ Z_{n1} = (A_{n1}, B_{n1}) & \cdots & Z_{nn} = (A_{nn}, B_{nn}) \end{pmatrix}$$

Let us formulate definitions of eigenvalue and eigenvectors of (Z_{ij}).

Definition 3 A Z-eigenvector of Z-valued matrix. A Z-valued eigenvalue of Z-valued square matrix (Z_{ij}) is such a Z-number $Z_\lambda = (A_\lambda, B_\lambda)$ that the following holds:

$$det(Z_{ij} - Z_\lambda I) = det \begin{pmatrix} Z_{11} - Z_\lambda & \cdots & Z_{1n} \\ . & \cdots & . \\ Z_{n1} & \cdots & Z_{nn} - Z_\lambda \end{pmatrix} = Z_0 \qquad (1)$$

where I is a traditional (non-fuzzy) identity matrix.

Thus, Z-valued eigenvalue.$Z_{\lambda j} = (A_{\lambda j}, B_{\lambda j})$ is a root of n-th order characteristic equation

$$Z_\lambda^n + Z_1 Z_\lambda^{n-1} + \ldots + Z_n = Z_0,$$

where $Z_l, l = 1, \ldots, n$ are coefficients induced by the elements of Z-valued matrix (Z_{ij}). So, n Z-valued eigenvalues
$Z_{\lambda j} = (A_{\lambda j}, B_{\lambda j})$ exist for a Z-valued square matrix (Z_{ij}).

Definition 4 A Z-eigenvector of Z-valued matrix. A vector of Z-numbers $(Z_Y) = (Z_{Y1} = (A_{Y1}, B_{Y1}), \ldots, Z_{Yn} = (A_{Yn}, B_{Yn}))$ is referred to as a Z-valued eigenvector of Z-valued square matrix (Z_{ij}) if it satisfies the following Z-valued linear system of equations:

$$(Z_{ij})(Z_Y) = Z_\lambda(Z_Y) \qquad (2)$$

where $Z_\lambda = (A_\lambda, B_\lambda)$ is Z-valued eigenvalue.

Thus, n Z-valued eigenvectors $(Z_{xj}), j = 1, \ldots, n$ exist, one for each Z-valued eigenvalue $Z_{\lambda j} = (A_{\lambda j}, B_{\lambda j})$

3 Statement of the Problem

Modeling proposed technique of the fuzzy AHP was applied for the software selection problem which it contains different and conflicting factors. Suppose that an MCDM problem involves 2 criteria: C_1-Reliability and C_2-Functionality. Every criteria is designed by Z number. Pairwise comparison of criteria C_1 and C_2 is given in Table 1.

Table 1. Pairwise comparison of criteria

Criteria	C_1	C_2
C_1	((0.93, 0.95, 1), (0.89, 0.95, 1))	((2, 2.5, 3), (0.7, 0.8, 1))
C_2	((0.33, 0.4, 0.5), (0.7, 0.8, 1))	((0.93, 0.94, 0.95), (0.89, 0.95, 1))

4 Solution of the Problem

The Z-matrix is:

$$(Z_{ij}) = \begin{pmatrix} ((0.93, 0.95, 1), \ (0.89, 0.95, 1)) & ((2, 2.5, 3), \ (0.7, 0.8, 1)) \\ ((0.33, 0.4, 0.5), \ (0.7, 0.8, 1)) & ((0.93, 0.93, 0.95), \ (0.89, 0.95, 1)) \end{pmatrix}$$

At the first step we consider the corresponding fuzzy matrix:

$$(A_{ij}) = \begin{pmatrix} (0.93, \ 0.95, \ 1) & (2, \ 2.5, \ 3) \\ (0.33, \ 0.4, \ 0.5) & (0.93, \ 0.93, \ 0.95) \end{pmatrix}$$

and the corresponding random matrix obtained by using the aggregation technique (3) (described in terms of mean and standard deviation of its elements):

$$(p_{ij}) = \begin{pmatrix} (m_{11}, \sigma_{11}) = (1, 0.01) & (m_{12}, \sigma_{12}) = (2.5, \ 0.45) \\ (m_{a21}, \sigma_{a13}) = (0.4, 1.8) & (m_{a22}, \sigma_{a22}) = (1, \ 0.01) \end{pmatrix}$$

At the second step, we determine a fuzzy eigenvalues A_λ of (A_{ij}) and probabilistic eigenvalues of (p_{ij}). The following crisp matrices are constructed based on (A_{ij}):

$$(A_{ijl}) = \begin{pmatrix} 0.93 & 2 \\ 0.33 & 0.93 \end{pmatrix}, \ (A_{ijm}) = \begin{pmatrix} 0.95 & 2.5 \\ 0.4 & 0.94 \end{pmatrix} \ (A_{iju}) = \begin{pmatrix} 1 & 3 \\ 0.5 & 0.95 \end{pmatrix}.$$

The related matrices $(\overline{A_{ijl}})$, $(\overline{A_{ijm}})$, $(\overline{A_{iju}})$, are

$$(\overline{A_{ijl}}) = \begin{pmatrix} 2.81 & 6.5 \\ 1.06 & 2.8 \end{pmatrix}, \ (\overline{A_{ijm}}) = \begin{pmatrix} 5.73 & 15 \\ 2.43 & 5.64 \end{pmatrix}, \ (\overline{A_{iju}}) = \begin{pmatrix} 2.95 & 8.5 \\ 1.4 & 2.84 \end{pmatrix}$$

The fuzzy eigenvalues $A_{\lambda 1} = (\lambda_{1l}, \lambda_{1m}, \lambda_{1u})$ computed by using matrices $(\overline{A_{ijl}})$, $(\overline{A_{ijm}})$, $(\overline{A_{iju}})$ are $A_{\lambda 1} = (1.7, 2, 2.17), A_{\lambda 2} = (-0.23 - 0.04, 0.8)$.

At the same time, we have to find $p_\lambda = (m_{\lambda, \lambda})$ for the random matrix (p_{ij}). The following probability distributions p_λ of eigenvalues are obtained:

$$p_\lambda = (2, 0.08), \ p_\lambda = (0, 0.01).$$

Thus, the obtained Z-eigenvalues are

$$Z_{\lambda 1} = (A_{\lambda 1}, p_{\lambda 1}) = ((1.7, 2, 2.17), (2, 0.08)),$$

$$Z_{\lambda 2} = (A_{\lambda 2}, p_{\lambda 2}) = ((-0.23 - 0.04, 0.8), (0, 0.01)).$$

At the third step, we compute eigenvectors $Z_{xj} = (A_{xj}, p_{xj})$ for the obtained eigenvalues $Z_\lambda = (A_\lambda, p_\lambda)$. Given fuzzy eigenvalues A_λ, the corresponding fuzzy eigenvectors obtained on the basis of (6)–(8) are given below:

$$(A_{x1}) = \begin{pmatrix} (0.61,\, 0.71,\, 0.77) \\ (0.24,\, 0.28,\, 0.31) \end{pmatrix}, \quad (A_{x2}) = \begin{pmatrix} (-14.4,\, 0.72,\, 4.11) \\ (-5.56,\, 0.28,\, 1.64) \end{pmatrix}$$

At the same time, for the probabilistic eigenvalues computed at the previous step, the corresponding eigenvectors are found as:

$$(p_{x1}) = \begin{pmatrix} (0.92,\, 0.025) \\ (0.38,\, 0.06) \end{pmatrix}, \quad (p_{x2}) = \begin{pmatrix} (-0.92,\, 0.025) \\ (0.38,\, 0.06) \end{pmatrix}$$

Thus, the obtained Z-eigenvectors $Z_{xj} = (A_{xj}, p_{xj}), j = 1, 2$ are

$$Z_{x1} = \begin{pmatrix} ((0.61,\, 0.71,\, 0.77),\, (0.92,\, 0.025)) \\ ((0.24,\, 0.28,\, 0.31),\, (0.38,\, 0.06)) \end{pmatrix},$$

$$Z_{x2} = \begin{pmatrix} ((-14.4,\, 0.72,\, 4.11),\, (-0.92,\, 0.025)) \\ ((-5.56,\, 0.28,\, 1.64),\, (0.38,\, 0.06)) \end{pmatrix}.$$

5 Conclusion

Estimation of eigenvalues and eigenvectors of pairwise comparison matrix is methodological challenger in multi-criterial decision problem under deep uncertainty is decision analysis is characterized with synergy of fuzzy and probabilistic uncertainty expressed by Z-number concept. Unfortunately, there are not research results on eigen solution of Z-valued matrices. In this paper we have suggested computation method of eigenvalues and eigenvectors of 2 by 2 matrix. The suggested method is applied to software selection problem under Z-information.

References

1. Buckley, B.: Fuzzy eigenvalue problems and input output analysis. Fuzzy Sets Syst. **34**, 187–195 (1998)
2. Chiao, C.: Generalized fuzzy eigenvalue problems. Tamsui Oxf. J. Math. Sci. **14**, 31–37 (1998)
3. Theodoroua, Y., Drossosb, C., Alevizosb, P.: Correspondence analysis with fuzzy data: the fuzzy eigenvalue problem. Fuzzy Sets Syst. **158**, 113–137 (2007)
4. Tian, Z.: Fuzzy eigenvectors of real matrix. J. Math. Res. **2**(3), 103 (2010)
5. Allahviranloo, T., Salahshour, S., Khezerloo, M.: Maximal- and minimal symmetric solutions of fully fuzzy linear systems. J. Comput. Appl. Math. **235**, 4652–4662 (2011)
6. Allahviranloo, T., Mikaelvand, N., Aftab Kiani, N., Mastani Shabestari, R.: Signed decomposition of fully fuzzy linear system. J. Comput. Appl. Math. **1**, 77–88 (2008)
7. Allahviranloo, T., Afshar Kermani, M.: Solution of a fuzzy system of linear equation. Appl. Math. Comput. **175**, 519–531 (2006)
8. Sevastjanov, P., Dymova, L.: A new method for solving interval and fuzzy equations: linear case. Inf. Sci. **179**, 925–937 (2009)

On Possibility of Using a Toothed Belt Drive in a Drive to Rocker-Machines

I. M. Kerimova[(⊠)]

20 Azadliq Avenue, Az1010 Baku, Azerbaijan
imkerimova@gmail.com

Abstract. A rocker-machine is an important type of oil and gas equipment and used for mechanical drive to oil well sucker-rod (plunger) pumps. The construction of a rocker-machine presents itself a balancer drive of sucker-rod pumps, consisting of an electric motor, a reducer and a two-set four linked joint mechanism. In this paper, the method of specifying of the dependency of the criterion of tensions of the belt sides on the angle of slide of the girth is designed.

Keywords: Rocker-machine · Belt drive · Reducer · Drive · Gear-ratio · Tension · Lifetime

1 Introduction

A rocker-machine is an important type of oil and gas equipment and used for mechanical drive to oil well sucker-rod (plunger) pumps. The construction of a rocker-machine presents itself a balancer drive of sucker-rod pumps, consisting of an electric motor, a reducer and a two-set four linked joint mechanism.

For decrease of the gear-ratio of the reducer in drives of rocker-machines they started to use flat-belt transmission, subsequently replaced with V-belt transmissions. However, because of series of the particularities such as, using of high-speed engines, a relative slip of belts increased when functioning and starting, and irregular preliminary extraction of V-belts and their wrong installation have brought to reduction of belts lifetime. Besides, using high-speed engines has required the reduction of the diameter of the driving pulley of a V-belt transmission that has brought to arising of great tensions in a belt because of a bend.

Constructions of low-speed rocker-machines with increased gear-ratio due to leading into the transmission an additional belt transmission are known. That allowed to reduce the oscillation frequency of the balancer to 0.8...1.7 per minute. For that a countershaft is assembled between an electric motor and a reducer with according small and big on diameter pulleys, installed console.

The warranty lifetime of V-belts, according to their ratings for the conditionally calculated length of different types of belts is found within 200...300 h. The calculated longevity does not exceed 500...600 operating hours. Herewith, a belt breakaway in the operational process is accompanied with its changing with a new one, which length often does not correspond to the length of other belts, earlier installed in the

R. A. Aliev et al. (Eds.): ICSCCW 2019, AISC 1095, pp. 763–767, 2020.
https://doi.org/10.1007/978-3-030-35249-3_99

transmission. It brings to disproportionate load distribution on belts and their irregular deterioration.

A frequent stopping of an operating process for the reason belts changing reduces efficiency of the use of the deep-well pumping unit. Replacement of V-belts with poly-V-belts, which have got broad spreading in drives of machines though can little enlarge the resource of the belt, but also does not bring to desired results. That is why some manufacturers, for instance Russian plants, have altered to production of rocker-machines, in drives of which belt transmissions are not used. However such problem solving has brought to increase of gear-ratio of the reducer, and so, to increase of the drive mass and to increase in cost of their production.

In the last decades toothed belt drives have been broadly started to be used in industry, their warranty lifetime according to passport data exceeds 2000 h. In this connection issues of toothed belt drives using in drives of rocker-machines are considered in this report. Methods of toothed belt selection and results of the longevity comparison of V-belts, poly-V-belts and toothed belts in drives of rocker-machines are given.

Using of results of problem solving on using of a toothed belt drive in drives of rocker-machines will allow to uncover their potential possibilities and, as effect, create high-performance drives of rocker-machines.

2   Statement of Problem

In [1] a determination method of the belt tighting force in its any point with provision for its own weight and hardness while transmissing of power by itself in power installation is designed. Besides following expressions are used in practice for determination of tighting in operating and down-leg sides of the flexible connection in process of mechanical work transmissing by itself [2].

$$
\left.
\begin{aligned}
P_1 &= \zeta Q \frac{e^{f\varphi_T}}{e^{f\varphi_T} - 1} + \frac{qV^2}{g} \\
P_2 &= \zeta Q \frac{1}{e^{f\varphi_T} - 1} + \frac{qV^2}{g}
\end{aligned}
\right\}. \tag{1}
$$

where $Q = 102 \, N/V_0$ is an effort given to the belt, N is a transmissing power, V_0 is a speed of the arc point of the pulley, $q = \delta g$ is a running weight of the flexible connection, ξ is a safety factor.

Converting expression (1) and bringing a dimensionless quantity $\lambda p = P_{B1}/P_{A1}$ we find the dependency of a tension criterion λ_p on an angle of slide φ_{as} of the girth, under $\alpha = 0$:

$$\frac{P_{A_1}(EF - \delta_{A_1}V_{A_1}^2)(EF - P_{A_1})}{(EF)^2}(\lambda_p - 1) + \frac{P_{A_1}^2(EF - \delta_{A_1}V_{A_1}^2)(EF - P_{A_1})}{2(EF)^2}(\lambda_p^2 - 1) -$$

$$-\frac{(P_{A_1} - \delta_{A_1}V_{A_1}^2)(P_{A_1} - EF)}{\delta_{A_1}V_{A_1}^2 - EF}(e^{f\varphi_T} - 1) - \frac{(P_{A_1} - \delta_{A_1}V_{A_1}^2)(P_{A_1} - EF)}{2(EF - \delta_{A_1}V_{A_1}^2)}(e^{f\varphi_T} - 1) -$$

$$- fg\delta_{A_1}R_2 Cos(\varphi_T - 1) + g\delta_{A_1}R_2 Sin\varphi_T = 0$$

$$(2)$$

3 Solution Method

Calculations were conducted under different values of initial tensions of the guided side and its modulus of elasticity for the concrete transmission. Namely, the interval of initial tension values is [1, 1.4], and the interval of modulus of elasticity values is [1, 1.125]. The further computations are based on these intervals. In the considered instance $\delta_{A1} = 0{,}18$ kg/m, $N = 6$ kWt, $E = 5 \cdot 10^8$ N/m^2, $F = 1.38$ sm^2, $R_2 = 470$ mm, $V_{A1} = 20$ m/sec, $f = 0.25$.

Numerical values of relationships of tensions of the belt sides are given in Table 1 depending on angles of slide of the girth under different values of initial tensions P_{A1}.

Table 1. Numerical values of relationship of tensions

φ_T	λ_p		
	λ_p ($P_1 = 30000N$ $E = 5.10^8$ N/M^2)	λ_p ($P_1 = 50000N$ $E = 5.10^8$ N/M^2)	λ_p ($P_1 = 30000N$ $E = 5.10^7$ N/M^2)
0	0,9999	1,00005	–
$\pi/4$	1,2116	1,2141	–
$3\pi/8$	1,3418	1,3443	–
$\pi/2$	1,4798	1,4805	–
$5\pi/8$	1,6322	1,633	1,639
$3\pi/4$	1,7982	1,7992	1,8062
$7\pi/8$	1,9861	1,9862	1,9932
π	2,19	2,19	2,19937

In Fig. 1 curves of the dependency of a tension criterion λ_p on an criterion of the girth angle $\lambda_\varphi = \varphi'_{as}/\varphi_{as}$ under different values of initial tensions P_{A1} of the belt guided side equal 30.000 N and 50.000 N are shown. The curve of that dependency under $P_{A1} = 30.000$ N and $E = 5 \cdot 10^7$ N/m^2 is also shown in this Fig. 1.

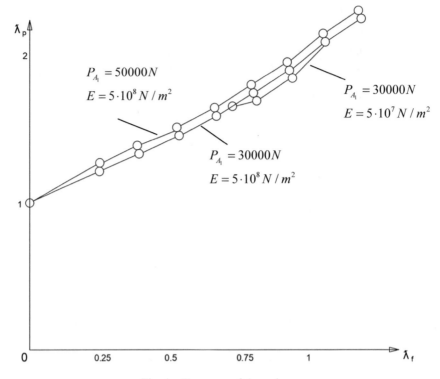

Fig. 1. The curve of dependency

An analysis of these curves shows that with increase of λ_ϑ relationships of the tensions of sides on the pulley λ_φ criterion of the angle of slide of the girth increases.

It is seen that under gradual growth of the tension criterion the angle of slide of the belt girth sharply increases. In the course of time as a result of friction and heating processes of the belt elements its modulus of elasticity decreases and the angle of slide of the girth increases. And this in its turn influences upon increase of the tension criterion. Hence, we get the important practical conclusion about that for normal operating of the flexible connection relationships of the tensions (the criterion of the tensions) must be less than 1.986 i.e. $\lambda_p \leq 1.986$. Besides from these curves it is seen that at reduction of the modulus of elasticity the belt loses capacity to work, for instance under $E = 5 \cdot 10^7$ N/m² the angle of slide of the girth $\varphi_{as} > \frac{3\pi}{4}$, which under this slide of the belt will be beyond the elastic sliding and belt skidding appears. As a result of analysis and selecting the equation on this curve, shown on Fig. 1, the shown dependency λ_p from λ_φ can be presented in the manner of the following power series:

$$\lambda_p = 1 + a_1\lambda_\varphi + a_2\lambda_\varphi^2 + a_3\lambda_\varphi^3, \tag{3}$$

where $a_1 = 0.229389\pi$, $a_2 = 0.0537789\pi^2$, $a_3 = -0.0033351\pi^3$.

Equation (3) presents itself a practical and generalized Euler formula. It makes possible to define from it tensions of the pulley for any value of the angle of the belt girth or on the contrary.

It follows to note that for designing and operating organizations alongside with calculation on friction and deterioration it means to calculate the drive belts on fatigue strength. Fatigue strength of the belt depends on those greatest tensions, which appear in its details in the process of its operating. In practice it is reasonable to produce them with a changeable distance between axes of pulleys. For the reason of tighting regulation of transmission sides they change the axle base by moving of one of the shafts.

Having selected the belt and having installed the safe load, it is possible to find the tensions and the angle of the belt girth according to the presented method and the belt safety margin is defined according to this tension. The proposed calculation method can be applied to high-speed transmissions. On the basis of aforesaid we come to a conclusion that the angle of slide of the girth rapidly increases and the criterion of the tensions of the belt sides slowly increases with reduction of the modulus of elasticity of the belt.

4 Conclusion

The method of specifying of the dependency of the criterion of tensions of the belt sides on the angle of slide of the girth is designed.

Dependency curves of the criterion of tensions on the angle of the belt girth are given and a practical formula for specifying of the tensions of the belt sides is derived.

Research results can be used at calculation of performance reliability of the belt transmission.

References

1. Gadirov, N.B., Bakhshaliyev, V.I.: The determination of tighting force of the flexible weighty V-belt applicable for the transmissing in power installation. In: Proceedings of Higher Schools "Oil and Gas", vol. 4, pp. 81–87 (1989)
2. Kolchin, N.I.: Machines Mechanics. Moscow (1972)

Calculation of Lyapunov Exponent of Time Series

Seving R. Mustafayeva$^{(\boxtimes)}$ (iD)

Azerbaijan State Oil and Industry University, Azadlig ave, 35, Baku, Azerbaijan
mustafayeva_81@mail.ru

Abstract. In this paper considered the problem of calculating the Lyapunov exponent of systems for forecasting time series of demand for marketing. Lyapunov index is one of the main characteristics of time series for forecasting problem. Because in now days the financial demand is problem with chaotic character.

Keywords: Time series · Lyapunov index · Forecasting

1 Introduction

The problem of analyzing and processing time series in the context of the principles of synergy is of particular relevance. Time series are the main result of experiments both full-scale and computational. The priorities usually indicate the tasks of meteorology, geophysics, and financial analysis. Recently, they were joined by physiology, medicine and social sciences. In the general case, the task of modeling by time series is known by the names of "system identification" and "system prediction". In this paper, the state and prospects for the implementation of the problem under study will be considered in the context of identification tasks and forecast tasks.

Identification task. When solving it, an attempt is made to answer the question of what are the parameters of the system that generated this time series. The parameters can be very different - statistical distributions, parameters of statistical models, spectral properties, etc. It is important that these parameters help to identify the system (process), that is, to distinguish it from others. The task of the forecast. It consists in predicting, by observational data, the future values of the measured characteristics or, more generally, the future state of the object being analyzed.

Thus, when processing time series, several types of problems are usually solved - problems of measuring the characteristics of dynamic systems that are invariant with respect to the change of variables (dimensions of attractors, Lyapunov exponents, entropies, approximations of the equations of motion from experimental data that relate to the identification problem). An important characteristic of the dynamical systems under consideration is the average growth rate of the state vector x(k), which for the semi ring with maximum and addition operations is defined in terms of this semi ring as the limit provided that this limit exists.

© Springer Nature Switzerland AG 2020
R. A. Aliev et al. (Eds.): ICSCCW 2019, AISC 1095, pp. 768–771, 2020.
https://doi.org/10.1007/978-3-030-35249-3_100

$$\lambda = \lim_{k \to \infty} kx(k)k_1/k \tag{1}$$

When analyzing systems with queues, the value of λ has the meaning of the average time of the service cycle, and the reciprocal of λ determines the capacity of the system. One of the features of the chaotic regimes is the instability of each trajectory belonging to the chaotic attractor. The quantitative measure of this instability was the Lyapunov characteristic exponents. In mathematics the Lyapunov exponent or Lyapunov characteristic exponent of a dynamical system is a quantity that characterizes the rate of separation of infinitely close trajectories. This sensitivity to initial conditions can be quantified as

$$\|\delta x(t)\| \approx e^{\lambda t}\|\delta x_0\| \tag{2}$$

where λ, the mean rate of separation of trajectories of the system, is called the leading Lyapunov exponent.

Rmax, × and Rmax, × , the results obtained for Rmax, +, can, as typically be extended to the remaining half rings. We will assume that the coordinates of the initial vector x(0) with probability 1 limited. Then with probability 1

$$c_1 \leq x(0) \leq c_2 \tag{3}$$

where c_1 and c_2 are some constants, $1 = (1, \ldots, 1)^T$

It is clear that under this condition the inequality holds

$$c_1\|A_k\| \leq \|x(k)\| \leq c_2\|A_k\| \tag{4}$$

It follows that the average growth rate λ of the state vector can be defined as

$$\lambda = \lim_{k \to \infty} \|x(k)\|^{1/k} \tag{5}$$

2 Lyapunov Exponent Determination

When analyzing systems with queues, the value of λ has the meaning of the average time of the service cycle, and the reciprocal of λ determines the capacity of the system. One of the features of the chaotic regimes is the instability of each trajectory belonging to the chaotic attractor. The quantitative measure of this instability was the Lyapunov characteristic exponents. In mathematics the Lyapunov exponent or Lyapunov characteristic exponent of a dynamical system is a quantity that characterizes the rate of separation of infinitesimally close trajectories. This sensitivity to initial conditions can be quantified as

$$\|\delta x(t)\| \approx e^{\lambda t}\|\delta x_0\| \tag{6}$$

where λ, the mean rate of separation of trajectories of the system, is called the leading Lyapunov exponent. In this paper we want to determine the Lyapunov index of time series for forecasting of the next year demand for the market according to the available data of recent years data. In particular, such a characteristic of the time series as the fractal dimension allows determining the moment when the system becomes unstable and ready to move to a new state. One of the main factors is the randomness of the fractal model, which is due to the exceptional interdependence of its input and output parameters. For this purpose we calculate the Lyapunov measure. The Lyapunov exponent, which is the (asymptotic) average growth rate of the state vector of the system, in semiring Rmax, + and Rmin, + is defined as the limit

$$\lambda = \lim_{k \to \infty} \|x(k)\|^{1/k} \tag{7}$$

3 Computer Simulation

For obtaining of Lyapunon index we were made all calculation were in Matlab environment.

1. We enter our input data of 2 years sales (see Table 1)
2. We determine Lyapunov index by using Matlab environment.

Table 1. Fragment of input data.

2012 year	2013 year
56	46
54	65
67	75
111	70
99	78
63	80
133	88
79	91
87	93
71	49
86	65
133	88
79	91
87	93
71	49
86	65
135	63

function zout = zlap(I, J, X, TAU)
% m file for function zlap used in lap.m for calculation of
% the first Lyapunov exponent using the method described by Wolf %
zout = X(I + (J–1) * TAU);
Lyapunov exponent
Lamda1 = 0.6609

4   Discussion and Conclusion

A positive Lyapunov index indicates the existence of chaotic motion in a dynamic system with limited trajectories. It is shown that, in contrast to the previously existing assumptions that the number of market participants is constant, this value also varies depending on the state of the market and the time interval. The number of market participants at short intervals (intraday) does not have any economic fundamental factors, and the market completely depends on the indicator of investor behavior. This conclusion allows us to explain the effectiveness of the use of technical analysis for intraday trading.

References

1. Holzfuss, J., Lauterborn, W.: Liapunov exponents from a time series of acoustic chaos. Phys. Rev. A **39**, 2146–2152 (1989)
2. Holzfuss, J., Parlitz, U.: Lyapunov exponents from time series. Lecture Notes in Mathematics, vol. 1486, pp. 263–270. Springer, Heidelberg (1991)
3. Eckmann, J.-P., Kamphorst, O.S., Ruelle, D., Cilliberto, S.: Liapunov exponents from time series. Phys. Rev. A **34**, 4971–4979 (1986)
4. Wolf, A., Swift, J.B., Swinney, H.L., Vastano, J.A.: Determining Lyapunov exponents from a time series. Physica D **16**, 285–317 (1985)
5. Aliev, R.A., Fazlollahi, B., Aliev, R.R.: Soft Computing and Its Applications in Business and Economics. Springer, Heidelberg (2004)
6. Bonfig, K.W., Mehdi, S.F., Gardashova, L.A.: Neural network based forecasting of time series and its application in the forecasting of oil prices in the world Market. ICAFS-2008, 8-th ICAFS and Soft Computing, Helsinki, Finland, pp. 275–281 (2008)

Application of Analytic Hierarchy Process Method for Ranking of Universities

Rashad R. Aliyev$^{(\boxtimes)}$ⓘ and Hasan Temizkanⓘ

Department of Mathematics, Faculty of Arts and Sciences, Eastern
Mediterranean University, Famagusta, North Cyprus, via Mersin 10, Turkey
{rashad.aliyev,hasan.temizkan}@emu.edu.tr

Abstract. Decision making is a process of choosing and identifying the best
alternative among available options to achieve the desired purpose in many
areas of human activity. Multiple Criteria Decision Making (MCDM) has been a
fast growing decision making tool for many years. It has taken its role in
different application areas due to being useful and attractive for solving complex
real-world problems. Analytic Hierarchy Process (AHP) is one of the mostly
used powerful MCDM techniques for formulating complex decisions. The aim
of this paper is to conduct AHP for ranking of five United Kingdom universities.
Four important criteria, and namely, teaching, research, citations and international outlook are used to rank the universities. The comparison matrix is used to
compare criteria as well as alternatives with respect to each criterion. Consistency index and consistency ratio are calculated by using eigenvalues to check
the consistency of comparison matrix.

Keywords: Decision making · Multiple Criteria Decision Making (MCDM) ·
Analytic Hierarchy Process (AHP) · Ranking · Consistency index · Consistency
ratio · Eigenvalue

1 Introduction

Decision making is a cognitive process that results in choosing an optimal decision
action. In decision making process, a finite set of alternatives as well as a finite set of
criteria for these alternatives are taken into consideration. The purpose is to make a best
choice which is suitable for preferences of a decision maker. In MCDM, a decision
maker should choose among available alternatives and rank them in preference order,
and the evaluation criteria which are weighted with respect to their importance are used
to achieve a desirable result. Analytic Hierarchy Process (AHP) is one of the effective
MCDM approaches, and was firstly presented by Thomas Saaty in 1980. AHP method
puts a problem into a hierarchical structure, and uses the pairwise comparisons of
alternatives to rank them. Prioritization, resource allocation, benchmarking, quality
management are some of decision cases to be defined by AHP method.

AHP technique provides a significant result to solve group decision making processes for a large number of academic evaluation cases [1]. Research papers have been
evaluated by using AHP method in Villanova University.

© Springer Nature Switzerland AG 2020
R. A. Aliev et al. (Eds.): ICSCCW 2019, AISC 1095, pp. 772–780, 2020.
https://doi.org/10.1007/978-3-030-35249-3_101

In [2], AHP approach is used to develop an effective academic staff promotion system in University of Kuala Lumpur. Some important criteria for staff evaluation are used to obtain the best alternative. AHP method is used to select the most convenient staff having enough qualifications for academic promotion.

Teaching evaluation is essential to improve the quality of teaching at the university. This may be possible by assessing the quality parameters of teachers. In [3], AHP method is used in to evaluate the quality of university teaching.

2 Comparison Matrix of AHP. Principle of Working with AHP

In AHP method, the comparison matrix (n × n) should be created for criteria as well as for alternatives with respect to each criterion. The form of the comparison matrix is as follows:

$$\begin{bmatrix} a_{11} & a_{12} & \cdots & a_{1n} \\ a_{21} & a_{22} & \cdots & a_{2n} \\ \cdot & & & \cdot \\ \cdot & & & \cdot \\ \cdot & & & \cdot \\ a_{n1} & a_{n2} & \cdots & a_{nn} \end{bmatrix}$$

The diagonal entries from a_{11} to a_{nn} take the value 1 ($[a_{ii}...a_{nn}] = 1$). The rest entries take the value between 1–9 describing numerical values of comparison matrix [4]. If a_{12} takes the value 3, then a_{21} becomes $1/a_{12}$ that is 1/3 ($[a_{ji} = 1/a_{ij}]$).

There are some steps to be followed to reach results using AHP method [4]. The principle of working with AHP method is stepwise explained below.

Step 1. The model is developed: the decision problem is broken down into a hierarchy structure, and the first, second, and third levels of the hierarchy consist of goal, criteria, and alternatives, respectively.

Step 2. Priorities (weights) for the criteria are obtained: for this reason, pairwise comparisons are executed between criteria. Then the consistency of judgments is checked to be sure about the proportionality and transitivity.

The pairwise comparisons between the criteria should be realized to obtain the priorities by considering a scale developed by Saaty [4]. The maximum numerical value 9 shows the extremely important property whereas the minimum numerical value 1 describes equally important property. The numerical values 2 and 3 describe moderately more important property; numerical values 4 and 5 describe strongly more important property. After defining the comparison matrix of criteria, it is required to calculate the normalized matrix which is carried out in the following order: (1) Sum the values in each column of the comparison matrix; (2) Divide each value by sum of related column to calculate the normalized matrix; (3) Obtain the priorities for the criteria by taking the average of each row from the normalized matrix.

The consistency of comparison matrix is checked as follows: (1) By using the comparison matrix, multiply each value in the first column with the first criterion priority, then multiply each value in the second column with the second criterion priority and continue this process for all columns; (2) Sum the values in each row to obtain the values that are called weighted sum; (3) Divide the values of weighted sum by related priority of each criterion. After division, take the average of the values to calculate λ_{max}; (4) Calculate the consistency index (CI) and consistency ratio (CR) by using the following formulas (n is a number of compared elements):

$$CI = \frac{\lambda_{max} - n}{n - 1} \tag{1}$$

$$CR = \frac{CI}{RI} \tag{2}$$

where RI (Random Index) gets available values from a randomly generated comparison matrix. The value of CR should be ≤ 0.1 to claim that it is consistent. For n = 4 the value of RI is 0.9; for n = 5 the value of RI is 1.12; for n = 6 the value of RI is 1.24 [4].

Step 3. Priorities for alternatives are obtained: the pairwise comparison is done between the alternatives with respect to each criterion. Then the consistency of the pairwise comparisons is checked: (1) The comparison matrix is done with respect to each criterion; (2) Calculate the normalized matrix and take the average of rows to obtain the priorities for each alternative; (3) Steps for consistency checking are same.

Step 4. Final (overall) priorities are obtained: the priorities of the alternatives with respect to each criterion are combined as weighted sum by considering the weight of each criterion to determine the final priorities. The alternative with the highest final priority is the best choice [4].

The priorities of alternatives for each criterion and priorities of criteria are considered. Final priorities of alternatives are calculated by multiplying each of alternative priorities (with respect to each criterion) with corresponding criteria weights, then by taking the summation of each row. So the best alternative is obtained.

3   Numerical Example on Ranking of Universities

In this paper, five UK universities are used for ranking process. Let's name these universities as A, B, C, D and E. The following criteria with their weights are used in performance evaluation of universities: (1) Teaching: %30; (2) Research: %30; (3) Citations: %30; (4) International outlook: %7.5; (5) Industry income: %2.5. The order of importance and the weights are taken from [5]. Since the weight of industry income criterion is too low to affect the calculation, this criterion is not considered in this paper. So, the above-mentioned first four criteria are used. The hierarchy structure for ranking of universities is depicted in Fig. 1.

The values of the comparison matrix for criteria are shown in Table 1. Table 2 shows the normalized matrix for Table 1. In order to have the criteria priorities, we use the average of each row from the normalized matrix (Table 2).

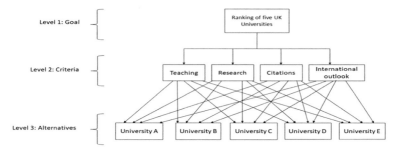

Fig. 1. Hierarchy structure for ranking of universities

Table 1. Comparison matrix for criteria

Criterion	Teaching	Research	Citations	International outlook
Teaching	1	1	1	4
Research	1	1	1	4
Citations	1	1	1	4
International outlook	0.25	0.25	0.25	1
Sum	3.25	3.25	3.25	13

Table 2. Normalized matrix and criteria priorities

Criterion	Teaching	Research	Citations	International outlook	Priority
Teaching	0.308	0.308	0.308	0.308	0.308
Research	0.308	0.308	0.308	0.308	0.308
Citations	0.308	0.308	0.308	0.308	0.308
International outlook	0.076	0.076	0.076	0.076	0.076

The consistency of comparison matrix should be checked. The comparison matrix is multiplied with the criteria priorities, and the weighted sum is calculated (Table 3).

Table 3. Results of weighted sum for criteria

Criterion	Teaching	Research	Citations	International outlook	Weighted sum
Criteria priorities	0.308	0.308	0.308	0.076	
Teaching	1	1	1	4	1.228
Research	1	1	1	4	1.228
Citations	1	1	1	4	1.228
International outlook	0.25	0.25	0.25	1	0.307

The weighted sum for each criterion is divided by respective priority to find eigenvalues λ's, then average is calculated to find the maximum eigenvalue λ_{max} (Table 4).

Table 4. Result for maximum eigenvalue λ_{max}

Weighted sum	Priority	λ	λ_{max}
1.228	0.308	3.987	16/4 = 4
1.228	0.308	3.987	
1.228	0.308	3.987	
0.307	0.076	4.039	
		Sum: 16	

The consistency index $CI = \frac{\lambda_{max}-n}{n-1} = \frac{4-4}{4-1} = 0$, and the consistency ratio $CR = \frac{CI}{RI} = \frac{0}{0.9} = 0 < 0.1$. So, the comparison matrix is perfectly consistent.

Figure 2 shows the computer simulation results for criteria priorities.

Fig. 2. Computer simulation results for criteria priorities

The same steps are done for each criterion by considering all alternatives. The weights for alternatives with respect to each criterion are taken from [6].

For teaching criterion priorities, comparison matrix is shown in Table 5, and the computer simulation results for teaching criterion priorities are shown in Fig. 3 ($\lambda_{max} = 5.0192, CI = 0.0048, CR = 0.0042 < 0.1$).

Table 5. Comparison matrix for teaching criterion

Teaching	A	B	C	D	E
A	1	1	2	3	1
B	1	1	2	2	1
C	0.5	0.5	1	1	0.5
D	0.333	0.5	1	1	0.5
E	1	1	2	2	1

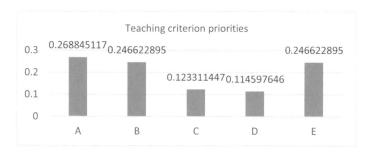

Fig. 3. Computer simulation results for teaching criterion priorities

For research criterion priorities, comparison matrix is shown in Table 6, and the computer simulation results for research criterion priorities are shown in Fig. 4 ($\lambda_{\max} = 5.093, \text{CI} = 0.02325, \text{CR} = 0.0207 < 0.1$).

Table 6. Comparison matrix for research criterion

Research	A	B	C	D	E
A	1	1	2	3	2
B	1	1	1	2	1
C	0.5	1	1	1	1
D	0.333	0.5	1	1	0.5
E	0.5	1	1	2	1

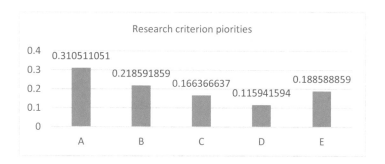

Fig. 4. Computer simulation results for research criterion priorities

For citations criterion priorities, the comparison matrix is shown in Table 7, and the computer simulation results for citations criterion priorities are represented in Fig. 5 ($\lambda_{\max} = 5.0172, \text{CI} = 0.0043, \text{CR} = 0.0038 < 0.1$).

For international outlook criterion priorities, the comparison matrix is shown in Table 8. Figure 6 shows the computer simulation results for international outlook priorities. Being very close to each other, the value 1 for weights of alternatives is assigned ($\lambda_{\max} = 5, \text{CI} = 0, \text{CR} = 0 < 0.1$).

Table 7. Comparison matrix for citations criterion

Citations	A	B	C	D	E
A	1	1	1	1	3
B	1	1	1	1	3
C	1	1	1	1	3
D	1	1	1	1	2
E	0.333	0.333	0.333	0.5	1

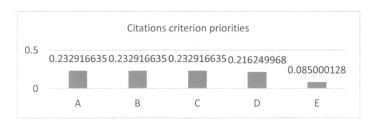

Fig. 5. Computer simulation results for citations criterion priorities

Table 8. Comparison matrix for international outlook criterion

International outlook	A	B	C	D	E
A	1	1	1	1	1
B	1	1	1	1	1
C	1	1	1	1	1
D	1	1	1	1	1
E	1	1	1	1	1

Fig. 6. Computer simulation results for international outlook criterion priorities

The priority of alternatives for each criterion is multiplied by criteria priorities and then each row is summed up to find final priorities of each alternative (Table 9).

The computer simulation results for final priorities of each alternative are shown in Fig. 7, and according to these results, the universities are ranked as $A > B > C > E > D$. So, the best alternative is obtained to be the university A.

Table 9. Results for final priorities of each alternative

	Teaching	Research	Citations	International outlook	Final Priorities
Criteria priorities	0.308	0.308	0.308	0.076	
A	0.269	0.310	0.233	0.2	0.2653
B	0.247	0.219	0.233	0.2	0.2305
C	0.123	0.166	0.233	0.2	0.1760
D	0.114	0.116	0.216	0.2	0.1526
E	0.247	0.189	0.085	0.2	0.1756

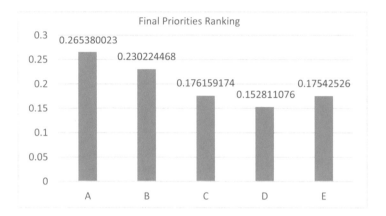

Fig. 7. Computer simulation results for final priorities

4 Conclusion

Ranking of five UK universities by applying AHP approach is studied in this paper. The criteria and the alternatives are mutually compared and the comparison matrices are formed by using the comparison scale. The computer simulation shows that the consistency ratio is 0 for criteria comparisons. The consistency ratios for alternatives comparison for teaching, research, citations, and international outlook criteria are 0.0042, 0.0207, 0.0038, and 0, respectively. Since all the consistency ratios of comparison matrices are less than 0.1, we decide that all comparison matrices are consistent; so the reliable ranking of universities is achieved.

References

1. Liberatore, M.J., Nydick, R.L.: Group decision making in higher education using the analytic hierarchy process. Res. High. Educ. **38**(5), 593–614 (1997)
2. Saaludin, N., Harun, S., Wahab, S.Z.A., Yahya, Y.: Academic staff promotion in higher education by using analytic hierarchy process (AHP). In: The 4th International Conference on Quantitative Sciences and Its Applications (ICOQSIA) (2016)

3. Sahito, M., Chandio, M.S.: An analytical hierarchy process model for the evaluation of university teaching quality. Sindh Univ. Res. J. Sci. Ser. **49**(004), 857–860 (2017)
4. Mu, E., Pereyra-Rojas, M.: Practical Decision Making. SOR. Springer, Cham (2017). https://doi.org/10.1007/978-3-319-33861-3
5. https://www.timeshighereducation.com/world-university-rankings/methodology-world-university-rankings-2019
6. https://www.timeshighereducation.com/student/best-universities/best-universities-uk

Toward a Derivative
of a Z-Number-Valued Function

O. H. Huseynov[(✉)] [iD]

Research Laboratory of Intelligent Control and Decision Making Systems in Industry and Economics, Azerbaijan State Oil and Industry University, Baku, Azerbaijan
oleg_huseynov@yahoo.com

Abstract. A study of Z-number-valued functions is important for formalization and processing of bimodal information. In particular, this requires formulation of differentiability and integrability of Z-number-valued functions. In this paper, we propose a result on relation between the concepts of derivative and integral for Z-number-valued functions.

Keywords: Z-number · Bimodal information · Derivative · Integral · Approximate limit

1 Introduction

The concept of a Z-number was introduced by Zadeh to deal with information characterized by fuzziness and partial reliability [1]. In recent years, important results were obtained in the field of computations with Z-numbers [2–10]. However, only a few works exist on analysis of Z-number-valued functions [5, 7, 8, 11–16]. In [5, 8] they formulate a concept of Z-number-valued function, propose the related definitions of limit and continuity. Also, in [5] the authors initiate the study of Hukuhara difference of Z-numbers and formulate definitions on differentiability of Z-number-valued function on this basis. The conditions of existence of Hukuhara difference of Z-numbers are analyzed in [7]. In [11] study of differentiability and integrability of Z-number-valued function based on Fuzzy Sumudu Transform is proposed. The results are applied to analysis of differential equations with Z-number-valued information.

Existing studies of Z-number-valued function are in an initial stage. Development of basics of differential and integral calculus for Z-number-valued functions is necessary for solving problems of forecasting, planning, control [17], decision making under fuzziness and partial reliability of information. Dealing with such type of information leads to principally new problems which cannot be solved by a simple generalization of calculus of fuzzy-valued functions. Particularly, these problems include formulation of an approximate limit, construction of joint probability, analysis of informativeness of computation results etc.

In this paper, we continue our study of Z-number-valued function initiated in [5, 8]. We introduce a theorem on relation between differentiability and integrability of Z-number-valued function on the basis of definitions formulated in [5, 8]. The paper is

© Springer Nature Switzerland AG 2020
R. A. Aliev et al. (Eds.): ICSCCW 2019, AISC 1095, pp. 781–786, 2020.
https://doi.org/10.1007/978-3-030-35249-3_102

structured as follows. Section 2 includes basic definitions of study of Z-number-valued functions. In Sect. 3 we formulate and prove the mentioned theorem. Section 4 is conclusion.

2 Preliminaries

Definition 1. A Z-number [1, 3–5]. A continuous Z-number is an ordered pair $Z = (A, B)$ where A is a continuous fuzzy number playing a role of a fuzzy constraint on values that a random variable X may take:

$$X \text{ is } A,$$

and B is a continuous fuzzy number with a membership function $\mu_B : [0, 1] \to [0, 1]$, playing a role of a fuzzy constraint on the probability measure of A:

$$P(A) \text{ is } B.$$

A discrete Z-number [4] is an ordered pair $Z = (A, B)$ of discrete fuzzy numbers A and B. A plays the same role as it does in a continuous Z-number. B is a discrete fuzzy number with a membership function $\mu_B : \{b_1, \ldots, b_n\} \to [0, 1], \{b_1, \ldots, b_n\} \subset [0, 1]$, playing a role of a fuzzy constraint on the probability measure of A: $P(A) = \sum_{i=1}^{n} \mu_A(x_i) p(x_i)$, $P(A) \in \text{supp}(B)$.

Let us denote by $\mathcal{Z}^n$ the space of elements which are n-vectors of discrete Z-numbers $\bar{Z} = (Z_1, Z_2, \ldots, Z_n)$. Denote $\mathcal{Z}_{[c,d]} = \left\{ (A, B) \middle| A \in \mathcal{D}_{[c,d]}, B \in \mathcal{D}_{[0,1]} \right\}$, $[c, d] \subset \mathcal{R}$, and $\mathcal{Z}_+ = \left\{ (A, B) \in \mathcal{Z} \middle| A \in \mathcal{D}_{[0,\infty)} \right\}$, $\mathcal{Z}_- = \mathcal{Z} \backslash \mathcal{Z}_+$ [8].

Let us consider a subset $\mathcal{A} \subset \mathcal{Z}$. Let $Z_1 = (A_1, B_1)$ and $Z_2 = (A_2, B_2)$ be two Z-numbers. Denote by $A_i^{\alpha_k}$ and $B_i^{\beta_k}$ k-th α-cuts of A_i and B_i respectively, $\alpha_k \in \{\alpha_1, \alpha_2, \ldots, \alpha_n\} \subset [0, 1]$, $\beta_k \in \{\beta_1, \beta_2, \ldots, \beta_n\} \subset [0, 1]$. Denote $a_{i\alpha_k}^L = \min A_i^\alpha$, $a_{i\alpha_k}^R = \max A_i^\alpha$ and $b_{i\beta_k}^L = \min B_i^{\beta_k}$, $b_{i\beta_k}^R = \max B_i^{\beta_k}$, $i = 1, 2$.

Definition 2. The Hamming distance based metrics on $\mathcal{Z}$ [8]. The Hamming distance based Z-metrics on $\mathcal{Z}$ is defined as

$$D(Z_1, Z_2) = \left(\frac{1}{n+1} \sum_{k=1}^{n} \left\{ \left| a_{1\alpha_k}^L - a_{2\alpha_k}^L \right| + \left| a_{1\alpha_k}^R - a_{2\alpha_k}^R \right| \right\} + \frac{1}{m+1} \sum_{k=1}^{m} \left\{ \left| b_{1\beta_k}^L - b_{2\beta_k}^L \right| + \left| b_{1\beta_k}^R - b_{2\beta_k}^R \right| \right\} \right)$$

Let us denote $\mathbf{Z} = \{Z_i | Z_i \in \mathcal{A}, i = 1, 2, \ldots, n, \ldots\}$ a sequence of Z-numbers.

Definition 3. An r-limit [5, 8]. An element $Z \in \mathcal{A}$ is called an r-limit of $\mathbf{Z}$ (denoted $Z = r - \lim \mathbf{Z}$) if for any $t \in R^+ \backslash \{0\}$ the inequality $D(Z, Z_i) \leq r + t$ is valid for almost all Z_i.

For the case of a sequence of continuous Z-numbers, one would have a definition of a limit where $r = 0$.

Definition 4. A discrete Z-number-valued function of discrete Z-numbers [5, 8]. A discrete Z-number-valued function of discrete Z-numbers is a mapping $f : \mathcal{A} \to \mathcal{A}$.

Analogously, in a continuous setting, the Z-function is defined as a mapping between spaces of continuous Z-numbers.

Definition 5. An r-limit of discrete Z-number-valued function of discrete Z-numbers [5, 8]. An element $Z_d \in f(Z)$ is called an r-limit of f at a point $Z_{a,i} \in Z$ and denoted $Z_d = r - \lim_{Z_x \to Z_a} f(Z_x)$ if for any sequence Z satisfying the condition $Z_a = \lim Z$, the equality $Z_d = r - \lim f(Z)$ is valid.

3   A Derivative of a Z-Number-Valued Function

In [5, 7] the authors formulated a concept of Hukuhara difference of Z-numbers $Z_1 -_h Z_2 = Z_{12}$. Let us shortly describe this concept.

Definition 6. Hukuhara difference of discrete Z-numbers [7]. For discrete Z-numbers $Z_1 = (A_1, B_1)$ and $Z_2 = (A_2, B_2)$ their Hukuhara difference denoted $Z_{12} = Z_1 -_h Z_2$ is the discrete Z-number $Z_{12} = (A_{12}, B_{12})$ such that $Z_1 = Z_2 + Z_{12}$.

The conditions on existence of Hukuhara difference $Z_{12} = Z_1 -_h Z_2$ include conditions on existence of Hukuhara difference of fuzzy numbers $A_{12} = A_1 -_h A_2$ and conditions on fuzzy sets of distributions underlying B_1 and B_2. Both these conditions inherit results of Minkowski set-valued arithmetic. These conditions are described in details in [7].

On the basis of the concept of Hukuhara difference of Z-numbers, the following definitions of derivative of Z-number-valued function were proposed.

Definition 7. Discrete derivative of a discrete Z-number-valued function at a crisp point [5]. $f : \Omega \to \mathcal{Z}$ may have the following discrete derivatives $\Delta f(n)$ at n:

1. the forward right-hand discrete derivative $\Delta_r f(n) = f(n+1) -_h f(n)$
2. the backward right-hand discrete derivative $\Delta_{r^-} f(n) = \frac{f(n) -_h f(n+1)}{-1}$
3. the forward left-hand discrete derivative $\Delta_l f(n) = f(n) -_h f(n-1)$.
4. the backward left-hand discrete derivative $\Delta_{l^-} f(n) = \frac{f(n-1) -_h f(n)}{-1}$.

An existence of $\Delta_r f(n)$ or $\Delta_l f(n)$ implies that the length of support of the first component A_n of a Z-number $f(n) = (A_n, B_n)$ is non-decreasing. Existence of the other two derivatives implies non-increasing support of the first component of a Z-number $f(n) = (A_n, B_n)$.

Based on Definition 6, the concept of a generalized Hukuhara differentiability is defined in [5] as follows.

Definition 8. Generalized Hukuhara differentiability [5]. Let $f : \Omega \to \mathcal{Z}$ and $n \in \Omega$. We say that f is differentiable at $n \in \Omega$ if

$\Delta_r f(n)$ and $\Delta_l f(n)$ exist

or

$\Delta_{r-} f(n)$ and $\Delta_{l-} f(n)$ exist

or

$\Delta_r f(n)$ and $\Delta_{l-} f(n)$ exist

or

$\Delta_{r-} f(n)$ and $\Delta_l f(n)$ exist

Definition 9. The definite integral. The definite integral I of Z-number-valued function $f : [a, b] \rightarrow Z$ is defined as an r-limit:

$$F(x) = \int_a^b f(t)dt = \sum_{i=1}^n f(x_i)\Delta x,$$

$$d(F(x), I) \leq r.$$

We have derived the following result on relation between differentiability and integrability.

Theorem. Let $f : [a, b] \rightarrow Z$ be a Z-valued function. Then the following is satisfied in terms of r-limit:

(1) A Z-valued function $F(x_k) = \int_a^x f(t)dt = \sum_{i=1}^k f(x_i)$ is differentiable function in terms of definition and $F'(x_k) = f(x)$

(2) A Z-valued function $G(x) = \int_x^b f(t)dt = \sum_{i=k}^n f(x_i)$ is differentiable function in terms of definition and $G'(x) = -f(x)$

Proof. Let us prove property (1), the proof for property (2) is analogous. We can write in terms of r-limit:

$$d(F(x_k), I) = d(F(x_{k-1}) + f(x_k), I) = r$$

and

$$d(F'(x_k), D) = d(f(x_k), D) = r.$$

Thus,

$$d(F(x_{k-1}) + f(x_k), I) = d(f(x_k), D),$$

$$d(F(x_{k-1}) + f(x_k) -_h f(x_k), I) = d(f(x_k) -_h f(x_k), D) \text{ and}$$

$$d(F(x_{k-1}), I) = d(Z_0, D).$$

On the other hand,

$$d(F(x_{k-1}) + f(x_k), I) = d(f(x_k), D)$$

$$d(F(x_{k-1}) -_h F(x_{k-1}) + f(x_k), I -_h F(x_{k-1})) = d(f(x_k), D),$$

$$d(f(x_k), I -_h F(x_{k-1})) = d(f(x_k), D).$$

This completes the proof of the theorem.

Note. In a special case one has:

$$I -_h F(x_{k-1}) = D.$$

Thus, an approximation level of derivative evaluation is equal to approximate evaluation of a definite integral.

4 Conclusion

In this paper we obtained a result on relation between differentiability and integrability of Z-number-valued function. The results is based on a concept of an approximate limit of Z-number-valued function. We analyze dependence between approximate evaluations of Z-number-valued function and derivative of its integral. Formulation of derivative and integral of Z-number-valued function as approximate limits correspond to linguistic approximation of information and may allow to reduce computational complexity of handling bimodal information. In this viewpoint, the proposed result is perspective from theoretical and practical viewpoints.

As compared to fuzzy functions, we have to deal with more complex granules, and new problems in theoretical, computational and interpretational viewpoints.

References

1. Zadeh, L.A.: A note on Z-numbers. Inform. Sci. **181**, 2923–2932 (2011)
2. Aliev, R.A., Alizadeh, A.V., Huseynov, O.H., Jabbarova, K.I.: Z-number based linear programming. Int. J. Intell. Syst. **30**, 563–589 (2015)
3. Aliev, R.A., Alizadeh, A.V., Huseynov, O.H.: The arithmetic of continuous Z-numbers. Inform. Sci. **373**, 441–460 (2016)
4. Aliev, R.A., Alizadeh, A.V., Huseynov, O.H.: The arithmetic of discrete Z-numbers. Inform. Sci. **290**, 134–155 (2015)
5. Aliev, R.A., Huseynov, O.H., Aliyev, R.R., Alizadeh, A.V.: The Arithmetic of Z-Numbers: Theory and Applications. World Scientific, Singapore (2015)
6. Aliev, R.A., Huseynov, O.H.: Decision Theory with Imperfect Information. World Scientific, Singapore (2014)
7. Aliev, R.A., Perdycz, W., Huseynov, O.H.: Hukuhara difference of Z-numbers. Inform. Sci. **466**, 13–24 (2018)

8. Aliev, R.A., Perdycz, W., Huseynov, O.H.: Functions defined on a set of Z-numbers. Inform. Sci. **423**, 353–375 (2018)
9. Aliev, R.A.: Uncertain Computation Based on Decision Theory. World Scientific Publishing, Singapore (2017)
10. Yager, R.: On Z-valuations using Zadeh's Z-numbers. Int. J. Intell. Syst. **27**, 259–278 (2012)
11. Jafari, R., Razvarz, S., Gegov, A.: Solving differential equations with Z-numbers by utilizing fuzzy sumudu transform. In: Arai, K., Kapoor, S., Bhatia, R. (eds.) IntelliSys 2018. AISC, vol. 869, pp. 1125–1138. Springer, Cham (2019). https://doi.org/10.1007/978-3-030-01057-7_82
12. Lorkowski, J., Aliev, R., Kreinovich, V.: Towards decision making under interval, set-valued, fuzzy, and Z-number uncertainty: a fair price approach. In: Proceedings of the IEEE International Conference on Fuzzy Systems, FUZZ-IEEE 2014, pp. 2244–2253. IEEE (2014)
13. Aliev, R.A., Kreinovich, V.: Z-numbers and type-2 fuzzy sets: a representation result. Intell. Autom. Soft Comput. (2017). https://doi.org/10.1080/10798587.2017.1330310
14. Jiang, W., Cao, Y., Deng, X.: A novel z-network model based on Bayesian network and Z-number. IEEE Trans. Fuzzy Syst. (2019). https://doi.org/10.1109/tfuzz.2019.2918999
15. Shen, K., Wang, J.: Z-VIKOR method based on a new comprehensive weighted distance measure of Z-number and its application. IEEE Trans. Fuzzy Syst. **26**(6), 3232–3245 (2018)
16. Kang, B., Deng, Y., Hewage, K., Sadiq, R.: A method of measuring uncertainty for Z-number. IEEE Trans. Fuzzy Syst. **27**, 731–738 (2018)
17. Aliev, R.A., Pedrycz, W.: Fundamentals of a fuzzy-logic-based generalized theory of stability. IEEE Trans. Syst. Man Cybern. Part B (Cybern.) **39**(4), 971–988 (2009)

Diagnostic Operation of Gas Pipelines Based on Artificial Neuron Technologies

E. K. Iskandarov$^{(\boxtimes)}$, G. G. Ismayilov, and F. B. Ismayilova

Azerbaijan State Oil and Industry University, Azadlıq 20,
AZ1010 Baku, Azerbaijan
e.iskenderov62@mail.ru, asi_zum@mail.ru,
fidan.ismayilova.2014@mail.ru

Abstract. The article deals with issues of diagnosing the functioning of gas pipeline systems based on one of the elements of modern information technology, the principles of artificial neuron networks. Artificial neuron networks (ANN) have been already applying in some areas of science and technology. Intellectual systems based on ANN allow solving a number of problems of pattern recognition, prediction, optimization, diagnostics and control. The technique for diagnosing the operation of gas pipelines proposed in the article based on the principles of ANN. In solving this problem, the principles for a simple pipeline operating on a squared friction mode are used. As a transfer function for the cases of presence and absence of gas leakage, was used the functional dependence of pressure loss. The analyzed changes in the operational parameters of the existing gas pipeline by applying artificial neuron networks are given in the article. The possibility of prompt and accurate determination of minor changes in the regime parameters are shown according to the dynamics of change at the output of the neuron network.

Keywords: Hydrocarbon losses · Gas leakage · Gas transport · Operating parameters · Neuron technology and pressure

1 Introduction

The gas industry of Azerbaijan is one of the leading industrial sectors making up the fuel and energy balance of the Republic. At present, the gas transmission network of Azerbaijan includes about 5,000 km of main gas pipelines, including five heavy-duty transit gas pipelines. Analysis of the pipeline system shows that the reliability and efficiency of the operation of gas pipelines depends on improving the system of their operation. The task of ensuring a high level of efficiency in the functioning of these systems is a complex problem. Its successful solution is laid at the stages of design and construction, carried out, as a rule, at the stage of operation. One of the main problems is the reduction of hydrocarbon losses throughout the natural gas transportation system.

The main causes of hydrocarbon losses during gas transportation are:

- Violation of the technological mode of gas transportation by the supplier;
- Poor control over the regime parameters (pressure, temperature and compositions) in the important nodes of the scheme;

R. A. Aliev et al. (Eds.): ICSCCW 2019, AISC 1095, pp. 787–791, 2020.
https://doi.org/10.1007/978-3-030-35249-3_103

- Violation of the technological mode of transportation due to poor control over gas distribution;
- Non-compliance with the requirements for ensuring the quality of the transported gas (dew points for water and for hydrocarbons at high humidity in the composition);
- Untimely removal of liquid products and mechanical impurities along the entire route;
- All possible gas leaks, accidents, etc.

To reduce the loss of hydrocarbons, it is necessary to comply with the technological regulations for gas transportation, compiled in accordance with international standards and norms.

2   Solution of the Problem

The possibility of diagnosing emergency gas leaks, based on the analysis of the mode of gas transportation through the Khachmaz-Mingachevir pipeline network by applying an artificial neuron network is shown below.

As the artificial neuron network operated incorporates new observation indicators of pipelines, these nets re-select models and determine the dependence of weight coefficients on the number of observations for selected models, including any changes in the system, and reflects various gas leakages. In the solution of the problem, the following equation for simple gas pipeline was used [1–5]:

$$P_b^2 - P_e^2 = A_2 \cdot Q^2 \tag{1}$$

Here, P_b and P_e - respectively, pressure at the beginning and end of the gas pipeline; $A_2 = \frac{zT\lambda L\Delta}{D^5}$; z-the gas compression coefficient and Δ-and relative density (according to the air); T - the average temperature on the gas pipeline; D and L - respectively, the diameter and length of the gas pipeline.

There should be a plurality of observation indicators (the initial data), in order to adapt artificial neuron networks to the processes in the pipelines, (for example, the pressure at the beginning and end of the pipeline and the daily flow rates of oil and gas over some months). Of course, the accuracy of the network's information is rising as the number of measured prices is high.

Natural gas entering the territory of Azerbaijan from Russia is distributed, also by appointment, at many points along the route from the Siyazan receiving point (base point) to the Mingachevir gas distribution station (final point). This circumstance gives the nature of complexity to the regime and that is why, the control over the state and regime parameters of the gas pipeline is complicated. Nevertheless, with a good organization of the work of the maintenance and monitoring of the pipeline, especially in the winter season, it is possible to prevent complications due to daily irregularity of consumption. This requires, in addition to maintenance and monitoring of the pipeline regime, regularly examine quality indicators, such as gas composition, water and hydrocarbon dew point, dew point pressure and loss of condensable hydrocarbons and process this information for developing measures to improve quality and reduce losses during gas transportation. However, it is difficult to cope with this task without experimental studies of the phase state of the transported gas, appropriate calculations on a computer and special programs. The processing of current practical information of observation with the same old methods has no basis and can lead to various kinds of complications and increase the loss of valuable products.

Thereby, we have used current daily measurements of pressure and temperature at the respective points of the gas distribution network (Table 1). According to the given data, the pressure dynamics at the beginning, end, and other points of gas transportation were determined (see Fig. 1). The pressure at the initial observation point varies widely. Therefore, in October 2002, the gas pipeline was also stopped (25.10.2002).

Table 1. Results of observation of the mode of gas transportation (October, 2002)

Data of measurement	Base point		43-km of line 1–2		Garasu 1 line		Garayazi 1–2 line		Mingachevir 1 line		Garamusa 1 line	
	P, bar	T, 0C	P, bar	T, 0C	P, bar	T, 0C	P, bar	T, 0C	P, bar	T, 0C	P, bar	T, 0C
01.10.02	42.07	21.20	15.08	16.00	12.39	22.80	15.40	21.80	9.19	23.00	7.71	21.60
02.10.02	41.70	21.10	16.61	15.80	12.20	22.10	14.94	21.60	10.94	22.90	8.02	20.70
03.10.02	40.90	19.20	16.94	15.60	11.93	21.60	15.24	21.50	11.79	22.60	7.82	19.40
04.10.02	40.80	18.80	16.88	15.50	12.00	21.90	15.16	21.30	12.03	22.40	8.18	19.20
05.10.02	34.30	18.30	16.24	15.30	12.86	22.30	14.57	21.30	12.11	22.30	7.61	18.70
06.10.02	24.40	19.00	14.89	15.20	12.45	22.60	13.90	21.10	11.51	22.40	7.22	18.80
07.10.02	24.70	19.00	14.25	15.20	11.94	22.70	12.80	21.10	10.53	22.20	7.15	20.30
08.10.02	30.40	20.20	15.26	15.10	11.77	22.30	17.10	21.10	10.02	22.30	7.84	20.30
09.10.02	36.10	20.30	16.60	15.00	14.52	23.00	14.82	21.10	10.99	22.40	8.54	20.30
10.10.02	34.60	19.70	17.37	14.90	15.53	23.00	16.10	21.10	11.92	22.40	8.83	20.20
11.10.02	36.00	19.90	18.11	14.80	15.57	22.60	16.65	21.00	12.17	22.30	9.62	19.50
12.10.02	36.10	20.10	18.32	14.70	17.07	23.00	16.96	20.90	14.50	22.10	9.34	19.40
11.10.02	31.00	18.30	17.67	14.50	16.62	23.00	16.17	20.70	14.68	22.40	9.43	19.60
14.10.02	28.30	17.10	16.76	14.40	15.55	21.00	14.89	20.60	16.80	22.00	8.19	19.60
15.10.02	28.30	17.50	16.78	14.30	15.41	22.80	14.85	20.50	11.87	21.80	8.04	18.90
16.10.02	34.40	18.80	17.77	14.30	12.53	21.00	15.98	20.30	11.84	21.80	9.19	19.20
17.10.02	41.10	18.80	17.72	14.10	14.04	21.60	15.98	20.40	12.88	21.90	9.24	19.20
18.10.02	41.50	19.00	17.62	14.00	14.72	21.90	15.85	20.30	11.20	22.00	9.70	19.50
19.10.02	42.70	19.20	17.89	14.00	15.84	22.30	16.21	20.30	12.94	21.90	9.58	19.30
20.10.02	41.00	18.70	17.80	19.00	16.54	22.60	16.18	20.30	14.10	21.90	9.69	19.30
21.10.02	41.60	16.70	18.28	18.00	17.03	22.40	16.90	20.30	12.40	21.80	10.26	17.60
22.10.02	39.50	16.70	17.74	16.00	15.90	21.80	16.75	20.10	14.56	21.90	9.20	16.90
23.10.02	38.60	17.70	14.53	13.00	11.62	17.90			14.68	21.50	7.79	17.80
24.10.02	28.00	24.20	7.89	11.00	8.60	18.20			11.71	21.30	6.41	18.40
25.10.02	0.00	0.00	6.54	12.00	7.18	19.50			7.37	21.00	5.80	19.40
26.10.02	36.60	18.40	6.29	19.00	6.85	19.20			6.16	21.10	5.03	19.20
27.10.02	40.40	17.60	19.90	14.20	14.18	19.50			5.33	21.10	8.14	19.20
28.10.02	38.70	16.40	18.06	18.00	18.41	15.60						
29.10.02	37.90	16.70	17.71	16.00								
30.10.02	35.20	15.60										
31.10.02	34.20	15.70										

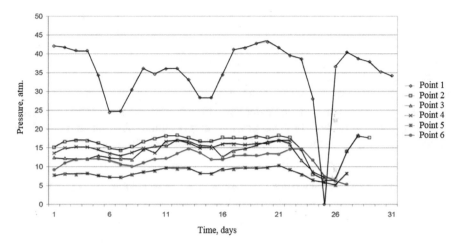

Fig. 1. Dynamics of pressure on points of the pipeline

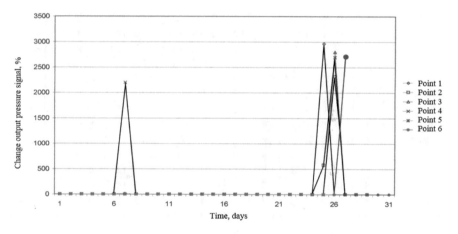

Fig. 2. The dynamics of the signal at output of the artificial neuron network

This indicates that the gas supply through this pipeline is irregular, i.e. irregular gas losses occur even during the month, and not the season. At the final point, it is relatively calm, apparently due to the forced regulation of consumption along the route of gas in Mingachevir. In other points, in contrast to the source point, there are also minor changes in the dynamics of pressure.

3 Conclusion

According to the dynamics of the pressure signal at the output of the artificial neuron network at all points after a pressure change occurred in the system (on the sixth day and on the 24th day), the analogue of the pressure signal increases 2200 and 3000 times, respectively (see Fig. 2). Abovementioned approves that, by an artificial neuron network, minor changes in the dynamics of the mode parameter, in particular the pressure at the beginning of the pipeline, can be detected quickly and with high accuracy.

References

1. Barsky, A.B.: Neuron networks and artificial intelligence. Inf. Technol. **1**, 17–25 (2003)
2. Ibishov, B.H., Ismailov, B.Q.: Diagnostics of oil leaks from pipelines to the environment. In: Materials of the International Conference "Azerbaijan - After Independence", Baku, pp. 134–135, 3–4 March 2003
3. Ibishov, B.G., Ismayilov, B.G.: Diagnosis of leakage of oil from offshore underwater pipelines based on artificial neuron networks. In: Reports of the Scientific and Practical Conference "Caspian Oil & Gas 2004", Baku, pp. 334–335 (2004)
4. Korotaeva, Y.P., Margulova, R.D.: Production, Preparation and Transportation of Natural Gas and Condensate, vol. II. Nedra, Moscow (1984)
5. Aliyev, R.A., Aliyev, R.R.: The theory of intellectual systems and its application Chashyogly, Baku (2001)

Modelling the Association Between Personality and Colour Preferences by Using Fuzzy Logic

Konul Memmedova[(⊠)] [iD]

Department of Psychological Counseling and Guidance, Near East University,
Lefkosa 98010, North Cyprus
konul.memmedova@neu.edu.tr

Abstract. Analysis of the relationship between personality traits and colour preferences is in the focus of environmental and social psychologist. However, in all published papers researchers made the assumption that the measurement of the personality factors is made precise. In this paper, it is proposed that the application of fuzzy logic approach for association of the personality traits with color preferences can be an emphatic alternative. The measurement instruments for this study are the Big-Five Factor Test to measure personality characters and the colour pallet to measure the colour preferences of individuals. Application of fuzzy logic approach allows to handle imprecision and uncertainty of measurement.

Keywords: Personality traits · Color preferences · Fuzzy logic · Big-Five Factor Test

1 Introduction

Modeling of the association between personality and colour preferences has attracted many researchers in the environmental social psychologies and physiology, [1–3]. Practical values of this study are also important for marketing (including advertising, branding and consumer shopping behavior), robotics, interior design, computer vision, etc. Warner in [1] gives the historical review of the published papers in personality and colour.

In [4] linkage is made between individuals' psychological characters and their preferred clotting colour. As shown in this study, "colour preferences are important, it affects individuals' presentations to others, and how other perceives individuals".

In all above mentioned papers a relationship between personality and color were derived by using the statistical approach. The statistical approach is based on the assumption that the collected data is precise and information obtained by measurement can be accurately quantified. In [5, 6] fuzzy logic approach was proposed to be providing useful tool for many applications. In [7] were considered application of fuzzy logic in psychological and educational research. Zhuxin et al. [8] provides a fuzzy logic model incorporating personality with heterogeneous pedestrians. In [9–12] is given impact of color on consumer shopping behavior, marketing and branding.

In [13], the application of neural network and fuzzy logic for colour recognition is considered. Correlation between Personality Types and Color Shade Preference is

© Springer Nature Switzerland AG 2020
R. A. Aliev et al. (Eds.): ICSCCW 2019, AISC 1095, pp. 792–799, 2020.
https://doi.org/10.1007/978-3-030-35249-3_104

presented in [14]. Paper [15] presents the primary color (RGB color) classification based on fuzzy logic approach, using data obtained from psychophysical experiment.

Application of fuzzy logic for colour verbalization presented in [16].

As explained in aforementioned review, this study is the first study concerning the association of personality traits with color preferences by using fuzzy logic.

2 Methodology

2.1 Measuring Personality Factors

The Big Five-50 Personality Questionnaire is used to measure the personality factors. This test consists of 50 questions grouped into five personality factors: Extraversion, Agreeableness, Conscientiousness, Emotional stability and Intellect or Imagination. The Turkish version of this Questionnaire and is reliability and validity are given in [17, 18].

Table 1. The big five-50 personality questionnaire

N^0	Item scale	VL	L	N	H	VH	
Extraversion							
1.	Don't talk a lot						(−)
2.	Keep in the background						(−)
3.	Have little to say						(−)
4.	Don't like to draw attention to myself						(−)
5.	I am quiet around strangers						(−)
6.	I am the life of the party						(+)
7.	Feel comfortable around people						(+)
8.	Start conversations						(+)
9.	Talk to a lot of different people at parties						(+)
10.	Don't mind being the center of attention						(+)
Agreeableness							
11.	Feel little concern for others						(−)
12.	Insult people						(−)
13.	I am not interested in other people's problems						(−)
14.	I am not really interested in others						(−)
15.	I am interested in people						(+)
16.	Sympathize with others' feelings						(+)
17.	Have a soft heart						(+)
18.	Take time out for others						(+)
19.	Feel others' emotions						(+)
20.	Make people feel at ease						(+)

(continued)

Table 1. (*continued*)

N⁰	Item scale	VL	L	N	H	VH	
Conscientiousness							
21.	Leave my belongings around						(−)
22.	Make a mess of things						(−)
23.	Often forget to put things back in their proper place						(−)
24.	Shirk my duties						(−)
25.	I am always prepared						(+)
26.	Pay attention to details						(+)
27.	Get chores done right away						(+)
28.	Like order						(+)
29.	Follow a schedule						(+)
30.	I am exacting in my work						(+)
Emotional stability							
31.	Get stressed out easily						(−)
32.	Worry about things						(−)
33.	I am easily disturbed						(−)
34.	Get upset easily						(−)
35.	Change my mood a lot						(−)
36.	Have frequent mood swings						(−)
37.	Get irritated easily						(−)
38.	Often feel blue						(−)
39.	I am relaxed most of the time						(+)
40.	Seldom feel blue						(+)
Intellect or Imagination							
41.	Have difficulty understanding abstract ideas						(−)
42.	I am not interested in abstract ideas						(−)
43.	Do not have a good imagination						(−)
44.	Have a rich vocabulary						(+)
45.	Have a vivid imagination						(+)
46.	Have excellent ideas						(+)
47.	I am quick to understand things						(+)
48.	Use difficult words						(+)
49.	Spend time reflecting on things						(+)
50.	I am full of ideas						

Where: VL-Very low, L-Low, N-Neutral, H-High, or VH-Very high.
The numbers parentheses after each item indicate direction of scoring (+ or −).

2.2 Colour and Colour Preferences Test

According to the color theory there are different models for representing the colors:

(a) RGB (Red, Green, Blue) model is additive model, meaning that any desired color can be obtained by the addition of three above mentioned colours (Fig. 1).

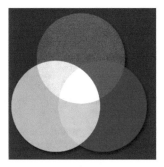

Fig. 1. RGB color model. Subtractive color model

(b) CMYB consists the following primary 4 colors: Cyan, Magneta, Yellow, Black (Fig. 2)

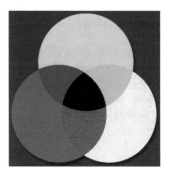

Fig. 2. CMYB model.

(c) HSL (hue, saturation, lightness) and HSV (hue, saturation, value) are alternative representations of the RGB colour model. According to the HSV theory there are 6 main visible colours: Red, Orange, Yellow, Green, Blue Purple (Fig. 3),

Fig. 3. HSL color model

The colours are determinant of human behaviour and emotion [2, 3]. The meaning of different colours was systemized and represented in the Table 2

Table 2. Meaning of different colours

Color	Determinant
Red	Happiness, agitation, stimulating, exiting, hostility, aggression, active, strong, masterful. *Red lover are sexually charged, extroverts, outgoing, outspoken, action driven*
Orange	Stimulating, pleasant, exiting, emotion, warmth, delight, distressed, upset, disturbed, happy. *Orange lover are curious, good natured, restless, socialize, energy driven*
Yellow	Stimulating, un pleasant, exiting, envy, hostility, aggression, cheerful, jovial, joyful *Yellow lover are in need of mathematical and logical order in their life, association with scientist and analytical thinker*
Green	Leisurely, controlled emotionality, youthful. *Green lover are need belonging, security and safety, community oriented, honest*
Blue	Dignity, sadness, cool, most pleasant, leisurely, drives toward control, social, secure Security drive, comfortable, tender, soothing. *Blue lover* are *compassionate, caring and highly introspective, do not like to change their point of view*
Purple	Depressing, vigorous, disagreeable, sad, deep but optimistic depression, dignified, stately, unhappy. *Purple lover put a big accent on emotion and perfection, are creative and caring*
Black	Sad, intense anxiety fears, depression, disagreeable, vague, evasion, fear, depression, powerful, strong, masterful. *Black lover intelligent, pro-active, impressive, dignified* *Do not like to show off and to be ostentatious*
White	Pure, spirited, solemn, empty, tender, soothing, *White lover have a need simplicity and independence, tend to meticulous and careful, self-sufficient*
Brown	Sad, disagreeable, secure, comfortable, full. *Brown lover want to live safety and comfort and they are mature, intelligent*

Color Pallet Preferences Test

The participants fill the 2nd row of the color pallet with the numbers 1, 2, 11 (1-the most preferred ,…, 11-least preferred)according to their color preferences, after completing the personality questionnaire (Table 1). The color palette consist of 11 basic colours: Violet (V), Blue (B), Ceyan (C), Green (G), Yellow(Y), Orange (O), Red (R), Brawn (Br), Black (Bl), Green (G) and White (W) (Fig. 4).

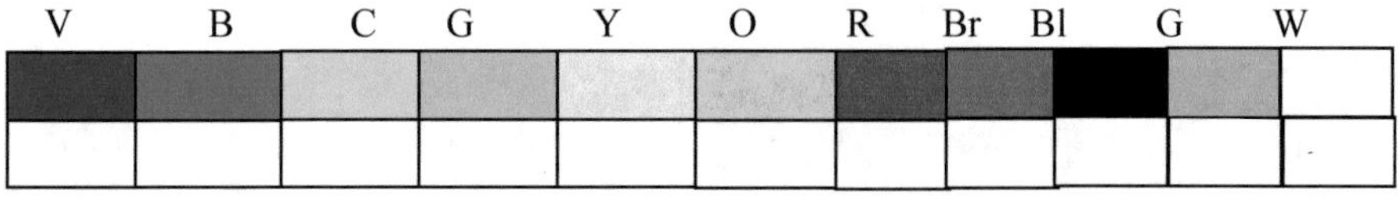

Fig. 4. Color pallet

3   Design of Fuzzy Logic Model

In Fig. 5 is given fuzzy-logic model to establish the relationship between the personality factors and individuals colour preferences.

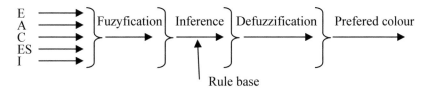

Fig. 5. Proposed fuzzy logic model

The personality factors: Extraversion (E), Agreeableness (A), Conscientiousness (C), Emotional Stability (ES) and Intellect/Imagination (I) of the individuals obtained from Table 1 is transformed into fuzzy membership functions by fuzzification. In Fig. 6 is given example of representation of ES by triangle membership functions.

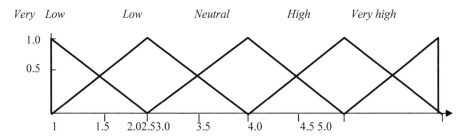

Fig. 6. Representation of Emotional stability by fuzzy membership function

Fuzzification of the others input factors (Extraversion, Agreeableness, Conscientiousness, Intellect/Imagination) can be performed by the same way.

In the study Takagi-Sugeno [18] inference is used to process the inputs and for mapping 5-inputs to the output which is represented by singleton spike. By weighted averaging crisp output (Preferred color) is generated as follows:

Although in the present research we are using a singleton variable (colour) as output, in general for the colour input Fuzzification (representation of color by membership function) can be realized regarding to the frequency spectrum (or wavelength) of colour (Fig. 7).

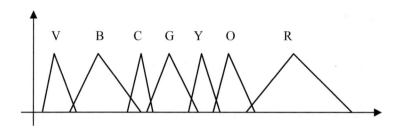

Fig. 7. Representation of colours by membership function: frequency band of color

Inference system selects the corresponding rule from the rule base regarding to the membership values. Rule base is a list of IF-THEN rules establishing input (personality factors) and output (colour) verbal relationship.

Each rule consists of two parts: antecedent and consequence parts.

The fragment of IF-THEN rules are given in Table 3:

Table 3. The fragment of IF-Then rules

N⁰	IF – THEN rules	
	Antecedent	Consequence
1	IF (E is Neutral) and (A is High) and (C is Low) and (ES is Very high) and I is High)	Then output is Blue
2	IF (E is High) and (A is Neutral) and (C is Very high) and (ES is High) and I is High)	Then output is Red
3	IF (E is Low) and (A is Neutral) and (C is High) and (ES is high) and I is Neutral)	The output is Green

4   Conclusion

In this paper a new model based on fuzzy logic theory that incorporates personality characters of individuals with their colour preferences is considered. The proposed model can handle imprecision and uncertainty of input data collected by Big-five-50 questionnaire. The fuzzy mapping of personality traits on colour preferences is made through the Inference system.

Simulation of the model using the variety of rules listed in Rule base demonstrated adequacy of the proposed approach.

In the future, the author intends to use the Z-number theory for prediction of personality of individuals regarding their colour preferences.

References

1. Warner, S.: On the relation of color and personality. J. Proj. Tech. Pers. Assess. **30**(6), 512–524 (1966)
2. Patricia Valdez, A., Mehrabian, A.: Effects of colour on emotions. J. Exp. Psychol. **123**(4), 394–409 (1994)
3. Lange, R., Testing, S., Dallas, Rentfrow, J.: Strong's Interest Inventory and Cattell's 16PF. Energia White Paper, DCS – 23. http://www.deweycolorsystem.com
4. Club, A.: The Relation Between Colors and Personality. Accessed 27 Mar 2015. https://attireclub.org/2015/03/27/colors-and-personality/
5. Zadeh, L.A.: Fuzzy sets. Inf. Comput. **8**, 338–353 (1965)
6. Aliev, R.: Soft Computing and Its Application, World Scientific, River Edge, NJ, USA (2001)
7. Kushwaha, G.S., Kumar, S.: Role of the fuzzy system in psychological research. Eur. J. Psychol. **5**(2), 123–134 (2009)
8. Zhuxin, X.Z., Xiangtao, F., Qingwen, J., Hongdeng, J., Jian, L.: Fuzzy logic based model that incorporates personality traits for heterogeneous pedestrians. Symmetry **9**(10), 239 (2017)
9. Seyed Fathollah, A.A., Honari, R.: Investigating the psychological impact of colors on process of consumer shopping behaviour. Rev. Manag. Bus. Res. **3**(2), 1244 (2014)
10. Ciotti, G.: The psychology of colours in marketing and branding. Accessed 5 April 2019. https://www.helpscout.com/blog/psychology-of-color/
11. D'Andrade, R., Egan, M.: The color of emotion. Am. Ethnologist **1**(1), 49–63 (1974)
12. Bakri, B., Noranis, M., Norwati, M., Noridayu, M.: Colour QR code recognition utilizing neural network and fuzzy logic techniques. J. Theor. Appl. Inf. Technol. **95**(15), 3703–3712 (2017)
13. Divya, G., Ravina, M.: Correlation between personality types and colour shade preference. Int. J. Indian Psychol. **01**(04), 70–79 (2014)
14. Mohd Alif, S., Bin, A., Nazrul, B.M., Yusman, Y., Mohd Fadzil, A.H.: Study RGB color classification using fuzzy logic. In: International Conference, Engineering Technology Empowerment via R&D, ETERD 2010, vol. 1, Universities Kuala, Malaysia France Institute, Lumpur (2010)
15. Abshire, C., Allebach, J., Gusev, D.A.: Psychophysical study of color verbalization using fuzzy logic. In: IS&T International Symposium on Electronic Imaging (2016)
16. Tatar, A.: Translation of big-five personality questionnaire(B5KT-50-Tr) into Turkish and comparing it with five factor personality inventory short form. Anat. J. Psychiatry **18**(1), 51–61 (2017)
17. Karamana, G., Dogana, T., Coban, A.: A study to adapt the big five inventory to Turkish. Proc. Soc. Behav. Sci. **2**, 2357–2359 (2010)
18. Takagi, T., Sugeno, M.: Fuzzy identification of systems and its applications to modeling and control. IEEE Trans. Syst. Man Cybern. **15**(1), 116–132 (1985)

Music Composition by Using Fuzzy Logic

Javanshir Guliyev[1] ⓘ and Konul Memmedova[2](✉) ⓘ

[1] Department of Acting, Near East University, Nicosia, North Cyprus
javanshir55@gmail.com
[2] Department of Psychological Counselling and Guidance, Near East University,
Nicosia, North Cyprus
konul.memmedova@neu.edu.tr

Abstract. Computer music composition has been recently used by the many musicians and scientists on artificial intelligence. One of main problem here is how to evaluate synthesized music by a computer. In this paper an approach is developed for estimation of the generated melodies by using fuzzy logic and human expertise. This music quality estimation is modeled using a fuzzy ranking approach.

Keywords: Fuzzy logic · Big data · Computer music composition

1 Introduction

In existing literature there are different methods for generating musical melodies with the aid of computer [1–8]. Music generation includes different aspects: melody generation, harmony, rhythm analysis etc. Music composition methods also can be divided into note-based and signal-based methods. Musicians are interested what kind of computer music is preferred by them. How to estimate the quality of the generated by computer music? Is computer compose new music as famous musician?

In this paper we consider the interaction between musician and computer to compose new melodies and to evaluate the generated melodies.

2 Preliminaries

In this section the basic notations used in fuzzy operations are introduced [2].

Definition 1. A popular fuzzy number is a triangular fuzzy number $\tilde{u} = (a, b, c)$ where b denotes a modal value. The membership function of a triangular fuzzy number is defined by:

$$
\mu_{\tilde{u}}(x) = \begin{cases} \frac{x-a}{b-a}, & a \leq x \leq b, \\ \frac{x-c}{b-c}, & b \leq x \leq c, \\ 0, & otherwise. \end{cases}
$$

Definition 2. A triangular fuzzy number (a, b, c) is said to be non-negative if $a \geq 0$.

© Springer Nature Switzerland AG 2020
R. A. Aliev et al. (Eds.): ICSCCW 2019, AISC 1095, pp. 800–804, 2020.
https://doi.org/10.1007/978-3-030-35249-3_105

Definition 3. Two triangular fuzzy number $\tilde{u} = (a, b, c)$ and $\tilde{v} = (e, f, g)$ are said to be equal if and only if $a = e$, $b = f$ and $c = g$.

The proposed fuzzy logic-based music composition system is shown in Fig. 1.

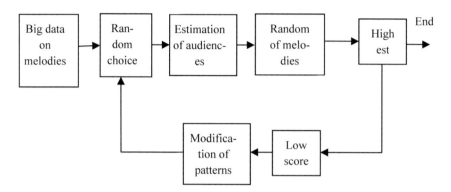

Fig. 1. Fuzzy logic-based music composition system

Big data on melodies includes different data set for example camping data set which contains hundred melodies. Example of the fragments of Azeri composer's music data set are shown in Fig. 2a, b, c.

Phase 2 performs random generation of desirable melodies from Big Data. Most pleasurable melodies are estimated in Phase 3 by audiences. Audiences give these melodies weights in [0–1] scale in accordance with codebook given in Fig. 3.

In this phase, aggregation of scores of audiences is performed using formula (1).

$$F_w = \sum_{k=1}^{n} w_k A_k \qquad (1)$$

where F_w is aggregated score of audiences, w_k is weight of A_k, A_k is fuzzy score of k-th audience.

Ranking of chosen in phase 3 melodies is performed in phase 4 by using fuzzy ranking of fuzzy numbers in accordance with formula (2).

$$\begin{aligned} &\text{if } g(A_i \geq A_j) > g(A_j \geq A_i) \text{ then } A_i \geq A_j, \\ &\text{else if } g(A_i \geq A_j) < g(A_j \geq A_i) \text{ then } A_j \geq A_i, \\ &\text{else } A_i = A_j, \end{aligned} \qquad (2)$$

where g is degree that A_i is greater than or equal to A_j.

Fig. 2. Fragments from Azeri music data-set

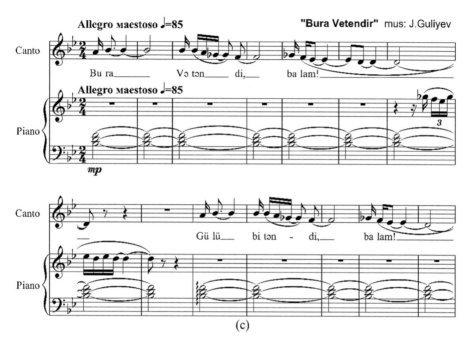

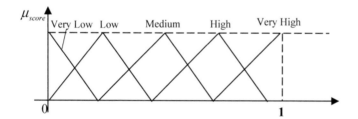

Fig. 3. Codebook on score of audiences.

If there is highest score (phase 5) generation of acceptable melody is ended. If no (phase 6) through phase 7 random choice of new patterns from Big Data is performed. In result a new acceptable melody is generated.

3 Conclusion

In this paper, fuzzy logic and Big Data-based music generation system is suggested. The proposed system is interactive human-computer system, which provides generation of pleasurable and meaningful melodies to help musicians.

References

1. Kaufmann, A., Gupta, M.M.: Introduction Fuzzy Arithmetic. Van Nostrand Reinhold, New York (1985)
2. Pinkerton, R.C.: Information theory and melody. Sci. Am. **194**(2), 77–87 (1956)
3. Brooks, F.P., Hopkins, A., Neumann, P.G., Wright, W.V.: An experiment in musical composition. IRE Trans. Electron. Comput. **3**, 175–182 (1957)
4. Pachet, F., Roy, P., Barbieri, G.: Finite-length Markov processes with constraints. In: Twenty-Second International Joint Conference on Artificial Intelligence (2011)
5. Davismoon, S., Eccles, J.: Combining musical constraints with markov transition probabilities to improve the generation of creative musical structures. In: European Conference on the Applications of Evolutionary Computation, pp. 361–370. Springer (2010)
6. McVicar, M., Fukayama, S., Goto, M.: AutoLeadGuitar: automatic generation of guitar solo phrases in the tablature space. In: 2014 12th International Conference on Signal Processing (ICSP), pp. 599–604. IEEE (2014)
7. Papadopoulos, A., Roy, P., Pachet, F.: Avoiding plagiarism in Markov sequence generation. In: Twenty-Eighth AAAI Conference on Artificial Intelligence (2014)
8. Agres, K., Herremans, D., Bigo, L., Conklin, D.: Harmonic structure predicts the enjoyment of uplifting trance music. Front. Psychol. **7**, 1999 (2017)

Estimation of Consumer Buying Behavior for Brand Choosing by Using Fuzzy IF-THEN Rules

Khatira J. Dovlatova(✉)

Joint MBA Program, Azerbaijan State Oil and Industry University,
20 Azadlig Avenue, Baku 1010, Azerbaijan
xdovlatova@gmail.com

Abstract. Consumer buying behavior is described by behavioral issues and some behavioral factors. It needs to consider consumer buying behavior in decision making that depends on consumer's attitudes, preferences and intentions. Analyzing of consumer behavior plays a main role in marketing. Current models of consumer behavior in decision making are not well appropriated to imprecise situation and uncertainty in real-life. For consumer buying behavior problems of casual modelling in marketing, the information is received by means of questionnaires. In this paper, we offer fuzzy logic based on 1- level hierarchical decision making model that contains main factors influence consumer buying behavior.

In the given model input variables such as fashion conscious, conservatism and cultural experience impact to output variables hedonism and shopping experience. Decision making process are characterized by using fuzzy aggregation methods and fuzzy reasoning on the bases of acquired fuzzy "IF-THEN" rules.

From obtained results we can say that hedonism and shopping experience is depend on how input variables influence them. Eventually, the numerical example is obtained for to show the reliability of the offered model.

Keywords: Decision making · Consumer buying behavior · Marketing modelling · Fuzzy IF-THEN rules · Fashion consciousness · Brand perception · Purchasing intention · Shopping experience

1 Introduction

The consumer behavior is an extensive and complex subject. Analyzing, understanding consumer behavior and "knowing consumers" are not ordinary [1]. It is not almost possible to forecast how consumers will act correctly in a specified situation. After analyzing consumer behavior this process allows to understand consumer's feeling, satisfaction and selection process [2]. By the way there are some cultural, psychological and individual factors influenced consumer buying behavior [3]. Marketing managers, academics and specialists have stressed the need for analyzing and explaining consumer buyer behavior based on input and output variables. Today's approach let describe relationship between input and output variables. During

© Springer Nature Switzerland AG 2020
R. A. Aliev et al. (Eds.): ICSCCW 2019, AISC 1095, pp. 805–812, 2020.
https://doi.org/10.1007/978-3-030-35249-3_106

purchasing brand clothes some input variables- fashion conscious, conservatism and cultural experience, output variables such as hedonism, shopping experience influence consumer buying behavior.

Fashion conscious give explanation of an individual's attitude toward, interests and ideas about fashion products during shopping orientations explains how to measure of variety, frequency and motivation [4].

Conservatives tend to differentiate themselves through products in order to show they are better than others by choosing products [5]. Conservatism is also connected with beliefs about what is moral and what is not [6].

Culture is the combination of essential values, needs, wants, demands and behaviors studied by a member of society from family and different significant institutions. Basically, culture is the main part of every civilization and the significant reason of personal wants and demands. Cultural influence on consumer buying behavior is different in each country marketing managers have to be very attentive in analyzing the culture of regions, different groups, social classes or even countries [7].

Hedonism can be described in brief as the pursuit of pleasure or a form of dedication to delight, particularly, sensorial delight or as a defending a behavioral style motivated by the desire to seek pleasure or avoid pain in the psychological sense. Hedonism also has been explained as a state of mind in which "pleasure is the highest beauty, pleasure-seeking is a doctrine that a life-style devoted to pleasure-seeking" [8].

It is general opinion between companies which a shopping experience starting point begins from customer's entering point to a shop or an e-commerce website and finish when this same customer leaves marketplace with its purchases. This is also what shoppers may think because they don't see behind the scenes. But a shopping experience is more common [7].

In some researches conceptual model is analyzed between X_1, X_2, Y_1, Y_2 and Z variables [9], but in this paper we research relationship between X_1, X_2, X_3 and Y_1 Y_2 variables.

2   Preliminaries

Definition 1 Fuzzy sets. A fuzzy set B in R (real line) is determined to be a set of ordered pairs $B = \{(x, \mu_B(x))/x \in R\}$, where $\mu_B(X)$ is named the membership function for the fuzzy set.

Definition 2 Fuzzy number. [10] A fuzzy number is a fuzzy set B on $\Re$ that owns the next characteristics: (a) B is normal fuzzy set; (b) B is a convex fuzzy set; (c) α -cut of B, B^α is a closed interval for every $\alpha \in (0, 1]$; (d) the support of B, sup(B) is bounded.

Definition 3 Linguistic IF-THEN rules. [10] Fuzzy sets and linguistic variables are commonly used for modelling nonlinear uncertain objects and for approximating functions. The conception of fuzzy modeling is composed of replacing the exact numerical relationship between object parameters by some qualitative relations indicated via linguistic IF-THEN rules.

The composition of Linguistic IF-THEN rules for objects and systems with multiply inputs (n) and outputs (m) (MIMO models) is indicated as below:

IF y_1 is C_{11} AND y_2 is C_{12} AND $\dots y_n$ is C_{1n} THEN

IF z_1 is D_{11} AND z_2 is D_{12} AND $\dots z_m$ is D_{1m}

AND

IF y_1 is C_{21} AND y_2 is C_{22} AND $\dots y_n$ is C_{2n} THEN

IF z_1 is D_{21} AND z_2 is D_{22} AND $\dots z_m$ is D_{2m}

AND

IF y_1 is C_{r1} AND y_2 is C_{r2} AND $\dots y_n$ is C_{rn} THEN

IF z_1 is D_{r1} AND z_2 is D_{r2} AND $\dots z_m$ is D_{rm}

In this condition $y_i(i = \overline{1,n})$, $z_j(j = \overline{1,m})$ are contain input and output linguistic variables of fuzzy system, and C_{ij}, D_{ij} are fuzzy sets. Every rule should be symbolized by IF-THEN fuzzy relation (linguistic implication).

Definition 4. [11] A distance between A1 and A2 fuzzy numbers is defined by using formula:

$$D(A_1, A_2) = \frac{1}{n+1} \sum_{k=1}^{n} \left\{ \left| a_{1\alpha k}^L - a_{2\alpha k}^L \right| + \left| a_{1\alpha k}^R - a_{2\alpha k}^R \right| \right\}$$

3 Statement of the Problem

The conceptual problem is modelling of consumer buying behavior by using fuzzy IF-THEN rules. The conceptual model of the consumer buying behavior for brand choosing in decision making process is shown in Fig. 1 [9].

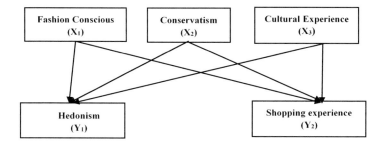

Fig. 1. Conceptual model

Fashion conscious, conservatism and cultural experience are input, hedonism and shopping experience are main output variables which affected brand clothes selection process. For calculation this conceptual model we should identify all factors and sub-factors influencing all variables. Fashion conscious (X_1), conservatism (X_2), cultural experience (X_3) influence hedonism (Y_1) and shopping experience (Y_2).

For modelling dependence between input (X_1, X_2, X_3) and output (Y_1, Y_2) variables it is necessary to arrange fuzzy logic rules in the statement of the problem. As a matter of fact we use experts' responses. If input variables are known we will find by using them output variables.

Measurement of input and output variables for the consumer buying behavior for brand choosing were acquired from questionnaire. This questionnaire consists of 27 questions and 23 experts answer these questions. Obtained information is analyzed and distance, similarity calculated by using codebook.

Codebook for fuzzy numbers is displayed in Table 1.

Table 1. Codebook for fuzzy numbers

1	High	(0.5 1 1)
2	Medium	(0 0.5 1)
3	Low	(0 0 0.5)

4 Solution of the Problem

We explain the solution algorithm of the problem which stated in the previous section. İnputs are differently described for 27 situation of consumer buying behavior. Y_1 and Y_2 are aggregated by using different experts' answers.

Step 1. Calculate fuzzy valued assessing of hedonism (Y_1) and as follows.

$$Y_1^j = \frac{\sum_{i=1}^{n} Y_{1i}^j}{n} \tag{1}$$

Here j = 1, 2, 3, ..., n n-number of experts i = 1, 2, ... n.

Step 2. Calculate fuzzy valued assessing of shopping experience (Y_2) and as follows.

$$Y_2^j = \frac{\sum_{i=1}^{n} Y_{2i}^j}{n} \tag{2}$$

Step 3. Set up fuzzy IF-THEN rule based models which define the relationship among fashion conscious (X_1), conservatism(X_2), cultural experience (X_3) influence hedonism (Y_1) and shopping experience (Y_2), as follows:

$$\textit{If } X_1^j \textit{ is } C_1^j, \; X_2^j \textit{ is } C_2^j, \; X_3^j \textit{ is } C_3^j \textit{ Then } Y_1^j \textit{ is } D_1^j \textit{ AND } Y_2^j \textit{ is } D_2^j$$

After aggregation process new fuzzy number is obtained. For measuring hedonism and shopping experience find distance, similarity between codebook and obtained fuzzy number.

5 Example

Assume that our questionnaire consist of 27 questions. According to the first situation hedonism is given as "low", "medium", "high" and 16 experts choose "low", 3 experts choose "medium" and 4 experts choose "high" measurement. Let us compute aggregated output based on answers according to (4)–(5). By this way hedonism (Y_1) and shopping experience (Y_2) are obtained. Aggregation process as follows for 27 rules:

$$Z_1^1 = \frac{low + medium + low + high\ldots\ldots\ldots + medium + high}{23}$$

1. $Y_1^1 = ((16 \times (0;\ 0;\ 0.5)) + (3 \times (0;\ 0.5;\ 1)) + (4 \times (0.5;\ 1;\ 1))/23)$
   $= (((0;\ 0;\ 8) + (0;\ 1.5;\ 3) + (2;\ 4;\ 4))/23) = (0.09;\ 0.24;\ 0.65)$
   2. $Y_1^2 = (0.02;\ 0.24;\ 0.65)$
   3. $Y_1^3 = (0.22;\ 0.54;\ 1.70)$
   .
   .
   .

27. $Y_1^{27} = (0.09;\ 0.26;\ 0.67)$

Aggregated answers are labelled by using similarity. For each rule, we compute similarity between current inputs. Consider similarity is computed by using Definition 4.

$$S(B_1, B_{1j}) = \frac{1}{1 + D(B_1, B_{1j})}, \quad j = 1, \ldots, 3.$$

Distance between first situation answers and "low" codebook

$$D((0.09; 0.24; 0.65), (0; 0; 0.5)) = (1/3 \times (|0.09 - 0| + |0.65 - 0.50| + |0.24 - 0| + |0.24 - 0|)) = 0.24$$

Similarity for the first situation is shown below:

$$S(A_1, A_{1J}) = \frac{1}{1 + 0.24} = 0.81$$

Given 27 rules are obtained from answers:

1. *If X_1^1 is low , X_2^1 is low and X_3^1 is low THEN Y_1^1 is low*   *AND* Y_2^1 is low
2. *If X_1^2 is low , X_2^2 is low and X_3^2 is medium THEN Y_1^2 is low*   *AND* Y_2^2 is low
3. *If X_1^3 is low , X_2^3 is low and X_3^3 is high THEN Y_1^3 is medium*   *AND* Y_2^3 is low

. . . .
. . . .
. . . .
. . . .

25. *If X_1^{25} is high , X_2^{25} is low and X_3^{25} is low THEN Y_1^{25} is high*   *AND* Y_2^{25} is high
26. *If X_1^{26} is high , X_2^{26} is low and X_3^{26} is low THEN Y_1^{26} is medium* *AND* Y_2^{26} is medium
27. *If X_1^{27} is high , X_2^{27} is low and X_3^{27} is low THEN Y_1^{27} is low*   *AND* Y_2^{27} is low

For new situation 2 given current inputs are evaluated and results shown below (Table 2):

Table 2. Evaluated current inputs and results

Current input 1	Low (0.1; 0.3; 0.5)	Medium (0.2; 0.6; 1)	High (0.7; 1; 1)
Current input 2	Low (0;0.5; 0.8)	Medium (0; 0.6; 1)	High (0.7; 1; 1)

Then we calculated the $S(B_1, B_{1J})$ of current input 1 and 2 with the inputs of the rules for hedonism:

$$1.\ S_1((0.48\,0.61\,0.81)),\ \text{High}) = 0.48$$
$$2.\ S_1((0.48\,0.61\,0.75)),\ \text{High}) = 0.48$$
$$3.\ S_1((0.48\,0.61\,0.75)),\ \text{High}) = 0.48$$
. . . . .
. . . . .
. . . . .
$$27.\ S_1((0.94\,0.61\,0.81)),\ \text{Medium}) = 0.61.$$

As follows we calculated minimum degrees of the rules for:
$S_1 = \min (0.48,\ 0.48,\ 0.48,\ \dots\ 0.61) = 0.48$ $S_2 = \min(0.48,\ 0.61,\ 0.64,\ \dots,$ $0.73) = 0.48$

After finding of the firing degrees we calculated the coefficients of linear interpolation based on similarity:

For current input 1:

$$w_1 = \frac{0.48}{15.59} = 0.03,\ldots, w_{25} = \frac{0.75}{15.59} = 0.05,\ldots, w_{27} = \frac{0.61}{15.59} = 0.04$$

For current input 2:

$$w_1 = \frac{0.48}{16.19} = 0.03,\ldots, w_5 = \frac{0.64}{16.19} = 0.04,\ldots, w_{27} = \frac{0.48}{16.19} = 0.03$$

The resulting output is calculated as follows:

$$Y_1 = 0.03 * low + 0.03 * low + 0.03 * medium + \ldots + 0.03 * low =$$
$$0.03 * (0\ 0\ 0.5) + 0.03 * (0\ 0.5\ 1) + \ldots + 0.03 * (0\ 0\ 0.5) = (0.1\ 0.5\ 0.9)$$

$$Y_1 = 0.03 * low + 0.03 * low + 0.04 * medium + \ldots + 0.05 * medium + 0.03 * low$$
$$= 0.03 * (0\ 0.5\ 0.8) + 0.03 * (0\ 0.5\ 0.8) + \ldots + 0.04 * (0\ 0.6\ 1) + 0.03 * (0\ 0.5\ 0.8)$$
$$= (0.13\ 0.53\ 0.9)$$

As stated by to the codebook this output can be named as Medium by using current input 1 and low by using current input 2 for hedonism.

Same calculation for shopping experience is obtained and codebook this output can be named as Medium.

6 Conclusion

The hierarchical decision-making method and using this system in marketing research were took into consideration in this paper. An information uses in marketing is obtained by use of a survey through that questionnaires were utilized. For uncertainty marketing data needs human-like judgment that can be accomplished by using fuzzy logic.

This paper is an important offering to quantitative marketing research. An offered analyzing process based on fuzzy logic can be used to find solution of many different problems in marketing research.

References

1. Durmaz, Y., Mucahit, C.: The impact of cultural factors on the consumer buying behaviors examined through an impirical study. Int. J. Bus. Soc. Sci. 2(5), 109–114 (2011)
2. Shiffman, L.G., Kanuk, L.L.: Consumer Behavior, 9th edn. Prentice Hall of India Private Limited, New Delhi (2008)
3. Varma, M.M., Aggarwal, R.K.: Consumer Behaviour, vol. 1, no. 5. King Books, Delhi (2014)

4. Rahma, L., Deasyana, D., Afrianty, S.: The influence of fashion consciousness and brand image on purchase decision: a survey on female consumer at house of ria Miranda. RJOAS **5** (77) (2018)

5. Ordabayeva, N.: How Liberals and Conservatives Shop Differently (2018)

6. Oyserman, D., Schwarz, N.: Conservatism as a situated identity: implications for consumer behavior. J. Consum. Psychol. (2017). https://doi.org/10.1016/j.jcps.2017.08.003

7. Retailnative Homepage. https://www.retailnative.com/defining-shopping-experience/

8. Kazakeviciute, A., Banyte, J.: The relationship of consumers perceived hedonic value and behavior. Inzinerine Ekonomika-Eng. Econ. **23**(5), 532–540 (2012)

9. Sadikoglu, G., Saner, T.: Fuzzy logic based modelling of decision buying process. In: ICAFS-2018. Springer (2018)

10. Aliev, R.A.: Uncertain Computation Based on Decision Theory. World Scientific Publishing (2017)

11. Aliev, R.A., Pedrycz, W., Huseynov, O.H., Eyupoglu, S.Z.: Approximate reasoning on a basis of Z-number-valued if–then rules. IEEE Trans. Fuzzy Syst. **25**(6), 1589–1600 (2017)

Studying of Wear Resistance of Piston Rings of Oilfield Compressors

Djahid A. Kerimov(✉) [iD]

Azerbaijan State Oil and Industry University, Azadlig 34,
Nasimi, Baku, Azerbaijan
haciyevanaila64@gmail.com

Abstract. The wear resistance of compressor rings made of AFQM material with graphite composition is studied experimentally. With the help of a reusable experiment, the dynamics of wear of piston rings at each compressor stage was studied. On this basis the correlation equations of these processes are worked out. As a result of these studies, the use of AFGM material in compressors of the oil and gas industry is recommended.

Keywords: Wear resistance · Compressor · Piston rings · Plastic materials · Correlation equations

1 Introduction

For the purpose of studying of wear resistance in exploitation conditions of the sealing elements of the compressors made of plastic materials, the process of wear of AFGM-type piston rings made of a special composition of plastic materials is considered. For this purpose, process of work of the piston rings made by a special technology, of the compressors of model 2SQ-50, 2SQ-6-50 and 20QD working in the oil and gas industry considered. After machining for each ring the permissible elasticity is defined and according to their size the value of permissible gap in a cylinder of the compressor is regulated.

The elasticity of the material of the piston ring is checked according to the method proposed by S.L. Anisimov's [1].

The elasticity of the material of the piston ring is determined by the following formula:

$$q_{\mathrm{w}} = \frac{I_{\mathrm{w}} \cdot E}{7,08 \cdot D_{\mathrm{w}}\left(\frac{D_c}{t} - 1\right)}, \tag{1}$$

where $I_w = a - k$ – elasticity of a fold of the lock; a – the gap arising in the lock in a free state; k – the gap arising because of warmth; t – radial thickness of a ring; Dc – diameter of a cylinder; E – elasticity module, $(9 \div 12) \times 10^3$ kgf/cm^2.

In industrial conditions, the test of piston rings carried out in three-stage compressors type 2SQ-50. The first-stage of this compressor is double acting, and the second and third stages are single-stage differential pistons.

© Springer Nature Switzerland AG 2020
R. A. Aliev et al. (Eds.): ICSCCW 2019, AISC 1095, pp. 813–817, 2020.
https://doi.org/10.1007/978-3-030-35249-3_107

Consolidation of the piston is provided with piston rings. Thus, the number of rings of the first stage is – 4, the diameter is 370/346 mm, the height of each is 12 mm; second – 5, diameter – 225/205 mm, height – 10 mm. The third stage is equipped with 8 piston rings (diameter 190/170 mm and height – 10 mm). All rings have a cross-section of the lock at an angle of 45°.

2 Studying of the Dynamics of Wear of Piston Rings Made of AFGM Material in Three-Stage Compressors of Type 2SQ-50

For the purpose of studying of nature of wear according to the stages of piston rings the sizes of rings of one compressor were measured each 1000 h in five different points of various cross-sections, the radial and axial sizes were checked. On the basis of the obtained experimental data of characteristic of wear of piston rings made of AFGM material in various operating conditions were represented by experimental curves. Here change of characteristic of curves shows that external piston rings of the first stage wear more, than average piston rings. The wear of all rings on time decreases from 0,15 to 0,05, and the intensity remains unchanged.

To learn properties of wear in the operating conditions of piston rings on the basis of the obtained data the correlation equations of dependence were defined.

Here the main task to determine the law of wears the piston rings in the different stages of the compressor by time. For this purpose by results of an experiment the correlation equations of dependence were calculated. For determining of type and stage of the correlation equation was mainly used numbers Chebyshev's [2]:

$$f(\tau) = \frac{\sum_j M_j}{n} + \frac{\sum_j M_j \psi_1(\tau_j)}{\sum_j \psi_1^2(\tau_j)} \cdot \psi_1(\tau_j) + \frac{\sum_j M_j \psi_2(\tau_j)}{\sum_j \psi_2^2(\tau_j)} \cdot \psi_2(\tau_j)$$

$$+ \ldots + \frac{\sum_j M_j \psi_m(\tau_j)}{\sum_j \psi_m^2(\tau_j)} \cdot \psi_m(\tau_j) \tag{2}$$

or

$$f(\tau) = k_0 + k_1\psi_1(\tau) + k_2\psi_2(\tau) + \ldots + k_m\psi_m(\tau). \tag{3}$$

In this row at each of coefficients km, ψm in different values Mj according to summing up value $\psi_m(\tau_j)$, function $\psi_m(\tau_j)$ satisfies the following condition

$$\psi_{m+1}(\tau) = \psi_1(\tau) \cdot \psi_m(\tau) - \frac{\lambda^2(n^2 - \lambda^2)}{4(2\lambda - 1) \cdot (2\lambda + 1)} \cdot \psi_{m-1}(\tau) \tag{4}$$

since

$$\psi_0(\tau) = 1$$

$$\psi_1(\tau) = \tau - \frac{n+1}{2}. \tag{5}$$

On the basis of these equations we find, in a special case

$$\left.\begin{array}{l} \psi_2(\tau) = \psi_1^2(\tau) - \dfrac{n^2 - 1}{12} \\[3mm] \psi_3(\tau) = \psi_1^3(\tau) - \dfrac{3n^2 - 7}{20} \cdot \psi_1(\tau) \end{array}\right\} \tag{6}$$

Considering the coefficient $\psi_m^1(\tau_j)$ the calculation is carried out according to the following formulas:

$$\left.\begin{array}{l} 2\psi_1 = -(n+1) + 2\tau \\[2mm] 3\psi_2 = \dfrac{3(2\psi_1)^2 - (n^2 - 1)}{4} \\[3mm] \dfrac{10}{3}\psi_3 = \dfrac{5 \cdot 3\psi_2 - (n^2 - 4)}{9} \cdot 2\psi_1 \end{array}\right\} \tag{7}$$

$\sum\limits_{j} \psi_m^2(\tau_j)$ – the sum is calculated by the following formula:

$$\sum_{j} \psi_m^2(\tau_j) = \frac{(m!)^2 \cdot n \cdot (n^2 - 1) \cdot (n^2 - 4)\ldots(n^2 - m^2)}{[(2m - 1)!!]^2 2^{2m}(2m + 1)} \tag{8}$$

For special case:

$$\left.\begin{array}{l} \sum\limits_{j} \psi_1^2(\tau_j) = \dfrac{n \cdot (n^2 - 1)}{2^2 \cdot 3} \\[4mm] \sum\limits_{j} \psi_2^2(\tau_j) = \dfrac{n \cdot (n^2 - 1) \cdot (n^2 - 4)}{2^2 \cdot 3^2 \cdot 5} \\[4mm] \sum\limits_{j} \psi_3^2(\tau_j) = \dfrac{n \cdot (n^2 - 1) \cdot (n^2 - 4) \cdot (n^2 - 9)}{2^4 \cdot 5^2 \cdot 7} \end{array}\right\} . \tag{9}$$

The correlation equations of the presented dependences calculated by Chebyshev's numbers for the material AFGM are given in the table.

The carried out calculations showed what for the specified dependences the correlation equations of the third order completely define and satisfy experimentally received results.

The above by means of the dispersive analysis are affirming.
Thus

$$F = \dfrac{k_3 \left[\dfrac{\sum\limits_{j} (c_3 \psi_3)^2}{c_3^2} \right]}{\sum\limits_{q=1}^{p} \sum\limits_{n=1}^{m} \left(\tau_{qn} - \tau_q \right)^2} \leq F_{tab}$$

the condition is satisfied.

In this case, the dispersions are relatively at the level of 5% $V_1 = 1$, $V_2 = 4$ and as a result are obtain $F_{tab} = 7{,}74$.

The analysis of the correlation equations shows that wear of the piston ring in the first stage submits to the law of a cubic parabola. 5000 h after exploitation linear wear in approximately an average 0,65 mm, and the wall of the cylinder, as well as the grooves of the piston ring, one can say do not wear out at all.

During exploitation in the second stage of piston rings are relatively more wear out and before reaching the maximum value becomes equal to 0,74 mm. At the same time the rings which are close located to the second stage little wear out (0,063 ÷ 0,690 mm), than rings which are close located to the third stage (0788 ÷ 0,804 mm).

3   Discussion and Conclusions

During exploitation the wear of rings according to time changes in accordance with the law of a cubic parabola. Wear of the piston rings working in the second stage was at the same time studied. The received results show that intensity of wear of the piston rings working in the second stage, one may say, that everywhere is identical. It once again confirms expediency of use of AFGM material. The piston rings working in the third stage in comparison with the first and second stages wear out more. 5000 h after exploitation the piston rings of the third stage wear out up to 0,944 mm, in this case rings of the first and second steps wear out respectively up to 0,680 mm and 0,692 mm. The received dependences in other rings repeated: in the third stage – 0,87 mm, in the first stage – 0,630 mm and in the second stage – 0,728 mm. It follows that the rings of the third stage are more loaded concerning the first and second stages. Consequently, in the third stage the temperature and pressure is high. For this reason piston rings of the third step wear out more.

References

1. Kerimov, D.A., Kurbanova, S.K.: Bases of Designing of Plastic Details and Press-Moulds. Publishing House "Elm", Baku (1997)
2. Kerimov, D.A., Gasanova, N.A.: Development of optimal regimes of cooling for plastic parts of thermosets. In: SAEQ Science & Applied Engineering Quarterly, 17–19 January-February-March (2017)
3. Gasanova, N.A.: Behavior of plastic working in oil-field equipment. Int. J. Innov. Res. Comput. Sci. Technol. (IJIRCST) 5(5) (2017)

Enhancing Students' Collaboration and Creative Thinking Skills by Using Online Encyclopedias

M. A. Salahli[1(✉)] ⑩, T. Gasimzadeh[2], F. Alasgarova[3],
and A. Guliyev[3]

[1] Department of Computer and Instructional Technologies Education,
Çanakkale Onsekiz Mart University, Çanakkale, Turkey
msalahli@comu.edu.tr
[2] Department of Instrument Making Engineering, Azerbaijan State Oil
and Industry University, 20 Azadliq Avenue, 1010 Baku, Azerbaijan
tgasimzade@silkwaywest.com
[3] Department of Information Technology and Natural Sciences,
Azerbaijan Tourism and Management University, Koroglu Rehimov Street,
822/23, 1072 Baku, Azerbaijan
flora.aleskerova@gmail.com, akber_guliyev@yahoo.com

Abstract. Cooperative learning and creative thinking skills play an important role in the skills that 21st century students should have. This study explores how these skills can be improved by using the opportunities provided by online encyclopedias. The proposed method aims develop cooperative learning skills in the process of writing research homework; developing creative thinking by conducting research on a given topic and writing a composition; integration of technology with education and training process. Tests conducted on student groups showed that students who participated in the writing of wiki articles improved their cooperative and creative thinking skills.

Keywords: Creative learning · Collaborative learning · Online encyclopedia · Wiki

1 Introduction

The rapid development of technology in the 21st century has led to the creation of new professions. In addition to other skills, the professionals who will do these professions must possess such skills as creative thinking and cooperative learning. For this reason, one of the main objectives of contemporary education system is to provide the students with the necessary knowledge and skills to perform these professions. In the standards of International Educational Technologies Association, they have determined the characteristics that students should have in order to be able to do their present and future professions [1]. Creative thinking and collaborative skills are among the features that students should possess in these standards.

Teaching creative thinking can enable students to produce original solutions to problem solving and develop various solutions to problems.

© Springer Nature Switzerland AG 2020
R. A. Aliev et al. (Eds.): ICSCCW 2019, AISC 1095, pp. 818–826, 2020.
https://doi.org/10.1007/978-3-030-35249-3_108

In creative thinking, reaching the right sources of information and using them for purpose are important conditions. In this context, encyclopedias stand out as the correct source of information.

The aim of this research is to improve the cooperative learning and creative thinking of high schools' students by using online encyclopedias.

The answers to the following questions were sought in order to achieve the aim:

- What is the level of students' creative thinking skills?
- What is the level of students' collaborative learning skills?
- What are the possibilities of using computer technologies to develop these two skills?
- Do online encyclopedias have a positive impact on the development of cooperative learning and creative thinking skills?

To answer these questions, an approach based on using online encyclopedia application has been developed.

The research was done on example of Turkish language and literature course. This study aims to investigate the effect of online encyclopedias on the cooperative learning and creative thinking of tenth grade students in Turkish language and literature course. In generally, it was aimed to achieve three goals:

- Students learn to use the MediaWiki tools and learn to write articles for online encyclopedias;
- Students gain collaborative learning skills during the article writing process;
- Research and writing articles on specific topics increase students' creative thinking skills.

2   The State of the Problem

2.1   Research on Creative and Collaborative Learning

One of the basic concepts of the contemporary education system is the development of creativity. In order to teach 21st century skills, educational innovations should be made, and different perspectives are required. We can incorporate creative thinking into students' lives by updating education to include 21st century skills, including creativity. In this way, an improvement can be achieved in terms of raising questioning generations that can use imagination, produce different ideas.

Laius and Rannikmae [2] studied the ways in which technological literacy education influences the creativity-thinking skills of ninth grade students in high school. Contradictory event test was applied to the students and three scales were used. These; giving an event and asking questions, specifying the reasons and predicting the outcome. Responses were examined at the levels of fluency, flexibility and complexity. As a result of the research, it was determined that scientific technological literacy teaching improved the students' creative thinking skills.

Law [3] determined that the experimental group applied jigsaw technique was more successful than the control group.

Thanh [4] examined the views of students and teachers about the applicability of collaborative learning in the classroom. At the end of the study, teachers stated that they wanted to use cooperative learning techniques but did not have enough information. Some teachers mentioned the lack of classroom materials. Thanh stated that cooperative learning was not sufficiently feasible as a workload and an excess of curriculum.

In [5] the problem of developing students' collective and creative thinking skills via Problem-Solving Activities in a Ubiquitous Learning Environment is studied.

Critical and creative thinking skills are essential for students who plan to work in the 21st-century workforce. This goal of the project reported in the article [13] was to define critical and creative thinking in a way that would be useful for classroom teachers charged with developing such skills in their students.

The work [6] reports on a pilot study which has investigated the creative impact of information and communication technology (ICT) in a rural primary school in Southwest England. A model of creativity is presented with three discrete but related modes of activity—problem solving, creative cognition, and social interaction.

2.2 Use of Wikis in Education

In terms of education and training, Wiki is the most important platform that provides support for the development of cooperative learning [7]. The contributions of wikis to the cooperation environment in the fields of education and training are given below.

(1) Creating a collection of pages produced and edited by groups,
(2) Ensuring that documents, pictures and assignments are kept on any subject,
(3) To perform group search scanning between documents,
(4) To monitor and manage the changes in the documents,
(5) The formation of online discussion environments.

In Wikis, if the student changes or deletes the content of the page as a result of an error, it provides the possibility to restore the page [8]. An important feature of Wikis is that it keeps all records. Thus, all changes made on the page are stored. The use of Wikis in the classroom provides many different designs for students. Students can easily use this when they want to search for the information they want and can be redirected to another page inside or outside the page thanks to the hyperlinks used in the Wiki. While users can only comment on blogs, they can also comment on wikis and change the content.

Students can collaborate with friends through this platform and help each other. The teacher can guide the student at any time, provides management on site, and can be involved in the process. All changes made by the students on the Wiki are tracked, including how changes are made, frequency of changes, the comparison between the previous page and the next page can be provided by reviewing [9].

Wikis are important tools that bring diversity in learning environments. The development of the content of the Wiki by many authors, provides a common working environment for learners, group production, providing opportunities for discussion and interaction with the students provides important opportunities [10].

Wikis are generally used in group studies in education and training. This environment allows users to create and edit pages, delete pages, such as increasing the interest in collaboration environments. With these features, students are provided to create pages related to course contents and participate in the collaboration process [11].

With the use of Wiki in education, the teacher is no longer the person who tells the information and becomes a guide teacher in the learning of the students. In this platform, the students close each other's gaps and enable each other to learn.

3 Theoretical Framework of Research

3.1 Use of Constructivist Approach in the Research

In our study, constructivist approach was applied. Constructivist approach includes individuals who are conscious, creative, researching, questioning, knowing what they are learning and who can develop the technology for themselves. In the constructivist approach, technology is used for effective learning, purposeful learning, original learning and cooperative learning. With the use of technology, in constructivism, the student can develop better ideas in identifying problems and generating appropriate ways to solve them. The constructivist approach should allow students to work in collaborative groups, develop and present projects and benefit from technology [12]. The use of technology plays an important role in providing an enriched learning environment to the student, increasing the interest and relevance of the course, and ensuring that the information is understood in a regular and understandable manner.

The constructivist approach contributes to the creation of a cooperative and creative teaching environment in instructional design. It assigns a sense of responsibility to the student in structuring information. This creates a flexible environment. The role of the teacher changes and supports the cooperative learning and creative thinking skills of the students and takes on the role of guidance. In constructivist instructional design, the use of technology is important in problem solving, structuring information in a collaborative environment, and linking learning with students' own experiences.

The use of the constructivist approach features in our study is summarized in Table 1.

3.2 The Methodology

The study group of the study consists of high schools' students who took Turkish Language and Literature course in Çanakkale Mehmet Akif Ersoy Vocational Technical and Industrial Vocational High School (N = 64). In the experimental group, courses were taught through the online encyclopedia, while in the control group the courses were taught with the traditional teaching method. The experimental procedure lasted for three weeks.

The SPSS 22.0 program was used in the analysis of the data obtained during the study. The creative thinking and cooperative learning skills of the groups were analyzed with independent sample t-test.

Table 1. Reflection of constructivist approach in the study

Features of constructivist approach	Reflection of constructivist approach in this study
1. Encourages students to use technology, teaches learning creatively	1. Through applications on pbworks, students learn how to use technology effectively and purposefully
2. Students search for information for problem solving	2. Students do research on "poems about the Dardanelles wars"
3. Students collaborate in group activities	3. Students write articles in collaborative environments
4. Encourages students to think creatively. The students discuss their ideas and compare their ideas	4. Students produce different ideas during the article writing process. On the Pbworks app, students compare their ideas with discussion pages

Important steps of implementation include:

- Experimental and control groups were determined; Collaboration Process Scale and Creative Thinking Scale pre-test were applied for both groups;
- Teachers and students were taught how to set up, run mediawiki, and how to write articles using the online encyclopedia editor;
- Experimental group students were assigned to write articles on a topic related to the course; Students were asked to write composition about Ferhat and Shirin's love in the first week. ("Ferhat and Shirin" is a popular folk epic about love; This topic has been chosen because it is very popular and very attractive for the age level of the target audience. It is aimed to make the experimental group more willing to participate in the research process);
- In the next two weeks, students were asked to write articles on "the poems about the Dardanelles wars" and "narration of social problems in folk poetry".

Students are told that online encyclopedias have great importance in sharing information and acquiring information. It is known that wiki articles are widely used all over the world and that these articles can be written in many languages of the world. By following the wiki rules, students will be able to write wiki articles and share them with the world. As a result, the students will be motivated to develop creative thinking by using technology.

Real-time co-editing system pbworks was used as editor for writing and design of Wiki articles. (https://en.wikipedia.org/wiki/PBworks). A collaboration system or collaboration editor is a form of collaboration software that allows several people to edit a computer file using different computers. This application is called general editing (https://en.wikipedia.org/wiki/Collaborative_real-time_editor). The reason for using Pbworks is that any student can access the wiki from your computer and phone connected to the Internet without any installation.

After Pbworks is explained to the students, the application stage of the research was started. In this stage, a class account for the students was created, and the accounts were distributed to the students. The students were divided into five groups. The experimental group was asked to write a collaborative composition on Pbworks about the emotion felt by Ferhat and Şhirin love. After reading the story, the students were asked to define love according to their imagination and develop new ideas. They shared their opinions about the ideas put forward through the application.

Real-time co-editing system Pbworks was used as editor for writing and design of Wiki articles. (https://en.wikipedia.org/wiki/PBworks). A collaboration system or collaboration editor is a form of collaboration software that allows several people to edit a computer file using different computers. This application is called general editing (https://en.wikipedia.org/wiki/Collaborative_real-time_editor). The reason for using Pbworks is that any student can access the wiki from the computer and phone connected to the Internet without any installation.

4   Analyses of the Test Results

To determine creative potential of the students, pre-test and post-test were conducted for the students of experimental and control groups. Whelton and Cameron's creativity scale "How creative are you" [14] was applied on these groups. According to this scale, the following fuzzy variables were used to determine the creativity level of the students:

non-creative; weak creative; medium creative; very creative; super creative.

The test results show that there is a significant difference between post-test creativity scale scores of the students in the experimental group and the control group (t (62) = 4.671, p = .00 < .05, d = .91). It was seen that the difference between the groups is in favor of the skill points of the students in the experimental group. The experimental and control groups creativity skills were found to be X = 4.50 and 3.70, respectively. It was observed that the creativity skills score of the experimental group students increased more. When these results are considered, it can be said that online encyclopedia practice improves students' creative thinking skills.

A collaborative working environment was created for the 30 students in the experimental group through the online encyclopedia application, and after the completion of all tasks, the results of the final test on cooperative learning were analyzed. From the answers of the students, it is seen that most of the students have "a high level" of cooperative learning. The student showing low level is only 1 person in the face to face and group process question groups and there is not any student showing "low level" in other question groups. It can be said that the students have shown great success for cooperative learning after practice. From these data, it is concluded that the application improves the students' cooperative learning skills. The use of online encyclopedias has a significant impact to improve students' cooperative learning skills.

5   Conclusions

In this study, an application system that deals with the process of preparing high school students' articles on certain subjects by using wiki software is described to improve the cooperative learning and creative thinking skills of high school students. Text editor Pbworks was used in the system. The application was used in 10th grade Turkish Language and Literature course. Pre-test and post-test measurements showed that students' cooperative learning and creative thinking skills changed positively.

Students found wiki environments fun, easy and useful. They stated that it would be better if the courses continued in this way. With the Wiki application, students were directed to do it again. With the Wiki application, students are able to conduct research, access to different information and interact. Students also had a learning environment outside the classroom.

Table 2. The creativity levels of the control group students based on the scores obtained from the pre-test.

Creativity levels	Score	F	%
Non-creative	Less than 15	–	–
Weak creative	7–25	–	–
Medium creative	18–45	14	41,2
Very creative	28–70	16	47
Süper creative	Greater than 65	4	11,8

Table 3. The creativity levels of the experimental group students based on the scores obtained from the pre-test

Creativity levels	Score	F	%
Non-creative	Less than 15	–	–
Weak creative	7–25	2	6,7
Medium creative	18–45	11	36,6
Very creative	28–70	15	50,0
Süper creative	Greater than 65	2	6,7

The creativity levels of the control and experimental group students based on the scores obtained from the pre-test are shown in Tables 2 and 3, respectively.

Table 4. The creativity levels of the control group students based on the scores obtained from the post-test

Creativity levels	Score	F	%
Non-creative	Less than 15	–	–
Weak creative	7–25	–	–
Medium creative	18–45	15	44,2
Very creative	28–70	14	41,2
Süper creative	Greater than 65	5	14,7

Table 5. The creativity levels of the experimental group students based on the scores obtained from the post-test.

Creativity levels	Score	F	%
Non-creative	Less than 15	–	–
Weak creative	7–25	–	–
Medium creative	18–45	2	6,7
Very creative	28–70	11	36,6
Süper creative	Greater than 65	17	56,7

References

1. ISTE: Uluslararası Eğitim Teknolojileri (2016). https://www.iste.org/standards. Accessed 10 Mar 2019
2. Laius, A., Ve Rannıkmae, M.: The influence of STL teaching on students' creative thinking, cresils contributions of research to enhancing students' interest in learning science. Esera, Barcelona, (2005)
3. Law, Y.K.: The effect of cooperative learning on enhancing Hong Kong fifth graders achievement goals, autonomous motivation and reading proficiency. J. Res. Read. **34**(4), 402–425 (2011)
4. Thanh, P.T.H.: An investigation of perceptions of Vietnamese teachers and students towards cooperative learning (CL). Int. Educ. Stud. **4**(1), 3–12 (2011)
5. Laisema, S., Wannapiroon, P.: Design of collaborative learning with creative problem-solving process learning activities in a ubiquitous learning environment to develop creative thinking skills. Procedia - Soc. Behav. Sci. **116**, 3921–3926 (2014)
6. Wheeler, S., Waite, S.J., Bromfield, C.: Promoting creative thinking through the use of ICT. J. Comput. Assist. Learn. **18**(3), 367–378 (2002)
7. Raman, M., Ryan, T., Olfman, L.: Designing knowledge management systems for teaching and learning with wiki technology. J. Inf. Syst. Educ. **16**(3), 311–321 (2005)
8. Engstrom, M.E., Jewett, D.: Collaborative learning the wiki way. TechTrends **49**, 12–15 (2007)
9. Morgan, B., Smith, R.D.: A wiki for classroom writing. Read. Teach. **62**(1), 80–82 (2008)
10. Bruns, A., Humphreys, S.: Building collaborative capacities in learners: the M/cyclopedia project revisited. In: Proceedings of the 2007 International Symposium on Wikis, pp. 1–10, 21–25 October, Montreal, Quebec, Canada (2007)

11. Frydenberg, M.: Wikis as a tool for collaborative course management. MERLOT J. Online Learn. Teach. **4**(2), 169–181 (2008)
12. Güneş, G., Asan, A.: Oluşturmacı yaklaşıma göre tasarlanan öğrenme ortamının matematik başarısına etkisi. J. Gazi Educ. Fac. **25**(1), 105–121 (2005)
13. Baum Combs, L., Cennamo, K.S., Newbill, P.L.: Developing critical and creative thinkers: toward a conceptual model of critical and creative thinking processes. Educ. Technol. **49**, 3–13 (2009)
14. Whelton, D.A., Cameron, K.S.: Developing Management Skills, 8th edn. Prentice Hall (2011)

Application of Fuzzy AHP-TOPSIS Method for Software Package Selection

Nihad Mehdiyev[(✉)] [iD]

Azerbaijan State Oil and Industry University, 20 Azadlig Avenue,
1010 Baku, Azerbaijan
nihadmehdi@gmail.com

Abstract. Selection the true software packages is very important for the growth or failure of any manufacturing or service company. There are a lot of influencing factors in the choosing of software for firm and to solve this problem was used multi-criteria decision making problem. In this paper, for choosing best software two methods- Analytic Hierarchy Process (AHP) and Technique for Order Preference by Similarity to Ideal Solution (TOPSIS) methods is utilized. Using AHP technique in software choosing we determine weights of the selected factors. Then, TOPSIS method is utilized to determine rating of the alternatives'. Finally, a software prototype for presenting both method's is implemented.

Keywords: Multiple criteria decision making · Software selection · AHP · TOPSIS · Fuzzy numbers

1 Introduction

The needs for suitable software packages in manufacturing and service systems is continuously increasing [1]. Software packages choosing may include hundreds of needs and products; thus, a structure is needed to adequately focus on the basic factors affecting the selection process [2, 3]. Simulation is very important tool recognizing designers to assume new systems and evaluation, observation system's actions. Presently, the market suggests a multiple simulation software packages [4]. So, an organized and standardized style for simulation software selection is needed: A lot of company's select set of tools without establishing formal estimation factors or exhaustively examining tools. It is very important to determine ways to consistently and objectively estimate a tool's utilization [5]. Tewoldeberhan et al. [6] described an exhaustive discrete-event simulation software selection technique for big foreign firms.

Dorado et al. [7] used the AHP technique for choosing software in education. The results of these works represented that some generic selection methods and AHP method is largely used as a decision tool for software selection. King and Newman [8] evaluated software for business simulation based on factors that are researching methods for estimating potential software and increasing the experience. Hlupic [9] represented a software tool that select simulation software given for the basic aspects. Hlupic et al. [10] developed a detailed framework for simulation software estimation where size is above 300 factors. Fuzzy-based methods make possible decision makers

© Springer Nature Switzerland AG 2020
R. A. Aliev et al. (Eds.): ICSCCW 2019, AISC 1095, pp. 827–834, 2020.
https://doi.org/10.1007/978-3-030-35249-3_109

to use linguistic language for estimating criteria easily and intuitively. It improves software selection process by placing the uncertainty and ambiguity occurred during human decision making [11]. Using fuzzy methodology in software selection process can also decries uncertainties specific in experts' judgments. In this paper, a methodology based on fuzzy AHP and fuzzy TOPSISis employed as MCDM methods.

The fuzzy analytical hierarchy process is used to define weights of the basic factors, and fuzzy technique for order preference by similarity to ideal solution is used for ranking of alternatives. AHP helps decision makers to state a selection problem into a hierarchy. It is a very important method for handling categorical and numerical multi-criteria problems. In addition, AHP procedures are appropriate to individual and group decision making. TOPSIS is a largely accepted multi-criteria method because its sound logic, simultaneous consideration of the ideal and the non-ideal solutions.

This paper is stated as follows: Sect. 2 represents proposed techniques, and conclusion of these methodology are discussed. In Sect. 3, the results of proposed techniques are tested through the sensitivity analysis. Results of this work are described in Sect. 4.

2 Preliminaries

Definition 1 A fuzzy number. A fuzzy number is a fuzzy set A on $\mathcal{R}$ which contain the following characteristics: (a) A is a normal fuzzy set; (b) A is a convex fuzzy set; (c) α-cut of A, A^α is a closed interval for every $\alpha \in (0, 1]$; (d) the support of A, supp(A) is bounded where $\underline{\mu}_A(\alpha) \leq \overline{\mu}_A(\alpha)$, $0 \leq \alpha \leq 1$.

Definition 2 Fuzzy triangle matrix. Any TFN A, is completely represented by a triplet $A = (a_1, a_2, a_3)$. A-is nonnegative symmetric TFN, if we have:
$A = (a_1, a_2, a_3)$ with $a_1, a_2, a_3 \in R^+$ and $a_2 - a_1 = a_3 - a_2$

Definition 3 Fuzzy eigenvector. Let $F = [a_{ij}]$ be a fuzzy square matrix. If there exist a

non-zero vector $X = \begin{bmatrix} x_1 \\ x_2 \\ \vdots \\ x_n \end{bmatrix}$, where $F \cdot X = \lambda X$, and vector X called eigenvector of F,

corresponding to the fuzzy eigenvalue λ.

Definition 4 Fuzzy eigenvalue. Let $F = [a_{ij}]$ be a square fuzzy matrix. The roots of the characteristic equation $|F - \lambda I| = 0$ is called fuzzy eigenvalues of F.

Definition 5 Consistency of fuzzy matrix. Let A is pairwise comparison matrix where $a_{ij}(i, j \in N)$ is assumed to reflect how many more times item/criterion j.

$$A_{n\times n} = (a_{ij})_{n\times n} = \begin{matrix} 1 \\ 2 \\ \vdots \\ n \end{matrix} \begin{bmatrix} a_{11} & a_{11} & \cdots & a_{11} \\ a_{11} & a_{11} & \cdots & a_{11} \\ \vdots & \vdots & \vdots & \vdots \\ a_{11} & a_{11} & \cdots & a_{11} \end{bmatrix}$$

Pairwise comparison matrices are consistent if the following rules:

(a) reciprocal - $a_{ij} = a_{ji}^{-1}; \forall i, j \in N$
(b) transitivity- $a_{ik} = a_{ji}^{-1} = a_{ij} \times a_{jk}; \forall i, k \in N | j \in N - \{i, k\}$

Definition 6. AHP. The Analytical Hierarchy Process is used for arraigning and examining complicated problems by mathematical calculations. Combination of separate performance indicators with one of main performance indicator are done so as to allocate different weights to each attribute. The process of AHP is mainly used to calculate weights. The inputs for AHP are alternatives and attributes.

3   Statement of the Problem

AHP and TOPSIS techniques was applied for the choice of the software packages. Three alternatives of software- S1, S2, S3, and four basic criteria -C1 – vendor, C2 – functionality, C3 – reliability, C4 – portability have been used. Decision matrix D, which consists of alternatives $S_i, i = \overline{1, 3}$ and criteria $C_j, j = \overline{1, 4}$ is given in Table 1. Elements of matrix D are TFN describing linguistic terms rating of the alternative S_i according C_j.

Table 1. Decision matrix D.

Attribute	C_1 Vendor	C_2 Functionality	C_3 Reliability	C_4 Portability
S_1-Software	Average 0.4, 0.5, 0.6, 0.7	High 0.6, 0.7, 0.8, 0.9	Very high 0.8, 0.9, 1, 1	High 0.6, 0.7, 0.8, 0.9
S_2-Software	Average 0.4, 0.5, 0.6, 0.7	Very high 0.8, 0.9, 1, 1	Average 0.4, 0.5, 0.6, 0.7	High 0.6, 0.7, 0.8, 0.9
S_3-Software	High 0.6, 0.7, 0.8, 0.9	Average 0.4, 0.5, 0.6, 0.7	Very high 0.8, 0.9, 1, 1	Very high 0.8, 0.9, 1, 1

The problem is to find best alternative S_i which fits are criteria optimally.

4   Solution of the Problem

Two methods, AHP and TOPSIS, are combined for ranking alternative software according to factors. AHP technique is utilized to calculate weights of factors, TOPSIS is used for ranking of alternatives.

4.1   Determining of Fuzzy Weights of Criteria

Fuzzy eigenvalues $\lambda(\lambda_l, \lambda_m, \lambda_u)$ and eigenvectors $w(w_l, w_m, w_u)$ are determined by (1)–(3) [13].

$$\overline{F_l}w_l + \overline{F_m}w_m + \overline{F_u}w_u - \overline{\lambda_l}w_l - \overline{\lambda_m}w_m - \overline{\lambda_u}w_u = 0 \qquad (1)$$

$$\overline{F_l}w_l = \overline{\lambda_l}w_l, \quad \overline{F_m}w_m = \overline{\lambda_m}w_m, \quad \overline{F_u}w_u = \overline{\lambda_u}w_u \qquad (2)$$

where,

$$\overline{\lambda_l} = 2\lambda_l + \lambda_m, \quad \overline{\lambda_m} = \lambda_l + 4\lambda_m, \quad \overline{\lambda_u} = \lambda_m + 2\lambda_u \qquad (3)$$

For determination of vectors of weights of criteria we use approach given in [13]. The comparison matrix for criteria is given in Table 2.

Table 2. The comparison matrix F for criteria's

Criteria's	C_1	C_2	C_3	C_4
C_1	(1, 1, 1)	(1.5, 2, 2.5)	(2, 2.5, 3)	(1, 1.5, 2)
C_2	(1/2.5, 1/2, 1/1.5)	(1, 1, 1)	(1, 1.5, 2)	(1.5, 2, 2.5)
C_3	(1/3, 1/2.5, 1/2)	(1/2, 1/1.5, 1)	(1, 1, 1)	(1/2.5, 1/2, 1/1.5)
C_4	(1/2, 1/1.5, 1)	(1/2.5, 1/2, 1/1.5)	(1.5, 2, 2.5)	(1, 1, 1)

So fuzzy matrix given in Table 2 can be expressed by three crisp matrices: F_l, F_m, F_u [14], where F_l, F_m and F_u are crisp reciprocal matrices of left, medium and upper parts of triangular fuzzy matrix F. The AHP provides a measure of the consistency of pairwise comparison judgments by computing a consistency ratio. AHP method uses a hierarchical structure to organize and control the complexity of decision involving many factors, and it uses expert opinion to measure the relative value or of these. Each alternative is described within fuzzy decision matrix (x_{ij}), where x_{ij} is a fuzzy estimation of i-th option by j-th criteria, $i = 1, 2, \ldots, m, j = 1, 2, \ldots, n$. (x_{ij}) is converted into (r_{ij}), where r_{ij} is defined as next:

$$r_{ij} = \frac{x_{ij}}{\sqrt{\sum_{i=1}^{M} x_{ij}^2}} \qquad (4)$$

For the fuzzy weighted normalized decision, a^* and a^- are defined:

$$a^* = \left\{ (\max_i v_{ij} | j \in J), (\min_i v_{ij} | j \in J | i = 1, 2, 3, \ldots, M) \right\} = \{v_{1^*}, v_{2^*}, \ldots v_{N^*}\} \qquad (5)$$

$$a^- = \left\{ (\min_i v_{ij} | j \in J), (\max_i v_{ij} | j \in J | i = 1, 2, 3, \ldots, M) \right\} = \{v_{1^-}, v_{2^-}, \ldots v_{N^-}\} \qquad (6)$$

For each option, fuzzy ideal solution and fuzzy negative ideal solution is determined by using Euclidean distance:

$$S_i^* = \sqrt{\sum (v_{ij} - v_j^*)^2}, \qquad i = 1, 2, 3, \ldots, M, \tag{7}$$

The similarity to the ideal solution a^*, denoted Ci, $0 \le Ci \le 1$, is computed for each alternative a_i:

$$C_i^* = \frac{S_{i-}}{S_i^* + S_{i-}}, \quad 0 \le C_i^* \le 1, \quad i = 1, 2, 3, \ldots, M, \tag{8}$$

AHP technique is a process that consists of the following steps: Step 1: State a set of all judgments in the comparison matrix by using the fundamental scale of pair-wise comparison. The comparison matrix is shown in Table 2. To investigate consistency of DM preferences given in Table 2 the decomposition approach is used. Matrices $\overline{F_l}, \overline{F_m}, \overline{F_u}$ are calculated by using (2) and is expressed in Table 3.

Table 3. $\overline{F_l}, \overline{F_m}, \overline{F_u}$ matrices.

$\overline{F_l}$				$\overline{F_m}$				$\overline{F_u}$			
3	5	6.5	3.5	3	7	8.5	5.5	6	12	15	9
0.9	3	3.5	5	1.83	3	5.5	7	3.06	6	6	12
1.3	1.66	3	1.3	1.4	2.66	3	1.83	2.43	4.16	6	3.06
1.66	1.3	5	3	2.66	1.83	7	3	4.16	4.16	12	6

Step 2: Using Eq. (1) $\lambda_l, \lambda_m, \lambda_u$ and w_l, w_m, w_u are obtained: $\overline{\lambda}_l = 10.9$, $\overline{\lambda}_m = 14.37$, $\overline{\lambda}_u = 24.9$, $\lambda_l = 3.41$, $\lambda_m = 4.08$, $\lambda_u = 5.17$
Using centroid method in defuzzification process we determine that $\lambda_{\max} = 4.22$.

$$\begin{aligned}
\overline{w_l} &= [0.33\,0.21\,0.19\,0.13] \\
\overline{w_m} &= [0.39\,0.24\,0.23\,0.16] \\
\overline{w_u} &= [0.49\,0.32\,0.29\,0.21]
\end{aligned} \tag{9}$$

Step 3: In this step Consistency Index (CI) and the Consistency Ratio (CR) are determined:

$$CI = \frac{\lambda_{\max} - k}{k - 1} = \frac{4.22 - 4}{3} = 0.073, \quad CR = \frac{0.073}{0.9} = 0.081$$

So decision preferences given by matrix F (Table 2) is consistent.
Now we present TOPSIS method to rank alternatives. This method consists of next steps:

Step 1: We describe the normalized decision matrix given in Table 4 by using the codebook in Table 1. Using centroid method in defuzzification process we determine that the relative weights of the criteria are $w_1 = 0.38$, $w_2 = 0.24$, $w_3 = 0.23$, $w_4 = 15$.

Table 4. Fuzzy decision matrix

	C_1-Vendor 0.38	C_2-Functionality 0.24	C_3-Reliability 0.23	C_4-Portability 0.15
S_1	0.4, 0.5, 0.6, 0.7	0.6, 0.7, 0.8, 0.9	0.8, 0.9, 1, 1	0.6, 0.7, 0.8, 0.9
S_2	0.4, 0.5, 0.6, 0.7	0.8, 0.9, 1, 1	0.4, 0.5, 0.6, 0.7	0.6, 0.7, 0.8, 0.9
S_3	0.6, 0.7, 0.8, 0.9	0.4, 0.5, 0.6, 0.7	0.8, 0.9, 1, 1	0.8, 0.9, 1, 1

Step 2: We construct the weighted fuzzy normalized decision matrix that is shown in Table 5:

Table 5. Fuzzy weighted normalized decision matrix

	C_1-Vendor	C_2-Functionality	C_3-Reliability	C_4-Portability
S_1	0.15, 0.19, 0.22, 0.26	0.14, 0.16, 0.19, 0.21	0.18, 0.2, 0.23, 0.23	0.09, 0.1, 0.12, 0.13
S_2	0.15, 0.19, 0.22, 0.26	0.19, 0.21, 0.24, 0.24	0.09, 0.11, 0.13, 0.16	0.09, 0.1, 0.12, 0.13
S_3	0.22, 0.26, 0.3, 0.34	0.09, 0.12, 0.14, 0.16	0.18, 0.2, 0.23, 0.23	0.12, 0.13, 0.15, 0.15

Step 3: At this step we determine the fuzzy ideal and the fuzzy negative ideal solutions S^* and S^- by using (5) and (6). The determined fuzzy ideal and negative ideal solutions are shown in Table 6.

Table 6. The Fuzzy Ideal and the Negative-ideal Solutions

	C_1-Vendor	C_2-Functionality	C_3-Reliability	C_4-Portability
S^*	0.22, 0.26, 0.3, 0.34	0.19, 0.21, 0.24, 0.24	0.18, 0.2, 0.23, 0.23	0.12, 0.13, 0.15, 0.15
S^-	0.15, 0.19, 0.22, 0.26	0.09, 0.12, 0.14, 0.16	0.09, 0.11, 0.13, 0.16	0.09, 0.1, 0.12, 0.13

Step 4. The separation of each alternative to the fuzzy ideal solution and fuzzy negative-ideal solutions is computed by using the Euclidean distance (7). The results are shown in Table 7:

Table 7. The separation distances

	C_1			C_2			C_3			C_4		
	S_1	S_2	S_3	S_1	S_2	S_3	S_1	S_2	S_3	S_1	S_2	S_3
S^*	0.15	0.15	0	0.09	0	0.13	0	0.15	0	0.06	0.06	0
S^-	0	0	0.15	0.13	0.09	0	0.15	0	0.15	0	0	0.06

Step 5. The likeness C_{i^*} to the fuzzy ideal solution is calculated for every alternative:

$$C_1^* = S_1^- / (S_1^* + S_1^-)$$
$$= [0/(0.15+0) + 0.13/(0.13+0.09) + 0.15/(0.15+0) + 0/(0+0.06)]/4$$
$$= 0.483,$$

$$C_2^* = [0/(0+0.15) + 0.09/0.09 + 0/0.15 + 0/0.05]/4 = 0.25,$$

$$C_3^* = [0.15/0.15 + 0/0.13 + 0.15/0.15 + 0.06/0.06]/4 = 0.75.$$

Step 6. Finally, we rank the alternatives with respect to C_{i^*}:

$$S_3 > S_1 > S_2$$

By using fuzzy integrated AHP and TOPSIS methodology, we determine that software S_3 is the best one and alternative S_2 is worst one in terms of enterprise software selection.

5 Conclusion

The article proposed a new procedure for enterprise software selection, to find the most suitable software among three alternatives in which based on predetermined selection factors. AHP was used to determine the weights of four criteria by pair-wise comparisons. Subsequently, TOPSIS was utilized to achieve final ranking preferences. Using a methodology based on fuzzy AHP and fuzzy TOPSIS approaches was determined that the best software is third software.

References

1. Lin, H.Y., Hsu, P.Y., Sheen, G.J.: A fuzzy-based decision-making procedure for data warehouse system selection. Expert Syst. Appl. 32(3), 939–953 (2007)
2. Bresnahan, J.: Mission impossible. CIO Magazine, October, p. 15 (1996)
3. Franch, X.: On the lightweight use of goal-oriented models for software package selection. In: Proceedings of the 17th Conference on Advanced Information Systems Engineering CAISE 2005. LNCS. Springer, vol. 3520, pp. 551–566 (2005)
4. Gupta, A., Verma, R., Singh, K.: Smart sim selector: a software for simulation software selection. Int. J. Eng. (IJE) 3(3), 175–185 (2009)
5. Chikofsky, E.J., Martin, D.E., Chang, H.: Assessing the state of tools assessment. IEEE Softw. 9(3), 18–21 (1992)
6. Tewoldeberhan, T.W., Verbraeck, A., Hlupic, V.: Implementing a discrete-event simulation software selection methodology for supporting decision making at Accenture. J. Oper. Res. Soc. 61(10), 1446–1458 (2010)
7. Dorado, R., Gómez-Moreno, A., Torres-Jiménez, E., López-Alba, E.: An AHP application to select software for engineering education. Comput. Appl. Eng. Educ. (2011). https://doi.org/10.1002/cae.20546
8. King, M., Newman, R.: Evaluating business simulation software: approach, tools and pedagogy. Horizon 17(4), 368–377 (2009)

9. Hlupic, V.: Simulation software selection using SimSelect. Simulation **69**(4), 231–239 (1997)
10. Hlupic, V., Irani, Z., Paul, R.: Evaluation framework for simulation software. Int. J. Adv. Manuf. Technol. **15**(5), 366–382 (1999)
11. Jadhav, A.S., Sonar, R.M.: Framework for evaluation and selection of the software packages: a hybrid knowledge based system approach. J. Syst. Softw. **84**(8), 1394–1407 (2011)
12. Saaty, T., Vargas, L.: Models, Methods, Concepts & Applications of the Analytic Hierarchy Process. International Series in Operations Research and Management Science (2001)
13. Prascevic, N., Prascevic, Z.: Application of fuzzy AHP method based on eigenvalues for decision making in construction industry. Tech. Gaz. **23**(1), 57–64 (2016)
14. Buckley, J.J.: Fuzzy hierarchical analysis. Fuzzy Sets Syst. **17**, 233–247 (1985)

Application of Fuzzy Linear Regression Model to Forecasting of Reservoir Rocks Properties

R. Y. Aliyarov⬤, A. B. Hasanov(✉), and G. A. Samadzada

Azerbaijan State Oil and Industry University, Azadlig Avenue, 20, Baku, Azerbaijan

r.aliyarov@asoiu.edu.az, adalathasanov@yahoo.com, gulus.szade@gmail.com

Abstract. Fuzzy linear regression has a great importance in developing mathematical models and it is widely used in different fields. In this paper we consider forecasting of reservoir rocks properties by using a linear regression with fuzzy coefficients. The forecasting results prove the validity of the proposed approach.

Keywords: Fuzzy regression · Reservoir rocks · Differential evolution

1 Introduction

In recent years, new views on the formation of oil and gas fields through the migration and accumulation of hydrocarbons in the upper crust have been actively discussed in the scientific literature. Moreover, the determining condition in this process is net volume and structure of the pore space. In the same time is believed that, in general, the natural potential of the productive capacity of terrigenous reservoirs is largely determined by their intergranular porosity and the nature of the grains packing. However, in addition, the shape of the grains, as well as the ratio of the content and distribution of grains of various sizes in the rock volume, also have a great influence on the porosity of the reservoirs. Attempts to simulate the common influencing effect on the multimodal distribution of intergranular porosity are widely known. Results of such attempts are well illustrated by studies [12] the authors of which, to identify the dependence of porosity on mechanical compaction. In these investigations, particles parameters averaging, such as particle size distribution of the rocks Md (grain size of the rocks), and the coefficients reflecting the sorting of the sediments (Ksort, Hr and the maximum content of any fraction Mf) are mainly used.

Regarding of above mentioned, the forecasting practice of reservoir rock properties is an important problem. This problem in one hand, is characterized by complexity of relationship between properties and from another hand, by uncertainty of relevant information. In view of this it is more effective to use fuzzy logic and soft computing methods for solving such problems. So in [5] multilag forecasting of five geological parameters was considered. But the approach used in [4, 5] is not effective to solve problems of extrapolation based forecasting. From the point of view of the representation of the polygons scattering experimental data and phase space parameters, fuzzy

R. A. Aliev et al. (Eds.): ICSCCW 2019, AISC 1095, pp. 835–840, 2020.
https://doi.org/10.1007/978-3-030-35249-3_110

petrophysical models were discussed in [4]. In [7] the authors consider to define the model of shale porosity in SCP. They constructed this model by using correlation-regression analysis method. To forecasting of petrophysical parameters based on actual log data is mentioned in [5]. The authors use the artificial intelligence method. [10] is dedicated forecasting of the oil and gas reservoir PVT parameters in Middle East Fields. Researchers used the combination of type-2 fuzzy logic system. The ability of type-2 fuzzy logic system to cope with uncertainty made the model more robust. In [8] was discussed applications of fuzzy logic in petrophysics and main concepts behind lithophiles are calculated using fuzzy logic methods. The uncertainty of the parameters is determined by the heterogeneity of the geological objects studied and expressed as quantitative estimates of reliability in [3]. These allow forecasting the reliability of graphical images based on the relationships between the parameters [8]. According to this, the forecasting is performed in the form of a fuzzy model of the distribution of possible values for each spatial point [5]. The disadvantage of this approach is the need to store and use large amounts of information. It is balanced by a high degree of visibility and the ability to achieve the reliability of each forecasting parameter to the expected values [9].

In this paper we apply fuzzy regression model [1] for forecasting of reservoir rock properties. This type of model is able to implement both extrapolation and interpolation based forecasting. For construction of this model Differential Evolution Optimization is used. The paper is structured as follows:

In Sect. 2 we provide necessary definitions. In Sect. 3 we describe the problem of construction of forecasting model on the basis of fuzzy regression. In Sect. 4 we illustrate application of this type of model to forecasting of reservoir rock properties in South Caspian Basin. Section 5 concludes.

2 Preliminaries

Definition 1 Fuzzy sets [2]. Let X be a classical set of objects, called the universe, whose generic elements are denoted x. Membership in a classical subset A of X is often viewed as a characteristic function $\mu_A(x)$ from X to $\{0,1\}$ such that

$$\mu_A(x) = \begin{cases} 1 & \text{iff} \quad x \in A \\ 0 & \text{iff} \quad x \notin A \end{cases}$$

where $\{0, 1\}$ is called a valuation set; 1 indicates membership while 0 - non-membership.

If the valuation set is allowed to be in the real interval $[0, 1]$, then A is called a fuzzy set.

Definition 2 Fuzzy numbers [2]. A fuzzy number is a fuzzy set A on R which possesses the following properties: (a) A is a normal fuzzy set; (b) A is a convex fuzzy set; (c) α-cuts of A A^α are a closed interval for every $\alpha \in (0,1]$; (d) the support of A A^{+0} is bounded.

Definition 3 Addition of fuzzy numbers [2]. Let A and B be two fuzzy numbers and A^α and B^α their α-cuts

$$A^\alpha = [a_1^\alpha, a_2^\alpha]; \ B^\alpha = [b_1^\alpha, b_2^\alpha]$$

Then we can write

$$A^\alpha + B^\alpha = [a_1^\alpha, a_2^\alpha] + [b_1^\alpha, b_2^\alpha] = [a_1^\alpha + b_1^\alpha, a_2^\alpha + b_2^\alpha],$$

where

$$A^\alpha = \{x/\mu_A(x) \geq \alpha\}; \ B^\alpha = \{x/\mu_B(x) \geq \alpha\}$$

Definition 4 Scalar multiplication of fuzzy number [2]. Scalar multiplication of fuzzy number A by ordinary numbers $k \in R^+$ is performed as follows

$$\forall A \subset R \ \ kA^\alpha = \left[ka_1^\alpha, ka_2^\alpha\right].$$

Definition 5 Differential Evolution Optimization. Differential Evolution (DE) is a stochastic optimization method [11]. It is effectively used for complex non-linear multiobjective problems, such as fuzzy optimization problems, neural networks learning problems and others.

The DE algorithm operators are differential mutation, probability crossover and selection. Mutation implements randomly selection from the population of 3 members (vectors) adding a weighted difference vector between two members to a third one. The crossover operation is applied to the mutated vector and another vector to generate new offspring vector. The selection process is done as follows. If the resulting vector yields a more optimal value of the objective function (usually, a lower one) than a predetermined population member does, the newly generated vector will replace the vector with which it was compared.

3   Statement of the Problem and Solution Method

The considered problem is forecasting of the properties of rocks in oil and gas fields. For this purpose, fuzzy linear regression-based forecasting model [1] is used to describe the interdependence of properties and the natural uncertainty of information. The following fuzzy regression model is considered:

$$\tilde{y}_k(x_1, x_2, \ldots, x_N) = \tilde{a}_{k0} + \sum_{i=1}^{N} \tilde{a}_{ki} x_i \tag{1}$$

x_i, $i = \overline{1, N}$ are values of parameters of rock properties (input data), y_k, $k = \overline{1, K}$ are forecasted values of parameters of rock properties (regression model output), and $\tilde{a}_{k0}$, $\tilde{a}_{ki}$ are coefficients of regression model described by triangular fuzzy numbers (TFNs).

For construction of model (1) we will use an approach proposed in [1]. According to this approach, evolutionary computation technique is used for determination of values of fuzzy coefficients $\tilde{a}_{k0}$, $\tilde{a}_{ki}$ which minimize the error between the model (1) and given data:

$$RMSE = \sqrt{\frac{\sum\limits_{h=1}^{n} (y_i(h) - y_i^*(h))^2}{n}} \rightarrow \min \qquad (2)$$

s.t.

$$\tilde{a}_{k0}, \tilde{a}_{ki} \in A, \qquad k = 1, \ldots, K. \qquad (3)$$

$y_i^*(h)$ actual values of parameters of rock properties at h depth, A is a search space. We intend to use the DE optimization technique to minimize error and obtain fuzzy coefficients. Given the constructed model (1) we will forecast the parameters of rock quality indicators y_k using the parameters values in the previous depth.

4   An Example

We consider the quality properties: clay content z_1, permeability z_2 and density of dry samples z_3. The data are given in the following Table 1:

Table 1. Reservoir rock properties data

h	z_1	z_2	z_3
$h_1 = h - 2$	0.83	15.65	2.5
$h_2 = h - 1$	0.86	18.21	2.55
$h_3 = h$	0.9	20.76	2.6
$h_4 = h + 1$	0.93	23.26	2.64
$h_5 = h + 2$	0.9	15.67	2.6
$h_6 = h + 3$	0.87	8.08	2.56
$h_7 = h + 4$	0.85	0.49	2.51
$h_8 = h + 5$	0.68	74.77	2.5
$h_9 = h + 6$	0.67	106.76	2.38
$h_{10} = h + 7$	0.37	101.26	2.14

According to the problem formulated in Sect. 3, we have to construct the following forecasting model (1):

$$\tilde{y}_1(x_1, x_2, \ldots, x_6) = \tilde{a}_{10} + \sum_{i=1}^{6} \tilde{a}_{1i} x_i,$$

$$\tilde{y}_2(x_1, x_2, \ldots, x_6) = \tilde{a}_{20} + \sum_{i=1}^{6} \tilde{a}_{2i} x_i,$$

$$\tilde{y}_3(x_1, x_2, \ldots, x_6) = \tilde{a}_{30} + \sum_{i=1}^{6} \tilde{a}_{3i} x_i.$$

where $\tilde{y}_k(x_1, x_2, \ldots, x_6) = z_k(h)$, $x_1 = z_1(h-1)$, $x_2 = z_1(h-2)$, $x_3 = z_2(h-1)$, $x_4 = z_2(h-2)$, $x_5 = z_3(h-1)$, $x_6 = z_3(h-2)$, $\tilde{a}_{k0}$, $\tilde{a}_{ki}$ are TFNs, $h-$ is depth, $k = 1, 2, .., 3$.

Construction of this model implies determination of fuzzy coefficients to minimize RMSE between the model and the data (Table 1). We used the DE optimization algorithm to determine fuzzy coefficients (by solving problem (1)–(2)). The following values of the parameters of the algorithm were used: mutation rate $F = 0.8$, crossover probability $CR = 0.7$, population size is $PN = 80$. The optimization converged with RMSE equal to 0.0006. The obtained fuzzy coefficients are as follows:

$$\tilde{a}_{10} = (-13.86, -12.6, -11.34), \quad \tilde{a}_{11} = (-2.53, -2.3, -2.07), \quad \tilde{a}_{12} = (0.49, 0.55, 0.61),$$
$$\tilde{a}_{13} = (2.10, 2.33, 2.57), \quad \tilde{a}_{14} = (-0.06, -0.05, -0.04), \quad \tilde{a}_{15} = (-0.54, -0.5, -0.44),$$
$$\tilde{a}_{16} = (-0.09, -0.08, -0.07);$$
$$\tilde{a}_{20} = (0.26, 0.29, 0.31), \quad \tilde{a}_{21} = (0.94, 1.04, 1.14), \quad \tilde{a}_{22} = (1.07, 1.18, 1.3),$$
$$\tilde{a}_{23} = (-0.6, -0.55, -0.5), \quad \tilde{a}_{24} = (-4.53, -4.12, -3.71), \quad \tilde{a}_{25} = (-0.21, -0.19, -0.17),$$
$$\tilde{a}_{26} = (-0.11, -0.1, -0.09);$$
$$\tilde{a}_{30} = (-1.66, -1.51, -1.35), \quad \tilde{a}_{31} = (-5.45, -4.96, -4.46), \quad \tilde{a}_{32} = (1.03, 1.15, 1.26),$$
$$\tilde{a}_{33} = (-2.61, -2.38, -2.14), \quad \tilde{a}_{34} = (-0.4, -0.37, -0.33), \quad \tilde{a}_{35} = (-1.14, -1.04, -0.93),$$
$$\tilde{a}_{36} = (-2.35, -2.14, -1.92).$$

Using these coefficients and given data we computed forecast the parameters of the $h = h_{11}$ depth:

$$\tilde{z}_1(h_{11}) = (0.164, 0.189, 0.214),$$

$$\tilde{z}_2(h_{11}) = (100.18, 100.21, 100.23),$$

$$\tilde{z}_3(h_{11}) = (2.29, 2.32, 2.34).$$

The obtained results of forecasting of behavior of rock properties may be considered as satisfactory.

5   Conclusion

We used fuzzy regression model for forecasting of reservoir rock properties. Model is able to implement both extrapolation and interpolation based forecasting. The obtained forecasting results show validity of the used approach.

References

1. Aliev, R.A., Fazlollahi, B., Vahidov, R.: Genetic algorithms-based fuzzy regression analysis. Soft. Comput. **6**, 470–475 (2002)
2. Aliev, R.A., Fazlollahi, B., Aliev, R.R.: Soft Computing and its Applications in Business and Economics. Springer, Heidelberg (2004)
3. Aliyarov, R.Y., Hasanov, A.B., et al.: Forecasting of qualitative characteristics of oil reservoirs. In: Materials of the Republican Scientific-Practical Conference devoted to the 95th Anniversary of H. Aliyev: Unity of Science, Education and Production at the Present Stage of Development, 7–8 May, pp. 22–31 (2018)
4. Aliyarov, R.Y, Hasanov, A.B., Ibrahimli, M.S, Ismayilova, Z.E., Jabiyeva, A.J.: Forecasting oil and gas reservoirs properties using of fuzzy-logic based methods. In: 13th International Conference on Theory and Application of Fuzzy Systems and Soft Computing, 27–28 August 2018, ICAFS-2018, pp. 769-773, Warsaw, Poland (2019)
5. Aliyarov, R.Y., Ramazanov, R.A.: Prediction of multivariable properties of reservoir rocks by using fuzzy clustering. In: 12th International Conference on Application of Fuzzy Systems and Soft Computing, 29–30 August 2016, ICAFS-2016, pp. 424–431, Vienna, Austria (2016)
6. Anifowose, F., Abdulraheem, A.: Prediction of porosity and permeability of oil and gas reservoirs using hybrid computational intelligence models. In: Proceedings of North Africa Technical Conference and Exhibition, 14–17 February, SPE 126649, Cairo, Egypt (2010)
7. Buryakovsky, L.A., Chilingar, G.V., Aminzadeh, F.: Petroleum Geology of the South Caspian Basin. Gulf Professional Publishing (2001)
8. Cuddy, S.: The application of the mathematics of fuzzy logic to petrophysics. In: 38th Annual Logging Symposium, SPWLA, Houston, Texas (1997)
9. Latyshova, M.G.: Practical guidance on the interpretation of geophysical research of wells. "Nedra", Moscow (1991)
10. Olatunji, S.O., Selamat, A., Abdurraheem, A.A.: Harnessing the power of type-2 fuzzy logic system in the prediction of reservoir properties. In: SPE Saudi Arabia Section Annual Technical Symposium and Exhibition, 21–23 April, SPE-178005-MS, Al-Khobar, Saudi Arabia (2015)
11. Price, K.V., Storm, R.M., Lampinen, J.A.: Differential Evolution - A Practical Approach to Global Optimization. Springer, New York (2005)
12. Romanovsky, S.I.: Application of information theory to solve some problems of lithology. In: Mathematical Methods in Geology, VSEGEI, pp. 75–92 (1968)
13. Wendelstein, B.Y., Rezvanov, R.A.: Geophysical methods for determining the parameters of oil and gas reservoirs. "Nedra", Moscow (1978)

Z-Matrix Consistency

Kamala Aliyeva[(✉)] [ORCID]

Azerbaijan State Oil and Industry University, Azadlig Ave.20,
AZ1010 Baku, Azerbaijan
kamalann64@gmail.com

Abstract. In this paper consistency of pairwise comparison Z matrix with elements is investigated. A consistency index of Z matrix with Z-elements will be considered as a perturbation of the elements of update matrix Z', where reciprocity and consistency are verified.

Keywords: Z-matrix · Reciprocity and consistency · Z-numbers · Consistency index

1 Introduction

Precedence correlations are very important submission of information used for solving decision making problems due to their efficiency in modeling. That precedence correlations are classified into multiple preference relations [5, 6], fuzzy preference relations [5–10] and linguistic preference relations [11–15]. Zhu et al. [16] determined that consistency analysis are based on ordinal consistency. Using analytic hierarchy process consistent matrix, Wang and Guo [17] define the priority vector by a model for solution the fuzzy matrix priority.

Zhang et al. [18] offered a technique due to the no-transitive rout number and no-transitive route contribution number for solution the fuzzy matrix without ordinal consistency. Wu [19] utilize the continual interval argument operator and the model for ranking, that very important for decision-makers in determining optimistic degree. Fan and Jiang [20] are used ranking technique of fuzzy matrix and offered an idea of consistency based on the new consistency approximation and ranking method to elimination incorrect conclusions. This feature simply the research of consistency among expert judgments.

Consistency index of a matrix using the average consistency index of the decision matrix is defined by Lamata and Pelaez [1]. Li and Ma [2] offered a method that help on decision making and used Gower charts for estimation the consistency by graphics. Basile and Dapuzzol [3] utilized the complex precise simple order to evaluate the consistency of matrix. Luo [4] developed the reviewing technique of matrix. Deriving consistent pairwise comparison matrices is considered in [21–23]. In [24] authors propose effective method for determining a basic weights from the DM preferences information included in a matrix without consistency and reciprocity properties. Unfortunately they reduce nonlinear optimization problem into linear one.

© Springer Nature Switzerland AG 2020
R. A. Aliev et al. (Eds.): ICSCCW 2019, AISC 1095, pp. 841–845, 2020.
https://doi.org/10.1007/978-3-030-35249-3_111

2 Preliminaries

Definition 1. A fuzzy set $\tilde{A}$ in X is represented by a membership function $\mu\tilde{A}(x)$ that relates with every element x in X a real number within interval [0, 1]. The value $\mu\tilde{A}(x)$ is called the rate of membership of x in $\tilde{A}$ [6].

Definition 2. A fuzzy number is a fuzzy set A on $\mathcal{R}$ and have next features: (a) A is a normal fuzzy set; (b) A is a bell shape fuzzy set; (c) α-cut of A, A^{α} is a enclosed interval for each $\alpha \in (0, 1]$; (d) the support of A, supp(A) is bounded where $\underline{\mu}_A(\alpha) \leq \overline{\mu_A}(\alpha), 0 \leq \alpha \leq 1$.

In all existing works consistency analysis is investigated with respect to pairwise comparison matrix with crisp or fuzzy elements without taking into reliability of preference information. For first time in this paper we consider consistency-driven approximation of pairwise composition matrix with Z-elements.

Definition 3. A Z-number is an ordered pair Z = (A, B) where A is a continuous fuzzy number that a random variable may take X is A and B is a continuous fuzzy number with a membership function μ_B: [0,1] → [0,1]

Definition 4. The elements of Z- matrix will be considered as a perturbation of the elements of update matrix Z′, where reciprocity and consistency are verified. Z- matrix is presented in next form:

$$Z = \begin{bmatrix} Z_{11} & Z_{12} & Z_{13} & \cdots & Z_{1n} \\ Z_{21} & Z_{22} & Z_{23} & \cdots & Z_{2n} \\ Z_{31} & Z_{32} & Z_{33} & \cdots & Z_{3n} \\ \vdots & \vdots & \vdots & & \vdots \\ Z_{n1} & Z_{n2} & Z_{n3} & \cdots & Z_{nn} \end{bmatrix} \tag{1}$$

3 Statement of the Problem

Z-matrix is given as (1) form. The problems to find update Z' with acceptable consistency index

$$Z' = \begin{bmatrix} Z'_{11} & Z'_{12} & Z'_{13} & \cdots & Z'_{1n} \\ Z'_{21} & Z'_{22} & Z'_{23} & \cdots & Z'_{2n} \\ Z'_{n1} & Z'_{n2} & Z'_{n3} & \cdots & Z'_{nn} \end{bmatrix} \tag{2}$$

The elements of matrix Z will be considered as a perturbation of the elements of update matrix Z′, where reciprocity and consistency are verified. A distance-based measure can be used to measure deviation between (1) and (2). For solution of this problem we use the next optimization problem

$$J = \sum_{i=1}^{n} \sum_{j=1}^{n} \left| Z_{ij} - Z'_{ij} \right| \rightarrow \min \tag{3}$$

$$Z'_{ij} \bullet Z'_{ji} \approx Z_{(1)} \qquad i = \overline{1,n}, \; j = \overline{1,n} \tag{4}$$

$$Z'_{ij} \bullet Z'_{jk} \approx Z'_{ik} \qquad \text{for all } i, j, k \tag{5}$$

$$Z'_{ij} \geq Z_{ij(0)} \tag{6}$$

Constraint (4) is related to conditions of reciprocity of Z' shown in (2). Constraint (5) concerns consistency conditions. As criterion (3) one can use classical Euclidian distance or different type of generalized distance.

4 Solution of the Problem

J can be represented as in (3). Assume that we investigate distance between Z_{ij} and Z'_{ij}.

$$Z_{ij} = (A_{ij}, B_{ij}) \text{ and } Z'_{ij} = (A'_{ij}, B'_{ij})$$

Distance between A_{ij} and A'_{ij} is calculated as

$$D_A^{\alpha}(A_{ij}, A'_{ij}) = \left| \frac{A_{ijL}(\alpha) + A_{ijR}(\alpha)}{2} - \frac{A'_{ijL}(\alpha) + A'_{ijR}(\alpha)}{2} \right| \tag{7}$$

Distance between B_{ij} and B'_{ij} are calculated as

$$D_B^{1\alpha}(B_{ij}, B'_{ij}) = \left| \frac{B_{ijL}(\alpha) + B_{ijR}(\alpha)}{2} - \frac{B'_{ijL}(\alpha) + B'_{ijR}(\alpha)}{2} \right| \tag{8}$$

and

$$D_B^{2\alpha}(B_{ij}, B'_{ij}) = \left| \left(B_{ijR}(\alpha) + B_{ijL}(\alpha) \right) - \left(B'_{ijR}(\alpha) + B'_{ijL}(\alpha) \right) \right| \tag{9}$$

Finally distance between B_{ij} and B'_{ij} is defined as

$$D_B(B_{ij}, B'_{ij}) = \frac{1}{2} \left(\int_0^1 \left(D_B^{1\alpha}\left(B_{ij}, B'_{ij} \right) + D_B^{2\alpha}\left(B_{ij}, B'_{ij} \right) \right) \right) d\alpha \tag{10}$$

Now we have to find distance between underlying probability distributions.
Let $P_{Z_{ij}}$ and $P'_{Z_{ij}}$ be the underlying probability distribution of Z_{ij} and Z'_{ij} respectively.

Let G_L be a set of probability distributions with conditions given below

$$G_l = \left\{ \begin{array}{c} P_Z : \int P_{Z_l}(x) dx = 1, \int P_{Z_l}(x) \bullet \mu_{A_l}(x) dx \text{ is } B_l \\ \int x \bullet P_{Z_l}(x) dx = \int x \bullet \mu_{A_l}(x) dx / \mu_{A_l}(x) dx \end{array} \right\} \qquad (11)$$

Then distance between $p_{Z_{ij}}$ and $p'_{Z_{ij}}$ can be expressed as

$$D_P\left(P_{Z_{ij}}, P'_{Z_{ij}}\right) = \inf\left\{ \left(1 - \int_R \left(\left(P_{Z_{ij}}(x) P'_{Z_{ij}}(x)\right)^{1/2} dx\right)^{1/2}\right)\right\} \qquad (12)$$

Here one can use Hellinger and other different types of probabilistic distances (total variation distance, Bhattacharyya distance, Mahalanobis distance, Jeffreys distance etc.).The proposed a comprehensive distance between Z_{ij} and Z'_{ij} is defined as

$$D^{total}(Z_{ij}, Z'_{ij}) = \left[\beta D_A^\alpha\left(A_{ij}, A'_{ij}\right)\right] + (1 - \beta) \times D_P\left(P_{Z_{ij}}, P'_{Z_{ij}}\right) \qquad (13)$$

To minimize D^{total} in (13) differential optimization method is used.

5   Conclusion

This study provides an alternative approach to the Saaty's classical eigensolution method. Saaty's consistency method gives ability to analysis of consistency of pairwise comparison matrices, suggested in this paper method provides the users tool to synthesis consistent matrix. Other distinguished characteristic of the proposed method is generalization of the classical consistent approximation to knowledge acquisition on DM preferences. Additionally, associated nonlinear optimization problem on consistency analysis is solved without transformation into a linear on due to use DEO method.

References

1. Lamata, M.T., Pelaez, J.I.: A method for improving the consistency of judgments. Int. J. Uncertainty Fuzziness knowl. Based Syst. **10**(6), 677–686 (2003)
2. Li, H.L., Ma, L.C.: Adjusting ordinal and cardinal inconsistencies in decision preferences based on Gower Plots. Asia-Pac. J. Oper. Res. **23**, 329–346 (2003)
3. Basile, L., Dapuzzol, L.: Weak consistency and quasilinear means imply the actual making. Int. J. Uncertainty Fuzziness Knowl.-Based Syst. **10**, 227–239 (2002)
4. Luo, Z.Q.: A new revising method for judgment matrix without consistency in AHP. Syst. Eng. Theor. Pract. **6**, 83–92 (2004)
5. Wang, Y.M., Fan, Z.P.: Fuzzy preference relations: aggregation and weight determination. Comput. Ind. Eng. **53**, 384–388 (2007)

6. Fan, Z.P., Jiang, Y.P., Mao, J.Y.: A method for repairing the inconsistency of fuzzy preference relations. Fuzzy Sets Syst. **157**(1), 20–33 (2006)
7. Chiclana, F., Herrera, F., Herrera-Viedma, E.: Integrating multiplicative preference relations in a multipurpose decision making model based on fuzzy preference relations. Fuzzy Sets Syst. **122**, 277–291 (2001)
8. Fan, Z.P., Ma, Y.P., Jiang, Y.H., Sun, L.M.: A goal programming approach to group decision making based on multiplicative preference relations and fuzzy preference relations. Eur. J. Oper. Res. **174**, 311–321 (2006)
9. Ma, J., Fan, Y.P., Jiang, J.Y., Mao, L.M.: A method of repairing the inconsistency of fuzzy preference relatio. Fuzzy Sets Syst. **157**, 20–33 (2006)
10. Wang, T.C., Chen, Y.N.: A new method on decision-making using fuzzy linguistic assessment variables and fuzzy preference relations. In: The Processing of the 9th World Multi-conference on Systemic Cybernetics and Informatics, Orlando, pp. 360–363 (2005)
11. Herrera, F., Herrera-Viedma, E.: Choice functions and mechanisms for linguistic preference relations. Eur. J. Oper. Res. **120**, 144–161 (2000)
12. Herrera, F., Herrera-Viedma, E., Chiclana, F.: Militiaperson decision-making based on multiplicative preference relations. Eur. J. Oper. Res. **129**, 372–385 (2001)
13. Herrera, F., Martinez, L.: A model based on linguistic 2-tuples of dealing with multi-granular hierarchical linguistic contexts in multi-expert decision making. IEEE Trans. Syst. Man Cybern. **31**, 227–234 (2001)
14. Xu, Z.S.: Goal programming models for obtaining the priority vector of incomplete fuzzy preference relation. Int. J. Approximate Reasoning **36**, 261–270 (2004)
15. Xu, Z.S.: Deviation measures of linguistic preference relations in group decision making. Omega **33**, 249–254 (2005)
16. Zhu, J.J., Wang, M.G., Liu, S.X.: Research on several problems for revising judgment matrix in AHP. Syst. Eng. Theor. Pract. **1**, 90–94 (2007)
17. Wang, X.G., Guo, Q.: The priority of fuzzy complementary judgment matrix and its application in procurement tenders for government project. In: 2010 17th International Conference on Industrial Engineering and Engineering Management, 29–31 October 2010, pp. 148–151 (2010)
18. Zhang, X., Yue, G., Liu, X., Yu, F.: The revising method for fuzzy judgment matrix without ordinal consistency. In: Proceeding of 2009 International Symposium on Information Processing, 21–23 August 2009, pp. 309–312 (2009)
19. Wu, J.: A new approach for priorities trapezoidal fuzzy number reciprocal judgment matrix. Chin. J. Manag. **18**(3), 95–100 (2010)
20. Fan, Z.P., Jiang, Y.P.: An overview on ranking method of fuzzy judgment matrix. Syst. Eng. **19**(5), 12–18 (2001). (in Chinese)
21. Saaty, T.L.: Fundamentals of Decision Making and Priority Theory with the Analytic Hierarchy Process. RWS Publications, Pittsburgh (1994)
22. Forman, E.H.: Random indices for incomplete pairwise comparison matrices. Eur. J. Oper. Res. **48**, 153–155 (1990)
23. Zhang, H., Sekhari, Y., Abdelaziz, O., Bouras, A.: Deriving consistent pairwise comparison matrices in decision making methodologies based on linear programming method. J. Intell. Fuzzy Syst. **27**(4), 1977–1989 (2014). https://doi.org/10.3233/ifs-141164
24. Esther, D., Jacinto, G.: Consistency-driven approximation of a pairwise comparison matrix. Kybernetika **39**(5), 561–568 (2003)

Deep Neural Networks in Semantic Analysis

Alexey Averkin[1,2]([⊠]) and Sergey Yarushev[2]

[1] Dorodnicyn Computing Centre, FRC CSC RAS, Moscow, Russia
averkin2003@inbox.ru
[2] Plekhanov Russian University of Economics, Moscow, Russia
sergey.yarushev@icloud.com

Abstract. This paper presents research of the possibilities of application deep neural networks in semantic analysis. This paper presents the current situation in this area and the prospects for application an artificial intelligence in semantic analysis and trend and tendencies of this science area. For better understanding future tendencies of researches in semantical area we present detailed review of the studies in semantic analysis with using artificial intelligence, studies about a human brain.

Keywords: Semantic analysis · Deep neural networks · Forecasting · Processing of natural language

1 Introduction

Please note that the first paragraph of a section or subsection is not indented. The first paragraphs that follows a table, figure, equation etc. does not have an indent, either. Machine learning is a central technology of artificial intelligence. In recent years, in this area all the attention was focused on the deep learning technology as a method for automatically extracting characteristic values necessary for the interpretation and evaluation of phenomena. Huge amounts of time series data collected from the devices, especially in the era of Internet of Things. Applying deep learning that data and classifying them with a high degree of accuracy, it is possible to carry out further analysis with a view to creating new products and solutions and open up new areas of business.

Deep learning [1] technology, which is seen as a breakthrough in the development of artificial intelligence, delivers the highest accuracy of image recognition and speech, but it is still applicable only to limited types of data. In particular, it has so far been difficult to classify accurately automatically variable time series data coming from the devices connected to the Internet of things.

The theory and practice of machine learning are experiencing this "deep revolution" caused by the successful application of methods Deep Learning (deep learning),

This work was supported by the Russian Foundation for Basic Research (Grant No. 17-07-01558) and the article was funded within the internal research of Plekhanov Russian University of Economics under the name "Development of methods to forecast prices for financial instruments based on neural networks".

R. A. Aliev et al. (Eds.): ICSCCW 2019, AISC 1095, pp. 846–853, 2020.
https://doi.org/10.1007/978-3-030-35249-3_112

representing the third generation of neural networks. In contrast to the classical (second generation) neural networks 80–90-ies of the last century, a new learning paradigm allowed to get rid of some of the problems that hindered the dissemination and successful application of traditional neural networks.

Networks trained by deep learning algorithms, not just beat the best in terms of accuracy, alternative approaches, but also in a number of tasks showed the beginnings of understanding the meaning of information supplied (for example, image recognition, analysis of textual information, and so on).

Most successful modern industrial methods of computer vision and speech recognition built on the use of deep networks, and IT-industry giants such as Apple, Google, Facebook, buying groups of researchers involved in deep neural networks.

Deep Learning - is part of a broader family of machine learning methods - learning concepts, which feature vectors are located directly on a variety of levels. These features are automatically determined and associated with each other, forming output data. At each level, abstract presented features based on attributes of the previous level. Thus, the deeper we move forward, the higher level of abstraction. The neural network is a plurality of layers of a plurality of levels with feature vectors that generate output data (Fig. 1).

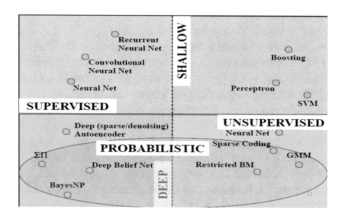

Fig. 1. Classification of deep architectures.

2   History of Deep Learning

The team of graduate students from the University of Toronto (Canada), led by Professor Jeffrey Hinton [2] (Hinton), won the competition held by the pharmaceutical company Merck. With access to a limited set of data describing the chemical structure of the molecules 15, a group of Hinton was able to create and use a special software system, which is to determine which of these molecules will be more effective than other work as drugs.

The peculiarity of this operation lies in the fact that system designers have used artificial neural network based on the so-called "deep learning" (Deep Learning). As a result, their system was able to conduct the necessary studies and calculations on the basis of very small set of input data: usually for training neural networks before using the required load in a vast array of information.

Achieving Hinton team looks particularly impressive when you consider that the application form has been submitted at the last moment. Moreover, the system of "deep learning" was created in the absence of concrete data on the interaction of molecules with the proposed target. The successful application of the technique of "deep learning" was another achievement in the development of artificial intelligence, which is very rich in the second half of 2012.

So, Jeff Dean this summer (Jeff Dean) and Andrew Ng (Andrew Y. Ng) from Google revealed a new system for image recognition to determine the accuracy of the cat in the picture 15.8%, where the cluster system for training of 16 thousand units to use the network ImageNet, contains 14 million images of 20,000 different objects. Last year, the Swiss scientists have demonstrated a system that is better recognized a person signs in the photos (the accuracy was 99.46% on a set of 50,000 images, with people maximum accuracy was 99.22%, and the average accuracy for a group of 32 people was "only" 98.84%). Finally, in October of this year, Richard Rashid (Richard F. Rashid) [3], coordinator of the Microsoft academic programs shown at a conference in Tianjin (China) technology live translation from English into Mandarin Chinese, maintaining the original voice.

All these technologies, demonstrating a breakthrough in the field of artificial intelligence, in one way or another based on the technique of "deep learning". The main contribution to the theory of in-depth training is now making just Professor Hinton, who, incidentally, is the great-grandson of George Boole, an English scientist, the inventor of Boolean algebra underlying modern computers.

The theory of deep learning complements conventional machine learning technology with special algorithms for the analysis of the input data at several levels of presentation. The peculiarity of the new approach is that the "deep learning" is studying the subject, until it finds enough informative presentation levels to take into account all the factors that could affect the characteristics of the studied subject.

Thus, a neural network based on this approach requires less input for training and trained network is able to analyze the information with much higher accuracy than the conventional neural network. Sam Hinton and colleagues say that their technology is particularly well suited for the search features in multidimensional, well-structured arrays of information.

Technology Artificial Intelligence (AI) in general, and in-depth training in particular, are now widely used in different systems, including voice assistant Apple Siri-based Nuance Communications technologies and recognition of addresses on Google Street View. However, scientists carefully evaluate progress in this area since the history of AI is replete with big promises and no less loud crunches.

Thus, in the 1960s, scientists have believed that before the establishment of full-fledged AI there are only about 10 years. Then, in the 1980s, there was a whole wave of young companies that offer a "ready-AI", then in this area has come true "Ice Age" - right up to the present time, when the enormous computing power available in the cloud services, opened a new level implementation of powerful neural networks using new theoretical and algorithmic basis.

New learning paradigm realizes the idea of training in two phases. In the first stage of a large array of un-partitioned data using avtoassotsiatorov (stratified by their learning without a teacher) extracted information about the internal structure of the input data. Then, using this information in a multi-layer neural network, trained with her teacher (tagged data) by known methods. The number of untagged data desirable to be as large as possible. Markup data may be much less. In our case, this is not very important.

Deep learning the most active and the most powerful tool in artificial intelligence area now. Based on the well-reviewed seen on many popular technologies of deep learning. Major corporations sent huge amounts of money in research in this area. The most popular field of application of deep learning today are the systems of computer vision, facial recognition, text, search engines and ontological system. To date, the use is gaining popularity depth training in forecasting time series. Based on this, in our further work we will carry out experiments in this direction. Today, deep learning technology based on the principles of service, which makes their use more convenient and faster in most areas.

Processing information in natural language analysis of the relationship between the collection of documents and terms presented in this document, understanding and determination of the direction and theme of the text - all tasks of semantic analysis. Latent semantic analysis are the search giant to find the text of one subject. A lot of work is carried out in the field of construction of semantic models for processing quality of the text, understanding the logical relationships, optimization of knowledge bases, as well as the widest range of applications.

How do children learn language? In particular, as they associate it with a sentence structure meaning? This issue is certainly associated with issue that is more global - how the brain connects the sequence of characters for building symbolic and sub-symbolic representations? Many scientists are conducting research to get the answers to these questions. One of the most important aspects in the natural language processing task is the speed with which this process occurs. Most clearly, this problem manifests itself in the research potential brain capacity. The most appropriate approach to the modeling of these processes is the use of neural networks. Proof of this can be a large number of works in this category.

In this paper, we consider a number in the field of semantic analysis using artificial neural networks applications.

3   Review of Text Classification Approaches

Chinese scientists [4] in their study presented a new model of text classification using a neural network and its learning by back propagation, as well as with the modified method. Use an effective method to reduce the characteristics of selecting the sample dimension, that of a bear increasing system performance. Standard learning algorithm backpropagation trained quite slow, so the researchers modified the algorithm to increase the speed of learning. Traditional word-matching based on the classification of text using vector space model. However, in this approach takes no account semantic relationships between terms, which can lead to deterioration of the classification accuracy. Latent semantic analysis can overcome the problems caused by the use of statistically derived conceptual indices and not just individual words. It creates kontsetualnye vector spaces, in which each term or a document is submitted as a vector in space. This not only significantly reduces the dimension, but also can detect important associative relations between terms. The researchers tested their model on a set of 20 news data, experimental results showed that the model with the modified method of backpropagation proposed in this paper, have surpassed the traditional teaching method. As well as the use of latent semantic analysis for this system allows you to dramatically reduce the dimension that can achieve good results classification.

Indian scientists compared the performance of common neuronal back propagation network with a combination of the neural network with the method of latent semantic indexing of text classification problem. In traditional neural networks with backpropagation error process of adjusting the scales locked in a local minimum, as well as the speed of learning network of this type is rather low, which entails a decrease in performance. In view of these shortcomings, scientists have decided to make a combination of latent semantic indexing and this neural network. Latent semantic representation of the structure of data in the low-dimensional space in which the documents, the terms and words sequences were also compared. The one-dimensional decomposition technique used in latent semantic analysis, multi-dimensional matrix in which the terms are broken down into a set of K orthogonal factors, which the original text data is changed to a smaller semantic space. The new vector documents can be found in the K-dimensional space. Also, the new coordinates are queries.

Performance tested a combination of these methods on the basis of the classification methodology 20 newsgroups from different categories, such as sports, medicine, business, politics, and others. As a result, these methods can significantly reduce the dimension and get the best results of the classification of the text.

One of the key figures in the study of human brain is Mauntkasl [5]. In this paper, he summarized his many years of research, he argues that, despite the diversity of their functions, all sections of the cerebral cortex are arranged, in principle, the same. This means that learning and pattern recognition takes place uniformly in the cortex, and the variety of its functions is a consequence of the diversity of signals processed by different parts of the cortex.

According Mauntkaslu bark has a two-dimensional honeycomb structure. The basic functional unit of the cortex is a mini-column of about 30 microns in diameter, consisting of about 100 neurons. Such mini-columns interconnected by positive and negative lateral connections. Moreover, the latter shall be sharp, but with a certain delay relative to the first. As a result, at the same time excited by a pool adjacent mini-column, involuntarily forcing the recall T. Kohonen self-organizing map [6]. As a result, throughout the cortex, we are seeing signs of self-organizing maps: detectors similar signals are located next to each other.

Experiments show that the elementary detectors in the area of these maps the order of 0.1 mm 2, i.e., they contain a mini-columns 102 or 104 neurons. Such functional units Mauntkasl calls the macro columns. It is they who determine the "resolution" of the cortex, and limit the number of signs that can remember people (only a few million). But the reliability of the memory is guaranteed by a large number of neurons that make up the macro column. So we keep his memory throughout life, even with the death of a substantial part of the neurons.

Thus, Kohonen maps are evidently the most appropriate tool for simulating the operation of the cortex. You just need to teach them how to work with dynamic patterns, with which only the brain works, because its main task - foresight.

As people acquire language, as well as two or more different languages with a nervous system remains an open question. To solve this problem, the French scientists, led by Peter blast furnace [7] proposed to build a model that will be able to learn any language from the very beginning. In this work, they offer the neural network approach, which handles proposals for like, word for word, without prior knowledge of the semantics of words. The proposed model does not "pre-bound" structure, but only random and trained compound model is based on Reservoir Computing technology. Scientists for robotic platforms through which users can teach the robot the basics of the English language, to later give him a different job, developed earlier model. In this paper, was added ability to handle rare words in order to be able to keep the dictionary size is quite small in the processing of natural language. Moreover, this approach was extended to the French language and shows that the neural network can learn two languages at the same time. Even with a small body language model is capable to learn and generalize knowledge in a monolingual or bilingual. This approach may be a more practical alternative for small shells of different languages than other training methods based on large data sets.

Many studies performed in the language processing by neural networks [8], and in recent years with the use of so-called Echo State Networks [9].

How the human brain processes the proposals that a person reads or hears? The task of understanding how the brain does it, occupies a central place in the research of scientists from the field. proposals processed in real time. Previous words in a sentence may affect the processing time in hundreds of milliseconds. Recent neurophysiological studies suggest that it is the frontal part of the brain plays a crucial role in this process. Hinaut [10] conducted a study that gives some insight into how certain aspects of the treatment of the proposals in real time occur, based on the dynamics of periodic cortical networks and plasticity in cortico-striatal system. They model the prefrontal area BA47 using recurrent neural network to obtain on-line input word categories in the processing of proposals. The system is trained on pairs of sentences in which the meaning is encoded as a function of the activation of the corresponding role of verbs and nouns in sentences. The model examines the expanded set of grammatical structures and demonstrates the ability to generate new designs. This shows how much early in the sentence a parallel set of predicates makes sense. The model shows how the online responses to the speech are influenced by the preceding words in a sentence and the previous proposals in the discourse, providing a new perspective on the neurophysiology of the brain cortex to recognize grammatical structure. The study found that recurrent neural network can decode the grammatical structure of the sentences in real-time in order to get an idea of the meaning of sentences.

Neural processing of natural language. The focus in cognitive science today is focused on the study of how the neural networks in the brain are used to read and understand the text. This issue explores a huge number of scientists in neuroscience along with recent studies that are designed to examine the brain processes involved in learning non-linguistic sequences or artificial grammar learning. Dominey [11] in their study attempted to combine data from several neurophysiological proposals processing models, through the specification of neural network model, the architecture of which is based on the well-known cortico-striatal-thalamo-cortical (KSTK) neuroanatomy system of human language. The challenge is to develop simulation models that take into account the limitations and neuroanatomical connections, and functional image data. In the proposed model, the structural cues encoded in a recurrent neural network in Kortikova BA47, activate circuit (KSTK) for modulating the flow lekskicheskoy semantic information in an integrated view on the meaning of the sentence level. The simulation results are shown in Caplan [12].

Modeling Language Authority spent SA Shumsky. In the work [13], the author puts forward three hypotheses: The first hypothesis is that the processing of the time series in the crust is done like modules, recognizing the typical temporal patterns, each with its own input. For example, the bark of the site responsible for the morphological analysis of words, recognize about 105 words and their constituent morphemes and syllables. Another cortical portion defining sentence structure, operates in the same way, but with a different primary alphabet, each character is no longer encodes a letter and a word unit. This area stores the characteristic patterns of the combination of words in a grammatically correct sentence. According to the second hypothesis, the input for the next cortical module, responsible for the analysis of the temporal structures of higher order thalamus serves as a compressed output signal from the previous module. According to the third hypothesis, in the "body language" there are two inter-channel "deep learning": the grammatical and semantic. Similarly, the dorsal (scene analysis) and ventral (object recognition) channels the analysis of visual information.

To test the hypotheses, the program complex "semantic processor Golem" was created, able to identify the hierarchy of language patterns in the training on large text arrays. Training was conducted on the text array of 6 GB, consisting of materials of Russian Internet media. In order to bring the conditions of the experiment to learn spoken language a child, all the words quoted in lowercase letters. The volume of training sample roughly corresponds to the linguistic experience of 20-year-old man (in the perception of $\sim$ 105 words per day). Education took about two months of work, a modern PC.

As a result, it developed a complex Shumsky pretty confident can recognize names, cities, countries and some other concepts. Also, an understanding has been achieved and what concepts in the proposal are related to what can be said that the Golem is able to adequately recognize and index the semantic content of the proposals.

4   Conclusions

In this paper, we present a review of recent works in the semantic analysis scientific area and natural language processing.

Based on the results obtained in each work, we can conclude that using of the deep neural network in semantic analysis task show high performance and extend the possibilities for analyzing text data, and indispensable technology for modeling cerebral activity, in particular simulation learning new languages and application Information technologies in building robots that can self-learn languages and understand their meaning.

References

1. Ranzato, M.A., Szummer, M.: Semi-supervised learning of compact document representations with deep networks. In: Proceedings of the 25th International Conference on Machine Learning, pp. 792–799. ACM (2008)
2. Hinton, G.E., Osindero, S., Teh, Y.W.: A fast learning algorithm for deep belief nets. Neural Comput. **18**(7), 1527–1554 (2012)
3. Deep Learning: Microsoft's Richard Rashid demos deep learning for speech recognition in China. http://deeplearning.net/2012/12/13/microsofts-richard-rashid-demos-deep-learning-for-speech-recognition-in-china/. Accessed 17 July 2017
4. Yu, B., Xu, Z., Li, C.: Latent semantic analysis for text categorization using neural network. Knowl.-Based Syst. **21**(8), 900–904 (2008)
5. Mountcastle, V.: The columnar organization of neocortex. Brain **120**, 701–722 (1997)
6. Kohonen, T.: Self-Organizing Maps. Springer, Heidelberg (2001)
7. Hinaut, X., et al.: A recurrent neural network for multiple language acquisition: starting with English and French. In: CoCo@ NIPS (2015)
8. Miikkulainen, R.: Subsymbolic case-role analysis of sentences with embedded clauses. Cogn. Sci. **20**, 47–73 (1996)
9. Hinaut, X., Dominey, P.F.: Real-time parallel processing of grammatical structure in the fronto-striatal system: a recurrent network simulation study using reservoir computing. PLoS ONE **8**(2), 52946 (2013)
10. Frank, S.L.: Strong systematicity in sentence processing by an echo state network. In: Kollias S.D., Stafylopatis A., Duch W., Oja E. (eds.) Artificial Neural Networks – ICANN 2006. ICANN. Lecture Notes in Computer Science, vol. 4131, pp. 505–514. Springer, Heidelberg (2006)
11. Dominey, P.F., Inui, T., Hoen, M.: Neural network processing of natural language: II. towards a unified model of corticostriatal function in learning sentence comprehension and non-linguistic sequencing. Brain Lang. **102**, 80–92 (2009)
12. Caplan, D., Baker, C., Dehaut, F.: Syntactic determinants of sentence comprehension in aphasia. Cognition **21**, 117–175 (1985)
13. Shumski, S.: Brain and language: hypotheses about structure of a natural language. HSE **2**(4), 15–23 (2017)

Strength of the Threaded Details Working in Oilfield Equipment

Djahid A. Kerimov$^{(\boxtimes)}$ (iD)

Azerbaijan State Oil and Industry University, Azadlig 34, Nasimi, Baku,
Azerbaijan
haciyevanaila64@gmail.com

Abstract. In practice of designing of machine details and the equipment for assessment the hazard of a tense situation usually use not criteria of strength, only the safety margin coefficient or equivalent stress. Methods of calculation of equivalent stress are considered on the basis of a method of calculation of criterion of strength depending on synchronous change of all three principal stresses. Obtained theoretical results are confirmed experimentally and the method of calculation of equivalent stress is developed that is necessary for practice.

Keywords: Safety margin coefficient · Equivalent stress · Main stress · Flat tense situation · Limiting stress

1 Introduction

In practice, the projecting a details of machine, instruments and installation for hazard assessment of tension use usually not criterion of strength, but a quantity of the safety margin coefficient of n or equivalent tension of σeq brought out of it.

For the safety margin coefficient and equivalent stress the criterion of strength is determined by the following formula [1]:

$$
\begin{aligned}
A(\sigma_{z_1} + \sigma_{z_2} + \sigma_{z_3}) + B\left(\sigma_{z_1}^2 + \sigma_{z_2}^2 + \sigma_{z_3}^2\right) + C(\sigma_{z_1}\sigma_{z_2} + \sigma_{z_1}\sigma_{z_3} + \sigma_{z_2}\sigma_{z_3}) + \\
+ D\sqrt{\sigma_{z_1}^2 + \sigma_{z_2}^2 + \sigma_{z_3}^2} + E\sqrt{(\sigma_{z_1} - \sigma_{z_2})^2 + (\sigma_{z_2} - \sigma_{z_3})^2 + (\sigma_{z_3} - \sigma_{z_1})^2} = 1.
\end{aligned}
\tag{1}
$$

For a flat case ($\sigma_{z_3} = 0$) temperature a criterion of strength, offered by I.N.Mirolyubov taking into account the theory Mohr's of strength for metal in a final version the formula (1) takes the following form [2]:

$$
\begin{aligned}
A(\sigma_{z_1} + \sigma_{z_2}) + B\left(\sigma_{z_1}^2 + \sigma_{z_2}^2\right) + C \cdot (\sigma_{z_1}\sigma_{z_2}) + D\sqrt{\sigma_{z_1}^2 + \sigma_{z_2}^2} + \\
+ E\sqrt{(\sigma_{z_1} - \sigma_{z_2})^2 + \sigma_{z_1}^2 + \sigma_{z_2}^2} = 1 \ .
\end{aligned}
\tag{2}
$$

© Springer Nature Switzerland AG 2020
R. A. Aliev et al. (Eds.): ICSCCW 2019, AISC 1095, pp. 854–858, 2020.
https://doi.org/10.1007/978-3-030-35249-3_113

2   The Method of Calculation of Equivalent Stress

Suppose that the sizes of the main normal tensions of $\sigma_1, \sigma_2, \sigma_3$ in a dangerous point of a detail be as a result received. Let us consider the limiting tense situation similar to given – $\sigma_{z1}, \sigma_{z2}, \sigma_{z3}$.

The limiting values of stress through the specified main stress and safety margin coefficient can be expressed:

$$\sigma_{z_1} = \sigma_1 n; \quad \sigma_{z_2} = \sigma_2 n; \quad \sigma_{z_3} = \sigma_3 n \tag{3}$$

Inserting the expression (1) to the Eq. (2), we will receive [3]

$$An^2(\sigma_1 + \sigma_2 + \sigma_3) + Bn^2(\sigma_1^2 + \sigma_2^2 + \sigma_3^2) + Cn^2(\sigma_1\sigma_2 + \sigma_2\sigma_3 + \sigma_3\sigma_1) + \\ + Dn^2\sqrt{\sigma_1^2 + \sigma_2^2 + \sigma_3^2} + En^2\sqrt{(\sigma_1 - \sigma_2)^2 + (\sigma_2 - \sigma_3)^2 + (\sigma_3 - \sigma_1)^2} - 1 = 0 \tag{4}$$

or

$$n^2\left[B(\sigma_1^2 + \sigma_2^2 + \sigma_3^2) + C(\sigma_1\sigma_2 + \sigma_2\sigma_3 + \sigma_3\sigma_1)\right] + n^2[A(\sigma_1 + \sigma_2 + \sigma_3) + \\ + D\sqrt{\sigma_1^2 + \sigma_2^2 + \sigma_3^2}] + n^2\left[E\sqrt{(\sigma_1 - \sigma_2)^2 + (\sigma_2 - \sigma_3)^2 + (\sigma_3 - \sigma_1)^2}\right] - 1 = 0. \tag{5}$$

From the quadratic Eq. (2) it is easy to find

$$n = \frac{1}{2\left[B(\sigma_1^2 + \sigma_2^2 + \sigma_3^2) + C(\sigma_1\sigma_2 + \sigma_2\sigma_3 + \sigma_3\sigma_1)\right]} \times$$

$$\times\{-A(\sigma_1 + \sigma_2 + \sigma_3) + D\sqrt{\sigma_1^2 + \sigma_2^2 + \sigma_3^2} + E\sqrt{(\sigma_1 - \sigma_2)^2 + (\sigma_2 - \sigma_3)^2 + (\sigma_3 - \sigma_1)^2} + \\ + \{[A(\sigma_1 + \sigma_2 + \sigma_3) + D\sqrt{\sigma_1^2 + \sigma_2^2 + \sigma_3^2} + \\ + E\sqrt{(\sigma_1 - \sigma_2)^2 + (\sigma_2 - \sigma_3)^2 + (\sigma_3 - \sigma_1)^2}]^2\}^{1/2} + \\ + 4[B(\sigma_1^2 + \sigma_2^2 + \sigma_3^2) + C(\sigma_1\sigma_2) + \sigma_2\sigma_3 + \sigma_3\sigma_1]\}. \tag{6}$$

In the case of a limiting tense situation ($\sigma_{z3} = 0$), the formula for the safety factor will take the form:

$$n = \frac{1}{2\left[B(\sigma_1^2 + \sigma_2^2 + \sigma_3^2) + C(\sigma_1\sigma_2)\right]} \cdot \{-[A(\sigma_1 + \sigma_2) + \\ + D\sqrt{\sigma_1^2 + \sigma_2^2} + E\sqrt{(\sigma_1 - \sigma_2)^2 + \sigma_2^2 + \sigma_1^2}] + \\ + \{[A(\sigma_1 + \sigma_2) + D\sqrt{\sigma_1^2 + \sigma_2^2} + E\sqrt{(\sigma_1 - \sigma_2)^2 + \sigma_2^2 + \sigma_1^2}]^2 \\ + 4B(\sigma_1^2 + \sigma_2^2 + C(\sigma_1\sigma_2))\}^{1/2}\}. \tag{7}$$

Using Eq. (3), we can derive formulas for equivalent stress. Usually, equivalent stress this main stress of an imaginary stretched element of the same material as the given element, and the stretched element is in the same dangerous tense situation. On condition that for brittle materials the limiting value of the main stress at tension is the temporary resistance of the material for tension σtem., then the equivalent stress must be equal to:

$$\sigma_{\text{ekv}} = \frac{\sigma_{\text{tem}}}{n} .$$ (8)

Multiply the left side of Eq. (4) by $\frac{\sigma_{\text{tem}}^2}{n^2}$:

$$
\begin{aligned}
& A\sigma_{\text{tem}} \cdot \frac{\sigma_{\text{tem}}}{n}(\sigma_1 + \sigma_2 + \sigma_3) + B\sigma_{\text{tem}}^2(\sigma_1^2 + \sigma_2^2 + \sigma_3^2) + \\
& + C\sigma_{\text{tem}}^2(\sigma_1\sigma_2 + \sigma_2\sigma_3 + \sigma_3\sigma_1) + D\sigma_{\text{tem}} \cdot \frac{\sigma_{\text{tem}}}{n}\sqrt{\sigma_1^2 + \sigma_2^2 + \sigma_3^2} + \\
& + E\sigma_{\text{tem}} \cdot \frac{\sigma_{\text{tem}}}{n}\sqrt{(\sigma_1 - \sigma_2)^2 + (\sigma_2 - \sigma_3)^2 + (\sigma_3 - \sigma_1)^2} - \frac{\sigma_{\text{tem}}^2}{n^2} = 0.
\end{aligned}
$$ (9)

We introduce new constant parameters:

$$a = A\sigma_{\text{tem}}; \quad b = B\sigma_{\text{tem}}^2; \quad c = C\sigma_{\text{tem}}^2; \quad d = D\sigma_{\text{tem}}; \quad e = E\sigma_{\text{tem}} .$$ (10)

Then the Eq. (10) with use of equality (9) assume the form

$$
\begin{aligned}
& \sigma_{\text{ekv}}^2 - \sigma_{\text{ekv}}\Big[a(\sigma_1 + \sigma_2 + \sigma_3) + d\sqrt{\sigma_1^2 + \sigma_2^2 + \sigma_3^2} + \\
& + e\sqrt{(\sigma_1 - \sigma_2)^2 + (\sigma_2 - \sigma_3)^2 + (\sigma_3 - \sigma_1)^2} + b\left(\sigma_1^2 + \sigma_2^2 + \sigma_3^2\right) + \\
& + c(\sigma_1\sigma_2 + \sigma_2\sigma_3 + \sigma_3\sigma_1)\Big] = 0.
\end{aligned}
$$ (11)

Equation (11) is the quadratic equation relatively to σekv, and the equation can be solved easily:

$$
\begin{aligned}
\sigma_{\text{ekv}} = & \tfrac{1}{2}\Big[a(\sigma_1 + \sigma_2 + \sigma_3) + d\sqrt{\sigma_1^2 + \sigma_2^2 + \sigma_3^2} + \\
& + e\sqrt{(\sigma_1 - \sigma_2)^2 + (\sigma_2 - \sigma_3)^2 + (\sigma_3 - \sigma_1)^2}\Big] + \\
& + \tfrac{1}{2}\Big\{\Big[a(\sigma_1 + \sigma_2 + \sigma_3) + d\sqrt{\sigma_1^2 + \sigma_2^2 + \sigma_3^2} + \\
& + e\sqrt{(\sigma_1 - \sigma_2)^2 + (\sigma_2 - \sigma_3)^2 + (\sigma_3 - \sigma_1)^2}\Big]\Big\}^{1/2} \dots \\
& + \tfrac{1}{2} + \{4[b\left(\sigma_1^2 + \sigma_2^2 + \sigma_3^2\right) + c(\sigma_1\sigma_2 + \sigma_2\sigma_3 + \sigma_3\sigma_1)]\}.
\end{aligned}
$$ (12)

The expression for the equivalent stress turned out to be somewhat simpler than the expression for the safety margin; therefore, practical calculations should be carried out using formula (12). The values of the coefficients a, b, c, d, e are given in the Table 1, in which for each coefficient the significant digits of 9 obtained during the determination of the parameters A, B, C on the computer and at calculations by formula (11) are stored.

Table 1. The values of the coefficients a, b, c, d, e

Parameters	K-18-2	K-18-2p
A	0,846153·10-3	0,944981·10-3
B	0,04409·10-6	0,19208·10-6
C	0,713242·10-6	0,809428·10-6
D	−1,31266·10-3	−1,85106·10-6
E	1,83274·10-3	2,08134·10-3
A	0,394307	0,444141
B	0,0095699	0,0424324
C	0,154885	0,178801
D	−0,611699	−0,8699
E	0,854059	0,978229

It is possible to reduce the values of the coefficients to 3 significant figures; however, the arithmetic error in the final result of formula (12) in this case will be about 10%. A further decrease of number of significant figures is undesirable, due to an increase in the summary arithmetic error, which is no difficult to verify by substituting the following values of principal stresses for any material in the formula (12): $\sigma_1 = 0$, $\sigma_2 = \sigma_{tem}$, $\sigma_3 = 0$. With a small number of signs at coefficients a, b, c, the value of σ_{ekv} may be different from the value of σ_{tem} due to large errors in the calculations.

For a case of flat tense situations ($\sigma_3 = 0$) the formula for equivalent stress drawn up by N.N. Davydenkov and A.N. Stavreqin has the following form:

$$\sigma_{ekv} = \frac{1}{2}\left[a(\sigma_1 + \sigma_2) + d\sqrt{\sigma_1^2 + \sigma_2^2} + e\sqrt{(\sigma_1 - \sigma_2)^2 + \sigma_1^2 + \sigma_2^2}\right] +$$

$$+ \frac{1}{2}\left\{\left[a(\sigma_1 + \sigma_2) + d\sqrt{\sigma_1^2 + \sigma_2^2} + e\sqrt{\left[(\sigma_1 - \sigma_2)^2 + \sigma_1^2 + \sigma_2^2\right]^2}\right]\right\}^{1/2} + \quad (13)$$

$$4\left[b\left(\sigma_1^2 + \sigma_2^2 + \sigma_3^2\right) + c\sigma_1\sigma_2\right].$$

If the detail is projected anew, then the condition of strength will be the following:

$$\sigma_{ekv} \leq [\sigma]_{ten}, \quad (14)$$

where $[\sigma]_{ten}$ – is the allowable stress at tension, equal:

$$[\sigma]_{ten} = \frac{\sigma_{tem}}{n}, \quad (15)$$

in which the safety margin coefficient is setting.

If check of ready details is made, then it is necessary to use a formula:

$$n = \frac{\sigma_{\text{tem}}}{\sigma_{\text{ekv}}}. \tag{16}$$

Of the formula (12) follows, that the coefficients a, b, c, d and e from table are dimensionless quantities. Then of the formula (11) it is easy to establish the dimensions of the parameters A, B, C, D and E:

$$[A] = [D] = [E] = \frac{cm^2}{kg^2}, \quad [B] = [C] = \frac{cm^2}{kg^2}.$$

To construct the limiting curves, formula (8) was programmed for a computer. Inserting in this formula any value $\sigma 1$ and $\sigma 2$, to find the safety margin coefficient for the tense situation with the ratio of main stress is equal $\sigma 2/\sigma 1$, consequently, and the limit values of stress by formula (9).

If at search of coordinates of points of a limiting curve, we take for one of stress $\sigma 1$ or $\sigma 2$ the value equal to unit, then the safety margin coefficient will be numerically equal to one of the limiting stress, for example, σ_{Z1}. Then it is easy to find the second limiting stress by formula:

$$\sigma_{z_2} = \sigma_{z_1}(\sigma_2/\sigma_1).$$

It should be noted that the derived formulas are suitable only for a case of simple loading when all three main stress change synchronously. As a rule, this case meets in the huge majority of details of machines.

3   Discussion and Conclusions

On the basis of a method of calculation of criterion of strength depending on synchronous change of all three principal stresses are considered methods of calculation of equivalent stress. Obtained theoretical results are confirmed experimentally and the method of calculation of equivalent stress is developed that is necessary for practice. It should be noted that the derived formulas are suitable only for a case of simple loading when all three main stress change synchronously. As a rule, this case meets in the huge majority of details of machines.

References

1. Kerimov, D.A.: Scientific bases and practical methods of optimization of parameters of quality of plastic details of the oil-field equipment. Dissertation of Doctorate Technical Sciences, Baku (1985)
2. Kerimov, D.A., Kurbanova, S.K.: Bases of Designing of Plastic Details and Press-Moulds. Elm Publishing House, Baku (1997)
3. Kerimov, D.A., Gasanova, N.A.: Development of optimal regimes of cooling for plastic parts of thermosets. SAEQ Sci. Appl. Eng. Q., 17–19, (2017)

Evaluation of Muscle Activities During Different Squat Variations Using Electromyography Signals

Erdag Deniz[1]($\boxtimes$) (iD) and Yavuz Hasan Ulas[2] (iD)

[1] Faculty of Sports Sciences, Near East University, Nicosia, North Cyprus
deniz.erdag@neu.edu.tr
[2] Department of Sports Medicine, Faculty of Medicine, Near East University,
Nicosia, North Cyprus
ulas.yavuz@neu.edu.tr

Abstract. The importance of Electromyography (EMG) signals analysis with advanced methodologies is increasing in biomedics, clinical diagnosis and biomechanics and it becomes a required practice for many scientists from both health and engineering fields. The squat is a very important exercise for improving athletic performance and for prevention and rehabilitation of injuries. It has many different variations which are supposedly focusing on different muscles. We aimed to compare vastus medialis, rectus femoris, vastus lateralis, gluteus maximus, semitendinosus, biceps femoris and erector spinae EMG activities during popular squat variations while the participants (14 healthy males, 23.7 $\pm$ 2.7 years-old) were performing 6 repetitions of front squat, back squat, hack squat, sumo squat and zercher squat with 60% of the 1 repetition maximum loading. Muscle EMG activities during different variations were compared by using Repeated measures ANOVA. The highest rectus femoris, vastus lateralis and vastus medialis EMG activities were observed during front squat with a significant difference with Zercher squat. EMG activities of Erector spinae and semitendinosus during Hack squat were significantly lower than all other squat variations ($p < 0.05$). These findings may suggest that front squat may be chosen to focus on quadriceps muscles while Hack squat may be a good choice for better knee and spinal stabilization.

Keywords: Strength training · Resistance training · Electromyography

1 Introduction

EMG signals arise from muscles electrical activity that is produced during contraction and under the control of nervous system. The signal represents the features of the anatomy and physiology of the related muscle. Briefly it is the electrical activity of the motor units of the tested muscle [1]. EMG signal can be measured from the skin surface during the contraction and relaxation of the muscle. Every movement of a muscle causes a specific activation pattern of motor units; thus multi-channel EMG signaling might be used for identification of the movement [2]. These signals are widely used electro-physiological signals in many different fields like medicine, engineering, biomechanics

© Springer Nature Switzerland AG 2020
R. A. Aliev et al. (Eds.): ICSCCW 2019, AISC 1095, pp. 859–865, 2020.
https://doi.org/10.1007/978-3-030-35249-3_114

etc. The squat has a very important place in all type of competitive sports and recreational exercises. It is regarded the "most important" exercise in strength training by coaches and sport scientists [3] or "king" [4] of strengthening exercises for the lower body [5–11]. During the squat, the force is expressed through the length of the leg while the feet are fixed to the ground [5, 12]. It is a multi-joint exercise activating many big and strong muscles and biomechanically similar to fundamental movements of athletics [10]. It is used by athletes to enhance performance and for prevention or rehabilitation of injuries. Gluteus maximus, hamstrings, quadriceps and gastrocnemius are the mainly active muscles during squat [6, 13, 14]. Ankle, knee and hip joints musculature are active with the contribution of the abdominals and spinal erectors [8, 14].

Considering the complexity of the exercise there are large number of different types of the loaded barbell squat that are used in many training programs for the lower-body development. During back squat the barbell is positioned right above deltoids, on trapezius muscle and subject crouches down till the femurs come parallel to the ground [13, 14]. Then he stands up to come back to the starting posture [13, 14]. Differently in front squat, the bar is placed athwart the clavicles and deltoid muscles and the subject holds the bar with the hands while the elbows are fully flexed and the humeri are parallel to the ground [13, 14]. Then he descents and ascends in the same way in back squat [7]. During the sumo squat the bar is carried across the back of the shoulders, feet are placed wider than shoulder width while the toes slightly externally rotated. In barbell hack squat the subject holds a barbell behind his back, stands straight while the elbows are fully extended and the in the zercher squat the bar is hold in the crook of the elbows and squat. Different applications of squat exercise have been investigated and/or compared in terms of kinetics, kinematics and muscle activation [7, 9–11, 15–18]. Gullet et al. evaluated knee biomechanics and muscle activities during back and front squat. They reported that 'the front squat was as effective as the back squat in terms of overall muscle recruitment, with significantly less compressive forces on the knee'. Yavuz et al. used maximal loading and reached the similar results with Gullet and emphasized that front squat can activate quadriceps as much as (if not more) back squat does despite the lower loads used [10]. Marcehtti et al. showed that gluteus maximus and quadriceps muscle activation could be affected by the knee position [18]. Eventually there have been many studies that compared different techniques, body and load positioning during squatting. However, none of these studies has compared back squat, front squat, sumo squat, hack squat and Zercher squat variations and their muscle activity patterns together. Understanding the muscle activity patterns of squat variations and their impact on lower-body musculature is very important to have optimal gains on muscles and allowing athletes and coaches to choose the effective technique for performance. It is also very important to choose the right technique for preventing injuries and planning the rehabilitation programs. Therefore, the purpose of this study was to compare the muscle EMG activity during popular squat variations.

2 Materials and Methods

2.1 Subjects

Fourteen healthy males (23.7 ± 2.7 years) who have been performing squats in their training routines (for 3.2 ± 0.6 years) took place in this study. The mean height and weight of the participants were 180.3 ± 6.4 cm and 86.8 ± 10.2 kg respectively. Right leg was dominant for all subjects. No subject had any orthopedic injury that may affect or prevent his squat performance. Informed consents were taken from the participants before participation. The procedure was approved by the Ethic Commission of Near East University (YDU/2017/43-361).

2.2   Instrumentation

We followed the methods of Yavuz et al. [11]. 8-channel portable EMG system (Myomonitor IV, Delsys) was used to evaluate the electromyographic (EMG) activity. EMG Works Acquisition 4.0.5 (Delsys) was used to collect data. The bandwidth frequency and data record were done according to procedure from Yavuz et al. [11]. Electrodes (41 mm $\times$ 20 mm $\times$ 5 mm, Delsys Inc.) were placed on the selected muscles on the dominant leg according to the procedures from Gullet et al. [7].

2.3   Procedures

The subjects attended 6 periods, about 50 min each, in approximately 2 weeks. Five sessions were pretesting sessions to determine the 1Repetetive Maximum (RM) (Maximum load that can be lifted by the subject with a proper technique) for each variation with two days off between each session in a random order for each subject. 1 RM's were taken according to the Kraemer and Fry [19]. Data collection was done 3 days after the pre-test sessions. All EMG data were obtained in the same testing session eliminating the possibility of day-to-day fluctuations and improper reapplication of EMG electrodes. After the EMG electrodes were placed maximum voluntary isometric contraction (MVIC) data were assessed with the method developed by Konrad et al. [20]. EMG data collection was simultaneously started with the initiation of each set of squats. For each variation, the subjects performed 6 repetitions of squats with 60% of their 1RM for each variation. The order of squat variations was random for each subject. Data obtained from the second, third, fourth and fifth repetitions were used for analyses. Ten minutes rest was given in between each set.

2.4   Data Processing

The procedure developed by International Society of Electrophysiology and Kinesiology [21] was used to analyze EMG data. The normalization of EMG data was done according the method of Yavuz et al. [11].

2.5   Statistical Analysis

The muscle EMG activities for different variations were given as a percentage of MVIC (MVIC%) and compared by repeated measures ANOVA. All EMG data were given as means with standard deviations.

3   Results

Figure 1 shows that the load for BS was significantly higher (103.5 ± 20.5, p = .000) compared to FS (79.7 ± 11.4, p = .000), HS (80.1 ± 18.3, p = .017) and ZS (74.1 ± 11.4, p = .000). Load for sumo squat (91.5 ± 12.9) was significantly higher compared to FS (73.1 ± 11.4, p = .006) and ZS (74.1 ± 11.4, p = .000).

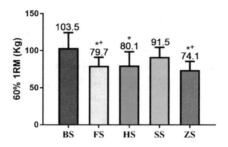

* BS found significantly higher compared to FS, HS and ZS p<0.05
+ Sumo squat found significantly higher compared to FS and ZS p<0.05

Fig. 1. Repeated measures ANOVA results for mean (±SD) 60% of 1 RM (kg) loads during BS, FS, HS, SS and ZS.

Table 1. I. The mean (± SD) EMG activities as MVIC% during back squats, front squats, hack squats, sumo squats and zercher squats performed with 60% of 1RM loads.

Muscle	Back squat	Front squat	Hack squat	Sumo squat	Zercher squat	p
Rectus femoris	31.8 ± 19.9	36.1 ± 17.7	32.1 ± 17.7	28.7 ± 18.2	29.2 ± 17.4[†]	**.005**
Vastus medialis	59.5 ± 21.6	61.6 ± 18.9	56.1 ± 22.4	59.6 ± 18.9	54.9 ± 18.1[†]	**.021**
Vastus lateralis	56.3 ± 19.5	58.4 ± 14.1	52.1 ± 15.4	56.5 ± 14.3	51.8 ± 15.3[†]	**.010**
Erector spinae	41.9 ± 22.2	36.8 ± 17.1	20.9 ± 9.1[*†ab]	39.2 ± 21.2	37.8 ± 14.9	**.002**
Gluteus maximus	19.5 ± 8.5	18.3 ± 6.7	14.2 ± 9.1	18.3 ± 7.5	18.1 ± 6.6	.119
Biceps femoris	7.1 ± 3.4	6.8 ± 6.7	6.5 ± 5.5	7.1 ± 3.1	7.3 ± 3.8	.769
Semitendinosus	10.1 ± 5.1	8.2 ± 3.1	5.1 ± 2.5[*†ab]	9.4 ± 3.7	9.9 ± 3.5	**.002**

*Back squat significantly higher than, †Front squat significantly higher than, [a]Sumo squat significantly higher than, [b]Zercher squat is significantly higher than p < 0.05

Table 1 shows that the front squat rectus femoris mean EMG activity (36.1 ± 17.7) was found as the highest and sumo squats rectus femoris mean EMG activity (28.7 ± 18.2) was found as the lowest compared to the other squat variations. The only statistically significant difference was that front squat (36.1 ± 17.7, P = .002) mean EMG activity was significantly higher than zercher squat (29.2 ± 17.4, P = .002). Similarly front squat vastus medialis (61.6 ± 18.9) and vastus lateralis (58.4 ± 14.1) mean EMG activities were found as the highest and zercher squat vastus medialis (54.9 ± 18.1) and vastus lateralis mean (51.8 ± 15.3) EMG activities were found as the lowest compared to other variations. During Hack squat, erector spinae mean EMG activity (20.9 ± 9.1) and semitendinosus mean EMG activity (5.1 ± 2.5) were found significantly lower than all other variations. There was no significant difference between any squat variations studied for biceps femoris and gluteus maximus muscles.

4 Discussion

The loads employed during the back squat were significantly higher than front squat, hack squat and zercher squat. Loads employed during sumo squat were significantly higher compared to front squat and zercher squat. Despite of the higher loads during back squat and sumo squat, we did not observe any effect of these higher loads on EMG activities. It may be because of, not the load, but the effort determines the amount of motor unit recruitment during muscle actions [22]; consequently, the EMG signaling. Since we used 60% of 1RM in all trials, the relative effort would be considered as equal or near equal.

Quadriceps femoris are the main active muscles during squat and they concentrically extent the knee while resisting the flexion eccentrically [9]. We observed that the overall quadriceps muscle EMG activities were the lowest during Zercher squat and it was significantly lower than front squat. No other difference was ob-served for any squat variations for quadriceps muscles. Gullett et al. [7] showed that the there was no difference between EMG activities for tested muscles during front and back squat despite of the lower loads during front squat. Our findings were in accordance with this study [6]. In our previous study we observed higher vastus medialis EMG activities during front squat with lower loads compared to back squat [10]. But the loads were %100 of 1RM for each squat variation. It was shown that the movement pattern and muscle activation could change with different loads [10, 11]. Gullet et al. [7] also observed increased compression in knee joint, which might hurt meniscus and cartilage during back squat. They emphasized that the high-er loads employed might be the reason of this increase. In the light of these findings it might be wise to use the front squat, which may equally activate the targeted muscle(s) with lower loads. GM and hamstrings are the primarily active hip muscles dur-ing squat. GM is primarily responsible for hip extension, controls descending eccentrically and overcomes the resistance during ascending concentrically. GM has also a very important role in knee and pelvis stabilization [9]. The hamstrings (semitendinosus, semimembranosus, biceps femoris) are antagonists of the quadriceps, respon-sible for knee flexion and opposing extensor moments. During squat, hamstrings and quadriceps contract together for stabilization of the knee joint [9]. We did not see any difference in EMG activities for GM and BF between any squat variations. It is known that squat depth is very important in activation of the GM. Caterisano et al. showed that GM activity increased with the increase of squat depth [23]. It was also shown that GM activity increased by increasing relative loads from 80% to 90 and 100% of 1 RM [11]. Since the squat depth and relative load were kept equal in all variations in our study, it might be reason that there was no difference between variations.

As an interesting finding of our study, we observed significantly lower se-mitendinosus and erector spinae muscle activities during Hack squat compared to all other variations. These muscles are important for stabilization of knee and hip while squatting. Although there are many muscles supporting the spine, the erector spinae have special importance since they help to protect spine against shear forces and keep integrity of spine. Eventually, they stabilize the spine [24]. Considering the im-portance of the knee and the spinal stabilization for preventing injuries, Hack squat might be a good choice with similar muscle activity and better knee and spinal stabi-lization. As a limitation of our study, we did not evaluate the kinematics of the squat variations. Further studies with kinematic analysis and muscle activity assessment are needed to be able to support our findings.

References

1. Chowdhury, R., Reaz, M., Ali, M., Bakar, A., Chellappan, K., Chang, T.: Surface electromyography signal processing and classification techniques. Sensors 13(9), 12431–12466 (2013)
2. Alkan, A., Günay, M.: Identification of EMG signals using discriminant analysis and SVM classifier. Expert Syst. Appl. 39(1), 0957–4174 (2012)
3. Comfort, P., Kasim, P.: Optimizing squat technique. Strength Cond. J. 1(29), 10–13 (2007)
4. O'shea, O.: Sports performance series: the parallel squat. Strength Cond. J. 7, 4–6 (1985)
5. Clark, D.R., Lambert, M.I., Hunter, A.M.: Muscle activation in the loaded free barbell squat: a brief review. J. Strength Cond. Res. 26, 1169–1178 (2012)
6. Escamilla, R.F.: Knee biomechanics of the dynamic squat exercise. Med. Sci. Sports Exerc. 33, 127–141 (2001)
7. Gullett, J.C., Tillman, M.D., Gutierrez, G.M., Chow, J.W.: A biomechanical comparison of back and front squats in healthy trained individuals. J. Strength Cond. Res. 23, 284–292 (2009)
8. Paoli, A., Marcolin, G., Petrone, N.: The effect of stance width on the electromyographical activity of eight superficial thigh muscles during back squat with different bar loads. J. Strength Cond. Res. 23, 246–250 (2009)
9. Schoenfeld, B.J.: Squatting kinematics and kinetics and their application to exercise performance. J. Strength Cond. Res. 24(12), 3497–3506 (2010)
10. Yavuz, H.U., Erdağ, D., Amca, A.M., Aritan, S.: Kinematic and EMG activities during front and back squat variations in maximum loads. J. Sports Sci. 33, 1058–1066 (2015)
11. Yavuz, H.U., Erdag, D.: Kinematic and electromyographic activity changes during back squat with submaximal and maximal loading. Appl. Bion. Biomech. Article ID 9084725, p. 8 (2017)
12. Escamilla, R.F., Fleisig, G.S., Lowry, T.M., Barrentine, S.W., Andrews, J.R.: A three-dimensional biomechanical analysis of the squat during varying stance widths. Med. Sci. Sports Exerc. 33, 984–998 (2001)
13. Baechle, T.R., Earle, R.: Essentials of Strength Training and Conditioning, 2nd edn. Human Kinetics, Champaign (2000)
14. Delavier, F.: Strength Training Anatomy. Human Kinetics, Champaign (2001)
15. Aspe, R.R., Swinton, P.A.: Electromyographic and kinetic comparison of the back squat and overhead squat. J. Strength Cond. Res. 28, 2827–2836 (2014)
16. Contreras, B., Vigotsky, A.D., Schoenfeld, B.J., Beardsley, C., Cronin, J.: A comparison of gluteus maximus, biceps femoris, and vastus lateralis electromyography amplitude in the parallel, full, and front squat variations in resistance-trained females. J. Appl. Biomech. 32, 16–22 (2016)
17. Dionisio, V.C., Almeida, G.L., Duarte, M., Hirata, R.P.: Kinematic, kinetic and EMG patterns during downward squatting. J. Electromyogr. Kinesiol. 18, 134–143 (2008)

18. Marchetti, P.H., Jarbas da Silva, J., Schoenfeld, J.B., Nardi, P.S., Pecoraro, S.L., D'Andréa Greve, J.M., Hartigan, E.: Muscle activation differs between three different knee joint-angle positions during a maximal isometric back squat exercise. J. Sports Med. Epub, Article ID 3846123, p. 6 (2016)
19. Kraemer, W.J., Fry, A.C., Ratamess, N., French, D.: Strength testing: development and evaluation of methodology. Physiol. Assess. Hum. Fitness 2, 119–150 (1995)
20. Konrad, P.: The ABC of EMG: A Practical Introduction to Kinesiological Electromyography, Noraxon Inc., USA (2005)
21. Merletti, R., Di Torino, P.: Standards for reporting EMG data. J. Electromyogr. Kinesiol. 9, 3–4 (1999)
22. Carpinelli, R.N.: The size principle and a critical analysis of the unsubstantiated heavier-is-better recommendation for resistance training. J. Exerc. Sci. Fitness 6(2), 67–86 (2008)
23. Caterisano, A., Moss, R.F., Pellinger, T.K., Woodruff, K., Lewis, V.C., Booth, W., Khadra, T.: The effect of back squat depth on the EMG activity of 4 superficial hip and thigh muscles. J Strength Cond. Res. 16, 428–432 (2002)
24. Toutoungi, D.E., Lu, T.W., Leardini, A., Catani, F., O'Connor, J.J.: Cruciate ligament forces in the human knee during rehabilitation exercises. Clin. Biomech. 15, 176–187 (2000)

Synthesis of "iron-cast iron-glass" Obsolete Powder Composite Materials Using Fuzzy Logic

T. G. Jabbarov[(⊠)] 🆔

Department of Mechanical and Materials Science Engineering,
Azerbaijan State Oil and Industry University, 20 Azadliq Avenue,
Az1010 Baku, Azerbaijan
tahir-cabbarov@mail.ru

Abstract. The beginning of the history of powder metallurgical development dates back to the 60s of the 20th century. All these years, the synthesis of obsolete composites has been based on experimental research. At the same time, heterogeneous structure, optimizing the composition of composite materials derived from the interaction of several components, is a requirement for determining the optimal composition of the material balance of computer applications. In this paper we consider construction of fuzzy IF-THEN rules from experimental data to describe relation between material composition and properties under uncertainty.

Keywords: Composite materials · Iron-cast iron-glass · Modelling · Phase · Material properties

1 Introduction

In order to improve inter-phase interaction between the iron-and-glass-bush powder composition materials and to increase the mechanical properties of the iron-glass oven composite materials and to increase the durability of the tribotehnical properties, it is necessary to add components such as cold-rolled ovulation material to the composition, they have. The use of shingles in the shingles makes it possible to achieve such effectiveness, as the amount of silica and maca is much higher in the composition of the chute. Which, in the process of heating, produce the corresponding oxidizations that are difficult to decompose themselves. Such oxides ensure the improvement of the heating of the metal carcass by glass in the cooking process.

The optical removal of the optical composition for the preparation of the "iron-cast iron-glass" oval pusher taken from the research object is considered as theoretically and experimentally important. Usually, such issues are solved on a case-by-case basis, with errors in the accuracy of the received prices. It is one of the actual requirements of the day to work on the synthesis of materials with computer programs to eliminate these errors.

The solution of the problem in this direction was based on the mathematical models and data obtained on the basis of the data gathered from the previous experiments. In this regard, Nosengo [7] has learned to use the computer and databases to prepare new

© Springer Nature Switzerland AG 2020
R. A. Aliev et al. (Eds.): ICSCCW 2019, AISC 1095, pp. 866–871, 2020.
https://doi.org/10.1007/978-3-030-35249-3_115

material in his article. He noted that the databases on existing materials and their properties were used by various academics. Such databases will help disseminate information about new materials with pre-existing features [2].

Raccuglia and his colleagues [8] used machine learning algorithms in the synthesis of new hybrid materials. According to the authors, this method can replace experimental methods.

Implementing fuzzy logic for modeling and planning of properties helps to better describe the development of material production and facilitate the results [3, 5, 6]. Particularly, it can be used to derive models from big data (huge set of structured and unstructured data from different sources [9]) on material composition and properties.

In this paper we use fuzzy If-Then rules to describe relation between material composition and related properties under uncertainty. The rules are derived by using fuzzy clustering of experimental data.

The paper is organized as follows. In Sect. 2 includes preliminary material used in the paper. In Sect. 3 we describe a problem of deriving a model of relation between material composition and related properties. The model is described in a form of fuzzy If-Then rules derived from experimental data by using fuzzy clustering. In Sect. 4 we consider application of rules construction for "iron-cast-glass" obsolete composite materials. Section 5 concludes.

2   Preliminaries

Definition 1. A fuzzy number [1] A fuzzy number is a fuzzy set A on R which possesses the following properties: (a) A is a normal fuzzy set; (b) A is a convex fuzzy set; (c) α-cut of A, A^{α} is a closed interval for every $\alpha \in (0, 1]$; (d) the support of A, supp (A) is bounded.

Definition 2. Fuzzy C-means (FCM) clustering [4]. FCM clustering problem is a problem of partitioning the dataset $X = \{x_1, \ldots, x_n\}$ into c fuzzy clusters formulated as follows:

$$J_m = \sum_{i=1}^{n} \sum_{j=1}^{c} u_{ij}^m \left\| x_i - v_j \right\|^2 \rightarrow \min$$

subject to

$$u_{ij}^m = \frac{1}{\sum_{j=1}^{c} \left(\frac{\left\| x_i - v_j \right\|}{\left\| x_i - v_k \right\|} \right)^{\frac{2}{m-1}}},$$

$$v_j = \frac{\sum_{i=1}^{n} u_{ij}^m x_i}{\sum_{i=1}^{n} u_{ij}^m},$$

$$1 \le m < \infty,$$

here u_{ij} is the degree of membership of data point x_i to the j-th fuzzy cluster, $\|\cdot\|$ is a norm, m is a fuzzifier which defines curvature of membership functions of obtained clusters, v_j is a center of the j-th fuzzy cluster.

3 Construction of Fuzzy Model of Relationship Between Material Composition and Properties

Suppose that big experimental data on relation of material composition and its properties exist (Table 1).

Table 1. Structure of experimental data.

Experiment	Material composition (mass in %)				Material properties		
#	Iron, y_1	...	Cast-iron, y_2	...	Glas, y_n	Characteristic1, z_1	...	Characteristic m, $z_{m,}$
1	y_{11}	...	y_{12}	...	y_{1n}	z_{11}	....	z_{1m}
.								
.								
.								
S	y_{sl}	...	y_{s2}	...	y_{sn}	z_{sl}	...	z_{sm}

Real-world experimental data are characterized by uncertainty. Our purpose is to construct a formal model of relation between material composition and its properties by that may account for uncertainty. In view of this, the considered problem is construction of fuzzy IF-THEN rules to summarize experimental data in Table 1. For deriving rules, we use fuzzy C-means clustering (Definition 2) of data in $n \times m$ dimensional space. As a result, data are partitioned into fuzzy clusters C_k, $k = 1,...,$ K. By projecting any fuzzy cluster on axes of multidimensional space, we uncover a fuzzy rule:

IF y_1 is A_{k1} and,..., and y_n is A_{kn} THEN z_1 is B_{k1} and,..., and z Medium is B_{km}, $k = 1,..., K$

$A_{k1},..., A_{kn}$ and $B_{k1},..., B_{km}$ are fuzzy terms (projections of C_k $n \times m$ on axes of dimensional space) describe imprecise values of percentage of components and material properties respectively, K is a number of rules. Such fuzzy rules provide a formal model that represent knowledge about dependence of material properties on material composition. Such a model provides an intuitive description of complex relationship under uncertainty.

The fuzzy IF-THEN rules based model can be used for material analysis (analysis of impact of individual components to properties etc.) and material synthesis (determination a composition that induce required property levels).

4   An Application

Let us consider a problem of modeling of "iron-cast-iron-glass" material properties dependence on material composition and test temperature. A fragment of related experimental data is shown in Table 2:

Table 2. A fragment of the big data on "iron-cast-iron-glass" obsolete powder composite materials.

Powder composition			Properties				
y_1 (Fe, mass, %)	y_2 (Cast-iron, mass, %)	y_3 (Glas, mass, %)	z_1 (conventional ultimate strength, MPa)	z_2 (conventional yield strength, MPa)	z_3 (hardness, HBB)	z_4 (resilience, kJ/m^2)	z_5 (relative density, %)
91	3	6	259	212	57	11	48
19	75	6	142	51	134	12	23
			.				
			.				
			.				
44	50	6	335	357	115	48	91
34	50	16	261	160	54	20	79

We applied FCM clustering to construct 5 fuzzy clusters for the considered experimental data. These fuzzy clusters are described by the following IF-THEN rules:

IF y_1 is *High* and y_2 is *Below Medium* and y_3 is *Medium* THEN z_1 is *High* and z_2 is *Very High* and z_3 is *High* and z_4 is *High* and z_5 is *High*,

IF y_1 is *Very High* and y_2 is *Very Low* and y_3 is *Medium* THEN z_1 is *Medium* and z_2 is *Medium* and z_3 is *Low* and z_4 is *L* and z_5 is *Medium*,

IF y_1 is *Very Low* and y_2 is *High* and y_3 is *Medium* THEN z_1 is *Low* and z_2 is *Very Low* and z_3 is *Very High* and z_4 is *L* and z_5 is *Low*,

IF y_1 is *Medium* and y_2 is *Medium* and y_3 is *Low* THEN z_1 is *High* and z_2 is *High* and z_3 is *Medium* and z_4 is *Medium* and z_5 is *High*,

IF y_1 is *Low* and y_2 is *Above Medium* and y_3 is *High* THEN z_1 is *Medium* and z_2 is *Land* z_3 is *Low* and z_4 is *Low* and z_3 is *Medium*.

The graphical description of these rules that includes membership functions of the antecedents (material composition) and consequents (properties) is shown in Fig. 1.

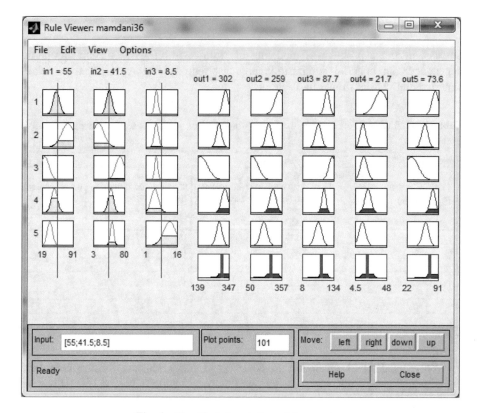

Fig. 1. Graphical description of fuzzy rules

5 Summary

We used FCM clustering of experimental data to derive fuzzy If-Then rules describing relation between material composition and properties. As an example, "iron-cast-iron-glass" is considered. This fuzzy model may be used for the purposes of material analysis and synthesis.

References

1. Aliev, R.A., Aliev, R.R.: Soft Computing and Its Application. World Scientific, New Jersey (2001)
2. Babanli, M.B.: Synthesis of new materials by using fuzzy and big data concepts. In: 9th International Conference on Theory and Application of Soft Computing with Wods and Perception, ICSCCW 2017, Budapest, Hungary, 22–23 August 2017, Procedia Computer Science, vol. 120, pp. 104–112 (2017)
3. Babanli, M.B., Huseynov, V.M.: Z-number-based material selection problem. In: 12th International Conference on Application of Fuzzy System sand Soft Computing, ICAFS 2016,

Vienna, Austria, 29–30 August 2016, Procedia Computer Science, vol. 102, pp. 183–189 (2016)

4. Bezdek, J.C.: Pattern Recognition with Fuzzy Objective Function Algoritms. Plenum Press, New York (1981)

5. Chen, D., Li, M., Wu, S.: Modeling of microstructure and constitutive relation during super plastic deformation by fuzzy-neural network. J. Mater. Process. Technol. **142**, 197–202 (2003)

6. Chen, S.-M.: A new method for tool steel materials selection under fuzzy environment. Fuzzy Sets Syst. **92**, 265–274 (1997)

7. Nosengo, N.: Can artificial intelligence create the next wonder material. Nature **533**, 22–25 (2016)

8. Raccuglia, P., Elbert, K.C., Adler, P.D.F., Falk, C., Wenny, M.B., Mollo, A., Zeller, M., Friedler, S.A., Schrier, J., Norquist, A.J.: Machine-learning-assisted materials discovery using failed experiments. Nature **533**, 73–76 (2016)

9. What is big data analytics? https://www.ibm.com/analytics/hadoop/big-data-analytics

Artificial Neural Networks for Predicting the Electrical Power of a New Configuration of Savonius Rotor

Youssef Kassem[1,2](✉) ⓘ, Hüseyin Gökçekuş[1], and Hüseyin Çamur[2]

[1] Faculty of Civil and Environmental Engineering, Near East University,
99138 Nicosia, North Cyprus, Turkey
{yousseuf.kassem, huseyin.gokcekus}@neu.edu.tr
[2] Faculty of Engineering, Mechanical Engineering Department,
Near East University, 99138 Nicosia, North Cyprus, Turkey
huseyin.camur@neu.edu.tr

Abstract. The electrical power of new configuration Savonius turbine is estimated using multilayer perceptron neural network (MLPNN) and radial basis function neural network (RBFNN) based on the experimental data which have been collected for 108 vertical axis wind turbine. In this work, the experiments were conducted at air velocities ranging from 3 to 12 m/s in front of a low-speed subsonic wind tunnel. Based on the experimental results, the new configuration of Savonius resulted in a noticeable improvement in the power compared to that of the classical Savonius turbine. This study reveals that the MLPNN has the potential of predicting the mechanical power of a Savonius wind turbine with minimal prediction error scores compared to the RBFNN model.

Keywords: Electrical power · Savonius turbine · MLPNN · RBFNN

1 Introduction

Renewable energy sources are considered clean alternatives to fossil fuels that can provide sustainable energy solutions [1, 2]. Renewable energies such as wind energy are recognized as alternative resources for generating electricity in the future [3]. A key advantage of wind energy is that they avoid carbon dioxide emissions [4].

Wind energy can be converted directly into electricity using wind turbines. Horizontal axis wind turbines are commonly used in energy production for grid-connected large utilities whereas vertical axis types are preferred for use in small-scale domestic applications. Throughout the years, researchers have given a lot of attention to the horizontal axis wind turbine with outstanding achievements in terms of further developing the technology. On the other hand, current conventional designs for the vertical axis wind turbine do not satisfactorily meet the requirements of users in cases of off-grid power generation at low wind speeds. Therefore, investigation of small-scale turbines for distributed energy systems has become popular [5, 6]. The Savonius wind turbine has the potential to fulfill the needs of users for such conditions. It has been reported that this type of wind turbine has a lower efficiency when compared to its

© Springer Nature Switzerland AG 2020
R. A. Aliev et al. (Eds.): ICSCCW 2019, AISC 1095, pp. 872–879, 2020.
https://doi.org/10.1007/978-3-030-35249-3_116

rivals. Savonius wind turbines are good for low wind speed and can be installed on the rooftop of the building or on top of communication towers [7].

Artificial neural network (ANN) is one alternative approach to predict the performance of the wind turbine. ANN has been used in energy systems application and renewable energy system in general.

Therefore, the objectives of the current study are: (1) design a new configuration Savonius wind turbine that can be used to meet the power for low demand applications, (2) investigate the geometries of wind turbine and number of blades on the electrical power of the new configuration of Savonius turbine and (3) predict the electrical power of the rotors using ANN approach and compare the estimated results with the experimental data to show the performance and accuracy of ANN approach for predicting the electrical power of the new configuration Savonius rotors.

2 Experimental Procedures

2.1 Rotor Design and Fabrication

In the present work, two-, three- and four-bladed rotors have been studied using semicircular blades. Top view and isometric view of the new Savonius-style rotors are shown in Fig. 1. The blades of rotors are manufactured from light plastic (PVC) tubes with different diameters and heights. The shaft of the rotor is made from stainless steel with 20 mm diameter and 700 mm long. An attempt has been made to study a variety of rotor configurations with various aspects and overlap ratio. In this study, the overlap ratio is the ratio between the external overlap (L') and the blade diameter (D). The external overlap is the distance between the desk and rotor blade and L is the distance between the center of the shaft and rotor blade as shown in Fig. 1. The wind turbine models were built with various external overlap. It is known that the large external overlap (L') leads to increase the torque and reduce the angular speed of the shaft. In the present investigations, two desks were made from fiberglass with a thickness of 5 mm and placed on the top and bottom of the model. In the present investigation, two desks are placed on the top and bottom of the model. The distance between the two desks depends on the height of the rotor blades. Dimensions of design parameters are shown in Table 1.

2.2 Test Facility

In this work, a low-speed wind tunnel with an open test section facility with a cross-sectional area of 700 mm × 700 mm was designed to evaluate the performance of the new configuration of a Savonius turbine as shown in Fig. 2. The rotor was placed at a distance of 200 mm from the exit of the tunnel. The air velocity was varied between 0–15 m/s and changed by the input voltage with the help of variac. Two Pitot tubes with an accuracy of ±0.1 m/s were used to measure the air velocity. RPM sensor was used to measure the rotational speed (RPM) of the rotor. In addition, the gearbox is used to increase the RPM delivered into the generator.

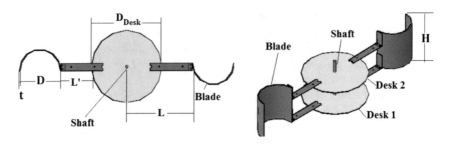

Fig. 1. The schematic shapes of the new Savonius-style rotors.

Table 1. Fixed and variable parameters of the design

Category	Design parameter	Value
Physical features	1. Blade	Semi-cylindrical
	2. Number of blades (N)	N = 2, 3 and 4 blades
	3. Blade material	Light plastic (PVC)
	4. Desk material	Fiberglass
	5. Shaft material	Stainless steel
Dimensional	6. Blade diameter (d)	d = 80, 100 and 160 mm
	7. Blade thickness (t)	t = 3 mm
	8. Desk diameter (D)	D = 300 mm
	9. Blade height (H)	H = 300, 400 and 500 mm
	10. External overlap (e)	e = 0, 50, 100 and 150 mm
	11. Aspect ratio (β = H/d)	β = 1.875, 2.5, 3, 3.125, 3.75, 4, 5, and 6.25
Operational	12. Rated wind speed (V)	V = 3, 4, 6, 8, 10 and 12 m/s

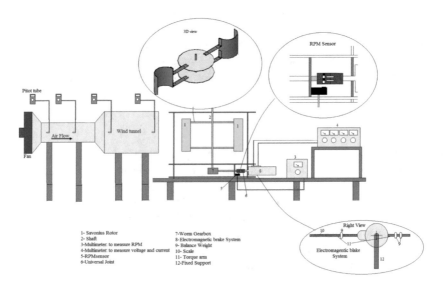

1- Savonius Rotor
2- Shaft
3-Multimeter: to measure RPM
4-Multimeter: to measure voltage and current
5-RPMsensor
6-Universal Joint
7-Worm Gearbox
8- Electromagnetic brake System
9- Balance Weight
10- Scale
11- Torque arm
12-Fixed Support

Fig. 2. Schematic diagram of the low-speed wind tunnel.

2.3 Experimental Setup

The experimental set-up shown in Fig. 2 consists of the wind tunnel, wind turbine model and measurement devices, namely the pitot tube, RPM sensor, and EMD. The Savonius rotor is placed at its proper position and is housed in a structure fabricated from mild steel plates. Two bearings (UC 204, NTN make) bolted to the mild steel plates supports the Savonius rotor. The seals are removed from the bearings and bearings are washed in petrol to remove any grease before mounting in order to reduce friction. The studs, nuts, and bolts used in constructing the housing facilitate the replacement of various test geometries of the Savonius rotor and also help to determine the proper position of the rotor axis at the center line of the wind tunnel. Furthermore, to increase the amount of voltage produced by the DC machine a gearbox was designed as shown in Fig. 2. The function of the gears is to increase the RPM input to the DC generator. Since the dimensions of the rotor are known and the wind speed can be measured, the amount of torque the rotor delivers under different conditions can be evaluated. The gear ratio of 1:10 was utilized in order to obtain optimum RPM needed for the particular generator. The rotors were attached to a gearbox, which was attached to the electromagnetic brakes system. An electromagnetic brakes system is used to measure force and torque from a rotating shaft and to measure the current and voltage produced by the rotor[1].

2.4 Experimental Methods

The mechanical power (P_m) is calculated by measuring the torque (T) on the rotating shaft and rotational speed (ω) at different wind speeds. The mechanical power can be estimated at each wind speed by:

$$P_m = T\omega \tag{1}$$

The force (F) shown on the scales of the electromagnetic brakes system (see Fig. 2) becomes torque when multiplied by the lever distance (d) from the center of rotation. The torque is obtained in (N.m) by

$$T = Fd \tag{2}$$

Electrical power (P_e) generated by the wind turbine model is estimated by multiplying the measured current (I), and voltage (V). These parameters were recorded by multi-meter device. The electrical power can be determined at each wind speed by:

$$P_e = VI \tag{3}$$

---

[1] The experimental setup for measuring the mechanical power and torque is explained in Ref. [8].

2.5   Modeling with ANN

The ANN is a mathematical modeling tool especially for complex systems and used in many different fields [9]. The multilayer perceptron neural network (MLPNN) and function neural network (RBFNN) are widely used in for solving engineering problems [10]. ANN model is developed to predict the electrical power of new configuration Savonius rotors. In the current work, 648 data were used to estimate the electrical power of the rotors. The air velocity (AV), rotational speeds (RPM), gap (G), number of blades (NB), aspect ratio (AR), mechanical power (MP) were the input of the models. The data were randomly split into 2 parts; 70% training, and 30% testing. Since the input and output variables have different magnitude, therefore, a normalization of them is required. The proposed models were developed using SPSS software.

3   Results and Discussions

3.1   Electrical Power of New Savonius-Style Rotors

Figure 3 illustrates the variation in electrical power with wind speed for the investigated rotors. The power for the 3- and 4-bladed rotors is almost the same and it is higher than the power obtained from the 2-bladed rotor. This may be because the net drag force on the rotor in the three- and four-blade cases is higher than that in the two-blade case. In addition, it is found that at 10 m/s wind speed, power production from the three-blade wind turbine increases steadily and the rotation produced can exceed the rotation of the four-blade wind turbine. The three-blade wind turbine model achieves better stability for similar power values than the four-blade wind turbine. It can be concluded from the experimental data that a three-bladed system has better overall performance than the other two models. In addition, as observed during testing, in the 4-blade turbine blade, at wind speeds higher than 10 m/s the blades started shaking. This shows that at high wind speeds, the turbine becomes unstable. Not only this may reduce the performance of the turbine, but it may also even break down.

3.2   Prediction the Electrical Power of New Configuration of Savonius Rotor Using ANN

The developed MLPNN and RBFNN model architectures for predicting the electrical power of the rotors are shown in Figs. 4 and 5, respectively. For the trained network for estimating the electrical power of the rotors, the sum squared error of MLPNN and RPFNN were 0.198 and 11.478, respectively. In addition, sum squared error of MLPNN and RPFNN for testing data were 0.088 and 4.211, respectively for testing data. Figure 6 compared the predicted value with the experimental data of the new configuration Savonius rotors. The results showed that the estimated values of the electrical power of rotors are closed to the experimental data.

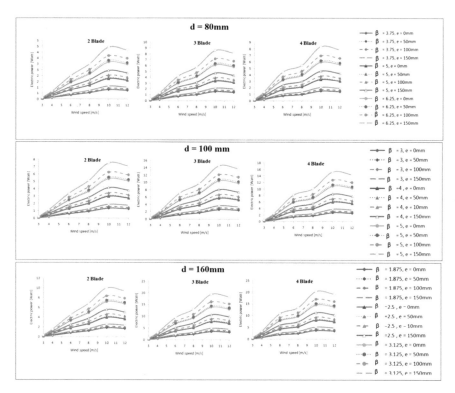

Fig. 3. Electrical power versus wind speed for various aspect ratios, overlaps, blade numbers, and blade diameters

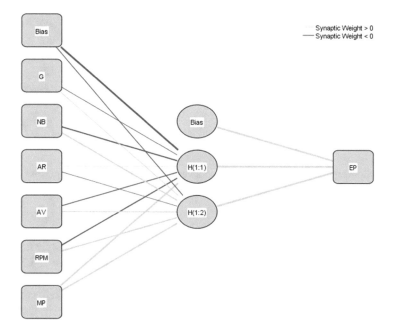

Fig. 4. The structure of MLPNN for predicting the electrical power of the rotors

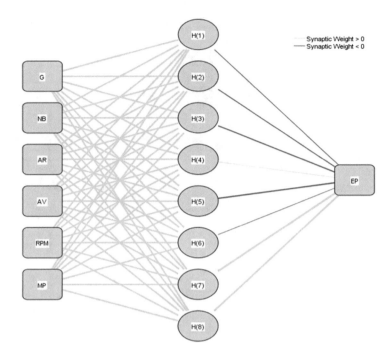

Fig. 5. The structure of RPFNN for predicting the electrical power of the rotors

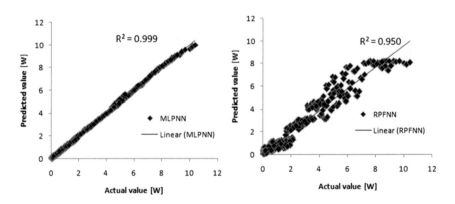

Fig. 6. Comparison between predicted and actual values of the electrical power of rotors.

4   Conclusions

In the present study, the performance of a newly developed Savonius-style turbine was evaluated experimentally and computationally at low wind velocities of 3–12 m/s. The results of the electrical power of the new configuration of the Savonius turbine are discussed and compared with ANN models. The experimental results indicated that the optimum value of wind speed is 10 m/s. In fact, it was observed during the tests that the turbine with the largest number of blades (4 blades) started shaking violently at higher wind speeds. This shows that at high wind speeds, turbines with four and larger numbers of blades are unstable. This may reduce the performance of the turbine, and it may break down. Moreover, the results confirmed that the MLPNN model was more robust and accurate in predicting the values of the electrical power of the rotors compared to the RBFNN model.

References

1. Zhou, Y., Wu, W., Liu, G.: Assessment of onshore wind energy resource and wind-generated electricity potential in Jiangsu, China. Energy Procedia **5**, 418–422 (2011)
2. Arreyndip, A., Joseph, E.: Small 500 kW onshore wind farm project in Kribi, Cameroon: sizing and checkers layout optimization model. Energy Rep. **4**, 528–535 (2018)
3. Dahlke, S.: Effects of wholesale electricity markets on wind generation in the Midwestern United States. Energy Policy **122**, 358–368 (2018)
4. Chadee, T., Clarke, M.: Wind resources and the levelized cost of wind generated electricity in the Caribbean islands of Trinidad and Tobago. Renew. Sustain. Energy Rev. **81**, 2526–2540 (2018)
5. Saeidi, D., Sedaghat, A., Alamdari, P., Alemrajabi, A.: Aerodynamic design and economical evaluation of site specific small vertical axis wind turbines. Appl. Energy **101**, 765–775 (2013)
6. Balduzzi, F., Bianchini, A., Carnevale, A., Ferrari, L., Magnani, S.: Feasibility analysis of a Darrieus vertical-axis wind turbine installation in the rooftop of a building. Appl. Energy **97**, 921–929 (2012)
7. Roy, S., Saha, U.K.: Review of experimental investigations into the design, performance and optimization of the Savonius rotor. Proc. Inst. Mech. Eng. Part A: J. Power Energy **227**(4), 528–542 (2013)
8. El-Ghazali, A.: The influence of turbine geometry on the performance of c-section vertical axis wind turbine. Master's thesis, Near East University (2016)
9. Sargolzaei, J., Kianifar, A.: Modeling and simulation of wind turbine Savonius rotors using artificial neural networks for estimation of the power ratio and torque. Simul. Model. Pract. Theory **17**(7), 1290–1298 (2009)
10. Jeong, H., Obaidat, S., Yen, Y., Park, J.: Advances in computer science and its applications: CSA 2013. Springer, Berlin (2014)

Prediction of Kinematic Viscosity and Density of Biodiesel Produced from Waste Sunflower and Canola Oils Using ANN and RSM: Comparative Study

Youssef Kassem[1,2](✉) ⓘ, Hüseyin Gökçekuş[1], and Hüseyin Çamur[2]

[1] Faculty of Civil and Environmental Engineering,
Department of Civil Engineering, Near East University,
99138 Nicosia (via Mersin 10, Turkey), Cyprus
{yousseuf.kassem, huseyin.gokcekus}@neu.edu.tr
[2] Faculty of Engineering, Department of Mechanical Engineering,
Near East University, 99138 Nicosia (via Mersin 10, Turkey), Cyprus
huseyin.camur@neu.edu.tr

Abstract. In this paper, the properties of two biodiesel obtained from waste cooking sunflower (WCSME) and waste cooking canola (WCCME) oils and their blends in the temperature range 20–300 °C is measured experimentally. This work focused on the application of Artificial Neural Network (ANN) as predictive tools for prediction the kinematic viscosity and density of the biodiesel. In the present study, temperature, the composition of methyl esters (wt%/100), the number of carbon atoms (NC), the number of hydrogen atoms (NH), molecular weight in g/mol, the number of double bond in the fatty acid chain and volume fraction of WCSME were used as input for the models. Moreover, Response surface methodology (RSM) was used to predict the effects of either temperature and volume fraction or volume fraction only on the selecting biodiesel properties using Two-variable or single variable model, respectively. Consequently, it was found that the single variable has more significant for predicting the kinematic viscosity and density compared to the two-variable model. Accordingly, the results indicated that the proposed ANN approach is able to provide a good agreement with the experimental data with the overall R^2 of 0.999 compared with the RSM models.

Keywords: ANN · Physicochemical properties · Waste cooking canola · Waste cooking sunflower

1   Introduction

The global energy demand is rapidly increased because of the consumption of fossil fuel. Therefore, the increases of energy demand have increased in recent years the significance of renewable energy as an alternative source to reduce greenhouse gas emissions (GHG). Biodiesel is considered an alternative fuel produced from vegetable oils or animal fats [1]. It is produced by transesterification with the aid of alcohol and a catalyst [1]. Biodiesel has a higher cetane number, combustion efficiency, viscosity,

© Springer Nature Switzerland AG 2020
R. A. Aliev et al. (Eds.): ICSCCW 2019, AISC 1095, pp. 880–887, 2020.
https://doi.org/10.1007/978-3-030-35249-3_117

density and so on [2]. The most important properties of biodiesel are the viscosity and density because they affect the engine performance and the injection system [3–5]. Viscosity and density of biodiesel are dependent on the type of vegetable oils [6]. Several scientific researchers have been measured viscosity and density experimentally for biodiesel fuels. Recently, various models were used to estimate the properties of biodiesel such as mathematical model [1, 2], Neuro-Fuzzy Inference System (ANFIS) [4], Artificial Neural Network (ANN) [7].

In the present study, the ANN approach was applied to estimate the kinematic viscosity and density of two biodiesel and their blends. The biodiesel samples were produced from waste cooking sunflower oil and waste canola oil through the transesterification reaction. Moreover, Two-variable and single-variable models of RSM were conducted to predict and compared the properties with each other as well as with the experimental values. Two-variable model of RSM is used to predict the kinematic viscosity and density as a function of temperature and volume fraction, whereas the single-variable model of RSM was carried out to evaluate the selecting properties as a function of temperature only. All models of ANNRSM were compared with each other and with the experimental data.

2   Experimental Procedures

2.1   Biodiesel Samples

Waste sunflower and waste canola cooking oils were used in this study and were collected from domestic restaurants. The two biodiesel used in this experimental work were transesterified fatty acid methyl ester of waste cooking sunflower (WCSME) and waste cooking canola (WCCME) vegetable oils and produced in the Mechanical Engineering laboratory. In this study, WCSME (waste cooking sunflower methyl ester) and WCCME (waste cooking canola methyl ester) were referred to the two pure biodiesel. The other three-biodiesel samples were termed as 25-WCSME, 50-WCSME, and 75-WCSME. Table 1 shows the fatty acid methyl ester composition of the biodiesel samples.

2.2   Kinematic Viscosity and Density

In the current study, Ubbelohde viscometer and Pycnometer with a bulb capacity of 25 ml were used to measure the kinematic viscosity and density of biodiesel samples, respectively. The kinematic viscosity and density were measured according to ASTM D-445 and ASTM D854, respectively. The properties of the two pure biodiesel and their blends were measured in the temperature range 20 °C to 300 °C in steps of 10 °C[1]. The measurements were repeated three times then the averaged of the results was taken.

---

[1] Reference [5] explained the experimental setup of the measuring the biodiesel properties from −10 to 300 °C.

Table 1. Fatty acid methyl ester composition (wt%)

Composition (wt%)	Methyl ester formula	M [g/mol]	N	Samples of current work				
				WCCME	25-WCSME	50-WCSME	75-WCSME	WCSME
C8:0	$C_9H_{18}O_2$	158.238	0	0	0.01	0.03	0.04	0.05
C10:0	$C_{11}H_{22}O_2$	186.2912	0	0	0.08	0.17	0.25	0.33
C12:0	$C_{13}H_{26}O_2$	214.3443	0	0.08	0.36	0.63	0.91	1.18
C14:0	$C_{15}H_{30}O_2$	242.3975	0	0	0.03	0.05	0.08	0.1
C16:0	$C_{17}H_{34}O_2$	270.4507	0	5.63	13.55	21.46	29.38	37.29
C18:0	$C_{19}H_{38}O_2$	298.5038	0	1.57	2.19	2.81	3.42	4.04
C18:1	$C_{19}H_{36}O_2$	296.4879	1	62.97	57.33	51.7	46.06	40.42
C18:2	$C_{19}H_{34}O_2$	294.4721	2	21.34	20.47	19.59	18.72	17.84
C18:3	$C_{19}H_{32}O_2$	292.4562	3	6.99	5.29	3.59	1.88	0.18
C20:0	$C_{21}H_{42}O_2$	326.557	0	0.46	0.35	0.23	0.12	0
C20:1	$C_{21}H_{40}O_2$	324.5411	1	1.04	0.78	0.52	0.26	0

2.3   Modeling with ANN

The ANN is a mathematical modeling tool especially for complex systems and used in many different fields [8]. ANN models are developed to predict the kinematic viscosity and density of biodiesel blended. In the current work, 290 data were used to estimate the biodiesel properties. The temperature, volume fraction, mass fraction (wt%/100), the number of carbon atoms (NC), the number of hydrogen atoms (NH), mass molecular weight in g/mol, and the number of a double bond in the fatty acid chain were the input of the models. Additionally, the output of the models was kinematic viscosity and density of biodiesel. The data were randomly split into 3 parts; 60% training, 20% validation and 20% testing. Since the input and output variables have different magnitude (Table 2), therefore, a normalization of them is required. The proposed models were developed using the toolbox in MATLAB R2015a.

3   Results and Discussions

3.1   Properties of Biodiesel Samples

Figure 1 illustrates the influence of the test temperature of the kinematic viscosity for all samples. Similarity, Fig. 2 shows the relationship between the temperature and the density of the samples. It is found that the A general trend of decreased in the kinematic viscosity and density of all biodiesel samples were observed with an increase in temperature.

Table 2. Minimum and maximum values of the input and output variables

Variable	Minimum-Maximum	Units
Input		
Temperature	20–300	°C
The volume fraction of WCSME	0–1	–
Mass fraction	0–62.92	wt%
Number of carbon atoms	9–21	–
Number of hydrogen atoms	18–42	–
Molecular weight	158.238–326.557	g/mol
Number of double bonds	0–3	–
Output		
Kinematic viscosity	0.404–9.75	mm^2/s
Density	653.68–887.48	kg/m^3

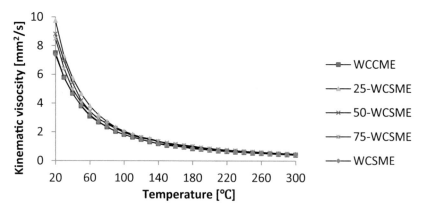

Fig. 1. Kinematic viscosity – temperature relationship for all samples.

3.2   Modeling with ANN

The developed ANN model architecture for predicting the engine performance and emission characteristics are shown in Fig. 3. Based on the training results, the ANN with seven hidden layers with twenty neurons has the best performance, i.e., 7:7:2 is the optimal selection of ANN for estimating the biodiesel properties. For the trained network for estimating the kinematic viscosity and density, the mean squared error (MSE) is equal to 1.95×10^{-6} and 3.69×10^{-6}, respectively. Figures 4 and 5 compared the predicted value with the experimental data for some selected test temperature. The results showed that the estimated values of viscosity and densities are closed to the experimental data.

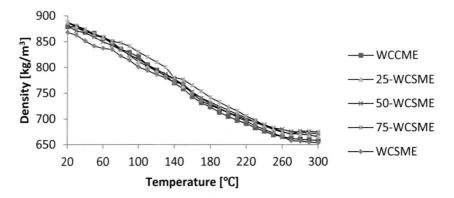

Fig. 2. Density – temperature relationship for all samples.

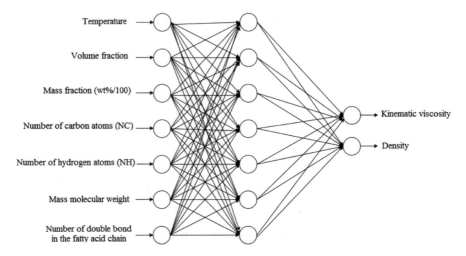

Fig. 3. The structure of the ANN for predicting the kinematic viscosity and density of the samples.

3.3 Modeling with RSM

In this study, Minitab software was used for Response surface methodology (RSM). Two models, namely; Single-variable model and Two-variable model were developed to determine the kinematic viscosity and densities of the samples. The mathematical equations used to estimate the biodiesel properties are shown in Table 3.

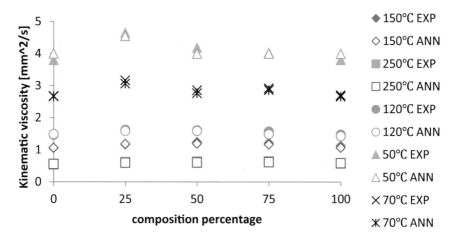

Fig. 4. A comparison between experimental and estimated data of kinematic viscosity vs percentage composition.

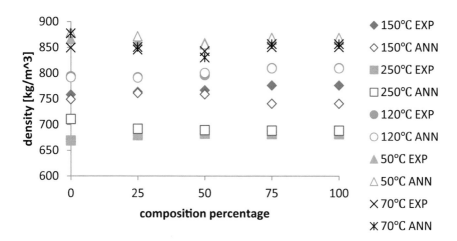

Fig. 5. A comparison between experimental and estimated data of density vs percentage composition.

Table 3. Correlations of present work to predict the biodiesel properties

Single-variable RSM	Eq. no.
$v(T) = 921.9e^{-0.00125T} - 49.35e^{-0.04448T}$	1
$\rho(T) = 11.66e^{-0.03731T} + 3.256e^{-0.00692T}$	2
Two-variable RSM	Eq. no.
$v(T, w_s) = 6.222 - 0.06838T + 1.154w_s + 0.000293T^2 - 0.00324Tw_s \\ \qquad - 1.244w_s^2 + 0.003998Tw_s^2$	3
$\rho(T, w_s) = 873.1 - 0.2512T + 48.03w_s - 0.00555T^2 - 0.05011Tw_s \\ \qquad - 25.94w_s^2 + 0.000138T^2w_s - 0.08195Tw_s^2$	4

3.4   Confirming ANN and RSM Models

The performance of the models is evaluated using R-squared (R^2) as shown in Table 4. It can be seen that the kinematic viscosity values obtained by ANN models have higher R-squared compared to RSM models. The results showed that the ANN model is highly capable of predicting the biodiesel properties compared to RSM models. Additionally, it is observed that both the individual single variable model (only temperature dependent) and the two-variable model (temperature and volume fraction) for density and kinematic viscosity prediction are provided a good agreement with the experimental data. It is found that the single-variable model has the higher performance to estimate the biodiesel properties compared to the Two-variable models due to the value of R-squared, i.e. R-squared values are within the range of 0.9836 to 0.9874.

Table 4. R-squared values for ANN and RSM models

Model	Properties	
	Kinematic viscosity	Density
ANN	0.9990	0.9991
Two-variable RSM	0.9370	0.9950
Single-variable RSM	0.9836	0.9874

4   Conclusions

In the present study, the kinematic viscosity and density of Two pure biodiesel produced from waste cooking sunflower and waste cooking canola oils and their blends were measured according to ASTM standards. Two predictive models are used to estimate the properties of the samples. After designing and testing the models, the predictive capability of the models was compared. According to the results, the following conclusions can be drawn.

1. The results indicate that the viscosity and density of the biodiesel sample decreased as the test temperature increases.
2. It is found that predicted values of biodiesel properties at different temperature using the ANN approach are closer to the experimental values.
3. The mathematical equations (RSM) can be used as a general equation to determine the viscosity and densities of biodiesel samples.
4. The results confirmed that the ANN models were more robust and accurate in predicting the values of viscosity and densities compared to the RSM models.

References

1. Kassem, Y., Çamur, H.: Effects of storage under different conditions on the fuel properties of biodiesel admixtures derived from waste frying and canola oils. Biomass Convers. Biorefin. **8**(4), 825–845 (2018)

2. Wakil, M., Kalam, M., Masjuki, H., Atabani, A., Fattah, R.: Influence of biodiesel blending on physicochemical properties and importance of mathematical model for predicting the properties of biodiesel blend. Energ. Convers. Manag. **94**, 51–67 (2015)
3. Kassem, Y., Çamur, H.: Prediction of biodiesel density for extended ranges of temperature and pressure using an adaptive neuro-fuzzy inference system (ANFIS) and radial basis function (RBF). Procedia Comput. Sci. **120**, 311–316 (2017)
4. Kassem, Y., Çamur, H., Esenel, E.: Adaptive neuro-fuzzy inference system (ANFIS) and response surface methodology (RSM) prediction of biodiesel dynamic viscosity at 313K. Procedia Comput. Sci. **120**, 521–528 (2017)
5. Kassem, Y., Çamur, H.: A laboratory study of the effects of wide range temperature on the properties of biodiesel produced from various waste vegetable oils. Waste Biomass Valoriz. **8**(6), 1995–2007 (2017)
6. Ayetor, K., Sunnu, A., Parbey, J.: Effect of biodiesel production parameters on viscosity and yield of methyl esters: Jatropha curcas, Elaeis guineensis, and Cocos nucifera. Alex. Eng. J. **54**(4), 1285–1290 (2015)
7. Tosun, E., Aydin, K., Bilgili, M.: Comparison of linear regression and artificial neural network model of a diesel engine fueled with biodiesel-alcohol mixtures. Alex. Eng. J. **55**(4), 3081–3089 (2016)
8. Parlak, A., Islamoglu, Y., Yasar, H., Egrisogut, A.: Application of artificial neural network to predict specific fuel consumption and exhaust temperature for a diesel engine. Appl. Therm. Eng. **26**(8–9), 824–828 (2006)

Thermodynamic Modeling of the Phase Diagram for Cu_2SnS_3-Cu_2SnSe_3 System

A. N. Mammadov[1,2], I. Dz. Alverdiev[3], Z. S. Aliev[4(✉)],
D. B. Tagiev[1], and M. B. Babanly[1]

[1] Nagiyev Institute of Catalysis and Inorganic Chemistry of ANAS,
Baku, Azerbaijan
asif.mammadov.47@mail.ru, dtagiyev@rambler.ru,
babanlymb@gmail.com
[2] Azerbaijan Technical University, Baku, Azerbaijan
[3] Ganja State University, Ganja, Azerbaijan
info@gdu.edu.az
[4] Azerbaijan State Oil and Industrial University, Baku, Azerbaijan
ziyasaliev@gmail.com

Abstract. The phase diagram of the Cu_2SnS_3-Cu_2SnSe_3 system was plotted by thermodynamic calculations. Experimental data from differential thermal (DTA) and X-ray diffraction (XRD) analyses have used for the calculation. From the fundamental principles of thermodynamics for heterogeneous equilibria, new equations have been obtained for the direct calculation of the coordinates for $(Cu_2SnS_3)_{1-x}(Cu_2SnSe_3)_x$ liquid and solid solutions. The parameters of the analytical dependencies of the Gibbs free energy within the asymmetric version of the model of regular solutions were determined by means the multipurpose genetic algorithm (MGA), whereas the boundaries of solid solutions are determined based on Gibbs function for the internal stability. Analytical dependencies between the variables and formation thermodynamic functions for the compounds allowed us to estimate the sensitivity of the calculated data with the input data. It was established that, the coordinates of the liquidus and solidus curves are insensitive to formation enthalpy of the Cu_2SnS_3 and Cu_2SnSe_3 compounds. At the same time, a high sensitivity of the liquidus and solidus coordinates to the excess free energy values of solutions was observed. A 3D model of the Gibbs energy dependences on compositions and temperatures was visualized.

Keywords: Phase diagrams · Cu_2SnS_3-Cu_2SnSe_3 system · Thermodynamic modeling · Multipurpose genetic algorithm

1 Introduction

Ternary thio- and selenostannates of copper and silver with general formulae Me_2SnX_3 (Me-Cu, Ag; X = S, Se) belong to diamond semiconductors and attracted much more attention as prospective functional materials for thermoelectric, photovoltaic and optical applications [1–5]. Search of optimal model by choosing the thermodynamic data to determine phase equilibria is a rather difficult problem in materials science due

© Springer Nature Switzerland AG 2020
R. A. Aliev et al. (Eds.): ICSCCW 2019, AISC 1095, pp. 888–895, 2020.
https://doi.org/10.1007/978-3-030-35249-3_118

to uncertainty of the determination of equations for phase equilibria. In recent years, many research works has been developed to optimize these phase equilibria using the multi-purpose genetic algorithm (MGA) in order to find the correct solutions for the thermodynamic equations for phase equilibria [6–10].

The purpose of this work is the thermodynamic modeling of the boundaries of liquid and solid solutions and the stability regions for solid solutions in the Cu_2SnS_3-Cu_2SnSe_3 system using a limited amount of experimental data from DTA and XRD. The phase diagram of the Cu_2SnS_3-Cu_2SnSe_3 system has not yet been studied so far and there is no data in the literature on the melting enthalpy and entropy of these compounds. Standard formations entropies, enthalpies and Gibbs formation energies for the Cu_2SnS_3 and Cu_2SnSe_3 were determined experimentally by the electromotive force measurements (EMF) [11].

2 Thermodynamic Equations

2.1 Equilibrium Solid Solution-Liquid Solution

Considering that the chemical potentials of each component in the liquid and solid phases, which are in heterogeneous equilibrium at constant temperature and pressure, are equal, we can write [12]:

$$\mu_i^l(x_i^l) = \mu_i^s(x_i^s), \text{ where } i = 1(Cu_2SnS_3),\ 2(Cu_2SnSe_3) \tag{1}$$

Here, l and s are liquid and solid phases, x_i^l and x_i^s are molar fractions of Cu_2SnS_3 (1) and Cu_2SnSe_3 (2) in liquid and solid solutions. Considering that the pure liquid copper, tin, sulfur and selenium are in standard state, then, for chemical potentials of Cu_2SnS_3 (1) and Cu_2SnSe_3 (2):

$$2\mu_{Cu}^{l,0} + \mu_{Sn}^{l,0} + 3\mu_{S}^{l,0} + RTlnx_1^l\gamma_1^l = 2\mu_{Cu}^{l,0} + \mu_{Sn}^{l,0} + 3\mu_{S}^{l,0} + \Delta G_1^{s,o} + RTlnx_1^s\gamma_1^s \tag{2}$$

$$2\mu_{Cu}^{l,0} + \mu_{Sn}^{l,0} + 3\mu_{Se}^{l,0} + RTlnx_2^l\gamma_2^l = 2\mu_{Cu}^{l,0} + \mu_{Sn}^{l,0} + 3\mu_{Se}^{l,0} + \Delta G_2^{s,o} + RTlnx_2^s\gamma_2^s \tag{3}$$

From here we can write

$$RTlnx_i^l\gamma_i^l = \Delta G_i^{s,o} + RTlnx_i^s\gamma_i^s \tag{4}$$

In (2) and (3), γ_i^l and γ_i^s are thermodynamic activities and activity coefficients of the component i in the liquid solution with respect to the pure liquid substance and in the solid solution with respect to the pure solid Cu_2SnS_3 (1) and Cu_2SnSe_3 (2), respectively. $\Delta G_i^{s,o}$ – Gibbs free energy formation of compounds Cu_2SnS_3 (1) and Cu_2SnSe_3 (2). According to the Gibbs-Helmholtz equation, we can write as:

$$\Delta G_i^{s,o} = \Delta H_i^{s,o} + T\Delta S_i^{s,o} \tag{5}$$

On the other hand, the Gibbs partial excess free energies are related to the activity coefficients by the following equations:

$$\Delta \bar{G}_i^{exs,l} = RT\ln\gamma_i^l; \quad \Delta \bar{G}_i^{exs,s} = RT\ln\gamma_i^s \tag{6}$$

Substituting (5) and (6) in (4), we obtain an equation for describing the liquidus and solidus systems, but taking into account the deviation of the properties of liquid and solid solutions from the ideal model:

$$\ln\left(\frac{x_i^l}{x_i^s}\right) = \Delta H_i^{s,o}/RT - \Delta S_i^{s,o}/R - \Delta \bar{G}_i^{exs,l/s}/RT \tag{7}$$

Here,

$$\Delta \bar{G}_i^{exs,l/s} = \left(\Delta \bar{G}_i^{exs,l} - \Delta \bar{G}_i^{exs,s}\right) \tag{8}$$

Denoting

$$F_i(T) = (\Delta S_i^{s,o} - \Delta H_i^{s,o}/T)/R + \Delta \bar{G}_i^{exs,l/s}/RT \tag{9}$$

from the Eq. (7), we obtain equations for approximating the concentration of the liquidus and solidus mole fractions for each fixed temperature:

$$x_2^l = \frac{1 - \exp(F_1(T))}{\exp(F_2(T)) - \exp(F_1(T))} \tag{10}$$

$$x_2^s = x_2^l \cdot \exp(F_2) \tag{11}$$

Here, $x_1^l = 1 - x_2^l; x_1^s = 1 - x_2^s$.

If the liquid and solid solutions are ideal $(\gamma_i^l = \gamma_i^s = 1)$, or pseudo-ideal $(\gamma_i^l/\gamma_i^s = 1)$, then the function $F_i(T)$ in (9) can be written as:

$$F_i(T) = (\Delta S_i^{s,o} - \Delta H_i^{s,o}/T)/R \tag{12}$$

2.2   Thermodynamic Stability of Solid Solutions

To calculate the free energy of formation for $(Cu_2SnS_3)_{1-x}(Cu_2SnSe_3)_x$ solid solutions, a modified version of the regular solutions model was used that has been successfully tested in [13]:

$$\Delta G_T^o = x(\Delta H_1^{s,0} - T\Delta S_1^{s,0}) + (1 - x)(\Delta H_2^{s,0} - T\Delta S_2^{s,0}) \\ + ax^m(1 - x)^n + pRT[x\ln x + (1 - x)\ln(1 - x)] \tag{13}$$

Here: 1 is Cu_2SnS_3; 2 is Cu_2SnSe_3; $x = x_2$.

Equation (13) is the free formation energy for $(Cu_2SnS_3)_{1-x}(Cu_2SnSe_3)_x$ solid solutions from pure liquid copper, tin, sulfur, and selenium. For the free energy of formation of $(Cu_2SnS_3)_{1-x}(Cu_2SnSe_3)_x$ solid solutions from compounds Cu_2SnS_3 and Cu_2SnSe_3, Eq. (13) can be written as:

$$\Delta G_T^o = ax^m(1-x)^n + pRT[x\ln x + (1-x)\ln(1-x)] \tag{14}$$

In Eq. (14), the first term represents the mixing enthalpy for solid solutions in an asymmetric version of the regular solutions model where solid solutions decomposing into two phases and the mixing parameter $a > 0$. The second term represents the configurational mixing entropy of solid solutions according to the non-molecular compounds model [13] where $P = 3$ and represents the number of different atoms in Cu_2SnS_3 and Cu_2SnSe_3. In all equations $R = 8.314$ J mol^{-1} K^{-1}.

The thermodynamic functions of the formation of Cu_2SnS_3 and Cu_2SnSe_3 compounds are listed in Table 1.

3 Methods for Solving Thermodynamic Equations by Using Multipurpose Genetic Algorithm (MGA)

In Eqs. (7, 9–11), the initial data are formation enthalpies and entropies of the Cu_2SnS_3 and Cu_2SnSe_3 compounds (Table 1).

Table 1. Enthalpies and entropies of formation of Cu_2SnS_3 and Cu_2SnSe_3 compounds from liquid tin, sulfur, selenium and super cooled copper, which were obtained by re-calculating data [11].

Variable	Unit	Value
$\Delta H_1^{s,0}(Cu_2SnS_3)$	J mol^{-1}	-280960 ± 2000
$\Delta H_2^{s,0}(Cu_2SnSe_3)$	J mol^{-1}	-260900 ± 2000
$\Delta S_1^{s,0}(Cu_2SnS_3)$	J mol^{-1}K	-58.64 ± 3
$\Delta S_2^{s,0}(Cu_2SnSe_3)$	J mol^{-1}K	-78.89 ± 4

To calculate an equilibrium compositions of liquid and solid solutions (Eqs. 7, 10, 11) and to optimize of the liquidus and solidus curves in the Cu_2SnS_3-Cu_2SnSe_3 system (Fig. 1), it is necessary to determine and approximate the value of $\Delta \bar{G}_i^{exs,l/s}$ in Eq. (8). These values were approximated in the form of functions:

$$\Delta \bar{G}_i^{exs,l/s}/RT = a + bT \tag{15}$$

Optimization of the liquidus and solidus curves was carried out according to the following scheme using the MGA [6]. Firstly, the range for each parameter is determined for the MGA based on the experimental data from DTA. Then, MGA changes the values of the variables depending on how these values generate the solidus and liquidus curves, which correspond to the present experimental data, taking into account the uncertainties of the experimental solidus and liquidus curves.

Fuzzy logic-weighting scheme looks at all the objective values of a particular member and rescales them to a value between 0 and 1. Zero if the value is the worst of the population. One if it falls within the experimental uncertainty. Once the objectives have been scaled, the average is taken over all objectives and that single number is the fitness for the member in the population. Using the fuzzy logic-weighting scheme, the GA is run until all the members of the population reach a fitness of 1 or at least reach a state of equilibrium where there is no more improvement. When this state is reached, the members of the final population are used to determine the uncertainty bounds on the model parameters. The population of final parameters can then be used to bound output of the model and show where the model is most uncertain and in need of more data. As a result of the iteration, the dependence (15) is obtained in the following form:

$$\Delta \bar{G}_i^{exs,l/s}/RT = -43.7158 + 0.01842T \tag{16}$$

Dependence (16) approximates the values of $\Delta \bar{G}_i^{exs,l/s}$ with high accuracy.

T, K	On Eq. (8)	On Eq. (16)
1017	−24.098	−24.9827
1055	−24.257	−24.2827
1075	−23.952	−23.9143
1125	−22.986	−22.9933

In Eqs. (13, 14), it is necessary to determine the parameter of the intermolecular interaction a and, the degrees of concentrations, m and n. For the model of strictly regular solutions, $m = n = 1$. It was found that, in order to approximate the thermodynamic function of formation, an asymmetric version of the regular solutions model should be used, according to which $m \neq n \neq 1$. To solve the equation containing two functional parameters (temperature and composition) and Gibbs excess free energy, the MGA was used [6]. The following conditions were used to carry out the iteration process:

$$x = 0 \div 1; \; a > 0; \; ; m > n > 1; \; 850K \leq T \leq 1150$$

As a result of the calculations, analytical expressions for Eqs. (13, 14) were determined. In particular, Eq. (14) has the form:

$$\Delta G_T^o(J/mol) = 5 \cdot 10^7 x^6 (1-x)^8 + 3RT[x \ln x + (1-x) \ln(1-x)] \tag{17}$$

Here, x is the mole fraction of Cu_2SnSe_3 in $(Cu_2SnS_3)_{1-x}(Cu_2SnSe_3)_x$ solid solution.

To determine the boundaries of the β_1- and β_2-solid solutions (Fig. 1), the thermodynamic condition for solutions with internal stability $(\partial^2 G/\partial x^2)_{P,T} > 0$ was used [12].

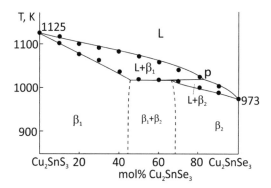

Fig. 1. Phase diagram of the system Cu₂SnS₃-Cu₂SnSe₃. Black points denote DTA data, curves obtained by thermodynamic calculation.

4 Results and Discussion

The phase diagram of the Cu_2SnS_3-Cu_2SnSe_3 system (Fig. 1), does not belong to a simple type. Along the section, there are wide areas of solid solutions based on both initial components. The homogeneity field of the β_2-solid solution based on Cu_2SnSe_3 is located below the peritectic line. The partial excess free energies of the equilibrium liquid and solutions are not significantly different:

$$\Delta \bar{G}_i^{exs,l/s} = \left(\Delta \bar{G}_i^{exs,l} - \Delta \bar{G}_i^{exs,s} \right) = -(23 \div 25)\,\mathrm{J/mol}$$

Although these values are commensurate with the errors of experimental determinations of thermodynamic parameters, the calculated data are very sensitive to these values. We believe that the simultaneous determination of the liquidus and solidus coordinates by calculation can lead to incorrect data. In this case, the coordinates of the liquidus should be determined experimentally. On the other hand, the coordinates of the solidus, which are difficult to determine experimentally, were calculated by the thermodynamic method. The 3D model of the analytical dependence (Fig. 2) of the free energy of formation of the $(Cu_2SnS_3)_{1-x}(Cu_2SnSe_3)_x$ solid solutions on the composition and temperatures indicates that there is a heterogeneous region over the entire temperature range of 850–1120K. This is due to the presence of the upper convexity on the surface of the free energy of solid solutions in Fig. 2.

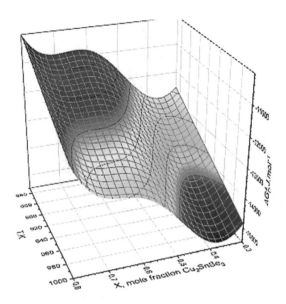

Fig. 2. Three-dimensional visualization of Eq. 17 for the dependence of the free energy of formation of $(Cu_2SnS_3)_{1-x}(Cu_2SnSe_3)_x$ solid solutions on composition and temperature.

5 Conclusions

In the exemplary case of the Cu_2SnS_3-Cu_2SnSe_3 system, it was found that in the absence of data on the melting enthalpy and entropy of the initial components of the system, the liquidus and solidus curves could be thermodynamically approximated based on the free energy, entropy and enthalpy of formation of compounds. It is established that the calculated data are very sensitive to the values of the partial thermodynamic functions of the equilibrium liquid and solid solutions. To describe the enthalpy component of the free energy of the formation energy of solid solutions, an asymmetric version of the regular solutions model should be used. Simultaneous determination of the coordinates of the liquidus and solidus by calculation can lead to incorrect data. This should be experimentally determine the coordinates of the liquidus, and the coordinates of the solidus, which are difficult to determine experimentally, to calculate the thermodynamic method.

Acknowledgments. This work was performed in the frame of a scientific program of the international laboratory between the Institute of Catalysis and Inorganic Chemistry of the National Academy of Sciences of Azerbaijan (Azerbaijan) and Centro de Fisica de Materials at Donostia (Spain).

References

1. Te, S.: Applications of Chalcogenides. Ed. by G.K. Ahluwalia. Springer (2016)
2. Yusibov, Y.A., Alverdiev, I.D., Mashadieva, L.F., Babanly, D.M., Mamedov, A.N., Babanly, M.B.: Experimental study and 3D modeling of the phase diagram of the Ag–Sn–Se

system. Russ. J. Inorg. Chem. **63**(12), 1622–1635 (2018). https://doi.org/10.1134/S0036023618120227

3. Bayazıt, T., Olgar, M.A., Küçükömeroğlu, T., Bacaksız, E., Tomakin, M.: Growth and characterization of Cu_2SnS_3 (CTS), Cu_2SnSe_3 (CTSe), and $Cu_2Sn(S,Se)_3$ (CTSSe) thin films using dip-coated Cu–Sn precursor. J. Mater. Sci. Mater. Electron. (2019). https://doi.org/10.1007/s10854-019-01622-4

4. Shigemi, A., Maeda, T., Wada, T.: First-principles calculation of Cu_2SnS_3 and related compounds. Phys. Status Solidi (b) **252**(6), 1230–1234 (2015). https://doi.org/10.1002/pssb.201400346

5. Choi, S.G., Kang, J., Li, J., Haneef, H., Podraza, N.J., Beall, C., Repins, I.L.: Optical function spectra and bandgap energy of Cu_2SnSe_3. Appl. Phys. Lett. **106**(4), 043902 (2015). https://doi.org/10.1063/1.4907202

6. Preuss, M., Wessing, S., Rudolph, G., Sadowski, G.: Solving phase equilibrium problems by means of avoidance-based multiobjectivization. In: Kacprzyk, J., Pedrycz, W. (eds.) Springer Handbook of Computational Intelligence, Part E.58, pp. 1159–1169. Springer, Heidelberg (2015). Evol. Comput. https://doi.org/10.1007/978-3-662-43505-2_58

7. Stan, M., Reardon, B.J.: A Bayesian approach to evaluating the uncertainty of thermodynamic data and phase diagrams. Calphad **27**, 319–323 (2003)

8. Duong, T.C., Hackenberg, R.E., Landa, A., Honarmandi, A., Talapatra, A.: Revisiting thermodynamics and kinetic diffusivities of uranium–niobium with Bayesian uncertainty analysis. Calphad **55**, 219–230 (2016)

9. Mammadov, A.N., Aliev, Z.S., Babanly, M.B.: Study of the uncertainty heterogeneous phase equilibria areas in the binary YbTe-SnTe alloy system. In: Aliev, R., Kacprzyk, J., Pedrycz, W., Jamshidi, M., Sadikoglu, F. (eds.) 13th International Conference on Theory and Application of Fuzzy Systems and Soft Computing - ICAFS-2018/Advances in Intelligent Systems and Computing, vol. 896, pp. 815–822. Springer, Cham (2019). https://doi.org/10.1007/978-3-030-04164-9_107

10. Mamedov, A.N., Akhmedova, N.Y., Asadov, C.M., Babanly, N.B., Mamedov, E.I.: Thermodynamic analysis and defection formation in alloys on the basis of lead selenide containing copper. Chem. Probl. **1**(17), 16–25 (2019). https://doi.org/10.32737/2221-8688-2019-1-16-25

11. Babanly, M.B., Yusibov, Y.A., Abishev, V.T.: Three-component chalcogenides based on copper and silver. BGU, Baku (1993). (in Russian)

12. Stolen, S., Grande, T.: Chemical Thermodynamics of Materials: Macroscopic and Microscopic Aspects. Wiley, Chichester (2004)

13. Mamedov, A.N., Tagiev, E.R., Aliev, Z.S., Babanly, M.B.: Phase boundaries of the $(YbTe)_x(PbTe)_{1-x}$ and $(YbTe)_x(SnTe)_{1-x}$ solid solutions series. Russ. J. Inorg. Mater. **52**(6), 543–545 (2016)

Toward a Definite Integral
of a Z-Number-Valued Function

O. H. Huseynov$^{(\boxtimes)}$ (iD)

Research Laboratory of Intelligent Control and Decision Making Systems in
Industry and Economics, Azerbaijan State Oil and Industry University, Baku,
Azerbaijan
oleg_huseynov@yahoo.com

Abstract. Z-number-valued function analysis is a new direction of studies on
processing of information characterized by fuzziness and probabilistic uncertainty. Differentiability and integrability are among the related key problems. In
this paper, we made a first step toward formulation of a definite integral of
Z-number-valued function. A definition of definite integral is proposed and its
relation with the concept of derivative is analyzed. In the proposed approach, the
formalism of approximate limits is used.

Keywords: Z-number · Bimodal information · Derivative · Integral ·
Approximate limit

1 Introduction

In order to describe fuzziness and partial reliability of information, Zadeh introduced
the concept of a Z-number [1]. A lot of papers on Z-numbers were published since then
and nowadays this field is of a high interest for researchers [2–11]. However, existing
works devoted to theoretical basics of Z-numbers, including fundamentals of
Z-number-valued functions, are scarce [5, 7, 8, 11–16]. In [5, 8] concepts of limit,
continuity, measurability and differentiability of Z-number-valued function. The proposed concept of derivative of Z-number-valued function is based on Hukuhara difference of Z-numbers [7]. In [11] another approach to study of differentiability and
integrability of Z-number-valued function is proposed. The results are applied to
analysis of differential equations with Z-number-valued information. However, the
proposed approach is related with transformation of Z-numbers to fuzzy numbers.

Unfortunately, there are no fundamental works on integration of a Z-number-valued
function. Research in this direction is necessary for development of a basis for
aggregation of Z-valued information, summarization of bimodal information and other
important problems of dealing with second order uncertainty. The potential applications can be found in data analysis, computing with words, intelligent systems design
and other fields. Let us mention that a problem of computation of a sum of a large
number of Z-numbers, addressed by L. Zadeh as an important practical problem, can be
theoretically formulated as a problem of computation of an integral of a Z-number-
valued function.

© Springer Nature Switzerland AG 2020
R. A. Aliev et al. (Eds.): ICSCCW 2019, AISC 1095, pp. 896–901, 2020.
https://doi.org/10.1007/978-3-030-35249-3_119

In this paper, we continue our study of Z-number-valued functions initiated in [5, 8]. We introduce a definition of a definite integral of a Z-number-valued function. A theorem on its relationship with derivative of a Z-number-valued function is formulated. The paper is structured as follows. Section 2 includes basic definitions of study of Z-number-valued functions. Section 3 is devoted to definite integral of a Z-number-valued function. Section 4 is conclusion.

2   Preliminaries

Definition 1. A Z-number [3–5]. A continuous Z-number is an ordered pair $Z = (A, B)$ where A is a continuous fuzzy number playing a role of a fuzzy constraint on values that a random variable X may take:

$$X \text{ is } A$$

and B is a continuous fuzzy number with a membership function $\mu_B : [0, 1] \to [0, 1]$, playing a role of a fuzzy constraint on the probability measure of A:

$$P(A) = \int_{\text{supp}A} \mu_A(x) p(x) dx \text{ is } B.$$

A discrete Z-number [4] is an ordered pair $Z = (A, B)$ of discrete fuzzy numbers A and B. A plays the same role as it does in a continuous Z-number. B is a discrete fuzzy number with a membership function $\mu_B : \{b_1, \ldots, b_n\} \to [0, 1]$, $\{b_1, \ldots, b_n\} \subset [0, 1]$, playing a role of a fuzzy constraint on the probability measure of A:

$$P(A) = \sum_{i=1}^{n} \mu_A(x_i) p(x_i), P(A) \in \text{supp}(B).$$

Let us denote by $\mathcal{Z}^n$ the space of elements which are n-vectors of discrete Z-numbers $\bar{Z} = (Z_1, Z_2, \ldots, Z_n)$. Denote $\mathcal{Z}_{[c,d]} = \left\{ (A, B) \middle| A \in \mathcal{D}_{[c,d]}, B \in \mathcal{D}_{[0,1]} \right\}$, $[c, d] \subset \mathcal{R}$, and $\mathcal{Z}_+ = \left\{ (A, B) \in \mathcal{Z} \middle| A \in \mathcal{D}_{[0,\infty)} \right\}$, $\mathcal{Z}_- = \mathcal{Z} \backslash \mathcal{Z}_+$ [8].

Let us consider a subset $\mathcal{A} \subset \mathcal{Z}$. Let $Z_1 = (A_1, B_1)$ and $Z_2 = (A_2, B_2)$ be two Z-numbers. Denote by $A_i^{\alpha_k}$ and $B_i^{\beta_k}$ k-th α-cuts of A_i and B_i respectively, $\alpha_k \in \{\alpha_1, \alpha_2, \ldots, \alpha_n\} \subset [0, 1]$, $\beta_k \in \{\beta_1, \beta_2, \ldots, \beta_n\} \subset [0, 1]$. Denote $a_{i\alpha_k}^L = \min A_i^\alpha$, $a_{i\alpha_k}^R = \max A_i^\alpha$ and $b_{i\beta_k}^L = \min B_i^{\beta_k}$, $b_{i\beta_k}^R = \max B_i^{\beta_k}$, $i = 1, 2$.

Definition 2. The Hamming distance based metrics on $\mathcal{Z}$ [8]. The Hamming distance based Z-metrics on $\mathcal{Z}$ is defined as

$$D(Z_1, Z_2) = \left(\frac{1}{n+1} \sum_{k=1}^{n} \left\{ \left| a_{1\alpha_k}^L - a_{2\alpha_k}^L \right| + \left| a_{1\alpha_k}^R - a_{2\alpha_k}^R \right| \right\} + \frac{1}{m+1} \sum_{k=1}^{m} \left\{ \left| b_{1\beta_k}^L - b_{2\beta_k}^L \right| + \left| b_{1\beta_k}^R - b_{2\beta_k}^R \right| \right\} \right)$$

Let $Z = \{ Z_i | Z_i \in \mathcal{A}, i = 1, 2, \ldots, n, \ldots \}$ be a sequence of Z-numbers.

Definition 3. An r-limit [5, 8]. An element $Z \in \mathcal{A}$ is called an r-limit of Z (denoted $Z = r\text{-lim } Z$) if for any $t \in R^+ \setminus \{0\}$ the inequality $D(Z, Z_i) \leq r + t$ is valid for almost all Z_i.

For the case of a sequence of continuous Z-numbers, one would have a definition of a limit where $r = 0$.

Definition 4. A discrete Z-number-valued function of discrete Z-numbers [5, 8]. A discrete Z-number-valued function of discrete Z-numbers is a mapping $f : \mathcal{A} \rightarrow \mathcal{A}$.

Analogously, in a continuous setting, the Z-function is defined as a mapping between spaces of continuous Z-numbers.

Definition 5. An r-limit of discrete Z-number-valued function of discrete Z-numbers [5, 8]. An element $Z_d \in f(Z)$ is called an r-limit of f at a point $Z_{a,i} \in Z$ and denoted $Z_d = r\text{-lim}_{Z_x \rightarrow Z_a} f(Z_x)$ if for any sequence Z satisfying the condition $Z_a = \lim Z$, the equality $Z_d = r\text{-lim } f(Z)$ is valid.

3 A Definite Integral of a Z-Number-Valued Function

The concepts of derivative and integral are closely related. In [5, 7] the authors considered a concept of Hukuhara difference of Z-numbers $Z_1 -_h Z_2 = Z_{12}$ formulated. Let us shortly describe this concept.

Definition 6. Hukuhara difference of discrete Z-numbers [7]. For discrete Z-numbers $Z_1 = (A_1, B_1)$ and $Z_2 = (A_2, B_2)$ their Hukuhara difference denoted $Z_{12} = Z_1 -_h Z_2$ is the discrete Z-number $Z_{12} = (A_{12}, B_{12})$ such that $Z_1 = Z_2 + Z_{12}$.

The conditions on existence of Hukuhara difference $Z_{12} = Z_1 -_h Z_2$ include conditions on existence of Hukuhara difference of fuzzy numbers $A_{12} = A_1 -_h A_2$ and conditions on fuzzy sets of distributions underlying B_1 and B_2. Both these conditions inherit results of Minkowski set-valued arithmetic. These conditions are described in details in [7].

On the basis of the concept of Hukuhara difference of Z-numbers, the following definitions of derivative of Z-number-valued function were proposed.

In view of this, we consider definition of a derivative of a discrete Z-number-valued function $f : \Omega \rightarrow \mathcal{Z}$, $\Omega \subseteq N$ (N is the set of natural numbers) [5]:

Definition 7. Discrete derivative of a discrete Z-number-valued function at a crisp point [5]. $f : \Omega \rightarrow \mathcal{Z}$ may have the following discrete derivatives $\Delta f(n)$ at n:

- the forward right-hand discrete derivative $\Delta_r f(n) = f(n+1) -_h f(n)$
- the backward right-hand discrete derivative $\Delta_{r^-} f(n) = \frac{f(n) -_h f(n+1)}{-1}$

- the forward left-hand discrete derivative $\Delta_l f(n) = f(n) -_h f(n-1)$.
- the backward left-hand discrete derivative $\Delta_{l-} f(n) = \frac{f(n-1) -_h f(n)}{-1}$.

An existence of $\Delta_r f(n)$ or $\Delta_l f(n)$ implies that the length of support of the first component A_n of a Z-number $f(n) = (A_n, B_n)$ is non-decreasing. Existence of the other two derivatives implies non-increasing support of the first component of a Z-number $f(n) = (A_n, B_n)$.

Based on Definition 6, the concept of a generalized Hukuhara differentiability is defined in [5] as follows.

Definition 8. Generalized Hukuhara differentiability [5]. Let $f : \Omega \to \mathcal{Z}$ and $n \in \Omega$. We say that f is differentiable at $n \in \Omega$ if

$\Delta_r f(n)$ and $\Delta_l f(n)$ exist

or

$\Delta_{r-} f(n)$ and $\Delta_{l-} f(n)$ exist

or

$\Delta_r f(n)$ and $\Delta_{l-} f(n)$ exist

or

$\Delta_{r-} f(n)$ and $\Delta_l f(n)$ exist

In the existing literature there is no definition of integral of Z-number-valued functions. We propose the following definition:

Definition 9. The definite integral. The definite integral I of Z-number-valued function $f : [a, b] \to \mathcal{Z}$ is defined as an r-limit:

$$F(x) = \int_a^b f(t)dt = \sum_{i=1}^n f(x_i)\Delta x,$$
$$d(F(x), I) \le r.$$

We derived the following result on relation between differentiability and integrability of a Z-number-valued function.

Theorem. If a Z-number-valued function $f : [a, b] \to \mathcal{Z}$ is differentiable on interval [a, b] then the following is satisfied in terms of r-limit:

$$\int_a^b f'(x)dx = \sum_{i=1}^{n-1} f'(x_i)\Delta x_i = f(b) -_h f(a).$$

Proof. We can write in terms of r-limit:

$$d(f'(x_i), D_i) = r_i, \, f'(x_i) = \frac{f(x_{i+1}) -_h f(x_i)}{\Delta x_i}, \, \Delta x_i = x_{i+1} - x_i, \text{ and}$$
$$= d(f(b) -_h f(a), J) = r.$$

The following is satisfied:

$$d\left(\sum_{i=1}^{n-1} f'(x_i)\Delta x_i, \sum_{i=1}^{n-1} D_i\Delta x_i\right) \leq \sum_{i=1}^{n-1} d(f'(x_i)\Delta x_i, D_i\Delta x_i) = \Delta x_i \sum_{i=1}^{n-1} d(f'(x_i), D_i) = \Delta x_i r_i$$

Then we have the following relation between r-limit J and r-limits D_i:

$$J \leq \sum_{i=1}^{n-1} D_i\Delta x_i.$$

This completes the proof of the theorem.

4 Conclusion

In this paper we made an attempt toward study of integration of a Z-number-valued function. A definition of a definite integral of Z-number-valued function in terms of an approximate limit is proposed and its relation with derivative is analyzed. Further development of derivative and integral of Z-number-valued function would provide a basis for bimodal information processing in decision making, forecasting, data summarization and other fields. Among the related problems, informativeness analysis, knowledge extraction, aggregation of bimodal information from a series of sources can be mentioned [17].

References

1. Zadeh, L.A.: A note on Z-numbers. Inf. Sci. **181**, 2923–2932 (2011)
2. Aliev, R.A., Alizadeh, A.V., Huseynov, O.H., Jabbarova, K.I.: Z-number based linear programming. Int. J. Intell. Syst. **30**, 563–589 (2015)
3. Aliev, R.A., Alizadeh, A.V., Huseynov, O.H.: The arithmetic of continuous Z-numbers. Inf. Sci. **373**, 441–460 (2016)
4. Aliev, R.A., Alizadeh, A.V., Huseynov, O.H.: The arithmetic of discrete Z-numbers. Inf. Sci. **290**, 134–155 (2015)
5. Aliev, R.A., Huseynov, O.H., Aliyev, R.R., Alizadeh, A.V.: The Arithmetic of Z-Numbers: Theory and Applications. World Scientific, Singapore (2015)
6. Aliev, R.A., Huseynov, O.H.: Decision Theory with Imperfect Information. World Scientific, Singapore (2014)
7. Aliev, R.A., Perdycz, W., Huseynov, O.H.: Hukuhara difference of Z-numbers. Inf. Sci. **466**, 13–24 (2018)
8. Aliev, R.A., Perdycz, W., Huseynov, O.H.: Functions defined on a set of Z-numbers. Inf. Sci. **423**, 353–375 (2018)
9. Aliev, R.A.: Uncertain Computation Based on Decision Theory. World Scientific Publishing, Singapore (2017)
10. Yager, R.: On Z-valuations using Zadeh's Z-numbers. Int. J. Intell. Syst. **27**, 259–278 (2012)

11. Jafari, R., Razvarz, S., Gegov, A.: Solving differential equations with Z-numbers by utilizing fuzzy Sumudu transform. In: Arai, K., Kapoor, S., Bhatia, R. (eds.) Intelligent Systems and Applications. IntelliSys 2018. Advances in Intelligent Systems and Computing, vol. 869, pp. 1125–1138. Springer, Cham (2019)
12. Lorkowski, J., Aliev, R., Kreinovich, V.: Towards decision making under interval, set-valued, fuzzy, and Z-number uncertainty: a fair price approach. In: Proceedings of the IEEE International Conference on Fuzzy Systems, FUZZ-IEEE 2014, pp. 2244–2253. IEEE (2014)
13. Aliev, R.A., Kreinovich, V.: Z-numbers and type-2 fuzzy sets: a representation result. Intell. Autom. Soft Comput. (2017). https://doi.org/10.1080/10798587.2017.1330310
14. Jiang, W., Cao, Y., Deng, X.: A novel Z-network model based on Bayesian network and Z-number. IEEE Trans. Fuzzy Syst. (2019). https://doi.org/10.1109/tfuzz.2019.2918999
15. Shen, K., Wang, J.: Z-VIKOR method based on a new comprehensive weighted distance measure of Z-number and its application. IEEE Trans. Fuzzy Syst. 26(6), 3232–3245 (2018)
16. Kang, B., Deng, Y., Hewage, K., Sadiq, R.: A method of measuring uncertainty for Z-number. IEEE Trans. Fuzzy Syst. 27, 731–738 (2018)
17. Aliev, R.A., Pedrycz, W.: Fundamentals of a fuzzy-logic-based generalized theory of stability. IEEE Trans. Syst. Man Cybern. Part B (Cybern.) 39(4), 971–988 (2009)

Sustainable Algorithms for the Synthesis of a Suboptimal Dynamic Object Management System

H. Z. Igamberdiev$^{1(\boxtimes)}$ (ID), E. A. Yusupov^{2} (ID), H. I. Sotvoldiev^{2} (ID), and B. S. Azamxonov^{3} (ID)

^{1} Tashkent State Technical University, Tashkent, Uzbekistan
ihz.tstu@gmail.com
^{2} Fergana Branch of the Tashkent University of Information Technologies, Fergana, Uzbekistan
tatuffdek@rambler.ru, h.sotvoldiyev@mail.ru
^{3} Fergana Polytechnic Institute, Fergana, Uzbekistan
b_azamxonov@mail.ru

Abstract. The article deals with the construction of sustainable algorithms for calculating the coefficients of the adaptive controller, identifying the parameters of the object and diagnosing its state within the objectives of synthesizing a system of suboptimal control of dynamic objects. When finding the coefficients of the regulator, a pseudo-inversion algorithm is used using skeletal decomposition of matrices. Identification algorithms are based on local optimization in accordance with the principle of separation. The solution of the problem of diagnosis is based on the analysis of the updating sequence of the Kalman filter.

Keywords: Dynamic object · Suboptimal control · Identification · Diagnosis · Robust synthesis algorithms

1 Introduction

High requirements to the quality of functioning of modern technical systems lead to the need of development of the adaptive management methods that allow optimizing management processes, ensure the efficiency of the management system when changing the static and dynamic characteristics of an object, and increase the reliability of its work. In implementing the various principles of adaptive control of an object, the question arises of how to choose the structure of the regulators of coordinate and parametric control and adaptation devices that change the parameters of the regulators and the observer.

One of the effective approaches to the problem of constructing suboptimal control systems for objects with incomplete information about their state, parameters and characteristics of external influences is the method of algorithmic design, which allows, from a single point of optimization of the system in the meaning of the designated quality functional approach the synthesis of regulators and adaptation devices. In this case, optimization algorithms are formed on the basis of necessary and sufficient

© Springer Nature Switzerland AG 2020
R. A. Aliev et al. (Eds.): ICSCCW 2019, AISC 1095, pp. 902–907, 2020.
https://doi.org/10.1007/978-3-030-35249-3_120

conditions of the optimality of the control system. The parameters of the algorithms are selected based on the requirements imposed on the stability of the motion of the system in any of its states corresponding to the peripheral values of the functional to the optimal, corresponding minimum of the quality of functional. The method of algorithmic design is also effectively used to solve problems of suboptimal control of linear non-stationary objects with Gaussian perturbations.

2 Problem Definition

Let's consider the observed and controlled object described by the equations

$$x(k+1) = A_0(k)x(k) + B_0(k)u(k) + w(k), \quad x(k_0) = x_0, \tag{1}$$

$$y(k) = Cx(k) + v(k),$$

where

$$M[x_0] = 0, \; M[x_0 x_0^T] = X_0,$$

$$M[w(k)] = M[v(k)] = 0, \; M[x_0 v^T(k)] = 0, \; M[x_0 w^T(k)] = 0,$$

$$M\left[\begin{pmatrix} w(k) \\ v(k) \end{pmatrix} (w^T(\tau) \; v^T(\tau)) \right] = \begin{bmatrix} W & G \\ G^T & V \end{bmatrix} \delta(k - \tau),$$

x_0 – Gaussian random vector; $A_0(k)$, $B_0(k)$, W, V and G – unknown matrices.
    It is necessary to construct the control $u(k)$, suboptimal in terms of minimum quality functionality:

$$J(x, \, u) = \frac{1}{2} M \left[x^T(t) F x(t) + \int_{t_0}^{T} (x^T(t)) \begin{pmatrix} Q & 0 \\ 0 & R \end{pmatrix} \begin{pmatrix} x(t) \\ u(t) \end{pmatrix} dt \right], \tag{2}$$

here Q – positive semidefinite matrix, F, R – positive definite matrices, $T - t_0$ set and far exceeds the object time constant.

3 Solution

We assume that the parameters of the object (1) $A_0(k)$ and $B_0(k)$ have the form:

$$A_0(k) = A^0 + a(k), \; B_0(k) = B^0 + b(k),$$

i.e. the reference values of the parameters A^0 and B^0 are preset, which, in the absence of parametric perturbations, should determine the dynamics of the object. In this case, it is proposed to optimize the control system by adaptive coordinate control of the object:

$$u(k) = H_1(k)\hat{x}(k) + [H_2(k) + E]u_1(k), \tag{3}$$

and adjusting filter gain matrices constructed for the object with the reference values of the parameters A^0 and B^0 and in this sense acting as an object model

$$\hat{x}(k+1) = A^0\hat{x}(k) + B^0 u(k) + K_\Phi(k)[y(k) - C\hat{x}(k)],$$

$$\hat{x}(k_0) = 0.$$

In expression (3) - matrices of adjustable forward and feedback coefficients, E is the identity matrix. The optimization of the control system and the selection of the control influence $u(k)$ is carried out due to the coordinate control of the object constructed according to the separation theorem for the reference values of the parameters A^0 and B^0.

The task of the adaptive regulator is to ensure implementation of the conditions at any time.

$$B_0(k)H_1(k) = \Delta a(k), \quad B_0(k)H_2(k) = \Delta b(k),$$

where $\Delta a(k)$ и $\Delta b(k)$ – are increments of parameters $a(k)$ and $b(k)$.

To implement the increments $\Delta a(k)$ and $\Delta b(k)$, it is necessary to fulfill adaptability conditions [1]. Then the coefficients of the adaptive controller are determined by the relations:

$$H_1(k) = B_0^+(k)\Delta a(k), \quad H_2(k) = B_0^+(k)\Delta b(k),$$

where $B_0^+(k)$ is the matrix pseudo-inverse to $B_0^+(k)$. Since $B_0(k)$ is unknown, for sufficiently small $\|B_0(k) - B^0\|$, approximate relations can be used

$$H_1(k) = (B^0)^+\Delta a(k), \quad H_2(k) = (B^0)^+\Delta b(k). \tag{4}$$

The task of determining $H_1(k)$ and $H_2(k)$ based on (4) is an incorrectly assigned task due to the use of a pseudo-inversion operation. In the practical implementation of relations (4) we will use the skeletal decomposition:

$$B^0 = QZ = \begin{Vmatrix} q_{11} & \cdots & q_{1r} \\ q_{21} & \cdots & q_{2r} \\ \cdots & \cdots & \cdots \\ q_{n1} & \cdots & q_{nr} \end{Vmatrix} \begin{Vmatrix} z_{11} & z_{12} & \cdots & z_{1m} \\ \cdots & \cdots & \cdots & \cdots \\ z_{r1} & z_{r2} & \cdots & z_{rm} \end{Vmatrix} \quad (r = r_{B^0}).$$

To give greater numerical stability in the implementation of algorithms (4) following, it is advisable to use expressions of the form:

$$(B^0)^+ = Z^+ Q^+ = Z^T g_\alpha(\hat{Z}) \, g_\beta(\hat{Q}) \, Q^T, \tag{5}$$

$$(B^0)^+ = \left(\frac{J^T}{P^T}\right) g_\alpha(\hat{V}) \, J \, g_\beta(\hat{W}) \, (J^T \Phi^T), \tag{6}$$

where

$$\hat{Z} = ZZ^T, \, \hat{Q} = Q^T Q, \, g_\alpha(\hat{Z}) = (\hat{Z} + \alpha I)^{-1},$$
$$g_\beta(\hat{Q}) = (\hat{Q} + \beta I)^{-1}, \, \hat{V} = JJ^T + PP^T, \, \hat{W} = J^T J + \Phi^T \Phi,$$
$$g_\alpha(\hat{V}) = (\hat{V} + \alpha I)^{-1}, \, g_\beta(\hat{W}) = (\hat{W} + \beta I)^{-1},$$

$\alpha > 0, \, \beta > 0$ – are regularization parameters.

The selection of the regularization parameters α and β in (5) and (6) should be carried out based on the method of model examples. The above given computational procedures allow to increase the accuracy of determining the control action when implementing the considered suboptimal control algorithm.

Here it is advisable to conduct identification of the object on the basis of local optimization in accordance with the principle of separation, when a custom model is assigned to object (1):

$$x(k+1) = A(k)x(k) + C(k)u(k) = B(k)\tilde{z}(k), \tag{7}$$

the parameters of which are refined using the recurrent least squares method:

$$B(k+1) = B(k) + (x(k+1) + B(k)z(k))z^T(k)\Gamma(k), \quad B_0 = B_0, \tag{8}$$

$$\Gamma(k) = \Gamma(k-1) - \frac{\Gamma(k-1)z(k)z^T(k)\Gamma(k-1)}{1 + z^T(k)\Gamma(k-1)z(k)}, \quad \Gamma_0 = \rho^{-1}I, \quad \rho > 0, \tag{9}$$

where $B(k) = \left(A(k)\vdots C(k)\right)$ is a composite matrix of parameter estimates, tunable at each iteration using algorithm (8), B_a is an a priori estimate of matrix B, $z^T(k) = \left(x^T(k)\vdots(u(k) + v(k))^T\right)$ is a generalized vector of disturbed inputs, e is a vector of independent random disturbances artificially entered into the control channel to improve the quality of the process identification, such, that $M\{v(k)\} = 0$, $M\left\{\|v(k)\|^2\right\} = \sigma_v^2 < \infty$, $M\{v(k)\,v^T(k)\} = P_v > 0$, $M\{v(k)\,\xi^T(k)\} = 0$.

During the operation of a suboptimal control system, it is necessary to diagnose its state [2]. To detect failures in such systems, you can use the updating sequence, which has the property that if the system functions normally, the sequence of updates $\Delta(k) = z(k) - H(k)\hat{x}(k/k-1)$ in the Kalman filter matched to the dynamics model is white Gaussian noise with zero mean and covariance matrix

$$P_\Delta(k) = H(k)P(k/k-1)H^T(k) + R(k),$$
$$P(k/k-1) = \Phi(k,k-1)P(k-1/k-1)\Phi^T(k,k-1) + G(k,k-1)Q(k-1)G^T(k,k-1),$$

where $P(k/k-1)$ is the covariance matrix of estimation errors at the previous step. The estimate of the state vector $\hat{x}(k/k)$ and the covariance error matrix of the estimates of $P(k/k)$ can be found using the Kalman filter:

$$\hat{x}(k/k) = \hat{x}(k/k-1) + K(k)\Delta(k),$$

$$K(k) = P(k/k-1)H^T(k)[H(k)P(k/k-1)H^T(k) + R(k)]^{-1},$$

$$P(k/k) = [I - K(k)H(k)]P(k/k-1),$$

where $K(k)$ is the matrix gain of the Kalman filter.

In order to detect failures, it is more convenient to use the normalized update sequence [2]:

$$\tilde{\Delta}(k) = [H(k)P(k/k-1)H^T(k) + R(k)]^{-1/2}\Delta(k),$$

because in this case $E[\tilde{\Delta}(k)\,\tilde{\Delta}^T(j)] = P_{\tilde{\Delta}} = I\delta(kj)$.

Failures resulting in changes in the system dynamics due to the appearance of jumps or shifts in the components of the state vector, sudden shifts appearing in the measurement channel will cause changes in sequence characteristics $\tilde{\Delta}(k)$ and make it different from white noise, shift the zero mean and change the unit covariance matrix. Therefore, the task of control in this case is reduced to the task of quickly detecting the fact that these characteristics differ from the nominal. Methods of checking the compliance of the updating sequence with white noise and identifying changes in its mathematical expectation are discussed in. In order to check the covariance matrix of the updating sequence $\tilde{\Delta}(k)$, in [1] it is proposed to use the trace of the selective covariance matrix

$$\widehat{S} = \frac{1}{M-1}\sum_{k=1}^{M}\left[\tilde{\Delta}(k) - \bar{\bar{\Delta}}\right]\left[\tilde{\Delta}(k) - \bar{\bar{\Delta}}\right]^T,$$

where $\bar{\bar{\Delta}} = \frac{1}{M}\sum_{k=1}^{M}\tilde{\Delta}(k)$ is the sample mean; M is the number of implementations in use, and χ^2 is the distribution.

4 Conclusion

The above given algorithms were used in solving problems of managing specific technological objects and showed their effectiveness.

References

1. Igamberdiev, H.Z., Abdurakhmanova, Yu.M., Mamirov, U.F.: Regular algorithms for state estimation of quasilinear control objects in apriori uncertainity. In: The 11th International Conference on Multimedia Information Technology and Applications (MITA 2015), Korea, pp. 196–198. IEEE Press (2015)
2. Himmelblau, D.M.: Fault Detection and Diagnosis in Chemical and Petrochemical Processes. Elsevier Scientific/Oxford, Amsterdam/New York (1978)

An Application of the Multi-attribute Decision Making Method to Car Selection Problem Under Imprecise Probabilities

Aynur I. Jabbarova[(✉)] [iD]

Azerbaijan State Economic University, Baku, Azerbaijan
stat_aynur@mail.ru

Abstract. Selection of a car is an important decision making problem. This article deals with the implementation of PROMETHEE technique for resolving such an issue under interval-valued weights of criteria is considered.

Keywords: Car selection · PROMETHEE method · Interval arithmetic

1 Introduction

Nowadays car selection is a difficult task for customers. It is related to various technical and operational characteristics of cars. In [1, 2], for solving this problem TOPSIS method is used. As important criteria lifespan, style, cost and fuel economy were taken. In [3] they used AHP and ANP methods. First, AHP and ANP methods were used to analyze customers' attention in buying a car, and the criteria have been determined. Finally, results of the application of AHP and AHP methods were compared.

The article consists of the following sections. In Sect. 2, primary information about interval arithmetic is given. Section 3 covers an application of PROMETHEE method in car selection problem. Section 4 includes conclusion.

2 Primary Information

Explanation 1. *Interval number* [4]. Let's $\underline{x}, \bar{x} \in R$ such that $\underline{x} \leq \bar{x}$. An interval number $[\underline{x}, \bar{x}]$ is a closed and bounded nonempty real interval, that is

$$[\underline{x}, \bar{x}] = \{x \in R | \underline{x} \leq x \leq \bar{x}\}.$$

Here $\underline{x} = \min([\underline{x}, \bar{x}])$ and $\bar{x} = \max([\underline{x}, \bar{x}])$ are the lower and upper endpoints of $[\underline{x}, \bar{x}]$. The equality of interval numbers is defined as

$$[\underline{x}, \bar{x}] = [\underline{y}, \bar{y}] \Leftrightarrow \underline{x} = \underline{y} \wedge \bar{x} = \bar{y}.$$

Let us describe basic operations over intervals.

Explanation 2. *Addition* [4]. Assume that two interval $[\underline{x}, \bar{x}]$ and $[\underline{y}, \bar{y}]$ are given. Interval addition is formulated as

© Springer Nature Switzerland AG 2020
R. A. Aliev et al. (Eds.): ICSCCW 2019, AISC 1095, pp. 908–913, 2020.
https://doi.org/10.1007/978-3-030-35249-3_121

$$[\underline{x}, \bar{x}] + [\underline{y}, \bar{y}] = [\underline{x} + \underline{y}, \ \bar{x} + \bar{y}].$$

Explanation 3. *Subtraction* [4]. Assume that interval numbers X and Y are given. Interval subtraction is defined by

$$X - Y = X + (-Y)$$

Explanation 4. Superiority of intervals [5]. Assume that interval numbers X and Y are given. Superiority of intervals is defined by

$$d(X,Y) = \begin{cases} \frac{\bar{X} - \bar{Y}}{|(\bar{X} - \bar{Y}) + (\underline{X} - \underline{Y})|} & \bar{X} > \bar{Y}, \ \underline{X} > \underline{Y} \\ 1 & \bar{X} = \bar{Y}, \ \underline{X} > \underline{Y} \ or \ \bar{X} > \bar{Y}, \ \underline{X} > \underline{Y} \ or \ \bar{X} = \bar{Y}, \ \underline{X} = \underline{Y} \\ 1 - d(Y,X) & otherwise \end{cases}$$

3   Problem Description

Let us reflect on the problem of deciding in car selection under interval-valued information. Assume that there are four cars: f_1 - Civic; f_2 - Saturn; f_3 - Ford; f_4 - Mazda. Each alternative is evaluated with following criteria: C_1 - cost, C_2 - reliability, C_3 - fuel eco., C_3 - style (Table 1).

Table 1. Decision matrix.

	Cost	Reliability	Fuel Eco.	Style
Civic (f_1)	8	9	9	7
Saturn (f_2)	7	7	8	8
Ford (f_3)	9	6	8	9
Mazda (f_4)	6	7	8	6

Importance weights of the criteria are interval-valued: $w_1 = [0.14 \quad 0.29]$, $w_2 = [0.38 \quad 0.42]$, $w_3 = [0.28 \quad 0.32]$, $w_4 = [0.05 \quad 0.12]$.

We apply the PROMETHEE method to solving the problem.

At the *first stage*, by the means of the following equation we normalize the decision matrix:

$$R_{ij} = [X_{ij} - \min(X_{ij})] / [\max(X_{ij}) - \min(X_{ij})] \ (i = 1, 2, \ldots, n : j = 1, 2, \ldots, m)$$

Where X_{ij} is the performance measure of ith alternative with respect to j^{th} criterion. For non-beneficial criteria, this equation can be rewritten as follows:

$$R_{ij} = \left[\max(X_{ij}) - X_{ij}\right] / \left[\max(X_{ij}) - \min(X_{ij})\right].$$

Table 2 indicates the results.

Table 2. Normalized decision matrix.

	Cost	Reliability	Fuel Eco.	Style
Civic	0.333333333	1	1	0.333333333
Saturn	0.666666667	0.333333333	0	0.666666667
Ford	0	0	0	1
Mazda	1	0.333333333	0	0

At the second stage, the evaluative alterations of ith alternative regarding other alternatives are calculated (Table 3).

Table 3. Evaluative alterations of ith alternative regarding the other alternatives.

	Cost	Reliability	Fuel Eco.	Style
$D(f_1 - f_2)$	−0.333333333	0.666666667	1	−0.333333333
$D(f_1 - f_3)$	0.333333333	1	1	−0.666666667
$D(f_1 - f_4)$	−0.666666667	0.666666667	1	0.333333333
$D(f_2 - f_1)$	0.333333333	−0.666666667	−1	0.333333333
$D(f_2 - f_3)$	0.666666667	0.333333333	0	−0.333333333
$D(f_2 - f_4)$	−0.333333333	0	0	0.666666667
$D(f_3 - f_1))$	−0.333333333	−1	−1	0.666666667
$D(f_3 - f_2)$	−0.666666667	−0.333333333	0	0.333333333
$D(f_3 - f_4)$	−1	−0.333333333	0	1
$D(f_4 - f_1)$	0.666666667	−0.666666667	−1	0.333333333
$D(f_4 - f_2)$	0.333333333	0	0	−0.666666667
$D(f_4 - f_3)$	1	0.333333333	0	−1

At the third stage, the preference function $P_j(i, i')$ is considered. The simplified preference function shown below is adopted here:

$$P_j(i, i') = 0 \text{ if } R_{ij} \le R_{i'j}$$

$$P_j(i, i') = \left(R_{ij} - R_{i'j}\right) \text{ if } R_{ij} > R_{i'j}.$$

Table 4 indicates the results.

Table 4. The preference function $P_j(i, i')$.

	Cost	Reliability	Fuel Eco.	Style
$P(f_1 - f_2)$	0	0.666666667	1	0
$P(f_1 - f_3)$	0.333333333	1	1	0
$P(f_1 - f_4)$	0	0.666666667	1	0.333333333
$P(f_2 - f_1)$	0.333333333	0	0	0.333333333
$P(f_2 - f_3)$	0.666666667	0.333333333	0	0
$P(f_2 - f_4)$	0	0	0	0.666666667
$P(f_3 - f_1))$	0	0	0	0.666666667
$P(f_3 - f_2)$	0	0	0	0.333333333
$P(f_3 - f_4)$	0	0	0	1
$P(f_4 - f_1)$	0.666666667	0	0	0
$P(f_4 - f_2)$	0.333333333	0	0	0
$P(f_4 - f_3)$	1	0.333333333	0	0

At the *fourth stage*, the following aggregated preference function estimating the attribute weights based on the interval-value is computed:

$$\pi(i, i') = \left[\sum_{j=1}^{m} W_j X P_j(i, i') \right] / \sum_{j=1}^{m} W_j$$

Where w_j is the relative importance (weight) of jth criterion. Table 5 indicates the results.

Table 5. Preference Aggregation function

	Civic		Saturn		Ford		Mazda	
Civic	–	–	0.463768	0.705882	0.614493	0.984314	0.478261	**0.75294**
Saturn	0.05507	0.16078	–	–	0.1913	0.39216	0.02899	0.0941
Ford	0.029	0.0942	0.0145	0.0471	–	–	0.0435	**0.1412**
Mazda	0.081	0.228	0.04058	0.11373	0.23188	0.50588	–	–

At the fifth step, we determine the leaving and entering outranking flows as follows:
Leaving (or positive) flow for i^{th} alternative, $\varphi^+(i) = \frac{1}{n-1} \sum_{i'=1}^{n} \pi(i, i')$, $i \neq i'$,
Entering (or negative) flow for i^{th} alternative, $\varphi^-(i) = \frac{1}{n-1} \sum_{i'=1}^{n} \pi(i', i)$, $i \neq i'$,
where n is the number of alternatives. Table 6 indicates the results.

Table 6. Leaving and entering flows

	$\varphi^+(i)$		$\varphi^-(i)$	
Civic	0.518841	0.814379	0.165217	0.482353
Saturn	0.091787	0.215686	0.518841	0.866667
Ford	0.028986	0.282353	1.037681	1.882353
Mazda	0.117874	0.282353	0.550725	0.988235

At the sixth stage, we compute the net outranking flow for each alternative as follows: $\varphi(i) = \varphi^+(i) - \varphi^-(i)$.

Table 7 indicates the results.

Table 7. Net outranking flow values

	$\varphi(i) = \varphi^+(i) - \varphi^-(i)$	
Civic	0.036488	0.649162
Saturn	−0.77488	−0.30315
Ford	−1.85337	−0.75533
Mazda	−0.87036	−0.26837

At the seventh stage we rank the alternatives by using Definition 4. The higher value of $\varphi(i)$ is, the better the alternative is. Table 8 indicates the results.

Table 8. Alternative ranks

	$\varphi(i) = \varphi^+(i) - \varphi^-(i)$		Rank
Civic	0.036488	0.649162	1
Saturn	−0.77488	−0.30315	2
Ford	−1.85337	−0.75533	4
Mazda	−0.87036	−0.26837	3

Thus, alternative f_1 (Civic) is the best one.

4   Conclusion

Car selection is multicriteria decision making problem which is characterized by interval-valued information on importance of criteria. In this paper, we apply PROMETHEE to selection of car under interval-valued weights of criteria. The obtained results show validity of application of the decision method.

References

1. Srikrishna, S., Reddy, S.M., Vani, S.: A new car selection in the market using TOPSIS technique. Int. J. Eng. Res. Gen. Sci. **2**(4), 177–181 (2014)
2. Yogesh, S., Nishant, Sh.: Application of multiple criteria decision making mathematical model for selecting best automobile. Int. J. Sci. Res. Dev. **5**(04), 1796–1801 (2017)
3. Yavaş, M., Ersöz, T., Kabak, M., Ersöz, F.: Proposal for multi-criteria approach to automobile selection. In: ISITES Conference 2014, Turkey, pp. 2060–2069 (2014)
4. Dawood, H.: Theories of Interval Arithmetic: Mathematical Foundations and Applications. LAP Lambert Academic Publishing, Germany (2011)
5. Magoc, T., Ceberio, M., Modave, F.: Interval-based multi-criteria decision making strategies to order intervals. Fuzzy Information Processing Society, pp. 1–6 (2008)

Exploring the Factors Affecting the Happy Marriages by Using Fuzzy TOPSIS

Farida Huseynova(✉)

Azerbaijan State Oil and Industry University, Baku, Azerbaijan
farida_hus@hotmail.com

Abstract. Different people characterize the experience of marriage differently, some consider that it may be based on arranged marriages, while, the others prove that there must be a real love. Even so, millions of people find relationships on Internet. Every person with full of hopes and full of a series of thoughts wishes succeeding everlasting happiness. However, the route to happiness is not always easy. We have regrettably to notice that, today's divorce statistics demonstrates sad reality- that many couples cannot complete their life journey to the end. The obscurity that stands on the road of a foundation of happy marriages is tried to be overviewed in this paper.

Keywords: Marriage · Real love · Arranged marriage · Love on internet · Happiness · Emotion

1 Introduction

The harmonious and happy family is the key to forming a stable and productive society. However, happiness of the family is quite different for everyone, besides the ways leading to happiness are different too.

Every person with full of hopes and series of thoughts portrays their future life and dreams about the experiencing a happy marriage. Many people consider that, the families should be established on appropriate principles and regulated accurately and only by this way happiness can be attained and be lifelong. Beyond doubt, there is no special model for happy perspectives.

Albanese [1] notes that "a value system reflects the culture and represents a higher level of superego functioning involving more abstract concepts that inform the person's life and provide guidance for the future but remain realistic, flexible, and widely shared by other members of society."

We have to stress that, happiness is a positive emotion experienced in a state of wellbeing. In Paul Ekman's six universal emotions happiness is the first basic one. The influence of affecting over person's emotion is composed of different parameters; mainly on values, perceptions, judgments, and beliefs.

As Lopez mentioned [2] a person with positive emotion can better control any situation, even avoid conflict and build meaningful relations. By their lofty ability of understanding the situations, they can easily meet the needs of those with whom they deal with.

© Springer Nature Switzerland AG 2020
R. A. Aliev et al. (Eds.): ICSCCW 2019, AISC 1095, pp. 914–922, 2020.
https://doi.org/10.1007/978-3-030-35249-3_122

Smith [3] argues that, when we are naturally involved to observe the emotions in others and to a certain extent, we witness them as the physical sensations. Understanding the companions' feelings, we search the ways to maximize their pleasures and minimize their pains, so that we may have a crucial role in their joys and enjoy their expressions of joyfulness and approval.

All the families are composed of human beings; with their differences on cultural values, personal values and own characteristics and personalities and naturally, these humans behave differently by making different decision in certain situations. Especially, with any new relationship, every person goes into the union with uncertain expectations, but as David Hume [4] underlines in his ''Theory of the Passions'' that reason can be accepted only as the slave of the passions, and there is no other way, than to serve and obey it.

After having had a look at the sociological aspects of family unions it would be important to provide our analyses on human's attitudes towards events and overview the road to happy marriages considering their main features. What are the main factors, fostering a happy family unity?

It is required to identify all type of marriages and avoid emotionally driven decisions in this path. Regrettably, for some people, money and beauty leading to recklessness is in the first place and people sacrifice most of their values upon it. Therefore, it would be useful do care, check, and examine on correct standards in spouse selection.

When the spouses are familiar with purpose and plans of the families in advance, they may accept it and prepare themselves for accompanying all future family situations. Therefore, it would be useful to provide a thorough and complete study of criterions for having a truthful information and overcoming the vague and surprising events. A feeling of expectation for a certain thing happens when there is no ambiguity referring to both future events and uncertain past or current events.

We tried to find the answer to the question "how to experience a happy marriage?" by providing survey among people of different ages, genders and occupations.

This paper aims to identify and clarify several factors that stand on the road of a foundation of happy marriages. Some people think that, it can be based on a real love, while, the others prove that arranged marriages are preferable, another ones hope to meet the expectations on Internet. For the purpose, fuzzy set theory is employed to deal with the key and crucial factors of success and advancement of the family according to above mentioned three alternatives. Hence, to deal with this problem fuzzy TOPSIS model was considered for calculating alternative preferences and weights of criteria.

This paper is organized as follows. Section 2 describes the analyses of arranged marriages, love, internet acquaintances, to see which of them is the most acceptable one for happiness.

Section 3 describes methods of analyses for the merits and excellence of marriages by using TOPSIS. In Sect. 4, finally, discussion and conclusions are presented.

2 Arranged Marriages

In arranged marriages, the acquaintance of young couples is performed mostly by parents, relatives or friends. The prediction by the families that the sides are more appropriate and coherent in alliances is the expected base. For some people of the world, love is hoped for after a couple marries, but not considered as a prerequisite to marriage. This can be the main disadvantage or problem in the family alliances.

However, the real life sometimes could be entirely inconsistent with the fundamental truth accordingly these marriages. However, these expectations can face many unpleasant results and lead to displeasures. The main problem here is that, without realizing whether they are compatible to each other or not, the couples are taking definitely the risk by exposing themselves to dangers in arranged marriages Many elder people of the world still consider that, the chances of happiness in arranged marriages succeeds high scores and more successes. They claim that love is not permanent, it is changeable.

That is why, in arranged marriage, respect, shared values, style of living and other factors are certainly considered as the main factors and the decisions of the families are mainly made according to those factors. Unfortunately, in most marriages based on love, the main values are not taken into consideration, as the most couples build their relationship based more on mutual attraction and friendship.

2.1 What Is Love? What Does Love Actually Feel Like?

It is considered, that the presence of love is an imperative element of a successful marriage.

Many people- philosophers, psychologists, and ordinary folk regard love as an emotion. Despite that, others argue that emotion cannot be accounted for romantic love.

Is the love feeling or emotion? Aristotle considers love as an emotion and as an intentional state [5].

It is natural to worship the one thing that millions of people cannot stop thinking about. People have different minds about Love. Some gauge it as craziness; the others appreciate it as beginning of a new life. You can spent hours upon hours wondering life questions about the feeling. What certain things do people need in love?

Scientific pioneer Minsky [6] offers an important model which allows to specify the work of a mind. According to him emotions, intuitions, and feelings cannot make our mind to think and to feel distinctly, therefore "When you fall in fall, you see the world quite differently".

Halvor Eifring [7] accepts love as an emotion too. He mentions that in the history of Western concept love is characterized as an emotion, feeling and sentiment and in a number of different languages it is acknowledged, such as in English and it as; basic instincts→emotions→positive feelings→love.

The latest research on love by Fredrickson [8] proves that "feelings of love is closely based and connected over a shared positive emotion".

There are some common aspects involved in love and emotional processes, whence emotion affect your behavior. Besides, human being is hardly able to make decision without emotion, as it is essential part of human intelligence. Future events and uncertain events can be ambiguous because of this factor.

Perhaps in love one would believe that a particular person has certain properties idealize the beloved, and judge them lovable. What might these properties be? Appearance, social status, communication ways, income etc.

Can being in love bring to happy marriages? Unfortunately, being in love not always lasts forevermore. It can be a fleeting state that either progresses into a long-term or it scatters, and the relationship disappears in a certain time. There are moments in the life, you can be in a mood when you feel happiness and show pleasure. There can be also times, when everything seems make us feeling weary or tedious and you can be described as being depressed by people around you.

People have such senses resulting all the weird effects as moods, feelings, emotions. What causes all these strange effects?

Today's problem is that, in many love stories, the couple spending a few happy hours together may think that they fall in love. Desirable consequences of an event based on beliefs, judgments are often thought of as reactive attitudes, and they can passively undergo. Can love be an emotion under this view? Emotions are not directly under our control. Therefore, we are overcome with joy, overwhelmed with fear, or gripped with anger. Intense feelings toward the individual, and their durability as the qualities of that individual change over time.

The level of idealization of the admiration will probably depend on the individual, that is, their upbringing and the culture surrounding their perception of love.

Though love appears to be universal, there are culture-specific norms and rules for it, and we tried to provide our analyses according to our environment.

2.2 Internet Acquaintances or Finding Love on the Internet

In today's modern world, millions of people are searching the ways of forming romantic relationships, their love stories on the Internet. However, it will be of great value to analyze the distinction between love on the internet from ordinary love and investigate their profound effects on the future success of the family. Internet love can be expressed as a verbal love or love through service not spending quality time together. It will also be important to examine if these types of love relationships prolong over time, and if the happiness is succeeded in them. Many people blame the high rate of marital failure on thoughts, as they did not spend enough quality of time together.

It is usually characterized as a much deeper problem, and the problem must be solved by quite different ways. The concept is simple - we all have different ways of communicating and there are certain types of interactions we value more than others.

Mendelsohn and Fiore [9] data scientists at Facebook also prove that, online dating provides an ecologically reasonable relationship on real environment, however it faces very big difficulties, as there are great deal of risks, vagueness and retributions of initiating in this kind family ties. According to a study led by Toma [10] "Many people misrepresent their height, weight or age in their profiles" in this relationships.

Therefore, it is important to have conscious effort in learning all the truth about the partners for overcoming the misadventures in building the relationship. Referring to the psychologists Mayer and Salovey's [11] advices, it is necessary to perceive, process and regulate information accurately and effectively in any environment according to one's capacity.

An intention of defining of required perspectives on marriage processes affect the future consequences and their respective functions provide important results.

3   Methods of Analyses for the Merits and Excellence of Marriages

We provided our survey amidst 200 people characterizing the experience of marriage differently, according to a real love, arranged marriages, or on Internet acquaintance. To the query "what kind of marriage| are preferable for you" different people responded quite differently, as it would be expected.

Our aim was defining these three ways and their strong impact to happiness. We are particularly interested to explore which family unions of above mentioned three alternatives could be the happiest.

The future of a happy family depends on the great extent on its form, if the loved ones are secure and satisfied with their life. The inner happiness of family can depend on criterions as; trust, income, morality, physical and mental health, beauty, knowledge and education, conditions of compatibility, equality in age, etc.

The one thing that any relationship needs to survive is balance. Couples can be happy not only for now, but forever. Consequently, we can forecast the behaviors and presume how they are able to adequate behave.

The most sensitive and important point of our discussion is that the merits and excellence of marriage depends not only of ways of alternatives, but also on some chosen criterions. In our research, three basic alternatives for happy marital life are chosen; trust, income, conditions of compatibility.

What is the way of identifying the harmonious spouse with these standards? The answers according to our criterions were certainly different how can we ascertain that our standards trust, income, conditions of compatibility can be ensured in all marriages?

By using fuzzy TOPSIS [12] the factors, affecting the happy marriages is explored.

We expect that, all three types of alternatives with the details of different criterions can be very interesting for comparison happiness in family problems. The decision matrix $\|x_{ij}\|$, where x_{ij} is a triangular fuzzy number $x_{ij} = (a_{ij}, b_{ij}, c_{ij})$ is shown in Table 1.

Table 1. Initial data

Weight	(0.2, 0.35, 0.45)	(0.1, 0.25, 0.65)	(0.2, 0.4, 0.4)
	Income	Trust	Compatibility
Love	(10.8, 12, 13.2)	(35.1, 39, 42.9)	(51.3, 57, 62.7)
Arranged marriage	(36.9, 41, 45.1)	(6.3, 7, 7.7)	(30.6, 34, 37.4)
Internet	(0.9, 1, 1.1)	(2.7, 3, 3.3)	(5.4, 6, 6.6)

In the first step, the normalized fuzzy decision matrix by using (1) is computed. The normalized fuzzy decision matrix

$$\tilde{r}_{ij} = \left(\frac{a_{ij}}{c_j^*}, \frac{b_{ij}}{c_j^*}, \frac{c_{ij}}{c_j^*}\right) \text{ and } c_j^* = \max_i\{c_{ij}\} \text{ (benefit criteria)}$$

In the second step, the weighted normalized fuzzy decision matrix is computed. The weighted normalized fuzzy decision matrix is

$$\tilde{V} = (\tilde{v}_{ij}), \text{ where } \tilde{v}_{ij} = \tilde{r}_{ij} \times w_j.$$

In the third step, the Fuzzy Positive Ideal Solution (FPIS) and Fuzzy Negative Ideal Solution (FNIS) is computed.

$$A^* = \tilde{v}_1^*, \tilde{v}_2^*, \ldots, \tilde{v}_n^*, \text{ where } \tilde{v}_j^* = \max_i\{v_{ij}\},$$

$$A^- = \tilde{v}_1^-, \tilde{v}_2^-, \ldots, \tilde{v}_n^-, \text{ where } \tilde{v}_j^* = \min_i\{v_{ij}\}.$$

In the fourth step, the distance from each alternative to the FPIS and to the FNIS is computed.

$$d_i^* = \sum_{j=1}^n d\left(\tilde{v}_{ij}, \tilde{v}_j^*\right),$$

$$d_i^- = \sum_{j=1}^n d\left(\tilde{v}_{ij}, \tilde{v}_j^-\right)$$

is the distance from each alternative A_i to the FPIS and to the FNIS, respectively.

In the fifth step, the closeness coefficient CC_i for each alternative is computed (Table 2).

$$CC_i = \frac{d_i^-}{d_i^- + d_i^*}.$$

Table 2. The closeness coefficient CC_i

Alternatives	CC_i
Love	(−2.1626, 0.6591, 3.0577)
Arranged marriage	(−3.2751, 0.5712, 4.4992)
Internet	(−0.9284, 0.0373, 0.8872)

Step 6. Rank the alternatives (Table 3).

First alternative- love is best one. Focused on survey - we have found in the data from which we have inferred our proposed model that, it is entirely possible to have happy marriages established mostly by love where criterions are also taken into account. We hypothesize that the more a member of a questioner assesses the income criterion then, they will be more satisfied with the relationship.

Table 3. Rank the alternatives

Alternatives	Grade	Rank
Love	0.96721	1
Arranged marriage	0.87274	2
Internet	0.720829	3

Undoubtedly, our interpretation is provided by our own way of thinking and analyses.

The work we have done is not provided as a kind of recipe to others in constructing and analyzing emotion models for achieving significant consequences. Undoubtedly, there are many other ways to do this.

4   Conclusion

The excellence of the marriage problems is overviewed in this paper.

However, it should be born in mind that a hundred percent homogeneity and harmony is not possible and a certain amount of distance and disharmony would definitely exist. It is assumed that, if the families are set upon correct foundations and are managed in the right way, then that 'love' and 'compassion' will continue and will remain life-long.

We keep moving on this problem until we reach a final decision (positive or negative). Our goal is to reach a clear-cut conclusion for the initial phase of research in our environment. And we are for ever lasting love in the families,

At a time when the standards and principles are in one's hand and one knows what one wants, then moving on the way is not very difficult (although it is a small thing to do and needs concentration).

It is difficult to make accurate generalizations of the factors that contribute to happiness of marriages. Some responses are too vague, whilst some are too specific. Thus we had a predesigned list containing 3 criterions trust, income, compatibility and let the interviewees to choose 1 answer. In this case, generalization would be more valid. Further, it is definitely not enough to construe the effects of boredom towards satisfaction and closeness of relationship by simply using three items. Therefore we should ask participants to choose the reasons of satisfaction from several choices including boredom immediately after asking the second item so as to double-check the validity of the correlation. Analyzing three criterions, we attempted to achieve a goal in identifying and clarifying respective ambiguities that stand in the way of a formalization of normal families.

These findings are particularly impactful given the need for efficient and effective representations. The problem is how we expect our partner to act and behave in family unions would be unrealistic without satisfaction.

Subsequently to our analyses of three criterions the aim was to identify and clarify respective ambiguities that stand in the way of a formalization of well- established families. It is evident, that there are individual differences, external events that are interpreted by different people differently and besides, emotional processes are comprehended inconsistently by people in all aspects.

The criterion is an integral part of forming and developing meaningful marriages. We could define that, over a series of surveys and observations, there were significant links between them and alternatives in successful interpersonal relations. Those participants who exhibited higher levels of EI also showed a greater propensity for empathic perspective taking, more satisfying relationships. Degree of satisfaction 7

Feeling of happiness 8 (responses) 22 in total Very/Somewhat Satisfied, Very/Somewhat Dissatisfied, Very Satisfied Sometimes/Often (12) 9 (75%), 3 (25%), 5 (41.7%), Rarely/Never (10) 9 (90%), 1 (10%), 9 (90%).

Understanding the causes and consequences of problems allows an individual to both manage the feeling and make an objective decision.

References

1. Albanese, P.J.: The Personality Continuum and Consumer Behavior. Quorum Books, Westport, American Psychiatric Association. Diagnostic and Statistical Manual of Mental Disorders. 4th ed. American Psychatric Association, Washington D.C (2002)
2. Ricky, W.G., Yvette, P.L.: Bad behavior in organizations: a review and typology for future research. J. Manag. **31**(6), 988–1005 (2005)
3. Smith, A.: The Theory of Moral Sentiments. (Glasgow Edition of the Works and Correspondence of Adam Smith), 1 (1985)
4. Selby-Bigge, L.A., Nidditch, P.H. (eds.): Treatise of Human Nature, Oxford (1978)
5. Nieuwenburg, P.: Emotions and perceptions in Aristotle's Rhetoric. Aust. J. Philos. **80**(1), 86–100 (2002)
6. Minsky, M.: The emotion machine: commonsense thinking, artificial intelligence, and the future of the human mind (draft), E-library. RoyalLib.Com (2006)
7. Emotions and the Conceptual History of Qíng. In: Love and Emotions in Traditional Chinese Literature. Leiden/Boston, Halvor Eifring University of Oslo (2018). https://www.researchgate.net/publication/325127435
8. Barbara, L.F.: The role of positive emotions in positive psychology. Broaden-and-build theory positive emotions. Am. Psychol. **56**(3), 218–226 (2001)
9. Mendelsohn, G.A., Fiore, A.T.: Black/White dating online: interracial courtship in the 21st century. American Psychological Association 3, 1, 2–18 (2014). https://www.researchgate.net/publication
10. Toma, C.L.: Examining the production, detection, and popular beliefs surrounding interpersonal deception in technologically-mediated environments. Palgrave Handbook of Deceptive Communication (2019, forthcoming)

11. Mayer, J.D., Salovey, P.: Yale Center for Emotional Intelligence. Sternberg 11 (2011)
12. Gardashova, L.A.: Z-number based TOPSIS method in multi-criteria decision making. In: Advances in Intelligent Systems and Computing, 896, Proceeding of Thirteenth International Conference on Applications of Fuzzy Systems and Soft Computing, pp. 42–50. Springer Nature Switzerland (2018)

Estimation of Ground Water Parameters by Developing a New Fuzzy Optimized Simulation Model

Nima Eini[1][(✉)] and Fidan Aslanova[2]

[1] Department of Civil Engineering, Faculty of Civil
and Environmental Engineering, Near East University,
P.O. Box: 99138 Nicosia, TRNC, Mersin 10, Turkey
nima.eini@neu.edu.tr
[2] Department of Environmental Engineering,
Faculty of Civil and Environmental Engineering, Near East University,
P.O. Box: 99138 Nicosia, TRNC, Mersin 10, Turkey
fidan.aslanova@neu.edu.tr

Abstract. Proper groundwater management is considered as one of the main sources of accurate groundwater parameters. Groundwater parameters are usually ignored in order to simplify the equations and uncertainty of seepage testing. In this research, an optimized simulation model has been developed to consider uncertainties in determining groundwater parameters. The optimized simulation model can accurately estimate the parameters of ground water due to the minimization of the difference between observation reduction and computational reduction loss. The proposed method is compared to the actual seepage test in the enclosed groundwater and its results are compared with the Thies method. Statistical error based on the results of proposed modeling simulation and the Thies method is solved by comparing the performance of two methods. For example, the average absolute relative error of the suggested simulation model and the Thies curve solution is respectively 0.69 and 1.13%, which indicates the accuracy of the proposed simulation model to the Thies solution. In the second section, due to the amount of discharge as an uncertain parameter, an optimized fuzzy simulated model is developed based on the fuzzy transformation method. Based on the results of this fuzzy method, the effect of this uncertainty on the optimal estimation of enclosed groundwater parameters has been investigated and the variation of groundwater parameters has been determined in different fuzzy sections. The results of two developed fuzzy methods show that the effect of discharge uncertainty on the estimation of the groundwater T parameter is higher than **S**.

Keywords: Groundwater parameters · Fuzzy transform method · Genetic algorithm · Seepage phenomenon

1 Introduction

Groundwater is one of the vital resources that play a significant role in providing water to agricultural, drinking and industrial sectors. The ability to store and transport of water in aquifers is controlled by their associated parameters such as storage coefficient

© Springer Nature Switzerland AG 2020
R. A. Aliev et al. (Eds.): ICSCCW 2019, AISC 1095, pp. 923–930, 2020.
https://doi.org/10.1007/978-3-030-35249-3_123

and transmissivity. Therefore, the optimum estimation of groundwater aquifers parameters is essential for optimal water resource management. Determining aquifer parameters using reverse methods is commonly used nowadays. It should be noted that in the above-mentioned methods, which are usually based on the matching of the curve by eyes, if the observation-time data are not consistent with the type curves, the results of the estimation of aquifer parameters won't be reliable; for this reason, new methods and optimization algorithms such as genetic algorithm (GA) are used to estimate optimal aquifer parameters. The purpose of using these algorithms is to minimize the difference between computational and observational losses. Evolutionary methods are one of the methods used in relatively optimal estimation problems for aquifer parameters like [1–5]. They used the method of curve fitting study, to model the specific simulation and SUMT comparison method. By comparing the total squared error of the three methods, the results indicated that the genetic optimization method was better and more suitable than the drawing method and SUMT method. Lu et al. used genetic optimization method to estimate horizontal and vertical hydraulic conductivity coefficients, storage factor, specific storage, delay index and thickness of the free characteristic aquifer [6]. The results indicate that the horizontal and vertical hydraulic conductivity coefficients and the anisotropic ratio[1] are strongly influenced by the aquifer thickness. The hypothesis is a generalized fuzzy set of the classical hypothesis, and a fuzzy set has set by its members with a degree of membership of 0 to 1 belong to it. In order to obtain reliable answers to real numbers, there should be exact values of the simulator model parameters, but in practice, these exact values are not present and the models are similar The instrument that has a degree of uncertainty. These uncertainties may be due to the lack or lack of a group of information, errors in measuring or simplifying complex issues [7–10]. The use of common fuzzy mathematics in nonlinear optimization problems is problematic due to the presence of several maximum and minimum points in different alpha sections for fuzzy inputs because it changes the shape of the fuzzy join function and its efficiency in practice [10–13]. Comparison of recently available articles suggests that uncertainties based on fuzzy logic have not been used for optimal estimation of enclosed aquifer parameters. In this research, a new fuzzy simulation-optimization method based on fuzzy and fuzzy mathematical modeling of fuzzy transformation and connection Simulation model-Optimization of genetic algorithm with the numerical solution of good function by Thies method for aquifer model has been developed [12].

2 Methodology

The developed method is based on the connection of the simulation model to optimize the genetic algorithm by numerical solution which using the Thies method and finally, simulation-optimization model shaped based on the fuzzy transformation method. The simulation-optimization is able to estimate the bounded aquifer parameters by minimizing the differences between the observed loss and computational loss. In the third

---

[1] Anisotropic ratio: $\left(\frac{k_r}{k_z}\right)$.

step, in order to examine the simulation-optimization model, its implementation for the probable values for The non-deterministic runoff test is performed and the objective function of the optimization simulation model is obtained corresponding to each of the above states, then the chart of changes in the value of the target function is plotted against the various values of the curve and the simulation-optimization model is analyzed. In the second step, the model needed to estimate the bounded aquifer parameters based on the genetic algorithm model which developed. To do this, decision variables, limitations, and the target function of the model should be determined. The simulation-optimized model is able to estimate the bounded aquifer parameters by minimizing the differences between the observed loss and computational loss. When the process of changes in the target function is consistent with the case, the fuzzy reduction method is used, otherwise, the general fuzzy transform method is used. In the last step, considering the unanimous or non-identical behavior of the optimized simulation model, the parameters Fuzzy optimized simulation model is decomposed with a suitable decay or general fuzzy transform method, and an optimized simulation model is implemented with a fuzzy transformation method. Then the conversion of the fuzzy intervals of the objective function of the optimized simulated model and the fuzzy function of the target and the bar of changes in decision variables are obtained in different fuzzy slices. Subsequent parts of the most important sections of the proposed method of identification include optimizing the simulation model and fuzzy transformation is described in detail.

3 Setting Optimized Simulated Model

The main objective of this study is to estimate the parameters of groundwater under uncertainty conditions using the genetic algorithm in order to minimize the difference between the loss of observation due to seepage or pumping and the calculated values, target function that purpose of this research is the following is:

$$Min(Z) = \left(\sum_{t=1}^{n} \frac{\left| s_0^t - s_c^t \right| / s_0^t}{n} \right), \tag{1}$$

$$sc = \frac{Q}{4\pi T} \int_{u}^{\infty} \frac{e^{-x}}{x} dx, \tag{2}$$

$$u = \frac{r^2 s}{4Tt} \tag{3}$$

s_0 is the observation loss in the observation well in time t, in terms of the meter, s_c is the computation loss using the this equation, n the total number, Q the equivalent of the discharge in cubic meters per day, t in terms of the day of the transfer discharge, T is transportation coefficient per square meter per day, r is the distance of observation well to discharge well, s is stored coefficient according to the above equations. S, T are selected as decision variables in an optimized simulation model.

3.1   Fuzzy Method

It can be said that one of the most important uncertain parameters in water penetration and seepage experiments performed to estimate groundwater parameters is discharge. This uncertainty is especially evident in cases where the measurement of the discharge has not been accurately taken. This research has been used to determine the uncertainty of the amount of discharge in an optimized simulation model. The method of fuzzy transformation is an uncertainty analysis method for non-deterministic models. FTM is a fuzzy simulation model of the complex nonlinear phenomenon in order to develop a fuzzy simulation model for estimating groundwater parameters in an appropriate optimization framework.

3.2   Fuzzy Transformation Method

Fuzzy transformation is one of the practical and useful tools for simulating systems that conclude non-deterministic parameters. This method can be used in two general and summary ways. The general method for simulating and analyzing non-uniform problems and equations and the summarized method could be used for phenomenon with uniform equations. Generalized and summary approach to fuzzy transformation is contain: In any issue, if we consider non-deterministic parameters $\tilde{p}_i = (i = 1, 2, \ldots, n)$ as fuzzy form, each particular parameter can be considered as P_i that set of series as $m + 1$ with $X_i^{(j)}(j = 0, 1, \ldots, m)$ would be decomposed into a multitude cluster [11]. The fuzzy transformation method considers all possible combinations of lower and upper bounds and additional quantities between them for all non-deterministic parameters. Therefore, this method can be very suitable for optimization and nonlinear simulation problems [10, 13].

4   Result and Discussion

Simulation model - optimized of the genetic algorithm based on the numerical solution of the good function in the software was prepared. To evaluate the proposed simulator model, the experimental data used in groundwater as detailed in Todd were used [11]. Then the simulation-optimization model was developed for this information and computational errors were determined. In the following, the calculated computational values of the model optimized simulation model were compared with the observed loss values and the graphical results of the Thies method. In order to measure the efficiency of the proposed method, mean absolute relative error (MARE), root means square error (RMSE), man absolute error (MAE) and correlation coefficient R^2 were calculated.

Table 1. Calculation of different statistical indices of the proposed simulation model in comparison with the approved Thies method

Method salutation	Statistical error indicators			
	MARE	RMSE	MAE	R^2
Curve Thies method	1.13	0.0085	0.0065	0.9994
Simulated optimized method	0.69	0.0055	0.0035	0.9997

In the present study, the MARE index is less than 10% acceptable, and the results presented in Table 1 show that the index for the simulated optimized proposed model is less than 1% and has a desirable accuracy. Also, it shows less error than Thies curve method. RMSE and MAE indicators are closer to zero, with higher accuracy. As shown in Table 1, the values of these errors are also lower for the proposed optimized simulation model than Thies method. R^2 the index in the groundwater which investigated in this study is much closer to 1, and it's greater than Thies method. In sum, it can be said that the performance of the optimized simulation model is better than Thies method. In Table 2, the results of estimating the applied parameters in the proposed optimization method are better than the curved this method.

Table 2. Calculation of groundwater parameters by proposed simulation model and comparison with Thies curve method

Parameters	Method/salutation		
	Thies method	Optimized simulation method	Error ratio
$T(m^2/day)$	1110	1145.68	3.11
S	0.000206	0.000188	9.57

After assessing the efficiency of the proposed optimized method based on the genetic algorithm, the main indeterminate parameters influencing the simulation model were determined. In this study, it seems that seepage is considered as an important non-linear parameter. This is especially convinced when discharge is not exactly measured. After the determination of the discharge as an uncertain parameter, the uniformity of the optimized simulated model has been investigated. Thus, for different possible values of discharge, for probable variations due to the measurement error of the model Optimized simulation has been implemented and the target function of the simulated optimized model is calibrated (Fig. 1).

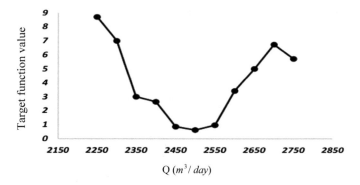

Fig. 1. Optimized simulated model target function variable chart for different values of the non-deterministic discharge parameter.

As shown in Fig. 1, the behavior of optimal simulation is observed. Regarding the non-uniformity of the simulation model, the uncertain parameter of the simulation model is analyzed based on the fuzzy transformation method, and the general fuzzy transformation method is optimized for the simulation model Used. Difference between the absolute contents of $|Q_{max} - Q_{min}|$ shows uncertainty in the measurement of discharge. The however greater size of $|Q_{max} - Q_{min}|$ lead us to the amount of uncertainty in the parameter. After designing a fuzzy optimized simulation model and executing it, target function of the simulated model is optimized by re-converting to the fuzzy intervals of simulation model in different alpha sections with increasing degrees of fuzzy variation limit of target function is reduced, which reflects the lesser effect of the uncertainty in the fuzzy section.

Table 3. The range of variation of fuzzy decision variables in different fuzzy sections

Fuzzy weight section	The range of variations of decision variables		
	–	$T(m^2/day)$	S
0	Minimum limit	1136.54	0.000183
	Maximum limit	1140.55	0.000188
0.25	Minimum limit	1125.15	0.000188
	Maximum limit	1152.62	0.000188
0.5	Minimum limit	1073.69	0.000188
	Maximum limit	1140.42	0.000192
0.75	Minimum limit	1123.93	0.000188
	Maximum limit	1165.68	0.000188
1	Minimum limit	1145.58	0.000019
	Maximum limit	1145.58	0.00019

The maximum changes for parameter T in the fuzzy section are 0.5.

5   Conclusion

In this research, the effectiveness of the proposed optimized simulated model with the connectable properties to a genetic algorithm based on Thies equations and the fuzzy transformation method was evaluated to estimate groundwater parameter. The target function of the optimized simulated model was based on minimizing the difference between observation reduction factors from the results and seepage test with computational reduction are taken from the Thies method. In order to evaluate the performance of this optimized simulation model, we compare this model with a curved method. The results show a better performance of the proposed method than the curved method. For example, the mean absolute relative error of the simulation model and curved Thies were respectively 0.69% and 1.13%. Therefore, an optimally model could be a suitable alternative to the Thies method. After evaluating the performance of the proposed simulation model by selecting discharge as non-deterministic parameter a fuzzy optimized simulation model was developed and implemented based on the fuzzy transformation method. Regarding the non-uniform behavior of the simulation model, the general fuzzy transformation method was used and the variation of groundwater parameters was determined in different fuzzy sections. Since the target function changes of the optimal simulation model are not uniform compared with the probable changes in the measured value of the discharge, and considering the nonlinearity of the optimal simulation model, the percentage of changes in groundwater parameters in different fuzzy sections does not follow the linear trend. This means that with increasing fuzzy degrees, the percentage of changes for groundwater parameters does not necessarily increase or decrease uniformly. As shown in Table 3, for the $\pm 10\%$ change in discharge, T has the most percentage change in phases of 0.5 and S has the highest percentage change in zero degrees. This means that if the flow rate changes from 2375 to 2625 m^3/day, the transmission coefficient parameters changes by 6.2%. Also, due to changes in the flow rate from 2250 to 2750 m^3/day, the coefficient of storage parameter changes by 2.7%.

References

1. Lingireddy, S.: Aquifer parameter estimation using genetic algorithms and neural networks. Civil Eng. Syst. **15**, 125–144 (1998)
2. Samuel, M.P., Jha, M.K.: Estimation of aquifer parameters from pumping test data by a genetic algorithm optimization technique. J. Irrig. Drainage Eng. **129**, 348–359 (2003)
3. Kerachian, R., Fallahnia, M., Bazargan-Lari, M.R., Mansoori, A., Sedghi, H.: A fuzzy game-theoretic approach for groundwater resources management. Application of Rubinstein bargaining theory. Resour. Conserv. Recycl. **54**, 673–682 (2010)
4. Abdel-Gawad, H.A.A.A., Elhadi, H.A.: Parameter estimation of pumping test data using a genetic algorithm. In: Thirteenth International Water Technology Conference, IWTC 13 (2009)
5. Rajesh, M., Kashyap, D., Hari Prasad, K.: Estimation of unconfined aquifer parameters by genetic algorithms. Hydrol. Sci. J. **55**, 403–413 (2010)
6. Lu, C., Shu, L., Chen, X., Cheng, C.: Parameter estimation for a karst aquifer with unknown thickness using the genetic algorithm method. Environ. Earth Sci. **63**(4), 797–807 (2011)

7. Bakhteni, S., Mortazavi-Naeini, M., Ataie-Ashtiani, B., Jeng, D., Khanbilvardi, R.: Evaluation of methods for estimating aquifer hydraulic parameters. Appl. Soft Comput. **28**, 541–549 (2015)

8. Zhuang, C., Zhou, Z., Zhan, H., Wang, G.: A new type curve method for estimating aquitard hydraulic parameters in a multi-layered aquifer system. J. Hydrol. **527**, 212–220 (2015)

9. Zadeh, L.A.: Fuzzy sets. Inf. Control **8**(3), 338–353 (1965)

10. Hanss, M.: The transformation method for the simulation and analysis of systems with uncertain parameters. Fuzzy Sets Syst. **130**, 277–289 (2002)

11. Hanss, M., Willner, K.: A fuzzy arithmetical approach to the solution of finite element problems with uncertain parameters. Mech. Res. Commun. **27**, 257–272 (2000)

12. Aramaki, T., Matsuo, T.: Evaluation model of policy scenarios for basin-wide water resources and quality management in the Tone River Japan. Water Sci. Technol. **38**, 59–67 (1998)

13. Sadegh, M., Kerachian, R.: Water resources allocation using solution concepts of fuzzy cooperative games: fuzzy least core and fuzzy weak least core. Water Resour. Manag. **25**, 2543–2573 (2011)

14. Nikoo, M.R., Kerachian, R., Karimi, A., Azadnia, A.A.: Optimal water and waste-load allocations in rivers using a fuzzy transformation technique: a case study. Environ. Monit. Assess. **185**, 2483–2502 (2013)

Fractal Dimension Determining for Demand Forecasting

Seving R. Mustafayeva$^{(\boxtimes)}$ (iD)

Azerbaijan State Oil and Industry University, Azadlig 35,
Nasimi, Baku, Azerbaijan
mustafayeva_81@mail.ru

Abstract. This paper analyzes the use of fractal dimensionality, and studies the conduct of complex dynamic systems, defining the criterion of instability by means of considering a value of the fractal dimension of an attractor. We propose fractal dimension determining for forecasting problem.

Keywords: Lyapunov index · Fractal dimension · Time series · Forecasting

1 Introduction

Modern economic theory has long proved the inconsistency and inadequacy of traditional linear models of market behavior. The practice shows that the dynamics of economic processes and phenomena are non-linear and often chaotic (unpredictable). This necessitates the search for alternative modeling methods using non-standard mathematical tools. The analysis of mathematical modeling in time series showed a number of interesting methods and findings, among which are: local and global methods; statistical processing methods; time series modeling using empirical Liouville equations and others. In recent decades, the set of traditional (linear) methods for studying time series has been significantly expanded by nonlinear methods derived from the theory of nonlinear dynamics and chaos; Many studies have been devoted to the evaluation of nonlinear characteristics and properties of natural and artificial systems. However, as the analysis showed, most nonlinear analysis methods require a significant amount of data or stationary time series, which is not always possible to obtain in real conditions.

Practice shows that the dynamics of economic processes and phenomena is non-linear and, often, chaotic (unpredictable) character. This necessitates the search for alternative modeling methods using non-standard mathematical tools. To date, there are quite a few areas in this field of economic and mathematical science. When analyzing socio-economic processes, such mathematical tools as fuzzy methods, neural networks, genetic algorithms, etc. are increasingly used.

Since the 1980s, many Western scientists have been actively involved in the introduction of the theory of fractals [1–4] into the economy, while domestic researchers began to consider this theory relatively recently. The use of fractal analysis in economics is described in the works of such eminent researchers as B. Mandelbrot, E. Peters, V. Arnold, P. Berger, I. Pomo, K. Vidal, G. Schuster, R. Manten, H. Stanley, V. C. The use of the mathematical apparatus of the theory of fractals opens up new

© Springer Nature Switzerland AG 2020
R. A. Aliev et al. (Eds.): ICSCCW 2019, AISC 1095, pp. 931–936, 2020.
https://doi.org/10.1007/978-3-030-35249-3_124

possibilities in the modeling of market processes. The key to this is the self-development of the fractal. This property characterizes a fractal as a mathematical object that most closely matches the systemic nature of social and economic processes occurring under the non-linear dynamics of a variety of factors of external and internal media.

In the real world, pure, ordered fractals, as a rule, do not exist, and one can speak only of fractal phenomena. They should be considered only as models that are approximately fractals in a statistical sense. How, D. Sornett, A. Yu. Loskutov, A.S. Mikhailov, N.V. Chumachenko, A.I. Lysenko and others. However, a well-constructed statistical fractal model makes it possible to obtain sufficiently accurate and adequate forecasts. Fractal analysis of markets, unlike the theory of efficient markets, postulates the dependence of future prices on their past changes. Thus, the pricing process in the markets is globally determined, dependent on "initial conditions", that is, past values. Locally, the pricing process is random, that is, in each case, the price has two options for development. The fractal analysis of the markets is directly based on the fractal theory and borrows the properties of fractals to obtain predictions.

Fractal computing is a new kind of soft computing. The concept of "soft computing" was introduced by the founder of fuzzy logic L. Zadeh at a seminar in 1994, as a consortium of computational methodologies that collectively provide the basis for understanding, designing and developing intelligent systems. Currently, soft computations include: fuzzy logic, neural computing, genetic computation, probabilistic computation, evidence-based reasoning, trust networks, sections of machine learning theory, fractal theory. In this paper, one kind of soft computing is introduced - calculations of fractal dimension time series.

2   Fractal Dimension Determination

Fractal dimension of a time series is calculated by measuring how jagged it is. We count the number of circles of a given fixed diameter and again count the number of circles required to cover the time series. If we continue this process we see that, the number of circles is related to the diameter of the circle according to the following relationship:

$$N \cdot d^D = 1,$$

where N - number of circles, d - diameter and D - fractal dimension. Equation can be transformed to find the fractal dimensions:

$$n = \frac{1}{s^D} \tag{1}$$

the familiar case.

$$\log n = \log\left(1/s^D\right) \tag{2}$$

take log of both sides.

$$\log n = D \log(1/s) \tag{3}$$

factor the exponent D out of the scale factor.

$$\frac{\log n}{\log 1/s} D \tag{4}$$

divided by $\log 1/s$ to set the equation equal to D.

We are left with the equation:

$$D = \frac{\log n}{\log 1/s} \tag{5}$$

Although there is no rule that dimension D has to have an integer value, this has been the convention in traditional geometry. Here we will carry out numeric calculations, for many different values of n and s, where D is not an integer dimension, but rather a "fractional" or "partial" dimension.

3 Computer Simulation

In this paper we want to determine the fractal dimension of time series for forecasting of the next year demand for the market according to the available data of recent years data. In particular, such a characteristic of the time series as the fractal dimension allows determining the moment when the system becomes unstable and ready to move to a new state (Fig. 1).

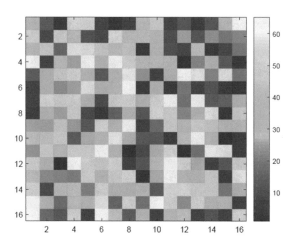

Fig. 1. Fragment of input data

Function zout = zlap(I, J, X, TAU)% m file for function zlap used in lap.m for calculation of % the first Lyapunov exponent using the method described by Wolf % zout = X(I + (J − 1)*TAU);

Lyapunov exponent
Lamda1 = 0.6609

A positive Lyapunov index indicates the existence of chaotic motion in a dynamic system with limited trajectories (Figs. 2 and 3).

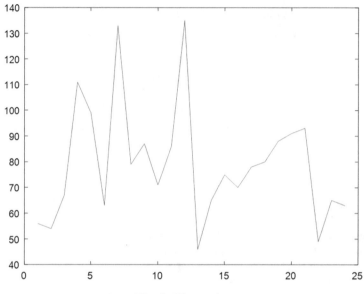

Fig. 2. Time series

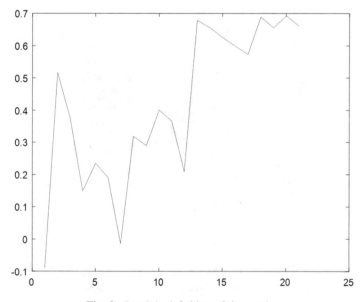

Fig. 3. Lambda definition of time series

As this dimension say about chaos of time series we can determine fractal dimension. With the using of MATLAB Simulink we obtain the fractal dimension of time series. Firstly we convert our data from linear form to 3d form. Then determine fractal dimension $D = 2.23$. Consequently that the time series has the effect of long-term memory, estimate its depth, reveal the presence of cycles (quasi-cycles). And we can say our result characterizes investments in the relevant financial instrument as "relatively risk-free" (Fig. 4).

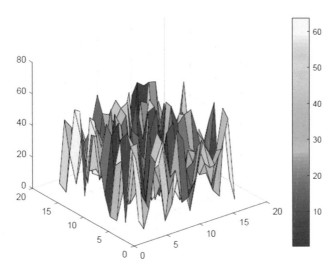

Fig. 4. Fractal dimension determination

4   Discussion and Conclusions

The study of the time series of an unstable market system made it possible to establish a connection between a state and an indicator of fractal dimension with fuzzy logic. For the market, the change in the fractal dimension is an indicator of a crisis state. Via this dimension is easy to forecast the demand of market. The main result of the work is the conclusion about the possibility of conducting a forecast analysis of financial time series by calculating a fractal measure. As a result of this analysis, we can conclude that the time series has the effect of long-term memory, estimate its depth, reveal the presence of cycles (quasi-cycles). At the same time, this analysis is not an exhaustive tool for the forecasting study of time series, since it does not always provide complete information about the behavior of the time series without the use of additional methods and algorithms.

References

1. Billingsley, P., et al.: Hausdorff dimension in probability theory. Ill. J. Math. **4**(2), 187–209 (1960)
2. Theiler, J.: Estimating fractal dimension. JOSA A **7**(6), 1055–1073 (1990)
3. Weron, A., Weron, R.: Fractal market hypothesis and two powerlaws. Chaos Solitons Fractals **11**(1), 289–296 (2000)
4. Zadeh, L.A.: Soft computing and fuzzy logic. IEEE Softw. Arch. **11**(6), 48–56 (1994)

Decision-Making on Steam Injection While High-Viscosity Oil Production Process Considering Uncertainty Conditions

D. A. Akhmetov[1], S. T. Zakenov[1], and O. G. Kirisenko[2($\boxtimes$)] (iD)

[1] Caspian State University of Technology and Engineering named after Sh. Yessenov, 32 Microdistricts, 130003 Aktau, Mangistau Region, Republic of Kazakhstan
aldee@list.ru, senbek@rambler.ru
[2] Oil and Gas Institute of the Azerbaijan National Academy of Sciences, F.Amirov 9, AZ1000 Baku, Azerbaijan
oleg.kirisenko@gmail.com

Abstract. The purpose of technological regimes during steam flooding should ensure the maximum increase in well flow rates for oil and a reduction in the flow rate of the working agent supplied. Based on the fact that the choice of the volume of steam occurs in a situation of uncertainty, as it is impossible to reliably predict the results of regime change, the application of the decision should be made taking into account this circumstance. In a situation when it is necessary to make a choice between modes with a minimum steam rate or a maximum liquid rate or some average between these modes, one of the decision-making criteria under conditions of uncertainty is used. This paper is devoted to the problem of optimizing the production of high-viscosity oil during steam flooding. The task in this case is two-criteria, that is, you need to choose the optimal mode, which should ensure the maximum yield with a minimum steam flowrate. In accordance with this, one or another method of making the optimal decision should be chosen. This problem was solved by the example of one of the fields in Kazakhstan with the application of the theory of fuzzy sets.

Keywords: Decision-making · Uncertainty · Steam-oil factor · Membership function · Oil production · Viscosity · Heavy oil

1 Introduction

The study shows that there are considerable numbers of fields with heavy and high-viscosity oils production. Regarding the geography of hard-to-recover oil reserves in terms of historical data, it can be noted that these hydrocarbons basins cover a wide geography. In this regard, various classifications of hard-to recover oils [1, 9–11] have been proposed, which allow the accurate select of methods on formation stimulation.

As a matter of fact, the oil in Kazakhstan fields is mostly heavy oil (ρ = 936 kg/m^3), high tarry oil composition (to 24%). The viscosity is high, therefor from the beginning of field reservoir exploitation (in the case under consideration Karazhanbasmunai), the project design decisions were made, according to this the thermal

© Springer Nature Switzerland AG 2020
R. A. Aliev et al. (Eds.): ICSCCW 2019, AISC 1095, pp. 937–942, 2020.
https://doi.org/10.1007/978-3-030-35249-3_125

methods were applied, the most important of which is steam injection [2–4]. Based on summaries of the experience of using thermal methods by stimulating formation [3], we performed an analysis of the thermal methods results has impact on it at the Karazhanbasmunai field [4, 5].

In this case, the task is that the assignment of technological regimes should ensure the maximum increase well production in oil flow rates and reduce the costs of supplied working agent.

Based on the fact that the selection of the steam volume occurs in an uncertainty situation, since it cannot be absolutely reliably predicted the results from regime changes, the decision implementation should be made taken into account this circumstance. In a situation when it is necessary to make a choice between regimes with minimum specific steam rate consumption or maximum fluid production rate or some average between these regimes, one of decision criteria under conditions of uncertainty is used.

In this case, the task of optimizing is two-criterial, that is, it needs to select the optimal regime, which should ensure the maximum flow rate with minimum specific steam rate consumption. In accordance with this, one or another method of making the optimal decision should be selected.

2 Task Assignment

The main point of the method of operating by steam injection is increasing the well production rates by using the necessary amount of working agent.

According to the studies review, presently, it is proposed the various methods for optimizing the steam thermal effect [6].

As shown by review, conventional technologies and methods of decision making that were previously developed do not allow finding straightforward decision under conditions of uncertainty.

Currently, there are about 1 million oil wells in the world [7]. The reservoir pressure in such wells, as a rule, does not allow oil to be produced by the flowing method, therefore more than 90% of them are operated using any method of artificial lift production to increase their productivity.

A large number of works are devoted to theoretical and experimental studies of steam thermal effects. These studies allowed a deeper study of the processes occurring with this method. Studies have shown the need for parallel field-theoretical studies, which complement each other, would allow following the dynamics of process indicators, to take a technological decision, taking into account the uncertainty of aims and constraints.

In recent investigations, research tasks are set for decision-making; however, setting the problem using two criteria requires the use of modern methods that take this circumstance into account and bring in fuzziness. Based on this, the task is to select the optimal combination of criteria, i.e. ensuring maximum production with the minimum amount of injected steam based on the analysis of field information using the provisions of the theory of fuzzy sets.

3   Research Results

To solve the problem, data were collected on the results of steam thermal effects, which were subjected to statistical processing.

Often, in practice, the production of the steam-thermal method uses the main characteristics - this is the dynamics of the specific steam consumption (steam-oil factor) and production rate of the total steam consumption. Such dependences in the form of graphs are shown in Fig. 1.

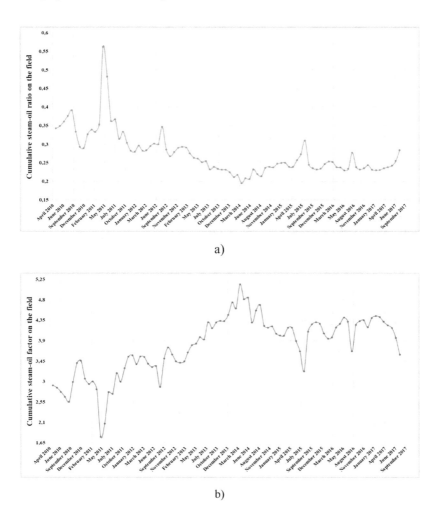

a)

b)

Fig. 1. Dynamics of steam-oil factor.

Such dependences allow monitoring changes in the parameters, to analyze the conditions, to execute calculations to substantiate and to determine the required quantity of injected steam. It enables the forecast assessment of the trend of oil production for the subsequent period using the graph of behavior of the steam-oil factor (hereinafter referred to as SOF). SOF is a ratio of the produced oil volume to the injected steam volume for a certain period of time. SOF is one of the main economic indices of steam treatment. Different modifications of this example are used in different cases in the literature and in the practice. Therefore, we show in Fig. 1 both these modifications, i.e. quantity of steam used for production of one ton of oil (ratio of the quantity of injected steam to the quantity of produced oil) - Fig. 1(a), and quantity of oil per ton of injected steam - Fig. 1(b).

It is clear that the both are one example; nevertheless, we show it in two figures for easy analysis and for clarity.

The above-mentioned uncertainty is caused by obscure wording of the purpose and restrictions in decision making. In this case, the objective is to obtain the maximum volume of oil with minimum volume of steam injected in a well. Usually, solving of such objectives cannot provide two mentioned conditions simultaneously, and in this case the methods allowing to make the trade-off decision are used. Such approach is an approach known from the fuzzy-set theory. Thereafter, it is necessary to assess the membership function for each selected criteria. Therefore, the appropriate calculations are executed subject to the provisions set forth in reference [8]. Then, such values of the membership function are used to find the least value of these two values, which is a value of the membership function of decision set (Table 1). The maximum value of this membership function corresponds to the optimal decision. The calculation results are presented in Table, which are, as an example, the sample from the total data set, where the optimal decision is marked.

Table 1. The results of calculation of membership function.

Date	Steam injection	Oil production owing to steam injection	μ_1	μ_2	μ_D
01.05.2010	263,176	90,255	0,614596	0,152333	0,152
01.06.2010	262,597	91,676	0,718065	0,170547	0,170547
—	—	—	—	—	—
01.04.2011	317,366	112,095	0,54871	0,729534	0,54871
01.05.2011	**207,684**	**116,704**	**0,927801**	**0,976167**	**0,9278**
01.06.2011	220,906	106,395	0,872623	0,500281	0,500281
—	—	—	—	—	—
01.06.2017	440,950	105,451	0,254695	0,469083	0,254695
01.07.2017	428,461	108,041	0,280381	0,559025	0,280381

Graphically, trends of membership function and optimal decision corresponding to the point of their crossing are shown in Fig. 2.

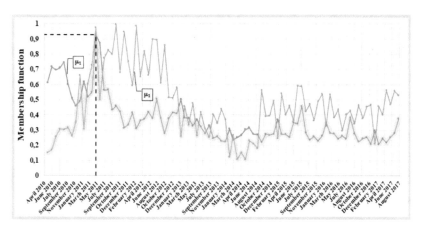

Fig. 2. Membership function of set of volumes of injected steam, produced oil and decisions (the latest is marked with yellow). Dashed lines show the optimal decision.

4   Conclusion

It is quietly clear that steam treatment methods of oil production are now one of the main technologies for development of high-viscosity oils and bitumen. All the while, the uncertainty of the processes in the reservoir system under impact of heat shall be noted. Thereat, decisions shall be made based on several criteria. The methods known from the fuzzy-set theory and designed for decision making under such conditions allow to increase efficiency of steam treatment owing to increase in oil production with minimum volumes of injected steam.

References

1. Akhmetov, D.A., Efendiyev, G.M., Karazhanova, M.K., Koylibaev, B.N.: Classification of hard-to-recover hydrocarbon reserves of Kazakhstan with the use of fuzzy cluster-analysis. In: 13th International Conference on Theory and Application of Fuzzy Systems and Soft Computing - ICAFS-2018, 27–28 August 2018, Warsaw, Poland, pp. 865–872. Springer Nature, Switzerland AG (2019)
2. Turkov, V.O.: Geological - Technical Information of Karazhanbas Field, April 2014
3. Esetov, Zh.A., Turdiyev, M.F., Kemalov, A.F., Abdrafikova, I.M.: Thermal-steam cyclic processing technology of development objects in Karazhanbas Kazakhstan field. Indian J. Sci. Technol. **9**(18), 1–7 (2016). https://doi.org/10.17485/ijst/2016/v9i18/93749
4. Akhmetov, D.A.: Analysis of results of steam treatment at Karazhanbas field. Azerbaijan Oil Facility **12**, 9–13 (2018)
5. Akhmetov, D.A., Efendiyev, G.M.: Experience of application of steam treatment at Karazhanbas Field. In: Materials of the International Scientific Practical Conference "Development of Science and Technology in Use of Subsoil of Kazakhstan", devoted to the 90-year anniversary of Academician Sh. Essenov, Aktau, pp. 170–173 (2017)
6. Zhdanov, S.A., Kryanev, D.Y., Simkin, E.M., Ursegov, S.O.: Method for Optimization of Steam Treatment during Development of Field with High-Viscosity Oils and Bitumen.

Patent 2445454 Russian Federation, IPC8 E 21 B 43/24. - No. 2010150291/03, announced on 09/12/2010, publ. 03/20/2012

7. Abraham, K.: High Prices, Instability Keep Activity High. World Oil 227, no. 9, September 2006. http://www.worldoil.com. Accessed 20 Dec 2006

8. Bellman, R.E., Zadeh, L.A.: Decision-making in a fuzzy environment. Manag. Sci. **17**, 141–164 (1970)

9. Classification of oils. https://studfiles.net/preview/1772355/page:2/

10. Muslimov, R.H.: The new classification of reserves and resources of oil and combustible gas – movement onward or backward? Georesursy = Georesources **18**(2), 80–87 (2016). https://doi.org/10.18599/grs.18.2.1

11. Lissovskiy, N.N., Khalimov, E.M.: About classification of hard to recover reserves. Bull. CKR Rosnedra **6**, 33–35 (2009)

Metabolic Syndrome Risk Evaluation Based on VDR Polymorphisms and Neural Networks

Adnan Khashman[1(✉)], Nedime Serakinci[2], and Meral Kizilkanat[3]

[1] European Centre for Research and Academic Affairs (ECRAA), Lefkosa,
Mersin 10, Turkey
adnan.khashman@ecraa.com
[2] Faculty of Medicine, Near East University, Lefkosa, Mersin 10, Turkey
nedime.serakinci@neu.edu.tr
[3] Polyclinic Laboratory, Nalbantoglu General Hospital, Lefkosa,
Mersin 10, Turkey
meral.karafistan@gmail.com

Abstract. The paper presents an intelligent implementation in medical genetics that supports clinical and laboratory practices by evaluating the risk of having metabolic syndrome (MetS) disorder based on its association with genetic variations or polymorphisms in Vitamin D Receptors (VDR). MetS is approximated in this work with irregularities in biochemical measurements of cholesterol and triglyceride levels in patients. The arbitration of this non-linear relation between VDR polymorphism and metabolic disorders is performed using a backpropagation neural network. The development of this risk evaluation system uses a dataset of biochemical and genetic data of 165 anonymous patients. The experimental results suggest that machine artificial neural networks can be successfully employed to evaluate the risk of metabolic syndrome using genetic and biochemical information.

Keywords: Neural networks · Back propagation · Risk evaluation · Metabolic syndrome · VDR polymorphism · DNA concentration · Genetic data

1 Introduction

Artificial intelligence applications are rapidly emerging in various fields with the aim of aiding our decision making and speeding up performance. The integration of advanced computational capabilities with artificial intelligence has encouraged many researchers to explore new application areas such as genetic medicine and health-care systems [1–9]. Intelligent system are often based on artificial neural networks; which are biologically inspired networks that have been successfully applied over the past three decades to solve a many tasks in various application areas such as in medicine, engineering, banking, military and so on [10–17].

Metabolic syndrome (MetS) is biochemically characterized with elevated levels of serum triglycerides, low-density lipoprotein (LDL), total cholesterol, reduced serum high-density lipoprotein (HDL) levels, and elevated glucose level together with increased insulin resistance [18]. Vitamin D Receptor (VDR) is a member of the steroid

© Springer Nature Switzerland AG 2020
R. A. Aliev et al. (Eds.): ICSCCW 2019, AISC 1095, pp. 943–949, 2020.
https://doi.org/10.1007/978-3-030-35249-3_126

hormone receptor family which acts as a transcriptional activator of many genes. The DNA polymorphisms which have been often reported for VDR gene are: BsmI (rs1544410 A > G), FokI (rs2228570 C > T), TaqI (rs731236 T > C), and ApaI (rs7975232 C > T) [19].

Recent studies reported that VDR polymorphisms are related to MetS [19–21]. For example, the work in [19] looked at the connection between VDR gene polymorphisms and MetS in North China. The authors suggested that VDR gene polymorphisms seem to be significantly linked with MetS, and that Allele B and BB genotype for BsmI were risk factors for MetS. They also reported that the BsmI polymorphism seemed to influence waist circumference, while the FokI polymorphism influenced BMI in subjects with MetS, thus, proposing that the BsmI and FokI polymorphisms of VDR can be potentially prognostic factors that could be used to predict the risk of developing MetS.

The use of artificially intelligence models to identify metabolic syndrome (MetS) has been suggested by several recent works [1–3, 7–9]. For example, in [1] the authors proposed the adoption of quantum particle swarm optimisation method to estimate the priority vector from the pairwise matrix of each MetS risk factors derived using the analytic hierarchy process. In [2], the authors employed decision tree and support vector machines to predict MetS and considered several factors in their investigation; including gender, age, smoking, triglycerides, total cholesterol, LDL, and HDL cholesterol. In [3], the authors suggested predicting individuals at high risk of MetS using a neural network model and based on physical signs. All these works, aimed at evaluating the risk of having metabolic syndrome, and for this purpose based their prediction systems on certain factors that are linked to MetS; such as LDH, HDL, etc.

In this work, we too aim at predicting metabolic syndrome (MetS) by examining four biochemical measurements (Triglyceride, total cholesterol, LDL, and HDL). For this purpose, we propose using the supervised back propagation (BPNN) neural model to evaluate the risk of having metabolic syndrome based on arbitrating the non-linear relation between Vitamin D Receptor genetic variations (polymorphisms) and the above four biochemical factors in 165 anonymous patients. We also consider other factors; such as the DNA concentration of the patient, their age and gender, and whether they smoke or not. To the best of our knowledge, this is very novel and has not been used in similar tasks before.

The paper is organization is as follows: in Sect. 2, the input and output attributes in the dataset are clarified and their coding into machine learning data is explained. In Sect. 3, the topology of the BPNN neural model; as well as the adopted learning scheme are explained. In Sect. 4, the achieved results are discussed and performance evaluation is presented. Finally, in Sect. 5 this work is concluded.

2   Input/Output Data Coding

This section explains the attributes of the dataset and their coding into binary digital data. The dataset contains information and biochemical measurements of 165 anonymous patients comprising 102 females and 63 males with a wide age span between 2 to 76 years old. We also consider the patient's smoking status which is often overlooked in such investigations since we believe it has an impact on the patient's health in general. The biomedical measurements include the four VDR gene polymorphisms (FokI, TaqI, BsmI, ApaI), the DNA concentration, Cholesterol, LDL, HDL, and Triglyceride levels.

The dataset had some missing values which were imputed by the average of the attribute values where a missing value occurs. An average value is worked out by dividing every entry in an attribute by the maximum recorded value within that attribute.

The dataset was organized into input and output attributes. In our hypothesis, there is a nonlinear relationship between the input and the output attributes and it is the task of the designed artificial intelligence system to model this relationship; in this case relating patients information and VDR polymorphisms to the levels of cholesterol, triglyceride, LDL and HDL. Thus, indicating the possibility of having metabolic syndrome (MetS). Table 1 lists the input and output attributes used in this work and shows their corresponding data coding which is applied prior to feeding the data to the prediction neural network model.

Table 1. Dataset input/output attributes and their machine coding.

	Input attributes	Information coding		
1.	Gender (102 female, 63 male)	Female {0}	Male {1}	
2.	Age (2–76) years old	Normalized via division by maximum value 76		
3.	Smoker	No {0}	Yes {1}	Unknown {0.5}
4.	DNA concentration(ng/ul)	Normalized via division by maximum value 650		
5.	VDR variant FokI	FF {1} normal	Ff {2} Carrier	ff {3} ill
6.	VDR variant TaqI	TT {1} normal	Tt {2} Carrier	tt {3} ill
7.	VDR variant BsmI	BB {1} normal	Bb {2} Carrier	bb {3} ill
8.	VDR variant ApaI	AA {1} normal	Aa {2} Carrier	aa {3} ill
	Output attributes	Information coding		
1.	Cholesterol (50–200) mg/dL	Within range Normal {1}	Out of range Disorder {0}	Greater than 2 normals is
2.	HDL (29–84) mg/dL	Within range Normal {1}	Out of range Disorder {0}	Low MetS Risk {1}
3.	LDL (60–100) mg/dL	Within range Normal {1}	Out of range Disorder {0}	Greater than 2 disorders is
4.	Triglyceride (<200) mg/dL	Within range Normal {1}	Out of range Disorder {0}	High MetS Risk {0}

3   The Risk Evaluation Neural Model

In This work we adopt the fully connected back propagation neural network (BPNN) as the risk evaluation model. Except for the neurons at the input layer, all other neurons are processing neurons which are activated using the sigmoid transfer function; and are aided out of local minima using extra bias neurons with true values of '1'. Synaptic weight updating during learning is performed using error minimization and the delta rule.

Figure 1 shows the topology of the proposed BPNN model with a total of three layers. The input layer has 8 input neurons receiving the coded input attributes data values which represent a patient's information and VDR polymorphisms. The number of neurons in the output layer is two, which is representative of the risk evaluation model's potential decisions of HIGH or LOW risk. The number of neurons in the hidden layer is initially set to one neuron and, thereafter, increased until minimal error and optimum learning values are obtained during training. The number of hidden neurons with the best performance was determined as 37 hidden processing neurons.

Training the BPNN neural network was executed utilizing a learning scheme with (51.5%:48.5%) training-to-testing data ratio, i.e. data of 85 patients was dedicated to training; and the remaining 80 patients for testing the risk evaluation model.

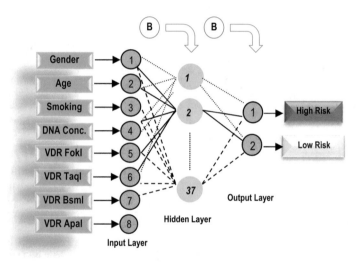

Fig. 1. Architecture of the BPNN risk evaluation model. B-neurons are bias neurons that help in triggering the processing neurons in hidden and output layers.

4   Experimental Results

The presented neural network model was simulated in C-language using the Dev-C^{++} 5.11 compiler running on Intel Core i3-7100 CPU 3.90 GHz, and 'Windows 10 Pro' 64-bit operating system, with 8 GB RAM.

The performance evaluation criteria in this work focusses on; *firstly*, the correct risk evaluation rate (CRER), and *secondly* the computational time. Table 2 shows the details of the experimental results. Using the training dataset, a CRER of 100% was achieved, whereas using the testing dataset yielded a CRER of 58.75%. The highest obtained overall combined CRER was 80%, which is considered reasonable considering the size of the dataset. The computational costs were as follows: training the neural model required 0.266 s whereas running the trained model with one forward pass required a fast 1.4×10^{-5} s.

Considering both performance criteria, the yielded results are considered successful and can be efficiently used in evaluating the risk of having metabolic syndrome (MetS).

Table 2. Optimized training parameters and evaluation results.

Neural Model	BPNN
Input neurons	8
Hidden neurons	37
Output neurons	2
Learning rate	0.0055
Momentum rate	0.47
Obtained error	0.0077
Iterations	1850
Training time (s)[1]	0.266
Testing time for run (s)[1]	1.4×10^{-5}
CRER[2] – Training	(85/85) 100%
CRER[2] – Testing	(47/80) 58.75%
CRER[2] – Overall	**(132/165) 80%**

[1]using the Dev-C++ 5.11 compiler running on Intel Core i3-7100 CPU 3.90 GHz, and 'Windows 10 Pro' 64-bit operating system, with 8 GB RAM. [2] CRER: Correct Risk Evaluation Rate.

5 Conclusion

This paper suggested a practical artificial intelligence implementation of neural networks in the field of medical genetics. The proposed neural model arbitrates the nonlinearity in relation between Vitamin D Receptors (VDR) genetic variations (Polymorphisms) and the risk of having metabolic syndrome (MetS). The importance of such an application is that it aids medical experts in their decision making by providing initial diagnosis via a fast risk evaluation system.

The MetS risk evaluation system is based on utilizing the popular and efficient backpropagation neural network model which is trained and tested using our own dataset. The dataset, which contains medical information of 165 anonymous patients, is organized into input attributes (four VDR polymorphisms, DNA concentration, age, gender, smoking status), and output attributes (Cholesterol, LDL, HDL, Triglyceride) which are biochemical measurement that are associated with MetS disorder.

The neural evaluation system performance was evaluated based on the overall correct risk evaluation rate (CRER) which was 80%, and the computational time which was very fast in training (0.266 s) as well as in running the trained model with one forward pass (1.4×10^{-5}). These successful results suggest that the developed neural network risk evaluation model, can be efficiently employed for similar medical applications. Further work will include exploring deep learning neural network models as risk evaluation models for similar tasks.

References

1. Kakudi, H.A., Loo, C.K., Pasupa, K.: Risk quantification of metabolic syndrome with quantum particle swarm optimisation. In: Proceedings of the 26th International Conference on World Wide Web Companion (2017)
2. Karimi-Alavijeh, F., Jalili, S., Sadeghi, M.: Predicting metabolic syndrome using decision tree and support vector machine methods. ARYA Atherosclerosis 12(3), 146–152 (2016)
3. Chen, H., Xiong, S., Ren, X.: Evaluating the risk of metabolic syndrome based on an artificial intelligence model. In: Abstract and Applied Analysis, pp. 1–12 (2014)
4. Khashman, A.: An emotional system with application to blood cell type identification. Trans. Inst. Meas. Control SAGE 34(2–3), 125–147 (2012)
5. Olaniyi, E.O., Khashman, A.: Onset diabetes diagnosis using artificial neural network. Int. J. Sci. Eng. Res. 5(10), 754–759 (2014)
6. Khashman, A., Dimililer, K.: Medical radiographs compression using neural networks and haar wavelet. In: EUROCON 2009, EUROCON 2009, pp. 1448–1453. IEEE (2009)
7. Christian, J.G., Byers, T.E., Christian, K.K., Goldstein, M.G., Bock, B.C., Prioreschi, B., Bessesen, D.H.: A computer support program that helps clinicians provide patients with metabolic syndrome tailored counseling to promote weight loss. J. Am. Diet. Assoc. 111(1), 75–83 (2011)
8. Baumgartner, C., Osl, M., Netzer, M., Baumgartner, D.: Bioinformatic-driven search for metabolic biomarkers in disease. J. Clin. Bioinf. 1(1), 2 (2011)
9. Lin, C.-C., Bai, Y.-M., Chen, J.-Y., Hwang, T.-J., Chen, T.-T., Chiu, H.-W., Li, Y.-C.: Easy and low-cost identification of metabolic syndrome in patients treated with second-generation antipsychotics. J. Clin. Psychiatry 71(03), 225–234 (2009)
10. Khashman, A., Sekeroglu, B.: Multi-banknote Identification using a single neural network. In: Blanc-Talon, J., Philips, W., Popescu, D., Scheunders, P. (eds.) Advanced Concepts for Intelligent Vision Systems (ACIVS2005). LNCS, vol. 3708, pp. 123–129. Springer, Heidelberg (2005)
11. Khashman, A., Sekeroglu, B.: A novel thresholding method for text separation and document enhancement. In: 11th Panhellenic conference on informatics, pp. 323–330. Greece (2007)
12. Khashman, A., Sekeroglu, B., Dimililer, K.: Intelligent coin identification system. In: IEEE Conference on Computer Aided Control System Design, 2006 IEEE International Conference on Control Applications, IEEE International Symposium on Intelligent Control, pp. 1226–1230 (2006)
13. Oyedotun, O.K., Khashman, A.: Document segmentation using textural features summarization and feedforward neural network. Appl. Intell. 45(1), 198–212 (2016)
14. Khashman, A.: Investigation of different neural models for blood cell type identification. Neural Comput. Appl. 21(6), 1177–1183 (2012)
15. Khashman, A.: Intelligent local face recognition. Recent Advances in Face Recognition, IntechOpen (2008)
16. Khashman, A.: Face recognition using neural networks and pattern averaging. In: International Symposium on Neural Networks, China, pp. 98–103 (2006)

17. Khashman, A.: Intelligent face recognition: local versus global pattern averaging. In: AI 2006: Advances in Artificial Intelligence, pp. 956–96 (2006)
18. Grundy, S.M.: Obesity, metabolic syndrome, and cardiovascular disease. J. Clin. Endocrinol. Metab. **89**(6), 2595–2600 (2004)
19. Zhao, Y., Liao, S., He, J., Jin, Y., Fu, H., Chen, X., Zhang, Y.: Association of vitamin D receptor gene polymorphisms with metabolic syndrome: a case–control design of population-based cross-sectional study in North China. Lipids Health Dis. **13**(1), 129 (2014)
20. Al-Daghri, N.M., Al-Attas, O.S., Alkharfy, K.M., Khan, N., Mohammed, A.K., Vinodson, B., Alokail, M.S.: Association of VDR-gene variants with factors related to the metabolic syndrome, type 2 diabetes and vitamin D deficiency. Gene **542**(2), 129–133 (2014)
21. Hasan, H.A., AbuOdeh, R.O., Muda, W.A.M.B.W., Mohamed, H.J.B.J., Samsudin, A.R.: Association of Vitamin D receptor gene polymorphisms with metabolic syndrome and its components among adult Arabs from the United Arab Emirates. Diabetes & Metabolic Syndrome: Clin. Res. Rev. **11**, S531–S537 (2017)

Intelligent Prediction of Initial Setting Time for Cement Pastes by Using Artificial Neural Network

Pinar Akpinar[1][(✉)] [iD] and Mariya A. Abubakar[2]

[1] Engineering Department, Near East University,
Nicosia, Lefkoşa, Mersin 10, Turkey
pinar.akpinar@neu.edu.tr
[2] Ministry of Works, Housing and Transport, Kano, Nigeria
aamariss02@gmail.com

Abstract. Concrete has become a major construction material all around the world with over ten billion tons consumed annually. One of the major issues to be kept monitored during manufacture of concrete is its initial setting time; that is to say, the time needed for the initiation of fresh concrete's solidification. This study aims to propose an intelligent model that will provide efficient prediction of setting time of cement pastes. An artificial Neural Network (ANN) model was proposed for the setting time predictions in this study; and its prediction performance was investigated systematically by using two training functions, under two different train:test data distributions together with five varying hidden neuron values. Setting time of cement pastes was predicted considering 12 input parameters. The results obtained indicates that the prediction accuracy of the employed ANN model is satisfactory; since it yielded remarkably high values of correlation coefficient and low mean square error such as 0.998 and 0.0003, respectively.

Keywords: Cement pastes · Initial setting time · Artificial neural network · Feedforward backpropagation · Train:test data distributions · Hidden neurons

1 Introduction

Cement is the fundamental material used in concrete manufacture. Setting is described as the solidifying of the cement paste. Setting mainly occurs through the 'hydration' reaction of cement compounds [1, 2]. Setting of fresh concrete is significant, since the fresh concrete should be maintained in the plastic stage for sufficient time period in order to ensure completion of mixing, placing and compaction processes of concrete in a feasible manner [1]. Information on the initial setting time of fresh concrete and/or cement paste are essential in construction scheduling, as well as in management of the haulage and placement of concrete. Conventional setting time determination tests are done with several trial and error batches of materials in the laboratory and each trial costs significant amount of materials, time and labor. Artificial Neural Networks (ANN) have been utilized as an effective device for several applications in different fields of civil engineering [3–7], as it is known to perform a vital role in imitation of complex and indirect procedures [3].

© Springer Nature Switzerland AG 2020
R. A. Aliev et al. (Eds.): ICSCCW 2019, AISC 1095, pp. 950–957, 2020.
https://doi.org/10.1007/978-3-030-35249-3_127

This study aims to propose an intelligent method for the prediction of initial setting time of cement pastes by using ANN that will be in a non-destructive manner; without consuming materials, time and labor. Even though ANN has been employed for several other aspects of civil engineering studies previously, it was observed that setting time predictions with ANN for cement pastes and concrete was not studied in detail in the related literature. Hence, another objective covered within this study is to provide insight on the changes of the effectiveness of proposed ANN model under the effects of varying train:test data distributions, as well as varying hidden neuron values. In this way, it is expected that future studies on the cement paste's setting time prediction in the related literature will be able to benefit the data that will be obtained from this study.

2 Methodology

2.1 Data Selection and Processing

In this study, the dataset used includes 174 cases having comparable parameters, that were extracted from 6 experimental studies published in the related literature [8–13]. The details of the input and output parameters are presented in Table 1.

Table 1. Input and output parameters used

Description		Parameters	Units
Inputs	1	Cement's CaO content	%
	2	Cement's Al_2O_3 content	%
	3	Cement's Fe_2O_3 content	%
	4	Cement's SiO_2 content	%
	5	water/cement ratio	%
	6	Cement content	kg/m^3
	7	Cement fineness	cm^2/g
	8	Temperature	°C
	9	Slag	%
	10	Fly ash content	%
	11	Silica fume content	%
	12	Cementitious materials' fineness	cm^2/g
Output	1	Initial setting time	min

The contents of cement's chemical compounds were considered as individual parameters for the very first time in this study, as they are known to hydrate and set at different rates [2]. The novel inclusion of each cement compound's content individually, enables the use of proposed ANN model for any kind of cement type of interest,

since the type of cements are known to be defined based on their chemical compositions. Other input parameters included were the water to cement ratio, temperature and
admixture used in the manufacturing of the cement paste. The target output considered
in this study was the cement paste's initial setting time. Since some of the inputs
presented in Table 1 are proposed and used for the very first time in this study, their
relationship with the defined output (i.e. setting time of cement paste) was verified by
multiple linear regression (MLR) that was used solely as an supplementary method in
this work. The coefficient of determination (R^2) was obtained as 0.80 as a result of
MLR analysis, which indicated that 80% of the variations in the output can be defined
by changes in the proposed set of multiple input parameters.

Normalization of input and output dataset is considered as an important data pre-
processing in training of an ANN model. The dataset was normalized between 0 and
1,considering its own minimum and maximum values, to ensure that each parameters
retains its importance within the network as suggested in the previous literature [3],
with using the below equation:

$$X_T = \frac{X_i - X_{min}}{X_{max} - X_{min}} \tag{1}$$

where X_{min} = minimum, X_{max} = maximum, X_T = normalized, X_i = actual values used.

2.2   ANN Architecture and Training

In this study, an ANN model employing feedforward backpropagation was proposed to
predict the setting time of cement pastes with carrying properties. The input layer of the
proposed model contains 12 nodes corresponding to the twelve input parameters
considered (See Fig. 1). Initial setting time is the only output parameter of concern. The
proposed architecture involved one hidden layer containing varying numbers of hidden
neurons as seen in Fig. 1.

The two considered data partitioning ratios considered in this study were; 70:30 and
50:50 for train:test data distribution. Five different hidden neuron values were selected
for network training. These values are 5, 12, 18, 23 and 30.

ANN trainings were developed using MATLAB® 2017a command line. For each
training, the networks were trained until an optimal performance, which was defined as
0.001 for mean square error, obtainable could be reached. In this study two training
functions were considered; Levenberg-Marquardt (LM) and Scaled Conjugate Gradient
(SCG). Levenberg-Marquardt has been a commonly used training algorithm in modelling in neural nets. The algorithm is favored due to its speed of convergence. Similarly, Scaled Conjugate Gradient was also selected as an alternative to LM, based on
the successful results and outcomes of the research with similar cementitious input
parameters presented in cement and concrete related literature [6]. The two referenced
training function trained respective networks with Sigmoid transfer function.

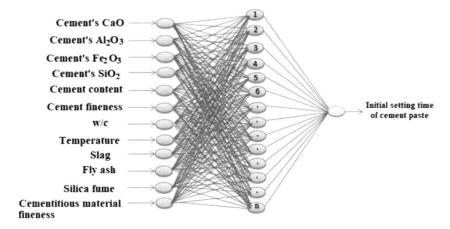

Fig. 1. ANN architecture used in this study

3   Results and Discussions

The results obtained for the prediction of initial setting time of cement pastes using the proposed ANN model with varying parameters are presented in detail in Tables 2 and 3. The prediction performance of each network training with different parameters was evaluated based on the values of correlation coefficient (R) and the mean square error (MSE).

Table 2. Prediction performance results with 50:50 data distribution

	H-N	R-train	R-test	R-overall	MSE-overall
LM	5	0.9977	0.9972	0.9966	0.0007
SCG		0.9909	0.9819	0.9883	0.0128
LM	12	0.9974	0.9716	0.9721	0.0073
SCG		0.9907	0.9287	0.9732	0.0020
LM	18	0.9967	0.9877	0.9933	0.0010
SCG		0.9905	0.8688	0.9633	0.0024
LM	23	0.9964	0.9841	0.9917	0.0010
SCG		0.9903	0.9644	0.9837	0.0036
LM	30	0.9974	0.9928	0.9945	0.0008
SCG		0.9866	0.8990	0.9498	0.0046

As seen in Tables 2 and 3, obtaining overall R values that are beyond 0.9498 in all cases, as well as obtaining overall MSE values less than 0.0056 in all cases, while employing a wide dataset including 174 experimental cases for the training and testing, indicates that the proposed ANN model have a high potential of serving satisfactorily as an alternative to conventional civil engineering destructive laboratory tests used to determine cement pastes' setting times.

Table 3. Prediction performance results with 70:30 data distribution

	H-N	R-Train	R-Test	R-Overall	MSE-Overall
LM	5	0.9986	0.9989	0.9986	0.0003
SCG		0.9898	0.9928	0.9901	0.0009
LM	12	0.9984	0.9757	0.9934	0.0007
SCG		0.9903	0.9850	0.9893	0.0008
LM	18	0.9987	0.9973	0.9985	0.0003
SCG		0.9893	0.9115	0.9682	0.0021
LM	23	0.9985	0.9975	0.9982	0.0005
SCG		0.9875	0.9850	0.9871	0.0011
LM	30	0.9972	0.9812	0.9935	0.0004
SCG		0.9892	0.9684	0.9806	0.0056

Figure 2a and b show the evolution of R (overall) and MSE (overall) respectively, for each hidden neuron value at both 50:50 and 70:30: train:test ratios, in a more comparative manner. With the aid of Fig. 2a, it can be clearly noted that 70:30 train:test distribution yielded a higher R value, indicating more accurate prediction of setting time, for all employed hidden neurons with both LM and SCG training functions.

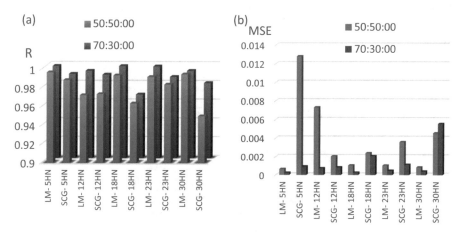

Fig. 2. Evolution of R-value (part a) and MSE (part b) with changing hidden neurons together with the data distributions

The increase in the R value when the train data set was increased from 50% to 70% is observed to be more significant at certain numbers of neurons in the hidden layer (e.g. at HN:12), and much less significant at other numbers of neurons in the hidden layer, as in the case of employing 5 neurons. Another noteworthy remark that can be seen from Fig. 2a is that; the effect of using increased train data distribution can be more significant depending on the training function employed for certain numbers of neurons in the hidden layer. This feature can be observed with increased R difference with SCG having 30 neurons in the hidden layer, rather than the much less significant R difference observed with LM when train data percentage is increased. A similar observation can be made with Fig. 2b, which shows the evolution of MSE for each selected numbers of neurons in the hidden layer *at both* 50:50 and 70:30: train:test ratios in a comparative manner. In Fig. 2b as well, the effect of yielding significantly higher accuracies for the case of increased train data set, seems to be training function-dependent also, occurring at certain numbers of neurons in the hidden layer. This can be observed with 5 neurons in the hidden layer in Fig. 2b; where for SCG, increasing the train data set from 50% to 70% made a clear difference in MSE, but this was not the case for LM at the same numbers of neurons in the hidden layer. Similarly, in the case of using 12 neurons; this time the difference in MSE for increasing train data set was more significant for LM but not for SCG. This finding implies that the selection of efficient train:test data distribution should be made carefully, considering the combinations of numbers of neurons in the hidden layer and training functions to be employed in each case.

In case of using 50:50 data distribution ratio (see Table 2 and Fig. 2a), it can be seen that the highest overall R-values for both LM and SCG were obtained with 5 neurons in the hidden layer; as 0.996 and 0.9883, respectively. In the case of the 70:30 train:test ratio (see the Table 3 and Fig. 2a), it can be seen that the highest overall R-values for both LM and SCG were obtained at hidden neuron 5; as 0.9989 and 0.9928 respectively, with a negligible difference of 0.0061.

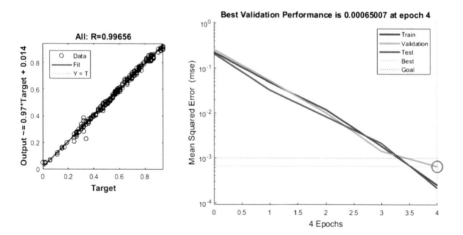

Fig. 3. Coefficient of correlation values (part a; on the left) and mean square error (part b; on the right) values for the best case of 50:50 data distribution; with LM at 5 Hidden neurons.

Figures 3 and 4 illustrate the correlation coefficient (part-a) and mean square error (part-b) for the best performance cases among all combinations having 50:50 data distribution presented in Table 2 and among all combinations having 70:30 data distribution presented in Table 3, respectively. In both data distributions, the best performance cases (i.e. highest R and lowest MSE) for both Levenberg-Marquardt

(LM) and Scaled Conjugate Gradient (SCG) functions occurred at 5 neurons in the hidden layer.

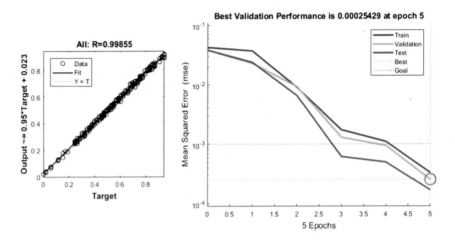

Fig. 4. Coefficient of correlation value (part a; on the left) and mean square error (part b; on the right) values for the best case of 70:30 data distribution; with LM at 5 Hidden neurons.

In both cases, LM was observed to yield R and MSE values indicating higher accuracies than the values yielded with SCG; even though the difference between the correlation functions yielded with LM and SCG become almost insignificant when the train data set percentage was increased to 70. As an overview, increasing the training dataset percentages clearly yielded an increase in the prediction accuracy; however, the level of this increases in the prediction accuracy is observed to be changing at each number of neurons sin the hidden layer, as well as for each training function used.

4   Conclusive Remarks

The study was carried out to propose an intelligent method for cement paste setting time predictions that could serve as an alternative to the 'time and resources con-suming' conventional civil engineering laboratory test methods. The employed ANN model yielded satisfactorily high R-values, such as 0.949 and above in all cases. Hence, ANN has the potential to serve efficiently as an alternative method for the determi-nation of cement setting time in related civil engineering applications.

    I. It was observed that greater training data distributions such as 70% used in this study, yielded a higher accuracy in setting time predictions for both training functions employed. The best prediction performance for both train:test data distributions were obtained with 5 neurons in the hidden layer, with both LM and SCG.

II. Higher prediction performances were observed to be obtained when Levenberg-Marquardt (LM) training function was employed, for all combinations of data distributions and hidden neurons used within this study. The highest particular value for coefficient of correlation was obtained as 0.9986 when LM was used with five neurons in the hidden layer and 70:30 train:test data distribution.

III. Detection of the most effective numbers of neurons in the hidden layer to be used is observed to be highly depending on the training function employed, as well as to the train:test data distribution value selected.

References

1. Neville, A.M., Brooks, J.J.: Concrete Technology. 2/E. Pearson Education, Canada (2010)
2. Neville, A.M.: Properties of Concrete, 5th edn. Pearson Education, Canada (2011)
3. Khademi, F.: Evaluation of concrete compressive strength using artificial neural network and multiple linear regression models. Iran Univ. Sci. Technol. **6**(3), 423–432 (2016)
4. Khashman, A., Akpinar, P.: Non-destructive prediction of concrete compressive strength using neural networks. Procedia Comput. Sci. **108**, 2358–2362 (2017)
5. Akpinar, P., Khashman, A.: Intelligent classification system for concrete compressive strength. Procedia Comput. Sci. **120**, 712–718 (2017)
6. Akpinar, P., Uwanuakwa, I.D.: Intelligent prediction of concrete carbonation depth using neural networks. Bull. Transilvania Univ. Brasov **9**(58), No. 2-2016, Series III:Mathematics, Informatics, Physiscs, pp. 99–108 (2016)
7. Brooks, J.J., Megat-Johari, M.A., Mazloom, M.: Effect of admixtures on the setting times of high-strength concrete. Cem. Concr. Compos. **22**(4), 293–301 (2000)
8. Gulbandilar, E., Kocak, Y.: Prediction of the effects of fly ash and silica fume on the setting time of Portland cement with fuzzy logic. Neural Comput. Appl. **22**(7–8), 1485–1491 (2013)
9. Khan, B., Ullah, M.: Effect of a retarding admixture on the setting time of cement pastes in hot weather. JKAU Eng. Sci. **15**(1), 63–79 (2004)
10. Tamas, F.D.: Acceleration and retardation of portland cement hydration by additives (8), 392–397 (1966)
11. Yurdakul, E.: Optimizing concrete mixtures with minimum cement content for performance and sustainability. Master thesis, Civil, Construction, and Environmental Engineering, Iowa State University, 112 p. (2010). https://doi.org/10.31274/etd-180810-1409
12. Güneyisi, E., Gesoglu, M., Özbay, E.: Evaluating and forecasting the initial and final setting times of self-compacting concretes containing mineral admixtures by neural network. Mater. Struct. Constr. **42**(4), 469–484 (2009)
13. Khademi, F., Jamal, S.M.: Predicting the 28 days compressive strength of concrete using artificial neural network. Managers J. Civ. Eng. **6**(2), 1–6 (2016)

Investigations on the Influence of Variations in Hidden Neurons and Training Data Percentage on the Efficiency of Concrete Carbonation Depth Prediction with ANN

Ikenna D. Uwanuakwa🆔 and Pinar Akpinar$^{(\boxtimes)}$🆔

Department of Civil Engineering, Near East University, Lefkoşa,
TRNC, Mersin 10, Turkey
ikenna.uwanuakwa@neu.edu.tr, pinar.akpinar@neu.edu.tr

Abstract. Concrete is undoubtedly one of the most popular construction materials. Carbonation is a well-known concrete durability problem that may negatively affect the performance of reinforced concrete buildings. In this study, efficient prediction of concrete carbonation depth was targetted by employing artificial neural networks with by One-step Secant method of optimization, as an alternative to conventionally used Levenberg-Marquardt algorithm. The effect of varying hidden neuron values and varying train:test data distribution on the evolution of network performance was aimed to be investigated in a sensitive and systematical way, as the scope of this study. Network training was carried out with combinations of 10 different hidden neurons and 11 data distribution ratios. For the task of predicting concrete carbonation depth as the output, the highest coefficient of correlation (R) obtained was 0.99. Results have shown that the variations of training dataset percentage within the range of 30–55% yielded a more significant improvement in the R-value than it is observed within the range of 60–80%. It was also observed that the variation of hidden neurons between the values 5–25 yielded relatively less significant changes on the prediction of accuracy, both in terms of R and MSE, for the range of training data percentages between 60–80%.

Keywords: Concrete carbonation · Artificial neural networks · One-step secant · Hidden neurons · Data distribution

1 Introduction

Carbonation in concrete buildings is a durability problem that is known to cause potentially severe alterations in concrete. The gradual decrease in the level of alkalinity in concrete due to the progress of carbonation not only changes concrete microstructure but also may lead to the initiation of corrosion in the steel reinforcing bars [1]. As a result of the increase in demand for having smarter buildings, the performance checks for concrete carbonation depth in structures must also become non-destructive, fast and efficient. ANN has proven within different fields of learning [2] and in concrete construction and durability [3, 4] to be effective in the analysis of the nonlinear separated

© Springer Nature Switzerland AG 2020
R. A. Aliev et al. (Eds.): ICSCCW 2019, AISC 1095, pp. 958–965, 2020.
https://doi.org/10.1007/978-3-030-35249-3_128

dataset. Previous studies carried out on the use of ANN for studying concrete carbonation depth yielded results emphasizing the critical importance of the selection of certain parameters, such as test:train data distribution and the value of hidden neurons, on the efficiency prediction results [5, 6]. Also, ANN models for carbonation studies in the related literature is observed to be limited with mainly on the use of the Levenberg-Marquardt (LM) algorithm only [7–10].The aim of this study is to provide insights on the effect of changing hidden neurons and training dataset percentages on the prediction performance of training networks that employ One-step Secant (OSS) as an alternative to conventional LM.

2   Trials and Methods

2.1   Variable Selection and Data Preparation

The detailed related-literature survey has been carried out carefully and the dataset used in this study was extracted from a total of 12 studies [11–22] with 378 fully comparable test cases on carbonation depth determination. Information on the used input and output parameters are presented in Table 1. Chemical compounds that present in both cement and fly ash (i.e. an additive used in concrete) such as CaO, SiO_2, Fe_2O_3, Al_2O_3 are known to be responsible for the production of $Ca(OH)_2$, which is the concrete compound that is susceptible to carbonation attack [23].

Table 1. Descriptive information on the used input and output variables.

	Variable [unit]	Min	Mean	Max		Variable [unit]	Min	Mean	Max
Input-1	CaO [%]	46	62.19	68	Input-10	Cement [kg/m^3]	67	300.08	486
Input-2	SiO_2 [%]	20	21.71	31	Input-11	Water [kg/m^3]	102	167.87	220
Input-3	Fe_2O_3 [%]	1	3.41	6	Input-12	w/b [%]	0.28	0.49	0.84
Input-4	Al_2O_3 [%]	3	4.71	9	Input-13	Curing time [days]	1	31.16	90
Input-5	Fly ash CaO [%]	0	1.43	5	Input-14	Curing RH [%]	50	89.38	100
Input-6	Fly ash SiO_2 [%]	0	23.98	55	Input-15	T [°C]	12	20.67	33
Input-7	Fly ash Fe_2O_3 [%]	0	4.75	14	Input-16	RH [%]	50	67.37	100
Input-8	Fly ash Al_2O_3 [%]	0	12.79	30	Input-17	CO_2 [%]	0.03	20.03	100
Input-9	Fly ash [kg/m^3]	0	50.19	280	Input-18	Age [days]	3	180.31	2070
Output-1	Depth [mm]	0	16.36	64					

Since these compounds' reactive behavior differ when they are contained by cement or by fly ash [24], their ultimate effect on the concrete's susceptibility to carbonation problem are studied by employing them as individual input parameters. Water/binder (i.e. water/(cement + fly ash)) ratio, curing period, as well as environmental variables like the ambient temperature (T), carbon dioxide (CO_2) content and the moisture content (RH) are the other important input parameters considered to be influencing concrete carbonation depth, which is the output parameter of interest, in this study (see Table 1). Data transformation was necessary for accurate prediction in neural nets. Normalization of variables which transformed all cases into having a zero mean was applied between 0 and 1 to conform with logistic Sigmoid transfer function used between the layers.

$$x_{ij}^{norm} = \frac{x_{ij} - x_{\min j}}{x_{\max j} - x_{\min j}} \qquad (1)$$

Where x_{ij}^{norm} is the normalized i value of the variable j, x_{ij} is the i original of the variable j, $x_{\min j}$ is the minimum value of the variable j and $x_{\max j}$ is the maximum value of the variable j.

2.2   Neural Network Architecture and Training Parameters

One-step secant (OSS) algorithm was used in this study to train the network as a proposed alternative to commonly used Levenberg-Marquardt Backpropagation (LM). Additional network training were also carried out with LM in order to validate the reliability of proposed OSS for the prediction of concrete carbonation. MATLAB 2015b script was used. Network random distribution was adopted and the early stopping was applied to the network, the maximum fall was set to 200 for both algorithms in order to reduce overfitting. The prediction model consisted of one hidden layer (including a varying number of hidden neurons), together with 18 input and one output parameters that are described in Table 1. The learning rate and max iterations were set to 0.001 and 20,000, respectively, while the activation function was defined as logsig. A feedforward backpropagation was trained with one-step secant as training function. Logistic Sigmund transfer function was used between layers.

As the scope of this study, 11 different partitions for train:test dataset were adopted as; 30:70%; 35:65%; 40:60%; 45:55%; 50:50%; 55:45%; 60:40%; 65:35%; 70:30%; 75:25% and 80:20%, respectively. The test dataset was further divided into test:validate in 50:50 ratio. In addition to the systematically increasing training dataset percentages, a wide range including 10 different hidden sizes were also used in this study with the aim of evaluating the changes in the behavior of neural nets with varying hidden neuron values. The values of 2, 5, 7, 10, 12, 15, 17, 20, 22 and 25 were selected to be used as hidden neurons in this work. No particular order was followed in selecting the hidden neuron values; however, both odd and even values were employed.

3   Results and Discussions

As a result of 110 sets of performed analyses with combinations of 11 data distribution and 10 hidden neurons, the highest R-value was obtained as **0.99** with **OSS** when the train:test data distribution was set as **80:20** (at the specific case of hidden neuron = 7). As a prior step to the investigation of the effect of varying parameters on prediction performance, LM was also employed to predict the concrete carbonation with the same data partitioning at all suggested hidden neuron values in order to validate the efficiency of OSS for this prediction task.

3.1   Comparison of OSS with the Conventional LM Model

Figure 1 presents the comparison of overall-R and MSE values obtained when LM and OSS were used with 80:20 train:test data partitioning (i.e. best result case).

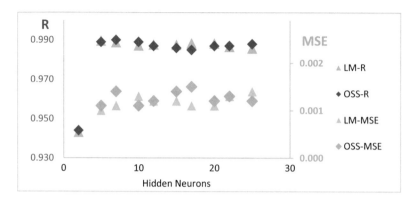

Fig. 1. Comparison of R and MSE values of OSS and LM at 80:20 data distribution.

It was observed that at this train:test data distribution, both algorithms provide significantly close accuracies, with almost matching R and almost matching MSE values. The convergence rate for LM was 7 s on average, while OSS had the highest training time of 266 s. Even 266 s recorded with OSS can be classified as insignificant when compared with laboratory experiments' duration needed to determine concrete carbonation depth, which might take several years in some cases. Hence, OSS could be used as an efficient alternative to LM algorithm for the application of concrete carbonation prediction, as it also provides even slightly higher R values.

3.2   The Effect of Varying Data Distribution on Network Performance

Figure 2 shows the evolution of R-value for the prediction of concrete carbonation with an increasing percentage of training data subset. The results show that the increasing percentage of training data percentage yields an increase in the R-value, indicating an improved prediction performance. It is observed that the improvement in R is more significant when the training data subset was increased from 30% to 55%. The improvement in R is less significant (i.e. only up to 0.01) when the training data percentage was increased beyond 60%.

Figure 2 also shows that hidden neuron value 2 was not sufficient to yield a relatively good network performance with all used train:test data distribution percentages. Hidden neuron values between 5–25 are observed to yield more comparable network performances, with changing peaks for each selected training data percentage. This implies that there is no single best hidden neuron value; instead, the optimum hidden neuron value and train:test data distribution combination should be verified in each case. The highest R-value was observed to be 0.99 with the combination of hidden neuron value-7 and 80:20 training:testing data distribution.

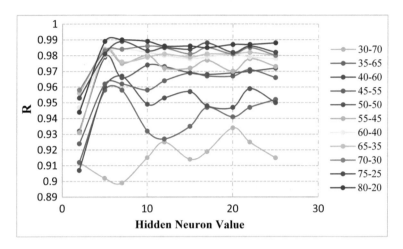

Fig. 2. Evolution of R-value with varying train:test data distribution percentages.

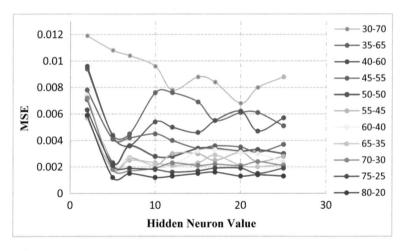

Fig. 3. Evolution of MSE with varying train:test data distribution percentages.

Figure 3 also indicates the positive effect of increasing training data percentage on the network performance, as MSE values are observed to merge to zero. Similar to the findings observed in Fig. 2, the inadequacy of hidden neuron value-2 in training the network for this prediction task, as well as the changing nature of "best hidden neurons" for each selected training percentage can also be seen in this Fig. 3. In general, training data percentages beyond 60% are observed to yield significantly low MSE values such as 0.003 and lower.

3.3 The Effect of Varying Hidden Neurons on the Network Performance

Figure 4 also confirms the previous findings: Hidden neuron value-2 was not a sufficient selection in any of the wide variety of train:test data distributions in this task, and also that the training data percentage of 60% and above yielded significantly higher R values, ranging basically between 0.98–0.99.

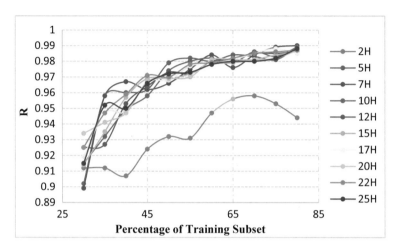

Fig. 4. Evolution of R value with varying hidden neuron values.

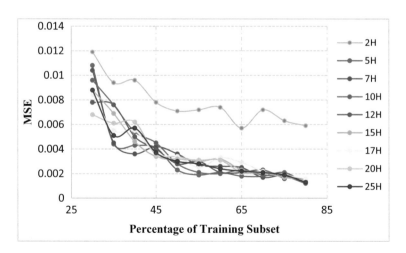

Fig. 5. Evolution of MSE value with varying hidden neuron values.

Figure 4 also shows that the change in hidden neuron values between 5–25 is much more critical (i.e. fluctuating more drastically) for relatively lower percentages of training data percentages such as 55% and lower. On the other hand, for training data percentages above 60%, the variations in hidden neuron values are not observed to yield critical changes in the prediction performance. In Fig. 5, it can be clearly observed that the MSE values were reduced significantly such as down to the value of 0.003 and lower, for training data percentages of 55% and more. For this range of relatively high training data percentages, it was observed that the variations in hidden neuron values became less significant.

4 Conclusions

The evaluation of the results obtained from a total of 110 network training with 11 different data distributions and 10 varying hidden neurons has been presented. The main conclusive remarks obtained from this study are as the followings:

I. Proposed ANN model for the concrete carbonation prediction using OSS algorithm yielded the highest accuracy with an R-value of 0.99 when hidden neuron value of 7 was used at a train:test data distribution of 80:20.
II. Employing OSS algorithm as an alternative to conventionally used LM algorithm is observed to be comparably highly efficient.
III. Generally, an increase in the training data set percentage yielded an increase in the carbonation depth prediction accuracy in terms of improved R values. The increase in the value of R is observed to be more significant when the training data percentage was increased from 30 to 55%, compared to the increase observed when the training data percentage was increased from 60 to 80%.
IV. The hidden neuron value-2 is observed to be insufficient in network training for all test:train data distributions. The effect of the variation of hidden neuron values between 5–25 was observed to be more critical for the prediction of accuracy, both in terms of R and MSE, for the lower range of training data; between 30–55%.

References

1. Thiery, M., Villain, G., Dangla, P., Platret, G.: Investigation of the carbonation front shape on cementitious materials: effects of the chemical kinetics. Cem. Concr. Res. **37**(7), 1047–1058 (2007)
2. Akpinar, P., Khashman, A.: Intelligent classification system for concrete compressive strength. Procedia Comput. Sci. **120**, 712–718 (2017)
3. Chandwani, V., Agrawal, V., Nagar, R.: Modeling slump of ready mix concrete using genetic algorithms assisted training of artificial neural networks. Expert Syst. Appl. **42**(2), 885–893 (2015)
4. Akpınar, P., Uwanuakwa, I.D.: Intelligent prediction of concrete carbonation depth using neural networks. Bull. Transilvania Univ. Brasov Ser. III Math. Inf. Phys. **9**(58), 99–107 (2016)
5. Panchal, G., Ganatra, A., Shah, P., Panchal, D.: Determination of over-learning and over-fitting problem in back propagation neural network. Int. J. Soft Comput. **2**(2), 40–51 (2011)
6. Khashman, A.: Neural networks for credit risk evaluation: investigation of different neural models and learning schemes. Expert Syst. Appl. **37**(9), 6233–6239 (2010)

7. Taffese, W.Z., Sistonen, E.: Neural network based hygrothermal prediction for deterioration risk analysis of surface-protected concrete façade element. Constr. Build. Mater. **113**, 34–48 (2016)
8. Güneyisi, E., Gesoğlu, M.: A study on durability properties of high-performance concretes incorporating high replacement levels of slag. Mater. Struct. **41**(3), 479–493 (2008)
9. Topçu, İ.B., Boğa, A.R., Hocaoğlu, F.O.: Modeling corrosion currents of reinforced concrete using ANN. Autom. Constr. **18**(2), 145–152 (2009)
10. Ghafoori, N., Najimi, M., Sobhani, J., Aqel, M.A.: Predicting rapid chloride permeability of self-consolidating concrete: a comparative study on statistical and neural network models. Constr. Build. Mater. **44**, 381–390 (2013)
11. Kari, O.P., Puttonen, J., Skantz, E.: Reactive transport modelling of long-term carbonation. Cem. Concr. Compos. **52**, 42–53 (2014)
12. Villain, G., Thiery, M., Platret, G.: Measurement methods of carbonation profiles in concrete: thermogravimetry, chemical analysis and gammadensimetry. Cem. Concr. Res. **37**(8), 1182–1192 (2007)
13. Atiş, C.D.: Accelerated carbonation and testing of concrete made with fly ash. Constr. Build. Mater. **17**(3), 147–152 (2003)
14. Rozière, E., Loukili, A., Cussigh, F.: A performance based approach for durability of concrete exposed to carbonation. Constr. Build. Mater. **23**(1), 190–199 (2009)
15. Hussain, S., Bhunia, D., Singh, S.B.: Comparative study of accelerated carbonation of plain cement and fly-ash concrete. J. Build. Eng. **10**, 26–31 (2017)
16. Villain, G., Thiery, M., Baroghel-Bouny, V., Platret, G.: Different methods to measure the carbonation profiles in concrete. In: International RILEM Workshop on Performance Based Evaluation and Indicators for Concrete Durability, pp. 89–98. RILEM Publications, Madrid (2006)
17. Younsi, A., Turcry, P., Aït-Mokhtar, A., Staquet, S.: Accelerated carbonation of concrete with high content of mineral additions: effect of interactions between hydration and drying. Cem. Concr. Res. **43**(1), 25–33 (2013)
18. Turcry, P., Oksri-Nelfia, L., Younsi, A., Aït-Mokhtar, A.: Analysis of an accelerated carbonation test with severe preconditioning. Cem. Concr. Res. **57**, 70–78 (2014)
19. Chang, C.F., Chen, J.W.: The experimental investigation of concrete carbonation depth. Cem. Concr. Res. **36**(9), 1760–1767 (2006)
20. Cui, H., Tang, W., Liu, W., Dong, Z., Xing, F.: Experimental study on effects of CO_2 concentrations on concrete carbonation and diffusion mechanisms. Constr. Build. Mater. **93**, 522–527 (2015)
21. Jiang, L., Lin, B., Cai, Y.: A model for predicting carbonation of high-volume fly ash concrete. Cem. Concr. Res. **30**(5), 699–702 (2000)
22. Balayssac, J.P., Détriché, C.H., Grandet, J.: Effects of curing upon carbonation of concrete. Constr. Build. Mater. **9**(2), 91–95 (1995)
23. Parrott, J.: Carbonation, moisture and empty pores. Adv. Cem. Res. **4**(15), 111–118 (1992)
24. Wang, A., Zhang, C., Sun, W.: Fly ash effects: II. The active effect of fly ash. Cem. Concr. Res. **34**(11), 2057–2060 (2004)

Author Index

Printed in the United States
By Bookmasters